Single Variable Calculus

Math 1A,B at UC Berkeley

8th Edition

Stewart

Australia • Brazil • Japan • Korea • Mexico • Singapore • Spain • United Kingdom • United States

Single Variable Calculus: Math 1A,B at UC Berkeley, 8th Edition

Calculus: Early Transcendentals, 8th Edition

© 2016, 2012 Cengage Learning. All rights reserved.

ALL RIGHTS RESERVED. No part of this work covered by the copyright herein may be reproduced, transmitted, stored or used in any form or by any means graphic, electronic, or mechanical, including but not limited to photocopying, recording, scanning, digitizing, taping, Web distribution, information networks, or information storage and retrieval systems, except as permitted under Section 107 or 108 of the 1976 United States Copyright Act, without the prior written permission of the publisher.

> For product information and technology assistance, contact us at
> **Cengage Learning Customer & Sales Support, 1-800-354-9706**
> For permission to use material from this text or product,
> submit all requests online at **cengage.com/permissions**
> Further permissions questions can be emailed to
> **permissionrequest@cengage.com**

This book contains select works from existing Cengage Learning resources and was produced by Cengage Learning Custom Solutions for collegiate use. As such, those adopting and/or contributing to this work are responsible for editorial content accuracy, continuity and completeness.

Compilation © 2015 Cengage Learning

ISBN: 978-1-305-76527-6

WCN: 01-100-101

Cengage Learning
20 Channel Center Street
Boston, MA 02210
USA

Cengage Learning is a leading provider of customized learning solutions with office locations around the globe, including Singapore, the United Kingdom, Australia, Mexico, Brazil, and Japan. Locate your local office at:
www.international.cengage.com/region.

Cengage Learning products are represented in Canada by Nelson Education, Ltd.

For your lifelong learning solutions, visit **www.cengage.com/custom.**

Visit our corporate website at **www.cengage.com.**

Printed at CLDPC, USA, 08-21

Contents

1 Functions and Models 9

- **1.1** Four Ways to Represent a Function 10
- **1.2** Mathematical Models: A Catalog of Essential Functions 23
- **1.3** New Functions from Old Functions 36
- **1.4** Exponential Functions 45
- **1.5** Inverse Functions and Logarithms 55
  Review 68

Principles of Problem Solving 71

2 Limits and Derivatives 77

- **2.1** The Tangent and Velocity Problems 78
- **2.2** The Limit of a Function 83
- **2.3** Calculating Limits Using the Limit Laws 95
- **2.4** The Precise Definition of a Limit 104
- **2.5** Continuity 114
- **2.6** Limits at Infinity; Horizontal Asymptotes 126
- **2.7** Derivatives and Rates of Change 140
  Writing Project • Early Methods for Finding Tangents 152
- **2.8** The Derivative as a Function 152
  Review 165

Problems Plus 169

3 Differentiation Rules 171

- **3.1** Derivatives of Polynomials and Exponential Functions 172
  - Applied Project • Building a Better Roller Coaster 182
- **3.2** The Product and Quotient Rules 183
- **3.3** Derivatives of Trigonometric Functions 190
- **3.4** The Chain Rule 197
  - Applied Project • Where Should a Pilot Start Descent? 208
- **3.5** Implicit Differentiation 208
  - Laboratory Project • Families of Implicit Curves 217
- **3.6** Derivatives of Logarithmic Functions 218
- **3.7** Rates of Change in the Natural and Social Sciences 224
- **3.8** Exponential Growth and Decay 237
  - Applied Project • Controlling Red Blood Cell Loss During Surgery 244
- **3.9** Related Rates 245
- **3.10** Linear Approximations and Differentials 251
  - Laboratory Project • Taylor Polynomials 258
- **3.11** Hyperbolic Functions 259
  - Review 266

Problems Plus 270

4 Applications of Differentiation 275

- **4.1** Maximum and Minimum Values 276
  - Applied Project • The Calculus of Rainbows 285
- **4.2** The Mean Value Theorem 287
- **4.3** How Derivatives Affect the Shape of a Graph 293
- **4.4** Indeterminate Forms and l'Hospital's Rule 304
  - Writing Project • The Origins of l'Hospital's Rule 314
- **4.5** Summary of Curve Sketching 315
- **4.6** Graphing with Calculus *and* Calculators 323
- **4.7** Optimization Problems 330
  - Applied Project • The Shape of a Can 343
  - Applied Project • Planes and Birds: Minimizing Energy 344
- **4.8** Newton's Method 345
- **4.9** Antiderivatives 350
  - Review 358

Problems Plus 363

5 Integrals 365

- **5.1** Areas and Distances 366
- **5.2** The Definite Integral 378
  - Discovery Project • Area Functions 391
- **5.3** The Fundamental Theorem of Calculus 392
- **5.4** Indefinite Integrals and the Net Change Theorem 402
  - Writing Project • Newton, Leibniz, and the Invention of Calculus 411
- **5.5** The Substitution Rule 412
- Review 421

Problems Plus 425

6 Applications of Integration 427

- **6.1** Areas Between Curves 428
  - Applied Project • The Gini Index 436
- **6.2** Volumes 438
- **6.3** Volumes by Cylindrical Shells 449
- **6.4** Work 455
- **6.5** Average Value of a Function 461
  - Applied Project • Calculus and Baseball 464
  - Applied Project • Where to Sit at the Movies 465
- Review 466

Problems Plus 468

7 Techniques of Integration 471

- **7.1** Integration by Parts 472
- **7.2** Trigonometric Integrals 479
- **7.3** Trigonometric Substitution 486
- **7.4** Integration of Rational Functions by Partial Fractions 493
- **7.5** Strategy for Integration 503
- **7.6** Integration Using Tables and Computer Algebra Systems 508
  - Discovery Project • Patterns in Integrals 513
- **7.7** Approximate Integration 514
- **7.8** Improper Integrals 527
- Review 537

Problems Plus 540

8 Further Applications of Integration   543

- **8.1** Arc Length   544
  - *Discovery Project* • Arc Length Contest   550
- **8.2** Area of a Surface of Revolution   551
  - *Discovery Project* • Rotating on a Slant   557
- **8.3** Applications to Physics and Engineering   558
  - *Discovery Project* • Complementary Coffee Cups   568
- **8.4** Applications to Economics and Biology   569
- **8.5** Probability   573
- Review   581

Problems Plus   583

9 Differential Equations   585

- **9.1** Modeling with Differential Equations   586
- **9.2** Direction Fields and Euler's Method   591
- **9.3** Separable Equations   599
  - *Applied Project* • How Fast Does a Tank Drain?   608
  - *Applied Project* • Which Is Faster, Going Up or Coming Down?   609
- **9.4** Models for Population Growth   610
- **9.5** Linear Equations   620
- **9.6** Predator-Prey Systems   627
- Review   634

Problems Plus   637

11 Infinite Sequences and Series   693

- **11.1** Sequences   694
  - *Laboratory Project* • Logistic Sequences   707
- **11.2** Series   707
- **11.3** The Integral Test and Estimates of Sums   719
- **11.4** The Comparison Tests   727
- **11.5** Alternating Series   732
- **11.6** Absolute Convergence and the Ratio and Root Tests   737
- **11.7** Strategy for Testing Series   744

11.8 Power Series 746

11.9 Representations of Functions as Power Series 752

11.10 Taylor and Maclaurin Series 759

Laboratory Project • An Elusive Limit 773

Writing Project • How Newton Discovered the Binomial Series 773

11.11 Applications of Taylor Polynomials 774

Applied Project • Radiation from the Stars 783

Review 784

Problems Plus 787

17 Second-Order Differential Equations 1153

17.1 Second-Order Linear Equations 1154

17.2 Nonhomogeneous Linear Equations 1160

17.3 Applications of Second-Order Differential Equations 1168

17.4 Series Solutions 1176

Review 1181

Appendixes A1

A Numbers, Inequalities, and Absolute Values A2

B Coordinate Geometry and Lines A10

C Graphs of Second-Degree Equations A16

D Trigonometry A24

E Sigma Notation A34

F Proofs of Theorems A39

G The Logarithm Defined as an Integral A50

H Complex Numbers A57

I Answers to Odd-Numbered Exercises A65

Index A139

1 Functions and Models

Pictura Collectus/Alamy

Often a graph is the best way to represent a function because it conveys so much information at a glance. Shown is a graph of the vertical ground acceleration created by the 2011 earthquake near Tohoku, Japan. The earthquake had a magnitude of 9.0 on the Richter scale and was so powerful that it moved northern Japan 8 feet closer to North America.

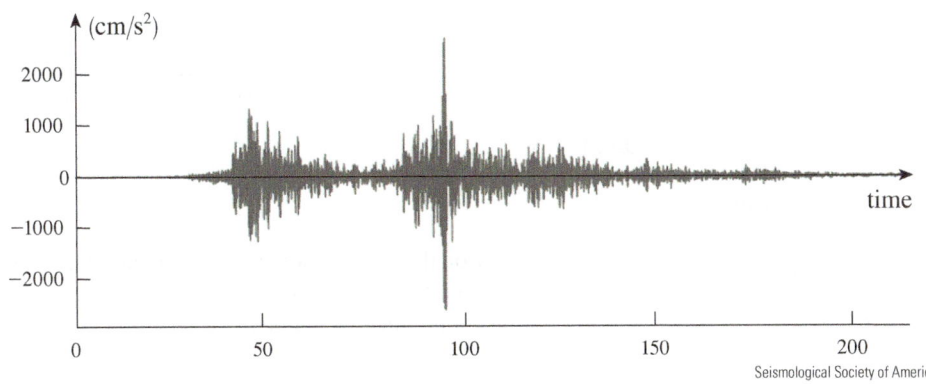

Seismological Society of America

THE FUNDAMENTAL OBJECTS THAT WE deal with in calculus are functions. This chapter prepares the way for calculus by discussing the basic ideas concerning functions, their graphs, and ways of transforming and combining them. We stress that a function can be represented in different ways: by an equation, in a table, by a graph, or in words. We look at the main types of functions that occur in calculus and describe the process of using these functions as mathematical models of real-world phenomena.

1.1 Four Ways to Represent a Function

Functions arise whenever one quantity depends on another. Consider the following four situations.

A. The area A of a circle depends on the radius r of the circle. The rule that connects r and A is given by the equation $A = \pi r^2$. With each positive number r there is associated one value of A, and we say that A is a *function* of r.

B. The human population of the world P depends on the time t. The table gives estimates of the world population $P(t)$ at time t, for certain years. For instance,

$$P(1950) \approx 2{,}560{,}000{,}000$$

But for each value of the time t there is a corresponding value of P, and we say that P is a function of t.

C. The cost C of mailing an envelope depends on its weight w. Although there is no simple formula that connects w and C, the post office has a rule for determining C when w is known.

D. The vertical acceleration a of the ground as measured by a seismograph during an earthquake is a function of the elapsed time t. Figure 1 shows a graph generated by seismic activity during the Northridge earthquake that shook Los Angeles in 1994. For a given value of t, the graph provides a corresponding value of a.

Year	Population (millions)
1900	1650
1910	1750
1920	1860
1930	2070
1940	2300
1950	2560
1960	3040
1970	3710
1980	4450
1990	5280
2000	6080
2010	6870

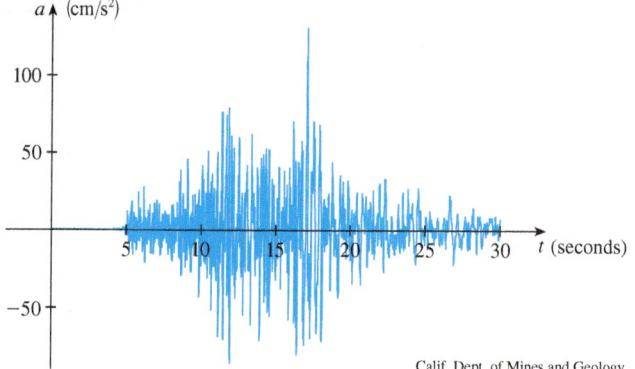

FIGURE 1
Vertical ground acceleration during the Northridge earthquake

Calif. Dept. of Mines and Geology

Each of these examples describes a rule whereby, given a number (r, t, w, or t), another number (A, P, C, or a) is assigned. In each case we say that the second number is a function of the first number.

> A **function** f is a rule that assigns to each element x in a set D exactly one element, called $f(x)$, in a set E.

We usually consider functions for which the sets D and E are sets of real numbers. The set D is called the **domain** of the function. The number $f(x)$ is the **value of f at x** and is read "f of x." The **range** of f is the set of all possible values of $f(x)$ as x varies throughout the domain. A symbol that represents an arbitrary number in the *domain* of a function f is called an **independent variable**. A symbol that represents a number in the *range* of f is called a **dependent variable**. In Example A, for instance, r is the independent variable and A is the dependent variable.

SECTION 1.1 Four Ways to Represent a Function   11

FIGURE 2
Machine diagram for a function f

It's helpful to think of a function as a **machine** (see Figure 2). If x is in the domain of the function f, then when x enters the machine, it's accepted as an input and the machine produces an output $f(x)$ according to the rule of the function. Thus we can think of the domain as the set of all possible inputs and the range as the set of all possible outputs.

The preprogrammed functions in a calculator are good examples of a function as a machine. For example, the square root key on your calculator computes such a function. You press the key labeled $\sqrt{}$ (or $\sqrt{x}$) and enter the input x. If $x < 0$, then x is not in the domain of this function; that is, x is not an acceptable input, and the calculator will indicate an error. If $x \geq 0$, then an *approximation* to $\sqrt{x}$ will appear in the display. Thus the $\sqrt{x}$ key on your calculator is not quite the same as the exact mathematical function f defined by $f(x) = \sqrt{x}$.

Another way to picture a function is by an **arrow diagram** as in Figure 3. Each arrow connects an element of D to an element of E. The arrow indicates that $f(x)$ is associated with x, $f(a)$ is associated with a, and so on.

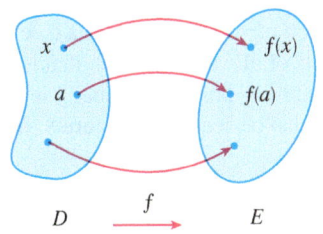

FIGURE 3
Arrow diagram for f

The most common method for visualizing a function is its graph. If f is a function with domain D, then its **graph** is the set of ordered pairs

$$\{(x, f(x)) \mid x \in D\}$$

(Notice that these are input-output pairs.) In other words, the graph of f consists of all points (x, y) in the coordinate plane such that $y = f(x)$ and x is in the domain of f.

The graph of a function f gives us a useful picture of the behavior or "life history" of a function. Since the y-coordinate of any point (x, y) on the graph is $y = f(x)$, we can read the value of $f(x)$ from the graph as being the height of the graph above the point x (see Figure 4). The graph of f also allows us to picture the domain of f on the x-axis and its range on the y-axis as in Figure 5.

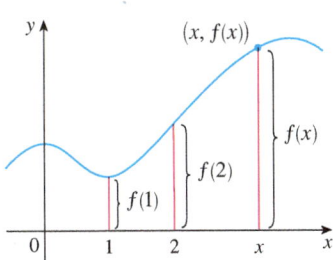

FIGURE 4

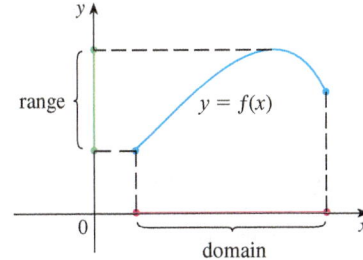

FIGURE 5

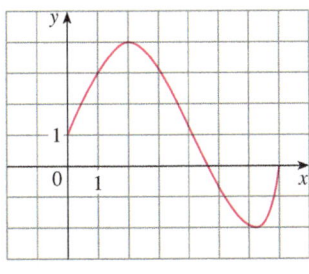

FIGURE 6

The notation for intervals is given in Appendix A.

EXAMPLE 1 The graph of a function f is shown in Figure 6.
(a) Find the values of $f(1)$ and $f(5)$.
(b) What are the domain and range of f?

SOLUTION
(a) We see from Figure 6 that the point $(1, 3)$ lies on the graph of f, so the value of f at 1 is $f(1) = 3$. (In other words, the point on the graph that lies above $x = 1$ is 3 units above the x-axis.)

When $x = 5$, the graph lies about 0.7 units below the x-axis, so we estimate that $f(5) \approx -0.7$.

(b) We see that $f(x)$ is defined when $0 \leq x \leq 7$, so the domain of f is the closed interval $[0, 7]$. Notice that f takes on all values from -2 to 4, so the range of f is

$$\{y \mid -2 \leq y \leq 4\} = [-2, 4]$$

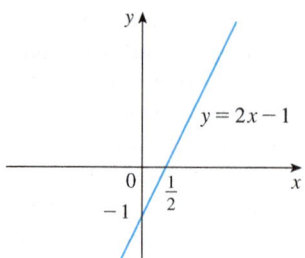

FIGURE 7

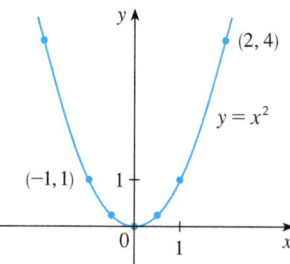

FIGURE 8

The expression
$$\frac{f(a+h) - f(a)}{h}$$
in Example 3 is called a **difference quotient** and occurs frequently in calculus. As we will see in Chapter 2, it represents the average rate of change of $f(x)$ between $x = a$ and $x = a + h$.

EXAMPLE 2 Sketch the graph and find the domain and range of each function.
(a) $f(x) = 2x - 1$ (b) $g(x) = x^2$

SOLUTION
(a) The equation of the graph is $y = 2x - 1$, and we recognize this as being the equation of a line with slope 2 and y-intercept -1. (Recall the slope-intercept form of the equation of a line: $y = mx + b$. See Appendix B.) This enables us to sketch a portion of the graph of f in Figure 7. The expression $2x - 1$ is defined for all real numbers, so the domain of f is the set of all real numbers, which we denote by $\mathbb{R}$. The graph shows that the range is also $\mathbb{R}$.

(b) Since $g(2) = 2^2 = 4$ and $g(-1) = (-1)^2 = 1$, we could plot the points $(2, 4)$ and $(-1, 1)$, together with a few other points on the graph, and join them to produce the graph (Figure 8). The equation of the graph is $y = x^2$, which represents a parabola (see Appendix C). The domain of g is $\mathbb{R}$. The range of g consists of all values of $g(x)$, that is, all numbers of the form x^2. But $x^2 \geq 0$ for all numbers x and any positive number y is a square. So the range of g is $\{y \mid y \geq 0\} = [0, \infty)$. This can also be seen from Figure 8. ■

EXAMPLE 3 If $f(x) = 2x^2 - 5x + 1$ and $h \neq 0$, evaluate $\dfrac{f(a+h) - f(a)}{h}$.

SOLUTION We first evaluate $f(a + h)$ by replacing x by $a + h$ in the expression for $f(x)$:

$$f(a + h) = 2(a + h)^2 - 5(a + h) + 1$$
$$= 2(a^2 + 2ah + h^2) - 5(a + h) + 1$$
$$= 2a^2 + 4ah + 2h^2 - 5a - 5h + 1$$

Then we substitute into the given expression and simplify:

$$\frac{f(a+h) - f(a)}{h} = \frac{(2a^2 + 4ah + 2h^2 - 5a - 5h + 1) - (2a^2 - 5a + 1)}{h}$$
$$= \frac{2a^2 + 4ah + 2h^2 - 5a - 5h + 1 - 2a^2 + 5a - 1}{h}$$
$$= \frac{4ah + 2h^2 - 5h}{h} = 4a + 2h - 5 \quad ■$$

■ Representations of Functions

There are four possible ways to represent a function:

- verbally   (by a description in words)
- numerically   (by a table of values)
- visually   (by a graph)
- algebraically   (by an explicit formula)

If a single function can be represented in all four ways, it's often useful to go from one representation to another to gain additional insight into the function. (In Example 2, for instance, we started with algebraic formulas and then obtained the graphs.) But certain functions are described more naturally by one method than by another. With this in mind, let's reexamine the four situations that we considered at the beginning of this section.

A. The most useful representation of the area of a circle as a function of its radius is probably the algebraic formula $A(r) = \pi r^2$, though it is possible to compile a table of values or to sketch a graph (half a parabola). Because a circle has to have a positive radius, the domain is $\{r \mid r > 0\} = (0, \infty)$, and the range is also $(0, \infty)$.

B. We are given a description of the function in words: $P(t)$ is the human population of the world at time t. Let's measure t so that $t = 0$ corresponds to the year 1900. The table of values of world population provides a convenient representation of this function. If we plot these values, we get the graph (called a *scatter plot*) in Figure 9. It too is a useful representation; the graph allows us to absorb all the data at once. What about a formula? Of course, it's impossible to devise an explicit formula that gives the exact human population $P(t)$ at any time t. But it is possible to find an expression for a function that *approximates* $P(t)$. In fact, using methods explained in Section 1.2, we obtain the approximation

$$P(t) \approx f(t) = (1.43653 \times 10^9) \cdot (1.01395)^t$$

Figure 10 shows that it is a reasonably good "fit." The function f is called a *mathematical model* for population growth. In other words, it is a function with an explicit formula that approximates the behavior of our given function. We will see, however, that the ideas of calculus can be applied to a table of values; an explicit formula is not necessary.

t (years since 1900)	Population (millions)
0	1650
10	1750
20	1860
30	2070
40	2300
50	2560
60	3040
70	3710
80	4450
90	5280
100	6080
110	6870

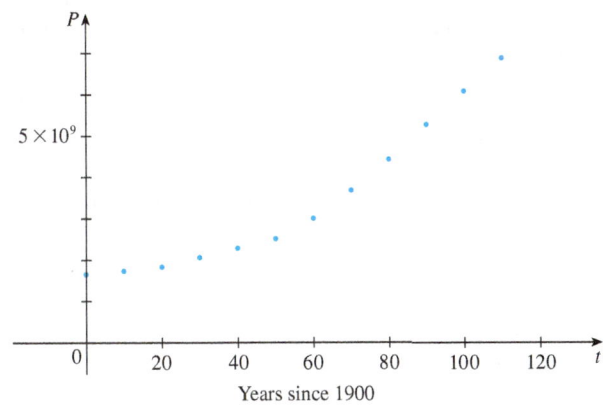

FIGURE 9

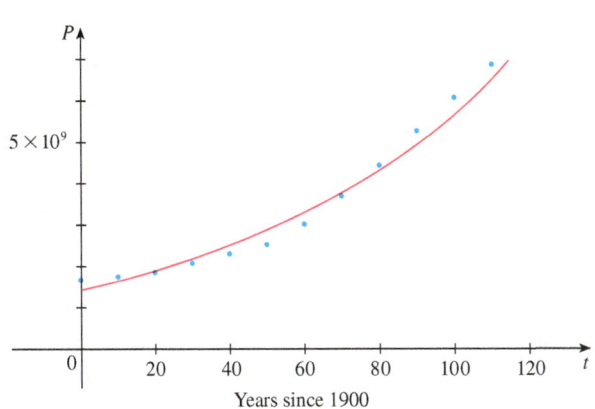

FIGURE 10

The function P is typical of the functions that arise whenever we attempt to apply calculus to the real world. We start with a verbal description of a function. Then we may be able to construct a table of values of the function, perhaps from instrument readings in a scientific experiment. Even though we don't have complete knowledge of the values of the function, we will see throughout the book that it is still possible to perform the operations of calculus on such a function.

C. Again the function is described in words: Let $C(w)$ be the cost of mailing a large envelope with weight w. The rule that the US Postal Service used as of 2015 is as follows: The cost is 98 cents for up to 1 oz, plus 21 cents for each additional ounce (or less) up to 13 oz. The table of values shown in the margin is the most convenient representation for this function, though it is possible to sketch a graph (see Example 10).

D. The graph shown in Figure 1 is the most natural representation of the vertical acceleration function $a(t)$. It's true that a table of values could be compiled, and it is even possible to devise an approximate formula. But everything a geologist needs to

A function defined by a table of values is called a *tabular* function.

w (ounces)	$C(w)$ (dollars)
$0 < w \leq 1$	0.98
$1 < w \leq 2$	1.19
$2 < w \leq 3$	1.40
$3 < w \leq 4$	1.61
$4 < w \leq 5$	1.82
⋮	⋮

know—amplitudes and patterns—can be seen easily from the graph. (The same is true for the patterns seen in electrocardiograms of heart patients and polygraphs for lie-detection.)

In the next example we sketch the graph of a function that is defined verbally.

EXAMPLE 4 When you turn on a hot-water faucet, the temperature T of the water depends on how long the water has been running. Draw a rough graph of T as a function of the time t that has elapsed since the faucet was turned on.

SOLUTION The initial temperature of the running water is close to room temperature because the water has been sitting in the pipes. When the water from the hot-water tank starts flowing from the faucet, T increases quickly. In the next phase, T is constant at the temperature of the heated water in the tank. When the tank is drained, T decreases to the temperature of the water supply. This enables us to make the rough sketch of T as a function of t in Figure 11. ∎

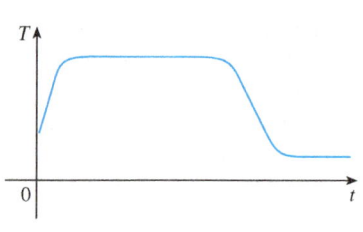

FIGURE 11

In the following example we start with a verbal description of a function in a physical situation and obtain an explicit algebraic formula. The ability to do this is a useful skill in solving calculus problems that ask for the maximum or minimum values of quantities.

EXAMPLE 5 A rectangular storage container with an open top has a volume of 10 m³. The length of its base is twice its width. Material for the base costs $10 per square meter; material for the sides costs $6 per square meter. Express the cost of materials as a function of the width of the base.

SOLUTION We draw a diagram as in Figure 12 and introduce notation by letting w and $2w$ be the width and length of the base, respectively, and h be the height.

The area of the base is $(2w)w = 2w^2$, so the cost, in dollars, of the material for the base is $10(2w^2)$. Two of the sides have area wh and the other two have area $2wh$, so the cost of the material for the sides is $6[2(wh) + 2(2wh)]$. The total cost is therefore

$$C = 10(2w^2) + 6[2(wh) + 2(2wh)] = 20w^2 + 36wh$$

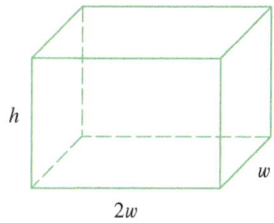

FIGURE 12

To express C as a function of w alone, we need to eliminate h and we do so by using the fact that the volume is 10 m³. Thus

$$w(2w)h = 10$$

which gives

$$h = \frac{10}{2w^2} = \frac{5}{w^2}$$

Substituting this into the expression for C, we have

$$C = 20w^2 + 36w\left(\frac{5}{w^2}\right) = 20w^2 + \frac{180}{w}$$

Therefore the equation

$$C(w) = 20w^2 + \frac{180}{w} \qquad w > 0$$

expresses C as a function of w. ∎

PS In setting up applied functions as in Example 5, it may be useful to review the principles of problem solving as discussed on page 71, particularly *Step 1: Understand the Problem.*

EXAMPLE 6 Find the domain of each function.

(a) $f(x) = \sqrt{x + 2}$ (b) $g(x) = \dfrac{1}{x^2 - x}$

Domain Convention

If a function is given by a formula and the domain is not stated explicitly, the convention is that the domain is the set of all numbers for which the formula makes sense and defines a real number.

SOLUTION

(a) Because the square root of a negative number is not defined (as a real number), the domain of f consists of all values of x such that $x + 2 \geq 0$. This is equivalent to $x \geq -2$, so the domain is the interval $[-2, \infty)$.

(b) Since

$$g(x) = \frac{1}{x^2 - x} = \frac{1}{x(x - 1)}$$

and division by 0 is not allowed, we see that $g(x)$ is not defined when $x = 0$ or $x = 1$. Thus the domain of g is

$$\{x \mid x \neq 0, x \neq 1\}$$

which could also be written in interval notation as

$$(-\infty, 0) \cup (0, 1) \cup (1, \infty)$$

∎

The graph of a function is a curve in the xy-plane. But the question arises: Which curves in the xy-plane are graphs of functions? This is answered by the following test.

> **The Vertical Line Test** A curve in the xy-plane is the graph of a function of x if and only if no vertical line intersects the curve more than once.

The reason for the truth of the Vertical Line Test can be seen in Figure 13. If each vertical line $x = a$ intersects a curve only once, at (a, b), then exactly one function value is defined by $f(a) = b$. But if a line $x = a$ intersects the curve twice, at (a, b) and (a, c), then the curve can't represent a function because a function can't assign two different values to a.

For example, the parabola $x = y^2 - 2$ shown in Figure 14(a) is not the graph of a function of x because, as you can see, there are vertical lines that intersect the parabola twice. The parabola, however, does contain the graphs of *two* functions of x. Notice that the equation $x = y^2 - 2$ implies $y^2 = x + 2$, so $y = \pm\sqrt{x + 2}$. Thus the upper and lower halves of the parabola are the graphs of the functions $f(x) = \sqrt{x + 2}$ [from Example 6(a)] and $g(x) = -\sqrt{x + 2}$. [See Figures 14(b) and (c).]

We observe that if we reverse the roles of x and y, then the equation $x = h(y) = y^2 - 2$ *does* define x as a function of y (with y as the independent variable and x as the dependent variable) and the parabola now appears as the graph of the function h.

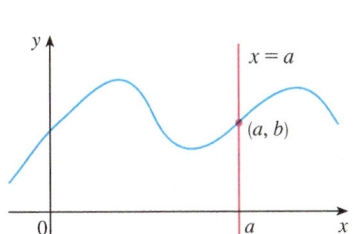

(a) This curve represents a function.

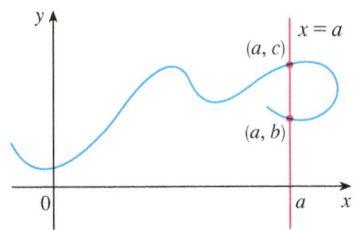

(b) This curve doesn't represent a function.

FIGURE 13

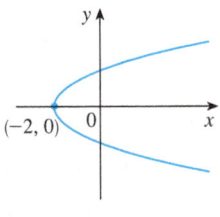

(a) $x = y^2 - 2$

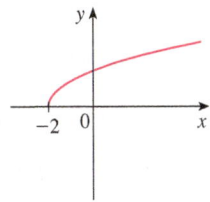

(b) $y = \sqrt{x + 2}$

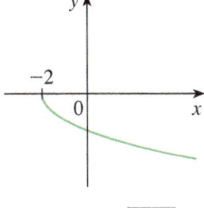

(c) $y = -\sqrt{x + 2}$

FIGURE 14

■ Piecewise Defined Functions

The functions in the following four examples are defined by different formulas in different parts of their domains. Such functions are called **piecewise defined functions**.

EXAMPLE 7 A function f is defined by

$$f(x) = \begin{cases} 1 - x & \text{if } x \leq -1 \\ x^2 & \text{if } x > -1 \end{cases}$$

Evaluate $f(-2)$, $f(-1)$, and $f(0)$ and sketch the graph.

SOLUTION Remember that a function is a rule. For this particular function the rule is the following: First look at the value of the input x. If it happens that $x \leq -1$, then the value of $f(x)$ is $1 - x$. On the other hand, if $x > -1$, then the value of $f(x)$ is x^2.

Since $-2 \leq -1$, we have $f(-2) = 1 - (-2) = 3$.

Since $-1 \leq -1$, we have $f(-1) = 1 - (-1) = 2$.

Since $0 > -1$, we have $f(0) = 0^2 = 0$.

How do we draw the graph of f? We observe that if $x \leq -1$, then $f(x) = 1 - x$, so the part of the graph of f that lies to the left of the vertical line $x = -1$ must coincide with the line $y = 1 - x$, which has slope -1 and y-intercept 1. If $x > -1$, then $f(x) = x^2$, so the part of the graph of f that lies to the right of the line $x = -1$ must coincide with the graph of $y = x^2$, which is a parabola. This enables us to sketch the graph in Figure 15. The solid dot indicates that the point $(-1, 2)$ is included on the graph; the open dot indicates that the point $(-1, 1)$ is excluded from the graph. ∎

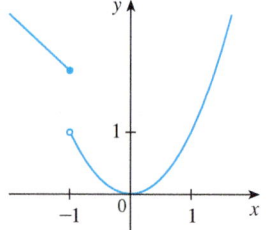

FIGURE 15

The next example of a piecewise defined function is the absolute value function. Recall that the **absolute value** of a number a, denoted by $|a|$, is the distance from a to 0 on the real number line. Distances are always positive or 0, so we have

$$|a| \geq 0 \quad \text{for every number } a$$

For a more extensive review of absolute values, see Appendix A.

For example,

$$|3| = 3 \quad |-3| = 3 \quad |0| = 0 \quad |\sqrt{2} - 1| = \sqrt{2} - 1 \quad |3 - \pi| = \pi - 3$$

In general, we have

$$\begin{aligned} |a| &= a \quad \text{if } a \geq 0 \\ |a| &= -a \quad \text{if } a < 0 \end{aligned}$$

(Remember that if a is negative, then $-a$ is positive.)

EXAMPLE 8 Sketch the graph of the absolute value function $f(x) = |x|$.

SOLUTION From the preceding discussion we know that

$$|x| = \begin{cases} x & \text{if } x \geq 0 \\ -x & \text{if } x < 0 \end{cases}$$

Using the same method as in Example 7, we see that the graph of f coincides with the line $y = x$ to the right of the y-axis and coincides with the line $y = -x$ to the left of the y-axis (see Figure 16). ∎

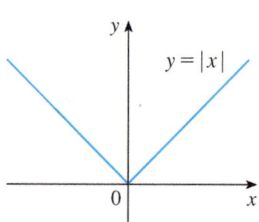

FIGURE 16

SECTION 1.1 Four Ways to Represent a Function 17

EXAMPLE 9 Find a formula for the function f graphed in Figure 17.

SOLUTION The line through $(0, 0)$ and $(1, 1)$ has slope $m = 1$ and y-intercept $b = 0$, so its equation is $y = x$. Thus, for the part of the graph of f that joins $(0, 0)$ to $(1, 1)$, we have

$$f(x) = x \quad \text{if } 0 \leq x \leq 1$$

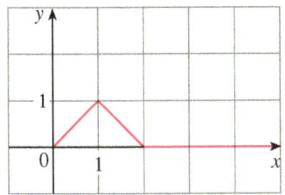

FIGURE 17

The line through $(1, 1)$ and $(2, 0)$ has slope $m = -1$, so its point-slope form is

$$y - 0 = (-1)(x - 2) \quad \text{or} \quad y = 2 - x$$

Point-slope form of the equation of a line:

$$y - y_1 = m(x - x_1)$$

See Appendix B.

So we have

$$f(x) = 2 - x \quad \text{if } 1 < x \leq 2$$

We also see that the graph of f coincides with the x-axis for $x > 2$. Putting this information together, we have the following three-piece formula for f:

$$f(x) = \begin{cases} x & \text{if } 0 \leq x \leq 1 \\ 2 - x & \text{if } 1 < x \leq 2 \\ 0 & \text{if } x > 2 \end{cases}$$

EXAMPLE 10 In Example C at the beginning of this section we considered the cost $C(w)$ of mailing a large envelope with weight w. In effect, this is a piecewise defined function because, from the table of values on page 13, we have

$$C(w) = \begin{cases} 0.98 & \text{if } 0 < w \leq 1 \\ 1.19 & \text{if } 1 < w \leq 2 \\ 1.40 & \text{if } 2 < w \leq 3 \\ 1.61 & \text{if } 3 < w \leq 4 \\ \vdots \end{cases}$$

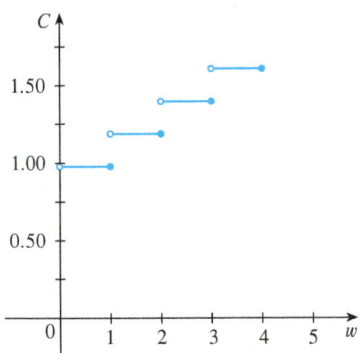

FIGURE 18

The graph is shown in Figure 18. You can see why functions similar to this one are called **step functions**—they jump from one value to the next. Such functions will be studied in Chapter 2.

■ **Symmetry**

If a function f satisfies $f(-x) = f(x)$ for every number x in its domain, then f is called an **even function**. For instance, the function $f(x) = x^2$ is even because

$$f(-x) = (-x)^2 = x^2 = f(x)$$

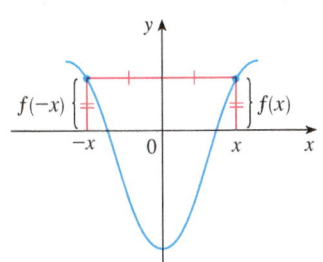

FIGURE 19
An even function

The geometric significance of an even function is that its graph is symmetric with respect to the y-axis (see Figure 19). This means that if we have plotted the graph of f for $x \geq 0$, we obtain the entire graph simply by reflecting this portion about the y-axis.

If f satisfies $f(-x) = -f(x)$ for every number x in its domain, then f is called an **odd function**. For example, the function $f(x) = x^3$ is odd because

$$f(-x) = (-x)^3 = -x^3 = -f(x)$$

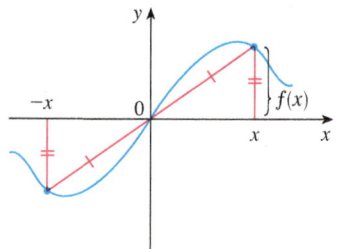

FIGURE 20
An odd function

The graph of an odd function is symmetric about the origin (see Figure 20). If we already have the graph of f for $x \geq 0$, we can obtain the entire graph by rotating this portion through 180° about the origin.

EXAMPLE 11 Determine whether each of the following functions is even, odd, or neither even nor odd.
(a) $f(x) = x^5 + x$ (b) $g(x) = 1 - x^4$ (c) $h(x) = 2x - x^2$

SOLUTION
(a)
$$f(-x) = (-x)^5 + (-x) = (-1)^5 x^5 + (-x)$$
$$= -x^5 - x = -(x^5 + x)$$
$$= -f(x)$$

Therefore f is an odd function.

(b) $$g(-x) = 1 - (-x)^4 = 1 - x^4 = g(x)$$

So g is even.

(c) $$h(-x) = 2(-x) - (-x)^2 = -2x - x^2$$

Since $h(-x) \neq h(x)$ and $h(-x) \neq -h(x)$, we conclude that h is neither even nor odd. ∎

The graphs of the functions in Example 11 are shown in Figure 21. Notice that the graph of h is symmetric neither about the y-axis nor about the origin.

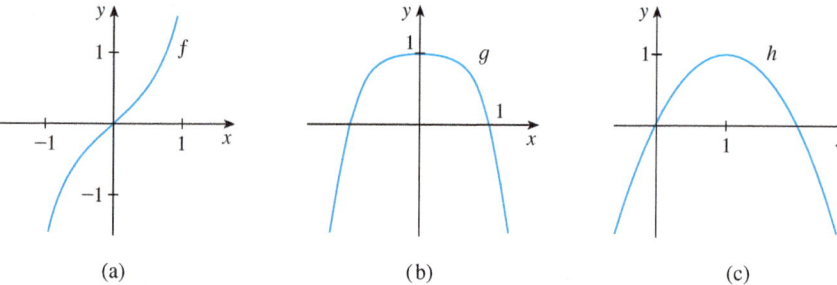

FIGURE 21   (a)   (b)   (c)

Increasing and Decreasing Functions

The graph shown in Figure 22 rises from A to B, falls from B to C, and rises again from C to D. The function f is said to be increasing on the interval $[a, b]$, decreasing on $[b, c]$, and increasing again on $[c, d]$. Notice that if x_1 and x_2 are any two numbers between a and b with $x_1 < x_2$, then $f(x_1) < f(x_2)$. We use this as the defining property of an increasing function.

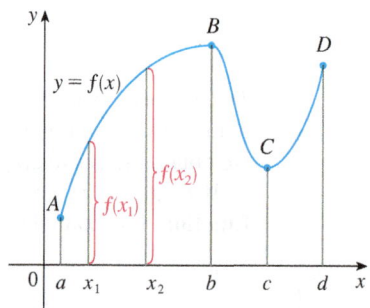

FIGURE 22

A function f is called **increasing** on an interval I if

$$f(x_1) < f(x_2) \qquad \text{whenever } x_1 < x_2 \text{ in } I$$

It is called **decreasing** on I if

$$f(x_1) > f(x_2) \qquad \text{whenever } x_1 < x_2 \text{ in } I$$

In the definition of an increasing function it is important to realize that the inequality $f(x_1) < f(x_2)$ must be satisfied for *every* pair of numbers x_1 and x_2 in I with $x_1 < x_2$.

You can see from Figure 23 that the function $f(x) = x^2$ is decreasing on the interval $(-\infty, 0]$ and increasing on the interval $[0, \infty)$.

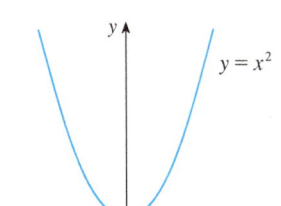

FIGURE 23

1.1 EXERCISES

1. If $f(x) = x + \sqrt{2 - x}$ and $g(u) = u + \sqrt{2 - u}$, is it true that $f = g$?

2. If
$$f(x) = \frac{x^2 - x}{x - 1} \quad \text{and} \quad g(x) = x$$
is it true that $f = g$?

3. The graph of a function f is given.
 (a) State the value of $f(1)$.
 (b) Estimate the value of $f(-1)$.
 (c) For what values of x is $f(x) = 1$?
 (d) Estimate the value of x such that $f(x) = 0$.
 (e) State the domain and range of f.
 (f) On what interval is f increasing?

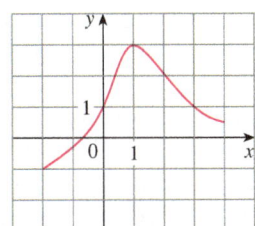

4. The graphs of f and g are given.

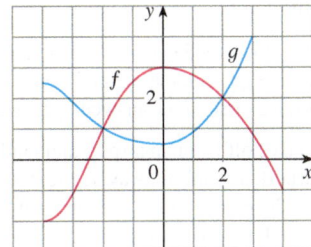

 (a) State the values of $f(-4)$ and $g(3)$.
 (b) For what values of x is $f(x) = g(x)$?

 (c) Estimate the solution of the equation $f(x) = -1$.
 (d) On what interval is f decreasing?
 (e) State the domain and range of f.
 (f) State the domain and range of g.

5. Figure 1 was recorded by an instrument operated by the California Department of Mines and Geology at the University Hospital of the University of Southern California in Los Angeles. Use it to estimate the range of the vertical ground acceleration function at USC during the Northridge earthquake.

6. In this section we discussed examples of ordinary, everyday functions: Population is a function of time, postage cost is a function of weight, water temperature is a function of time. Give three other examples of functions from everyday life that are described verbally. What can you say about the domain and range of each of your functions? If possible, sketch a rough graph of each function.

7–10 Determine whether the curve is the graph of a function of x. If it is, state the domain and range of the function.

7.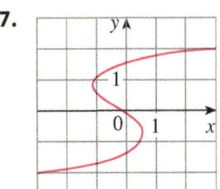

8.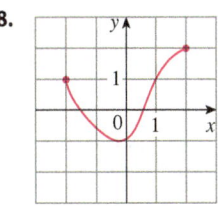

9.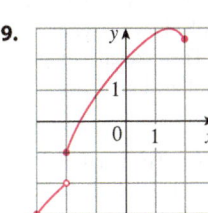

10.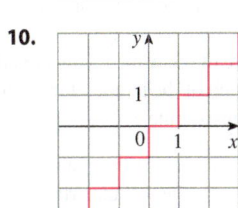

11. Shown is a graph of the global average temperature T during the 20th century. Estimate the following.
   (a) The global average temperature in 1950
   (b) The year when the average temperature was 14.2°C
   (c) The year when the temperature was smallest? Largest?
   (d) The range of T

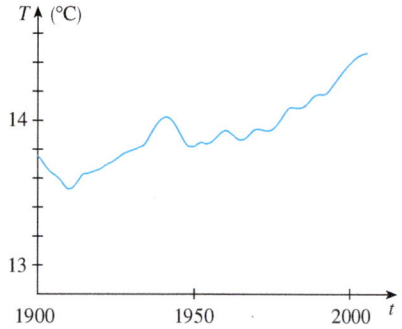

Source: Adapted from *Globe and Mail* [Toronto], 5 Dec. 2009. Print.

12. Trees grow faster and form wider rings in warm years and grow more slowly and form narrower rings in cooler years. The figure shows ring widths of a Siberian pine from 1500 to 2000.
   (a) What is the range of the ring width function?
   (b) What does the graph tend to say about the temperature of the earth? Does the graph reflect the volcanic eruptions of the mid-19th century?

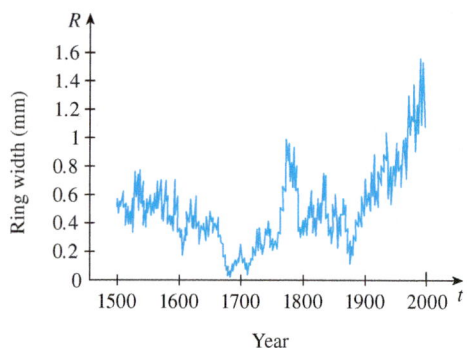

Source: Adapted from G. Jacoby et al., "Mongolian Tree Rings and 20th-Century Warming," *Science* 273 (1996): 771–73.

13. You put some ice cubes in a glass, fill the glass with cold water, and then let the glass sit on a table. Describe how the temperature of the water changes as time passes. Then sketch a rough graph of the temperature of the water as a function of the elapsed time.

14. Three runners compete in a 100-meter race. The graph depicts the distance run as a function of time for each runner. Describe in words what the graph tells you about this race. Who won the race? Did each runner finish the race?

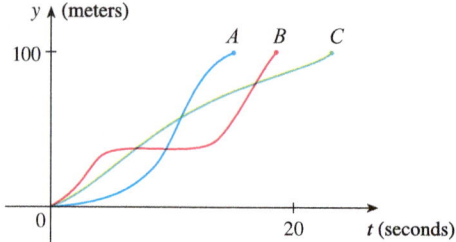

15. The graph shows the power consumption for a day in September in San Francisco. (P is measured in megawatts; t is measured in hours starting at midnight.)
   (a) What was the power consumption at 6 AM? At 6 PM?
   (b) When was the power consumption the lowest? When was it the highest? Do these times seem reasonable?

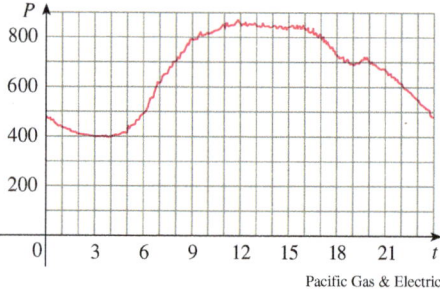

Pacific Gas & Electric

16. Sketch a rough graph of the number of hours of daylight as a function of the time of year.

17. Sketch a rough graph of the outdoor temperature as a function of time during a typical spring day.

18. Sketch a rough graph of the market value of a new car as a function of time for a period of 20 years. Assume the car is well maintained.

19. Sketch the graph of the amount of a particular brand of coffee sold by a store as a function of the price of the coffee.

20. You place a frozen pie in an oven and bake it for an hour. Then you take it out and let it cool before eating it. Describe how the temperature of the pie changes as time passes. Then sketch a rough graph of the temperature of the pie as a function of time.

21. A homeowner mows the lawn every Wednesday afternoon. Sketch a rough graph of the height of the grass as a function of time over the course of a four-week period.

22. An airplane takes off from an airport and lands an hour later at another airport, 400 miles away. If t represents the time in minutes since the plane has left the terminal building, let $x(t)$ be the horizontal distance traveled and $y(t)$ be the altitude of the plane.
   (a) Sketch a possible graph of $x(t)$.
   (b) Sketch a possible graph of $y(t)$.

(c) Sketch a possible graph of the ground speed.
(d) Sketch a possible graph of the vertical velocity.

23. Temperature readings T (in °F) were recorded every two hours from midnight to 2:00 PM in Atlanta on June 4, 2013. The time t was measured in hours from midnight.

t	0	2	4	6	8	10	12	14
T	74	69	68	66	70	78	82	86

(a) Use the readings to sketch a rough graph of T as a function of t.
(b) Use your graph to estimate the temperature at 9:00 AM.

24. Researchers measured the blood alcohol concentration (BAC) of eight adult male subjects after rapid consumption of 30 mL of ethanol (corresponding to two standard alcoholic drinks). The table shows the data they obtained by averaging the BAC (in mg/mL) of the eight men.
(a) Use the readings to sketch the graph of the BAC as a function of t.
(b) Use your graph to describe how the effect of alcohol varies with time.

t (hours)	BAC	t (hours)	BAC
0	0	1.75	0.22
0.2	0.25	2.0	0.18
0.5	0.41	2.25	0.15
0.75	0.40	2.5	0.12
1.0	0.33	3.0	0.07
1.25	0.29	3.5	0.03
1.5	0.24	4.0	0.01

Source: Adapted from P. Wilkinson et al., "Pharmacokinetics of Ethanol after Oral Administration in the Fasting State," *Journal of Pharmacokinetics and Biopharmaceutics* 5 (1977): 207–24.

25. If $f(x) = 3x^2 - x + 2$, find $f(2)$, $f(-2)$, $f(a)$, $f(-a)$, $f(a + 1)$, $2f(a)$, $f(2a)$, $f(a^2)$, $[f(a)]^2$, and $f(a + h)$.

26. A spherical balloon with radius r inches has volume $V(r) = \frac{4}{3}\pi r^3$. Find a function that represents the amount of air required to inflate the balloon from a radius of r inches to a radius of $r + 1$ inches.

27–30 Evaluate the difference quotient for the given function. Simplify your answer.

27. $f(x) = 4 + 3x - x^2$, $\dfrac{f(3 + h) - f(3)}{h}$

28. $f(x) = x^3$, $\dfrac{f(a + h) - f(a)}{h}$

29. $f(x) = \dfrac{1}{x}$, $\dfrac{f(x) - f(a)}{x - a}$

30. $f(x) = \dfrac{x + 3}{x + 1}$, $\dfrac{f(x) - f(1)}{x - 1}$

31–37 Find the domain of the function.

31. $f(x) = \dfrac{x + 4}{x^2 - 9}$

32. $f(x) = \dfrac{2x^3 - 5}{x^2 + x - 6}$

33. $f(t) = \sqrt[3]{2t - 1}$

34. $g(t) = \sqrt{3 - t} - \sqrt{2 + t}$

35. $h(x) = \dfrac{1}{\sqrt[4]{x^2 - 5x}}$

36. $f(u) = \dfrac{u + 1}{1 + \dfrac{1}{u + 1}}$

37. $F(p) = \sqrt{2 - \sqrt{p}}$

38. Find the domain and range and sketch the graph of the function $h(x) = \sqrt{4 - x^2}$.

39–40 Find the domain and sketch the graph of the function.

39. $f(x) = 1.6x - 2.4$

40. $g(t) = \dfrac{t^2 - 1}{t + 1}$

41–44 Evaluate $f(-3)$, $f(0)$, and $f(2)$ for the piecewise defined function. Then sketch the graph of the function.

41. $f(x) = \begin{cases} x + 2 & \text{if } x < 0 \\ 1 - x & \text{if } x \geq 0 \end{cases}$

42. $f(x) = \begin{cases} 3 - \frac{1}{2}x & \text{if } x < 2 \\ 2x - 5 & \text{if } x \geq 2 \end{cases}$

43. $f(x) = \begin{cases} x + 1 & \text{if } x \leq -1 \\ x^2 & \text{if } x > -1 \end{cases}$

44. $f(x) = \begin{cases} -1 & \text{if } x \leq 1 \\ 7 - 2x & \text{if } x > 1 \end{cases}$

45–50 Sketch the graph of the function.

45. $f(x) = x + |x|$

46. $f(x) = |x + 2|$

47. $g(t) = |1 - 3t|$

48. $h(t) = |t| + |t + 1|$

49. $f(x) = \begin{cases} |x| & \text{if } |x| \leq 1 \\ 1 & \text{if } |x| > 1 \end{cases}$

50. $g(x) = ||x| - 1|$

51–56 Find an expression for the function whose graph is the given curve.

51. The line segment joining the points $(1, -3)$ and $(5, 7)$

52. The line segment joining the points $(-5, 10)$ and $(7, -10)$

53. The bottom half of the parabola $x + (y - 1)^2 = 0$

54. The top half of the circle $x^2 + (y - 2)^2 = 4$

55. 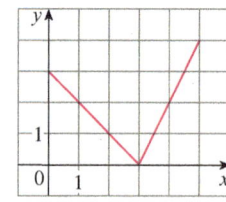 **56.**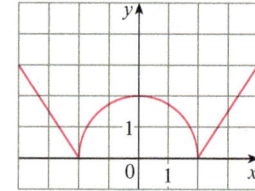

57–61 Find a formula for the described function and state its domain.

57. A rectangle has perimeter 20 m. Express the area of the rectangle as a function of the length of one of its sides.

58. A rectangle has area 16 m². Express the perimeter of the rectangle as a function of the length of one of its sides.

59. Express the area of an equilateral triangle as a function of the length of a side.

60. A closed rectangular box with volume 8 ft³ has length twice the width. Express the height of the box as a function of the width.

61. An open rectangular box with volume 2 m³ has a square base. Express the surface area of the box as a function of the length of a side of the base.

62. A Norman window has the shape of a rectangle surmounted by a semicircle. If the perimeter of the window is 30 ft, express the area A of the window as a function of the width x of the window.

63. A box with an open top is to be constructed from a rectangular piece of cardboard with dimensions 12 in. by 20 in. by cutting out equal squares of side x at each corner and then folding up the sides as in the figure. Express the volume V of the box as a function of x.

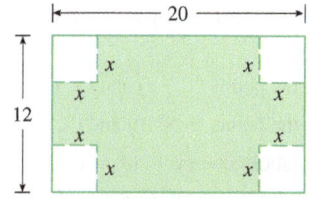 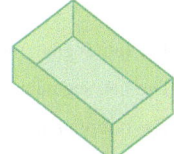

64. A cell phone plan has a basic charge of $35 a month. The plan includes 400 free minutes and charges 10 cents for each additional minute of usage. Write the monthly cost C as a function of the number x of minutes used and graph C as a function of x for $0 \leq x \leq 600$.

65. In a certain state the maximum speed permitted on freeways is 65 mi/h and the minimum speed is 40 mi/h. The fine for violating these limits is $15 for every mile per hour above the maximum speed or below the minimum speed. Express the amount of the fine F as a function of the driving speed x and graph $F(x)$ for $0 \leq x \leq 100$.

66. An electricity company charges its customers a base rate of $10 a month, plus 6 cents per kilowatt-hour (kWh) for the first 1200 kWh and 7 cents per kWh for all usage over 1200 kWh. Express the monthly cost E as a function of the amount x of electricity used. Then graph the function E for $0 \leq x \leq 2000$.

67. In a certain country, income tax is assessed as follows. There is no tax on income up to $10,000. Any income over $10,000 is taxed at a rate of 10%, up to an income of $20,000. Any income over $20,000 is taxed at 15%.
(a) Sketch the graph of the tax rate R as a function of the income I.
(b) How much tax is assessed on an income of $14,000? On $26,000?
(c) Sketch the graph of the total assessed tax T as a function of the income I.

68. The functions in Example 10 and Exercise 67 are called *step functions* because their graphs look like stairs. Give two other examples of step functions that arise in everyday life.

69–70 Graphs of f and g are shown. Decide whether each function is even, odd, or neither. Explain your reasoning.

69. 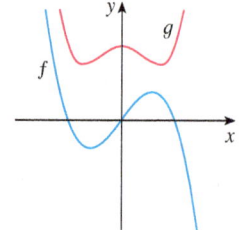 **70.**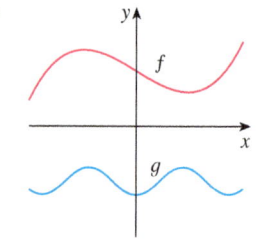

71. (a) If the point $(5, 3)$ is on the graph of an even function, what other point must also be on the graph?
(b) If the point $(5, 3)$ is on the graph of an odd function, what other point must also be on the graph?

72. A function f has domain $[-5, 5]$ and a portion of its graph is shown.
(a) Complete the graph of f if it is known that f is even.
(b) Complete the graph of f if it is known that f is odd.

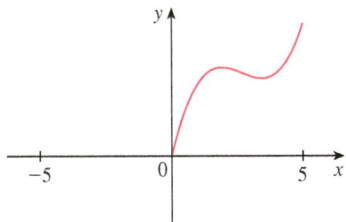

73–78 Determine whether f is even, odd, or neither. If you have a graphing calculator, use it to check your answer visually.

73. $f(x) = \dfrac{x}{x^2 + 1}$

74. $f(x) = \dfrac{x^2}{x^4 + 1}$

75. $f(x) = \dfrac{x}{x + 1}$

76. $f(x) = x|x|$

77. $f(x) = 1 + 3x^2 - x^4$

78. $f(x) = 1 + 3x^3 - x^5$

79. If f and g are both even functions, is $f + g$ even? If f and g are both odd functions, is $f + g$ odd? What if f is even and g is odd? Justify your answers.

80. If f and g are both even functions, is the product fg even? If f and g are both odd functions, is fg odd? What if f is even and g is odd? Justify your answers.

1.2 Mathematical Models: A Catalog of Essential Functions

A **mathematical model** is a mathematical description (often by means of a function or an equation) of a real-world phenomenon such as the size of a population, the demand for a product, the speed of a falling object, the concentration of a product in a chemical reaction, the life expectancy of a person at birth, or the cost of emission reductions. The purpose of the model is to understand the phenomenon and perhaps to make predictions about future behavior.

Figure 1 illustrates the process of mathematical modeling. Given a real-world problem, our first task is to formulate a mathematical model by identifying and naming the independent and dependent variables and making assumptions that simplify the phenomenon enough to make it mathematically tractable. We use our knowledge of the physical situation and our mathematical skills to obtain equations that relate the variables. In situations where there is no physical law to guide us, we may need to collect data (either from a library or the Internet or by conducting our own experiments) and examine the data in the form of a table in order to discern patterns. From this numerical representation of a function we may wish to obtain a graphical representation by plotting the data. The graph might even suggest a suitable algebraic formula in some cases.

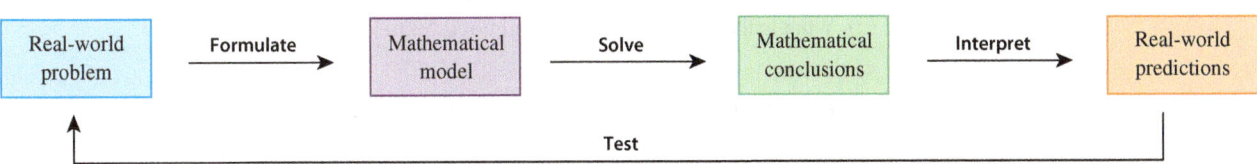

FIGURE 1
The modeling process

The second stage is to apply the mathematics that we know (such as the calculus that will be developed throughout this book) to the mathematical model that we have formulated in order to derive mathematical conclusions. Then, in the third stage, we take those mathematical conclusions and interpret them as information about the original real-world phenomenon by way of offering explanations or making predictions. The final step is to test our predictions by checking against new real data. If the predictions don't compare well with reality, we need to refine our model or to formulate a new model and start the cycle again.

A mathematical model is never a completely accurate representation of a physical situation—it is an *idealization*. A good model simplifies reality enough to permit mathematical calculations but is accurate enough to provide valuable conclusions. It is important to realize the limitations of the model. In the end, Mother Nature has the final say.

There are many different types of functions that can be used to model relationships observed in the real world. In what follows, we discuss the behavior and graphs of these functions and give examples of situations appropriately modeled by such functions.

■ Linear Models

The coordinate geometry of lines is reviewed in Appendix B.

When we say that y is a **linear function** of x, we mean that the graph of the function is a line, so we can use the slope-intercept form of the equation of a line to write a formula for the function as

$$y = f(x) = mx + b$$

where m is the slope of the line and b is the y-intercept.

A characteristic feature of linear functions is that they grow at a constant rate. For instance, Figure 2 shows a graph of the linear function $f(x) = 3x - 2$ and a table of sample values. Notice that whenever x increases by 0.1, the value of $f(x)$ increases by 0.3. So $f(x)$ increases three times as fast as x. Thus the slope of the graph $y = 3x - 2$, namely 3, can be interpreted as the rate of change of y with respect to x.

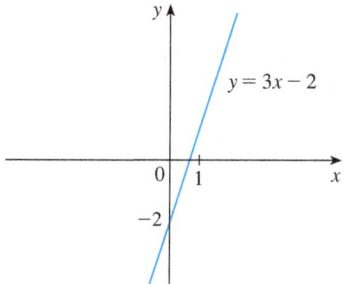

x	$f(x) = 3x - 2$
1.0	1.0
1.1	1.3
1.2	1.6
1.3	1.9
1.4	2.2
1.5	2.5

FIGURE 2

EXAMPLE 1

(a) As dry air moves upward, it expands and cools. If the ground temperature is 20°C and the temperature at a height of 1 km is 10°C, express the temperature T (in °C) as a function of the height h (in kilometers), assuming that a linear model is appropriate.
(b) Draw the graph of the function in part (a). What does the slope represent?
(c) What is the temperature at a height of 2.5 km?

SOLUTION

(a) Because we are assuming that T is a linear function of h, we can write

$$T = mh + b$$

We are given that $T = 20$ when $h = 0$, so

$$20 = m \cdot 0 + b = b$$

In other words, the y-intercept is $b = 20$.

We are also given that $T = 10$ when $h = 1$, so

$$10 = m \cdot 1 + 20$$

The slope of the line is therefore $m = 10 - 20 = -10$ and the required linear function is

$$T = -10h + 20$$

(b) The graph is sketched in Figure 3. The slope is $m = -10°C/km$, and this represents the rate of change of temperature with respect to height.

(c) At a height of $h = 2.5$ km, the temperature is

$$T = -10(2.5) + 20 = -5°C$$

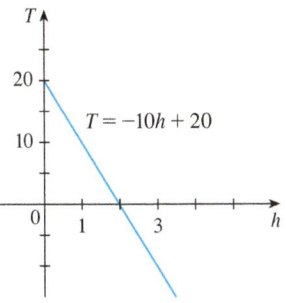

FIGURE 3

If there is no physical law or principle to help us formulate a model, we construct an **empirical model**, which is based entirely on collected data. We seek a curve that "fits" the data in the sense that it captures the basic trend of the data points.

EXAMPLE 2 Table 1 lists the average carbon dioxide level in the atmosphere, measured in parts per million at Mauna Loa Observatory from 1980 to 2012. Use the data in Table 1 to find a model for the carbon dioxide level.

SOLUTION We use the data in Table 1 to make the scatter plot in Figure 4, where t represents time (in years) and C represents the CO_2 level (in parts per million, ppm).

Table 1

Year	CO_2 level (in ppm)	Year	CO_2 level (in ppm)
1980	338.7	1998	366.5
1982	341.2	2000	369.4
1984	344.4	2002	373.2
1986	347.2	2004	377.5
1988	351.5	2006	381.9
1990	354.2	2008	385.6
1992	356.3	2010	389.9
1994	358.6	2012	393.8
1996	362.4		

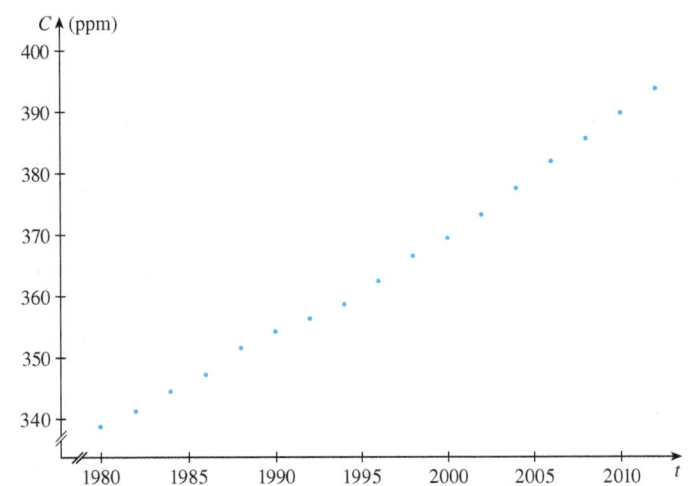

FIGURE 4 Scatter plot for the average CO_2 level

Notice that the data points appear to lie close to a straight line, so it's natural to choose a linear model in this case. But there are many possible lines that approximate these data points, so which one should we use? One possibility is the line that passes through the first and last data points. The slope of this line is

$$\frac{393.8 - 338.7}{2012 - 1980} = \frac{55.1}{32} = 1.721875 \approx 1.722$$

We write its equation as

$$C - 338.7 = 1.722(t - 1980)$$

or

$$\boxed{1} \qquad C = 1.722t - 3070.86$$

Equation 1 gives one possible linear model for the carbon dioxide level; it is graphed in Figure 5.

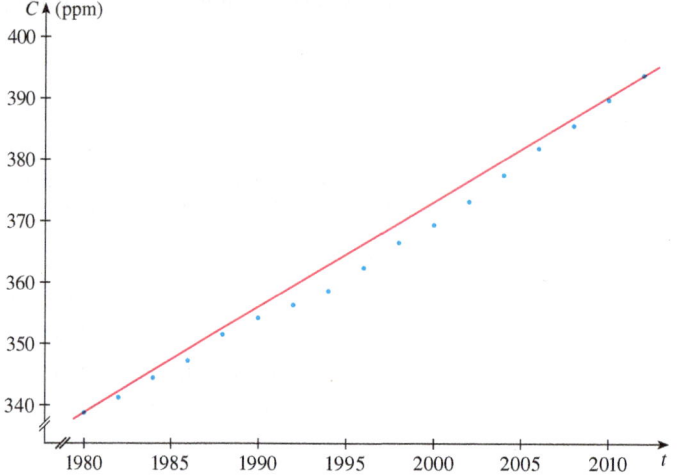

FIGURE 5
Linear model through first and last data points

A computer or graphing calculator finds the regression line by the method of **least squares**, which is to minimize the sum of the squares of the vertical distances between the data points and the line. The details are explained in Section 14.7.

Notice that our model gives values higher than most of the actual CO_2 levels. A better linear model is obtained by a procedure from statistics called *linear regression*. If we use a graphing calculator, we enter the data from Table 1 into the data editor and choose the linear regression command. (With Maple we use the fit[leastsquare] command in the stats package; with Mathematica we use the Fit command.) The machine gives the slope and y-intercept of the regression line as

$$m = 1.71262 \qquad b = -3054.14$$

So our least squares model for the CO_2 level is

$$\boxed{2} \qquad C = 1.71262t - 3054.14$$

In Figure 6 we graph the regression line as well as the data points. Comparing with Figure 5, we see that it gives a better fit than our previous linear model.

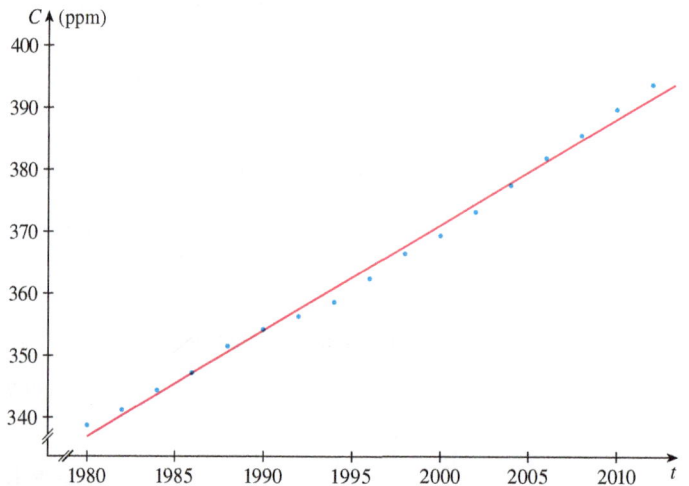

FIGURE 6
The regression line

EXAMPLE 3 Use the linear model given by Equation 2 to estimate the average CO_2 level for 1987 and to predict the level for the year 2020. According to this model, when will the CO_2 level exceed 420 parts per million?

SOLUTION Using Equation 2 with $t = 1987$, we estimate that the average CO_2 level in 1987 was

$$C(1987) = (1.71262)(1987) - 3054.14 \approx 348.84$$

This is an example of *interpolation* because we have estimated a value *between* observed values. (In fact, the Mauna Loa Observatory reported that the average CO_2 level in 1987 was 348.93 ppm, so our estimate is quite accurate.)

With $t = 2020$, we get

$$C(2020) = (1.71262)(2020) - 3054.14 \approx 405.35$$

So we predict that the average CO_2 level in the year 2020 will be 405.4 ppm. This is an example of *extrapolation* because we have predicted a value *outside* the time frame of observations. Consequently, we are far less certain about the accuracy of our prediction.

Using Equation 2, we see that the CO_2 level exceeds 420 ppm when

$$1.71262t - 3054.14 > 420$$

Solving this inequality, we get

$$t > \frac{3474.14}{1.71262} \approx 2028.55$$

We therefore predict that the CO_2 level will exceed 420 ppm by the year 2029. This prediction is risky because it involves a time quite remote from our observations. In fact, we see from Figure 6 that the trend has been for CO_2 levels to increase rather more rapidly in recent years, so the level might exceed 420 ppm well before 2029. ∎

■ Polynomials

A function P is called a **polynomial** if

$$P(x) = a_n x^n + a_{n-1} x^{n-1} + \cdots + a_2 x^2 + a_1 x + a_0$$

where n is a nonnegative integer and the numbers $a_0, a_1, a_2, \ldots, a_n$ are constants called the **coefficients** of the polynomial. The domain of any polynomial is $\mathbb{R} = (-\infty, \infty)$. If the leading coefficient $a_n \neq 0$, then the **degree** of the polynomial is n. For example, the function

$$P(x) = 2x^6 - x^4 + \tfrac{2}{5}x^3 + \sqrt{2}$$

is a polynomial of degree 6.

A polynomial of degree 1 is of the form $P(x) = mx + b$ and so it is a linear function. A polynomial of degree 2 is of the form $P(x) = ax^2 + bx + c$ and is called a **quadratic function**. Its graph is always a parabola obtained by shifting the parabola $y = ax^2$, as we will see in the next section. The parabola opens upward if $a > 0$ and downward if $a < 0$. (See Figure 7.)

A polynomial of degree 3 is of the form

$$P(x) = ax^3 + bx^2 + cx + d \qquad a \neq 0$$

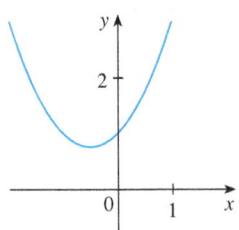

(a) $y = x^2 + x + 1$

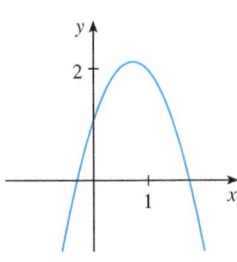

(b) $y = -2x^2 + 3x + 1$

FIGURE 7
The graphs of quadratic functions are parabolas.

and is called a **cubic function**. Figure 8 shows the graph of a cubic function in part (a) and graphs of polynomials of degrees 4 and 5 in parts (b) and (c). We will see later why the graphs have these shapes.

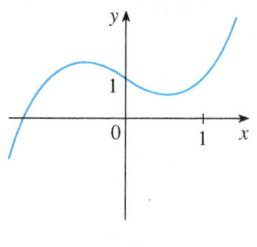
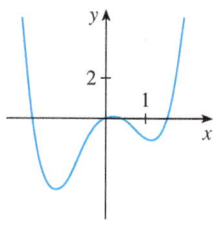
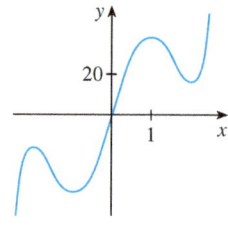

FIGURE 8

(a) $y = x^3 - x + 1$ (b) $y = x^4 - 3x^2 + x$ (c) $y = 3x^5 - 25x^3 + 60x$

Polynomials are commonly used to model various quantities that occur in the natural and social sciences. For instance, in Section 3.7 we will explain why economists often use a polynomial $P(x)$ to represent the cost of producing x units of a commodity. In the following example we use a quadratic function to model the fall of a ball.

Table 2

Time (seconds)	Height (meters)
0	450
1	445
2	431
3	408
4	375
5	332
6	279
7	216
8	143
9	61

EXAMPLE 4 A ball is dropped from the upper observation deck of the CN Tower, 450 m above the ground, and its height h above the ground is recorded at 1-second intervals in Table 2. Find a model to fit the data and use the model to predict the time at which the ball hits the ground.

SOLUTION We draw a scatter plot of the data in Figure 9 and observe that a linear model is inappropriate. But it looks as if the data points might lie on a parabola, so we try a quadratic model instead. Using a graphing calculator or computer algebra system (which uses the least squares method), we obtain the following quadratic model:

$$\boxed{3} \qquad h = 449.36 + 0.96t - 4.90t^2$$

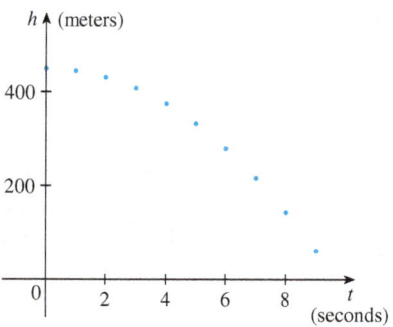

FIGURE 9
Scatter plot for a falling ball

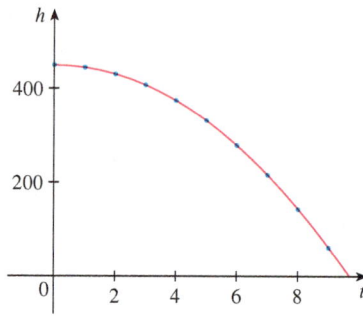

FIGURE 10
Quadratic model for a falling ball

In Figure 10 we plot the graph of Equation 3 together with the data points and see that the quadratic model gives a very good fit.

The ball hits the ground when $h = 0$, so we solve the quadratic equation

$$-4.90t^2 + 0.96t + 449.36 = 0$$

The quadratic formula gives

$$t = \frac{-0.96 \pm \sqrt{(0.96)^2 - 4(-4.90)(449.36)}}{2(-4.90)}$$

The positive root is $t \approx 9.67$, so we predict that the ball will hit the ground after about 9.7 seconds. ∎

Power Functions

A function of the form $f(x) = x^a$, where a is a constant, is called a **power function**. We consider several cases.

(i) $a = n$, **where n is a positive integer**

The graphs of $f(x) = x^n$ for $n = 1, 2, 3, 4,$ and 5 are shown in Figure 11. (These are polynomials with only one term.) We already know the shape of the graphs of $y = x$ (a line through the origin with slope 1) and $y = x^2$ [a parabola, see Example 1.1.2(b)].

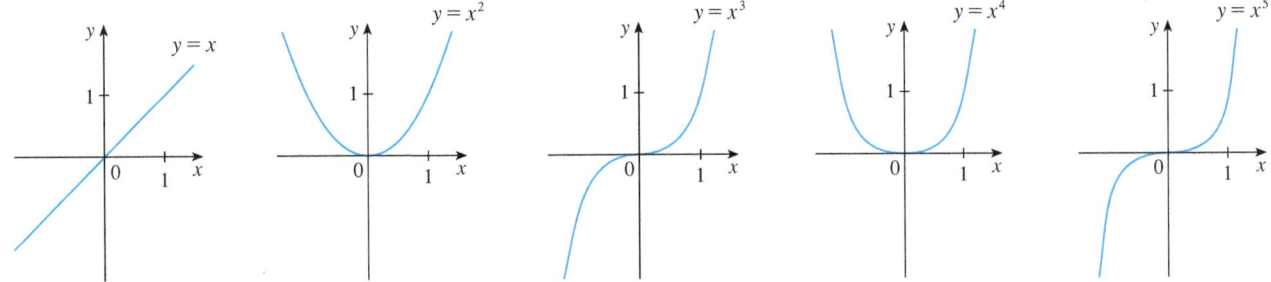

FIGURE 11 Graphs of $f(x) = x^n$ for $n = 1, 2, 3, 4, 5$

The general shape of the graph of $f(x) = x^n$ depends on whether n is even or odd. If n is even, then $f(x) = x^n$ is an even function and its graph is similar to the parabola $y = x^2$. If n is odd, then $f(x) = x^n$ is an odd function and its graph is similar to that of $y = x^3$. Notice from Figure 12, however, that as n increases, the graph of $y = x^n$ becomes flatter near 0 and steeper when $|x| \geq 1$. (If x is small, then x^2 is smaller, x^3 is even smaller, x^4 is smaller still, and so on.)

A **family of functions** is a collection of functions whose equations are related. Figure 12 shows two families of power functions, one with even powers and one with odd powers.

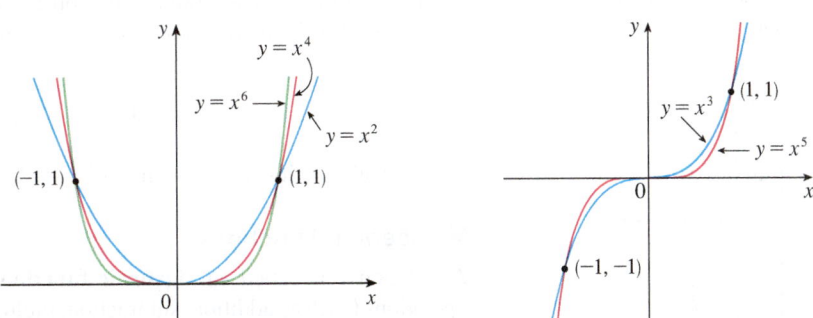

FIGURE 12

(ii) $a = 1/n$, **where n is a positive integer**

The function $f(x) = x^{1/n} = \sqrt[n]{x}$ is a **root function**. For $n = 2$ it is the square root function $f(x) = \sqrt{x}$, whose domain is $[0, \infty)$ and whose graph is the upper half of the parabola $x = y^2$. [See Figure 13(a).] For other even values of n, the graph of $y = \sqrt[n]{x}$ is similar to that of $y = \sqrt{x}$. For $n = 3$ we have the cube root function $f(x) = \sqrt[3]{x}$ whose domain is $\mathbb{R}$ (recall that every real number has a cube root) and whose graph is shown in Figure 13(b). The graph of $y = \sqrt[n]{x}$ for n odd ($n > 3$) is similar to that of $y = \sqrt[3]{x}$.

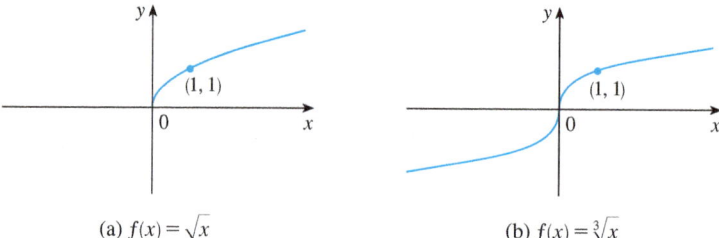

FIGURE 13
Graphs of root functions

(a) $f(x) = \sqrt{x}$ (b) $f(x) = \sqrt[3]{x}$

(iii) $a = -1$

The graph of the **reciprocal function** $f(x) = x^{-1} = 1/x$ is shown in Figure 14. Its graph has the equation $y = 1/x$, or $xy = 1$, and is a hyperbola with the coordinate axes as its asymptotes. This function arises in physics and chemistry in connection with Boyle's Law, which says that, when the temperature is constant, the volume V of a gas is inversely proportional to the pressure P:

$$V = \frac{C}{P}$$

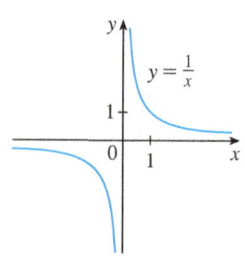

FIGURE 14
The reciprocal function

where C is a constant. Thus the graph of V as a function of P (see Figure 15) has the same general shape as the right half of Figure 14.

Power functions are also used to model species-area relationships (Exercises 30–31), illumination as a function of distance from a light source (Exercise 29), and the period of revolution of a planet as a function of its distance from the sun (Exercise 32).

■ **Rational Functions**

A **rational function** f is a ratio of two polynomials:

$$f(x) = \frac{P(x)}{Q(x)}$$

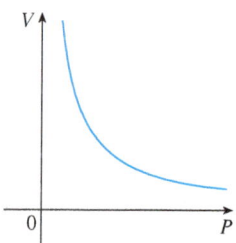

FIGURE 15
Volume as a function of pressure at constant temperature

where P and Q are polynomials. The domain consists of all values of x such that $Q(x) \neq 0$. A simple example of a rational function is the function $f(x) = 1/x$, whose domain is $\{x \mid x \neq 0\}$; this is the reciprocal function graphed in Figure 14. The function

$$f(x) = \frac{2x^4 - x^2 + 1}{x^2 - 4}$$

is a rational function with domain $\{x \mid x \neq \pm 2\}$. Its graph is shown in Figure 16.

■ **Algebraic Functions**

A function f is called an **algebraic function** if it can be constructed using algebraic operations (such as addition, subtraction, multiplication, division, and taking roots) starting with polynomials. Any rational function is automatically an algebraic function. Here are two more examples:

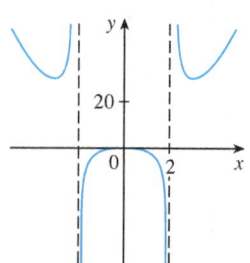

FIGURE 16
$f(x) = \dfrac{2x^4 - x^2 + 1}{x^2 - 4}$

$$f(x) = \sqrt{x^2 + 1} \qquad g(x) = \frac{x^4 - 16x^2}{x + \sqrt{x}} + (x - 2)\sqrt[3]{x + 1}$$

When we sketch algebraic functions in Chapter 4, we will see that their graphs can assume a variety of shapes. Figure 17 illustrates some of the possibilities.

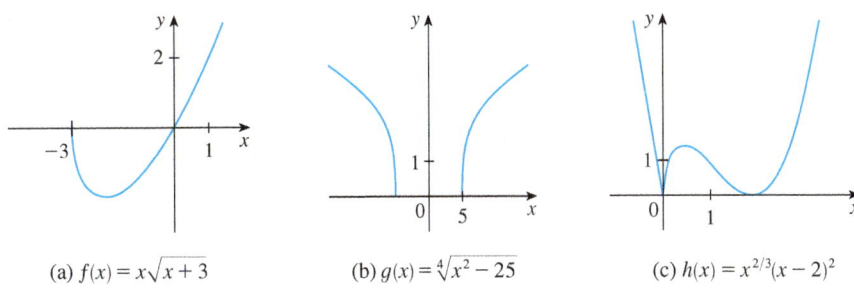

FIGURE 17 (a) $f(x) = x\sqrt{x+3}$ (b) $g(x) = \sqrt[4]{x^2 - 25}$ (c) $h(x) = x^{2/3}(x-2)^2$

An example of an algebraic function occurs in the theory of relativity. The mass of a particle with velocity v is

$$m = f(v) = \frac{m_0}{\sqrt{1 - v^2/c^2}}$$

where m_0 is the rest mass of the particle and $c = 3.0 \times 10^5$ km/s is the speed of light in a vacuum.

■ Trigonometric Functions

The Reference Pages are located at the back of the book.

Trigonometry and the trigonometric functions are reviewed on Reference Page 2 and also in Appendix D. In calculus the convention is that radian measure is always used (except when otherwise indicated). For example, when we use the function $f(x) = \sin x$, it is understood that $\sin x$ means the sine of the angle whose radian measure is x. Thus the graphs of the sine and cosine functions are as shown in Figure 18.

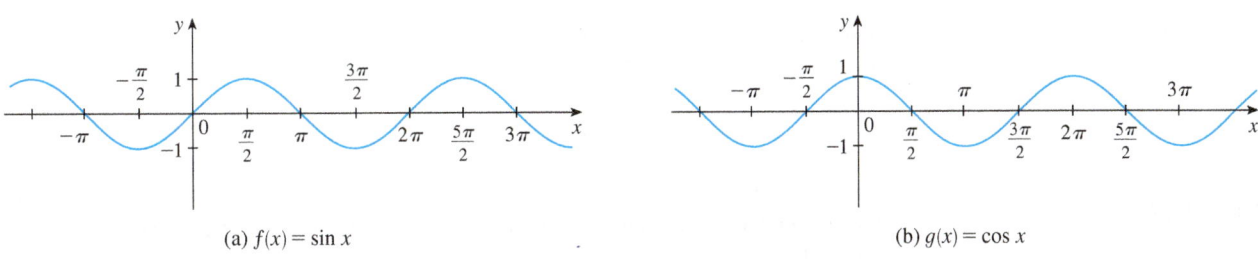

(a) $f(x) = \sin x$ (b) $g(x) = \cos x$

FIGURE 18

Notice that for both the sine and cosine functions the domain is $(-\infty, \infty)$ and the range is the closed interval $[-1, 1]$. Thus, for all values of x, we have

$$-1 \leq \sin x \leq 1 \qquad -1 \leq \cos x \leq 1$$

or, in terms of absolute values,

$$|\sin x| \leq 1 \qquad |\cos x| \leq 1$$

Also, the zeros of the sine function occur at the integer multiples of π; that is,

$$\sin x = 0 \quad \text{when} \quad x = n\pi \quad n \text{ an integer}$$

An important property of the sine and cosine functions is that they are periodic functions and have period 2π. This means that, for all values of x,

$$\sin(x + 2\pi) = \sin x \qquad \cos(x + 2\pi) = \cos x$$

The periodic nature of these functions makes them suitable for modeling repetitive phenomena such as tides, vibrating springs, and sound waves. For instance, in Example 1.3.4 we will see that a reasonable model for the number of hours of daylight in Philadelphia t days after January 1 is given by the function

$$L(t) = 12 + 2.8 \sin\left[\frac{2\pi}{365}(t - 80)\right]$$

EXAMPLE 5 What is the domain of the function $f(x) = \dfrac{1}{1 - 2\cos x}$?

SOLUTION This function is defined for all values of x except for those that make the denominator 0. But

$$1 - 2\cos x = 0 \iff \cos x = \frac{1}{2} \iff x = \frac{\pi}{3} + 2n\pi \text{ or } x = \frac{5\pi}{3} + 2n\pi$$

where n is any integer (because the cosine function has period 2π). So the domain of f is the set of all real numbers except for the ones noted above. ∎

The tangent function is related to the sine and cosine functions by the equation

$$\tan x = \frac{\sin x}{\cos x}$$

and its graph is shown in Figure 19. It is undefined whenever $\cos x = 0$, that is, when $x = \pm\pi/2, \pm 3\pi/2, \ldots$. Its range is $(-\infty, \infty)$. Notice that the tangent function has period π:

$$\tan(x + \pi) = \tan x \qquad \text{for all } x$$

The remaining three trigonometric functions (cosecant, secant, and cotangent) are the reciprocals of the sine, cosine, and tangent functions. Their graphs are shown in Appendix D.

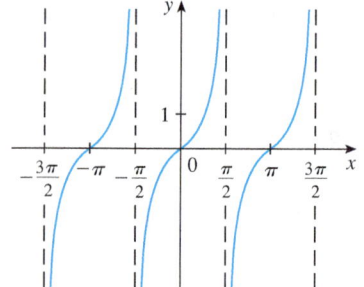

FIGURE 19
$y = \tan x$

Exponential Functions

The **exponential function**s are the functions of the form $f(x) = b^x$, where the base b is a positive constant. The graphs of $y = 2^x$ and $y = (0.5)^x$ are shown in Figure 20. In both cases the domain is $(-\infty, \infty)$ and the range is $(0, \infty)$.

Exponential functions will be studied in detail in Section 1.4, and we will see that they are useful for modeling many natural phenomena, such as population growth (if $b > 1$) and radioactive decay (if $b < 1$).

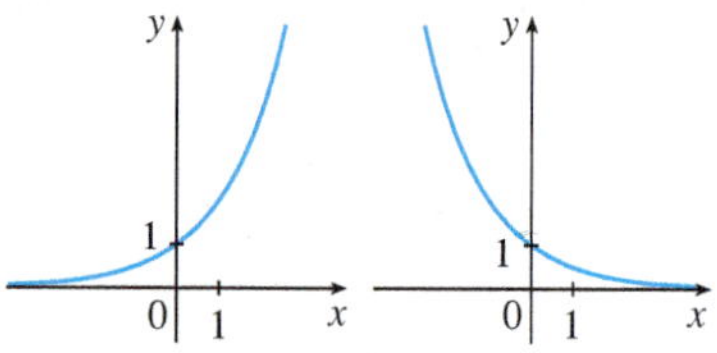

(a) $y = 2^x$ (b) $y = (0.5)^x$

FIGURE 20

Logarithmic Functions

The **logarithmic functions** $f(x) = \log_b x$, where the base b is a positive constant, are the inverse functions of the exponential functions. They will be studied in Section 1.5. Figure

SECTION 1.2 Mathematical Models: A Catalog of Essential Functions

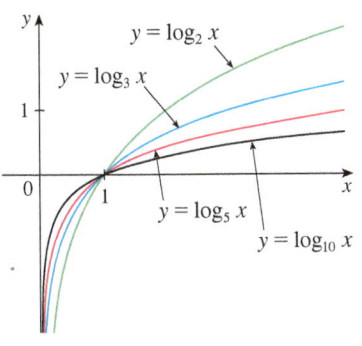

FIGURE 21

21 shows the graphs of four logarithmic functions with various bases. In each case the domain is $(0, \infty)$, the range is $(-\infty, \infty)$, and the function increases slowly when $x > 1$.

EXAMPLE 6 Classify the following functions as one of the types of functions that we have discussed.

(a) $f(x) = 5^x$
(b) $g(x) = x^5$
(c) $h(x) = \dfrac{1 + x}{1 - \sqrt{x}}$
(d) $u(t) = 1 - t + 5t^4$

SOLUTION

(a) $f(x) = 5^x$ is an exponential function. (The x is the exponent.)

(b) $g(x) = x^5$ is a power function. (The x is the base.) We could also consider it to be a polynomial of degree 5.

(c) $h(x) = \dfrac{1 + x}{1 - \sqrt{x}}$ is an algebraic function.

(d) $u(t) = 1 - t + 5t^4$ is a polynomial of degree 4. ∎

1.2 EXERCISES

1–2 Classify each function as a power function, root function, polynomial (state its degree), rational function, algebraic function, trigonometric function, exponential function, or logarithmic function.

1. (a) $f(x) = \log_2 x$
(b) $g(x) = \sqrt[4]{x}$
(c) $h(x) = \dfrac{2x^3}{1 - x^2}$
(d) $u(t) = 1 - 1.1t + 2.54t^2$
(e) $v(t) = 5^t$
(f) $w(\theta) = \sin \theta \, \cos^2 \theta$

2. (a) $y = \pi^x$
(b) $y = x^\pi$
(c) $y = x^2(2 - x^3)$
(d) $y = \tan t - \cos t$
(e) $y = \dfrac{s}{1 + s}$
(f) $y = \dfrac{\sqrt{x^3 - 1}}{1 + \sqrt[3]{x}}$

3–4 Match each equation with its graph. Explain your choices. (Don't use a computer or graphing calculator.)

3. (a) $y = x^2$ (b) $y = x^5$ (c) $y = x^8$

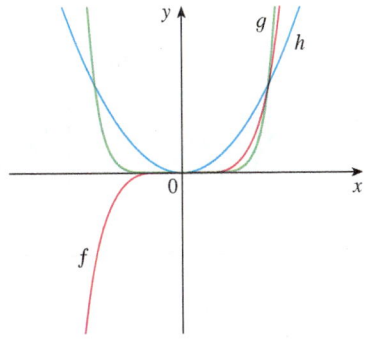

4. (a) $y = 3x$ (b) $y = 3^x$ (c) $y = x^3$ (d) $y = \sqrt[3]{x}$

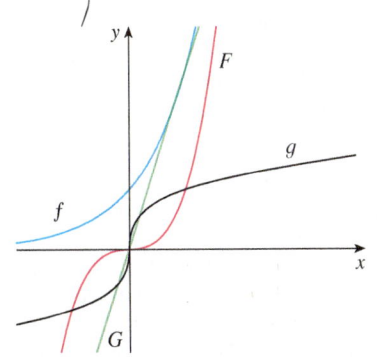

5–6 Find the domain of the function.

5. $f(x) = \dfrac{\cos x}{1 - \sin x}$
6. $g(x) = \dfrac{1}{1 - \tan x}$

7. (a) Find an equation for the family of linear functions with slope 2 and sketch several members of the family.
(b) Find an equation for the family of linear functions such that $f(2) = 1$ and sketch several members of the family.
(c) Which function belongs to both families?

8. What do all members of the family of linear functions $f(x) = 1 + m(x + 3)$ have in common? Sketch several members of the family.

9. What do all members of the family of linear functions $f(x) = c - x$ have in common? Sketch several members of the family.

10. Find expressions for the quadratic functions whose graphs are shown.

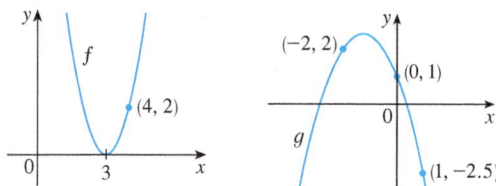

11. Find an expression for a cubic function f if $f(1) = 6$ and $f(-1) = f(0) = f(2) = 0$.

12. Recent studies indicate that the average surface temperature of the earth has been rising steadily. Some scientists have modeled the temperature by the linear function $T = 0.02t + 8.50$, where T is temperature in °C and t represents years since 1900.
  (a) What do the slope and T-intercept represent?
  (b) Use the equation to predict the average global surface temperature in 2100.

13. If the recommended adult dosage for a drug is D (in mg), then to determine the appropriate dosage c for a child of age a, pharmacists use the equation $c = 0.0417D(a + 1)$. Suppose the dosage for an adult is 200 mg.
  (a) Find the slope of the graph of c. What does it represent?
  (b) What is the dosage for a newborn?

14. The manager of a weekend flea market knows from past experience that if he charges x dollars for a rental space at the market, then the number y of spaces he can rent is given by the equation $y = 200 - 4x$.
  (a) Sketch a graph of this linear function. (Remember that the rental charge per space and the number of spaces rented can't be negative quantities.)
  (b) What do the slope, the y-intercept, and the x-intercept of the graph represent?

15. The relationship between the Fahrenheit (F) and Celsius (C) temperature scales is given by the linear function $F = \frac{9}{5}C + 32$.
  (a) Sketch a graph of this function.
  (b) What is the slope of the graph and what does it represent? What is the F-intercept and what does it represent?

16. Jason leaves Detroit at 2:00 PM and drives at a constant speed west along I-94. He passes Ann Arbor, 40 mi from Detroit, at 2:50 PM.
  (a) Express the distance traveled in terms of the time elapsed.
  (b) Draw the graph of the equation in part (a).
  (c) What is the slope of this line? What does it represent?

17. Biologists have noticed that the chirping rate of crickets of a certain species is related to temperature, and the relationship appears to be very nearly linear. A cricket produces 113 chirps per minute at 70°F and 173 chirps per minute at 80°F.
  (a) Find a linear equation that models the temperature T as a function of the number of chirps per minute N.
  (b) What is the slope of the graph? What does it represent?
  (c) If the crickets are chirping at 150 chirps per minute, estimate the temperature.

18. The manager of a furniture factory finds that it costs $2200 to manufacture 100 chairs in one day and $4800 to produce 300 chairs in one day.
  (a) Express the cost as a function of the number of chairs produced, assuming that it is linear. Then sketch the graph.
  (b) What is the slope of the graph and what does it represent?
  (c) What is the y-intercept of the graph and what does it represent?

19. At the surface of the ocean, the water pressure is the same as the air pressure above the water, 15 lb/in². Below the surface, the water pressure increases by 4.34 lb/in² for every 10 ft of descent.
  (a) Express the water pressure as a function of the depth below the ocean surface.
  (b) At what depth is the pressure 100 lb/in²?

20. The monthly cost of driving a car depends on the number of miles driven. Lynn found that in May it cost her $380 to drive 480 mi and in June it cost her $460 to drive 800 mi.
  (a) Express the monthly cost C as a function of the distance driven d, assuming that a linear relationship gives a suitable model.
  (b) Use part (a) to predict the cost of driving 1500 miles per month.
  (c) Draw the graph of the linear function. What does the slope represent?
  (d) What does the C-intercept represent?
  (e) Why does a linear function give a suitable model in this situation?

21–22 For each scatter plot, decide what type of function you might choose as a model for the data. Explain your choices.

21.

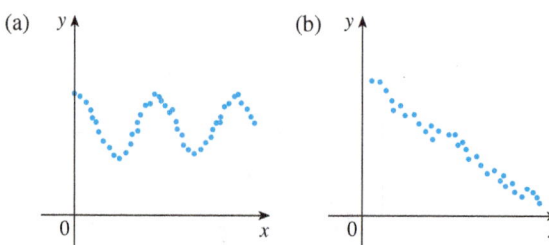

22.

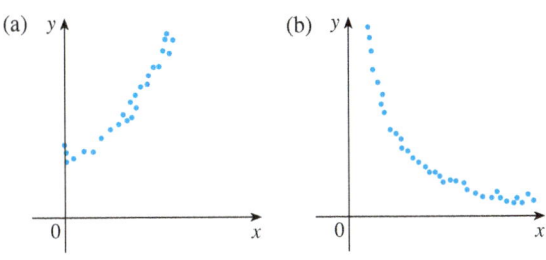

23. The table shows (lifetime) peptic ulcer rates (per 100 population) for various family incomes as reported by the National Health Interview Survey.

Income	Ulcer rate (per 100 population)
$4,000	14.1
$6,000	13.0
$8,000	13.4
$12,000	12.5
$16,000	12.0
$20,000	12.4
$30,000	10.5
$45,000	9.4
$60,000	8.2

(a) Make a scatter plot of these data and decide whether a linear model is appropriate.
(b) Find and graph a linear model using the first and last data points.
(c) Find and graph the least squares regression line.
(d) Use the linear model in part (c) to estimate the ulcer rate for an income of $25,000.
(e) According to the model, how likely is someone with an income of $80,000 to suffer from peptic ulcers?
(f) Do you think it would be reasonable to apply the model to someone with an income of $200,000?

24. Biologists have observed that the chirping rate of crickets of a certain species appears to be related to temperature. The table shows the chirping rates for various temperatures.
(a) Make a scatter plot of the data.
(b) Find and graph the regression line.
(c) Use the linear model in part (b) to estimate the chirping rate at 100°F.

Temperature (°F)	Chirping rate (chirps/min)	Temperature (°F)	Chirping rate (chirps/min)
50	20	75	140
55	46	80	173
60	79	85	198
65	91	90	211
70	113		

25. Anthropologists use a linear model that relates human femur (thighbone) length to height. The model allows an anthropologist to determine the height of an individual when only a partial skeleton (including the femur) is found. Here we find the model by analyzing the data on femur length and height for the eight males given in the following table.
(a) Make a scatter plot of the data.
(b) Find and graph the regression line that models the data.
(c) An anthropologist finds a human femur of length 53 cm. How tall was the person?

Femur length (cm)	Height (cm)	Femur length (cm)	Height (cm)
50.1	178.5	44.5	168.3
48.3	173.6	42.7	165.0
45.2	164.8	39.5	155.4
44.7	163.7	38.0	155.8

26. When laboratory rats are exposed to asbestos fibers, some of them develop lung tumors. The table lists the results of several experiments by different scientists.
(a) Find the regression line for the data.
(b) Make a scatter plot and graph the regression line. Does the regression line appear to be a suitable model for the data?
(c) What does the y-intercept of the regression line represent?

Asbestos exposure (fibers/mL)	Percent of mice that develop lung tumors	Asbestos exposure (fibers/mL)	Percent of mice that develop lung tumors
50	2	1600	42
400	6	1800	37
500	5	2000	38
900	10	3000	50
1100	26		

27. The table shows world average daily oil consumption from 1985 to 2010 measured in thousands of barrels per day.
(a) Make a scatter plot and decide whether a linear model is appropriate.
(b) Find and graph the regression line.
(c) Use the linear model to estimate the oil consumption in 2002 and 2012.

Years since 1985	Thousands of barrels of oil per day
0	60,083
5	66,533
10	70,099
15	76,784
20	84,077
25	87,302

Source: US Energy Information Administration

28. The table shows average US retail residential prices of electricity from 2000 to 2012, measured in cents per kilowatt hour.
   (a) Make a scatter plot. Is a linear model appropriate?
   (b) Find and graph the regression line.
   (c) Use your linear model from part (b) to estimate the average retail price of electricity in 2005 and 2013.

Years since 2000	Cents/kWh
0	8.24
2	8.44
4	8.95
6	10.40
8	11.26
10	11.54
12	11.58

Source: US Energy Information Administration

29. Many physical quantities are connected by *inverse square laws*, that is, by power functions of the form $f(x) = kx^{-2}$. In particular, the illumination of an object by a light source is inversely proportional to the square of the distance from the source. Suppose that after dark you are in a room with just one lamp and you are trying to read a book. The light is too dim and so you move halfway to the lamp. How much brighter is the light?

30. It makes sense that the larger the area of a region, the larger the number of species that inhabit the region. Many ecologists have modeled the species-area relation with a power function and, in particular, the number of species S of bats living in caves in central Mexico has been related to the surface area A of the caves by the equation $S = 0.7A^{0.3}$.
   (a) The cave called *Misión Imposible* near Puebla, Mexico, has a surface area of $A = 60$ m². How many species of bats would you expect to find in that cave?
   (b) If you discover that four species of bats live in a cave, estimate the area of the cave.

31. The table shows the number N of species of reptiles and amphibians inhabiting Caribbean islands and the area A of the island in square miles.
   (a) Use a power function to model N as a function of A.
   (b) The Caribbean island of Dominica has area 291 mi². How many species of reptiles and amphibians would you expect to find on Dominica?

Island	A	N
Saba	4	5
Monserrat	40	9
Puerto Rico	3,459	40
Jamaica	4,411	39
Hispaniola	29,418	84
Cuba	44,218	76

32. The table shows the mean (average) distances d of the planets from the sun (taking the unit of measurement to be the distance from the earth to the sun) and their periods T (time of revolution in years).
   (a) Fit a power model to the data.
   (b) Kepler's Third Law of Planetary Motion states that "The square of the period of revolution of a planet is proportional to the cube of its mean distance from the sun."
   Does your model corroborate Kepler's Third Law?

Planet	d	T
Mercury	0.387	0.241
Venus	0.723	0.615
Earth	1.000	1.000
Mars	1.523	1.881
Jupiter	5.203	11.861
Saturn	9.541	29.457
Uranus	19.190	84.008
Neptune	30.086	164.784

1.3 New Functions from Old Functions

In this section we start with the basic functions we discussed in Section 1.2 and obtain new functions by shifting, stretching, and reflecting their graphs. We also show how to combine pairs of functions by the standard arithmetic operations and by composition.

■ Transformations of Functions

By applying certain transformations to the graph of a given function we can obtain the graphs of related functions. This will give us the ability to sketch the graphs of many functions quickly by hand. It will also enable us to write equations for given graphs.

Let's first consider **translations**. If c is a positive number, then the graph of $y = f(x) + c$ is just the graph of $y = f(x)$ shifted upward a distance of c units (because each y-coordinate is increased by the same number c). Likewise, if $g(x) = f(x - c)$, where $c > 0$, then the value of g at x is the same as the value of f at $x - c$ (c units to the left of x). Therefore the graph of $y = f(x - c)$ is just the graph of $y = f(x)$ shifted c units to the right (see Figure 1).

> **Vertical and Horizontal Shifts** Suppose $c > 0$. To obtain the graph of
>
> $y = f(x) + c$, shift the graph of $y = f(x)$ a distance c units upward
> $y = f(x) - c$, shift the graph of $y = f(x)$ a distance c units downward
> $y = f(x - c)$, shift the graph of $y = f(x)$ a distance c units to the right
> $y = f(x + c)$, shift the graph of $y = f(x)$ a distance c units to the left

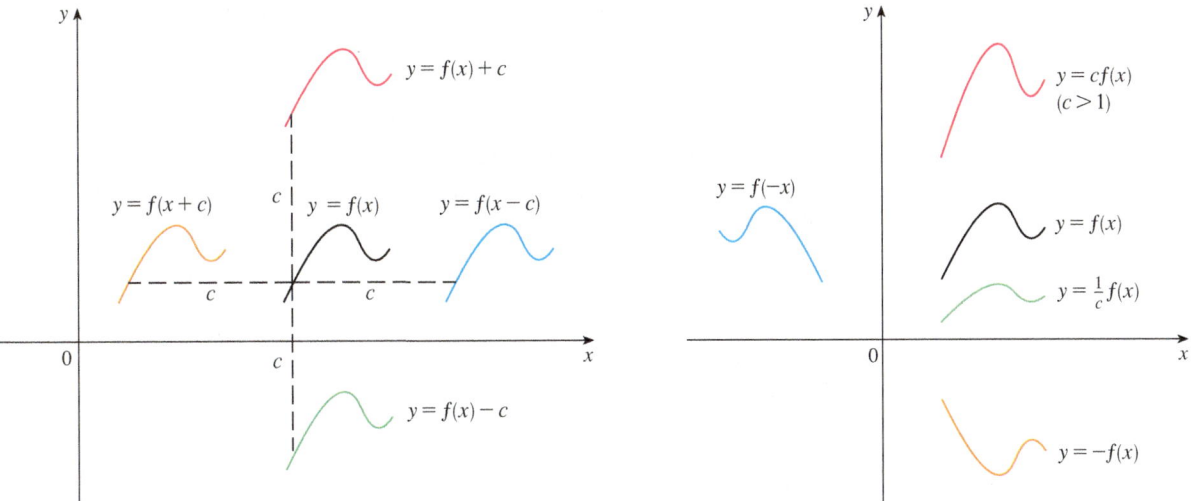

FIGURE 1 Translating the graph of f

FIGURE 2 Stretching and reflecting the graph of f

Now let's consider the **stretching** and **reflecting** transformations. If $c > 1$, then the graph of $y = cf(x)$ is the graph of $y = f(x)$ stretched by a factor of c in the vertical direction (because each y-coordinate is multiplied by the same number c). The graph of $y = -f(x)$ is the graph of $y = f(x)$ reflected about the x-axis because the point (x, y) is replaced by the point $(x, -y)$. (See Figure 2 and the following chart, where the results of other stretching, shrinking, and reflecting transformations are also given.)

> **Vertical and Horizontal Stretching and Reflecting** Suppose $c > 1$. To obtain the graph of
>
> $y = cf(x)$, stretch the graph of $y = f(x)$ vertically by a factor of c
> $y = (1/c)f(x)$, shrink the graph of $y = f(x)$ vertically by a factor of c
> $y = f(cx)$, shrink the graph of $y = f(x)$ horizontally by a factor of c
> $y = f(x/c)$, stretch the graph of $y = f(x)$ horizontally by a factor of c
> $y = -f(x)$, reflect the graph of $y = f(x)$ about the x-axis
> $y = f(-x)$, reflect the graph of $y = f(x)$ about the y-axis

Figure 3 illustrates these stretching transformations when applied to the cosine function with $c = 2$. For instance, in order to get the graph of $y = 2 \cos x$ we multiply the y-coordinate of each point on the graph of $y = \cos x$ by 2. This means that the graph of $y = \cos x$ gets stretched vertically by a factor of 2.

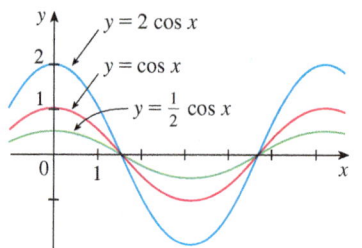
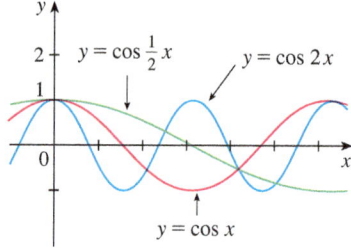

FIGURE 3

EXAMPLE 1 Given the graph of $y = \sqrt{x}$, use transformations to graph $y = \sqrt{x} - 2$, $y = \sqrt{x - 2}$, $y = -\sqrt{x}$, $y = 2\sqrt{x}$, and $y = \sqrt{-x}$.

SOLUTION The graph of the square root function $y = \sqrt{x}$, obtained from Figure 1.2.13(a), is shown in Figure 4(a). In the other parts of the figure we sketch $y = \sqrt{x} - 2$ by shifting 2 units downward, $y = \sqrt{x - 2}$ by shifting 2 units to the right, $y = -\sqrt{x}$ by reflecting about the x-axis, $y = 2\sqrt{x}$ by stretching vertically by a factor of 2, and $y = \sqrt{-x}$ by reflecting about the y-axis.

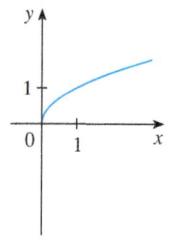
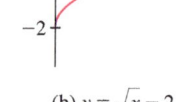
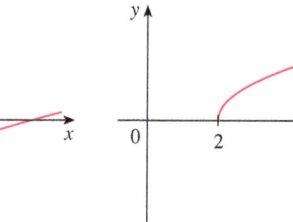
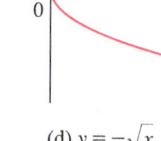
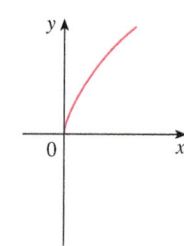
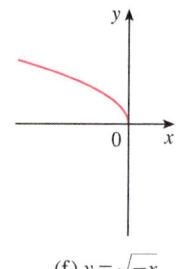

(a) $y = \sqrt{x}$ (b) $y = \sqrt{x} - 2$ (c) $y = \sqrt{x - 2}$ (d) $y = -\sqrt{x}$ (e) $y = 2\sqrt{x}$ (f) $y = \sqrt{-x}$

FIGURE 4

EXAMPLE 2 Sketch the graph of the function $f(x) = x^2 + 6x + 10$.

SOLUTION Completing the square, we write the equation of the graph as

$$y = x^2 + 6x + 10 = (x + 3)^2 + 1$$

This means we obtain the desired graph by starting with the parabola $y = x^2$ and shifting 3 units to the left and then 1 unit upward (see Figure 5).

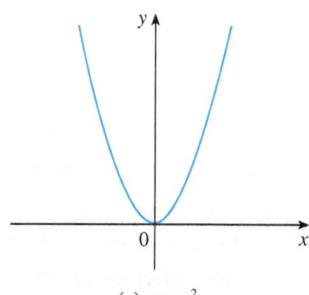
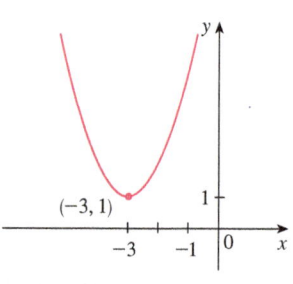

FIGURE 5   (a) $y = x^2$ (b) $y = (x + 3)^2 + 1$

SECTION 1.3 New Functions from Old Functions 39

EXAMPLE 3 Sketch the graphs of the following functions.
(a) $y = \sin 2x$ (b) $y = 1 - \sin x$

SOLUTION
(a) We obtain the graph of $y = \sin 2x$ from that of $y = \sin x$ by compressing horizontally by a factor of 2. (See Figures 6 and 7.) Thus, whereas the period of $y = \sin x$ is 2π, the period of $y = \sin 2x$ is $2\pi/2 = \pi$.

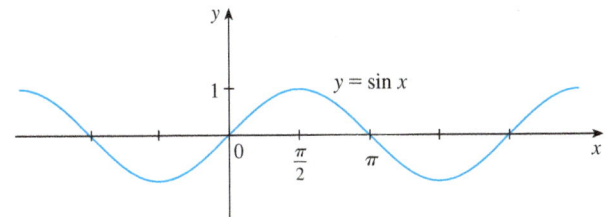

FIGURE 6

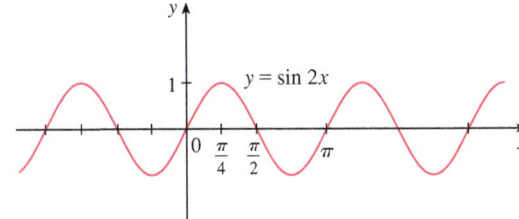

FIGURE 7

(b) To obtain the graph of $y = 1 - \sin x$, we again start with $y = \sin x$. We reflect about the x-axis to get the graph of $y = -\sin x$ and then we shift 1 unit upward to get $y = 1 - \sin x$. (See Figure 8.)

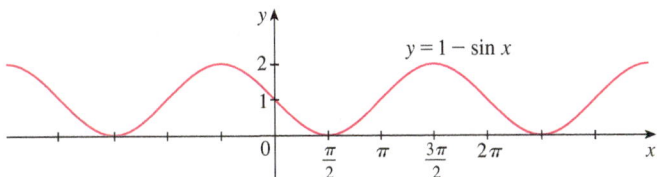

FIGURE 8

EXAMPLE 4 Figure 9 shows graphs of the number of hours of daylight as functions of the time of the year at several latitudes. Given that Philadelphia is located at approximately 40°N latitude, find a function that models the length of daylight at Philadelphia.

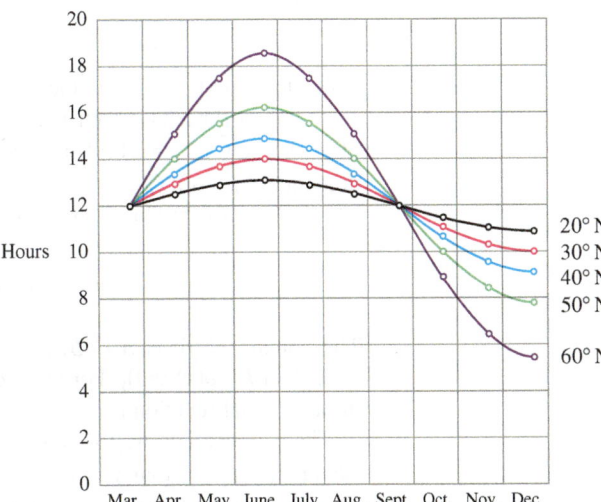

FIGURE 9
Graph of the length of daylight from March 21 through December 21 at various latitudes

Source: Adapted from L. Harrison, *Daylight, Twilight, Darkness and Time* (New York: Silver, Burdett, 1935), 40.

SOLUTION Notice that each curve resembles a shifted and stretched sine function. By looking at the blue curve we see that, at the latitude of Philadelphia, daylight lasts about 14.8 hours on June 21 and 9.2 hours on December 21, so the amplitude of the curve (the factor by which we have to stretch the sine curve vertically) is $\frac{1}{2}(14.8 - 9.2) = 2.8$.

By what factor do we need to stretch the sine curve horizontally if we measure the time t in days? Because there are about 365 days in a year, the period of our model should be 365. But the period of $y = \sin t$ is 2π, so the horizontal stretching factor is $2\pi/365$.

We also notice that the curve begins its cycle on March 21, the 80th day of the year, so we have to shift the curve 80 units to the right. In addition, we shift it 12 units upward. Therefore we model the length of daylight in Philadelphia on the tth day of the year by the function

$$L(t) = 12 + 2.8 \sin\left[\frac{2\pi}{365}(t - 80)\right]$$

Another transformation of some interest is taking the *absolute value* of a function. If $y = |f(x)|$, then according to the definition of absolute value, $y = f(x)$ when $f(x) \geq 0$ and $y = -f(x)$ when $f(x) < 0$. This tells us how to get the graph of $y = |f(x)|$ from the graph of $y = f(x)$: The part of the graph that lies above the x-axis remains the same; the part that lies below the x-axis is reflected about the x-axis.

EXAMPLE 5 Sketch the graph of the function $y = |x^2 - 1|$.

SOLUTION We first graph the parabola $y = x^2 - 1$ in Figure 10(a) by shifting the parabola $y = x^2$ downward 1 unit. We see that the graph lies below the x-axis when $-1 < x < 1$, so we reflect that part of the graph about the x-axis to obtain the graph of $y = |x^2 - 1|$ in Figure 10(b).

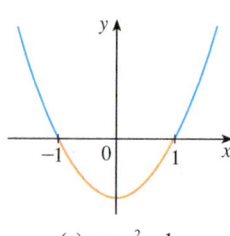

(a) $y = x^2 - 1$

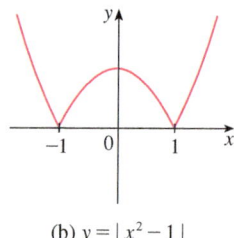

(b) $y = |x^2 - 1|$

FIGURE 10

Combinations of Functions

Two functions f and g can be combined to form new functions $f + g$, $f - g$, fg, and f/g in a manner similar to the way we add, subtract, multiply, and divide real numbers. The sum and difference functions are defined by

$$(f + g)(x) = f(x) + g(x) \qquad (f - g)(x) = f(x) - g(x)$$

If the domain of f is A and the domain of g is B, then the domain of $f + g$ is the intersection $A \cap B$ because both $f(x)$ and $g(x)$ have to be defined. For example, the domain of $f(x) = \sqrt{x}$ is $A = [0, \infty)$ and the domain of $g(x) = \sqrt{2 - x}$ is $B = (-\infty, 2]$, so the domain of $(f + g)(x) = \sqrt{x} + \sqrt{2 - x}$ is $A \cap B = [0, 2]$.

Similarly, the product and quotient functions are defined by

$$(fg)(x) = f(x)g(x) \qquad \left(\frac{f}{g}\right)(x) = \frac{f(x)}{g(x)}$$

The domain of fg is $A \cap B$, but we can't divide by 0 and so the domain of f/g is $\{x \in A \cap B \mid g(x) \neq 0\}$. For instance, if $f(x) = x^2$ and $g(x) = x - 1$, then the domain of the rational function $(f/g)(x) = x^2/(x - 1)$ is $\{x \mid x \neq 1\}$, or $(-\infty, 1) \cup (1, \infty)$.

There is another way of combining two functions to obtain a new function. For example, suppose that $y = f(u) = \sqrt{u}$ and $u = g(x) = x^2 + 1$. Since y is a function of u and u is, in turn, a function of x, it follows that y is ultimately a function of x.

We compute this by substitution:

$$y = f(u) = f(g(x)) = f(x^2 + 1) = \sqrt{x^2 + 1}$$

The procedure is called *composition* because the new function is *composed* of the two given functions f and g.

In general, given any two functions f and g, we start with a number x in the domain of g and calculate $g(x)$. If this number $g(x)$ is in the domain of f, then we can calculate the value of $f(g(x))$. Notice that the output of one function is used as the input to the next function. The result is a new function $h(x) = f(g(x))$ obtained by substituting g into f. It is called the *composition* (or *composite*) of f and g and is denoted by $f \circ g$ ("f circle g").

Definition Given two functions f and g, the **composite function** $f \circ g$ (also called the **composition** of f and g) is defined by

$$(f \circ g)(x) = f(g(x))$$

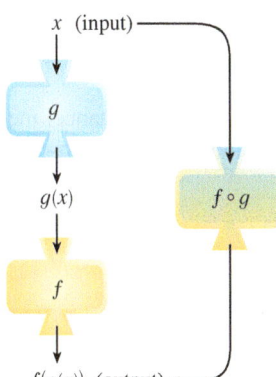

FIGURE 11
The $f \circ g$ machine is composed of the g machine (first) and then the f machine.

The domain of $f \circ g$ is the set of all x in the domain of g such that $g(x)$ is in the domain of f. In other words, $(f \circ g)(x)$ is defined whenever both $g(x)$ and $f(g(x))$ are defined. Figure 11 shows how to picture $f \circ g$ in terms of machines.

EXAMPLE 6 If $f(x) = x^2$ and $g(x) = x - 3$, find the composite functions $f \circ g$ and $g \circ f$.

SOLUTION We have

$$(f \circ g)(x) = f(g(x)) = f(x - 3) = (x - 3)^2$$

$$(g \circ f)(x) = g(f(x)) = g(x^2) = x^2 - 3$$

NOTE You can see from Example 6 that, in general, $f \circ g \neq g \circ f$. Remember, the notation $f \circ g$ means that the function g is applied first and then f is applied second. In Example 6, $f \circ g$ is the function that *first* subtracts 3 and *then* squares; $g \circ f$ is the function that *first* squares and *then* subtracts 3.

EXAMPLE 7 If $f(x) = \sqrt{x}$ and $g(x) = \sqrt{2 - x}$, find each of the following functions and their domains.
(a) $f \circ g$ (b) $g \circ f$ (c) $f \circ f$ (d) $g \circ g$

SOLUTION
(a) $$(f \circ g)(x) = f(g(x)) = f(\sqrt{2 - x}) = \sqrt{\sqrt{2 - x}} = \sqrt[4]{2 - x}$$

The domain of $f \circ g$ is $\{x \mid 2 - x \geq 0\} = \{x \mid x \leq 2\} = (-\infty, 2]$.

(b) $$(g \circ f)(x) = g(f(x)) = g(\sqrt{x}) = \sqrt{2 - \sqrt{x}}$$

For $\sqrt{x}$ to be defined we must have $x \geq 0$. For $\sqrt{2 - \sqrt{x}}$ to be defined we must have $2 - \sqrt{x} \geq 0$, that is, $\sqrt{x} \leq 2$, or $x \leq 4$. Thus we have $0 \leq x \leq 4$, so the domain of $g \circ f$ is the closed interval $[0, 4]$.

If $0 \leq a \leq b$, then $a^2 \leq b^2$.

(c) $$(f \circ f)(x) = f(f(x)) = f(\sqrt{x}) = \sqrt{\sqrt{x}} = \sqrt[4]{x}$$

The domain of $f \circ f$ is $[0, \infty)$.

(d) $\quad (g \circ g)(x) = g(g(x)) = g(\sqrt{2-x}) = \sqrt{2 - \sqrt{2-x}}$

This expression is defined when both $2 - x \geq 0$ and $2 - \sqrt{2-x} \geq 0$. The first inequality means $x \leq 2$, and the second is equivalent to $\sqrt{2-x} \leq 2$, or $2 - x \leq 4$, or $x \geq -2$. Thus $-2 \leq x \leq 2$, so the domain of $g \circ g$ is the closed interval $[-2, 2]$. ■

It is possible to take the composition of three or more functions. For instance, the composite function $f \circ g \circ h$ is found by first applying h, then g, and then f as follows:

$$(f \circ g \circ h)(x) = f(g(h(x)))$$

EXAMPLE 8 Find $f \circ g \circ h$ if $f(x) = x/(x + 1)$, $g(x) = x^{10}$, and $h(x) = x + 3$.

SOLUTION
$$(f \circ g \circ h)(x) = f(g(h(x))) = f(g(x + 3))$$
$$= f((x + 3)^{10}) = \frac{(x + 3)^{10}}{(x + 3)^{10} + 1}$$
■

So far we have used composition to build complicated functions from simpler ones. But in calculus it is often useful to be able to *decompose* a complicated function into simpler ones, as in the following example.

EXAMPLE 9 Given $F(x) = \cos^2(x + 9)$, find functions f, g, and h such that $F = f \circ g \circ h$.

SOLUTION Since $F(x) = [\cos(x + 9)]^2$, the formula for F says: First add 9, then take the cosine of the result, and finally square. So we let

$$h(x) = x + 9 \qquad g(x) = \cos x \qquad f(x) = x^2$$

Then
$$(f \circ g \circ h)(x) = f(g(h(x))) = f(g(x + 9)) = f(\cos(x + 9))$$
$$= [\cos(x + 9)]^2 = F(x)$$
■

1.3 EXERCISES

1. Suppose the graph of f is given. Write equations for the graphs that are obtained from the graph of f as follows.
   (a) Shift 3 units upward.
   (b) Shift 3 units downward.
   (c) Shift 3 units to the right.
   (d) Shift 3 units to the left.
   (e) Reflect about the x-axis.
   (f) Reflect about the y-axis.
   (g) Stretch vertically by a factor of 3.
   (h) Shrink vertically by a factor of 3.

2. Explain how each graph is obtained from the graph of $y = f(x)$.
   (a) $y = f(x) + 8$
   (b) $y = f(x + 8)$
   (c) $y = 8f(x)$
   (d) $y = f(8x)$
   (e) $y = -f(x) - 1$
   (f) $y = 8f(\frac{1}{8}x)$

3. The graph of $y = f(x)$ is given. Match each equation with its graph and give reasons for your choices.
   (a) $y = f(x - 4)$
   (b) $y = f(x) + 3$
   (c) $y = \frac{1}{3}f(x)$
   (d) $y = -f(x + 4)$
   (e) $y = 2f(x + 6)$

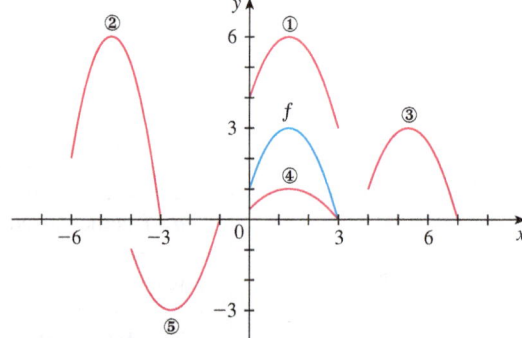

4. The graph of f is given. Draw the graphs of the following functions.
(a) $y = f(x) - 3$
(b) $y = f(x + 1)$
(c) $y = \frac{1}{2} f(x)$
(d) $y = -f(x)$

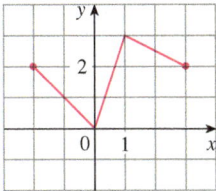

5. The graph of f is given. Use it to graph the following functions.
(a) $y = f(2x)$
(b) $y = f(\frac{1}{2}x)$
(c) $y = f(-x)$
(d) $y = -f(-x)$

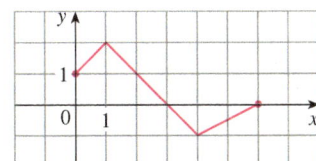

6–7 The graph of $y = \sqrt{3x - x^2}$ is given. Use transformations to create a function whose graph is as shown.

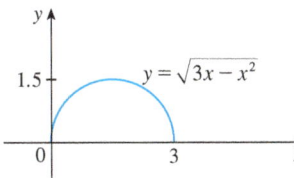

6. 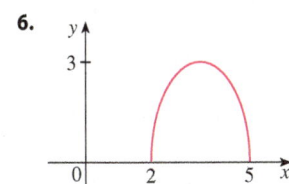 **7.**

8. (a) How is the graph of $y = 2 \sin x$ related to the graph of $y = \sin x$? Use your answer and Figure 6 to sketch the graph of $y = 2 \sin x$.
(b) How is the graph of $y = 1 + \sqrt{x}$ related to the graph of $y = \sqrt{x}$? Use your answer and Figure 4(a) to sketch the graph of $y = 1 + \sqrt{x}$.

9–24 Graph the function by hand, not by plotting points, but by starting with the graph of one of the standard functions given in Section 1.2, and then applying the appropriate transformations.

9. $y = -x^2$
10. $y = (x - 3)^2$
11. $y = x^3 + 1$
12. $y = 1 - \frac{1}{x}$
13. $y = 2 \cos 3x$
14. $y = 2\sqrt{x + 1}$
15. $y = x^2 - 4x + 5$
16. $y = 1 + \sin \pi x$
17. $y = 2 - \sqrt{x}$
18. $y = 3 - 2 \cos x$
19. $y = \sin(\frac{1}{2}x)$
20. $y = |x| - 2$
21. $y = |x - 2|$
22. $y = \frac{1}{4} \tan(x - \frac{\pi}{4})$
23. $y = |\sqrt{x} - 1|$
24. $y = |\cos \pi x|$

25. The city of New Orleans is located at latitude 30°N. Use Figure 9 to find a function that models the number of hours of daylight at New Orleans as a function of the time of year. To check the accuracy of your model, use the fact that on March 31 the sun rises at 5:51 AM and sets at 6:18 PM in New Orleans.

26. A variable star is one whose brightness alternately increases and decreases. For the most visible variable star, Delta Cephei, the time between periods of maximum brightness is 5.4 days, the average brightness (or magnitude) of the star is 4.0, and its brightness varies by ±0.35 magnitude. Find a function that models the brightness of Delta Cephei as a function of time.

27. Some of the highest tides in the world occur in the Bay of Fundy on the Atlantic Coast of Canada. At Hopewell Cape the water depth at low tide is about 2.0 m and at high tide it is about 12.0 m. The natural period of oscillation is about 12 hours and on June 30, 2009, high tide occurred at 6:45 AM. Find a function involving the cosine function that models the water depth $D(t)$ (in meters) as a function of time t (in hours after midnight) on that day.

28. In a normal respiratory cycle the volume of air that moves into and out of the lungs is about 500 mL. The reserve and residue volumes of air that remain in the lungs occupy about 2000 mL and a single respiratory cycle for an average human takes about 4 seconds. Find a model for the total volume of air $V(t)$ in the lungs as a function of time.

29. (a) How is the graph of $y = f(|x|)$ related to the graph of f?
(b) Sketch the graph of $y = \sin |x|$.
(c) Sketch the graph of $y = \sqrt{|x|}$.

30. Use the given graph of f to sketch the graph of $y = 1/f(x)$. Which features of f are the most important in sketching $y = 1/f(x)$? Explain how they are used.

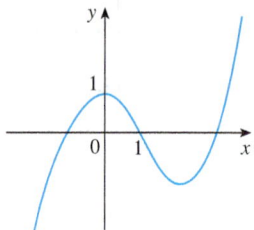

31–32 Find (a) $f + g$, (b) $f - g$, (c) fg, and (d) f/g and state their domains.

31. $f(x) = x^3 + 2x^2$, $g(x) = 3x^2 - 1$

32. $f(x) = \sqrt{3-x}$, $g(x) = \sqrt{x^2 - 1}$

33–38 Find the functions (a) $f \circ g$, (b) $g \circ f$, (c) $f \circ f$, and (d) $g \circ g$ and their domains.

33. $f(x) = 3x + 5$, $g(x) = x^2 + x$

34. $f(x) = x^3 - 2$, $g(x) = 1 - 4x$

35. $f(x) = \sqrt{x+1}$, $g(x) = 4x - 3$

36. $f(x) = \sin x$, $g(x) = x^2 + 1$

37. $f(x) = x + \dfrac{1}{x}$, $g(x) = \dfrac{x+1}{x+2}$

38. $f(x) = \dfrac{x}{1+x}$, $g(x) = \sin 2x$

39–42 Find $f \circ g \circ h$.

39. $f(x) = 3x - 2$, $g(x) = \sin x$, $h(x) = x^2$

40. $f(x) = |x - 4|$, $g(x) = 2^x$, $h(x) = \sqrt{x}$

41. $f(x) = \sqrt{x-3}$, $g(x) = x^2$, $h(x) = x^3 + 2$

42. $f(x) = \tan x$, $g(x) = \dfrac{x}{x-1}$, $h(x) = \sqrt[3]{x}$

43–48 Express the function in the form $f \circ g$.

43. $F(x) = (2x + x^2)^4$

44. $F(x) = \cos^2 x$

45. $F(x) = \dfrac{\sqrt[3]{x}}{1 + \sqrt[3]{x}}$

46. $G(x) = \sqrt[3]{\dfrac{x}{1+x}}$

47. $v(t) = \sec(t^2) \tan(t^2)$

48. $u(t) = \dfrac{\tan t}{1 + \tan t}$

49–51 Express the function in the form $f \circ g \circ h$.

49. $R(x) = \sqrt{\sqrt{x} - 1}$

50. $H(x) = \sqrt[8]{2 + |x|}$

51. $S(t) = \sin^2(\cos t)$

52. Use the table to evaluate each expression.
(a) $f(g(1))$ (b) $g(f(1))$ (c) $f(f(1))$
(d) $g(g(1))$ (e) $(g \circ f)(3)$ (f) $(f \circ g)(6)$

x	1	2	3	4	5	6
$f(x)$	3	1	4	2	2	5
$g(x)$	6	3	2	1	2	3

53. Use the given graphs of f and g to evaluate each expression, or explain why it is undefined.
(a) $f(g(2))$ (b) $g(f(0))$ (c) $(f \circ g)(0)$
(d) $(g \circ f)(6)$ (e) $(g \circ g)(-2)$ (f) $(f \circ f)(4)$

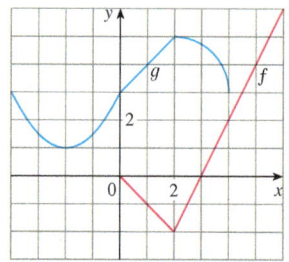

54. Use the given graphs of f and g to estimate the value of $f(g(x))$ for $x = -5, -4, -3, \ldots, 5$. Use these estimates to sketch a rough graph of $f \circ g$.

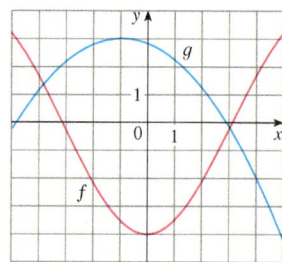

55. A stone is dropped into a lake, creating a circular ripple that travels outward at a speed of 60 cm/s.
(a) Express the radius r of this circle as a function of the time t (in seconds).
(b) If A is the area of this circle as a function of the radius, find $A \circ r$ and interpret it.

56. A spherical balloon is being inflated and the radius of the balloon is increasing at a rate of 2 cm/s.
(a) Express the radius r of the balloon as a function of the time t (in seconds).
(b) If V is the volume of the balloon as a function of the radius, find $V \circ r$ and interpret it.

57. A ship is moving at a speed of 30 km/h parallel to a straight shoreline. The ship is 6 km from shore and it passes a lighthouse at noon.
(a) Express the distance s between the lighthouse and the ship

as a function of d, the distance the ship has traveled since noon; that is, find f so that $s = f(d)$.
(b) Express d as a function of t, the time elapsed since noon; that is, find g so that $d = g(t)$.
(c) Find $f \circ g$. What does this function represent?

58. An airplane is flying at a speed of 350 mi/h at an altitude of one mile and passes directly over a radar station at time $t = 0$.
(a) Express the horizontal distance d (in miles) that the plane has flown as a function of t.
(b) Express the distance s between the plane and the radar station as a function of d.
(c) Use composition to express s as a function of t.

59. The **Heaviside function** H is defined by

$$H(t) = \begin{cases} 0 & \text{if } t < 0 \\ 1 & \text{if } t \geq 0 \end{cases}$$

It is used in the study of electric circuits to represent the sudden surge of electric current, or voltage, when a switch is instantaneously turned on.
(a) Sketch the graph of the Heaviside function.
(b) Sketch the graph of the voltage $V(t)$ in a circuit if the switch is turned on at time $t = 0$ and 120 volts are applied instantaneously to the circuit. Write a formula for $V(t)$ in terms of $H(t)$.
(c) Sketch the graph of the voltage $V(t)$ in a circuit if the switch is turned on at time $t = 5$ seconds and 240 volts are applied instantaneously to the circuit. Write a formula for $V(t)$ in terms of $H(t)$. (Note that starting at $t = 5$ corresponds to a translation.)

60. The Heaviside function defined in Exercise 59 can also be used to define the **ramp function** $y = ctH(t)$, which represents a gradual increase in voltage or current in a circuit.
(a) Sketch the graph of the ramp function $y = tH(t)$.
(b) Sketch the graph of the voltage $V(t)$ in a circuit if the switch is turned on at time $t = 0$ and the voltage is gradually increased to 120 volts over a 60-second time interval. Write a formula for $V(t)$ in terms of $H(t)$ for $t \leq 60$.
(c) Sketch the graph of the voltage $V(t)$ in a circuit if the switch is turned on at time $t = 7$ seconds and the voltage is gradually increased to 100 volts over a period of 25 seconds. Write a formula for $V(t)$ in terms of $H(t)$ for $t \leq 32$.

61. Let f and g be linear functions with equations $f(x) = m_1 x + b_1$ and $g(x) = m_2 x + b_2$. Is $f \circ g$ also a linear function? If so, what is the slope of its graph?

62. If you invest x dollars at 4% interest compounded annually, then the amount $A(x)$ of the investment after one year is $A(x) = 1.04x$. Find $A \circ A$, $A \circ A \circ A$, and $A \circ A \circ A \circ A$. What do these compositions represent? Find a formula for the composition of n copies of A.

63. (a) If $g(x) = 2x + 1$ and $h(x) = 4x^2 + 4x + 7$, find a function f such that $f \circ g = h$. (Think about what operations you would have to perform on the formula for g to end up with the formula for h.)
(b) If $f(x) = 3x + 5$ and $h(x) = 3x^2 + 3x + 2$, find a function g such that $f \circ g = h$.

64. If $f(x) = x + 4$ and $h(x) = 4x - 1$, find a function g such that $g \circ f = h$.

65. Suppose g is an even function and let $h = f \circ g$. Is h always an even function?

66. Suppose g is an odd function and let $h = f \circ g$. Is h always an odd function? What if f is odd? What if f is even?

1.4 Exponential Functions

The function $f(x) = 2^x$ is called an *exponential function* because the variable, x, is the exponent. It should not be confused with the power function $g(x) = x^2$, in which the variable is the base.

In general, an **exponential function** is a function of the form

$$f(x) = b^x$$

where b is a positive constant. Let's recall what this means.
If $x = n$, a positive integer, then

$$b^n = \underbrace{b \cdot b \cdot \ \cdots \ \cdot b}_{n \text{ factors}}$$

If $x = 0$, then $b^0 = 1$, and if $x = -n$, where n is a positive integer, then

$$b^{-n} = \frac{1}{b^n}$$

In Appendix G we present an alternative approach to the exponential and logarithmic functions using integral calculus.

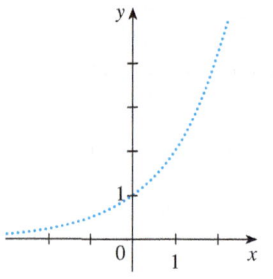

FIGURE 1
Representation of $y = 2^x$, x rational

If x is a rational number, $x = p/q$, where p and q are integers and $q > 0$, then

$$b^x = b^{p/q} = \sqrt[q]{b^p} = \left(\sqrt[q]{b}\right)^p$$

But what is the meaning of b^x if x is an irrational number? For instance, what is meant by $2^{\sqrt{3}}$ or 5^π?

To help us answer this question we first look at the graph of the function $y = 2^x$, where x is rational. A representation of this graph is shown in Figure 1. We want to enlarge the domain of $y = 2^x$ to include both rational and irrational numbers.

There are holes in the graph in Figure 1 corresponding to irrational values of x. We want to fill in the holes by defining $f(x) = 2^x$, where $x \in \mathbb{R}$, so that f is an increasing function. In particular, since the irrational number $\sqrt{3}$ satisfies

$$1.7 < \sqrt{3} < 1.8$$

we must have

$$2^{1.7} < 2^{\sqrt{3}} < 2^{1.8}$$

and we know what $2^{1.7}$ and $2^{1.8}$ mean because 1.7 and 1.8 are rational numbers. Similarly, if we use better approximations for $\sqrt{3}$, we obtain better approximations for $2^{\sqrt{3}}$:

$$1.73 < \sqrt{3} < 1.74 \quad \Rightarrow \quad 2^{1.73} < 2^{\sqrt{3}} < 2^{1.74}$$

$$1.732 < \sqrt{3} < 1.733 \quad \Rightarrow \quad 2^{1.732} < 2^{\sqrt{3}} < 2^{1.733}$$

$$1.7320 < \sqrt{3} < 1.7321 \quad \Rightarrow \quad 2^{1.7320} < 2^{\sqrt{3}} < 2^{1.7321}$$

$$1.73205 < \sqrt{3} < 1.73206 \quad \Rightarrow \quad 2^{1.73205} < 2^{\sqrt{3}} < 2^{1.73206}$$

$$\vdots \qquad\qquad\qquad\qquad \vdots$$

A proof of this fact is given in J. Marsden and A. Weinstein, *Calculus Unlimited* (Menlo Park, CA: Benjamin/Cummings, 1981).

It can be shown that there is exactly one number that is greater than all of the numbers

$$2^{1.7}, \quad 2^{1.73}, \quad 2^{1.732}, \quad 2^{1.7320}, \quad 2^{1.73205}, \quad \ldots$$

and less than all of the numbers

$$2^{1.8}, \quad 2^{1.74}, \quad 2^{1.733}, \quad 2^{1.7321}, \quad 2^{1.73206}, \quad \ldots$$

We define $2^{\sqrt{3}}$ to be this number. Using the preceding approximation process we can compute it correct to six decimal places:

$$2^{\sqrt{3}} \approx 3.321997$$

Similarly, we can define 2^x (or b^x, if $b > 0$) where x is any irrational number. Figure 2 shows how all the holes in Figure 1 have been filled to complete the graph of the function $f(x) = 2^x$, $x \in \mathbb{R}$.

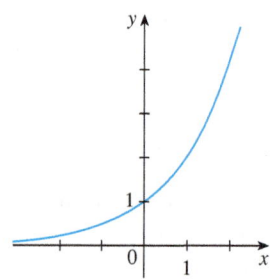

FIGURE 2
$y = 2^x$, x real

The graphs of members of the family of functions $y = b^x$ are shown in Figure 3 for various values of the base b. Notice that all of these graphs pass through the same point $(0, 1)$ because $b^0 = 1$ for $b \neq 0$. Notice also that as the base b gets larger, the exponential function grows more rapidly (for $x > 0$).

If $0 < b < 1$, then b^x approaches 0 as x becomes large. If $b > 1$, then b^x approaches 0 as x decreases through negative values. In both cases the x-axis is a horizontal asymptote. These matters are discussed in Section 2.6.

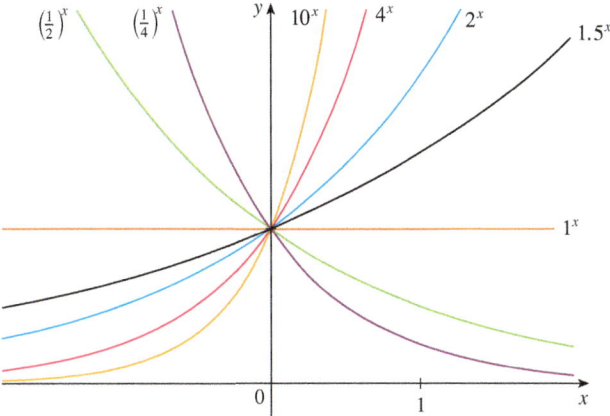

FIGURE 3

You can see from Figure 3 that there are basically three kinds of exponential functions $y = b^x$. If $0 < b < 1$, the exponential function decreases; if $b = 1$, it is a constant; and if $b > 1$, it increases. These three cases are illustrated in Figure 4. Observe that if $b \neq 1$, then the exponential function $y = b^x$ has domain $\mathbb{R}$ and range $(0, \infty)$. Notice also that, since $(1/b)^x = 1/b^x = b^{-x}$, the graph of $y = (1/b)^x$ is just the reflection of the graph of $y = b^x$ about the y-axis.

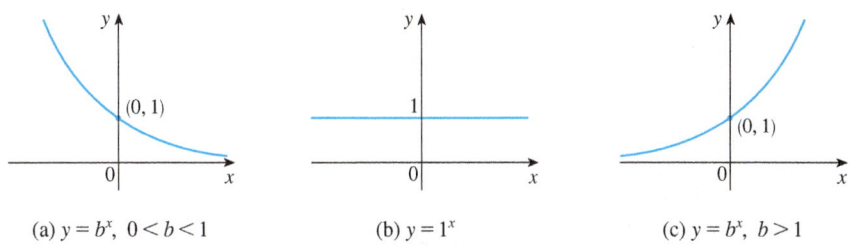

FIGURE 4 (a) $y = b^x$, $0 < b < 1$ (b) $y = 1^x$ (c) $y = b^x$, $b > 1$

One reason for the importance of the exponential function lies in the following properties. If x and y are rational numbers, then these laws are well known from elementary algebra. It can be proved that they remain true for arbitrary real numbers x and y.

www.stewartcalculus.com
For review and practice using the Laws of Exponents, click on *Review of Algebra*.

Laws of Exponents If a and b are positive numbers and x and y are any real numbers, then

1. $b^{x+y} = b^x b^y$ 2. $b^{x-y} = \dfrac{b^x}{b^y}$ 3. $(b^x)^y = b^{xy}$ 4. $(ab)^x = a^x b^x$

EXAMPLE 1 Sketch the graph of the function $y = 3 - 2^x$ and determine its domain and range.

For a review of reflecting and shifting graphs, see Section 1.3.

SOLUTION First we reflect the graph of $y = 2^x$ [shown in Figures 2 and 5(a)] about the x-axis to get the graph of $y = -2^x$ in Figure 5(b). Then we shift the graph of $y = -2^x$

upward 3 units to obtain the graph of $y = 3 - 2^x$ in Figure 5(c). The domain is $\mathbb{R}$ and the range is $(-\infty, 3)$.

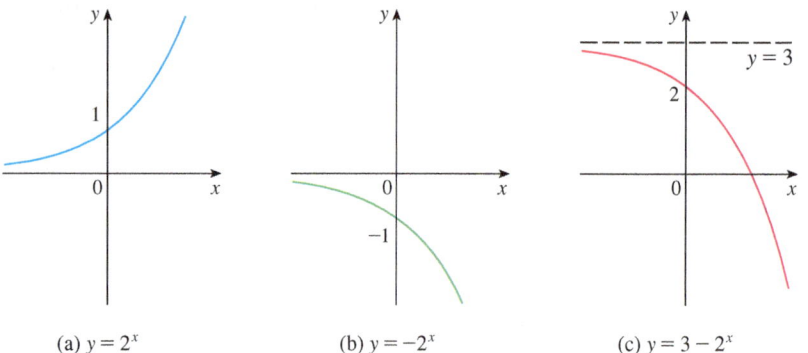

FIGURE 5 (a) $y = 2^x$ (b) $y = -2^x$ (c) $y = 3 - 2^x$

EXAMPLE 2 Use a graphing device to compare the exponential function $f(x) = 2^x$ and the power function $g(x) = x^2$. Which function grows more quickly when x is large?

SOLUTION Figure 6 shows both functions graphed in the viewing rectangle $[-2, 6]$ by $[0, 40]$. We see that the graphs intersect three times, but for $x > 4$ the graph of $f(x) = 2^x$ stays above the graph of $g(x) = x^2$. Figure 7 gives a more global view and shows that for large values of x, the exponential function $y = 2^x$ grows far more rapidly than the power function $y = x^2$.

Example 2 shows that $y = 2^x$ increases more quickly than $y = x^2$. To demonstrate just how quickly $f(x) = 2^x$ increases, let's perform the following thought experiment. Suppose we start with a piece of paper a thousandth of an inch thick and we fold it in half 50 times. Each time we fold the paper in half, the thickness of the paper doubles, so the thickness of the resulting paper would be $2^{50}/1000$ inches. How thick do you think that is? It works out to be more than 17 million miles!

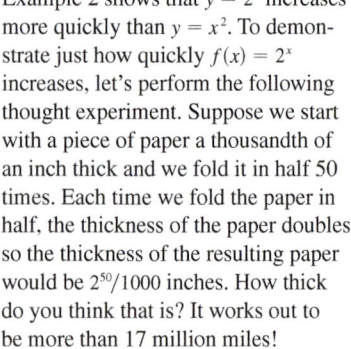

FIGURE 6

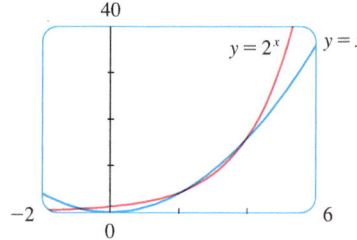

FIGURE 7

Applications of Exponential Functions

The exponential function occurs very frequently in mathematical models of nature and society. Here we indicate briefly how it arises in the description of population growth and radioactive decay. In later chapters we will pursue these and other applications in greater detail.

First we consider a population of bacteria in a homogeneous nutrient medium. Suppose that by sampling the population at certain intervals it is determined that the population doubles every hour. If the number of bacteria at time t is $p(t)$, where t is measured in hours, and the initial population is $p(0) = 1000$, then we have

$$p(1) = 2p(0) = 2 \times 1000$$

$$p(2) = 2p(1) = 2^2 \times 1000$$

$$p(3) = 2p(2) = 2^3 \times 1000$$

It seems from this pattern that, in general,

$$p(t) = 2^t \times 1000 = (1000)2^t$$

This population function is a constant multiple of the exponential function $y = 2^t$, so it exhibits the rapid growth that we observed in Figures 2 and 7. Under ideal conditions (unlimited space and nutrition and absence of disease) this exponential growth is typical of what actually occurs in nature.

What about the human population? Table 1 shows data for the population of the world in the 20th century and Figure 8 shows the corresponding scatter plot.

Table 1

t (years since 1900)	Population (millions)
0	1650
10	1750
20	1860
30	2070
40	2300
50	2560
60	3040
70	3710
80	4450
90	5280
100	6080
110	6870

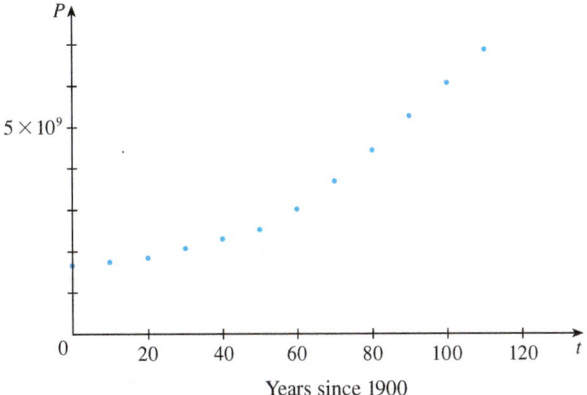

FIGURE 8
Scatter plot for world population growth

The pattern of the data points in Figure 8 suggests exponential growth, so we use a graphing calculator with exponential regression capability to apply the method of least squares and obtain the exponential model

$$P = (1436.53) \cdot (1.01395)^t$$

where $t = 0$ corresponds to 1900. Figure 9 shows the graph of this exponential function together with the original data points. We see that the exponential curve fits the data reasonably well. The period of relatively slow population growth is explained by the two world wars and the Great Depression of the 1930s.

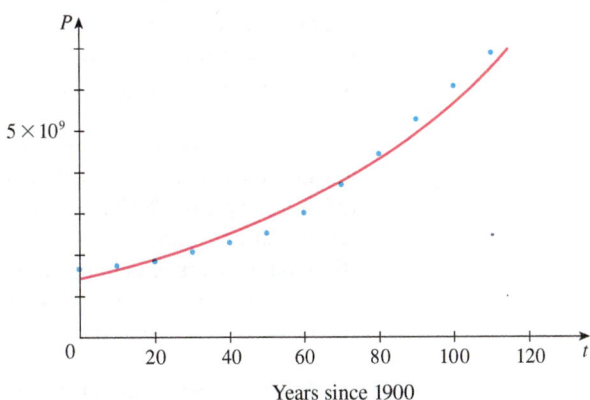

FIGURE 9
Exponential model for population growth

In 1995 a paper appeared detailing the effect of the protease inhibitor ABT-538 on the human immunodeficiency virus HIV-1.[1] Table 2 shows values of the plasma viral load $V(t)$ of patient 303, measured in RNA copies per mL, t days after ABT-538 treatment was begun. The corresponding scatter plot is shown in Figure 10.

Table 2

t (days)	$V(t)$
1	76.0
4	53.0
8	18.0
11	9.4
15	5.2
22	3.6

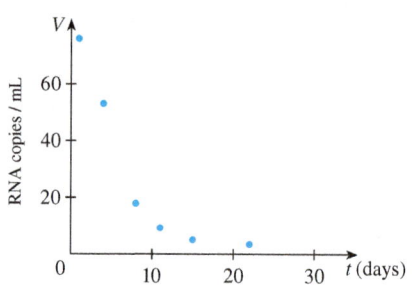

FIGURE 10 Plasma viral load in patient 303

The rather dramatic decline of the viral load that we see in Figure 10 reminds us of the graphs of the exponential function $y = b^x$ in Figures 3 and 4(a) for the case where the base b is less than 1. So let's model the function $V(t)$ by an exponential function. Using a graphing calculator or computer to fit the data in Table 2 with an exponential function of the form $y = a \cdot b^t$, we obtain the model

$$V = 96.39785 \cdot (0.818656)^t$$

In Figure 11 we graph this exponential function with the data points and see that the model represents the viral load reasonably well for the first month of treatment.

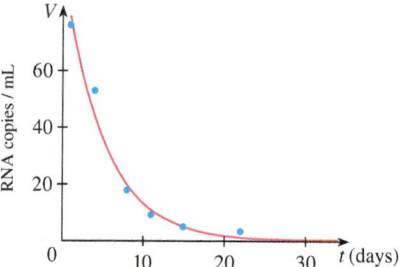

FIGURE 11
Exponential model for viral load

We could use the graph in Figure 11 to estimate the **half-life** of V, that is, the time required for the viral load to be reduced to half its initial value (see Exercise 33). In the next example we are given the half-life of a radioactive element and asked to find the mass of a sample at any time.

EXAMPLE 3 The half-life of strontium-90, ^{90}Sr, is 25 years. This means that half of any given quantity of ^{90}Sr will disintegrate in 25 years.
(a) If a sample of ^{90}Sr has a mass of 24 mg, find an expression for the mass $m(t)$ that remains after t years.
(b) Find the mass remaining after 40 years, correct to the nearest milligram.
(c) Use a graphing device to graph $m(t)$ and use the graph to estimate the time required for the mass to be reduced to 5 mg.

---

1. D. Ho et al., "Rapid Turnover of Plasma Virions and CD4 Lymphocytes in HIV-1 Infection," *Nature* 373 (1995): 123–26.

SOLUTION

(a) The mass is initially 24 mg and is halved during each 25-year period, so

$$m(0) = 24$$

$$m(25) = \frac{1}{2}(24)$$

$$m(50) = \frac{1}{2} \cdot \frac{1}{2}(24) = \frac{1}{2^2}(24)$$

$$m(75) = \frac{1}{2} \cdot \frac{1}{2^2}(24) = \frac{1}{2^3}(24)$$

$$m(100) = \frac{1}{2} \cdot \frac{1}{2^3}(24) = \frac{1}{2^4}(24)$$

From this pattern, it appears that the mass remaining after t years is

$$m(t) = \frac{1}{2^{t/25}}(24) = 24 \cdot 2^{-t/25} = 24 \cdot (2^{-1/25})^t$$

This is an exponential function with base $b = 2^{-1/25} = 1/2^{1/25}$.

(b) The mass that remains after 40 years is

$$m(40) = 24 \cdot 2^{-40/25} \approx 7.9 \text{ mg}$$

(c) We use a graphing calculator or computer to graph the function $m(t) = 24 \cdot 2^{-t/25}$ in Figure 12. We also graph the line $m = 5$ and use the cursor to estimate that $m(t) = 5$ when $t \approx 57$. So the mass of the sample will be reduced to 5 mg after about 57 years. ∎

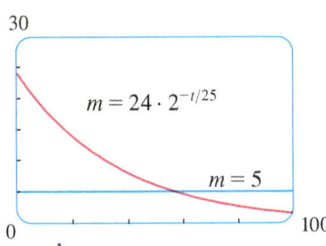

FIGURE 12

The Number e

Of all possible bases for an exponential function, there is one that is most convenient for the purposes of calculus. The choice of a base b is influenced by the way the graph of $y = b^x$ crosses the y-axis. Figures 13 and 14 show the tangent lines to the graphs of $y = 2^x$ and $y = 3^x$ at the point $(0, 1)$. (Tangent lines will be defined precisely in Section 2.7. For present purposes, you can think of the tangent line to an exponential graph at a point as the line that touches the graph only at that point.) If we measure the slopes of these tangent lines at $(0, 1)$, we find that $m \approx 0.7$ for $y = 2^x$ and $m \approx 1.1$ for $y = 3^x$.

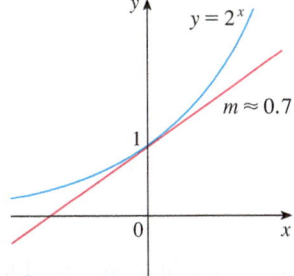
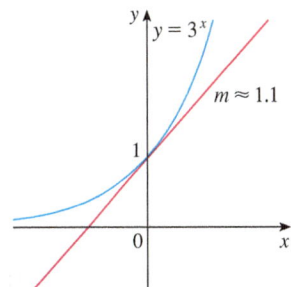

FIGURE 13   **FIGURE 14**

It turns out, as we will see in Chapter 3, that some of the formulas of calculus will be greatly simplified if we choose the base b so that the slope of the tangent line to $y = b^x$

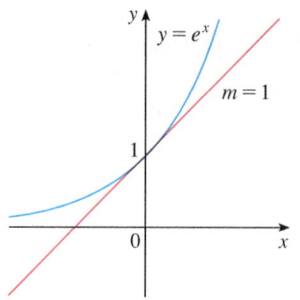

FIGURE 15
The natural exponential function crosses the y-axis with a slope of 1.

TEC Module 1.4 enables you to graph exponential functions with various bases and their tangent lines in order to estimate more closely the value of b for which the tangent has slope 1.

at $(0, 1)$ is *exactly* 1. (See Figure 15.) In fact, there *is* such a number and it is denoted by the letter e. (This notation was chosen by the Swiss mathematician Leonhard Euler in 1727, probably because it is the first letter of the word *exponential*.) In view of Figures 13 and 14, it comes as no surprise that the number e lies between 2 and 3 and the graph of $y = e^x$ lies between the graphs of $y = 2^x$ and $y = 3^x$. (See Figure 16.) In Chapter 3 we will see that the value of e, correct to five decimal places, is

$$e \approx 2.71828$$

We call the function $f(x) = e^x$ the **natural exponential function**.

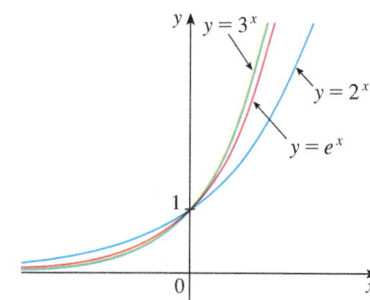

FIGURE 16

EXAMPLE 4 Graph the function $y = \frac{1}{2}e^{-x} - 1$ and state the domain and range.

SOLUTION We start with the graph of $y = e^x$ from Figures 15 and 17(a) and reflect about the y-axis to get the graph of $y = e^{-x}$ in Figure 17(b). (Notice that the graph crosses the y-axis with a slope of -1). Then we compress the graph vertically by a factor of 2 to obtain the graph of $y = \frac{1}{2}e^{-x}$ in Figure 17(c). Finally, we shift the graph downward one unit to get the desired graph in Figure 17(d). The domain is $\mathbb{R}$ and the range is $(-1, \infty)$.

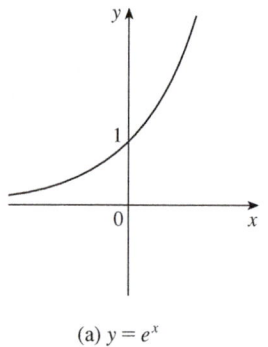 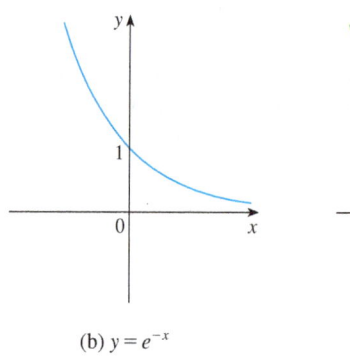 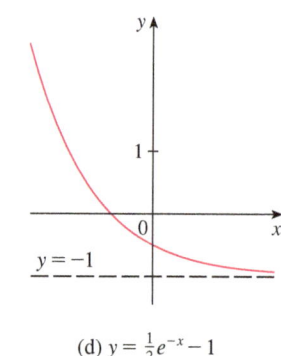

(a) $y = e^x$ (b) $y = e^{-x}$ (c) $y = \frac{1}{2}e^{-x}$ (d) $y = \frac{1}{2}e^{-x} - 1$

FIGURE 17

How far to the right do you think we would have to go for the height of the graph of $y = e^x$ to exceed a million? The next example demonstrates the rapid growth of this function by providing an answer that might surprise you.

EXAMPLE 5 Use a graphing device to find the values of x for which $e^x > 1{,}000{,}000$.

SOLUTION In Figure 18 we graph both the function $y = e^x$ and the horizontal line $y = 1,000,000$. We see that these curves intersect when $x \approx 13.8$. Thus $e^x > 10^6$ when $x > 13.8$. It is perhaps surprising that the values of the exponential function have already surpassed a million when x is only 14.

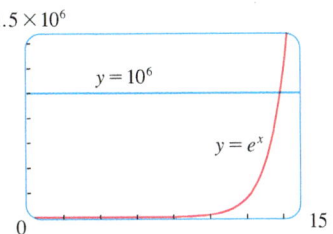

FIGURE 18

1.4 EXERCISES

1–4 Use the Law of Exponents to rewrite and simplify the expression.

1. (a) $\dfrac{4^{-3}}{2^{-8}}$ (b) $\dfrac{1}{\sqrt[3]{x^4}}$

2. (a) $8^{4/3}$ (b) $x(3x^2)^3$

3. (a) $b^8(2b)^4$ (b) $\dfrac{(6y^3)^4}{2y^5}$

4. (a) $\dfrac{x^{2n} \cdot x^{3n-1}}{x^{n+2}}$ (b) $\dfrac{\sqrt{a\sqrt{b}}}{\sqrt[3]{ab}}$

5. (a) Write an equation that defines the exponential function with base $b > 0$.
(b) What is the domain of this function?
(c) If $b \neq 1$, what is the range of this function?
(d) Sketch the general shape of the graph of the exponential function for each of the following cases.
   (i) $b > 1$
   (ii) $b = 1$
   (iii) $0 < b < 1$

6. (a) How is the number e defined?
(b) What is an approximate value for e?
(c) What is the natural exponential function?

7–10 Graph the given functions on a common screen. How are these graphs related?

7. $y = 2^x$, $y = e^x$, $y = 5^x$, $y = 20^x$

8. $y = e^x$, $y = e^{-x}$, $y = 8^x$, $y = 8^{-x}$

9. $y = 3^x$, $y = 10^x$, $y = \left(\tfrac{1}{3}\right)^x$, $y = \left(\tfrac{1}{10}\right)^x$

10. $y = 0.9^x$, $y = 0.6^x$, $y = 0.3^x$, $y = 0.1^x$

11–16 Make a rough sketch of the graph of the function. Do not use a calculator. Just use the graphs given in Figures 3 and 13 and, if necessary, the transformations of Section 1.3.

11. $y = 4^x - 1$ **12.** $y = (0.5)^{x-1}$

13. $y = -2^{-x}$ **14.** $y = e^{|x|}$

15. $y = 1 - \tfrac{1}{2}e^{-x}$ **16.** $y = 2(1 - e^x)$

17. Starting with the graph of $y = e^x$, write the equation of the graph that results from
(a) shifting 2 units downward.
(b) shifting 2 units to the right.
(c) reflecting about the x-axis.
(d) reflecting about the y-axis.
(e) reflecting about the x-axis and then about the y-axis.

18. Starting with the graph of $y = e^x$, find the equation of the graph that results from
(a) reflecting about the line $y = 4$.
(b) reflecting about the line $x = 2$.

19–20 Find the domain of each function.

19. (a) $f(x) = \dfrac{1 - e^{x^2}}{1 - e^{1-x^2}}$ (b) $f(x) = \dfrac{1 + x}{e^{\cos x}}$

20. (a) $g(t) = \sqrt{10^t - 100}$ (b) $g(t) = \sin(e^t - 1)$

21–22 Find the exponential function $f(x) = Cb^x$ whose graph is given.

21.

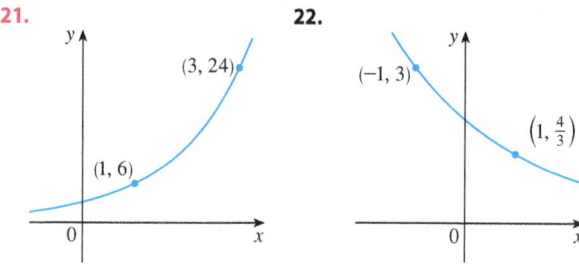

22.

23. If $f(x) = 5^x$, show that
$$\frac{f(x+h) - f(x)}{h} = 5^x \left(\frac{5^h - 1}{h} \right)$$

24. Suppose you are offered a job that lasts one month. Which of the following methods of payment do you prefer?
   I. One million dollars at the end of the month.
   II. One cent on the first day of the month, two cents on the second day, four cents on the third day, and, in general, 2^{n-1} cents on the nth day.

25. Suppose the graphs of $f(x) = x^2$ and $g(x) = 2^x$ are drawn on a coordinate grid where the unit of measurement is 1 inch. Show that, at a distance 2 ft to the right of the origin, the height of the graph of f is 48 ft but the height of the graph of g is about 265 mi.

26. Compare the functions $f(x) = x^5$ and $g(x) = 5^x$ by graphing both functions in several viewing rectangles. Find all points of intersection of the graphs correct to one decimal place. Which function grows more rapidly when x is large?

27. Compare the functions $f(x) = x^{10}$ and $g(x) = e^x$ by graphing both f and g in several viewing rectangles. When does the graph of g finally surpass the graph of f?

28. Use a graph to estimate the values of x such that $e^x > 1{,}000{,}000{,}000$.

29. A researcher is trying to determine the doubling time for a population of the bacterium *Giardia lamblia*. He starts a culture in a nutrient solution and estimates the bacteria count every four hours. His data are shown in the table.

Time (hours)	0	4	8	12	16	20	24
Bacteria count (CFU/mL)	37	47	63	78	105	130	173

(a) Make a scatter plot of the data.
(b) Use a graphing calculator to find an exponential curve $f(t) = a \cdot b^t$ that models the bacteria population t hours later.
(c) Graph the model from part (b) together with the scatter plot in part (a). Use the TRACE feature to determine how long it takes for the bacteria count to double.

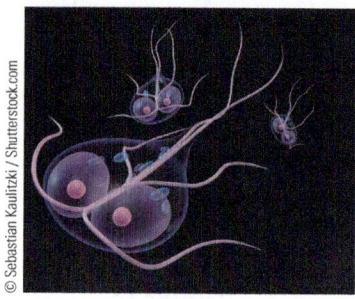

G. lamblia

30. A bacteria culture starts with 500 bacteria and doubles in size every half hour.
   (a) How many bacteria are there after 3 hours?
   (b) How many bacteria are there after t hours?
   (c) How many bacteria are there after 40 minutes?
   (d) Graph the population function and estimate the time for the population to reach 100,000.

31. The half-life of bismuth-210, ^{210}Bi, is 5 days.
   (a) If a sample has a mass of 200 mg, find the amount remaining after 15 days.
   (b) Find the amount remaining after t days.
   (c) Estimate the amount remaining after 3 weeks.
   (d) Use a graph to estimate the time required for the mass to be reduced to 1 mg.

32. An isotope of sodium, ^{24}Na, has a half-life of 15 hours. A sample of this isotope has mass 2 g.
   (a) Find the amount remaining after 60 hours.
   (b) Find the amount remaining after t hours.
   (c) Estimate the amount remaining after 4 days.
   (d) Use a graph to estimate the time required for the mass to be reduced to 0.01 g.

33. Use the graph of V in Figure 11 to estimate the half-life of the viral load of patient 303 during the first month of treatment.

34. After alcohol is fully absorbed into the body, it is metabolized with a half-life of about 1.5 hours. Suppose you have had three alcoholic drinks and an hour later, at midnight, your blood alcohol concentration (BAC) is 0.6 mg/mL.
   (a) Find an exponential decay model for your BAC t hours after midnight.
   (b) Graph your BAC and use the graph to determine when you can drive home if the legal limit is 0.08 mg/mL.

Source: Adapted from P. Wilkinson et al., "Pharmacokinetics of Ethanol after Oral Administration in the Fasting State," *Journal of Pharmacokinetics and Biopharmaceutics* 5 (1977): 207–24.

35. Use a graphing calculator with exponential regression capability to model the population of the world with the data from 1950 to 2010 in Table 1 on page 49. Use the model to estimate the population in 1993 and to predict the population in the year 2020.

36. The table gives the population of the United States, in millions, for the years 1900–2010. Use a graphing calculator with exponential regression capability to model the US population since 1900. Use the model to estimate the population in 1925 and to predict the population in the year 2020.

Year	Population	Year	Population
1900	76	1960	179
1910	92	1970	203
1920	106	1980	227
1930	123	1990	250
1940	131	2000	281
1950	150	2010	310

37. If you graph the function
$$f(x) = \frac{1 - e^{1/x}}{1 + e^{1/x}}$$
you'll see that f appears to be an odd function. Prove it.

38. Graph several members of the family of functions
$$f(x) = \frac{1}{1 + ae^{bx}}$$
where $a > 0$. How does the graph change when b changes? How does it change when a changes?

1.5 Inverse Functions and Logarithms

Table 1 gives data from an experiment in which a bacteria culture started with 100 bacteria in a limited nutrient medium; the size of the bacteria population was recorded at hourly intervals. The number of bacteria N is a function of the time t: $N = f(t)$.

Suppose, however, that the biologist changes her point of view and becomes interested in the time required for the population to reach various levels. In other words, she is thinking of t as a function of N. This function is called the *inverse function* of f, denoted by f^{-1}, and read "f inverse." Thus $t = f^{-1}(N)$ is the time required for the population level to reach N. The values of f^{-1} can be found by reading Table 1 from right to left or by consulting Table 2. For instance, $f^{-1}(550) = 6$ because $f(6) = 550$.

Table 1 N as a function of t

t (hours)	$N = f(t)$ = population at time t
0	100
1	168
2	259
3	358
4	445
5	509
6	550
7	573
8	586

Table 2 t as a function of N

N	$t = f^{-1}(N)$ = time to reach N bacteria
100	0
168	1
259	2
358	3
445	4
509	5
550	6
573	7
586	8

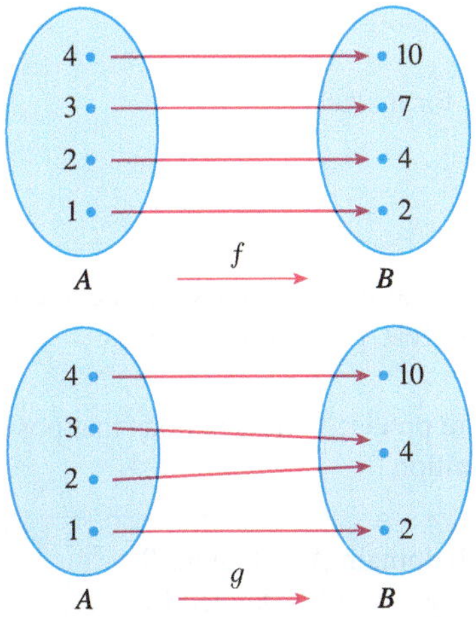

Not all functions possess inverses. Let's compare the functions f and g whose arrow diagrams are shown in Figure 1. Note that f never takes on the same value twice (any two inputs in A have different outputs), whereas g does take on the same value twice (both 2 and 3 have the same output, 4). In symbols,
$$g(2) = g(3)$$
but
$$f(x_1) \neq f(x_2) \quad \text{whenever } x_1 \neq x_2$$

Functions that share this property with f are called *one-to-one functions*.

FIGURE 1
f is one-to-one; g is not.

In the language of inputs and outputs, this definition says that f is one-to-one if each output corresponds to only one input.

> **[1] Definition** A function f is called a **one-to-one function** if it never takes on the same value twice; that is,
> $$f(x_1) \neq f(x_2) \quad \text{whenever } x_1 \neq x_2$$

If a horizontal line intersects the graph of f in more than one point, then we see from Figure 2 that there are numbers x_1 and x_2 such that $f(x_1) = f(x_2)$. This means that f is not one-to-one.

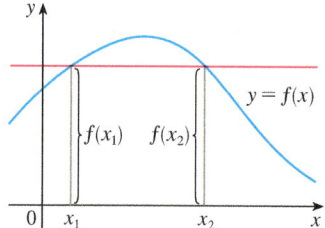

FIGURE 2
This function is not one-to-one because $f(x_1) = f(x_2)$.

Therefore we have the following geometric method for determining whether a function is one-to-one.

> **Horizontal Line Test** A function is one-to-one if and only if no horizontal line intersects its graph more than once.

EXAMPLE 1 Is the function $f(x) = x^3$ one-to-one?

SOLUTION 1 If $x_1 \neq x_2$, then $x_1^3 \neq x_2^3$ (two different numbers can't have the same cube). Therefore, by Definition 1, $f(x) = x^3$ is one-to-one.

SOLUTION 2 From Figure 3 we see that no horizontal line intersects the graph of $f(x) = x^3$ more than once. Therefore, by the Horizontal Line Test, f is one-to-one. ∎

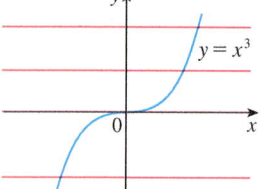

FIGURE 3
$f(x) = x^3$ is one-to-one.

EXAMPLE 2 Is the function $g(x) = x^2$ one-to-one?

SOLUTION 1 This function is not one-to-one because, for instance,
$$g(1) = 1 = g(-1)$$
and so 1 and -1 have the same output.

SOLUTION 2 From Figure 4 we see that there are horizontal lines that intersect the graph of g more than once. Therefore, by the Horizontal Line Test, g is not one-to-one. ∎

One-to-one functions are important because they are precisely the functions that possess inverse functions according to the following definition.

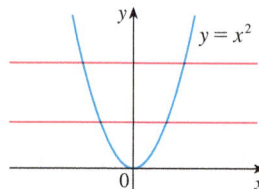

FIGURE 4
$g(x) = x^2$ is not one-to-one.

> **[2] Definition** Let f be a one-to-one function with domain A and range B. Then its **inverse function** f^{-1} has domain B and range A and is defined by
> $$f^{-1}(y) = x \iff f(x) = y$$
> for any y in B.

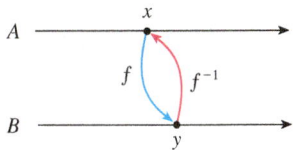

FIGURE 5

This definition says that if f maps x into y, then f^{-1} maps y back into x. (If f were not one-to-one, then f^{-1} would not be uniquely defined.) The arrow diagram in Figure 5 indicates that f^{-1} reverses the effect of f. Note that

$$\text{domain of } f^{-1} = \text{range of } f$$
$$\text{range of } f^{-1} = \text{domain of } f$$

For example, the inverse function of $f(x) = x^3$ is $f^{-1}(x) = x^{1/3}$ because if $y = x^3$, then

$$f^{-1}(y) = f^{-1}(x^3) = (x^3)^{1/3} = x$$

CAUTION Do not mistake the -1 in f^{-1} for an exponent. Thus

$$f^{-1}(x) \quad \text{does } not \text{ mean} \quad \frac{1}{f(x)}$$

The reciprocal $1/f(x)$ could, however, be written as $[f(x)]^{-1}$.

EXAMPLE 3 If $f(1) = 5$, $f(3) = 7$, and $f(8) = -10$, find $f^{-1}(7)$, $f^{-1}(5)$, and $f^{-1}(-10)$.

SOLUTION From the definition of f^{-1} we have

$$f^{-1}(7) = 3 \quad \text{because} \quad f(3) = 7$$
$$f^{-1}(5) = 1 \quad \text{because} \quad f(1) = 5$$
$$f^{-1}(-10) = 8 \quad \text{because} \quad f(8) = -10$$

The diagram in Figure 6 makes it clear how f^{-1} reverses the effect of f in this case. ■

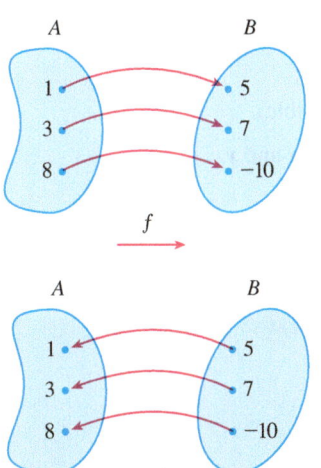

FIGURE 6
The inverse function reverses inputs and outputs.

The letter x is traditionally used as the independent variable, so when we concentrate on f^{-1} rather than on f, we usually reverse the roles of x and y in Definition 2 and write

$$\boxed{3} \qquad f^{-1}(x) = y \iff f(y) = x$$

By substituting for y in Definition 2 and substituting for x in (3), we get the following **cancellation equations**:

$$\boxed{4} \qquad f^{-1}(f(x)) = x \quad \text{for every } x \text{ in } A$$
$$f(f^{-1}(x)) = x \quad \text{for every } x \text{ in } B$$

The first cancellation equation says that if we start with x, apply f, and then apply f^{-1}, we arrive back at x, where we started (see the machine diagram in Figure 7). Thus f^{-1} undoes what f does. The second equation says that f undoes what f^{-1} does.

FIGURE 7

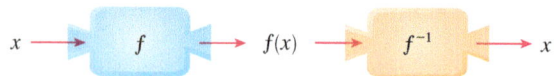

For example, if $f(x) = x^3$, then $f^{-1}(x) = x^{1/3}$ and so the cancellation equations become

$$f^{-1}(f(x)) = (x^3)^{1/3} = x$$

$$f(f^{-1}(x)) = (x^{1/3})^3 = x$$

These equations simply say that the cube function and the cube root function cancel each other when applied in succession.

Now let's see how to compute inverse functions. If we have a function $y = f(x)$ and are able to solve this equation for x in terms of y, then according to Definition 2 we must have $x = f^{-1}(y)$. If we want to call the independent variable x, we then interchange x and y and arrive at the equation $y = f^{-1}(x)$.

---

5 How to Find the Inverse Function of a One-to-One Function f

STEP 1 Write $y = f(x)$.

STEP 2 Solve this equation for x in terms of y (if possible).

STEP 3 To express f^{-1} as a function of x, interchange x and y. The resulting equation is $y = f^{-1}(x)$.

EXAMPLE 4 Find the inverse function of $f(x) = x^3 + 2$.

SOLUTION According to (5) we first write

$$y = x^3 + 2$$

Then we solve this equation for x:

$$x^3 = y - 2$$

$$x = \sqrt[3]{y - 2}$$

Finally, we interchange x and y:

$$y = \sqrt[3]{x - 2}$$

Therefore the inverse function is $f^{-1}(x) = \sqrt[3]{x - 2}$. ∎

In Example 4, notice how f^{-1} reverses the effect of f. The function f is the rule "Cube, then add 2"; f^{-1} is the rule "Subtract 2, then take the cube root."

The principle of interchanging x and y to find the inverse function also gives us the method for obtaining the graph of f^{-1} from the graph of f. Since $f(a) = b$ if and only if $f^{-1}(b) = a$, the point (a, b) is on the graph of f if and only if the point (b, a) is on the graph of f^{-1}. But we get the point (b, a) from (a, b) by reflecting about the line $y = x$. (See Figure 8.)

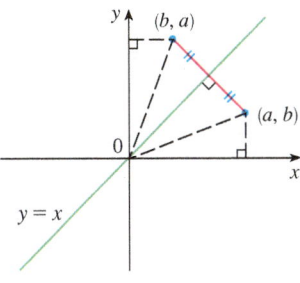

FIGURE 8

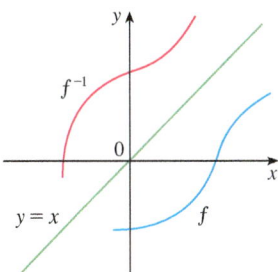

FIGURE 9

Therefore, as illustrated by Figure 9:

> The graph of f^{-1} is obtained by reflecting the graph of f about the line $y = x$.

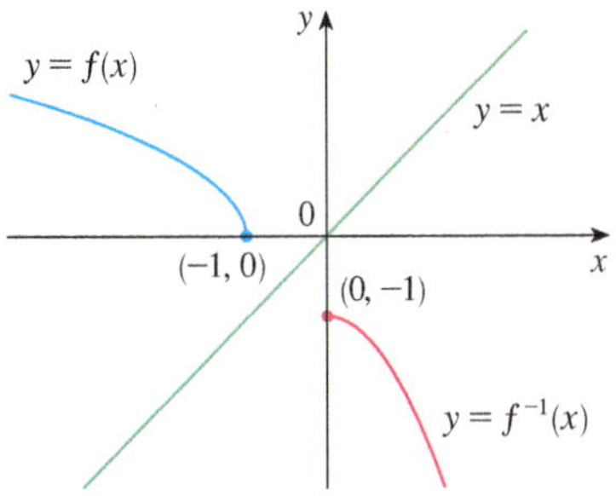

EXAMPLE 5 Sketch the graphs of $f(x) = \sqrt{-1 - x}$ and its inverse function using the same coordinate axes.

SOLUTION First we sketch the curve $y = \sqrt{-1 - x}$ (the top half of the parabola $y^2 = -1 - x$, or $x = -y^2 - 1$) and then we reflect about the line $y = x$ to get the graph of f^{-1}. (See Figure 10.) As a check on our graph, notice that the expression for f^{-1} is $f^{-1}(x) = -x^2 - 1$, $x \geq 0$. So the graph of f^{-1} is the right half of the parabola $y = -x^2 - 1$ and this seems reasonable from Figure 10. ■

FIGURE 10

■ Logarithmic Functions

If $b > 0$ and $b \neq 1$, the exponential function $f(x) = b^x$ is either increasing or decreasing and so it is one-to-one by the Horizontal Line Test. It therefore has an inverse function f^{-1}, which is called the **logarithmic function with base b** and is denoted by $\log_b$. If we use the formulation of an inverse function given by (3),

$$f^{-1}(x) = y \iff f(y) = x$$

then we have

$$\boxed{6} \qquad \log_b x = y \iff b^y = x$$

Thus, if $x > 0$, then $\log_b x$ is the exponent to which the base b must be raised to give x. For example, $\log_{10} 0.001 = -3$ because $10^{-3} = 0.001$.

The cancellation equations (4), when applied to the functions $f(x) = b^x$ and $f^{-1}(x) = \log_b x$, become

$$\boxed{7} \qquad \begin{aligned} \log_b(b^x) &= x \quad \text{for every } x \in \mathbb{R} \\ b^{\log_b x} &= x \quad \text{for every } x > 0 \end{aligned}$$

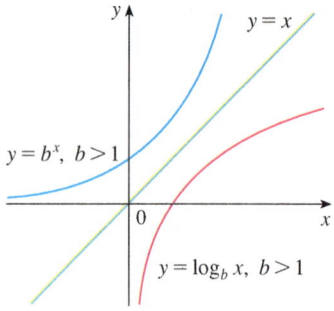

FIGURE 11

The logarithmic function $\log_b$ has domain $(0, \infty)$ and range $\mathbb{R}$. Its graph is the reflection of the graph of $y = b^x$ about the line $y = x$.

Figure 11 shows the case where $b > 1$. (The most important logarithmic functions have base $b > 1$.) The fact that $y = b^x$ is a very rapidly increasing function for $x > 0$ is reflected in the fact that $y = \log_b x$ is a very slowly increasing function for $x > 1$.

Figure 12 shows the graphs of $y = \log_b x$ with various values of the base $b > 1$. Since $\log_b 1 = 0$, the graphs of all logarithmic functions pass through the point $(1, 0)$.

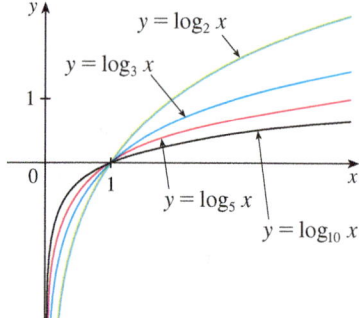

FIGURE 12

The following properties of logarithmic functions follow from the corresponding properties of exponential functions given in Section 1.4.

> **Laws of Logarithms** If x and y are positive numbers, then
>
> **1.** $\log_b(xy) = \log_b x + \log_b y$
>
> **2.** $\log_b\left(\dfrac{x}{y}\right) = \log_b x - \log_b y$
>
> **3.** $\log_b(x^r) = r \log_b x$ (where r is any real number)

EXAMPLE 6 Use the laws of logarithms to evaluate $\log_2 80 - \log_2 5$.

SOLUTION Using Law 2, we have

$$\log_2 80 - \log_2 5 = \log_2\left(\frac{80}{5}\right) = \log_2 16 = 4$$

because $2^4 = 16$. ∎

◼ Natural Logarithms

Of all possible bases b for logarithms, we will see in Chapter 3 that the most convenient choice of a base is the number e, which was defined in Section 1.4. The logarithm with base e is called the **natural logarithm** and has a special notation:

$$\log_e x = \ln x$$

Notation for Logarithms Most textbooks in calculus and the sciences, as well as calculators, use the notation $\ln x$ for the natural logarithm and $\log x$ for the "common logarithm," $\log_{10} x$. In the more advanced mathematical and scientific literature and in computer languages, however, the notation $\log x$ usually denotes the natural logarithm.

If we put $b = e$ and replace $\log_e$ with "ln" in (6) and (7), then the defining properties of the natural logarithm function become

SECTION 1.5 Inverse Functions and Logarithms

8
$$\ln x = y \iff e^y = x$$

9
$$\ln(e^x) = x \quad x \in \mathbb{R}$$
$$e^{\ln x} = x \quad x > 0$$

In particular, if we set $x = 1$, we get

$$\ln e = 1$$

EXAMPLE 7 Find x if $\ln x = 5$.

SOLUTION 1 From (8) we see that

$$\ln x = 5 \quad \text{means} \quad e^5 = x$$

Therefore $x = e^5$.

(If you have trouble working with the "ln" notation, just replace it by $\log_e$. Then the equation becomes $\log_e x = 5$; so, by the definition of logarithm, $e^5 = x$.)

SOLUTION 2 Start with the equation

$$\ln x = 5$$

and apply the exponential function to both sides of the equation:

$$e^{\ln x} = e^5$$

But the second cancellation equation in (9) says that $e^{\ln x} = x$. Therefore $x = e^5$. ∎

EXAMPLE 8 Solve the equation $e^{5-3x} = 10$.

SOLUTION We take natural logarithms of both sides of the equation and use (9):

$$\ln(e^{5-3x}) = \ln 10$$
$$5 - 3x = \ln 10$$
$$3x = 5 - \ln 10$$
$$x = \tfrac{1}{3}(5 - \ln 10)$$

Since the natural logarithm is found on scientific calculators, we can approximate the solution: to four decimal places, $x \approx 0.8991$. ∎

EXAMPLE 9 Express $\ln a + \tfrac{1}{2} \ln b$ as a single logarithm.

SOLUTION Using Laws 3 and 1 of logarithms, we have

$$\ln a + \tfrac{1}{2} \ln b = \ln a + \ln b^{1/2}$$
$$= \ln a + \ln \sqrt{b}$$
$$= \ln(a\sqrt{b})$$

∎

The following formula shows that logarithms with any base can be expressed in terms of the natural logarithm.

> **[10] Change of Base Formula** For any positive number b ($b \neq 1$), we have
> $$\log_b x = \frac{\ln x}{\ln b}$$

PROOF Let $y = \log_b x$. Then, from (6), we have $b^y = x$. Taking natural logarithms of both sides of this equation, we get $y \ln b = \ln x$. Therefore

$$y = \frac{\ln x}{\ln b}$$

Scientific calculators have a key for natural logarithms, so Formula 10 enables us to use a calculator to compute a logarithm with any base (as shown in the following example). Similarly, Formula 10 allows us to graph any logarithmic function on a graphing calculator or computer (see Exercises 43 and 44).

EXAMPLE 10 Evaluate $\log_8 5$ correct to six decimal places.

SOLUTION Formula 10 gives

$$\log_8 5 = \frac{\ln 5}{\ln 8} \approx 0.773976$$

Graph and Growth of the Natural Logarithm

The graphs of the exponential function $y = e^x$ and its inverse function, the natural logarithm function, are shown in Figure 13. Because the curve $y = e^x$ crosses the y-axis with a slope of 1, it follows that the reflected curve $y = \ln x$ crosses the x-axis with a slope of 1.

In common with all other logarithmic functions with base greater than 1, the natural logarithm is an increasing function defined on $(0, \infty)$ and the y-axis is a vertical asymptote. (This means that the values of $\ln x$ become very large negative as x approaches 0.)

EXAMPLE 11 Sketch the graph of the function $y = \ln(x - 2) - 1$.

SOLUTION We start with the graph of $y = \ln x$ as given in Figure 13. Using the transformations of Section 1.3, we shift it 2 units to the right to get the graph of $y = \ln(x - 2)$ and then we shift it 1 unit downward to get the graph of $y = \ln(x - 2) - 1$. (See Figure 14.)

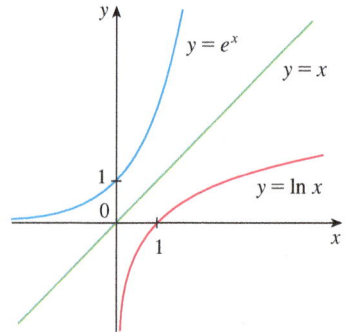

FIGURE 13
The graph of $y = \ln x$ is the reflection of the graph of $y = e^x$ about the line $y = x$.

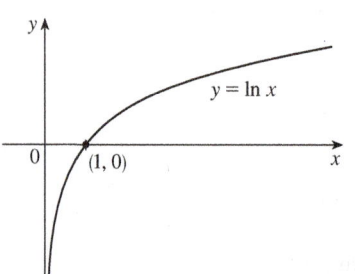
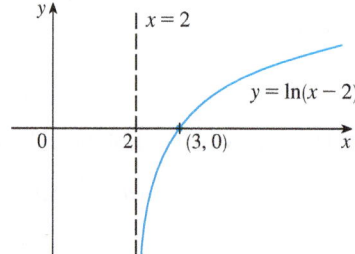
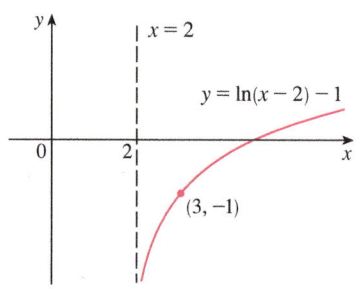

FIGURE 14

Although ln x is an increasing function, it grows *very* slowly when $x > 1$. In fact, ln x grows more slowly than any positive power of x. To illustrate this fact, we compare approximate values of the functions $y = \ln x$ and $y = x^{1/2} = \sqrt{x}$ in the following table and we graph them in Figures 15 and 16. You can see that initially the graphs of $y = \sqrt{x}$ and $y = \ln x$ grow at comparable rates, but eventually the root function far surpasses the logarithm.

x	1	2	5	10	50	100	500	1000	10,000	100,000
$\ln x$	0	0.69	1.61	2.30	3.91	4.6	6.2	6.9	9.2	11.5
$\sqrt{x}$	1	1.41	2.24	3.16	7.07	10.0	22.4	31.6	100	316
$\dfrac{\ln x}{\sqrt{x}}$	0	0.49	0.72	0.73	0.55	0.46	0.28	0.22	0.09	0.04

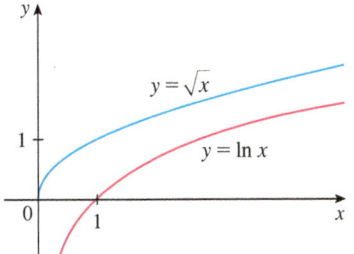

FIGURE 15

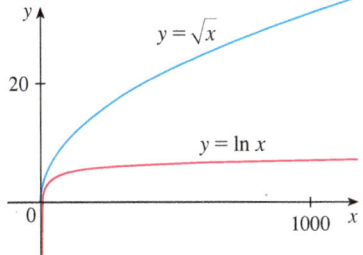

FIGURE 16

■ Inverse Trigonometric Functions

When we try to find the inverse trigonometric functions, we have a slight difficulty: Because the trigonometric functions are not one-to-one, they don't have inverse functions. The difficulty is overcome by restricting the domains of these functions so that they become one-to-one.

You can see from Figure 17 that the sine function $y = \sin x$ is not one-to-one (use the Horizontal Line Test). But the function $f(x) = \sin x$, $-\pi/2 \leq x \leq \pi/2$, is one-to-one (see Figure 18). The inverse function of this restricted sine function f exists and is denoted by $\sin^{-1}$ or arcsin. It is called the **inverse sine function** or the **arcsine function**.

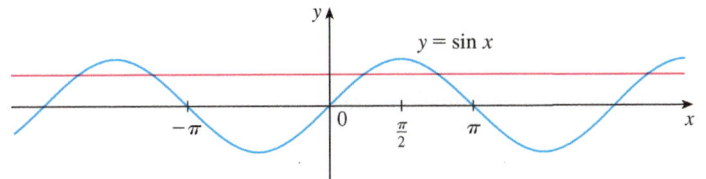

FIGURE 17

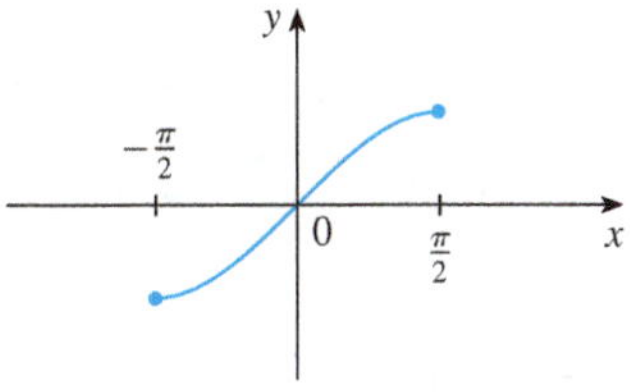

FIGURE 18
$y = \sin x$, $-\frac{\pi}{2} \leq x \leq \frac{\pi}{2}$

Since the definition of an inverse function says that

$$f^{-1}(x) = y \iff f(y) = x$$

we have

$$\sin^{-1}x = y \iff \sin y = x \quad \text{and} \quad -\frac{\pi}{2} \leq y \leq \frac{\pi}{2}$$

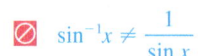

Thus, if $-1 \leq x \leq 1$, $\sin^{-1}x$ is the number between $-\pi/2$ and $\pi/2$ whose sine is x.

EXAMPLE 12 Evaluate (a) $\sin^{-1}\left(\frac{1}{2}\right)$ and (b) $\tan(\arcsin \frac{1}{3})$.

SOLUTION

(a) We have

$$\sin^{-1}\left(\frac{1}{2}\right) = \frac{\pi}{6}$$

because $\sin(\pi/6) = \frac{1}{2}$ and $\pi/6$ lies between $-\pi/2$ and $\pi/2$.

(b) Let $\theta = \arcsin \frac{1}{3}$, so $\sin \theta = \frac{1}{3}$. Then we can draw a right triangle with angle θ as in Figure 19 and deduce from the Pythagorean Theorem that the third side has length $\sqrt{9-1} = 2\sqrt{2}$. This enables us to read from the triangle that

$$\tan(\arcsin \tfrac{1}{3}) = \tan \theta = \frac{1}{2\sqrt{2}}$$

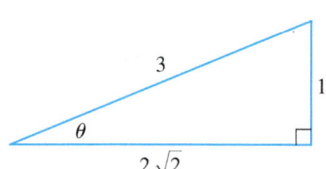

FIGURE 19

The cancellation equations for inverse functions become, in this case,

$$\sin^{-1}(\sin x) = x \quad \text{for } -\frac{\pi}{2} \leq x \leq \frac{\pi}{2}$$

$$\sin(\sin^{-1}x) = x \quad \text{for } -1 \leq x \leq 1$$

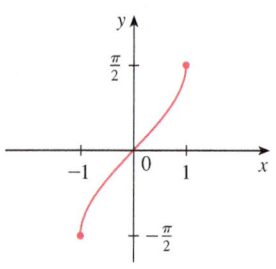

FIGURE 20
$y = \sin^{-1}x = \arcsin x$

The inverse sine function, $\sin^{-1}$, has domain $[-1, 1]$ and range $[-\pi/2, \pi/2]$, and its graph, shown in Figure 20, is obtained from that of the restricted sine function (Figure 18) by reflection about the line $y = x$.

The **inverse cosine function** is handled similarly. The restricted cosine function $f(x) = \cos x$, $0 \leq x \leq \pi$, is one-to-one (see Figure 21) and so it has an inverse function denoted by $\cos^{-1}$ or arccos.

$$\cos^{-1}x = y \iff \cos y = x \quad \text{and} \quad 0 \leq y \leq \pi$$

The cancellation equations are

$$\cos^{-1}(\cos x) = x \quad \text{for } 0 \leq x \leq \pi$$

$$\cos(\cos^{-1}x) = x \quad \text{for } -1 \leq x \leq 1$$

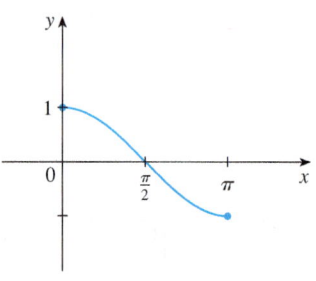

FIGURE 21
$y = \cos x, 0 \leq x \leq \pi$

The inverse cosine function, $\cos^{-1}$, has domain $[-1, 1]$ and range $[0, \pi]$. Its graph is shown in Figure 22.

The tangent function can be made one-to-one by restricting it to the interval $(-\pi/2, \pi/2)$. Thus the **inverse tangent function** is defined as the inverse of the function $f(x) = \tan x$, $-\pi/2 < x < \pi/2$. (See Figure 23.) It is denoted by $\tan^{-1}$ or arctan.

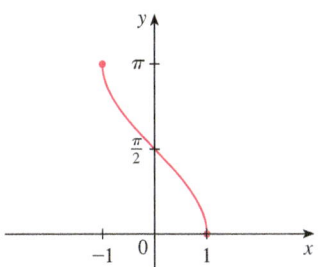

FIGURE 22
$y = \cos^{-1} x = \arccos x$

$$\tan^{-1} x = y \quad \Longleftrightarrow \quad \tan y = x \quad \text{and} \quad -\frac{\pi}{2} < y < \frac{\pi}{2}$$

EXAMPLE 13 Simplify the expression $\cos(\tan^{-1} x)$.

SOLUTION 1 Let $y = \tan^{-1} x$. Then $\tan y = x$ and $-\pi/2 < y < \pi/2$. We want to find $\cos y$ but, since $\tan y$ is known, it is easier to find $\sec y$ first:

$$\sec^2 y = 1 + \tan^2 y = 1 + x^2$$

$$\sec y = \sqrt{1 + x^2} \qquad \text{(since } \sec y > 0 \text{ for } -\pi/2 < y < \pi/2\text{)}$$

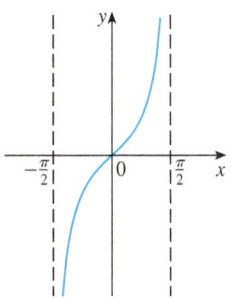

FIGURE 23
$y = \tan x$, $-\frac{\pi}{2} < x < \frac{\pi}{2}$

Thus
$$\cos(\tan^{-1} x) = \cos y = \frac{1}{\sec y} = \frac{1}{\sqrt{1 + x^2}}$$

SOLUTION 2 Instead of using trigonometric identities as in Solution 1, it is perhaps easier to use a diagram. If $y = \tan^{-1} x$, then $\tan y = x$, and we can read from Figure 24 (which illustrates the case $y > 0$) that

$$\cos(\tan^{-1} x) = \cos y = \frac{1}{\sqrt{1 + x^2}}$$

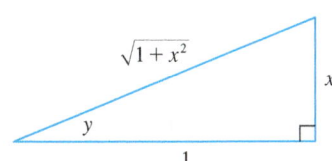

FIGURE 24

The inverse tangent function, $\tan^{-1} = \arctan$, has domain $\mathbb{R}$ and range $(-\pi/2, \pi/2)$. Its graph is shown in Figure 25.

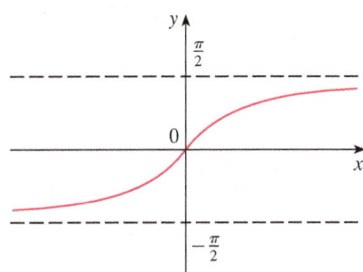

FIGURE 25
$y = \tan^{-1} x = \arctan x$

We know that the lines $x = \pm\pi/2$ are vertical asymptotes of the graph of tan. Since the graph of $\tan^{-1}$ is obtained by reflecting the graph of the restricted tangent function about the line $y = x$, it follows that the lines $y = \pi/2$ and $y = -\pi/2$ are horizontal asymptotes of the graph of $\tan^{-1}$.

The remaining inverse trigonometric functions are not used as frequently and are summarized here.

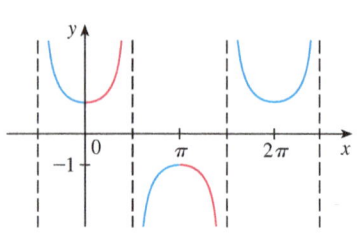

FIGURE 26
$y = \sec x$

$$\boxed{11} \quad \begin{aligned} y = \csc^{-1} x \; (|x| \geq 1) &\iff \csc y = x \text{ and } y \in (0, \pi/2] \cup (\pi, 3\pi/2] \\ y = \sec^{-1} x \; (|x| \geq 1) &\iff \sec y = x \text{ and } y \in [0, \pi/2) \cup [\pi, 3\pi/2) \\ y = \cot^{-1} x \; (x \in \mathbb{R}) &\iff \cot y = x \text{ and } y \in (0, \pi) \end{aligned}$$

The choice of intervals for y in the definitions of $\csc^{-1}$ and $\sec^{-1}$ is not universally agreed upon. For instance, some authors use $y \in [0, \pi/2) \cup (\pi/2, \pi]$ in the definition of $\sec^{-1}$. [You can see from the graph of the secant function in Figure 26 that both this choice and the one in (11) will work.]

1.5 EXERCISES

1. (a) What is a one-to-one function?
(b) How can you tell from the graph of a function whether it is one-to-one?

2. (a) Suppose f is a one-to-one function with domain A and range B. How is the inverse function f^{-1} defined? What is the domain of f^{-1}? What is the range of f^{-1}?
(b) If you are given a formula for f, how do you find a formula for f^{-1}?
(c) If you are given the graph of f, how do you find the graph of f^{-1}?

3–14 A function is given by a table of values, a graph, a formula, or a verbal description. Determine whether it is one-to-one.

3.

x	1	2	3	4	5	6
$f(x)$	1.5	2.0	3.6	5.3	2.8	2.0

4.

x	1	2	3	4	5	6
$f(x)$	1.0	1.9	2.8	3.5	3.1	2.9

5.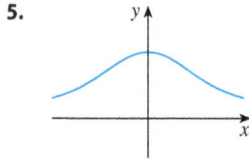

6.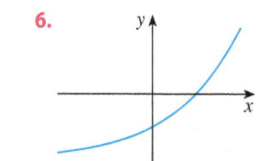

7.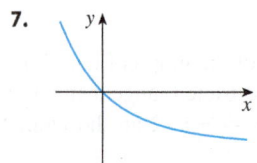

8.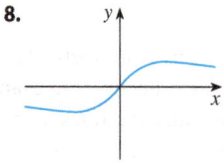

9. $f(x) = 2x - 3$

10. $f(x) = x^4 - 16$

11. $g(x) = 1 - \sin x$

12. $g(x) = \sqrt[3]{x}$

13. $f(t)$ is the height of a football t seconds after kickoff.

14. $f(t)$ is your height at age t.

15. Assume that f is a one-to-one function.
(a) If $f(6) = 17$, what is $f^{-1}(17)$?
(b) If $f^{-1}(3) = 2$, what is $f(2)$?

16. If $f(x) = x^5 + x^3 + x$, find $f^{-1}(3)$ and $f(f^{-1}(2))$.

17. If $g(x) = 3 + x + e^x$, find $g^{-1}(4)$.

18. The graph of f is given.
(a) Why is f one-to-one?
(b) What are the domain and range of f^{-1}?
(c) What is the value of $f^{-1}(2)$?
(d) Estimate the value of $f^{-1}(0)$.

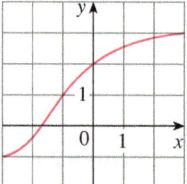

19. The formula $C = \frac{5}{9}(F - 32)$, where $F \geq -459.67$, expresses the Celsius temperature C as a function of the Fahrenheit temperature F. Find a formula for the inverse function and interpret it. What is the domain of the inverse function?

20. In the theory of relativity, the mass of a particle with speed v is

$$m = f(v) = \frac{m_0}{\sqrt{1 - v^2/c^2}}$$

where m_0 is the rest mass of the particle and c is the speed of light in a vacuum. Find the inverse function of f and explain its meaning.

21–26 Find a formula for the inverse of the function.

21. $f(x) = 1 + \sqrt{2 + 3x}$ **22.** $f(x) = \dfrac{4x - 1}{2x + 3}$

23. $f(x) = e^{2x-1}$ **24.** $y = x^2 - x$, $x \geq \frac{1}{2}$

25. $y = \ln(x + 3)$ **26.** $y = \dfrac{1 - e^{-x}}{1 + e^{-x}}$

27–28 Find an explicit formula for f^{-1} and use it to graph f^{-1}, f, and the line $y = x$ on the same screen. To check your work, see whether the graphs of f and f^{-1} are reflections about the line.

27. $f(x) = \sqrt{4x + 3}$ **28.** $f(x) = 1 + e^{-x}$

29–30 Use the given graph of f to sketch the graph of f^{-1}.

29. 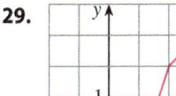   **30.**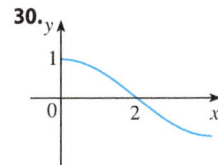

31. Let $f(x) = \sqrt{1 - x^2}$, $0 \leq x \leq 1$.
  (a) Find f^{-1}. How is it related to f?
  (b) Identify the graph of f and explain your answer to part (a).

32. Let $g(x) = \sqrt[3]{1 - x^3}$.
  (a) Find g^{-1}. How is it related to g?
  (b) Graph g. How do you explain your answer to part (a)?

33. (a) How is the logarithmic function $y = \log_b x$ defined?
  (b) What is the domain of this function?
  (c) What is the range of this function?
  (d) Sketch the general shape of the graph of the function $y = \log_b x$ if $b > 1$.

34. (a) What is the natural logarithm?
  (b) What is the common logarithm?
  (c) Sketch the graphs of the natural logarithm function and the natural exponential function with a common set of axes.

35–38 Find the exact value of each expression.

35. (a) $\log_2 32$ (b) $\log_8 2$

36. (a) $\log_5 \frac{1}{125}$ (b) $\ln(1/e^2)$

37. (a) $\log_{10} 40 + \log_{10} 2.5$
  (b) $\log_8 60 - \log_8 3 - \log_8 5$

38. (a) $e^{-\ln 2}$ (b) $e^{\ln(\ln e^3)}$

39–41 Express the given quantity as a single logarithm.

39. $\ln 10 + 2 \ln 5$ **40.** $\ln b + 2 \ln c - 3 \ln d$

41. $\frac{1}{3} \ln(x + 2)^3 + \frac{1}{2} [\ln x - \ln(x^2 + 3x + 2)^2]$

42. Use Formula 10 to evaluate each logarithm correct to six decimal places.
  (a) $\log_5 10$ (b) $\log_3 57$

43–44 Use Formula 10 to graph the given functions on a common screen. How are these graphs related?

43. $y = \log_{1.5} x$, $y = \ln x$, $y = \log_{10} x$, $y = \log_{50} x$

44. $y = \ln x$, $y = \log_{10} x$, $y = e^x$, $y = 10^x$

45. Suppose that the graph of $y = \log_2 x$ is drawn on a coordinate grid where the unit of measurement is an inch. How many miles to the right of the origin do we have to move before the height of the curve reaches 3 ft?

46. Compare the functions $f(x) = x^{0.1}$ and $g(x) = \ln x$ by graphing both f and g in several viewing rectangles. When does the graph of f finally surpass the graph of g?

47–48 Make a rough sketch of the graph of each function. Do not use a calculator. Just use the graphs given in Figures 12 and 13 and, if necessary, the transformations of Section 1.3.

47. (a) $y = \log_{10}(x + 5)$ (b) $y = -\ln x$

48. (a) $y = \ln(-x)$ (b) $y = \ln|x|$

49–50 (a) What are the domain and range of f?
(b) What is the x-intercept of the graph of f?
(c) Sketch the graph of f.

49. $f(x) = \ln x + 2$ **50.** $f(x) = \ln(x - 1) - 1$

51–54 Solve each equation for x.

51. (a) $e^{7-4x} = 6$ (b) $\ln(3x - 10) = 2$

52. (a) $\ln(x^2 - 1) = 3$ (b) $e^{2x} - 3e^x + 2 = 0$

53. (a) $2^{x-5} = 3$ (b) $\ln x + \ln(x - 1) = 1$

54. (a) $\ln(\ln x) = 1$ (b) $e^{ax} = Ce^{bx}$, where $a \neq b$

55–56 Solve each inequality for x.

55. (a) $\ln x < 0$ (b) $e^x > 5$

56. (a) $1 < e^{3x-1} < 2$ (b) $1 - 2 \ln x < 3$

57. (a) Find the domain of $f(x) = \ln(e^x - 3)$.
  (b) Find f^{-1} and its domain.

58. (a) What are the values of $e^{\ln 300}$ and $\ln(e^{300})$?
(b) Use your calculator to evaluate $e^{\ln 300}$ and $\ln(e^{300})$. What do you notice? Can you explain why the calculator has trouble?

CAS 59. Graph the function $f(x) = \sqrt{x^3 + x^2 + x + 1}$ and explain why it is one-to-one. Then use a computer algebra system to find an explicit expression for $f^{-1}(x)$. (Your CAS will produce three possible expressions. Explain why two of them are irrelevant in this context.)

CAS 60. (a) If $g(x) = x^6 + x^4$, $x \geq 0$, use a computer algebra system to find an expression for $g^{-1}(x)$.
(b) Use the expression in part (a) to graph $y = g(x)$, $y = x$, and $y = g^{-1}(x)$ on the same screen.

61. If a bacteria population starts with 100 bacteria and doubles every three hours, then the number of bacteria after t hours is $n = f(t) = 100 \cdot 2^{t/3}$.
(a) Find the inverse of this function and explain its meaning.
(b) When will the population reach 50,000?

62. When a camera flash goes off, the batteries immediately begin to recharge the flash's capacitor, which stores electric charge given by
$$Q(t) = Q_0(1 - e^{-t/a})$$
(The maximum charge capacity is Q_0 and t is measured in seconds.)
(a) Find the inverse of this function and explain its meaning.
(b) How long does it take to recharge the capacitor to 90% of capacity if $a = 2$?

63–68 Find the exact value of each expression.

63. (a) $\cos^{-1}(-1)$ (b) $\sin^{-1}(0.5)$

64. (a) $\tan^{-1}\sqrt{3}$ (b) $\arctan(-1)$

65. (a) $\csc^{-1}\sqrt{2}$ (b) $\arcsin 1$

66. (a) $\sin^{-1}(-1/\sqrt{2})$ (b) $\cos^{-1}(\sqrt{3}/2)$

67. (a) $\cot^{-1}(-\sqrt{3})$ (b) $\sec^{-1} 2$

68. (a) $\arcsin(\sin(5\pi/4))$ (b) $\cos(2\sin^{-1}(\frac{5}{13}))$

69. Prove that $\cos(\sin^{-1} x) = \sqrt{1 - x^2}$.

70–72 Simplify the expression.

70. $\tan(\sin^{-1} x)$ **71.** $\sin(\tan^{-1} x)$ **72.** $\sin(2 \arccos x)$

73–74 Graph the given functions on the same screen. How are these graphs related?

73. $y = \sin x$, $-\pi/2 \leq x \leq \pi/2$; $y = \sin^{-1} x$; $y = x$

74. $y = \tan x$, $-\pi/2 < x < \pi/2$; $y = \tan^{-1} x$; $y = x$

75. Find the domain and range of the function
$$g(x) = \sin^{-1}(3x + 1)$$

76. (a) Graph the function $f(x) = \sin(\sin^{-1} x)$ and explain the appearance of the graph.
(b) Graph the function $g(x) = \sin^{-1}(\sin x)$. How do you explain the appearance of this graph?

77. (a) If we shift a curve to the left, what happens to its reflection about the line $y = x$? In view of this geometric principle, find an expression for the inverse of $g(x) = f(x + c)$, where f is a one-to-one function.
(b) Find an expression for the inverse of $h(x) = f(cx)$, where $c \neq 0$.

1 REVIEW

CONCEPT CHECK

Answers to the Concept Check can be found on the back endpapers.

1. (a) What is a function? What are its domain and range?
(b) What is the graph of a function?
(c) How can you tell whether a given curve is the graph of a function?

2. Discuss four ways of representing a function. Illustrate your discussion with examples.

3. (a) What is an even function? How can you tell if a function is even by looking at its graph? Give three examples of an even function.
(b) What is an odd function? How can you tell if a function is odd by looking at its graph? Give three examples of an odd function.

4. What is an increasing function?

5. What is a mathematical model?

6. Give an example of each type of function.
(a) Linear function   (b) Power function
(c) Exponential function   (d) Quadratic function
(e) Polynomial of degree 5   (f) Rational function

7. Sketch by hand, on the same axes, the graphs of the following functions.
(a) $f(x) = x$ (b) $g(x) = x^2$
(c) $h(x) = x^3$ (d) $j(x) = x^4$

8. Draw, by hand, a rough sketch of the graph of each function.
   (a) $y = \sin x$ (b) $y = \tan x$ (c) $y = e^x$
   (d) $y = \ln x$ (e) $y = 1/x$ (f) $y = |x|$
   (g) $y = \sqrt{x}$ (h) $y = \tan^{-1} x$

9. Suppose that f has domain A and g has domain B.
   (a) What is the domain of $f + g$?
   (b) What is the domain of fg?
   (c) What is the domain of f/g?

10. How is the composite function $f \circ g$ defined? What is its domain?

11. Suppose the graph of f is given. Write an equation for each of the graphs that are obtained from the graph of f as follows.
    (a) Shift 2 units upward.   (b) Shift 2 units downward.
    (c) Shift 2 units to the right.   (d) Shift 2 units to the left.
    (e) Reflect about the x-axis.
    (f) Reflect about the y-axis.
    (g) Stretch vertically by a factor of 2.
    (h) Shrink vertically by a factor of 2.
    (i) Stretch horizontally by a factor of 2.
    (j) Shrink horizontally by a factor of 2.

12. (a) What is a one-to-one function? How can you tell if a function is one-to-one by looking at its graph?
    (b) If f is a one-to-one function, how is its inverse function f^{-1} defined? How do you obtain the graph of f^{-1} from the graph of f?

13. (a) How is the inverse sine function $f(x) = \sin^{-1} x$ defined? What are its domain and range?
    (b) How is the inverse cosine function $f(x) = \cos^{-1} x$ defined? What are its domain and range?
    (c) How is the inverse tangent function $f(x) = \tan^{-1} x$ defined? What are its domain and range?

TRUE-FALSE QUIZ

Determine whether the statement is true or false. If it is true, explain why. If it is false, explain why or give an example that disproves the statement.

1. If f is a function, then $f(s + t) = f(s) + f(t)$.
2. If $f(s) = f(t)$, then $s = t$.
3. If f is a function, then $f(3x) = 3f(x)$.
4. If $x_1 < x_2$ and f is a decreasing function, then $f(x_1) > f(x_2)$.
5. A vertical line intersects the graph of a function at most once.
6. If f and g are functions, then $f \circ g = g \circ f$.
7. If f is one-to-one, then $f^{-1}(x) = \dfrac{1}{f(x)}$.
8. You can always divide by e^x.
9. If $0 < a < b$, then $\ln a < \ln b$.
10. If $x > 0$, then $(\ln x)^6 = 6 \ln x$.
11. If $x > 0$ and $a > 1$, then $\dfrac{\ln x}{\ln a} = \ln \dfrac{x}{a}$.
12. $\tan^{-1}(-1) = 3\pi/4$
13. $\tan^{-1} x = \dfrac{\sin^{-1} x}{\cos^{-1} x}$
14. If x is any real number, then $\sqrt{x^2} = x$.

EXERCISES

1. Let f be the function whose graph is given.

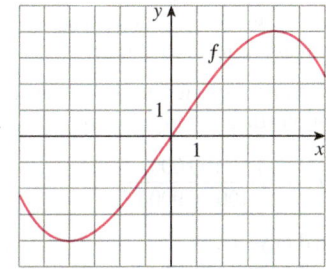

   (a) Estimate the value of $f(2)$.
   (b) Estimate the values of x such that $f(x) = 3$.
   (c) State the domain of f.
   (d) State the range of f.
   (e) On what interval is f increasing?
   (f) Is f one-to-one? Explain.
   (g) Is f even, odd, or neither even nor odd? Explain.

2. The graph of g is given.

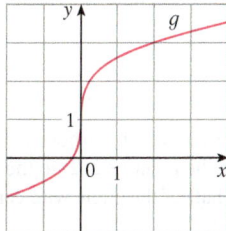

   (a) State the value of $g(2)$.
   (b) Why is g one-to-one?
   (c) Estimate the value of $g^{-1}(2)$.
   (d) Estimate the domain of g^{-1}.
   (e) Sketch the graph of g^{-1}.

3. If $f(x) = x^2 - 2x + 3$, evaluate the difference quotient
$$\frac{f(a + h) - f(a)}{h}$$

4. Sketch a rough graph of the yield of a crop as a function of the amount of fertilizer used.

5–8 Find the domain and range of the function. Write your answer in interval notation.

5. $f(x) = 2/(3x - 1)$
6. $g(x) = \sqrt{16 - x^4}$
7. $h(x) = \ln(x + 6)$
8. $F(t) = 3 + \cos 2t$

9. Suppose that the graph of f is given. Describe how the graphs of the following functions can be obtained from the graph of f.
(a) $y = f(x) + 8$
(b) $y = f(x + 8)$
(c) $y = 1 + 2f(x)$
(d) $y = f(x - 2) - 2$
(e) $y = -f(x)$
(f) $y = f^{-1}(x)$

10. The graph of f is given. Draw the graphs of the following functions.
(a) $y = f(x - 8)$
(b) $y = -f(x)$
(c) $y = 2 - f(x)$
(d) $y = \frac{1}{2}f(x) - 1$
(e) $y = f^{-1}(x)$
(f) $y = f^{-1}(x + 3)$

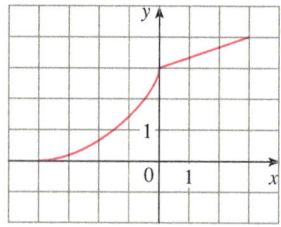

11–16 Use transformations to sketch the graph of the function.

11. $y = (x - 2)^3$
12. $y = 2\sqrt{x}$
13. $y = x^2 - 2x + 2$
14. $y = \ln(x + 1)$
15. $f(x) = -\cos 2x$
16. $f(x) = \begin{cases} -x & \text{if } x < 0 \\ e^x - 1 & \text{if } x \geq 0 \end{cases}$

17. Determine whether f is even, odd, or neither even nor odd.
(a) $f(x) = 2x^5 - 3x^2 + 2$
(b) $f(x) = x^3 - x^7$
(c) $f(x) = e^{-x^2}$
(d) $f(x) = 1 + \sin x$

18. Find an expression for the function whose graph consists of the line segment from the point $(-2, 2)$ to the point $(-1, 0)$ together with the top half of the circle with center the origin and radius 1.

19. If $f(x) = \ln x$ and $g(x) = x^2 - 9$, find the functions (a) $f \circ g$, (b) $g \circ f$, (c) $f \circ f$, (d) $g \circ g$, and their domains.

20. Express the function $F(x) = 1/\sqrt{x + \sqrt{x}}$ as a composition of three functions.

21. Life expectancy improved dramatically in the 20th century. The table gives the life expectancy at birth (in years) of males born in the United States. Use a scatter plot to choose an appropriate type of model. Use your model to predict the life span of a male born in the year 2010.

Birth year	Life expectancy	Birth year	Life expectancy
1900	48.3	1960	66.6
1910	51.1	1970	67.1
1920	55.2	1980	70.0
1930	57.4	1990	71.8
1940	62.5	2000	73.0
1950	65.6		

22. A small-appliance manufacturer finds that it costs $9000 to produce 1000 toaster ovens a week and $12,000 to produce 1500 toaster ovens a week.
(a) Express the cost as a function of the number of toaster ovens produced, assuming that it is linear. Then sketch the graph.
(b) What is the slope of the graph and what does it represent?
(c) What is the y-intercept of the graph and what does it represent?

23. If $f(x) = 2x + \ln x$, find $f^{-1}(2)$.

24. Find the inverse function of $f(x) = \dfrac{x + 1}{2x + 1}$.

25. Find the exact value of each expression.
(a) $e^{2\ln 3}$
(b) $\log_{10} 25 + \log_{10} 4$
(c) $\tan(\arcsin \frac{1}{2})$
(d) $\sin(\cos^{-1}(\frac{4}{5}))$

26. Solve each equation for x.
(a) $e^x = 5$
(b) $\ln x = 2$
(c) $e^{e^x} = 2$
(d) $\tan^{-1} x = 1$

27. The half-life of palladium-100, ^{100}Pd, is four days. (So half of any given quantity of ^{100}Pd will disintegrate in four days.) The initial mass of a sample is one gram.
(a) Find the mass that remains after 16 days.
(b) Find the mass $m(t)$ that remains after t days.
(c) Find the inverse of this function and explain its meaning.
(d) When will the mass be reduced to 0.01g?

28. The population of a certain species in a limited environment with initial population 100 and carrying capacity 1000 is

$$P(t) = \frac{100{,}000}{100 + 900e^{-t}}$$

where t is measured in years.
(a) Graph this function and estimate how long it takes for the population to reach 900.
(b) Find the inverse of this function and explain its meaning.
(c) Use the inverse function to find the time required for the population to reach 900. Compare with the result of part (a).

Principles of Problem Solving

There are no hard and fast rules that will ensure success in solving problems. However, it is possible to outline some general steps in the problem-solving process and to give some principles that may be useful in the solution of certain problems. These steps and principles are just common sense made explicit. They have been adapted from George Polya's book *How To Solve It*.

1 UNDERSTAND THE PROBLEM

The first step is to read the problem and make sure that you understand it clearly. Ask yourself the following questions:

What is the unknown?

What are the given quantities?

What are the given conditions?

For many problems it is useful to

draw a diagram

and identify the given and required quantities on the diagram.

Usually it is necessary to

introduce suitable notation

In choosing symbols for the unknown quantities we often use letters such as a, b, c, m, n, x, and y, but in some cases it helps to use initials as suggestive symbols; for instance, V for volume or t for time.

2 THINK OF A PLAN

Find a connection between the given information and the unknown that will enable you to calculate the unknown. It often helps to ask yourself explicitly: "How can I relate the given to the unknown?" If you don't see a connection immediately, the following ideas may be helpful in devising a plan.

Try to Recognize Something Familiar Relate the given situation to previous knowledge. Look at the unknown and try to recall a more familiar problem that has a similar unknown.

Try to Recognize Patterns Some problems are solved by recognizing that some kind of pattern is occurring. The pattern could be geometric, or numerical, or algebraic. If you can see regularity or repetition in a problem, you might be able to guess what the continuing pattern is and then prove it.

Use Analogy Try to think of an analogous problem, that is, a similar problem, a related problem, but one that is easier than the original problem. If you can solve the similar, simpler problem, then it might give you the clues you need to solve the original, more difficult problem. For instance, if a problem involves very large numbers, you could first try a similar problem with smaller numbers. Or if the problem involves three-dimensional geometry, you could look for a similar problem in two-dimensional geometry. Or if the problem you start with is a general one, you could first try a special case.

Introduce Something Extra It may sometimes be necessary to introduce something new, an auxiliary aid, to help make the connection between the given and the unknown. For instance, in a problem where a diagram is useful the auxiliary aid could be a new line drawn in a diagram. In a more algebraic problem it could be a new unknown that is related to the original unknown.

Take Cases We may sometimes have to split a problem into several cases and give a different argument for each of the cases. For instance, we often have to use this strategy in dealing with absolute value.

Work Backward Sometimes it is useful to imagine that your problem is solved and work backward, step by step, until you arrive at the given data. Then you may be able to reverse your steps and thereby construct a solution to the original problem. This procedure is commonly used in solving equations. For instance, in solving the equation $3x - 5 = 7$, we suppose that x is a number that satisfies $3x - 5 = 7$ and work backward. We add 5 to each side of the equation and then divide each side by 3 to get $x = 4$. Since each of these steps can be reversed, we have solved the problem.

Establish Subgoals In a complex problem it is often useful to set subgoals (in which the desired situation is only partially fulfilled). If we can first reach these subgoals, then we may be able to build on them to reach our final goal.

Indirect Reasoning Sometimes it is appropriate to attack a problem indirectly. In using proof by contradiction to prove that P implies Q, we assume that P is true and Q is false and try to see why this can't happen. Somehow we have to use this information and arrive at a contradiction to what we absolutely know is true.

Mathematical Induction In proving statements that involve a positive integer n, it is frequently helpful to use the following principle.

Principle of Mathematical Induction Let S_n be a statement about the positive integer n. Suppose that

1. S_1 is true.
2. S_{k+1} is true whenever S_k is true.

Then S_n is true for all positive integers n.

This is reasonable because, since S_1 is true, it follows from condition 2 (with $k = 1$) that S_2 is true. Then, using condition 2 with $k = 2$, we see that S_3 is true. Again using condition 2, this time with $k = 3$, we have that S_4 is true. This procedure can be followed indefinitely.

3 CARRY OUT THE PLAN

In Step 2 a plan was devised. In carrying out that plan we have to check each stage of the plan and write the details that prove that each stage is correct.

4 LOOK BACK

Having completed our solution, it is wise to look back over it, partly to see if we have made errors in the solution and partly to see if we can think of an easier way to solve the problem. Another reason for looking back is that it will familiarize us with the method of solution and this may be useful for solving a future problem. Descartes said, "Every problem that I solved became a rule which served afterwards to solve other problems."

These principles of problem solving are illustrated in the following examples. Before you look at the solutions, try to solve these problems yourself, referring to these Principles of Problem Solving if you get stuck. You may find it useful to refer to this section from time to time as you solve the exercises in the remaining chapters of this book.

EXAMPLE 1 Express the hypotenuse h of a right triangle with area 25 m² as a function of its perimeter P.

PS Understand the problem

SOLUTION Let's first sort out the information by identifying the unknown quantity and the data:

                  Unknown: hypotenuse h

                  Given quantities: perimeter P, area 25 m²

PS Draw a diagram

It helps to draw a diagram and we do so in Figure 1.

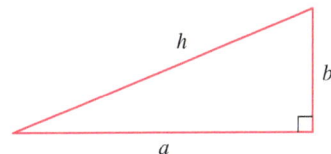

FIGURE 1

PS Connect the given with the unknown
PS Introduce something extra

In order to connect the given quantities to the unknown, we introduce two extra variables a and b, which are the lengths of the other two sides of the triangle. This enables us to express the given condition, which is that the triangle is right-angled, by the Pythagorean Theorem:

$$h^2 = a^2 + b^2$$

The other connections among the variables come by writing expressions for the area and perimeter:

$$25 = \tfrac{1}{2}ab \qquad P = a + b + h$$

Since P is given, notice that we now have three equations in the three unknowns a, b, and h:

$$\boxed{1} \qquad h^2 = a^2 + b^2$$
$$\boxed{2} \qquad 25 = \tfrac{1}{2}ab$$
$$\boxed{3} \qquad P = a + b + h$$

PS Relate to the familiar

Although we have the correct number of equations, they are not easy to solve in a straightforward fashion. But if we use the problem-solving strategy of trying to recognize something familiar, then we can solve these equations by an easier method. Look at the right sides of Equations 1, 2, and 3. Do these expressions remind you of anything familiar? Notice that they contain the ingredients of a familiar formula:

$$(a + b)^2 = a^2 + 2ab + b^2$$

Using this idea, we express $(a + b)^2$ in two ways. From Equations 1 and 2 we have

$$(a + b)^2 = (a^2 + b^2) + 2ab = h^2 + 4(25)$$

From Equation 3 we have

$$(a + b)^2 = (P - h)^2 = P^2 - 2Ph + h^2$$

Thus
$$h^2 + 100 = P^2 - 2Ph + h^2$$
$$2Ph = P^2 - 100$$
$$h = \frac{P^2 - 100}{2P}$$

This is the required expression for h as a function of P.

As the next example illustrates, it is often necessary to use the problem-solving principle of *taking cases* when dealing with absolute values.

EXAMPLE 2 Solve the inequality $|x - 3| + |x + 2| < 11$.

SOLUTION Recall the definition of absolute value:

$$|x| = \begin{cases} x & \text{if } x \geq 0 \\ -x & \text{if } x < 0 \end{cases}$$

It follows that

$$|x-3| = \begin{cases} x-3 & \text{if } x-3 \geq 0 \\ -(x-3) & \text{if } x-3 < 0 \end{cases}$$

$$= \begin{cases} x-3 & \text{if } x \geq 3 \\ -x+3 & \text{if } x < 3 \end{cases}$$

Similarly

$$|x+2| = \begin{cases} x+2 & \text{if } x+2 \geq 0 \\ -(x+2) & \text{if } x+2 < 0 \end{cases}$$

$$= \begin{cases} x+2 & \text{if } x \geq -2 \\ -x-2 & \text{if } x < -2 \end{cases}$$

PS Take cases

These expressions show that we must consider three cases:

$$x < -2 \qquad -2 \leq x < 3 \qquad x \geq 3$$

CASE I If $x < -2$, we have

$$|x-3| + |x+2| < 11$$

$$-x + 3 - x - 2 < 11$$

$$-2x < 10$$

$$x > -5$$

CASE II If $-2 \leq x < 3$, the given inequality becomes

$$-x + 3 + x + 2 < 11$$

$$5 < 11 \qquad \text{(always true)}$$

CASE III If $x \geq 3$, the inequality becomes

$$x - 3 + x + 2 < 11$$

$$2x < 12$$

$$x < 6$$

Combining cases I, II, and III, we see that the inequality is satisfied when $-5 < x < 6$. So the solution is the interval $(-5, 6)$. ■

In the following example we first guess the answer by looking at special cases and recognizing a pattern. Then we prove our conjecture by mathematical induction.

In using the Principle of Mathematical Induction, we follow three steps:

Step 1 Prove that S_n is true when $n = 1$.

Step 2 Assume that S_n is true when $n = k$ and deduce that S_n is true when $n = k + 1$.

Step 3 Conclude that S_n is true for all n by the Principle of Mathematical Induction.

EXAMPLE 3 If $f_0(x) = x/(x+1)$ and $f_{n+1} = f_0 \circ f_n$ for $n = 0, 1, 2, \ldots$, find a formula for $f_n(x)$.

PS Analogy: Try a similar, simpler problem

SOLUTION We start by finding formulas for $f_n(x)$ for the special cases $n = 1, 2,$ and 3.

$$f_1(x) = (f_0 \circ f_0)(x) = f_0(f_0(x)) = f_0\left(\frac{x}{x+1}\right)$$

$$= \frac{\frac{x}{x+1}}{\frac{x}{x+1} + 1} = \frac{\frac{x}{x+1}}{\frac{2x+1}{x+1}} = \frac{x}{2x+1}$$

$$f_2(x) = (f_0 \circ f_1)(x) = f_0(f_1(x)) = f_0\left(\frac{x}{2x+1}\right)$$

$$= \frac{\frac{x}{2x+1}}{\frac{x}{2x+1} + 1} = \frac{\frac{x}{2x+1}}{\frac{3x+1}{2x+1}} = \frac{x}{3x+1}$$

$$f_3(x) = (f_0 \circ f_2)(x) = f_0(f_2(x)) = f_0\left(\frac{x}{3x+1}\right)$$

$$= \frac{\frac{x}{3x+1}}{\frac{x}{3x+1} + 1} = \frac{\frac{x}{3x+1}}{\frac{4x+1}{3x+1}} = \frac{x}{4x+1}$$

PS Look for a pattern

We notice a pattern: The coefficient of x in the denominator of $f_n(x)$ is $n+1$ in the three cases we have computed. So we make the guess that, in general,

$$\boxed{4} \qquad f_n(x) = \frac{x}{(n+1)x + 1}$$

To prove this, we use the Principle of Mathematical Induction. We have already verified that (4) is true for $n = 1$. Assume that it is true for $n = k$, that is,

$$f_k(x) = \frac{x}{(k+1)x + 1}$$

Then $\qquad f_{k+1}(x) = (f_0 \circ f_k)(x) = f_0(f_k(x)) = f_0\left(\frac{x}{(k+1)x + 1}\right)$

$$= \frac{\frac{x}{(k+1)x+1}}{\frac{x}{(k+1)x+1} + 1} = \frac{\frac{x}{(k+1)x+1}}{\frac{(k+2)x+1}{(k+1)x+1}} = \frac{x}{(k+2)x+1}$$

This expression shows that (4) is true for $n = k + 1$. Therefore, by mathematical induction, it is true for all positive integers n.

Problems

1. One of the legs of a right triangle has length 4 cm. Express the length of the altitude perpendicular to the hypotenuse as a function of the length of the hypotenuse.

2. The altitude perpendicular to the hypotenuse of a right triangle is 12 cm. Express the length of the hypotenuse as a function of the perimeter.

3. Solve the equation $|2x - 1| - |x + 5| = 3$.

4. Solve the inequality $|x - 1| - |x - 3| \geq 5$.

5. Sketch the graph of the function $f(x) = |x^2 - 4|x| + 3|$.

6. Sketch the graph of the function $g(x) = |x^2 - 1| - |x^2 - 4|$.

7. Draw the graph of the equation $x + |x| = y + |y|$.

8. Sketch the region in the plane consisting of all points (x, y) such that
$$|x - y| + |x| - |y| \leq 2$$

9. The notation $\max\{a, b, \ldots\}$ means the largest of the numbers $a, b, \ldots$. Sketch the graph of each function.
   (a) $f(x) = \max\{x, 1/x\}$
   (b) $f(x) = \max\{\sin x, \cos x\}$
   (c) $f(x) = \max\{x^2, 2 + x, 2 - x\}$

10. Sketch the region in the plane defined by each of the following equations or inequalities.
    (a) $\max\{x, 2y\} = 1$
    (b) $-1 \leq \max\{x, 2y\} \leq 1$
    (c) $\max\{x, y^2\} = 1$

11. Evaluate $(\log_2 3)(\log_3 4)(\log_4 5) \cdots (\log_{31} 32)$.

12. (a) Show that the function $f(x) = \ln(x + \sqrt{x^2 + 1})$ is an odd function.
    (b) Find the inverse function of f.

13. Solve the inequality $\ln(x^2 - 2x - 2) \leq 0$.

14. Use indirect reasoning to prove that $\log_2 5$ is an irrational number.

15. A driver sets out on a journey. For the first half of the distance she drives at the leisurely pace of 30 mi/h; she drives the second half at 60 mi/h. What is her average speed on this trip?

16. Is it true that $f \circ (g + h) = f \circ g + f \circ h$?

17. Prove that if n is a positive integer, then $7^n - 1$ is divisible by 6.

18. Prove that $1 + 3 + 5 + \cdots + (2n - 1) = n^2$.

19. If $f_0(x) = x^2$ and $f_{n+1}(x) = f_0(f_n(x))$ for $n = 0, 1, 2, \ldots$, find a formula for $f_n(x)$.

20. (a) If $f_0(x) = \dfrac{1}{2 - x}$ and $f_{n+1} = f_0 \circ f_n$ for $n = 0, 1, 2, \ldots$, find an expression for $f_n(x)$ and use mathematical induction to prove it.
    (b) Graph f_0, f_1, f_2, f_3 on the same screen and describe the effects of repeated composition.

2 Limits and Derivatives

The maximum sustainable swimming speed S of salmon depends on the water temperature T. Exercise 58 in Section 2.7 asks you to analyze how S varies as T changes by estimating the derivative of S with respect to T.

© Jody Ann / Shutterstock.com

IN *A PREVIEW OF CALCULUS* (page 1) we saw how the idea of a limit underlies the various branches of calculus. It is therefore appropriate to begin our study of calculus by investigating limits and their properties. The special type of limit that is used to find tangents and velocities gives rise to the central idea in differential calculus, the derivative.

2.1 The Tangent and Velocity Problems

In this section we see how limits arise when we attempt to find the tangent to a curve or the velocity of an object.

■ The Tangent Problem

The word *tangent* is derived from the Latin word *tangens*, which means "touching." Thus a tangent to a curve is a line that touches the curve. In other words, a tangent line should have the same direction as the curve at the point of contact. How can this idea be made precise?

For a circle we could simply follow Euclid and say that a tangent is a line that intersects the circle once and only once, as in Figure 1(a). For more complicated curves this definition is inadequate. Figure 1(b) shows two lines l and t passing through a point P on a curve C. The line l intersects C only once, but it certainly does not look like what we think of as a tangent. The line t, on the other hand, looks like a tangent but it intersects C twice.

To be specific, let's look at the problem of trying to find a tangent line t to the parabola $y = x^2$ in the following example.

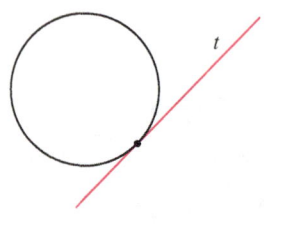

FIGURE 1

EXAMPLE 1 Find an equation of the tangent line to the parabola $y = x^2$ at the point $P(1, 1)$.

SOLUTION We will be able to find an equation of the tangent line t as soon as we know its slope m. The difficulty is that we know only one point, P, on t, whereas we need two points to compute the slope. But observe that we can compute an approximation to m by choosing a nearby point $Q(x, x^2)$ on the parabola (as in Figure 2) and computing the slope m_{PQ} of the secant line PQ. [A **secant line**, from the Latin word *secans*, meaning cutting, is a line that cuts (intersects) a curve more than once.]

We choose $x \neq 1$ so that $Q \neq P$. Then

$$m_{PQ} = \frac{x^2 - 1}{x - 1}$$

For instance, for the point $Q(1.5, 2.25)$ we have

$$m_{PQ} = \frac{2.25 - 1}{1.5 - 1} = \frac{1.25}{0.5} = 2.5$$

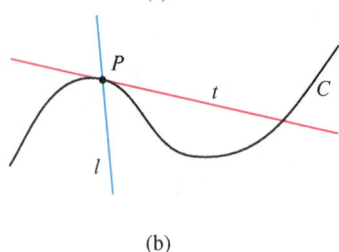

FIGURE 2

The tables in the margin show the values of m_{PQ} for several values of x close to 1. The closer Q is to P, the closer x is to 1 and, it appears from the tables, the closer m_{PQ} is to 2. This suggests that the slope of the tangent line t should be $m = 2$.

We say that the slope of the tangent line is the *limit* of the slopes of the secant lines, and we express this symbolically by writing

$$\lim_{Q \to P} m_{PQ} = m \quad \text{and} \quad \lim_{x \to 1} \frac{x^2 - 1}{x - 1} = 2$$

Assuming that the slope of the tangent line is indeed 2, we use the point-slope form of the equation of a line [$y - y_1 = m(x - x_1)$, see Appendix B] to write the equation of the tangent line through $(1, 1)$ as

$$y - 1 = 2(x - 1) \quad \text{or} \quad y = 2x - 1$$

x	m_{PQ}
2	3
1.5	2.5
1.1	2.1
1.01	2.01
1.001	2.001

x	m_{PQ}
0	1
0.5	1.5
0.9	1.9
0.99	1.99
0.999	1.999

Figure 3 illustrates the limiting process that occurs in this example. As Q approaches P along the parabola, the corresponding secant lines rotate about P and approach the tangent line t.

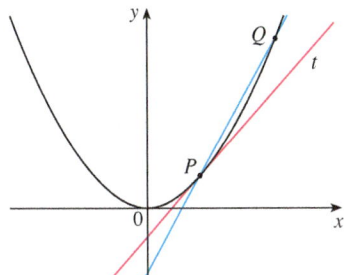
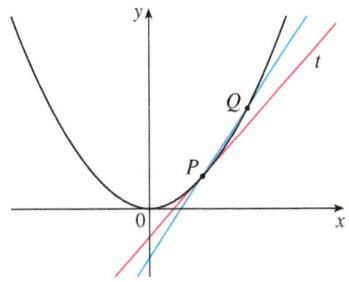
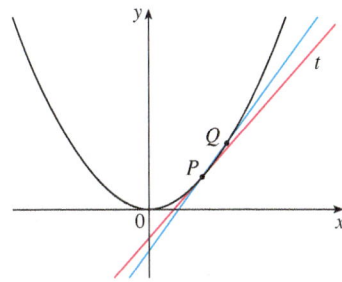

Q approaches P from the right

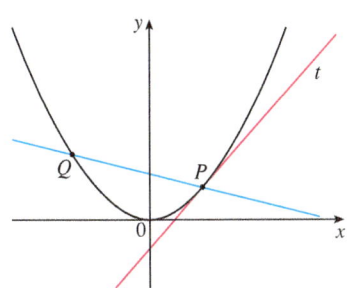
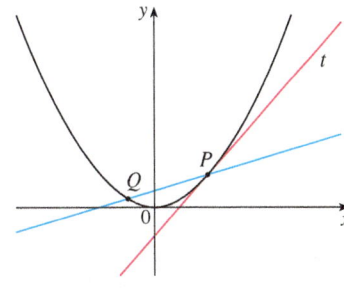
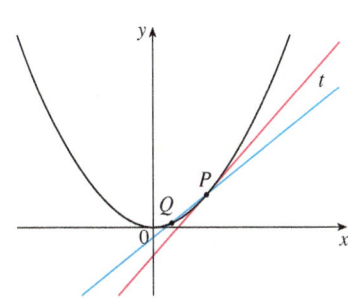

Q approaches P from the left

FIGURE 3

TEC In Visual 2.1 you can see how the process in Figure 3 works for additional functions.

t	Q
0.00	100.00
0.02	81.87
0.04	67.03
0.06	54.88
0.08	44.93
0.10	36.76

Many functions that occur in science are not described by explicit equations; they are defined by experimental data. The next example shows how to estimate the slope of the tangent line to the graph of such a function.

EXAMPLE 2 The flash unit on a camera operates by storing charge on a capacitor and releasing it suddenly when the flash is set off. The data in the table describe the charge Q remaining on the capacitor (measured in microcoulombs) at time t (measured in seconds after the flash goes off). Use the data to draw the graph of this function and estimate the slope of the tangent line at the point where $t = 0.04$. [*Note:* The slope of the tangent line represents the electric current flowing from the capacitor to the flash bulb (measured in microamperes).]

SOLUTION In Figure 4 we plot the given data and use them to sketch a curve that approximates the graph of the function.

FIGURE 4

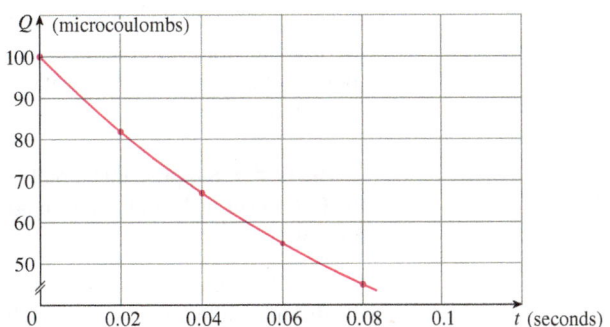

Given the points $P(0.04, 67.03)$ and $R(0.00, 100.00)$ on the graph, we find that the slope of the secant line PR is

$$m_{PR} = \frac{100.00 - 67.03}{0.00 - 0.04} = -824.25$$

R	m_{PR}
(0.00, 100.00)	−824.25
(0.02, 81.87)	−742.00
(0.06, 54.88)	−607.50
(0.08, 44.93)	−552.50
(0.10, 36.76)	−504.50

The table at the left shows the results of similar calculations for the slopes of other secant lines. From this table we would expect the slope of the tangent line at $t = 0.04$ to lie somewhere between -742 and -607.5. In fact, the average of the slopes of the two closest secant lines is

$$\tfrac{1}{2}(-742 - 607.5) = -674.75$$

So, by this method, we estimate the slope of the tangent line to be about -675.

Another method is to draw an approximation to the tangent line at P and measure the sides of the triangle ABC, as in Figure 5.

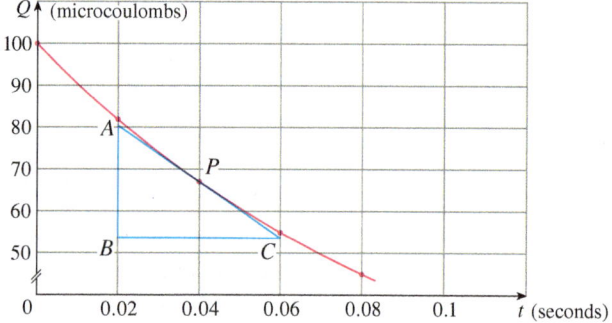

FIGURE 5

This gives an estimate of the slope of the tangent line as

$$-\frac{|AB|}{|BC|} \approx -\frac{80.4 - 53.6}{0.06 - 0.02} = -670 \quad \blacksquare$$

The physical meaning of the answer in Example 2 is that the electric current flowing from the capacitor to the flash bulb after 0.04 seconds is about -670 microamperes.

■ The Velocity Problem

If you watch the speedometer of a car as you travel in city traffic, you see that the speed doesn't stay the same for very long; that is, the velocity of the car is not constant. We assume from watching the speedometer that the car has a definite velocity at each moment, but how is the "instantaneous" velocity defined? Let's investigate the example of a falling ball.

EXAMPLE 3 Suppose that a ball is dropped from the upper observation deck of the CN Tower in Toronto, 450 m above the ground. Find the velocity of the ball after 5 seconds.

SOLUTION Through experiments carried out four centuries ago, Galileo discovered that the distance fallen by any freely falling body is proportional to the square of the time it has been falling. (This model for free fall neglects air resistance.) If the distance fallen

SECTION 2.1 The Tangent and Velocity Problems

after t seconds is denoted by $s(t)$ and measured in meters, then Galileo's law is expressed by the equation

$$s(t) = 4.9t^2$$

The difficulty in finding the velocity after 5 seconds is that we are dealing with a single instant of time ($t = 5$), so no time interval is involved. However, we can approximate the desired quantity by computing the average velocity over the brief time interval of a tenth of a second from $t = 5$ to $t = 5.1$:

$$\text{average velocity} = \frac{\text{change in position}}{\text{time elapsed}}$$

$$= \frac{s(5.1) - s(5)}{0.1}$$

$$= \frac{4.9(5.1)^2 - 4.9(5)^2}{0.1} = 49.49 \text{ m/s}$$

The following table shows the results of similar calculations of the average velocity over successively smaller time periods.

Time interval	Average velocity (m/s)
$5 \leq t \leq 6$	53.9
$5 \leq t \leq 5.1$	49.49
$5 \leq t \leq 5.05$	49.245
$5 \leq t \leq 5.01$	49.049
$5 \leq t \leq 5.001$	49.0049

The CN Tower in Toronto was the tallest freestanding building in the world for 32 years.

It appears that as we shorten the time period, the average velocity is becoming closer to 49 m/s. The **instantaneous velocity** when $t = 5$ is defined to be the limiting value of these average velocities over shorter and shorter time periods that start at $t = 5$. Thus it appears that the (instantaneous) velocity after 5 seconds is

$$v = 49 \text{ m/s} \qquad \blacksquare$$

You may have the feeling that the calculations used in solving this problem are very similar to those used earlier in this section to find tangents. In fact, there is a close connection between the tangent problem and the problem of finding velocities. If we draw the graph of the distance function of the ball (as in Figure 6) and we consider the points $P(a, 4.9a^2)$ and $Q(a + h, 4.9(a + h)^2)$ on the graph, then the slope of the secant line PQ is

$$m_{PQ} = \frac{4.9(a + h)^2 - 4.9a^2}{(a + h) - a}$$

which is the same as the average velocity over the time interval $[a, a + h]$. Therefore the velocity at time $t = a$ (the limit of these average velocities as h approaches 0) must be equal to the slope of the tangent line at P (the limit of the slopes of the secant lines).

Examples 1 and 3 show that in order to solve tangent and velocity problems we must be able to find limits. After studying methods for computing limits in the next five sections, we will return to the problems of finding tangents and velocities in Section 2.7.

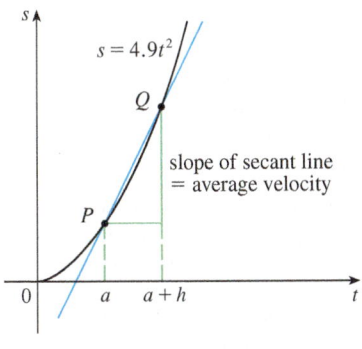
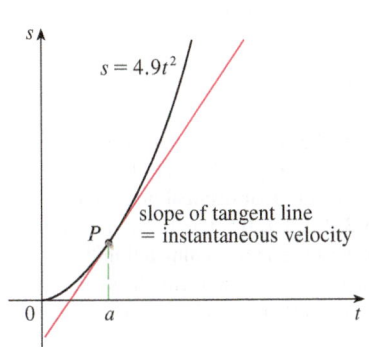

FIGURE 6

2.1 EXERCISES

1. A tank holds 1000 gallons of water, which drains from the bottom of the tank in half an hour. The values in the table show the volume V of water remaining in the tank (in gallons) after t minutes.

t (min)	5	10	15	20	25	30
V (gal)	694	444	250	111	28	0

   (a) If P is the point $(15, 250)$ on the graph of V, find the slopes of the secant lines PQ when Q is the point on the graph with $t = 5, 10, 20, 25,$ and 30.
   (b) Estimate the slope of the tangent line at P by averaging the slopes of two secant lines.
   (c) Use a graph of the function to estimate the slope of the tangent line at P. (This slope represents the rate at which the water is flowing from the tank after 15 minutes.)

2. A cardiac monitor is used to measure the heart rate of a patient after surgery. It compiles the number of heartbeats after t minutes. When the data in the table are graphed, the slope of the tangent line represents the heart rate in beats per minute.

t (min)	36	38	40	42	44
Heartbeats	2530	2661	2806	2948	3080

   The monitor estimates this value by calculating the slope of a secant line. Use the data to estimate the patient's heart rate after 42 minutes using the secant line between the points with the given values of t.
   (a) $t = 36$ and $t = 42$ (b) $t = 38$ and $t = 42$
   (c) $t = 40$ and $t = 42$ (d) $t = 42$ and $t = 44$
   What are your conclusions?

3. The point $P(2, -1)$ lies on the curve $y = 1/(1 - x)$.
   (a) If Q is the point $(x, 1/(1 - x))$, use your calculator to find the slope of the secant line PQ (correct to six decimal places) for the following values of x:
   (i) 1.5   (ii) 1.9   (iii) 1.99   (iv) 1.999
   (v) 2.5   (vi) 2.1   (vii) 2.01   (viii) 2.001
   (b) Using the results of part (a), guess the value of the slope of the tangent line to the curve at $P(2, -1)$.
   (c) Using the slope from part (b), find an equation of the tangent line to the curve at $P(2, -1)$.

4. The point $P(0.5, 0)$ lies on the curve $y = \cos \pi x$.
   (a) If Q is the point $(x, \cos \pi x)$, use your calculator to find the slope of the secant line PQ (correct to six decimal places) for the following values of x:
   (i) 0   (ii) 0.4   (iii) 0.49
   (iv) 0.499   (v) 1   (vi) 0.6
   (vii) 0.51   (viii) 0.501
   (b) Using the results of part (a), guess the value of the slope of the tangent line to the curve at $P(0.5, 0)$.
   (c) Using the slope from part (b), find an equation of the tangent line to the curve at $P(0.5, 0)$.
   (d) Sketch the curve, two of the secant lines, and the tangent line.

5. If a ball is thrown into the air with a velocity of 40 ft/s, its height in feet t seconds later is given by $y = 40t - 16t^2$.
   (a) Find the average velocity for the time period beginning when $t = 2$ and lasting
   (i) 0.5 seconds   (ii) 0.1 seconds
   (iii) 0.05 seconds   (iv) 0.01 seconds
   (b) Estimate the instantaneous velocity when $t = 2$.

6. If a rock is thrown upward on the planet Mars with a velocity of 10 m/s, its height in meters t seconds later is given by $y = 10t - 1.86t^2$.
   (a) Find the average velocity over the given time intervals:
   (i) [1, 2]   (ii) [1, 1.5]
   (iii) [1, 1.1]   (iv) [1, 1.01]
   (v) [1, 1.001]
   (b) Estimate the instantaneous velocity when $t = 1$.

7. The table shows the position of a motorcyclist after accelerating from rest.

t (seconds)	0	1	2	3	4	5	6
s (feet)	0	4.9	20.6	46.5	79.2	124.8	176.7

   (a) Find the average velocity for each time period:
   (i) [2, 4]   (ii) [3, 4]   (iii) [4, 5]   (iv) [4, 6]
   (b) Use the graph of s as a function of t to estimate the instantaneous velocity when $t = 3$.

8. The displacement (in centimeters) of a particle moving back and forth along a straight line is given by the equation of motion $s = 2 \sin \pi t + 3 \cos \pi t$, where t is measured in seconds.
   (a) Find the average velocity during each time period:
   (i) [1, 2]   (ii) [1, 1.1]
   (iii) [1, 1.01]   (iv) [1, 1.001]
   (b) Estimate the instantaneous velocity of the particle when $t = 1$.

9. The point $P(1, 0)$ lies on the curve $y = \sin(10\pi/x)$.
   (a) If Q is the point $(x, \sin(10\pi/x))$, find the slope of the secant line PQ (correct to four decimal places) for $x = 2, 1.5, 1.4, 1.3, 1.2, 1.1, 0.5, 0.6, 0.7, 0.8,$ and 0.9. Do the slopes appear to be approaching a limit?
   (b) Use a graph of the curve to explain why the slopes of the secant lines in part (a) are not close to the slope of the tangent line at P.
   (c) By choosing appropriate secant lines, estimate the slope of the tangent line at P.

2.2 The Limit of a Function

Having seen in the preceding section how limits arise when we want to find the tangent to a curve or the velocity of an object, we now turn our attention to limits in general and numerical and graphical methods for computing them.

Let's investigate the behavior of the function f defined by $f(x) = x^2 - x + 2$ for values of x near 2. The following table gives values of $f(x)$ for values of x close to 2 but not equal to 2.

x	$f(x)$	x	$f(x)$
1.0	2.000000	3.0	8.000000
1.5	2.750000	2.5	5.750000
1.8	3.440000	2.2	4.640000
1.9	3.710000	2.1	4.310000
1.95	3.852500	2.05	4.152500
1.99	3.970100	2.01	4.030100
1.995	3.985025	2.005	4.015025
1.999	3.997001	2.001	4.003001

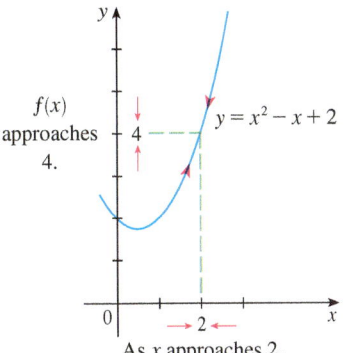

As x approaches 2,

FIGURE 1

From the table and the graph of f (a parabola) shown in Figure 1 we see that the closer x is to 2 (on either side of 2), the closer $f(x)$ is to 4. In fact, it appears that we can make the values of $f(x)$ as close as we like to 4 by taking x sufficiently close to 2. We express this by saying "the limit of the function $f(x) = x^2 - x + 2$ as x approaches 2 is equal to 4." The notation for this is

$$\lim_{x \to 2} (x^2 - x + 2) = 4$$

In general, we use the following notation.

1 Intuitive Definition of a Limit Suppose $f(x)$ is defined when x is near the number a. (This means that f is defined on some open interval that contains a, except possibly at a itself.) Then we write

$$\lim_{x \to a} f(x) = L$$

and say   "the limit of $f(x)$, as x approaches a, equals L"

if we can make the values of $f(x)$ arbitrarily close to L (as close to L as we like) by restricting x to be sufficiently close to a (on either side of a) but not equal to a.

Roughly speaking, this says that the values of $f(x)$ approach L as x approaches a. In other words, the values of $f(x)$ tend to get closer and closer to the number L as x gets closer and closer to the number a (from either side of a) but $x \ne a$. (A more precise definition will be given in Section 2.4.)

An alternative notation for

$$\lim_{x \to a} f(x) = L$$

is $f(x) \to L$ as $x \to a$

which is usually read "$f(x)$ approaches L as x approaches a."

Notice the phrase "but $x \neq a$" in the definition of limit. This means that in finding the limit of $f(x)$ as x approaches a, we never consider $x = a$. In fact, $f(x)$ need not even be defined when $x = a$. The only thing that matters is how f is defined *near a*.

Figure 2 shows the graphs of three functions. Note that in part (c), $f(a)$ is not defined and in part (b), $f(a) \neq L$. But in each case, regardless of what happens at a, it is true that $\lim_{x \to a} f(x) = L$.

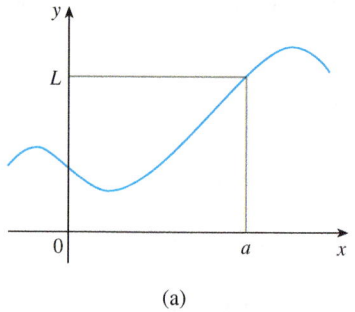

(a)

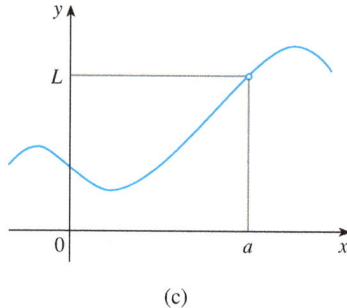

(b) (c)

FIGURE 2 $\lim_{x \to a} f(x) = L$ in all three cases

EXAMPLE 1 Guess the value of $\lim_{x \to 1} \dfrac{x - 1}{x^2 - 1}$.

SOLUTION Notice that the function $f(x) = (x - 1)/(x^2 - 1)$ is not defined when $x = 1$, but that doesn't matter because the definition of $\lim_{x \to a} f(x)$ says that we consider values of x that are close to a but not equal to a.

The tables at the left give values of $f(x)$ (correct to six decimal places) for values of x that approach 1 (but are not equal to 1). On the basis of the values in the tables, we make the guess that

$$\lim_{x \to 1} \frac{x - 1}{x^2 - 1} = 0.5$$

$x < 1$	$f(x)$
0.5	0.666667
0.9	0.526316
0.99	0.502513
0.999	0.500250
0.9999	0.500025

$x > 1$	$f(x)$
1.5	0.400000
1.1	0.476190
1.01	0.497512
1.001	0.499750
1.0001	0.499975

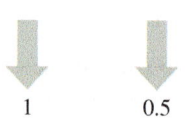

1   0.5

Example 1 is illustrated by the graph of f in Figure 3. Now let's change f slightly by giving it the value 2 when $x = 1$ and calling the resulting function g:

$$g(x) = \begin{cases} \dfrac{x - 1}{x^2 - 1} & \text{if } x \neq 1 \\ 2 & \text{if } x = 1 \end{cases}$$

This new function g still has the same limit as x approaches 1. (See Figure 4.)

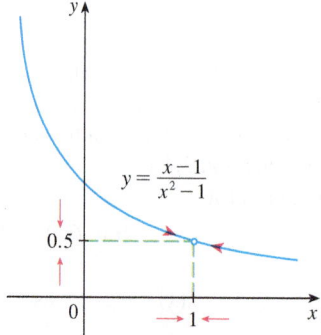

FIGURE 3

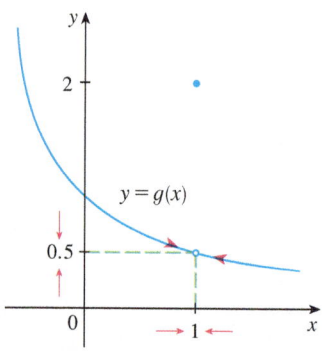

FIGURE 4

EXAMPLE 2 Estimate the value of $\lim_{t \to 0} \dfrac{\sqrt{t^2 + 9} - 3}{t^2}$.

SOLUTION The table lists values of the function for several values of t near 0.

t	$\dfrac{\sqrt{t^2 + 9} - 3}{t^2}$
±1.0	0.162277...
±0.5	0.165525...
±0.1	0.166620...
±0.05	0.166655...
±0.01	0.166666...

As t approaches 0, the values of the function seem to approach 0.1666666... and so we guess that

$$\lim_{t \to 0} \dfrac{\sqrt{t^2 + 9} - 3}{t^2} = \dfrac{1}{6}$$

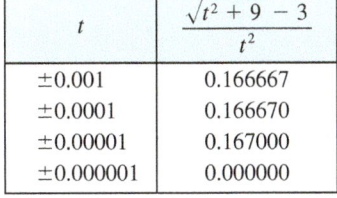

In Example 2 what would have happened if we had taken even smaller values of t? The table in the margin shows the results from one calculator; you can see that something strange seems to be happening.

t	$\dfrac{\sqrt{t^2 + 9} - 3}{t^2}$
±0.001	0.166667
±0.0001	0.166670
±0.00001	0.167000
±0.000001	0.000000

If you try these calculations on your own calculator you might get different values, but eventually you will get the value 0 if you make t sufficiently small. Does this mean that the answer is really 0 instead of $\tfrac{1}{6}$? No, the value of the limit is $\tfrac{1}{6}$, as we will show in the next section. The problem is that the calculator gave false values because $\sqrt{t^2 + 9}$ is very close to 3 when t is small. (In fact, when t is sufficiently small, a calculator's value for $\sqrt{t^2 + 9}$ is 3.000... to as many digits as the calculator is capable of carrying.)

www.stewartcalculus.com
For a further explanation of why calculators sometimes give false values, click on *Lies My Calculator and Computer Told Me*. In particular, see the section called *The Perils of Subtraction*.

Something similar happens when we try to graph the function

$$f(t) = \dfrac{\sqrt{t^2 + 9} - 3}{t^2}$$

of Example 2 on a graphing calculator or computer. Parts (a) and (b) of Figure 5 show quite accurate graphs of f, and when we use the trace mode (if available) we can estimate easily that the limit is about $\tfrac{1}{6}$. But if we zoom in too much, as in parts (c) and (d), then we get inaccurate graphs, again because of rounding errors from the subtraction.

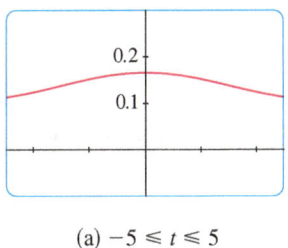

(a) $-5 \leq t \leq 5$

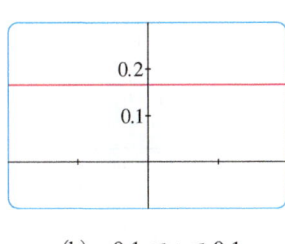

(b) $-0.1 \leq t \leq 0.1$

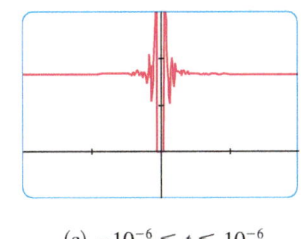

(c) $-10^{-6} \leq t \leq 10^{-6}$

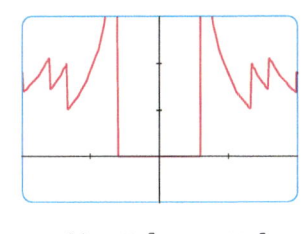

(d) $-10^{-7} \leq t \leq 10^{-7}$

FIGURE 5

86 CHAPTER 2 Limits and Derivatives

EXAMPLE 3 Guess the value of $\lim_{x \to 0} \dfrac{\sin x}{x}$.

SOLUTION The function $f(x) = (\sin x)/x$ is not defined when $x = 0$. Using a calculator (and remembering that, if $x \in \mathbb{R}$, $\sin x$ means the sine of the angle whose *radian* measure is x), we construct a table of values correct to eight decimal places. From the table at the left and the graph in Figure 6 we guess that

$$\lim_{x \to 0} \frac{\sin x}{x} = 1$$

This guess is in fact correct, as will be proved in Chapter 3 using a geometric argument.

x	$\dfrac{\sin x}{x}$
±1.0	0.84147098
±0.5	0.95885108
±0.4	0.97354586
±0.3	0.98506736
±0.2	0.99334665
±0.1	0.99833417
±0.05	0.99958339
±0.01	0.99998333
±0.005	0.99999583
±0.001	0.99999983

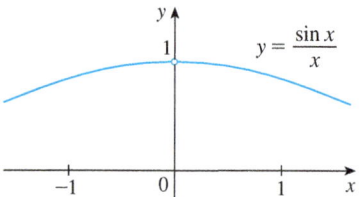

FIGURE 6

EXAMPLE 4 Investigate $\lim_{x \to 0} \sin \dfrac{\pi}{x}$.

SOLUTION Again the function $f(x) = \sin(\pi/x)$ is undefined at 0. Evaluating the function for some small values of x, we get

$$f(1) = \sin \pi = 0 \qquad f(\tfrac{1}{2}) = \sin 2\pi = 0$$
$$f(\tfrac{1}{3}) = \sin 3\pi = 0 \qquad f(\tfrac{1}{4}) = \sin 4\pi = 0$$
$$f(0.1) = \sin 10\pi = 0 \qquad f(0.01) = \sin 100\pi = 0$$

Similarly, $f(0.001) = f(0.0001) = 0$. On the basis of this information we might be tempted to guess that

$$\lim_{x \to 0} \sin \frac{\pi}{x} = 0$$

Computer Algebra Systems
Computer algebra systems (CAS) have commands that compute limits. In order to avoid the types of pitfalls demonstrated in Examples 2, 4, and 5, they don't find limits by numerical experimentation. Instead, they use more sophisticated techniques such as computing infinite series. If you have access to a CAS, use the limit command to compute the limits in the examples of this section and to check your answers in the exercises of this chapter.

but this time our guess is wrong. Note that although $f(1/n) = \sin n\pi = 0$ for any integer n, it is also true that $f(x) = 1$ for infinitely many values of x (such as 2/5 or 2/101) that approach 0. You can see this from the graph of f shown in Figure 7.

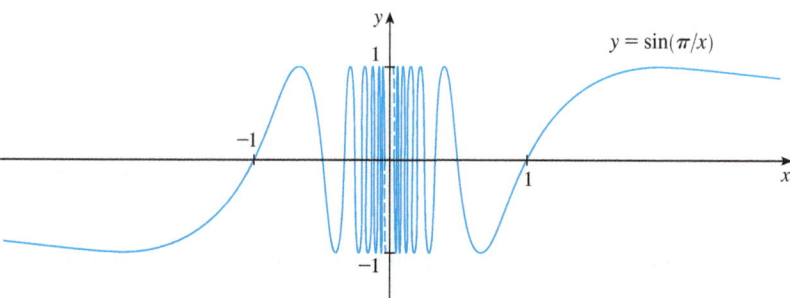

FIGURE 7

The dashed lines near the y-axis indicate that the values of $\sin(\pi/x)$ oscillate between 1 and -1 infinitely often as x approaches 0. (See Exercise 51.)

Since the values of $f(x)$ do not approach a fixed number as x approaches 0,

$$\lim_{x \to 0} \sin \frac{\pi}{x} \text{ does not exist}$$

EXAMPLE 5 Find $\lim_{x \to 0} \left(x^3 + \dfrac{\cos 5x}{10{,}000} \right)$.

SOLUTION As before, we construct a table of values. From the first table in the margin it appears that

$$\lim_{x \to 0} \left(x^3 + \frac{\cos 5x}{10{,}000} \right) = 0$$

x	$x^3 + \dfrac{\cos 5x}{10{,}000}$
1	1.000028
0.5	0.124920
0.1	0.001088
0.05	0.000222
0.01	0.000101

x	$x^3 + \dfrac{\cos 5x}{10{,}000}$
0.005	0.00010009
0.001	0.00010000

But if we persevere with smaller values of x, the second table suggests that

$$\lim_{x \to 0} \left(x^3 + \frac{\cos 5x}{10{,}000} \right) = 0.000100 = \frac{1}{10{,}000}$$

Later we will see that $\lim_{x \to 0} \cos 5x = 1$; then it follows that the limit is 0.0001.

Examples 4 and 5 illustrate some of the pitfalls in guessing the value of a limit. It is easy to guess the wrong value if we use inappropriate values of x, but it is difficult to know when to stop calculating values. And, as the discussion after Example 2 shows, sometimes calculators and computers give the wrong values. In the next section, however, we will develop foolproof methods for calculating limits.

■ One-Sided Limits

EXAMPLE 6 The Heaviside function H is defined by

$$H(t) = \begin{cases} 0 & \text{if } t < 0 \\ 1 & \text{if } t \geq 0 \end{cases}$$

[This function is named after the electrical engineer Oliver Heaviside (1850–1925) and can be used to describe an electric current that is switched on at time $t = 0$.] Its graph is shown in Figure 8.

As t approaches 0 from the left, $H(t)$ approaches 0. As t approaches 0 from the right, $H(t)$ approaches 1. There is no single number that $H(t)$ approaches as t approaches 0. Therefore $\lim_{t \to 0} H(t)$ does not exist.

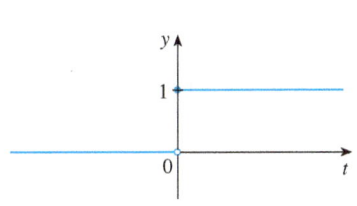

FIGURE 8
The Heaviside function

We noticed in Example 6 that $H(t)$ approaches 0 as t approaches 0 from the left and $H(t)$ approaches 1 as t approaches 0 from the right. We indicate this situation symbolically by writing

$$\lim_{t \to 0^-} H(t) = 0 \quad \text{and} \quad \lim_{t \to 0^+} H(t) = 1$$

The notation $t \to 0^-$ indicates that we consider only values of t that are less than 0. Likewise, $t \to 0^+$ indicates that we consider only values of t that are greater than 0.

2 Definition of One-Sided Limits We write

$$\lim_{x \to a^-} f(x) = L$$

and say the **left-hand limit of $f(x)$ as x approaches a** [or the **limit of $f(x)$ as x approaches a from the left**] is equal to L if we can make the values of $f(x)$ arbitrarily close to L by taking x to be sufficiently close to a with x less than a.

Notice that Definition 2 differs from Definition 1 only in that we require x to be less than a. Similarly, if we require that x be greater than a, we get "the **right-hand limit of $f(x)$ as x approaches a** is equal to L" and we write

$$\lim_{x \to a^+} f(x) = L$$

Thus the notation $x \to a^+$ means that we consider only x greater than a. These definitions are illustrated in Figure 9.

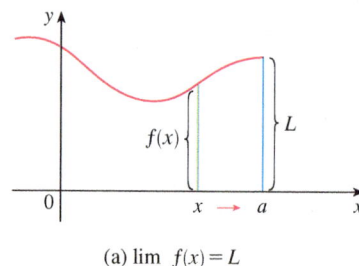
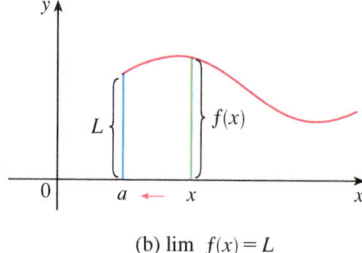

FIGURE 9 (a) $\lim_{x \to a^-} f(x) = L$ (b) $\lim_{x \to a^+} f(x) = L$

By comparing Definition 1 with the definitions of one-sided limits, we see that the following is true.

3 $\lim_{x \to a} f(x) = L$ if and only if $\lim_{x \to a^-} f(x) = L$ and $\lim_{x \to a^+} f(x) = L$

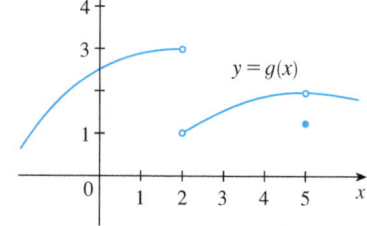

FIGURE 10

EXAMPLE 7 The graph of a function g is shown in Figure 10. Use it to state the values (if they exist) of the following:

(a) $\lim_{x \to 2^-} g(x)$ (b) $\lim_{x \to 2^+} g(x)$ (c) $\lim_{x \to 2} g(x)$

(d) $\lim_{x \to 5^-} g(x)$ (e) $\lim_{x \to 5^+} g(x)$ (f) $\lim_{x \to 5} g(x)$

SOLUTION From the graph we see that the values of $g(x)$ approach 3 as x approaches 2 from the left, but they approach 1 as x approaches 2 from the right. Therefore

(a) $\lim_{x \to 2^-} g(x) = 3$ and   (b) $\lim_{x \to 2^+} g(x) = 1$

(c) Since the left and right limits are different, we conclude from (3) that $\lim_{x \to 2} g(x)$ does not exist.

The graph also shows that

(d) $\lim_{x \to 5^-} g(x) = 2$ and   (e) $\lim_{x \to 5^+} g(x) = 2$

(f) This time the left and right limits are the same and so, by (3), we have

$$\lim_{x \to 5} g(x) = 2$$

Despite this fact, notice that $g(5) \neq 2$.

Infinite Limits

EXAMPLE 8 Find $\lim_{x \to 0} \dfrac{1}{x^2}$ if it exists.

SOLUTION As x becomes close to 0, x^2 also becomes close to 0, and $1/x^2$ becomes very large. (See the table in the margin.) In fact, it appears from the graph of the function $f(x) = 1/x^2$ shown in Figure 11 that the values of $f(x)$ can be made arbitrarily large by taking x close enough to 0. Thus the values of $f(x)$ do not approach a number, so $\lim_{x \to 0} (1/x^2)$ does not exist.

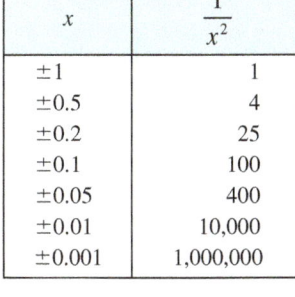

x	$\dfrac{1}{x^2}$
±1	1
±0.5	4
±0.2	25
±0.1	100
±0.05	400
±0.01	10,000
±0.001	1,000,000

To indicate the kind of behavior exhibited in Example 8, we use the notation

$$\lim_{x \to 0} \dfrac{1}{x^2} = \infty$$

This does not mean that we are regarding ∞ as a number. Nor does it mean that the limit exists. It simply expresses the particular way in which the limit does not exist: $1/x^2$ can be made as large as we like by taking x close enough to 0.

In general, we write symbolically

$$\lim_{x \to a} f(x) = \infty$$

to indicate that the values of $f(x)$ tend to become larger and larger (or "increase without bound") as x becomes closer and closer to a.

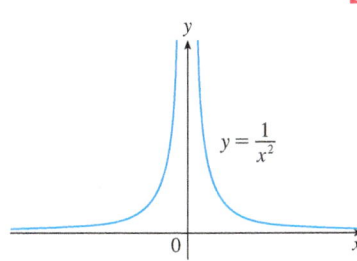

FIGURE 11

> **4** **Intuitive Definition of an Infinite Limit** Let f be a function defined on both sides of a, except possibly at a itself. Then
>
> $$\lim_{x \to a} f(x) = \infty$$
>
> means that the values of $f(x)$ can be made arbitrarily large (as large as we please) by taking x sufficiently close to a, but not equal to a.

Another notation for $\lim_{x \to a} f(x) = \infty$ is

$$f(x) \to \infty \qquad \text{as} \qquad x \to a$$

Again, the symbol ∞ is not a number, but the expression $\lim_{x \to a} f(x) = \infty$ is often read as

"the limit of $f(x)$, as x approaches a, is infinity"

or   "$f(x)$ becomes infinite as x approaches a"

or   "$f(x)$ increases without bound as x approaches a"

This definition is illustrated graphically in Figure 12.

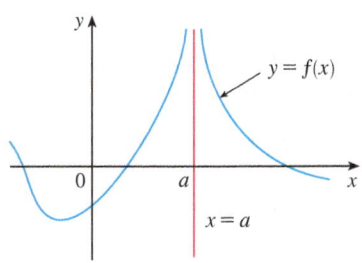

FIGURE 12
$\lim_{x \to a} f(x) = \infty$

90   CHAPTER 2   Limits and Derivatives

When we say a number is "large negative," we mean that it is negative but its magnitude (absolute value) is large.

A similar sort of limit, for functions that become large negative as x gets close to a, is defined in Definition 5 and is illustrated in Figure 13.

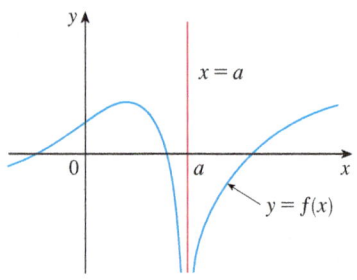

FIGURE 13
$\lim_{x \to a} f(x) = -\infty$

> **5 Definition** Let f be a function defined on both sides of a, except possibly at a itself. Then
> $$\lim_{x \to a} f(x) = -\infty$$
> means that the values of $f(x)$ can be made arbitrarily large negative by taking x sufficiently close to a, but not equal to a.

The symbol $\lim_{x \to a} f(x) = -\infty$ can be read as "the limit of $f(x)$, as x approaches a, is negative infinity" or "$f(x)$ decreases without bound as x approaches a." As an example we have

$$\lim_{x \to 0}\left(-\frac{1}{x^2}\right) = -\infty$$

Similar definitions can be given for the one-sided infinite limits

$$\lim_{x \to a^-} f(x) = \infty \qquad \lim_{x \to a^+} f(x) = \infty$$

$$\lim_{x \to a^-} f(x) = -\infty \qquad \lim_{x \to a^+} f(x) = -\infty$$

remembering that $x \to a^-$ means that we consider only values of x that are less than a, and similarly $x \to a^+$ means that we consider only $x > a$. Illustrations of these four cases are given in Figure 14.

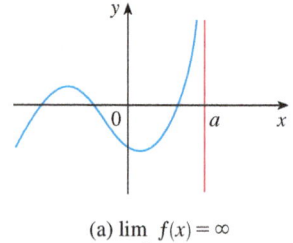 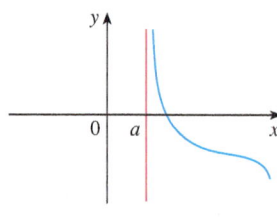 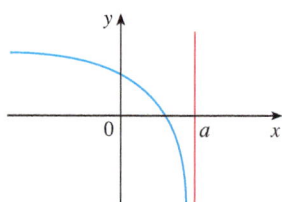 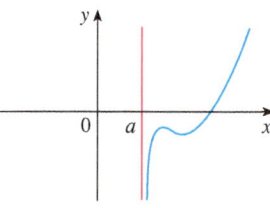

(a) $\lim_{x \to a^-} f(x) = \infty$ (b) $\lim_{x \to a^+} f(x) = \infty$ (c) $\lim_{x \to a^-} f(x) = -\infty$ (d) $\lim_{x \to a^+} f(x) = -\infty$

FIGURE 14

> **6 Definition** The vertical line $x = a$ is called a **vertical asymptote** of the curve $y = f(x)$ if at least one of the following statements is true:
> $$\lim_{x \to a} f(x) = \infty \qquad \lim_{x \to a^-} f(x) = \infty \qquad \lim_{x \to a^+} f(x) = \infty$$
> $$\lim_{x \to a} f(x) = -\infty \qquad \lim_{x \to a^-} f(x) = -\infty \qquad \lim_{x \to a^+} f(x) = -\infty$$

For instance, the y-axis is a vertical asymptote of the curve $y = 1/x^2$ because $\lim_{x \to 0}(1/x^2) = \infty$. In Figure 14 the line $x = a$ is a vertical asymptote in each of the four cases shown. In general, knowledge of vertical asymptotes is very useful in sketching graphs.

EXAMPLE 9 Find $\lim_{x \to 3^+} \dfrac{2x}{x-3}$ and $\lim_{x \to 3^-} \dfrac{2x}{x-3}$.

SOLUTION If x is close to 3 but larger than 3, then the denominator $x - 3$ is a small positive number and $2x$ is close to 6. So the quotient $2x/(x - 3)$ is a large *positive* number. [For instance, if $x = 3.01$ then $2x/(x - 3) = 6.02/0.01 = 602$.] Thus, intuitively, we see that

$$\lim_{x \to 3^+} \frac{2x}{x-3} = \infty$$

Likewise, if x is close to 3 but smaller than 3, then $x - 3$ is a small negative number but $2x$ is still a positive number (close to 6). So $2x/(x - 3)$ is a numerically large *negative* number. Thus

$$\lim_{x \to 3^-} \frac{2x}{x-3} = -\infty$$

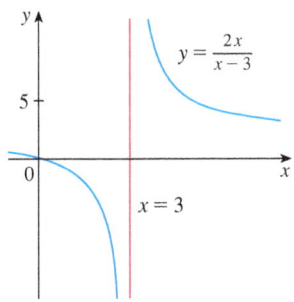

FIGURE 15

The graph of the curve $y = 2x/(x - 3)$ is given in Figure 15. The line $x = 3$ is a vertical asymptote. ∎

EXAMPLE 10 Find the vertical asymptotes of $f(x) = \tan x$.

SOLUTION Because

$$\tan x = \frac{\sin x}{\cos x}$$

there are potential vertical asymptotes where $\cos x = 0$. In fact, since $\cos x \to 0^+$ as $x \to (\pi/2)^-$ and $\cos x \to 0^-$ as $x \to (\pi/2)^+$, whereas $\sin x$ is positive (near 1) when x is near $\pi/2$, we have

$$\lim_{x \to (\pi/2)^-} \tan x = \infty \quad \text{and} \quad \lim_{x \to (\pi/2)^+} \tan x = -\infty$$

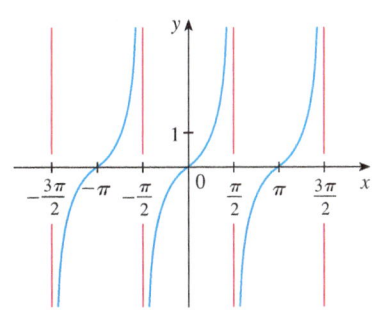

FIGURE 16
$y = \tan x$

This shows that the line $x = \pi/2$ is a vertical asymptote. Similar reasoning shows that the lines $x = \pi/2 + n\pi$, where n is an integer, are all vertical asymptotes of $f(x) = \tan x$. The graph in Figure 16 confirms this. ∎

Another example of a function whose graph has a vertical asymptote is the natural logarithmic function $y = \ln x$. From Figure 17 we see that

$$\lim_{x \to 0^+} \ln x = -\infty$$

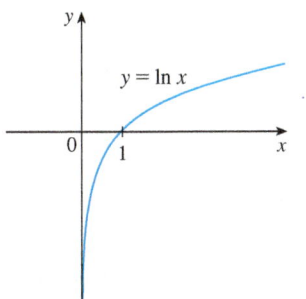

FIGURE 17
The y-axis is a vertical asymptote of the natural logarithmic function.

and so the line $x = 0$ (the y-axis) is a vertical asymptote. In fact, the same is true for $y = \log_b x$ provided that $b > 1$. (See Figures 1.5.11 and 1.5.12.)

2.2 EXERCISES

1. Explain in your own words what is meant by the equation
$$\lim_{x \to 2} f(x) = 5$$
Is it possible for this statement to be true and yet $f(2) = 3$? Explain.

2. Explain what it means to say that
$$\lim_{x \to 1^-} f(x) = 3 \quad \text{and} \quad \lim_{x \to 1^+} f(x) = 7$$
In this situation is it possible that $\lim_{x \to 1} f(x)$ exists? Explain.

3. Explain the meaning of each of the following.
   (a) $\lim_{x \to -3} f(x) = \infty$
   (b) $\lim_{x \to 4^+} f(x) = -\infty$

4. Use the given graph of f to state the value of each quantity, if it exists. If it does not exist, explain why.
   (a) $\lim_{x \to 2^-} f(x)$
   (b) $\lim_{x \to 2^+} f(x)$
   (c) $\lim_{x \to 2} f(x)$
   (d) $f(2)$
   (e) $\lim_{x \to 4} f(x)$
   (f) $f(4)$

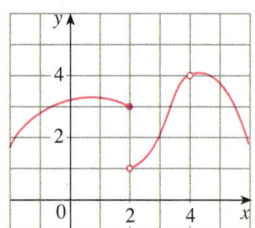

5. For the function f whose graph is given, state the value of each quantity, if it exists. If it does not exist, explain why.
   (a) $\lim_{x \to 1} f(x)$
   (b) $\lim_{x \to 3^-} f(x)$
   (c) $\lim_{x \to 3^+} f(x)$
   (d) $\lim_{x \to 3} f(x)$
   (e) $f(3)$

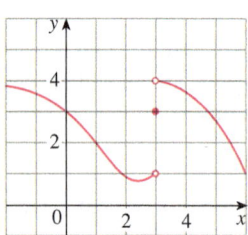

6. For the function h whose graph is given, state the value of each quantity, if it exists. If it does not exist, explain why.
   (a) $\lim_{x \to -3^-} h(x)$
   (b) $\lim_{x \to -3^+} h(x)$
   (c) $\lim_{x \to -3} h(x)$
   (d) $h(-3)$
   (e) $\lim_{x \to 0^-} h(x)$
   (f) $\lim_{x \to 0^+} h(x)$
   (g) $\lim_{x \to 0} h(x)$
   (h) $h(0)$
   (i) $\lim_{x \to 2} h(x)$
   (j) $h(2)$
   (k) $\lim_{x \to 5^+} h(x)$
   (l) $\lim_{x \to 5^-} h(x)$

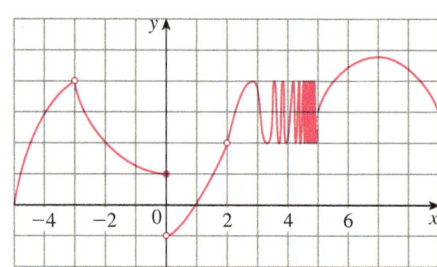

7. For the function g whose graph is given, state the value of each quantity, if it exists. If it does not exist, explain why.
   (a) $\lim_{t \to 0^-} g(t)$
   (b) $\lim_{t \to 0^+} g(t)$
   (c) $\lim_{t \to 0} g(t)$
   (d) $\lim_{t \to 2^-} g(t)$
   (e) $\lim_{t \to 2^+} g(t)$
   (f) $\lim_{t \to 2} g(t)$
   (g) $g(2)$
   (h) $\lim_{t \to 4} g(t)$

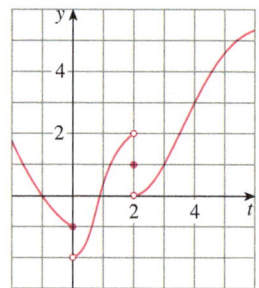

8. For the function A whose graph is shown, state the following.
   (a) $\lim_{x \to -3} A(x)$
   (b) $\lim_{x \to 2^-} A(x)$
   (c) $\lim_{x \to 2^+} A(x)$
   (d) $\lim_{x \to -1} A(x)$
   (e) The equations of the vertical asymptotes

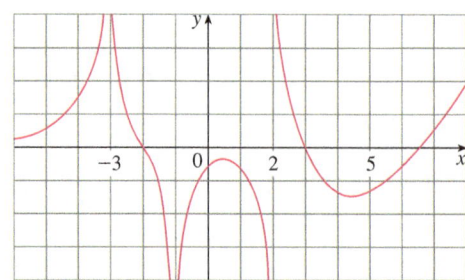

9. For the function f whose graph is shown, state the following.
   (a) $\lim_{x \to -7} f(x)$
   (b) $\lim_{x \to -3} f(x)$
   (c) $\lim_{x \to 0} f(x)$
   (d) $\lim_{x \to 6^-} f(x)$
   (e) $\lim_{x \to 6^+} f(x)$

(f) The equations of the vertical asymptotes.

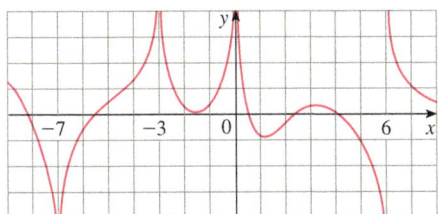

10. A patient receives a 150-mg injection of a drug every 4 hours. The graph shows the amount $f(t)$ of the drug in the bloodstream after t hours. Find
$$\lim_{t \to 12^-} f(t) \quad \text{and} \quad \lim_{t \to 12^+} f(t)$$
and explain the significance of these one-sided limits.

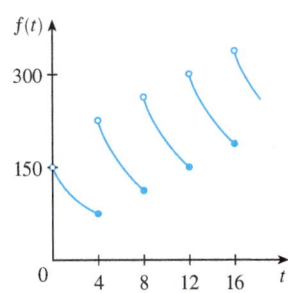

11–12 Sketch the graph of the function and use it to determine the values of a for which $\lim_{x \to a} f(x)$ exists.

11. $f(x) = \begin{cases} 1 + x & \text{if } x < -1 \\ x^2 & \text{if } -1 \leq x < 1 \\ 2 - x & \text{if } x \geq 1 \end{cases}$

12. $f(x) = \begin{cases} 1 + \sin x & \text{if } x < 0 \\ \cos x & \text{if } 0 \leq x \leq \pi \\ \sin x & \text{if } x > \pi \end{cases}$

13–14 Use the graph of the function f to state the value of each limit, if it exists. If it does not exist, explain why.
(a) $\lim_{x \to 0^-} f(x)$ (b) $\lim_{x \to 0^+} f(x)$ (c) $\lim_{x \to 0} f(x)$

13. $f(x) = \dfrac{1}{1 + e^{1/x}}$ **14.** $f(x) = \dfrac{x^2 + x}{\sqrt{x^3 + x^2}}$

15–18 Sketch the graph of an example of a function f that satisfies all of the given conditions.

15. $\lim_{x \to 0^-} f(x) = -1$, $\lim_{x \to 0^+} f(x) = 2$, $f(0) = 1$

16. $\lim_{x \to 0} f(x) = 1$, $\lim_{x \to 3^-} f(x) = -2$, $\lim_{x \to 3^+} f(x) = 2$,
$f(0) = -1$, $f(3) = 1$

17. $\lim_{x \to 3^+} f(x) = 4$, $\lim_{x \to 3^-} f(x) = 2$, $\lim_{x \to -2} f(x) = 2$,
$f(3) = 3$, $f(-2) = 1$

18. $\lim_{x \to 0^-} f(x) = 2$, $\lim_{x \to 0^+} f(x) = 0$, $\lim_{x \to 4^-} f(x) = 3$,
$\lim_{x \to 4^+} f(x) = 0$, $f(0) = 2$, $f(4) = 1$

19–22 Guess the value of the limit (if it exists) by evaluating the function at the given numbers (correct to six decimal places).

19. $\lim_{x \to 3} \dfrac{x^2 - 3x}{x^2 - 9}$,
$x = 3.1, 3.05, 3.01, 3.001, 3.0001,$
$2.9, 2.95, 2.99, 2.999, 2.9999$

20. $\lim_{x \to -3} \dfrac{x^2 - 3x}{x^2 - 9}$,
$x = -2.5, -2.9, -2.95, -2.99, -2.999, -2.9999,$
$-3.5, -3.1, -3.05, -3.01, -3.001, -3.0001$

21. $\lim_{t \to 0} \dfrac{e^{5t} - 1}{t}$, $t = \pm 0.5, \pm 0.1, \pm 0.01, \pm 0.001, \pm 0.0001$

22. $\lim_{h \to 0} \dfrac{(2 + h)^5 - 32}{h}$,
$h = \pm 0.5, \pm 0.1, \pm 0.01, \pm 0.001, \pm 0.0001$

23–28 Use a table of values to estimate the value of the limit. If you have a graphing device, use it to confirm your result graphically.

23. $\lim_{x \to 4} \dfrac{\ln x - \ln 4}{x - 4}$ **24.** $\lim_{p \to -1} \dfrac{1 + p^9}{1 + p^{15}}$

25. $\lim_{\theta \to 0} \dfrac{\sin 3\theta}{\tan 2\theta}$ **26.** $\lim_{t \to 0} \dfrac{5^t - 1}{t}$

27. $\lim_{x \to 0^+} x^x$ **28.** $\lim_{x \to 0^+} x^2 \ln x$

29. (a) By graphing the function $f(x) = (\cos 2x - \cos x)/x^2$ and zooming in toward the point where the graph crosses the y-axis, estimate the value of $\lim_{x \to 0} f(x)$.
(b) Check your answer in part (a) by evaluating $f(x)$ for values of x that approach 0.

30. (a) Estimate the value of
$$\lim_{x \to 0} \dfrac{\sin x}{\sin \pi x}$$
by graphing the function $f(x) = (\sin x)/(\sin \pi x)$. State your answer correct to two decimal places.
(b) Check your answer in part (a) by evaluating $f(x)$ for values of x that approach 0.

31–43 Determine the infinite limit.

31. $\lim\limits_{x \to 5^+} \dfrac{x+1}{x-5}$

32. $\lim\limits_{x \to 5^-} \dfrac{x+1}{x-5}$

33. $\lim\limits_{x \to 1} \dfrac{2-x}{(x-1)^2}$

34. $\lim\limits_{x \to 3^-} \dfrac{\sqrt{x}}{(x-3)^5}$

35. $\lim\limits_{x \to 3^+} \ln(x^2 - 9)$

36. $\lim\limits_{x \to 0^+} \ln(\sin x)$

37. $\lim\limits_{x \to (\pi/2)^+} \dfrac{1}{x} \sec x$

38. $\lim\limits_{x \to \pi^-} \cot x$

39. $\lim\limits_{x \to 2\pi^-} x \csc x$

40. $\lim\limits_{x \to 2^-} \dfrac{x^2 - 2x}{x^2 - 4x + 4}$

41. $\lim\limits_{x \to 2^+} \dfrac{x^2 - 2x - 8}{x^2 - 5x + 6}$

42. $\lim\limits_{x \to 0^+} \left(\dfrac{1}{x} - \ln x \right)$

43. $\lim\limits_{x \to 0} (\ln x^2 - x^{-2})$

44. (a) Find the vertical asymptotes of the function
$$y = \dfrac{x^2 + 1}{3x - 2x^2}$$
(b) Confirm your answer to part (a) by graphing the function.

45. Determine $\lim\limits_{x \to 1^-} \dfrac{1}{x^3 - 1}$ and $\lim\limits_{x \to 1^+} \dfrac{1}{x^3 - 1}$

(a) by evaluating $f(x) = 1/(x^3 - 1)$ for values of x that approach 1 from the left and from the right,
(b) by reasoning as in Example 9, and
(c) from a graph of f.

46. (a) By graphing the function $f(x) = (\tan 4x)/x$ and zooming in toward the point where the graph crosses the y-axis, estimate the value of $\lim_{x \to 0} f(x)$.
(b) Check your answer in part (a) by evaluating $f(x)$ for values of x that approach 0.

47. (a) Estimate the value of the limit $\lim_{x \to 0} (1 + x)^{1/x}$ to five decimal places. Does this number look familiar?
(b) Illustrate part (a) by graphing the function $y = (1 + x)^{1/x}$.

48. (a) Graph the function $f(x) = e^x + \ln|x - 4|$ for $0 \le x \le 5$. Do you think the graph is an accurate representation of f?
(b) How would you get a graph that represents f better?

49. (a) Evaluate the function $f(x) = x^2 - (2^x/1000)$ for $x = 1, 0.8, 0.6, 0.4, 0.2, 0.1,$ and 0.05, and guess the value of
$$\lim\limits_{x \to 0} \left(x^2 - \dfrac{2^x}{1000} \right)$$

(b) Evaluate $f(x)$ for $x = 0.04, 0.02, 0.01, 0.005, 0.003,$ and 0.001. Guess again.

50. (a) Evaluate $h(x) = (\tan x - x)/x^3$ for $x = 1, 0.5, 0.1, 0.05, 0.01,$ and 0.005.
(b) Guess the value of $\lim\limits_{x \to 0} \dfrac{\tan x - x}{x^3}$.
(c) Evaluate $h(x)$ for successively smaller values of x until you finally reach a value of 0 for $h(x)$. Are you still confident that your guess in part (b) is correct? Explain why you eventually obtained 0 values. (In Section 4.4 a method for evaluating this limit will be explained.)
(d) Graph the function h in the viewing rectangle $[-1, 1]$ by $[0, 1]$. Then zoom in toward the point where the graph crosses the y-axis to estimate the limit of $h(x)$ as x approaches 0. Continue to zoom in until you observe distortions in the graph of h. Compare with the results of part (c).

51. Graph the function $f(x) = \sin(\pi/x)$ of Example 4 in the viewing rectangle $[-1, 1]$ by $[-1, 1]$. Then zoom in toward the origin several times. Comment on the behavior of this function.

52. Consider the function $f(x) = \tan \dfrac{1}{x}$.

(a) Show that $f(x) = 0$ for $x = \dfrac{1}{\pi}, \dfrac{1}{2\pi}, \dfrac{1}{3\pi}, \ldots$

(b) Show that $f(x) = 1$ for $x = \dfrac{4}{\pi}, \dfrac{4}{5\pi}, \dfrac{4}{9\pi}, \ldots$

(c) What can you conclude about $\lim\limits_{x \to 0^+} \tan \dfrac{1}{x}$?

53. Use a graph to estimate the equations of all the vertical asymptotes of the curve
$$y = \tan(2 \sin x) \qquad -\pi \le x \le \pi$$
Then find the exact equations of these asymptotes.

54. In the theory of relativity, the mass of a particle with velocity v is
$$m = \dfrac{m_0}{\sqrt{1 - v^2/c^2}}$$
where m_0 is the mass of the particle at rest and c is the speed of light. What happens as $v \to c^-$?

55. (a) Use numerical and graphical evidence to guess the value of the limit
$$\lim\limits_{x \to 1} \dfrac{x^3 - 1}{\sqrt{x} - 1}$$
(b) How close to 1 does x have to be to ensure that the function in part (a) is within a distance 0.5 of its limit?

2.3 Calculating Limits Using the Limit Laws

In Section 2.2 we used calculators and graphs to guess the values of limits, but we saw that such methods don't always lead to the correct answer. In this section we use the following properties of limits, called the *Limit Laws*, to calculate limits.

> **Limit Laws** Suppose that c is a constant and the limits
> $$\lim_{x \to a} f(x) \quad \text{and} \quad \lim_{x \to a} g(x)$$
> exist. Then
>
> 1. $\lim_{x \to a} [f(x) + g(x)] = \lim_{x \to a} f(x) + \lim_{x \to a} g(x)$
> 2. $\lim_{x \to a} [f(x) - g(x)] = \lim_{x \to a} f(x) - \lim_{x \to a} g(x)$
> 3. $\lim_{x \to a} [cf(x)] = c \lim_{x \to a} f(x)$
> 4. $\lim_{x \to a} [f(x) g(x)] = \lim_{x \to a} f(x) \cdot \lim_{x \to a} g(x)$
> 5. $\lim_{x \to a} \dfrac{f(x)}{g(x)} = \dfrac{\lim_{x \to a} f(x)}{\lim_{x \to a} g(x)}$ if $\lim_{x \to a} g(x) \neq 0$

These five laws can be stated verbally as follows:

Sum Law
1. The limit of a sum is the sum of the limits.

Difference Law
2. The limit of a difference is the difference of the limits.

Constant Multiple Law
3. The limit of a constant times a function is the constant times the limit of the function.

Product Law
4. The limit of a product is the product of the limits.

Quotient Law
5. The limit of a quotient is the quotient of the limits (provided that the limit of the denominator is not 0).

It is easy to believe that these properties are true. For instance, if $f(x)$ is close to L and $g(x)$ is close to M, it is reasonable to conclude that $f(x) + g(x)$ is close to $L + M$. This gives us an intuitive basis for believing that Law 1 is true. In Section 2.4 we give a precise definition of a limit and use it to prove this law. The proofs of the remaining laws are given in Appendix F.

EXAMPLE 1 Use the Limit Laws and the graphs of f and g in Figure 1 to evaluate the following limits, if they exist.

(a) $\lim_{x \to -2} [f(x) + 5g(x)]$ (b) $\lim_{x \to 1} [f(x) g(x)]$ (c) $\lim_{x \to 2} \dfrac{f(x)}{g(x)}$

SOLUTION

(a) From the graphs of f and g we see that

$$\lim_{x \to -2} f(x) = 1 \quad \text{and} \quad \lim_{x \to -2} g(x) = -1$$

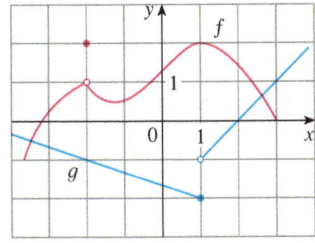

FIGURE 1

Therefore we have

$$\lim_{x \to -2} [f(x) + 5g(x)] = \lim_{x \to -2} f(x) + \lim_{x \to -2} [5g(x)] \quad \text{(by Limit Law 1)}$$

$$= \lim_{x \to -2} f(x) + 5 \lim_{x \to -2} g(x) \quad \text{(by Limit Law 3)}$$

$$= 1 + 5(-1) = -4$$

(b) We see that $\lim_{x \to 1} f(x) = 2$. But $\lim_{x \to 1} g(x)$ does not exist because the left and right limits are different:

$$\lim_{x \to 1^-} g(x) = -2 \qquad \lim_{x \to 1^+} g(x) = -1$$

So we can't use Law 4 for the desired limit. But we *can* use Law 4 for the one-sided limits:

$$\lim_{x \to 1^-} [f(x)g(x)] = \lim_{x \to 1^-} f(x) \cdot \lim_{x \to 1^-} g(x) = 2 \cdot (-2) = -4$$

$$\lim_{x \to 1^+} [f(x)g(x)] = \lim_{x \to 1^+} f(x) \cdot \lim_{x \to 1^+} g(x) = 2 \cdot (-1) = -2$$

The left and right limits aren't equal, so $\lim_{x \to 1} [f(x)g(x)]$ does not exist.

(c) The graphs show that

$$\lim_{x \to 2} f(x) \approx 1.4 \quad \text{and} \quad \lim_{x \to 2} g(x) = 0$$

Because the limit of the denominator is 0, we can't use Law 5. The given limit does not exist because the denominator approaches 0 while the numerator approaches a nonzero number. ∎

If we use the Product Law repeatedly with $g(x) = f(x)$, we obtain the following law.

Power Law

6. $\lim_{x \to a} [f(x)]^n = \left[\lim_{x \to a} f(x)\right]^n \qquad$ where n is a positive integer

In applying these six limit laws, we need to use two special limits:

7. $\lim_{x \to a} c = c \qquad$ **8.** $\lim_{x \to a} x = a$

These limits are obvious from an intuitive point of view (state them in words or draw graphs of $y = c$ and $y = x$), but proofs based on the precise definition are requested in the exercises for Section 2.4.

If we now put $f(x) = x$ in Law 6 and use Law 8, we get another useful special limit.

9. $\lim_{x \to a} x^n = a^n \qquad$ where n is a positive integer

A similar limit holds for roots as follows. (For square roots the proof is outlined in Exercise 2.4.37.)

10. $\lim_{x \to a} \sqrt[n]{x} = \sqrt[n]{a} \qquad$ where n is a positive integer

(If n is even, we assume that $a > 0$.)

More generally, we have the following law, which is proved in Section 2.5 as a consequence of Law 10.

Root Law

11. $\lim_{x \to a} \sqrt[n]{f(x)} = \sqrt[n]{\lim_{x \to a} f(x)}$ where n is a positive integer

$\left[\text{If } n \text{ is even, we assume that } \lim_{x \to a} f(x) > 0.\right]$

EXAMPLE 2 Evaluate the following limits and justify each step.

(a) $\lim_{x \to 5} (2x^2 - 3x + 4)$

(b) $\lim_{x \to -2} \dfrac{x^3 + 2x^2 - 1}{5 - 3x}$

SOLUTION

(a)
$$\lim_{x \to 5} (2x^2 - 3x + 4) = \lim_{x \to 5} (2x^2) - \lim_{x \to 5} (3x) + \lim_{x \to 5} 4 \quad \text{(by Laws 2 and 1)}$$
$$= 2 \lim_{x \to 5} x^2 - 3 \lim_{x \to 5} x + \lim_{x \to 5} 4 \quad \text{(by 3)}$$
$$= 2(5^2) - 3(5) + 4 \quad \text{(by 9, 8, and 7)}$$
$$= 39$$

(b) We start by using Law 5, but its use is fully justified only at the final stage when we see that the limits of the numerator and denominator exist and the limit of the denominator is not 0.

$$\lim_{x \to -2} \frac{x^3 + 2x^2 - 1}{5 - 3x} = \frac{\lim_{x \to -2} (x^3 + 2x^2 - 1)}{\lim_{x \to -2} (5 - 3x)} \quad \text{(by Law 5)}$$
$$= \frac{\lim_{x \to -2} x^3 + 2 \lim_{x \to -2} x^2 - \lim_{x \to -2} 1}{\lim_{x \to -2} 5 - 3 \lim_{x \to -2} x} \quad \text{(by 1, 2, and 3)}$$
$$= \frac{(-2)^3 + 2(-2)^2 - 1}{5 - 3(-2)} \quad \text{(by 9, 8, and 7)}$$
$$= -\frac{1}{11}$$

NOTE If we let $f(x) = 2x^2 - 3x + 4$, then $f(5) = 39$. In other words, we would have gotten the correct answer in Example 2(a) by substituting 5 for x. Similarly, direct substitution provides the correct answer in part (b). The functions in Example 2 are a polynomial and a rational function, respectively, and similar use of the Limit Laws proves that direct substitution always works for such functions (see Exercises 57 and 58). We state this fact as follows.

Direct Substitution Property If f is a polynomial or a rational function and a is in the domain of f, then
$$\lim_{x \to a} f(x) = f(a)$$

Newton and Limits

Isaac Newton was born on Christmas Day in 1642, the year of Galileo's death. When he entered Cambridge University in 1661 Newton didn't know much mathematics, but he learned quickly by reading Euclid and Descartes and by attending the lectures of Isaac Barrow. Cambridge was closed because of the plague in 1665 and 1666, and Newton returned home to reflect on what he had learned. Those two years were amazingly productive for at that time he made four of his major discoveries: (1) his representation of functions as sums of infinite series, including the binomial theorem; (2) his work on differential and integral calculus; (3) his laws of motion and law of universal gravitation; and (4) his prism experiments on the nature of light and color. Because of a fear of controversy and criticism, he was reluctant to publish his discoveries and it wasn't until 1687, at the urging of the astronomer Halley, that Newton published *Principia Mathematica*. In this work, the greatest scientific treatise ever written, Newton set forth his version of calculus and used it to investigate mechanics, fluid dynamics, and wave motion, and to explain the motion of planets and comets.

The beginnings of calculus are found in the calculations of areas and volumes by ancient Greek scholars such as Eudoxus and Archimedes. Although aspects of the idea of a limit are implicit in their "method of exhaustion," Eudoxus and Archimedes never explicitly formulated the concept of a limit. Likewise, mathematicians such as Cavalieri, Fermat, and Barrow, the immediate precursors of Newton in the development of calculus, did not actually use limits. It was Isaac Newton who was the first to talk explicitly about limits. He explained that the main idea behind limits is that quantities "approach nearer than by any given difference." Newton stated that the limit was the basic concept in calculus, but it was left to later mathematicians like Cauchy to clarify his ideas about limits.

98   CHAPTER 2   Limits and Derivatives

Functions with the Direct Substitution Property are called *continuous at a* and will be studied in Section 2.5. However, not all limits can be evaluated by direct substitution, as the following examples show.

EXAMPLE 3 Find $\displaystyle\lim_{x \to 1} \frac{x^2 - 1}{x - 1}$.

SOLUTION Let $f(x) = (x^2 - 1)/(x - 1)$. We can't find the limit by substituting $x = 1$ because $f(1)$ isn't defined. Nor can we apply the Quotient Law, because the limit of the denominator is 0. Instead, we need to do some preliminary algebra. We factor the numerator as a difference of squares:

$$\frac{x^2 - 1}{x - 1} = \frac{(x - 1)(x + 1)}{x - 1}$$

Notice that in Example 3 we do not have an infinite limit even though the denominator approaches 0 as $x \to 1$. When both numerator and denominator approach 0, the limit may be infinite or it may be some finite value.

The numerator and denominator have a common factor of $x - 1$. When we take the limit as x approaches 1, we have $x \neq 1$ and so $x - 1 \neq 0$. Therefore we can cancel the common factor and then compute the limit by direct substitution as follows:

$$\lim_{x \to 1} \frac{x^2 - 1}{x - 1} = \lim_{x \to 1} \frac{(x - 1)(x + 1)}{x - 1}$$

$$= \lim_{x \to 1} (x + 1)$$

$$= 1 + 1 = 2$$

The limit in this example arose in Example 2.1.1 when we were trying to find the tangent to the parabola $y = x^2$ at the point $(1, 1)$. ∎

NOTE In Example 3 we were able to compute the limit by replacing the given function $f(x) = (x^2 - 1)/(x - 1)$ by a simpler function, $g(x) = x + 1$, with the same limit. This is valid because $f(x) = g(x)$ except when $x = 1$, and in computing a limit as x approaches 1 we don't consider what happens when x is actually *equal* to 1. In general, we have the following useful fact.

> If $f(x) = g(x)$ when $x \neq a$, then $\displaystyle\lim_{x \to a} f(x) = \lim_{x \to a} g(x)$, provided the limits exist.

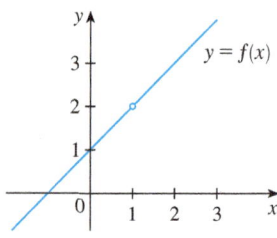

EXAMPLE 4 Find $\displaystyle\lim_{x \to 1} g(x)$ where

$$g(x) = \begin{cases} x + 1 & \text{if } x \neq 1 \\ \pi & \text{if } x = 1 \end{cases}$$

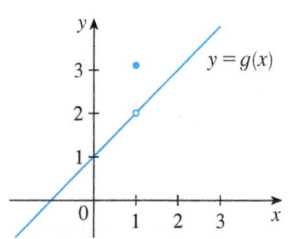

SOLUTION Here g is defined at $x = 1$ and $g(1) = \pi$, but the value of a limit as x approaches 1 does not depend on the value of the function *at* 1. Since $g(x) = x + 1$ for $x \neq 1$, we have

$$\lim_{x \to 1} g(x) = \lim_{x \to 1} (x + 1) = 2$$ ∎

FIGURE 2
The graphs of the functions f (from Example 3) and g (from Example 4)

Note that the values of the functions in Examples 3 and 4 are identical except when $x = 1$ (see Figure 2) and so they have the same limit as x approaches 1.

EXAMPLE 5 Evaluate $\lim_{h \to 0} \dfrac{(3+h)^2 - 9}{h}$.

SOLUTION If we define

$$F(h) = \dfrac{(3+h)^2 - 9}{h}$$

then, as in Example 3, we can't compute $\lim_{h \to 0} F(h)$ by letting $h = 0$ since $F(0)$ is undefined. But if we simplify $F(h)$ algebraically, we find that

$$F(h) = \dfrac{(9 + 6h + h^2) - 9}{h} = \dfrac{6h + h^2}{h} = \dfrac{h(6+h)}{h} = 6 + h$$

(Recall that we consider only $h \neq 0$ when letting h approach 0.) Thus

$$\lim_{h \to 0} \dfrac{(3+h)^2 - 9}{h} = \lim_{h \to 0} (6+h) = 6 \qquad \blacksquare$$

EXAMPLE 6 Find $\lim_{t \to 0} \dfrac{\sqrt{t^2 + 9} - 3}{t^2}$.

SOLUTION We can't apply the Quotient Law immediately, since the limit of the denominator is 0. Here the preliminary algebra consists of rationalizing the numerator:

$$\lim_{t \to 0} \dfrac{\sqrt{t^2 + 9} - 3}{t^2} = \lim_{t \to 0} \dfrac{\sqrt{t^2 + 9} - 3}{t^2} \cdot \dfrac{\sqrt{t^2 + 9} + 3}{\sqrt{t^2 + 9} + 3}$$

$$= \lim_{t \to 0} \dfrac{(t^2 + 9) - 9}{t^2 (\sqrt{t^2 + 9} + 3)}$$

$$= \lim_{t \to 0} \dfrac{t^2}{t^2 (\sqrt{t^2 + 9} + 3)}$$

$$= \lim_{t \to 0} \dfrac{1}{\sqrt{t^2 + 9} + 3}$$

$$= \dfrac{1}{\sqrt{\lim_{t \to 0} (t^2 + 9)} + 3}$$

Here we use several properties of limits (5, 1, 10, 7, 9).

$$= \dfrac{1}{3 + 3} = \dfrac{1}{6}$$

This calculation confirms the guess that we made in Example 2.2.2. $\qquad \blacksquare$

Some limits are best calculated by first finding the left- and right-hand limits. The following theorem is a reminder of what we discovered in Section 2.2. It says that a two-sided limit exists if and only if both of the one-sided limits exist and are equal.

1 Theorem $\lim_{x \to a} f(x) = L$ if and only if $\lim_{x \to a^-} f(x) = L = \lim_{x \to a^+} f(x)$

When computing one-sided limits, we use the fact that the Limit Laws also hold for one-sided limits.

EXAMPLE 7 Show that $\lim_{x \to 0} |x| = 0$.

SOLUTION Recall that

$$|x| = \begin{cases} x & \text{if } x \geq 0 \\ -x & \text{if } x < 0 \end{cases}$$

Since $|x| = x$ for $x > 0$, we have

$$\lim_{x \to 0^+} |x| = \lim_{x \to 0^+} x = 0$$

For $x < 0$ we have $|x| = -x$ and so

$$\lim_{x \to 0^-} |x| = \lim_{x \to 0^-} (-x) = 0$$

Therefore, by Theorem 1,

$$\lim_{x \to 0} |x| = 0$$

The result of Example 7 looks plausible from Figure 3.

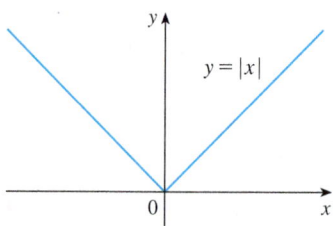

FIGURE 3

EXAMPLE 8 Prove that $\lim_{x \to 0} \dfrac{|x|}{x}$ does not exist.

SOLUTION Using the facts that $|x| = x$ when $x > 0$ and $|x| = -x$ when $x < 0$, we have

$$\lim_{x \to 0^+} \frac{|x|}{x} = \lim_{x \to 0^+} \frac{x}{x} = \lim_{x \to 0^+} 1 = 1$$

$$\lim_{x \to 0^-} \frac{|x|}{x} = \lim_{x \to 0^-} \frac{-x}{x} = \lim_{x \to 0^-} (-1) = -1$$

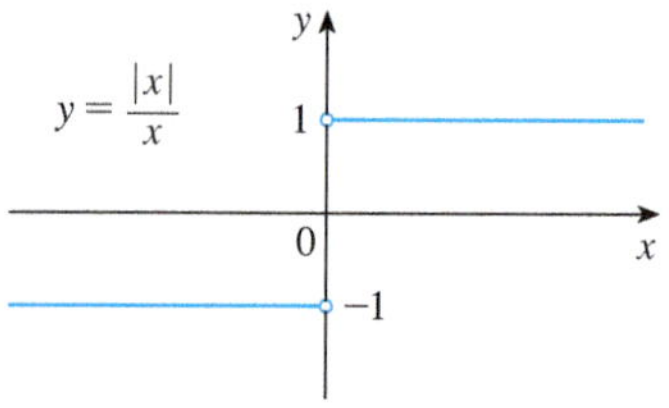

Since the right- and left-hand limits are different, it follows from Theorem 1 that $\lim_{x \to 0} |x|/x$ does not exist. The graph of the function $f(x) = |x|/x$ is shown in Figure 4 and supports the one-sided limits that we found.

FIGURE 4

EXAMPLE 9 If

$$f(x) = \begin{cases} \sqrt{x - 4} & \text{if } x > 4 \\ 8 - 2x & \text{if } x < 4 \end{cases}$$

determine whether $\lim_{x \to 4} f(x)$ exists.

SOLUTION Since $f(x) = \sqrt{x - 4}$ for $x > 4$, we have

$$\lim_{x \to 4^+} f(x) = \lim_{x \to 4^+} \sqrt{x - 4} = \sqrt{4 - 4} = 0$$

Since $f(x) = 8 - 2x$ for $x < 4$, we have

$$\lim_{x \to 4^-} f(x) = \lim_{x \to 4^-} (8 - 2x) = 8 - 2 \cdot 4 = 0$$

The right- and left-hand limits are equal. Thus the limit exists and

$$\lim_{x \to 4} f(x) = 0$$

It is shown in Example 2.4.3 that $\lim_{x \to 0^+} \sqrt{x} = 0$.

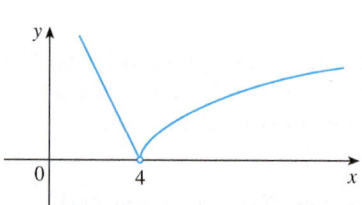

FIGURE 5

The graph of f is shown in Figure 5.

Other notations for $[\![x]\!]$ are $[x]$ and $\lfloor x \rfloor$. The greatest integer function is sometimes called the *floor function*.

EXAMPLE 10 The **greatest integer function** is defined by $[\![x]\!]$ = the largest integer that is less than or equal to x. (For instance, $[\![4]\!] = 4$, $[\![4.8]\!] = 4$, $[\![\pi]\!] = 3$, $[\![\sqrt{2}]\!] = 1$, $[\![-\frac{1}{2}]\!] = -1$.) Show that $\lim_{x \to 3} [\![x]\!]$ does not exist.

SOLUTION The graph of the greatest integer function is shown in Figure 6. Since $[\![x]\!] = 3$ for $3 \leq x < 4$, we have

$$\lim_{x \to 3^+} [\![x]\!] = \lim_{x \to 3^+} 3 = 3$$

Since $[\![x]\!] = 2$ for $2 \leq x < 3$, we have

$$\lim_{x \to 3^-} [\![x]\!] = \lim_{x \to 3^-} 2 = 2$$

Because these one-sided limits are not equal, $\lim_{x \to 3} [\![x]\!]$ does not exist by Theorem 1. ∎

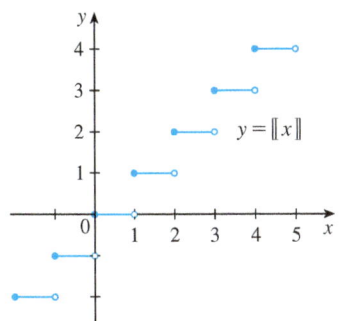

FIGURE 6
Greatest integer function

The next two theorems give two additional properties of limits. Their proofs can be found in Appendix F.

2 Theorem If $f(x) \leq g(x)$ when x is near a (except possibly at a) and the limits of f and g both exist as x approaches a, then

$$\lim_{x \to a} f(x) \leq \lim_{x \to a} g(x)$$

3 The Squeeze Theorem If $f(x) \leq g(x) \leq h(x)$ when x is near a (except possibly at a) and

$$\lim_{x \to a} f(x) = \lim_{x \to a} h(x) = L$$

then

$$\lim_{x \to a} g(x) = L$$

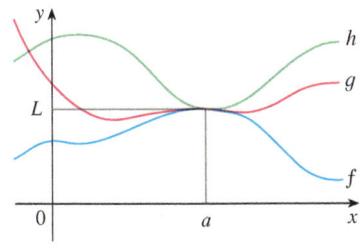

FIGURE 7

The Squeeze Theorem, which is sometimes called the Sandwich Theorem or the Pinching Theorem, is illustrated by Figure 7. It says that if $g(x)$ is squeezed between $f(x)$ and $h(x)$ near a, and if f and h have the same limit L at a, then g is forced to have the same limit L at a.

EXAMPLE 11 Show that $\lim_{x \to 0} x^2 \sin \dfrac{1}{x} = 0$.

SOLUTION First note that we **cannot** use

$$\lim_{x \to 0} x^2 \sin \frac{1}{x} = \lim_{x \to 0} x^2 \cdot \lim_{x \to 0} \sin \frac{1}{x}$$

because $\lim_{x \to 0} \sin(1/x)$ does not exist (see Example 2.2.4).

Instead we apply the Squeeze Theorem, and so we need to find a function f smaller than $g(x) = x^2 \sin(1/x)$ and a function h bigger than g such that both $f(x)$ and $h(x)$ approach 0. To do this we use our knowledge of the sine function. Because the sine of any number lies between -1 and 1, we can write.

$$\boxed{4} \qquad -1 \leq \sin \frac{1}{x} \leq 1$$

Any inequality remains true when multiplied by a positive number. We know that $x^2 \geq 0$ for all x and so, multiplying each side of the inequalities in (4) by x^2, we get

$$-x^2 \leq x^2 \sin \frac{1}{x} \leq x^2$$

as illustrated by Figure 8. We know that

$$\lim_{x \to 0} x^2 = 0 \quad \text{and} \quad \lim_{x \to 0} (-x^2) = 0$$

Taking $f(x) = -x^2$, $g(x) = x^2 \sin(1/x)$, and $h(x) = x^2$ in the Squeeze Theorem, we obtain

$$\lim_{x \to 0} x^2 \sin \frac{1}{x} = 0$$

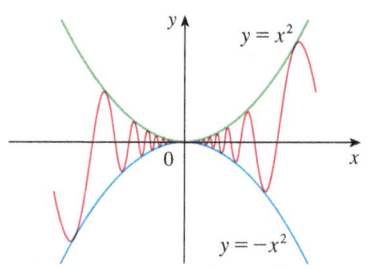

FIGURE 8
$y = x^2 \sin(1/x)$

2.3 EXERCISES

1. Given that

$$\lim_{x \to 2} f(x) = 4 \qquad \lim_{x \to 2} g(x) = -2 \qquad \lim_{x \to 2} h(x) = 0$$

find the limits that exist. If the limit does not exist, explain why.

(a) $\lim_{x \to 2} [f(x) + 5g(x)]$ (b) $\lim_{x \to 2} [g(x)]^3$

(c) $\lim_{x \to 2} \sqrt{f(x)}$ (d) $\lim_{x \to 2} \frac{3f(x)}{g(x)}$

(e) $\lim_{x \to 2} \frac{g(x)}{h(x)}$ (f) $\lim_{x \to 2} \frac{g(x)h(x)}{f(x)}$

2. The graphs of f and g are given. Use them to evaluate each limit, if it exists. If the limit does not exist, explain why.

(a) $\lim_{x \to 2} [f(x) + g(x)]$ (b) $\lim_{x \to 0} [f(x) - g(x)]$

(c) $\lim_{x \to -1} [f(x)g(x)]$ (d) $\lim_{x \to 3} \frac{f(x)}{g(x)}$

(e) $\lim_{x \to 2} [x^2 f(x)]$ (f) $f(-1) + \lim_{x \to -1} g(x)$

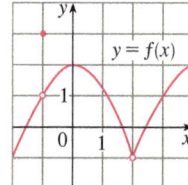
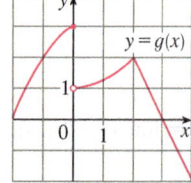

3–9 Evaluate the limit and justify each step by indicating the appropriate Limit Law(s).

3. $\lim_{x \to 3} (5x^3 - 3x^2 + x - 6)$

4. $\lim_{x \to -1} (x^4 - 3x)(x^2 + 5x + 3)$

5. $\lim_{t \to -2} \frac{t^4 - 2}{2t^2 - 3t + 2}$

6. $\lim_{u \to -2} \sqrt{u^4 + 3u + 6}$

7. $\lim_{x \to 8} (1 + \sqrt[3]{x})(2 - 6x^2 + x^3)$

8. $\lim_{t \to 2} \left(\frac{t^2 - 2}{t^3 - 3t + 5} \right)^2$

9. $\lim_{x \to 2} \sqrt{\frac{2x^2 + 1}{3x - 2}}$

10. (a) What is wrong with the following equation?

$$\frac{x^2 + x - 6}{x - 2} = x + 3$$

(b) In view of part (a), explain why the equation

$$\lim_{x \to 2} \frac{x^2 + x - 6}{x - 2} = \lim_{x \to 2} (x + 3)$$

is correct.

11–32 Evaluate the limit, if it exists.

11. $\lim_{x \to 5} \frac{x^2 - 6x + 5}{x - 5}$

12. $\lim_{x \to -3} \frac{x^2 + 3x}{x^2 - x - 12}$

13. $\lim_{x \to 5} \frac{x^2 - 5x + 6}{x - 5}$

14. $\lim_{x \to 4} \frac{x^2 + 3x}{x^2 - x - 12}$

15. $\lim_{t \to -3} \frac{t^2 - 9}{2t^2 + 7t + 3}$

16. $\lim_{x \to -1} \frac{2x^2 + 3x + 1}{x^2 - 2x - 3}$

17. $\lim_{h \to 0} \frac{(-5 + h)^2 - 25}{h}$

18. $\lim_{h \to 0} \frac{(2 + h)^3 - 8}{h}$

19. $\lim\limits_{x \to -2} \dfrac{x+2}{x^3+8}$

20. $\lim\limits_{t \to 1} \dfrac{t^4-1}{t^3-1}$

21. $\lim\limits_{h \to 0} \dfrac{\sqrt{9+h}-3}{h}$

22. $\lim\limits_{u \to 2} \dfrac{\sqrt{4u+1}-3}{u-2}$

23. $\lim\limits_{x \to 3} \dfrac{\frac{1}{x}-\frac{1}{3}}{x-3}$

24. $\lim\limits_{h \to 0} \dfrac{(3+h)^{-1}-3^{-1}}{h}$

25. $\lim\limits_{t \to 0} \dfrac{\sqrt{1+t}-\sqrt{1-t}}{t}$

26. $\lim\limits_{t \to 0} \left(\dfrac{1}{t}-\dfrac{1}{t^2+t}\right)$

27. $\lim\limits_{x \to 16} \dfrac{4-\sqrt{x}}{16x-x^2}$

28. $\lim\limits_{x \to 2} \dfrac{x^2-4x+4}{x^4-3x^2-4}$

29. $\lim\limits_{t \to 0} \left(\dfrac{1}{t\sqrt{1+t}}-\dfrac{1}{t}\right)$

30. $\lim\limits_{x \to -4} \dfrac{\sqrt{x^2+9}-5}{x+4}$

31. $\lim\limits_{h \to 0} \dfrac{(x+h)^3-x^3}{h}$

32. $\lim\limits_{h \to 0} \dfrac{\frac{1}{(x+h)^2}-\frac{1}{x^2}}{h}$

33. (a) Estimate the value of
$$\lim_{x \to 0} \dfrac{x}{\sqrt{1+3x}-1}$$
by graphing the function $f(x) = x/(\sqrt{1+3x}-1)$.
(b) Make a table of values of $f(x)$ for x close to 0 and guess the value of the limit.
(c) Use the Limit Laws to prove that your guess is correct.

34. (a) Use a graph of
$$f(x) = \dfrac{\sqrt{3+x}-\sqrt{3}}{x}$$
to estimate the value of $\lim_{x \to 0} f(x)$ to two decimal places.
(b) Use a table of values of $f(x)$ to estimate the limit to four decimal places.
(c) Use the Limit Laws to find the exact value of the limit.

35. Use the Squeeze Theorem to show that $\lim_{x \to 0} (x^2 \cos 20\pi x) = 0$. Illustrate by graphing the functions $f(x) = -x^2$, $g(x) = x^2 \cos 20\pi x$, and $h(x) = x^2$ on the same screen.

36. Use the Squeeze Theorem to show that
$$\lim_{x \to 0} \sqrt{x^3+x^2}\, \sin \dfrac{\pi}{x} = 0$$
Illustrate by graphing the functions f, g, and h (in the notation of the Squeeze Theorem) on the same screen.

37. If $4x - 9 \leq f(x) \leq x^2 - 4x + 7$ for $x \geq 0$, find $\lim\limits_{x \to 4} f(x)$.

38. If $2x \leq g(x) \leq x^4 - x^2 + 2$ for all x, evaluate $\lim\limits_{x \to 1} g(x)$.

39. Prove that $\lim\limits_{x \to 0} x^4 \cos \dfrac{2}{x} = 0$.

40. Prove that $\lim\limits_{x \to 0^+} \sqrt{x}\, e^{\sin(\pi/x)} = 0$.

41–46 Find the limit, if it exists. If the limit does not exist, explain why.

41. $\lim\limits_{x \to 3} (2x + |x-3|)$

42. $\lim\limits_{x \to -6} \dfrac{2x+12}{|x+6|}$

43. $\lim\limits_{x \to 0.5^-} \dfrac{2x-1}{|2x^3-x^2|}$

44. $\lim\limits_{x \to -2} \dfrac{2-|x|}{2+x}$

45. $\lim\limits_{x \to 0^-} \left(\dfrac{1}{x}-\dfrac{1}{|x|}\right)$

46. $\lim\limits_{x \to 0^+} \left(\dfrac{1}{x}-\dfrac{1}{|x|}\right)$

47. The *signum* (or *sign*) *function*, denoted by sgn, is defined by
$$\operatorname{sgn} x = \begin{cases} -1 & \text{if } x < 0 \\ 0 & \text{if } x = 0 \\ 1 & \text{if } x > 0 \end{cases}$$
(a) Sketch the graph of this function.
(b) Find each of the following limits or explain why it does not exist.
  (i) $\lim\limits_{x \to 0^+} \operatorname{sgn} x$ (ii) $\lim\limits_{x \to 0^-} \operatorname{sgn} x$
  (iii) $\lim\limits_{x \to 0} \operatorname{sgn} x$ (iv) $\lim\limits_{x \to 0} |\operatorname{sgn} x|$

48. Let $g(x) = \operatorname{sgn}(\sin x)$.
(a) Find each of the following limits or explain why it does not exist.
  (i) $\lim\limits_{x \to 0^+} g(x)$ (ii) $\lim\limits_{x \to 0^-} g(x)$ (iii) $\lim\limits_{x \to 0} g(x)$
  (iv) $\lim\limits_{x \to \pi^+} g(x)$ (v) $\lim\limits_{x \to \pi^-} g(x)$ (vi) $\lim\limits_{x \to \pi} g(x)$
(b) For which values of a does $\lim_{x \to a} g(x)$ not exist?
(c) Sketch a graph of g.

49. Let $g(x) = \dfrac{x^2+x-6}{|x-2|}$.
(a) Find
  (i) $\lim\limits_{x \to 2^+} g(x)$ (ii) $\lim\limits_{x \to 2^-} g(x)$
(b) Does $\lim_{x \to 2} g(x)$ exist?
(c) Sketch the graph of g.

50. Let
$$f(x) = \begin{cases} x^2+1 & \text{if } x < 1 \\ (x-2)^2 & \text{if } x \geq 1 \end{cases}$$
(a) Find $\lim_{x \to 1^-} f(x)$ and $\lim_{x \to 1^+} f(x)$.
(b) Does $\lim_{x \to 1} f(x)$ exist?
(c) Sketch the graph of f.

51. Let
$$B(t) = \begin{cases} 4 - \tfrac{1}{2}t & \text{if } t < 2 \\ \sqrt{t+c} & \text{if } t \geq 2 \end{cases}$$
Find the value of c so that $\lim\limits_{t \to 2} B(t)$ exists.

52. Let
$$g(x) = \begin{cases} x & \text{if } x < 1 \\ 3 & \text{if } x = 1 \\ 2 - x^2 & \text{if } 1 < x \leq 2 \\ x - 3 & \text{if } x > 2 \end{cases}$$

(a) Evaluate each of the following, if it exists.
(i) $\lim_{x \to 1^-} g(x)$ (ii) $\lim_{x \to 1} g(x)$ (iii) $g(1)$
(iv) $\lim_{x \to 2^-} g(x)$ (v) $\lim_{x \to 2^+} g(x)$ (vi) $\lim_{x \to 2} g(x)$

(b) Sketch the graph of g.

53. (a) If the symbol $[\![\ ]\!]$ denotes the greatest integer function defined in Example 10, evaluate
(i) $\lim_{x \to -2^+} [\![x]\!]$ (ii) $\lim_{x \to -2} [\![x]\!]$ (iii) $\lim_{x \to -2.4} [\![x]\!]$

(b) If n is an integer, evaluate
(i) $\lim_{x \to n^-} [\![x]\!]$ (ii) $\lim_{x \to n^+} [\![x]\!]$

(c) For what values of a does $\lim_{x \to a} [\![x]\!]$ exist?

54. Let $f(x) = [\![\cos x]\!]$, $-\pi \leq x \leq \pi$.
(a) Sketch the graph of f.
(b) Evaluate each limit, if it exists.
(i) $\lim_{x \to 0} f(x)$ (ii) $\lim_{x \to (\pi/2)^-} f(x)$
(iii) $\lim_{x \to (\pi/2)^+} f(x)$ (iv) $\lim_{x \to \pi/2} f(x)$
(c) For what values of a does $\lim_{x \to a} f(x)$ exist?

55. If $f(x) = [\![x]\!] + [\![-x]\!]$, show that $\lim_{x \to 2} f(x)$ exists but is not equal to $f(2)$.

56. In the theory of relativity, the Lorentz contraction formula
$$L = L_0 \sqrt{1 - v^2/c^2}$$
expresses the length L of an object as a function of its velocity v with respect to an observer, where L_0 is the length of the object at rest and c is the speed of light. Find $\lim_{v \to c^-} L$ and interpret the result. Why is a left-hand limit necessary?

57. If p is a polynomial, show that $\lim_{x \to a} p(x) = p(a)$.

58. If r is a rational function, use Exercise 57 to show that $\lim_{x \to a} r(x) = r(a)$ for every number a in the domain of r.

59. If $\lim_{x \to 1} \dfrac{f(x) - 8}{x - 1} = 10$, find $\lim_{x \to 1} f(x)$.

60. If $\lim_{x \to 0} \dfrac{f(x)}{x^2} = 5$, find the following limits.
(a) $\lim_{x \to 0} f(x)$ (b) $\lim_{x \to 0} \dfrac{f(x)}{x}$

61. If
$$f(x) = \begin{cases} x^2 & \text{if } x \text{ is rational} \\ 0 & \text{if } x \text{ is irrational} \end{cases}$$
prove that $\lim_{x \to 0} f(x) = 0$.

62. Show by means of an example that $\lim_{x \to a} [f(x) + g(x)]$ may exist even though neither $\lim_{x \to a} f(x)$ nor $\lim_{x \to a} g(x)$ exists.

63. Show by means of an example that $\lim_{x \to a} [f(x) g(x)]$ may exist even though neither $\lim_{x \to a} f(x)$ nor $\lim_{x \to a} g(x)$ exists.

64. Evaluate $\lim_{x \to 2} \dfrac{\sqrt{6 - x} - 2}{\sqrt{3 - x} - 1}$.

65. Is there a number a such that
$$\lim_{x \to -2} \frac{3x^2 + ax + a + 3}{x^2 + x - 2}$$
exists? If so, find the value of a and the value of the limit.

66. The figure shows a fixed circle C_1 with equation $(x - 1)^2 + y^2 = 1$ and a shrinking circle C_2 with radius r and center the origin. P is the point $(0, r)$, Q is the upper point of intersection of the two circles, and R is the point of intersection of the line PQ and the x-axis. What happens to R as C_2 shrinks, that is, as $r \to 0^+$?

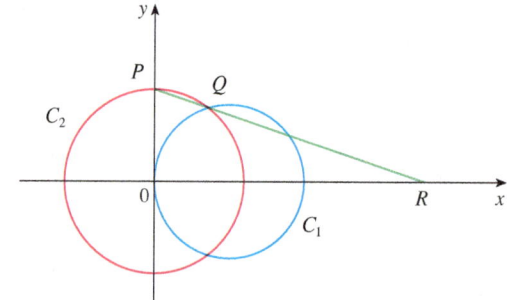

2.4 The Precise Definition of a Limit

The intuitive definition of a limit given in Section 2.2 is inadequate for some purposes because such phrases as "x is close to 2" and "$f(x)$ gets closer and closer to L" are vague. In order to be able to prove conclusively that

$$\lim_{x \to 0} \left(x^3 + \frac{\cos 5x}{10{,}000} \right) = 0.0001 \quad \text{or} \quad \lim_{x \to 0} \frac{\sin x}{x} = 1$$

we must make the definition of a limit precise.

SECTION 2.4 The Precise Definition of a Limit

To motivate the precise definition of a limit, let's consider the function

$$f(x) = \begin{cases} 2x - 1 & \text{if } x \neq 3 \\ 6 & \text{if } x = 3 \end{cases}$$

Intuitively, it is clear that when x is close to 3 but $x \neq 3$, then $f(x)$ is close to 5, and so $\lim_{x \to 3} f(x) = 5$.

To obtain more detailed information about how $f(x)$ varies when x is close to 3, we ask the following question:

How close to 3 does x have to be so that $f(x)$ differs from 5 by less than 0.1?

The distance from x to 3 is $|x - 3|$ and the distance from $f(x)$ to 5 is $|f(x) - 5|$, so our problem is to find a number δ such that

It is traditional to use the Greek letter δ (delta) in this situation.

$$|f(x) - 5| < 0.1 \quad \text{if} \quad |x - 3| < \delta \text{ but } x \neq 3$$

If $|x - 3| > 0$, then $x \neq 3$, so an equivalent formulation of our problem is to find a number δ such that

$$|f(x) - 5| < 0.1 \quad \text{if} \quad 0 < |x - 3| < \delta$$

Notice that if $0 < |x - 3| < (0.1)/2 = 0.05$, then

$$|f(x) - 5| = |(2x - 1) - 5| = |2x - 6| = 2|x - 3| < 2(0.05) = 0.1$$

that is, $\quad |f(x) - 5| < 0.1 \quad \text{if} \quad 0 < |x - 3| < 0.05$

Thus an answer to the problem is given by $\delta = 0.05$; that is, if x is within a distance of 0.05 from 3, then $f(x)$ will be within a distance of 0.1 from 5.

If we change the number 0.1 in our problem to the smaller number 0.01, then by using the same method we find that $f(x)$ will differ from 5 by less than 0.01 provided that x differs from 3 by less than $(0.01)/2 = 0.005$:

$$|f(x) - 5| < 0.01 \quad \text{if} \quad 0 < |x - 3| < 0.005$$

Similarly,

$$|f(x) - 5| < 0.001 \quad \text{if} \quad 0 < |x - 3| < 0.0005$$

The numbers 0.1, 0.01, and 0.001 that we have considered are *error tolerances* that we might allow. For 5 to be the precise limit of $f(x)$ as x approaches 3, we must not only be able to bring the difference between $f(x)$ and 5 below each of these three numbers; we must be able to bring it below *any* positive number. And, by the same reasoning, we can! If we write ε (the Greek letter epsilon) for an arbitrary positive number, then we find as before that

$$\boxed{1} \qquad |f(x) - 5| < \varepsilon \quad \text{if} \quad 0 < |x - 3| < \delta = \frac{\varepsilon}{2}$$

This is a precise way of saying that $f(x)$ is close to 5 when x is close to 3 because (1) says that we can make the values of $f(x)$ within an arbitrary distance ε from 5 by restricting the values of x to be within a distance $\varepsilon/2$ from 3 (but $x \neq 3$).

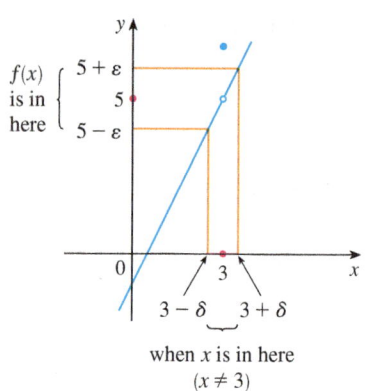

FIGURE 1

Note that (1) can be rewritten as follows:

if $3 - \delta < x < 3 + \delta$ $(x \neq 3)$ then $5 - \varepsilon < f(x) < 5 + \varepsilon$

and this is illustrated in Figure 1. By taking the values of x ($\neq 3$) to lie in the interval $(3 - \delta, 3 + \delta)$ we can make the values of $f(x)$ lie in the interval $(5 - \varepsilon, 5 + \varepsilon)$. Using (1) as a model, we give a precise definition of a limit.

2 Precise Definition of a Limit Let f be a function defined on some open interval that contains the number a, except possibly at a itself. Then we say that the **limit of $f(x)$ as x approaches a is L**, and we write

$$\lim_{x \to a} f(x) = L$$

if for every number $\varepsilon > 0$ there is a number $\delta > 0$ such that

if $0 < |x - a| < \delta$ then $|f(x) - L| < \varepsilon$

Since $|x - a|$ is the distance from x to a and $|f(x) - L|$ is the distance from $f(x)$ to L, and since ε can be arbitrarily small, the definition of a limit can be expressed in words as follows:

$\lim_{x \to a} f(x) = L$ means that the distance between $f(x)$ and L can be made arbitrarily small by requiring that the distance from x to a be sufficiently small (but not 0).

Alternatively,

$\lim_{x \to a} f(x) = L$ means that the values of $f(x)$ can be made as close as we please to L by requiring x to be close enough to a (but not equal to a).

We can also reformulate Definition 2 in terms of intervals by observing that the inequality $|x - a| < \delta$ is equivalent to $-\delta < x - a < \delta$, which in turn can be written as $a - \delta < x < a + \delta$. Also $0 < |x - a|$ is true if and only if $x - a \neq 0$, that is, $x \neq a$. Similarly, the inequality $|f(x) - L| < \varepsilon$ is equivalent to the pair of inequalities $L - \varepsilon < f(x) < L + \varepsilon$. Therefore, in terms of intervals, Definition 2 can be stated as follows:

$\lim_{x \to a} f(x) = L$ means that for every $\varepsilon > 0$ (no matter how small ε is) we can find $\delta > 0$ such that if x lies in the open interval $(a - \delta, a + \delta)$ and $x \neq a$, then $f(x)$ lies in the open interval $(L - \varepsilon, L + \varepsilon)$.

We interpret this statement geometrically by representing a function by an arrow diagram as in Figure 2, where f maps a subset of $\mathbb{R}$ onto another subset of $\mathbb{R}$.

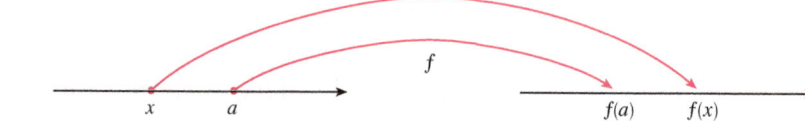

FIGURE 2

The definition of limit says that if any small interval $(L - \varepsilon, L + \varepsilon)$ is given around L, then we can find an interval $(a - \delta, a + \delta)$ around a such that f maps all the points in $(a - \delta, a + \delta)$ (except possibly a) into the interval $(L - \varepsilon, L + \varepsilon)$. (See Figure 3.)

SECTION 2.4 The Precise Definition of a Limit

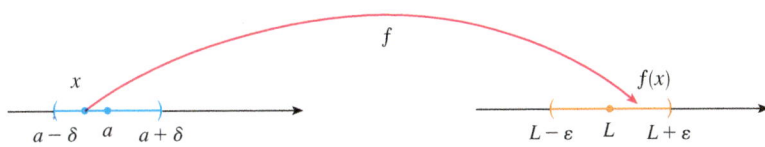

FIGURE 3

Another geometric interpretation of limits can be given in terms of the graph of a function. If $\varepsilon > 0$ is given, then we draw the horizontal lines $y = L + \varepsilon$ and $y = L - \varepsilon$ and the graph of f. (See Figure 4.) If $\lim_{x \to a} f(x) = L$, then we can find a number $\delta > 0$ such that if we restrict x to lie in the interval $(a - \delta, a + \delta)$ and take $x \neq a$, then the curve $y = f(x)$ lies between the lines $y = L - \varepsilon$ and $y = L + \varepsilon$. (See Figure 5.) You can see that if such a δ has been found, then any smaller δ will also work.

It is important to realize that the process illustrated in Figures 4 and 5 must work for *every* positive number ε, no matter how small it is chosen. Figure 6 shows that if a smaller ε is chosen, then a smaller δ may be required.

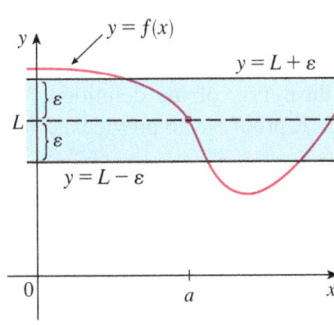

FIGURE 4

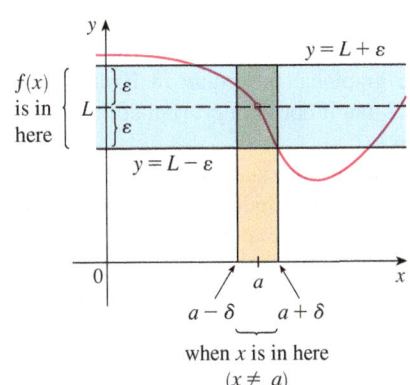

FIGURE 5

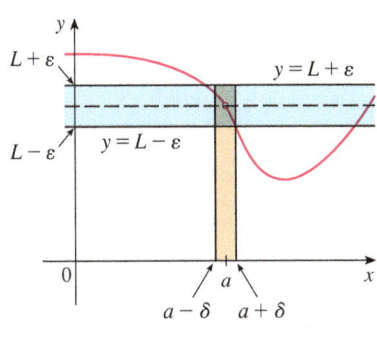

FIGURE 6

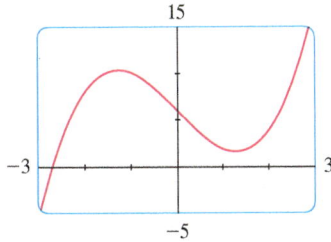

FIGURE 7

EXAMPLE 1 Since $f(x) = x^3 - 5x + 6$ is a polynomial, we know from the Direct Substitution Property that $\lim_{x \to 1} f(x) = f(1) = 1^3 - 5(1) + 6 = 2$. Use a graph to find a number δ such that if x is within δ of 1, then y is within 0.2 of 2, that is,

$$\text{if} \quad |x - 1| < \delta \quad \text{then} \quad |(x^3 - 5x + 6) - 2| < 0.2$$

In other words, find a number δ that corresponds to $\varepsilon = 0.2$ in the definition of a limit for the function $f(x) = x^3 - 5x + 6$ with $a = 1$ and $L = 2$.

SOLUTION A graph of f is shown in Figure 7; we are interested in the region near the point $(1, 2)$. Notice that we can rewrite the inequality

$$|(x^3 - 5x + 6) - 2| < 0.2$$

as

$$-0.2 < (x^3 - 5x + 6) - 2 < 0.2$$

or equivalently

$$1.8 < x^3 - 5x + 6 < 2.2$$

So we need to determine the values of x for which the curve $y = x^3 - 5x + 6$ lies between the horizontal lines $y = 1.8$ and $y = 2.2$. Therefore we graph the curves $y = x^3 - 5x + 6$, $y = 1.8$, and $y = 2.2$ near the point $(1, 2)$ in Figure 8. Then we

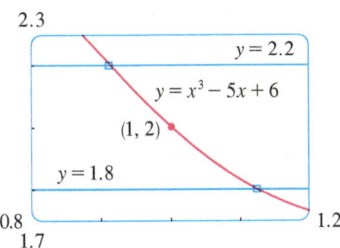

FIGURE 8

use the cursor to estimate that the x-coordinate of the point of intersection of the line $y = 2.2$ and the curve $y = x^3 - 5x + 6$ is about 0.911. Similarly, $y = x^3 - 5x + 6$ intersects the line $y = 1.8$ when $x \approx 1.124$. So, rounding toward 1 to be safe, we can say that

$$\text{if} \quad 0.92 < x < 1.12 \quad \text{then} \quad 1.8 < x^3 - 5x + 6 < 2.2$$

This interval (0.92, 1.12) is not symmetric about $x = 1$. The distance from $x = 1$ to the left endpoint is $1 - 0.92 = 0.08$ and the distance to the right endpoint is 0.12. We can choose δ to be the smaller of these numbers, that is, $\delta = 0.08$. Then we can rewrite our inequalities in terms of distances as follows:

$$\text{if} \quad |x - 1| < 0.08 \quad \text{then} \quad |(x^3 - 5x + 6) - 2| < 0.2$$

This just says that by keeping x within 0.08 of 1, we are able to keep $f(x)$ within 0.2 of 2.

Although we chose $\delta = 0.08$, any smaller positive value of δ would also have worked. ■

The graphical procedure in Example 1 gives an illustration of the definition for $\varepsilon = 0.2$, but it does not *prove* that the limit is equal to 2. A proof has to provide a δ for *every* ε.

In proving limit statements it may be helpful to think of the definition of limit as a challenge. First it challenges you with a number ε. Then you must be able to produce a suitable δ. You have to be able to do this for *every* $\varepsilon > 0$, not just a particular ε.

Imagine a contest between two people, A and B, and imagine yourself to be B. Person A stipulates that the fixed number L should be approximated by the values of $f(x)$ to within a degree of accuracy ε (say, 0.01). Person B then responds by finding a number δ such that if $0 < |x - a| < \delta$, then $|f(x) - L| < \varepsilon$. Then A may become more exacting and challenge B with a smaller value of ε (say, 0.0001). Again B has to respond by finding a corresponding δ. Usually the smaller the value of ε, the smaller the corresponding value of δ must be. If B always wins, no matter how small A makes ε, then $\lim_{x \to a} f(x) = L$.

TEC In Module 2.4/2.6 you can explore the precise definition of a limit both graphically and numerically.

EXAMPLE 2 Prove that $\lim_{x \to 3} (4x - 5) = 7$.

SOLUTION

1. *Preliminary analysis of the problem (guessing a value for δ).* Let ε be a given positive number. We want to find a number δ such that

$$\text{if} \quad 0 < |x - 3| < \delta \quad \text{then} \quad |(4x - 5) - 7| < \varepsilon$$

But $|(4x - 5) - 7| = |4x - 12| = |4(x - 3)| = 4|x - 3|$. Therefore we want δ such that

$$\text{if} \quad 0 < |x - 3| < \delta \quad \text{then} \quad 4|x - 3| < \varepsilon$$

that is,

$$\text{if} \quad 0 < |x - 3| < \delta \quad \text{then} \quad |x - 3| < \frac{\varepsilon}{4}$$

This suggests that we should choose $\delta = \varepsilon/4$.

2. *Proof (showing that this δ works).* Given $\varepsilon > 0$, choose $\delta = \varepsilon/4$. If $0 < |x - 3| < \delta$, then

$$|(4x - 5) - 7| = |4x - 12| = 4|x - 3| < 4\delta = 4\left(\frac{\varepsilon}{4}\right) = \varepsilon$$

Thus

$$\text{if} \quad 0 < |x - 3| < \delta \quad \text{then} \quad |(4x - 5) - 7| < \varepsilon$$

Therefore, by the definition of a limit,

$$\lim_{x \to 3} (4x - 5) = 7$$

This example is illustrated by Figure 9.

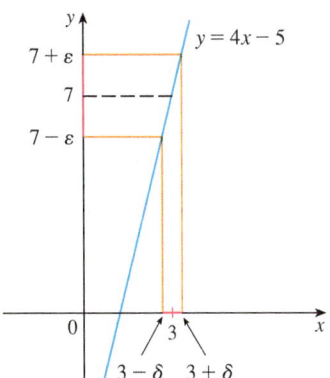

FIGURE 9

Cauchy and Limits

After the invention of calculus in the 17th century, there followed a period of free development of the subject in the 18th century. Mathematicians like the Bernoulli brothers and Euler were eager to exploit the power of calculus and boldly explored the consequences of this new and wonderful mathematical theory without worrying too much about whether their proofs were completely correct.

The 19th century, by contrast, was the Age of Rigor in mathematics. There was a movement to go back to the foundations of the subject—to provide careful definitions and rigorous proofs. At the forefront of this movement was the French mathematician Augustin-Louis Cauchy (1789–1857), who started out as a military engineer before becoming a mathematics professor in Paris. Cauchy took Newton's idea of a limit, which was kept alive in the 18th century by the French mathematician Jean d'Alembert, and made it more precise. His definition of a limit reads as follows: "When the successive values attributed to a variable approach indefinitely a fixed value so as to end by differing from it by as little as one wishes, this last is called the *limit* of all the others." But when Cauchy used this definition in examples and proofs, he often employed delta-epsilon inequalities similar to the ones in this section. A typical Cauchy proof starts with: "Designate by δ and ε two very small numbers; . . ." He used ε because of the correspondence between epsilon and the French word *erreur* and δ because delta corresponds to *différence*. Later, the German mathematician Karl Weierstrass (1815–1897) stated the definition of a limit exactly as in our Definition 2.

Note that in the solution of Example 2 there were two stages—guessing and proving. We made a preliminary analysis that enabled us to guess a value for δ. But then in the second stage we had to go back and prove in a careful, logical fashion that we had made a correct guess. This procedure is typical of much of mathematics. Sometimes it is necessary to first make an intelligent guess about the answer to a problem and then prove that the guess is correct.

The intuitive definitions of one-sided limits that were given in Section 2.2 can be precisely reformulated as follows.

3 Definition of Left-Hand Limit

$$\lim_{x \to a^-} f(x) = L$$

if for every number $\varepsilon > 0$ there is a number $\delta > 0$ such that

$$\text{if} \quad a - \delta < x < a \quad \text{then} \quad |f(x) - L| < \varepsilon$$

4 Definition of Right-Hand Limit

$$\lim_{x \to a^+} f(x) = L$$

if for every number $\varepsilon > 0$ there is a number $\delta > 0$ such that

$$\text{if} \quad a < x < a + \delta \quad \text{then} \quad |f(x) - L| < \varepsilon$$

Notice that Definition 3 is the same as Definition 2 except that x is restricted to lie in the *left* half $(a - \delta, a)$ of the interval $(a - \delta, a + \delta)$. In Definition 4, x is restricted to lie in the *right* half $(a, a + \delta)$ of the interval $(a - \delta, a + \delta)$.

EXAMPLE 3 Use Definition 4 to prove that $\lim_{x \to 0^+} \sqrt{x} = 0$.

SOLUTION

1. *Guessing a value for δ.* Let ε be a given positive number. Here $a = 0$ and $L = 0$, so we want to find a number δ such that

$$\text{if} \quad 0 < x < \delta \quad \text{then} \quad |\sqrt{x} - 0| < \varepsilon$$

that is,

$$\text{if} \quad 0 < x < \delta \quad \text{then} \quad \sqrt{x} < \varepsilon$$

or, squaring both sides of the inequality $\sqrt{x} < \varepsilon$, we get

$$\text{if} \quad 0 < x < \delta \quad \text{then} \quad x < \varepsilon^2$$

This suggests that we should choose $\delta = \varepsilon^2$.

2. *Showing that this δ works.* Given $\varepsilon > 0$, let $\delta = \varepsilon^2$. If $0 < x < \delta$, then

$$\sqrt{x} < \sqrt{\delta} = \sqrt{\varepsilon^2} = \varepsilon$$

so

$$|\sqrt{x} - 0| < \varepsilon$$

According to Definition 4, this shows that $\lim_{x \to 0^+} \sqrt{x} = 0$. ∎

EXAMPLE 4 Prove that $\lim_{x \to 3} x^2 = 9$.

SOLUTION

1. *Guessing a value for δ.* Let $\varepsilon > 0$ be given. We have to find a number $\delta > 0$ such that

$$\text{if} \quad 0 < |x - 3| < \delta \quad \text{then} \quad |x^2 - 9| < \varepsilon$$

To connect $|x^2 - 9|$ with $|x - 3|$ we write $|x^2 - 9| = |(x + 3)(x - 3)|$. Then we want

$$\text{if} \quad 0 < |x - 3| < \delta \quad \text{then} \quad |x + 3||x - 3| < \varepsilon$$

Notice that if we can find a positive constant C such that $|x + 3| < C$, then

$$|x + 3||x - 3| < C|x - 3|$$

and we can make $C|x - 3| < \varepsilon$ by taking $|x - 3| < \varepsilon/C$, so we could choose $\delta = \varepsilon/C$.

We can find such a number C if we restrict x to lie in some interval centered at 3. In fact, since we are interested only in values of x that are close to 3, it is reasonable to assume that x is within a distance 1 from 3, that is, $|x - 3| < 1$. Then $2 < x < 4$, so $5 < x + 3 < 7$. Thus we have $|x + 3| < 7$, and so $C = 7$ is a suitable choice for the constant.

But now there are two restrictions on $|x - 3|$, namely

$$|x - 3| < 1 \quad \text{and} \quad |x - 3| < \frac{\varepsilon}{C} = \frac{\varepsilon}{7}$$

To make sure that both of these inequalities are satisfied, we take δ to be the smaller of the two numbers 1 and $\varepsilon/7$. The notation for this is $\delta = \min\{1, \varepsilon/7\}$.

2. *Showing that this δ works.* Given $\varepsilon > 0$, let $\delta = \min\{1, \varepsilon/7\}$. If $0 < |x - 3| < \delta$, then $|x - 3| < 1 \Rightarrow 2 < x < 4 \Rightarrow |x + 3| < 7$ (as in part 1). We also have $|x - 3| < \varepsilon/7$, so

$$|x^2 - 9| = |x + 3||x - 3| < 7 \cdot \frac{\varepsilon}{7} = \varepsilon$$

This shows that $\lim_{x \to 3} x^2 = 9$. ∎

As Example 4 shows, it is not always easy to prove that limit statements are true using the ε, δ definition. In fact, if we had been given a more complicated function such as $f(x) = (6x^2 - 8x + 9)/(2x^2 - 1)$, a proof would require a great deal of ingenuity. Fortunately this is unnecessary because the Limit Laws stated in Section 2.3 can be proved using Definition 2, and then the limits of complicated functions can be found rigorously from the Limit Laws without resorting to the definition directly.

For instance, we prove the Sum Law: If $\lim_{x \to a} f(x) = L$ and $\lim_{x \to a} g(x) = M$ both exist, then

$$\lim_{x \to a} [f(x) + g(x)] = L + M$$

The remaining laws are proved in the exercises and in Appendix F.

PROOF OF THE SUM LAW Let $\varepsilon > 0$ be given. We must find $\delta > 0$ such that

if $\quad 0 < |x - a| < \delta \quad$ then $\quad |f(x) + g(x) - (L + M)| < \varepsilon$

Triangle Inequality:
$$|a + b| \leq |a| + |b|$$
(See Appendix A).

Using the Triangle Inequality we can write

$$\boxed{5} \quad |f(x) + g(x) - (L + M)| = |(f(x) - L) + (g(x) - M)|$$
$$\leq |f(x) - L| + |g(x) - M|$$

We make $|f(x) + g(x) - (L + M)|$ less than ε by making each of the terms $|f(x) - L|$ and $|g(x) - M|$ less than $\varepsilon/2$.

Since $\varepsilon/2 > 0$ and $\lim_{x \to a} f(x) = L$, there exists a number $\delta_1 > 0$ such that

if $\quad 0 < |x - a| < \delta_1 \quad$ then $\quad |f(x) - L| < \frac{\varepsilon}{2}$

Similarly, since $\lim_{x \to a} g(x) = M$, there exists a number $\delta_2 > 0$ such that

if $\quad 0 < |x - a| < \delta_2 \quad$ then $\quad |g(x) - M| < \frac{\varepsilon}{2}$

Let $\delta = \min\{\delta_1, \delta_2\}$, the smaller of the numbers δ_1 and δ_2. Notice that

if $\quad 0 < |x - a| < \delta \quad$ then $\quad 0 < |x - a| < \delta_1 \quad$ and $\quad 0 < |x - a| < \delta_2$

and so $\quad |f(x) - L| < \frac{\varepsilon}{2} \quad$ and $\quad |g(x) - M| < \frac{\varepsilon}{2}$

Therefore, by (5),

$$|f(x) + g(x) - (L + M)| \leq |f(x) - L| + |g(x) - M|$$
$$< \frac{\varepsilon}{2} + \frac{\varepsilon}{2} = \varepsilon$$

To summarize,

$$\text{if} \quad 0 < |x - a| < \delta \quad \text{then} \quad |f(x) + g(x) - (L + M)| < \varepsilon$$

Thus, by the definition of a limit,

$$\lim_{x \to a} [f(x) + g(x)] = L + M$$

■ Infinite Limits

Infinite limits can also be defined in a precise way. The following is a precise version of Definition 2.2.4.

6 **Precise Definition of an Infinite Limit** Let f be a function defined on some open interval that contains the number a, except possibly at a itself. Then

$$\lim_{x \to a} f(x) = \infty$$

means that for every positive number M there is a positive number δ such that

$$\text{if} \quad 0 < |x - a| < \delta \quad \text{then} \quad f(x) > M$$

This says that the values of $f(x)$ can be made arbitrarily large (larger than any given number M) by requiring x to be close enough to a (within a distance δ, where δ depends on M, but with $x \neq a$). A geometric illustration is shown in Figure 10.

Given any horizontal line $y = M$, we can find a number $\delta > 0$ such that if we restrict x to lie in the interval $(a - \delta, a + \delta)$ but $x \neq a$, then the curve $y = f(x)$ lies above the line $y = M$. You can see that if a larger M is chosen, then a smaller δ may be required.

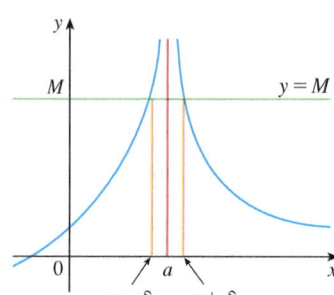

FIGURE 10

EXAMPLE 5 Use Definition 6 to prove that $\lim\limits_{x \to 0} \dfrac{1}{x^2} = \infty$.

SOLUTION Let M be a given positive number. We want to find a number δ such that

$$\text{if} \quad 0 < |x| < \delta \quad \text{then} \quad 1/x^2 > M$$

But $\dfrac{1}{x^2} > M \iff x^2 < \dfrac{1}{M} \iff \sqrt{x^2} < \sqrt{\dfrac{1}{M}} \iff |x| < \dfrac{1}{\sqrt{M}}$

So if we choose $\delta = 1/\sqrt{M}$ and $0 < |x| < \delta = 1/\sqrt{M}$, then $1/x^2 > M$. This shows that $1/x^2 \to \infty$ as $x \to 0$. ■

Similarly, the following is a precise version of Definition 2.2.5. It is illustrated by Figure 11.

7 **Definition** Let f be a function defined on some open interval that contains the number a, except possibly at a itself. Then

$$\lim_{x \to a} f(x) = -\infty$$

means that for every negative number N there is a positive number δ such that

$$\text{if} \quad 0 < |x - a| < \delta \quad \text{then} \quad f(x) < N$$

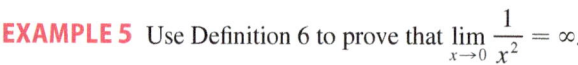

FIGURE 11

2.4 EXERCISES

1. Use the given graph of f to find a number δ such that
   if $\quad |x - 1| < \delta \quad$ then $\quad |f(x) - 1| < 0.2$

   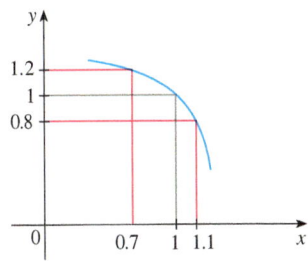

2. Use the given graph of f to find a number δ such that
   if $\quad 0 < |x - 3| < \delta \quad$ then $\quad |f(x) - 2| < 0.5$

   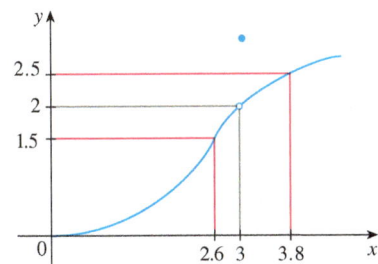

3. Use the given graph of $f(x) = \sqrt{x}$ to find a number δ such that
   if $\quad |x - 4| < \delta \quad$ then $\quad |\sqrt{x} - 2| < 0.4$

   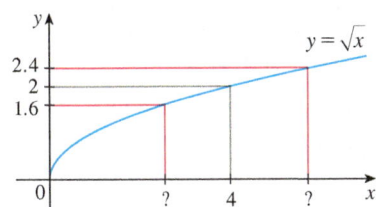

4. Use the given graph of $f(x) = x^2$ to find a number δ such that
   if $\quad |x - 1| < \delta \quad$ then $\quad |x^2 - 1| < \frac{1}{2}$

   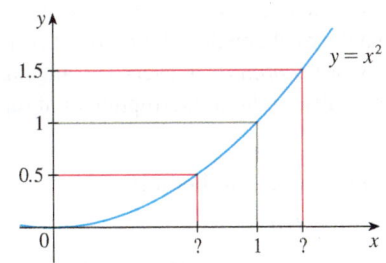

5. Use a graph to find a number δ such that
   if $\quad \left|x - \dfrac{\pi}{4}\right| < \delta \quad$ then $\quad |\tan x - 1| < 0.2$

6. Use a graph to find a number δ such that
   if $\quad |x - 1| < \delta \quad$ then $\quad \left|\dfrac{2x}{x^2 + 4} - 0.4\right| < 0.1$

7. For the limit
   $$\lim_{x \to 2} (x^3 - 3x + 4) = 6$$
   illustrate Definition 2 by finding values of δ that correspond to $\varepsilon = 0.2$ and $\varepsilon = 0.1$.

8. For the limit
   $$\lim_{x \to 0} \frac{e^{2x} - 1}{x} = 2$$
   illustrate Definition 2 by finding values of δ that correspond to $\varepsilon = 0.5$ and $\varepsilon = 0.1$.

9. (a) Use a graph to find a number δ such that
   if $\quad 2 < x < 2 + \delta \quad$ then $\quad \dfrac{1}{\ln(x - 1)} > 100$

   (b) What limit does part (a) suggest is true?

10. Given that $\lim_{x \to \pi} \csc^2 x = \infty$, illustrate Definition 6 by finding values of δ that correspond to (a) $M = 500$ and (b) $M = 1000$.

11. A machinist is required to manufacture a circular metal disk with area 1000 cm².
    (a) What radius produces such a disk?
    (b) If the machinist is allowed an error tolerance of ± 5 cm² in the area of the disk, how close to the ideal radius in part (a) must the machinist control the radius?
    (c) In terms of the ε, δ definition of $\lim_{x \to a} f(x) = L$, what is x? What is $f(x)$? What is a? What is L? What value of ε is given? What is the corresponding value of δ?

12. A crystal growth furnace is used in research to determine how best to manufacture crystals used in electronic components for the space shuttle. For proper growth of the crystal, the temperature must be controlled accurately by adjusting the input power. Suppose the relationship is given by
    $$T(w) = 0.1w^2 + 2.155w + 20$$
    where T is the temperature in degrees Celsius and w is the power input in watts.
    (a) How much power is needed to maintain the temperature at 200°C?

(b) If the temperature is allowed to vary from 200°C by up to ±1°C, what range of wattage is allowed for the input power?

(c) In terms of the ε, δ definition of $\lim_{x \to a} f(x) = L$, what is x? What is $f(x)$? What is a? What is L? What value of ε is given? What is the corresponding value of δ?

13. (a) Find a number δ such that if $|x - 2| < \delta$, then $|4x - 8| < \varepsilon$, where $\varepsilon = 0.1$.
(b) Repeat part (a) with $\varepsilon = 0.01$.

14. Given that $\lim_{x \to 2}(5x - 7) = 3$, illustrate Definition 2 by finding values of δ that correspond to $\varepsilon = 0.1$, $\varepsilon = 0.05$, and $\varepsilon = 0.01$.

15–18 Prove the statement using the ε, δ definition of a limit and illustrate with a diagram like Figure 9.

15. $\lim_{x \to 3} \left(1 + \tfrac{1}{3}x\right) = 2$

16. $\lim_{x \to 4}(2x - 5) = 3$

17. $\lim_{x \to -3}(1 - 4x) = 13$

18. $\lim_{x \to -2}(3x + 5) = -1$

19–32 Prove the statement using the ε, δ definition of a limit.

19. $\lim_{x \to 1} \dfrac{2 + 4x}{3} = 2$

20. $\lim_{x \to 10}\left(3 - \tfrac{4}{5}x\right) = -5$

21. $\lim_{x \to 4} \dfrac{x^2 - 2x - 8}{x - 4} = 6$

22. $\lim_{x \to -1.5} \dfrac{9 - 4x^2}{3 + 2x} = 6$

23. $\lim_{x \to a} x = a$

24. $\lim_{x \to a} c = c$

25. $\lim_{x \to 0} x^2 = 0$

26. $\lim_{x \to 0} x^3 = 0$

27. $\lim_{x \to 0} |x| = 0$

28. $\lim_{x \to -6^+} \sqrt[8]{6 + x} = 0$

29. $\lim_{x \to 2}(x^2 - 4x + 5) = 1$

30. $\lim_{x \to 2}(x^2 + 2x - 7) = 1$

31. $\lim_{x \to -2}(x^2 - 1) = 3$

32. $\lim_{x \to 2} x^3 = 8$

33. Verify that another possible choice of δ for showing that $\lim_{x \to 3} x^2 = 9$ in Example 4 is $\delta = \min\{2, \varepsilon/8\}$.

34. Verify, by a geometric argument, that the largest possible choice of δ for showing that $\lim_{x \to 3} x^2 = 9$ is $\delta = \sqrt{9 + \varepsilon} - 3$.

CAS 35. (a) For the limit $\lim_{x \to 1}(x^3 + x + 1) = 3$, use a graph to find a value of δ that corresponds to $\varepsilon = 0.4$.
(b) By using a computer algebra system to solve the cubic equation $x^3 + x + 1 = 3 + \varepsilon$, find the largest possible value of δ that works for any given $\varepsilon > 0$.
(c) Put $\varepsilon = 0.4$ in your answer to part (b) and compare with your answer to part (a).

36. Prove that $\lim_{x \to 2} \dfrac{1}{x} = \dfrac{1}{2}$.

37. Prove that $\lim_{x \to a} \sqrt{x} = \sqrt{a}$ if $a > 0$.

$\left[\text{Hint: Use } |\sqrt{x} - \sqrt{a}| = \dfrac{|x - a|}{\sqrt{x} + \sqrt{a}}.\right]$

38. If H is the Heaviside function defined in Example 2.2.6, prove, using Definition 2, that $\lim_{t \to 0} H(t)$ does not exist. [*Hint:* Use an indirect proof as follows. Suppose that the limit is L. Take $\varepsilon = \tfrac{1}{2}$ in the definition of a limit and try to arrive at a contradiction.]

39. If the function f is defined by

$$f(x) = \begin{cases} 0 & \text{if } x \text{ is rational} \\ 1 & \text{if } x \text{ is irrational} \end{cases}$$

prove that $\lim_{x \to 0} f(x)$ does not exist.

40. By comparing Definitions 2, 3, and 4, prove Theorem 2.3.1.

41. How close to -3 do we have to take x so that

$$\dfrac{1}{(x + 3)^4} > 10{,}000$$

42. Prove, using Definition 6, that $\lim_{x \to -3} \dfrac{1}{(x + 3)^4} = \infty$.

43. Prove that $\lim_{x \to 0^+} \ln x = -\infty$.

44. Suppose that $\lim_{x \to a} f(x) = \infty$ and $\lim_{x \to a} g(x) = c$, where c is a real number. Prove each statement.
(a) $\lim_{x \to a}[f(x) + g(x)] = \infty$
(b) $\lim_{x \to a}[f(x)g(x)] = \infty$ if $c > 0$
(c) $\lim_{x \to a}[f(x)g(x)] = -\infty$ if $c < 0$

2.5 Continuity

We noticed in Section 2.3 that the limit of a function as x approaches a can often be found simply by calculating the value of the function at a. Functions with this property are called *continuous at a*. We will see that the mathematical definition of continuity corresponds closely with the meaning of the word *continuity* in everyday language. (A continuous process is one that takes place gradually, without interruption or abrupt change.)

1 Definition A function f is **continuous at a number** a if

$$\lim_{x \to a} f(x) = f(a)$$

As illustrated in Figure 1, if f is continuous, then the points $(x, f(x))$ on the graph of f approach the point $(a, f(a))$ on the graph. So there is no gap in the curve.

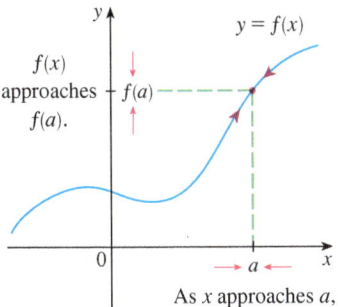

FIGURE 1

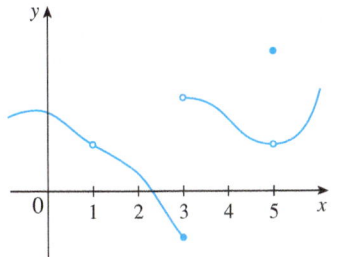

FIGURE 2

Notice that Definition 1 implicitly requires three things if f is continuous at a:

1. $f(a)$ is defined (that is, a is in the domain of f)
2. $\lim_{x \to a} f(x)$ exists
3. $\lim_{x \to a} f(x) = f(a)$

The definition says that f is continuous at a if $f(x)$ approaches $f(a)$ as x approaches a. Thus a continuous function f has the property that a small change in x produces only a small change in $f(x)$. In fact, the change in $f(x)$ can be kept as small as we please by keeping the change in x sufficiently small.

If f is defined near a (in other words, f is defined on an open interval containing a, except perhaps at a), we say that f is **discontinuous at a** (or f has a **discontinuity** at a) if f is not continuous at a.

Physical phenomena are usually continuous. For instance, the displacement or velocity of a vehicle varies continuously with time, as does a person's height. But discontinuities do occur in such situations as electric currents. [See Example 2.2.6, where the Heaviside function is discontinuous at 0 because $\lim_{t \to 0} H(t)$ does not exist.]

Geometrically, you can think of a function that is continuous at every number in an interval as a function whose graph has no break in it: the graph can be drawn without removing your pen from the paper.

EXAMPLE 1 Figure 2 shows the graph of a function f. At which numbers is f discontinuous? Why?

SOLUTION It looks as if there is a discontinuity when $a = 1$ because the graph has a break there. The official reason that f is discontinuous at 1 is that $f(1)$ is not defined.

The graph also has a break when $a = 3$, but the reason for the discontinuity is different. Here, $f(3)$ is defined, but $\lim_{x \to 3} f(x)$ does not exist (because the left and right limits are different). So f is discontinuous at 3.

What about $a = 5$? Here, $f(5)$ is defined and $\lim_{x \to 5} f(x)$ exists (because the left and right limits are the same). But

$$\lim_{x \to 5} f(x) \neq f(5)$$

So f is discontinuous at 5. ∎

Now let's see how to detect discontinuities when a function is defined by a formula.

EXAMPLE 2 Where are each of the following functions discontinuous?

(a) $f(x) = \dfrac{x^2 - x - 2}{x - 2}$

(b) $f(x) = \begin{cases} \dfrac{1}{x^2} & \text{if } x \neq 0 \\ 1 & \text{if } x = 0 \end{cases}$

(c) $f(x) = \begin{cases} \dfrac{x^2 - x - 2}{x - 2} & \text{if } x \neq 2 \\ 1 & \text{if } x = 2 \end{cases}$

(d) $f(x) = [\![x]\!]$

SOLUTION
(a) Notice that $f(2)$ is not defined, so f is discontinuous at 2. Later we'll see why f is continuous at all other numbers.

(b) Here $f(0) = 1$ is defined but

$$\lim_{x \to 0} f(x) = \lim_{x \to 0} \frac{1}{x^2}$$

does not exist. (See Example 2.2.8.) So f is discontinuous at 0.

(c) Here $f(2) = 1$ is defined and

$$\lim_{x \to 2} f(x) = \lim_{x \to 2} \frac{x^2 - x - 2}{x - 2} = \lim_{x \to 2} \frac{(x - 2)(x + 1)}{x - 2} = \lim_{x \to 2} (x + 1) = 3$$

exists. But

$$\lim_{x \to 2} f(x) \neq f(2)$$

so f is not continuous at 2.

(d) The greatest integer function $f(x) = [\![x]\!]$ has discontinuities at all of the integers because $\lim_{x \to n} [\![x]\!]$ does not exist if n is an integer. (See Example 2.3.10 and Exercise 2.3.53.) ∎

Figure 3 shows the graphs of the functions in Example 2. In each case the graph can't be drawn without lifting the pen from the paper because a hole or break or jump occurs in the graph. The kind of discontinuity illustrated in parts (a) and (c) is called **removable** because we could remove the discontinuity by redefining f at just the single number 2. [The function $g(x) = x + 1$ is continuous.] The discontinuity in part (b) is called an **infinite discontinuity**. The discontinuities in part (d) are called **jump discontinuities** because the function "jumps" from one value to another.

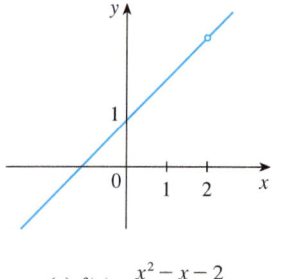
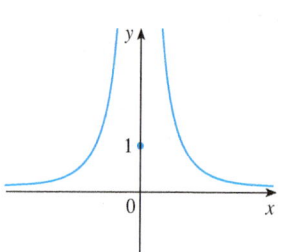
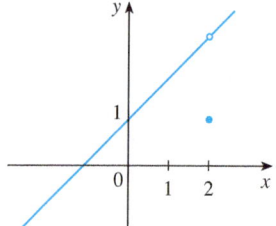
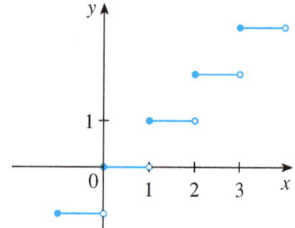

(a) $f(x) = \dfrac{x^2 - x - 2}{x - 2}$

(b) $f(x) = \begin{cases} \dfrac{1}{x^2} & \text{if } x \neq 0 \\ 1 & \text{if } x = 0 \end{cases}$

(c) $f(x) = \begin{cases} \dfrac{x^2 - x - 2}{x - 2} & \text{if } x \neq 2 \\ 1 & \text{if } x = 2 \end{cases}$

(d) $f(x) = [\![x]\!]$

FIGURE 3
Graphs of the functions in Example 2

> **2 Definition** A function f is **continuous from the right at a number** a if
>
> $$\lim_{x \to a^+} f(x) = f(a)$$
>
> and f is **continuous from the left at** a if
>
> $$\lim_{x \to a^-} f(x) = f(a)$$

EXAMPLE 3 At each integer n, the function $f(x) = [\![x]\!]$ [see Figure 3(d)] is continuous from the right but discontinuous from the left because

$$\lim_{x \to n^+} f(x) = \lim_{x \to n^+} [\![x]\!] = n = f(n)$$

but
$$\lim_{x \to n^-} f(x) = \lim_{x \to n^-} [\![x]\!] = n - 1 \neq f(n)$$

> **3** **Definition** A function f is **continuous on an interval** if it is continuous at every number in the interval. (If f is defined only on one side of an endpoint of the interval, we understand *continuous* at the endpoint to mean *continuous from the right* or *continuous from the left*.)

EXAMPLE 4 Show that the function $f(x) = 1 - \sqrt{1 - x^2}$ is continuous on the interval $[-1, 1]$.

SOLUTION If $-1 < a < 1$, then using the Limit Laws, we have

$$\lim_{x \to a} f(x) = \lim_{x \to a} \left(1 - \sqrt{1 - x^2}\right)$$
$$= 1 - \lim_{x \to a} \sqrt{1 - x^2} \quad \text{(by Laws 2 and 7)}$$
$$= 1 - \sqrt{\lim_{x \to a} (1 - x^2)} \quad \text{(by 11)}$$
$$= 1 - \sqrt{1 - a^2} \quad \text{(by 2, 7, and 9)}$$
$$= f(a)$$

Thus, by Definition 1, f is continuous at a if $-1 < a < 1$. Similar calculations show that

$$\lim_{x \to -1^+} f(x) = 1 = f(-1) \quad \text{and} \quad \lim_{x \to 1^-} f(x) = 1 = f(1)$$

so f is continuous from the right at -1 and continuous from the left at 1. Therefore, according to Definition 3, f is continuous on $[-1, 1]$.

The graph of f is sketched in Figure 4. It is the lower half of the circle

$$x^2 + (y - 1)^2 = 1$$

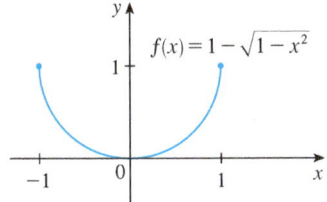

FIGURE 4

Instead of always using Definitions 1, 2, and 3 to verify the continuity of a function as we did in Example 4, it is often convenient to use the next theorem, which shows how to build up complicated continuous functions from simple ones.

> **4** **Theorem** If f and g are continuous at a and c is a constant, then the following functions are also continuous at a:
> 1. $f + g$
> 2. $f - g$
> 3. cf
> 4. fg
> 5. $\dfrac{f}{g}$ if $g(a) \neq 0$

PROOF Each of the five parts of this theorem follows from the corresponding Limit Law in Section 2.3. For instance, we give the proof of part 1. Since f and g are continuous at a, we have

$$\lim_{x \to a} f(x) = f(a) \quad \text{and} \quad \lim_{x \to a} g(x) = g(a)$$

Therefore
$$\lim_{x \to a} (f + g)(x) = \lim_{x \to a} [f(x) + g(x)]$$
$$= \lim_{x \to a} f(x) + \lim_{x \to a} g(x) \quad \text{(by Law 1)}$$
$$= f(a) + g(a)$$
$$= (f + g)(a)$$

This shows that $f + g$ is continuous at a. ■

It follows from Theorem 4 and Definition 3 that if f and g are continuous on an interval, then so are the functions $f + g$, $f - g$, cf, fg, and (if g is never 0) f/g. The following theorem was stated in Section 2.3 as the Direct Substitution Property.

5 Theorem
(a) Any polynomial is continuous everywhere; that is, it is continuous on $\mathbb{R} = (-\infty, \infty)$.
(b) Any rational function is continuous wherever it is defined; that is, it is continuous on its domain.

PROOF
(a) A polynomial is a function of the form
$$P(x) = c_n x^n + c_{n-1} x^{n-1} + \cdots + c_1 x + c_0$$
where $c_0, c_1, \ldots, c_n$ are constants. We know that
$$\lim_{x \to a} c_0 = c_0 \quad \text{(by Law 7)}$$
and
$$\lim_{x \to a} x^m = a^m \quad m = 1, 2, \ldots, n \quad \text{(by 9)}$$

This equation is precisely the statement that the function $f(x) = x^m$ is a continuous function. Thus, by part 3 of Theorem 4, the function $g(x) = cx^m$ is continuous. Since P is a sum of functions of this form and a constant function, it follows from part 1 of Theorem 4 that P is continuous.

(b) A rational function is a function of the form
$$f(x) = \frac{P(x)}{Q(x)}$$
where P and Q are polynomials. The domain of f is $D = \{x \in \mathbb{R} \mid Q(x) \neq 0\}$. We know from part (a) that P and Q are continuous everywhere. Thus, by part 5 of Theorem 4, f is continuous at every number in D. ■

As an illustration of Theorem 5, observe that the volume of a sphere varies continuously with its radius because the formula $V(r) = \frac{4}{3}\pi r^3$ shows that V is a polynomial function of r. Likewise, if a ball is thrown vertically into the air with a velocity of 50 ft/s, then the height of the ball in feet t seconds later is given by the formula $h = 50t - 16t^2$. Again this is a polynomial function, so the height is a continuous function of the elapsed time, as we might expect.

Knowledge of which functions are continuous enables us to evaluate some limits very quickly, as the following example shows. Compare it with Example 2.3.2(b).

EXAMPLE 5 Find $\lim_{x \to -2} \dfrac{x^3 + 2x^2 - 1}{5 - 3x}$.

SOLUTION The function

$$f(x) = \frac{x^3 + 2x^2 - 1}{5 - 3x}$$

is rational, so by Theorem 5 it is continuous on its domain, which is $\{x \mid x \neq \tfrac{5}{3}\}$. Therefore

$$\lim_{x \to -2} \frac{x^3 + 2x^2 - 1}{5 - 3x} = \lim_{x \to -2} f(x) = f(-2)$$

$$= \frac{(-2)^3 + 2(-2)^2 - 1}{5 - 3(-2)} = -\frac{1}{11}$$

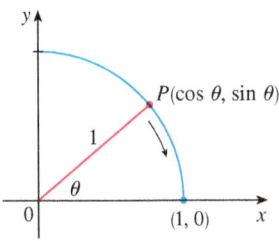

FIGURE 5

It turns out that most of the familiar functions are continuous at every number in their domains. For instance, Limit Law 10 (page 96) is exactly the statement that root functions are continuous.

From the appearance of the graphs of the sine and cosine functions (Figure 1.2.18), we would certainly guess that they are continuous. We know from the definitions of $\sin \theta$ and $\cos \theta$ that the coordinates of the point P in Figure 5 are $(\cos \theta, \sin \theta)$. As $\theta \to 0$, we see that P approaches the point $(1, 0)$ and so $\cos \theta \to 1$ and $\sin \theta \to 0$. Thus

$$\boxed{6} \qquad \lim_{\theta \to 0} \cos \theta = 1 \qquad \lim_{\theta \to 0} \sin \theta = 0$$

Another way to establish the limits in (6) is to use the Squeeze Theorem with the inequality $\sin \theta < \theta$ (for $\theta > 0$), which is proved in Section 3.3.

Since $\cos 0 = 1$ and $\sin 0 = 0$, the equations in (6) assert that the cosine and sine functions are continuous at 0. The addition formulas for cosine and sine can then be used to deduce that these functions are continuous everywhere (see Exercises 64 and 65).

It follows from part 5 of Theorem 4 that

$$\tan x = \frac{\sin x}{\cos x}$$

is continuous except where $\cos x = 0$. This happens when x is an odd integer multiple of $\pi/2$, so $y = \tan x$ has infinite discontinuities when $x = \pm \pi/2, \pm 3\pi/2, \pm 5\pi/2,$ and so on (see Figure 6).

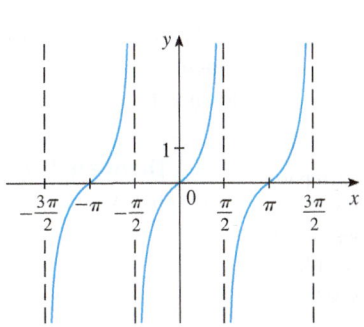

FIGURE 6
$y = \tan x$

The inverse function of any continuous one-to-one function is also continuous. (This fact is proved in Appendix F, but our geometric intuition makes it seem plausible: The graph of f^{-1} is obtained by reflecting the graph of f about the line $y = x$. So if the graph of f has no break in it, neither does the graph of f^{-1}.) Thus the inverse trigonometric functions are continuous.

In Section 1.4 we defined the exponential function $y = b^x$ so as to fill in the holes in the graph of $y = b^x$ where x is rational. In other words, the very definition of $y = b^x$ makes it a continuous function on $\mathbb{R}$. Therefore its inverse function $y = \log_b x$ is continuous on $(0, \infty)$.

The inverse trigonometric functions are reviewed in Section 1.5.

7 Theorem The following types of functions are continuous at every number in their domains:

- polynomials
- rational functions
- root functions
- trigonometric functions
- inverse trigonometric functions
- exponential functions
- logarithmic functions

EXAMPLE 6 Where is the function $f(x) = \dfrac{\ln x + \tan^{-1} x}{x^2 - 1}$ continuous?

SOLUTION We know from Theorem 7 that the function $y = \ln x$ is continuous for $x > 0$ and $y = \tan^{-1} x$ is continuous on $\mathbb{R}$. Thus, by part 1 of Theorem 4, $y = \ln x + \tan^{-1} x$ is continuous on $(0, \infty)$. The denominator, $y = x^2 - 1$, is a polynomial, so it is continuous everywhere. Therefore, by part 5 of Theorem 4, f is continuous at all positive numbers x except where $x^2 - 1 = 0 \iff x = \pm 1$. So f is continuous on the intervals $(0, 1)$ and $(1, \infty)$. ∎

EXAMPLE 7 Evaluate $\lim\limits_{x \to \pi} \dfrac{\sin x}{2 + \cos x}$.

SOLUTION Theorem 7 tells us that $y = \sin x$ is continuous. The function in the denominator, $y = 2 + \cos x$, is the sum of two continuous functions and is therefore continuous. Notice that this function is never 0 because $\cos x \geq -1$ for all x and so $2 + \cos x > 0$ everywhere. Thus the ratio

$$f(x) = \frac{\sin x}{2 + \cos x}$$

is continuous everywhere. Hence, by the definition of a continuous function,

$$\lim_{x \to \pi} \frac{\sin x}{2 + \cos x} = \lim_{x \to \pi} f(x) = f(\pi) = \frac{\sin \pi}{2 + \cos \pi} = \frac{0}{2 - 1} = 0$$
∎

Another way of combining continuous functions f and g to get a new continuous function is to form the composite function $f \circ g$. This fact is a consequence of the following theorem.

This theorem says that a limit symbol can be moved through a function symbol if the function is continuous and the limit exists. In other words, the order of these two symbols can be reversed.

8 Theorem If f is continuous at b and $\lim\limits_{x \to a} g(x) = b$, then $\lim\limits_{x \to a} f(g(x)) = f(b)$. In other words,

$$\lim_{x \to a} f(g(x)) = f\left(\lim_{x \to a} g(x)\right)$$

Intuitively, Theorem 8 is reasonable because if x is close to a, then $g(x)$ is close to b, and since f is continuous at b, if $g(x)$ is close to b, then $f(g(x))$ is close to $f(b)$. A proof of Theorem 8 is given in Appendix F.

EXAMPLE 8 Evaluate $\lim\limits_{x \to 1} \arcsin\left(\dfrac{1 - \sqrt{x}}{1 - x}\right)$.

SOLUTION Because arcsin is a continuous function, we can apply Theorem 8:

$$\lim_{x \to 1} \arcsin\left(\frac{1 - \sqrt{x}}{1 - x}\right) = \arcsin\left(\lim_{x \to 1} \frac{1 - \sqrt{x}}{1 - x}\right)$$

$$= \arcsin\left(\lim_{x \to 1} \frac{1 - \sqrt{x}}{(1 - \sqrt{x})(1 + \sqrt{x})}\right)$$

$$= \arcsin\left(\lim_{x \to 1} \frac{1}{1 + \sqrt{x}}\right)$$

$$= \arcsin \frac{1}{2} = \frac{\pi}{6} \qquad \blacksquare$$

Let's now apply Theorem 8 in the special case where $f(x) = \sqrt[n]{x}$, with n being a positive integer. Then

$$f(g(x)) = \sqrt[n]{g(x)}$$

and

$$f\left(\lim_{x \to a} g(x)\right) = \sqrt[n]{\lim_{x \to a} g(x)}$$

If we put these expressions into Theorem 8, we get

$$\lim_{x \to a} \sqrt[n]{g(x)} = \sqrt[n]{\lim_{x \to a} g(x)}$$

and so Limit Law 11 has now been proved. (We assume that the roots exist.)

9 Theorem If g is continuous at a and f is continuous at $g(a)$, then the composite function $f \circ g$ given by $(f \circ g)(x) = f(g(x))$ is continuous at a.

This theorem is often expressed informally by saying "a continuous function of a continuous function is a continuous function."

PROOF Since g is continuous at a, we have

$$\lim_{x \to a} g(x) = g(a)$$

Since f is continuous at $b = g(a)$, we can apply Theorem 8 to obtain

$$\lim_{x \to a} f(g(x)) = f(g(a))$$

which is precisely the statement that the function $h(x) = f(g(x))$ is continuous at a; that is, $f \circ g$ is continuous at a. ∎

EXAMPLE 9 Where are the following functions continuous?
(a) $h(x) = \sin(x^2)$ (b) $F(x) = \ln(1 + \cos x)$

SOLUTION
(a) We have $h(x) = f(g(x))$, where

$$g(x) = x^2 \quad \text{and} \quad f(x) = \sin x$$

Now g is continuous on $\mathbb{R}$ since it is a polynomial, and f is also continuous everywhere. Thus $h = f \circ g$ is continuous on $\mathbb{R}$ by Theorem 9.

(b) We know from Theorem 7 that $f(x) = \ln x$ is continuous and $g(x) = 1 + \cos x$ is continuous (because both $y = 1$ and $y = \cos x$ are continuous). Therefore, by Theorem 9, $F(x) = f(g(x))$ is continuous wherever it is defined. Now $\ln(1 + \cos x)$ is defined when $1 + \cos x > 0$. So it is undefined when $\cos x = -1$, and this happens when $x = \pm\pi, \pm 3\pi, \ldots$. Thus F has discontinuities when x is an odd multiple of π and is continuous on the intervals between these values (see Figure 7). ∎

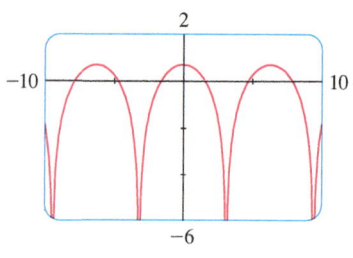

FIGURE 7
$y = \ln(1 + \cos x)$

An important property of continuous functions is expressed by the following theorem, whose proof is found in more advanced books on calculus.

> **10 The Intermediate Value Theorem** Suppose that f is continuous on the closed interval $[a, b]$ and let N be any number between $f(a)$ and $f(b)$, where $f(a) \neq f(b)$. Then there exists a number c in (a, b) such that $f(c) = N$.

The Intermediate Value Theorem states that a continuous function takes on every intermediate value between the function values $f(a)$ and $f(b)$. It is illustrated by Figure 8. Note that the value N can be taken on once [as in part (a)] or more than once [as in part (b)].

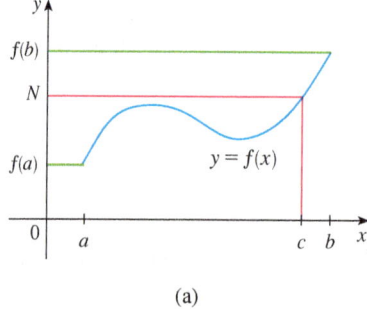
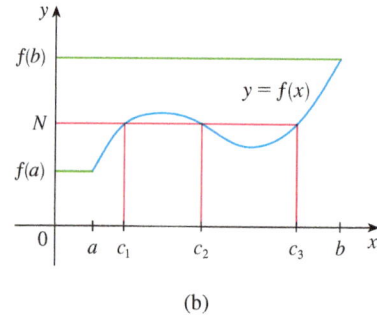

(a)   (b)

FIGURE 8

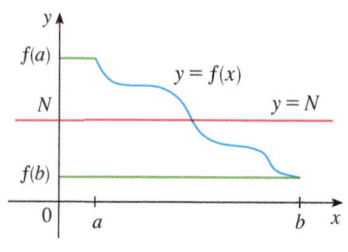

FIGURE 9

If we think of a continuous function as a function whose graph has no hole or break, then it is easy to believe that the Intermediate Value Theorem is true. In geometric terms it says that if any horizontal line $y = N$ is given between $y = f(a)$ and $y = f(b)$ as in Figure 9, then the graph of f can't jump over the line. It must intersect $y = N$ somewhere.

It is important that the function f in Theorem 10 be continuous. The Intermediate Value Theorem is not true in general for discontinuous functions (see Exercise 50).

One use of the Intermediate Value Theorem is in locating roots of equations as in the following example.

EXAMPLE 10 Show that there is a root of the equation
$$4x^3 - 6x^2 + 3x - 2 = 0$$
between 1 and 2.

SOLUTION Let $f(x) = 4x^3 - 6x^2 + 3x - 2$. We are looking for a solution of the given equation, that is, a number c between 1 and 2 such that $f(c) = 0$. Therefore we take $a = 1$, $b = 2$, and $N = 0$ in Theorem 10. We have
$$f(1) = 4 - 6 + 3 - 2 = -1 < 0$$
and
$$f(2) = 32 - 24 + 6 - 2 = 12 > 0$$

Thus $f(1) < 0 < f(2)$; that is, $N = 0$ is a number between $f(1)$ and $f(2)$. Now f is continuous since it is a polynomial, so the Intermediate Value Theorem says there is a number c between 1 and 2 such that $f(c) = 0$. In other words, the equation $4x^3 - 6x^2 + 3x - 2 = 0$ has at least one root c in the interval $(1, 2)$.

In fact, we can locate a root more precisely by using the Intermediate Value Theorem again. Since
$$f(1.2) = -0.128 < 0 \quad \text{and} \quad f(1.3) = 0.548 > 0$$
a root must lie between 1.2 and 1.3. A calculator gives, by trial and error,
$$f(1.22) = -0.007008 < 0 \quad \text{and} \quad f(1.23) = 0.056068 > 0$$
so a root lies in the interval $(1.22, 1.23)$. ■

We can use a graphing calculator or computer to illustrate the use of the Intermediate Value Theorem in Example 10. Figure 10 shows the graph of f in the viewing rectangle $[-1, 3]$ by $[-3, 3]$ and you can see that the graph crosses the x-axis between 1 and 2. Figure 11 shows the result of zooming in to the viewing rectangle $[1.2, 1.3]$ by $[-0.2, 0.2]$.

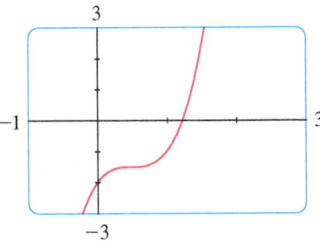

FIGURE 10

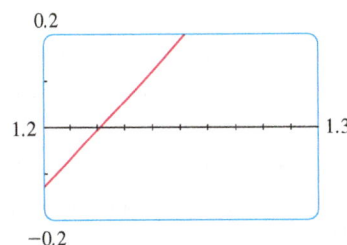

FIGURE 11

In fact, the Intermediate Value Theorem plays a role in the very way these graphing devices work. A computer calculates a finite number of points on the graph and turns on the pixels that contain these calculated points. It assumes that the function is continuous and takes on all the intermediate values between two consecutive points. The computer therefore "connects the dots" by turning on the intermediate pixels.

2.5 EXERCISES

1. Write an equation that expresses the fact that a function f is continuous at the number 4.

2. If f is continuous on $(-\infty, \infty)$, what can you say about its graph?

3. (a) From the graph of f, state the numbers at which f is discontinuous and explain why.
   (b) For each of the numbers stated in part (a), determine whether f is continuous from the right, or from the left, or neither.

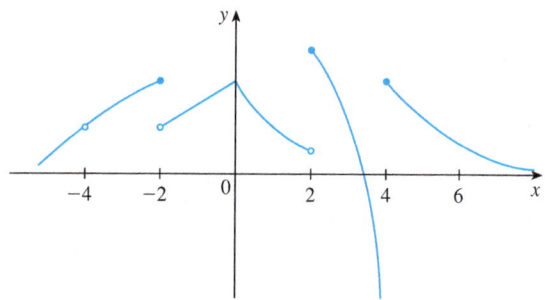

4. From the graph of g, state the intervals on which g is continuous.

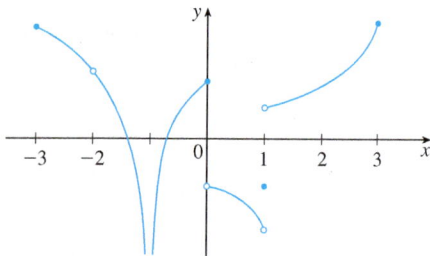

5–8 Sketch the graph of a function f that is continuous except for the stated discontinuity.

5. Discontinuous, but continuous from the right, at 2

6. Discontinuities at -1 and 4, but continuous from the left at -1 and from the right at 4

7. Removable discontinuity at 3, jump discontinuity at 5

8. Neither left nor right continuous at -2, continuous only from the left at 2

9. The toll T charged for driving on a certain stretch of a toll road is \$5 except during rush hours (between 7 AM and 10 AM and between 4 PM and 7 PM) when the toll is \$7.
   (a) Sketch a graph of T as a function of the time t, measured in hours past midnight.
   (b) Discuss the discontinuities of this function and their significance to someone who uses the road.

10. Explain why each function is continuous or discontinuous.
    (a) The temperature at a specific location as a function of time
    (b) The temperature at a specific time as a function of the distance due west from New York City
    (c) The altitude above sea level as a function of the distance due west from New York City
    (d) The cost of a taxi ride as a function of the distance traveled
    (e) The current in the circuit for the lights in a room as a function of time

11–14 Use the definition of continuity and the properties of limits to show that the function is continuous at the given number a.

11. $f(x) = (x + 2x^3)^4$, $\quad a = -1$

12. $g(t) = \dfrac{t^2 + 5t}{2t + 1}$, $\quad a = 2$

13. $p(v) = 2\sqrt{3v^2 + 1}$, $\quad a = 1$

14. $f(x) = 3x^4 - 5x + \sqrt[3]{x^2 + 4}$, $\quad a = 2$

15–16 Use the definition of continuity and the properties of limits to show that the function is continuous on the given interval.

15. $f(x) = x + \sqrt{x - 4}$, $\quad [4, \infty)$

16. $g(x) = \dfrac{x - 1}{3x + 6}$, $\quad (-\infty, -2)$

17–22 Explain why the function is discontinuous at the given number a. Sketch the graph of the function.

17. $f(x) = \dfrac{1}{x + 2}$ $\quad a = -2$

18. $f(x) = \begin{cases} \dfrac{1}{x+2} & \text{if } x \neq -2 \\ 1 & \text{if } x = -2 \end{cases}$ $\quad a = -2$

19. $f(x) = \begin{cases} x + 3 & \text{if } x \leq -1 \\ 2^x & \text{if } x > -1 \end{cases}$ $\quad a = -1$

20. $f(x) = \begin{cases} \dfrac{x^2 - x}{x^2 - 1} & \text{if } x \neq 1 \\ 1 & \text{if } x = 1 \end{cases}$ $\quad a = 1$

21. $f(x) = \begin{cases} \cos x & \text{if } x < 0 \\ 0 & \text{if } x = 0 \\ 1 - x^2 & \text{if } x > 0 \end{cases}$ $\quad a = 0$

22. $f(x) = \begin{cases} \dfrac{2x^2 - 5x - 3}{x - 3} & \text{if } x \neq 3 \\ 6 & \text{if } x = 3 \end{cases}$ $\quad a = 3$

23–24 How would you "remove the discontinuity" of f? In other words, how would you define $f(2)$ in order to make f continuous at 2?

23. $f(x) = \dfrac{x^2 - x - 2}{x - 2}$

24. $f(x) = \dfrac{x^3 - 8}{x^2 - 4}$

25–32 Explain, using Theorems 4, 5, 7, and 9, why the function is continuous at every number in its domain. State the domain.

25. $F(x) = \dfrac{2x^2 - x - 1}{x^2 + 1}$

26. $G(x) = \dfrac{x^2 + 1}{2x^2 - x - 1}$

27. $Q(x) = \dfrac{\sqrt[3]{x - 2}}{x^3 - 2}$

28. $R(t) = \dfrac{e^{\sin t}}{2 + \cos \pi t}$

29. $A(t) = \arcsin(1 + 2t)$

30. $B(x) = \dfrac{\tan x}{\sqrt{4 - x^2}}$

31. $M(x) = \sqrt{1 + \dfrac{1}{x}}$

32. $N(r) = \tan^{-1}(1 + e^{-r^2})$

33–34 Locate the discontinuities of the function and illustrate by graphing.

33. $y = \dfrac{1}{1 + e^{1/x}}$

34. $y = \ln(\tan^2 x)$

35–38 Use continuity to evaluate the limit.

35. $\lim\limits_{x \to 2} x\sqrt{20 - x^2}$

36. $\lim\limits_{x \to \pi} \sin(x + \sin x)$

37. $\lim\limits_{x \to 1} \ln\left(\dfrac{5 - x^2}{1 + x}\right)$

38. $\lim\limits_{x \to 4} 3^{\sqrt{x^2 - 2x - 4}}$

39–40 Show that f is continuous on $(-\infty, \infty)$.

39. $f(x) = \begin{cases} 1 - x^2 & \text{if } x \leq 1 \\ \ln x & \text{if } x > 1 \end{cases}$

40. $f(x) = \begin{cases} \sin x & \text{if } x < \pi/4 \\ \cos x & \text{if } x \geq \pi/4 \end{cases}$

41–43 Find the numbers at which f is discontinuous. At which of these numbers is f continuous from the right, from the left, or neither? Sketch the graph of f.

41. $f(x) = \begin{cases} x^2 & \text{if } x < -1 \\ x & \text{if } -1 \leq x < 1 \\ 1/x & \text{if } x \geq 1 \end{cases}$

42. $f(x) = \begin{cases} 2^x & \text{if } x \leq 1 \\ 3 - x & \text{if } 1 < x \leq 4 \\ \sqrt{x} & \text{if } x > 4 \end{cases}$

43. $f(x) = \begin{cases} x + 2 & \text{if } x < 0 \\ e^x & \text{if } 0 \leq x \leq 1 \\ 2 - x & \text{if } x > 1 \end{cases}$

44. The gravitational force exerted by the planet Earth on a unit mass at a distance r from the center of the planet is

$$F(r) = \begin{cases} \dfrac{GMr}{R^3} & \text{if } r < R \\ \dfrac{GM}{r^2} & \text{if } r \geq R \end{cases}$$

where M is the mass of Earth, R is its radius, and G is the gravitational constant. Is F a continuous function of r?

45. For what value of the constant c is the function f continuous on $(-\infty, \infty)$?

$$f(x) = \begin{cases} cx^2 + 2x & \text{if } x < 2 \\ x^3 - cx & \text{if } x \geq 2 \end{cases}$$

46. Find the values of a and b that make f continuous everywhere.

$$f(x) = \begin{cases} \dfrac{x^2 - 4}{x - 2} & \text{if } x < 2 \\ ax^2 - bx + 3 & \text{if } 2 \leq x < 3 \\ 2x - a + b & \text{if } x \geq 3 \end{cases}$$

47. Suppose f and g are continuous functions such that $g(2) = 6$ and $\lim\limits_{x \to 2}[3f(x) + f(x)g(x)] = 36$. Find $f(2)$.

48. Let $f(x) = 1/x$ and $g(x) = 1/x^2$.
(a) Find $(f \circ g)(x)$.
(b) Is $f \circ g$ continuous everywhere? Explain.

49. Which of the following functions f has a removable discontinuity at a? If the discontinuity is removable, find a function g that agrees with f for $x \neq a$ and is continuous at a.
(a) $f(x) = \dfrac{x^4 - 1}{x - 1}, \quad a = 1$
(b) $f(x) = \dfrac{x^3 - x^2 - 2x}{x - 2}, \quad a = 2$
(c) $f(x) = [\![\sin x]\!], \quad a = \pi$

50. Suppose that a function f is continuous on $[0, 1]$ except at 0.25 and that $f(0) = 1$ and $f(1) = 3$. Let $N = 2$. Sketch two possible graphs of f, one showing that f might not satisfy the conclusion of the Intermediate Value Theorem and one showing that f might still satisfy the conclusion of the Intermediate Value Theorem (even though it doesn't satisfy the hypothesis).

51. If $f(x) = x^2 + 10 \sin x$, show that there is a number c such that $f(c) = 1000$.

52. Suppose f is continuous on $[1, 5]$ and the only solutions of the equation $f(x) = 6$ are $x = 1$ and $x = 4$. If $f(2) = 8$, explain why $f(3) > 6$.

53–56 Use the Intermediate Value Theorem to show that there is a root of the given equation in the specified interval.

53. $x^4 + x - 3 = 0$, $(1, 2)$

54. $\ln x = x - \sqrt{x}$, $(2, 3)$

55. $e^x = 3 - 2x$, $(0, 1)$

56. $\sin x = x^2 - x$, $(1, 2)$

57–58 (a) Prove that the equation has at least one real root.
(b) Use your calculator to find an interval of length 0.01 that contains a root.

57. $\cos x = x^3$ **58.** $\ln x = 3 - 2x$

59–60 (a) Prove that the equation has at least one real root.
(b) Use your graphing device to find the root correct to three decimal places.

59. $100e^{-x/100} = 0.01x^2$

60. $\arctan x = 1 - x$

61–62 Prove, without graphing, that the graph of the function has at least two x-intercepts in the specified interval.

61. $y = \sin x^3$, $(1, 2)$

62. $y = x^2 - 3 + 1/x$, $(0, 2)$

63. Prove that f is continuous at a if and only if
$$\lim_{h \to 0} f(a + h) = f(a)$$

64. To prove that sine is continuous, we need to show that $\lim_{x \to a} \sin x = \sin a$ for every real number a. By Exercise 63 an equivalent statement is that
$$\lim_{h \to 0} \sin(a + h) = \sin a$$
Use (6) to show that this is true.

65. Prove that cosine is a continuous function.

66. (a) Prove Theorem 4, part 3.
(b) Prove Theorem 4, part 5.

67. For what values of x is f continuous?
$$f(x) = \begin{cases} 0 & \text{if } x \text{ is rational} \\ 1 & \text{if } x \text{ is irrational} \end{cases}$$

68. For what values of x is g continuous?
$$g(x) = \begin{cases} 0 & \text{if } x \text{ is rational} \\ x & \text{if } x \text{ is irrational} \end{cases}$$

69. Is there a number that is exactly 1 more than its cube?

70. If a and b are positive numbers, prove that the equation
$$\frac{a}{x^3 + 2x^2 - 1} + \frac{b}{x^3 + x - 2} = 0$$
has at least one solution in the interval $(-1, 1)$.

71. Show that the function
$$f(x) = \begin{cases} x^4 \sin(1/x) & \text{if } x \neq 0 \\ 0 & \text{if } x = 0 \end{cases}$$
is continuous on $(-\infty, \infty)$.

72. (a) Show that the absolute value function $F(x) = |x|$ is continuous everywhere.
(b) Prove that if f is a continuous function on an interval, then so is $|f|$.
(c) Is the converse of the statement in part (b) also true? In other words, if $|f|$ is continuous, does it follow that f is continuous? If so, prove it. If not, find a counterexample.

73. A Tibetan monk leaves the monastery at 7:00 AM and takes his usual path to the top of the mountain, arriving at 7:00 PM. The following morning, he starts at 7:00 AM at the top and takes the same path back, arriving at the monastery at 7:00 PM. Use the Intermediate Value Theorem to show that there is a point on the path that the monk will cross at exactly the same time of day on both days.

2.6 Limits at Infinity; Horizontal Asymptotes

In Sections 2.2 and 2.4 we investigated infinite limits and vertical asymptotes. There we let x approach a number and the result was that the values of y became arbitrarily large (positive or negative). In this section we let x become arbitrarily large (positive or negative) and see what happens to y.

Let's begin by investigating the behavior of the function f defined by
$$f(x) = \frac{x^2 - 1}{x^2 + 1}$$

SECTION 2.6 Limits at Infinity; Horizontal Asymptotes 127

x	$f(x)$
0	-1
± 1	0
± 2	0.600000
± 3	0.800000
± 4	0.882353
± 5	0.923077
± 10	0.980198
± 50	0.999200
± 100	0.999800
± 1000	0.999998

as x becomes large. The table at the left gives values of this function correct to six decimal places, and the graph of f has been drawn by a computer in Figure 1.

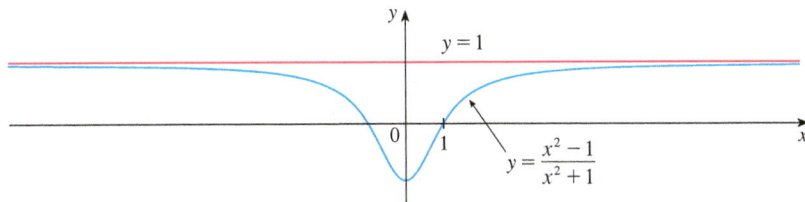

FIGURE 1

As x grows larger and larger you can see that the values of $f(x)$ get closer and closer to 1. (The graph of f approaches the horizontal line $y = 1$ as we look to the right.) In fact, it seems that we can make the values of $f(x)$ as close as we like to 1 by taking x sufficiently large. This situation is expressed symbolically by writing

$$\lim_{x \to \infty} \frac{x^2 - 1}{x^2 + 1} = 1$$

In general, we use the notation

$$\lim_{x \to \infty} f(x) = L$$

to indicate that the values of $f(x)$ approach L as x becomes larger and larger.

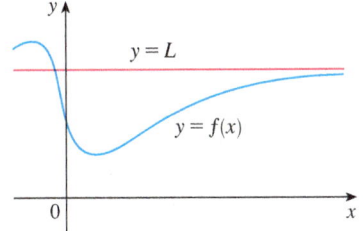

1 Intuitive Definition of a Limit at Infinity Let f be a function defined on some interval (a, ∞). Then

$$\lim_{x \to \infty} f(x) = L$$

means that the values of $f(x)$ can be made arbitrarily close to L by requiring x to be sufficiently large.

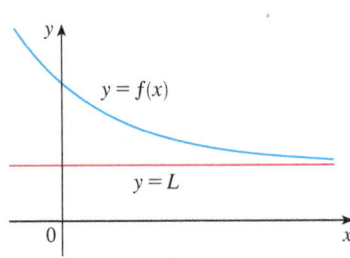

Another notation for $\lim_{x \to \infty} f(x) = L$ is

$$f(x) \to L \quad \text{as} \quad x \to \infty$$

The symbol ∞ does not represent a number. Nonetheless, the expression $\lim_{x \to \infty} f(x) = L$ is often read as

"the limit of $f(x)$, as x approaches infinity, is L"

or   "the limit of $f(x)$, as x becomes infinite, is L"

or   "the limit of $f(x)$, as x increases without bound, is L"

The meaning of such phrases is given by Definition 1. A more precise definition, similar to the ε, δ definition of Section 2.4, is given at the end of this section.

Geometric illustrations of Definition 1 are shown in Figure 2. Notice that there are many ways for the graph of f to approach the line $y = L$ (which is called a *horizontal asymptote*) as we look to the far right of each graph.

Referring back to Figure 1, we see that for numerically large negative values of x, the values of $f(x)$ are close to 1. By letting x decrease through negative values without bound, we can make $f(x)$ as close to 1 as we like. This is expressed by writing

$$\lim_{x \to -\infty} \frac{x^2 - 1}{x^2 + 1} = 1$$

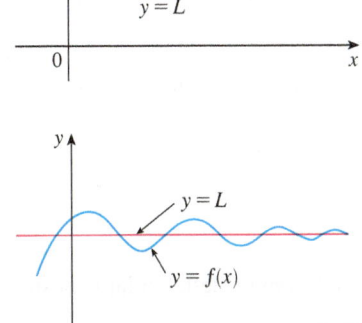

FIGURE 2
Examples illustrating $\lim_{x \to \infty} f(x) = L$

128 CHAPTER 2 Limits and Derivatives

The general definition is as follows.

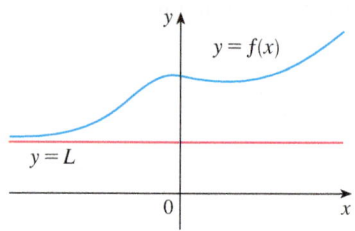

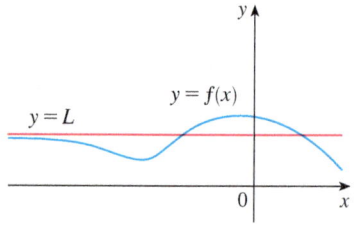

FIGURE 3
Examples illustrating $\lim_{x \to -\infty} f(x) = L$

[2] Definition Let f be a function defined on some interval $(-\infty, a)$. Then
$$\lim_{x \to -\infty} f(x) = L$$
means that the values of $f(x)$ can be made arbitrarily close to L by requiring x to be sufficiently large negative.

Again, the symbol $-\infty$ does not represent a number, but the expression $\lim_{x \to -\infty} f(x) = L$ is often read as

"the limit of $f(x)$, as x approaches negative infinity, is L"

Definition 2 is illustrated in Figure 3. Notice that the graph approaches the line $y = L$ as we look to the far left of each graph.

[3] Definition The line $y = L$ is called a **horizontal asymptote** of the curve $y = f(x)$ if either
$$\lim_{x \to \infty} f(x) = L \quad \text{or} \quad \lim_{x \to -\infty} f(x) = L$$

For instance, the curve illustrated in Figure 1 has the line $y = 1$ as a horizontal asymptote because
$$\lim_{x \to \infty} \frac{x^2 - 1}{x^2 + 1} = 1$$

An example of a curve with two horizontal asymptotes is $y = \tan^{-1}x$. (See Figure 4.) In fact,

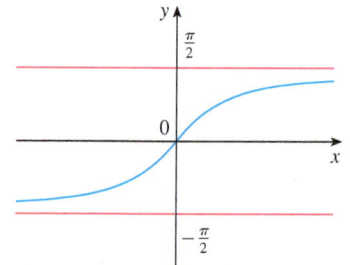

FIGURE 4
$y = \tan^{-1}x$

[4]
$$\lim_{x \to -\infty} \tan^{-1}x = -\frac{\pi}{2} \qquad \lim_{x \to \infty} \tan^{-1}x = \frac{\pi}{2}$$

so both of the lines $y = -\pi/2$ and $y = \pi/2$ are horizontal asymptotes. (This follows from the fact that the lines $x = \pm\pi/2$ are vertical asymptotes of the graph of the tangent function.)

EXAMPLE 1 Find the infinite limits, limits at infinity, and asymptotes for the function f whose graph is shown in Figure 5.

SOLUTION We see that the values of $f(x)$ become large as $x \to -1$ from both sides, so
$$\lim_{x \to -1} f(x) = \infty$$

Notice that $f(x)$ becomes large negative as x approaches 2 from the left, but large positive as x approaches 2 from the right. So
$$\lim_{x \to 2^-} f(x) = -\infty \quad \text{and} \quad \lim_{x \to 2^+} f(x) = \infty$$

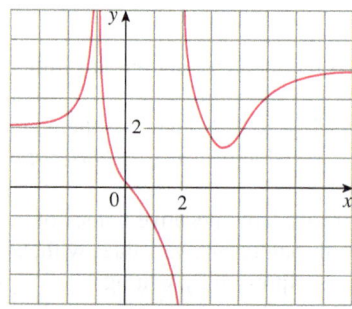

FIGURE 5

Thus both of the lines $x = -1$ and $x = 2$ are vertical asymptotes.

As x becomes large, it appears that $f(x)$ approaches 4. But as x decreases through negative values, $f(x)$ approaches 2. So

$$\lim_{x \to \infty} f(x) = 4 \quad \text{and} \quad \lim_{x \to -\infty} f(x) = 2$$

This means that both $y = 4$ and $y = 2$ are horizontal asymptotes.

EXAMPLE 2 Find $\lim_{x \to \infty} \dfrac{1}{x}$ and $\lim_{x \to -\infty} \dfrac{1}{x}$.

SOLUTION Observe that when x is large, $1/x$ is small. For instance,

$$\frac{1}{100} = 0.01 \qquad \frac{1}{10,000} = 0.0001 \qquad \frac{1}{1,000,000} = 0.000001$$

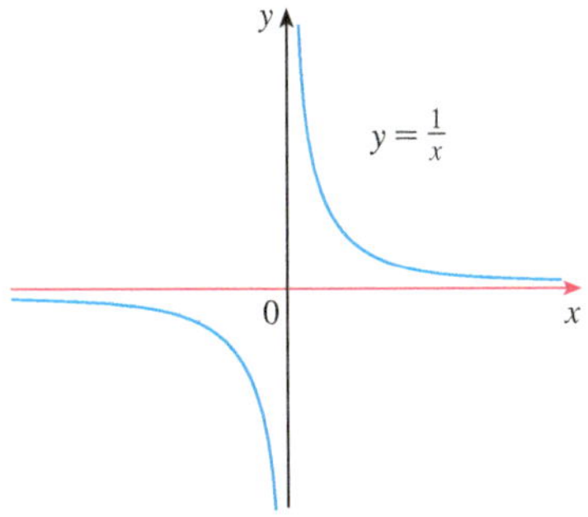

In fact, by taking x large enough, we can make $1/x$ as close to 0 as we please. Therefore, according to Definition 1, we have

$$\lim_{x \to \infty} \frac{1}{x} = 0$$

Similar reasoning shows that when x is large negative, $1/x$ is small negative, so we also have

$$\lim_{x \to -\infty} \frac{1}{x} = 0$$

It follows that the line $y = 0$ (the x-axis) is a horizontal asymptote of the curve $y = 1/x$. (This is an equilateral hyperbola; see Figure 6.)

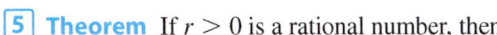

FIGURE 6
$\lim_{x \to \infty} \dfrac{1}{x} = 0$, $\lim_{x \to -\infty} \dfrac{1}{x} = 0$

Most of the Limit Laws that were given in Section 2.3 also hold for limits at infinity. It can be proved that *the Limit Laws listed in Section 2.3 (with the exception of Laws 9 and 10) are also valid if "$x \to a$" is replaced by "$x \to \infty$" or "$x \to -\infty$."* In particular, if we combine Laws 6 and 11 with the results of Example 2, we obtain the following important rule for calculating limits.

5 Theorem If $r > 0$ is a rational number, then

$$\lim_{x \to \infty} \frac{1}{x^r} = 0$$

If $r > 0$ is a rational number such that x^r is defined for all x, then

$$\lim_{x \to -\infty} \frac{1}{x^r} = 0$$

EXAMPLE 3 Evaluate

$$\lim_{x \to \infty} \frac{3x^2 - x - 2}{5x^2 + 4x + 1}$$

and indicate which properties of limits are used at each stage.

SOLUTION As x becomes large, both numerator and denominator become large, so it isn't obvious what happens to their ratio. We need to do some preliminary algebra.

130 CHAPTER 2 Limits and Derivatives

To evaluate the limit at infinity of any rational function, we first divide both the numerator and denominator by the highest power of x that occurs in the denominator. (We may assume that $x \neq 0$, since we are interested only in large values of x.) In this case the highest power of x in the denominator is x^2, so we have

$$\lim_{x \to \infty} \frac{3x^2 - x - 2}{5x^2 + 4x + 1} = \lim_{x \to \infty} \frac{\dfrac{3x^2 - x - 2}{x^2}}{\dfrac{5x^2 + 4x + 1}{x^2}} = \lim_{x \to \infty} \frac{3 - \dfrac{1}{x} - \dfrac{2}{x^2}}{5 + \dfrac{4}{x} + \dfrac{1}{x^2}}$$

$$= \frac{\lim\limits_{x \to \infty} \left(3 - \dfrac{1}{x} - \dfrac{2}{x^2}\right)}{\lim\limits_{x \to \infty} \left(5 + \dfrac{4}{x} + \dfrac{1}{x^2}\right)} \quad \text{(by Limit Law 5)}$$

$$= \frac{\lim\limits_{x \to \infty} 3 - \lim\limits_{x \to \infty} \dfrac{1}{x} - 2 \lim\limits_{x \to \infty} \dfrac{1}{x^2}}{\lim\limits_{x \to \infty} 5 + 4 \lim\limits_{x \to \infty} \dfrac{1}{x} + \lim\limits_{x \to \infty} \dfrac{1}{x^2}} \quad \text{(by 1, 2, and 3)}$$

$$= \frac{3 - 0 - 0}{5 + 0 + 0} \quad \text{(by 7 and Theorem 5)}$$

$$= \frac{3}{5}$$

A similar calculation shows that the limit as $x \to -\infty$ is also $\frac{3}{5}$. Figure 7 illustrates the results of these calculations by showing how the graph of the given rational function approaches the horizontal asymptote $y = \frac{3}{5} = 0.6$. ∎

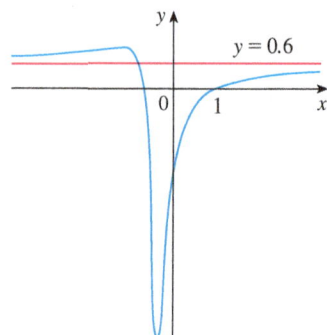

FIGURE 7
$y = \dfrac{3x^2 - x - 2}{5x^2 + 4x + 1}$

EXAMPLE 4 Find the horizontal and vertical asymptotes of the graph of the function

$$f(x) = \frac{\sqrt{2x^2 + 1}}{3x - 5}$$

SOLUTION Dividing both numerator and denominator by x and using the properties of limits, we have

$$\lim_{x \to \infty} \frac{\sqrt{2x^2 + 1}}{3x - 5} = \lim_{x \to \infty} \frac{\dfrac{\sqrt{2x^2 + 1}}{x}}{\dfrac{3x - 5}{x}} = \lim_{x \to \infty} \frac{\sqrt{\dfrac{2x^2 + 1}{x^2}}}{\dfrac{3x - 5}{x}} \quad \text{(since } \sqrt{x^2} = x \text{ for } x > 0\text{)}$$

$$= \frac{\lim\limits_{x \to \infty} \sqrt{2 + \dfrac{1}{x^2}}}{\lim\limits_{x \to \infty} \left(3 - \dfrac{5}{x}\right)} = \frac{\sqrt{\lim\limits_{x \to \infty} 2 + \lim\limits_{x \to \infty} \dfrac{1}{x^2}}}{\lim\limits_{x \to \infty} 3 - 5 \lim\limits_{x \to \infty} \dfrac{1}{x}} = \frac{\sqrt{2 + 0}}{3 - 5 \cdot 0} = \frac{\sqrt{2}}{3}$$

Therefore the line $y = \sqrt{2}/3$ is a horizontal asymptote of the graph of f.

In computing the limit as $x \to -\infty$, we must remember that for $x < 0$, we have $\sqrt{x^2} = |x| = -x$. So when we divide the numerator by x, for $x < 0$ we get

$$\frac{\sqrt{2x^2 + 1}}{x} = \frac{\sqrt{2x^2 + 1}}{-\sqrt{x^2}} = -\sqrt{\frac{2x^2 + 1}{x^2}} = -\sqrt{2 + \frac{1}{x^2}}$$

Therefore

$$\lim_{x \to -\infty} \frac{\sqrt{2x^2 + 1}}{3x - 5} = \lim_{x \to -\infty} \frac{-\sqrt{2 + \frac{1}{x^2}}}{3 - \frac{5}{x}} = \frac{-\sqrt{2 + \lim_{x \to -\infty} \frac{1}{x^2}}}{3 - 5 \lim_{x \to -\infty} \frac{1}{x}} = -\frac{\sqrt{2}}{3}$$

Thus the line $y = -\sqrt{2}/3$ is also a horizontal asymptote.

A vertical asymptote is likely to occur when the denominator, $3x - 5$, is 0, that is, when $x = \frac{5}{3}$. If x is close to $\frac{5}{3}$ and $x > \frac{5}{3}$, then the denominator is close to 0 and $3x - 5$ is positive. The numerator $\sqrt{2x^2 + 1}$ is always positive, so $f(x)$ is positive. Therefore

$$\lim_{x \to (5/3)^+} \frac{\sqrt{2x^2 + 1}}{3x - 5} = \infty$$

(Notice that the numerator does *not* approach 0 as $x \to 5/3$.)

If x is close to $\frac{5}{3}$ but $x < \frac{5}{3}$, then $3x - 5 < 0$ and so $f(x)$ is large negative. Thus

$$\lim_{x \to (5/3)^-} \frac{\sqrt{2x^2 + 1}}{3x - 5} = -\infty$$

The vertical asymptote is $x = \frac{5}{3}$. All three asymptotes are shown in Figure 8. ∎

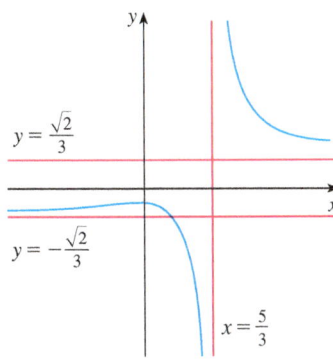

FIGURE 8
$y = \dfrac{\sqrt{2x^2 + 1}}{3x - 5}$

EXAMPLE 5 Compute $\lim_{x \to \infty} \left(\sqrt{x^2 + 1} - x \right)$.

SOLUTION Because both $\sqrt{x^2 + 1}$ and x are large when x is large, it's difficult to see what happens to their difference, so we use algebra to rewrite the function. We first multiply numerator and denominator by the conjugate radical:

We can think of the given function as having a denominator of 1.

$$\lim_{x \to \infty} \left(\sqrt{x^2 + 1} - x \right) = \lim_{x \to \infty} \left(\sqrt{x^2 + 1} - x \right) \cdot \frac{\sqrt{x^2 + 1} + x}{\sqrt{x^2 + 1} + x}$$

$$= \lim_{x \to \infty} \frac{(x^2 + 1) - x^2}{\sqrt{x^2 + 1} + x} = \lim_{x \to \infty} \frac{1}{\sqrt{x^2 + 1} + x}$$

Notice that the denominator of this last expression $\left(\sqrt{x^2 + 1} + x \right)$ becomes large as $x \to \infty$ (it's bigger than x). So

$$\lim_{x \to \infty} \left(\sqrt{x^2 + 1} - x \right) = \lim_{x \to \infty} \frac{1}{\sqrt{x^2 + 1} + x} = 0$$

Figure 9 illustrates this result. ∎

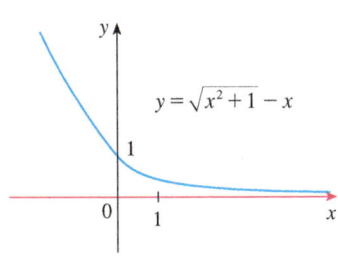

FIGURE 9

EXAMPLE 6 Evaluate $\lim_{x \to 2^+} \arctan\left(\dfrac{1}{x-2}\right)$.

SOLUTION If we let $t = 1/(x-2)$, we know that $t \to \infty$ as $x \to 2^+$. Therefore, by the second equation in (4), we have

$$\lim_{x \to 2^+} \arctan\left(\dfrac{1}{x-2}\right) = \lim_{t \to \infty} \arctan t = \dfrac{\pi}{2}$$ ∎

The graph of the natural exponential function $y = e^x$ has the line $y = 0$ (the x-axis) as a horizontal asymptote. (The same is true of any exponential function with base $b > 1$.) In fact, from the graph in Figure 10 and the corresponding table of values, we see that

$$\boxed{6 \qquad \lim_{x \to -\infty} e^x = 0}$$

Notice that the values of e^x approach 0 very rapidly.

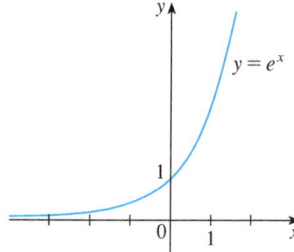

x	e^x
0	1.00000
−1	0.36788
−2	0.13534
−3	0.04979
−5	0.00674
−8	0.00034
−10	0.00005

FIGURE 10

EXAMPLE 7 Evaluate $\lim_{x \to 0^-} e^{1/x}$.

SOLUTION If we let $t = 1/x$, we know that $t \to -\infty$ as $x \to 0^-$. Therefore, by (6),

$$\lim_{x \to 0^-} e^{1/x} = \lim_{t \to -\infty} e^t = 0$$

(See Exercise 81.) ∎

PS The problem-solving strategy for Examples 6 and 7 is *introducing something extra* (see page 71). Here, the something extra, the auxiliary aid, is the new variable t.

EXAMPLE 8 Evaluate $\lim_{x \to \infty} \sin x$.

SOLUTION As x increases, the values of $\sin x$ oscillate between 1 and −1 infinitely often and so they don't approach any definite number. Thus $\lim_{x \to \infty} \sin x$ does not exist. ∎

■ Infinite Limits at Infinity

The notation

$$\lim_{x \to \infty} f(x) = \infty$$

is used to indicate that the values of $f(x)$ become large as x becomes large. Similar meanings are attached to the following symbols:

$$\lim_{x \to -\infty} f(x) = \infty \qquad \lim_{x \to \infty} f(x) = -\infty \qquad \lim_{x \to -\infty} f(x) = -\infty$$

EXAMPLE 9 Find $\lim_{x \to \infty} x^3$ and $\lim_{x \to -\infty} x^3$.

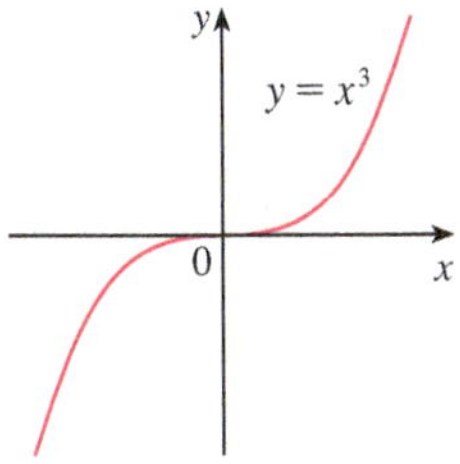

SOLUTION When x becomes large, x^3 also becomes large. For instance,

$$10^3 = 1000 \qquad 100^3 = 1{,}000{,}000 \qquad 1000^3 = 1{,}000{,}000{,}000$$

In fact, we can make x^3 as big as we like by requiring x to be large enough. Therefore we can write

$$\lim_{x \to \infty} x^3 = \infty$$

Similarly, when x is large negative, so is x^3. Thus

$$\lim_{x \to -\infty} x^3 = -\infty$$

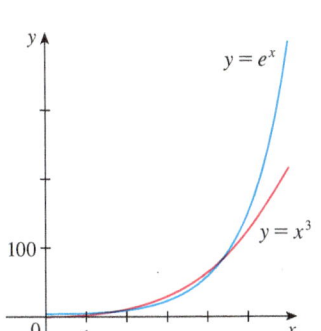

FIGURE 11
$\lim_{x \to \infty} x^3 = \infty$, $\lim_{x \to -\infty} x^3 = -\infty$

These limit statements can also be seen from the graph of $y = x^3$ in Figure 11. ■

Looking at Figure 10 we see that

$$\lim_{x \to \infty} e^x = \infty$$

but, as Figure 12 demonstrates, $y = e^x$ becomes large as $x \to \infty$ at a much faster rate than $y = x^3$.

FIGURE 12
e^x is much larger than x^3 when x is large.

EXAMPLE 10 Find $\lim_{x \to \infty} (x^2 - x)$.

SOLUTION It would be **wrong** to write

$$\lim_{x \to \infty} (x^2 - x) = \lim_{x \to \infty} x^2 - \lim_{x \to \infty} x = \infty - \infty$$

The Limit Laws can't be applied to infinite limits because ∞ is not a number ($\infty - \infty$ can't be defined). However, we *can* write

$$\lim_{x \to \infty} (x^2 - x) = \lim_{x \to \infty} x(x - 1) = \infty$$

because both x and $x - 1$ become arbitrarily large and so their product does too. ■

EXAMPLE 11 Find $\lim_{x \to \infty} \dfrac{x^2 + x}{3 - x}$.

SOLUTION As in Example 3, we divide the numerator and denominator by the highest power of x in the denominator, which is just x:

$$\lim_{x \to \infty} \frac{x^2 + x}{3 - x} = \lim_{x \to \infty} \frac{x + 1}{\dfrac{3}{x} - 1} = -\infty$$

because $x + 1 \to \infty$ and $3/x - 1 \to 0 - 1 = -1$ as $x \to \infty$. ■

The next example shows that by using infinite limits at infinity, together with intercepts, we can get a rough idea of the graph of a polynomial without having to plot a large number of points.

EXAMPLE 12 Sketch the graph of $y = (x - 2)^4(x + 1)^3(x - 1)$ by finding its intercepts and its limits as $x \to \infty$ and as $x \to -\infty$.

SOLUTION The y-intercept is $f(0) = (-2)^4(1)^3(-1) = -16$ and the x-intercepts are found by setting $y = 0$: $x = 2, -1, 1$. Notice that since $(x - 2)^4$ is never negative, the function doesn't change sign at 2; thus the graph doesn't cross the x-axis at 2. The graph crosses the axis at -1 and 1.

When x is large positive, all three factors are large, so

$$\lim_{x \to \infty} (x - 2)^4(x + 1)^3(x - 1) = \infty$$

When x is large negative, the first factor is large positive and the second and third factors are both large negative, so

$$\lim_{x \to -\infty} (x - 2)^4(x + 1)^3(x - 1) = \infty$$

Combining this information, we give a rough sketch of the graph in Figure 13. ∎

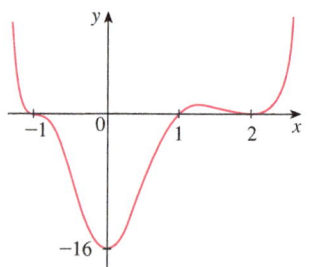

FIGURE 13
$y = (x - 2)^4(x + 1)^3(x - 1)$

Precise Definitions

Definition 1 can be stated precisely as follows.

7 Precise Definition of a Limit at Infinity Let f be a function defined on some interval (a, ∞). Then
$$\lim_{x \to \infty} f(x) = L$$
means that for every $\varepsilon > 0$ there is a corresponding number N such that
$$\text{if} \quad x > N \quad \text{then} \quad |f(x) - L| < \varepsilon$$

In words, this says that the values of $f(x)$ can be made arbitrarily close to L (within a distance ε, where ε is any positive number) by requiring x to be sufficiently large (larger than N, where N depends on ε). Graphically it says that by keeping x large enough (larger than some number N) we can make the graph of f lie between the given horizontal lines $y = L - \varepsilon$ and $y = L + \varepsilon$ as in Figure 14. This must be true no matter how small we choose ε.

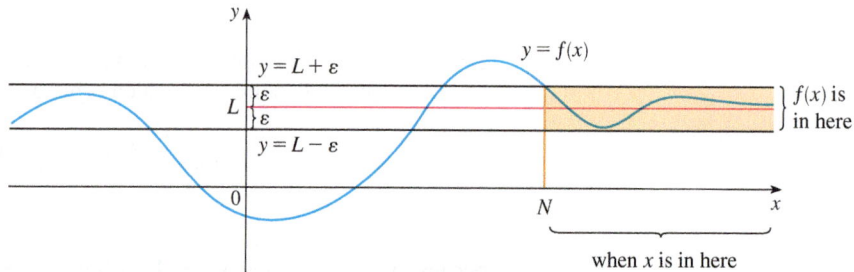

FIGURE 14
$\lim_{x \to \infty} f(x) = L$

Figure 15 shows that if a smaller value of ε is chosen, then a larger value of N may be required.

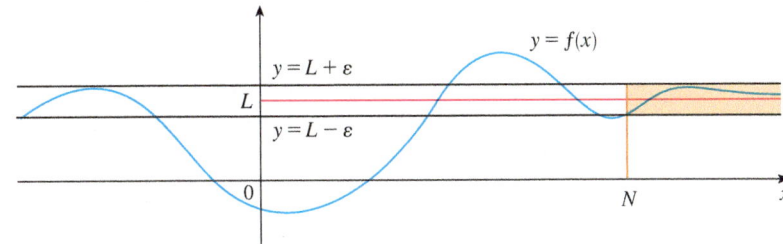

FIGURE 15
$\lim_{x \to \infty} f(x) = L$

Similarly, a precise version of Definition 2 is given by Definition 8, which is illustrated in Figure 16.

8 Definition Let f be a function defined on some interval $(-\infty, a)$. Then
$$\lim_{x \to -\infty} f(x) = L$$
means that for every $\varepsilon > 0$ there is a corresponding number N such that
$$\text{if} \quad x < N \quad \text{then} \quad |f(x) - L| < \varepsilon$$

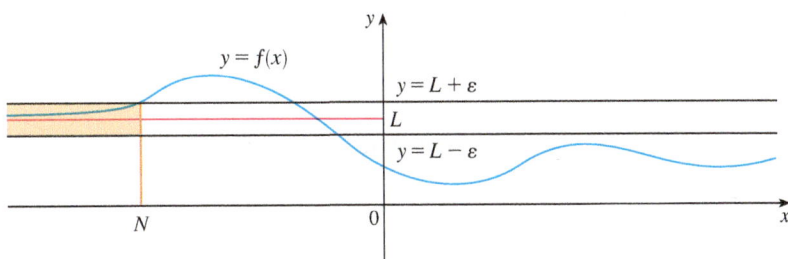

FIGURE 16
$\lim_{x \to -\infty} f(x) = L$

In Example 3 we calculated that
$$\lim_{x \to \infty} \frac{3x^2 - x - 2}{5x^2 + 4x + 1} = \frac{3}{5}$$

In the next example we use a graphing device to relate this statement to Definition 7 with $L = \frac{3}{5} = 0.6$ and $\varepsilon = 0.1$.

TEC In Module 2.4/2.6 you can explore the precise definition of a limit both graphically and numerically.

EXAMPLE 13 Use a graph to find a number N such that
$$\text{if} \quad x > N \quad \text{then} \quad \left| \frac{3x^2 - x - 2}{5x^2 + 4x + 1} - 0.6 \right| < 0.1$$

136   **CHAPTER 2**   Limits and Derivatives

SOLUTION We rewrite the given inequality as

$$0.5 < \frac{3x^2 - x - 2}{5x^2 + 4x + 1} < 0.7$$

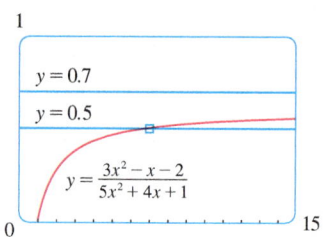

We need to determine the values of x for which the given curve lies between the horizontal lines $y = 0.5$ and $y = 0.7$. So we graph the curve and these lines in Figure 17. Then we use the cursor to estimate that the curve crosses the line $y = 0.5$ when $x \approx 6.7$. To the right of this number it seems that the curve stays between the lines $y = 0.5$ and $y = 0.7$. Rounding up to be safe, we can say that

$$\text{if} \quad x > 7 \quad \text{then} \quad \left| \frac{3x^2 - x - 2}{5x^2 + 4x + 1} - 0.6 \right| < 0.1$$

FIGURE 17

In other words, for $\varepsilon = 0.1$ we can choose $N = 7$ (or any larger number) in Definition 7. ∎

EXAMPLE 14 Use Definition 7 to prove that $\displaystyle\lim_{x \to \infty} \frac{1}{x} = 0$.

SOLUTION Given $\varepsilon > 0$, we want to find N such that

$$\text{if} \quad x > N \quad \text{then} \quad \left| \frac{1}{x} - 0 \right| < \varepsilon$$

In computing the limit we may assume that $x > 0$. Then $1/x < \varepsilon \iff x > 1/\varepsilon$. Let's choose $N = 1/\varepsilon$. So

$$\text{if} \quad x > N = \frac{1}{\varepsilon} \quad \text{then} \quad \left| \frac{1}{x} - 0 \right| = \frac{1}{x} < \varepsilon$$

Therefore, by Definition 7,

$$\lim_{x \to \infty} \frac{1}{x} = 0$$

Figure 18 illustrates the proof by showing some values of ε and the corresponding values of N.

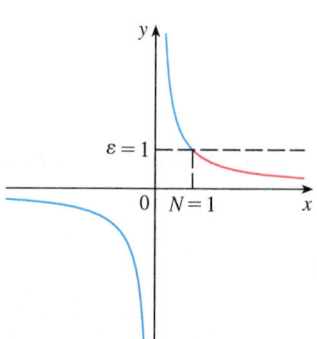
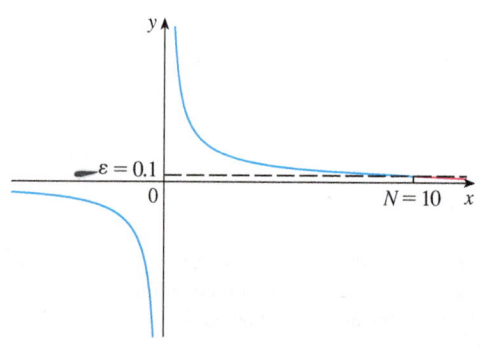

FIGURE 18 ∎

SECTION 2.6 Limits at Infinity; Horizontal Asymptotes   137

Finally we note that an infinite limit at infinity can be defined as follows. The geometric illustration is given in Figure 19.

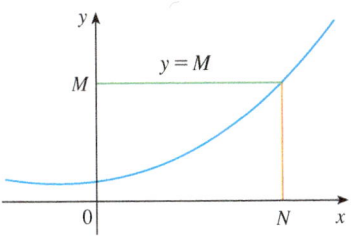

FIGURE 19
$\lim_{x \to \infty} f(x) = \infty$

9 Definition of an Infinite Limit at Infinity Let f be a function defined on some interval (a, ∞). Then

$$\lim_{x \to \infty} f(x) = \infty$$

means that for every positive number M there is a corresponding positive number N such that

$$\text{if} \quad x > N \quad \text{then} \quad f(x) > M$$

Similar definitions apply when the symbol ∞ is replaced by $-\infty$. (See Exercise 80.)

2.6 EXERCISES

1. Explain in your own words the meaning of each of the following.
   (a) $\lim_{x \to \infty} f(x) = 5$
   (b) $\lim_{x \to -\infty} f(x) = 3$

2. (a) Can the graph of $y = f(x)$ intersect a vertical asymptote? Can it intersect a horizontal asymptote? Illustrate by sketching graphs.
   (b) How many horizontal asymptotes can the graph of $y = f(x)$ have? Sketch graphs to illustrate the possibilities.

3. For the function f whose graph is given, state the following.
   (a) $\lim_{x \to \infty} f(x)$
   (b) $\lim_{x \to -\infty} f(x)$
   (c) $\lim_{x \to 1} f(x)$
   (d) $\lim_{x \to 3} f(x)$
   (e) The equations of the asymptotes

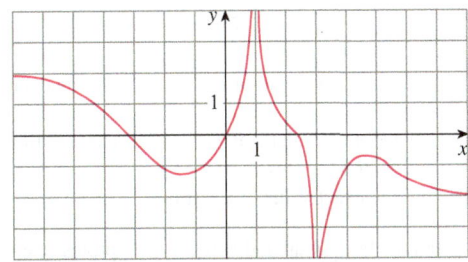

4. For the function g whose graph is given, state the following.
   (a) $\lim_{x \to \infty} g(x)$
   (b) $\lim_{x \to -\infty} g(x)$
   (c) $\lim_{x \to 0} g(x)$
   (d) $\lim_{x \to 2^-} g(x)$
   (e) $\lim_{x \to 2^+} g(x)$
   (f) The equations of the asymptotes

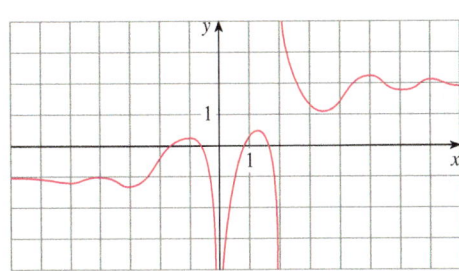

5–10 Sketch the graph of an example of a function f that satisfies all of the given conditions.

5. $\lim_{x \to 0} f(x) = -\infty$, $\quad \lim_{x \to -\infty} f(x) = 5$, $\quad \lim_{x \to \infty} f(x) = -5$

6. $\lim_{x \to 2} f(x) = \infty$, $\quad \lim_{x \to -2^+} f(x) = \infty$, $\quad \lim_{x \to -2^-} f(x) = -\infty$,
   $\lim_{x \to -\infty} f(x) = 0$, $\quad \lim_{x \to \infty} f(x) = 0$, $\quad f(0) = 0$

7. $\lim_{x \to 2} f(x) = -\infty$, $\quad \lim_{x \to \infty} f(x) = \infty$, $\quad \lim_{x \to -\infty} f(x) = 0$,
   $\lim_{x \to 0^+} f(x) = \infty$, $\quad \lim_{x \to 0^-} f(x) = -\infty$

8. $\lim_{x \to \infty} f(x) = 3$, $\lim_{x \to 2^-} f(x) = \infty$, $\lim_{x \to 2^+} f(x) = -\infty$, f is odd

9. $f(0) = 3$, $\quad \lim_{x \to 0^-} f(x) = 4$, $\quad \lim_{x \to 0^+} f(x) = 2$,
   $\lim_{x \to -\infty} f(x) = -\infty$, $\quad \lim_{x \to 4^-} f(x) = -\infty$, $\quad \lim_{x \to 4^+} f(x) = \infty$,
   $\lim_{x \to \infty} f(x) = 3$

10. $\lim_{x \to 3} f(x) = -\infty$, $\quad \lim_{x \to \infty} f(x) = 2$, $\quad f(0) = 0$, $\quad f$ is even

11. Guess the value of the limit

$$\lim_{x\to\infty} \frac{x^2}{2^x}$$

by evaluating the function $f(x) = x^2/2^x$ for $x = 0, 1, 2, 3, 4, 5, 6, 7, 8, 9, 10, 20, 50,$ and 100. Then use a graph of f to support your guess.

12. (a) Use a graph of

$$f(x) = \left(1 - \frac{2}{x}\right)^x$$

to estimate the value of $\lim_{x\to\infty} f(x)$ correct to two decimal places.
(b) Use a table of values of $f(x)$ to estimate the limit to four decimal places.

13–14 Evaluate the limit and justify each step by indicating the appropriate properties of limits.

13. $\lim_{x\to\infty} \frac{2x^2 - 7}{5x^2 + x - 3}$

14. $\lim_{x\to\infty} \sqrt{\frac{9x^3 + 8x - 4}{3 - 5x + x^3}}$

15–42 Find the limit or show that it does not exist.

15. $\lim_{x\to\infty} \frac{3x - 2}{2x + 1}$

16. $\lim_{x\to\infty} \frac{1 - x^2}{x^3 - x + 1}$

17. $\lim_{x\to-\infty} \frac{x - 2}{x^2 + 1}$

18. $\lim_{x\to-\infty} \frac{4x^3 + 6x^2 - 2}{2x^3 - 4x + 5}$

19. $\lim_{t\to\infty} \frac{\sqrt{t} + t^2}{2t - t^2}$

20. $\lim_{t\to\infty} \frac{t - t\sqrt{t}}{2t^{3/2} + 3t - 5}$

21. $\lim_{x\to\infty} \frac{(2x^2 + 1)^2}{(x - 1)^2(x^2 + x)}$

22. $\lim_{x\to\infty} \frac{x^2}{\sqrt{x^4 + 1}}$

23. $\lim_{x\to\infty} \frac{\sqrt{1 + 4x^6}}{2 - x^3}$

24. $\lim_{x\to-\infty} \frac{\sqrt{1 + 4x^6}}{2 - x^3}$

25. $\lim_{x\to\infty} \frac{\sqrt{x + 3x^2}}{4x - 1}$

26. $\lim_{x\to\infty} \frac{x + 3x^2}{4x - 1}$

27. $\lim_{x\to\infty} \left(\sqrt{9x^2 + x} - 3x\right)$

28. $\lim_{x\to-\infty} \left(\sqrt{4x^2 + 3x} + 2x\right)$

29. $\lim_{x\to\infty} \left(\sqrt{x^2 + ax} - \sqrt{x^2 + bx}\right)$

30. $\lim_{x\to\infty} \sqrt{x^2 + 1}$

31. $\lim_{x\to\infty} \frac{x^4 - 3x^2 + x}{x^3 - x + 2}$

32. $\lim_{x\to\infty} (e^{-x} + 2\cos 3x)$

33. $\lim_{x\to-\infty} (x^2 + 2x^7)$

34. $\lim_{x\to-\infty} \frac{1 + x^6}{x^4 + 1}$

35. $\lim_{x\to\infty} \arctan(e^x)$

36. $\lim_{x\to\infty} \frac{e^{3x} - e^{-3x}}{e^{3x} + e^{-3x}}$

37. $\lim_{x\to\infty} \frac{1 - e^x}{1 + 2e^x}$

38. $\lim_{x\to\infty} \frac{\sin^2 x}{x^2 + 1}$

39. $\lim_{x\to\infty} (e^{-2x} \cos x)$

40. $\lim_{x\to 0^+} \tan^{-1}(\ln x)$

41. $\lim_{x\to\infty} [\ln(1 + x^2) - \ln(1 + x)]$

42. $\lim_{x\to\infty} [\ln(2 + x) - \ln(1 + x)]$

43. (a) For $f(x) = \dfrac{x}{\ln x}$ find each of the following limits.
(i) $\lim_{x\to 0^+} f(x)$ (ii) $\lim_{x\to 1^-} f(x)$ (iii) $\lim_{x\to 1^+} f(x)$
(b) Use a table of values to estimate $\lim_{x\to\infty} f(x)$.
(c) Use the information from parts (a) and (b) to make a rough sketch of the graph of f.

44. For $f(x) = \dfrac{2}{x} - \dfrac{1}{\ln x}$ find each of the following limits.
(a) $\lim_{x\to\infty} f(x)$ (b) $\lim_{x\to 0^+} f(x)$
(c) $\lim_{x\to 1^-} f(x)$ (d) $\lim_{x\to 1^+} f(x)$
(e) Use the information from parts (a)–(d) to make a rough sketch of the graph of f.

45. (a) Estimate the value of

$$\lim_{x\to-\infty} \left(\sqrt{x^2 + x + 1} + x\right)$$

by graphing the function $f(x) = \sqrt{x^2 + x + 1} + x$.
(b) Use a table of values of $f(x)$ to guess the value of the limit.
(c) Prove that your guess is correct.

46. (a) Use a graph of

$$f(x) = \sqrt{3x^2 + 8x + 6} - \sqrt{3x^2 + 3x + 1}$$

to estimate the value of $\lim_{x\to\infty} f(x)$ to one decimal place.
(b) Use a table of values of $f(x)$ to estimate the limit to four decimal places.
(c) Find the exact value of the limit.

47–52 Find the horizontal and vertical asymptotes of each curve. If you have a graphing device, check your work by graphing the curve and estimating the asymptotes.

47. $y = \dfrac{5 + 4x}{x + 3}$

48. $y = \dfrac{2x^2 + 1}{3x^2 + 2x - 1}$

49. $y = \dfrac{2x^2 + x - 1}{x^2 + x - 2}$

50. $y = \dfrac{1 + x^4}{x^2 - x^4}$

51. $y = \dfrac{x^3 - x}{x^2 - 6x + 5}$ **52.** $y = \dfrac{2e^x}{e^x - 5}$

53. Estimate the horizontal asymptote of the function
$$f(x) = \dfrac{3x^3 + 500x^2}{x^3 + 500x^2 + 100x + 2000}$$
by graphing f for $-10 \leq x \leq 10$. Then calculate the equation of the asymptote by evaluating the limit. How do you explain the discrepancy?

54. (a) Graph the function
$$f(x) = \dfrac{\sqrt{2x^2 + 1}}{3x - 5}$$
How many horizontal and vertical asymptotes do you observe? Use the graph to estimate the values of the limits
$$\lim_{x \to \infty} \dfrac{\sqrt{2x^2 + 1}}{3x - 5} \quad \text{and} \quad \lim_{x \to -\infty} \dfrac{\sqrt{2x^2 + 1}}{3x - 5}$$
(b) By calculating values of $f(x)$, give numerical estimates of the limits in part (a).
(c) Calculate the exact values of the limits in part (a). Did you get the same value or different values for these two limits? [In view of your answer to part (a), you might have to check your calculation for the second limit.]

55. Let P and Q be polynomials. Find
$$\lim_{x \to \infty} \dfrac{P(x)}{Q(x)}$$
if the degree of P is (a) less than the degree of Q and (b) greater than the degree of Q.

56. Make a rough sketch of the curve $y = x^n$ (n an integer) for the following five cases:
  (i) $n = 0$ (ii) $n > 0$, n odd
  (iii) $n > 0$, n even   (iv) $n < 0$, n odd
  (v) $n < 0$, n even

Then use these sketches to find the following limits.
(a) $\lim_{x \to 0^+} x^n$ (b) $\lim_{x \to 0^-} x^n$
(c) $\lim_{x \to \infty} x^n$ (d) $\lim_{x \to -\infty} x^n$

57. Find a formula for a function f that satisfies the following conditions:
$$\lim_{x \to \pm\infty} f(x) = 0, \quad \lim_{x \to 0} f(x) = -\infty, \quad f(2) = 0,$$
$$\lim_{x \to 3^-} f(x) = \infty, \quad \lim_{x \to 3^+} f(x) = -\infty$$

58. Find a formula for a function that has vertical asymptotes $x = 1$ and $x = 3$ and horizontal asymptote $y = 1$.

59. A function f is a ratio of quadratic functions and has a vertical asymptote $x = 4$ and just one x-intercept, $x = 1$.
It is known that f has a removable discontinuity at $x = -1$ and $\lim_{x \to -1} f(x) = 2$. Evaluate
(a) $f(0)$ (b) $\lim_{x \to \infty} f(x)$

60–64 Find the limits as $x \to \infty$ and as $x \to -\infty$. Use this information, together with intercepts, to give a rough sketch of the graph as in Example 12.

60. $y = 2x^3 - x^4$ **61.** $y = x^4 - x^6$
62. $y = x^3(x + 2)^2(x - 1)$
63. $y = (3 - x)(1 + x)^2(1 - x)^4$
64. $y = x^2(x^2 - 1)^2(x + 2)$

65. (a) Use the Squeeze Theorem to evaluate $\lim_{x \to \infty} \dfrac{\sin x}{x}$.
(b) Graph $f(x) = (\sin x)/x$. How many times does the graph cross the asymptote?

66. By the *end behavior* of a function we mean the behavior of its values as $x \to \infty$ and as $x \to -\infty$.
(a) Describe and compare the end behavior of the functions
$$P(x) = 3x^5 - 5x^3 + 2x \qquad Q(x) = 3x^5$$
by graphing both functions in the viewing rectangles $[-2, 2]$ by $[-2, 2]$ and $[-10, 10]$ by $[-10{,}000, 10{,}000]$.
(b) Two functions are said to have the *same end behavior* if their ratio approaches 1 as $x \to \infty$. Show that P and Q have the same end behavior.

67. Find $\lim_{x \to \infty} f(x)$ if, for all $x > 1$,
$$\dfrac{10e^x - 21}{2e^x} < f(x) < \dfrac{5\sqrt{x}}{\sqrt{x - 1}}$$

68. (a) A tank contains 5000 L of pure water. Brine that contains 30 g of salt per liter of water is pumped into the tank at a rate of 25 L/min. Show that the concentration of salt after t minutes (in grams per liter) is
$$C(t) = \dfrac{30t}{200 + t}$$
(b) What happens to the concentration as $t \to \infty$?

69. In Chapter 9 we will be able to show, under certain assumptions, that the velocity $v(t)$ of a falling raindrop at time t is
$$v(t) = v^*(1 - e^{-gt/v^*})$$
where g is the acceleration due to gravity and v^* is the *terminal velocity* of the raindrop.
(a) Find $\lim_{t \to \infty} v(t)$.
(b) Graph $v(t)$ if $v^* = 1$ m/s and $g = 9.8$ m/s². How long does it take for the velocity of the raindrop to reach 99% of its terminal velocity?

70. (a) By graphing $y = e^{-x/10}$ and $y = 0.1$ on a common screen, discover how large you need to make x so that $e^{-x/10} < 0.1$.
(b) Can you solve part (a) without using a graphing device?

71. Use a graph to find a number N such that

if $\quad x > N \quad$ then $\quad \left| \dfrac{3x^2 + 1}{2x^2 + x + 1} - 1.5 \right| < 0.05$

72. For the limit

$$\lim_{x \to \infty} \frac{1 - 3x}{\sqrt{x^2 + 1}} = -3$$

illustrate Definition 7 by finding values of N that correspond to $\varepsilon = 0.1$ and $\varepsilon = 0.05$.

73. For the limit

$$\lim_{x \to -\infty} \frac{1 - 3x}{\sqrt{x^2 + 1}} = 3$$

illustrate Definition 8 by finding values of N that correspond to $\varepsilon = 0.1$ and $\varepsilon = 0.05$.

74. For the limit

$$\lim_{x \to \infty} \sqrt{x \ln x} = \infty$$

illustrate Definition 9 by finding a value of N that corresponds to $M = 100$.

75. (a) How large do we have to take x so that $1/x^2 < 0.0001$?
(b) Taking $r = 2$ in Theorem 5, we have the statement

$$\lim_{x \to \infty} \frac{1}{x^2} = 0$$

Prove this directly using Definition 7.

76. (a) How large do we have to take x so that $1/\sqrt{x} < 0.0001$?
(b) Taking $r = \tfrac{1}{2}$ in Theorem 5, we have the statement

$$\lim_{x \to \infty} \frac{1}{\sqrt{x}} = 0$$

Prove this directly using Definition 7.

77. Use Definition 8 to prove that $\displaystyle\lim_{x \to -\infty} \frac{1}{x} = 0$.

78. Prove, using Definition 9, that $\displaystyle\lim_{x \to \infty} x^3 = \infty$.

79. Use Definition 9 to prove that $\displaystyle\lim_{x \to \infty} e^x = \infty$.

80. Formulate a precise definition of

$$\lim_{x \to -\infty} f(x) = -\infty$$

Then use your definition to prove that

$$\lim_{x \to -\infty} (1 + x^3) = -\infty$$

81. (a) Prove that

$$\lim_{x \to \infty} f(x) = \lim_{t \to 0^+} f(1/t)$$

and

$$\lim_{x \to -\infty} f(x) = \lim_{t \to 0^-} f(1/t)$$

if these limits exist.
(b) Use part (a) and Exercise 65 to find

$$\lim_{x \to 0^+} x \sin \frac{1}{x}$$

2.7 Derivatives and Rates of Change

The problem of finding the tangent line to a curve and the problem of finding the velocity of an object both involve finding the same type of limit, as we saw in Section 2.1. This special type of limit is called a *derivative* and we will see that it can be interpreted as a rate of change in any of the natural or social sciences or engineering.

■ Tangents

If a curve C has equation $y = f(x)$ and we want to find the tangent line to C at the point $P(a, f(a))$, then we consider a nearby point $Q(x, f(x))$, where $x \neq a$, and compute the slope of the secant line PQ:

$$m_{PQ} = \frac{f(x) - f(a)}{x - a}$$

Then we let Q approach P along the curve C by letting x approach a. If m_{PQ} approaches

SECTION 2.7 Derivatives and Rates of Change   **141**

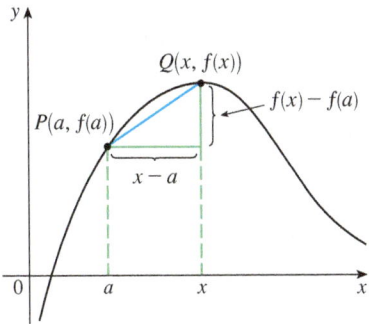

a number m, then we define the *tangent* t to be the line through P with slope m. (This amounts to saying that the tangent line is the limiting position of the secant line PQ as Q approaches P. See Figure 1.)

> **1 Definition** The **tangent line** to the curve $y = f(x)$ at the point $P(a, f(a))$ is the line through P with slope
> $$m = \lim_{x \to a} \frac{f(x) - f(a)}{x - a}$$
> provided that this limit exists.

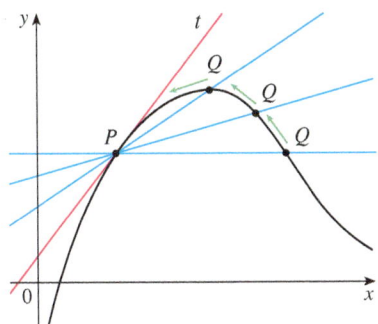

In our first example we confirm the guess we made in Example 2.1.1.

EXAMPLE 1 Find an equation of the tangent line to the parabola $y = x^2$ at the point $P(1, 1)$.

SOLUTION Here we have $a = 1$ and $f(x) = x^2$, so the slope is

$$m = \lim_{x \to 1} \frac{f(x) - f(1)}{x - 1} = \lim_{x \to 1} \frac{x^2 - 1}{x - 1}$$

$$= \lim_{x \to 1} \frac{(x - 1)(x + 1)}{x - 1}$$

$$= \lim_{x \to 1} (x + 1) = 1 + 1 = 2$$

FIGURE 1

Point-slope form for a line through the point (x_1, y_1) with slope m:
$$y - y_1 = m(x - x_1)$$

Using the point-slope form of the equation of a line, we find that an equation of the tangent line at $(1, 1)$ is

$$y - 1 = 2(x - 1) \quad \text{or} \quad y = 2x - 1 \quad \blacksquare$$

TEC Visual 2.7 shows an animation of Figure 2.

We sometimes refer to the slope of the tangent line to a curve at a point as the **slope of the curve** at the point. The idea is that if we zoom in far enough toward the point, the curve looks almost like a straight line. Figure 2 illustrates this procedure for the curve $y = x^2$ in Example 1. The more we zoom in, the more the parabola looks like a line. In other words, the curve becomes almost indistinguishable from its tangent line.

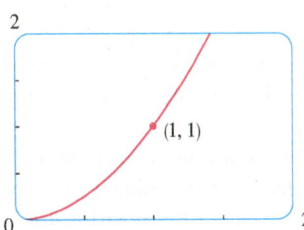 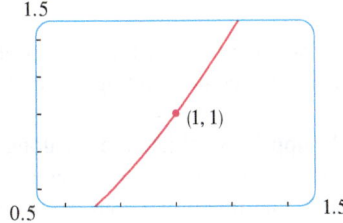 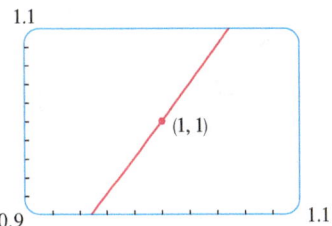

FIGURE 2 Zooming in toward the point $(1, 1)$ on the parabola $y = x^2$

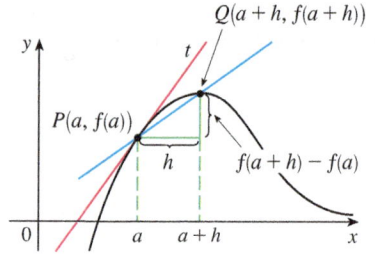

FIGURE 3

There is another expression for the slope of a tangent line that is sometimes easier to use. If $h = x - a$, then $x = a + h$ and so the slope of the secant line PQ is

$$m_{PQ} = \frac{f(a+h) - f(a)}{h}$$

(See Figure 3 where the case $h > 0$ is illustrated and Q is to the right of P. If it happened that $h < 0$, however, Q would be to the left of P.)

Notice that as x approaches a, h approaches 0 (because $h = x - a$) and so the expression for the slope of the tangent line in Definition 1 becomes

$$\boxed{2 \qquad m = \lim_{h \to 0} \frac{f(a+h) - f(a)}{h}}$$

EXAMPLE 2 Find an equation of the tangent line to the hyperbola $y = 3/x$ at the point $(3, 1)$.

SOLUTION Let $f(x) = 3/x$. Then, by Equation 2, the slope of the tangent at $(3, 1)$ is

$$m = \lim_{h \to 0} \frac{f(3+h) - f(3)}{h}$$

$$= \lim_{h \to 0} \frac{\dfrac{3}{3+h} - 1}{h} = \lim_{h \to 0} \dfrac{\dfrac{3 - (3+h)}{3+h}}{h}$$

$$= \lim_{h \to 0} \frac{-h}{h(3+h)} = \lim_{h \to 0} -\frac{1}{3+h} = -\frac{1}{3}$$

Therefore an equation of the tangent at the point $(3, 1)$ is

$$y - 1 = -\tfrac{1}{3}(x - 3)$$

which simplifies to $\qquad x + 3y - 6 = 0$

The hyperbola and its tangent are shown in Figure 4.

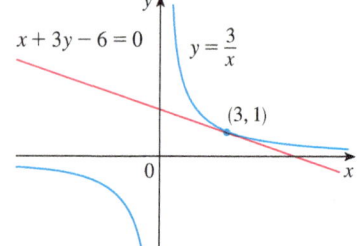

FIGURE 4

■ Velocities

In Section 2.1 we investigated the motion of a ball dropped from the CN Tower and defined its velocity to be the limiting value of average velocities over shorter and shorter time periods.

In general, suppose an object moves along a straight line according to an equation of motion $s = f(t)$, where s is the displacement (directed distance) of the object from the origin at time t. The function f that describes the motion is called the **position function** of the object. In the time interval from $t = a$ to $t = a + h$ the change in position is $f(a + h) - f(a)$. (See Figure 5.)

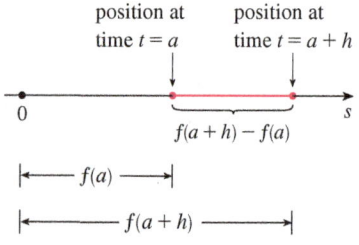

FIGURE 5

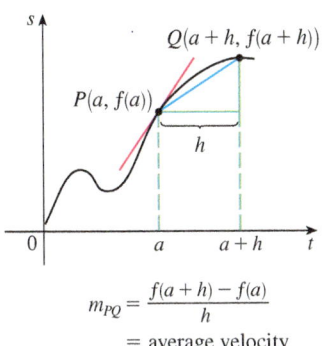

$$m_{PQ} = \frac{f(a+h) - f(a)}{h}$$
= average velocity

FIGURE 6

The average velocity over this time interval is

$$\text{average velocity} = \frac{\text{displacement}}{\text{time}} = \frac{f(a+h) - f(a)}{h}$$

which is the same as the slope of the secant line PQ in Figure 6.

Now suppose we compute the average velocities over shorter and shorter time intervals $[a, a + h]$. In other words, we let h approach 0. As in the example of the falling ball, we define the **velocity** (or **instantaneous velocity**) $v(a)$ at time $t = a$ to be the limit of these average velocities:

$$\boxed{3 \quad v(a) = \lim_{h \to 0} \frac{f(a+h) - f(a)}{h}}$$

This means that the velocity at time $t = a$ is equal to the slope of the tangent line at P (compare Equations 2 and 3).

Now that we know how to compute limits, let's reconsider the problem of the falling ball.

EXAMPLE 3 Suppose that a ball is dropped from the upper observation deck of the CN Tower, 450 m above the ground.
(a) What is the velocity of the ball after 5 seconds?
(b) How fast is the ball traveling when it hits the ground?

Recall from Section 2.1: The distance (in meters) fallen after t seconds is $4.9t^2$.

SOLUTION We will need to find the velocity both when $t = 5$ and when the ball hits the ground, so it's efficient to start by finding the velocity at a general time t. Using the equation of motion $s = f(t) = 4.9t^2$, we have

$$v(t) = \lim_{h \to 0} \frac{f(t+h) - f(t)}{h} = \lim_{h \to 0} \frac{4.9(t+h)^2 - 4.9t^2}{h}$$

$$= \lim_{h \to 0} \frac{4.9(t^2 + 2th + h^2 - t^2)}{h} = \lim_{h \to 0} \frac{4.9(2th + h^2)}{h}$$

$$= \lim_{h \to 0} \frac{4.9h(2t + h)}{h} = \lim_{h \to 0} 4.9(2t + h) = 9.8t$$

(a) The velocity after 5 seconds is $v(5) = (9.8)(5) = 49$ m/s.

(b) Since the observation deck is 450 m above the ground, the ball will hit the ground at the time t when $s(t) = 450$, that is,

$$4.9t^2 = 450$$

This gives

$$t^2 = \frac{450}{4.9} \quad \text{and} \quad t = \sqrt{\frac{450}{4.9}} \approx 9.6 \text{ s}$$

The velocity of the ball as it hits the ground is therefore

$$v\left(\sqrt{\frac{450}{4.9}}\right) = 9.8\sqrt{\frac{450}{4.9}} \approx 94 \text{ m/s}$$

■ Derivatives

We have seen that the same type of limit arises in finding the slope of a tangent line (Equation 2) or the velocity of an object (Equation 3). In fact, limits of the form

$$\lim_{h \to 0} \frac{f(a+h) - f(a)}{h}$$

arise whenever we calculate a rate of change in any of the sciences or engineering, such as a rate of reaction in chemistry or a marginal cost in economics. Since this type of limit occurs so widely, it is given a special name and notation.

> **4 Definition** The **derivative of a function f at a number a**, denoted by $f'(a)$, is
>
> $$f'(a) = \lim_{h \to 0} \frac{f(a+h) - f(a)}{h}$$
>
> if this limit exists.

$f'(a)$ is read "f prime of a."

If we write $x = a + h$, then we have $h = x - a$ and h approaches 0 if and only if x approaches a. Therefore an equivalent way of stating the definition of the derivative, as we saw in finding tangent lines, is

> **5**
> $$f'(a) = \lim_{x \to a} \frac{f(x) - f(a)}{x - a}$$

EXAMPLE 4 Find the derivative of the function $f(x) = x^2 - 8x + 9$ at the number a.

SOLUTION From Definition 4 we have

Definitions 4 and 5 are equivalent, so we can use either one to compute the derivative. In practice, Definition 4 often leads to simpler computations.

$$f'(a) = \lim_{h \to 0} \frac{f(a+h) - f(a)}{h}$$

$$= \lim_{h \to 0} \frac{[(a+h)^2 - 8(a+h) + 9] - [a^2 - 8a + 9]}{h}$$

$$= \lim_{h \to 0} \frac{a^2 + 2ah + h^2 - 8a - 8h + 9 - a^2 + 8a - 9}{h}$$

$$= \lim_{h \to 0} \frac{2ah + h^2 - 8h}{h} = \lim_{h \to 0} (2a + h - 8)$$

$$= 2a - 8$$

We defined the tangent line to the curve $y = f(x)$ at the point $P(a, f(a))$ to be the line that passes through P and has slope m given by Equation 1 or 2. Since, by Definition 4, this is the same as the derivative $f'(a)$, we can now say the following.

> The tangent line to $y = f(x)$ at $(a, f(a))$ is the line through $(a, f(a))$ whose slope is equal to $f'(a)$, the derivative of f at a.

If we use the point-slope form of the equation of a line, we can write an equation of the tangent line to the curve $y = f(x)$ at the point $(a, f(a))$:

$$y - f(a) = f'(a)(x - a)$$

EXAMPLE 5 Find an equation of the tangent line to the parabola $y = x^2 - 8x + 9$ at the point $(3, -6)$.

SOLUTION From Example 4 we know that the derivative of $f(x) = x^2 - 8x + 9$ at the number a is $f'(a) = 2a - 8$. Therefore the slope of the tangent line at $(3, -6)$ is $f'(3) = 2(3) - 8 = -2$. Thus an equation of the tangent line, shown in Figure 7, is

$$y - (-6) = (-2)(x - 3) \quad \text{or} \quad y = -2x$$

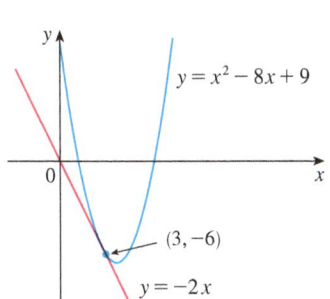

FIGURE 7

Rates of Change

Suppose y is a quantity that depends on another quantity x. Thus y is a function of x and we write $y = f(x)$. If x changes from x_1 to x_2, then the change in x (also called the **increment** of x) is

$$\Delta x = x_2 - x_1$$

and the corresponding change in y is

$$\Delta y = f(x_2) - f(x_1)$$

The difference quotient

$$\frac{\Delta y}{\Delta x} = \frac{f(x_2) - f(x_1)}{x_2 - x_1}$$

is called the **average rate of change of y with respect to x** over the interval $[x_1, x_2]$ and can be interpreted as the slope of the secant line PQ in Figure 8.

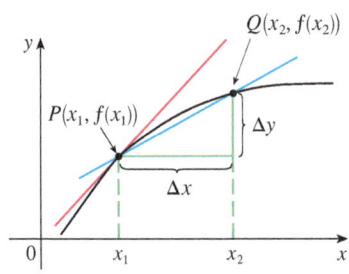

average rate of change $= m_{PQ}$

instantaneous rate of change $=$ slope of tangent at P

FIGURE 8

By analogy with velocity, we consider the average rate of change over smaller and smaller intervals by letting x_2 approach x_1 and therefore letting Δx approach 0. The limit of these average rates of change is called the **(instantaneous) rate of change of y with respect to x** at $x = x_1$, which (as in the case of velocity) is interpreted as the slope of the tangent to the curve $y = f(x)$ at $P(x_1, f(x_1))$:

> **6** $\quad$ instantaneous rate of change $= \lim\limits_{\Delta x \to 0} \dfrac{\Delta y}{\Delta x} = \lim\limits_{x_2 \to x_1} \dfrac{f(x_2) - f(x_1)}{x_2 - x_1}$

We recognize this limit as being the derivative $f'(x_1)$.

We know that one interpretation of the derivative $f'(a)$ is as the slope of the tangent line to the curve $y = f(x)$ when $x = a$. We now have a second interpretation:

> The derivative $f'(a)$ is the instantaneous rate of change of $y = f(x)$ with respect to x when $x = a$.

The connection with the first interpretation is that if we sketch the curve $y = f(x)$, then the instantaneous rate of change is the slope of the tangent to this curve at the point where $x = a$. This means that when the derivative is large (and therefore the curve is steep, as at the point P in Figure 9), the y-values change rapidly. When the derivative is small, the curve is relatively flat (as at point Q) and the y-values change slowly.

In particular, if $s = f(t)$ is the position function of a particle that moves along a straight line, then $f'(a)$ is the rate of change of the displacement s with respect to the time t. In other words, $f'(a)$ *is the velocity of the particle at time $t = a$*. The **speed** of the particle is the absolute value of the velocity, that is, $|f'(a)|$.

In the next example we discuss the meaning of the derivative of a function that is defined verbally.

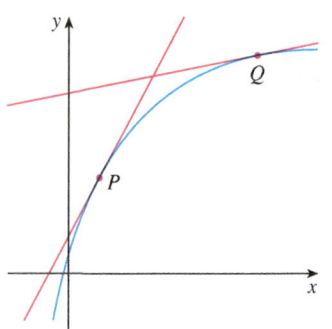

FIGURE 9
The y-values are changing rapidly at P and slowly at Q.

EXAMPLE 6 A manufacturer produces bolts of a fabric with a fixed width. The cost of producing x yards of this fabric is $C = f(x)$ dollars.
(a) What is the meaning of the derivative $f'(x)$? What are its units?
(b) In practical terms, what does it mean to say that $f'(1000) = 9$?
(c) Which do you think is greater, $f'(50)$ or $f'(500)$? What about $f'(5000)$?

SOLUTION
(a) The derivative $f'(x)$ is the instantaneous rate of change of C with respect to x; that is, $f'(x)$ means the rate of change of the production cost with respect to the number of yards produced. (Economists call this rate of change the *marginal cost*. This idea is discussed in more detail in Sections 3.7 and 4.7.)
Because
$$f'(x) = \lim_{\Delta x \to 0} \frac{\Delta C}{\Delta x}$$

the units for $f'(x)$ are the same as the units for the difference quotient $\Delta C/\Delta x$. Since ΔC is measured in dollars and Δx in yards, it follows that the units for $f'(x)$ are dollars per yard.

(b) The statement that $f'(1000) = 9$ means that, after 1000 yards of fabric have been manufactured, the rate at which the production cost is increasing is \$9/yard. (When $x = 1000$, C is increasing 9 times as fast as x.)

Since $\Delta x = 1$ is small compared with $x = 1000$, we could use the approximation

$$f'(1000) \approx \frac{\Delta C}{\Delta x} = \frac{\Delta C}{1} = \Delta C$$

Here we are assuming that the cost function is well behaved; in other words, $C(x)$ doesn't oscillate rapidly near $x = 1000$.

and say that the cost of manufacturing the 1000th yard (or the 1001st) is about \$9.

(c) The rate at which the production cost is increasing (per yard) is probably lower when $x = 500$ than when $x = 50$ (the cost of making the 500th yard is less than the cost of the 50th yard) because of economies of scale. (The manufacturer makes more efficient use of the fixed costs of production.) So

$$f'(50) > f'(500)$$

But, as production expands, the resulting large-scale operation might become inefficient and there might be overtime costs. Thus it is possible that the rate of increase of costs will eventually start to rise. So it may happen that

$$f'(5000) > f'(500)$$ ∎

In the following example we estimate the rate of change of the national debt with respect to time. Here the function is defined not by a formula but by a table of values.

EXAMPLE 7 Let $D(t)$ be the US national debt at time t. The table in the margin gives approximate values of this function by providing end of year estimates, in billions of dollars, from 1985 to 2010. Interpret and estimate the value of $D'(2000)$.

SOLUTION The derivative $D'(2000)$ means the rate of change of D with respect to t when $t = 2000$, that is, the rate of increase of the national debt in 2000.

According to Equation 5,

$$D'(2000) = \lim_{t \to 2000} \frac{D(t) - D(2000)}{t - 2000}$$

So we compute and tabulate values of the difference quotient (the average rates of change) as follows.

t	D(t)
1985	1945.9
1990	3364.8
1995	4988.7
2000	5662.2
2005	8170.4
2010	14,025.2

Source: US Dept. of the Treasury

t	Time interval	Average rate of change $= \dfrac{D(t) - D(2000)}{t - 2000}$
1985	[1985, 2000]	247.75
1990	[1990, 2000]	229.74
1995	[1995, 2000]	134.70
2005	[2000, 2005]	501.64
2010	[2000, 2010]	836.30

From this table we see that $D'(2000)$ lies somewhere between 134.70 and 501.64 billion dollars per year. [Here we are making the reasonable assumption that the debt didn't fluctuate wildly between 1995 and 2005.] We estimate that the rate of increase of the national debt of the United States in 2000 was the average of these two numbers, namely

$$D'(2000) \approx 318 \text{ billion dollars per year}$$

Another method would be to plot the debt function and estimate the slope of the tangent line when $t = 2000$. ∎

A Note On Units

The units for the average rate of change $\Delta D/\Delta t$ are the units for ΔD divided by the units for Δt, namely billions of dollars per year. The instantaneous rate of change is the limit of the average rates of change, so it is measured in the same units: billions of dollars per year.

In Examples 3, 6, and 7 we saw three specific examples of rates of change: the velocity of an object is the rate of change of displacement with respect to time; marginal cost is the rate of change of production cost with respect to the number of items produced; the rate of change of the debt with respect to time is of interest in economics. Here is a small sample of other rates of change: In physics, the rate of change of work with respect to time is called *power*. Chemists who study a chemical reaction are interested in the rate of change in the concentration of a reactant with respect to time (called the *rate of reaction*).

A biologist is interested in the rate of change of the population of a colony of bacteria with respect to time. In fact, the computation of rates of change is important in all of the natural sciences, in engineering, and even in the social sciences. Further examples will be given in Section 3.7.

All these rates of change are derivatives and can therefore be interpreted as slopes of tangents. This gives added significance to the solution of the tangent problem. Whenever we solve a problem involving tangent lines, we are not just solving a problem in geometry. We are also implicitly solving a great variety of problems involving rates of change in science and engineering.

2.7 EXERCISES

1. A curve has equation $y = f(x)$.
   (a) Write an expression for the slope of the secant line through the points $P(3, f(3))$ and $Q(x, f(x))$.
   (b) Write an expression for the slope of the tangent line at P.

2. Graph the curve $y = e^x$ in the viewing rectangles $[-1, 1]$ by $[0, 2]$, $[-0.5, 0.5]$ by $[0.5, 1.5]$, and $[-0.1, 0.1]$ by $[0.9, 1.1]$. What do you notice about the curve as you zoom in toward the point $(0, 1)$?

3. (a) Find the slope of the tangent line to the parabola $y = 4x - x^2$ at the point $(1, 3)$
   (i) using Definition 1   (ii) using Equation 2
   (b) Find an equation of the tangent line in part (a).
   (c) Graph the parabola and the tangent line. As a check on your work, zoom in toward the point $(1, 3)$ until the parabola and the tangent line are indistinguishable.

4. (a) Find the slope of the tangent line to the curve $y = x - x^3$ at the point $(1, 0)$
   (i) using Definition 1   (ii) using Equation 2
   (b) Find an equation of the tangent line in part (a).
   (c) Graph the curve and the tangent line in successively smaller viewing rectangles centered at $(1, 0)$ until the curve and the line appear to coincide.

5–8 Find an equation of the tangent line to the curve at the given point.

5. $y = 4x - 3x^2$, $(2, -4)$
6. $y = x^3 - 3x + 1$, $(2, 3)$
7. $y = \sqrt{x}$, $(1, 1)$
8. $y = \dfrac{2x + 1}{x + 2}$, $(1, 1)$

9. (a) Find the slope of the tangent to the curve $y = 3 + 4x^2 - 2x^3$ at the point where $x = a$.
   (b) Find equations of the tangent lines at the points $(1, 5)$ and $(2, 3)$.
   (c) Graph the curve and both tangents on a common screen.

10. (a) Find the slope of the tangent to the curve $y = 1/\sqrt{x}$ at the point where $x = a$.
    (b) Find equations of the tangent lines at the points $(1, 1)$ and $(4, \tfrac{1}{2})$.
    (c) Graph the curve and both tangents on a common screen.

11. (a) A particle starts by moving to the right along a horizontal line; the graph of its position function is shown in the figure. When is the particle moving to the right? Moving to the left? Standing still?
    (b) Draw a graph of the velocity function.

12. Shown are graphs of the position functions of two runners, A and B, who run a 100-meter race and finish in a tie.

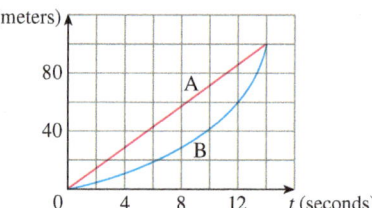

    (a) Describe and compare how the runners run the race.
    (b) At what time is the distance between the runners the greatest?
    (c) At what time do they have the same velocity?

13. If a ball is thrown into the air with a velocity of 40 ft/s, its height (in feet) after t seconds is given by $y = 40t - 16t^2$. Find the velocity when $t = 2$.

14. If a rock is thrown upward on the planet Mars with a velocity of 10 m/s, its height (in meters) after t seconds is given by $H = 10t - 1.86t^2$.
(a) Find the velocity of the rock after one second.
(b) Find the velocity of the rock when $t = a$.
(c) When will the rock hit the surface?
(d) With what velocity will the rock hit the surface?

15. The displacement (in meters) of a particle moving in a straight line is given by the equation of motion $s = 1/t^2$, where t is measured in seconds. Find the velocity of the particle at times $t = a$, $t = 1$, $t = 2$, and $t = 3$.

16. The displacement (in feet) of a particle moving in a straight line is given by $s = \frac{1}{2}t^2 - 6t + 23$, where t is measured in seconds.
(a) Find the average velocity over each time interval:
  (i) [4, 8]   (ii) [6, 8]
  (iii) [8, 10]   (iv) [8, 12]
(b) Find the instantaneous velocity when $t = 8$.
(c) Draw the graph of s as a function of t and draw the secant lines whose slopes are the average velocities in part (a). Then draw the tangent line whose slope is the instantaneous velocity in part (b).

17. For the function g whose graph is given, arrange the following numbers in increasing order and explain your reasoning:

$$0 \quad g'(-2) \quad g'(0) \quad g'(2) \quad g'(4)$$

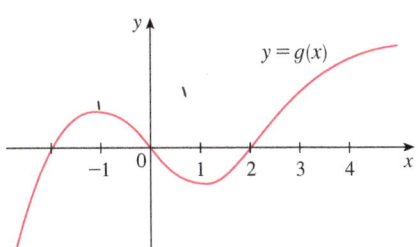

18. The graph of a function f is shown.
(a) Find the average rate of change of f on the interval [20, 60].
(b) Identify an interval on which the average rate of change of f is 0.
(c) Which interval gives a larger average rate of change, [40, 60] or [40, 70]?
(d) Compute $\dfrac{f(40) - f(10)}{40 - 10}$; what does this value represent geometrically?

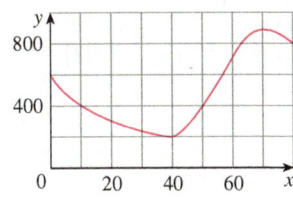

19. For the function f graphed in Exercise 18:
(a) Estimate the value of $f'(50)$.
(b) Is $f'(10) > f'(30)$?
(c) Is $f'(60) > \dfrac{f(80) - f(40)}{80 - 40}$? Explain.

20. Find an equation of the tangent line to the graph of $y = g(x)$ at $x = 5$ if $g(5) = -3$ and $g'(5) = 4$.

21. If an equation of the tangent line to the curve $y = f(x)$ at the point where $a = 2$ is $y = 4x - 5$, find $f(2)$ and $f'(2)$.

22. If the tangent line to $y = f(x)$ at (4, 3) passes through the point (0, 2), find $f(4)$ and $f'(4)$.

23. Sketch the graph of a function f for which $f(0) = 0$, $f'(0) = 3$, $f'(1) = 0$, and $f'(2) = -1$.

24. Sketch the graph of a function g for which
$g(0) = g(2) = g(4) = 0$, $g'(1) = g'(3) = 0$,
$g'(0) = g'(4) = 1$, $g'(2) = -1$, $\lim_{x \to \infty} g(x) = \infty$, and $\lim_{x \to -\infty} g(x) = -\infty$.

25. Sketch the graph of a function g that is continuous on its domain $(-5, 5)$ and where $g(0) = 1$, $g'(0) = 1$, $g'(-2) = 0$, $\lim_{x \to -5^+} g(x) = \infty$, and $\lim_{x \to 5^-} g(x) = 3$.

26. Sketch the graph of a function f where the domain is $(-2, 2)$, $f'(0) = -2$, $\lim_{x \to 2^-} f(x) = \infty$, f is continuous at all numbers in its domain except ± 1, and f is odd.

27. If $f(x) = 3x^2 - x^3$, find $f'(1)$ and use it to find an equation of the tangent line to the curve $y = 3x^2 - x^3$ at the point (1, 2).

28. If $g(x) = x^4 - 2$, find $g'(1)$ and use it to find an equation of the tangent line to the curve $y = x^4 - 2$ at the point (1, −1).

29. (a) If $F(x) = 5x/(1 + x^2)$, find $F'(2)$ and use it to find an equation of the tangent line to the curve $y = 5x/(1 + x^2)$ at the point (2, 2).
(b) Illustrate part (a) by graphing the curve and the tangent line on the same screen.

30. (a) If $G(x) = 4x^2 - x^3$, find $G'(a)$ and use it to find equations of the tangent lines to the curve $y = 4x^2 - x^3$ at the points (2, 8) and (3, 9).
(b) Illustrate part (a) by graphing the curve and the tangent lines on the same screen.

31–36 Find $f'(a)$.

31. $f(x) = 3x^2 - 4x + 1$

32. $f(t) = 2t^3 + t$

33. $f(t) = \dfrac{2t + 1}{t + 3}$

34. $f(x) = x^{-2}$

35. $f(x) = \sqrt{1 - 2x}$

36. $f(x) = \dfrac{4}{\sqrt{1 - x}}$

37–42 Each limit represents the derivative of some function f at some number a. State such an f and a in each case.

37. $\lim\limits_{h \to 0} \dfrac{\sqrt{9 + h} - 3}{h}$

38. $\lim\limits_{h \to 0} \dfrac{e^{-2+h} - e^{-2}}{h}$

39. $\lim\limits_{x \to 2} \dfrac{x^6 - 64}{x - 2}$

40. $\lim\limits_{x \to 1/4} \dfrac{\frac{1}{x} - 4}{x - \frac{1}{4}}$

41. $\lim\limits_{h \to 0} \dfrac{\cos(\pi + h) + 1}{h}$

42. $\lim\limits_{\theta \to \pi/6} \dfrac{\sin\theta - \frac{1}{2}}{\theta - \pi/6}$

43–44 A particle moves along a straight line with equation of motion $s = f(t)$, where s is measured in meters and t in seconds. Find the velocity and the speed when $t = 4$.

43. $f(t) = 80t - 6t^2$

44. $f(t) = 10 + \dfrac{45}{t + 1}$

45. A warm can of soda is placed in a cold refrigerator. Sketch the graph of the temperature of the soda as a function of time. Is the initial rate of change of temperature greater or less than the rate of change after an hour?

46. A roast turkey is taken from an oven when its temperature has reached 185°F and is placed on a table in a room where the temperature is 75°F. The graph shows how the temperature of the turkey decreases and eventually approaches room temperature. By measuring the slope of the tangent, estimate the rate of change of the temperature after an hour.

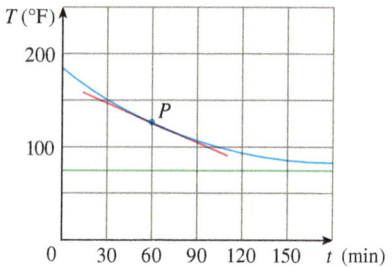

47. Researchers measured the average blood alcohol concentration $C(t)$ of eight men starting one hour after consumption of 30 mL of ethanol (corresponding to two alcoholic drinks).

t (hours)	1.0	1.5	2.0	2.5	3.0
$C(t)$ (mg/mL)	0.33	0.24	0.18	0.12	0.07

(a) Find the average rate of change of C with respect to t over each time interval:
   (i) [1.0, 2.0]   (ii) [1.5, 2.0]
   (iii) [2.0, 2.5]   (iv) [2.0, 3.0]
   In each case, include the units.
(b) Estimate the instantaneous rate of change at $t = 2$ and interpret your result. What are the units?

Source: Adapted from P. Wilkinson et al., "Pharmacokinetics of Ethanol after Oral Administration in the Fasting State," *Journal of Pharmacokinetics and Biopharmaceutics* 5 (1977): 207–24.

48. The number N of locations of a popular coffeehouse chain is given in the table. (The numbers of locations as of October 1 are given.)

Year	2004	2006	2008	2010	2012
N	8569	12,440	16,680	16,858	18,066

(a) Find the average rate of growth
   (i) from 2006 to 2008
   (ii) from 2008 to 2010
   In each case, include the units. What can you conclude?
(b) Estimate the instantaneous rate of growth in 2010 by taking the average of two average rates of change. What are its units?
(c) Estimate the instantaneous rate of growth in 2010 by measuring the slope of a tangent.

49. The table shows world average daily oil consumption from 1985 to 2010 measured in thousands of barrels per day.
(a) Compute and interpret the average rate of change from 1990 to 2005. What are the units?
(b) Estimate the instantaneous rate of change in 2000 by taking the average of two average rates of change. What are its units?

Years since 1985	Thousands of barrels of oil per day
0	60,083
5	66,533
10	70,099
15	76,784
20	84,077
25	87,302

Source: US Energy Information Administration

50. The table shows values of the viral load $V(t)$ in HIV patient 303, measured in RNA copies/mL, t days after ABT-538 treatment was begun.

t	4	8	11	15	22
$V(t)$	53	18	9.4	5.2	3.6

(a) Find the average rate of change of V with respect to t over each time interval:
   (i) [4, 11]   (ii) [8, 11]
   (iii) [11, 15]   (iv) [11, 22]
   What are the units?
(b) Estimate and interpret the value of the derivative $V'(11)$.

Source: Adapted from D. Ho et al., "Rapid Turnover of Plasma Virions and CD4 Lymphocytes in HIV-1 Infection," *Nature* 373 (1995): 123–26.

51. The cost (in dollars) of producing x units of a certain commodity is $C(x) = 5000 + 10x + 0.05x^2$.
  (a) Find the average rate of change of C with respect to x when the production level is changed
    (i) from $x = 100$ to $x = 105$
    (ii) from $x = 100$ to $x = 101$
  (b) Find the instantaneous rate of change of C with respect to x when $x = 100$. (This is called the *marginal cost*. Its significance will be explained in Section 3.7.)

52. If a cylindrical tank holds 100,000 gallons of water, which can be drained from the bottom of the tank in an hour, then Torricelli's Law gives the volume V of water remaining in the tank after t minutes as

$$V(t) = 100,000\left(1 - \tfrac{1}{60}t\right)^2 \qquad 0 \le t \le 60$$

Find the rate at which the water is flowing out of the tank (the instantaneous rate of change of V with respect to t) as a function of t. What are its units? For times $t = 0, 10, 20, 30, 40, 50,$ and 60 min, find the flow rate and the amount of water remaining in the tank. Summarize your findings in a sentence or two. At what time is the flow rate the greatest? The least?

53. The cost of producing x ounces of gold from a new gold mine is $C = f(x)$ dollars.
  (a) What is the meaning of the derivative $f'(x)$? What are its units?
  (b) What does the statement $f'(800) = 17$ mean?
  (c) Do you think the values of $f'(x)$ will increase or decrease in the short term? What about the long term? Explain.

54. The number of bacteria after t hours in a controlled laboratory experiment is $n = f(t)$.
  (a) What is the meaning of the derivative $f'(5)$? What are its units?
  (b) Suppose there is an unlimited amount of space and nutrients for the bacteria. Which do you think is larger, $f'(5)$ or $f'(10)$? If the supply of nutrients is limited, would that affect your conclusion? Explain.

55. Let $H(t)$ be the daily cost (in dollars) to heat an office building when the outside temperature is t degrees Fahrenheit.
  (a) What is the meaning of $H'(58)$? What are its units?
  (b) Would you expect $H'(58)$ to be positive or negative? Explain.

56. The quantity (in pounds) of a gourmet ground coffee that is sold by a coffee company at a price of p dollars per pound is $Q = f(p)$.
  (a) What is the meaning of the derivative $f'(8)$? What are its units?
  (b) Is $f'(8)$ positive or negative? Explain.

57. The quantity of oxygen that can dissolve in water depends on the temperature of the water. (So thermal pollution influences the oxygen content of water.) The graph shows how oxygen solubility S varies as a function of the water temperature T.
  (a) What is the meaning of the derivative $S'(T)$? What are its units?
  (b) Estimate the value of $S'(16)$ and interpret it.

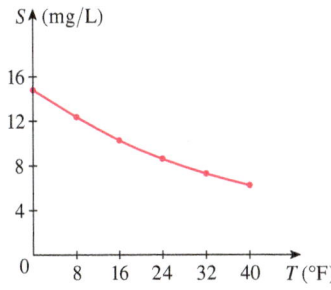

Source: C. Kupchella et al., *Environmental Science: Living Within the System of Nature*, 2d ed. (Boston: Allyn and Bacon, 1989).

58. The graph shows the influence of the temperature T on the maximum sustainable swimming speed S of Coho salmon.
  (a) What is the meaning of the derivative $S'(T)$? What are its units?
  (b) Estimate the values of $S'(15)$ and $S'(25)$ and interpret them.

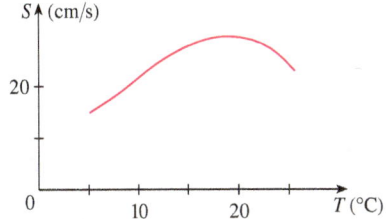

59–60 Determine whether $f'(0)$ exists.

59. $f(x) = \begin{cases} x \sin \dfrac{1}{x} & \text{if } x \ne 0 \\ 0 & \text{if } x = 0 \end{cases}$

60. $f(x) = \begin{cases} x^2 \sin \dfrac{1}{x} & \text{if } x \ne 0 \\ 0 & \text{if } x = 0 \end{cases}$

61. (a) Graph the function $f(x) = \sin x - \tfrac{1}{1000}\sin(1000x)$ in the viewing rectangle $[-2\pi, 2\pi]$ by $[-4, 4]$. What slope does the graph appear to have at the origin?
  (b) Zoom in to the viewing window $[-0.4, 0.4]$ by $[-0.25, 0.25]$ and estimate the value of $f'(0)$. Does this agree with your answer from part (a)?
  (c) Now zoom in to the viewing window $[-0.008, 0.008]$ by $[-0.005, 0.005]$. Do you wish to revise your estimate for $f'(0)$?

WRITING PROJECT EARLY METHODS FOR FINDING TANGENTS

The first person to formulate explicitly the ideas of limits and derivatives was Sir Isaac Newton in the 1660s. But Newton acknowledged that "If I have seen further than other men, it is because I have stood on the shoulders of giants." Two of those giants were Pierre Fermat (1601–1665) and Newton's mentor at Cambridge, Isaac Barrow (1630–1677). Newton was familiar with the methods that these men used to find tangent lines, and their methods played a role in Newton's eventual formulation of calculus.

The following references contain explanations of these methods. Read one or more of the references and write a report comparing the methods of either Fermat or Barrow to modern methods. In particular, use the method of Section 2.7 to find an equation of the tangent line to the curve $y = x^3 + 2x$ at the point (1, 3) and show how either Fermat or Barrow would have solved the same problem. Although you used derivatives and they did not, point out similarities between the methods.

1. Carl Boyer and Uta Merzbach, *A History of Mathematics* (New York: Wiley, 1989), pp. 389, 432.

2. C. H. Edwards, *The Historical Development of the Calculus* (New York: Springer-Verlag, 1979), pp. 124, 132.

3. Howard Eves, *An Introduction to the History of Mathematics*, 6th ed. (New York: Saunders, 1990), pp. 391, 395.

4. Morris Kline, *Mathematical Thought from Ancient to Modern Times* (New York: Oxford University Press, 1972), pp. 344, 346.

2.8 The Derivative as a Function

In the preceding section we considered the derivative of a function f at a fixed number a:

$$\boxed{1} \qquad f'(a) = \lim_{h \to 0} \frac{f(a + h) - f(a)}{h}$$

Here we change our point of view and let the number a vary. If we replace a in Equation 1 by a variable x, we obtain

$$\boxed{2} \qquad f'(x) = \lim_{h \to 0} \frac{f(x + h) - f(x)}{h}$$

Given any number x for which this limit exists, we assign to x the number $f'(x)$. So we can regard f' as a new function, called the **derivative of f** and defined by Equation 2. We know that the value of f' at x, $f'(x)$, can be interpreted geometrically as the slope of the tangent line to the graph of f at the point $(x, f(x))$.

SECTION 2.8 The Derivative as a Function 153

The function f' is called the derivative of f because it has been "derived" from f by the limiting operation in Equation 2. The domain of f' is the set $\{x \mid f'(x) \text{ exists}\}$ and may be smaller than the domain of f.

EXAMPLE 1 The graph of a function f is given in Figure 1. Use it to sketch the graph of the derivative f'.

SOLUTION We can estimate the value of the derivative at any value of x by drawing the tangent at the point $(x, f(x))$ and estimating its slope. For instance, for $x = 5$ we draw the tangent at P in Figure 2(a) and estimate its slope to be about $\frac{3}{2}$, so $f'(5) \approx 1.5$. This allows us to plot the point $P'(5, 1.5)$ on the graph of f' directly beneath P. (The slope of the graph of f becomes the y-value on the graph of f'.) Repeating this procedure at several points, we get the graph shown in Figure 2(b). Notice that the tangents at A, B, and C are horizontal, so the derivative is 0 there and the graph of f' crosses the x-axis (where $y = 0$) at the points A', B', and C', directly beneath A, B, and C. Between A and B the tangents have positive slope, so $f'(x)$ is positive there. (The graph is above the x-axis.) But between B and C the tangents have negative slope, so $f'(x)$ is negative there.

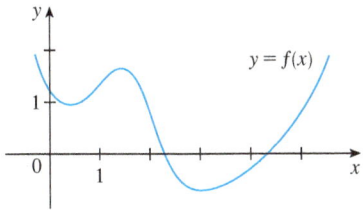

FIGURE 1

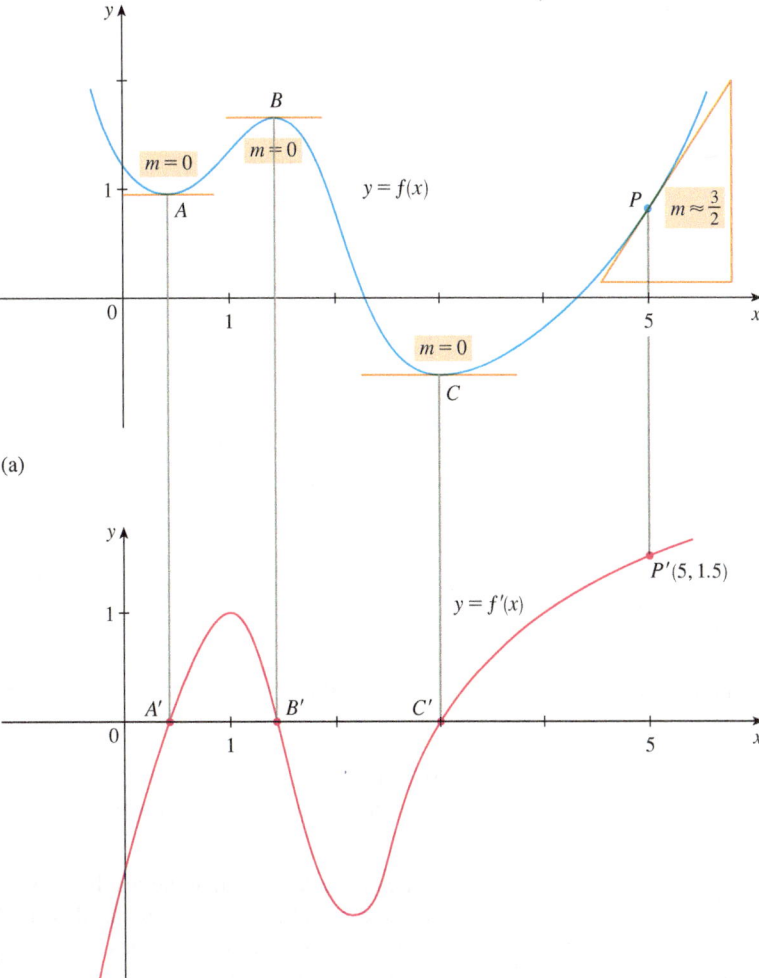

TEC Visual 2.8 shows an animation of Figure 2 for several functions.

FIGURE 2

EXAMPLE 2
(a) If $f(x) = x^3 - x$, find a formula for $f'(x)$.
(b) Illustrate this formula by comparing the graphs of f and f'.

SOLUTION
(a) When using Equation 2 to compute a derivative, we must remember that the variable is h and that x is temporarily regarded as a constant during the calculation of the limit.

$$f'(x) = \lim_{h \to 0} \frac{f(x+h) - f(x)}{h} = \lim_{h \to 0} \frac{[(x+h)^3 - (x+h)] - [x^3 - x]}{h}$$

$$= \lim_{h \to 0} \frac{x^3 + 3x^2h + 3xh^2 + h^3 - x - h - x^3 + x}{h}$$

$$= \lim_{h \to 0} \frac{3x^2h + 3xh^2 + h^3 - h}{h}$$

$$= \lim_{h \to 0} (3x^2 + 3xh + h^2 - 1) = 3x^2 - 1$$

(b) We use a graphing device to graph f and f' in Figure 3. Notice that $f'(x) = 0$ when f has horizontal tangents and $f'(x)$ is positive when the tangents have positive slope. So these graphs serve as a check on our work in part (a). ∎

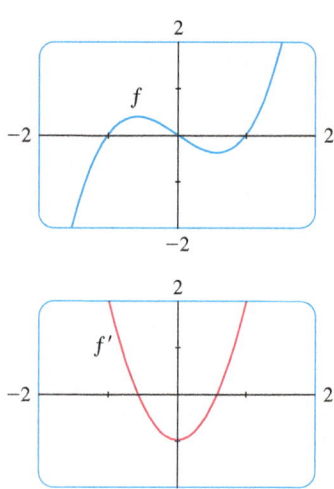

FIGURE 3

EXAMPLE 3
If $f(x) = \sqrt{x}$, find the derivative of f. State the domain of f'.

SOLUTION

$$f'(x) = \lim_{h \to 0} \frac{f(x+h) - f(x)}{h}$$

$$= \lim_{h \to 0} \frac{\sqrt{x+h} - \sqrt{x}}{h}$$

$$= \lim_{h \to 0} \left(\frac{\sqrt{x+h} - \sqrt{x}}{h} \cdot \frac{\sqrt{x+h} + \sqrt{x}}{\sqrt{x+h} + \sqrt{x}} \right) \quad \text{(Rationalize the numerator.)}$$

$$= \lim_{h \to 0} \frac{(x+h) - x}{h(\sqrt{x+h} + \sqrt{x})} = \lim_{h \to 0} \frac{h}{h(\sqrt{x+h} + \sqrt{x})}$$

$$= \lim_{h \to 0} \frac{1}{\sqrt{x+h} + \sqrt{x}} = \frac{1}{\sqrt{x} + \sqrt{x}} = \frac{1}{2\sqrt{x}}$$

We see that $f'(x)$ exists if $x > 0$, so the domain of f' is $(0, \infty)$. This is slightly smaller than the domain of f, which is $[0, \infty)$. ∎

Let's check to see that the result of Example 3 is reasonable by looking at the graphs of f and f' in Figure 4. When x is close to 0, $\sqrt{x}$ is also close to 0, so $f'(x) = 1/(2\sqrt{x})$ is very large and this corresponds to the steep tangent lines near $(0, 0)$ in Figure 4(a) and the large values of $f'(x)$ just to the right of 0 in Figure 4(b). When x is large, $f'(x)$ is very small and this corresponds to the flatter tangent lines at the far right of the graph of f and the horizontal asymptote of the graph of f'.

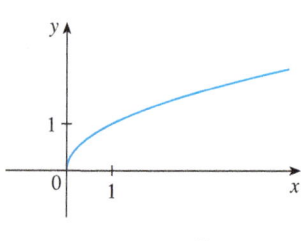

(a) $f(x) = \sqrt{x}$

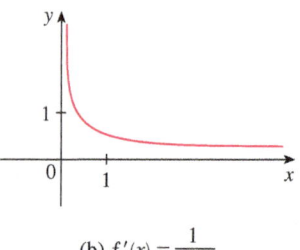

(b) $f'(x) = \dfrac{1}{2\sqrt{x}}$

FIGURE 4

EXAMPLE 4 Find f' if $f(x) = \dfrac{1-x}{2+x}$.

SOLUTION

$$f'(x) = \lim_{h \to 0} \frac{f(x+h) - f(x)}{h}$$

$$= \lim_{h \to 0} \frac{\dfrac{1-(x+h)}{2+(x+h)} - \dfrac{1-x}{2+x}}{h}$$

$$= \lim_{h \to 0} \frac{(1-x-h)(2+x) - (1-x)(2+x+h)}{h(2+x+h)(2+x)}$$

$$= \lim_{h \to 0} \frac{(2-x-2h-x^2-xh) - (2-x+h-x^2-xh)}{h(2+x+h)(2+x)}$$

$$= \lim_{h \to 0} \frac{-3h}{h(2+x+h)(2+x)} = \lim_{h \to 0} \frac{-3}{(2+x+h)(2+x)} = -\frac{3}{(2+x)^2} \blacksquare$$

$$\frac{\dfrac{a}{b} - \dfrac{c}{d}}{e} = \frac{ad - bc}{bd} \cdot \frac{1}{e}$$

■ Other Notations

If we use the traditional notation $y = f(x)$ to indicate that the independent variable is x and the dependent variable is y, then some common alternative notations for the derivative are as follows:

$$f'(x) = y' = \frac{dy}{dx} = \frac{df}{dx} = \frac{d}{dx}f(x) = Df(x) = D_x f(x)$$

The symbols D and d/dx are called **differentiation operators** because they indicate the operation of **differentiation**, which is the process of calculating a derivative.

The symbol dy/dx, which was introduced by Leibniz, should not be regarded as a ratio (for the time being); it is simply a synonym for $f'(x)$. Nonetheless, it is a very useful and suggestive notation, especially when used in conjunction with increment notation. Referring to Equation 2.7.6, we can rewrite the definition of derivative in Leibniz notation in the form

$$\frac{dy}{dx} = \lim_{\Delta x \to 0} \frac{\Delta y}{\Delta x}$$

If we want to indicate the value of a derivative dy/dx in Leibniz notation at a specific number a, we use the notation

$$\left. \frac{dy}{dx} \right|_{x=a} \quad \text{or} \quad \left. \frac{dy}{dx} \right]_{x=a}$$

which is a synonym for $f'(a)$. The vertical bar means "evaluate at."

3 **Definition** A function f is **differentiable at a** if $f'(a)$ exists. It is **differentiable on an open interval** (a, b) [or (a, ∞) or $(-\infty, a)$ or $(-\infty, \infty)$] if it is differentiable at every number in the interval.

Leibniz

Gottfried Wilhelm Leibniz was born in Leipzig in 1646 and studied law, theology, philosophy, and mathematics at the university there, graduating with a bachelor's degree at age 17. After earning his doctorate in law at age 20, Leibniz entered the diplomatic service and spent most of his life traveling to the capitals of Europe on political missions. In particular, he worked to avert a French military threat against Germany and attempted to reconcile the Catholic and Protestant churches.

His serious study of mathematics did not begin until 1672 while he was on a diplomatic mission in Paris. There he built a calculating machine and met scientists, like Huygens, who directed his attention to the latest developments in mathematics and science. Leibniz sought to develop a symbolic logic and system of notation that would simplify logical reasoning. In particular, the version of calculus that he published in 1684 established the notation and the rules for finding derivatives that we use today.

Unfortunately, a dreadful priority dispute arose in the 1690s between the followers of Newton and those of Leibniz as to who had invented calculus first. Leibniz was even accused of plagiarism by members of the Royal Society in England. The truth is that each man invented calculus independently. Newton arrived at his version of calculus first but, because of his fear of controversy, did not publish it immediately. So Leibniz's 1684 account of calculus was the first to be published.

EXAMPLE 5 Where is the function $f(x) = |x|$ differentiable?

SOLUTION If $x > 0$, then $|x| = x$ and we can choose h small enough that $x + h > 0$ and hence $|x + h| = x + h$. Therefore, for $x > 0$, we have

$$f'(x) = \lim_{h \to 0} \frac{|x+h| - |x|}{h} = \lim_{h \to 0} \frac{(x+h) - x}{h}$$

$$= \lim_{h \to 0} \frac{h}{h} = \lim_{h \to 0} 1 = 1$$

and so f is differentiable for any $x > 0$.

Similarly, for $x < 0$ we have $|x| = -x$ and h can be chosen small enough that $x + h < 0$ and so $|x + h| = -(x + h)$. Therefore, for $x < 0$,

$$f'(x) = \lim_{h \to 0} \frac{|x+h| - |x|}{h} = \lim_{h \to 0} \frac{-(x+h) - (-x)}{h}$$

$$= \lim_{h \to 0} \frac{-h}{h} = \lim_{h \to 0} (-1) = -1$$

and so f is differentiable for any $x < 0$.

For $x = 0$ we have to investigate

$$f'(0) = \lim_{h \to 0} \frac{f(0+h) - f(0)}{h}$$

$$= \lim_{h \to 0} \frac{|0+h| - |0|}{h} = \lim_{h \to 0} \frac{|h|}{h} \quad \text{(if it exists)}$$

Let's compute the left and right limits separately:

$$\lim_{h \to 0^+} \frac{|h|}{h} = \lim_{h \to 0^+} \frac{h}{h} = \lim_{h \to 0^+} 1 = 1$$

and

$$\lim_{h \to 0^-} \frac{|h|}{h} = \lim_{h \to 0^-} \frac{-h}{h} = \lim_{h \to 0^-} (-1) = -1$$

Since these limits are different, $f'(0)$ does not exist. Thus f is differentiable at all x except 0.

A formula for f' is given by

$$f'(x) = \begin{cases} 1 & \text{if } x > 0 \\ -1 & \text{if } x < 0 \end{cases}$$

and its graph is shown in Figure 5(b). The fact that $f'(0)$ does not exist is reflected geometrically in the fact that the curve $y = |x|$ does not have a tangent line at $(0, 0)$. [See Figure 5(a).]

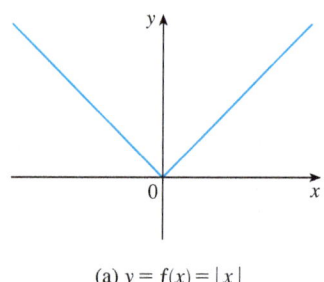

(a) $y = f(x) = |x|$

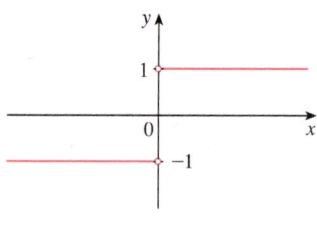

(b) $y = f'(x)$

FIGURE 5

Both continuity and differentiability are desirable properties for a function to have. The following theorem shows how these properties are related.

4 Theorem If f is differentiable at a, then f is continuous at a.

PROOF To prove that f is continuous at a, we have to show that $\lim_{x \to a} f(x) = f(a)$. We do this by showing that the difference $f(x) - f(a)$ approaches 0.

The given information is that f is differentiable at a, that is,

$$f'(a) = \lim_{x \to a} \frac{f(x) - f(a)}{x - a}$$

PS An important aspect of problem solving is trying to find a connection between the given and the unknown. See Step 2 (Think of a Plan) in *Principles of Problem Solving* on page 71.

exists (see Equation 2.7.5). To connect the given and the unknown, we divide and multiply $f(x) - f(a)$ by $x - a$ (which we can do when $x \neq a$):

$$f(x) - f(a) = \frac{f(x) - f(a)}{x - a}(x - a)$$

Thus, using the Product Law and (2.7.5), we can write

$$\lim_{x \to a} [f(x) - f(a)] = \lim_{x \to a} \frac{f(x) - f(a)}{x - a}(x - a)$$

$$= \lim_{x \to a} \frac{f(x) - f(a)}{x - a} \cdot \lim_{x \to a} (x - a)$$

$$= f'(a) \cdot 0 = 0$$

To use what we have just proved, we start with $f(x)$ and add and subtract $f(a)$:

$$\lim_{x \to a} f(x) = \lim_{x \to a} [f(a) + (f(x) - f(a))]$$

$$= \lim_{x \to a} f(a) + \lim_{x \to a} [f(x) - f(a)]$$

$$= f(a) + 0 = f(a)$$

Therefore f is continuous at a. ∎

NOTE The converse of Theorem 4 is false; that is, there are functions that are continuous but not differentiable. For instance, the function $f(x) = |x|$ is continuous at 0 because

$$\lim_{x \to 0} f(x) = \lim_{x \to 0} |x| = 0 = f(0)$$

(See Example 2.3.7.) But in Example 5 we showed that f is not differentiable at 0.

■ How Can a Function Fail To Be Differentiable?

We saw that the function $y = |x|$ in Example 5 is not differentiable at 0 and Figure 5(a) shows that its graph changes direction abruptly when $x = 0$. In general, if the graph of a function f has a "corner" or "kink" in it, then the graph of f has no tangent at this point and f is not differentiable there. [In trying to compute $f'(a)$, we find that the left and right limits are different.]

Theorem 4 gives another way for a function not to have a derivative. It says that if f is not continuous at a, then f is not differentiable at a. So at any discontinuity (for instance, a jump discontinuity) f fails to be differentiable.

158   **CHAPTER 2**   Limits and Derivatives

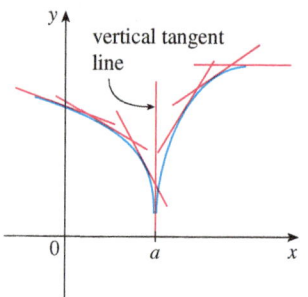

FIGURE 6

A third possibility is that the curve has a **vertical tangent line** when $x = a$; that is, f is continuous at a and

$$\lim_{x \to a} |f'(x)| = \infty$$

This means that the tangent lines become steeper and steeper as $x \to a$. Figure 6 shows one way that this can happen; Figure 7(c) shows another. Figure 7 illustrates the three possibilities that we have discussed.

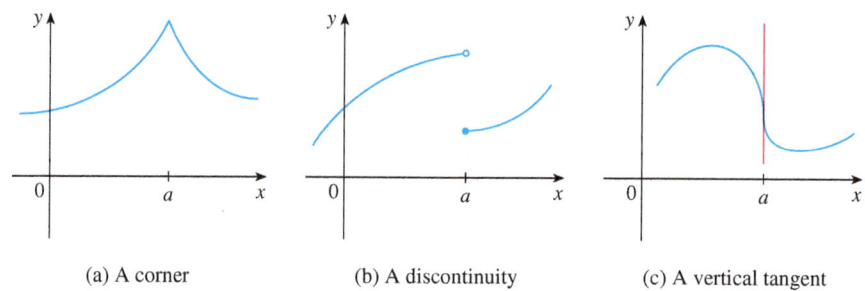

FIGURE 7
Three ways for f not to be differentiable at a

(a) A corner   (b) A discontinuity   (c) A vertical tangent

A graphing calculator or computer provides another way of looking at differentiability. If f is differentiable at a, then when we zoom in toward the point $(a, f(a))$ the graph straightens out and appears more and more like a line. (See Figure 8. We saw a specific example of this in Figure 2.7.2.) But no matter how much we zoom in toward a point like the ones in Figures 6 and 7(a), we can't eliminate the sharp point or corner (see Figure 9).

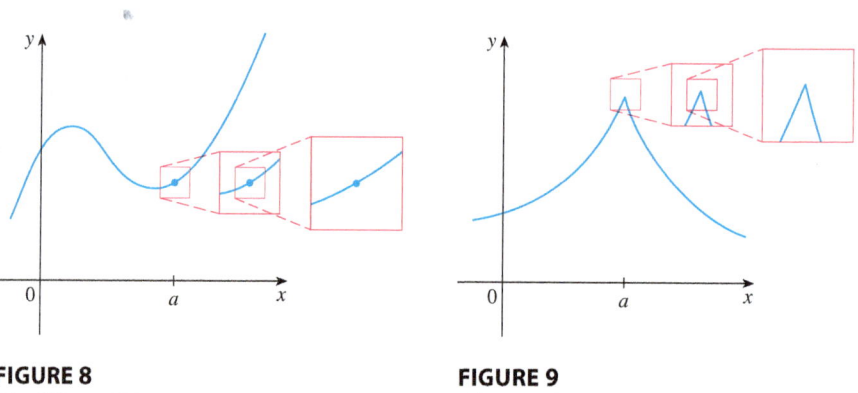

FIGURE 8
f is differentiable at a.

FIGURE 9
f is not differentiable at a.

■ Higher Derivatives

If f is a differentiable function, then its derivative f' is also a function, so f' may have a derivative of its own, denoted by $(f')' = f''$. This new function f'' is called the **second derivative** of f because it is the derivative of the derivative of f. Using Leibniz notation, we write the second derivative of $y = f(x)$ as

$$\underbrace{\frac{d}{dx}}_{\text{derivative of}} \underbrace{\left(\frac{dy}{dx} \right)}_{\text{first derivative}} = \underbrace{\frac{d^2y}{dx^2}}_{\text{second derivative}}$$

EXAMPLE 6 If $f(x) = x^3 - x$, find and interpret $f''(x)$.

SOLUTION In Example 2 we found that the first derivative is $f'(x) = 3x^2 - 1$. So the second derivative is

$$f''(x) = (f')'(x) = \lim_{h \to 0} \frac{f'(x+h) - f'(x)}{h}$$

$$= \lim_{h \to 0} \frac{[3(x+h)^2 - 1] - [3x^2 - 1]}{h}$$

$$= \lim_{h \to 0} \frac{3x^2 + 6xh + 3h^2 - 1 - 3x^2 + 1}{h}$$

$$= \lim_{h \to 0} (6x + 3h) = 6x$$

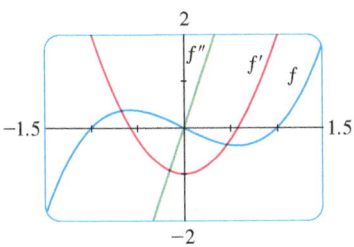

FIGURE 10

TEC In Module 2.8 you can see how changing the coefficients of a polynomial f affects the appearance of the graphs of f, f', and f''.

The graphs of f, f', and f'' are shown in Figure 10.

We can interpret $f''(x)$ as the slope of the curve $y = f'(x)$ at the point $(x, f'(x))$. In other words, it is the rate of change of the slope of the original curve $y = f(x)$.

Notice from Figure 10 that $f''(x)$ is negative when $y = f'(x)$ has negative slope and positive when $y = f'(x)$ has positive slope. So the graphs serve as a check on our calculations.

In general, we can interpret a second derivative as a rate of change of a rate of change. The most familiar example of this is *acceleration*, which we define as follows.

If $s = s(t)$ is the position function of an object that moves in a straight line, we know that its first derivative represents the velocity $v(t)$ of the object as a function of time:

$$v(t) = s'(t) = \frac{ds}{dt}$$

The instantaneous rate of change of velocity with respect to time is called the **acceleration** $a(t)$ of the object. Thus the acceleration function is the derivative of the velocity function and is therefore the second derivative of the position function:

$$a(t) = v'(t) = s''(t)$$

or, in Leibniz notation,

$$a = \frac{dv}{dt} = \frac{d^2s}{dt^2}$$

Acceleration is the change in velocity you feel when speeding up or slowing down in a car.

The **third derivative** f''' is the derivative of the second derivative: $f''' = (f'')'$. So $f'''(x)$ can be interpreted as the slope of the curve $y = f''(x)$ or as the rate of change of $f''(x)$. If $y = f(x)$, then alternative notations for the third derivative are

$$y''' = f'''(x) = \frac{d}{dx}\left(\frac{d^2y}{dx^2}\right) = \frac{d^3y}{dx^3}$$

We can also interpret the third derivative physically in the case where the function is the position function $s = s(t)$ of an object that moves along a straight line. Because $s''' = (s'')' = a'$, the third derivative of the position function is the derivative of the acceleration function and is called the **jerk**:

$$j = \frac{da}{dt} = \frac{d^3s}{dt^3}$$

Thus the jerk j is the rate of change of acceleration. It is aptly named because a large jerk means a sudden change in acceleration, which causes an abrupt movement in a vehicle.

The differentiation process can be continued. The fourth derivative f'''' is usually denoted by $f^{(4)}$. In general, the nth derivative of f is denoted by $f^{(n)}$ and is obtained from f by differentiating n times. If $y = f(x)$, we write

$$y^{(n)} = f^{(n)}(x) = \frac{d^n y}{dx^n}$$

EXAMPLE 7 If $f(x) = x^3 - x$, find $f'''(x)$ and $f^{(4)}(x)$.

SOLUTION In Example 6 we found that $f''(x) = 6x$. The graph of the second derivative has equation $y = 6x$ and so it is a straight line with slope 6. Since the derivative $f'''(x)$ is the slope of $f''(x)$, we have

$$f'''(x) = 6$$

for all values of x. So f''' is a constant function and its graph is a horizontal line. Therefore, for all values of x,

$$f^{(4)}(x) = 0$$

We have seen that one application of second and third derivatives occurs in analyzing the motion of objects using acceleration and jerk. We will investigate another application of second derivatives in Section 4.3, where we show how knowledge of f'' gives us information about the shape of the graph of f. In Chapter 11 we will see how second and higher derivatives enable us to represent functions as sums of infinite series.

2.8 EXERCISES

1–2 Use the given graph to estimate the value of each derivative. Then sketch the graph of f'.

1. (a) $f'(-3)$ (b) $f'(-2)$ (c) $f'(-1)$ (d) $f'(0)$
    (e) $f'(1)$ (f) $f'(2)$ (g) $f'(3)$

2. (a) $f'(0)$ (b) $f'(1)$ (c) $f'(2)$ (d) $f'(3)$
   (e) $f'(4)$ (f) $f'(5)$ (g) $f'(6)$ (h) $f'(7)$

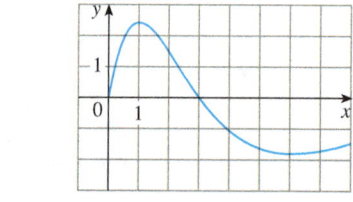

3. Match the graph of each function in (a)–(d) with the graph of its derivative in I–IV. Give reasons for your choices.

(a)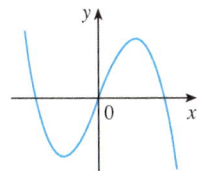

(b)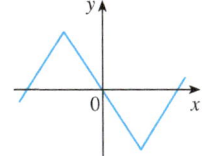

(c)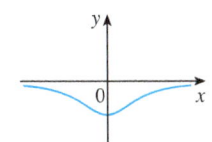

(d)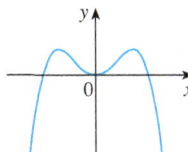

I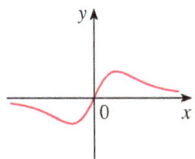

II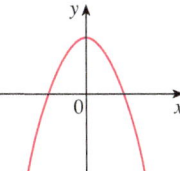

III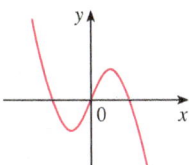

IV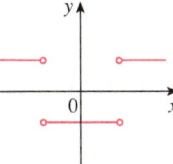

4–11 Trace or copy the graph of the given function f. (Assume that the axes have equal scales.) Then use the method of Example 1 to sketch the graph of f' below it.

4.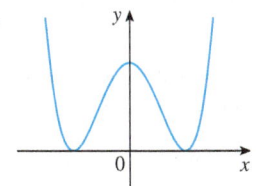

5.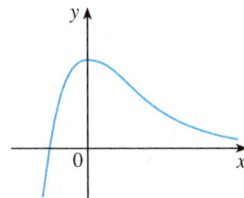

6.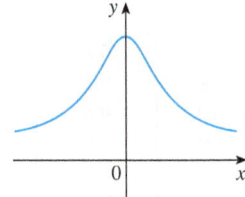

7.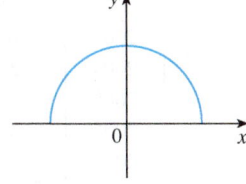

8.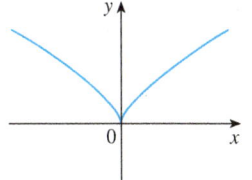

9.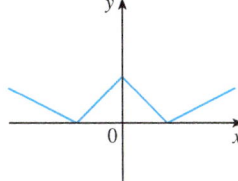

10.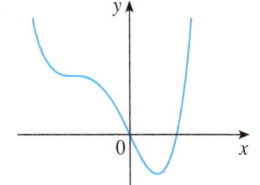

11.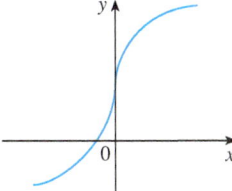

12. Shown is the graph of the population function $P(t)$ for yeast cells in a laboratory culture. Use the method of Example 1 to graph the derivative $P'(t)$. What does the graph of P' tell us about the yeast population?

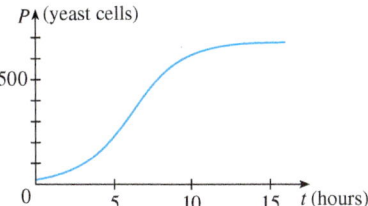

13. A rechargeable battery is plugged into a charger. The graph shows $C(t)$, the percentage of full capacity that the battery reaches as a function of time t elapsed (in hours).
    (a) What is the meaning of the derivative $C'(t)$?
    (b) Sketch the graph of $C'(t)$. What does the graph tell you?

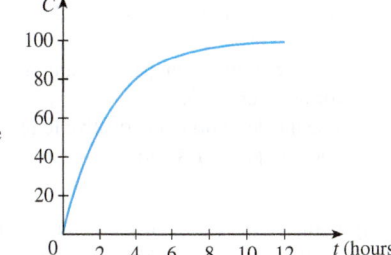

14. The graph (from the US Department of Energy) shows how driving speed affects gas mileage. Fuel economy F is measured in miles per gallon and speed v is measured in miles per hour.
(a) What is the meaning of the derivative $F'(v)$?
(b) Sketch the graph of $F'(v)$.
(c) At what speed should you drive if you want to save on gas?

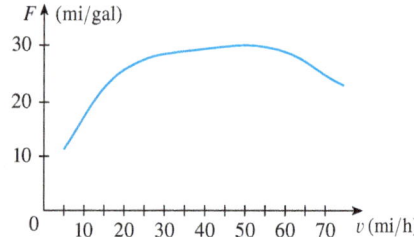

15. The graph shows how the average age of first marriage of Japanese men varied in the last half of the 20th century. Sketch the graph of the derivative function $M'(t)$. During which years was the derivative negative?

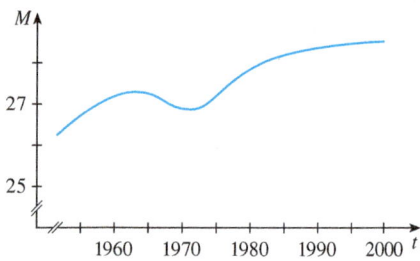

16–18 Make a careful sketch of the graph of f and below it sketch the graph of f' in the same manner as in Exercises 4–11. Can you guess a formula for $f'(x)$ from its graph?

16. $f(x) = \sin x$ **17.** $f(x) = e^x$ **18.** $f(x) = \ln x$

19. Let $f(x) = x^2$.
(a) Estimate the values of $f'(0)$, $f'(\frac{1}{2})$, $f'(1)$, and $f'(2)$ by using a graphing device to zoom in on the graph of f.
(b) Use symmetry to deduce the values of $f'(-\frac{1}{2})$, $f'(-1)$, and $f'(-2)$.
(c) Use the results from parts (a) and (b) to guess a formula for $f'(x)$.
(d) Use the definition of derivative to prove that your guess in part (c) is correct.

20. Let $f(x) = x^3$.
(a) Estimate the values of $f'(0)$, $f'(\frac{1}{2})$, $f'(1)$, $f'(2)$, and $f'(3)$ by using a graphing device to zoom in on the graph of f.

(b) Use symmetry to deduce the values of $f'(-\frac{1}{2})$, $f'(-1)$, $f'(-2)$, and $f'(-3)$.
(c) Use the values from parts (a) and (b) to graph f'.
(d) Guess a formula for $f'(x)$.
(e) Use the definition of derivative to prove that your guess in part (d) is correct.

21–31 Find the derivative of the function using the definition of derivative. State the domain of the function and the domain of its derivative.

21. $f(x) = 3x - 8$ **22.** $f(x) = mx + b$

23. $f(t) = 2.5t^2 + 6t$ **24.** $f(x) = 4 + 8x - 5x^2$

25. $f(x) = x^2 - 2x^3$ **26.** $g(t) = \dfrac{1}{\sqrt{t}}$

27. $g(x) = \sqrt{9 - x}$ **28.** $f(x) = \dfrac{x^2 - 1}{2x - 3}$

29. $G(t) = \dfrac{1 - 2t}{3 + t}$ **30.** $f(x) = x^{3/2}$

31. $f(x) = x^4$

---

32. (a) Sketch the graph of $f(x) = \sqrt{6 - x}$ by starting with the graph of $y = \sqrt{x}$ and using the transformations of Section 1.3.
(b) Use the graph from part (a) to sketch the graph of f'.
(c) Use the definition of a derivative to find $f'(x)$. What are the domains of f and f'?
(d) Use a graphing device to graph f' and compare with your sketch in part (b).

33. (a) If $f(x) = x^4 + 2x$, find $f'(x)$.
(b) Check to see that your answer to part (a) is reasonable by comparing the graphs of f and f'.

34. (a) If $f(x) = x + 1/x$, find $f'(x)$.
(b) Check to see that your answer to part (a) is reasonable by comparing the graphs of f and f'.

35. The unemployment rate $U(t)$ varies with time. The table gives the percentage of unemployed in the US labor force from 2003 to 2012.
(a) What is the meaning of $U'(t)$? What are its units?
(b) Construct a table of estimated values for $U'(t)$.

t	$U(t)$	t	$U(t)$
2003	6.0	2008	5.8
2004	5.5	2009	9.3
2005	5.1	2010	9.6
2006	4.6	2011	8.9
2007	4.6	2012	8.1

Source: US Bureau of Labor Statistics

36. The table gives the number $N(t)$, measured in thousands, of minimally invasive cosmetic surgery procedures performed in the United States for various years t.

t	$N(t)$ (thousands)
2000	5,500
2002	4,897
2004	7,470
2006	9,138
2008	10,897
2010	11,561
2012	13,035

Source: American Society of Plastic Surgeons

(a) What is the meaning of $N'(t)$? What are its units?
(b) Construct a table of estimated values for $N'(t)$.
(c) Graph N and N'.
(d) How would it be possible to get more accurate values for $N'(t)$?

37. The table gives the height as time passes of a typical pine tree grown for lumber at a managed site.

Tree age (years)	14	21	28	35	42	49
Height (feet)	41	54	64	72	78	83

Source: Arkansas Forestry Commission

If $H(t)$ is the height of the tree after t years, construct a table of estimated values for H' and sketch its graph.

38. Water temperature affects the growth rate of brook trout. The table shows the amount of weight gained by brook trout after 24 days in various water temperatures.

Temperature (°C)	15.5	17.7	20.0	22.4	24.4
Weight gained (g)	37.2	31.0	19.8	9.7	−9.8

If $W(x)$ is the weight gain at temperature x, construct a table of estimated values for W' and sketch its graph. What are the units for $W'(x)$?

Source: Adapted from J. Chadwick Jr., "Temperature Effects on Growth and Stress Physiology of Brook Trout: Implications for Climate Change Impacts on an Iconic Cold-Water Fish." *Masters Theses.* Paper 897. 2012. scholarworks.umass.edu/theses/897.

39. Let P represent the percentage of a city's electrical power that is produced by solar panels t years after January 1, 2000.
(a) What does dP/dt represent in this context?
(b) Interpret the statement
$$\left.\frac{dP}{dt}\right|_{t=2} = 3.5$$

40. Suppose N is the number of people in the United States who travel by car to another state for a vacation this year when the average price of gasoline is p dollars per gallon. Do you expect dN/dp to be positive or negative? Explain.

41–44 The graph of f is given. State, with reasons, the numbers at which f is *not* differentiable.

41.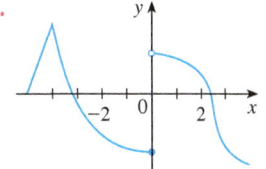

42.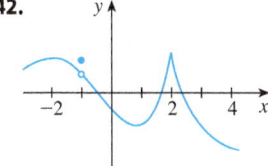

43.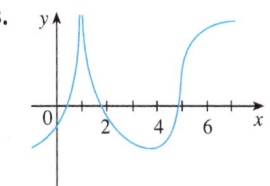

44.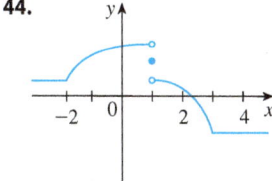

45. Graph the function $f(x) = x + \sqrt{|x|}$. Zoom in repeatedly, first toward the point $(-1, 0)$ and then toward the origin. What is different about the behavior of f in the vicinity of these two points? What do you conclude about the differentiability of f?

46. Zoom in toward the points $(1, 0)$, $(0, 1)$, and $(-1, 0)$ on the graph of the function $g(x) = (x^2 - 1)^{2/3}$. What do you notice? Account for what you see in terms of the differentiability of g.

47–48 The graphs of a function f and its derivative f' are shown. Which is bigger, $f'(-1)$ or $f''(1)$?

47.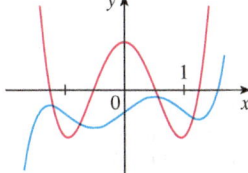

48.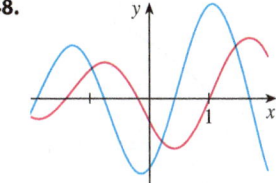

49. The figure shows the graphs of f, f', and f''. Identify each curve, and explain your choices.

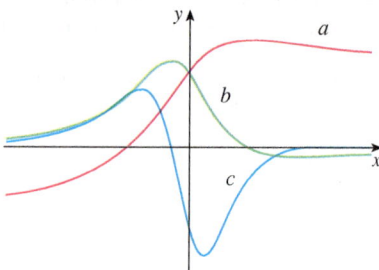

50. The figure shows graphs of f, f', f'', and f'''. Identify each curve, and explain your choices.

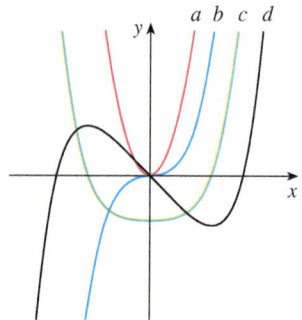

51. The figure shows the graphs of three functions. One is the position function of a car, one is the velocity of the car, and one is its acceleration. Identify each curve, and explain your choices.

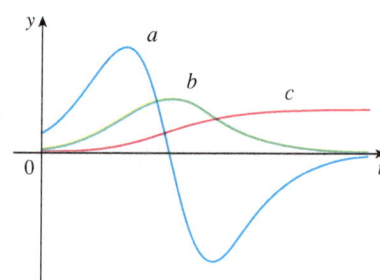

52. The figure shows the graphs of four functions. One is the position function of a car, one is the velocity of the car, one is its acceleration, and one is its jerk. Identify each curve, and explain your choices.

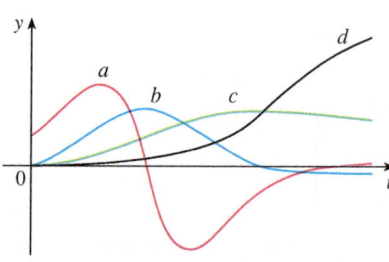

53–54 Use the definition of a derivative to find $f'(x)$ and $f''(x)$. Then graph f, f', and f'' on a common screen and check to see if your answers are reasonable.

53. $f(x) = 3x^2 + 2x + 1$ **54.** $f(x) = x^3 - 3x$

55. If $f(x) = 2x^2 - x^3$, find $f'(x)$, $f''(x)$, $f'''(x)$, and $f^{(4)}(x)$. Graph f, f', f'', and f''' on a common screen. Are the graphs consistent with the geometric interpretations of these derivatives?

56. (a) The graph of a position function of a car is shown, where s is measured in feet and t in seconds. Use it to graph the velocity and acceleration of the car. What is the acceleration at $t = 10$ seconds?

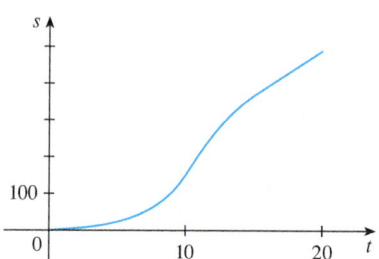

(b) Use the acceleration curve from part (a) to estimate the jerk at $t = 10$ seconds. What are the units for jerk?

57. Let $f(x) = \sqrt[3]{x}$.
(a) If $a \neq 0$, use Equation 2.7.5 to find $f'(a)$.
(b) Show that $f'(0)$ does not exist.
(c) Show that $y = \sqrt[3]{x}$ has a vertical tangent line at $(0, 0)$. (Recall the shape of the graph of f. See Figure 1.2.13.)

58. (a) If $g(x) = x^{2/3}$, show that $g'(0)$ does not exist.
(b) If $a \neq 0$, find $g'(a)$.
(c) Show that $y = x^{2/3}$ has a vertical tangent line at $(0, 0)$.
(d) Illustrate part (c) by graphing $y = x^{2/3}$.

59. Show that the function $f(x) = |x - 6|$ is not differentiable at 6. Find a formula for f' and sketch its graph.

60. Where is the greatest integer function $f(x) = [\![x]\!]$ not differentiable? Find a formula for f' and sketch its graph.

61. (a) Sketch the graph of the function $f(x) = x|x|$.
(b) For what values of x is f differentiable?
(c) Find a formula for f'.

62. (a) Sketch the graph of the function $g(x) = x + |x|$.
(b) For what values of x is g differentiable?
(c) Find a formula for g'.

63. Recall that a function f is called *even* if $f(-x) = f(x)$ for all x in its domain and *odd* if $f(-x) = -f(x)$ for all such x. Prove each of the following.
(a) The derivative of an even function is an odd function.
(b) The derivative of an odd function is an even function.

64. The **left-hand** and **right-hand derivatives** of f at a are defined by

$$f'_-(a) = \lim_{h \to 0^-} \frac{f(a+h) - f(a)}{h}$$

and

$$f'_+(a) = \lim_{h \to 0^+} \frac{f(a+h) - f(a)}{h}$$

if these limits exist. Then $f'(a)$ exists if and only if these one-sided derivatives exist and are equal.
(a) Find $f'_-(4)$ and $f'_+(4)$ for the function

$$f(x) = \begin{cases} 0 & \text{if } x \leq 0 \\ 5 - x & \text{if } 0 < x < 4 \\ \dfrac{1}{5-x} & \text{if } x \geq 4 \end{cases}$$

(b) Sketch the graph of f.
(c) Where is f discontinuous?
(d) Where is f not differentiable?

65. Nick starts jogging and runs faster and faster for 3 mintues, then he walks for 5 minutes. He stops at an intersection for 2 minutes, runs fairly quickly for 5 minutes, then walks for 4 minutes.
(a) Sketch a possible graph of the distance s Nick has covered after t minutes.
(b) Sketch a graph of ds/dt.

66. When you turn on a hot-water faucet, the temperature T of the water depends on how long the water has been running.
(a) Sketch a possible graph of T as a function of the time t that has elapsed since the faucet was turned on.
(b) Describe how the rate of change of T with respect to t varies as t increases.
(c) Sketch a graph of the derivative of T.

67. Let ℓ be the tangent line to the parabola $y = x^2$ at the point $(1, 1)$. The *angle of inclination* of ℓ is the angle ϕ that ℓ makes with the positive direction of the x-axis. Calculate ϕ correct to the nearest degree.

2 REVIEW

CONCEPT CHECK

Answers to the Concept Check can be found on the back endpapers.

1. Explain what each of the following means and illustrate with a sketch.
   (a) $\lim_{x \to a} f(x) = L$ (b) $\lim_{x \to a^+} f(x) = L$ (c) $\lim_{x \to a^-} f(x) = L$
   (d) $\lim_{x \to a} f(x) = \infty$ (e) $\lim_{x \to \infty} f(x) = L$

2. Describe several ways in which a limit can fail to exist. Illustrate with sketches.

3. State the following Limit Laws.
   (a) Sum Law
   (b) Difference Law
   (c) Constant Multiple Law
   (d) Product Law
   (e) Quotient Law
   (f) Power Law
   (g) Root Law

4. What does the Squeeze Theorem say?

5. (a) What does it mean to say that the line $x = a$ is a vertical asymptote of the curve $y = f(x)$? Draw curves to illustrate the various possibilities.
   (b) What does it mean to say that the line $y = L$ is a horizontal asymptote of the curve $y = f(x)$? Draw curves to illustrate the various possibilities.

6. Which of the following curves have vertical asymptotes? Which have horizontal asymptotes?
   (a) $y = x^4$ (b) $y = \sin x$ (c) $y = \tan x$
   (d) $y = \tan^{-1} x$ (e) $y = e^x$ (f) $y = \ln x$
   (g) $y = 1/x$ (h) $y = \sqrt{x}$

7. (a) What does it mean for f to be continuous at a?
   (b) What does it mean for f to be continuous on the interval $(-\infty, \infty)$? What can you say about the graph of such a function?

8. (a) Give examples of functions that are continuous on $[-1, 1]$.
   (b) Give an example of a function that is not continuous on $[0, 1]$.

9. What does the Intermediate Value Theorem say?

10. Write an expression for the slope of the tangent line to the curve $y = f(x)$ at the point $(a, f(a))$.

11. Suppose an object moves along a straight line with position $f(t)$ at time t. Write an expression for the instantaneous velocity of the object at time $t = a$. How can you interpret this velocity in terms of the graph of f?

12. If $y = f(x)$ and x changes from x_1 to x_2, write expressions for the following.
    (a) The average rate of change of y with respect to x over the interval $[x_1, x_2]$.
    (b) The instantaneous rate of change of y with respect to x at $x = x_1$.

13. Define the derivative $f'(a)$. Discuss two ways of interpreting this number.

14. Define the second derivative of f. If $f(t)$ is the position function of a particle, how can you interpret the second derivative?

15. (a) What does it mean for f to be differentiable at a?
(b) What is the relation between the differentiability and continuity of a function?
(c) Sketch the graph of a function that is continuous but not differentiable at $a = 2$.

16. Describe several ways in which a function can fail to be differentiable. Illustrate with sketches.

TRUE-FALSE QUIZ

Determine whether the statement is true or false. If it is true, explain why. If it is false, explain why or give an example that disproves the statement.

1. $\lim_{x \to 4} \left(\dfrac{2x}{x-4} - \dfrac{8}{x-4} \right) = \lim_{x \to 4} \dfrac{2x}{x-4} - \lim_{x \to 4} \dfrac{8}{x-4}$

2. $\lim_{x \to 1} \dfrac{x^2 + 6x - 7}{x^2 + 5x - 6} = \dfrac{\lim_{x \to 1}(x^2 + 6x - 7)}{\lim_{x \to 1}(x^2 + 5x - 6)}$

3. $\lim_{x \to 1} \dfrac{x - 3}{x^2 + 2x - 4} = \dfrac{\lim_{x \to 1}(x - 3)}{\lim_{x \to 1}(x^2 + 2x - 4)}$

4. $\dfrac{x^2 - 9}{x - 3} = x + 3$

5. $\lim_{x \to 3} \dfrac{x^2 - 9}{x - 3} = \lim_{x \to 3}(x + 3)$

6. If $\lim_{x \to 5} f(x) = 2$ and $\lim_{x \to 5} g(x) = 0$, then $\lim_{x \to 5}[f(x)/g(x)]$ does not exist.

7. If $\lim_{x \to 5} f(x) = 0$ and $\lim_{x \to 5} g(x) = 0$, then $\lim_{x \to 5}[f(x)/g(x)]$ does not exist.

8. If neither $\lim_{x \to a} f(x)$ nor $\lim_{x \to a} g(x)$ exists, then $\lim_{x \to a}[f(x) + g(x)]$ does not exist.

9. If $\lim_{x \to a} f(x)$ exists but $\lim_{x \to a} g(x)$ does not exist, then $\lim_{x \to a}[f(x) + g(x)]$ does not exist.

10. If $\lim_{x \to 6}[f(x) g(x)]$ exists, then the limit must be $f(6) g(6)$.

11. If p is a polynomial, then $\lim_{x \to b} p(x) = p(b)$.

12. If $\lim_{x \to 0} f(x) = \infty$ and $\lim_{x \to 0} g(x) = \infty$, then $\lim_{x \to 0}[f(x) - g(x)] = 0$.

13. A function can have two different horizontal asymptotes.

14. If f has domain $[0, \infty)$ and has no horizontal asymptote, then $\lim_{x \to \infty} f(x) = \infty$ or $\lim_{x \to \infty} f(x) = -\infty$.

15. If the line $x = 1$ is a vertical asymptote of $y = f(x)$, then f is not defined at 1.

16. If $f(1) > 0$ and $f(3) < 0$, then there exists a number c between 1 and 3 such that $f(c) = 0$.

17. If f is continuous at 5 and $f(5) = 2$ and $f(4) = 3$, then $\lim_{x \to 2} f(4x^2 - 11) = 2$.

18. If f is continuous on $[-1, 1]$ and $f(-1) = 4$ and $f(1) = 3$, then there exists a number r such that $|r| < 1$ and $f(r) = \pi$.

19. Let f be a function such that $\lim_{x \to 0} f(x) = 6$. Then there exists a positive number δ such that if $0 < |x| < \delta$, then $|f(x) - 6| < 1$.

20. If $f(x) > 1$ for all x and $\lim_{x \to 0} f(x)$ exists, then $\lim_{x \to 0} f(x) > 1$.

21. If f is continuous at a, then f is differentiable at a.

22. If $f'(r)$ exists, then $\lim_{x \to r} f(x) = f(r)$.

23. $\dfrac{d^2 y}{dx^2} = \left(\dfrac{dy}{dx} \right)^2$

24. The equation $x^{10} - 10x^2 + 5 = 0$ has a root in the interval $(0, 2)$.

25. If f is continuous at a, so is $|f|$.

26. If $|f|$ is continuous at a, so is f.

EXERCISES

1. The graph of f is given.

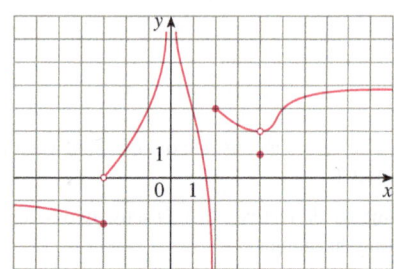

(a) Find each limit, or explain why it does not exist.
(i) $\lim_{x \to 2^+} f(x)$ (ii) $\lim_{x \to -3^+} f(x)$ (iii) $\lim_{x \to -3} f(x)$
(iv) $\lim_{x \to 4} f(x)$ (v) $\lim_{x \to 0} f(x)$ (vi) $\lim_{x \to 2^-} f(x)$
(vii) $\lim_{x \to \infty} f(x)$ (viii) $\lim_{x \to -\infty} f(x)$

(b) State the equations of the horizontal asymptotes.
(c) State the equations of the vertical asymptotes.
(d) At what numbers is f discontinuous? Explain.

2. Sketch the graph of a function f that satisfies all of the following conditions:

$\lim_{x \to -\infty} f(x) = -2$, $\lim_{x \to \infty} f(x) = 0$, $\lim_{x \to -3} f(x) = \infty$,

$\lim_{x \to 3^-} f(x) = -\infty$, $\lim_{x \to 3^+} f(x) = 2$,

f is continuous from the right at 3

3–20 Find the limit.

3. $\lim\limits_{x \to 1} e^{x^3 - x}$

4. $\lim\limits_{x \to 3} \dfrac{x^2 - 9}{x^2 + 2x - 3}$

5. $\lim\limits_{x \to -3} \dfrac{x^2 - 9}{x^2 + 2x - 3}$

6. $\lim\limits_{x \to 1^+} \dfrac{x^2 - 9}{x^2 + 2x - 3}$

7. $\lim\limits_{h \to 0} \dfrac{(h - 1)^3 + 1}{h}$

8. $\lim\limits_{t \to 2} \dfrac{t^2 - 4}{t^3 - 8}$

9. $\lim\limits_{r \to 9} \dfrac{\sqrt{r}}{(r - 9)^4}$

10. $\lim\limits_{v \to 4^+} \dfrac{4 - v}{|4 - v|}$

11. $\lim\limits_{u \to 1} \dfrac{u^4 - 1}{u^3 + 5u^2 - 6u}$

12. $\lim\limits_{x \to 3} \dfrac{\sqrt{x + 6} - x}{x^3 - 3x^2}$

13. $\lim\limits_{x \to \infty} \dfrac{\sqrt{x^2 - 9}}{2x - 6}$

14. $\lim\limits_{x \to -\infty} \dfrac{\sqrt{x^2 - 9}}{2x - 6}$

15. $\lim\limits_{x \to \pi^-} \ln(\sin x)$

16. $\lim\limits_{x \to -\infty} \dfrac{1 - 2x^2 - x^4}{5 + x - 3x^4}$

17. $\lim\limits_{x \to \infty} \left(\sqrt{x^2 + 4x + 1} - x\right)$

18. $\lim\limits_{x \to \infty} e^{x - x^2}$

19. $\lim\limits_{x \to 0^+} \tan^{-1}(1/x)$

20. $\lim\limits_{x \to 1} \left(\dfrac{1}{x - 1} + \dfrac{1}{x^2 - 3x + 2}\right)$

21–22 Use graphs to discover the asymptotes of the curve. Then prove what you have discovered.

21. $y = \dfrac{\cos^2 x}{x^2}$

22. $y = \sqrt{x^2 + x + 1} - \sqrt{x^2 - x}$

23. If $2x - 1 \leq f(x) \leq x^2$ for $0 < x < 3$, find $\lim\limits_{x \to 1} f(x)$.

24. Prove that $\lim\limits_{x \to 0} x^2 \cos(1/x^2) = 0$.

25–28 Prove the statement using the precise definition of a limit.

25. $\lim\limits_{x \to 2} (14 - 5x) = 4$

26. $\lim\limits_{x \to 0} \sqrt[3]{x} = 0$

27. $\lim\limits_{x \to 2} (x^2 - 3x) = -2$

28. $\lim\limits_{x \to 4^+} \dfrac{2}{\sqrt{x - 4}} = \infty$

29. Let
$$f(x) = \begin{cases} \sqrt{-x} & \text{if } x < 0 \\ 3 - x & \text{if } 0 \leq x < 3 \\ (x - 3)^2 & \text{if } x > 3 \end{cases}$$

(a) Evaluate each limit, if it exists.
 (i) $\lim\limits_{x \to 0^+} f(x)$ (ii) $\lim\limits_{x \to 0^-} f(x)$ (iii) $\lim\limits_{x \to 0} f(x)$
 (iv) $\lim\limits_{x \to 3^-} f(x)$ (v) $\lim\limits_{x \to 3^+} f(x)$ (vi) $\lim\limits_{x \to 3} f(x)$

(b) Where is f discontinuous?
(c) Sketch the graph of f.

30. Let
$$g(x) = \begin{cases} 2x - x^2 & \text{if } 0 \leq x \leq 2 \\ 2 - x & \text{if } 2 < x \leq 3 \\ x - 4 & \text{if } 3 < x < 4 \\ \pi & \text{if } x \geq 4 \end{cases}$$

(a) For each of the numbers 2, 3, and 4, discover whether g is continuous from the left, continuous from the right, or continuous at the number.
(b) Sketch the graph of g.

31–32 Show that the function is continuous on its domain. State the domain.

31. $h(x) = xe^{\sin x}$

32. $g(x) = \dfrac{\sqrt{x^2 - 9}}{x^2 - 2}$

33–34 Use the Intermediate Value Theorem to show that there is a root of the equation in the given interval.

33. $x^5 - x^3 + 3x - 5 = 0$, $(1, 2)$

34. $\cos \sqrt{x} = e^x - 2$, $(0, 1)$

35. (a) Find the slope of the tangent line to the curve $y = 9 - 2x^2$ at the point $(2, 1)$.
 (b) Find an equation of this tangent line.

36. Find equations of the tangent lines to the curve
$$y = \dfrac{2}{1 - 3x}$$
at the points with x-coordinates 0 and -1.

37. The displacement (in meters) of an object moving in a straight line is given by $s = 1 + 2t + \frac{1}{4}t^2$, where t is measured in seconds.
 (a) Find the average velocity over each time period.
  (i) $[1, 3]$ (ii) $[1, 2]$ (iii) $[1, 1.5]$ (iv) $[1, 1.1]$
 (b) Find the instantaneous velocity when $t = 1$.

38. According to Boyle's Law, if the temperature of a confined gas is held fixed, then the product of the pressure P and the volume V is a constant. Suppose that, for a certain gas, $PV = 800$, where P is measured in pounds per square inch and V is measured in cubic inches.
 (a) Find the average rate of change of P as V increases from 200 in³ to 250 in³.
 (b) Express V as a function of P and show that the instantaneous rate of change of V with respect to P is inversely proportional to the square of P.

39. (a) Use the definition of a derivative to find $f'(2)$, where $f(x) = x^3 - 2x$.
 (b) Find an equation of the tangent line to the curve $y = x^3 - 2x$ at the point $(2, 4)$.
 (c) Illustrate part (b) by graphing the curve and the tangent line on the same screen.

40. Find a function f and a number a such that
$$\lim\limits_{h \to 0} \dfrac{(2 + h)^6 - 64}{h} = f'(a)$$

41. The total cost of repaying a student loan at an interest rate of $r\%$ per year is $C = f(r)$.
 (a) What is the meaning of the derivative $f'(r)$? What are its units?

(b) What does the statement $f'(10) = 1200$ mean?
(c) Is $f'(r)$ always positive or does it change sign?

42–44 Trace or copy the graph of the function. Then sketch a graph of its derivative directly beneath.

42.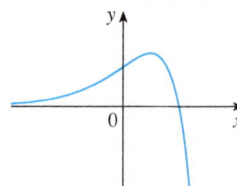

43.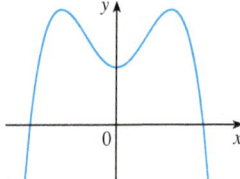

44.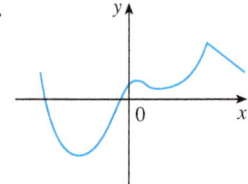

---

45. (a) If $f(x) = \sqrt{3 - 5x}$, use the definition of a derivative to find $f'(x)$.
(b) Find the domains of f and f'.
(c) Graph f and f' on a common screen. Compare the graphs to see whether your answer to part (a) is reasonable.

46. (a) Find the asymptotes of the graph of $f(x) = \dfrac{4 - x}{3 + x}$ and use them to sketch the graph.
(b) Use your graph from part (a) to sketch the graph of f'.
(c) Use the definition of a derivative to find $f'(x)$.
(d) Use a graphing device to graph f' and compare with your sketch in part (b).

47. The graph of f is shown. State, with reasons, the numbers at which f is not differentiable.

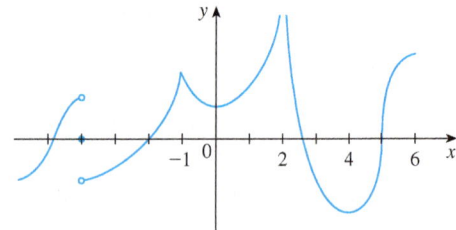

48. The figure shows the graphs of f, f', and f''. Identify each curve, and explain your choices.

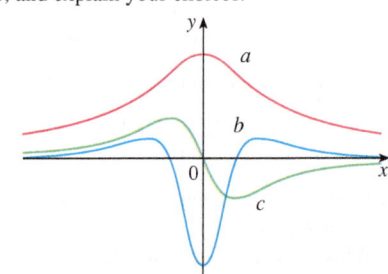

49. Sketch the graph of a function f that satisfies all of the following conditions: The domain of f is all real numbers except 0, $\lim_{x \to 0^-} f(x) = 1$, $\lim_{x \to 0^+} f(x) = 0$, $f'(x) > 0$ for all x in the domain of f, $\lim_{x \to -\infty} f'(x) = 0$, $\lim_{x \to \infty} f'(x) = 1$.

50. Let $P(t)$ be the percentage of Americans under the age of 18 at time t. The table gives values of this function in census years from 1950 to 2010.

t	$P(t)$	t	$P(t)$
1950	31.1	1990	25.7
1960	35.7	2000	25.7
1970	34.0	2010	24.0
1980	28.0		

(a) What is the meaning of $P'(t)$? What are its units?
(b) Construct a table of estimated values for $P'(t)$.
(c) Graph P and P'.
(d) How would it be possible to get more accurate values for $P'(t)$?

51. Let $B(t)$ be the number of US $20 bills in circulation at time t. The table gives values of this function from 1990 to 2010, as of December 31, in billions. Interpret and estimate the value of $B'(2000)$.

t	1990	1995	2000	2005	2010
$B(t)$	3.45	4.21	4.93	5.77	6.53

52. The *total fertility rate* at time t, denoted by $F(t)$, is an estimate of the average number of children born to each woman (assuming that current birth rates remain constant). The graph of the total fertility rate in the United States shows the fluctuations from 1940 to 2010.
(a) Estimate the values of $F'(1950)$, $F'(1965)$, and $F'(1987)$.
(b) What are the meanings of these derivatives?
(c) Can you suggest reasons for the values of these derivatives?

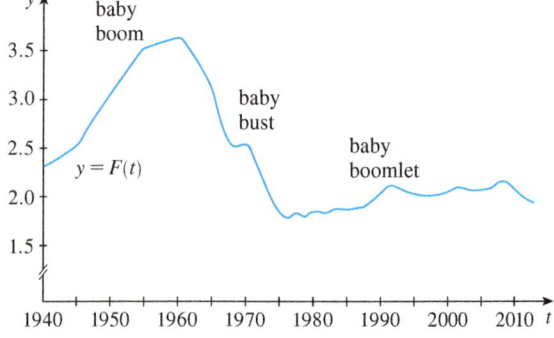

53. Suppose that $|f(x)| \leq g(x)$ for all x, where $\lim_{x \to a} g(x) = 0$. Find $\lim_{x \to a} f(x)$.

54. Let $f(x) = [\![x]\!] + [\![-x]\!]$.
(a) For what values of a does $\lim_{x \to a} f(x)$ exist?
(b) At what numbers is f discontinuous?

Problems Plus

In our discussion of the principles of problem solving we considered the problem-solving strategy of *introducing something extra* (see page 71). In the following example we show how this principle is sometimes useful when we evaluate limits. The idea is to change the variable—to introduce a new variable that is related to the original variable—in such a way as to make the problem simpler. Later, in Section 5.5, we will make more extensive use of this general idea.

EXAMPLE 1 Evaluate $\lim\limits_{x \to 0} \dfrac{\sqrt[3]{1 + cx} - 1}{x}$, where c is a constant.

SOLUTION As it stands, this limit looks challenging. In Section 2.3 we evaluated several limits in which both numerator and denominator approached 0. There our strategy was to perform some sort of algebraic manipulation that led to a simplifying cancellation, but here it's not clear what kind of algebra is necessary.

So we introduce a new variable t by the equation

$$t = \sqrt[3]{1 + cx}$$

We also need to express x in terms of t, so we solve this equation:

$$t^3 = 1 + cx \qquad x = \frac{t^3 - 1}{c} \quad (\text{if } c \neq 0)$$

Notice that $x \to 0$ is equivalent to $t \to 1$. This allows us to convert the given limit into one involving the variable t:

$$\lim_{x \to 0} \frac{\sqrt[3]{1 + cx} - 1}{x} = \lim_{t \to 1} \frac{t - 1}{(t^3 - 1)/c}$$

$$= \lim_{t \to 1} \frac{c(t - 1)}{t^3 - 1}$$

The change of variable allowed us to replace a relatively complicated limit by a simpler one of a type that we have seen before. Factoring the denominator as a difference of cubes, we get

$$\lim_{t \to 1} \frac{c(t - 1)}{t^3 - 1} = \lim_{t \to 1} \frac{c(t - 1)}{(t - 1)(t^2 + t + 1)}$$

$$= \lim_{t \to 1} \frac{c}{t^2 + t + 1} = \frac{c}{3}$$

In making the change of variable we had to rule out the case $c = 0$. But if $c = 0$, the function is 0 for all nonzero x and so its limit is 0. Therefore, in all cases, the limit is $c/3$. ∎

The following problems are meant to test and challenge your problem-solving skills. Some of them require a considerable amount of time to think through, so don't be discouraged if you can't solve them right away. If you get stuck, you might find it helpful to refer to the discussion of the principles of problem solving on page 71.

Problems

1. Evaluate $\lim\limits_{x \to 1} \dfrac{\sqrt[3]{x} - 1}{\sqrt{x} - 1}$.

2. Find numbers a and b such that $\lim\limits_{x \to 0} \dfrac{\sqrt{ax + b} - 2}{x} = 1$.

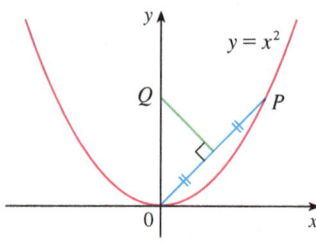

FIGURE FOR PROBLEM 4

3. Evaluate $\displaystyle\lim_{x \to 0} \frac{|2x-1| - |2x+1|}{x}$.

4. The figure shows a point P on the parabola $y = x^2$ and the point Q where the perpendicular bisector of OP intersects the y-axis. As P approaches the origin along the parabola, what happens to Q? Does it have a limiting position? If so, find it.

5. Evaluate the following limits, if they exist, where $[\![x]\!]$ denotes the greatest integer function.
   (a) $\displaystyle\lim_{x \to 0} \frac{[\![x]\!]}{x}$
   (b) $\displaystyle\lim_{x \to 0} x [\![1/x]\!]$

6. Sketch the region in the plane defined by each of the following equations.
   (a) $[\![x]\!]^2 + [\![y]\!]^2 = 1$
   (b) $[\![x]\!]^2 - [\![y]\!]^2 = 3$
   (c) $[\![x + y]\!]^2 = 1$
   (d) $[\![x]\!] + [\![y]\!] = 1$

7. Find all values of a such that f is continuous on $\mathbb{R}$:

$$f(x) = \begin{cases} x + 1 & \text{if } x \leq a \\ x^2 & \text{if } x > a \end{cases}$$

8. A **fixed point** of a function f is a number c in its domain such that $f(c) = c$. (The function doesn't move c; it stays fixed.)
   (a) Sketch the graph of a continuous function with domain $[0, 1]$ whose range also lies in $[0, 1]$. Locate a fixed point of f.
   (b) Try to draw the graph of a continuous function with domain $[0, 1]$ and range in $[0, 1]$ that does *not* have a fixed point. What is the obstacle?
   (c) Use the Intermediate Value Theorem to prove that any continuous function with domain $[0, 1]$ and range in $[0, 1]$ must have a fixed point.

9. If $\lim_{x \to a} [f(x) + g(x)] = 2$ and $\lim_{x \to a} [f(x) - g(x)] = 1$, find $\lim_{x \to a} [f(x) g(x)]$.

10. (a) The figure shows an isosceles triangle ABC with $\angle B = \angle C$. The bisector of angle B intersects the side AC at the point P. Suppose that the base BC remains fixed but the altitude $|AM|$ of the triangle approaches 0, so A approaches the midpoint M of BC. What happens to P during this process? Does it have a limiting position? If so, find it.
    (b) Try to sketch the path traced out by P during this process. Then find an equation of this curve and use this equation to sketch the curve.

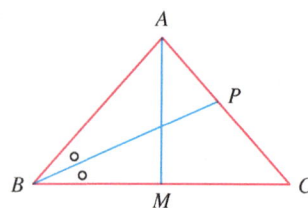

FIGURE FOR PROBLEM 10

11. (a) If we start from 0° latitude and proceed in a westerly direction, we can let $T(x)$ denote the temperature at the point x at any given time. Assuming that T is a continuous function of x, show that at any fixed time there are at least two diametrically opposite points on the equator that have exactly the same temperature.
    (b) Does the result in part (a) hold for points lying on any circle on the earth's surface?
    (c) Does the result in part (a) hold for barometric pressure and for altitude above sea level?

12. If f is a differentiable function and $g(x) = xf(x)$, use the definition of a derivative to show that $g'(x) = xf'(x) + f(x)$.

13. Suppose f is a function that satisfies the equation

$$f(x + y) = f(x) + f(y) + x^2 y + xy^2$$

for all real numbers x and y. Suppose also that

$$\lim_{x \to 0} \frac{f(x)}{x} = 1$$

(a) Find $f(0)$.
(b) Find $f'(0)$.
(c) Find $f'(x)$.

14. Suppose f is a function with the property that $|f(x)| \leq x^2$ for all x. Show that $f(0) = 0$. Then show that $f'(0) = 0$.

3 Differentiation Rules

In the project on page 208 you will calculate the distance from an airport runway at which a pilot should start descent for a smooth landing.

WE HAVE SEEN HOW TO INTERPRET derivatives as slopes and rates of change. We have seen how to estimate derivatives of functions given by tables of values. We have learned how to graph derivatives of functions that are defined graphically. We have used the definition of a derivative to calculate the derivatives of functions defined by formulas. But it would be tedious if we always had to use the definition, so in this chapter we develop rules for finding derivatives without having to use the definition directly. These differentiation rules enable us to calculate with relative ease the derivatives of polynomials, rational functions, algebraic functions, exponential and logarithmic functions, and trigonometric and inverse trigonometric functions. We then use these rules to solve problems involving rates of change and the approximation of functions.

3.1 Derivatives of Polynomials and Exponential Functions

In this section we learn how to differentiate constant functions, power functions, polynomials, and exponential functions.

Let's start with the simplest of all functions, the constant function $f(x) = c$. The graph of this function is the horizontal line $y = c$, which has slope 0, so we must have $f'(x) = 0$. (See Figure 1.) A formal proof, from the definition of a derivative, is also easy:

$$f'(x) = \lim_{h \to 0} \frac{f(x+h) - f(x)}{h} = \lim_{h \to 0} \frac{c - c}{h} = \lim_{h \to 0} 0 = 0$$

In Leibniz notation, we write this rule as follows.

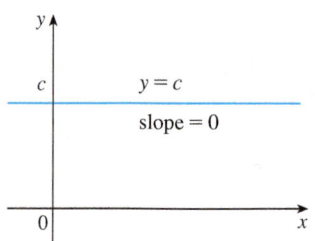

FIGURE 1
The graph of $f(x) = c$ is the line $y = c$, so $f'(x) = 0$.

Derivative of a Constant Function

$$\frac{d}{dx}(c) = 0$$

Power Functions

We next look at the functions $f(x) = x^n$, where n is a positive integer. If $n = 1$, the graph of $f(x) = x$ is the line $y = x$, which has slope 1. (See Figure 2.) So

[1] $$\frac{d}{dx}(x) = 1$$

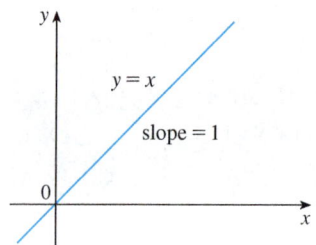

FIGURE 2
The graph of $f(x) = x$ is the line $y = x$, so $f'(x) = 1$.

(You can also verify Equation 1 from the definition of a derivative.) We have already investigated the cases $n = 2$ and $n = 3$. In fact, in Section 2.8 (Exercises 19 and 20) we found that

[2] $$\frac{d}{dx}(x^2) = 2x \qquad \frac{d}{dx}(x^3) = 3x^2$$

For $n = 4$ we find the derivative of $f(x) = x^4$ as follows:

$$f'(x) = \lim_{h \to 0} \frac{f(x+h) - f(x)}{h} = \lim_{h \to 0} \frac{(x+h)^4 - x^4}{h}$$

$$= \lim_{h \to 0} \frac{x^4 + 4x^3h + 6x^2h^2 + 4xh^3 + h^4 - x^4}{h}$$

$$= \lim_{h \to 0} \frac{4x^3h + 6x^2h^2 + 4xh^3 + h^4}{h}$$

$$= \lim_{h \to 0} (4x^3 + 6x^2h + 4xh^2 + h^3) = 4x^3$$

Thus

[3] $$\frac{d}{dx}(x^4) = 4x^3$$

Comparing the equations in (1), (2), and (3), we see a pattern emerging. It seems to be a reasonable guess that, when n is a positive integer, $(d/dx)(x^n) = nx^{n-1}$. This turns out to be true.

The Power Rule If n is a positive integer, then
$$\frac{d}{dx}(x^n) = nx^{n-1}$$

FIRST PROOF The formula
$$x^n - a^n = (x - a)(x^{n-1} + x^{n-2}a + \cdots + xa^{n-2} + a^{n-1})$$

can be verified simply by multiplying out the right-hand side (or by summing the second factor as a geometric series). If $f(x) = x^n$, we can use Equation 2.7.5 for $f'(a)$ and the equation above to write

$$f'(a) = \lim_{x \to a} \frac{f(x) - f(a)}{x - a} = \lim_{x \to a} \frac{x^n - a^n}{x - a}$$

$$= \lim_{x \to a} (x^{n-1} + x^{n-2}a + \cdots + xa^{n-2} + a^{n-1})$$

$$= a^{n-1} + a^{n-2}a + \cdots + aa^{n-2} + a^{n-1}$$

$$= na^{n-1}$$

SECOND PROOF

$$f'(x) = \lim_{h \to 0} \frac{f(x + h) - f(x)}{h} = \lim_{h \to 0} \frac{(x + h)^n - x^n}{h}$$

In finding the derivative of x^4 we had to expand $(x + h)^4$. Here we need to expand $(x + h)^n$ and we use the Binomial Theorem to do so:

The Binomial Theorem is given on Reference Page 1.

$$f'(x) = \lim_{h \to 0} \frac{\left[x^n + nx^{n-1}h + \frac{n(n-1)}{2}x^{n-2}h^2 + \cdots + nxh^{n-1} + h^n \right] - x^n}{h}$$

$$= \lim_{h \to 0} \frac{nx^{n-1}h + \frac{n(n-1)}{2}x^{n-2}h^2 + \cdots + nxh^{n-1} + h^n}{h}$$

$$= \lim_{h \to 0} \left[nx^{n-1} + \frac{n(n-1)}{2}x^{n-2}h + \cdots + nxh^{n-2} + h^{n-1} \right]$$

$$= nx^{n-1}$$

because every term except the first has h as a factor and therefore approaches 0. ∎

We illustrate the Power Rule using various notations in Example 1.

EXAMPLE 1
(a) If $f(x) = x^6$, then $f'(x) = 6x^5$.
(b) If $y = x^{1000}$, then $y' = 1000x^{999}$.
(c) If $y = t^4$, then $\dfrac{dy}{dt} = 4t^3$.
(d) $\dfrac{d}{dr}(r^3) = 3r^2$ ∎

What about power functions with negative integer exponents? In Exercise 65 we ask you to verify from the definition of a derivative that

$$\frac{d}{dx}\left(\frac{1}{x}\right) = -\frac{1}{x^2}$$

We can rewrite this equation as

$$\frac{d}{dx}(x^{-1}) = (-1)x^{-2}$$

and so the Power Rule is true when $n = -1$. In fact, we will show in the next section [Exercise 3.2.64(c)] that it holds for all negative integers.

What if the exponent is a fraction? In Example 2.8.3 we found that

$$\frac{d}{dx}\sqrt{x} = \frac{1}{2\sqrt{x}}$$

which can be written as

$$\frac{d}{dx}(x^{1/2}) = \tfrac{1}{2}x^{-1/2}$$

This shows that the Power Rule is true even when $n = \tfrac{1}{2}$. In fact, we will show in Section 3.6 that it is true for all real numbers n.

> **The Power Rule (General Version)** If n is any real number, then
> $$\frac{d}{dx}(x^n) = nx^{n-1}$$

EXAMPLE 2 Differentiate:
(a) $f(x) = \dfrac{1}{x^2}$
(b) $y = \sqrt[3]{x^2}$

SOLUTION In each case we rewrite the function as a power of x.
(a) Since $f(x) = x^{-2}$, we use the Power Rule with $n = -2$:

$$f'(x) = \frac{d}{dx}(x^{-2}) = -2x^{-2-1} = -2x^{-3} = -\frac{2}{x^3}$$

(b) $\dfrac{dy}{dx} = \dfrac{d}{dx}\left(\sqrt[3]{x^2}\right) = \dfrac{d}{dx}(x^{2/3}) = \tfrac{2}{3}x^{(2/3)-1} = \tfrac{2}{3}x^{-1/3}$ ∎

Observe from Figure 3 that the function y in Example 2(b) is increasing when y' is positive and is decreasing when y' is negative. In Chapter 4 we will prove that, in general, *a function increases when its derivative is positive and decreases when its derivative is negative.*

Figure 3 shows the function y in Example 2(b) and its derivative y'. Notice that y is not differentiable at 0 (y' is not defined there).

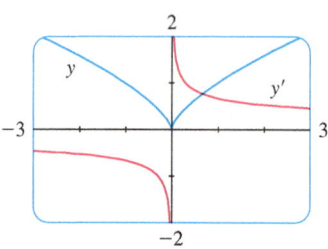

FIGURE 3
$y = \sqrt[3]{x^2}$

The Power Rule enables us to find tangent lines without having to resort to the definition of a derivative. It also enables us to find *normal lines*. The **normal line** to a curve C at a point P is the line through P that is perpendicular to the tangent line at P. (In the study of optics, one needs to consider the angle between a light ray and the normal line to a lens.)

EXAMPLE 3 Find equations of the tangent line and normal line to the curve $y = x\sqrt{x}$ at the point $(1, 1)$. Illustrate by graphing the curve and these lines.

SOLUTION The derivative of $f(x) = x\sqrt{x} = xx^{1/2} = x^{3/2}$ is

$$f'(x) = \tfrac{3}{2} x^{(3/2)-1} = \tfrac{3}{2} x^{1/2} = \tfrac{3}{2}\sqrt{x}$$

So the slope of the tangent line at $(1, 1)$ is $f'(1) = \tfrac{3}{2}$. Therefore an equation of the tangent line is

$$y - 1 = \tfrac{3}{2}(x - 1) \quad \text{or} \quad y = \tfrac{3}{2}x - \tfrac{1}{2}$$

The normal line is perpendicular to the tangent line, so its slope is the negative reciprocal of $\tfrac{3}{2}$, that is, $-\tfrac{2}{3}$. Thus an equation of the normal line is

$$y - 1 = -\tfrac{2}{3}(x - 1) \quad \text{or} \quad y = -\tfrac{2}{3}x + \tfrac{5}{3}$$

We graph the curve and its tangent line and normal line in Figure 4.

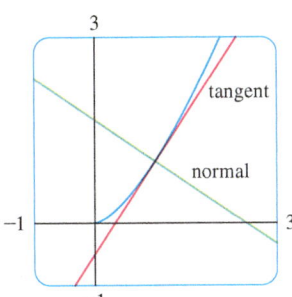

FIGURE 4
$y = x\sqrt{x}$

New Derivatives from Old

When new functions are formed from old functions by addition, subtraction, or multiplication by a constant, their derivatives can be calculated in terms of derivatives of the old functions. In particular, the following formula says that *the derivative of a constant times a function is the constant times the derivative of the function.*

Geometric Interpretation of the Constant Multiple Rule

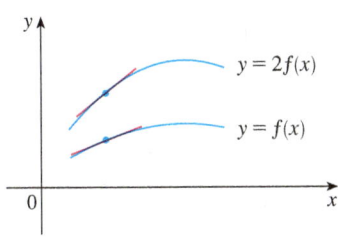

Multiplying by $c = 2$ stretches the graph vertically by a factor of 2. All the rises have been doubled but the runs stay the same. So the slopes are doubled too.

The Constant Multiple Rule If c is a constant and f is a differentiable function, then

$$\frac{d}{dx}[cf(x)] = c \frac{d}{dx} f(x)$$

PROOF Let $g(x) = cf(x)$. Then

$$g'(x) = \lim_{h \to 0} \frac{g(x + h) - g(x)}{h} = \lim_{h \to 0} \frac{cf(x + h) - cf(x)}{h}$$

$$= \lim_{h \to 0} c \left[\frac{f(x + h) - f(x)}{h} \right]$$

$$= c \lim_{h \to 0} \frac{f(x + h) - f(x)}{h} \quad \text{(by Limit Law 3)}$$

$$= cf'(x)$$

EXAMPLE 4

(a) $\dfrac{d}{dx}(3x^4) = 3\dfrac{d}{dx}(x^4) = 3(4x^3) = 12x^3$

(b) $\dfrac{d}{dx}(-x) = \dfrac{d}{dx}[(-1)x] = (-1)\dfrac{d}{dx}(x) = -1(1) = -1$ ∎

The next rule tells us that *the derivative of a sum of functions is the sum of the derivatives.*

Using prime notation, we can write the Sum Rule as
$(f + g)' = f' + g'$

The Sum Rule If f and g are both differentiable, then

$$\dfrac{d}{dx}[f(x) + g(x)] = \dfrac{d}{dx}f(x) + \dfrac{d}{dx}g(x)$$

PROOF Let $F(x) = f(x) + g(x)$. Then

$$F'(x) = \lim_{h \to 0} \dfrac{F(x + h) - F(x)}{h}$$

$$= \lim_{h \to 0} \dfrac{[f(x + h) + g(x + h)] - [f(x) + g(x)]}{h}$$

$$= \lim_{h \to 0} \left[\dfrac{f(x + h) - f(x)}{h} + \dfrac{g(x + h) - g(x)}{h}\right]$$

$$= \lim_{h \to 0} \dfrac{f(x + h) - f(x)}{h} + \lim_{h \to 0} \dfrac{g(x + h) - g(x)}{h} \quad \text{(by Limit Law 1)}$$

$$= f'(x) + g'(x) \qquad \blacksquare$$

The Sum Rule can be extended to the sum of any number of functions. For instance, using this theorem twice, we get

$$(f + g + h)' = [(f + g) + h]' = (f + g)' + h' = f' + g' + h'$$

By writing $f - g$ as $f + (-1)g$ and applying the Sum Rule and the Constant Multiple Rule, we get the following formula.

The Difference Rule If f and g are both differentiable, then

$$\dfrac{d}{dx}[f(x) - g(x)] = \dfrac{d}{dx}f(x) - \dfrac{d}{dx}g(x)$$

The Constant Multiple Rule, the Sum Rule, and the Difference Rule can be combined with the Power Rule to differentiate any polynomial, as the following examples demonstrate.

EXAMPLE 5

$$\frac{d}{dx}(x^8 + 12x^5 - 4x^4 + 10x^3 - 6x + 5)$$

$$= \frac{d}{dx}(x^8) + 12\frac{d}{dx}(x^5) - 4\frac{d}{dx}(x^4) + 10\frac{d}{dx}(x^3) - 6\frac{d}{dx}(x) + \frac{d}{dx}(5)$$

$$= 8x^7 + 12(5x^4) - 4(4x^3) + 10(3x^2) - 6(1) + 0$$

$$= 8x^7 + 60x^4 - 16x^3 + 30x^2 - 6$$

EXAMPLE 6 Find the points on the curve $y = x^4 - 6x^2 + 4$ where the tangent line is horizontal.

SOLUTION Horizontal tangents occur where the derivative is zero. We have

$$\frac{dy}{dx} = \frac{d}{dx}(x^4) - 6\frac{d}{dx}(x^2) + \frac{d}{dx}(4)$$

$$= 4x^3 - 12x + 0 = 4x(x^2 - 3)$$

Thus $dy/dx = 0$ if $x = 0$ or $x^2 - 3 = 0$, that is, $x = \pm\sqrt{3}$. So the given curve has horizontal tangents when $x = 0, \sqrt{3},$ and $-\sqrt{3}$. The corresponding points are $(0, 4)$, $(\sqrt{3}, -5)$, and $(-\sqrt{3}, -5)$. (See Figure 5.)

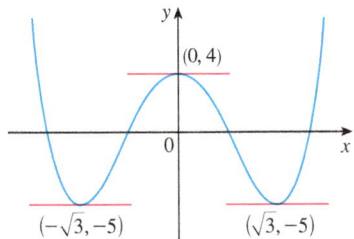

FIGURE 5
The curve $y = x^4 - 6x^2 + 4$ and its horizontal tangents

EXAMPLE 7 The equation of motion of a particle is $s = 2t^3 - 5t^2 + 3t + 4$, where s is measured in centimeters and t in seconds. Find the acceleration as a function of time. What is the acceleration after 2 seconds?

SOLUTION The velocity and acceleration are

$$v(t) = \frac{ds}{dt} = 6t^2 - 10t + 3$$

$$a(t) = \frac{dv}{dt} = 12t - 10$$

The acceleration after 2 s is $a(2) = 14$ cm/s².

Exponential Functions

Let's try to compute the derivative of the exponential function $f(x) = b^x$ using the definition of a derivative:

$$f'(x) = \lim_{h \to 0} \frac{f(x + h) - f(x)}{h} = \lim_{h \to 0} \frac{b^{x+h} - b^x}{h}$$

$$= \lim_{h \to 0} \frac{b^x b^h - b^x}{h} = \lim_{h \to 0} \frac{b^x(b^h - 1)}{h}$$

The factor b^x doesn't depend on h, so we can take it in front of the limit:

$$f'(x) = b^x \lim_{h \to 0} \frac{b^h - 1}{h}$$

Notice that the limit is the value of the derivative of f at 0, that is,

$$\lim_{h \to 0} \frac{b^h - 1}{h} = f'(0)$$

Therefore we have shown that if the exponential function $f(x) = b^x$ is differentiable at 0, then it is differentiable everywhere and

4 $$f'(x) = f'(0)b^x$$

This equation says that *the rate of change of any exponential function is proportional to the function itself.* (The slope is proportional to the height.)

Numerical evidence for the existence of $f'(0)$ is given in the table at the left for the cases $b = 2$ and $b = 3$. (Values are stated correct to four decimal places.) It appears that the limits exist and

h	$\dfrac{2^h - 1}{h}$	$\dfrac{3^h - 1}{h}$
0.1	0.7177	1.1612
0.01	0.6956	1.1047
0.001	0.6934	1.0992
0.0001	0.6932	1.0987

$$\text{for } b = 2, \quad f'(0) = \lim_{h \to 0} \frac{2^h - 1}{h} \approx 0.69$$

$$\text{for } b = 3, \quad f'(0) = \lim_{h \to 0} \frac{3^h - 1}{h} \approx 1.10$$

In fact, it can be proved that these limits exist and, correct to six decimal places, the values are

$$\frac{d}{dx}(2^x)\bigg|_{x=0} \approx 0.693147 \qquad \frac{d}{dx}(3^x)\bigg|_{x=0} \approx 1.098612$$

Thus, from Equation 4, we have

5 $$\frac{d}{dx}(2^x) \approx (0.69)2^x \qquad \frac{d}{dx}(3^x) \approx (1.10)3^x$$

Of all possible choices for the base b in Equation 4, the simplest differentiation formula occurs when $f'(0) = 1$. In view of the estimates of $f'(0)$ for $b = 2$ and $b = 3$, it seems reasonable that there is a number b between 2 and 3 for which $f'(0) = 1$. It is traditional to denote this value by the letter e. (In fact, that is how we introduced e in Section 1.4.) Thus we have the following definition.

In Exercise 1 we will see that e lies between 2.7 and 2.8. Later we will be able to show that, correct to five decimal places,
$$e \approx 2.71828$$

Definition of the Number e

e is the number such that $\quad \displaystyle\lim_{h \to 0} \frac{e^h - 1}{h} = 1$

Geometrically, this means that of all the possible exponential functions $y = b^x$, the function $f(x) = e^x$ is the one whose tangent line at $(0, 1)$ has a slope $f'(0)$ that is exactly 1. (See Figures 6 and 7.)

SECTION 3.1 Derivatives of Polynomials and Exponential Functions 179

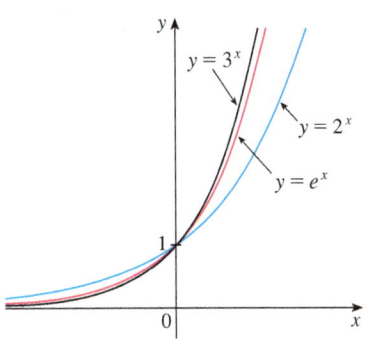

FIGURE 6

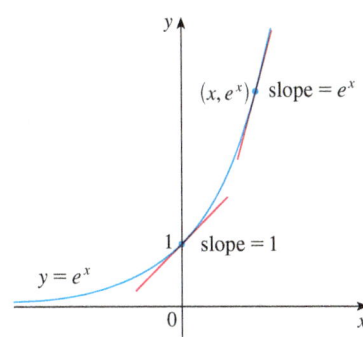

FIGURE 7

If we put $b = e$ and, therefore, $f'(0) = 1$ in Equation 4, it becomes the following important differentiation formula.

Derivative of the Natural Exponential Function
$$\frac{d}{dx}(e^x) = e^x$$

TEC Visual 3.1 uses the slope-a-scope to illustrate this formula.

Thus the exponential function $f(x) = e^x$ has the property that it is its own derivative. The geometrical significance of this fact is that the slope of a tangent line to the curve $y = e^x$ is equal to the y-coordinate of the point (see Figure 7).

EXAMPLE 8 If $f(x) = e^x - x$, find f' and f''. Compare the graphs of f and f'.

SOLUTION Using the Difference Rule, we have

$$f'(x) = \frac{d}{dx}(e^x - x) = \frac{d}{dx}(e^x) - \frac{d}{dx}(x) = e^x - 1$$

In Section 2.8 we defined the second derivative as the derivative of f', so

$$f''(x) = \frac{d}{dx}(e^x - 1) = \frac{d}{dx}(e^x) - \frac{d}{dx}(1) = e^x$$

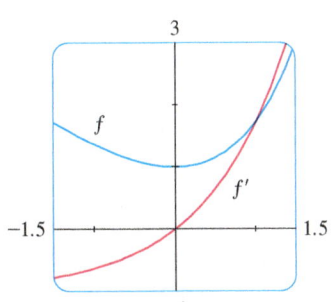

FIGURE 8

The function f and its derivative f' are graphed in Figure 8. Notice that f has a horizontal tangent when $x = 0$; this corresponds to the fact that $f'(0) = 0$. Notice also that, for $x > 0$, $f'(x)$ is positive and f is increasing. When $x < 0$, $f'(x)$ is negative and f is decreasing.

EXAMPLE 9 At what point on the curve $y = e^x$ is the tangent line parallel to the line $y = 2x$?

SOLUTION Since $y = e^x$, we have $y' = e^x$. Let the x-coordinate of the point in question be a. Then the slope of the tangent line at that point is e^a. This tangent line will be parallel to the line $y = 2x$ if it has the same slope, that is, 2. Equating slopes, we get

$$e^a = 2 \qquad a = \ln 2$$

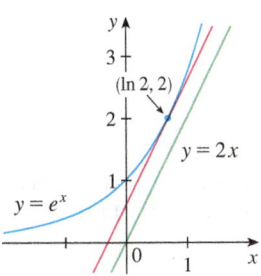

FIGURE 9

Therefore the required point is $(a, e^a) = (\ln 2, 2)$. (See Figure 9.)

3.1 EXERCISES

1. (a) How is the number e defined?
(b) Use a calculator to estimate the values of the limits

$$\lim_{h \to 0} \frac{2.7^h - 1}{h} \quad \text{and} \quad \lim_{h \to 0} \frac{2.8^h - 1}{h}$$

correct to two decimal places. What can you conclude about the value of e?

2. (a) Sketch, by hand, the graph of the function $f(x) = e^x$, paying particular attention to how the graph crosses the y-axis. What fact allows you to do this?
(b) What types of functions are $f(x) = e^x$ and $g(x) = x^e$? Compare the differentiation formulas for f and g.
(c) Which of the two functions in part (b) grows more rapidly when x is large?

3–32 Differentiate the function.

3. $f(x) = 2^{40}$

4. $f(x) = e^5$

5. $f(x) = 5.2x + 2.3$

6. $g(x) = \frac{7}{4}x^2 - 3x + 12$

7. $f(t) = 2t^3 - 3t^2 - 4t$

8. $f(t) = 1.4t^5 - 2.5t^2 + 6.7$

9. $g(x) = x^2(1 - 2x)$

10. $H(u) = (3u - 1)(u + 2)$

11. $g(t) = 2t^{-3/4}$

12. $B(y) = cy^{-6}$

13. $F(r) = \dfrac{5}{r^3}$

14. $y = x^{5/3} - x^{2/3}$

15. $R(a) = (3a + 1)^2$

16. $h(t) = \sqrt[4]{t} - 4e^t$

17. $S(p) = \sqrt{p} - p$

18. $y = \sqrt[3]{x}(2 + x)$

19. $y = 3e^x + \dfrac{4}{\sqrt[3]{x}}$

20. $S(R) = 4\pi R^2$

21. $h(u) = Au^3 + Bu^2 + Cu$

22. $y = \dfrac{\sqrt{x} + x}{x^2}$

23. $y = \dfrac{x^2 + 4x + 3}{\sqrt{x}}$

24. $G(t) = \sqrt{5t} + \dfrac{\sqrt{7}}{t}$

25. $j(x) = x^{2.4} + e^{2.4}$

26. $k(r) = e^r + r^e$

27. $G(q) = (1 + q^{-1})^2$

28. $F(z) = \dfrac{A + Bz + Cz^2}{z^2}$

29. $f(v) = \dfrac{\sqrt[3]{v} - 2ve^v}{v}$

30. $D(t) = \dfrac{1 + 16t^2}{(4t)^3}$

31. $z = \dfrac{A}{y^{10}} + Be^y$

32. $y = e^{x+1} + 1$

33–36 Find an equation of the tangent line to the curve at the given point.

33. $y = 2x^3 - x^2 + 2$, $(1, 3)$

34. $y = 2e^x + x$, $(0, 2)$

35. $y = x + \dfrac{2}{x}$, $(2, 3)$

36. $y = \sqrt[4]{x} - x$, $(1, 0)$

37–38 Find equations of the tangent line and normal line to the curve at the given point.

37. $y = x^4 + 2e^x$, $(0, 2)$

38. $y^2 = x^3$, $(1, 1)$

39–40 Find an equation of the tangent line to the curve at the given point. Illustrate by graphing the curve and the tangent line on the same screen.

39. $y = 3x^2 - x^3$, $(1, 2)$

40. $y = x - \sqrt{x}$, $(1, 0)$

41–42 Find $f'(x)$. Compare the graphs of f and f' and use them to explain why your answer is reasonable.

41. $f(x) = x^4 - 2x^3 + x^2$

42. $f(x) = x^5 - 2x^3 + x - 1$

43. (a) Graph the function
$$f(x) = x^4 - 3x^3 - 6x^2 + 7x + 30$$
in the viewing rectangle $[-3, 5]$ by $[-10, 50]$.
(b) Using the graph in part (a) to estimate slopes, make a rough sketch, by hand, of the graph of f'. (See Example 2.8.1.)
(c) Calculate $f'(x)$ and use this expression, with a graphing device, to graph f'. Compare with your sketch in part (b).

44. (a) Graph the function $g(x) = e^x - 3x^2$ in the viewing rectangle $[-1, 4]$ by $[-8, 8]$.
(b) Using the graph in part (a) to estimate slopes, make a rough sketch, by hand, of the graph of g'. (See Example 2.8.1.)
(c) Calculate $g'(x)$ and use this expression, with a graphing device, to graph g'. Compare with your sketch in part (b).

45–46 Find the first and second derivatives of the function.

45. $f(x) = 0.001x^5 - 0.02x^3$

46. $G(r) = \sqrt{r} + \sqrt[3]{r}$

47–48 Find the first and second derivatives of the function. Check to see that your answers are reasonable by comparing the graphs of f, f', and f''.

47. $f(x) = 2x - 5x^{3/4}$

48. $f(x) = e^x - x^3$

49. The equation of motion of a particle is $s = t^3 - 3t$, where s is in meters and t is in seconds. Find
(a) the velocity and acceleration as functions of t,
(b) the acceleration after 2 s, and
(c) the acceleration when the velocity is 0.

50. The equation of motion of a particle is $s = t^4 - 2t^3 + t^2 - t$, where s is in meters and t is in seconds.
(a) Find the velocity and acceleration as functions of t.
(b) Find the acceleration after 1 s.
(c) Graph the position, velocity, and acceleration functions on the same screen.

51. Biologists have proposed a cubic polynomial to model the length L of Alaskan rockfish at age A:

$$L = 0.0155A^3 - 0.372A^2 + 3.95A + 1.21$$

where L is measured in inches and A in years. Calculate

$$\left.\frac{dL}{dA}\right|_{A=12}$$

and interpret your answer.

52. The number of tree species S in a given area A in the Pasoh Forest Reserve in Malaysia has been modeled by the power function

$$S(A) = 0.882A^{0.842}$$

where A is measured in square meters. Find $S'(100)$ and interpret your answer.

Source: Adapted from K. Kochummen et al., "Floristic Composition of Pasoh Forest Reserve, A Lowland Rain Forest in Peninsular Malaysia," *Journal of Tropical Forest Science* 3 (1991):1–13.

53. Boyle's Law states that when a sample of gas is compressed at a constant temperature, the pressure P of the gas is inversely proportional to the volume V of the gas.
(a) Suppose that the pressure of a sample of air that occupies 0.106 m^3 at $25°C$ is 50 kPa. Write V as a function of P.
(b) Calculate dV/dP when $P = 50$ kPa. What is the meaning of the derivative? What are its units?

54. Car tires need to be inflated properly because overinflation or underinflation can cause premature tread wear. The data in the table show tire life L (in thousands of miles) for a certain type of tire at various pressures P (in lb/in²).

P	26	28	31	35	38	42	45
L	50	66	78	81	74	70	59

(a) Use a calculator to model tire life with a quadratic function of the pressure.
(b) Use the model to estimate dL/dP when $P = 30$ and when $P = 40$. What is the meaning of the derivative? What are the units? What is the significance of the signs of the derivatives?

55. Find the points on the curve $y = 2x^3 + 3x^2 - 12x + 1$ where the tangent is horizontal.

56. For what value of x does the graph of $f(x) = e^x - 2x$ have a horizontal tangent?

57. Show that the curve $y = 2e^x + 3x + 5x^3$ has no tangent line with slope 2.

58. Find an equation of the tangent line to the curve $y = x^4 + 1$ that is parallel to the line $32x - y = 15$.

59. Find equations of both lines that are tangent to the curve $y = x^3 - 3x^2 + 3x - 3$ and are parallel to the line $3x - y = 15$.

60. At what point on the curve $y = 1 + 2e^x - 3x$ is the tangent line parallel to the line $3x - y = 5$? Illustrate by graphing the curve and both lines.

61. Find an equation of the normal line to the curve $y = \sqrt{x}$ that is parallel to the line $2x + y = 1$.

62. Where does the normal line to the parabola $y = x^2 - 1$ at the point $(-1, 0)$ intersect the parabola a second time? Illustrate with a sketch.

63. Draw a diagram to show that there are two tangent lines to the parabola $y = x^2$ that pass through the point $(0, -4)$. Find the coordinates of the points where these tangent lines intersect the parabola.

64. (a) Find equations of both lines through the point $(2, -3)$ that are tangent to the parabola $y = x^2 + x$.
(b) Show that there is no line through the point $(2, 7)$ that is tangent to the parabola. Then draw a diagram to see why.

65. Use the definition of a derivative to show that if $f(x) = 1/x$, then $f'(x) = -1/x^2$. (This proves the Power Rule for the case $n = -1$.)

66. Find the nth derivative of each function by calculating the first few derivatives and observing the pattern that occurs.
(a) $f(x) = x^n$ (b) $f(x) = 1/x$

67. Find a second-degree polynomial P such that $P(2) = 5$, $P'(2) = 3$, and $P''(2) = 2$.

68. The equation $y'' + y' - 2y = x^2$ is called a **differential equation** because it involves an unknown function y and its derivatives y' and y''. Find constants A, B, and C such that the function $y = Ax^2 + Bx + C$ satisfies this equation. (Differential equations will be studied in detail in Chapter 9.)

69. Find a cubic function $y = ax^3 + bx^2 + cx + d$ whose graph has horizontal tangents at the points $(-2, 6)$ and $(2, 0)$.

70. Find a parabola with equation $y = ax^2 + bx + c$ that has slope 4 at $x = 1$, slope -8 at $x = -1$, and passes through the point $(2, 15)$.

71. Let
$$f(x) = \begin{cases} x^2 + 1 & \text{if } x < 1 \\ x + 1 & \text{if } x \geq 1 \end{cases}$$

Is f differentiable at 1? Sketch the graphs of f and f'.

72. At what numbers is the following function g differentiable?
$$g(x) = \begin{cases} 2x & \text{if } x \leq 0 \\ 2x - x^2 & \text{if } 0 < x < 2 \\ 2 - x & \text{if } x \geq 2 \end{cases}$$

Give a formula for g' and sketch the graphs of g and g'.

73. (a) For what values of x is the function $f(x) = |x^2 - 9|$ differentiable? Find a formula for f'.
(b) Sketch the graphs of f and f'.

74. Where is the function $h(x) = |x - 1| + |x + 2|$ differentiable? Give a formula for h' and sketch the graphs of h and h'.

75. Find the parabola with equation $y = ax^2 + bx$ whose tangent line at (1, 1) has equation $y = 3x - 2$.

76. Suppose the curve $y = x^4 + ax^3 + bx^2 + cx + d$ has a tangent line when $x = 0$ with equation $y = 2x + 1$ and a tangent line when $x = 1$ with equation $y = 2 - 3x$. Find the values of a, b, c, and d.

77. For what values of a and b is the line $2x + y = b$ tangent to the parabola $y = ax^2$ when $x = 2$?

78. Find the value of c such that the line $y = \frac{3}{2}x + 6$ is tangent to the curve $y = c\sqrt{x}$.

79. What is the value of c such that the line $y = 2x + 3$ is tangent to the parabola $y = cx^2$?

80. The graph of any quadratic function $f(x) = ax^2 + bx + c$ is a parabola. Prove that the average of the slopes of the tangent lines to the parabola at the endpoints of any interval $[p, q]$ equals the slope of the tangent line at the midpoint of the interval.

81. Let
$$f(x) = \begin{cases} x^2 & \text{if } x \leq 2 \\ mx + b & \text{if } x > 2 \end{cases}$$

Find the values of m and b that make f differentiable everywhere.

82. A tangent line is drawn to the hyperbola $xy = c$ at a point P.
(a) Show that the midpoint of the line segment cut from this tangent line by the coordinate axes is P.
(b) Show that the triangle formed by the tangent line and the coordinate axes always has the same area, no matter where P is located on the hyperbola.

83. Evaluate $\lim\limits_{x \to 1} \dfrac{x^{1000} - 1}{x - 1}$.

84. Draw a diagram showing two perpendicular lines that intersect on the y-axis and are both tangent to the parabola $y = x^2$. Where do these lines intersect?

85. If $c > \frac{1}{2}$, how many lines through the point $(0, c)$ are normal lines to the parabola $y = x^2$? What if $c \leq \frac{1}{2}$?

86. Sketch the parabolas $y = x^2$ and $y = x^2 - 2x + 2$. Do you think there is a line that is tangent to both curves? If so, find its equation. If not, why not?

APPLIED PROJECT — BUILDING A BETTER ROLLER COASTER

Suppose you are asked to design the first ascent and drop for a new roller coaster. By studying photographs of your favorite coasters, you decide to make the slope of the ascent 0.8 and the slope of the drop -1.6. You decide to connect these two straight stretches $y = L_1(x)$ and $y = L_2(x)$ with part of a parabola $y = f(x) = ax^2 + bx + c$, where x and $f(x)$ are measured in feet. For the track to be smooth there can't be abrupt changes in direction, so you want the linear segments L_1 and L_2 to be tangent to the parabola at the transition points P and Q. (See the figure.) To simplify the equations, you decide to place the origin at P.

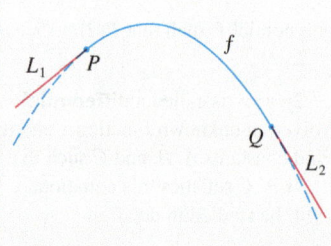

1. (a) Suppose the horizontal distance between P and Q is 100 ft. Write equations in a, b, and c that will ensure that the track is smooth at the transition points.
(b) Solve the equations in part (a) for a, b, and c to find a formula for $f(x)$.
(c) Plot L_1, f, and L_2 to verify graphically that the transitions are smooth.
(d) Find the difference in elevation between P and Q.

2. The solution in Problem 1 might *look* smooth, but it might not *feel* smooth because the piecewise defined function [consisting of $L_1(x)$ for $x < 0$, $f(x)$ for $0 \leq x \leq 100$, and

$L_2(x)$ for $x > 100$] doesn't have a continuous second derivative. So you decide to improve the design by using a quadratic function $q(x) = ax^2 + bx + c$ only on the interval $10 \leq x \leq 90$ and connecting it to the linear functions by means of two cubic functions:

$$g(x) = kx^3 + lx^2 + mx + n \qquad 0 \leq x < 10$$

$$h(x) = px^3 + qx^2 + rx + s \qquad 90 < x \leq 100$$

(a) Write a system of equations in 11 unknowns that ensure that the functions and their first two derivatives agree at the transition points.

(b) Solve the equations in part (a) with a computer algebra system to find formulas for $q(x)$, $g(x)$, and $h(x)$.

(c) Plot L_1, g, q, h, and L_2, and compare with the plot in Problem 1(c).

3.2 The Product and Quotient Rules

The formulas of this section enable us to differentiate new functions formed from old functions by multiplication or division.

The Product Rule

By analogy with the Sum and Difference Rules, one might be tempted to guess, as Leibniz did three centuries ago, that the derivative of a product is the product of the derivatives. We can see, however, that this guess is wrong by looking at a particular example. Let $f(x) = x$ and $g(x) = x^2$. Then the Power Rule gives $f'(x) = 1$ and $g'(x) = 2x$. But $(fg)(x) = x^3$, so $(fg)'(x) = 3x^2$. Thus $(fg)' \neq f'g'$. The correct formula was discovered by Leibniz (soon after his false start) and is called the Product Rule.

Before stating the Product Rule, let's see how we might discover it. We start by assuming that $u = f(x)$ and $v = g(x)$ are both positive differentiable functions. Then we can interpret the product uv as an area of a rectangle (see Figure 1). If x changes by an amount Δx, then the corresponding changes in u and v are

$$\Delta u = f(x + \Delta x) - f(x) \qquad \Delta v = g(x + \Delta x) - g(x)$$

and the new value of the product, $(u + \Delta u)(v + \Delta v)$, can be interpreted as the area of the large rectangle in Figure 1 (provided that Δu and Δv happen to be positive). The change in the area of the rectangle is

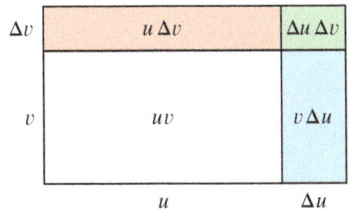

FIGURE 1
The geometry of the Product Rule

$$\boxed{1} \qquad \Delta(uv) = (u + \Delta u)(v + \Delta v) - uv = u\,\Delta v + v\,\Delta u + \Delta u\,\Delta v$$

$$= \text{the sum of the three shaded areas}$$

If we divide by Δx, we get

$$\frac{\Delta(uv)}{\Delta x} = u\,\frac{\Delta v}{\Delta x} + v\,\frac{\Delta u}{\Delta x} + \Delta u\,\frac{\Delta v}{\Delta x}$$

184   **CHAPTER 3**   Differentiation Rules

Recall that in Leibniz notation the definition of a derivative can be written as

$$\frac{dy}{dx} = \lim_{\Delta x \to 0} \frac{\Delta y}{\Delta x}$$

If we now let $\Delta x \to 0$, we get the derivative of uv:

$$\frac{d}{dx}(uv) = \lim_{\Delta x \to 0} \frac{\Delta(uv)}{\Delta x} = \lim_{\Delta x \to 0} \left(u \frac{\Delta v}{\Delta x} + v \frac{\Delta u}{\Delta x} + \Delta u \frac{\Delta v}{\Delta x} \right)$$

$$= u \lim_{\Delta x \to 0} \frac{\Delta v}{\Delta x} + v \lim_{\Delta x \to 0} \frac{\Delta u}{\Delta x} + \left(\lim_{\Delta x \to 0} \Delta u \right) \left(\lim_{\Delta x \to 0} \frac{\Delta v}{\Delta x} \right)$$

$$= u \frac{dv}{dx} + v \frac{du}{dx} + 0 \cdot \frac{dv}{dx}$$

$$\boxed{2} \quad \frac{d}{dx}(uv) = u \frac{dv}{dx} + v \frac{du}{dx}$$

(Notice that $\Delta u \to 0$ as $\Delta x \to 0$ since f is differentiable and therefore continuous.)

Although we started by assuming (for the geometric interpretation) that all the quantities are positive, we notice that Equation 1 is always true. (The algebra is valid whether u, v, Δu, and Δv are positive or negative.) So we have proved Equation 2, known as the Product Rule, for all differentiable functions u and v.

In prime notation:

$$(fg)' = fg' + gf'$$

> **The Product Rule**   If f and g are both differentiable, then
>
> $$\frac{d}{dx}[f(x)g(x)] = f(x)\frac{d}{dx}[g(x)] + g(x)\frac{d}{dx}[f(x)]$$

In words, the Product Rule says that *the derivative of a product of two functions is the first function times the derivative of the second function plus the second function times the derivative of the first function.*

EXAMPLE 1
(a) If $f(x) = xe^x$, find $f'(x)$.
(b) Find the nth derivative, $f^{(n)}(x)$.

SOLUTION
(a) By the Product Rule, we have

$$f'(x) = \frac{d}{dx}(xe^x)$$

$$= x\frac{d}{dx}(e^x) + e^x \frac{d}{dx}(x)$$

$$= xe^x + e^x \cdot 1 = (x + 1)e^x$$

(b) Using the Product Rule a second time, we get

$$f''(x) = \frac{d}{dx}[(x + 1)e^x]$$

$$= (x + 1)\frac{d}{dx}(e^x) + e^x \frac{d}{dx}(x + 1)$$

$$= (x + 1)e^x + e^x \cdot 1 = (x + 2)e^x$$

Figure 2 shows the graphs of the function f of Example 1 and its derivative f'. Notice that $f'(x)$ is positive when f is increasing and negative when f is decreasing.

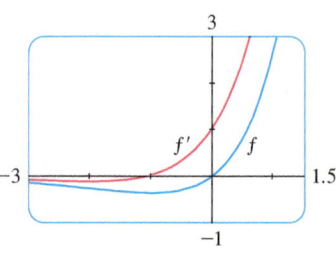

FIGURE 2

Further applications of the Product Rule give

$$f'''(x) = (x+3)e^x \quad f^{(4)}(x) = (x+4)e^x$$

In fact, each successive differentiation adds another term e^x, so

$$f^{(n)}(x) = (x+n)e^x \quad \blacksquare$$

EXAMPLE 2 Differentiate the function $f(t) = \sqrt{t}\,(a + bt)$.

SOLUTION 1 Using the Product Rule, we have

$$f'(t) = \sqrt{t}\,\frac{d}{dt}(a + bt) + (a + bt)\frac{d}{dt}\left(\sqrt{t}\right)$$

$$= \sqrt{t} \cdot b + (a + bt) \cdot \tfrac{1}{2}t^{-1/2}$$

$$= b\sqrt{t} + \frac{a + bt}{2\sqrt{t}} = \frac{a + 3bt}{2\sqrt{t}}$$

SOLUTION 2 If we first use the laws of exponents to rewrite $f(t)$, then we can proceed directly without using the Product Rule.

$$f(t) = a\sqrt{t} + bt\sqrt{t} = at^{1/2} + bt^{3/2}$$

$$f'(t) = \tfrac{1}{2}at^{-1/2} + \tfrac{3}{2}bt^{1/2}$$

which is equivalent to the answer given in Solution 1. $\quad\blacksquare$

Example 2 shows that it is sometimes easier to simplify a product of functions before differentiating than to use the Product Rule. In Example 1, however, the Product Rule is the only possible method.

EXAMPLE 3 If $f(x) = \sqrt{x}\,g(x)$, where $g(4) = 2$ and $g'(4) = 3$, find $f'(4)$.

SOLUTION Applying the Product Rule, we get

$$f'(x) = \frac{d}{dx}\left[\sqrt{x}\,g(x)\right] = \sqrt{x}\,\frac{d}{dx}[g(x)] + g(x)\frac{d}{dx}\left[\sqrt{x}\right]$$

$$= \sqrt{x}\,g'(x) + g(x) \cdot \tfrac{1}{2}x^{-1/2}$$

$$= \sqrt{x}\,g'(x) + \frac{g(x)}{2\sqrt{x}}$$

So

$$f'(4) = \sqrt{4}\,g'(4) + \frac{g(4)}{2\sqrt{4}} = 2 \cdot 3 + \frac{2}{2 \cdot 2} = 6.5 \quad \blacksquare$$

In Example 2, a and b are constants. It is customary in mathematics to use letters near the beginning of the alphabet to represent constants and letters near the end of the alphabet to represent variables.

■ The Quotient Rule

We find a rule for differentiating the quotient of two differentiable functions $u = f(x)$ and $v = g(x)$ in much the same way that we found the Product Rule. If x, u, and v change by amounts Δx, Δu, and Δv, then the corresponding change in the quotient u/v is

$$\Delta\left(\frac{u}{v}\right) = \frac{u + \Delta u}{v + \Delta v} - \frac{u}{v} = \frac{(u + \Delta u)v - u(v + \Delta v)}{v(v + \Delta v)}$$

$$= \frac{v\Delta u - u\Delta v}{v(v + \Delta v)}$$

so

$$\frac{d}{dx}\left(\frac{u}{v}\right) = \lim_{\Delta x \to 0} \frac{\Delta(u/v)}{\Delta x} = \lim_{\Delta x \to 0} \frac{v\dfrac{\Delta u}{\Delta x} - u\dfrac{\Delta v}{\Delta x}}{v(v + \Delta v)}$$

As $\Delta x \to 0$, $\Delta v \to 0$ also, because $v = g(x)$ is differentiable and therefore continuous. Thus, using the Limit Laws, we get

$$\frac{d}{dx}\left(\frac{u}{v}\right) = \frac{v \lim\limits_{\Delta x \to 0} \dfrac{\Delta u}{\Delta x} - u \lim\limits_{\Delta x \to 0} \dfrac{\Delta v}{\Delta x}}{v \lim\limits_{\Delta x \to 0}(v + \Delta v)} = \frac{v \dfrac{du}{dx} - u \dfrac{dv}{dx}}{v^2}$$

In prime notation:

$$\left(\frac{f}{g}\right)' = \frac{gf' - fg'}{g^2}$$

The Quotient Rule If f and g are differentiable, then

$$\frac{d}{dx}\left[\frac{f(x)}{g(x)}\right] = \frac{g(x)\dfrac{d}{dx}[f(x)] - f(x)\dfrac{d}{dx}[g(x)]}{[g(x)]^2}$$

In words, the Quotient Rule says that the *derivative of a quotient is the denominator times the derivative of the numerator minus the numerator times the derivative of the denominator, all divided by the square of the denominator.*

The Quotient Rule and the other differentiation formulas enable us to compute the derivative of any rational function, as the next example illustrates.

We can use a graphing device to check that the answer to Example 4 is plausible. Figure 3 shows the graphs of the function of Example 4 and its derivative. Notice that when y grows rapidly (near -2), y' is large. And when y grows slowly, y' is near 0.

EXAMPLE 4 Let $y = \dfrac{x^2 + x - 2}{x^3 + 6}$. Then

$$y' = \frac{(x^3 + 6)\dfrac{d}{dx}(x^2 + x - 2) - (x^2 + x - 2)\dfrac{d}{dx}(x^3 + 6)}{(x^3 + 6)^2}$$

$$= \frac{(x^3 + 6)(2x + 1) - (x^2 + x - 2)(3x^2)}{(x^3 + 6)^2}$$

$$= \frac{(2x^4 + x^3 + 12x + 6) - (3x^4 + 3x^3 - 6x^2)}{(x^3 + 6)^2}$$

$$= \frac{-x^4 - 2x^3 + 6x^2 + 12x + 6}{(x^3 + 6)^2}$$

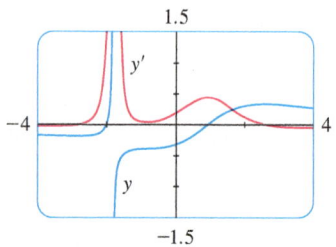

FIGURE 3

SECTION 3.2 The Product and Quotient Rules

EXAMPLE 5 Find an equation of the tangent line to the curve $y = e^x/(1 + x^2)$ at the point $\left(1, \tfrac{1}{2}e\right)$.

SOLUTION According to the Quotient Rule, we have

$$\frac{dy}{dx} = \frac{(1 + x^2)\dfrac{d}{dx}(e^x) - e^x \dfrac{d}{dx}(1 + x^2)}{(1 + x^2)^2}$$

$$= \frac{(1 + x^2)e^x - e^x(2x)}{(1 + x^2)^2} = \frac{e^x(1 - 2x + x^2)}{(1 + x^2)^2}$$

$$= \frac{e^x(1 - x)^2}{(1 + x^2)^2}$$

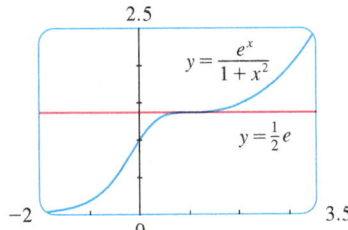

FIGURE 4

So the slope of the tangent line at $\left(1, \tfrac{1}{2}e\right)$ is

$$\left.\frac{dy}{dx}\right|_{x=1} = 0$$

This means that the tangent line at $\left(1, \tfrac{1}{2}e\right)$ is horizontal and its equation is $y = \tfrac{1}{2}e$. [See Figure 4. Notice that the function is increasing and crosses its tangent line at $\left(1, \tfrac{1}{2}e\right)$.]

NOTE Don't use the Quotient Rule *every* time you see a quotient. Sometimes it's easier to rewrite a quotient first to put it in a form that is simpler for the purpose of differentiation. For instance, although it is possible to differentiate the function

$$F(x) = \frac{3x^2 + 2\sqrt{x}}{x}$$

using the Quotient Rule, it is much easier to perform the division first and write the function as

$$F(x) = 3x + 2x^{-1/2}$$

before differentiating.

We summarize the differentiation formulas we have learned so far as follows.

Table of Differentiation Formulas

$$\frac{d}{dx}(c) = 0 \qquad \frac{d}{dx}(x^n) = nx^{n-1} \qquad \frac{d}{dx}(e^x) = e^x$$

$$(cf)' = cf' \qquad (f + g)' = f' + g' \qquad (f - g)' = f' - g'$$

$$(fg)' = fg' + gf' \qquad \left(\frac{f}{g}\right)' = \frac{gf' - fg'}{g^2}$$

3.2 EXERCISES

1. Find the derivative of $f(x) = (1 + 2x^2)(x - x^2)$ in two ways: by using the Product Rule and by performing the multiplication first. Do your answers agree?

2. Find the derivative of the function

$$F(x) = \frac{x^4 - 5x^3 + \sqrt{x}}{x^2}$$

in two ways: by using the Quotient Rule and by simplifying first. Show that your answers are equivalent. Which method do you prefer?

3–26 Differentiate.

3. $f(x) = (3x^2 - 5x)e^x$

4. $g(x) = (x + 2\sqrt{x})e^x$

5. $y = \dfrac{x}{e^x}$

6. $y = \dfrac{e^x}{1 - e^x}$

7. $g(x) = \dfrac{1 + 2x}{3 - 4x}$

8. $G(x) = \dfrac{x^2 - 2}{2x + 1}$

9. $H(u) = (u - \sqrt{u})(u + \sqrt{u})$

10. $J(v) = (v^3 - 2v)(v^{-4} + v^{-2})$

11. $F(y) = \left(\dfrac{1}{y^2} - \dfrac{3}{y^4}\right)(y + 5y^3)$

12. $f(z) = (1 - e^z)(z + e^z)$

13. $y = \dfrac{x^2 + 1}{x^3 - 1}$

14. $y = \dfrac{\sqrt{x}}{2 + x}$

15. $y = \dfrac{t^3 + 3t}{t^2 - 4t + 3}$

16. $y = \dfrac{1}{t^3 + 2t^2 - 1}$

17. $y = e^p(p + p\sqrt{p})$

18. $h(r) = \dfrac{ae^r}{b + e^r}$

19. $y = \dfrac{s - \sqrt{s}}{s^2}$

20. $y = (z^2 + e^z)\sqrt{z}$

21. $f(t) = \dfrac{\sqrt[3]{t}}{t - 3}$

22. $V(t) = \dfrac{4 + t}{te^t}$

23. $f(x) = \dfrac{x^2 e^x}{x^2 + e^x}$

24. $F(t) = \dfrac{At}{Bt^2 + Ct^3}$

25. $f(x) = \dfrac{x}{x + \dfrac{c}{x}}$

26. $f(x) = \dfrac{ax + b}{cx + d}$

27–30 Find $f'(x)$ and $f''(x)$.

27. $f(x) = (x^3 + 1)e^x$

28. $f(x) = \sqrt{x}e^x$

29. $f(x) = \dfrac{x^2}{1 + e^x}$

30. $f(x) = \dfrac{x}{x^2 - 1}$

31–32 Find an equation of the tangent line to the given curve at the specified point.

31. $y = \dfrac{x^2 - 1}{x^2 + x + 1}$, $(1, 0)$

32. $y = \dfrac{1 + x}{1 + e^x}$, $\left(0, \frac{1}{2}\right)$

33–34 Find equations of the tangent line and normal line to the given curve at the specified point.

33. $y = 2xe^x$, $(0, 0)$

34. $y = \dfrac{2x}{x^2 + 1}$, $(1, 1)$

35. (a) The curve $y = 1/(1 + x^2)$ is called a **witch of Maria Agnesi**. Find an equation of the tangent line to this curve at the point $\left(-1, \frac{1}{2}\right)$.
(b) Illustrate part (a) by graphing the curve and the tangent line on the same screen.

36. (a) The curve $y = x/(1 + x^2)$ is called a **serpentine**. Find an equation of the tangent line to this curve at the point $(3, 0.3)$.
(b) Illustrate part (a) by graphing the curve and the tangent line on the same screen.

37. (a) If $f(x) = (x^3 - x)e^x$, find $f'(x)$.
(b) Check to see that your answer to part (a) is reasonable by comparing the graphs of f and f'.

38. (a) If $f(x) = e^x/(2x^2 + x + 1)$, find $f'(x)$.
(b) Check to see that your answer to part (a) is reasonable by comparing the graphs of f and f'.

39. (a) If $f(x) = (x^2 - 1)/(x^2 + 1)$, find $f'(x)$ and $f''(x)$.
(b) Check to see that your answers to part (a) are reasonable by comparing the graphs of f, f', and f''.

40. (a) If $f(x) = (x^2 - 1)e^x$, find $f'(x)$ and $f''(x)$.
(b) Check to see that your answers to part (a) are reasonable by comparing the graphs of f, f', and f''.

41. If $f(x) = x^2/(1 + x)$, find $f''(1)$.

42. If $g(x) = x/e^x$, find $g^{(n)}(x)$.

43. Suppose that $f(5) = 1$, $f'(5) = 6$, $g(5) = -3$, and $g'(5) = 2$. Find the following values.
(a) $(fg)'(5)$
(b) $(f/g)'(5)$
(c) $(g/f)'(5)$

44. Suppose that $f(4) = 2$, $g(4) = 5$, $f'(4) = 6$, and $g'(4) = -3$. Find $h'(4)$.
(a) $h(x) = 3f(x) + 8g(x)$
(b) $h(x) = f(x)g(x)$
(c) $h(x) = \dfrac{f(x)}{g(x)}$
(d) $h(x) = \dfrac{g(x)}{f(x) + g(x)}$

45. If $f(x) = e^x g(x)$, where $g(0) = 2$ and $g'(0) = 5$, find $f'(0)$.

46. If $h(2) = 4$ and $h'(2) = -3$, find

$$\frac{d}{dx}\left(\frac{h(x)}{x}\right)\bigg|_{x=2}$$

47. If $g(x) = xf(x)$, where $f(3) = 4$ and $f'(3) = -2$, find an equation of the tangent line to the graph of g at the point where $x = 3$.

48. If $f(2) = 10$ and $f'(x) = x^2 f(x)$ for all x, find $f''(2)$.

49. If f and g are the functions whose graphs are shown, let $u(x) = f(x)g(x)$ and $v(x) = f(x)/g(x)$.
(a) Find $u'(1)$. (b) Find $v'(5)$.

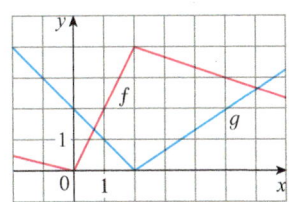

50. Let $P(x) = F(x)G(x)$ and $Q(x) = F(x)/G(x)$, where F and G are the functions whose graphs are shown.
(a) Find $P'(2)$. (b) Find $Q'(7)$.

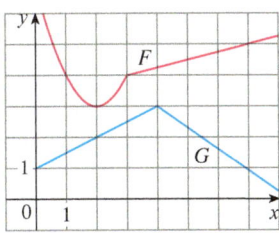

51. If g is a differentiable function, find an expression for the derivative of each of the following functions.
(a) $y = xg(x)$ (b) $y = \dfrac{x}{g(x)}$
(c) $y = \dfrac{g(x)}{x}$

52. If f is a differentiable function, find an expression for the derivative of each of the following functions.
(a) $y = x^2 f(x)$ (b) $y = \dfrac{f(x)}{x^2}$
(c) $y = \dfrac{x^2}{f(x)}$ (d) $y = \dfrac{1 + xf(x)}{\sqrt{x}}$

53. How many tangent lines to the curve $y = x/(x + 1)$ pass through the point $(1, 2)$? At which points do these tangent lines touch the curve?

54. Find equations of the tangent lines to the curve

$$y = \frac{x - 1}{x + 1}$$

that are parallel to the line $x - 2y = 2$.

55. Find $R'(0)$, where

$$R(x) = \frac{x - 3x^3 + 5x^5}{1 + 3x^3 + 6x^6 + 9x^9}$$

Hint: Instead of finding $R'(x)$ first, let $f(x)$ be the numerator and $g(x)$ the denominator of $R(x)$ and compute $R'(0)$ from $f(0)$, $f'(0)$, $g(0)$, and $g'(0)$.

56. Use the method of Exercise 55 to compute $Q'(0)$, where

$$Q(x) = \frac{1 + x + x^2 + xe^x}{1 - x + x^2 - xe^x}$$

57. In this exercise we estimate the rate at which the total personal income is rising in the Richmond-Petersburg, Virginia, metropolitan area. In 1999, the population of this area was 961,400, and the population was increasing at roughly 9200 people per year. The average annual income was $30,593 per capita, and this average was increasing at about $1400 per year (a little above the national average of about $1225 yearly). Use the Product Rule and these figures to estimate the rate at which total personal income was rising in the Richmond-Petersburg area in 1999. Explain the meaning of each term in the Product Rule.

58. A manufacturer produces bolts of a fabric with a fixed width. The quantity q of this fabric (measured in yards) that is sold is a function of the selling price p (in dollars per yard), so we can write $q = f(p)$. Then the total revenue earned with selling price p is $R(p) = pf(p)$.
(a) What does it mean to say that $f(20) = 10,000$ and $f'(20) = -350$?
(b) Assuming the values in part (a), find $R'(20)$ and interpret your answer.

59. The Michaelis-Menten equation for the enzyme chymotrypsin is

$$v = \frac{0.14[S]}{0.015 + [S]}$$

where v is the rate of an enzymatic reaction and $[S]$ is the concentration of a substrate S. Calculate $dv/d[S]$ and interpret it.

60. The *biomass* $B(t)$ of a fish population is the total mass of the members of the population at time t. It is the product of the number of individuals $N(t)$ in the population and the average mass $M(t)$ of a fish at time t. In the case of guppies, breeding occurs continually. Suppose that at time $t = 4$ weeks the population is 820 guppies and is growing at a rate of 50 guppies per week, while the average mass is 1.2 g and is increasing at a rate of 0.14 g/week. At what rate is the biomass increasing when $t = 4$?

61. (a) Use the Product Rule twice to prove that if f, g, and h are differentiable, then $(fgh)' = f'gh + fg'h + fgh'$.
(b) Taking $f = g = h$ in part (a), show that
$$\frac{d}{dx}[f(x)]^3 = 3[f(x)]^2 f'(x)$$
(c) Use part (b) to differentiate $y = e^{3x}$.

62. (a) If $F(x) = f(x)g(x)$, where f and g have derivatives of all orders, show that $F'' = f''g + 2f'g' + fg''$.
(b) Find similar formulas for F''' and $F^{(4)}$.
(c) Guess a formula for $F^{(n)}$.

63. Find expressions for the first five derivatives of $f(x) = x^2 e^x$. Do you see a pattern in these expressions? Guess a formula for $f^{(n)}(x)$ and prove it using mathematical induction.

64. (a) If g is differentiable, the **Reciprocal Rule** says that
$$\frac{d}{dx}\left[\frac{1}{g(x)}\right] = -\frac{g'(x)}{[g(x)]^2}$$
Use the Quotient Rule to prove the Reciprocal Rule.
(b) Use the Reciprocal Rule to differentiate the function in Exercise 16.
(c) Use the Reciprocal Rule to verify that the Power Rule is valid for negative integers, that is,
$$\frac{d}{dx}(x^{-n}) = -nx^{-n-1}$$
for all positive integers n.

3.3 Derivatives of Trigonometric Functions

A review of the trigonometric functions is given in Appendix D.

Before starting this section, you might need to review the trigonometric functions. In particular, it is important to remember that when we talk about the function f defined for all real numbers x by
$$f(x) = \sin x$$
it is understood that $\sin x$ means the sine of the angle whose *radian* measure is x. A similar convention holds for the other trigonometric functions cos, tan, csc, sec, and cot. Recall from Section 2.5 that all of the trigonometric functions are continuous at every number in their domains.

If we sketch the graph of the function $f(x) = \sin x$ and use the interpretation of $f'(x)$ as the slope of the tangent to the sine curve in order to sketch the graph of f' (see Exercise 2.8.16), then it looks as if the graph of f' may be the same as the cosine curve (see Figure 1).

TEC Visual 3.3 shows an animation of Figure 1.

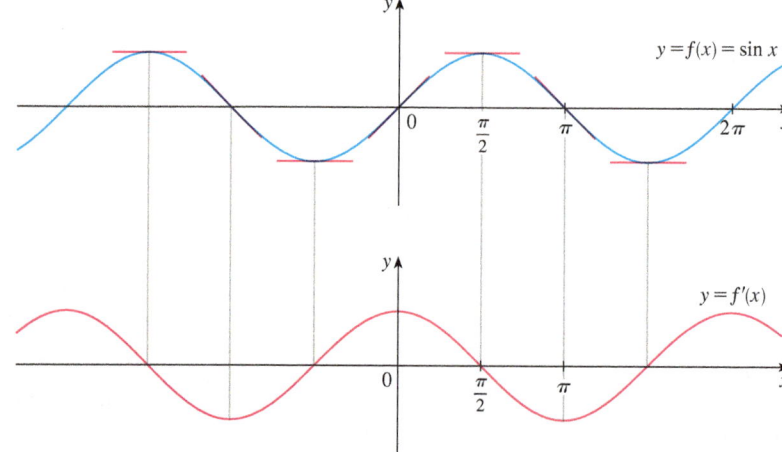

FIGURE 1

Let's try to confirm our guess that if $f(x) = \sin x$, then $f'(x) = \cos x$. From the definition of a derivative, we have

$$f'(x) = \lim_{h \to 0} \frac{f(x+h) - f(x)}{h} = \lim_{h \to 0} \frac{\sin(x+h) - \sin x}{h}$$

We have used the addition formula for sine. See Appendix D.

$$= \lim_{h \to 0} \frac{\sin x \cos h + \cos x \sin h - \sin x}{h}$$

$$= \lim_{h \to 0} \left[\frac{\sin x \cos h - \sin x}{h} + \frac{\cos x \sin h}{h} \right]$$

$$= \lim_{h \to 0} \left[\sin x \left(\frac{\cos h - 1}{h} \right) + \cos x \left(\frac{\sin h}{h} \right) \right]$$

1 $$= \lim_{h \to 0} \sin x \cdot \lim_{h \to 0} \frac{\cos h - 1}{h} + \lim_{h \to 0} \cos x \cdot \lim_{h \to 0} \frac{\sin h}{h}$$

Two of these four limits are easy to evaluate. Since we regard x as a constant when computing a limit as $h \to 0$, we have

$$\lim_{h \to 0} \sin x = \sin x \quad \text{and} \quad \lim_{h \to 0} \cos x = \cos x$$

The limit of $(\sin h)/h$ is not so obvious. In Example 2.2.3 we made the guess, on the basis of numerical and graphical evidence, that

2 $$\lim_{\theta \to 0} \frac{\sin \theta}{\theta} = 1$$

We now use a geometric argument to prove Equation 2. Assume first that θ lies between 0 and $\pi/2$. Figure 2(a) shows a sector of a circle with center O, central angle θ, and radius 1. BC is drawn perpendicular to OA. By the definition of radian measure, we have arc $AB = \theta$. Also $|BC| = |OB| \sin \theta = \sin \theta$. From the diagram we see that

$$|BC| < |AB| < \text{arc } AB$$

Therefore $\qquad \sin \theta < \theta \qquad$ so $\qquad \dfrac{\sin \theta}{\theta} < 1$

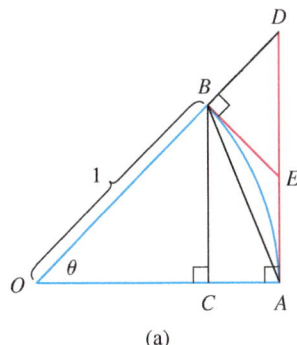

(a)

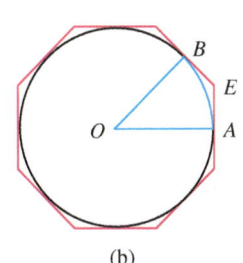

(b)

FIGURE 2

Let the tangent lines at A and B intersect at E. You can see from Figure 2(b) that the circumference of a circle is smaller than the length of a circumscribed polygon, and so arc $AB < |AE| + |EB|$. Thus

$$\theta = \text{arc } AB < |AE| + |EB|$$
$$< |AE| + |ED|$$
$$= |AD| = |OA| \tan \theta$$
$$= \tan \theta$$

(In Appendix F the inequality $\theta \leq \tan \theta$ is proved directly from the definition of the length of an arc without resorting to geometric intuition as we did here.) Therefore we have

$$\theta < \frac{\sin\theta}{\cos\theta}$$

so

$$\cos\theta < \frac{\sin\theta}{\theta} < 1$$

We know that $\lim_{\theta\to 0} 1 = 1$ and $\lim_{\theta\to 0} \cos\theta = 1$, so by the Squeeze Theorem, we have

$$\lim_{\theta\to 0^+} \frac{\sin\theta}{\theta} = 1$$

But the function $(\sin\theta)/\theta$ is an even function, so its right and left limits must be equal. Hence, we have

$$\lim_{\theta\to 0} \frac{\sin\theta}{\theta} = 1$$

so we have proved Equation 2.

We can deduce the value of the remaining limit in (1) as follows:

We multiply numerator and denominator by $\cos\theta + 1$ in order to put the function in a form in which we can use the limits we know.

$$\lim_{\theta\to 0} \frac{\cos\theta - 1}{\theta} = \lim_{\theta\to 0} \left(\frac{\cos\theta - 1}{\theta} \cdot \frac{\cos\theta + 1}{\cos\theta + 1}\right) = \lim_{\theta\to 0} \frac{\cos^2\theta - 1}{\theta(\cos\theta + 1)}$$

$$= \lim_{\theta\to 0} \frac{-\sin^2\theta}{\theta(\cos\theta + 1)} = -\lim_{\theta\to 0}\left(\frac{\sin\theta}{\theta} \cdot \frac{\sin\theta}{\cos\theta + 1}\right)$$

$$= -\lim_{\theta\to 0} \frac{\sin\theta}{\theta} \cdot \lim_{\theta\to 0} \frac{\sin\theta}{\cos\theta + 1}$$

$$= -1 \cdot \left(\frac{0}{1 + 1}\right) = 0 \quad \text{(by Equation 2)}$$

3
$$\lim_{\theta\to 0} \frac{\cos\theta - 1}{\theta} = 0$$

If we now put the limits (2) and (3) in (1), we get

$$f'(x) = \lim_{h\to 0} \sin x \cdot \lim_{h\to 0} \frac{\cos h - 1}{h} + \lim_{h\to 0} \cos x \cdot \lim_{h\to 0} \frac{\sin h}{h}$$

$$= (\sin x) \cdot 0 + (\cos x) \cdot 1 = \cos x$$

So we have proved the formula for the derivative of the sine function:

4
$$\frac{d}{dx}(\sin x) = \cos x$$

Figure 3 shows the graphs of the function of Example 1 and its derivative. Notice that $y' = 0$ whenever y has a horizontal tangent.

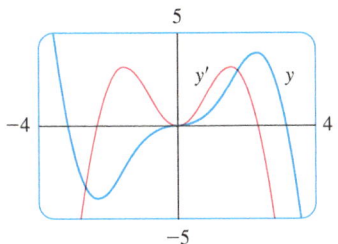

FIGURE 3

EXAMPLE 1 Differentiate $y = x^2 \sin x$.

SOLUTION Using the Product Rule and Formula 4, we have

$$\frac{dy}{dx} = x^2 \frac{d}{dx}(\sin x) + \sin x \frac{d}{dx}(x^2)$$

$$= x^2 \cos x + 2x \sin x \qquad \blacksquare$$

Using the same methods as in the proof of Formula 4, one can prove (see Exercise 20) that

$$\boxed{5} \qquad \frac{d}{dx}(\cos x) = -\sin x$$

The tangent function can also be differentiated by using the definition of a derivative, but it is easier to use the Quotient Rule together with Formulas 4 and 5:

$$\frac{d}{dx}(\tan x) = \frac{d}{dx}\left(\frac{\sin x}{\cos x}\right)$$

$$= \frac{\cos x \frac{d}{dx}(\sin x) - \sin x \frac{d}{dx}(\cos x)}{\cos^2 x}$$

$$= \frac{\cos x \cdot \cos x - \sin x (-\sin x)}{\cos^2 x}$$

$$= \frac{\cos^2 x + \sin^2 x}{\cos^2 x}$$

$$= \frac{1}{\cos^2 x} = \sec^2 x$$

$$\boxed{6} \qquad \frac{d}{dx}(\tan x) = \sec^2 x$$

The derivatives of the remaining trigonometric functions, csc, sec, and cot, can also be found easily using the Quotient Rule (see Exercises 17–19). We collect all the differentiation formulas for trigonometric functions in the following table. Remember that they are valid only when x is measured in radians.

When you memorize this table, it is helpful to notice that the minus signs go with the derivatives of the "cofunctions," that is, cosine, cosecant, and cotangent.

Derivatives of Trigonometric Functions

$$\frac{d}{dx}(\sin x) = \cos x \qquad\qquad \frac{d}{dx}(\csc x) = -\csc x \cot x$$

$$\frac{d}{dx}(\cos x) = -\sin x \qquad\qquad \frac{d}{dx}(\sec x) = \sec x \tan x$$

$$\frac{d}{dx}(\tan x) = \sec^2 x \qquad\qquad \frac{d}{dx}(\cot x) = -\csc^2 x$$

EXAMPLE 2 Differentiate $f(x) = \dfrac{\sec x}{1 + \tan x}$. For what values of x does the graph of f have a horizontal tangent?

SOLUTION The Quotient Rule gives

$$f'(x) = \frac{(1 + \tan x)\dfrac{d}{dx}(\sec x) - \sec x \dfrac{d}{dx}(1 + \tan x)}{(1 + \tan x)^2}$$

$$= \frac{(1 + \tan x)\sec x \tan x - \sec x \cdot \sec^2 x}{(1 + \tan x)^2}$$

$$= \frac{\sec x\,(\tan x + \tan^2 x - \sec^2 x)}{(1 + \tan x)^2}$$

$$= \frac{\sec x\,(\tan x - 1)}{(1 + \tan x)^2}$$

In simplifying the answer we have used the identity $\tan^2 x + 1 = \sec^2 x$.

Since $\sec x$ is never 0, we see that $f'(x) = 0$ when $\tan x = 1$, and this occurs when $x = n\pi + \pi/4$, where n is an integer (see Figure 4).

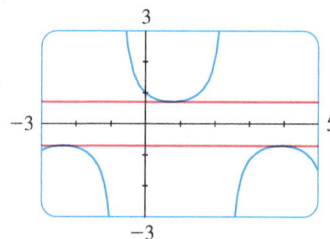

FIGURE 4
The horizontal tangents in Example 2

Trigonometric functions are often used in modeling real-world phenomena. In particular, vibrations, waves, elastic motions, and other quantities that vary in a periodic manner can be described using trigonometric functions. In the following example we discuss an instance of simple harmonic motion.

EXAMPLE 3 An object at the end of a vertical spring is stretched 4 cm beyond its rest position and released at time $t = 0$. (See Figure 5 and note that the downward direction is positive.) Its position at time t is

$$s = f(t) = 4 \cos t$$

Find the velocity and acceleration at time t and use them to analyze the motion of the object.

SOLUTION The velocity and acceleration are

$$v = \frac{ds}{dt} = \frac{d}{dt}(4 \cos t) = 4\frac{d}{dt}(\cos t) = -4 \sin t$$

$$a = \frac{dv}{dt} = \frac{d}{dt}(-4 \sin t) = -4\frac{d}{dt}(\sin t) = -4 \cos t$$

The object oscillates from the lowest point ($s = 4$ cm) to the highest point ($s = -4$ cm). The period of the oscillation is 2π, the period of $\cos t$.

The speed is $|v| = 4|\sin t|$, which is greatest when $|\sin t| = 1$, that is, when $\cos t = 0$. So the object moves fastest as it passes through its equilibrium position ($s = 0$). Its speed is 0 when $\sin t = 0$, that is, at the high and low points.

The acceleration $a = -4 \cos t = 0$ when $s = 0$. It has greatest magnitude at the high and low points. See the graphs in Figure 6.

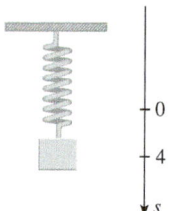

FIGURE 5

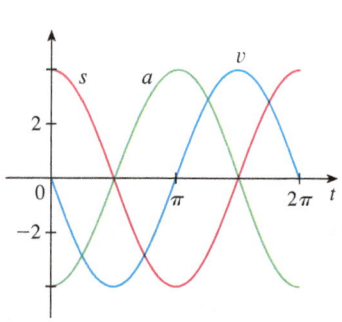

FIGURE 6

EXAMPLE 4 Find the 27th derivative of $\cos x$.

SOLUTION The first few derivatives of $f(x) = \cos x$ are as follows:

$$f'(x) = -\sin x$$
$$f''(x) = -\cos x$$
$$f'''(x) = \sin x$$
$$f^{(4)}(x) = \cos x$$
$$f^{(5)}(x) = -\sin x$$

We see that the successive derivatives occur in a cycle of length 4 and, in particular, $f^{(n)}(x) = \cos x$ whenever n is a multiple of 4. Therefore

$$f^{(24)}(x) = \cos x$$

and, differentiating three more times, we have

$$f^{(27)}(x) = \sin x$$

∎

Our main use for the limit in Equation 2 has been to prove the differentiation formula for the sine function. But this limit is also useful in finding certain other trigonometric limits, as the following two examples show.

EXAMPLE 5 Find $\displaystyle\lim_{x \to 0} \frac{\sin 7x}{4x}$.

SOLUTION In order to apply Equation 2, we first rewrite the function by multiplying and dividing by 7:

$$\frac{\sin 7x}{4x} = \frac{7}{4}\left(\frac{\sin 7x}{7x}\right)$$

If we let $\theta = 7x$, then $\theta \to 0$ as $x \to 0$, so by Equation 2 we have

$$\lim_{x \to 0} \frac{\sin 7x}{4x} = \frac{7}{4} \lim_{x \to 0} \left(\frac{\sin 7x}{7x}\right)$$
$$= \frac{7}{4} \lim_{\theta \to 0} \frac{\sin \theta}{\theta} = \frac{7}{4} \cdot 1 = \frac{7}{4}$$

∎

EXAMPLE 6 Calculate $\displaystyle\lim_{x \to 0} x \cot x$.

SOLUTION Here we divide numerator and denominator by x:

$$\lim_{x \to 0} x \cot x = \lim_{x \to 0} \frac{x \cos x}{\sin x}$$
$$= \lim_{x \to 0} \frac{\cos x}{\frac{\sin x}{x}} = \frac{\displaystyle\lim_{x \to 0} \cos x}{\displaystyle\lim_{x \to 0} \frac{\sin x}{x}}$$
$$= \frac{\cos 0}{1} \qquad \text{(by the continuity of cosine and Equation 2)}$$
$$= 1$$

∎

Look for a pattern.

Note that $\sin 7x \neq 7 \sin x$.

3.3 EXERCISES

1–16 Differentiate.

1. $f(x) = x^2 \sin x$
2. $f(x) = x \cos x + 2 \tan x$
3. $f(x) = e^x \cos x$
4. $y = 2 \sec x - \csc x$
5. $y = \sec \theta \tan \theta$
6. $g(\theta) = e^\theta (\tan \theta - \theta)$
7. $y = c \cos t + t^2 \sin t$
8. $f(t) = \dfrac{\cot t}{e^t}$
9. $y = \dfrac{x}{2 - \tan x}$
10. $y = \sin \theta \cos \theta$
11. $f(\theta) = \dfrac{\sin \theta}{1 + \cos \theta}$
12. $y = \dfrac{\cos x}{1 - \sin x}$
13. $y = \dfrac{t \sin t}{1 + t}$
14. $y = \dfrac{\sin t}{1 + \tan t}$
15. $f(\theta) = \theta \cos \theta \sin \theta$
16. $f(t) = t e^t \cot t$

17. Prove that $\dfrac{d}{dx} (\csc x) = -\csc x \cot x$.

18. Prove that $\dfrac{d}{dx} (\sec x) = \sec x \tan x$.

19. Prove that $\dfrac{d}{dx} (\cot x) = -\csc^2 x$.

20. Prove, using the definition of derivative, that if $f(x) = \cos x$, then $f'(x) = -\sin x$.

21–24 Find an equation of the tangent line to the curve at the given point.

21. $y = \sin x + \cos x$, $(0, 1)$
22. $y = e^x \cos x$, $(0, 1)$
23. $y = \cos x - \sin x$, $(\pi, -1)$
24. $y = x + \tan x$, (π, π)

25. (a) Find an equation of the tangent line to the curve $y = 2x \sin x$ at the point $(\pi/2, \pi)$.
    (b) Illustrate part (a) by graphing the curve and the tangent line on the same screen.

26. (a) Find an equation of the tangent line to the curve $y = 3x + 6 \cos x$ at the point $(\pi/3, \pi + 3)$.
    (b) Illustrate part (a) by graphing the curve and the tangent line on the same screen.

27. (a) If $f(x) = \sec x - x$, find $f'(x)$.
    (b) Check to see that your answer to part (a) is reasonable by graphing both f and f' for $|x| < \pi/2$.

28. (a) If $f(x) = e^x \cos x$, find $f'(x)$ and $f''(x)$.
    (b) Check to see that your answers to part (a) are reasonable by graphing f, f', and f''.

29. If $H(\theta) = \theta \sin \theta$, find $H'(\theta)$ and $H''(\theta)$.

30. If $f(t) = \sec t$, find $f''(\pi/4)$.

31. (a) Use the Quotient Rule to differentiate the function
$$f(x) = \dfrac{\tan x - 1}{\sec x}$$
    (b) Simplify the expression for $f(x)$ by writing it in terms of $\sin x$ and $\cos x$, and then find $f'(x)$.
    (c) Show that your answers to parts (a) and (b) are equivalent.

32. Suppose $f(\pi/3) = 4$ and $f'(\pi/3) = -2$, and let $g(x) = f(x) \sin x$ and $h(x) = (\cos x)/f(x)$. Find
    (a) $g'(\pi/3)$
    (b) $h'(\pi/3)$

33–34 For what values of x does the graph of f have a horizontal tangent?

33. $f(x) = x + 2 \sin x$
34. $f(x) = e^x \cos x$

35. A mass on a spring vibrates horizontally on a smooth level surface (see the figure). Its equation of motion is $x(t) = 8 \sin t$, where t is in seconds and x in centimeters.
    (a) Find the velocity and acceleration at time t.
    (b) Find the position, velocity, and acceleration of the mass at time $t = 2\pi/3$. In what direction is it moving at that time?

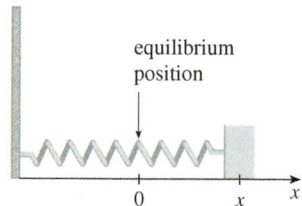

36. An elastic band is hung on a hook and a mass is hung on the lower end of the band. When the mass is pulled downward and then released, it vibrates vertically. The equation of motion is $s = 2 \cos t + 3 \sin t$, $t \geq 0$, where s is measured in centimeters and t in seconds. (Take the positive direction to be downward.)
    (a) Find the velocity and acceleration at time t.
    (b) Graph the velocity and acceleration functions.
    (c) When does the mass pass through the equilibrium position for the first time?
    (d) How far from its equilibrium position does the mass travel?
    (e) When is the speed the greatest?

37. A ladder 10 ft long rests against a vertical wall. Let θ be the angle between the top of the ladder and the wall and let x be the distance from the bottom of the ladder to the wall. If the bottom of the ladder slides away from the wall, how fast does x change with respect to θ when $\theta = \pi/3$?

38. An object with weight W is dragged along a horizontal plane by a force acting along a rope attached to the object.

If the rope makes an angle θ with the plane, then the magnitude of the force is

$$F = \frac{\mu W}{\mu \sin \theta + \cos \theta}$$

where μ is a constant called the *coefficient of friction*.
(a) Find the rate of change of F with respect to θ.
(b) When is this rate of change equal to 0?
(c) If $W = 50$ lb and $\mu = 0.6$, draw the graph of F as a function of θ and use it to locate the value of θ for which $dF/d\theta = 0$. Is the value consistent with your answer to part (b)?

39–50 Find the limit.

39. $\lim\limits_{x \to 0} \dfrac{\sin 5x}{3x}$

40. $\lim\limits_{x \to 0} \dfrac{\sin x}{\sin \pi x}$

41. $\lim\limits_{t \to 0} \dfrac{\tan 6t}{\sin 2t}$

42. $\lim\limits_{\theta \to 0} \dfrac{\cos \theta - 1}{\sin \theta}$

43. $\lim\limits_{x \to 0} \dfrac{\sin 3x}{5x^3 - 4x}$

44. $\lim\limits_{x \to 0} \dfrac{\sin 3x \sin 5x}{x^2}$

45. $\lim\limits_{\theta \to 0} \dfrac{\sin \theta}{\theta + \tan \theta}$

46. $\lim\limits_{x \to 0} \csc x \sin(\sin x)$

47. $\lim\limits_{\theta \to 0} \dfrac{\cos \theta - 1}{2\theta^2}$

48. $\lim\limits_{x \to 0} \dfrac{\sin(x^2)}{x}$

49. $\lim\limits_{x \to \pi/4} \dfrac{1 - \tan x}{\sin x - \cos x}$

50. $\lim\limits_{x \to 1} \dfrac{\sin(x - 1)}{x^2 + x - 2}$

51–52 Find the given derivative by finding the first few derivatives and observing the pattern that occurs.

51. $\dfrac{d^{99}}{dx^{99}}(\sin x)$

52. $\dfrac{d^{35}}{dx^{35}}(x \sin x)$

53. Find constants A and B such that the function $y = A \sin x + B \cos x$ satisfies the differential equation $y'' + y' - 2y = \sin x$.

54. (a) Evaluate $\lim\limits_{x \to \infty} x \sin \dfrac{1}{x}$.

(b) Evaluate $\lim\limits_{x \to 0} x \sin \dfrac{1}{x}$.

(c) Illustrate parts (a) and (b) by graphing $y = x \sin(1/x)$.

55. Differentiate each trigonometric identity to obtain a new (or familiar) identity.

(a) $\tan x = \dfrac{\sin x}{\cos x}$

(b) $\sec x = \dfrac{1}{\cos x}$

(c) $\sin x + \cos x = \dfrac{1 + \cot x}{\csc x}$

56. A semicircle with diameter PQ sits on an isosceles triangle PQR to form a region shaped like a two-dimensional ice-cream cone, as shown in the figure. If $A(\theta)$ is the area of the semicircle and $B(\theta)$ is the area of the triangle, find

$$\lim_{\theta \to 0^+} \frac{A(\theta)}{B(\theta)}$$

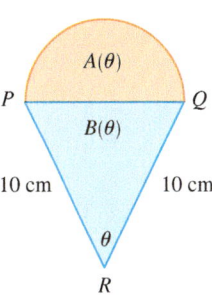

57. The figure shows a circular arc of length s and a chord of length d, both subtended by a central angle θ. Find

$$\lim_{\theta \to 0^+} \frac{s}{d}$$

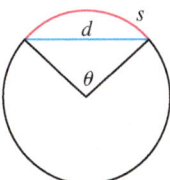

58. Let $f(x) = \dfrac{x}{\sqrt{1 - \cos 2x}}$.

(a) Graph f. What type of discontinuity does it appear to have at 0?
(b) Calculate the left and right limits of f at 0. Do these values confirm your answer to part (a)?

3.4 The Chain Rule

Suppose you are asked to differentiate the function

$$F(x) = \sqrt{x^2 + 1}$$

The differentiation formulas you learned in the previous sections of this chapter do not enable you to calculate $F'(x)$.

See Section 1.3 for a review of composite functions.

Observe that F is a composite function. In fact, if we let $y = f(u) = \sqrt{u}$ and let $u = g(x) = x^2 + 1$, then we can write $y = F(x) = f(g(x))$, that is, $F = f \circ g$. We know how to differentiate both f and g, so it would be useful to have a rule that tells us how to find the derivative of $F = f \circ g$ in terms of the derivatives of f and g.

It turns out that the derivative of the composite function $f \circ g$ is the product of the derivatives of f and g. This fact is one of the most important of the differentiation rules and is called the *Chain Rule*. It seems plausible if we interpret derivatives as rates of change. Regard du/dx as the rate of change of u with respect to x, dy/du as the rate of change of y with respect to u, and dy/dx as the rate of change of y with respect to x. If u changes twice as fast as x and y changes three times as fast as u, then it seems reasonable that y changes six times as fast as x, and so we expect that

$$\frac{dy}{dx} = \frac{dy}{du} \frac{du}{dx}$$

The Chain Rule If g is differentiable at x and f is differentiable at $g(x)$, then the composite function $F = f \circ g$ defined by $F(x) = f(g(x))$ is differentiable at x and F' is given by the product

$$F'(x) = f'(g(x)) \cdot g'(x)$$

In Leibniz notation, if $y = f(u)$ and $u = g(x)$ are both differentiable functions, then

$$\frac{dy}{dx} = \frac{dy}{du} \frac{du}{dx}$$

James Gregory
The first person to formulate the Chain Rule was the Scottish mathematician James Gregory (1638–1675), who also designed the first practical reflecting telescope. Gregory discovered the basic ideas of calculus at about the same time as Newton. He became the first Professor of Mathematics at the University of St. Andrews and later held the same position at the University of Edinburgh. But one year after accepting that position he died at the age of 36.

COMMENTS ON THE PROOF OF THE CHAIN RULE Let Δu be the change in u corresponding to a change of Δx in x, that is,

$$\Delta u = g(x + \Delta x) - g(x)$$

Then the corresponding change in y is

$$\Delta y = f(u + \Delta u) - f(u)$$

It is tempting to write

$$\frac{dy}{dx} = \lim_{\Delta x \to 0} \frac{\Delta y}{\Delta x}$$

$$\boxed{1} \qquad = \lim_{\Delta x \to 0} \frac{\Delta y}{\Delta u} \cdot \frac{\Delta u}{\Delta x}$$

$$= \lim_{\Delta x \to 0} \frac{\Delta y}{\Delta u} \cdot \lim_{\Delta x \to 0} \frac{\Delta u}{\Delta x}$$

$$= \lim_{\Delta u \to 0} \frac{\Delta y}{\Delta u} \cdot \lim_{\Delta x \to 0} \frac{\Delta u}{\Delta x} \qquad \text{(Note that } \Delta u \to 0 \text{ as } \Delta x \to 0 \text{ since } g \text{ is continuous.)}$$

$$= \frac{dy}{du} \frac{du}{dx}$$

The only flaw in this reasoning is that in (1) it might happen that $\Delta u = 0$ (even when $\Delta x \neq 0$) and, of course, we can't divide by 0. Nonetheless, this reasoning does at least *suggest* that the Chain Rule is true. A full proof of the Chain Rule is given at the end of this section. ∎

The Chain Rule can be written either in the prime notation

$$\boxed{2} \qquad (f \circ g)'(x) = f'(g(x)) \cdot g'(x)$$

or, if $y = f(u)$ and $u = g(x)$, in Leibniz notation:

$$\boxed{3} \qquad \frac{dy}{dx} = \frac{dy}{du} \frac{du}{dx}$$

Equation 3 is easy to remember because if dy/du and du/dx were quotients, then we could cancel du. Remember, however, that du has not been defined and du/dx should not be thought of as an actual quotient.

EXAMPLE 1 Find $F'(x)$ if $F(x) = \sqrt{x^2 + 1}$.

SOLUTION 1 (using Equation 2): At the beginning of this section we expressed F as $F(x) = (f \circ g)(x) = f(g(x))$ where $f(u) = \sqrt{u}$ and $g(x) = x^2 + 1$. Since

$$f'(u) = \tfrac{1}{2} u^{-1/2} = \frac{1}{2\sqrt{u}} \qquad \text{and} \qquad g'(x) = 2x$$

we have

$$F'(x) = f'(g(x)) \cdot g'(x)$$

$$= \frac{1}{2\sqrt{x^2 + 1}} \cdot 2x = \frac{x}{\sqrt{x^2 + 1}}$$

SOLUTION 2 (using Equation 3): If we let $u = x^2 + 1$ and $y = \sqrt{u}$, then

$$F'(x) = \frac{dy}{du} \frac{du}{dx} = \frac{1}{2\sqrt{u}}(2x) = \frac{1}{2\sqrt{x^2 + 1}}(2x) = \frac{x}{\sqrt{x^2 + 1}}$$ ∎

When using Formula 3 we should bear in mind that dy/dx refers to the derivative of y when y is considered as a function of x (called the *derivative of y with respect to x*), whereas dy/du refers to the derivative of y when considered as a function of u (the derivative of y with respect to u). For instance, in Example 1, y can be considered as a function of x $\left(y = \sqrt{x^2 + 1}\right)$ and also as a function of u $\left(y = \sqrt{u}\right)$. Note that

$$\frac{dy}{dx} = F'(x) = \frac{x}{\sqrt{x^2 + 1}} \qquad \text{whereas} \qquad \frac{dy}{du} = f'(u) = \frac{1}{2\sqrt{u}}$$

NOTE In using the Chain Rule we work from the outside to the inside. Formula 2 says that *we differentiate the outer function f [at the inner function $g(x)$] and then we multiply by the derivative of the inner function.*

$$\frac{d}{dx} \underbrace{f}_{\substack{\text{outer} \\ \text{function}}} \underbrace{(g(x))}_{\substack{\text{evaluated} \\ \text{at inner} \\ \text{function}}} = \underbrace{f'}_{\substack{\text{derivative} \\ \text{of outer} \\ \text{function}}} \underbrace{(g(x))}_{\substack{\text{evaluated} \\ \text{at inner} \\ \text{function}}} \cdot \underbrace{g'(x)}_{\substack{\text{derivative} \\ \text{of inner} \\ \text{function}}}$$

EXAMPLE 2 Differentiate (a) $y = \sin(x^2)$ and (b) $y = \sin^2 x$.

SOLUTION

(a) If $y = \sin(x^2)$, then the outer function is the sine function and the inner function is the squaring function, so the Chain Rule gives

$$\frac{dy}{dx} = \frac{d}{dx} \underbrace{\sin}_{\substack{\text{outer} \\ \text{function}}} \underbrace{(x^2)}_{\substack{\text{evaluated} \\ \text{at inner} \\ \text{function}}} = \underbrace{\cos}_{\substack{\text{derivative} \\ \text{of outer} \\ \text{function}}} \underbrace{(x^2)}_{\substack{\text{evaluated} \\ \text{at inner} \\ \text{function}}} \cdot \underbrace{2x}_{\substack{\text{derivative} \\ \text{of inner} \\ \text{function}}}$$

$$= 2x \cos(x^2)$$

(b) Note that $\sin^2 x = (\sin x)^2$. Here the outer function is the squaring function and the inner function is the sine function. So

$$\frac{dy}{dx} = \frac{d}{dx} \underbrace{(\sin x)^2}_{\substack{\text{inner} \\ \text{function}}} = \underbrace{2}_{\substack{\text{derivative} \\ \text{of outer} \\ \text{function}}} \cdot \underbrace{(\sin x)}_{\substack{\text{evaluated} \\ \text{at inner} \\ \text{function}}} \cdot \underbrace{\cos x}_{\substack{\text{derivative} \\ \text{of inner} \\ \text{function}}}$$

The answer can be left as $2 \sin x \cos x$ or written as $\sin 2x$ (by a trigonometric identity known as the double-angle formula).

See Reference Page 2 or Appendix D.

In Example 2(a) we combined the Chain Rule with the rule for differentiating the sine function. In general, if $y = \sin u$, where u is a differentiable function of x, then, by the Chain Rule,

$$\frac{dy}{dx} = \frac{dy}{du}\frac{du}{dx} = \cos u \frac{du}{dx}$$

Thus
$$\frac{d}{dx}(\sin u) = \cos u \frac{du}{dx}$$

In a similar fashion, all of the formulas for differentiating trigonometric functions can be combined with the Chain Rule.

Let's make explicit the special case of the Chain Rule where the outer function f is a power function. If $y = [g(x)]^n$, then we can write $y = f(u) = u^n$ where $u = g(x)$. By using the Chain Rule and then the Power Rule, we get

$$\frac{dy}{dx} = \frac{dy}{du}\frac{du}{dx} = nu^{n-1}\frac{du}{dx} = n[g(x)]^{n-1}g'(x)$$

---

4 **The Power Rule Combined with the Chain Rule** If n is any real number and $u = g(x)$ is differentiable, then

$$\frac{d}{dx}(u^n) = nu^{n-1}\frac{du}{dx}$$

Alternatively,
$$\frac{d}{dx}[g(x)]^n = n[g(x)]^{n-1} \cdot g'(x)$$

Notice that the derivative in Example 1 could be calculated by taking $n = \frac{1}{2}$ in Rule 4.

EXAMPLE 3 Differentiate $y = (x^3 - 1)^{100}$.

SOLUTION Taking $u = g(x) = x^3 - 1$ and $n = 100$ in (4), we have

$$\frac{dy}{dx} = \frac{d}{dx}(x^3 - 1)^{100} = 100(x^3 - 1)^{99} \frac{d}{dx}(x^3 - 1)$$

$$= 100(x^3 - 1)^{99} \cdot 3x^2 = 300x^2(x^3 - 1)^{99}$$

EXAMPLE 4 Find $f'(x)$ if $f(x) = \dfrac{1}{\sqrt[3]{x^2 + x + 1}}$.

SOLUTION First rewrite f: $\quad f(x) = (x^2 + x + 1)^{-1/3}$

Thus $\quad f'(x) = -\tfrac{1}{3}(x^2 + x + 1)^{-4/3} \dfrac{d}{dx}(x^2 + x + 1)$

$$= -\tfrac{1}{3}(x^2 + x + 1)^{-4/3}(2x + 1)$$

EXAMPLE 5 Find the derivative of the function

$$g(t) = \left(\frac{t-2}{2t+1}\right)^9$$

SOLUTION Combining the Power Rule, Chain Rule, and Quotient Rule, we get

$$g'(t) = 9\left(\frac{t-2}{2t+1}\right)^8 \frac{d}{dt}\left(\frac{t-2}{2t+1}\right)$$

$$= 9\left(\frac{t-2}{2t+1}\right)^8 \frac{(2t+1) \cdot 1 - 2(t-2)}{(2t+1)^2} = \frac{45(t-2)^8}{(2t+1)^{10}}$$

EXAMPLE 6 Differentiate $y = (2x + 1)^5(x^3 - x + 1)^4$.

SOLUTION In this example we must use the Product Rule before using the Chain Rule:

$$\frac{dy}{dx} = (2x+1)^5 \frac{d}{dx}(x^3 - x + 1)^4 + (x^3 - x + 1)^4 \frac{d}{dx}(2x+1)^5$$

$$= (2x+1)^5 \cdot 4(x^3 - x + 1)^3 \frac{d}{dx}(x^3 - x + 1)$$

$$+ (x^3 - x + 1)^4 \cdot 5(2x+1)^4 \frac{d}{dx}(2x+1)$$

$$= 4(2x+1)^5(x^3 - x + 1)^3(3x^2 - 1) + 5(x^3 - x + 1)^4(2x+1)^4 \cdot 2$$

Noticing that each term has the common factor $2(2x + 1)^4(x^3 - x + 1)^3$, we could factor it out and write the answer as

$$\frac{dy}{dx} = 2(2x+1)^4(x^3 - x + 1)^3(17x^3 + 6x^2 - 9x + 3)$$

The graphs of the functions y and y' in Example 6 are shown in Figure 1. Notice that y' is large when y increases rapidly and $y' = 0$ when y has a horizontal tangent. So our answer appears to be reasonable.

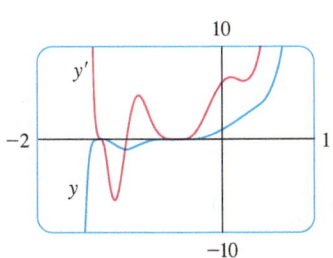

FIGURE 1

More generally, the Chain Rule gives

$$\frac{d}{dx}(e^u) = e^u \frac{du}{dx}$$

Don't confuse Formula 5 (where x is the *exponent*) with the Power Rule (where x is the *base*):

$$\frac{d}{dx}(x^n) = nx^{n-1}$$

EXAMPLE 7 Differentiate $y = e^{\sin x}$.

SOLUTION Here the inner function is $g(x) = \sin x$ and the outer function is the exponential function $f(x) = e^x$. So, by the Chain Rule,

$$\frac{dy}{dx} = \frac{d}{dx}(e^{\sin x}) = e^{\sin x}\frac{d}{dx}(\sin x) = e^{\sin x}\cos x \qquad \blacksquare$$

We can use the Chain Rule to differentiate an exponential function with any base $b > 0$. Recall from Section 1.5 that $b = e^{\ln b}$. So

$$b^x = (e^{\ln b})^x = e^{(\ln b)x}$$

and the Chain Rule gives

$$\frac{d}{dx}(b^x) = \frac{d}{dx}(e^{(\ln b)x}) = e^{(\ln b)x}\frac{d}{dx}(\ln b)x$$

$$= e^{(\ln b)x} \cdot \ln b = b^x \ln b$$

because $\ln b$ is a constant. So we have the formula

$$\boxed{\frac{d}{dx}(b^x) = b^x \ln b} \qquad \boxed{5}$$

In particular, if $b = 2$, we get

$$\frac{d}{dx}(2^x) = 2^x \ln 2 \qquad \boxed{6}$$

In Section 3.1 we gave the estimate

$$\frac{d}{dx}(2^x) \approx (0.69)2^x$$

This is consistent with the exact formula (6) because $\ln 2 \approx 0.693147$.

The reason for the name "Chain Rule" becomes clear when we make a longer chain by adding another link. Suppose that $y = f(u)$, $u = g(x)$, and $x = h(t)$, where f, g, and h are differentiable functions. Then, to compute the derivative of y with respect to t, we use the Chain Rule twice:

$$\frac{dy}{dt} = \frac{dy}{dx}\frac{dx}{dt} = \frac{dy}{du}\frac{du}{dx}\frac{dx}{dt}$$

EXAMPLE 8 If $f(x) = \sin(\cos(\tan x))$, then

$$f'(x) = \cos(\cos(\tan x))\frac{d}{dx}\cos(\tan x)$$

$$= \cos(\cos(\tan x))[-\sin(\tan x)]\frac{d}{dx}(\tan x)$$

$$= -\cos(\cos(\tan x))\sin(\tan x)\sec^2 x$$

Notice that we used the Chain Rule twice. $\qquad \blacksquare$

EXAMPLE 9 Differentiate $y = e^{\sec 3\theta}$.

SOLUTION The outer function is the exponential function, the middle function is the secant function, and the inner function is the tripling function. So we have

$$\frac{dy}{d\theta} = e^{\sec 3\theta} \frac{d}{d\theta}(\sec 3\theta)$$

$$= e^{\sec 3\theta} \sec 3\theta \tan 3\theta \frac{d}{d\theta}(3\theta)$$

$$= 3e^{\sec 3\theta} \sec 3\theta \tan 3\theta \qquad \blacksquare$$

■ How to Prove the Chain Rule

Recall that if $y = f(x)$ and x changes from a to $a + \Delta x$, we define the increment of y as

$$\Delta y = f(a + \Delta x) - f(a)$$

According to the definition of a derivative, we have

$$\lim_{\Delta x \to 0} \frac{\Delta y}{\Delta x} = f'(a)$$

So if we denote by ε the difference between the difference quotient and the derivative, we obtain

$$\lim_{\Delta x \to 0} \varepsilon = \lim_{\Delta x \to 0} \left(\frac{\Delta y}{\Delta x} - f'(a) \right) = f'(a) - f'(a) = 0$$

But $\qquad \varepsilon = \dfrac{\Delta y}{\Delta x} - f'(a) \quad \Rightarrow \quad \Delta y = f'(a) \Delta x + \varepsilon \Delta x$

If we define ε to be 0 when $\Delta x = 0$, then ε becomes a continuous function of Δx. Thus, for a differentiable function f, we can write

$$\boxed{7} \qquad \Delta y = f'(a) \Delta x + \varepsilon \Delta x \quad \text{where} \quad \varepsilon \to 0 \text{ as } \Delta x \to 0$$

and ε is a continuous function of Δx. This property of differentiable functions is what enables us to prove the Chain Rule.

PROOF OF THE CHAIN RULE Suppose $u = g(x)$ is differentiable at a and $y = f(u)$ is differentiable at $b = g(a)$. If Δx is an increment in x and Δu and Δy are the corresponding increments in u and y, then we can use Equation 7 to write

$$\boxed{8} \qquad \Delta u = g'(a) \Delta x + \varepsilon_1 \Delta x = [g'(a) + \varepsilon_1] \Delta x$$

where $\varepsilon_1 \to 0$ as $\Delta x \to 0$. Similarly

$$\boxed{9} \qquad \Delta y = f'(b) \Delta u + \varepsilon_2 \Delta u = [f'(b) + \varepsilon_2] \Delta u$$

where $\varepsilon_2 \to 0$ as $\Delta u \to 0$. If we now substitute the expression for Δu from Equation 8 into Equation 9, we get

$$\Delta y = [f'(b) + \varepsilon_2][g'(a) + \varepsilon_1] \Delta x$$

so
$$\frac{\Delta y}{\Delta x} = [f'(b) + \varepsilon_2][g'(a) + \varepsilon_1]$$

As $\Delta x \to 0$, Equation 8 shows that $\Delta u \to 0$. So both $\varepsilon_1 \to 0$ and $\varepsilon_2 \to 0$ as $\Delta x \to 0$. Therefore

$$\frac{dy}{dx} = \lim_{\Delta x \to 0} \frac{\Delta y}{\Delta x} = \lim_{\Delta x \to 0} [f'(b) + \varepsilon_2][g'(a) + \varepsilon_1]$$

$$= f'(b)g'(a) = f'(g(a))g'(a)$$

This proves the Chain Rule. ∎

3.4 EXERCISES

1–6 Write the composite function in the form $f(g(x))$. [Identify the inner function $u = g(x)$ and the outer function $y = f(u)$.] Then find the derivative dy/dx.

1. $y = \sqrt[3]{1 + 4x}$
2. $y = (2x^3 + 5)^4$
3. $y = \tan \pi x$
4. $y = \sin(\cot x)$
5. $y = e^{\sqrt{x}}$
6. $y = \sqrt{2 - e^x}$

7–46 Find the derivative of the function.

7. $F(x) = (5x^6 + 2x^3)^4$
8. $F(x) = (1 + x + x^2)^{99}$
9. $f(x) = \sqrt{5x + 1}$
10. $f(x) = \dfrac{1}{\sqrt[3]{x^2 - 1}}$
11. $f(\theta) = \cos(\theta^2)$
12. $g(\theta) = \cos^2\theta$
13. $y = x^2 e^{-3x}$
14. $f(t) = t \sin \pi t$
15. $f(t) = e^{at} \sin bt$
16. $g(x) = e^{x^2 - x}$
17. $f(x) = (2x - 3)^4(x^2 + x + 1)^5$
18. $g(x) = (x^2 + 1)^3(x^2 + 2)^6$
19. $h(t) = (t + 1)^{2/3}(2t^2 - 1)^3$
20. $F(t) = (3t - 1)^4(2t + 1)^{-3}$
21. $y = \sqrt{\dfrac{x}{x + 1}}$
22. $y = \left(x + \dfrac{1}{x}\right)^5$
23. $y = e^{\tan \theta}$
24. $f(t) = 2^{t^3}$
25. $g(u) = \left(\dfrac{u^3 - 1}{u^3 + 1}\right)^8$
26. $s(t) = \sqrt{\dfrac{1 + \sin t}{1 + \cos t}}$
27. $r(t) = 10^{2\sqrt{t}}$
28. $f(z) = e^{z/(z-1)}$
29. $H(r) = \dfrac{(r^2 - 1)^3}{(2r + 1)^5}$
30. $J(\theta) = \tan^2(n\theta)$
31. $F(t) = e^{t \sin 2t}$
32. $F(t) = \dfrac{t^2}{\sqrt{t^3 + 1}}$
33. $G(x) = 4^{C/x}$
34. $U(y) = \left(\dfrac{y^4 + 1}{y^2 + 1}\right)^5$
35. $y = \cos\left(\dfrac{1 - e^{2x}}{1 + e^{2x}}\right)$
36. $y = x^2 e^{-1/x}$
37. $y = \cot^2(\sin \theta)$
38. $y = \sqrt{1 + xe^{-2x}}$
39. $f(t) = \tan(\sec(\cos t))$
40. $y = e^{\sin 2x} + \sin(e^{2x})$
41. $f(t) = \sin^2(e^{\sin^2 t})$
42. $y = \sqrt{x + \sqrt{x + \sqrt{x}}}$
43. $g(x) = (2ra^{rx} + n)^p$
44. $y = 2^{3^{4^x}}$
45. $y = \cos\sqrt{\sin(\tan \pi x)}$
46. $y = [x + (x + \sin^2 x)^3]^4$

47–50 Find y' and y''.

47. $y = \cos(\sin 3\theta)$
48. $y = \dfrac{1}{(1 + \tan x)^2}$
49. $y = \sqrt{1 - \sec t}$
50. $y = e^{e^x}$

51–54 Find an equation of the tangent line to the curve at the given point.

51. $y = 2^x$, $(0, 1)$
52. $y = \sqrt{1 + x^3}$, $(2, 3)$
53. $y = \sin(\sin x)$, $(\pi, 0)$
54. $y = xe^{-x^2}$, $(0, 0)$

55. (a) Find an equation of the tangent line to the curve $y = 2/(1 + e^{-x})$ at the point $(0, 1)$.
    (b) Illustrate part (a) by graphing the curve and the tangent line on the same screen.

56. (a) The curve $y = |x|/\sqrt{2 - x^2}$ is called a *bullet-nose curve*. Find an equation of the tangent line to this curve at the point $(1, 1)$.
(b) Illustrate part (a) by graphing the curve and the tangent line on the same screen.

57. (a) If $f(x) = x\sqrt{2 - x^2}$, find $f'(x)$.
(b) Check to see that your answer to part (a) is reasonable by comparing the graphs of f and f'.

58. The function $f(x) = \sin(x + \sin 2x)$, $0 \leq x \leq \pi$, arises in applications to frequency modulation (FM) synthesis.
(a) Use a graph of f produced by a calculator to make a rough sketch of the graph of f'.
(b) Calculate $f'(x)$ and use this expression, with a calculator, to graph f'. Compare with your sketch in part (a).

59. Find all points on the graph of the function $f(x) = 2 \sin x + \sin^2 x$ at which the tangent line is horizontal.

60. At what point on the curve $y = \sqrt{1 + 2x}$ is the tangent line perpendicular to the line $6x + 2y = 1$?

61. If $F(x) = f(g(x))$, where $f(-2) = 8$, $f'(-2) = 4$, $f'(5) = 3$, $g(5) = -2$, and $g'(5) = 6$, find $F'(5)$.

62. If $h(x) = \sqrt{4 + 3f(x)}$, where $f(1) = 7$ and $f'(1) = 4$, find $h'(1)$.

63. A table of values for f, g, f', and g' is given.

x	$f(x)$	$g(x)$	$f'(x)$	$g'(x)$
1	3	2	4	6
2	1	8	5	7
3	7	2	7	9

(a) If $h(x) = f(g(x))$, find $h'(1)$.
(b) If $H(x) = g(f(x))$, find $H'(1)$.

64. Let f and g be the functions in Exercise 63.
(a) If $F(x) = f(f(x))$, find $F'(2)$.
(b) If $G(x) = g(g(x))$, find $G'(3)$.

65. If f and g are the functions whose graphs are shown, let $u(x) = f(g(x))$, $v(x) = g(f(x))$, and $w(x) = g(g(x))$. Find each derivative, if it exists. If it does not exist, explain why.
(a) $u'(1)$ (b) $v'(1)$ (c) $w'(1)$

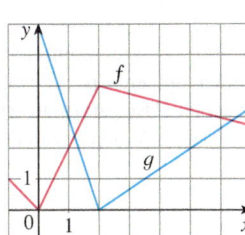

66. If f is the function whose graph is shown, let $h(x) = f(f(x))$ and $g(x) = f(x^2)$. Use the graph of f to estimate the value of each derivative.
(a) $h'(2)$ (b) $g'(2)$

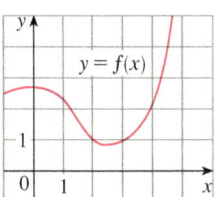

67. If $g(x) = \sqrt{f(x)}$, where the graph of f is shown, evaluate $g'(3)$.

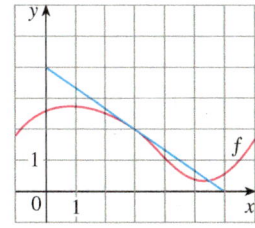

68. Suppose f is differentiable on $\mathbb{R}$ and α is a real number. Let $F(x) = f(x^\alpha)$ and $G(x) = [f(x)]^\alpha$. Find expressions for (a) $F'(x)$ and (b) $G'(x)$.

69. Suppose f is differentiable on $\mathbb{R}$. Let $F(x) = f(e^x)$ and $G(x) = e^{f(x)}$. Find expressions for (a) $F'(x)$ and (b) $G'(x)$.

70. Let $g(x) = e^{cx} + f(x)$ and $h(x) = e^{kx} f(x)$, where $f(0) = 3$, $f'(0) = 5$, and $f''(0) = -2$.
(a) Find $g'(0)$ and $g''(0)$ in terms of c.
(b) In terms of k, find an equation of the tangent line to the graph of h at the point where $x = 0$.

71. Let $r(x) = f(g(h(x)))$, where $h(1) = 2$, $g(2) = 3$, $h'(1) = 4$, $g'(2) = 5$, and $f'(3) = 6$. Find $r'(1)$.

72. If g is a twice differentiable function and $f(x) = xg(x^2)$, find f'' in terms of g, g', and g''.

73. If $F(x) = f(3f(4f(x)))$, where $f(0) = 0$ and $f'(0) = 2$, find $F'(0)$.

74. If $F(x) = f(xf(xf(x)))$, where $f(1) = 2$, $f(2) = 3$, $f'(1) = 4$, $f'(2) = 5$, and $f'(3) = 6$, find $F'(1)$.

75. Show that the function $y = e^{2x}(A \cos 3x + B \sin 3x)$ satisfies the differential equation $y'' - 4y' + 13y = 0$.

76. For what values of r does the function $y = e^{rx}$ satisfy the differential equation $y'' - 4y' + y = 0$?

77. Find the 50th derivative of $y = \cos 2x$.

78. Find the 1000th derivative of $f(x) = xe^{-x}$.

79. The displacement of a particle on a vibrating string is given by the equation $s(t) = 10 + \frac{1}{4}\sin(10\pi t)$ where s is measured in centimeters and t in seconds. Find the velocity of the particle after t seconds.

80. If the equation of motion of a particle is given by $s = A\cos(\omega t + \delta)$, the particle is said to undergo *simple harmonic motion*.
(a) Find the velocity of the particle at time t.
(b) When is the velocity 0?

81. A Cepheid variable star is a star whose brightness alternately increases and decreases. The most easily visible such star is Delta Cephei, for which the interval between times of maximum brightness is 5.4 days. The average brightness of this star is 4.0 and its brightness changes by ± 0.35. In view of these data, the brightness of Delta Cephei at time t, where t is measured in days, has been modeled by the function

$$B(t) = 4.0 + 0.35\sin\left(\frac{2\pi t}{5.4}\right)$$

(a) Find the rate of change of the brightness after t days.
(b) Find, correct to two decimal places, the rate of increase after one day.

82. In Example 1.3.4 we arrived at a model for the length of daylight (in hours) in Philadelphia on the tth day of the year:

$$L(t) = 12 + 2.8\sin\left[\frac{2\pi}{365}(t - 80)\right]$$

Use this model to compare how the number of hours of daylight is increasing in Philadelphia on March 21 and May 21.

83. The motion of a spring that is subject to a frictional force or a damping force (such as a shock absorber in a car) is often modeled by the product of an exponential function and a sine or cosine function. Suppose the equation of motion of a point on such a spring is

$$s(t) = 2e^{-1.5t}\sin 2\pi t$$

where s is measured in centimeters and t in seconds. Find the velocity after t seconds and graph both the position and velocity functions for $0 \le t \le 2$.

84. Under certain circumstances a rumor spreads according to the equation

$$p(t) = \frac{1}{1 + ae^{-kt}}$$

where $p(t)$ is the proportion of the population that has heard the rumor at time t and a and k are positive constants. [In Section 9.4 we will see that this is a reasonable equation for $p(t)$.]
(a) Find $\lim_{t\to\infty} p(t)$.
(b) Find the rate of spread of the rumor.

(c) Graph p for the case $a = 10$, $k = 0.5$ with t measured in hours. Use the graph to estimate how long it will take for 80% of the population to hear the rumor.

85. The average blood alcohol concentration (BAC) of eight male subjects was measured after consumption of 15 mL of ethanol (corresponding to one alcoholic drink). The resulting data were modeled by the concentration function

$$C(t) = 0.0225te^{-0.0467t}$$

where t is measured in minutes after consumption and C is measured in mg/mL.
(a) How rapidly was the BAC increasing after 10 minutes?
(b) How rapidly was it decreasing half an hour later?

Source: Adapted from P. Wilkinson et al., "Pharmacokinetics of Ethanol after Oral Administration in the Fasting State," *Journal of Pharmacokinetics and Biopharmaceutics* 5 (1977): 207–24.

86. In Section 1.4 we modeled the world population from 1900 to 2010 with the exponential function

$$P(t) = (1436.53) \cdot (1.01395)^t$$

where $t = 0$ corresponds to the year 1900 and $P(t)$ is measured in millions. According to this model, what was the rate of increase of world population in 1920? In 1950? In 2000?

87. A particle moves along a straight line with displacement $s(t)$, velocity $v(t)$, and acceleration $a(t)$. Show that

$$a(t) = v(t)\frac{dv}{ds}$$

Explain the difference between the meanings of the derivatives dv/dt and dv/ds.

88. Air is being pumped into a spherical weather balloon. At any time t, the volume of the balloon is $V(t)$ and its radius is $r(t)$.
(a) What do the derivatives dV/dr and dV/dt represent?
(b) Express dV/dt in terms of dr/dt.

89. The flash unit on a camera operates by storing charge on a capacitor and releasing it suddenly when the flash is set off. The following data describe the charge Q remaining on the capacitor (measured in microcoulombs, μC) at time t (measured in seconds).

t	0.00	0.02	0.04	0.06	0.08	0.10
Q	100.00	81.87	67.03	54.88	44.93	36.76

(a) Use a graphing calculator or computer to find an exponential model for the charge.
(b) The derivative $Q'(t)$ represents the electric current (measured in microamperes, μA) flowing from the capacitor to the flash bulb. Use part (a) to estimate the current when $t = 0.04$ s. Compare with the result of Example 2.1.2.

90. The table gives the US population from 1790 to 1860.

Year	Population	Year	Population
1790	3,929,000	1830	12,861,000
1800	5,308,000	1840	17,063,000
1810	7,240,000	1850	23,192,000
1820	9,639,000	1860	31,443,000

(a) Use a graphing calculator or computer to fit an exponential function to the data. Graph the data points and the exponential model. How good is the fit?
(b) Estimate the rates of population growth in 1800 and 1850 by averaging slopes of secant lines.
(c) Use the exponential model in part (a) to estimate the rates of growth in 1800 and 1850. Compare these estimates with the ones in part (b).
(d) Use the exponential model to predict the population in 1870. Compare with the actual population of 38,558,000. Can you explain the discrepancy?

91. Computer algebra systems have commands that differentiate functions, but the form of the answer may not be convenient and so further commands may be necessary to simplify the answer.
(a) Use a CAS to find the derivative in Example 5 and compare with the answer in that example. Then use the simplify command and compare again.
(b) Use a CAS to find the derivative in Example 6. What happens if you use the simplify command? What happens if you use the factor command? Which form of the answer would be best for locating horizontal tangents?

92. (a) Use a CAS to differentiate the function

$$f(x) = \sqrt{\frac{x^4 - x + 1}{x^4 + x + 1}}$$

and to simplify the result.
(b) Where does the graph of f have horizontal tangents?
(c) Graph f and f' on the same screen. Are the graphs consistent with your answer to part (b)?

93. Use the Chain Rule to prove the following.
(a) The derivative of an even function is an odd function.
(b) The derivative of an odd function is an even function.

94. Use the Chain Rule and the Product Rule to give an alternative proof of the Quotient Rule.
[*Hint:* Write $f(x)/g(x) = f(x)[g(x)]^{-1}$.]

95. (a) If n is a positive integer, prove that

$$\frac{d}{dx}(\sin^n x \cos nx) = n \sin^{n-1} x \cos(n+1)x$$

(b) Find a formula for the derivative of $y = \cos^n x \cos nx$ that is similar to the one in part (a).

96. Suppose $y = f(x)$ is a curve that always lies above the x-axis and never has a horizontal tangent, where f is differentiable everywhere. For what value of y is the rate of change of y^5 with respect to x eighty times the rate of change of y with respect to x?

97. Use the Chain Rule to show that if θ is measured in degrees, then

$$\frac{d}{d\theta}(\sin \theta) = \frac{\pi}{180} \cos \theta$$

(This gives one reason for the convention that radian measure is always used when dealing with trigonometric functions in calculus: the differentiation formulas would not be as simple if we used degree measure.)

98. (a) Write $|x| = \sqrt{x^2}$ and use the Chain Rule to show that

$$\frac{d}{dx}|x| = \frac{x}{|x|}$$

(b) If $f(x) = |\sin x|$, find $f'(x)$ and sketch the graphs of f and f'. Where is f not differentiable?
(c) If $g(x) = \sin |x|$, find $g'(x)$ and sketch the graphs of g and g'. Where is g not differentiable?

99. If $y = f(u)$ and $u = g(x)$, where f and g are twice differentiable functions, show that

$$\frac{d^2y}{dx^2} = \frac{d^2y}{du^2}\left(\frac{du}{dx}\right)^2 + \frac{dy}{du}\frac{d^2u}{dx^2}$$

100. If $y = f(u)$ and $u = g(x)$, where f and g possess third derivatives, find a formula for d^3y/dx^3 similar to the one given in Exercise 99.

APPLIED PROJECT WHERE SHOULD A PILOT START DESCENT?

An approach path for an aircraft landing is shown in the figure and satisfies the following conditions:

(i) The cruising altitude is h when descent starts at a horizontal distance ℓ from touchdown at the origin.

(ii) The pilot must maintain a constant horizontal speed v throughout descent.

(iii) The absolute value of the vertical acceleration should not exceed a constant k (which is much less than the acceleration due to gravity).

1. Find a cubic polynomial $P(x) = ax^3 + bx^2 + cx + d$ that satisfies condition (i) by imposing suitable conditions on $P(x)$ and $P'(x)$ at the start of descent and at touchdown.

2. Use conditions (ii) and (iii) to show that
$$\frac{6hv^2}{\ell^2} \leq k$$

3. Suppose that an airline decides not to allow vertical acceleration of a plane to exceed $k = 860$ mi/h². If the cruising altitude of a plane is 35,000 ft and the speed is 300 mi/h, how far away from the airport should the pilot start descent?

4. Graph the approach path if the conditions stated in Problem 3 are satisfied.

3.5 Implicit Differentiation

The functions that we have met so far can be described by expressing one variable explicitly in terms of another variable—for example,

$$y = \sqrt{x^3 + 1} \quad \text{or} \quad y = x \sin x$$

or, in general, $y = f(x)$. Some functions, however, are defined implicitly by a relation between x and y such as

$$x^2 + y^2 = 25$$

or

$$x^3 + y^3 = 6xy$$

In some cases it is possible to solve such an equation for y as an explicit function (or several functions) of x. For instance, if we solve Equation 1 for y, we get $y = \pm\sqrt{25 - x^2}$, so two of the functions determined by the implicit Equation 1 are $f(x) = \sqrt{25 - x^2}$ and $g(x) = -\sqrt{25 - x^2}$. The graphs of f and g are the upper and lower semicircles of the circle $x^2 + y^2 = 25$. (See Figure 1.)

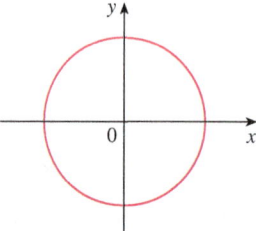
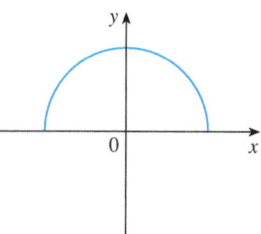
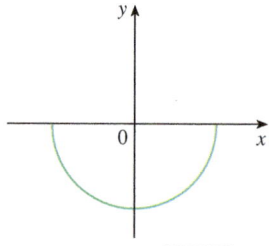

FIGURE 1 (a) $x^2 + y^2 = 25$ (b) $f(x) = \sqrt{25 - x^2}$ (c) $g(x) = -\sqrt{25 - x^2}$

It's not easy to solve Equation 2 for y explicitly as a function of x by hand. (A computer algebra system has no trouble, but the expressions it obtains are very complicated.) Nonetheless, (2) is the equation of a curve called the **folium of Descartes** shown in Figure 2 and it implicitly defines y as several functions of x. The graphs of three such functions are shown in Figure 3. When we say that f is a function defined implicitly by Equation 2, we mean that the equation

$$x^3 + [f(x)]^3 = 6xf(x)$$

is true for all values of x in the domain of f.

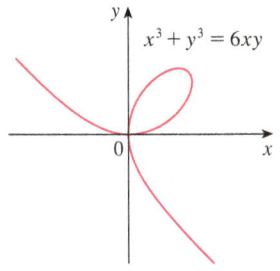

FIGURE 2 The folium of Descartes

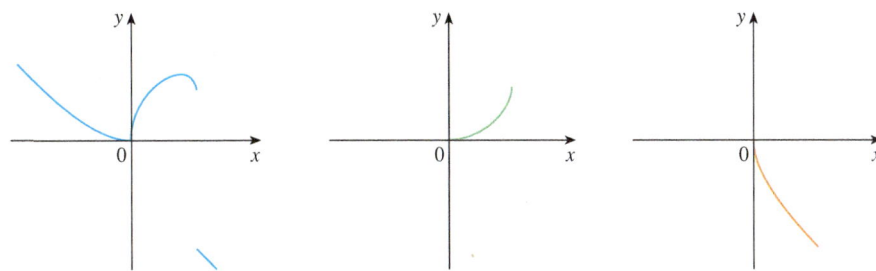

FIGURE 3 Graphs of three functions defined by the folium of Descartes

Fortunately, we don't need to solve an equation for y in terms of x in order to find the derivative of y. Instead we can use the method of **implicit differentiation**. This consists of differentiating both sides of the equation with respect to x and then solving the resulting equation for y'. In the examples and exercises of this section it is always assumed that the given equation determines y implicitly as a differentiable function of x so that the method of implicit differentiation can be applied.

EXAMPLE 1

(a) If $x^2 + y^2 = 25$, find $\dfrac{dy}{dx}$.

(b) Find an equation of the tangent to the circle $x^2 + y^2 = 25$ at the point $(3, 4)$.

SOLUTION 1

(a) Differentiate both sides of the equation $x^2 + y^2 = 25$:

$$\frac{d}{dx}(x^2 + y^2) = \frac{d}{dx}(25)$$

$$\frac{d}{dx}(x^2) + \frac{d}{dx}(y^2) = 0$$

Remembering that y is a function of x and using the Chain Rule, we have

$$\frac{d}{dx}(y^2) = \frac{d}{dy}(y^2)\frac{dy}{dx} = 2y\frac{dy}{dx}$$

Thus
$$2x + 2y\frac{dy}{dx} = 0$$

Now we solve this equation for dy/dx:

$$\frac{dy}{dx} = -\frac{x}{y}$$

(b) At the point (3, 4) we have $x = 3$ and $y = 4$, so

$$\frac{dy}{dx} = -\frac{3}{4}$$

An equation of the tangent to the circle at (3, 4) is therefore

$$y - 4 = -\tfrac{3}{4}(x - 3) \quad \text{or} \quad 3x + 4y = 25$$

SOLUTION 2

(b) Solving the equation $x^2 + y^2 = 25$ for y, we get $y = \pm\sqrt{25 - x^2}$. The point (3, 4) lies on the upper semicircle $y = \sqrt{25 - x^2}$ and so we consider the function $f(x) = \sqrt{25 - x^2}$. Differentiating f using the Chain Rule, we have

$$f'(x) = \tfrac{1}{2}(25 - x^2)^{-1/2} \frac{d}{dx}(25 - x^2)$$

$$= \tfrac{1}{2}(25 - x^2)^{-1/2}(-2x) = -\frac{x}{\sqrt{25 - x^2}}$$

So

$$f'(3) = -\frac{3}{\sqrt{25 - 3^2}} = -\frac{3}{4}$$

and, as in Solution 1, an equation of the tangent is $3x + 4y = 25$. ∎

Example 1 illustrates that even when it is possible to solve an equation explicitly for y in terms of x, it may be easier to use implicit differentiation.

NOTE 1 The expression $dy/dx = -x/y$ in Solution 1 gives the derivative in terms of both x and y. It is correct no matter which function y is determined by the given equation. For instance, for $y = f(x) = \sqrt{25 - x^2}$ we have

$$\frac{dy}{dx} = -\frac{x}{y} = -\frac{x}{\sqrt{25 - x^2}}$$

whereas for $y = g(x) = -\sqrt{25 - x^2}$ we have

$$\frac{dy}{dx} = -\frac{x}{y} = -\frac{x}{-\sqrt{25 - x^2}} = \frac{x}{\sqrt{25 - x^2}}$$

EXAMPLE 2

(a) Find y' if $x^3 + y^3 = 6xy$.
(b) Find the tangent to the folium of Descartes $x^3 + y^3 = 6xy$ at the point (3, 3).
(c) At what point in the first quadrant is the tangent line horizontal?

SOLUTION

(a) Differentiating both sides of $x^3 + y^3 = 6xy$ with respect to x, regarding y as a function of x, and using the Chain Rule on the term y^3 and the Product Rule on the term $6xy$, we get

$$3x^2 + 3y^2 y' = 6xy' + 6y$$

or

$$x^2 + y^2 y' = 2xy' + 2y$$

We now solve for y':

$$y^2 y' - 2xy' = 2y - x^2$$

$$(y^2 - 2x)y' = 2y - x^2$$

$$y' = \frac{2y - x^2}{y^2 - 2x}$$

(b) When $x = y = 3$,

$$y' = \frac{2 \cdot 3 - 3^2}{3^2 - 2 \cdot 3} = -1$$

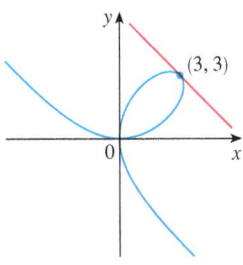

FIGURE 4

and a glance at Figure 4 confirms that this is a reasonable value for the slope at $(3, 3)$. So an equation of the tangent to the folium at $(3, 3)$ is

$$y - 3 = -1(x - 3) \quad \text{or} \quad x + y = 6$$

(c) The tangent line is horizontal if $y' = 0$. Using the expression for y' from part (a), we see that $y' = 0$ when $2y - x^2 = 0$ (provided that $y^2 - 2x \neq 0$). Substituting $y = \tfrac{1}{2}x^2$ in the equation of the curve, we get

$$x^3 + \left(\tfrac{1}{2}x^2\right)^3 = 6x\left(\tfrac{1}{2}x^2\right)$$

which simplifies to $x^6 = 16x^3$. Since $x \neq 0$ in the first quadrant, we have $x^3 = 16$. If $x = 16^{1/3} = 2^{4/3}$, then $y = \tfrac{1}{2}(2^{8/3}) = 2^{5/3}$. Thus the tangent is horizontal at $(2^{4/3}, 2^{5/3})$, which is approximately $(2.5198, 3.1748)$. Looking at Figure 5, we see that our answer is reasonable. ∎

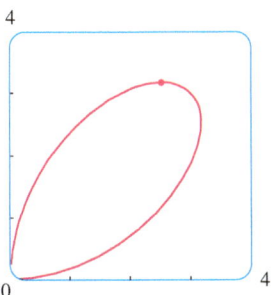

FIGURE 5

NOTE 2 There is a formula for the three roots of a cubic equation that is like the quadratic formula but much more complicated. If we use this formula (or a computer algebra system) to solve the equation $x^3 + y^3 = 6xy$ for y in terms of x, we get three functions determined by the equation:

$$y = f(x) = \sqrt[3]{-\tfrac{1}{2}x^3 + \sqrt{\tfrac{1}{4}x^6 - 8x^3}} + \sqrt[3]{-\tfrac{1}{2}x^3 - \sqrt{\tfrac{1}{4}x^6 - 8x^3}}$$

and

$$y = \tfrac{1}{2}\left[-f(x) \pm \sqrt{-3}\left(\sqrt[3]{-\tfrac{1}{2}x^3 + \sqrt{\tfrac{1}{4}x^6 - 8x^3}} - \sqrt[3]{-\tfrac{1}{2}x^3 - \sqrt{\tfrac{1}{4}x^6 - 8x^3}}\right)\right]$$

(These are the three functions whose graphs are shown in Figure 3.) You can see that the method of implicit differentiation saves an enormous amount of work in cases such as this. Moreover, implicit differentiation works just as easily for equations such as

$$y^5 + 3x^2y^2 + 5x^4 = 12$$

for which it is *impossible* to find a similar expression for y in terms of x.

Abel and Galois

The Norwegian mathematician Niels Abel proved in 1824 that no general formula can be given for the roots of a fifth-degree equation in terms of radicals. Later the French mathematician Evariste Galois proved that it is impossible to find a general formula for the roots of an nth-degree equation (in terms of algebraic operations on the coefficients) if n is any integer larger than 4.

EXAMPLE 3 Find y' if $\sin(x + y) = y^2 \cos x$.

SOLUTION Differentiating implicitly with respect to x and remembering that y is a function of x, we get

$$\cos(x + y) \cdot (1 + y') = y^2(-\sin x) + (\cos x)(2yy')$$

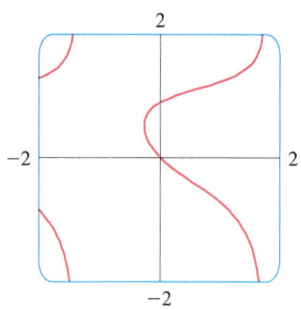

FIGURE 6

(Note that we have used the Chain Rule on the left side and the Product Rule and Chain Rule on the right side.) If we collect the terms that involve y', we get

$$\cos(x+y) + y^2 \sin x = (2y\cos x)y' - \cos(x+y) \cdot y'$$

So

$$y' = \frac{y^2 \sin x + \cos(x+y)}{2y\cos x - \cos(x+y)}$$

Figure 6, drawn with the implicit-plotting command of a computer algebra system, shows part of the curve $\sin(x+y) = y^2 \cos x$. As a check on our calculation, notice that $y' = -1$ when $x = y = 0$ and it appears from the graph that the slope is approximately -1 at the origin. ∎

Figures 7, 8, and 9 show three more curves produced by a computer algebra system with an implicit-plotting command. In Exercises 41–42 you will have an opportunity to create and examine unusual curves of this nature.

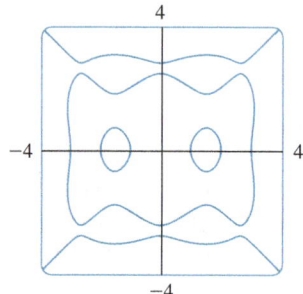

FIGURE 7
$(x^2 - 1)(x^2 - 4)(x^2 - 9)$
$= y^2(y^2 - 4)(y^2 - 9)$

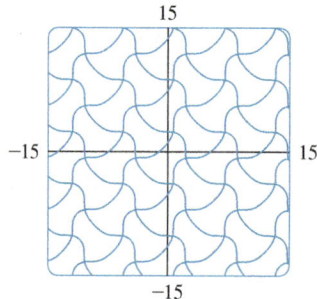

FIGURE 8
$\cos(x - \sin y) = \sin(y - \sin x)$

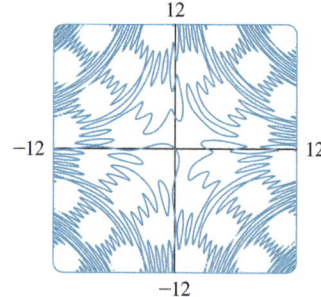

FIGURE 9
$\sin(xy) = \sin x + \sin y$

The following example shows how to find the second derivative of a function that is defined implicitly.

EXAMPLE 4 Find y'' if $x^4 + y^4 = 16$.

SOLUTION Differentiating the equation implicitly with respect to x, we get

$$4x^3 + 4y^3 y' = 0$$

Solving for y' gives

$$\boxed{3} \qquad y' = -\frac{x^3}{y^3}$$

To find y'' we differentiate this expression for y' using the Quotient Rule and remembering that y is a function of x:

$$y'' = \frac{d}{dx}\left(-\frac{x^3}{y^3}\right) = -\frac{y^3 (d/dx)(x^3) - x^3 (d/dx)(y^3)}{(y^3)^2}$$

$$= -\frac{y^3 \cdot 3x^2 - x^3(3y^2 y')}{y^6}$$

If we now substitute Equation 3 into this expression, we get

$$y'' = -\frac{3x^2y^3 - 3x^3y^2\left(-\dfrac{x^3}{y^3}\right)}{y^6}$$

$$= -\frac{3(x^2y^4 + x^6)}{y^7} = -\frac{3x^2(y^4 + x^4)}{y^7}$$

But the values of x and y must satisfy the original equation $x^4 + y^4 = 16$. So the answer simplifies to

$$y'' = -\frac{3x^2(16)}{y^7} = -48\,\frac{x^2}{y^7}$$

Figure 10 shows the graph of the curve $x^4 + y^4 = 16$ of Example 4. Notice that it's a stretched and flattened version of the circle $x^2 + y^2 = 4$. For this reason it's sometimes called a *fat circle*. It starts out very steep on the left but quickly becomes very flat. This can be seen from the expression

$$y' = -\frac{x^3}{y^3} = -\left(\frac{x}{y}\right)^3$$

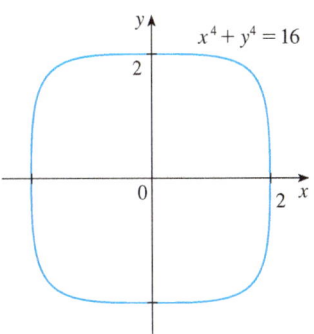

FIGURE 10

Derivatives of Inverse Trigonometric Functions

The inverse trigonometric functions were reviewed in Section 1.5. We discussed their continuity in Section 2.5 and their asymptotes in Section 2.6. Here we use implicit differentiation to find the derivatives of the inverse trigonometric functions, assuming that these functions are differentiable. [In fact, if f is any one-to-one differentiable function, it can be proved that its inverse function f^{-1} is also differentiable, except where its tangents are vertical. This is plausible because the graph of a differentiable function has no corner or kink and so if we reflect it about $y = x$, the graph of its inverse function also has no corner or kink.]

Recall the definition of the arcsine function:

$$y = \sin^{-1}x \quad \text{means} \quad \sin y = x \quad \text{and} \quad -\frac{\pi}{2} \leq y \leq \frac{\pi}{2}$$

Differentiating $\sin y = x$ implicitly with respect to x, we obtain

$$\cos y\,\frac{dy}{dx} = 1 \quad \text{or} \quad \frac{dy}{dx} = \frac{1}{\cos y}$$

Now $\cos y \geq 0$, since $-\pi/2 \leq y \leq \pi/2$, so

$$\cos y = \sqrt{1 - \sin^2 y} = \sqrt{1 - x^2}$$

The same method can be used to find a formula for the derivative of *any* inverse function. See Exercise 77.

Therefore
$$\frac{dy}{dx} = \frac{1}{\cos y} = \frac{1}{\sqrt{1-x^2}}$$

$$\frac{d}{dx}(\sin^{-1}x) = \frac{1}{\sqrt{1-x^2}}$$

Figure 11 shows the graph of $f(x) = \tan^{-1}x$ and its derivative $f'(x) = 1/(1+x^2)$. Notice that f is increasing and $f'(x)$ is always positive. The fact that $\tan^{-1}x \to \pm\pi/2$ as $x \to \pm\infty$ is reflected in the fact that $f'(x) \to 0$ as $x \to \pm\infty$.

The formula for the derivative of the arctangent function is derived in a similar way. If $y = \tan^{-1}x$, then $\tan y = x$. Differentiating this latter equation implicitly with respect to x, we have

$$\sec^2 y \frac{dy}{dx} = 1$$

$$\frac{dy}{dx} = \frac{1}{\sec^2 y} = \frac{1}{1+\tan^2 y} = \frac{1}{1+x^2}$$

$$\frac{d}{dx}(\tan^{-1}x) = \frac{1}{1+x^2}$$

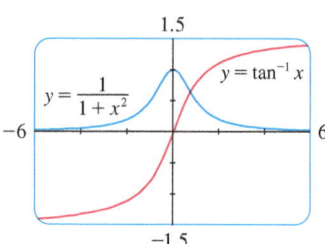

FIGURE 11

EXAMPLE 5 Differentiate (a) $y = \dfrac{1}{\sin^{-1}x}$ and (b) $f(x) = x \arctan\sqrt{x}$.

SOLUTION

(a) $\dfrac{dy}{dx} = \dfrac{d}{dx}(\sin^{-1}x)^{-1} = -(\sin^{-1}x)^{-2}\dfrac{d}{dx}(\sin^{-1}x)$

$= -\dfrac{1}{(\sin^{-1}x)^2\sqrt{1-x^2}}$

Recall that arctan x is an alternative notation for $\tan^{-1}x$.

(b) $f'(x) = x\dfrac{1}{1+(\sqrt{x})^2}\left(\tfrac{1}{2}x^{-1/2}\right) + \arctan\sqrt{x}$

$= \dfrac{\sqrt{x}}{2(1+x)} + \arctan\sqrt{x}$ ∎

The inverse trigonometric functions that occur most frequently are the ones that we have just discussed. The derivatives of the remaining four are given in the following table. The proofs of the formulas are left as exercises.

The formulas for the derivatives of $\csc^{-1}x$ and $\sec^{-1}x$ depend on the definitions that are used for these functions. See Exercise 64.

Derivatives of Inverse Trigonometric Functions

$$\frac{d}{dx}(\sin^{-1}x) = \frac{1}{\sqrt{1-x^2}} \qquad \frac{d}{dx}(\csc^{-1}x) = -\frac{1}{x\sqrt{x^2-1}}$$

$$\frac{d}{dx}(\cos^{-1}x) = -\frac{1}{\sqrt{1-x^2}} \qquad \frac{d}{dx}(\sec^{-1}x) = \frac{1}{x\sqrt{x^2-1}}$$

$$\frac{d}{dx}(\tan^{-1}x) = \frac{1}{1+x^2} \qquad \frac{d}{dx}(\cot^{-1}x) = -\frac{1}{1+x^2}$$

3.5 EXERCISES

1–4
(a) Find y' by implicit differentiation.
(b) Solve the equation explicitly for y and differentiate to get y' in terms of x.
(c) Check that your solutions to parts (a) and (b) are consistent by substituting the expression for y into your solution for part (a).

1. $9x^2 - y^2 = 1$ **2.** $2x^2 + x + xy = 1$

3. $\sqrt{x} + \sqrt{y} = 1$ **4.** $\dfrac{2}{x} - \dfrac{1}{y} = 4$

5–20 Find dy/dx by implicit differentiation.

5. $x^2 - 4xy + y^2 = 4$ **6.** $2x^2 + xy - y^2 = 2$

7. $x^4 + x^2y^2 + y^3 = 5$ **8.** $x^3 - xy^2 + y^3 = 1$

9. $\dfrac{x^2}{x + y} = y^2 + 1$ **10.** $xe^y = x - y$

11. $y \cos x = x^2 + y^2$ **12.** $\cos(xy) = 1 + \sin y$

13. $\sqrt{x + y} = x^4 + y^4$ **14.** $e^y \sin x = x + xy$

15. $e^{x/y} = x - y$ **16.** $xy = \sqrt{x^2 + y^2}$

17. $\tan^{-1}(x^2 y) = x + xy^2$ **18.** $x \sin y + y \sin x = 1$

19. $\sin(xy) = \cos(x + y)$ **20.** $\tan(x - y) = \dfrac{y}{1 + x^2}$

21. If $f(x) + x^2[f(x)]^3 = 10$ and $f(1) = 2$, find $f'(1)$.

22. If $g(x) + x \sin g(x) = x^2$, find $g'(0)$.

23–24 Regard y as the independent variable and x as the dependent variable and use implicit differentiation to find dx/dy.

23. $x^4 y^2 - x^3 y + 2xy^3 = 0$ **24.** $y \sec x = x \tan y$

25–32 Use implicit differentiation to find an equation of the tangent line to the curve at the given point.

25. $y \sin 2x = x \cos 2y$, $(\pi/2,\ \pi/4)$

26. $\sin(x + y) = 2x - 2y$, $(\pi,\ \pi)$

27. $x^2 - xy - y^2 = 1$, $(2, 1)$ (hyperbola)

28. $x^2 + 2xy + 4y^2 = 12$, $(2, 1)$ (ellipse)

29. $x^2 + y^2 = (2x^2 + 2y^2 - x)^2$, $(0, \tfrac{1}{2})$, (cardioid)

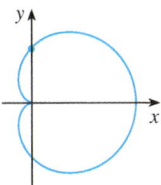

30. $x^{2/3} + y^{2/3} = 4$, $(-3\sqrt{3},\ 1)$, (astroid)

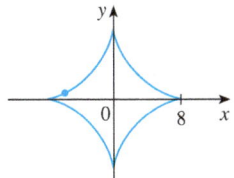

31. $2(x^2 + y^2)^2 = 25(x^2 - y^2)$, $(3, 1)$, (lemniscate)

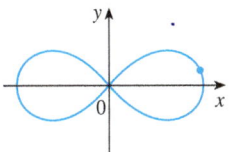

32. $y^2(y^2 - 4) = x^2(x^2 - 5)$, $(0, -2)$, (devil's curve)

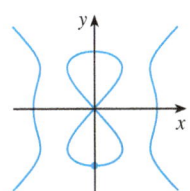

33. (a) The curve with equation $y^2 = 5x^4 - x^2$ is called a **kampyle of Eudoxus**. Find an equation of the tangent line to this curve at the point $(1, 2)$.
(b) Illustrate part (a) by graphing the curve and the tangent line on a common screen. (If your graphing device will graph implicitly defined curves, then use that capability. If not, you can still graph this curve by graphing its upper and lower halves separately.)

34. (a) The curve with equation $y^2 = x^3 + 3x^2$ is called the **Tschirnhausen cubic**. Find an equation of the tangent line to this curve at the point $(1, -2)$.
(b) At what points does this curve have horizontal tangents?
(c) Illustrate parts (a) and (b) by graphing the curve and the tangent lines on a common screen.

35–38 Find y'' by implicit differentiation.

35. $x^2 + 4y^2 = 4$ **36.** $x^2 + xy + y^2 = 3$

37. $\sin y + \cos x = 1$ **38.** $x^3 - y^3 = 7$

39. If $xy + e^y = e$, find the value of y'' at the point where $x = 0$.

40. If $x^2 + xy + y^3 = 1$, find the value of y''' at the point where $x = 1$.

CAS 41. Fanciful shapes can be created by using the implicit plotting capabilities of computer algebra systems.
  (a) Graph the curve with equation
$$y(y^2 - 1)(y - 2) = x(x - 1)(x - 2)$$
  At how many points does this curve have horizontal tangents? Estimate the x-coordinates of these points.
  (b) Find equations of the tangent lines at the points $(0, 1)$ and $(0, 2)$.
  (c) Find the exact x-coordinates of the points in part (a).
  (d) Create even more fanciful curves by modifying the equation in part (a).

CAS 42. (a) The curve with equation
$$2y^3 + y^2 - y^5 = x^4 - 2x^3 + x^2$$
  has been likened to a bouncing wagon. Use a computer algebra system to graph this curve and discover why.
  (b) At how many points does this curve have horizontal tangent lines? Find the x-coordinates of these points.

43. Find the points on the lemniscate in Exercise 31 where the tangent is horizontal.

44. Show by implicit differentiation that the tangent to the ellipse
$$\frac{x^2}{a^2} + \frac{y^2}{b^2} = 1$$
at the point (x_0, y_0) is
$$\frac{x_0 x}{a^2} + \frac{y_0 y}{b^2} = 1$$

45. Find an equation of the tangent line to the hyperbola
$$\frac{x^2}{a^2} - \frac{y^2}{b^2} = 1$$
at the point (x_0, y_0).

46. Show that the sum of the x- and y-intercepts of any tangent line to the curve $\sqrt{x} + \sqrt{y} = \sqrt{c}$ is equal to c.

47. Show, using implicit differentiation, that any tangent line at a point P to a circle with center O is perpendicular to the radius OP.

48. The Power Rule can be proved using implicit differentiation for the case where n is a rational number, $n = p/q$, and $y = f(x) = x^n$ is assumed beforehand to be a differentiable function. If $y = x^{p/q}$, then $y^q = x^p$. Use implicit differentiation to show that
$$y' = \frac{p}{q} x^{(p/q)-1}$$

49–60 Find the derivative of the function. Simplify where possible.

49. $y = (\tan^{-1} x)^2$ **50.** $y = \tan^{-1}(x^2)$

51. $y = \sin^{-1}(2x + 1)$ **52.** $g(x) = \arccos\sqrt{x}$

53. $F(x) = x \sec^{-1}(x^3)$

54. $y = \tan^{-1}(x - \sqrt{1 + x^2})$

55. $h(t) = \cot^{-1}(t) + \cot^{-1}(1/t)$ **56.** $R(t) = \arcsin(1/t)$

57. $y = x \sin^{-1} x + \sqrt{1 - x^2}$ **58.** $y = \cos^{-1}(\sin^{-1} t)$

59. $y = \arccos\left(\dfrac{b + a \cos x}{a + b \cos x}\right)$, $0 \leq x \leq \pi, \ a > b > 0$

60. $y = \arctan\sqrt{\dfrac{1 - x}{1 + x}}$

61–62 Find $f'(x)$. Check that your answer is reasonable by comparing the graphs of f and f'.

61. $f(x) = \sqrt{1 - x^2} \arcsin x$ **62.** $f(x) = \arctan(x^2 - x)$

63. Prove the formula for $(d/dx)(\cos^{-1} x)$ by the same method as for $(d/dx)(\sin^{-1} x)$.

64. (a) One way of defining $\sec^{-1} x$ is to say that
$y = \sec^{-1} x \iff \sec y = x$ and $0 \leq y < \pi/2$ or $\pi \leq y < 3\pi/2$. Show that, with this definition,
$$\frac{d}{dx}(\sec^{-1} x) = \frac{1}{x\sqrt{x^2 - 1}}$$
  (b) Another way of defining $\sec^{-1} x$ that is sometimes used is to say that $y = \sec^{-1} x \iff \sec y = x$ and $0 \leq y \leq \pi, y \neq \pi/2$. Show that, with this definition,
$$\frac{d}{dx}(\sec^{-1} x) = \frac{1}{|x|\sqrt{x^2 - 1}}$$

65–68 Two curves are **orthogonal** if their tangent lines are perpendicular at each point of intersection. Show that the given families of curves are **orthogonal trajectories** of each other; that is, every curve in one family is orthogonal to every curve in the other family. Sketch both families of curves on the same axes.

65. $x^2 + y^2 = r^2$, $ax + by = 0$

66. $x^2 + y^2 = ax$, $x^2 + y^2 = by$

67. $y = cx^2$, $x^2 + 2y^2 = k$

68. $y = ax^3$, $x^2 + 3y^2 = b$

69. Show that the ellipse $x^2/a^2 + y^2/b^2 = 1$ and the hyperbola $x^2/A^2 - y^2/B^2 = 1$ are orthogonal trajectories if $A^2 < a^2$ and $a^2 - b^2 = A^2 + B^2$ (so the ellipse and hyperbola have the same foci).

70. Find the value of the number a such that the families of curves $y = (x + c)^{-1}$ and $y = a(x + k)^{1/3}$ are orthogonal trajectories.

71. (a) The *van der Waals equation* for n moles of a gas is
$$\left(P + \frac{n^2 a}{V^2}\right)(V - nb) = nRT$$
where P is the pressure, V is the volume, and T is the temperature of the gas. The constant R is the universal gas constant and a and b are positive constants that are characteristic of a particular gas. If T remains constant, use implicit differentiation to find dV/dP.

(b) Find the rate of change of volume with respect to pressure of 1 mole of carbon dioxide at a volume of $V = 10$ L and a pressure of $P = 2.5$ atm. Use $a = 3.592$ L^2-atm/mole^2 and $b = 0.04267$ L/mole.

72. (a) Use implicit differentiation to find y' if
$$x^2 + xy + y^2 + 1 = 0$$

(b) Plot the curve in part (a). What do you see? Prove that what you see is correct.
(c) In view of part (b), what can you say about the expression for y' that you found in part (a)?

73. The equation $x^2 - xy + y^2 = 3$ represents a "rotated ellipse," that is, an ellipse whose axes are not parallel to the coordinate axes. Find the points at which this ellipse crosses the x-axis and show that the tangent lines at these points are parallel.

74. (a) Where does the normal line to the ellipse $x^2 - xy + y^2 = 3$ at the point $(-1, 1)$ intersect the ellipse a second time?

(b) Illustrate part (a) by graphing the ellipse and the normal line.

75. Find all points on the curve $x^2 y^2 + xy = 2$ where the slope of the tangent line is -1.

76. Find equations of both the tangent lines to the ellipse $x^2 + 4y^2 = 36$ that pass through the point $(12, 3)$.

77. (a) Suppose f is a one-to-one differentiable function and its inverse function f^{-1} is also differentiable. Use implicit differentiation to show that
$$(f^{-1})'(x) = \frac{1}{f'(f^{-1}(x))}$$
provided that the denominator is not 0.
(b) If $f(4) = 5$ and $f'(4) = \frac{2}{3}$, find $(f^{-1})'(5)$.

78. (a) Show that $f(x) = x + e^x$ is one-to-one.
(b) What is the value of $f^{-1}(1)$?
(c) Use the formula from Exercise 77(a) to find $(f^{-1})'(1)$.

79. The **Bessel function** of order 0, $y = J(x)$, satisfies the differential equation $xy'' + y' + xy = 0$ for all values of x and its value at 0 is $J(0) = 1$.
(a) Find $J'(0)$.
(b) Use implicit differentiation to find $J''(0)$.

80. The figure shows a lamp located three units to the right of the y-axis and a shadow created by the elliptical region $x^2 + 4y^2 \leq 5$. If the point $(-5, 0)$ is on the edge of the shadow, how far above the x-axis is the lamp located?

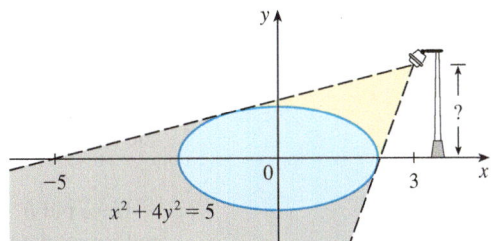

LABORATORY PROJECT FAMILIES OF IMPLICIT CURVES

In this project you will explore the changing shapes of implicitly defined curves as you vary the constants in a family, and determine which features are common to all members of the family.

1. Consider the family of curves
$$y^2 - 2x^2(x + 8) = c[(y + 1)^2(y + 9) - x^2]$$

(a) By graphing the curves with $c = 0$ and $c = 2$, determine how many points of intersection there are. (You might have to zoom in to find all of them.)
(b) Now add the curves with $c = 5$ and $c = 10$ to your graphs in part (a). What do you notice? What about other values of c?

2. (a) Graph several members of the family of curves
$$x^2 + y^2 + cx^2y^2 = 1$$
Describe how the graph changes as you change the value of c.
(b) What happens to the curve when $c = -1$? Describe what appears on the screen. Can you prove it algebraically?
(c) Find y' by implicit differentiation. For the case $c = -1$, is your expression for y' consistent with what you discovered in part (b)?

3.6 Derivatives of Logarithmic Functions

In this section we use implicit differentiation to find the derivatives of the logarithmic functions $y = \log_b x$ and, in particular, the natural logarithmic function $y = \ln x$. [It can be proved that logarithmic functions are differentiable; this is certainly plausible from their graphs (see Figure 1.5.12).]

$$\boxed{1} \qquad \frac{d}{dx}(\log_b x) = \frac{1}{x \ln b}$$

PROOF Let $y = \log_b x$. Then
$$b^y = x$$

Formula 3.4.5 says that
$$\frac{d}{dx}(b^x) = b^x \ln b$$

Differentiating this equation implicitly with respect to x, using Formula 3.4.5, we get
$$b^y (\ln b) \frac{dy}{dx} = 1$$

and so
$$\frac{dy}{dx} = \frac{1}{b^y \ln b} = \frac{1}{x \ln b} \qquad \blacksquare$$

If we put $b = e$ in Formula 1, then the factor $\ln b$ on the right side becomes $\ln e = 1$ and we get the formula for the derivative of the natural logarithmic function $\log_e x = \ln x$:

$$\boxed{2} \qquad \frac{d}{dx}(\ln x) = \frac{1}{x}$$

By comparing Formulas 1 and 2, we see one of the main reasons that natural logarithms (logarithms with base e) are used in calculus: The differentiation formula is simplest when $b = e$ because $\ln e = 1$.

EXAMPLE 1 Differentiate $y = \ln(x^3 + 1)$.

SOLUTION To use the Chain Rule, we let $u = x^3 + 1$. Then $y = \ln u$, so
$$\frac{dy}{dx} = \frac{dy}{du} \frac{du}{dx} = \frac{1}{u} \frac{du}{dx}$$
$$= \frac{1}{x^3 + 1}(3x^2) = \frac{3x^2}{x^3 + 1} \qquad \blacksquare$$

In general, if we combine Formula 2 with the Chain Rule as in Example 1, we get

$$\boxed{\frac{d}{dx}(\ln u) = \frac{1}{u}\frac{du}{dx}} \quad \text{or} \quad \boxed{\frac{d}{dx}[\ln g(x)] = \frac{g'(x)}{g(x)}}$$

EXAMPLE 2 Find $\dfrac{d}{dx}\ln(\sin x)$.

SOLUTION Using (3), we have

$$\frac{d}{dx}\ln(\sin x) = \frac{1}{\sin x}\frac{d}{dx}(\sin x) = \frac{1}{\sin x}\cos x = \cot x$$

EXAMPLE 3 Differentiate $f(x) = \sqrt{\ln x}$.

SOLUTION This time the logarithm is the inner function, so the Chain Rule gives

$$f'(x) = \tfrac{1}{2}(\ln x)^{-1/2}\frac{d}{dx}(\ln x) = \frac{1}{2\sqrt{\ln x}}\cdot\frac{1}{x} = \frac{1}{2x\sqrt{\ln x}}$$

EXAMPLE 4 Differentiate $f(x) = \log_{10}(2 + \sin x)$.

SOLUTION Using Formula 1 with $b = 10$, we have

$$f'(x) = \frac{d}{dx}\log_{10}(2 + \sin x)$$

$$= \frac{1}{(2 + \sin x)\ln 10}\frac{d}{dx}(2 + \sin x)$$

$$= \frac{\cos x}{(2 + \sin x)\ln 10}$$

EXAMPLE 5 Find $\dfrac{d}{dx}\ln\dfrac{x + 1}{\sqrt{x - 2}}$.

SOLUTION 1

$$\frac{d}{dx}\ln\frac{x + 1}{\sqrt{x - 2}} = \frac{1}{\dfrac{x + 1}{\sqrt{x - 2}}}\frac{d}{dx}\frac{x + 1}{\sqrt{x - 2}}$$

$$= \frac{\sqrt{x - 2}}{x + 1}\cdot\frac{\sqrt{x - 2}\cdot 1 - (x + 1)(\tfrac{1}{2})(x - 2)^{-1/2}}{x - 2}$$

$$= \frac{x - 2 - \tfrac{1}{2}(x + 1)}{(x + 1)(x - 2)}$$

$$= \frac{x - 5}{2(x + 1)(x - 2)}$$

Figure 1 shows the graph of the function f of Example 5 together with the graph of its derivative. It gives a visual check on our calculation. Notice that $f'(x)$ is large negative when f is rapidly decreasing.

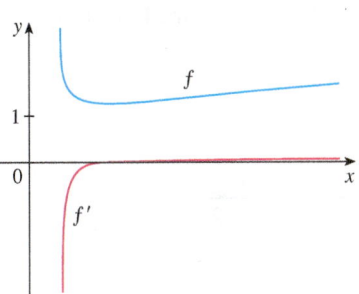

FIGURE 1

SOLUTION 2 If we first simplify the given function using the laws of logarithms, then the differentiation becomes easier:

$$\frac{d}{dx} \ln \frac{x+1}{\sqrt{x-2}} = \frac{d}{dx}\left[\ln(x+1) - \tfrac{1}{2}\ln(x-2)\right]$$

$$= \frac{1}{x+1} - \frac{1}{2}\left(\frac{1}{x-2}\right)$$

(This answer can be left as written, but if we used a common denominator we would see that it gives the same answer as in Solution 1.) ■

EXAMPLE 6 Find $f'(x)$ if $f(x) = \ln|x|$.

SOLUTION Since

$$f(x) = \begin{cases} \ln x & \text{if } x > 0 \\ \ln(-x) & \text{if } x < 0 \end{cases}$$

it follows that

$$f'(x) = \begin{cases} \dfrac{1}{x} & \text{if } x > 0 \\ \dfrac{1}{-x}(-1) = \dfrac{1}{x} & \text{if } x < 0 \end{cases}$$

Thus $f'(x) = 1/x$ for all $x \neq 0$. ■

Figure 2 shows the graph of the function $f(x) = \ln|x|$ in Example 6 and its derivative $f'(x) = 1/x$. Notice that when x is small, the graph of $y = \ln|x|$ is steep and so $f'(x)$ is large (positive or negative).

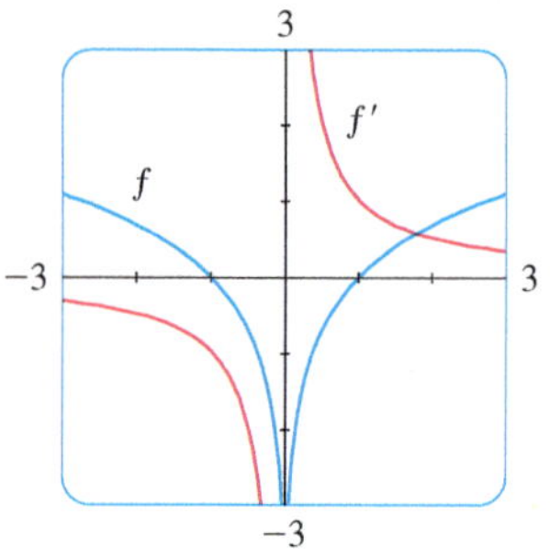

FIGURE 2

The result of Example 6 is worth remembering:

4
$$\boxed{\frac{d}{dx} \ln|x| = \frac{1}{x}}$$

■ Logarithmic Differentiation

The calculation of derivatives of complicated functions involving products, quotients, or powers can often be simplified by taking logarithms. The method used in the following example is called **logarithmic differentiation**.

EXAMPLE 7 Differentiate $y = \dfrac{x^{3/4}\sqrt{x^2+1}}{(3x+2)^5}$.

SOLUTION We take logarithms of both sides of the equation and use the Laws of Logarithms to simplify:

$$\ln y = \tfrac{3}{4}\ln x + \tfrac{1}{2}\ln(x^2+1) - 5\ln(3x+2)$$

Differentiating implicitly with respect to x gives

$$\frac{1}{y}\frac{dy}{dx} = \frac{3}{4}\cdot\frac{1}{x} + \frac{1}{2}\cdot\frac{2x}{x^2+1} - 5\cdot\frac{3}{3x+2}$$

Solving for dy/dx, we get

$$\frac{dy}{dx} = y\left(\frac{3}{4x} + \frac{x}{x^2+1} - \frac{15}{3x+2}\right)$$

If we hadn't used logarithmic differentiation in Example 7, we would have had to use both the Quotient Rule and the Product Rule. The resulting calculation would have been horrendous.

Because we have an explicit expression for y, we can substitute and write

$$\frac{dy}{dx} = \frac{x^{3/4}\sqrt{x^2+1}}{(3x+2)^5}\left(\frac{3}{4x} + \frac{x}{x^2+1} - \frac{15}{3x+2}\right) \qquad \blacksquare$$

> **Steps in Logarithmic Differentiation**
>
> 1. Take natural logarithms of both sides of an equation $y = f(x)$ and use the Laws of Logarithms to simplify.
> 2. Differentiate implicitly with respect to x.
> 3. Solve the resulting equation for y'.

If $f(x) < 0$ for some values of x, then $\ln f(x)$ is not defined, but we can write $|y| = |f(x)|$ and use Equation 4. We illustrate this procedure by proving the general version of the Power Rule, as promised in Section 3.1.

> **The Power Rule** If n is any real number and $f(x) = x^n$, then
>
> $$f'(x) = nx^{n-1}$$

If $x = 0$, we can show that $f'(0) = 0$ for $n > 1$ directly from the definition of a derivative.

PROOF Let $y = x^n$ and use logarithmic differentiation:

$$\ln|y| = \ln|x|^n = n\ln|x| \qquad x \neq 0$$

Therefore
$$\frac{y'}{y} = \frac{n}{x}$$

Hence
$$y' = n\frac{y}{x} = n\frac{x^n}{x} = nx^{n-1} \qquad \blacksquare$$

 You should distinguish carefully between the Power Rule $[(x^n)' = nx^{n-1}]$, where the base is variable and the exponent is constant, and the rule for differentiating exponential functions $[(b^x)' = b^x \ln b]$, where the base is constant and the exponent is variable.

In general there are four cases for exponents and bases:

Constant base, constant exponent
1. $\dfrac{d}{dx}(b^n) = 0 \qquad$ (b and n are constants)

Variable base, constant exponent
2. $\dfrac{d}{dx}[f(x)]^n = n[f(x)]^{n-1}f'(x)$

Constant base, variable exponent
3. $\dfrac{d}{dx}[b^{g(x)}] = b^{g(x)}(\ln b)g'(x)$

Variable base, variable exponent
4. To find $(d/dx)[f(x)]^{g(x)}$, logaritmic differentiation can be used, as in the next example.

Figure 3 illustrates Example 8 by showing the graphs of $f(x) = x^{\sqrt{x}}$ and its derivative.

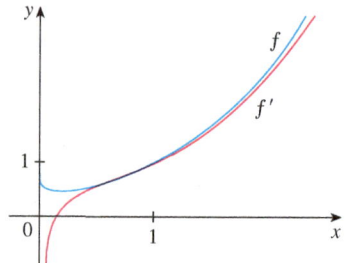

FIGURE 3

EXAMPLE 8 Differentiate $y = x^{\sqrt{x}}$.

SOLUTION 1 Since both the base and the exponent are variable, we use logarithmic differentiation:

$$\ln y = \ln x^{\sqrt{x}} = \sqrt{x} \ln x$$

$$\frac{y'}{y} = \sqrt{x} \cdot \frac{1}{x} + (\ln x) \frac{1}{2\sqrt{x}}$$

$$y' = y\left(\frac{1}{\sqrt{x}} + \frac{\ln x}{2\sqrt{x}}\right) = x^{\sqrt{x}} \left(\frac{2 + \ln x}{2\sqrt{x}}\right)$$

SOLUTION 2 Another method is to write $x^{\sqrt{x}} = (e^{\ln x})^{\sqrt{x}}$:

$$\frac{d}{dx}\left(x^{\sqrt{x}}\right) = \frac{d}{dx}\left(e^{\sqrt{x} \ln x}\right) = e^{\sqrt{x} \ln x} \frac{d}{dx}\left(\sqrt{x} \ln x\right)$$

$$= x^{\sqrt{x}} \left(\frac{2 + \ln x}{2\sqrt{x}}\right) \quad \text{(as in Solution 1)} \quad \blacksquare$$

The Number e as a Limit

We have shown that if $f(x) = \ln x$, then $f'(x) = 1/x$. Thus $f'(1) = 1$. We now use this fact to express the number e as a limit.

From the definition of a derivative as a limit, we have

$$f'(1) = \lim_{h \to 0} \frac{f(1 + h) - f(1)}{h} = \lim_{x \to 0} \frac{f(1 + x) - f(1)}{x}$$

$$= \lim_{x \to 0} \frac{\ln(1 + x) - \ln 1}{x} = \lim_{x \to 0} \frac{1}{x} \ln(1 + x)$$

$$= \lim_{x \to 0} \ln(1 + x)^{1/x}$$

Because $f'(1) = 1$, we have

$$\lim_{x \to 0} \ln(1 + x)^{1/x} = 1$$

Then, by Theorem 2.5.8 and the continuity of the exponential function, we have

$$e = e^1 = e^{\lim_{x \to 0} \ln(1+x)^{1/x}} = \lim_{x \to 0} e^{\ln(1+x)^{1/x}} = \lim_{x \to 0} (1 + x)^{1/x}$$

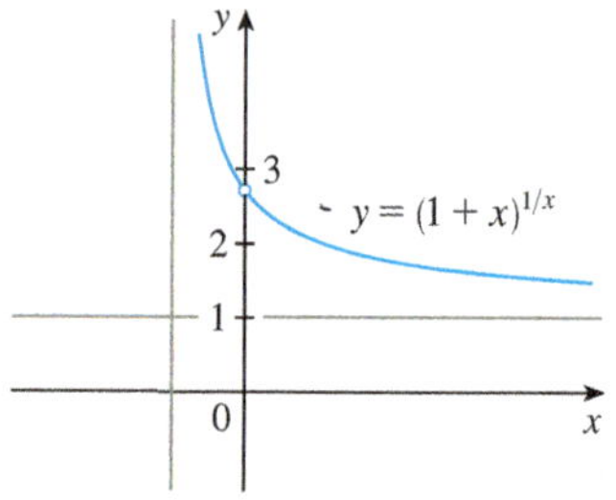

$$\boxed{5} \qquad e = \lim_{x \to 0} (1 + x)^{1/x}$$

Formula 5 is illustrated by the graph of the function $y = (1 + x)^{1/x}$ in Figure 4 and a table of values for small values of x. This illustrates the fact that, correct to seven decimal places,

$$e \approx 2.7182818$$

If we put $n = 1/x$ in Formula 5, then $n \to \infty$ as $x \to 0^+$ and so an alternative expression for e is

$$\boxed{6} \qquad e = \lim_{n \to \infty} \left(1 + \frac{1}{n}\right)^n$$

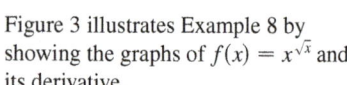

FIGURE 4

x	$(1 + x)^{1/x}$
0.1	2.59374246
0.01	2.70481383
0.001	2.71692393
0.0001	2.71814593
0.00001	2.71826824
0.000001	2.71828047
0.0000001	2.71828169
0.00000001	2.71828181

3.6 EXERCISES

1. Explain why the natural logarithmic function $y = \ln x$ is used much more frequently in calculus than the other logarithmic functions $y = \log_b x$.

2–22 Differentiate the function.

2. $f(x) = x \ln x - x$

3. $f(x) = \sin(\ln x)$

4. $f(x) = \ln(\sin^2 x)$

5. $f(x) = \ln \dfrac{1}{x}$

6. $y = \dfrac{1}{\ln x}$

7. $f(x) = \log_{10}(1 + \cos x)$

8. $f(x) = \log_{10} \sqrt{x}$

9. $g(x) = \ln(xe^{-2x})$

10. $g(t) = \sqrt{1 + \ln t}$

11. $F(t) = (\ln t)^2 \sin t$

12. $h(x) = \ln(x + \sqrt{x^2 - 1})$

13. $G(y) = \ln \dfrac{(2y + 1)^5}{\sqrt{y^2 + 1}}$

14. $P(v) = \dfrac{\ln v}{1 - v}$

15. $F(s) = \ln \ln s$

16. $y = \ln|1 + t - t^3|$

17. $T(z) = 2^z \log_2 z$

18. $y = \ln(\csc x - \cot x)$

19. $y = \ln(e^{-x} + xe^{-x})$

20. $H(z) = \ln \sqrt{\dfrac{a^2 - z^2}{a^2 + z^2}}$

21. $y = \tan[\ln(ax + b)]$

22. $y = \log_2(x \log_5 x)$

23–26 Find y' and y''.

23. $y = \sqrt{x} \ln x$

24. $y = \dfrac{\ln x}{1 + \ln x}$

25. $y = \ln|\sec x|$

26. $y = \ln(1 + \ln x)$

27–30 Differentiate f and find the domain of f.

27. $f(x) = \dfrac{x}{1 - \ln(x - 1)}$

28. $f(x) = \sqrt{2 + \ln x}$

29. $f(x) = \ln(x^2 - 2x)$

30. $f(x) = \ln \ln \ln x$

31. If $f(x) = \ln(x + \ln x)$, find $f'(1)$.

32. If $f(x) = \cos(\ln x^2)$, find $f'(1)$.

33–34 Find an equation of the tangent line to the curve at the given point.

33. $y = \ln(x^2 - 3x + 1)$, $(3, 0)$

34. $y = x^2 \ln x$, $(1, 0)$

35. If $f(x) = \sin x + \ln x$, find $f'(x)$. Check that your answer is reasonable by comparing the graphs of f and f'.

36. Find equations of the tangent lines to the curve $y = (\ln x)/x$ at the points $(1, 0)$ and $(e, 1/e)$. Illustrate by graphing the curve and its tangent lines.

37. Let $f(x) = cx + \ln(\cos x)$. For what value of c is $f'(\pi/4) = 6$?

38. Let $f(x) = \log_b(3x^2 - 2)$. For what value of b is $f'(1) = 3$?

39–50 Use logarithmic differentiation to find the derivative of the function.

39. $y = (x^2 + 2)^2(x^4 + 4)^4$

40. $y = \dfrac{e^{-x} \cos^2 x}{x^2 + x + 1}$

41. $y = \sqrt{\dfrac{x - 1}{x^4 + 1}}$

42. $y = \sqrt{x} e^{x^2 - x}(x + 1)^{2/3}$

43. $y = x^x$

44. $y = x^{\cos x}$

45. $y = x^{\sin x}$

46. $y = \sqrt{x}^x$

47. $y = (\cos x)^x$

48. $y = (\sin x)^{\ln x}$

49. $y = (\tan x)^{1/x}$

50. $y = (\ln x)^{\cos x}$

51. Find y' if $y = \ln(x^2 + y^2)$.

52. Find y' if $x^y = y^x$.

53. Find a formula for $f^{(n)}(x)$ if $f(x) = \ln(x - 1)$.

54. Find $\dfrac{d^9}{dx^9}(x^8 \ln x)$.

55. Use the definition of derivative to prove that

$$\lim_{x \to 0} \dfrac{\ln(1 + x)}{x} = 1$$

56. Show that $\displaystyle\lim_{n \to \infty} \left(1 + \dfrac{x}{n}\right)^n = e^x$ for any $x > 0$.

3.7 Rates of Change in the Natural and Social Sciences

We know that if $y = f(x)$, then the derivative dy/dx can be interpreted as the rate of change of y with respect to x. In this section we examine some of the applications of this idea to physics, chemistry, biology, economics, and other sciences.

Let's recall from Section 2.7 the basic idea behind rates of change. If x changes from x_1 to x_2, then the change in x is

$$\Delta x = x_2 - x_1$$

and the corresponding change in y is

$$\Delta y = f(x_2) - f(x_1)$$

The difference quotient

$$\frac{\Delta y}{\Delta x} = \frac{f(x_2) - f(x_1)}{x_2 - x_1}$$

is **the average rate of change of y with respect to x** over the interval $[x_1, x_2]$ and can be interpreted as the slope of the secant line PQ in Figure 1. Its limit as $\Delta x \to 0$ is the derivative $f'(x_1)$, which can therefore be interpreted as the **instantaneous rate of change of y with respect to x** or the slope of the tangent line at $P(x_1, f(x_1))$. Using Leibniz notation, we write the process in the form

$$\frac{dy}{dx} = \lim_{\Delta x \to 0} \frac{\Delta y}{\Delta x}$$

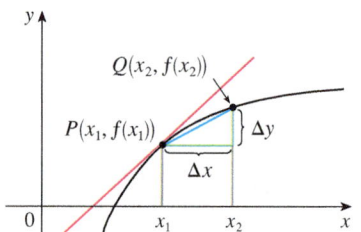

m_{PQ} = average rate of change
$m = f'(x_1)$ = instantaneous rate of change

FIGURE 1

Whenever the function $y = f(x)$ has a specific interpretation in one of the sciences, its derivative will have a specific interpretation as a rate of change. (As we discussed in Section 2.7, the units for dy/dx are the units for y divided by the units for x.) We now look at some of these interpretations in the natural and social sciences.

■ Physics

If $s = f(t)$ is the position function of a particle that is moving in a straight line, then $\Delta s/\Delta t$ represents the average velocity over a time period Δt, and $v = ds/dt$ represents the instantaneous **velocity** (the rate of change of displacement with respect to time). The instantaneous rate of change of velocity with respect to time is **acceleration**: $a(t) = v'(t) = s''(t)$. This was discussed in Sections 2.7 and 2.8, but now that we know the differentiation formulas, we are able to solve problems involving the motion of objects more easily.

EXAMPLE 1 The position of a particle is given by the equation

$$s = f(t) = t^3 - 6t^2 + 9t$$

where t is measured in seconds and s in meters.
(a) Find the velocity at time t.
(b) What is the velocity after 2 s? After 4 s?
(c) When is the particle at rest?
(d) When is the particle moving forward (that is, in the positive direction)?
(e) Draw a diagram to represent the motion of the particle.
(f) Find the total distance traveled by the particle during the first five seconds.
(g) Find the acceleration at time t and after 4 s.

(h) Graph the position, velocity, and acceleration functions for $0 \le t \le 5$.
(i) When is the particle speeding up? When is it slowing down?

SOLUTION

(a) The velocity function is the derivative of the position function.

$$s = f(t) = t^3 - 6t^2 + 9t$$

$$v(t) = \frac{ds}{dt} = 3t^2 - 12t + 9$$

(b) The velocity after 2 s means the instantaneous velocity when $t = 2$, that is,

$$v(2) = \frac{ds}{dt}\bigg|_{t=2} = 3(2)^2 - 12(2) + 9 = -3 \text{ m/s}$$

The velocity after 4 s is

$$v(4) = 3(4)^2 - 12(4) + 9 = 9 \text{ m/s}$$

(c) The particle is at rest when $v(t) = 0$, that is,

$$3t^2 - 12t + 9 = 3(t^2 - 4t + 3) = 3(t-1)(t-3) = 0$$

and this is true when $t = 1$ or $t = 3$. Thus the particle is at rest after 1 s and after 3 s.

(d) The particle moves in the positive direction when $v(t) > 0$, that is,

$$3t^2 - 12t + 9 = 3(t-1)(t-3) > 0$$

This inequality is true when both factors are positive ($t > 3$) or when both factors are negative ($t < 1$). Thus the particle moves in the positive direction in the time intervals $t < 1$ and $t > 3$. It moves backward (in the negative direction) when $1 < t < 3$.

(e) Using the information from part (d) we make a schematic sketch in Figure 2 of the motion of the particle back and forth along a line (the s-axis).

(f) Because of what we learned in parts (d) and (e), we need to calculate the distances traveled during the time intervals [0, 1], [1, 3], and [3, 5] separately.
The distance traveled in the first second is

$$|f(1) - f(0)| = |4 - 0| = 4 \text{ m}$$

From $t = 1$ to $t = 3$ the distance traveled is

$$|f(3) - f(1)| = |0 - 4| = 4 \text{ m}$$

From $t = 3$ to $t = 5$ the distance traveled is

$$|f(5) - f(3)| = |20 - 0| = 20 \text{ m}$$

The total distance is $4 + 4 + 20 = 28$ m.

(g) The acceleration is the derivative of the velocity function:

$$a(t) = \frac{d^2s}{dt^2} = \frac{dv}{dt} = 6t - 12$$

$$a(4) = 6(4) - 12 = 12 \text{ m/s}^2$$

(h) Figure 3 shows the graphs of s, v, and a.

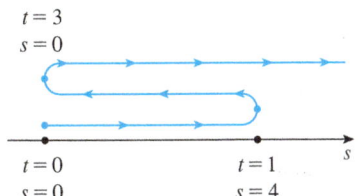

FIGURE 2

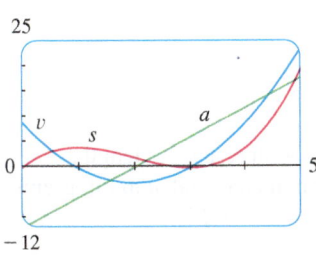

FIGURE 3

(i) The particle speeds up when the velocity is positive and increasing (v and a are both positive) and also when the velocity is negative and decreasing (v and a are both negative). In other words, the particle speeds up when the velocity and acceleration have the same sign. (The particle is pushed in the same direction it is moving.) From Figure 3 we see that this happens when $1 < t < 2$ and when $t > 3$. The particle slows down when v and a have opposite signs, that is, when $0 \leq t < 1$ and when $2 < t < 3$. Figure 4 summarizes the motion of the particle.

TEC In Module 3.7 you can see an animation of Figure 4 with an expression for s that you can choose yourself.

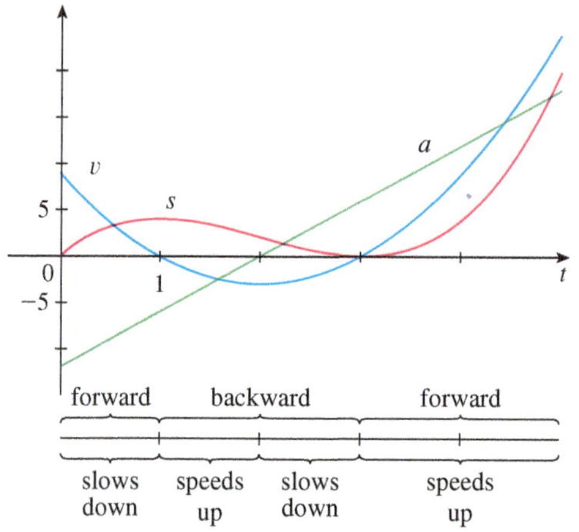

FIGURE 4

EXAMPLE 2 If a rod or piece of wire is homogeneous, then its linear density is uniform and is defined as the mass per unit length ($\rho = m/l$) and measured in kilograms per meter. Suppose, however, that the rod is not homogeneous but that its mass measured from its left end to a point x is $m = f(x)$, as shown in Figure 5.

FIGURE 5 This part of the rod has mass $f(x)$.

The mass of the part of the rod that lies between $x = x_1$ and $x = x_2$ is given by $\Delta m = f(x_2) - f(x_1)$, so the average density of that part of the rod is

$$\text{average density} = \frac{\Delta m}{\Delta x} = \frac{f(x_2) - f(x_1)}{x_2 - x_1}$$

If we now let $\Delta x \to 0$ (that is, $x_2 \to x_1$), we are computing the average density over smaller and smaller intervals. The **linear density** ρ at x_1 is the limit of these average densities as $\Delta x \to 0$; that is, the linear density is the rate of change of mass with respect to length. Symbolically,

$$\rho = \lim_{\Delta x \to 0} \frac{\Delta m}{\Delta x} = \frac{dm}{dx}$$

Thus the linear density of the rod is the derivative of mass with respect to length.

For instance, if $m = f(x) = \sqrt{x}$, where x is measured in meters and m in kilograms, then the average density of the part of the rod given by $1 \leq x \leq 1.2$ is

$$\frac{\Delta m}{\Delta x} = \frac{f(1.2) - f(1)}{1.2 - 1} = \frac{\sqrt{1.2} - 1}{0.2} \approx 0.48 \text{ kg/m}$$

while the density right at $x = 1$ is

$$\rho = \frac{dm}{dx}\bigg|_{x=1} = \frac{1}{2\sqrt{x}}\bigg|_{x=1} = 0.50 \text{ kg/m}$$

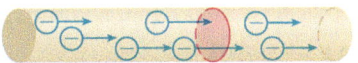

FIGURE 6

EXAMPLE 3 A current exists whenever electric charges move. Figure 6 shows part of a wire and electrons moving through a plane surface, shaded red. If ΔQ is the net charge that passes through this surface during a time period Δt, then the average current during this time interval is defined as

$$\text{average current} = \frac{\Delta Q}{\Delta t} = \frac{Q_2 - Q_1}{t_2 - t_1}$$

If we take the limit of this average current over smaller and smaller time intervals, we get what is called the **current** I at a given time t_1:

$$I = \lim_{\Delta t \to 0} \frac{\Delta Q}{\Delta t} = \frac{dQ}{dt}$$

Thus the current is the rate at which charge flows through a surface. It is measured in units of charge per unit time (often coulombs per second, called amperes).

Velocity, density, and current are not the only rates of change that are important in physics. Others include power (the rate at which work is done), the rate of heat flow, temperature gradient (the rate of change of temperature with respect to position), and the rate of decay of a radioactive substance in nuclear physics.

■ Chemistry

EXAMPLE 4 A chemical reaction results in the formation of one or more substances (called *products*) from one or more starting materials (called *reactants*). For instance, the "equation"

$$2\text{H}_2 + \text{O}_2 \to 2\text{H}_2\text{O}$$

indicates that two molecules of hydrogen and one molecule of oxygen form two molecules of water. Let's consider the reaction

$$\text{A} + \text{B} \to \text{C}$$

where A and B are the reactants and C is the product. The **concentration** of a reactant A is the number of moles (1 mole = 6.022×10^{23} molecules) per liter and is denoted by [A]. The concentration varies during a reaction, so [A], [B], and [C] are all functions of time (t). The average rate of reaction of the product C over a time interval $t_1 \leq t \leq t_2$ is

$$\frac{\Delta[\text{C}]}{\Delta t} = \frac{[\text{C}](t_2) - [\text{C}](t_1)}{t_2 - t_1}$$

But chemists are more interested in the **instantaneous rate of reaction**, which is obtained by taking the limit of the average rate of reaction as the time interval Δt approaches 0:

$$\text{rate of reaction} = \lim_{\Delta t \to 0} \frac{\Delta[\text{C}]}{\Delta t} = \frac{d[\text{C}]}{dt}$$

Since the concentration of the product increases as the reaction proceeds, the derivative $d[\text{C}]/dt$ will be positive, and so the rate of reaction of C is positive. The concentrations of the reactants, however, decrease during the reaction, so, to make the rates of reaction of A and B positive numbers, we put minus signs in front of the derivatives $d[\text{A}]/dt$ and $d[\text{B}]/dt$. Since [A] and [B] each decrease at the same rate that [C] increases, we have

$$\text{rate of reaction} = \frac{d[\text{C}]}{dt} = -\frac{d[\text{A}]}{dt} = -\frac{d[\text{B}]}{dt}$$

More generally, it turns out that for a reaction of the form

$$a\text{A} + b\text{B} \rightarrow c\text{C} + d\text{D}$$

we have

$$-\frac{1}{a}\frac{d[\text{A}]}{dt} = -\frac{1}{b}\frac{d[\text{B}]}{dt} = \frac{1}{c}\frac{d[\text{C}]}{dt} = \frac{1}{d}\frac{d[\text{D}]}{dt}$$

The rate of reaction can be determined from data and graphical methods. In some cases there are explicit formulas for the concentrations as functions of time, which enable us to compute the rate of reaction (see Exercise 24). ■

EXAMPLE 5 One of the quantities of interest in thermodynamics is compressibility. If a given substance is kept at a constant temperature, then its volume V depends on its pressure P. We can consider the rate of change of volume with respect to pressure—namely, the derivative dV/dP. As P increases, V decreases, so $dV/dP < 0$. The **compressibility** is defined by introducing a minus sign and dividing this derivative by the volume V:

$$\text{isothermal compressibility} = \beta = -\frac{1}{V}\frac{dV}{dP}$$

Thus β measures how fast, per unit volume, the volume of a substance decreases as the pressure on it increases at constant temperature.

For instance, the volume V (in cubic meters) of a sample of air at 25°C was found to be related to the pressure P (in kilopascals) by the equation

$$V = \frac{5.3}{P}$$

The rate of change of V with respect to P when $P = 50$ kPa is

$$\left.\frac{dV}{dP}\right|_{P=50} = \left.-\frac{5.3}{P^2}\right|_{P=50}$$

$$= -\frac{5.3}{2500} = -0.00212 \text{ m}^3/\text{kPa}$$

The compressibility at that pressure is

$$\beta = -\frac{1}{V}\left.\frac{dV}{dP}\right|_{P=50} = \frac{0.00212}{\frac{5.3}{50}} = 0.02 \text{ (m}^3/\text{kPa)/m}^3 \quad ■$$

Biology

EXAMPLE 6 Let $n = f(t)$ be the number of individuals in an animal or plant population at time t. The change in the population size between the times $t = t_1$ and $t = t_2$ is $\Delta n = f(t_2) - f(t_1)$, and so the average rate of growth during the time period $t_1 \leq t \leq t_2$ is

$$\text{average rate of growth} = \frac{\Delta n}{\Delta t} = \frac{f(t_2) - f(t_1)}{t_2 - t_1}$$

The **instantaneous rate of growth** is obtained from this average rate of growth by letting the time period Δt approach 0:

$$\text{growth rate} = \lim_{\Delta t \to 0} \frac{\Delta n}{\Delta t} = \frac{dn}{dt}$$

Strictly speaking, this is not quite accurate because the actual graph of a population function $n = f(t)$ would be a step function that is discontinuous whenever a birth or death occurs and therefore not differentiable. However, for a large animal or plant population, we can replace the graph by a smooth approximating curve as in Figure 7.

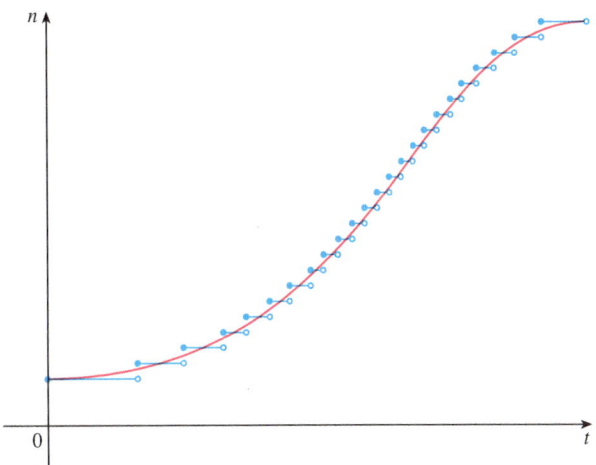

FIGURE 7
A smooth curve approximating a growth function

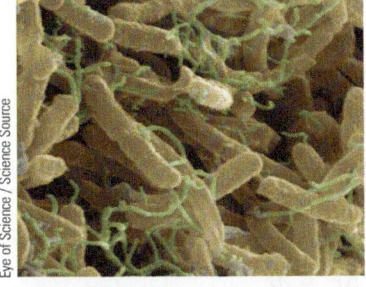

E. coli bacteria are about 2 micrometers (μm) long and 0.75 μm wide. The image was produced with a scanning electron microscope.

To be more specific, consider a population of bacteria in a homogeneous nutrient medium. Suppose that by sampling the population at certain intervals it is determined that the population doubles every hour. If the initial population is n_0 and the time t is measured in hours, then

$$f(1) = 2f(0) = 2n_0$$

$$f(2) = 2f(1) = 2^2 n_0$$

$$f(3) = 2f(2) = 2^3 n_0$$

and, in general,

$$f(t) = 2^t n_0$$

The population function is $n = n_0 2^t$.

In Section 3.4 we showed that

$$\frac{d}{dx}(b^x) = b^x \ln b$$

So the rate of growth of the bacteria population at time t is

$$\frac{dn}{dt} = \frac{d}{dt}(n_0 2^t) = n_0 2^t \ln 2$$

For example, suppose that we start with an initial population of $n_0 = 100$ bacteria. Then the rate of growth after 4 hours is

$$\left.\frac{dn}{dt}\right|_{t=4} = 100 \cdot 2^4 \ln 2 = 1600 \ln 2 \approx 1109$$

This means that, after 4 hours, the bacteria population is growing at a rate of about 1109 bacteria per hour. ■

EXAMPLE 7 When we consider the flow of blood through a blood vessel, such as a vein or artery, we can model the shape of the blood vessel by a cylindrical tube with radius R and length l as illustrated in Figure 8.

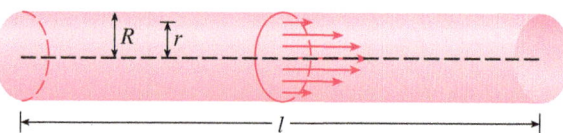

FIGURE 8
Blood flow in an artery

Because of friction at the walls of the tube, the velocity v of the blood is greatest along the central axis of the tube and decreases as the distance r from the axis increases until v becomes 0 at the wall. The relationship between v and r is given by the **law of laminar flow** discovered by the French physician Jean-Louis-Marie Poiseuille in 1840. This law states that

$$\boxed{1} \qquad v = \frac{P}{4\eta l}(R^2 - r^2)$$

For more detailed information, see W. Nichols and M. O'Rourke (eds.), *McDonald's Blood Flow in Arteries: Theoretical, Experimental, and Clinical Principles*, 5th ed. (New York, 2005).

where η is the viscosity of the blood and P is the pressure difference between the ends of the tube. If P and l are constant, then v is a function of r with domain $[0, R]$.

The average rate of change of the velocity as we move from $r = r_1$ outward to $r = r_2$ is given by

$$\frac{\Delta v}{\Delta r} = \frac{v(r_2) - v(r_1)}{r_2 - r_1}$$

and if we let $\Delta r \to 0$, we obtain the **velocity gradient**, that is, the instantaneous rate of change of velocity with respect to r:

$$\text{velocity gradient} = \lim_{\Delta r \to 0} \frac{\Delta v}{\Delta r} = \frac{dv}{dr}$$

Using Equation 1, we obtain

$$\frac{dv}{dr} = \frac{P}{4\eta l}(0 - 2r) = -\frac{Pr}{2\eta l}$$

For one of the smaller human arteries we can take $\eta = 0.027$, $R = 0.008$ cm, $l = 2$ cm, and $P = 4000$ dynes/cm², which gives

$$v = \frac{4000}{4(0.027)2}(0.000064 - r^2)$$

$$\approx 1.85 \times 10^4(6.4 \times 10^{-5} - r^2)$$

At $r = 0.002$ cm the blood is flowing at a speed of

$$v(0.002) \approx 1.85 \times 10^4(64 \times 10^{-6} - 4 \times 10^{-6})$$

$$= 1.11 \text{ cm/s}$$

and the velocity gradient at that point is

$$\left.\frac{dv}{dr}\right|_{r=0.002} = -\frac{4000(0.002)}{2(0.027)2} \approx -74 \text{ (cm/s)/cm}$$

To get a feeling for what this statement means, let's change our units from centimeters to micrometers (1 cm = 10,000 μm). Then the radius of the artery is 80 μm. The velocity at the central axis is 11,850 μm/s, which decreases to 11,110 μm/s at a distance of $r = 20$ μm. The fact that $dv/dr = -74$ (μm/s)/μm means that, when $r = 20$ μm, the velocity is decreasing at a rate of about 74 μm/s for each micrometer that we proceed away from the center. ∎

■ Economics

EXAMPLE 8 Suppose $C(x)$ is the total cost that a company incurs in producing x units of a certain commodity. The function C is called a **cost function**. If the number of items produced is increased from x_1 to x_2, then the additional cost is $\Delta C = C(x_2) - C(x_1)$, and the average rate of change of the cost is

$$\frac{\Delta C}{\Delta x} = \frac{C(x_2) - C(x_1)}{x_2 - x_1} = \frac{C(x_1 + \Delta x) - C(x_1)}{\Delta x}$$

The limit of this quantity as $\Delta x \to 0$, that is, the instantaneous rate of change of cost with respect to the number of items produced, is called the **marginal cost** by economists:

$$\text{marginal cost} = \lim_{\Delta x \to 0} \frac{\Delta C}{\Delta x} = \frac{dC}{dx}$$

[Since x often takes on only integer values, it may not make literal sense to let Δx approach 0, but we can always replace $C(x)$ by a smooth approximating function as in Example 6.]

Taking $\Delta x = 1$ and n large (so that Δx is small compared to n), we have

$$C'(n) \approx C(n+1) - C(n)$$

Thus the marginal cost of producing n units is approximately equal to the cost of producing one more unit [the $(n+1)$st unit].

It is often appropriate to represent a total cost function by a polynomial

$$C(x) = a + bx + cx^2 + dx^3$$

where a represents the overhead cost (rent, heat, maintenance) and the other terms represent the cost of raw materials, labor, and so on. (The cost of raw materials may be proportional to x, but labor costs might depend partly on higher powers of x because of overtime costs and inefficiencies involved in large-scale operations.)

For instance, suppose a company has estimated that the cost (in dollars) of producing x items is

$$C(x) = 10{,}000 + 5x + 0.01x^2$$

Then the marginal cost function is

$$C'(x) = 5 + 0.02x$$

The marginal cost at the production level of 500 items is

$$C'(500) = 5 + 0.02(500) = \$15/\text{item}$$

This gives the rate at which costs are increasing with respect to the production level when $x = 500$ and predicts the cost of the 501st item.

The actual cost of producing the 501st item is

$$C(501) - C(500) = [10{,}000 + 5(501) + 0.01(501)^2]$$
$$- [10{,}000 + 5(500) + 0.01(500)^2]$$
$$= \$15.01$$

Notice that $C'(500) \approx C(501) - C(500)$. ■

Economists also study marginal demand, marginal revenue, and marginal profit, which are the derivatives of the demand, revenue, and profit functions. These will be considered in Chapter 4 after we have developed techniques for finding the maximum and minimum values of functions.

■ Other Sciences

Rates of change occur in all the sciences. A geologist is interested in knowing the rate at which an intruded body of molten rock cools by conduction of heat into surrounding rocks. An engineer wants to know the rate at which water flows into or out of a reservoir. An urban geographer is interested in the rate of change of the population density in a city as the distance from the city center increases. A meteorologist is concerned with the rate of change of atmospheric pressure with respect to height (see Exercise 3.8.19).

In psychology, those interested in learning theory study the so-called learning curve, which graphs the performance $P(t)$ of someone learning a skill as a function of the training time t. Of particular interest is the rate at which performance improves as time passes, that is, dP/dt.

In sociology, differential calculus is used in analyzing the spread of rumors (or innovations or fads or fashions). If $p(t)$ denotes the proportion of a population that knows a rumor by time t, then the derivative dp/dt represents the rate of spread of the rumor (see Exercise 3.4.84).

■ A Single Idea, Many Interpretations

Velocity, density, current, power, and temperature gradient in physics; rate of reaction and compressibility in chemistry; rate of growth and blood velocity gradient in biology;

3.7 EXERCISES

1–4 A particle moves according to a law of motion $s = f(t)$, $t \geq 0$, where t is measured in seconds and s in feet.
(a) Find the velocity at time t.
(b) What is the velocity after 1 second?
(c) When is the particle at rest?
(d) When is the particle moving in the positive direction?
(e) Find the total distance traveled during the first 6 seconds.
(f) Draw a diagram like Figure 2 to illustrate the motion of the particle.
(g) Find the acceleration at time t and after 1 second.
(h) Graph the position, velocity, and acceleration functions for $0 \leq t \leq 6$.
(i) When is the particle speeding up? When is it slowing down?

1. $f(t) = t^3 - 8t^2 + 24t$

2. $f(t) = \dfrac{9t}{t^2 + 9}$

3. $f(t) = \sin(\pi t/2)$

4. $f(t) = t^2 e^{-t}$

5. Graphs of the *velocity* functions of two particles are shown, where t is measured in seconds. When is each particle speeding up? When is it slowing down? Explain.

(a) 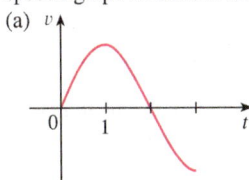 (b)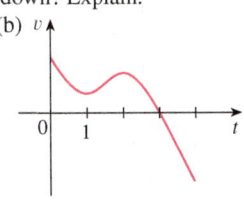

6. Graphs of the *position* functions of two particles are shown, where t is measured in seconds. When is each particle speeding up? When is it slowing down? Explain.

(a) 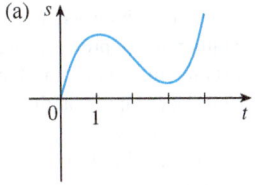 (b)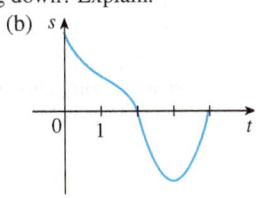

7. The height (in meters) of a projectile shot vertically upward from a point 2 m above ground level with an initial velocity of 24.5 m/s is $h = 2 + 24.5t - 4.9t^2$ after t seconds.
(a) Find the velocity after 2 s and after 4 s.
(b) When does the projectile reach its maximum height?
(c) What is the maximum height?
(d) When does it hit the ground?
(e) With what velocity does it hit the ground?

8. If a ball is thrown vertically upward with a velocity of 80 ft/s, then its height after t seconds is $s = 80t - 16t^2$.
(a) What is the maximum height reached by the ball?
(b) What is the velocity of the ball when it is 96 ft above the ground on its way up? On its way down?

9. If a rock is thrown vertically upward from the surface of Mars with velocity 15 m/s, its height after t seconds is $h = 15t - 1.86t^2$.
(a) What is the velocity of the rock after 2 s?
(b) What is the velocity of the rock when its height is 25 m on its way up? On its way down?

10. A particle moves with position function
$$s = t^4 - 4t^3 - 20t^2 + 20t \qquad t \geq 0$$
(a) At what time does the particle have a velocity of 20 m/s?
(b) At what time is the acceleration 0? What is the significance of this value of t?

11. (a) A company makes computer chips from square wafers of silicon. It wants to keep the side length of a wafer very close to 15 mm and it wants to know how the area $A(x)$ of a wafer changes when the side length x changes. Find $A'(15)$ and explain its meaning in this situation.
(b) Show that the rate of change of the area of a square with respect to its side length is half its perimeter. Try to explain geometrically why this is true by drawing a square whose side length x is increased by an amount Δx. How can you approximate the resulting change in area ΔA if Δx is small?

12. (a) Sodium chlorate crystals are easy to grow in the shape of cubes by allowing a solution of water and sodium chlorate to evaporate slowly. If V is the volume of such a cube with side length x, calculate dV/dx when $x = 3$ mm and explain its meaning.
(b) Show that the rate of change of the volume of a cube with respect to its edge length is equal to half the surface area of the cube. Explain geometrically why this result is true by arguing by analogy with Exercise 11(b).

13. (a) Find the average rate of change of the area of a circle with respect to its radius r as r changes from
    (i) 2 to 3   (ii) 2 to 2.5   (iii) 2 to 2.1
(b) Find the instantaneous rate of change when $r = 2$.
(c) Show that the rate of change of the area of a circle with respect to its radius (at any r) is equal to the circumference of the circle. Try to explain geometrically why this is true by drawing a circle whose radius is increased by an amount Δr. How can you approximate the resulting change in area ΔA if Δr is small?

14. A stone is dropped into a lake, creating a circular ripple that travels outward at a speed of 60 cm/s. Find the rate at which the area within the circle is increasing after (a) 1 s, (b) 3 s, and (c) 5 s. What can you conclude?

15. A spherical balloon is being inflated. Find the rate of increase of the surface area ($S = 4\pi r^2$) with respect to the radius r when r is (a) 1 ft, (b) 2 ft, and (c) 3 ft. What conclusion can you make?

16. (a) The volume of a growing spherical cell is $V = \frac{4}{3}\pi r^3$, where the radius r is measured in micrometers (1 μm $= 10^{-6}$ m). Find the average rate of change of V with respect to r when r changes from
    (i) 5 to 8 μm   (ii) 5 to 6 μm   (iii) 5 to 5.1 μm
(b) Find the instantaneous rate of change of V with respect to r when $r = 5$ μm.
(c) Show that the rate of change of the volume of a sphere with respect to its radius is equal to its surface area. Explain geometrically why this result is true. Argue by analogy with Exercise 13(c).

17. The mass of the part of a metal rod that lies between its left end and a point x meters to the right is $3x^2$ kg. Find the linear density (see Example 2) when x is (a) 1 m, (b) 2 m, and (c) 3 m. Where is the density the highest? The lowest?

18. If a tank holds 5000 gallons of water, which drains from the bottom of the tank in 40 minutes, then Torricelli's Law gives the volume V of water remaining in the tank after t minutes as
$$V = 5000\left(1 - \tfrac{1}{40}t\right)^2 \qquad 0 \le t \le 40$$
Find the rate at which water is draining from the tank after (a) 5 min, (b) 10 min, (c) 20 min, and (d) 40 min. At what time is the water flowing out the fastest? The slowest? Summarize your findings.

19. The quantity of charge Q in coulombs (C) that has passed through a point in a wire up to time t (measured in seconds) is given by $Q(t) = t^3 - 2t^2 + 6t + 2$. Find the current when (a) $t = 0.5$ s and (b) $t = 1$ s. [See Example 3. The unit of current is an ampere (1 A = 1 C/s).] At what time is the current lowest?

20. Newton's Law of Gravitation says that the magnitude F of the force exerted by a body of mass m on a body of mass M is
$$F = \frac{GmM}{r^2}$$
where G is the gravitational constant and r is the distance between the bodies.
(a) Find dF/dr and explain its meaning. What does the minus sign indicate?
(b) Suppose it is known that the earth attracts an object with a force that decreases at the rate of 2 N/km when $r = 20{,}000$ km. How fast does this force change when $r = 10{,}000$ km?

21. The force F acting on a body with mass m and velocity v is the rate of change of momentum: $F = (d/dt)(mv)$. If m is constant, this becomes $F = ma$, where $a = dv/dt$ is the acceleration. But in the theory of relativity the mass of a particle varies with v as follows: $m = m_0/\sqrt{1 - v^2/c^2}$, where m_0 is the mass of the particle at rest and c is the speed of light. Show that
$$F = \frac{m_0 a}{(1 - v^2/c^2)^{3/2}}$$

22. Some of the highest tides in the world occur in the Bay of Fundy on the Atlantic Coast of Canada. At Hopewell Cape the water depth at low tide is about 2.0 m and at high tide it is about 12.0 m. The natural period of oscillation is a little more than 12 hours and on June 30, 2009, high tide occurred at 6:45 AM. This helps explain the following model for the water depth D (in meters) as a function of the time t (in hours after midnight) on that day:
$$D(t) = 7 + 5\cos[0.503(t - 6.75)]$$
How fast was the tide rising (or falling) at the following times?
(a) 3:00 AM   (b) 6:00 AM
(c) 9:00 AM   (d) Noon

23. Boyle's Law states that when a sample of gas is compressed at a constant temperature, the product of the pressure and the volume remains constant: $PV = C$.
(a) Find the rate of change of volume with respect to pressure.
(b) A sample of gas is in a container at low pressure and is steadily compressed at constant temperature for 10 minutes. Is the volume decreasing more rapidly at the beginning or the end of the 10 minutes? Explain.
(c) Prove that the isothermal compressibility (see Example 5) is given by $\beta = 1/P$.

24. If, in Example 4, one molecule of the product C is formed from one molecule of the reactant A and one molecule of the reactant B, and the initial concentrations of A and B have a common value $[A] = [B] = a$ moles/L, then

$$[C] = a^2kt/(akt + 1)$$

where k is a constant.
 (a) Find the rate of reaction at time t.
 (b) Show that if $x = [C]$, then

$$\frac{dx}{dt} = k(a - x)^2$$

 (c) What happens to the concentration as $t \to \infty$?
 (d) What happens to the rate of reaction as $t \to \infty$?
 (e) What do the results of parts (c) and (d) mean in practical terms?

25. In Example 6 we considered a bacteria population that doubles every hour. Suppose that another population of bacteria triples every hour and starts with 400 bacteria. Find an expression for the number n of bacteria after t hours and use it to estimate the rate of growth of the bacteria population after 2.5 hours.

26. The number of yeast cells in a laboratory culture increases rapidly initially but levels off eventually. The population is modeled by the function

$$n = f(t) = \frac{a}{1 + be^{-0.7t}}$$

where t is measured in hours. At time $t = 0$ the population is 20 cells and is increasing at a rate of 12 cells/hour. Find the values of a and b. According to this model, what happens to the yeast population in the long run?

27. The table gives the population of the world $P(t)$, in millions, where t is measured in years and $t = 0$ corresponds to the year 1900.

t	Population (millions)	t	Population (millions)
0	1650	60	3040
10	1750	70	3710
20	1860	80	4450
30	2070	90	5280
40	2300	100	6080
50	2560	110	6870

 (a) Estimate the rate of population growth in 1920 and in 1980 by averaging the slopes of two secant lines.
 (b) Use a graphing device to find a cubic function (a third-degree polynomial) that models the data.
 (c) Use your model in part (b) to find a model for the rate of population growth.
 (d) Use part (c) to estimate the rates of growth in 1920 and 1980. Compare with your estimates in part (a).
 (e) In Section 1.1 we modeled $P(t)$ with the exponential function

$$f(t) = (1.43653 \times 10^9) \cdot (1.01395)^t$$

Use this model to find a model for the rate of population growth.
 (f) Use your model in part (e) to estimate the rate of growth in 1920 and 1980. Compare with your estimates in parts (a) and (d).
 (g) Estimate the rate of growth in 1985.

28. The table shows how the average age of first marriage of Japanese women has varied since 1950.

t	$A(t)$	t	$A(t)$
1950	23.0	1985	25.5
1955	23.8	1990	25.9
1960	24.4	1995	26.3
1965	24.5	2000	27.0
1970	24.2	2005	28.0
1975	24.7	2010	28.8
1980	25.2		

 (a) Use a graphing calculator or computer to model these data with a fourth-degree polynomial.
 (b) Use part (a) to find a model for $A'(t)$.
 (c) Estimate the rate of change of marriage age for women in 1990.
 (d) Graph the data points and the models for A and A'.

29. Refer to the law of laminar flow given in Example 7. Consider a blood vessel with radius 0.01 cm, length 3 cm, pressure difference 3000 dynes/cm², and viscosity $\eta = 0.027$.
 (a) Find the velocity of the blood along the centerline $r = 0$, at radius $r = 0.005$ cm, and at the wall $r = R = 0.01$ cm.
 (b) Find the velocity gradient at $r = 0$, $r = 0.005$, and $r = 0.01$.
 (c) Where is the velocity the greatest? Where is the velocity changing most?

30. The frequency of vibrations of a vibrating violin string is given by

$$f = \frac{1}{2L}\sqrt{\frac{T}{\rho}}$$

where L is the length of the string, T is its tension, and ρ is its linear density. [See Chapter 11 in D. E. Hall, *Musical Acoustics*, 3rd ed. (Pacific Grove, CA: Brooks/Cole, 2002).]
 (a) Find the rate of change of the frequency with respect to
  (i) the length (when T and ρ are constant),
  (ii) the tension (when L and ρ are constant), and
  (iii) the linear density (when L and T are constant).
 (b) The pitch of a note (how high or low the note sounds) is determined by the frequency f. (The higher the frequency, the higher the pitch.) Use the signs of the

derivatives in part (a) to determine what happens to the pitch of a note
  (i) when the effective length of a string is decreased by placing a finger on the string so a shorter portion of the string vibrates,
  (ii) when the tension is increased by turning a tuning peg,
  (iii) when the linear density is increased by switching to another string.

31. Suppose that the cost (in dollars) for a company to produce x pairs of a new line of jeans is
$$C(x) = 2000 + 3x + 0.01x^2 + 0.0002x^3$$
(a) Find the marginal cost function.
(b) Find $C'(100)$ and explain its meaning. What does it predict?
(c) Compare $C'(100)$ with the cost of manufacturing the 101st pair of jeans.

32. The cost function for a certain commodity is
$$C(q) = 84 + 0.16q - 0.0006q^2 + 0.000003q^3$$
(a) Find and interpret $C'(100)$.
(b) Compare $C'(100)$ with the cost of producing the 101st item.

33. If $p(x)$ is the total value of the production when there are x workers in a plant, then the *average productivity* of the workforce at the plant is
$$A(x) = \frac{p(x)}{x}$$
(a) Find $A'(x)$. Why does the company want to hire more workers if $A'(x) > 0$?
(b) Show that $A'(x) > 0$ if $p'(x)$ is greater than the average productivity.

34. If R denotes the reaction of the body to some stimulus of strength x, the *sensitivity* S is defined to be the rate of change of the reaction with respect to x. A particular example is that when the brightness x of a light source is increased, the eye reacts by decreasing the area R of the pupil. The experimental formula
$$R = \frac{40 + 24x^{0.4}}{1 + 4x^{0.4}}$$
has been used to model the dependence of R on x when R is measured in square millimeters and x is measured in appropriate units of brightness.
(a) Find the sensitivity.
(b) Illustrate part (a) by graphing both R and S as functions of x. Comment on the values of R and S at low levels of brightness. Is this what you would expect?

35. Patients undergo dialysis treatment to remove urea from their blood when their kidneys are not functioning properly. Blood is diverted from the patient through a machine that filters out urea. Under certain conditions, the duration of dialysis required, given that the initial urea concentration is $c > 1$, is given by the equation
$$t = \ln\left(\frac{3c + \sqrt{9c^2 - 8c}}{2}\right)$$
Calculate the derivative of t with respect to c and interpret it.

36. Invasive species often display a wave of advance as they colonize new areas. Mathematical models based on random dispersal and reproduction have demonstrated that the speed with which such waves move is given by the function $f(r) = 2\sqrt{Dr}$, where r is the reproductive rate of individuals and D is a parameter quantifying dispersal. Calculate the derivative of the wave speed with respect to the reproductive rate r and explain its meaning.

37. The gas law for an ideal gas at absolute temperature T (in kelvins), pressure P (in atmospheres), and volume V (in liters) is $PV = nRT$, where n is the number of moles of the gas and $R = 0.0821$ is the gas constant. Suppose that, at a certain instant, $P = 8.0$ atm and is increasing at a rate of 0.10 atm/min and $V = 10$ L and is decreasing at a rate of 0.15 L/min. Find the rate of change of T with respect to time at that instant if $n = 10$ mol.

38. In a fish farm, a population of fish is introduced into a pond and harvested regularly. A model for the rate of change of the fish population is given by the equation
$$\frac{dP}{dt} = r_0\left(1 - \frac{P(t)}{P_c}\right)P(t) - \beta P(t)$$
where r_0 is the birth rate of the fish, P_c is the maximum population that the pond can sustain (called the *carrying capacity*), and β is the percentage of the population that is harvested.
(a) What value of dP/dt corresponds to a stable population?
(b) If the pond can sustain 10,000 fish, the birth rate is 5%, and the harvesting rate is 4%, find the stable population level.
(c) What happens if β is raised to 5%?

39. In the study of ecosystems, *predator-prey models* are often used to study the interaction between species. Consider populations of tundra wolves, given by $W(t)$, and caribou, given by $C(t)$, in northern Canada. The interaction has been modeled by the equations
$$\frac{dC}{dt} = aC - bCW \qquad \frac{dW}{dt} = -cW + dCW$$
(a) What values of dC/dt and dW/dt correspond to stable populations?
(b) How would the statement "The caribou go extinct" be represented mathematically?
(c) Suppose that $a = 0.05$, $b = 0.001$, $c = 0.05$, and $d = 0.0001$. Find all population pairs (C, W) that lead to stable populations. According to this model, is it possible for the two species to live in balance or will one or both species become extinct?

3.8 Exponential Growth and Decay

In many natural phenomena, quantities grow or decay at a rate proportional to their size. For instance, if $y = f(t)$ is the number of individuals in a population of animals or bacteria at time t, then it seems reasonable to expect that the rate of growth $f'(t)$ is proportional to the population $f(t)$; that is, $f'(t) = kf(t)$ for some constant k. Indeed, under ideal conditions (unlimited environment, adequate nutrition, immunity to disease) the mathematical model given by the equation $f'(t) = kf(t)$ predicts what actually happens fairly accurately. Another example occurs in nuclear physics where the mass of a radioactive substance decays at a rate proportional to the mass. In chemistry, the rate of a unimolecular first-order reaction is proportional to the concentration of the substance. In finance, the value of a savings account with continuously compounded interest increases at a rate proportional to that value.

In general, if $y(t)$ is the value of a quantity y at time t and if the rate of change of y with respect to t is proportional to its size $y(t)$ at any time, then

$$\boxed{1} \qquad \frac{dy}{dt} = ky$$

where k is a constant. Equation 1 is sometimes called the **law of natural growth** (if $k > 0$) or the **law of natural decay** (if $k < 0$). It is called a **differential equation** because it involves an unknown function y and its derivative dy/dt.

It's not hard to think of a solution of Equation 1. This equation asks us to find a function whose derivative is a constant multiple of itself. We have met such functions in this chapter. Any exponential function of the form $y(t) = Ce^{kt}$, where C is a constant, satisfies

$$y'(t) = C(ke^{kt}) = k(Ce^{kt}) = ky(t)$$

We will see in Section 9.4 that *any* function that satisfies $dy/dt = ky$ must be of the form $y = Ce^{kt}$. To see the significance of the constant C, we observe that

$$y(0) = Ce^{k \cdot 0} = C$$

Therefore C is the initial value of the function.

$\boxed{2}$ **Theorem** The only solutions of the differential equation $dy/dt = ky$ are the exponential functions

$$y(t) = y(0)e^{kt}$$

■ Population Growth

What is the significance of the proportionality constant k? In the context of population growth, where $P(t)$ is the size of a population at time t, we can write

$$\boxed{3} \qquad \frac{dP}{dt} = kP \qquad \text{or} \qquad \frac{1}{P}\frac{dP}{dt} = k$$

The quantity

$$\frac{1}{P}\frac{dP}{dt}$$

is the growth rate divided by the population size; it is called the **relative growth rate**.

According to (3), instead of saying "the growth rate is proportional to population size" we could say "the relative growth rate is constant." Then (2) says that a population with constant relative growth rate must grow exponentially. Notice that the relative growth rate k appears as the coefficient of t in the exponential function Ce^{kt}. For instance, if

$$\frac{dP}{dt} = 0.02P$$

and t is measured in years, then the relative growth rate is $k = 0.02$ and the population grows at a relative rate of 2% per year. If the population at time 0 is P_0, then the expression for the population is

$$P(t) = P_0 e^{0.02t}$$

EXAMPLE 1 Use the fact that the world population was 2560 million in 1950 and 3040 million in 1960 to model the population of the world in the second half of the 20th century. (Assume that the growth rate is proportional to the population size.) What is the relative growth rate? Use the model to estimate the world population in 1993 and to predict the population in the year 2020.

SOLUTION We measure the time t in years and let $t = 0$ in the year 1950. We measure the population $P(t)$ in millions of people. Then $P(0) = 2560$ and $P(10) = 3040$. Since we are assuming that $dP/dt = kP$, Theorem 2 gives

$$P(t) = P(0)e^{kt} = 2560e^{kt}$$

$$P(10) = 2560e^{10k} = 3040$$

$$k = \frac{1}{10} \ln \frac{3040}{2560} \approx 0.017185$$

The relative growth rate is about 1.7% per year and the model is

$$P(t) = 2560e^{0.017185t}$$

We estimate that the world population in 1993 was

$$P(43) = 2560e^{0.017185(43)} \approx 5360 \text{ million}$$

The model predicts that the population in 2020 will be

$$P(70) = 2560e^{0.017185(70)} \approx 8524 \text{ million}$$

The graph in Figure 1 shows that the model is fairly accurate to the end of the 20th century (the dots represent the actual population), so the estimate for 1993 is quite reliable. But the prediction for 2020 is riskier.

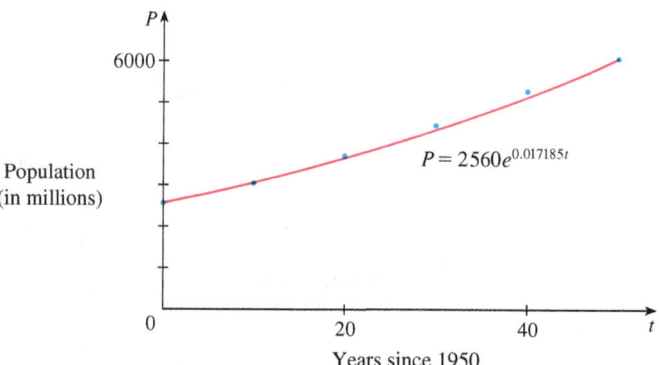

FIGURE 1
A model for world population growth in the second half of the 20th century

Radioactive Decay

Radioactive substances decay by spontaneously emitting radiation. If $m(t)$ is the mass remaining from an initial mass m_0 of the substance after time t, then the relative decay rate

$$-\frac{1}{m}\frac{dm}{dt}$$

has been found experimentally to be constant. (Since dm/dt is negative, the relative decay rate is positive.) It follows that

$$\frac{dm}{dt} = km$$

where k is a negative constant. In other words, radioactive substances decay at a rate proportional to the remaining mass. This means that we can use (2) to show that the mass decays exponentially:

$$m(t) = m_0 e^{kt}$$

Physicists express the rate of decay in terms of **half-life**, the time required for half of any given quantity to decay.

EXAMPLE 2 The half-life of radium-226 is 1590 years.
(a) A sample of radium-226 has a mass of 100 mg. Find a formula for the mass of the sample that remains after t years.
(b) Find the mass after 1000 years correct to the nearest milligram.
(c) When will the mass be reduced to 30 mg?

SOLUTION
(a) Let $m(t)$ be the mass of radium-226 (in milligrams) that remains after t years. Then $dm/dt = km$ and $m(0) = 100$, so (2) gives

$$m(t) = m(0)e^{kt} = 100e^{kt}$$

In order to determine the value of k, we use the fact that $m(1590) = \tfrac{1}{2}(100)$. Thus

$$100e^{1590k} = 50 \quad \text{so} \quad e^{1590k} = \tfrac{1}{2}$$

and

$$1590k = \ln \tfrac{1}{2} = -\ln 2$$

$$k = -\frac{\ln 2}{1590}$$

Therefore

$$m(t) = 100e^{-(\ln 2)t/1590}$$

We could use the fact that $e^{\ln 2} = 2$ to write the expression for $m(t)$ in the alternative form

$$m(t) = 100 \times 2^{-t/1590}$$

(b) The mass after 1000 years is

$$m(1000) = 100e^{-(\ln 2)1000/1590} \approx 65 \text{ mg}$$

(c) We want to find the value of t such that $m(t) = 30$, that is,

$$100e^{-(\ln 2)t/1590} = 30 \quad \text{or} \quad e^{-(\ln 2)t/1590} = 0.3$$

We solve this equation for t by taking the natural logarithm of both sides:

$$-\frac{\ln 2}{1590} t = \ln 0.3$$

Thus
$$t = -1590 \frac{\ln 0.3}{\ln 2} \approx 2762 \text{ years} \quad \blacksquare$$

As a check on our work in Example 2, we use a graphing device to draw the graph of $m(t)$ in Figure 2 together with the horizontal line $m = 30$. These curves intersect when $t \approx 2800$, and this agrees with the answer to part (c).

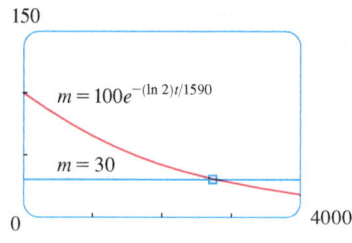

FIGURE 2

■ Newton's Law of Cooling

Newton's Law of Cooling states that the rate of cooling of an object is proportional to the temperature difference between the object and its surroundings, provided that this difference is not too large. (This law also applies to warming.) If we let $T(t)$ be the temperature of the object at time t and T_s be the temperature of the surroundings, then we can formulate Newton's Law of Cooling as a differential equation:

$$\frac{dT}{dt} = k(T - T_s)$$

where k is a constant. This equation is not quite the same as Equation 1, so we make the change of variable $y(t) = T(t) - T_s$. Because T_s is constant, we have $y'(t) = T'(t)$ and so the equation becomes

$$\frac{dy}{dt} = ky$$

We can then use (2) to find an expression for y, from which we can find T.

EXAMPLE 3 A bottle of soda pop at room temperature (72°F) is placed in a refrigerator where the temperature is 44°F. After half an hour the soda pop has cooled to 61°F.
(a) What is the temperature of the soda pop after another half hour?
(b) How long does it take for the soda pop to cool to 50°F?

SOLUTION
(a) Let $T(t)$ be the temperature of the soda after t minutes. The surrounding temperature is $T_s = 44°F$, so Newton's Law of Cooling states that

$$\frac{dT}{dt} = k(T - 44)$$

If we let $y = T - 44$, then $y(0) = T(0) - 44 = 72 - 44 = 28$, so y satisfies

$$\frac{dy}{dt} = ky \qquad y(0) = 28$$

and by (2) we have

$$y(t) = y(0)e^{kt} = 28e^{kt}$$

We are given that $T(30) = 61$, so $y(30) = 61 - 44 = 17$ and

$$28e^{30k} = 17 \qquad e^{30k} = \tfrac{17}{28}$$

Taking logarithms, we have

$$k = \frac{\ln\left(\frac{17}{28}\right)}{30} \approx -0.01663$$

Thus

$$y(t) = 28e^{-0.01663t}$$

$$T(t) = 44 + 28e^{-0.01663t}$$

$$T(60) = 44 + 28e^{-0.01663(60)} \approx 54.3$$

So after another half hour the pop has cooled to about 54°F.

(b) We have $T(t) = 50$ when

$$44 + 28e^{-0.01663t} = 50$$

$$e^{-0.01663t} = \frac{6}{28}$$

$$t = \frac{\ln\left(\frac{6}{28}\right)}{-0.01663} \approx 92.6$$

The pop cools to 50°F after about 1 hour 33 minutes.

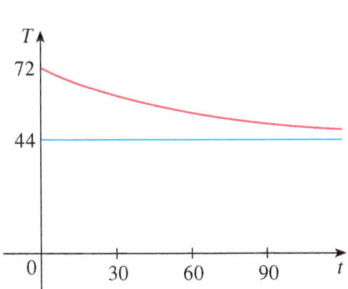

FIGURE 3

Notice that in Example 3, we have

$$\lim_{t \to \infty} T(t) = \lim_{t \to \infty} (44 + 28e^{-0.01663t}) = 44 + 28 \cdot 0 = 44$$

which is to be expected. The graph of the temperature function is shown in Figure 3.

Continuously Compounded Interest

EXAMPLE 4 If $1000 is invested at 6% interest, compounded annually, then after 1 year the investment is worth $1000(1.06) = $1060, after 2 years it's worth $[1000(1.06)]1.06 = $1123.60, and after t years it's worth $1000(1.06)^t$. In general, if an amount A_0 is invested at an interest rate r ($r = 0.06$ in this example), then after t years it's worth $A_0(1 + r)^t$. Usually, however, interest is compounded more frequently, say, n times a year. Then in each compounding period the interest rate is r/n and there are nt compounding periods in t years, so the value of the investment is

$$A_0\left(1 + \frac{r}{n}\right)^{nt}$$

For instance, after 3 years at 6% interest a $1000 investment will be worth

$$\$1000(1.06)^3 = \$1191.02 \quad \text{with annual compounding}$$

$$\$1000(1.03)^6 = \$1194.05 \quad \text{with semiannual compounding}$$

$$\$1000(1.015)^{12} = \$1195.62 \quad \text{with quarterly compounding}$$

$$\$1000(1.005)^{36} = \$1196.68 \quad \text{with monthly compounding}$$

$$\$1000\left(1 + \frac{0.06}{365}\right)^{365 \cdot 3} = \$1197.20 \quad \text{with daily compounding}$$

You can see that the interest paid increases as the number of compounding periods (n) increases. If we let $n \to \infty$, then we will be compounding the interest **continuously** and the value of the investment will be

$$A(t) = \lim_{n \to \infty} A_0 \left(1 + \frac{r}{n}\right)^{nt}$$

$$= \lim_{n \to \infty} A_0 \left[\left(1 + \frac{r}{n}\right)^{n/r}\right]^{rt}$$

$$= A_0 \left[\lim_{n \to \infty} \left(1 + \frac{r}{n}\right)^{n/r}\right]^{rt}$$

$$= A_0 \left[\lim_{m \to \infty} \left(1 + \frac{1}{m}\right)^{m}\right]^{rt} \qquad \text{(where } m = n/r\text{)}$$

But the limit in this expression is equal to the number e (see Equation 3.6.6). So with continuous compounding of interest at interest rate r, the amount after t years is

$$A(t) = A_0 e^{rt}$$

If we differentiate this equation, we get

$$\frac{dA}{dt} = rA_0 e^{rt} = rA(t)$$

which says that, with continuous compounding of interest, the rate of increase of an investment is proportional to its size.

Returning to the example of $1000 invested for 3 years at 6% interest, we see that with continuous compounding of interest the value of the investment will be

$$A(3) = \$1000 e^{(0.06)3} = \$1197.22$$

Notice how close this is to the amount we calculated for daily compounding, $1197.20. But the amount is easier to compute if we use continuous compounding. ■

3.8 EXERCISES

1. A population of protozoa develops with a constant relative growth rate of 0.7944 per member per day. On day zero the population consists of two members. Find the population size after six days.

2. A common inhabitant of human intestines is the bacterium *Escherichia coli*, named after the German pediatrician Theodor Escherich, who identified it in 1885. A cell of this bacterium in a nutrient-broth medium divides into two cells every 20 minutes. The initial population of a culture is 50 cells.
   (a) Find the relative growth rate.
   (b) Find an expression for the number of cells after t hours.
   (c) Find the number of cells after 6 hours.
   (d) Find the rate of growth after 6 hours.
   (e) When will the population reach a million cells?

3. A bacteria culture initially contains 100 cells and grows at a rate proportional to its size. After an hour the population has increased to 420.
   (a) Find an expression for the number of bacteria after t hours.
   (b) Find the number of bacteria after 3 hours.
   (c) Find the rate of growth after 3 hours.
   (d) When will the population reach 10,000?

4. A bacteria culture grows with constant relative growth rate. The bacteria count was 400 after 2 hours and 25,600 after 6 hours.
   (a) What is the relative growth rate? Express your answer as a percentage.
   (b) What was the intitial size of the culture?
   (c) Find an expression for the number of bacteria after t hours.
   (d) Find the number of cells after 4.5 hours.
   (e) Find the rate of growth after 4.5 hours.
   (f) When will the population reach 50,000?

5. The table gives estimates of the world population, in millions, from 1750 to 2000.

Year	Population	Year	Population
1750	790	1900	1650
1800	980	1950	2560
1850	1260	2000	6080

   (a) Use the exponential model and the population figures for 1750 and 1800 to predict the world population in 1900 and 1950. Compare with the actual figures.
   (b) Use the exponential model and the population figures for 1850 and 1900 to predict the world population in 1950. Compare with the actual population.
   (c) Use the exponential model and the population figures for 1900 and 1950 to predict the world population in 2000. Compare with the actual population and try to explain the discrepancy.

6. The table gives the population of Indonesia, in millions, for the second half of the 20th century.

Year	Population
1950	83
1960	100
1970	122
1980	150
1990	182
2000	214

   (a) Assuming the population grows at a rate proportional to its size, use the census figures for 1950 and 1960 to predict the population in 1980. Compare with the actual figure.
   (b) Use the census figures for 1960 and 1980 to predict the population in 2000. Compare with the actual population.
   (c) Use the census figures for 1980 and 2000 to predict the population in 2010 and compare with the actual population of 243 million.
   (d) Use the model in part (c) to predict the population in 2020. Do you think the prediction will be too high or too low? Why?

7. Experiments show that if the chemical reaction
   $$N_2O_5 \to 2NO_2 + \tfrac{1}{2}O_2$$
   takes place at 45°C, the rate of reaction of dinitrogen pentoxide is proportional to its concentration as follows:
   $$-\frac{d[N_2O_5]}{dt} = 0.0005[N_2O_5]$$
   (See Example 3.7.4.)
   (a) Find an expression for the concentration $[N_2O_5]$ after t seconds if the initial concentration is C.
   (b) How long will the reaction take to reduce the concentration of N_2O_5 to 90% of its original value?

8. Strontium-90 has a half-life of 28 days.
   (a) A sample has a mass of 50 mg initially. Find a formula for the mass remaining after t days.
   (b) Find the mass remaining after 40 days.
   (c) How long does it take the sample to decay to a mass of 2 mg?
   (d) Sketch the graph of the mass function.

9. The half-life of cesium-137 is 30 years. Suppose we have a 100-mg sample.
   (a) Find the mass that remains after t years.
   (b) How much of the sample remains after 100 years?
   (c) After how long will only 1 mg remain?

10. A sample of tritium-3 decayed to 94.5% of its original amount after a year.
    (a) What is the half-life of tritium-3?
    (b) How long would it take the sample to decay to 20% of its original amount?

11. Scientists can determine the age of ancient objects by the method of *radiocarbon dating*. The bombardment of the upper atmosphere by cosmic rays converts nitrogen to a radioactive isotope of carbon, ^{14}C, with a half-life of about 5730 years. Vegetation absorbs carbon dioxide through the atmosphere and animal life assimilates ^{14}C through food chains. When a plant or animal dies, it stops replacing its carbon and the amount of ^{14}C begins to decrease through radioactive decay. Therefore the level of radioactivity must also decay exponentially.

    A discovery revealed a parchment fragment that had about 74% as much ^{14}C radioactivity as does plant material on the earth today. Estimate the age of the parchment.

12. Dinosaur fossils are too old to be reliably dated using carbon-14. (See Exercise 11.) Suppose we had a 68-million-year-old dinosaur fossil. What fraction of the living dinosaur's ^{14}C would be remaining today? Suppose the minimum detectable amount is 0.1%. What is the maximum age of a fossil that we could date using ^{14}C?

13. Dinosaur fossils are often dated by using an element other than carbon, such as potassium-40, that has a longer half-life (in this case, approximately 1.25 billion years). Suppose the minimum detectable amount is 0.1% and a dinosaur is dated

with ^{40}K to be 68 million years old. Is such a dating possible? In other words, what is the maximum age of a fossil that we could date using ^{40}K?

14. A curve passes through the point (0, 5) and has the property that the slope of the curve at every point P is twice the y-coordinate of P. What is the equation of the curve?

15. A roast turkey is taken from an oven when its temperature has reached 185°F and is placed on a table in a room where the temperature is 75°F.
(a) If the temperature of the turkey is 150°F after half an hour, what is the temperature after 45 minutes?
(b) When will the turkey have cooled to 100°F?

16. In a murder investigation, the temperature of the corpse was 32.5°C at 1:30 PM and 30.3°C an hour later. Normal body temperature is 37.0°C and the temperature of the surroundings was 20.0°C. When did the murder take place?

17. When a cold drink is taken from a refrigerator, its temperature is 5°C. After 25 minutes in a 20°C room its temperature has increased to 10°C.
(a) What is the temperature of the drink after 50 minutes?
(b) When will its temperature be 15°C?

18. A freshly brewed cup of coffee has temperature 95°C in a 20°C room. When its temperature is 70°C, it is cooling at a rate of 1°C per minute. When does this occur?

19. The rate of change of atmospheric pressure P with respect to altitude h is proportional to P, provided that the temperature is constant. At 15°C the pressure is 101.3 kPa at sea level and 87.14 kPa at $h = 1000$ m.
(a) What is the pressure at an altitude of 3000 m?
(b) What is the pressure at the top of Mount McKinley, at an altitude of 6187 m?

20. (a) If $1000 is borrowed at 8% interest, find the amounts due at the end of 3 years if the interest is compounded (i) annually, (ii) quarterly, (iii) monthly, (iv) weekly, (v) daily, (vi) hourly, and (vii) continuously.
(b) Suppose $1000 is borrowed and the interest is compounded continuously. If $A(t)$ is the amount due after t years, where $0 \le t \le 3$, graph $A(t)$ for each of the interest rates 6%, 8%, and 10% on a common screen.

21. (a) If $3000 is invested at 5% interest, find the value of the investment at the end of 5 years if the interest is compounded (i) annually, (ii) semiannually, (iii) monthly, (iv) weekly, (v) daily, and (vi) continuously.
(b) If $A(t)$ is the amount of the investment at time t for the case of continuous compounding, write a differential equation and an initial condition satisfied by $A(t)$.

22. (a) How long will it take an investment to double in value if the interest rate is 6% compounded continuously?
(b) What is the equivalent annual interest rate?

APPLIED PROJECT

CONTROLLING RED BLOOD CELL LOSS DURING SURGERY

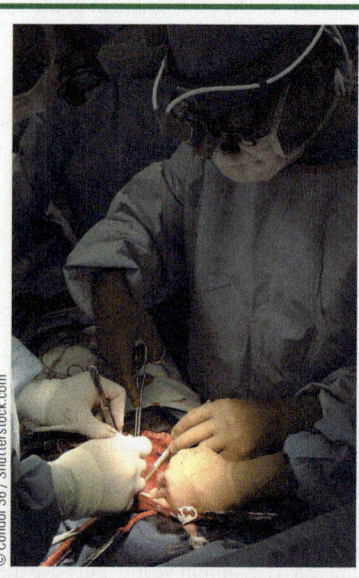

A typical volume of blood in the human body is about 5 L. A certain percentage of that volume (called the *hematocrit*) consists of red blood cells (RBCs); typically the hematocrit is about 45% in males. Suppose that a surgery takes four hours and a male patient bleeds 2.5 L of blood. During surgery the patient's blood volume is maintained at 5 L by injection of saline solution, which mixes quickly with the blood but dilutes it so that the hematocrit decreases as time passes.

1. Assuming that the rate of RBC loss is proportional to the volume of RBCs, determine the patient's volume of RBCs by the end of the operation.

2. A procedure called *acute normovolemic hemodilution* (ANH) has been developed to minimize RBC loss during surgery. In this procedure blood is extracted from the patient before the operation and replaced with saline solution. This dilutes the patient's blood, resulting in fewer RBCs being lost during the bleeding. The extracted blood is then returned to the patient after surgery. Only a certain amount of blood can be extracted, however, because the RBC concentration can never be allowed to drop below 25% during surgery. What is the maximum amount of blood that can be extracted in the ANH procedure for the surgery described in this project?

3. What is the RBC loss without the ANH procedure? What is the loss if the procedure is carried out with the volume calculated in Problem 2?

3.9 Related Rates

If we are pumping air into a balloon, both the volume and the radius of the balloon are increasing and their rates of increase are related to each other. But it is much easier to measure directly the rate of increase of the volume than the rate of increase of the radius.

In a related rates problem the idea is to compute the rate of change of one quantity in terms of the rate of change of another quantity (which may be more easily measured). The procedure is to find an equation that relates the two quantities and then use the Chain Rule to differentiate both sides with respect to time.

EXAMPLE 1 Air is being pumped into a spherical balloon so that its volume increases at a rate of 100 cm³/s. How fast is the radius of the balloon increasing when the diameter is 50 cm?

SOLUTION We start by identifying two things:

the *given information:*

the rate of increase of the volume of air is 100 cm³/s

and the *unknown:*

the rate of increase of the radius when the diameter is 50 cm

In order to express these quantities mathematically, we introduce some suggestive *notation:*

Let V be the volume of the balloon and let r be its radius.

The key thing to remember is that rates of change are derivatives. In this problem, the volume and the radius are both functions of the time t. The rate of increase of the volume with respect to time is the derivative dV/dt, and the rate of increase of the radius is dr/dt. We can therefore restate the given and the unknown as follows:

Given: $\dfrac{dV}{dt} = 100 \text{ cm}^3/\text{s}$

Unknown: $\dfrac{dr}{dt}$ when $r = 25$ cm

In order to connect dV/dt and dr/dt, we first relate V and r by the formula for the volume of a sphere:

$$V = \tfrac{4}{3}\pi r^3$$

In order to use the given information, we differentiate each side of this equation with respect to t. To differentiate the right side, we need to use the Chain Rule:

$$\frac{dV}{dt} = \frac{dV}{dr}\frac{dr}{dt} = 4\pi r^2 \frac{dr}{dt}$$

Now we solve for the unknown quantity:

$$\frac{dr}{dt} = \frac{1}{4\pi r^2}\frac{dV}{dt}$$

If we put $r = 25$ and $dV/dt = 100$ in this equation, we obtain

$$\frac{dr}{dt} = \frac{1}{4\pi(25)^2} 100 = \frac{1}{25\pi}$$

The radius of the balloon is increasing at the rate of $1/(25\pi) \approx 0.0127$ cm/s. ∎

EXAMPLE 2 A ladder 10 ft long rests against a vertical wall. If the bottom of the ladder slides away from the wall at a rate of 1 ft/s, how fast is the top of the ladder sliding down the wall when the bottom of the ladder is 6 ft from the wall?

SOLUTION We first draw a diagram and label it as in Figure 1. Let x feet be the distance from the bottom of the ladder to the wall and y feet the distance from the top of the ladder to the ground. Note that x and y are both functions of t (time, measured in seconds).

We are given that $dx/dt = 1$ ft/s and we are asked to find dy/dt when $x = 6$ ft (see Figure 2). In this problem, the relationship between x and y is given by the Pythagorean Theorem:

$$x^2 + y^2 = 100$$

Differentiating each side with respect to t using the Chain Rule, we have

$$2x \frac{dx}{dt} + 2y \frac{dy}{dt} = 0$$

and solving this equation for the desired rate, we obtain

$$\frac{dy}{dt} = -\frac{x}{y} \frac{dx}{dt}$$

When $x = 6$, the Pythagorean Theorem gives $y = 8$ and so, substituting these values and $dx/dt = 1$, we have

$$\frac{dy}{dt} = -\frac{6}{8}(1) = -\frac{3}{4} \text{ ft/s}$$

The fact that dy/dt is negative means that the distance from the top of the ladder to the ground is *decreasing* at a rate of $\frac{3}{4}$ ft/s. In other words, the top of the ladder is sliding down the wall at a rate of $\frac{3}{4}$ ft/s. ∎

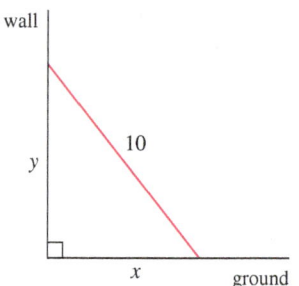

FIGURE 1

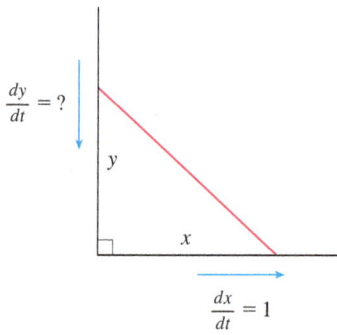

FIGURE 2

EXAMPLE 3 A water tank has the shape of an inverted circular cone with base radius 2 m and height 4 m. If water is being pumped into the tank at a rate of 2 m³/min, find the rate at which the water level is rising when the water is 3 m deep.

SOLUTION We first sketch the cone and label it as in Figure 3. Let V, r, and h be the volume of the water, the radius of the surface, and the height of the water at time t, where t is measured in minutes.

We are given that $dV/dt = 2$ m³/min and we are asked to find dh/dt when h is 3 m. The quantities V and h are related by the equation

$$V = \tfrac{1}{3}\pi r^2 h$$

but it is very useful to express V as a function of h alone. In order to eliminate r, we use

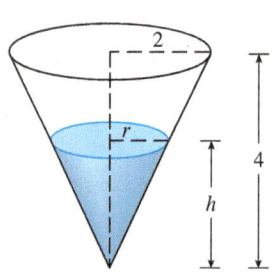

FIGURE 3

the similar triangles in Figure 3 to write

$$\frac{r}{h} = \frac{2}{4} \qquad r = \frac{h}{2}$$

and the expression for V becomes

$$V = \frac{1}{3}\pi\left(\frac{h}{2}\right)^2 h = \frac{\pi}{12}h^3$$

Now we can differentiate each side with respect to t:

$$\frac{dV}{dt} = \frac{\pi}{4}h^2 \frac{dh}{dt}$$

so

$$\frac{dh}{dt} = \frac{4}{\pi h^2} \frac{dV}{dt}$$

Substituting $h = 3$ m and $dV/dt = 2$ m³/min, we have

$$\frac{dh}{dt} = \frac{4}{\pi(3)^2} \cdot 2 = \frac{8}{9\pi}$$

The water level is rising at a rate of $8/(9\pi) \approx 0.28$ m/min. ∎

PS Look back: What have we learned from Examples 1–3 that will help us solve future problems?

Problem Solving Strategy It is useful to recall some of the problem-solving principles from page 71 and adapt them to related rates in light of our experience in Examples 1–3:

1. Read the problem carefully.
2. Draw a diagram if possible.
3. Introduce notation. Assign symbols to all quantities that are functions of time.
4. Express the given information and the required rate in terms of derivatives.
5. Write an equation that relates the various quantities of the problem. If necessary, use the geometry of the situation to eliminate one of the variables by substitution (as in Example 3).
6. Use the Chain Rule to differentiate both sides of the equation with respect to t.
7. Substitute the given information into the resulting equation and solve for the unknown rate.

⊘ **WARNING** A common error is to substitute the given numerical information (for quantities that vary with time) too early. This should be done only *after* the differentiation. (Step 7 follows Step 6.) For instance, in Example 3 we dealt with general values of h until we finally substituted $h = 3$ at the last stage. (If we had put $h = 3$ earlier, we would have gotten $dV/dt = 0$, which is clearly wrong.)

The following examples are further illustrations of the strategy.

EXAMPLE 4 Car A is traveling west at 50 mi/h and car B is traveling north at 60 mi/h. Both are headed for the intersection of the two roads. At what rate are the cars approaching each other when car A is 0.3 mi and car B is 0.4 mi from the intersection?

SOLUTION We draw Figure 4, where C is the intersection of the roads. At a given time t, let x be the distance from car A to C, let y be the distance from car B to C, and let z be the distance between the cars, where x, y, and z are measured in miles.

We are given that $dx/dt = -50$ mi/h and $dy/dt = -60$ mi/h. (The derivatives are negative because x and y are decreasing.) We are asked to find dz/dt. The equation that

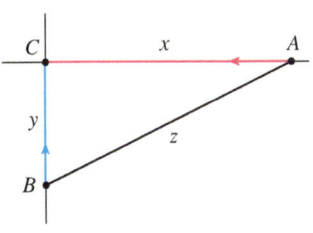

FIGURE 4

relates x, y, and z is given by the Pythagorean Theorem:

$$z^2 = x^2 + y^2$$

Differentiating each side with respect to t, we have

$$2z \frac{dz}{dt} = 2x \frac{dx}{dt} + 2y \frac{dy}{dt}$$

$$\frac{dz}{dt} = \frac{1}{z}\left(x \frac{dx}{dt} + y \frac{dy}{dt}\right)$$

When $x = 0.3$ mi and $y = 0.4$ mi, the Pythagorean Theorem gives $z = 0.5$ mi, so

$$\frac{dz}{dt} = \frac{1}{0.5}[0.3(-50) + 0.4(-60)]$$

$$= -78 \text{ mi/h}$$

The cars are approaching each other at a rate of 78 mi/h. ∎

EXAMPLE 5 A man walks along a straight path at a speed of 4 ft/s. A searchlight is located on the ground 20 ft from the path and is kept focused on the man. At what rate is the searchlight rotating when the man is 15 ft from the point on the path closest to the searchlight?

SOLUTION We draw Figure 5 and let x be the distance from the man to the point on the path closest to the searchlight. We let θ be the angle between the beam of the searchlight and the perpendicular to the path.

We are given that $dx/dt = 4$ ft/s and are asked to find $d\theta/dt$ when $x = 15$. The equation that relates x and θ can be written from Figure 5:

$$\frac{x}{20} = \tan\theta \qquad x = 20\tan\theta$$

Differentiating each side with respect to t, we get

$$\frac{dx}{dt} = 20\sec^2\theta \frac{d\theta}{dt}$$

so

$$\frac{d\theta}{dt} = \frac{1}{20}\cos^2\theta \frac{dx}{dt}$$

$$= \frac{1}{20}\cos^2\theta \, (4) = \frac{1}{5}\cos^2\theta$$

When $x = 15$, the length of the beam is 25, so $\cos\theta = \frac{4}{5}$ and

$$\frac{d\theta}{dt} = \frac{1}{5}\left(\frac{4}{5}\right)^2 = \frac{16}{125} = 0.128$$

The searchlight is rotating at a rate of 0.128 rad/s. ∎

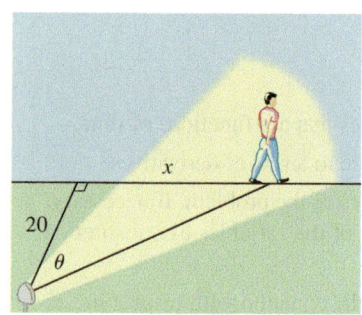

FIGURE 5

3.9 EXERCISES

1. If V is the volume of a cube with edge length x and the cube expands as time passes, find dV/dt in terms of dx/dt.

2. (a) If A is the area of a circle with radius r and the circle expands as time passes, find dA/dt in terms of dr/dt.
   (b) Suppose oil spills from a ruptured tanker and spreads in a circular pattern. If the radius of the oil spill increases at a constant rate of 1 m/s, how fast is the area of the spill increasing when the radius is 30 m?

3. Each side of a square is increasing at a rate of 6 cm/s. At what rate is the area of the square increasing when the area of the square is 16 cm²?

4. The length of a rectangle is increasing at a rate of 8 cm/s and its width is increasing at a rate of 3 cm/s. When the length is 20 cm and the width is 10 cm, how fast is the area of the rectangle increasing?

5. A cylindrical tank with radius 5 m is being filled with water at a rate of 3 m³/min. How fast is the height of the water increasing?

6. The radius of a sphere is increasing at a rate of 4 mm/s. How fast is the volume increasing when the diameter is 80 mm?

7. The radius of a spherical ball is increasing at a rate of 2 cm/min. At what rate is the surface area of the ball increasing when the radius is 8 cm?

8. The area of a triangle with sides of lengths a and b and contained angle θ is
$$A = \tfrac{1}{2} ab \sin \theta$$
   (a) If $a = 2$ cm, $b = 3$ cm, and θ increases at a rate of 0.2 rad/min, how fast is the area increasing when $\theta = \pi/3$?
   (b) If $a = 2$ cm, b increases at a rate of 1.5 cm/min, and θ increases at a rate of 0.2 rad/min, how fast is the area increasing when $b = 3$ cm and $\theta = \pi/3$?
   (c) If a increases at a rate of 2.5 cm/min, b increases at a rate of 1.5 cm/min, and θ increases at a rate of 0.2 rad/min, how fast is the area increasing when $a = 2$ cm, $b = 3$ cm, and $\theta = \pi/3$?

9. Suppose $y = \sqrt{2x + 1}$, where x and y are functions of t.
   (a) If $dx/dt = 3$, find dy/dt when $x = 4$.
   (b) If $dy/dt = 5$, find dx/dt when $x = 12$.

10. Suppose $4x^2 + 9y^2 = 36$, where x and y are functions of t.
    (a) If $dy/dt = \tfrac{1}{3}$, find dx/dt when $x = 2$ and $y = \tfrac{2}{3}\sqrt{5}$.
    (b) If $dx/dt = 3$, find dy/dt when $x = -2$ and $y = \tfrac{2}{3}\sqrt{5}$.

11. If $x^2 + y^2 + z^2 = 9$, $dx/dt = 5$, and $dy/dt = 4$, find dz/dt when $(x, y, z) = (2, 2, 1)$.

12. A particle is moving along a hyperbola $xy = 8$. As it reaches the point (4, 2), the y-coordinate is decreasing at a rate of 3 cm/s. How fast is the x-coordinate of the point changing at that instant?

13–16
(a) What quantities are given in the problem?
(b) What is the unknown?
(c) Draw a picture of the situation for any time t.
(d) Write an equation that relates the quantities.
(e) Finish solving the problem.

13. A plane flying horizontally at an altitude of 1 mi and a speed of 500 mi/h passes directly over a radar station. Find the rate at which the distance from the plane to the station is increasing when it is 2 mi away from the station.

14. If a snowball melts so that its surface area decreases at a rate of 1 cm²/min, find the rate at which the diameter decreases when the diameter is 10 cm.

15. A street light is mounted at the top of a 15-ft-tall pole. A man 6 ft tall walks away from the pole with a speed of 5 ft/s along a straight path. How fast is the tip of his shadow moving when he is 40 ft from the pole?

16. At noon, ship A is 150 km west of ship B. Ship A is sailing east at 35 km/h and ship B is sailing north at 25 km/h. How fast is the distance between the ships changing at 4:00 PM?

17. Two cars start moving from the same point. One travels south at 60 mi/h and the other travels west at 25 mi/h. At what rate is the distance between the cars increasing two hours later?

18. A spotlight on the ground shines on a wall 12 m away. If a man 2 m tall walks from the spotlight toward the building at a speed of 1.6 m/s, how fast is the length of his shadow on the building decreasing when he is 4 m from the building?

19. A man starts walking north at 4 ft/s from a point P. Five minutes later a woman starts walking south at 5 ft/s from a point 500 ft due east of P. At what rate are the people moving apart 15 min after the woman starts walking?

20. A baseball diamond is a square with side 90 ft. A batter hits the ball and runs toward first base with a speed of 24 ft/s.
    (a) At what rate is his distance from second base decreasing when he is halfway to first base?
    (b) At what rate is his distance from third base increasing at the same moment?

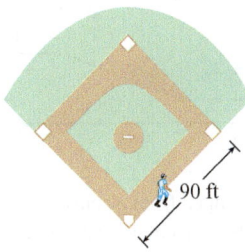

21. The altitude of a triangle is increasing at a rate of 1 cm/min while the area of the triangle is increasing at a rate of 2 cm²/min. At what rate is the base of the triangle changing when the altitude is 10 cm and the area is 100 cm²?

22. A boat is pulled into a dock by a rope attached to the bow of the boat and passing through a pulley on the dock that is 1 m higher than the bow of the boat. If the rope is pulled in at a rate of 1 m/s, how fast is the boat approaching the dock when it is 8 m from the dock?

23. At noon, ship A is 100 km west of ship B. Ship A is sailing south at 35 km/h and ship B is sailing north at 25 km/h. How fast is the distance between the ships changing at 4:00 PM?

24. A particle moves along the curve $y = 2 \sin(\pi x/2)$. As the particle passes through the point $(\frac{1}{3}, 1)$, its x-coordinate increases at a rate of $\sqrt{10}$ cm/s. How fast is the distance from the particle to the origin changing at this instant?

25. Water is leaking out of an inverted conical tank at a rate of 10,000 cm³/min at the same time that water is being pumped into the tank at a constant rate. The tank has height 6 m and the diameter at the top is 4 m. If the water level is rising at a rate of 20 cm/min when the height of the water is 2 m, find the rate at which water is being pumped into the tank.

26. A trough is 10 ft long and its ends have the shape of isosceles triangles that are 3 ft across at the top and have a height of 1 ft. If the trough is being filled with water at a rate of 12 ft³/min, how fast is the water level rising when the water is 6 inches deep?

27. A water trough is 10 m long and a cross-section has the shape of an isosceles trapezoid that is 30 cm wide at the bottom, 80 cm wide at the top, and has height 50 cm. If the trough is being filled with water at the rate of 0.2 m³/min, how fast is the water level rising when the water is 30 cm deep?

28. A swimming pool is 20 ft wide, 40 ft long, 3 ft deep at the shallow end, and 9 ft deep at its deepest point. A cross-section is shown in the figure. If the pool is being filled at a rate of 0.8 ft³/min, how fast is the water level rising when the depth at the deepest point is 5 ft?

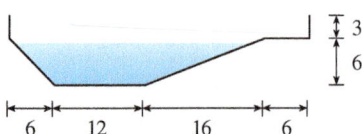

29. Gravel is being dumped from a conveyor belt at a rate of 30 ft³/min, and its coarseness is such that it forms a pile in the shape of a cone whose base diameter and height are always equal. How fast is the height of the pile increasing when the pile is 10 ft high?

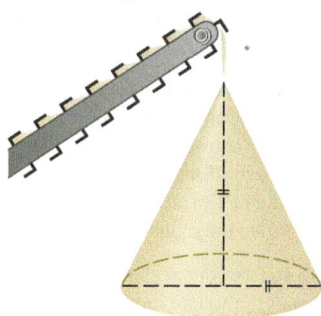

30. A kite 100 ft above the ground moves horizontally at a speed of 8 ft/s. At what rate is the angle between the string and the horizontal decreasing when 200 ft of string has been let out?

31. The sides of an equilateral triangle are increasing at a rate of 10 cm/min. At what rate is the area of the triangle increasing when the sides are 30 cm long?

32. How fast is the angle between the ladder and the ground changing in Example 2 when the bottom of the ladder is 6 ft from the wall?

33. The top of a ladder slides down a vertical wall at a rate of 0.15 m/s. At the moment when the bottom of the ladder is 3 m from the wall, it slides away from the wall at a rate of 0.2 m/s. How long is the ladder?

34. According to the model we used to solve Example 2, what happens as the top of the ladder approaches the ground? Is the model appropriate for small values of y?

35. If the minute hand of a clock has length r (in centimeters), find the rate at which it sweeps out area as a function of r.

36. A faucet is filling a hemispherical basin of diameter 60 cm with water at a rate of 2 L/min. Find the rate at which the water is rising in the basin when it is half full. [Use the following facts: 1 L is 1000 cm³. The volume of the portion of a sphere with radius r from the bottom to a height h is $V = \pi(rh^2 - \frac{1}{3}h^3)$, as we will show in Chapter 6.]

37. Boyle's Law states that when a sample of gas is compressed at a constant temperature, the pressure P and volume V satisfy the equation $PV = C$, where C is a constant. Suppose that at a certain instant the volume is 600 cm³, the pressure is 150 kPa, and the pressure is increasing at a rate of 20 kPa/min. At what rate is the volume decreasing at this instant?

38. When air expands adiabatically (without gaining or losing heat), its pressure P and volume V are related by the equation $PV^{1.4} = C$, where C is a constant. Suppose that at a certain instant the volume is 400 cm³ and the pressure is 80 kPa and is decreasing at a rate of 10 kPa/min. At what rate is the volume increasing at this instant?

39. If two resistors with resistances R_1 and R_2 are connected in parallel, as in the figure, then the total resistance R, measured in ohms (Ω), is given by

$$\frac{1}{R} = \frac{1}{R_1} + \frac{1}{R_2}$$

If R_1 and R_2 are increasing at rates of $0.3 \; \Omega/s$ and $0.2 \; \Omega/s$, respectively, how fast is R changing when $R_1 = 80 \; \Omega$ and $R_2 = 100 \; \Omega$?

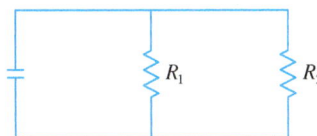

40. Brain weight B as a function of body weight W in fish has been modeled by the power function $B = 0.007W^{2/3}$, where B and W are measured in grams. A model for body weight as a function of body length L (measured in centimeters) is $W = 0.12L^{2.53}$. If, over 10 million years, the average length of a certain species of fish evolved from 15 cm to 20 cm at a constant rate, how fast was this species' brain growing when the average length was 18 cm?

41. Two sides of a triangle have lengths 12 m and 15 m. The angle between them is increasing at a rate of $2°/min$. How fast is the length of the third side increasing when the angle between the sides of fixed length is $60°$?

42. Two carts, A and B, are connected by a rope 39 ft long that passes over a pulley P (see the figure). The point Q is on the floor 12 ft directly beneath P and between the carts. Cart A is being pulled away from Q at a speed of 2 ft/s. How fast is cart B moving toward Q at the instant when cart A is 5 ft from Q?

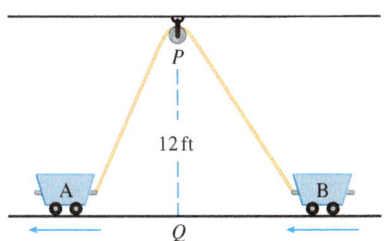

43. A television camera is positioned 4000 ft from the base of a rocket launching pad. The angle of elevation of the camera has to change at the correct rate in order to keep the rocket in sight. Also, the mechanism for focusing the camera has to take into account the increasing distance from the camera to the rising rocket. Let's assume the rocket rises vertically and its speed is 600 ft/s when it has risen 3000 ft.
(a) How fast is the distance from the television camera to the rocket changing at that moment?
(b) If the television camera is always kept aimed at the rocket, how fast is the camera's angle of elevation changing at that same moment?

44. A lighthouse is located on a small island 3 km away from the nearest point P on a straight shoreline and its light makes four revolutions per minute. How fast is the beam of light moving along the shoreline when it is 1 km from P?

45. A plane flies horizontally at an altitude of 5 km and passes directly over a tracking telescope on the ground. When the angle of elevation is $\pi/3$, this angle is decreasing at a rate of $\pi/6$ rad/min. How fast is the plane traveling at that time?

46. A Ferris wheel with a radius of 10 m is rotating at a rate of one revolution every 2 minutes. How fast is a rider rising when his seat is 16 m above ground level?

47. A plane flying with a constant speed of 300 km/h passes over a ground radar station at an altitude of 1 km and climbs at an angle of $30°$. At what rate is the distance from the plane to the radar station increasing a minute later?

48. Two people start from the same point. One walks east at 3 mi/h and the other walks northeast at 2 mi/h. How fast is the distance between the people changing after 15 minutes?

49. A runner sprints around a circular track of radius 100 m at a constant speed of 7 m/s. The runner's friend is standing at a distance 200 m from the center of the track. How fast is the distance between the friends changing when the distance between them is 200 m?

50. The minute hand on a watch is 8 mm long and the hour hand is 4 mm long. How fast is the distance between the tips of the hands changing at one o'clock?

3.10 Linear Approximations and Differentials

We have seen that a curve lies very close to its tangent line near the point of tangency. In fact, by zooming in toward a point on the graph of a differentiable function, we noticed that the graph looks more and more like its tangent line. (See Figure 2.7.2.) This observation is the basis for a method of finding approximate values of functions.

The idea is that it might be easy to calculate a value $f(a)$ of a function, but difficult (or even impossible) to compute nearby values of f. So we settle for the easily computed

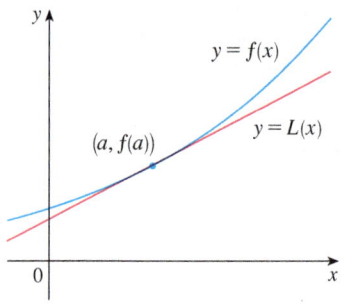

FIGURE 1

values of the linear function L whose graph is the tangent line of f at $(a, f(a))$. (See Figure 1.)

In other words, we use the tangent line at $(a, f(a))$ as an approximation to the curve $y = f(x)$ when x is near a. An equation of this tangent line is

$$y = f(a) + f'(a)(x - a)$$

and the approximation

$$\boxed{1 \quad f(x) \approx f(a) + f'(a)(x - a)}$$

is called the **linear approximation** or **tangent line approximation** of f at a. The linear function whose graph is this tangent line, that is,

$$\boxed{2 \quad L(x) = f(a) + f'(a)(x - a)}$$

is called the **linearization** of f at a.

EXAMPLE 1 Find the linearization of the function $f(x) = \sqrt{x + 3}$ at $a = 1$ and use it to approximate the numbers $\sqrt{3.98}$ and $\sqrt{4.05}$. Are these approximations overestimates or underestimates?

SOLUTION The derivative of $f(x) = (x + 3)^{1/2}$ is

$$f'(x) = \tfrac{1}{2}(x + 3)^{-1/2} = \frac{1}{2\sqrt{x + 3}}$$

and so we have $f(1) = 2$ and $f'(1) = \tfrac{1}{4}$. Putting these values into Equation 2, we see that the linearization is

$$L(x) = f(1) + f'(1)(x - 1) = 2 + \tfrac{1}{4}(x - 1) = \frac{7}{4} + \frac{x}{4}$$

The corresponding linear approximation (1) is

$$\sqrt{x + 3} \approx \frac{7}{4} + \frac{x}{4} \qquad \text{(when } x \text{ is near 1)}$$

In particular, we have

$$\sqrt{3.98} \approx \tfrac{7}{4} + \tfrac{0.98}{4} = 1.995 \quad \text{and} \quad \sqrt{4.05} \approx \tfrac{7}{4} + \tfrac{1.05}{4} = 2.0125$$

The linear approximation is illustrated in Figure 2. We see that, indeed, the tangent line approximation is a good approximation to the given function when x is near 1. We also see that our approximations are overestimates because the tangent line lies above the curve.

Of course, a calculator could give us approximations for $\sqrt{3.98}$ and $\sqrt{4.05}$, but the linear approximation gives an approximation *over an entire interval*.

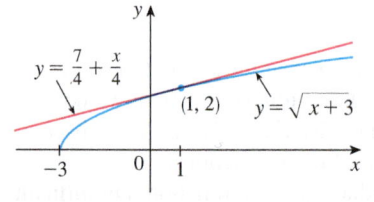

FIGURE 2

In the following table we compare the estimates from the linear approximation in Example 1 with the true values. Notice from this table, and also from Figure 2, that the tangent line approximation gives good estimates when x is close to 1 but the accuracy of the approximation deteriorates when x is farther away from 1.

	x	From $L(x)$	Actual value
$\sqrt{3.9}$	0.9	1.975	1.97484176...
$\sqrt{3.98}$	0.98	1.995	1.99499373...
$\sqrt{4}$	1	2	2.00000000...
$\sqrt{4.05}$	1.05	2.0125	2.01246117...
$\sqrt{4.1}$	1.1	2.025	2.02484567...
$\sqrt{5}$	2	2.25	2.23606797...
$\sqrt{6}$	3	2.5	2.44948974...

How good is the approximation that we obtained in Example 1? The next example shows that by using a graphing calculator or computer we can determine an interval throughout which a linear approximation provides a specified accuracy.

EXAMPLE 2 For what values of x is the linear approximation

$$\sqrt{x+3} \approx \frac{7}{4} + \frac{x}{4}$$

accurate to within 0.5? What about accuracy to within 0.1?

SOLUTION Accuracy to within 0.5 means that the functions should differ by less than 0.5:

$$\left| \sqrt{x+3} - \left(\frac{7}{4} + \frac{x}{4} \right) \right| < 0.5$$

Equivalently, we could write

$$\sqrt{x+3} - 0.5 < \frac{7}{4} + \frac{x}{4} < \sqrt{x+3} + 0.5$$

This says that the linear approximation should lie between the curves obtained by shifting the curve $y = \sqrt{x+3}$ upward and downward by an amount 0.5. Figure 3 shows the tangent line $y = (7 + x)/4$ intersecting the upper curve $y = \sqrt{x+3} + 0.5$ at P and Q. Zooming in and using the cursor, we estimate that the x-coordinate of P is about -2.66 and the x-coordinate of Q is about 8.66. Thus we see from the graph that the approximation

$$\sqrt{x+3} \approx \frac{7}{4} + \frac{x}{4}$$

is accurate to within 0.5 when $-2.6 < x < 8.6$. (We have rounded to be safe.)

Similarly, from Figure 4 we see that the approximation is accurate to within 0.1 when $-1.1 < x < 3.9$.

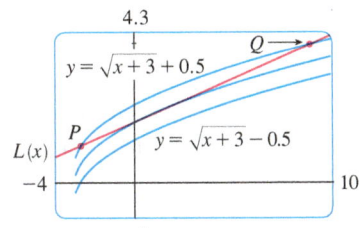

FIGURE 3

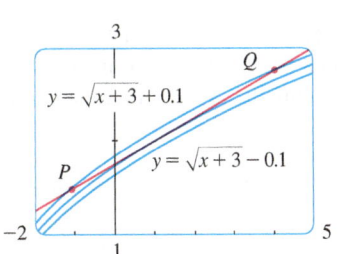

FIGURE 4

Applications to Physics

Linear approximations are often used in physics. In analyzing the consequences of an equation, a physicist sometimes needs to simplify a function by replacing it with its linear approximation. For instance, in deriving a formula for the period of a pendulum, physics textbooks obtain the expression $a_T = -g \sin \theta$ for tangential acceleration and then replace $\sin \theta$ by θ with the remark that $\sin \theta$ is very close to θ if θ is not too large. [See, for example, *Physics: Calculus*, 2d ed., by Eugene Hecht (Pacific Grove, CA: Brooks/Cole, 2000), p. 431.] You can verify that the linearization of the function $f(x) = \sin x$ at $a = 0$ is $L(x) = x$ and so the linear approximation at 0 is

$$\sin x \approx x$$

(see Exercise 42). So, in effect, the derivation of the formula for the period of a pendulum uses the tangent line approximation for the sine function.

Another example occurs in the theory of optics, where light rays that arrive at shallow angles relative to the optical axis are called *paraxial rays.* In paraxial (or Gaussian) optics, both $\sin \theta$ and $\cos \theta$ are replaced by their linearizations. In other words, the linear approximations

$$\sin \theta \approx \theta \qquad \text{and} \qquad \cos \theta \approx 1$$

are used because θ is close to 0. The results of calculations made with these approximations became the basic theoretical tool used to design lenses. [See *Optics,* 4th ed., by Eugene Hecht (San Francisco, 2002), p. 154.]

In Section 11.11 we will present several other applications of the idea of linear approximations to physics and engineering.

■ Differentials

The ideas behind linear approximations are sometimes formulated in the terminology and notation of *differentials*. If $y = f(x)$, where f is a differentiable function, then the **differential** dx is an independent variable; that is, dx can be given the value of any real number. The **differential** dy is then defined in terms of dx by the equation

3 $$dy = f'(x)\, dx$$

So dy is a dependent variable; it depends on the values of x and dx. If dx is given a specific value and x is taken to be some specific number in the domain of f, then the numerical value of dy is determined.

The geometric meaning of differentials is shown in Figure 5. Let $P(x, f(x))$ and $Q(x + \Delta x, f(x + \Delta x))$ be points on the graph of f and let $dx = \Delta x$. The corresponding change in y is

$$\Delta y = f(x + \Delta x) - f(x)$$

The slope of the tangent line PR is the derivative $f'(x)$. Thus the directed distance from S to R is $f'(x)\, dx = dy$. Therefore dy represents the amount that the tangent line rises or falls (the change in the linearization), whereas Δy represents the amount that the curve $y = f(x)$ rises or falls when x changes by an amount dx.

If $dx \neq 0$, we can divide both sides of Equation 3 by dx to obtain

$$\frac{dy}{dx} = f'(x)$$

We have seen similar equations before, but now the left side can genuinely be interpreted as a ratio of differentials.

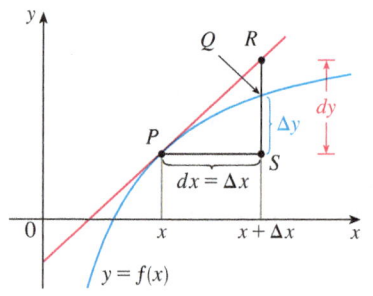

FIGURE 5

EXAMPLE 3 Compare the values of Δy and dy if $y = f(x) = x^3 + x^2 - 2x + 1$ and x changes (a) from 2 to 2.05 and (b) from 2 to 2.01.

SOLUTION
(a) We have
$$f(2) = 2^3 + 2^2 - 2(2) + 1 = 9$$

$$f(2.05) = (2.05)^3 + (2.05)^2 - 2(2.05) + 1 = 9.717625$$

$$\Delta y = f(2.05) - f(2) = 0.717625$$

Figure 6 shows the function in Example 3 and a comparison of dy and Δy when $a = 2$. The viewing rectangle is [1.8, 2.5] by [6, 18].

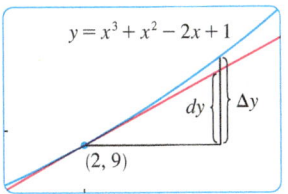

FIGURE 6

In general, $$dy = f'(x)\,dx = (3x^2 + 2x - 2)\,dx$$

When $x = 2$ and $dx = \Delta x = 0.05$, this becomes
$$dy = [3(2)^2 + 2(2) - 2]0.05 = 0.7$$

(b) $$f(2.01) = (2.01)^3 + (2.01)^2 - 2(2.01) + 1 = 9.140701$$
$$\Delta y = f(2.01) - f(2) = 0.140701$$

When $dx = \Delta x = 0.01$,
$$dy = [3(2)^2 + 2(2) - 2]0.01 = 0.14$$

Notice that the approximation $\Delta y \approx dy$ becomes better as Δx becomes smaller in Example 3. Notice also that dy was easier to compute than Δy. For more complicated functions it may be impossible to compute Δy exactly. In such cases the approximation by differentials is especially useful.

In the notation of differentials, the linear approximation (1) can be written as

$$f(a + dx) \approx f(a) + dy$$

For instance, for the function $f(x) = \sqrt{x + 3}$ in Example 1, we have

$$dy = f'(x)\,dx = \frac{dx}{2\sqrt{x + 3}}$$

If $a = 1$ and $dx = \Delta x = 0.05$, then

$$dy = \frac{0.05}{2\sqrt{1 + 3}} = 0.0125$$

and $$\sqrt{4.05} = f(1.05) \approx f(1) + dy = 2.0125$$

just as we found in Example 1.

Our final example illustrates the use of differentials in estimating the errors that occur because of approximate measurements.

EXAMPLE 4 The radius of a sphere was measured and found to be 21 cm with a possible error in measurement of at most 0.05 cm. What is the maximum error in using this value of the radius to compute the volume of the sphere?

SOLUTION If the radius of the sphere is r, then its volume is $V = \frac{4}{3}\pi r^3$. If the error in the measured value of r is denoted by $dr = \Delta r$, then the corresponding error in the calculated value of V is ΔV, which can be approximated by the differential

$$dV = 4\pi r^2\,dr$$

When $r = 21$ and $dr = 0.05$, this becomes

$$dV = 4\pi(21)^2 0.05 \approx 277$$

The maximum error in the calculated volume is about 277 cm^3.

256   **CHAPTER 3**   Differentiation Rules

NOTE Although the possible error in Example 4 may appear to be rather large, a better picture of the error is given by the **relative error**, which is computed by dividing the error by the total volume:

$$\frac{\Delta V}{V} \approx \frac{dV}{V} = \frac{4\pi r^2\, dr}{\frac{4}{3}\pi r^3} = 3\,\frac{dr}{r}$$

Thus the relative error in the volume is about three times the relative error in the radius. In Example 4 the relative error in the radius is approximately $dr/r = 0.05/21 \approx 0.0024$ and it produces a relative error of about 0.007 in the volume. The errors could also be expressed as **percentage errors** of 0.24% in the radius and 0.7% in the volume.

3.10 EXERCISES

1–4 Find the linearization $L(x)$ of the function at a.

1. $f(x) = x^3 - x^2 + 3$, $a = -2$

2. $f(x) = \sin x$, $a = \pi/6$

3. $f(x) = \sqrt{x}$, $a = 4$

4. $f(x) = 2^x$, $a = 0$

5. Find the linear approximation of the function $f(x) = \sqrt{1-x}$ at $a = 0$ and use it to approximate the numbers $\sqrt{0.9}$ and $\sqrt{0.99}$. Illustrate by graphing f and the tangent line.

6. Find the linear approximation of the function $g(x) = \sqrt[3]{1+x}$ at $a = 0$ and use it to approximate the numbers $\sqrt[3]{0.95}$ and $\sqrt[3]{1.1}$. Illustrate by graphing g and the tangent line.

7–10 Verify the given linear approximation at $a = 0$. Then determine the values of x for which the linear approximation is accurate to within 0.1.

7. $\ln(1+x) \approx x$ **8.** $(1+x)^{-3} \approx 1 - 3x$

9. $\sqrt[4]{1+2x} \approx 1 + \frac{1}{2}x$ **10.** $e^x \cos x \approx 1 + x$

11–14 Find the differential of each function.

11. (a) $y = xe^{-4x}$ (b) $y = \sqrt{1 - t^4}$

12. (a) $y = \dfrac{1+2u}{1+3u}$ (b) $y = \theta^2 \sin 2\theta$

13. (a) $y = \tan \sqrt{t}$ (b) $y = \dfrac{1-v^2}{1+v^2}$

14. (a) $y = \ln(\sin \theta)$ (b) $y = \dfrac{e^x}{1-e^x}$

15–18 (a) Find the differential dy and (b) evaluate dy for the given values of x and dx.

15. $y = e^{x/10}$, $x = 0$, $dx = 0.1$

16. $y = \cos \pi x$, $x = \frac{1}{3}$, $dx = -0.02$

17. $y = \sqrt{3 + x^2}$, $x = 1$, $dx = -0.1$

18. $y = \dfrac{x+1}{x-1}$, $x = 2$, $dx = 0.05$

19–22 Compute Δy and dy for the given values of x and $dx = \Delta x$. Then sketch a diagram like Figure 5 showing the line segments with lengths dx, dy, and Δy.

19. $y = x^2 - 4x$, $x = 3$, $\Delta x = 0.5$

20. $y = x - x^3$, $x = 0$, $\Delta x = -0.3$

21. $y = \sqrt{x-2}$, $x = 3$, $\Delta x = 0.8$

22. $y = e^x$, $x = 0$, $\Delta x = 0.5$

23–28 Use a linear approximation (or differentials) to estimate the given number.

23. $(1.999)^4$ **24.** $1/4.002$

25. $\sqrt[3]{1001}$ **26.** $\sqrt{100.5}$

27. $e^{0.1}$ **28.** $\cos 29°$

29–31 Explain, in terms of linear approximations or differentials, why the approximation is reasonable.

29. $\sec 0.08 \approx 1$ **30.** $\sqrt{4.02} \approx 2.005$

31. $\dfrac{1}{9.98} \approx 0.1002$

32. Let $f(x) = (x-1)^2 \quad g(x) = e^{-2x}$

and $\quad h(x) = 1 + \ln(1-2x)$

(a) Find the linearizations of f, g, and h at $a = 0$. What do you notice? How do you explain what happened?

(b) Graph f, g, and h and their linear approximations. For which function is the linear approximation best? For which is it worst? Explain.

33. The edge of a cube was found to be 30 cm with a possible error in measurement of 0.1 cm. Use differentials to estimate the maximum possible error, relative error, and percentage error in computing (a) the volume of the cube and (b) the surface area of the cube.

34. The radius of a circular disk is given as 24 cm with a maximum error in measurement of 0.2 cm.
(a) Use differentials to estimate the maximum error in the calculated area of the disk.
(b) What is the relative error? What is the percentage error?

35. The circumference of a sphere was measured to be 84 cm with a possible error of 0.5 cm.
(a) Use differentials to estimate the maximum error in the calculated surface area. What is the relative error?
(b) Use differentials to estimate the maximum error in the calculated volume. What is the relative error?

36. Use differentials to estimate the amount of paint needed to apply a coat of paint 0.05 cm thick to a hemispherical dome with diameter 50 m.

37. (a) Use differentials to find a formula for the approximate volume of a thin cylindrical shell with height h, inner radius r, and thickness Δr.
(b) What is the error involved in using the formula from part (a)?

38. One side of a right triangle is known to be 20 cm long and the opposite angle is measured as $30°$, with a possible error of $\pm 1°$.
(a) Use differentials to estimate the error in computing the length of the hypotenuse.
(b) What is the percentage error?

39. If a current I passes through a resistor with resistance R, Ohm's Law states that the voltage drop is $V = RI$. If V is constant and R is measured with a certain error, use differentials to show that the relative error in calculating I is approximately the same (in magnitude) as the relative error in R.

40. When blood flows along a blood vessel, the flux F (the volume of blood per unit time that flows past a given point) is proportional to the fourth power of the radius R of the blood vessel:

$$F = kR^4$$

(This is known as Poiseuille's Law; we will show why it is true in Section 8.4.) A partially clogged artery can be expanded by an operation called angioplasty, in which a balloon-tipped catheter is inflated inside the artery in order to widen it and restore the normal blood flow.

Show that the relative change in F is about four times the relative change in R. How will a 5% increase in the radius affect the flow of blood?

41. Establish the following rules for working with differentials (where c denotes a constant and u and v are functions of x).
(a) $dc = 0$ (b) $d(cu) = c\, du$
(c) $d(u + v) = du + dv$ (d) $d(uv) = u\, dv + v\, du$
(e) $d\left(\dfrac{u}{v}\right) = \dfrac{v\, du - u\, dv}{v^2}$ (f) $d(x^n) = nx^{n-1}\, dx$

42. On page 431 of *Physics: Calculus*, 2d ed., by Eugene Hecht (Pacific Grove, CA: Brooks/Cole, 2000), in the course of deriving the formula $T = 2\pi\sqrt{L/g}$ for the period of a pendulum of length L, the author obtains the equation $a_T = -g\sin\theta$ for the tangential acceleration of the bob of the pendulum. He then says, "for small angles, the value of θ in radians is very nearly the value of $\sin\theta$; they differ by less than 2% out to about $20°$."
(a) Verify the linear approximation at 0 for the sine function:

$$\sin x \approx x$$

(b) Use a graphing device to determine the values of x for which $\sin x$ and x differ by less than 2%. Then verify Hecht's statement by converting from radians to degrees.

43. Suppose that the only information we have about a function f is that $f(1) = 5$ and the graph of its *derivative* is as shown.
(a) Use a linear approximation to estimate $f(0.9)$ and $f(1.1)$.
(b) Are your estimates in part (a) too large or too small? Explain.

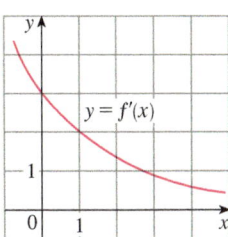

44. Suppose that we don't have a formula for $g(x)$ but we know that $g(2) = -4$ and $g'(x) = \sqrt{x^2 + 5}$ for all x.
(a) Use a linear approximation to estimate $g(1.95)$ and $g(2.05)$.
(b) Are your estimates in part (a) too large or too small? Explain.

LABORATORY PROJECT TAYLOR POLYNOMIALS

The tangent line approximation $L(x)$ is the best first-degree (linear) approximation to $f(x)$ near $x = a$ because $f(x)$ and $L(x)$ have the same rate of change (derivative) at a. For a better approximation than a linear one, let's try a second-degree (quadratic) approximation $P(x)$. In other words, we approximate a curve by a parabola instead of by a straight line. To make sure that the approximation is a good one, we stipulate the following:

(i) $P(a) = f(a)$ (P and f should have the same value at a.)
(ii) $P'(a) = f'(a)$ (P and f should have the same rate of change at a.)
(iii) $P''(a) = f''(a)$ (The slopes of P and f should change at the same rate at a.)

1. Find the quadratic approximation $P(x) = A + Bx + Cx^2$ to the function $f(x) = \cos x$ that satisfies conditions (i), (ii), and (iii) with $a = 0$. Graph P, f, and the linear approximation $L(x) = 1$ on a common screen. Comment on how well the functions P and L approximate f.

2. Determine the values of x for which the quadratic approximation $f(x) \approx P(x)$ in Problem 1 is accurate to within 0.1. [*Hint:* Graph $y = P(x)$, $y = \cos x - 0.1$, and $y = \cos x + 0.1$ on a common screen.]

3. To approximate a function f by a quadratic function P near a number a, it is best to write P in the form
$$P(x) = A + B(x - a) + C(x - a)^2$$
Show that the quadratic function that satisfies conditions (i), (ii), and (iii) is
$$P(x) = f(a) + f'(a)(x - a) + \tfrac{1}{2}f''(a)(x - a)^2$$

4. Find the quadratic approximation to $f(x) = \sqrt{x + 3}$ near $a = 1$. Graph f, the quadratic approximation, and the linear approximation from Example 3.10.2 on a common screen. What do you conclude?

5. Instead of being satisfied with a linear or quadratic approximation to $f(x)$ near $x = a$, let's try to find better approximations with higher-degree polynomials. We look for an nth-degree polynomial
$$T_n(x) = c_0 + c_1(x - a) + c_2(x - a)^2 + c_3(x - a)^3 + \cdots + c_n(x - a)^n$$
such that T_n and its first n derivatives have the same values at $x = a$ as f and its first n derivatives. By differentiating repeatedly and setting $x = a$, show that these conditions are satisfied if $c_0 = f(a)$, $c_1 = f'(a)$, $c_2 = \tfrac{1}{2}f''(a)$, and in general
$$c_k = \frac{f^{(k)}(a)}{k!}$$
where $k! = 1 \cdot 2 \cdot 3 \cdot 4 \cdots \cdot k$. The resulting polynomial
$$T_n(x) = f(a) + f'(a)(x - a) + \frac{f''(a)}{2!}(x - a)^2 + \cdots + \frac{f^{(n)}(a)}{n!}(x - a)^n$$
is called the **nth-degree Taylor polynomial of f centered at a**.

6. Find the 8th-degree Taylor polynomial centered at $a = 0$ for the function $f(x) = \cos x$. Graph f together with the Taylor polynomials T_2, T_4, T_6, T_8 in the viewing rectangle $[-5, 5]$ by $[-1.4, 1.4]$ and comment on how well they approximate f.

3.11 Hyperbolic Functions

Certain even and odd combinations of the exponential functions e^x and e^{-x} arise so frequently in mathematics and its applications that they deserve to be given special names. In many ways they are analogous to the trigonometric functions, and they have the same relationship to the hyperbola that the trigonometric functions have to the circle. For this reason they are collectively called **hyperbolic functions** and individually called **hyperbolic sine**, **hyperbolic cosine**, and so on.

Definition of the Hyperbolic Functions

$$\sinh x = \frac{e^x - e^{-x}}{2} \qquad \operatorname{csch} x = \frac{1}{\sinh x}$$

$$\cosh x = \frac{e^x + e^{-x}}{2} \qquad \operatorname{sech} x = \frac{1}{\cosh x}$$

$$\tanh x = \frac{\sinh x}{\cosh x} \qquad \coth x = \frac{\cosh x}{\sinh x}$$

The graphs of hyperbolic sine and cosine can be sketched using graphical addition as in Figures 1 and 2.

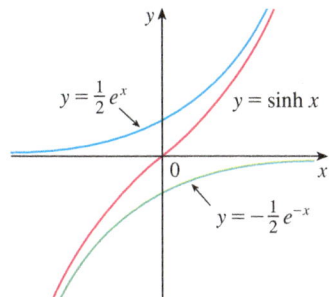

FIGURE 1
$y = \sinh x = \frac{1}{2}e^x - \frac{1}{2}e^{-x}$

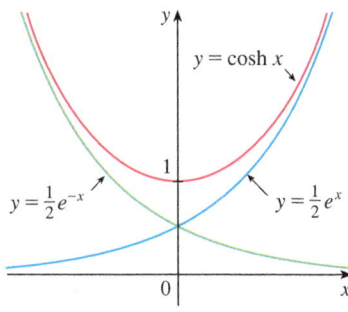

FIGURE 2
$y = \cosh x = \frac{1}{2}e^x + \frac{1}{2}e^{-x}$

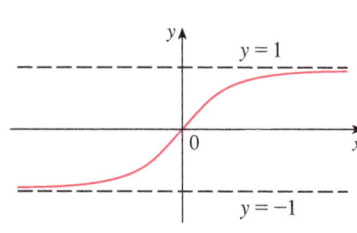

FIGURE 3
$y = \tanh x$

Note that sinh has domain $\mathbb{R}$ and range $\mathbb{R}$, while cosh has domain $\mathbb{R}$ and range $[1, \infty)$. The graph of tanh is shown in Figure 3. It has the horizontal asymptotes $y = \pm 1$. (See Exercise 23.)

Some of the mathematical uses of hyperbolic functions will be seen in Chapter 7. Applications to science and engineering occur whenever an entity such as light, velocity, electricity, or radioactivity is gradually absorbed or extinguished, for the decay can be represented by hyperbolic functions. The most famous application is the use of hyperbolic cosine to describe the shape of a hanging wire. It can be proved that if a heavy flexible cable (such as a telephone or power line) is suspended between two points at the same height, then it takes the shape of a curve with equation $y = c + a \cosh(x/a)$ called a *catenary* (see Figure 4). (The Latin word *catena* means "chain.")

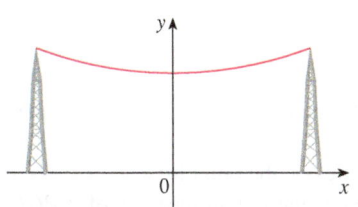

FIGURE 4
A catenary $y = c + a \cosh(x/a)$

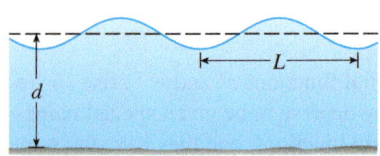

FIGURE 5
Idealized ocean wave

Another application of hyperbolic functions occurs in the description of ocean waves: The velocity of a water wave with length L moving across a body of water with depth d is modeled by the function

$$v = \sqrt{\frac{gL}{2\pi} \tanh\left(\frac{2\pi d}{L}\right)}$$

where g is the acceleration due to gravity. (See Figure 5 and Exercise 49.)

The hyperbolic functions satisfy a number of identities that are similar to well-known trigonometric identities. We list some of them here and leave most of the proofs to the exercises.

Hyperbolic Identities

$$\sinh(-x) = -\sinh x \qquad \cosh(-x) = \cosh x$$

$$\cosh^2 x - \sinh^2 x = 1 \qquad 1 - \tanh^2 x = \text{sech}^2 x$$

$$\sinh(x + y) = \sinh x \cosh y + \cosh x \sinh y$$

$$\cosh(x + y) = \cosh x \cosh y + \sinh x \sinh y$$

The Gateway Arch in St. Louis was designed using a hyperbolic cosine function (see Exercise 48).

EXAMPLE 1 Prove (a) $\cosh^2 x - \sinh^2 x = 1$ and (b) $1 - \tanh^2 x = \text{sech}^2 x$.

SOLUTION

(a) $$\cosh^2 x - \sinh^2 x = \left(\frac{e^x + e^{-x}}{2}\right)^2 - \left(\frac{e^x - e^{-x}}{2}\right)^2$$

$$= \frac{e^{2x} + 2 + e^{-2x}}{4} - \frac{e^{2x} - 2 + e^{-2x}}{4}$$

$$= \frac{4}{4} = 1$$

(b) We start with the identity proved in part (a):

$$\cosh^2 x - \sinh^2 x = 1$$

If we divide both sides by $\cosh^2 x$, we get

$$1 - \frac{\sinh^2 x}{\cosh^2 x} = \frac{1}{\cosh^2 x}$$

or $$1 - \tanh^2 x = \text{sech}^2 x \qquad \blacksquare$$

The identity proved in Example 1(a) gives a clue to the reason for the name "hyperbolic" functions:

If t is any real number, then the point $P(\cos t, \sin t)$ lies on the unit circle $x^2 + y^2 = 1$ because $\cos^2 t + \sin^2 t = 1$. In fact, t can be interpreted as the radian measure of $\angle POQ$ in Figure 6. For this reason the trigonometric functions are sometimes called *circular* functions.

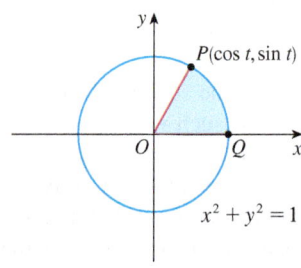

FIGURE 6

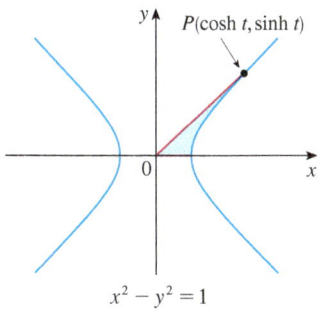

FIGURE 7

Likewise, if t is any real number, then the point $P(\cosh t, \sinh t)$ lies on the right branch of the hyperbola $x^2 - y^2 = 1$ because $\cosh^2 t - \sinh^2 t = 1$ and $\cosh t \geq 1$. This time, t does not represent the measure of an angle. However, it turns out that t represents twice the area of the shaded hyperbolic sector in Figure 7, just as in the trigonometric case t represents twice the area of the shaded circular sector in Figure 6.

The derivatives of the hyperbolic functions are easily computed. For example,

$$\frac{d}{dx}(\sinh x) = \frac{d}{dx}\left(\frac{e^x - e^{-x}}{2}\right) = \frac{e^x + e^{-x}}{2} = \cosh x$$

We list the differentiation formulas for the hyperbolic functions as Table 1. The remaining proofs are left as exercises. Note the analogy with the differentiation formulas for trigonometric functions, but beware that the signs are different in some cases.

1 Derivatives of Hyperbolic Functions

$$\frac{d}{dx}(\sinh x) = \cosh x \qquad \frac{d}{dx}(\operatorname{csch} x) = -\operatorname{csch} x \coth x$$

$$\frac{d}{dx}(\cosh x) = \sinh x \qquad \frac{d}{dx}(\operatorname{sech} x) = -\operatorname{sech} x \tanh x$$

$$\frac{d}{dx}(\tanh x) = \operatorname{sech}^2 x \qquad \frac{d}{dx}(\coth x) = -\operatorname{csch}^2 x$$

EXAMPLE 2 Any of these differentiation rules can be combined with the Chain Rule. For instance,

$$\frac{d}{dx}\left(\cosh \sqrt{x}\right) = \sinh \sqrt{x} \cdot \frac{d}{dx}\sqrt{x} = \frac{\sinh \sqrt{x}}{2\sqrt{x}} \qquad \blacksquare$$

■ Inverse Hyperbolic Functions

You can see from Figures 1 and 3 that sinh and tanh are one-to-one functions and so they have inverse functions denoted by $\sinh^{-1}$ and $\tanh^{-1}$. Figure 2 shows that cosh is not one-to-one, but when restricted to the domain $[0, \infty)$ it becomes one-to-one. The inverse hyperbolic cosine function is defined as the inverse of this restricted function.

2
$$y = \sinh^{-1} x \iff \sinh y = x$$
$$y = \cosh^{-1} x \iff \cosh y = x \text{ and } y \geq 0$$
$$y = \tanh^{-1} x \iff \tanh y = x$$

The remaining inverse hyperbolic functions are defined similarly (see Exercise 28).

We can sketch the graphs of $\sinh^{-1}$, $\cosh^{-1}$, and $\tanh^{-1}$ in Figures 8, 9, and 10 by using Figures 1, 2, and 3.

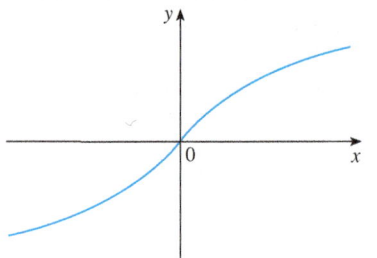

FIGURE 8 $y = \sinh^{-1} x$
domain $= \mathbb{R}$ range $= \mathbb{R}$

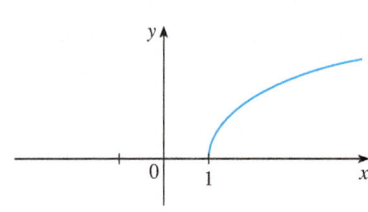

FIGURE 9 $y = \cosh^{-1} x$
domain $= [1, \infty)$ range $= [0, \infty)$

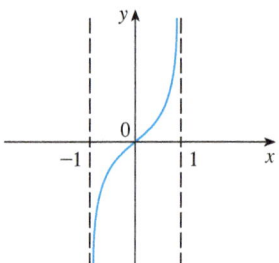

FIGURE 10 $y = \tanh^{-1} x$
domain $= (-1, 1)$ range $= \mathbb{R}$

Since the hyperbolic functions are defined in terms of exponential functions, it's not surprising to learn that the inverse hyperbolic functions can be expressed in terms of logarithms. In particular, we have:

Formula 3 is proved in Example 3. The proofs of Formulas 4 and 5 are requested in Exercises 26 and 27.

$$\boxed{3} \qquad \sinh^{-1} x = \ln\left(x + \sqrt{x^2 + 1}\right) \qquad x \in \mathbb{R}$$

$$\boxed{4} \qquad \cosh^{-1} x = \ln\left(x + \sqrt{x^2 - 1}\right) \qquad x \geq 1$$

$$\boxed{5} \qquad \tanh^{-1} x = \tfrac{1}{2} \ln\left(\frac{1 + x}{1 - x}\right) \qquad -1 < x < 1$$

EXAMPLE 3 Show that $\sinh^{-1} x = \ln\left(x + \sqrt{x^2 + 1}\right)$.

SOLUTION Let $y = \sinh^{-1} x$. Then

$$x = \sinh y = \frac{e^y - e^{-y}}{2}$$

so

$$e^y - 2x - e^{-y} = 0$$

or, multiplying by e^y,

$$e^{2y} - 2xe^y - 1 = 0$$

This is really a quadratic equation in e^y:

$$(e^y)^2 - 2x(e^y) - 1 = 0$$

Solving by the quadratic formula, we get

$$e^y = \frac{2x \pm \sqrt{4x^2 + 4}}{2} = x \pm \sqrt{x^2 + 1}$$

Note that $e^y > 0$, but $x - \sqrt{x^2 + 1} < 0$ (because $x < \sqrt{x^2 + 1}$). Thus the minus sign is inadmissible and we have

$$e^y = x + \sqrt{x^2 + 1}$$

Therefore $\quad y = \ln(e^y) = \ln(x + \sqrt{x^2 + 1})$

This shows that $\quad \sinh^{-1}x = \ln(x + \sqrt{x^2 + 1})$

(See Exercise 25 for another method.)

---

6 Derivatives of Inverse Hyperbolic Functions

$$\frac{d}{dx}(\sinh^{-1}x) = \frac{1}{\sqrt{1 + x^2}} \qquad \frac{d}{dx}(\operatorname{csch}^{-1}x) = -\frac{1}{|x|\sqrt{x^2 + 1}}$$

$$\frac{d}{dx}(\cosh^{-1}x) = \frac{1}{\sqrt{x^2 - 1}} \qquad \frac{d}{dx}(\operatorname{sech}^{-1}x) = -\frac{1}{x\sqrt{1 - x^2}}$$

$$\frac{d}{dx}(\tanh^{-1}x) = \frac{1}{1 - x^2} \qquad \frac{d}{dx}(\coth^{-1}x) = \frac{1}{1 - x^2}$$

Notice that the formulas for the derivatives of $\tanh^{-1}x$ and $\coth^{-1}x$ appear to be identical. But the domains of these functions have no numbers in common: $\tanh^{-1}x$ is defined for $|x| < 1$, whereas $\coth^{-1}x$ is defined for $|x| > 1$.

The inverse hyperbolic functions are all differentiable because the hyperbolic functions are differentiable. The formulas in Table 6 can be proved either by the method for inverse functions or by differentiating Formulas 3, 4, and 5.

EXAMPLE 4 Prove that $\dfrac{d}{dx}(\sinh^{-1}x) = \dfrac{1}{\sqrt{1 + x^2}}$.

SOLUTION 1 Let $y = \sinh^{-1}x$. Then $\sinh y = x$. If we differentiate this equation implicitly with respect to x, we get

$$\cosh y \frac{dy}{dx} = 1$$

Since $\cosh^2 y - \sinh^2 y = 1$ and $\cosh y \geq 0$, we have $\cosh y = \sqrt{1 + \sinh^2 y}$, so

$$\frac{dy}{dx} = \frac{1}{\cosh y} = \frac{1}{\sqrt{1 + \sinh^2 y}} = \frac{1}{\sqrt{1 + x^2}}$$

SOLUTION 2 From Equation 3 (proved in Example 3), we have

$$\frac{d}{dx}(\sinh^{-1}x) = \frac{d}{dx}\ln(x + \sqrt{x^2 + 1})$$

$$= \frac{1}{x + \sqrt{x^2 + 1}} \frac{d}{dx}(x + \sqrt{x^2 + 1})$$

$$= \frac{1}{x + \sqrt{x^2 + 1}}\left(1 + \frac{x}{\sqrt{x^2 + 1}}\right)$$

$$= \frac{\sqrt{x^2 + 1} + x}{(x + \sqrt{x^2 + 1})\sqrt{x^2 + 1}}$$

$$= \frac{1}{\sqrt{x^2 + 1}}$$

EXAMPLE 5 Find $\dfrac{d}{dx}[\tanh^{-1}(\sin x)]$.

SOLUTION Using Table 6 and the Chain Rule, we have

$$\frac{d}{dx}[\tanh^{-1}(\sin x)] = \frac{1}{1-(\sin x)^2}\frac{d}{dx}(\sin x)$$

$$= \frac{1}{1-\sin^2 x}\cos x = \frac{\cos x}{\cos^2 x} = \sec x$$

3.11 EXERCISES

1–6 Find the numerical value of each expression.

1. (a) sinh 0 (b) cosh 0
2. (a) tanh 0 (b) tanh 1
3. (a) cosh(ln 5) (b) cosh 5
4. (a) sinh 4 (b) sinh(ln 4)
5. (a) sech 0 (b) $\cosh^{-1} 1$
6. (a) sinh 1 (b) $\sinh^{-1} 1$

7–19 Prove the identity.

7. $\sinh(-x) = -\sinh x$
   (This shows that sinh is an odd function.)
8. $\cosh(-x) = \cosh x$
   (This shows that cosh is an even function.)
9. $\cosh x + \sinh x = e^x$
10. $\cosh x - \sinh x = e^{-x}$
11. $\sinh(x+y) = \sinh x \cosh y + \cosh x \sinh y$
12. $\cosh(x+y) = \cosh x \cosh y + \sinh x \sinh y$
13. $\coth^2 x - 1 = \csch^2 x$
14. $\tanh(x+y) = \dfrac{\tanh x + \tanh y}{1 + \tanh x \tanh y}$
15. $\sinh 2x = 2 \sinh x \cosh x$
16. $\cosh 2x = \cosh^2 x + \sinh^2 x$
17. $\tanh(\ln x) = \dfrac{x^2 - 1}{x^2 + 1}$
18. $\dfrac{1 + \tanh x}{1 - \tanh x} = e^{2x}$
19. $(\cosh x + \sinh x)^n = \cosh nx + \sinh nx$
    (n any real number)

20. If $\tanh x = \tfrac{12}{13}$, find the values of the other hyperbolic functions at x.

21. If $\cosh x = \tfrac{5}{3}$ and $x > 0$, find the values of the other hyperbolic functions at x.

22. (a) Use the graphs of sinh, cosh, and tanh in Figures 1–3 to draw the graphs of csch, sech, and coth.
    (b) Check the graphs that you sketched in part (a) by using a graphing device to produce them.

23. Use the definitions of the hyperbolic functions to find each of the following limits.
    (a) $\lim\limits_{x \to \infty} \tanh x$
    (b) $\lim\limits_{x \to -\infty} \tanh x$
    (c) $\lim\limits_{x \to \infty} \sinh x$
    (d) $\lim\limits_{x \to -\infty} \sinh x$
    (e) $\lim\limits_{x \to \infty} \sech x$
    (f) $\lim\limits_{x \to \infty} \coth x$
    (g) $\lim\limits_{x \to 0^+} \coth x$
    (h) $\lim\limits_{x \to 0^-} \coth x$
    (i) $\lim\limits_{x \to -\infty} \csch x$
    (j) $\lim\limits_{x \to \infty} \dfrac{\sinh x}{e^x}$

24. Prove the formulas given in Table 1 for the derivatives of the functions (a) cosh, (b) tanh, (c) csch, (d) sech, and (e) coth.

25. Give an alternative solution to Example 3 by letting $y = \sinh^{-1} x$ and then using Exercise 9 and Example 1(a) with x replaced by y.

26. Prove Equation 4.

27. Prove Equation 5 using (a) the method of Example 3 and (b) Exercise 18 with x replaced by y.

28. For each of the following functions (i) give a definition like those in (2), (ii) sketch the graph, and (iii) find a formula similar to Equation 3.
    (a) $\csch^{-1}$ (b) $\sech^{-1}$ (c) $\coth^{-1}$

29. Prove the formulas given in Table 6 for the derivatives of the following functions.
    (a) $\cosh^{-1}$ (b) $\tanh^{-1}$ (c) $\csch^{-1}$
    (d) $\sech^{-1}$ (e) $\coth^{-1}$

30–45 Find the derivative. Simplify where possible.

30. $f(x) = e^x \cosh x$
31. $f(x) = \tanh \sqrt{x}$
32. $g(x) = \sinh^2 x$

33. $h(x) = \sinh(x^2)$ **34.** $F(t) = \ln(\sinh t)$

35. $G(t) = \sinh(\ln t)$

36. $y = \text{sech}\, x\,(1 + \ln \text{sech}\, x)$

37. $y = e^{\cosh 3x}$ **38.** $f(t) = \dfrac{1 + \sinh t}{1 - \sinh t}$

39. $g(t) = t\coth\sqrt{t^2 + 1}$ **40.** $y = \sinh^{-1}(\tan x)$

41. $y = \cosh^{-1}\sqrt{x}$

42. $y = x\tanh^{-1}x + \ln\sqrt{1 - x^2}$

43. $y = x\sinh^{-1}(x/3) - \sqrt{9 + x^2}$

44. $y = \text{sech}^{-1}(e^{-x})$

45. $y = \coth^{-1}(\sec x)$

46. Show that $\dfrac{d}{dx}\sqrt[4]{\dfrac{1 + \tanh x}{1 - \tanh x}} = \tfrac{1}{2}e^{x/2}$.

47. Show that $\dfrac{d}{dx}\arctan(\tanh x) = \text{sech}\, 2x$.

48. The Gateway Arch in St. Louis was designed by Eero Saarinen and was constructed using the equation
$$y = 211.49 - 20.96 \cosh 0.03291765x$$
for the central curve of the arch, where x and y are measured in meters and $|x| \leq 91.20$.
(a) Graph the central curve.
(b) What is the height of the arch at its center?
(c) At what points is the height 100 m?
(d) What is the slope of the arch at the points in part (c)?

49. If a water wave with length L moves with velocity v in a body of water with depth d, then
$$v = \sqrt{\dfrac{gL}{2\pi}\tanh\left(\dfrac{2\pi d}{L}\right)}$$
where g is the acceleration due to gravity. (See Figure 5.) Explain why the approximation
$$v \approx \sqrt{\dfrac{gL}{2\pi}}$$
is appropriate in deep water.

50. A flexible cable always hangs in the shape of a catenary $y = c + a\cosh(x/a)$, where c and a are constants and $a > 0$ (see Figure 4 and Exercise 52). Graph several members of the family of functions $y = a\cosh(x/a)$. How does the graph change as a varies?

51. A telephone line hangs between two poles 14 m apart in the shape of the catenary $y = 20\cosh(x/20) - 15$, where x and y are measured in meters.
(a) Find the slope of this curve where it meets the right pole.
(b) Find the angle θ between the line and the pole.

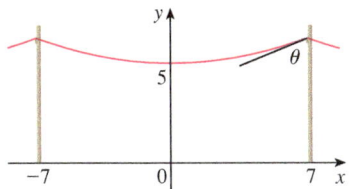

52. Using principles from physics it can be shown that when a cable is hung between two poles, it takes the shape of a curve $y = f(x)$ that satisfies the differential equation
$$\dfrac{d^2y}{dx^2} = \dfrac{\rho g}{T}\sqrt{1 + \left(\dfrac{dy}{dx}\right)^2}$$
where ρ is the linear density of the cable, g is the acceleration due to gravity, T is the tension in the cable at its lowest point, and the coordinate system is chosen appropriately. Verify that the function
$$y = f(x) = \dfrac{T}{\rho g}\cosh\left(\dfrac{\rho g x}{T}\right)$$
is a solution of this differential equation.

53. A cable with linear density $\rho = 2$ kg/m is strung from the tops of two poles that are 200 m apart.
(a) Use Exercise 52 to find the tension T so that the cable is 60 m above the ground at its lowest point. How tall are the poles?
(b) If the tension is doubled, what is the new low point of the cable? How tall are the poles now?

54. A model for the velocity of a falling object after time t is
$$v(t) = \sqrt{\dfrac{mg}{k}}\tanh\left(t\sqrt{\dfrac{gk}{m}}\right)$$
where m is the mass of the object, $g = 9.8$ m/s^2 is the acceleration due to gravity, k is a constant, t is measured in seconds, and v in m/s.
(a) Calculate the terminal velocity of the object, that is, $\lim_{t\to\infty} v(t)$.
(b) If a person falls from a building, the value of the constant k depends on his or her position. For a "belly-to-earth" position, $k = 0.515$ kg/s, but for a "feet-first" position, $k = 0.067$ kg/s. If a 60-kg person falls in belly-to-earth position, what is the terminal velocity? What about feet-first?

Source: L. Long et al., "How Terminal Is Terminal Velocity?" *American Mathematical Monthly* 113 (2006): 752–55.

55. (a) Show that any function of the form
$$y = A\sinh mx + B\cosh mx$$
satisfies the differential equation $y'' = m^2y$.
(b) Find $y = y(x)$ such that $y'' = 9y$, $y(0) = -4$, and $y'(0) = 6$.

56. If $x = \ln(\sec\theta + \tan\theta)$, show that $\sec\theta = \cosh x$.

266   CHAPTER 3   Differentiation Rules

57. At what point of the curve $y = \cosh x$ does the tangent have slope 1?

58. Investigate the family of functions

$$f_n(x) = \tanh(n \sin x)$$

where n is a positive integer. Describe what happens to the graph of f_n when n becomes large.

59. Show that if $a \neq 0$ and $b \neq 0$, then there exist numbers α and β such that $ae^x + be^{-x}$ equals either

$$\alpha \sinh(x + \beta) \quad \text{or} \quad \alpha \cosh(x + \beta)$$

In other words, almost every function of the form $f(x) = ae^x + be^{-x}$ is a shifted and stretched hyperbolic sine or cosine function.

3   REVIEW

CONCEPT CHECK

Answers to the Concept Check can be found on the back endpapers.

1. State each differentiation rule both in symbols and in words.
  (a) The Power Rule   (b) The Constant Multiple Rule
  (c) The Sum Rule   (d) The Difference Rule
  (e) The Product Rule (f) The Quotient Rule
  (g) The Chain Rule

2. State the derivative of each function.
  (a) $y = x^n$ (b) $y = e^x$ (c) $y = b^x$
  (d) $y = \ln x$ (e) $y = \log_b x$ (f) $y = \sin x$
  (g) $y = \cos x$ (h) $y = \tan x$ (i) $y = \csc x$
  (j) $y = \sec x$ (k) $y = \cot x$ (l) $y = \sin^{-1} x$
  (m) $y = \cos^{-1} x$ (n) $y = \tan^{-1} x$ (o) $y = \sinh x$
  (p) $y = \cosh x$ (q) $y = \tanh x$ (r) $y = \sinh^{-1} x$
  (s) $y = \cosh^{-1} x$ (t) $y = \tanh^{-1} x$

3. (a) How is the number e defined?
  (b) Express e as a limit.
  (c) Why is the natural exponential function $y = e^x$ used more often in calculus than the other exponential functions $y = b^x$?

  (d) Why is the natural logarithmic function $y = \ln x$ used more often in calculus than the other logarithmic functions $y = \log_b x$?

4. (a) Explain how implicit differentiation works.
  (b) Explain how logarithmic differentiation works.

5. Give several examples of how the derivative can be interpreted as a rate of change in physics, chemistry, biology, economics, or other sciences.

6. (a) Write a differential equation that expresses the law of natural growth.
  (b) Under what circumstances is this an appropriate model for population growth?
  (c) What are the solutions of this equation?

7. (a) Write an expression for the linearization of f at a.
  (b) If $y = f(x)$, write an expression for the differential dy.
  (c) If $dx = \Delta x$, draw a picture showing the geometric meanings of Δy and dy.

TRUE-FALSE QUIZ

Determine whether the statement is true or false. If it is true, explain why. If it is false, explain why or give an example that disproves the statement.

1. If f and g are differentiable, then

$$\frac{d}{dx}[f(x) + g(x)] = f'(x) + g'(x)$$

2. If f and g are differentiable, then

$$\frac{d}{dx}[f(x)g(x)] = f'(x)g'(x)$$

3. If f and g are differentiable, then

$$\frac{d}{dx}[f(g(x))] = f'(g(x))g'(x)$$

4. If f is differentiable, then $\dfrac{d}{dx}\sqrt{f(x)} = \dfrac{f'(x)}{2\sqrt{f(x)}}$.

5. If f is differentiable, then $\dfrac{d}{dx} f(\sqrt{x}) = \dfrac{f'(x)}{2\sqrt{x}}$.

6. If $y = e^2$, then $y' = 2e$.

7. $\dfrac{d}{dx}(10^x) = x10^{x-1}$

8. $\dfrac{d}{dx}(\ln 10) = \dfrac{1}{10}$

9. $\dfrac{d}{dx}(\tan^2 x) = \dfrac{d}{dx}(\sec^2 x)$

10. $\dfrac{d}{dx}|x^2 + x| = |2x + 1|$

11. The derivative of a polynomial is a polynomial.

12. If $f(x) = (x^6 - x^4)^5$, then $f^{(31)}(x) = 0$.

13. The derivative of a rational function is a rational function.

14. An equation of the tangent line to the parabola $y = x^2$ at $(-2, 4)$ is $y - 4 = 2x(x + 2)$.

15. If $g(x) = x^5$, then $\lim_{x \to 2} \dfrac{g(x) - g(2)}{x - 2} = 80$

EXERCISES

1–50 Calculate y'.

1. $y = (x^2 + x^3)^4$

2. $y = \dfrac{1}{\sqrt{x}} - \dfrac{1}{\sqrt[5]{x^3}}$

3. $y = \dfrac{x^2 - x + 2}{\sqrt{x}}$

4. $y = \dfrac{\tan x}{1 + \cos x}$

5. $y = x^2 \sin \pi x$

6. $y = x \cos^{-1} x$

7. $y = \dfrac{t^4 - 1}{t^4 + 1}$

8. $xe^y = y \sin x$

9. $y = \ln(x \ln x)$

10. $y = e^{mx} \cos nx$

11. $y = \sqrt{x} \cos \sqrt{x}$

12. $y = (\arcsin 2x)^2$

13. $y = \dfrac{e^{1/x}}{x^2}$

14. $y = \ln \sec x$

15. $y + x \cos y = x^2 y$

16. $y = \left(\dfrac{u - 1}{u^2 + u + 1}\right)^4$

17. $y = \sqrt{\arctan x}$

18. $y = \cot(\csc x)$

19. $y = \tan\left(\dfrac{t}{1 + t^2}\right)$

20. $y = e^{x \sec x}$

21. $y = 3^{x \ln x}$

22. $y = \sec(1 + x^2)$

23. $y = (1 - x^{-1})^{-1}$

24. $y = 1/\sqrt[3]{x + \sqrt{x}}$

25. $\sin(xy) = x^2 - y$

26. $y = \sqrt{\sin \sqrt{x}}$

27. $y = \log_5(1 + 2x)$

28. $y = (\cos x)^x$

29. $y = \ln \sin x - \tfrac{1}{2} \sin^2 x$

30. $y = \dfrac{(x^2 + 1)^4}{(2x + 1)^3 (3x - 1)^5}$

31. $y = x \tan^{-1}(4x)$

32. $y = e^{\cos x} + \cos(e^x)$

33. $y = \ln |\sec 5x + \tan 5x|$

34. $y = 10^{\tan \pi \theta}$

35. $y = \cot(3x^2 + 5)$

36. $y = \sqrt{t \ln(t^4)}$

37. $y = \sin(\tan \sqrt{1 + x^3})$

38. $y = \arctan(\arcsin \sqrt{x})$

39. $y = \tan^2(\sin \theta)$

40. $xe^y = y - 1$

41. $y = \dfrac{\sqrt{x + 1}\,(2 - x)^5}{(x + 3)^7}$

42. $y = \dfrac{(x + \lambda)^4}{x^4 + \lambda^4}$

43. $y = x \sinh(x^2)$

44. $y = \dfrac{\sin mx}{x}$

45. $y = \ln(\cosh 3x)$

46. $y = \ln \left| \dfrac{x^2 - 4}{2x + 5} \right|$

47. $y = \cosh^{-1}(\sinh x)$

48. $y = x \tanh^{-1} \sqrt{x}$

49. $y = \cos\left(e^{\sqrt{\tan 3x}}\right)$

50. $y = \sin^2(\cos \sqrt{\sin \pi x})$

51. If $f(t) = \sqrt{4t + 1}$, find $f''(2)$.

52. If $g(\theta) = \theta \sin \theta$, find $g''(\pi/6)$.

53. Find y'' if $x^6 + y^6 = 1$.

54. Find $f^{(n)}(x)$ if $f(x) = 1/(2 - x)$.

55. Use mathematical induction (page 72) to show that if $f(x) = xe^x$, then $f^{(n)}(x) = (x + n)e^x$.

56. Evaluate $\lim_{t \to 0} \dfrac{t^3}{\tan^3(2t)}$.

57–59 Find an equation of the tangent to the curve at the given point.

57. $y = 4 \sin^2 x$, $(\pi/6, 1)$

58. $y = \dfrac{x^2 - 1}{x^2 + 1}$, $(0, -1)$

59. $y = \sqrt{1 + 4 \sin x}$, $(0, 1)$

60–61 Find equations of the tangent line and normal line to the curve at the given point.

60. $x^2 + 4xy + y^2 = 13$, $(2, 1)$

61. $y = (2 + x)e^{-x}$, $(0, 2)$

62. If $f(x) = xe^{\sin x}$, find $f'(x)$. Graph f and f' on the same screen and comment.

63. (a) If $f(x) = x\sqrt{5 - x}$, find $f'(x)$.
(b) Find equations of the tangent lines to the curve $y = x\sqrt{5 - x}$ at the points $(1, 2)$ and $(4, 4)$.
(c) Illustrate part (b) by graphing the curve and tangent lines on the same screen.
(d) Check to see that your answer to part (a) is reasonable by comparing the graphs of f and f'.

64. (a) If $f(x) = 4x - \tan x$, $-\pi/2 < x < \pi/2$, find f' and f''.
(b) Check to see that your answers to part (a) are reasonable by comparing the graphs of f, f', and f''.

65. At what points on the curve $y = \sin x + \cos x$, $0 \leq x \leq 2\pi$, is the tangent line horizontal?

66. Find the points on the ellipse $x^2 + 2y^2 = 1$ where the tangent line has slope 1.

67. If $f(x) = (x-a)(x-b)(x-c)$, show that
$$\frac{f'(x)}{f(x)} = \frac{1}{x-a} + \frac{1}{x-b} + \frac{1}{x-c}$$

68. (a) By differentiating the double-angle formula
$$\cos 2x = \cos^2 x - \sin^2 x$$
obtain the double-angle formula for the sine function.
(b) By differentiating the addition formula
$$\sin(x+a) = \sin x \cos a + \cos x \sin a$$
obtain the addition formula for the cosine function.

69. Suppose that
$$f(1) = 2 \quad f'(1) = 3 \quad f(2) = 1 \quad f'(2) = 2$$
$$g(1) = 3 \quad g'(1) = 1 \quad g(2) = 1 \quad g'(2) = 4$$
(a) If $S(x) = f(x) + g(x)$, find $S'(1)$.
(b) If $P(x) = f(x)g(x)$, find $P'(2)$.
(c) If $Q(x) = f(x)/g(x)$, find $Q'(1)$.
(d) If $C(x) = f(g(x))$, find $C'(2)$.

70. If f and g are the functions whose graphs are shown, let $P(x) = f(x)g(x)$, $Q(x) = f(x)/g(x)$, and $C(x) = f(g(x))$. Find (a) $P'(2)$, (b) $Q'(2)$, and (c) $C'(2)$.

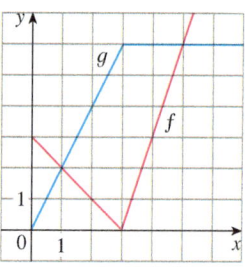

71–78 Find f' in terms of g'.

71. $f(x) = x^2 g(x)$

72. $f(x) = g(x^2)$

73. $f(x) = [g(x)]^2$

74. $f(x) = g(g(x))$

75. $f(x) = g(e^x)$

76. $f(x) = e^{g(x)}$

77. $f(x) = \ln|g(x)|$

78. $f(x) = g(\ln x)$

79–81 Find h' in terms of f' and g'.

79. $h(x) = \dfrac{f(x)g(x)}{f(x) + g(x)}$

80. $h(x) = \sqrt{\dfrac{f(x)}{g(x)}}$

81. $h(x) = f(g(\sin 4x))$

82. (a) Graph the function $f(x) = x - 2\sin x$ in the viewing rectangle $[0, 8]$ by $[-2, 8]$.
(b) On which interval is the average rate of change larger: $[1, 2]$ or $[2, 3]$?
(c) At which value of x is the instantaneous rate of change larger: $x = 2$ or $x = 5$?
(d) Check your visual estimates in part (c) by computing $f'(x)$ and comparing the numerical values of $f'(2)$ and $f'(5)$.

83. At what point on the curve $y = [\ln(x+4)]^2$ is the tangent horizontal?

84. (a) Find an equation of the tangent to the curve $y = e^x$ that is parallel to the line $x - 4y = 1$.
(b) Find an equation of the tangent to the curve $y = e^x$ that passes through the origin.

85. Find a parabola $y = ax^2 + bx + c$ that passes through the point $(1, 4)$ and whose tangent lines at $x = -1$ and $x = 5$ have slopes 6 and -2, respectively.

86. The function $C(t) = K(e^{-at} - e^{-bt})$, where a, b, and K are positive constants and $b > a$, is used to model the concentration at time t of a drug injected into the bloodstream.
(a) Show that $\lim_{t \to \infty} C(t) = 0$.
(b) Find $C'(t)$, the rate of change of drug concentration in the blood.
(c) When is this rate equal to 0?

87. An equation of motion of the form $s = Ae^{-ct}\cos(\omega t + \delta)$ represents damped oscillation of an object. Find the velocity and acceleration of the object.

88. A particle moves along a horizontal line so that its coordinate at time t is $x = \sqrt{b^2 + c^2 t^2}$, $t \geq 0$, where b and c are positive constants.
(a) Find the velocity and acceleration functions.
(b) Show that the particle always moves in the positive direction.

89. A particle moves on a vertical line so that its coordinate at time t is $y = t^3 - 12t + 3$, $t \geq 0$.
(a) Find the velocity and acceleration functions.
(b) When is the particle moving upward and when is it moving downward?
(c) Find the distance that the particle travels in the time interval $0 \leq t \leq 3$.
(d) Graph the position, velocity, and acceleration functions for $0 \leq t \leq 3$.
(e) When is the particle speeding up? When is it slowing down?

90. The volume of a right circular cone is $V = \frac{1}{3}\pi r^2 h$, where r is the radius of the base and h is the height.
(a) Find the rate of change of the volume with respect to the height if the radius is constant.

(b) Find the rate of change of the volume with respect to the radius if the height is constant.

91. The mass of part of a wire is $x(1 + \sqrt{x})$ kilograms, where x is measured in meters from one end of the wire. Find the linear density of the wire when $x = 4$ m.

92. The cost, in dollars, of producing x units of a certain commodity is
$$C(x) = 920 + 2x - 0.02x^2 + 0.00007x^3$$
(a) Find the marginal cost function.
(b) Find $C'(100)$ and explain its meaning.
(c) Compare $C'(100)$ with the cost of producing the 101st item.

93. A bacteria culture contains 200 cells initially and grows at a rate proportional to its size. After half an hour the population has increased to 360 cells.
(a) Find the number of bacteria after t hours.
(b) Find the number of bacteria after 4 hours.
(c) Find the rate of growth after 4 hours.
(d) When will the population reach 10,000?

94. Cobalt-60 has a half-life of 5.24 years.
(a) Find the mass that remains from a 100-mg sample after 20 years.
(b) How long would it take for the mass to decay to 1 mg?

95. Let $C(t)$ be the concentration of a drug in the bloodstream. As the body eliminates the drug, $C(t)$ decreases at a rate that is proportional to the amount of the drug that is present at the time. Thus $C'(t) = -kC(t)$, where k is a positive number called the *elimination constant* of the drug.
(a) If C_0 is the concentration at time $t = 0$, find the concentration at time t.
(b) If the body eliminates half the drug in 30 hours, how long does it take to eliminate 90% of the drug?

96. A cup of hot chocolate has temperature 80°C in a room kept at 20°C. After half an hour the hot chocolate cools to 60°C.
(a) What is the temperature of the chocolate after another half hour?
(b) When will the chocolate have cooled to 40°C?

97. The volume of a cube is increasing at a rate of 10 cm³/min. How fast is the surface area increasing when the length of an edge is 30 cm?

98. A paper cup has the shape of a cone with height 10 cm and radius 3 cm (at the top). If water is poured into the cup at a rate of 2 cm³/s, how fast is the water level rising when the water is 5 cm deep?

99. A balloon is rising at a constant speed of 5 ft/s. A boy is cycling along a straight road at a speed of 15 ft/s. When he passes under the balloon, it is 45 ft above him. How fast is the distance between the boy and the balloon increasing 3 s later?

100. A waterskier skis over the ramp shown in the figure at a speed of 30 ft/s. How fast is she rising as she leaves the ramp?

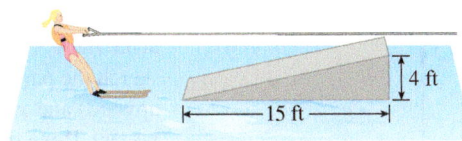

101. The angle of elevation of the sun is decreasing at a rate of 0.25 rad/h. How fast is the shadow cast by a 400-ft-tall building increasing when the angle of elevation of the sun is $\pi/6$?

102. (a) Find the linear approximation to $f(x) = \sqrt{25 - x^2}$ near 3.
(b) Illustrate part (a) by graphing f and the linear approximation.
(c) For what values of x is the linear approximation accurate to within 0.1?

103. (a) Find the linearization of $f(x) = \sqrt[3]{1 + 3x}$ at $a = 0$. State the corresponding linear approximation and use it to give an approximate value for $\sqrt[3]{1.03}$.
(b) Determine the values of x for which the linear approximation given in part (a) is accurate to within 0.1.

104. Evaluate dy if $y = x^3 - 2x^2 + 1$, $x = 2$, and $dx = 0.2$.

105. A window has the shape of a square surmounted by a semicircle. The base of the window is measured as having width 60 cm with a possible error in measurement of 0.1 cm. Use differentials to estimate the maximum error possible in computing the area of the window.

106–108 Express the limit as a derivative and evaluate.

106. $\displaystyle\lim_{x \to 1} \frac{x^{17} - 1}{x - 1}$

107. $\displaystyle\lim_{h \to 0} \frac{\sqrt[4]{16 + h} - 2}{h}$

108. $\displaystyle\lim_{\theta \to \pi/3} \frac{\cos \theta - 0.5}{\theta - \pi/3}$

109. Evaluate $\displaystyle\lim_{x \to 0} \frac{\sqrt{1 + \tan x} - \sqrt{1 + \sin x}}{x^3}$.

110. Suppose f is a differentiable function such that $f(g(x)) = x$ and $f'(x) = 1 + [f(x)]^2$. Show that $g'(x) = 1/(1 + x^2)$.

111. Find $f'(x)$ if it is known that
$$\frac{d}{dx}[f(2x)] = x^2$$

112. Show that the length of the portion of any tangent line to the astroid $x^{2/3} + y^{2/3} = a^{2/3}$ cut off by the coordinate axes is constant.

Problems Plus

Before you look at the examples, cover up the solutions and try them yourself first.

EXAMPLE 1 How many lines are tangent to both of the parabolas $y = -1 - x^2$ and $y = 1 + x^2$? Find the coordinates of the points at which these tangents touch the parabolas.

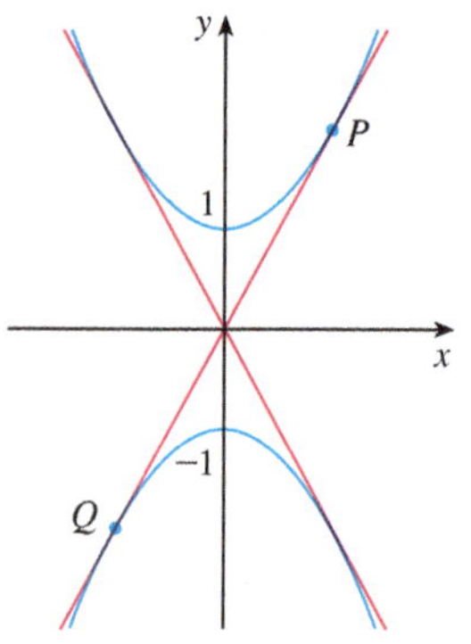

SOLUTION To gain insight into this problem, it is essential to draw a diagram. So we sketch the parabolas $y = 1 + x^2$ (which is the standard parabola $y = x^2$ shifted 1 unit upward) and $y = -1 - x^2$ (which is obtained by reflecting the first parabola about the x-axis). If we try to draw a line tangent to both parabolas, we soon discover that there are only two possibilities, as illustrated in Figure 1.

Let P be a point at which one of these tangents touches the upper parabola and let a be its x-coordinate. (The choice of notation for the unknown is important. Of course we could have used b or c or x_0 or x_1 instead of a. However, it's not advisable to use x in place of a because that x could be confused with the variable x in the equation of the parabola.) Then, since P lies on the parabola $y = 1 + x^2$, its y-coordinate must be $1 + a^2$. Because of the symmetry shown in Figure 1, the coordinates of the point Q where the tangent touches the lower parabola must be $(-a, -(1 + a^2))$.

To use the given information that the line is a tangent, we equate the slope of the line PQ to the slope of the tangent line at P. We have

$$m_{PQ} = \frac{1 + a^2 - (-1 - a^2)}{a - (-a)} = \frac{1 + a^2}{a}$$

If $f(x) = 1 + x^2$, then the slope of the tangent line at P is $f'(a) = 2a$. Thus the condition that we need to use is that

$$\frac{1 + a^2}{a} = 2a$$

Solving this equation, we get $1 + a^2 = 2a^2$, so $a^2 = 1$ and $a = \pm 1$. Therefore the points are $(1, 2)$ and $(-1, -2)$. By symmetry, the two remaining points are $(-1, 2)$ and $(1, -2)$. ■

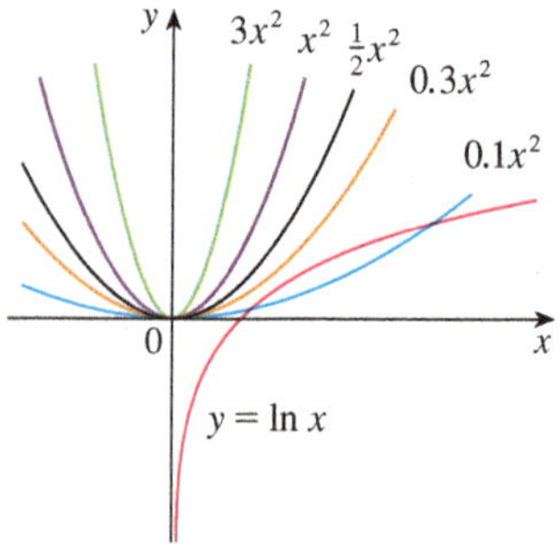

EXAMPLE 2 For what values of c does the equation $\ln x = cx^2$ have exactly one solution?

SOLUTION One of the most important principles of problem solving is to draw a diagram, even if the problem as stated doesn't explicitly mention a geometric situation. Our present problem can be reformulated geometrically as follows: For what values of c does the curve $y = \ln x$ intersect the curve $y = cx^2$ in exactly one point?

Let's start by graphing $y = \ln x$ and $y = cx^2$ for various values of c. We know that, for $c \neq 0$, $y = cx^2$ is a parabola that opens upward if $c > 0$ and downward if $c < 0$. Figure 2 shows the parabolas $y = cx^2$ for several positive values of c. Most of them don't intersect $y = \ln x$ at all and one intersects twice. We have the feeling that there must be a value of c (somewhere between 0.1 and 0.3) for which the curves intersect exactly once, as in Figure 3.

To find that particular value of c, we let a be the x-coordinate of the single point of intersection. In other words, $\ln a = ca^2$, so a is the unique solution of the given equation. We see from Figure 3 that the curves just touch, so they have a common tangent line when $x = a$. That means the curves $y = \ln x$ and $y = cx^2$ have the same slope when $x = a$. Therefore

$$\frac{1}{a} = 2ca$$

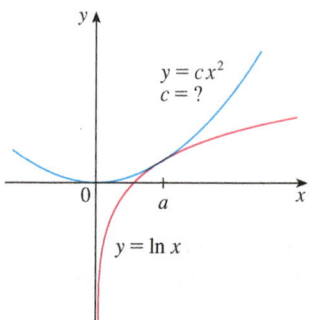

FIGURE 1

FIGURE 2

FIGURE 3

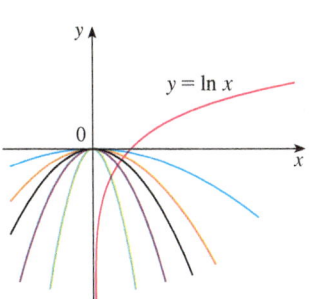

FIGURE 4

Solving the equations $\ln a = ca^2$ and $1/a = 2ca$, we get

$$\ln a = ca^2 = c \cdot \frac{1}{2c} = \frac{1}{2}$$

Thus $a = e^{1/2}$ and

$$c = \frac{\ln a}{a^2} = \frac{\ln e^{1/2}}{e} = \frac{1}{2e}$$

For negative values of c we have the situation illustrated in Figure 4: All parabolas $y = cx^2$ with negative values of c intersect $y = \ln x$ exactly once. And let's not forget about $c = 0$: The curve $y = 0x^2 = 0$ is just the x-axis, which intersects $y = \ln x$ exactly once.

To summarize, the required values of c are $c = 1/(2e)$ and $c \leq 0$. ∎

Problems

1. Find points P and Q on the parabola $y = 1 - x^2$ so that the triangle ABC formed by the x-axis and the tangent lines at P and Q is an equilateral triangle. (See the figure.)

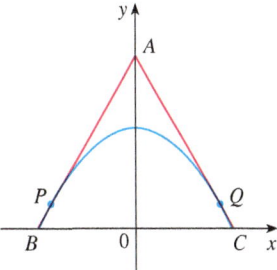

2. Find the point where the curves $y = x^3 - 3x + 4$ and $y = 3(x^2 - x)$ are tangent to each other, that is, have a common tangent line. Illustrate by sketching both curves and the common tangent.

3. Show that the tangent lines to the parabola $y = ax^2 + bx + c$ at any two points with x-coordinates p and q must intersect at a point whose x-coordinate is halfway between p and q.

4. Show that

$$\frac{d}{dx}\left(\frac{\sin^2 x}{1 + \cot x} + \frac{\cos^2 x}{1 + \tan x}\right) = -\cos 2x$$

5. If $f(x) = \lim\limits_{t \to x} \dfrac{\sec t - \sec x}{t - x}$, find the value of $f'(\pi/4)$.

6. Find the values of the constants a and b such that

$$\lim_{x \to 0} \frac{\sqrt[3]{ax + b} - 2}{x} = \frac{5}{12}$$

7. Show that $\sin^{-1}(\tanh x) = \tan^{-1}(\sinh x)$.

8. A car is traveling at night along a highway shaped like a parabola with its vertex at the origin (see the figure). The car starts at a point 100 m west and 100 m north of the origin and travels in an easterly direction. There is a statue located 100 m east and 50 m north of the origin. At what point on the highway will the car's headlights illuminate the statue?

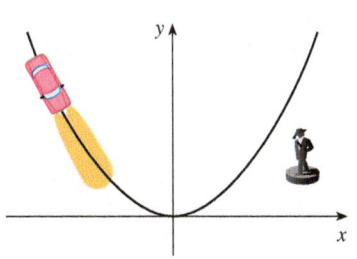

FIGURE FOR PROBLEM 8

9. Prove that $\dfrac{d^n}{dx^n}(\sin^4 x + \cos^4 x) = 4^{n-1}\cos(4x + n\pi/2)$.

10. If f is differentiable at a, where $a > 0$, evaluate the following limit in terms of $f'(a)$:

$$\lim_{x \to a} \frac{f(x) - f(a)}{\sqrt{x} - \sqrt{a}}$$

11. The figure shows a circle with radius 1 inscribed in the parabola $y = x^2$. Find the center of the circle.

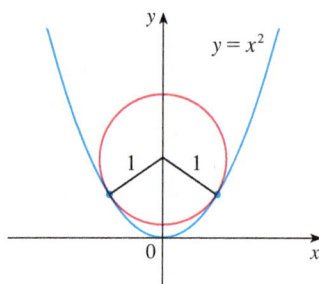

12. Find all values of c such that the parabolas $y = 4x^2$ and $x = c + 2y^2$ intersect each other at right angles.

13. How many lines are tangent to both of the circles $x^2 + y^2 = 4$ and $x^2 + (y-3)^2 = 1$? At what points do these tangent lines touch the circles?

14. If $f(x) = \dfrac{x^{46} + x^{45} + 2}{1 + x}$, calculate $f^{(46)}(3)$. Express your answer using factorial notation: $n! = 1 \cdot 2 \cdot 3 \cdot \cdots \cdot (n-1) \cdot n$.

15. The figure shows a rotating wheel with radius 40 cm and a connecting rod AP with length 1.2 m. The pin P slides back and forth along the x-axis as the wheel rotates counter-clockwise at a rate of 360 revolutions per minute.
(a) Find the angular velocity of the connecting rod, $d\alpha/dt$, in radians per second, when $\theta = \pi/3$.
(b) Express the distance $x = |OP|$ in terms of θ.
(c) Find an expression for the velocity of the pin P in terms of θ.

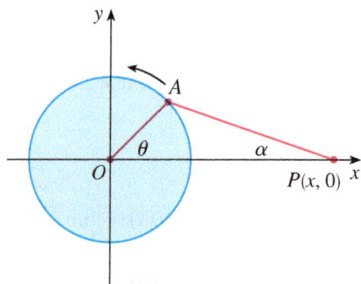

16. Tangent lines T_1 and T_2 are drawn at two points P_1 and P_2 on the parabola $y = x^2$ and they intersect at a point P. Another tangent line T is drawn at a point between P_1 and P_2; it intersects T_1 at Q_1 and T_2 at Q_2. Show that

$$\frac{|PQ_1|}{|PP_1|} + \frac{|PQ_2|}{|PP_2|} = 1$$

17. Show that

$$\frac{d^n}{dx^n}(e^{ax} \sin bx) = r^n e^{ax} \sin(bx + n\theta)$$

where a and b are positive numbers, $r^2 = a^2 + b^2$, and $\theta = \tan^{-1}(b/a)$.

18. Evaluate $\lim\limits_{x \to \pi} \dfrac{e^{\sin x} - 1}{x - \pi}$.

19. Let T and N be the tangent and normal lines to the ellipse $x^2/9 + y^2/4 = 1$ at any point P on the ellipse in the first quadrant. Let x_T and y_T be the x- and y-intercepts of T and x_N and y_N be the intercepts of N. As P moves along the ellipse in the first quadrant (but not on the axes), what values can x_T, y_T, x_N, and y_N take on? First try to guess the answers just by looking at the figure. Then use calculus to solve the problem and see how good your intuition is.

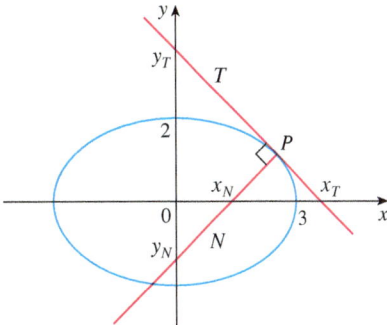

20. Evaluate $\lim\limits_{x \to 0} \dfrac{\sin(3 + x)^2 - \sin 9}{x}$.

21. (a) Use the identity for $\tan(x - y)$ (see Equation 14b in Appendix D) to show that if two lines L_1 and L_2 intersect at an angle α, then

$$\tan \alpha = \frac{m_2 - m_1}{1 + m_1 m_2}$$

where m_1 and m_2 are the slopes of L_1 and L_2, respectively.

(b) The **angle between the curves** C_1 and C_2 at a point of intersection P is defined to be the angle between the tangent lines to C_1 and C_2 at P (if these tangent lines exist). Use part (a) to find, correct to the nearest degree, the angle between each pair of curves at each point of intersection.
 (i) $y = x^2$ and $y = (x - 2)^2$
 (ii) $x^2 - y^2 = 3$ and $x^2 - 4x + y^2 + 3 = 0$

22. Let $P(x_1, y_1)$ be a point on the parabola $y^2 = 4px$ with focus $F(p, 0)$. Let α be the angle between the parabola and the line segment FP, and let β be the angle between the horizontal line $y = y_1$ and the parabola as in the figure. Prove that $\alpha = \beta$. (Thus, by a principle of geometrical optics, light from a source placed at F will be reflected along a line parallel to the x-axis. This explains why *paraboloids,* the surfaces obtained by rotating parabolas about their axes, are used as the shape of some automobile headlights and mirrors for telescopes.)

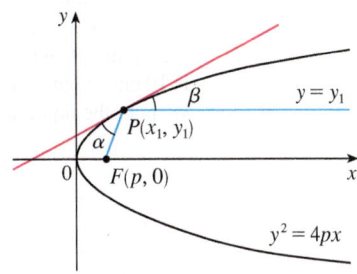

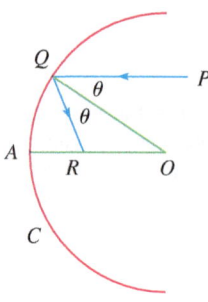

FIGURE FOR PROBLEM 23

23. Suppose that we replace the parabolic mirror of Problem 22 by a spherical mirror. Although the mirror has no focus, we can show the existence of an *approximate* focus. In the figure, C is a semicircle with center O. A ray of light coming in toward the mirror parallel to the axis along the line PQ will be reflected to the point R on the axis so that $\angle PQO = \angle OQR$ (the angle of incidence is equal to the angle of reflection). What happens to the point R as P is taken closer and closer to the axis?

24. If f and g are differentiable functions with $f(0) = g(0) = 0$ and $g'(0) \neq 0$, show that
$$\lim_{x \to 0} \frac{f(x)}{g(x)} = \frac{f'(0)}{g'(0)}$$

25. Evaluate $\displaystyle\lim_{x \to 0} \frac{\sin(a + 2x) - 2\sin(a + x) + \sin a}{x^2}$.

26. (a) The cubic function $f(x) = x(x - 2)(x - 6)$ has three distinct zeros: 0, 2, and 6. Graph f and its tangent lines at the *average* of each pair of zeros. What do you notice?
    (b) Suppose the cubic function $f(x) = (x - a)(x - b)(x - c)$ has three distinct zeros: a, b, and c. Prove, with the help of a computer algebra system, that a tangent line drawn at the average of the zeros a and b intersects the graph of f at the third zero.

27. For what value of k does the equation $e^{2x} = k\sqrt{x}$ have exactly one solution?

28. For which positive numbers a is it true that $a^x \geq 1 + x$ for all x?

29. If
$$y = \frac{x}{\sqrt{a^2 - 1}} - \frac{2}{\sqrt{a^2 - 1}} \arctan \frac{\sin x}{a + \sqrt{a^2 - 1} + \cos x}$$
show that $y' = \dfrac{1}{a + \cos x}$.

30. Given an ellipse $x^2/a^2 + y^2/b^2 = 1$, where $a \neq b$, find the equation of the set of all points from which there are two tangents to the curve whose slopes are (a) reciprocals and (b) negative reciprocals.

31. Find the two points on the curve $y = x^4 - 2x^2 - x$ that have a common tangent line.

32. Suppose that three points on the parabola $y = x^2$ have the property that their normal lines intersect at a common point. Show that the sum of their x-coordinates is 0.

33. A *lattice point* in the plane is a point with integer coordinates. Suppose that circles with radius r are drawn using all lattice points as centers. Find the smallest value of r such that any line with slope $\frac{2}{5}$ intersects some of these circles.

34. A cone of radius r centimeters and height h centimeters is lowered point first at a rate of 1 cm/s into a tall cylinder of radius R centimeters that is partially filled with water. How fast is the water level rising at the instant the cone is completely submerged?

35. A container in the shape of an inverted cone has height 16 cm and radius 5 cm at the top. It is partially filled with a liquid that oozes through the sides at a rate proportional to the area of the container that is in contact with the liquid. (The surface area of a cone is $\pi r l$, where r is the radius and l is the slant height.) If we pour the liquid into the container at a rate of 2 cm³/min, then the height of the liquid decreases at a rate of 0.3 cm/min when the height is 10 cm. If our goal is to keep the liquid at a constant height of 10 cm, at what rate should we pour the liquid into the container?

4 Applications of Differentiation

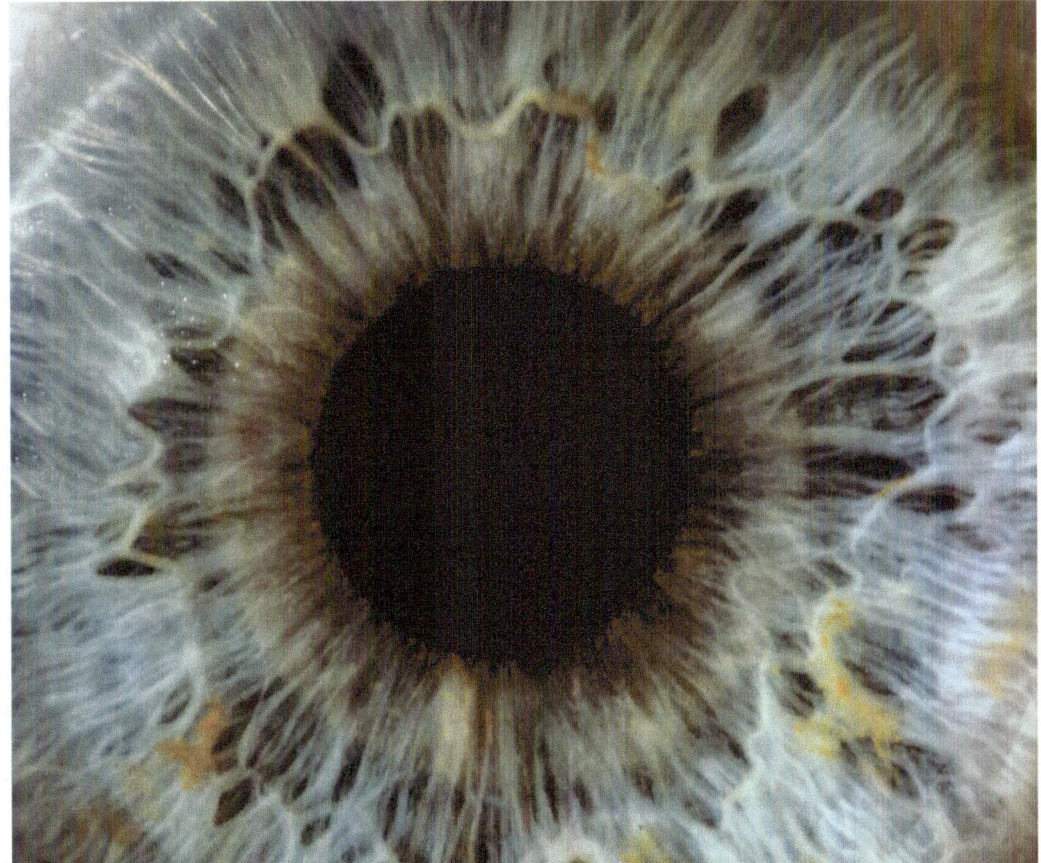

When we view the world around us, the light entering the eye near the center of the pupil is perceived brighter than light entering closer to the edges of the pupil. This phenomenon, known as the *Stiles–Crawford effect*, is explored as the pupil changes in radius in Exercise 80 on page 313.

WE HAVE ALREADY INVESTIGATED SOME of the applications of derivatives, but now that we know the differentiation rules we are in a better position to pursue the applications of differentiation in greater depth. Here we learn how derivatives affect the shape of a graph of a function and, in particular, how they help us locate maximum and minimum values of functions. Many practical problems require us to minimize a cost or maximize an area or somehow find the best possible outcome of a situation. In particular, we will be able to investigate the optimal shape of a can and to explain the location of rainbows in the sky.

4.1 Maximum and Minimum Values

Some of the most important applications of differential calculus are *optimization problems*, in which we are required to find the optimal (best) way of doing something. Here are examples of such problems that we will solve in this chapter:

- What is the shape of a can that minimizes manufacturing costs?
- What is the maximum acceleration of a space shuttle? (This is an important question to the astronauts who have to withstand the effects of acceleration.)
- What is the radius of a contracted windpipe that expels air most rapidly during a cough?
- At what angle should blood vessels branch so as to minimize the energy expended by the heart in pumping blood?

These problems can be reduced to finding the maximum or minimum values of a function. Let's first explain exactly what we mean by maximum and minimum values.

We see that the highest point on the graph of the function f shown in Figure 1 is the point $(3, 5)$. In other words, the largest value of f is $f(3) = 5$. Likewise, the smallest value is $f(6) = 2$. We say that $f(3) = 5$ is the *absolute maximum* of f and $f(6) = 2$ is the *absolute minimum*. In general, we use the following definition.

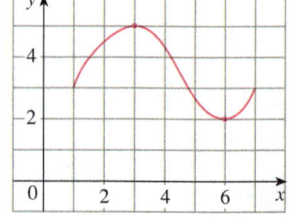

FIGURE 1

1 Definition Let c be a number in the domain D of a function f. Then $f(c)$ is the

- **absolute maximum** value of f on D if $f(c) \geq f(x)$ for all x in D.
- **absolute minimum** value of f on D if $f(c) \leq f(x)$ for all x in D.

An absolute maximum or minimum is sometimes called a **global** maximum or minimum. The maximum and minimum values of f are called **extreme values** of f.

Figure 2 shows the graph of a function f with absolute maximum at d and absolute minimum at a. Note that $(d, f(d))$ is the highest point on the graph and $(a, f(a))$ is the lowest point. In Figure 2, if we consider only values of x near b [for instance, if we restrict our attention to the interval (a, c)], then $f(b)$ is the largest of those values of $f(x)$ and is called a *local maximum value* of f. Likewise, $f(c)$ is called a *local minimum value* of f because $f(c) \leq f(x)$ for x near c [in the interval (b, d), for instance]. The function f also has a local minimum at e. In general, we have the following definition.

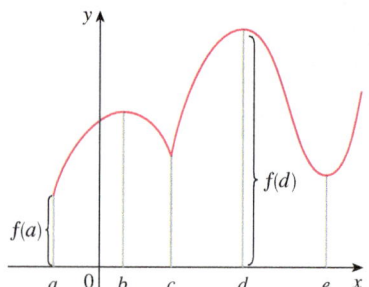

FIGURE 2
Abs min $f(a)$, abs max $f(d)$,
loc min $f(c), f(e)$, loc max $f(b), f(d)$

2 Definition The number $f(c)$ is a

- **local maximum** value of f if $f(c) \geq f(x)$ when x is near c.
- **local minimum** value of f if $f(c) \leq f(x)$ when x is near c.

In Definition 2 (and elsewhere), if we say that something is true **near** c, we mean that it is true on some open interval containing c. For instance, in Figure 3 we see that $f(4) = 5$ is a local minimum because it's the smallest value of f on the interval I. It's not the absolute minimum because $f(x)$ takes smaller values when x is near 12 (in the interval K, for instance). In fact $f(12) = 3$ is both a local minimum and the absolute minimum. Similarly, $f(8) = 7$ is a local maximum, but not the absolute maximum because f takes larger values near 1.

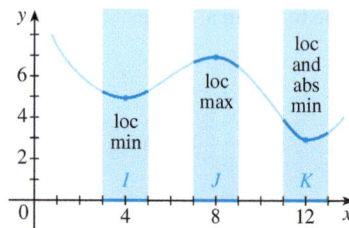

FIGURE 3

SECTION 4.1 Maximum and Minimum Values 277

EXAMPLE 1 The function $f(x) = \cos x$ takes on its (local and absolute) maximum value of 1 infinitely many times, since $\cos 2n\pi = 1$ for any integer n and $-1 \leq \cos x \leq 1$ for all x. (See Figure 4.) Likewise, $\cos(2n + 1)\pi = -1$ is its minimum value, where n is any integer.

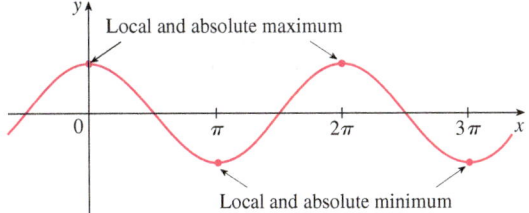

FIGURE 4
$y = \cos x$

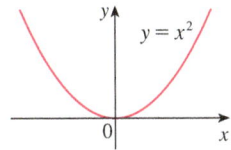

FIGURE 5
Mimimum value 0, no maximum

EXAMPLE 2 If $f(x) = x^2$, then $f(x) \geq f(0)$ because $x^2 \geq 0$ for all x. Therefore $f(0) = 0$ is the absolute (and local) minimum value of f. This corresponds to the fact that the origin is the lowest point on the parabola $y = x^2$. (See Figure 5.) However, there is no highest point on the parabola and so this function has no maximum value.

EXAMPLE 3 From the graph of the function $f(x) = x^3$, shown in Figure 6, we see that this function has neither an absolute maximum value nor an absolute minimum value. In fact, it has no local extreme values either.

EXAMPLE 4 The graph of the function

$$f(x) = 3x^4 - 16x^3 + 18x^2 \qquad -1 \leq x \leq 4$$

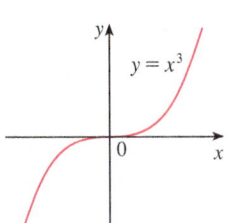

FIGURE 6
No mimimum, no maximum

is shown in Figure 7. You can see that $f(1) = 5$ is a local maximum, whereas the absolute maximum is $f(-1) = 37$. (This absolute maximum is not a local maximum because it occurs at an endpoint.) Also, $f(0) = 0$ is a local minimum and $f(3) = -27$ is both a local and an absolute minimum. Note that f has neither a local nor an absolute maximum at $x = 4$.

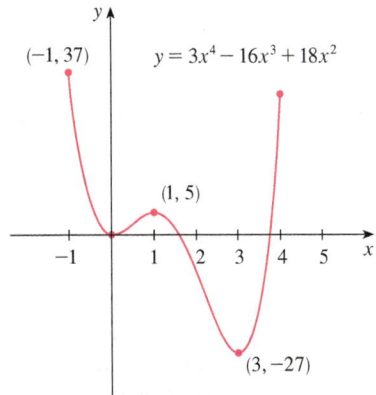

FIGURE 7

We have seen that some functions have extreme values, whereas others do not. The following theorem gives conditions under which a function is guaranteed to possess extreme values.

3 The Extreme Value Theorem If f is continuous on a closed interval $[a, b]$, then f attains an absolute maximum value $f(c)$ and an absolute minimum value $f(d)$ at some numbers c and d in $[a, b]$.

The Extreme Value Theorem is illustrated in Figure 8. Note that an extreme value can be taken on more than once. Although the Extreme Value Theorem is intuitively very plausible, it is difficult to prove and so we omit the proof.

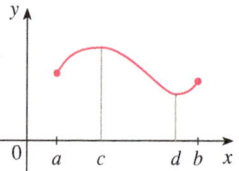 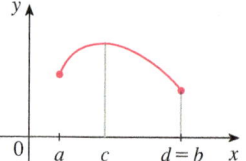 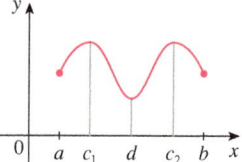

FIGURE 8 Functions continuous on a closed interval always attain extreme values.

Figures 9 and 10 show that a function need not possess extreme values if either hypothesis (continuity or closed interval) is omitted from the Extreme Value Theorem.

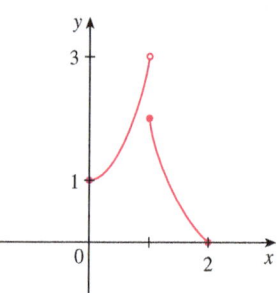 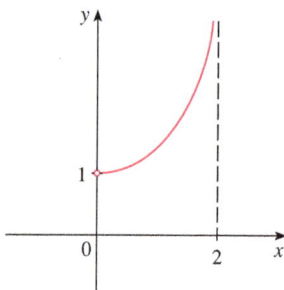

FIGURE 9
This function has minimum value $f(2) = 0$, but no maximum value.

FIGURE 10
This continuous function g has no maximum or minimum.

The function f whose graph is shown in Figure 9 is defined on the closed interval $[0, 2]$ but has no maximum value. (Notice that the range of f is $[0, 3)$. The function takes on values arbitrarily close to 3, but never actually attains the value 3.) This does not contradict the Extreme Value Theorem because f is not continuous. [Nonetheless, a discontinuous function *could* have maximum and minimum values. See Exercise 13(b).]

The function g shown in Figure 10 is continuous on the open interval $(0, 2)$ but has neither a maximum nor a minimum value. [The range of g is $(1, \infty)$. The function takes on arbitrarily large values.] This does not contradict the Extreme Value Theorem because the interval $(0, 2)$ is not closed.

The Extreme Value Theorem says that a continuous function on a closed interval has a maximum value and a minimum value, but it does not tell us how to find these extreme values. Notice in Figure 8 that the absolute maximum and minimum values that are *between* a and b occur at local maximum or minimum values, so we start by looking for local extreme values.

Figure 11 shows the graph of a function f with a local maximum at c and a local minimum at d. It appears that at the maximum and minimum points the tangent lines are horizontal and therefore each has slope 0. We know that the derivative is the slope of the

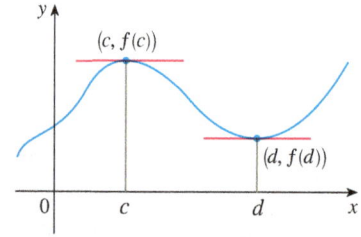

FIGURE 11

tangent line, so it appears that $f'(c) = 0$ and $f'(d) = 0$. The following theorem says that this is always true for differentiable functions.

> **[4] Fermat's Theorem** If f has a local maximum or minimum at c, and if $f'(c)$ exists, then $f'(c) = 0$.

Fermat

Fermat's Theorem is named after Pierre Fermat (1601–1665), a French lawyer who took up mathematics as a hobby. Despite his amateur status, Fermat was one of the two inventors of analytic geometry (Descartes was the other). His methods for finding tangents to curves and maximum and minimum values (before the invention of limits and derivatives) made him a forerunner of Newton in the creation of differential calculus.

PROOF Suppose, for the sake of definiteness, that f has a local maximum at c. Then, according to Definition 2, $f(c) \geq f(x)$ if x is sufficiently close to c. This implies that if h is sufficiently close to 0, with h being positive or negative, then

$$f(c) \geq f(c + h)$$

and therefore

[5]
$$f(c + h) - f(c) \leq 0$$

We can divide both sides of an inequality by a positive number. Thus, if $h > 0$ and h is sufficiently small, we have

$$\frac{f(c + h) - f(c)}{h} \leq 0$$

Taking the right-hand limit of both sides of this inequality (using Theorem 2.3.2), we get

$$\lim_{h \to 0^+} \frac{f(c + h) - f(c)}{h} \leq \lim_{h \to 0^+} 0 = 0$$

But since $f'(c)$ exists, we have

$$f'(c) = \lim_{h \to 0} \frac{f(c + h) - f(c)}{h} = \lim_{h \to 0^+} \frac{f(c + h) - f(c)}{h}$$

and so we have shown that $f'(c) \leq 0$.

If $h < 0$, then the direction of the inequality (5) is reversed when we divide by h:

$$\frac{f(c + h) - f(c)}{h} \geq 0 \qquad h < 0$$

So, taking the left-hand limit, we have

$$f'(c) = \lim_{h \to 0} \frac{f(c + h) - f(c)}{h} = \lim_{h \to 0^-} \frac{f(c + h) - f(c)}{h} \geq 0$$

We have shown that $f'(c) \geq 0$ and also that $f'(c) \leq 0$. Since both of these inequalities must be true, the only possibility is that $f'(c) = 0$.

We have proved Fermat's Theorem for the case of a local maximum. The case of a local minimum can be proved in a similar manner, or we could use Exercise 78 to deduce it from the case we have just proved (see Exercise 79). ∎

The following examples caution us against reading too much into Fermat's Theorem: We can't expect to locate extreme values simply by setting $f'(x) = 0$ and solving for x.

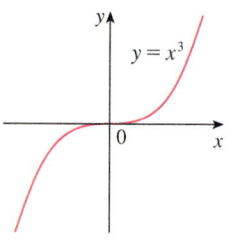

FIGURE 12
If $f(x) = x^3$, then $f'(0) = 0$, but f has no maximum or minimum.

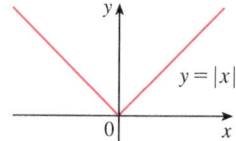

FIGURE 13
If $f(x) = |x|$, then $f(0) = 0$ is a minimum value, but $f'(0)$ does not exist.

Figure 14 shows a graph of the function f in Example 7. It supports our answer because there is a horizontal tangent when $x = 1.5$ [where $f'(x) = 0$] and a vertical tangent when $x = 0$ [where $f'(x)$ is undefined].

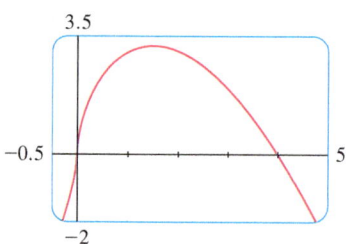

FIGURE 14

EXAMPLE 5 If $f(x) = x^3$, then $f'(x) = 3x^2$, so $f'(0) = 0$. But f has no maximum or minimum at 0, as you can see from its graph in Figure 12. (Or observe that $x^3 > 0$ for $x > 0$ but $x^3 < 0$ for $x < 0$.) The fact that $f'(0) = 0$ simply means that the curve $y = x^3$ has a horizontal tangent at $(0, 0)$. Instead of having a maximum or minimum at $(0, 0)$, the curve crosses its horizontal tangent there. ∎

EXAMPLE 6 The function $f(x) = |x|$ has its (local and absolute) minimum value at 0, but that value can't be found by setting $f'(x) = 0$ because, as was shown in Example 2.8.5, $f'(0)$ does not exist. (See Figure 13.) ∎

⊘ **WARNING** Examples 5 and 6 show that we must be careful when using Fermat's Theorem. Example 5 demonstrates that *even when $f'(c) = 0$ there need not be a maximum or minimum at c.* (In other words, the converse of Fermat's Theorem is false in general.) Furthermore, *there may be an extreme value even when $f'(c)$ does not exist* (as in Example 6).

Fermat's Theorem does suggest that we should at least *start* looking for extreme values of f at the numbers c where $f'(c) = 0$ or where $f'(c)$ does not exist. Such numbers are given a special name.

6 Definition A **critical number** of a function f is a number c in the domain of f such that either $f'(c) = 0$ or $f'(c)$ does not exist.

EXAMPLE 7 Find the critical numbers of $f(x) = x^{3/5}(4 - x)$.

SOLUTION The Product Rule gives

$$f'(x) = x^{3/5}(-1) + (4 - x)(\tfrac{3}{5}x^{-2/5}) = -x^{3/5} + \frac{3(4 - x)}{5x^{2/5}}$$

$$= \frac{-5x + 3(4 - x)}{5x^{2/5}} = \frac{12 - 8x}{5x^{2/5}}$$

[The same result could be obtained by first writing $f(x) = 4x^{3/5} - x^{8/5}$.] Therefore $f'(x) = 0$ if $12 - 8x = 0$, that is, $x = \tfrac{3}{2}$, and $f'(x)$ does not exist when $x = 0$. Thus the critical numbers are $\tfrac{3}{2}$ and 0. ∎

In terms of critical numbers, Fermat's Theorem can be rephrased as follows (compare Definition 6 with Theorem 4):

7 If f has a local maximum or minimum at c, then c is a critical number of f.

To find an absolute maximum or minimum of a continuous function on a closed interval, we note that either it is local [in which case it occurs at a critical number by (7)] or it occurs at an endpoint of the interval, as we see from the examples in Figure 8. Thus the following three-step procedure always works.

SECTION 4.1 Maximum and Minimum Values

The Closed Interval Method To find the *absolute* maximum and minimum values of a continuous function f on a closed interval $[a, b]$:

1. Find the values of f at the critical numbers of f in (a, b).
2. Find the values of f at the endpoints of the interval.
3. The largest of the values from Steps 1 and 2 is the absolute maximum value; the smallest of these values is the absolute minimum value.

EXAMPLE 8 Find the absolute maximum and minimum values of the function
$$f(x) = x^3 - 3x^2 + 1 \qquad -\tfrac{1}{2} \leq x \leq 4$$

SOLUTION Since f is continuous on $\left[-\tfrac{1}{2}, 4\right]$, we can use the Closed Interval Method:
$$f(x) = x^3 - 3x^2 + 1$$
$$f'(x) = 3x^2 - 6x = 3x(x - 2)$$

Since $f'(x)$ exists for all x, the only critical numbers of f occur when $f'(x) = 0$, that is, $x = 0$ or $x = 2$. Notice that each of these critical numbers lies in the interval $\left(-\tfrac{1}{2}, 4\right)$. The values of f at these critical numbers are
$$f(0) = 1 \qquad f(2) = -3$$

The values of f at the endpoints of the interval are
$$f\left(-\tfrac{1}{2}\right) = \tfrac{1}{8} \qquad f(4) = 17$$

Comparing these four numbers, we see that the absolute maximum value is $f(4) = 17$ and the absolute minimum value is $f(2) = -3$.

Note that in this example the absolute maximum occurs at an endpoint, whereas the absolute minimum occurs at a critical number. The graph of f is sketched in Figure 15. ∎

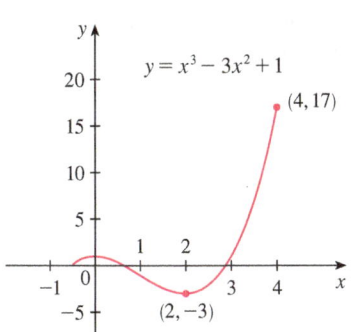

FIGURE 15

If you have a graphing calculator or a computer with graphing software, it is possible to estimate maximum and minimum values very easily. But, as the next example shows, calculus is needed to find the *exact* values.

EXAMPLE 9
(a) Use a graphing device to estimate the absolute minimum and maximum values of the function $f(x) = x - 2\sin x$, $0 \leq x \leq 2\pi$.
(b) Use calculus to find the exact minimum and maximum values.

SOLUTION
(a) Figure 16 shows a graph of f in the viewing rectangle $[0, 2\pi]$ by $[-1, 8]$. By moving the cursor close to the maximum point, we see that the y-coordinates don't change very much in the vicinity of the maximum. The absolute maximum value is about 6.97 and it occurs when $x \approx 5.2$. Similarly, by moving the cursor close to the minimum point, we see that the absolute minimum value is about -0.68 and it occurs when $x \approx 1.0$. It is possible to get more accurate estimates by zooming in toward the

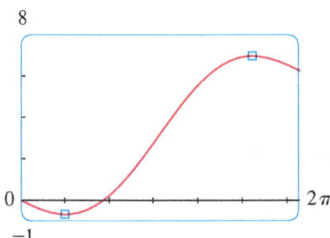

FIGURE 16

maximum and minimum points (or using a built-in maximum or minimum feature), but instead let's use calculus.

(b) The function $f(x) = x - 2 \sin x$ is continuous on $[0, 2\pi]$. Since $f'(x) = 1 - 2 \cos x$, we have $f'(x) = 0$ when $\cos x = \frac{1}{2}$ and this occurs when $x = \pi/3$ or $5\pi/3$. The values of f at these critical numbers are

$$f(\pi/3) = \frac{\pi}{3} - 2 \sin \frac{\pi}{3} = \frac{\pi}{3} - \sqrt{3} \approx -0.684853$$

and

$$f(5\pi/3) = \frac{5\pi}{3} - 2 \sin \frac{5\pi}{3} = \frac{5\pi}{3} + \sqrt{3} \approx 6.968039$$

The values of f at the endpoints are

$$f(0) = 0 \quad \text{and} \quad f(2\pi) = 2\pi \approx 6.28$$

Comparing these four numbers and using the Closed Interval Method, we see that the absolute minimum value is $f(\pi/3) = \pi/3 - \sqrt{3}$ and the absolute maximum value is $f(5\pi/3) = 5\pi/3 + \sqrt{3}$. The values from part (a) serve as a check on our work. ∎

EXAMPLE 10 The Hubble Space Telescope was deployed on April 24, 1990, by the space shuttle *Discovery*. A model for the velocity of the shuttle during this mission, from liftoff at $t = 0$ until the solid rocket boosters were jettisoned at $t = 126$ seconds, is given by

$$v(t) = 0.001302t^3 - 0.09029t^2 + 23.61t - 3.083$$

(in feet per second). Using this model, estimate the absolute maximum and minimum values of the *acceleration* of the shuttle between liftoff and the jettisoning of the boosters.

SOLUTION We are asked for the extreme values not of the given velocity function, but rather of the acceleration function. So we first need to differentiate to find the acceleration:

$$a(t) = v'(t) = \frac{d}{dt}(0.001302t^3 - 0.09029t^2 + 23.61t - 3.083)$$

$$= 0.003906t^2 - 0.18058t + 23.61$$

We now apply the Closed Interval Method to the continuous function a on the interval $0 \leq t \leq 126$. Its derivative is

$$a'(t) = 0.007812t - 0.18058$$

The only critical number occurs when $a'(t) = 0$:

$$t_1 = \frac{0.18058}{0.007812} \approx 23.12$$

Evaluating $a(t)$ at the critical number and at the endpoints, we have

$$a(0) = 23.61 \qquad a(t_1) \approx 21.52 \qquad a(126) \approx 62.87$$

So the maximum acceleration is about 62.87 ft/s^2 and the minimum acceleration is about 21.52 ft/s^2. ∎

4.1 EXERCISES

1. Explain the difference between an absolute minimum and a local minimum.

2. Suppose f is a continuous function defined on a closed interval $[a, b]$.
   (a) What theorem guarantees the existence of an absolute maximum value and an absolute minimum value for f?
   (b) What steps would you take to find those maximum and minimum values?

3–4 For each of the numbers a, b, c, d, r, and s, state whether the function whose graph is shown has an absolute maximum or minimum, a local maximum or minimum, or neither a maximum nor a minimum.

3.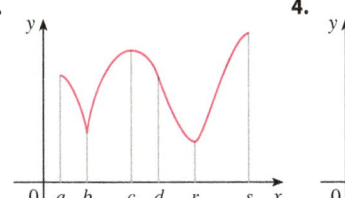
4.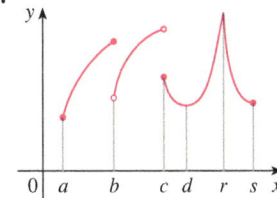

5–6 Use the graph to state the absolute and local maximum and minimum values of the function.

5.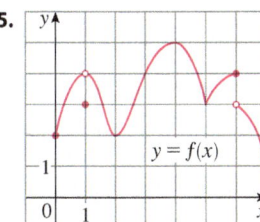
6.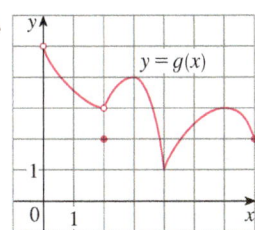

7–10 Sketch the graph of a function f that is continuous on $[1, 5]$ and has the given properties.

7. Absolute maximum at 5, absolute minimum at 2, local maximum at 3, local minima at 2 and 4

8. Absolute maximum at 4, absolute minimum at 5, local maximum at 2, local minimum at 3

9. Absolute minimum at 3, absolute maximum at 4, local maximum at 2

10. Absolute maximum at 2, absolute minimum at 5, 4 is a critical number but there is no local maximum or minimum there.

11. (a) Sketch the graph of a function that has a local maximum at 2 and is differentiable at 2.
    (b) Sketch the graph of a function that has a local maximum at 2 and is continuous but not differentiable at 2.
    (c) Sketch the graph of a function that has a local maximum at 2 and is not continuous at 2.

12. (a) Sketch the graph of a function on $[-1, 2]$ that has an absolute maximum but no local maximum.
    (b) Sketch the graph of a function on $[-1, 2]$ that has a local maximum but no absolute maximum.

13. (a) Sketch the graph of a function on $[-1, 2]$ that has an absolute maximum but no absolute minimum.
    (b) Sketch the graph of a function on $[-1, 2]$ that is discontinuous but has both an absolute maximum and an absolute minimum.

14. (a) Sketch the graph of a function that has two local maxima, one local minimum, and no absolute minimum.
    (b) Sketch the graph of a function that has three local minima, two local maxima, and seven critical numbers.

15–28 Sketch the graph of f by hand and use your sketch to find the absolute and local maximum and minimum values of f. (Use the graphs and transformations of Sections 1.2 and 1.3.)

15. $f(x) = \frac{1}{2}(3x - 1), \quad x \leq 3$

16. $f(x) = 2 - \frac{1}{3}x, \quad x \geq -2$

17. $f(x) = 1/x, \quad x \geq 1$

18. $f(x) = 1/x, \quad 1 < x < 3$

19. $f(x) = \sin x, \quad 0 \leq x < \pi/2$

20. $f(x) = \sin x, \quad 0 < x \leq \pi/2$

21. $f(x) = \sin x, \quad -\pi/2 \leq x \leq \pi/2$

22. $f(t) = \cos t, \quad -3\pi/2 \leq t \leq 3\pi/2$

23. $f(x) = \ln x, \quad 0 < x \leq 2$

24. $f(x) = |x|$

25. $f(x) = 1 - \sqrt{x}$

26. $f(x) = e^x$

27. $f(x) = \begin{cases} x^2 & \text{if } -1 \leq x \leq 0 \\ 2 - 3x & \text{if } 0 < x \leq 1 \end{cases}$

28. $f(x) = \begin{cases} 2x + 1 & \text{if } 0 \leq x < 1 \\ 4 - 2x & \text{if } 1 \leq x \leq 3 \end{cases}$

29–44 Find the critical numbers of the function.

29. $f(x) = 4 + \frac{1}{3}x - \frac{1}{2}x^2$

30. $f(x) = x^3 + 6x^2 - 15x$

31. $f(x) = 2x^3 - 3x^2 - 36x$

32. $f(x) = 2x^3 + x^2 + 2x$

33. $g(t) = t^4 + t^3 + t^2 + 1$

34. $g(t) = |3t - 4|$

35. $g(y) = \dfrac{y - 1}{y^2 - y + 1}$

36. $h(p) = \dfrac{p - 1}{p^2 + 4}$

37. $h(t) = t^{3/4} - 2t^{1/4}$

38. $g(x) = \sqrt[3]{4 - x^2}$

39. $F(x) = x^{4/5}(x - 4)^2$

40. $g(\theta) = 4\theta - \tan\theta$

41. $f(\theta) = 2\cos\theta + \sin^2\theta$

42. $h(t) = 3t - \arcsin t$

43. $f(x) = x^2 e^{-3x}$

44. $f(x) = x^{-2}\ln x$

45–46 A formula for the *derivative* of a function f is given. How many critical numbers does f have?

45. $f'(x) = 5e^{-0.1|x|}\sin x - 1$

46. $f'(x) = \dfrac{100\cos^2 x}{10 + x^2} - 1$

47–62 Find the absolute maximum and absolute minimum values of f on the given interval.

47. $f(x) = 12 + 4x - x^2$, $[0, 5]$

48. $f(x) = 5 + 54x - 2x^3$, $[0, 4]$

49. $f(x) = 2x^3 - 3x^2 - 12x + 1$, $[-2, 3]$

50. $f(x) = x^3 - 6x^2 + 5$, $[-3, 5]$

51. $f(x) = 3x^4 - 4x^3 - 12x^2 + 1$, $[-2, 3]$

52. $f(t) = (t^2 - 4)^3$, $[-2, 3]$

53. $f(x) = x + \dfrac{1}{x}$, $[0.2, 4]$

54. $f(x) = \dfrac{x}{x^2 - x + 1}$, $[0, 3]$

55. $f(t) = t - \sqrt[3]{t}$, $[-1, 4]$

56. $f(t) = \dfrac{\sqrt{t}}{1 + t^2}$, $[0, 2]$

57. $f(t) = 2\cos t + \sin 2t$, $[0, \pi/2]$

58. $f(t) = t + \cot(t/2)$, $[\pi/4, 7\pi/4]$

59. $f(x) = x^{-2}\ln x$, $\left[\tfrac{1}{2}, 4\right]$

60. $f(x) = xe^{x/2}$, $[-3, 1]$

61. $f(x) = \ln(x^2 + x + 1)$, $[-1, 1]$

62. $f(x) = x - 2\tan^{-1}x$, $[0, 4]$

63. If a and b are positive numbers, find the maximum value of $f(x) = x^a(1 - x)^b$, $0 \leq x \leq 1$.

64. Use a graph to estimate the critical numbers of $f(x) = |1 + 5x - x^3|$ correct to one decimal place.

65–68
(a) Use a graph to estimate the absolute maximum and minimum values of the function to two decimal places.
(b) Use calculus to find the exact maximum and minimum values.

65. $f(x) = x^5 - x^3 + 2$, $-1 \leq x \leq 1$

66. $f(x) = e^x + e^{-2x}$, $0 \leq x \leq 1$

67. $f(x) = x\sqrt{x - x^2}$

68. $f(x) = x - 2\cos x$, $-2 \leq x \leq 0$

69. After the consumption of an alcoholic beverage, the concentration of alcohol in the bloodstream (blood alcohol concentration, or BAC) surges as the alcohol is absorbed, followed by a gradual decline as the alcohol is metabolized. The function

$$C(t) = 1.35te^{-2.802t}$$

models the average BAC, measured in mg/mL, of a group of eight male subjects t hours after rapid consumption of 15 mL of ethanol (corresponding to one alcoholic drink). What is the maximum average BAC during the first 3 hours? When does it occur?

Source: Adapted from P. Wilkinson et al., "Pharmacokinetics of Ethanol after Oral Administration in the Fasting State," *Journal of Pharmacokinetics and Biopharmaceutics* 5 (1977): 207–24.

70. After an antibiotic tablet is taken, the concentration of the antibiotic in the bloodstream is modeled by the function

$$C(t) = 8(e^{-0.4t} - e^{-0.6t})$$

where the time t is measured in hours and C is measured in μg/mL. What is the maximum concentration of the antibiotic during the first 12 hours?

71. Between 0°C and 30°C, the volume V (in cubic centimeters) of 1 kg of water at a temperature T is given approximately by the formula

$$V = 999.87 - 0.06426T + 0.0085043T^2 - 0.0000679T^3$$

Find the temperature at which water has its maximum density.

72. An object with weight W is dragged along a horizontal plane by a force acting along a rope attached to the object. If the rope makes an angle θ with the plane, then the magnitude of the force is

$$F = \dfrac{\mu W}{\mu \sin\theta + \cos\theta}$$

where μ is a positive constant called the *coefficient of friction* and where $0 \leq \theta \leq \pi/2$. Show that F is minimized when $\tan\theta = \mu$.

73. The water level, measured in feet above mean sea level, of Lake Lanier in Georgia, USA, during 2012 can be modeled by the function

$$L(t) = 0.01441t^3 - 0.4177t^2 + 2.703t + 1060.1$$

where t is measured in months since January 1, 2012. Estimate when the water level was highest during 2012.

74. On May 7, 1992, the space shuttle *Endeavour* was launched on mission STS-49, the purpose of which was to install a new perigee kick motor in an Intelsat communications satellite. The table gives the velocity data for the shuttle between liftoff and the jettisoning of the solid rocket boosters.
(a) Use a graphing calculator or computer to find the cubic polynomial that best models the velocity of the shuttle for

the time interval $t \in [0, 125]$. Then graph this polynomial.
(b) Find a model for the acceleration of the shuttle and use it to estimate the maximum and minimum values of the acceleration during the first 125 seconds.

Event	Time (s)	Velocity (ft/s)
Launch	0	0
Begin roll maneuver	10	185
End roll maneuver	15	319
Throttle to 89%	20	447
Throttle to 67%	32	742
Throttle to 104%	59	1325
Maximum dynamic pressure	62	1445
Solid rocket booster separation	125	4151

75. When a foreign object lodged in the trachea (windpipe) forces a person to cough, the diaphragm thrusts upward causing an increase in pressure in the lungs. This is accompanied by a contraction of the trachea, making a narrower channel for the expelled air to flow through. For a given amount of air to escape in a fixed time, it must move faster through the narrower channel than the wider one. The greater the velocity of the airstream, the greater the force on the foreign object. X rays show that the radius of the circular tracheal tube contracts to about two-thirds of its normal radius during a cough. According to a mathematical model of coughing, the velocity v of the airstream is related to the radius r of the trachea by the equation

$$v(r) = k(r_0 - r)r^2 \qquad \tfrac{1}{2}r_0 \leq r \leq r_0$$

where k is a constant and r_0 is the normal radius of the trachea.

The restriction on r is due to the fact that the tracheal wall stiffens under pressure and a contraction greater than $\tfrac{1}{2}r_0$ is prevented (otherwise the person would suffocate).
(a) Determine the value of r in the interval $[\tfrac{1}{2}r_0, r_0]$ at which v has an absolute maximum. How does this compare with experimental evidence?
(b) What is the absolute maximum value of v on the interval?
(c) Sketch the graph of v on the interval $[0, r_0]$.

76. Show that 5 is a critical number of the function

$$g(x) = 2 + (x - 5)^3$$

but g does not have a local extreme value at 5.

77. Prove that the function

$$f(x) = x^{101} + x^{51} + x + 1$$

has neither a local maximum nor a local minimum.

78. If f has a local minimum value at c, show that the function $g(x) = -f(x)$ has a local maximum value at c.

79. Prove Fermat's Theorem for the case in which f has a local minimum at c.

80. A cubic function is a polynomial of degree 3; that is, it has the form $f(x) = ax^3 + bx^2 + cx + d$, where $a \neq 0$.
(a) Show that a cubic function can have two, one, or no critical number(s). Give examples and sketches to illustrate the three possibilities.
(b) How many local extreme values can a cubic function have?

APPLIED PROJECT

THE CALCULUS OF RAINBOWS

Rainbows are created when raindrops scatter sunlight. They have fascinated mankind since ancient times and have inspired attempts at scientific explanation since the time of Aristotle. In this project we use the ideas of Descartes and Newton to explain the shape, location, and colors of rainbows.

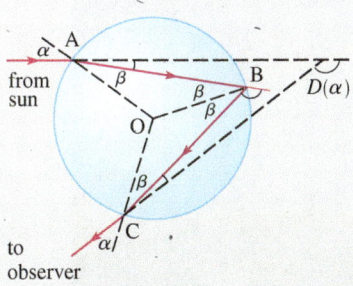

Formation of the primary rainbow

1. The figure shows a ray of sunlight entering a spherical raindrop at A. Some of the light is reflected, but the line AB shows the path of the part that enters the drop. Notice that the light is refracted toward the normal line AO and in fact Snell's Law says that $\sin \alpha = k \sin \beta$, where α is the angle of incidence, β is the angle of refraction, and $k \approx \tfrac{4}{3}$ is the index of refraction for water. At B some of the light passes through the drop and is refracted into the air, but the line BC shows the part that is reflected. (The angle of incidence equals the angle of reflection.) When the ray reaches C, part of it is reflected, but for the time being we are more interested in the part that leaves the raindrop at C. (Notice that it is refracted away from the normal line.) The *angle of deviation* $D(\alpha)$ is the amount of clockwise rotation that the ray has undergone during this three-stage process. Thus

$$D(\alpha) = (\alpha - \beta) + (\pi - 2\beta) + (\alpha - \beta) = \pi + 2\alpha - 4\beta$$

Show that the minimum value of the deviation is $D(\alpha) \approx 138°$ and occurs when $\alpha \approx 59.4°$.

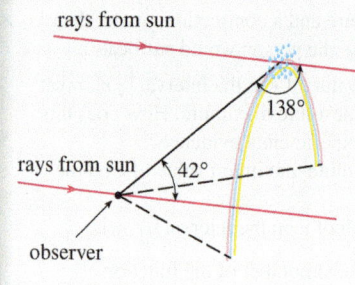

The significance of the minimum deviation is that when $\alpha \approx 59.4°$ we have $D'(\alpha) \approx 0$, so $\Delta D/\Delta \alpha \approx 0$. This means that many rays with $\alpha \approx 59.4°$ become deviated by approximately the same amount. It is the *concentration* of rays coming from near the direction of minimum deviation that creates the brightness of the primary rainbow. The figure at the left shows that the angle of elevation from the observer up to the highest point on the rainbow is $180° - 138° = 42°$. (This angle is called the *rainbow angle*.)

2. Problem 1 explains the location of the primary rainbow, but how do we explain the colors? Sunlight comprises a range of wavelengths, from the red range through orange, yellow, green, blue, indigo, and violet. As Newton discovered in his prism experiments of 1666, the index of refraction is different for each color. (The effect is called *dispersion*.) For red light the refractive index is $k \approx 1.3318$, whereas for violet light it is $k \approx 1.3435$. By repeating the calculation of Problem 1 for these values of k, show that the rainbow angle is about $42.3°$ for the red bow and $40.6°$ for the violet bow. So the rainbow really consists of seven individual bows corresponding to the seven colors.

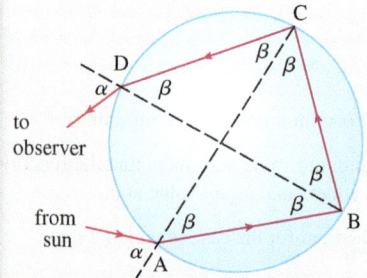

Formation of the secondary rainbow

3. Perhaps you have seen a fainter secondary rainbow above the primary bow. That results from the part of a ray that enters a raindrop and is refracted at A, reflected twice (at B and C), and refracted as it leaves the drop at D (see the figure at the left). This time the deviation angle $D(\alpha)$ is the total amount of counterclockwise rotation that the ray undergoes in this four-stage process. Show that

$$D(\alpha) = 2\alpha - 6\beta + 2\pi$$

and $D(\alpha)$ has a minimum value when

$$\cos \alpha = \sqrt{\frac{k^2 - 1}{8}}$$

Taking $k = \frac{4}{3}$, show that the minimum deviation is about $129°$ and so the rainbow angle for the secondary rainbow is about $51°$, as shown in the figure at the left.

4. Show that the colors in the secondary rainbow appear in the opposite order from those in the primary rainbow.

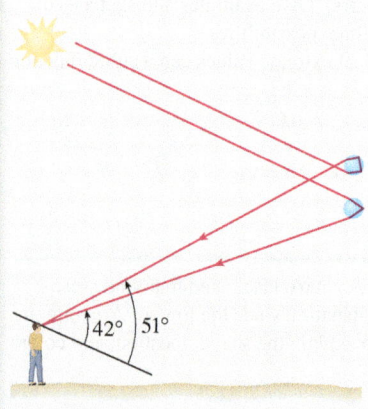

4.2 The Mean Value Theorem

We will see that many of the results of this chapter depend on one central fact, which is called the Mean Value Theorem. But to arrive at the Mean Value Theorem we first need the following result.

Rolle's Theorem Let f be a function that satisfies the following three hypotheses:

1. f is continuous on the closed interval $[a, b]$.
2. f is differentiable on the open interval (a, b).
3. $f(a) = f(b)$

Then there is a number c in (a, b) such that $f'(c) = 0$.

Rolle
Rolle's Theorem was first published in 1691 by the French mathematician Michel Rolle (1652–1719) in a book entitled *Méthode pour resoudre les Egalitez*. He was a vocal critic of the methods of his day and attacked calculus as being a "collection of ingenious fallacies." Later, however, he became convinced of the essential correctness of the methods of calculus.

Before giving the proof let's take a look at the graphs of some typical functions that satisfy the three hypotheses. Figure 1 shows the graphs of four such functions. In each case it appears that there is at least one point $(c, f(c))$ on the graph where the tangent is horizontal and therefore $f'(c) = 0$. Thus Rolle's Theorem is plausible.

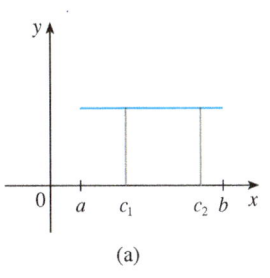
(a)

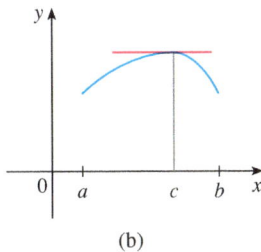
(b)

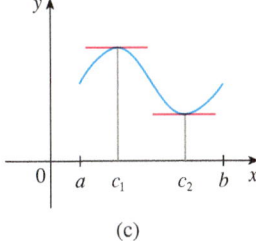
(c)

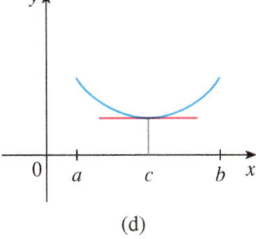
(d)

FIGURE 1

PS Take cases

PROOF There are three cases:

CASE I $f(x) = k$, **a constant**
Then $f'(x) = 0$, so the number c can be taken to be *any* number in (a, b).

CASE II $f(x) > f(a)$ **for some x in (a, b)** [as in Figure 1(b) or (c)]
By the Extreme Value Theorem (which we can apply by hypothesis 1), f has a maximum value somewhere in $[a, b]$. Since $f(a) = f(b)$, it must attain this maximum value at a number c in the open interval (a, b). Then f has a *local* maximum at c and, by hypothesis 2, f is differentiable at c. Therefore $f'(c) = 0$ by Fermat's Theorem.

CASE III $f(x) < f(a)$ **for some x in (a, b)** [as in Figure 1(c) or (d)]
By the Extreme Value Theorem, f has a minimum value in $[a, b]$ and, since $f(a) = f(b)$, it attains this minimum value at a number c in (a, b). Again $f'(c) = 0$ by Fermat's Theorem. ■

EXAMPLE 1 Let's apply Rolle's Theorem to the position function $s = f(t)$ of a moving object. If the object is in the same place at two different instants $t = a$ and $t = b$, then $f(a) = f(b)$. Rolle's Theorem says that there is some instant of time $t = c$ between a and b when $f'(c) = 0$; that is, the velocity is 0. (In particular, you can see that this is true when a ball is thrown directly upward.) ■

EXAMPLE 2 Prove that the equation $x^3 + x - 1 = 0$ has exactly one real root.

SOLUTION First we use the Intermediate Value Theorem (2.5.10) to show that a root exists. Let $f(x) = x^3 + x - 1$. Then $f(0) = -1 < 0$ and $f(1) = 1 > 0$. Since f is a

Figure 2 shows a graph of the function $f(x) = x^3 + x - 1$ discussed in Example 2. Rolle's Theorem shows that, no matter how much we enlarge the viewing rectangle, we can never find a second x-intercept.

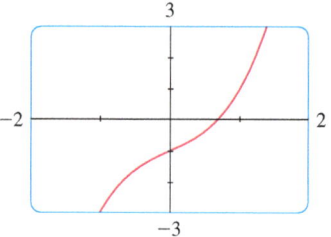

FIGURE 2

The Mean Value Theorem is an example of what is called an existence theorem. Like the Intermediate Value Theorem, the Extreme Value Theorem, and Rolle's Theorem, it guarantees that there *exists* a number with a certain property, but it doesn't tell us how to find the number.

polynomial, it is continuous, so the Intermediate Value Theorem states that there is a number c between 0 and 1 such that $f(c) = 0$. Thus the given equation has a root.

To show that the equation has no other real root, we use Rolle's Theorem and argue by contradiction. Suppose that it had two roots a and b. Then $f(a) = 0 = f(b)$ and, since f is a polynomial, it is differentiable on (a, b) and continuous on $[a, b]$. Thus, by Rolle's Theorem, there is a number c between a and b such that $f'(c) = 0$. But

$$f'(x) = 3x^2 + 1 \geq 1 \qquad \text{for all } x$$

(since $x^2 \geq 0$) so $f'(x)$ can never be 0. This gives a contradiction. Therefore the equation can't have two real roots. ∎

Our main use of Rolle's Theorem is in proving the following important theorem, which was first stated by another French mathematician, Joseph-Louis Lagrange.

The Mean Value Theorem Let f be a function that satisfies the following hypotheses:

1. f is continuous on the closed interval $[a, b]$.
2. f is differentiable on the open interval (a, b).

Then there is a number c in (a, b) such that

$$\boxed{1} \qquad f'(c) = \frac{f(b) - f(a)}{b - a}$$

or, equivalently,

$$\boxed{2} \qquad f(b) - f(a) = f'(c)(b - a)$$

Before proving this theorem, we can see that it is reasonable by interpreting it geometrically. Figures 3 and 4 show the points $A(a, f(a))$ and $B(b, f(b))$ on the graphs of two differentiable functions. The slope of the secant line AB is

$$\boxed{3} \qquad m_{AB} = \frac{f(b) - f(a)}{b - a}$$

which is the same expression as on the right side of Equation 1. Since $f'(c)$ is the slope of the tangent line at the point $(c, f(c))$, the Mean Value Theorem, in the form given by Equation 1, says that there is at least one point $P(c, f(c))$ on the graph where the slope of the tangent line is the same as the slope of the secant line AB. In other words, there is a point P where the tangent line is parallel to the secant line AB. (Imagine a line far away that stays parallel to AB while moving toward AB until it touches the graph for the first time.)

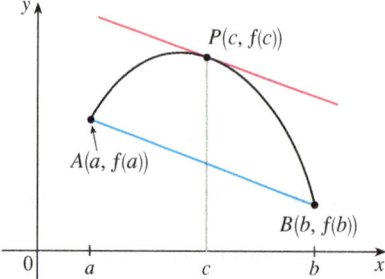

FIGURE 3

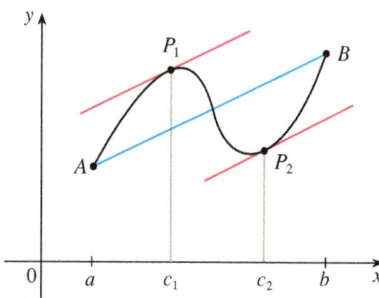

FIGURE 4

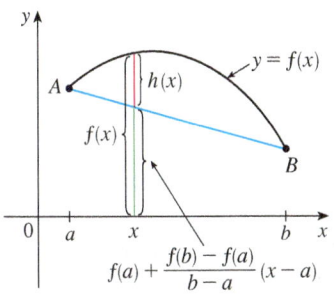

FIGURE 5

Lagrange and the Mean Value Theorem

The Mean Value Theorem was first formulated by Joseph-Louis Lagrange (1736–1813), born in Italy of a French father and an Italian mother. He was a child prodigy and became a professor in Turin at the tender age of 19. Lagrange made great contributions to number theory, theory of functions, theory of equations, and analytical and celestial mechanics. In particular, he applied calculus to the analysis of the stability of the solar system. At the invitation of Frederick the Great, he succeeded Euler at the Berlin Academy and, when Frederick died, Lagrange accepted King Louis XVI's invitation to Paris, where he was given apartments in the Louvre and became a professor at the Ecole Polytechnique. Despite all the trappings of luxury and fame, he was a kind and quiet man, living only for science.

PROOF We apply Rolle's Theorem to a new function h defined as the difference between f and the function whose graph is the secant line AB. Using Equation 3 and the point-slope equation of a line, we see that the equation of the line AB can be written as

$$y - f(a) = \frac{f(b) - f(a)}{b - a}(x - a)$$

or as

$$y = f(a) + \frac{f(b) - f(a)}{b - a}(x - a)$$

So, as shown in Figure 5,

$$\boxed{4} \quad h(x) = f(x) - f(a) - \frac{f(b) - f(a)}{b - a}(x - a)$$

First we must verify that h satisfies the three hypotheses of Rolle's Theorem.

1. The function h is continuous on $[a, b]$ because it is the sum of f and a first-degree polynomial, both of which are continuous.

2. The function h is differentiable on (a, b) because both f and the first-degree polynomial are differentiable. In fact, we can compute h' directly from Equation 4:

$$h'(x) = f'(x) - \frac{f(b) - f(a)}{b - a}$$

(Note that $f(a)$ and $[f(b) - f(a)]/(b - a)$ are constants.)

3. $$h(a) = f(a) - f(a) - \frac{f(b) - f(a)}{b - a}(a - a) = 0$$

$$h(b) = f(b) - f(a) - \frac{f(b) - f(a)}{b - a}(b - a)$$

$$= f(b) - f(a) - [f(b) - f(a)] = 0$$

Therefore $h(a) = h(b)$.

Since h satisfies the hypotheses of Rolle's Theorem, that theorem says there is a number c in (a, b) such that $h'(c) = 0$. Therefore

$$0 = h'(c) = f'(c) - \frac{f(b) - f(a)}{b - a}$$

and so

$$f'(c) = \frac{f(b) - f(a)}{b - a}$$

∎

EXAMPLE 3 To illustrate the Mean Value Theorem with a specific function, let's consider $f(x) = x^3 - x$, $a = 0$, $b = 2$. Since f is a polynomial, it is continuous and differentiable for all x, so it is certainly continuous on $[0, 2]$ and differentiable on $(0, 2)$. Therefore, by the Mean Value Theorem, there is a number c in $(0, 2)$ such that

$$f(2) - f(0) = f'(c)(2 - 0)$$

Now $f(2) = 6$, $f(0) = 0$, and $f'(x) = 3x^2 - 1$, so this equation becomes

$$6 = (3c^2 - 1)2 = 6c^2 - 2$$

which gives $c^2 = \frac{4}{3}$, that is, $c = \pm 2/\sqrt{3}$. But c must lie in $(0, 2)$, so $c = 2/\sqrt{3}$.

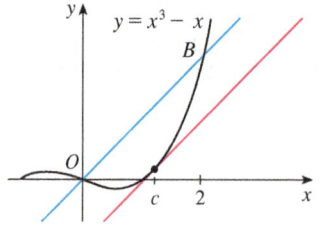

FIGURE 6

Figure 6 illustrates this calculation: The tangent line at this value of c is parallel to the secant line OB. ∎

EXAMPLE 4 If an object moves in a straight line with position function $s = f(t)$, then the average velocity between $t = a$ and $t = b$ is

$$\frac{f(b) - f(a)}{b - a}$$

and the velocity at $t = c$ is $f'(c)$. Thus the Mean Value Theorem (in the form of Equation 1) tells us that at some time $t = c$ between a and b the instantaneous velocity $f'(c)$ is equal to that average velocity. For instance, if a car traveled 180 km in 2 hours, then the speedometer must have read 90 km/h at least once.

In general, the Mean Value Theorem can be interpreted as saying that there is a number at which the instantaneous rate of change is equal to the average rate of change over an interval. ∎

The main significance of the Mean Value Theorem is that it enables us to obtain information about a function from information about its derivative. The next example provides an instance of this principle.

EXAMPLE 5 Suppose that $f(0) = -3$ and $f'(x) \leq 5$ for all values of x. How large can $f(2)$ possibly be?

SOLUTION We are given that f is differentiable (and therefore continuous) everywhere. In particular, we can apply the Mean Value Theorem on the interval $[0, 2]$. There exists a number c such that

$$f(2) - f(0) = f'(c)(2 - 0)$$

so

$$f(2) = f(0) + 2f'(c) = -3 + 2f'(c)$$

We are given that $f'(x) \leq 5$ for all x, so in particular we know that $f'(c) \leq 5$. Multiplying both sides of this inequality by 2, we have $2f'(c) \leq 10$, so

$$f(2) = -3 + 2f'(c) \leq -3 + 10 = 7$$

The largest possible value for $f(2)$ is 7. ∎

The Mean Value Theorem can be used to establish some of the basic facts of differential calculus. One of these basic facts is the following theorem. Others will be found in the following sections.

> **5 Theorem** If $f'(x) = 0$ for all x in an interval (a, b), then f is constant on (a, b).

PROOF Let x_1 and x_2 be any two numbers in (a, b) with $x_1 < x_2$. Since f is differentiable on (a, b), it must be differentiable on (x_1, x_2) and continuous on $[x_1, x_2]$. By applying the Mean Value Theorem to f on the interval $[x_1, x_2]$, we get a number c such that $x_1 < c < x_2$ and

$$\boxed{6} \qquad f(x_2) - f(x_1) = f'(c)(x_2 - x_1)$$

Since $f'(x) = 0$ for all x, we have $f'(c) = 0$, and so Equation 6 becomes

$$f(x_2) - f(x_1) = 0 \quad \text{or} \quad f(x_2) = f(x_1)$$

Therefore f has the same value at *any* two numbers x_1 and x_2 in (a, b). This means that f is constant on (a, b). ∎

Corollary 7 says that if two functions have the same derivatives on an interval, then their graphs must be vertical translations of each other there. In other words, the graphs have the same shape, but could be shifted up or down.

7 Corollary If $f'(x) = g'(x)$ for all x in an interval (a, b), then $f - g$ is constant on (a, b); that is, $f(x) = g(x) + c$ where c is a constant.

PROOF Let $F(x) = f(x) - g(x)$. Then

$$F'(x) = f'(x) - g'(x) = 0$$

for all x in (a, b). Thus, by Theorem 5, F is constant; that is, $f - g$ is constant. ∎

NOTE Care must be taken in applying Theorem 5. Let

$$f(x) = \frac{x}{|x|} = \begin{cases} 1 & \text{if } x > 0 \\ -1 & \text{if } x < 0 \end{cases}$$

The domain of f is $D = \{x \mid x \neq 0\}$ and $f'(x) = 0$ for all x in D. But f is obviously not a constant function. This does not contradict Theorem 5 because D is not an interval. Notice that f is constant on the interval $(0, \infty)$ and also on the interval $(-\infty, 0)$.

EXAMPLE 6 Prove the identity $\tan^{-1}x + \cot^{-1}x = \pi/2$.

SOLUTION Although calculus isn't needed to prove this identity, the proof using calculus is quite simple. If $f(x) = \tan^{-1}x + \cot^{-1}x$, then

$$f'(x) = \frac{1}{1 + x^2} - \frac{1}{1 + x^2} = 0$$

for all values of x. Therefore $f(x) = C$, a constant. To determine the value of C, we put $x = 1$ [because we can evaluate $f(1)$ exactly]. Then

$$C = f(1) = \tan^{-1}1 + \cot^{-1}1 = \frac{\pi}{4} + \frac{\pi}{4} = \frac{\pi}{2}$$

Thus $\tan^{-1}x + \cot^{-1}x = \pi/2$. ∎

4.2 EXERCISES

1. The graph of a function f is shown. Verify that f satisfies the hypotheses of Rolle's Theorem on the interval $[0, 8]$. Then estimate the value(s) of c that satisfy the conclusion of Rolle's Theorem on that interval.

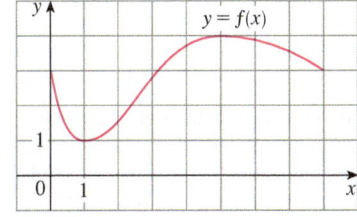

2. Draw the graph of a function defined on $[0, 8]$ such that $f(0) = f(8) = 3$ and the function does not satisfy the conclusion of Rolle's Theorem on $[0, 8]$.

3. The graph of a function g is shown.

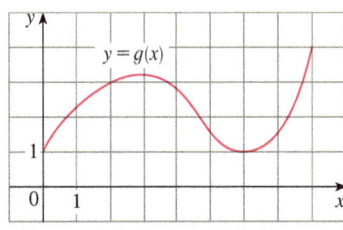

(a) Verify that g satisfies the hypotheses of the Mean Value Theorem on the interval $[0, 8]$.
(b) Estimate the value(s) of c that satisfy the conclusion of the Mean Value Theorem on the interval $[0, 8]$.
(c) Estimate the value(s) of c that satisfy the conclusion of the Mean Value Theorem on the interval $[2, 6]$.

4. Draw the graph of a function that is continuous on $[0, 8]$ where $f(0) = 1$ and $f(8) = 4$ and that does not satisfy the conclusion of the Mean Value Theorem on $[0, 8]$.

5–8 Verify that the function satisfies the three hypotheses of Rolle's Theorem on the given interval. Then find all numbers c that satisfy the conclusion of Rolle's Theorem.

5. $f(x) = 2x^2 - 4x + 5$, $\quad [-1, 3]$

6. $f(x) = x^3 - 2x^2 - 4x + 2$, $\quad [-2, 2]$

7. $f(x) = \sin(x/2)$, $\quad [\pi/2, 3\pi/2]$

8. $f(x) = x + 1/x$, $\quad [\tfrac{1}{2}, 2]$

9. Let $f(x) = 1 - x^{2/3}$. Show that $f(-1) = f(1)$ but there is no number c in $(-1, 1)$ such that $f'(c) = 0$. Why does this not contradict Rolle's Theorem?

10. Let $f(x) = \tan x$. Show that $f(0) = f(\pi)$ but there is no number c in $(0, \pi)$ such that $f'(c) = 0$. Why does this not contradict Rolle's Theorem?

11–14 Verify that the function satisfies the hypotheses of the Mean Value Theorem on the given interval. Then find all numbers c that satisfy the conclusion of the Mean Value Theorem.

11. $f(x) = 2x^2 - 3x + 1$, $\quad [0, 2]$

12. $f(x) = x^3 - 3x + 2$, $\quad [-2, 2]$

13. $f(x) = \ln x$, $\quad [1, 4]$ $\qquad$ **14.** $f(x) = 1/x$, $\quad [1, 3]$

15–16 Find the number c that satisfies the conclusion of the Mean Value Theorem on the given interval. Graph the function, the secant line through the endpoints, and the tangent line at $(c, f(c))$. Are the secant line and the tangent line parallel?

15. $f(x) = \sqrt{x}$, $\quad [0, 4]$ $\qquad$ **16.** $f(x) = e^{-x}$, $\quad [0, 2]$

17. Let $f(x) = (x - 3)^{-2}$. Show that there is no value of c in $(1, 4)$ such that $f(4) - f(1) = f'(c)(4 - 1)$. Why does this not contradict the Mean Value Theorem?

18. Let $f(x) = 2 - |2x - 1|$. Show that there is no value of c such that $f(3) - f(0) = f'(c)(3 - 0)$. Why does this not contradict the Mean Value Theorem?

19–20 Show that the equation has exactly one real root.

19. $2x + \cos x = 0$ $\qquad$ **20.** $x^3 + e^x = 0$

21. Show that the equation $x^3 - 15x + c = 0$ has at most one root in the interval $[-2, 2]$.

22. Show that the equation $x^4 + 4x + c = 0$ has at most two real roots.

23. (a) Show that a polynomial of degree 3 has at most three real roots.

(b) Show that a polynomial of degree n has at most n real roots.

24. (a) Suppose that f is differentiable on $\mathbb{R}$ and has two roots. Show that f' has at least one root.
(b) Suppose f is twice differentiable on $\mathbb{R}$ and has three roots. Show that f'' has at least one real root.
(c) Can you generalize parts (a) and (b)?

25. If $f(1) = 10$ and $f'(x) \geq 2$ for $1 \leq x \leq 4$, how small can $f(4)$ possibly be?

26. Suppose that $3 \leq f'(x) \leq 5$ for all values of x. Show that $18 \leq f(8) - f(2) \leq 30$.

27. Does there exist a function f such that $f(0) = -1$, $f(2) = 4$, and $f'(x) \leq 2$ for all x?

28. Suppose that f and g are continuous on $[a, b]$ and differentiable on (a, b). Suppose also that $f(a) = g(a)$ and $f'(x) < g'(x)$ for $a < x < b$. Prove that $f(b) < g(b)$. [Hint: Apply the Mean Value Theorem to the function $h = f - g$.]

29. Show that $\sin x < x$ if $0 < x < 2\pi$.

30. Suppose f is an odd function and is differentiable everywhere. Prove that for every positive number b, there exists a number c in $(-b, b)$ such that $f'(c) = f(b)/b$.

31. Use the Mean Value Theorem to prove the inequality

$$|\sin a - \sin b| \leq |a - b| \quad \text{for all } a \text{ and } b$$

32. If $f'(x) = c$ (c a constant) for all x, use Corollary 7 to show that $f(x) = cx + d$ for some constant d.

33. Let $f(x) = 1/x$ and

$$g(x) = \begin{cases} \dfrac{1}{x} & \text{if } x > 0 \\ 1 + \dfrac{1}{x} & \text{if } x < 0 \end{cases}$$

Show that $f'(x) = g'(x)$ for all x in their domains. Can we conclude from Corollary 7 that $f - g$ is constant?

34. Use the method of Example 6 to prove the identity

$$2 \sin^{-1} x = \cos^{-1}(1 - 2x^2) \qquad x \geq 0$$

35. Prove the identity $\arcsin \dfrac{x - 1}{x + 1} = 2 \arctan \sqrt{x} - \dfrac{\pi}{2}$.

36. At 2:00 PM a car's speedometer reads 30 mi/h. At 2:10 PM it reads 50 mi/h. Show that at some time between 2:00 and 2:10 the acceleration is exactly 120 mi/h^2.

37. Two runners start a race at the same time and finish in a tie. Prove that at some time during the race they have the same speed. [Hint: Consider $f(t) = g(t) - h(t)$, where g and h are the position functions of the two runners.]

38. A number a is called a **fixed point** of a function f if $f(a) = a$. Prove that if $f'(x) \neq 1$ for all real numbers x, then f has at most one fixed point.

4.3 How Derivatives Affect the Shape of a Graph

Many of the applications of calculus depend on our ability to deduce facts about a function f from information concerning its derivatives. Because $f'(x)$ represents the slope of the curve $y = f(x)$ at the point $(x, f(x))$, it tells us the direction in which the curve proceeds at each point. So it is reasonable to expect that information about $f'(x)$ will provide us with information about $f(x)$.

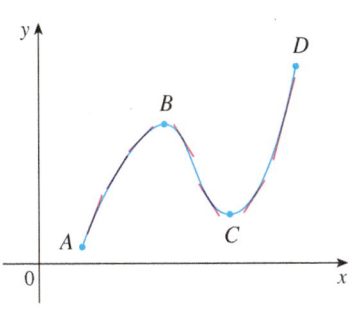

FIGURE 1

■ What Does f' Say About f?

To see how the derivative of f can tell us where a function is increasing or decreasing, look at Figure 1. (Increasing functions and decreasing functions were defined in Section 1.1.) Between A and B and between C and D, the tangent lines have positive slope and so $f'(x) > 0$. Between B and C, the tangent lines have negative slope and so $f'(x) < 0$. Thus it appears that f increases when $f'(x)$ is positive and decreases when $f'(x)$ is negative. To prove that this is always the case, we use the Mean Value Theorem.

Let's abbreviate the name of this test to the I/D Test.

> **Increasing/Decreasing Test**
> (a) If $f'(x) > 0$ on an interval, then f is increasing on that interval.
> (b) If $f'(x) < 0$ on an interval, then f is decreasing on that interval.

PROOF
(a) Let x_1 and x_2 be any two numbers in the interval with $x_1 < x_2$. According to the definition of an increasing function (page 19), we have to show that $f(x_1) < f(x_2)$.

Because we are given that $f'(x) > 0$, we know that f is differentiable on $[x_1, x_2]$. So, by the Mean Value Theorem, there is a number c between x_1 and x_2 such that

$$\boxed{1} \qquad f(x_2) - f(x_1) = f'(c)(x_2 - x_1)$$

Now $f'(c) > 0$ by assumption and $x_2 - x_1 > 0$ because $x_1 < x_2$. Thus the right side of Equation 1 is positive, and so

$$f(x_2) - f(x_1) > 0 \quad \text{or} \quad f(x_1) < f(x_2)$$

This shows that f is increasing.
Part (b) is proved similarly. ∎

EXAMPLE 1 Find where the function $f(x) = 3x^4 - 4x^3 - 12x^2 + 5$ is increasing and where it is decreasing.

SOLUTION We start by differentiating f:

$$f'(x) = 12x^3 - 12x^2 - 24x = 12x(x - 2)(x + 1)$$

To use the I/D Test we have to know where $f'(x) > 0$ and where $f'(x) < 0$. To solve these inequalities we first find where $f'(x) = 0$, namely at $x = 0, 2,$ and -1. These are the critical numbers of f, and they divide the domain into four intervals (see the number line at the left). Within each interval, $f'(x)$ must be always positive or always negative. (See Examples 3 and 4 in Appendix A.) We can determine which is the case for each interval from the signs of the three factors of $f'(x)$, namely, $12x, x - 2,$ and $x + 1$, as shown in the following chart. A plus sign indicates that the given expression is positive, and a minus sign indicates that it is negative. The last col-

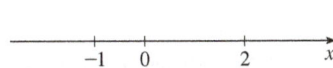

umn of the chart gives the conclusion based on the I/D Test. For instance, $f'(x) < 0$ for $0 < x < 2$, so f is decreasing on $(0, 2)$. (It would also be true to say that f is decreasing on the closed interval $[0, 2]$.)

Interval	$12x$	$x - 2$	$x + 1$	$f'(x)$	f
$x < -1$	$-$	$-$	$-$	$-$	decreasing on $(-\infty, -1)$
$-1 < x < 0$	$-$	$-$	$+$	$+$	increasing on $(-1, 0)$
$0 < x < 2$	$+$	$-$	$+$	$-$	decreasing on $(0, 2)$
$x > 2$	$+$	$+$	$+$	$+$	increasing on $(2, \infty)$

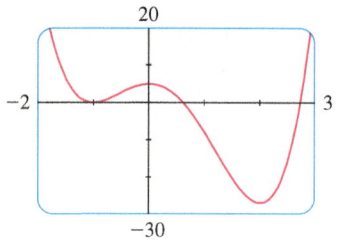

FIGURE 2

The graph of f shown in Figure 2 confirms the information in the chart. ∎

Local Extreme Values

Recall from Section 4.1 that if f has a local maximum or minimum at c, then c must be a critical number of f (by Fermat's Theorem), but not every critical number gives rise to a maximum or a minimum. We therefore need a test that will tell us whether or not f has a local maximum or minimum at a critical number.

You can see from Figure 2 that $f(0) = 5$ is a local maximum value of f because f increases on $(-1, 0)$ and decreases on $(0, 2)$. Or, in terms of derivatives, $f'(x) > 0$ for $-1 < x < 0$ and $f'(x) < 0$ for $0 < x < 2$. In other words, the sign of $f'(x)$ changes from positive to negative at 0. This observation is the basis of the following test.

> **The First Derivative Test** Suppose that c is a critical number of a continuous function f.
> (a) If f' changes from positive to negative at c, then f has a local maximum at c.
> (b) If f' changes from negative to positive at c, then f has a local minimum at c.
> (c) If f' is positive to the left and right of c, or negative to the left and right of c, then f has no local maximum or minimum at c.

The First Derivative Test is a consequence of the I/D Test. In part (a), for instance, since the sign of $f'(x)$ changes from positive to negative at c, f is increasing to the left of c and decreasing to the right of c. It follows that f has a local maximum at c.

It is easy to remember the First Derivative Test by visualizing diagrams such as those in Figure 3.

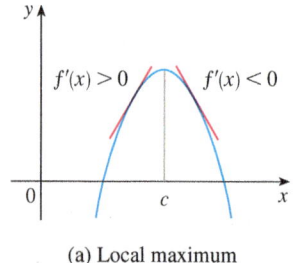
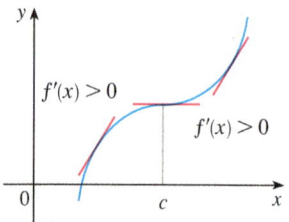
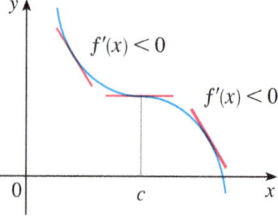

(a) Local maximum (b) Local minimum (c) No maximum or minimum (d) No maximum or minimum

FIGURE 3

EXAMPLE 2 Find the local minimum and maximum values of the function f in Example 1.

SOLUTION From the chart in the solution to Example 1 we see that $f'(x)$ changes from negative to positive at -1, so $f(-1) = 0$ is a local minimum value by the First Derivative Test. Similarly, f' changes from negative to positive at 2, so $f(2) = -27$ is also a local minimum value. As noted previously, $f(0) = 5$ is a local maximum value because $f'(x)$ changes from positive to negative at 0. ∎

EXAMPLE 3 Find the local maximum and minimum values of the function
$$g(x) = x + 2 \sin x \qquad 0 \leq x \leq 2\pi$$

SOLUTION As in Example 1, we start by finding the critical numbers. The derivative is:
$$g'(x) = 1 + 2 \cos x$$

so $g'(x) = 0$ when $\cos x = -\frac{1}{2}$. The solutions of this equation are $2\pi/3$ and $4\pi/3$. Because g is differentiable everywhere, the only critical numbers are $2\pi/3$ and $4\pi/3$. We split the domain into intervals according to the critical numbers. Within each interval, $g'(x)$ is either always positive or always negative and so we analyze g in the following chart.

The $+$ signs in the chart come from the fact that $g'(x) > 0$ when $\cos x > -\frac{1}{2}$. From the graph of $y = \cos x$, this is true in the indicated intervals.

Interval	$g'(x) = 1 + 2\cos x$	g
$0 < x < 2\pi/3$	$+$	increasing on $(0, 2\pi/3)$
$2\pi/3 < x < 4\pi/3$	$-$	decreasing on $(2\pi/3, 4\pi/3)$
$4\pi/3 < x < 2\pi$	$+$	increasing on $(4\pi/3, 2\pi)$

Because $g'(x)$ changes from positive to negative at $2\pi/3$, the First Derivative Test tells us that there is a local maximum at $2\pi/3$ and the local maximum value is

$$g(2\pi/3) = \frac{2\pi}{3} + 2 \sin \frac{2\pi}{3} = \frac{2\pi}{3} + 2\left(\frac{\sqrt{3}}{2}\right) = \frac{2\pi}{3} + \sqrt{3} \approx 3.83$$

Likewise, $g'(x)$ changes from negative to positive at $4\pi/3$ and so

$$g(4\pi/3) = \frac{4\pi}{3} + 2 \sin \frac{4\pi}{3} = \frac{4\pi}{3} + 2\left(-\frac{\sqrt{3}}{2}\right) = \frac{4\pi}{3} - \sqrt{3} \approx 2.46$$

is a local minimum value. The graph of g in Figure 4 supports our conclusion. ∎

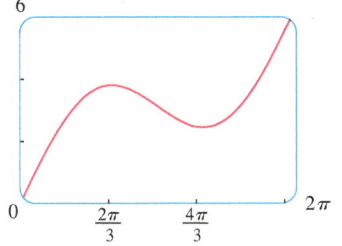

FIGURE 4
$g(x) = x + 2\sin x$

■ What Does f'' Say About f?

Figure 5 shows the graphs of two increasing functions on (a, b). Both graphs join point A to point B but they look different because they bend in different directions. How can we distinguish between these two types of behavior?

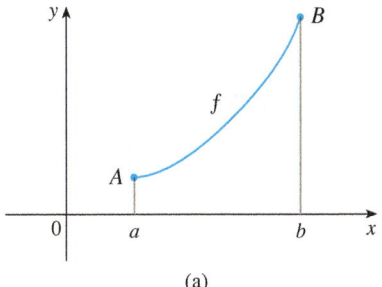
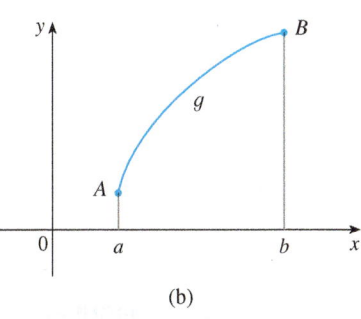

FIGURE 5 (a) (b)

296 CHAPTER 4 Applications of Differentiation

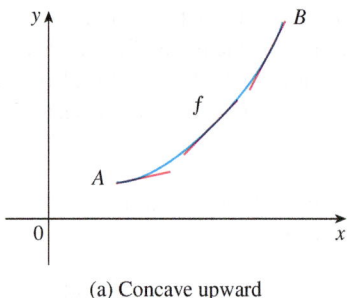

(a) Concave upward

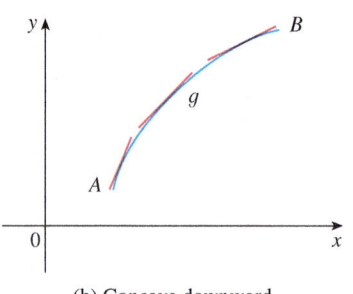

(b) Concave downward

FIGURE 6

In Figure 6 tangents to these curves have been drawn at several points. In (a) the curve lies above the tangents and f is called *concave upward* on (a, b). In (b) the curve lies below the tangents and g is called *concave downward* on (a, b).

> **Definition** If the graph of f lies above all of its tangents on an interval I, then it is called **concave upward** on I. If the graph of f lies below all of its tangents on I, it is called **concave downward** on I.

Figure 7 shows the graph of a function that is concave upward (abbreviated CU) on the intervals (b, c), (d, e), and (e, p) and concave downward (CD) on the intervals (a, b), (c, d), and (p, q).

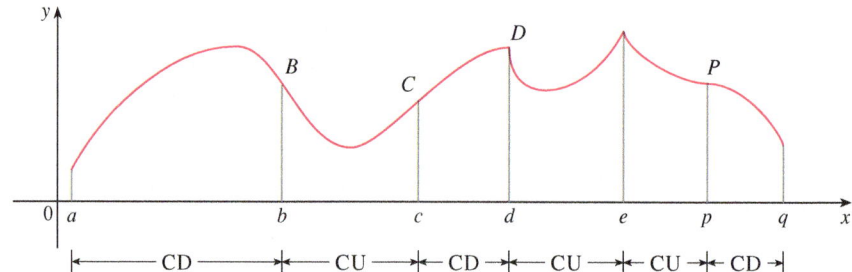

FIGURE 7

Let's see how the second derivative helps determine the intervals of concavity. Looking at Figure 6(a), you can see that, going from left to right, the slope of the tangent increases. This means that the derivative f' is an increasing function and therefore its derivative f'' is positive. Likewise, in Figure 6(b) the slope of the tangent decreases from left to right, so f' decreases and therefore f'' is negative. This reasoning can be reversed and suggests that the following theorem is true. A proof is given in Appendix F with the help of the Mean Value Theorem.

> **Concavity Test**
> (a) If $f''(x) > 0$ for all x in I, then the graph of f is concave upward on I.
> (b) If $f''(x) < 0$ for all x in I, then the graph of f is concave downward on I.

EXAMPLE 4 Figure 8 shows a population graph for Cyprian honeybees raised in an apiary. How does the rate of population increase change over time? When is this rate highest? Over what intervals is P concave upward or concave downward?

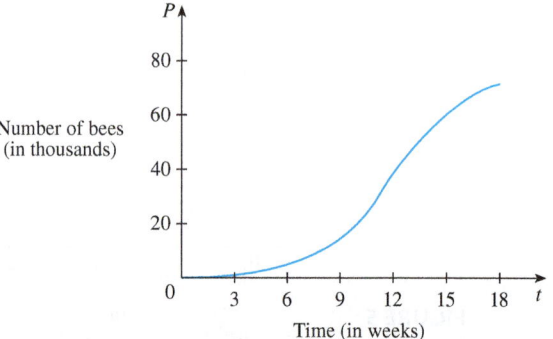

FIGURE 8

SOLUTION By looking at the slope of the curve as t increases, we see that the rate of increase of the population is initially very small, then gets larger until it reaches a maximum at about $t = 12$ weeks, and decreases as the population begins to level off. As the population approaches its maximum value of about 75,000 (called the *carrying capacity*), the rate of increase, $P'(t)$, approaches 0. The curve appears to be concave upward on $(0, 12)$ and concave downward on $(12, 18)$. ∎

In Example 4, the population curve changed from concave upward to concave downward at approximately the point (12, 38,000). This point is called an *inflection point* of the curve. The significance of this point is that the rate of population increase has its maximum value there. In general, an inflection point is a point where a curve changes its direction of concavity.

> **Definition** A point P on a curve $y = f(x)$ is called an **inflection point** if f is continuous there and the curve changes from concave upward to concave downward or from concave downward to concave upward at P.

For instance, in Figure 7, B, C, D, and P are the points of inflection. Notice that if a curve has a tangent at a point of inflection, then the curve crosses its tangent there.

In view of the Concavity Test, there is a point of inflection at any point where the second derivative changes sign.

EXAMPLE 5 Sketch a possible graph of a function f that satisfies the following conditions:

(i) $f'(x) > 0$ on $(-\infty, 1)$, $f'(x) < 0$ on $(1, \infty)$
(ii) $f''(x) > 0$ on $(-\infty, -2)$ and $(2, \infty)$, $f''(x) < 0$ on $(-2, 2)$
(iii) $\lim_{x \to -\infty} f(x) = -2$, $\lim_{x \to \infty} f(x) = 0$

SOLUTION Condition (i) tells us that f is increasing on $(-\infty, 1)$ and decreasing on $(1, \infty)$. Condition (ii) says that f is concave upward on $(-\infty, -2)$ and $(2, \infty)$, and concave downward on $(-2, 2)$. From condition (iii) we know that the graph of f has two horizontal asymptotes: $y = -2$ (to the left) and $y = 0$ (to the right).

We first draw the horizontal asymptote $y = -2$ as a dashed line (see Figure 9). We then draw the graph of f approaching this asymptote at the far left, increasing to its maximum point at $x = 1$, and decreasing toward the x-axis as at the far right. We also make sure that the graph has inflection points when $x = -2$ and 2. Notice that we made the curve bend upward for $x < -2$ and $x > 2$, and bend downward when x is between -2 and 2. ∎

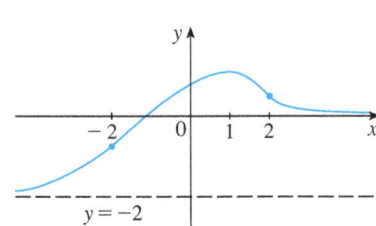

FIGURE 9

Another application of the second derivative is the following test for identifying local maximum and minimum values. It is a consequence of the Concavity Test, and serves as an alternative to the First Derivative Test.

> **The Second Derivative Test** Suppose f'' is continuous near c.
> (a) If $f'(c) = 0$ and $f''(c) > 0$, then f has a local minimum at c.
> (b) If $f'(c) = 0$ and $f''(c) < 0$, then f has a local maximum at c.

For instance, part (a) is true because $f''(x) > 0$ near c and so f is concave upward near c. This means that the graph of f lies *above* its horizontal tangent at c and so f has a local minimum at c. (See Figure 10.)

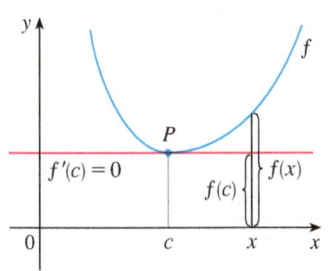

FIGURE 10
$f''(c) > 0$, f is concave upward

EXAMPLE 6 Discuss the curve $y = x^4 - 4x^3$ with respect to concavity, points of inflection, and local maxima and minima. Use this information to sketch the curve.

SOLUTION If $f(x) = x^4 - 4x^3$, then

$$f'(x) = 4x^3 - 12x^2 = 4x^2(x - 3)$$

$$f''(x) = 12x^2 - 24x = 12x(x - 2)$$

To find the critical numbers we set $f'(x) = 0$ and obtain $x = 0$ and $x = 3$. (Note that f' is a polynomial and hence defined everywhere.) To use the Second Derivative Test we evaluate f'' at these critical numbers:

$$f''(0) = 0 \qquad f''(3) = 36 > 0$$

Since $f'(3) = 0$ and $f''(3) > 0$, $f(3) = -27$ is a local minimum. [In fact, the expression for $f'(x)$ shows that f decreases to the left of 3 and increases to the right of 3.] Since $f''(0) = 0$, the Second Derivative Test gives no information about the critical number 0. But since $f'(x) < 0$ for $x < 0$ and also for $0 < x < 3$, the First Derivative Test tells us that f does not have a local maximum or minimum at 0.

Since $f''(x) = 0$ when $x = 0$ or 2, we divide the real line into intervals with these numbers as endpoints and complete the following chart.

Interval	$f''(x) = 12x(x - 2)$	Concavity
$(-\infty, 0)$	+	upward
$(0, 2)$	−	downward
$(2, \infty)$	+	upward

The point $(0, 0)$ is an inflection point since the curve changes from concave upward to concave downward there. Also $(2, -16)$ is an inflection point since the curve changes from concave downward to concave upward there.

Using the local minimum, the intervals of concavity, and the inflection points, we sketch the curve in Figure 11.

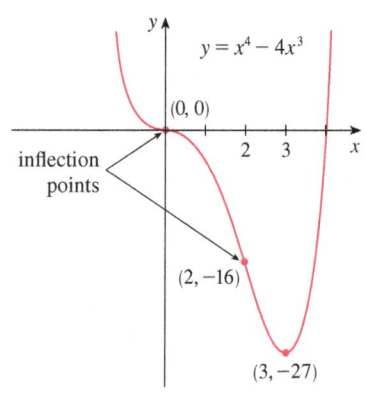

FIGURE 11

NOTE The Second Derivative Test is inconclusive when $f''(c) = 0$. In other words, at such a point there might be a maximum, there might be a minimum, or there might be neither (as in Example 6). This test also fails when $f''(c)$ does not exist. In such cases the First Derivative Test must be used. In fact, even when both tests apply, the First Derivative Test is often the easier one to use.

EXAMPLE 7 Sketch the graph of the function $f(x) = x^{2/3}(6 - x)^{1/3}$.

SOLUTION Calculation of the first two derivatives gives

Use the differentiation rules to check these calculations.

$$f'(x) = \frac{4 - x}{x^{1/3}(6 - x)^{2/3}} \qquad f''(x) = \frac{-8}{x^{4/3}(6 - x)^{5/3}}$$

Since $f'(x) = 0$ when $x = 4$ and $f'(x)$ does not exist when $x = 0$ or $x = 6$, the critical numbers are 0, 4, and 6.

Try reproducing the graph in Figure 12 with a graphing calculator or computer. Some machines produce the complete graph, some produce only the portion to the right of the y-axis, and some produce only the portion between $x = 0$ and $x = 6$. For an explanation and cure, see Example 7 in "Graphing Calculators and Computers" at www.stewartcalculus.com. An equivalent expression that gives the correct graph is

$$y = (x^2)^{1/3} \cdot \frac{6-x}{|6-x|} |6-x|^{1/3}$$

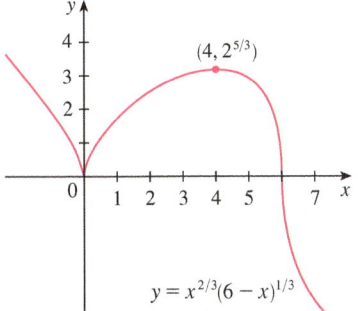

FIGURE 12

TEC In Module 4.3 you can practice using information about f', f'', and asymptotes to determine the shape of the graph of f.

Interval	$4-x$	$x^{1/3}$	$(6-x)^{2/3}$	$f'(x)$	f
$x < 0$	+	−	+	−	decreasing on $(-\infty, 0)$
$0 < x < 4$	+	+	+	+	increasing on $(0, 4)$
$4 < x < 6$	−	+	+	−	decreasing on $(4, 6)$
$x > 6$	−	+	+	−	decreasing on $(6, \infty)$

To find the local extreme values we use the First Derivative Test. Since f' changes from negative to positive at 0, $f(0) = 0$ is a local minimum. Since f' changes from positive to negative at 4, $f(4) = 2^{5/3}$ is a local maximum. The sign of f' does not change at 6, so there is no minimum or maximum there. (The Second Derivative Test could be used at 4 but not at 0 or 6 since f'' does not exist at either of these numbers.)

Looking at the expression for $f''(x)$ and noting that $x^{4/3} \geq 0$ for all x, we have $f''(x) < 0$ for $x < 0$ and for $0 < x < 6$ and $f''(x) > 0$ for $x > 6$. So f is concave downward on $(-\infty, 0)$ and $(0, 6)$ and concave upward on $(6, \infty)$, and the only inflection point is $(6, 0)$. The graph is sketched in Figure 12. Note that the curve has vertical tangents at $(0, 0)$ and $(6, 0)$ because $|f'(x)| \to \infty$ as $x \to 0$ and as $x \to 6$. ■

EXAMPLE 8 Use the first and second derivatives of $f(x) = e^{1/x}$, together with asymptotes, to sketch its graph.

SOLUTION Notice that the domain of f is $\{x \mid x \neq 0\}$, so we check for vertical asymptotes by computing the left and right limits as $x \to 0$. As $x \to 0^+$, we know that $t = 1/x \to \infty$, so

$$\lim_{x \to 0^+} e^{1/x} = \lim_{t \to \infty} e^t = \infty$$

and this shows that $x = 0$ is a vertical asymptote. As $x \to 0^-$, we have $t = 1/x \to -\infty$, so

$$\lim_{x \to 0^-} e^{1/x} = \lim_{t \to -\infty} e^t = 0$$

As $x \to \pm\infty$, we have $1/x \to 0$ and so

$$\lim_{x \to \pm\infty} e^{1/x} = e^0 = 1$$

This shows that $y = 1$ is a horizontal asymptote (both to the left and right).

Now let's compute the derivative. The Chain Rule gives

$$f'(x) = -\frac{e^{1/x}}{x^2}$$

Since $e^{1/x} > 0$ and $x^2 > 0$ for all $x \neq 0$, we have $f'(x) < 0$ for all $x \neq 0$. Thus f is decreasing on $(-\infty, 0)$ and on $(0, \infty)$. There is no critical number, so the function has no local maximum or minimum. The second derivative is

$$f''(x) = -\frac{x^2 e^{1/x}(-1/x^2) - e^{1/x}(2x)}{x^4} = \frac{e^{1/x}(2x+1)}{x^4}$$

Since $e^{1/x} > 0$ and $x^4 > 0$, we have $f''(x) > 0$ when $x > -\frac{1}{2}$ ($x \neq 0$) and $f''(x) < 0$ when $x < -\frac{1}{2}$. So the curve is concave downward on $\left(-\infty, -\frac{1}{2}\right)$ and concave upward on $\left(-\frac{1}{2}, 0\right)$ and on $(0, \infty)$. The inflection point is $\left(-\frac{1}{2}, e^{-2}\right)$.

To sketch the graph of f we first draw the horizontal asymptote $y = 1$ (as a dashed line), together with the parts of the curve near the asymptotes in a preliminary sketch

[Figure 13(a)]. These parts reflect the information concerning limits and the fact that f is decreasing on both $(-\infty, 0)$ and $(0, \infty)$. Notice that we have indicated that $f(x) \to 0$ as $x \to 0^-$ even though $f(0)$ does not exist. In Figure 13(b) we finish the sketch by incorporating the information concerning concavity and the inflection point. In Figure 13(c) we check our work with a graphing device.

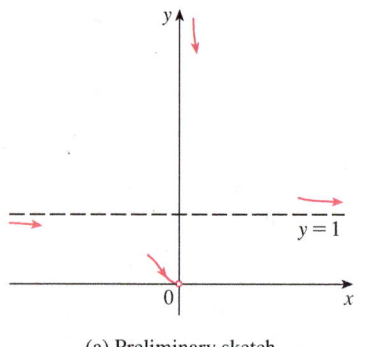
(a) Preliminary sketch

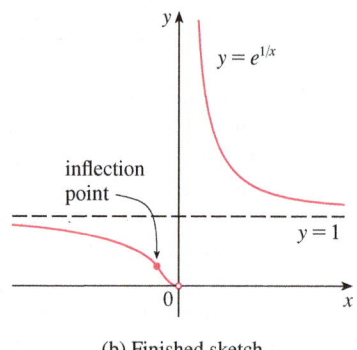
(b) Finished sketch

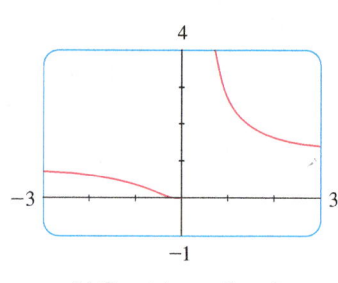
(c) Computer confirmation

FIGURE 13

4.3 EXERCISES

1–2 Use the given graph of f to find the following.
(a) The open intervals on which f is increasing.
(b) The open intervals on which f is decreasing.
(c) The open intervals on which f is concave upward.
(d) The open intervals on which f is concave downward.
(e) The coordinates of the points of inflection.

1. 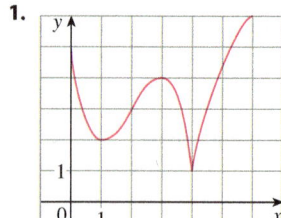   **2.**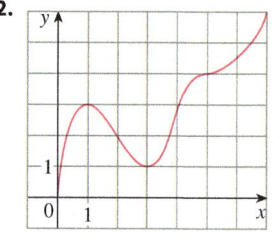

3. Suppose you are given a formula for a function f.
(a) How do you determine where f is increasing or decreasing?
(b) How do you determine where the graph of f is concave upward or concave downward?
(c) How do you locate inflection points?

4. (a) State the First Derivative Test.
(b) State the Second Derivative Test. Under what circumstances is it inconclusive? What do you do if it fails?

5–6 The graph of the *derivative* f' of a function f is shown.
(a) On what intervals is f increasing or decreasing?
(b) At what values of x does f have a local maximum or minimum?

5.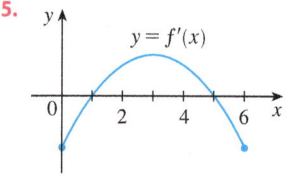

6.

7. In each part state the x-coordinates of the inflection points of f. Give reasons for your answers.
(a) The curve is the graph of f.
(b) The curve is the graph of f'.
(c) The curve is the graph of f''.

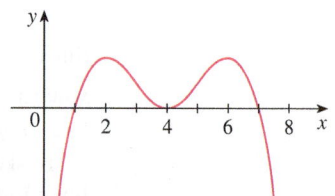

8. The graph of the first derivative f' of a function f is shown.
   (a) On what intervals is f increasing? Explain.
   (b) At what values of x does f have a local maximum or minimum? Explain.
   (c) On what intervals is f concave upward or concave downward? Explain.
   (d) What are the x-coordinates of the inflection points of f? Why?

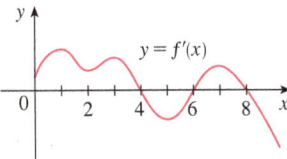

9–18
(a) Find the intervals on which f is increasing or decreasing.
(b) Find the local maximum and minimum values of f.
(c) Find the intervals of concavity and the inflection points.

9. $f(x) = x^3 - 3x^2 - 9x + 4$
10. $f(x) = 2x^3 - 9x^2 + 12x - 3$
11. $f(x) = x^4 - 2x^2 + 3$
12. $f(x) = \dfrac{x}{x^2 + 1}$
13. $f(x) = \sin x + \cos x, \quad 0 \le x \le 2\pi$
14. $f(x) = \cos^2 x - 2 \sin x, \quad 0 \le x \le 2\pi$
15. $f(x) = e^{2x} + e^{-x}$
16. $f(x) = x^2 \ln x$
17. $f(x) = x^2 - x - \ln x$
18. $f(x) = x^4 e^{-x}$

19–21 Find the local maximum and minimum values of f using both the First and Second Derivative Tests. Which method do you prefer?

19. $f(x) = 1 + 3x^2 - 2x^3$
20. $f(x) = \dfrac{x^2}{x - 1}$
21. $f(x) = \sqrt{x} - \sqrt[4]{x}$

22. (a) Find the critical numbers of $f(x) = x^4(x - 1)^3$.
    (b) What does the Second Derivative Test tell you about the behavior of f at these critical numbers?
    (c) What does the First Derivative Test tell you?

23. Suppose f'' is continuous on $(-\infty, \infty)$.
    (a) If $f'(2) = 0$ and $f''(2) = -5$, what can you say about f?
    (b) If $f'(6) = 0$ and $f''(6) = 0$, what can you say about f?

24–31 Sketch the graph of a function that satisfies all of the given conditions.

24. (a) $f'(x) < 0$ and $f''(x) < 0$ for all x
    (b) $f'(x) > 0$ and $f''(x) > 0$ for all x

25. (a) $f'(x) > 0$ and $f''(x) < 0$ for all x
    (b) $f'(x) < 0$ and $f''(x) > 0$ for all x

26. Vertical asymptote $x = 0$, $f'(x) > 0$ if $x < -2$,
    $f'(x) < 0$ if $x > -2$ $(x \ne 0)$,
    $f''(x) < 0$ if $x < 0$, $f''(x) > 0$ if $x > 0$

27. $f'(0) = f'(2) = f'(4) = 0$,
    $f'(x) > 0$ if $x < 0$ or $2 < x < 4$,
    $f'(x) < 0$ if $0 < x < 2$ or $x > 4$,
    $f''(x) > 0$ if $1 < x < 3$, $f''(x) < 0$ if $x < 1$ or $x > 3$

28. $f'(x) > 0$ for all $x \ne 1$, vertical asymptote $x = 1$,
    $f''(x) > 0$ if $x < 1$ or $x > 3$, $f''(x) < 0$ if $1 < x < 3$

29. $f'(5) = 0$, $f'(x) < 0$ when $x < 5$,
    $f'(x) > 0$ when $x > 5$, $f''(2) = 0$, $f''(8) = 0$,
    $f''(x) < 0$ when $x < 2$ or $x > 8$,
    $f''(x) > 0$ for $2 < x < 8$, $\lim\limits_{x \to \infty} f(x) = 3$, $\lim\limits_{x \to -\infty} f(x) = 3$

30. $f'(0) = f'(4) = 0$, $f'(x) = 1$ if $x < -1$,
    $f'(x) > 0$ if $0 < x < 2$,
    $f'(x) < 0$ if $-1 < x < 0$ or $2 < x < 4$ or $x > 4$,
    $\lim\limits_{x \to 2^-} f'(x) = \infty$, $\lim\limits_{x \to 2^+} f'(x) = -\infty$,
    $f''(x) > 0$ if $-1 < x < 2$ or $2 < x < 4$,
    $f''(x) < 0$ if $x > 4$

31. $f'(x) > 0$ if $x \ne 2$, $f''(x) > 0$ if $x < 2$,
    $f''(x) < 0$ if $x > 2$, f has inflection point $(2, 5)$,
    $\lim\limits_{x \to \infty} f(x) = 8$, $\lim\limits_{x \to -\infty} f(x) = 0$

32. Suppose $f(3) = 2$, $f'(3) = \frac{1}{2}$, and $f'(x) > 0$ and $f''(x) < 0$ for all x.
    (a) Sketch a possible graph for f.
    (b) How many solutions does the equation $f(x) = 0$ have? Why?
    (c) Is it possible that $f'(2) = \frac{1}{3}$? Why?

33. Suppose f is a continuous function where $f(x) > 0$ for all x, $f(0) = 4$, $f'(x) > 0$ if $x < 0$ or $x > 2$, $f'(x) < 0$ if $0 < x < 2$, $f''(-1) = f''(1) = 0$, $f''(x) > 0$ if $x < -1$ or $x > 1$, $f''(x) < 0$ if $-1 < x < 1$.
    (a) Can f have an absolute maximum? If so, sketch a possible graph of f. If not, explain why.
    (b) Can f have an absolute minimum? If so, sketch a possible graph of f. If not, explain why.
    (c) Sketch a possible graph for f that does *not* achieve an absolute minimum.

34. The graph of a function $y = f(x)$ is shown. At which point(s) are the following true?
    (a) $\dfrac{dy}{dx}$ and $\dfrac{d^2 y}{dx^2}$ are both positive.
    (b) $\dfrac{dy}{dx}$ and $\dfrac{d^2 y}{dx^2}$ are both negative.
    (c) $\dfrac{dy}{dx}$ is negative but $\dfrac{d^2 y}{dx^2}$ is positive.

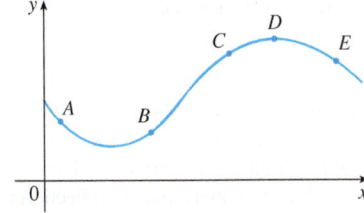

35–36 The graph of the derivative f' of a continuous function f is shown.
(a) On what intervals is f increasing? Decreasing?
(b) At what values of x does f have a local maximum? Local minimum?
(c) On what intervals is f concave upward? Concave downward?
(d) State the x-coordinate(s) of the point(s) of inflection.
(e) Assuming that $f(0) = 0$, sketch a graph of f.

35.
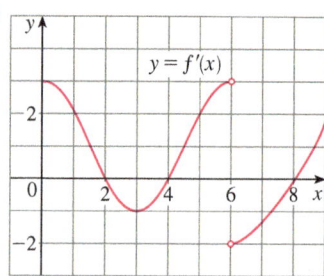

36.
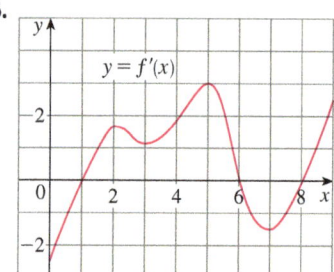

37–48
(a) Find the intervals of increase or decrease.
(b) Find the local maximum and minimum values.
(c) Find the intervals of concavity and the inflection points.
(d) Use the information from parts (a)–(c) to sketch the graph. Check your work with a graphing device if you have one.

37. $f(x) = x^3 - 12x + 2$
38. $f(x) = 36x + 3x^2 - 2x^3$
39. $f(x) = \frac{1}{2}x^4 - 4x^2 + 3$
40. $g(x) = 200 + 8x^3 + x^4$
41. $h(x) = (x + 1)^5 - 5x - 2$
42. $h(x) = 5x^3 - 3x^5$
43. $F(x) = x\sqrt{6 - x}$
44. $G(x) = 5x^{2/3} - 2x^{5/3}$
45. $C(x) = x^{1/3}(x + 4)$
46. $f(x) = \ln(x^2 + 9)$
47. $f(\theta) = 2\cos\theta + \cos^2\theta$, $0 \le \theta \le 2\pi$
48. $S(x) = x - \sin x$, $0 \le x \le 4\pi$

49–56
(a) Find the vertical and horizontal asymptotes.
(b) Find the intervals of increase or decrease.
(c) Find the local maximum and minimum values.
(d) Find the intervals of concavity and the inflection points.
(e) Use the information from parts (a)–(d) to sketch the graph of f.

49. $f(x) = 1 + \dfrac{1}{x} - \dfrac{1}{x^2}$
50. $f(x) = \dfrac{x^2 - 4}{x^2 + 4}$
51. $f(x) = \sqrt{x^2 + 1} - x$
52. $f(x) = \dfrac{e^x}{1 - e^x}$
53. $f(x) = e^{-x^2}$
54. $f(x) = x - \frac{1}{6}x^2 - \frac{2}{3}\ln x$
55. $f(x) = \ln(1 - \ln x)$
56. $f(x) = e^{\arctan x}$

57. Suppose the derivative of a function f is $f'(x) = (x + 1)^2(x - 3)^5(x - 6)^4$. On what interval is f increasing?

58. Use the methods of this section to sketch the curve $y = x^3 - 3a^2x + 2a^3$, where a is a positive constant. What do the members of this family of curves have in common? How do they differ from each other?

59–60
(a) Use a graph of f to estimate the maximum and minimum values. Then find the exact values.
(b) Estimate the value of x at which f increases most rapidly. Then find the exact value.

59. $f(x) = \dfrac{x + 1}{\sqrt{x^2 + 1}}$
60. $f(x) = x^2 e^{-x}$

61–62
(a) Use a graph of f to give a rough estimate of the intervals of concavity and the coordinates of the points of inflection.
(b) Use a graph of f'' to give better estimates.

61. $f(x) = \sin 2x + \sin 4x$, $0 \le x \le \pi$
62. $f(x) = (x - 1)^2(x + 1)^3$

CAS 63–64 Estimate the intervals of concavity to one decimal place by using a computer algebra system to compute and graph f''.

63. $f(x) = \dfrac{x^4 + x^3 + 1}{\sqrt{x^2 + x + 1}}$
64. $f(x) = \dfrac{x^2 \tan^{-1} x}{1 + x^3}$

65. A graph of a population of yeast cells in a new laboratory culture as a function of time is shown.

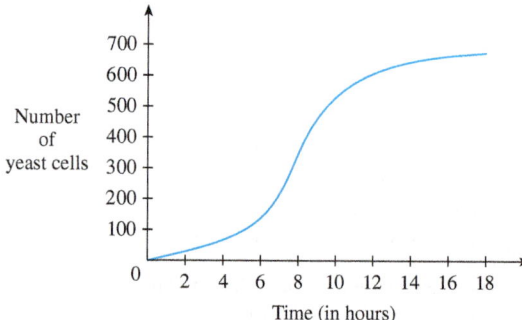

(a) Describe how the rate of population increase varies.

(b) When is this rate highest?
(c) On what intervals is the population function concave upward or downward?
(d) Estimate the coordinates of the inflection point.

66. In an episode of *The Simpsons* television show, Homer reads from a newspaper and announces "Here's good news! According to this eye-catching article, SAT scores are declining at a slower rate." Interpret Homer's statement in terms of a function and its first and second derivatives.

67. The president announces that the national deficit is increasing, but at a decreasing rate. Interpret this statement in terms of a function and its first and second derivatives.

68. Let $f(t)$ be the temperature at time t where you live and suppose that at time $t = 3$ you feel uncomfortably hot. How do you feel about the given data in each case?
(a) $f'(3) = 2$, $f''(3) = 4$
(b) $f'(3) = 2$, $f''(3) = -4$
(c) $f'(3) = -2$, $f''(3) = 4$
(d) $f'(3) = -2$, $f''(3) = -4$

69. Let $K(t)$ be a measure of the knowledge you gain by studying for a test for t hours. Which do you think is larger, $K(8) - K(7)$ or $K(3) - K(2)$? Is the graph of K concave upward or concave downward? Why?

70. Coffee is being poured into the mug shown in the figure at a constant rate (measured in volume per unit time). Sketch a rough graph of the depth of the coffee in the mug as a function of time. Account for the shape of the graph in terms of concavity. What is the significance of the inflection point?

71. A *drug response curve* describes the level of medication in the bloodstream after a drug is administered. A surge function $S(t) = At^p e^{-kt}$ is often used to model the response curve, reflecting an initial surge in the drug level and then a more gradual decline. If, for a particular drug, $A = 0.01$, $p = 4$, $k = 0.07$, and t is measured in minutes, estimate the times corresponding to the inflection points and explain their significance. If you have a graphing device, use it to graph the drug response curve.

72. The family of bell-shaped curves
$$y = \frac{1}{\sigma\sqrt{2\pi}} e^{-(x-\mu)^2/(2\sigma^2)}$$
occurs in probability and statistics, where it is called the *normal density function*. The constant μ is called the *mean* and the positive constant σ is called the *standard deviation*. For simplicity, let's scale the function so as to remove the factor $1/(\sigma\sqrt{2\pi})$ and let's analyze the special case where $\mu = 0$. So we study the function
$$f(x) = e^{-x^2/(2\sigma^2)}$$
(a) Find the asymptote, maximum value, and inflection points of f.
(b) What role does σ play in the shape of the curve?
(c) Illustrate by graphing four members of this family on the same screen.

73. Find a cubic function $f(x) = ax^3 + bx^2 + cx + d$ that has a local maximum value of 3 at $x = -2$ and a local minimum value of 0 at $x = 1$.

74. For what values of the numbers a and b does the function
$$f(x) = axe^{bx^2}$$
have the maximum value $f(2) = 1$?

75. (a) If the function $f(x) = x^3 + ax^2 + bx$ has the local minimum value $-\frac{2}{9}\sqrt{3}$ at $x = 1/\sqrt{3}$, what are the values of a and b?
(b) Which of the tangent lines to the curve in part (a) has the smallest slope?

76. For what values of a and b is $(2, 2.5)$ an inflection point of the curve $x^2y + ax + by = 0$? What additional inflection points does the curve have?

77. Show that the curve $y = (1 + x)/(1 + x^2)$ has three points of inflection and they all lie on one straight line.

78. Show that the curves $y = e^{-x}$ and $y = -e^{-x}$ touch the curve $y = e^{-x} \sin x$ at its inflection points.

79. Show that the inflection points of the curve $y = x \sin x$ lie on the curve $y^2(x^2 + 4) = 4x^2$.

80–82 Assume that all of the functions are twice differentiable and the second derivatives are never 0.

80. (a) If f and g are concave upward on I, show that $f + g$ is concave upward on I.
(b) If f is positive and concave upward on I, show that the function $g(x) = [f(x)]^2$ is concave upward on I.

81. (a) If f and g are positive, increasing, concave upward functions on I, show that the product function fg is concave upward on I.
(b) Show that part (a) remains true if f and g are both decreasing.
(c) Suppose f is increasing and g is decreasing. Show, by giving three examples, that fg may be concave upward, concave downward, or linear. Why doesn't the argument in parts (a) and (b) work in this case?

82. Suppose f and g are both concave upward on $(-\infty, \infty)$. Under what condition on f will the composite function $h(x) = f(g(x))$ be concave upward?

83. Show that $\tan x > x$ for $0 < x < \pi/2$. [*Hint:* Show that $f(x) = \tan x - x$ is increasing on $(0, \pi/2)$.]

84. (a) Show that $e^x \geq 1 + x$ for $x \geq 0$.
(b) Deduce that $e^x \geq 1 + x + \frac{1}{2}x^2$ for $x \geq 0$.
(c) Use mathematical induction to prove that for $x \geq 0$ and any positive integer n,
$$e^x \geq 1 + x + \frac{x^2}{2!} + \cdots + \frac{x^n}{n!}$$

85. Show that a cubic function (a third-degree polynomial) always has exactly one point of inflection. If its graph has three x-intercepts x_1, x_2, and x_3, show that the x-coordinate of the inflection point is $(x_1 + x_2 + x_3)/3$.

86. For what values of c does the polynomial $P(x) = x^4 + cx^3 + x^2$ have two inflection points? One inflection point? None? Illustrate by graphing P for several values of c. How does the graph change as c decreases?

87. Prove that if $(c, f(c))$ is a point of inflection of the graph of f and f'' exists in an open interval that contains c, then $f''(c) = 0$. [*Hint:* Apply the First Derivative Test and Fermat's Theorem to the function $g = f'$.]

88. Show that if $f(x) = x^4$, then $f''(0) = 0$, but $(0, 0)$ is not an inflection point of the graph of f.

89. Show that the function $g(x) = x|x|$ has an inflection point at $(0, 0)$ but $g''(0)$ does not exist.

90. Suppose that f''' is continuous and $f'(c) = f''(c) = 0$, but $f'''(c) > 0$. Does f have a local maximum or minimum at c? Does f have a point of inflection at c?

91. Suppose f is differentiable on an interval I and $f'(x) > 0$ for all numbers x in I except for a single number c. Prove that f is increasing on the entire interval I.

92. For what values of c is the function
$$f(x) = cx + \frac{1}{x^2 + 3}$$
increasing on $(-\infty, \infty)$?

93. The three cases in the First Derivative Test cover the situations one commonly encounters but do not exhaust all possibilities. Consider the functions f, g, and h whose values at 0 are all 0 and, for $x \neq 0$,
$$f(x) = x^4 \sin \frac{1}{x} \qquad g(x) = x^4\left(2 + \sin \frac{1}{x}\right)$$
$$h(x) = x^4\left(-2 + \sin \frac{1}{x}\right)$$
(a) Show that 0 is a critical number of all three functions but their derivatives change sign infinitely often on both sides of 0.
(b) Show that f has neither a local maximum nor a local minimum at 0, g has a local minimum, and h has a local maximum.

4.4 Indeterminate Forms and l'Hospital's Rule

Suppose we are trying to analyze the behavior of the function
$$F(x) = \frac{\ln x}{x - 1}$$

Although F is not defined when $x = 1$, we need to know how F behaves *near* 1. In particular, we would like to know the value of the limit

$$\lim_{x \to 1} \frac{\ln x}{x - 1}$$

In computing this limit we can't apply Law 5 of limits (the limit of a quotient is the quotient of the limits, see Section 2.3) because the limit of the denominator is 0. In fact, although the limit in (1) exists, its value is not obvious because both numerator and denominator approach 0 and $\frac{0}{0}$ is not defined.

In general, if we have a limit of the form
$$\lim_{x \to a} \frac{f(x)}{g(x)}$$
where both $f(x) \to 0$ and $g(x) \to 0$ as $x \to a$, then this limit may or may not exist and is called an **indeterminate form of type $\frac{0}{0}$**. We met some limits of this type in Chapter 2. For rational functions, we can cancel common factors:
$$\lim_{x \to 1} \frac{x^2 - x}{x^2 - 1} = \lim_{x \to 1} \frac{x(x - 1)}{(x + 1)(x - 1)} = \lim_{x \to 1} \frac{x}{x + 1} = \frac{1}{2}$$

We used a geometric argument to show that

$$\lim_{x \to 0} \frac{\sin x}{x} = 1$$

But these methods do not work for limits such as (1), so in this section we introduce a systematic method, known as *l'Hospital's Rule*, for the evaluation of indeterminate forms.

Another situation in which a limit is not obvious occurs when we look for a horizontal asymptote of F and need to evaluate the limit

$$\boxed{2} \qquad \lim_{x \to \infty} \frac{\ln x}{x - 1}$$

It isn't obvious how to evaluate this limit because both numerator and denominator become large as $x \to \infty$. There is a struggle between numerator and denominator. If the numerator wins, the limit will be ∞ (the numerator was increasing significantly faster than the denominator); if the denominator wins, the answer will be 0. Or there may be some compromise, in which case the answer will be some finite positive number.

In general, if we have a limit of the form

$$\lim_{x \to a} \frac{f(x)}{g(x)}$$

where both $f(x) \to \infty$ (or $-\infty$) and $g(x) \to \infty$ (or $-\infty$), then the limit may or may not exist and is called an **indeterminate form of type** ∞/∞. We saw in Section 2.6 that this type of limit can be evaluated for certain functions, including rational functions, by dividing numerator and denominator by the highest power of x that occurs in the denominator. For instance,

$$\lim_{x \to \infty} \frac{x^2 - 1}{2x^2 + 1} = \lim_{x \to \infty} \frac{1 - \frac{1}{x^2}}{2 + \frac{1}{x^2}} = \frac{1 - 0}{2 + 0} = \frac{1}{2}$$

This method does not work for limits such as (2), but l'Hospital's Rule also applies to this type of indeterminate form.

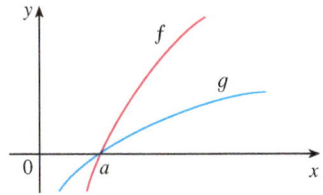

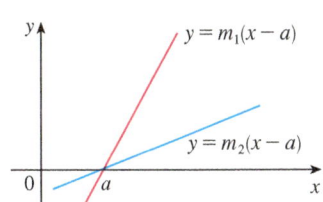

FIGURE 1

Figure 1 suggests visually why l'Hospital's Rule might be true. The first graph shows two differentiable functions f and g, each of which approaches 0 as $x \to a$. If we were to zoom in toward the point $(a, 0)$, the graphs would start to look almost linear. But if the functions actually *were* linear, as in the second graph, then their ratio would be

$$\frac{m_1(x - a)}{m_2(x - a)} = \frac{m_1}{m_2}$$

which is the ratio of their derivatives. This suggests that

$$\lim_{x \to a} \frac{f(x)}{g(x)} = \lim_{x \to a} \frac{f'(x)}{g'(x)}$$

L'Hospital's Rule Suppose f and g are differentiable and $g'(x) \neq 0$ on an open interval I that contains a (except possibly at a). Suppose that

$$\lim_{x \to a} f(x) = 0 \quad \text{and} \quad \lim_{x \to a} g(x) = 0$$

or that

$$\lim_{x \to a} f(x) = \pm\infty \quad \text{and} \quad \lim_{x \to a} g(x) = \pm\infty$$

(In other words, we have an indeterminate form of type $\frac{0}{0}$ or ∞/∞.) Then

$$\lim_{x \to a} \frac{f(x)}{g(x)} = \lim_{x \to a} \frac{f'(x)}{g'(x)}$$

if the limit on the right side exists (or is ∞ or $-\infty$).

NOTE 1 L'Hospital's Rule says that the limit of a quotient of functions is equal to the limit of the quotient of their derivatives, provided that the given conditions are satisfied.

306   **CHAPTER 4**   Applications of Differentiation

It is especially important to verify the conditions regarding the limits of f and g before using l'Hospital's Rule.

NOTE 2   L'Hospital's Rule is also valid for one-sided limits and for limits at infinity or negative infinity; that is, "$x \to a$" can be replaced by any of the symbols $x \to a^+$, $x \to a^-$, $x \to \infty$, or $x \to -\infty$.

NOTE 3   For the special case in which $f(a) = g(a) = 0$, f' and g' are continuous, and $g'(a) \neq 0$, it is easy to see why l'Hospital's Rule is true. In fact, using the alternative form of the definition of a derivative, we have

$$\lim_{x \to a} \frac{f'(x)}{g'(x)} = \frac{f'(a)}{g'(a)} = \frac{\lim_{x \to a} \dfrac{f(x) - f(a)}{x - a}}{\lim_{x \to a} \dfrac{g(x) - g(a)}{x - a}} = \lim_{x \to a} \frac{\dfrac{f(x) - f(a)}{x - a}}{\dfrac{g(x) - g(a)}{x - a}}$$

$$= \lim_{x \to a} \frac{f(x) - f(a)}{g(x) - g(a)} = \lim_{x \to a} \frac{f(x)}{g(x)} \qquad [\text{since } f(a) = g(a) = 0]$$

It is more difficult to prove the general version of l'Hospital's Rule. See Appendix F.

> **L'Hospital**
> L'Hospital's Rule is named after a French nobleman, the Marquis de l'Hospital (1661–1704), but was discovered by a Swiss mathematician, John Bernoulli (1667–1748). You might sometimes see l'Hospital spelled as l'Hôpital, but he spelled his own name l'Hospital, as was common in the 17th century. See Exercise 83 for the example that the Marquis used to illustrate his rule. See the project on page 314 for further historical details.

EXAMPLE 1   Find $\displaystyle\lim_{x \to 1} \frac{\ln x}{x - 1}$.

SOLUTION   Since

$$\lim_{x \to 1} \ln x = \ln 1 = 0 \qquad \text{and} \qquad \lim_{x \to 1} (x - 1) = 0$$

the limit is an indeterminate form of type $\tfrac{0}{0}$, so we can apply l'Hospital's Rule:

$$\lim_{x \to 1} \frac{\ln x}{x - 1} = \lim_{x \to 1} \frac{\dfrac{d}{dx}(\ln x)}{\dfrac{d}{dx}(x - 1)} = \lim_{x \to 1} \frac{1/x}{1}$$

$$= \lim_{x \to 1} \frac{1}{x} = 1 \qquad \blacksquare$$

> ⊘ Notice that when using l'Hospital's Rule we differentiate the numerator and denominator *separately*. We do *not* use the Quotient Rule.

EXAMPLE 2   Calculate $\displaystyle\lim_{x \to \infty} \frac{e^x}{x^2}$.

SOLUTION   We have $\lim_{x \to \infty} e^x = \infty$ and $\lim_{x \to \infty} x^2 = \infty$, so the limit is an indeterminate form of type ∞/∞, and l'Hospital's Rule gives

$$\lim_{x \to \infty} \frac{e^x}{x^2} = \lim_{x \to \infty} \frac{\dfrac{d}{dx}(e^x)}{\dfrac{d}{dx}(x^2)} = \lim_{x \to \infty} \frac{e^x}{2x}$$

Since $e^x \to \infty$ and $2x \to \infty$ as $x \to \infty$, the limit on the right side is also indeterminate, but a second application of l'Hospital's Rule gives

$$\lim_{x \to \infty} \frac{e^x}{x^2} = \lim_{x \to \infty} \frac{e^x}{2x} = \lim_{x \to \infty} \frac{e^x}{2} = \infty \qquad \blacksquare$$

> The graph of the function of Example 2 is shown in Figure 2. We have noticed previously that exponential functions grow far more rapidly than power functions, so the result of Example 2 is not unexpected. See also Exercise 73.

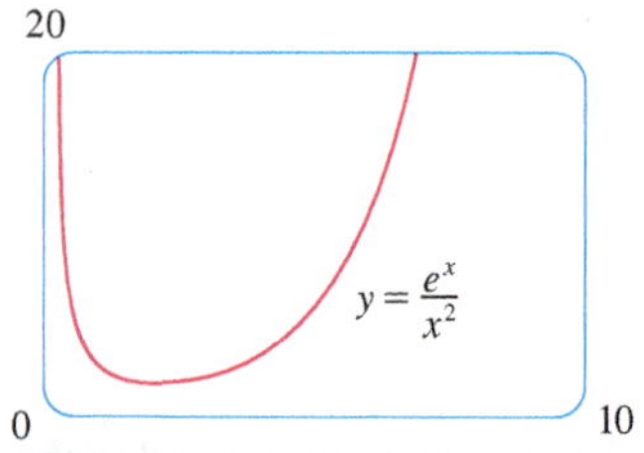

FIGURE 2

SECTION 4.4 Indeterminate Forms and l'Hospital's Rule 307

The graph of the function of Example 3 is shown in Figure 3. We have discussed previously the slow growth of logarithms, so it isn't surprising that this ratio approaches 0 as $x \to \infty$. See also Exercise 74.

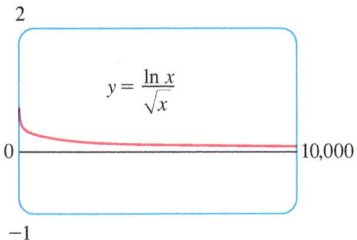

FIGURE 3

EXAMPLE 3 Calculate $\lim_{x \to \infty} \dfrac{\ln x}{\sqrt{x}}$.

SOLUTION Since $\ln x \to \infty$ and $\sqrt{x} \to \infty$ as $x \to \infty$, l'Hospital's Rule applies:

$$\lim_{x \to \infty} \frac{\ln x}{\sqrt{x}} = \lim_{x \to \infty} \frac{1/x}{\frac{1}{2}x^{-1/2}} = \lim_{x \to \infty} \frac{1/x}{1/(2\sqrt{x})}$$

Notice that the limit on the right side is now indeterminate of type $\frac{0}{0}$. But instead of applying l'Hospital's Rule a second time as we did in Example 2, we simplify the expression and see that a second application is unnecessary:

$$\lim_{x \to \infty} \frac{\ln x}{\sqrt{x}} = \lim_{x \to \infty} \frac{1/x}{1/(2\sqrt{x})} = \lim_{x \to \infty} \frac{2}{\sqrt{x}} = 0 \quad \blacksquare$$

In both Examples 2 and 3 we evaluated limits of type ∞/∞, but we got two different results. In Example 2, the infinite limit tells us that the numerator e^x increases significantly faster than the denominator x^2, resulting in larger and larger ratios. In fact, $y = e^x$ grows more quickly than all the power functions $y = x^n$ (see Exercise 73). In Example 3 we have the opposite situation; the limit of 0 means that the denominator outpaces the numerator, and the ratio eventually approaches 0.

EXAMPLE 4 Find $\lim_{x \to 0} \dfrac{\tan x - x}{x^3}$. (See Exercise 2.2.50.)

SOLUTION Noting that both $\tan x - x \to 0$ and $x^3 \to 0$ as $x \to 0$, we use l'Hospital's Rule:

$$\lim_{x \to 0} \frac{\tan x - x}{x^3} = \lim_{x \to 0} \frac{\sec^2 x - 1}{3x^2}$$

Since the limit on the right side is still indeterminate of type $\frac{0}{0}$, we apply l'Hospital's Rule again:

$$\lim_{x \to 0} \frac{\sec^2 x - 1}{3x^2} = \lim_{x \to 0} \frac{2 \sec^2 x \tan x}{6x}$$

The graph in Figure 4 gives visual confirmation of the result of Example 4. If we were to zoom in too far, however, we would get an inaccurate graph because $\tan x$ is close to x when x is small. See Exercise 2.2.50(d).

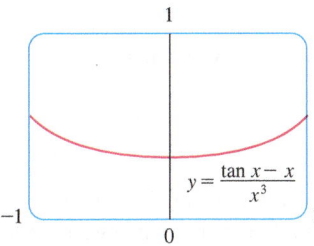

FIGURE 4

Because $\lim_{x \to 0} \sec^2 x = 1$, we simplify the calculation by writing

$$\lim_{x \to 0} \frac{2 \sec^2 x \tan x}{6x} = \frac{1}{3} \lim_{x \to 0} \sec^2 x \cdot \lim_{x \to 0} \frac{\tan x}{x} = \frac{1}{3} \lim_{x \to 0} \frac{\tan x}{x}$$

We can evaluate this last limit either by using l'Hospital's Rule a third time or by writing $\tan x$ as $(\sin x)/(\cos x)$ and making use of our knowledge of trigonometric limits. Putting together all the steps, we get

$$\lim_{x \to 0} \frac{\tan x - x}{x^3} = \lim_{x \to 0} \frac{\sec^2 x - 1}{3x^2} = \lim_{x \to 0} \frac{2 \sec^2 x \tan x}{6x}$$

$$= \frac{1}{3} \lim_{x \to 0} \frac{\tan x}{x} = \frac{1}{3} \lim_{x \to 0} \frac{\sec^2 x}{1} = \frac{1}{3} \quad \blacksquare$$

EXAMPLE 5 Find $\lim\limits_{x \to \pi^-} \dfrac{\sin x}{1 - \cos x}$.

SOLUTION If we blindly attempted to use l'Hospital's Rule, we would get

$$\lim_{x \to \pi^-} \frac{\sin x}{1 - \cos x} = \lim_{x \to \pi^-} \frac{\cos x}{\sin x} = -\infty$$

This is wrong! Although the numerator $\sin x \to 0$ as $x \to \pi^-$, notice that the denominator $(1 - \cos x)$ does not approach 0, so l'Hospital's Rule can't be applied here.

The required limit is, in fact, easy to find because the function is continuous at π and the denominator is nonzero there:

$$\lim_{x \to \pi^-} \frac{\sin x}{1 - \cos x} = \frac{\sin \pi}{1 - \cos \pi} = \frac{0}{1 - (-1)} = 0$$

Example 5 shows what can go wrong if you use l'Hospital's Rule without thinking. Other limits *can* be found using l'Hospital's Rule but are more easily found by other methods. (See Examples 2.3.3, 2.3.5, and 2.6.3, and the discussion at the beginning of this section.) So when evaluating any limit, you should consider other methods before using l'Hospital's Rule.

■ Indeterminate Products

If $\lim_{x \to a} f(x) = 0$ and $\lim_{x \to a} g(x) = \infty$ (or $-\infty$), then it isn't clear what the value of $\lim_{x \to a} [f(x) g(x)]$, if any, will be. There is a struggle between f and g. If f wins, the answer will be 0; if g wins, the answer will be ∞ (or $-\infty$). Or there may be a compromise where the answer is a finite nonzero number. This kind of limit is called an **indeterminate form of type $0 \cdot \infty$**. We can deal with it by writing the product fg as a quotient:

$$fg = \frac{f}{1/g} \quad \text{or} \quad fg = \frac{g}{1/f}$$

This converts the given limit into an indeterminate form of type $\tfrac{0}{0}$ or ∞/∞ so that we can use l'Hospital's Rule.

EXAMPLE 6 Evaluate $\lim\limits_{x \to 0^+} x \ln x$.

SOLUTION The given limit is indeterminate because, as $x \to 0^+$, the first factor (x) approaches 0 while the second factor $(\ln x)$ approaches $-\infty$. Writing $x = 1/(1/x)$, we have $1/x \to \infty$ as $x \to 0^+$, so l'Hospital's Rule gives

$$\lim_{x \to 0^+} x \ln x = \lim_{x \to 0^+} \frac{\ln x}{1/x} = \lim_{x \to 0^+} \frac{1/x}{-1/x^2} = \lim_{x \to 0^+} (-x) = 0$$

NOTE In solving Example 6 another possible option would have been to write

$$\lim_{x \to 0^+} x \ln x = \lim_{x \to 0^+} \frac{x}{1/\ln x}$$

This gives an indeterminate form of the type $\tfrac{0}{0}$, but if we apply l'Hospital's Rule we get a more complicated expression than the one we started with. In general, when we rewrite an indeterminate product, we try to choose the option that leads to the simpler limit.

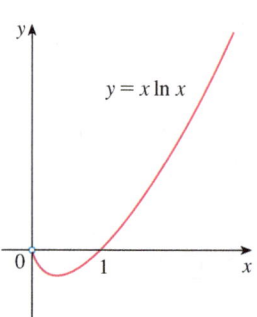

Figure 5 shows the graph of the function in Example 6. Notice that the function is undefined at $x = 0$; the graph approaches the origin but never quite reaches it.

FIGURE 5

Indeterminate Differences

If $\lim_{x \to a} f(x) = \infty$ and $\lim_{x \to a} g(x) = \infty$, then the limit

$$\lim_{x \to a} [f(x) - g(x)]$$

is called an **indeterminate form of type $\infty - \infty$**. Again there is a contest between f and g. Will the answer be ∞ (f wins) or will it be $-\infty$ (g wins) or will they compromise on a finite number? To find out, we try to convert the difference into a quotient (for instance, by using a common denominator, or rationalization, or factoring out a common factor) so that we have an indeterminate form of type $\frac{0}{0}$ or ∞/∞.

EXAMPLE 7 Compute $\lim_{x \to 1^+} \left(\dfrac{1}{\ln x} - \dfrac{1}{x-1} \right)$.

SOLUTION First notice that $1/(\ln x) \to \infty$ and $1/(x-1) \to \infty$ as $x \to 1^+$, so the limit is indeterminate of type $\infty - \infty$. Here we can start with a common denominator:

$$\lim_{x \to 1^+} \left(\dfrac{1}{\ln x} - \dfrac{1}{x-1} \right) = \lim_{x \to 1^+} \dfrac{x - 1 - \ln x}{(x-1) \ln x}$$

Both numerator and denominator have a limit of 0, so l'Hospital's Rule applies, giving

$$\lim_{x \to 1^+} \dfrac{x - 1 - \ln x}{(x-1) \ln x} = \lim_{x \to 1^+} \dfrac{1 - \dfrac{1}{x}}{(x-1) \cdot \dfrac{1}{x} + \ln x} = \lim_{x \to 1^+} \dfrac{x-1}{x - 1 + x \ln x}$$

Again we have an indeterminate limit of type $\frac{0}{0}$, so we apply l'Hospital's Rule a second time:

$$\lim_{x \to 1^+} \dfrac{x-1}{x - 1 + x \ln x} = \lim_{x \to 1^+} \dfrac{1}{1 + x \cdot \dfrac{1}{x} + \ln x}$$

$$= \lim_{x \to 1^+} \dfrac{1}{2 + \ln x} = \dfrac{1}{2}$$

EXAMPLE 8 Calculate $\lim_{x \to \infty} (e^x - x)$.

SOLUTION This is an indeterminate difference because both e^x and x approach infinity. We would expect the limit to be infinity because $e^x \to \infty$ much faster than x. But we can verify this by factoring out x:

$$e^x - x = x \left(\dfrac{e^x}{x} - 1 \right)$$

The term $e^x/x \to \infty$ as $x \to \infty$ by l'Hospital's Rule and so we now have a product in which both factors grow large:

$$\lim_{x \to \infty} (e^x - x) = \lim_{x \to \infty} \left[x \left(\dfrac{e^x}{x} - 1 \right) \right] = \infty$$

■ Indeterminate Powers

Several indeterminate forms arise from the limit

$$\lim_{x \to a} [f(x)]^{g(x)}$$

1. $\lim_{x \to a} f(x) = 0$ and $\lim_{x \to a} g(x) = 0$ type 0^0
2. $\lim_{x \to a} f(x) = \infty$ and $\lim_{x \to a} g(x) = 0$ type ∞^0
3. $\lim_{x \to a} f(x) = 1$ and $\lim_{x \to a} g(x) = \pm\infty$ type 1^∞

Each of these three cases can be treated either by taking the natural logarithm:

$$\text{let} \quad y = [f(x)]^{g(x)}, \quad \text{then} \quad \ln y = g(x) \ln f(x)$$

or by writing the function as an exponential:

$$[f(x)]^{g(x)} = e^{g(x) \ln f(x)}$$

(Recall that both of these methods were used in differentiating such functions.) In either method we are led to the indeterminate product $g(x) \ln f(x)$, which is of type $0 \cdot \infty$.

> Although forms of the type 0^0, ∞^0, and 1^∞ are indeterminate, the form 0^∞ is not indeterminate. (See Exercise 86.)

EXAMPLE 9 Calculate $\lim_{x \to 0^+} (1 + \sin 4x)^{\cot x}$.

SOLUTION First notice that as $x \to 0^+$, we have $1 + \sin 4x \to 1$ and $\cot x \to \infty$, so the given limit is indeterminate (type 1^∞). Let

$$y = (1 + \sin 4x)^{\cot x}$$

Then $\quad \ln y = \ln[(1 + \sin 4x)^{\cot x}] = \cot x \ln(1 + \sin 4x) = \dfrac{\ln(1 + \sin 4x)}{\tan x}$

so l'Hospital's Rule gives

$$\lim_{x \to 0^+} \ln y = \lim_{x \to 0^+} \frac{\ln(1 + \sin 4x)}{\tan x} = \lim_{x \to 0^+} \frac{\dfrac{4 \cos 4x}{1 + \sin 4x}}{\sec^2 x} = 4$$

So far we have computed the limit of $\ln y$, but what we want is the limit of y. To find this we use the fact that $y = e^{\ln y}$:

$$\lim_{x \to 0^+} (1 + \sin 4x)^{\cot x} = \lim_{x \to 0^+} y = \lim_{x \to 0^+} e^{\ln y} = e^4 \qquad \blacksquare$$

EXAMPLE 10 Find $\lim_{x \to 0^+} x^x$.

SOLUTION Notice that this limit is indeterminate since $0^x = 0$ for any $x > 0$ but $x^0 = 1$ for any $x \neq 0$. (Recall that 0^0 is undefined.) We could proceed as in Example 9 or by writing the function as an exponential:

$$x^x = (e^{\ln x})^x = e^{x \ln x}$$

In Example 6 we used l'Hospital's Rule to show that

$$\lim_{x \to 0^+} x \ln x = 0$$

> The graph of the function $y = x^x$, $x > 0$, is shown in Figure 6. Notice that although 0^0 is not defined, the values of the function approach 1 as $x \to 0^+$. This confirms the result of Example 10.

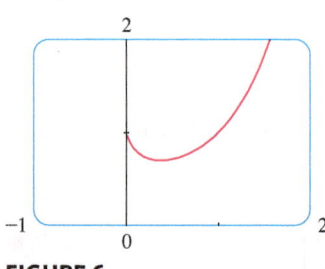

FIGURE 6

Therefore

$$\lim_{x \to 0^+} x^x = \lim_{x \to 0^+} e^{x \ln x} = e^0 = 1$$

4.4 EXERCISES

1–4 Given that

$$\lim_{x \to a} f(x) = 0 \quad \lim_{x \to a} g(x) = 0 \quad \lim_{x \to a} h(x) = 1$$

$$\lim_{x \to a} p(x) = \infty \quad \lim_{x \to a} q(x) = \infty$$

which of the following limits are indeterminate forms? For those that are not an indeterminate form, evaluate the limit where possible.

1. (a) $\lim_{x \to a} \dfrac{f(x)}{g(x)}$ (b) $\lim_{x \to a} \dfrac{f(x)}{p(x)}$

(c) $\lim_{x \to a} \dfrac{h(x)}{p(x)}$ (d) $\lim_{x \to a} \dfrac{p(x)}{f(x)}$

(e) $\lim_{x \to a} \dfrac{p(x)}{q(x)}$

2. (a) $\lim_{x \to a} [f(x)p(x)]$ (b) $\lim_{x \to a} [h(x)p(x)]$

(c) $\lim_{x \to a} [p(x)q(x)]$

3. (a) $\lim_{x \to a} [f(x) - p(x)]$ (b) $\lim_{x \to a} [p(x) - q(x)]$

(c) $\lim_{x \to a} [p(x) + q(x)]$

4. (a) $\lim_{x \to a} [f(x)]^{g(x)}$ (b) $\lim_{x \to a} [f(x)]^{p(x)}$

(c) $\lim_{x \to a} [h(x)]^{p(x)}$ (d) $\lim_{x \to a} [p(x)]^{f(x)}$

(e) $\lim_{x \to a} [p(x)]^{q(x)}$ (f) $\lim_{x \to a} \sqrt[q(x)]{p(x)}$

5–6 Use the graphs of f and g and their tangent lines at $(2, 0)$ to find $\lim_{x \to 2} \dfrac{f(x)}{g(x)}$.

5. 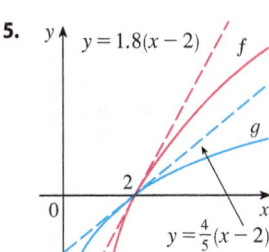   **6.**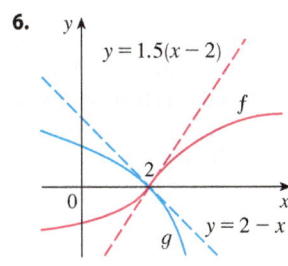

7. The graph of a function f and its tangent line at 0 are shown. What is the value of $\lim_{x \to 0} \dfrac{f(x)}{e^x - 1}$?

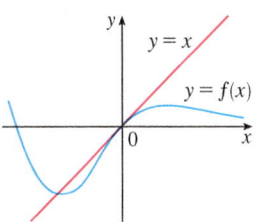

8–68 Find the limit. Use l'Hospital's Rule where appropriate. If there is a more elementary method, consider using it. If l'Hospital's Rule doesn't apply, explain why.

8. $\lim_{x \to 3} \dfrac{x - 3}{x^2 - 9}$

9. $\lim_{x \to 4} \dfrac{x^2 - 2x - 8}{x - 4}$

10. $\lim_{x \to -2} \dfrac{x^3 + 8}{x + 2}$

11. $\lim_{x \to 1} \dfrac{x^3 - 2x^2 + 1}{x^3 - 1}$

12. $\lim_{x \to 1/2} \dfrac{6x^2 + 5x - 4}{4x^2 + 16x - 9}$

13. $\lim_{x \to (\pi/2)^+} \dfrac{\cos x}{1 - \sin x}$

14. $\lim_{x \to 0} \dfrac{\tan 3x}{\sin 2x}$

15. $\lim_{t \to 0} \dfrac{e^{2t} - 1}{\sin t}$

16. $\lim_{x \to 0} \dfrac{x^2}{1 - \cos x}$

17. $\lim_{\theta \to \pi/2} \dfrac{1 - \sin \theta}{1 + \cos 2\theta}$

18. $\lim_{\theta \to \pi} \dfrac{1 + \cos \theta}{1 - \cos \theta}$

19. $\lim_{x \to \infty} \dfrac{\ln x}{\sqrt{x}}$

20. $\lim_{x \to \infty} \dfrac{x + x^2}{1 - 2x^2}$

21. $\lim_{x \to 0^+} \dfrac{\ln x}{x}$

22. $\lim_{x \to \infty} \dfrac{\ln \sqrt{x}}{x^2}$

23. $\lim_{t \to 1} \dfrac{t^8 - 1}{t^5 - 1}$

24. $\lim_{t \to 0} \dfrac{8^t - 5^t}{t}$

25. $\lim_{x \to 0} \dfrac{\sqrt{1 + 2x} - \sqrt{1 - 4x}}{x}$

26. $\lim_{u \to \infty} \dfrac{e^{u/10}}{u^3}$

27. $\lim_{x \to 0} \dfrac{e^x - 1 - x}{x^2}$

28. $\lim_{x \to 0} \dfrac{\sinh x - x}{x^3}$

29. $\lim_{x \to 0} \dfrac{\tanh x}{\tan x}$

30. $\lim_{x \to 0} \dfrac{x - \sin x}{x - \tan x}$

31. $\lim_{x \to 0} \dfrac{\sin^{-1} x}{x}$

32. $\lim_{x \to \infty} \dfrac{(\ln x)^2}{x}$

33. $\lim_{x \to 0} \dfrac{x 3^x}{3^x - 1}$

34. $\lim_{x \to 0} \dfrac{\cos mx - \cos nx}{x^2}$

35. $\lim_{x \to 0} \dfrac{\ln(1 + x)}{\cos x + e^x - 1}$

36. $\lim_{x \to 1} \dfrac{x \sin(x - 1)}{2x^2 - x - 1}$

37. $\lim_{x \to 0^+} \dfrac{\arctan(2x)}{\ln x}$

38. $\lim_{x \to 0^+} \dfrac{x^x - 1}{\ln x + x - 1}$

39. $\lim_{x \to 1} \dfrac{x^a - 1}{x^b - 1}, \; b \ne 0$

40. $\lim_{x \to 0} \dfrac{e^x - e^{-x} - 2x}{x - \sin x}$

41. $\lim_{x \to 0} \dfrac{\cos x - 1 + \frac{1}{2}x^2}{x^4}$

42. $\lim_{x \to a^+} \dfrac{\cos x \ln(x - a)}{\ln(e^x - e^a)}$

43. $\lim_{x \to \infty} x \sin(\pi/x)$

44. $\lim_{x \to \infty} \sqrt{x}\, e^{-x/2}$

45. $\lim_{x \to 0} \sin 5x \csc 3x$

46. $\lim_{x \to -\infty} x \ln\left(1 - \dfrac{1}{x}\right)$

47. $\lim_{x \to \infty} x^3 e^{-x^2}$

48. $\lim_{x \to \infty} x^{3/2} \sin(1/x)$

49. $\lim_{x \to 1^+} \ln x \tan(\pi x/2)$

50. $\lim_{x \to (\pi/2)^-} \cos x \sec 5x$

51. $\lim_{x \to 1} \left(\dfrac{x}{x - 1} - \dfrac{1}{\ln x}\right)$

52. $\lim_{x \to 0} (\csc x - \cot x)$

53. $\lim_{x \to 0^+} \left(\dfrac{1}{x} - \dfrac{1}{e^x - 1}\right)$

54. $\lim_{x \to 0^+} \left(\dfrac{1}{x} - \dfrac{1}{\tan^{-1} x}\right)$

55. $\lim_{x \to \infty} (x - \ln x)$

56. $\lim_{x \to 1^+} [\ln(x^7 - 1) - \ln(x^5 - 1)]$

57. $\lim_{x \to 0^+} x^{\sqrt{x}}$

58. $\lim_{x \to 0^+} (\tan 2x)^x$

59. $\lim_{x \to 0} (1 - 2x)^{1/x}$

60. $\lim_{x \to \infty} \left(1 + \dfrac{a}{x}\right)^{bx}$

61. $\lim_{x \to 1^+} x^{1/(1-x)}$

62. $\lim_{x \to \infty} x^{(\ln 2)/(1 + \ln x)}$

63. $\lim_{x \to \infty} x^{1/x}$

64. $\lim_{x \to \infty} x^{e^{-x}}$

65. $\lim_{x \to 0^+} (4x + 1)^{\cot x}$

66. $\lim_{x \to 1} (2 - x)^{\tan(\pi x/2)}$

67. $\lim_{x \to 0^+} (1 + \sin 3x)^{1/x}$

68. $\lim_{x \to \infty} \left(\dfrac{2x - 3}{2x + 5}\right)^{2x+1}$

69–70 Use a graph to estimate the value of the limit. Then use l'Hospital's Rule to find the exact value.

69. $\lim_{x \to \infty} \left(1 + \dfrac{2}{x}\right)^x$

70. $\lim_{x \to 0} \dfrac{5^x - 4^x}{3^x - 2^x}$

71–72 Illustrate l'Hospital's Rule by graphing both $f(x)/g(x)$ and $f'(x)/g'(x)$ near $x = 0$ to see that these ratios have the same limit as $x \to 0$. Also, calculate the exact value of the limit.

71. $f(x) = e^x - 1, \quad g(x) = x^3 + 4x$

72. $f(x) = 2x \sin x, \quad g(x) = \sec x - 1$

73. Prove that
$$\lim_{x \to \infty} \dfrac{e^x}{x^n} = \infty$$
for any positive integer n. This shows that the exponential function approaches infinity faster than any power of x.

74. Prove that
$$\lim_{x \to \infty} \dfrac{\ln x}{x^p} = 0$$
for any number $p > 0$. This shows that the logarithmic function approaches infinity more slowly than any power of x.

75–76 What happens if you try to use l'Hospital's Rule to find the limit? Evaluate the limit using another method.

75. $\lim_{x \to \infty} \dfrac{x}{\sqrt{x^2 + 1}}$

76. $\lim_{x \to (\pi/2)^-} \dfrac{\sec x}{\tan x}$

77. Investigate the family of curves $f(x) = e^x - cx$. In particular, find the limits as $x \to \pm\infty$ and determine the values of c for which f has an absolute minimum. What happens to the minimum points as c increases?

78. If an object with mass m is dropped from rest, one model for its speed v after t seconds, taking air resistance into account, is
$$v = \dfrac{mg}{c}\left(1 - e^{-ct/m}\right)$$
where g is the acceleration due to gravity and c is a positive constant. (In Chapter 9 we will be able to deduce this equation from the assumption that the air resistance is proportional to the speed of the object; c is the proportionality constant.)
(a) Calculate $\lim_{t \to \infty} v$. What is the meaning of this limit?
(b) For fixed t, use l'Hospital's Rule to calculate $\lim_{c \to 0^+} v$. What can you conclude about the velocity of a falling object in a vacuum?

79. If an initial amount A_0 of money is invested at an interest rate r compounded n times a year, the value of the investment after t years is

$$A = A_0\left(1 + \frac{r}{n}\right)^{nt}$$

If we let $n \to \infty$, we refer to the *continuous compounding* of interest. Use l'Hospital's Rule to show that if interest is compounded continuously, then the amount after t years is

$$A = A_0 e^{rt}$$

80. Light enters the eye through the pupil and strikes the retina, where photoreceptor cells sense light and color. W. Stanley Stiles and B. H. Crawford studied the phenomenon in which measured brightness decreases as light enters farther from the center of the pupil. (See the figure.)

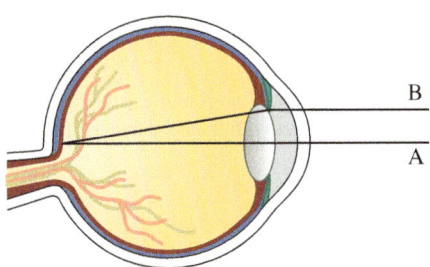

A light beam A that enters through the center of the pupil measures brighter than a beam B entering near the edge of the pupil.

They detailed their findings of this phenomenon, known as the *Stiles–Crawford effect of the first kind*, in an important paper published in 1933. In particular, they observed that the amount of luminance sensed was *not* proportional to the area of the pupil as they expected. The percentage P of the total luminance entering a pupil of radius r mm that is sensed at the retina can be described by

$$P = \frac{1 - 10^{-\rho r^2}}{\rho r^2 \ln 10}$$

where ρ is an experimentally determined constant, typically about 0.05.
(a) What is the percentage of luminance sensed by a pupil of radius 3 mm? Use $\rho = 0.05$.
(b) Compute the percentage of luminance sensed by a pupil of radius 2 mm. Does it make sense that it is larger than the answer to part (a)?
(c) Compute $\lim_{r \to 0^+} P$. Is the result what you would expect? Is this result physically possible?

Source: Adapted from W. Stiles and B. Crawford, "The Luminous Efficiency of Rays Entering the Eye Pupil at Different Points." *Proceedings of the Royal Society of London, Series B: Biological Sciences* 112 (1933): 428–50.

81. Some populations initally grow exponentially but eventually level off. Equations of the form

$$P(t) = \frac{M}{1 + Ae^{-kt}}$$

where M, A, and k are positive constants, are called *logistic equations* and are often used to model such populations. (We will investigate these in detail in Chapter 9.) Here M is called the *carrying capacity* and represents the maximum population size that can be supported, and $A = \dfrac{M - P_0}{P_0}$, where P_0 is the initial population.
(a) Compute $\lim_{t \to \infty} P(t)$. Explain why your answer is to be expected.
(b) Compute $\lim_{M \to \infty} P(t)$. (Note that A is defined in terms of M.) What kind of function is your result?

82. A metal cable has radius r and is covered by insulation so that the distance from the center of the cable to the exterior of the insulation is R. The velocity v of an electrical impulse in the cable is

$$v = -c\left(\frac{r}{R}\right)^2 \ln\left(\frac{r}{R}\right)$$

where c is a positive constant. Find the following limits and interpret your answers.
(a) $\lim_{R \to r^+} v$
(b) $\lim_{r \to 0^+} v$

83. The first appearance in print of l'Hospital's Rule was in the book *Analyse des Infiniment Petits* published by the Marquis de l'Hospital in 1696. This was the first calculus textbook ever published and the example that the Marquis used in that book to illustrate his rule was to find the limit of the function

$$y = \frac{\sqrt{2a^3x - x^4} - a\sqrt[3]{aax}}{a - \sqrt[4]{ax^3}}$$

as x approaches a, where $a > 0$. (At that time it was common to write aa instead of a^2.) Solve this problem.

84. The figure shows a sector of a circle with central angle θ. Let $A(\theta)$ be the area of the segment between the chord PR and the arc PR. Let $B(\theta)$ be the area of the triangle PQR. Find $\lim_{\theta \to 0^+} A(\theta)/B(\theta)$.

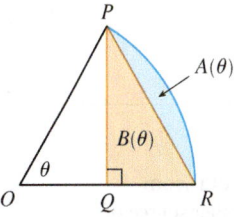

85. Evaluate

$$\lim_{x \to \infty}\left[x - x^2\ln\left(\frac{1 + x}{x}\right)\right].$$

86. Suppose f is a positive function. If $\lim_{x \to a} f(x) = 0$ and $\lim_{x \to a} g(x) = \infty$, show that

$$\lim_{x \to a} [f(x)]^{g(x)} = 0$$

This shows that 0^∞ is not an indeterminate form.

87. If f' is continuous, $f(2) = 0$, and $f'(2) = 7$, evaluate

$$\lim_{x \to 0} \frac{f(2 + 3x) + f(2 + 5x)}{x}$$

88. For what values of a and b is the following equation true?

$$\lim_{x \to 0} \left(\frac{\sin 2x}{x^3} + a + \frac{b}{x^2} \right) = 0$$

89. If f' is continuous, use l'Hospital's Rule to show that

$$\lim_{h \to 0} \frac{f(x + h) - f(x - h)}{2h} = f'(x)$$

Explain the meaning of this equation with the aid of a diagram.

90. If f'' is continuous, show that

$$\lim_{h \to 0} \frac{f(x + h) - 2f(x) + f(x - h)}{h^2} = f''(x)$$

91. Let

$$f(x) = \begin{cases} e^{-1/x^2} & \text{if } x \neq 0 \\ 0 & \text{if } x = 0 \end{cases}$$

(a) Use the definition of derivative to compute $f'(0)$.
(b) Show that f has derivatives of all orders that are defined on $\mathbb{R}$. [*Hint:* First show by induction that there is a polynomial $p_n(x)$ and a nonnegative integer k_n such that $f^{(n)}(x) = p_n(x)f(x)/x^{k_n}$ for $x \neq 0$.]

92. Let

$$f(x) = \begin{cases} |x|^x & \text{if } x \neq 0 \\ 1 & \text{if } x = 0 \end{cases}$$

(a) Show that f is continuous at 0.
(b) Investigate graphically whether f is differentiable at 0 by zooming in several times toward the point (0, 1) on the graph of f.
(c) Show that f is not differentiable at 0. How can you reconcile this fact with the appearance of the graphs in part (b)?

WRITING PROJECT THE ORIGINS OF L'HOSPITAL'S RULE

L'Hospital's Rule was first published in 1696 in the Marquis de l'Hospital's calculus textbook *Analyse des Infiniment Petits*, but the rule was discovered in 1694 by the Swiss mathematician John (Johann) Bernoulli. The explanation is that these two mathematicians had entered into a curious business arrangement whereby the Marquis de l'Hospital bought the rights to Bernoulli's mathematical discoveries. The details, including a translation of l'Hospital's letter to Bernoulli proposing the arrangement, can be found in the book by Eves [1].

Write a report on the historical and mathematical origins of l'Hospital's Rule. Start by providing brief biographical details of both men (the dictionary edited by Gillispie [2] is a good source) and outline the business deal between them. Then give l'Hospital's statement of his rule, which is found in Struik's sourcebook [4] and more briefly in the book of Katz [3]. Notice that l'Hospital and Bernoulli formulated the rule geometrically and gave the answer in terms of differentials. Compare their statement with the version of l'Hospital's Rule given in Section 4.4 and show that the two statements are essentially the same.

1. Howard Eves, *In Mathematical Circles (Volume 2: Quadrants III and IV)* (Boston: Prindle, Weber and Schmidt, 1969), pp. 20–22.

2. C. C. Gillispie, ed., *Dictionary of Scientific Biography* (New York: Scribner's, 1974). See the article on Johann Bernoulli by E. A. Fellmann and J. O. Fleckenstein in Volume II and the article on the Marquis de l'Hospital by Abraham Robinson in Volume VIII.

3. Victor Katz, *A History of Mathematics: An Introduction* (New York: HarperCollins, 1993), p. 484.

4. D. J. Struik, ed., *A Sourcebook in Mathematics, 1200–1800* (Princeton, NJ: Princeton University Press, 1969), pp. 315–16.

www.stewartcalculus.com
The Internet is another source of information for this project. Click on *History of Mathematics* for a list of reliable websites.

4.5 Summary of Curve Sketching

So far we have been concerned with some particular aspects of curve sketching: domain, range, and symmetry in Chapter 1; limits, continuity, and asymptotes in Chapter 2; derivatives and tangents in Chapters 2 and 3; and extreme values, intervals of increase and decrease, concavity, points of inflection, and l'Hospital's Rule in this chapter. It is now time to put all of this information together to sketch graphs that reveal the important features of functions.

You might ask: Why don't we just use a graphing calculator or computer to graph a curve? Why do we need to use calculus?

It's true that modern technology is capable of producing very accurate graphs. But even the best graphing devices have to be used intelligently. It is easy to arrive at a misleading graph, or to miss important details of a curve, when relying solely on technology. (See "Graphing Calculators and Computers" at www.stewartcalculus.com, especially Examples 1, 3, 4, and 5. See also Section 4.6.) The use of calculus enables us to discover the most interesting aspects of graphs and in many cases to calculate maximum and minimum points and inflection points *exactly* instead of approximately.

For instance, Figure 1 shows the graph of $f(x) = 8x^3 - 21x^2 + 18x + 2$. At first glance it seems reasonable: It has the same shape as cubic curves like $y = x^3$, and it appears to have no maximum or minimum point. But if you compute the derivative, you will see that there is a maximum when $x = 0.75$ and a minimum when $x = 1$. Indeed, if we zoom in to this portion of the graph, we see that behavior exhibited in Figure 2. Without calculus, we could easily have overlooked it.

In the next section we will graph functions by using the interaction between calculus and graphing devices. In this section we draw graphs by first considering the following information. We don't assume that you have a graphing device, but if you do have one you should use it as a check on your work.

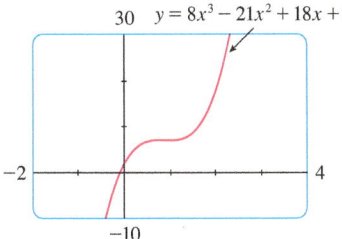

FIGURE 1

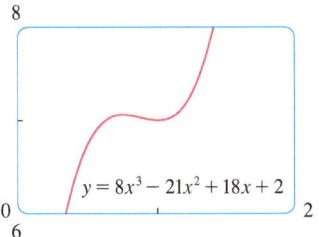

FIGURE 2

■ Guidelines for Sketching a Curve

The following checklist is intended as a guide to sketching a curve $y = f(x)$ by hand. Not every item is relevant to every function. (For instance, a given curve might not have an asymptote or possess symmetry.) But the guidelines provide all the information you need to make a sketch that displays the most important aspects of the function.

A. Domain It's often useful to start by determining the domain D of f, that is, the set of values of x for which $f(x)$ is defined.

B. Intercepts The y-intercept is $f(0)$ and this tells us where the curve intersects the y-axis. To find the x-intercepts, we set $y = 0$ and solve for x. (You can omit this step if the equation is difficult to solve.)

C. Symmetry

(i) If $f(-x) = f(x)$ for all x in D, that is, the equation of the curve is unchanged when x is replaced by $-x$, then f is an **even function** and the curve is symmetric about the y-axis. This means that our work is cut in half. If we know what the curve looks like for $x \geq 0$, then we need only reflect about the y-axis to obtain the complete curve [see Figure 3(a)]. Here are some examples: $y = x^2$, $y = x^4$, $y = |x|$, and $y = \cos x$.

(ii) If $f(-x) = -f(x)$ for all x in D, then f is an **odd function** and the curve is symmetric about the origin. Again we can obtain the complete curve if we know what it looks like for $x \geq 0$. [Rotate 180° about the origin; see Figure 3(b).] Some simple examples of odd functions are $y = x$, $y = x^3$, $y = x^5$, and $y = \sin x$.

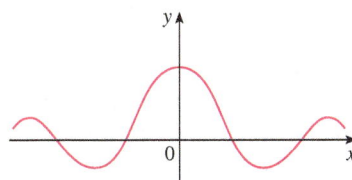

(a) Even function: reflectional symmetry

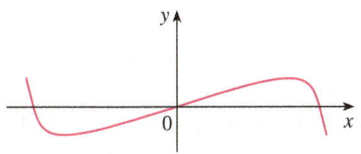

(b) Odd function: rotational symmetry

FIGURE 3

(iii) If $f(x + p) = f(x)$ for all x in D, where p is a positive constant, then f is called a **periodic function** and the smallest such number p is called the **period**. For instance, $y = \sin x$ has period 2π and $y = \tan x$ has period π. If we know what the graph looks like in an interval of length p, then we can use translation to sketch the entire graph (see Figure 4).

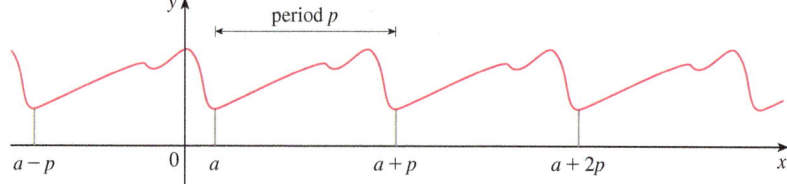

FIGURE 4
Periodic function: translational symmetry

D. Asymptotes

(i) *Horizontal Asymptotes.* Recall from Section 2.6 that if either $\lim_{x \to \infty} f(x) = L$ or $\lim_{x \to -\infty} f(x) = L$, then the line $y = L$ is a horizontal asymptote of the curve $y = f(x)$. If it turns out that $\lim_{x \to \infty} f(x) = \infty$ (or $-\infty$), then we do not have an asymptote to the right, but this fact is still useful information for sketching the curve.

(ii) *Vertical Asymptotes.* Recall from Section 2.2 that the line $x = a$ is a vertical asymptote if at least one of the following statements is true:

$$\boxed{1} \qquad \lim_{x \to a^+} f(x) = \infty \qquad \lim_{x \to a^-} f(x) = \infty$$

$$\lim_{x \to a^+} f(x) = -\infty \qquad \lim_{x \to a^-} f(x) = -\infty$$

(For rational functions you can locate the vertical asymptotes by equating the denominator to 0 after canceling any common factors. But for other functions this method does not apply.) Furthermore, in sketching the curve it is very useful to know exactly which of the statements in (1) is true. If $f(a)$ is not defined but a is an endpoint of the domain of f, then you should compute $\lim_{x \to a^-} f(x)$ or $\lim_{x \to a^+} f(x)$, whether or not this limit is infinite.

(iii) *Slant Asymptotes.* These are discussed at the end of this section.

E. Intervals of Increase or Decrease Use the I/D Test. Compute $f'(x)$ and find the intervals on which $f'(x)$ is positive (f is increasing) and the intervals on which $f'(x)$ is negative (f is decreasing).

F. Local Maximum and Minimum Values Find the critical numbers of f [the numbers c where $f'(c) = 0$ or $f'(c)$ does not exist]. Then use the First Derivative Test. If f' changes from positive to negative at a critical number c, then $f(c)$ is a local maximum. If f' changes from negative to positive at c, then $f(c)$ is a local minimum. Although it is usually preferable to use the First Derivative Test, you can use the Second Derivative Test if $f'(c) = 0$ and $f''(c) \neq 0$. Then $f''(c) > 0$ implies that $f(c)$ is a local minimum, whereas $f''(c) < 0$ implies that $f(c)$ is a local maximum.

G. Concavity and Points of Inflection Compute $f''(x)$ and use the Concavity Test. The curve is concave upward where $f''(x) > 0$ and concave downward where $f''(x) < 0$. Inflection points occur where the direction of concavity changes.

H. Sketch the Curve Using the information in items A–G, draw the graph. Sketch the asymptotes as dashed lines. Plot the intercepts, maximum and minimum points, and inflection points. Then make the curve pass through these points, rising and falling according to E, with concavity according to G, and approaching the asymptotes.

SECTION 4.5 Summary of Curve Sketching   317

If additional accuracy is desired near any point, you can compute the value of the derivative there. The tangent indicates the direction in which the curve proceeds.

EXAMPLE 1 Use the guidelines to sketch the curve $y = \dfrac{2x^2}{x^2 - 1}$.

A. The domain is
$$\{x \mid x^2 - 1 \neq 0\} = \{x \mid x \neq \pm 1\} = (-\infty, -1) \cup (-1, 1) \cup (1, \infty)$$

B. The x- and y-intercepts are both 0.
C. Since $f(-x) = f(x)$, the function f is even. The curve is symmetric about the y-axis.

D.
$$\lim_{x \to \pm\infty} \frac{2x^2}{x^2 - 1} = \lim_{x \to \pm\infty} \frac{2}{1 - 1/x^2} = 2$$

Therefore the line $y = 2$ is a horizontal asymptote.
Since the denominator is 0 when $x = \pm 1$, we compute the following limits:

$$\lim_{x \to 1^+} \frac{2x^2}{x^2 - 1} = \infty \qquad \lim_{x \to 1^-} \frac{2x^2}{x^2 - 1} = -\infty$$

$$\lim_{x \to -1^+} \frac{2x^2}{x^2 - 1} = -\infty \qquad \lim_{x \to -1^-} \frac{2x^2}{x^2 - 1} = \infty$$

Therefore the lines $x = 1$ and $x = -1$ are vertical asymptotes. This information about limits and asymptotes enables us to draw the preliminary sketch in Figure 5, showing the parts of the curve near the asymptotes.

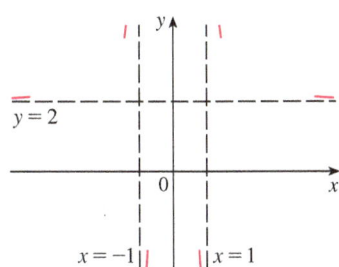

FIGURE 5
Preliminary sketch

We have shown the curve approaching its horizontal asymptote from above in Figure 5. This is confirmed by the intervals of increase and decrease.

E.
$$f'(x) = \frac{(x^2 - 1)(4x) - 2x^2 \cdot 2x}{(x^2 - 1)^2} = \frac{-4x}{(x^2 - 1)^2}$$

Since $f'(x) > 0$ when $x < 0$ ($x \neq -1$) and $f'(x) < 0$ when $x > 0$ ($x \neq 1$), f is increasing on $(-\infty, -1)$ and $(-1, 0)$ and decreasing on $(0, 1)$ and $(1, \infty)$.

F. The only critical number is $x = 0$. Since f' changes from positive to negative at 0, $f(0) = 0$ is a local maximum by the First Derivative Test.

G.
$$f''(x) = \frac{(x^2 - 1)^2(-4) + 4x \cdot 2(x^2 - 1)2x}{(x^2 - 1)^4} = \frac{12x^2 + 4}{(x^2 - 1)^3}$$

Since $12x^2 + 4 > 0$ for all x, we have
$$f''(x) > 0 \iff x^2 - 1 > 0 \iff |x| > 1$$

and $f''(x) < 0 \iff |x| < 1$. Thus the curve is concave upward on the intervals $(-\infty, -1)$ and $(1, \infty)$ and concave downward on $(-1, 1)$. It has no point of inflection since 1 and -1 are not in the domain of f.

H. Using the information in E–G, we finish the sketch in Figure 6.

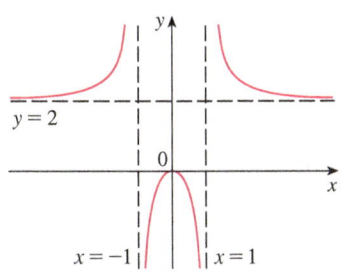

FIGURE 6
Finished sketch of $y = \dfrac{2x^2}{x^2 - 1}$

EXAMPLE 2 Sketch the graph of $f(x) = \dfrac{x^2}{\sqrt{x + 1}}$.

A. Domain $= \{x \mid x + 1 > 0\} = \{x \mid x > -1\} = (-1, \infty)$

B. The x- and y-intercepts are both 0.
C. Symmetry: None
D. Since
$$\lim_{x \to \infty} \frac{x^2}{\sqrt{x+1}} = \infty$$
there is no horizontal asymptote. Since $\sqrt{x+1} \to 0$ as $x \to -1^+$ and $f(x)$ is always positive, we have
$$\lim_{x \to -1^+} \frac{x^2}{\sqrt{x+1}} = \infty$$
and so the line $x = -1$ is a vertical asymptote.

E. $f'(x) = \dfrac{\sqrt{x+1}\,(2x) - x^2 \cdot 1/(2\sqrt{x+1})}{x+1} = \dfrac{3x^2 + 4x}{2(x+1)^{3/2}} = \dfrac{x(3x+4)}{2(x+1)^{3/2}}$

We see that $f'(x) = 0$ when $x = 0$ (notice that $-\frac{4}{3}$ is not in the domain of f), so the only critical number is 0. Since $f'(x) < 0$ when $-1 < x < 0$ and $f'(x) > 0$ when $x > 0$, f is decreasing on $(-1, 0)$ and increasing on $(0, \infty)$.

F. Since $f'(0) = 0$ and f' changes from negative to positive at 0, $f(0) = 0$ is a local (and absolute) minimum by the First Derivative Test.

G. $f''(x) = \dfrac{2(x+1)^{3/2}(6x+4) - (3x^2 + 4x)3(x+1)^{1/2}}{4(x+1)^3} = \dfrac{3x^2 + 8x + 8}{4(x+1)^{5/2}}$

Note that the denominator is always positive. The numerator is the quadratic $3x^2 + 8x + 8$, which is always positive because its discriminant is $b^2 - 4ac = -32$, which is negative, and the coefficient of x^2 is positive. Thus $f''(x) > 0$ for all x in the domain of f, which means that f is concave upward on $(-1, \infty)$ and there is no point of inflection.

H. The curve is sketched in Figure 7.

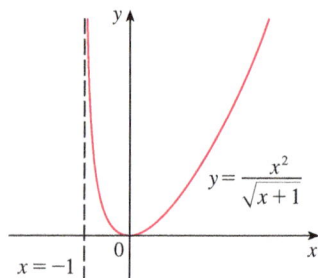

FIGURE 7

EXAMPLE 3 Sketch the graph of $f(x) = xe^x$.

A. The domain is $\mathbb{R}$.
B. The x- and y-intercepts are both 0.
C. Symmetry: None
D. Because both x and e^x become large as $x \to \infty$, we have $\lim_{x \to \infty} xe^x = \infty$. As $x \to -\infty$, however, $e^x \to 0$ and so we have an indeterminate product that requires the use of l'Hospital's Rule:
$$\lim_{x \to -\infty} xe^x = \lim_{x \to -\infty} \frac{x}{e^{-x}} = \lim_{x \to -\infty} \frac{1}{-e^{-x}} = \lim_{x \to -\infty} (-e^x) = 0$$
Thus the x-axis is a horizontal asymptote.

E. $$f'(x) = xe^x + e^x = (x+1)e^x$$

Since e^x is always positive, we see that $f'(x) > 0$ when $x + 1 > 0$, and $f'(x) < 0$ when $x + 1 < 0$. So f is increasing on $(-1, \infty)$ and decreasing on $(-\infty, -1)$.

F. Because $f'(-1) = 0$ and f' changes from negative to positive at $x = -1$, $f(-1) = -e^{-1} \approx -0.37$ is a local (and absolute) minimum.

G. $$f''(x) = (x+1)e^x + e^x = (x+2)e^x$$

Since $f''(x) > 0$ if $x > -2$ and $f''(x) < 0$ if $x < -2$, f is concave upward on

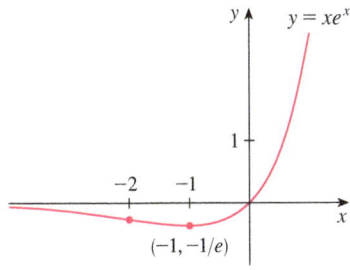

FIGURE 8

(−2, ∞) and concave downward on (−∞, −2). The inflection point is (−2, −2e^{−2}) ≈ (−2, −0.27).

H. We use this information to sketch the curve in Figure 8.

EXAMPLE 4 Sketch the graph of $f(x) = \dfrac{\cos x}{2 + \sin x}$.

A. The domain is $\mathbb{R}$.

B. The y-intercept is $f(0) = \tfrac{1}{2}$. The x-intercepts occur when $\cos x = 0$, that is, $x = (\pi/2) + n\pi$, where n is an integer.

C. f is neither even nor odd, but $f(x + 2\pi) = f(x)$ for all x and so f is periodic and has period 2π. Thus, in what follows, we need to consider only $0 \leq x \leq 2\pi$ and then extend the curve by translation in part H.

D. Asymptotes: None

E. $$f'(x) = \frac{(2 + \sin x)(-\sin x) - \cos x\,(\cos x)}{(2 + \sin x)^2} = -\frac{2 \sin x + 1}{(2 + \sin x)^2}$$

The denominator is always positive, so $f'(x) > 0$ when $2 \sin x + 1 < 0 \iff \sin x < -\tfrac{1}{2} \iff 7\pi/6 < x < 11\pi/6$. So f is increasing on $(7\pi/6, 11\pi/6)$ and decreasing on $(0, 7\pi/6)$ and $(11\pi/6, 2\pi)$.

F. From part E and the First Derivative Test, we see that the local minimum value is $f(7\pi/6) = -1/\sqrt{3}$ and the local maximum value is $f(11\pi/6) = 1/\sqrt{3}$.

G. If we use the Quotient Rule again and simplify, we get

$$f''(x) = -\frac{2 \cos x\,(1 - \sin x)}{(2 + \sin x)^3}$$

Because $(2 + \sin x)^3 > 0$ and $1 - \sin x \geq 0$ for all x, we know that $f''(x) > 0$ when $\cos x < 0$, that is, $\pi/2 < x < 3\pi/2$. So f is concave upward on $(\pi/2, 3\pi/2)$ and concave downward on $(0, \pi/2)$ and $(3\pi/2, 2\pi)$. The inflection points are $(\pi/2, 0)$ and $(3\pi/2, 0)$.

H. The graph of the function restricted to $0 \leq x \leq 2\pi$ is shown in Figure 9. Then we extend it, using periodicity, to the complete graph in Figure 10.

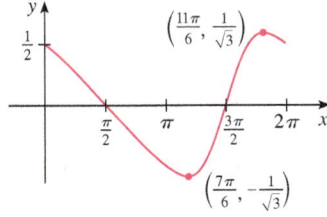

FIGURE 9

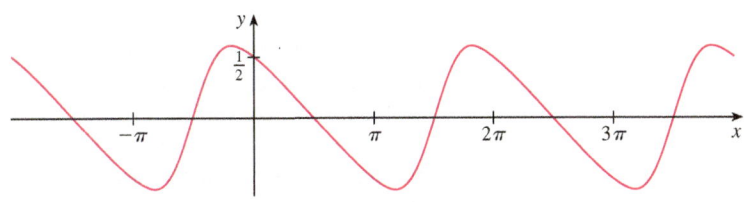

FIGURE 10

EXAMPLE 5 Sketch the graph of $y = \ln(4 - x^2)$.

A. The domain is

$$\{x \mid 4 - x^2 > 0\} = \{x \mid x^2 < 4\} = \{x \mid |x| < 2\} = (-2, 2)$$

B. The y-intercept is $f(0) = \ln 4$. To find the x-intercept we set
$$y = \ln(4 - x^2) = 0$$
We know that $\ln 1 = 0$, so we have $4 - x^2 = 1 \;\Rightarrow\; x^2 = 3$ and therefore the x-intercepts are $\pm\sqrt{3}$.

C. Since $f(-x) = f(x)$, f is even and the curve is symmetric about the y-axis.

D. We look for vertical asymptotes at the endpoints of the domain. Since $4 - x^2 \to 0^+$ as $x \to 2^-$ and also as $x \to -2^+$, we have
$$\lim_{x \to 2^-} \ln(4 - x^2) = -\infty \qquad \lim_{x \to -2^+} \ln(4 - x^2) = -\infty$$
Thus the lines $x = 2$ and $x = -2$ are vertical asymptotes.

E.
$$f'(x) = \frac{-2x}{4 - x^2}$$

Since $f'(x) > 0$ when $-2 < x < 0$ and $f'(x) < 0$ when $0 < x < 2$, f is increasing on $(-2, 0)$ and decreasing on $(0, 2)$.

F. The only critical number is $x = 0$. Since f' changes from positive to negative at 0, $f(0) = \ln 4$ is a local maximum by the First Derivative Test.

G.
$$f''(x) = \frac{(4 - x^2)(-2) + 2x(-2x)}{(4 - x^2)^2} = \frac{-8 - 2x^2}{(4 - x^2)^2}$$

Since $f''(x) < 0$ for all x, the curve is concave downward on $(-2, 2)$ and has no inflection point.

H. Using this information, we sketch the curve in Figure 11. ∎

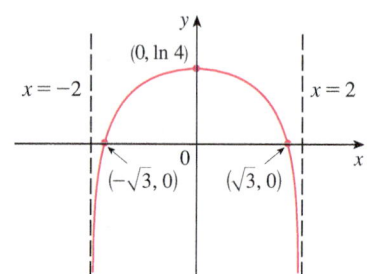

FIGURE 11
$y = \ln(4 - x^2)$

Slant Asymptotes

Some curves have asymptotes that are *oblique*, that is, neither horizontal nor vertical. If
$$\lim_{x \to \infty} [f(x) - (mx + b)] = 0$$
where $m \neq 0$, then the line $y = mx + b$ is called a **slant asymptote** because the vertical distance between the curve $y = f(x)$ and the line $y = mx + b$ approaches 0, as in Figure 12. (A similar situation exists if we let $x \to -\infty$.) For rational functions, slant asymptotes occur when the degree of the numerator is one more than the degree of the denominator. In such a case the equation of the slant asymptote can be found by long division as in the following example.

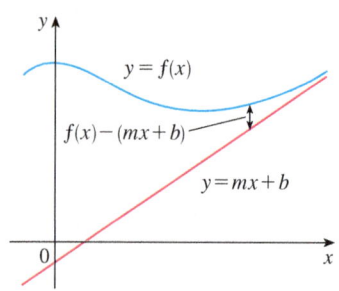

FIGURE 12

EXAMPLE 6 Sketch the graph of $f(x) = \dfrac{x^3}{x^2 + 1}$.

A. The domain is $\mathbb{R} = (-\infty, \infty)$.

B. The x- and y-intercepts are both 0.

C. Since $f(-x) = -f(x)$, f is odd and its graph is symmetric about the origin.

D. Since $x^2 + 1$ is never 0, there is no vertical asymptote. Since $f(x) \to \infty$ as $x \to \infty$ and $f(x) \to -\infty$ as $x \to -\infty$, there is no horizontal asymptote. But long division gives

$$f(x) = \frac{x^3}{x^2 + 1} = x - \frac{x}{x^2 + 1}$$

This equation suggests that $y = x$ is a candidate for a slant asymptote. In fact,

$$f(x) - x = -\frac{x}{x^2 + 1} = -\frac{\frac{1}{x}}{1 + \frac{1}{x^2}} \to 0 \quad \text{as} \quad x \to \pm\infty$$

So the line $y = x$ is a slant asymptote.

E. $$f'(x) = \frac{(x^2 + 1)(3x^2) - x^3 \cdot 2x}{(x^2 + 1)^2} = \frac{x^2(x^2 + 3)}{(x^2 + 1)^2}$$

Since $f'(x) > 0$ for all x (except 0), f is increasing on $(-\infty, \infty)$.

F. Although $f'(0) = 0$, f' does not change sign at 0, so there is no local maximum or minimum.

G. $$f''(x) = \frac{(x^2 + 1)^2(4x^3 + 6x) - (x^4 + 3x^2) \cdot 2(x^2 + 1)2x}{(x^2 + 1)^4} = \frac{2x(3 - x^2)}{(x^2 + 1)^3}$$

Since $f''(x) = 0$ when $x = 0$ or $x = \pm\sqrt{3}$, we set up the following chart:

Interval	x	$3 - x^2$	$(x^2 + 1)^3$	$f''(x)$	f
$x < -\sqrt{3}$	−	−	+	+	CU on $(-\infty, -\sqrt{3})$
$-\sqrt{3} < x < 0$	−	+	+	−	CD on $(-\sqrt{3}, 0)$
$0 < x < \sqrt{3}$	+	+	+	+	CU on $(0, \sqrt{3})$
$x > \sqrt{3}$	+	−	+	−	CD on $(\sqrt{3}, \infty)$

The points of inflection are $\left(-\sqrt{3}, -\tfrac{3}{4}\sqrt{3}\right)$, $(0, 0)$, and $\left(\sqrt{3}, \tfrac{3}{4}\sqrt{3}\right)$.

H. The graph of f is sketched in Figure 13.

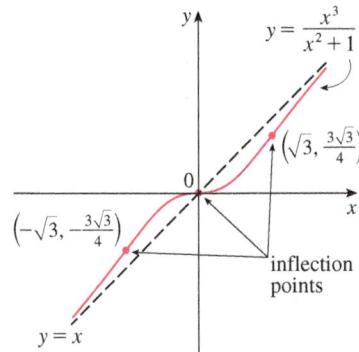

FIGURE 13

4.5 EXERCISES

1–54 Use the guidelines of this section to sketch the curve.

1. $y = x^3 + 3x^2$
2. $y = 2 + 3x^2 - x^3$
3. $y = x^4 - 4x$
4. $y = x^4 - 8x^2 + 8$
5. $y = x(x - 4)^3$
6. $y = x^5 - 5x$
7. $y = \tfrac{1}{5}x^5 - \tfrac{8}{3}x^3 + 16x$
8. $y = (4 - x^2)^5$
9. $y = \dfrac{x}{x - 1}$
10. $y = \dfrac{x^2 + 5x}{25 - x^2}$
11. $y = \dfrac{x - x^2}{2 - 3x + x^2}$
12. $y = 1 + \dfrac{1}{x} + \dfrac{1}{x^2}$
13. $y = \dfrac{x}{x^2 - 4}$
14. $y = \dfrac{1}{x^2 - 4}$
15. $y = \dfrac{x^2}{x^2 + 3}$
16. $y = \dfrac{(x - 1)^2}{x^2 + 1}$
17. $y = \dfrac{x - 1}{x^2}$
18. $y = \dfrac{x}{x^3 - 1}$

19. $y = \dfrac{x^3}{x^3 + 1}$

20. $y = \dfrac{x^3}{x - 2}$

21. $y = (x - 3)\sqrt{x}$

22. $y = (x - 4)\sqrt[3]{x}$

23. $y = \sqrt{x^2 + x - 2}$

24. $y = \sqrt{x^2 + x} - x$

25. $y = \dfrac{x}{\sqrt{x^2 + 1}}$

26. $y = x\sqrt{2 - x^2}$

27. $y = \dfrac{\sqrt{1 - x^2}}{x}$

28. $y = \dfrac{x}{\sqrt{x^2 - 1}}$

29. $y = x - 3x^{1/3}$

30. $y = x^{5/3} - 5x^{2/3}$

31. $y = \sqrt[3]{x^2 - 1}$

32. $y = \sqrt[3]{x^3 + 1}$

33. $y = \sin^3 x$

34. $y = x + \cos x$

35. $y = x \tan x$, $\ -\pi/2 < x < \pi/2$

36. $y = 2x - \tan x$, $\ -\pi/2 < x < \pi/2$

37. $y = \sin x + \sqrt{3}\cos x$, $\ -2\pi \leq x \leq 2\pi$

38. $y = \csc x - 2\sin x$, $\ 0 < x < \pi$

39. $y = \dfrac{\sin x}{1 + \cos x}$

40. $y = \dfrac{\sin x}{2 + \cos x}$

41. $y = \arctan(e^x)$

42. $y = (1 - x)e^x$

43. $y = 1/(1 + e^{-x})$

44. $y = e^{-x} \sin x$, $\ 0 \leq x \leq 2\pi$

45. $y = \dfrac{1}{x} + \ln x$

46. $y = e^{2x} - e^x$

47. $y = (1 + e^x)^{-2}$

48. $y = e^x/x^2$

49. $y = \ln(\sin x)$

50. $y = \ln(1 + x^3)$

51. $y = xe^{-1/x}$

52. $y = \dfrac{\ln x}{x^2}$

53. $y = e^{\arctan x}$

54. $y = \tan^{-1}\left(\dfrac{x - 1}{x + 1}\right)$

55. In the theory of relativity, the mass of a particle is

$$m = \dfrac{m_0}{\sqrt{1 - v^2/c^2}}$$

where m_0 is the rest mass of the particle, m is the mass when the particle moves with speed v relative to the observer, and c is the speed of light. Sketch the graph of m as a function of v.

56. In the theory of relativity, the energy of a particle is

$$E = \sqrt{m_0^2 c^4 + h^2 c^2/\lambda^2}$$

where m_0 is the rest mass of the particle, λ is its wave length, and h is Planck's constant. Sketch the graph of E as a function of λ. What does the graph say about the energy?

57. A model for the spread of a rumor is given by the equation

$$p(t) = \dfrac{1}{1 + ae^{-kt}}$$

where $p(t)$ is the proportion of the population that knows the rumor at time t and a and k are positive constants.
(a) When will half the population have heard the rumor?
(b) When is the rate of spread of the rumor greatest?
(c) Sketch the graph of p.

58. A model for the concentration at time t of a drug injected into the bloodstream is

$$C(t) = K(e^{-at} - e^{-bt})$$

where a, b, and K are positive constants and $b > a$. Sketch the graph of the concentration function. What does the graph tell us about how the concentration varies as time passes?

59. The figure shows a beam of length L embedded in concrete walls. If a constant load W is distributed evenly along its length, the beam takes the shape of the deflection curve

$$y = -\dfrac{W}{24EI}x^4 + \dfrac{WL}{12EI}x^3 - \dfrac{WL^2}{24EI}x^2$$

where E and I are positive constants. (E is Young's modulus of elasticity and I is the moment of inertia of a cross-section of the beam.) Sketch the graph of the deflection curve.

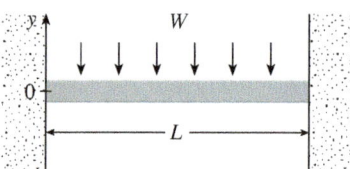

60. Coulomb's Law states that the force of attraction between two charged particles is directly proportional to the product of the charges and inversely proportional to the square of the distance between them. The figure shows particles with charge 1 located at positions 0 and 2 on a coordinate line and a particle with charge -1 at a position x between them. It follows from Coulomb's Law that the net force acting on the middle particle is

$$F(x) = -\dfrac{k}{x^2} + \dfrac{k}{(x - 2)^2} \qquad 0 < x < 2$$

where k is a positive constant. Sketch the graph of the net force function. What does the graph say about the force?

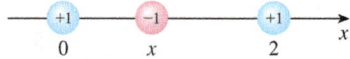

61–64 Find an equation of the slant asymptote. Do not sketch the curve.

61. $y = \dfrac{x^2 + 1}{x + 1}$

62. $y = \dfrac{4x^3 - 10x^2 - 11x + 1}{x^2 - 3x}$

63. $y = \dfrac{2x^3 - 5x^2 + 3x}{x^2 - x - 2}$

64. $y = \dfrac{-6x^4 + 2x^3 + 3}{2x^3 - x}$

65–70 Use the guidelines of this section to sketch the curve. In guideline D find an equation of the slant asymptote.

65. $y = \dfrac{x^2}{x - 1}$

66. $y = \dfrac{1 + 5x - 2x^2}{x - 2}$

67. $y = \dfrac{x^3 + 4}{x^2}$

68. $y = \dfrac{x^3}{(x + 1)^2}$

69. $y = 1 + \tfrac{1}{2}x + e^{-x}$

70. $y = 1 - x + e^{1 + x/3}$

71. Show that the curve $y = x - \tan^{-1} x$ has two slant asymptotes: $y = x + \pi/2$ and $y = x - \pi/2$. Use this fact to help sketch the curve.

72. Show that the curve $y = \sqrt{x^2 + 4x}$ has two slant asymptotes: $y = x + 2$ and $y = -x - 2$. Use this fact to help sketch the curve.

73. Show that the lines $y = (b/a)x$ and $y = -(b/a)x$ are slant asymptotes of the hyperbola $(x^2/a^2) - (y^2/b^2) = 1$.

74. Let $f(x) = (x^3 + 1)/x$. Show that

$$\lim_{x \to \pm\infty} [f(x) - x^2] = 0$$

This shows that the graph of f approaches the graph of $y = x^2$, and we say that the curve $y = f(x)$ is *asymptotic* to the parabola $y = x^2$. Use this fact to help sketch the graph of f.

75. Discuss the asymptotic behavior of $f(x) = (x^4 + 1)/x$ in the same manner as in Exercise 74. Then use your results to help sketch the graph of f.

76. Use the asymptotic behavior of $f(x) = \sin x + e^{-x}$ to sketch its graph without going through the curve-sketching procedure of this section.

4.6 Graphing with Calculus *and* Calculators

You may want to read "Graphing Calculators and Computers" at www.stewartcalculus.com if you haven't already. In particular, it explains how to avoid some of the pitfalls of graphing devices by choosing appropriate viewing rectangles.

The method we used to sketch curves in the preceding section was a culmination of much of our study of differential calculus. The graph was the final object that we produced. In this section our point of view is completely different. Here we *start* with a graph produced by a graphing calculator or computer and then we refine it. We use calculus to make sure that we reveal all the important aspects of the curve. And with the use of graphing devices we can tackle curves that would be far too complicated to consider without technology. The theme is the *interaction* between calculus and calculators.

EXAMPLE 1 Graph the polynomial $f(x) = 2x^6 + 3x^5 + 3x^3 - 2x^2$. Use the graphs of f' and f'' to estimate all maximum and minimum points and intervals of concavity.

SOLUTION If we specify a domain but not a range, many graphing devices will deduce a suitable range from the values computed. Figure 1 shows the plot from one such device if we specify that $-5 \le x \le 5$. Although this viewing rectangle is useful for showing that the asymptotic behavior (or end behavior) is the same as for $y = 2x^6$, it is obviously hiding some finer detail. So we change to the viewing rectangle $[-3, 2]$ by $[-50, 100]$ shown in Figure 2.

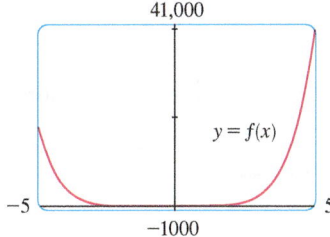

FIGURE 1

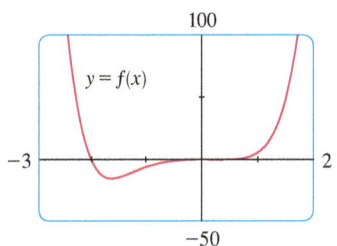

FIGURE 2

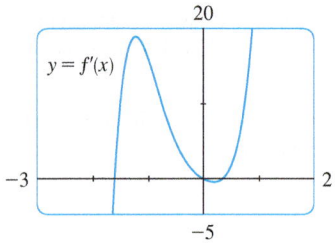

FIGURE 3

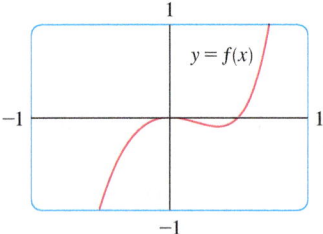

FIGURE 4

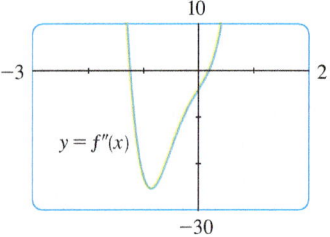

FIGURE 5

From this graph it appears that there is an absolute minimum value of about -15.33 when $x \approx -1.62$ (by using the cursor) and f is decreasing on $(-\infty, -1.62)$ and increasing on $(-1.62, \infty)$. Also there appears to be a horizontal tangent at the origin and inflection points when $x = 0$ and when x is somewhere between -2 and -1.

Now let's try to confirm these impressions using calculus. We differentiate and get

$$f'(x) = 12x^5 + 15x^4 + 9x^2 - 4x$$

$$f''(x) = 60x^4 + 60x^3 + 18x - 4$$

When we graph f' in Figure 3 we see that $f'(x)$ changes from negative to positive when $x \approx -1.62$; this confirms (by the First Derivative Test) the minimum value that we found earlier. But, perhaps to our surprise, we also notice that $f'(x)$ changes from positive to negative when $x = 0$ and from negative to positive when $x \approx 0.35$. This means that f has a local maximum at 0 and a local minimum when $x \approx 0.35$, but these were hidden in Figure 2. Indeed, if we now zoom in toward the origin in Figure 4, we see what we missed before: a local maximum value of 0 when $x = 0$ and a local minimum value of about -0.1 when $x \approx 0.35$.

What about concavity and inflection points? From Figures 2 and 4 there appear to be inflection points when x is a little to the left of -1 and when x is a little to the right of 0. But it's difficult to determine inflection points from the graph of f, so we graph the second derivative f'' in Figure 5. We see that f'' changes from positive to negative when $x \approx -1.23$ and from negative to positive when $x \approx 0.19$. So, correct to two decimal places, f is concave upward on $(-\infty, -1.23)$ and $(0.19, \infty)$ and concave downward on $(-1.23, 0.19)$. The inflection points are $(-1.23, -10.18)$ and $(0.19, -0.05)$.

We have discovered that no single graph reveals all the important features of this polynomial. But Figures 2 and 4, when taken together, do provide an accurate picture. ■

EXAMPLE 2 Draw the graph of the function

$$f(x) = \frac{x^2 + 7x + 3}{x^2}$$

in a viewing rectangle that contains all the important features of the function. Estimate the maximum and minimum values and the intervals of concavity. Then use calculus to find these quantities exactly.

SOLUTION Figure 6, produced by a computer with automatic scaling, is a disaster. Some graphing calculators use $[-10, 10]$ by $[-10, 10]$ as the default viewing rectangle, so let's try it. We get the graph shown in Figure 7; it's a major improvement.

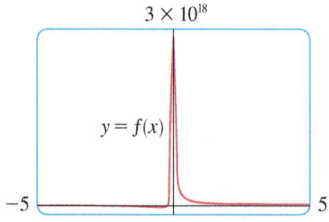

FIGURE 6

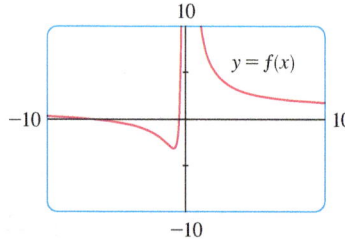

FIGURE 7

The y-axis appears to be a vertical asymptote and indeed it is because

$$\lim_{x \to 0} \frac{x^2 + 7x + 3}{x^2} = \infty$$

Figure 7 also allows us to estimate the x-intercepts: about -0.5 and -6.5. The exact values are obtained by using the quadratic formula to solve the equation $x^2 + 7x + 3 = 0$; we get $x = (-7 \pm \sqrt{37})/2$.

To get a better look at horizontal asymptotes, we change to the viewing rectangle $[-20, 20]$ by $[-5, 10]$ in Figure 8. It appears that $y = 1$ is the horizontal asymptote and this is easily confirmed:

$$\lim_{x \to \pm\infty} \frac{x^2 + 7x + 3}{x^2} = \lim_{x \to \pm\infty} \left(1 + \frac{7}{x} + \frac{3}{x^2}\right) = 1$$

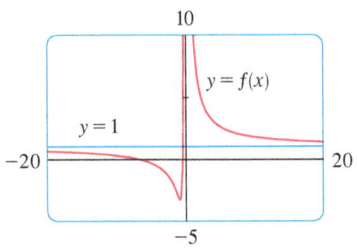

FIGURE 8

To estimate the minimum value we zoom in to the viewing rectangle $[-3, 0]$ by $[-4, 2]$ in Figure 9. The cursor indicates that the absolute minimum value is about -3.1 when $x \approx -0.9$, and we see that the function decreases on $(-\infty, -0.9)$ and $(0, \infty)$ and increases on $(-0.9, 0)$. The exact values are obtained by differentiating:

$$f'(x) = -\frac{7}{x^2} - \frac{6}{x^3} = -\frac{7x + 6}{x^3}$$

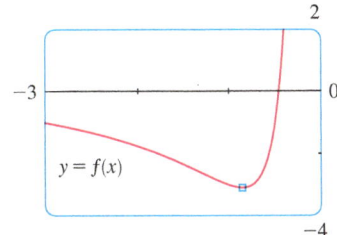

FIGURE 9

This shows that $f'(x) > 0$ when $-\frac{6}{7} < x < 0$ and $f'(x) < 0$ when $x < -\frac{6}{7}$ and when $x > 0$. The exact minimum value is $f(-\frac{6}{7}) = -\frac{37}{12} \approx -3.08$.

Figure 9 also shows that an inflection point occurs somewhere between $x = -1$ and $x = -2$. We could estimate it much more accurately using the graph of the second derivative, but in this case it's just as easy to find exact values. Since

$$f''(x) = \frac{14}{x^3} + \frac{18}{x^4} = \frac{2(7x + 9)}{x^4}$$

we see that $f''(x) > 0$ when $x > -\frac{9}{7}$ ($x \neq 0$). So f is concave upward on $\left(-\frac{9}{7}, 0\right)$ and $(0, \infty)$ and concave downward on $\left(-\infty, -\frac{9}{7}\right)$. The inflection point is $\left(-\frac{9}{7}, -\frac{71}{27}\right)$.

The analysis using the first two derivatives shows that Figure 8 displays all the major aspects of the curve. ∎

EXAMPLE 3 Graph the function $f(x) = \dfrac{x^2(x + 1)^3}{(x - 2)^2(x - 4)^4}$.

SOLUTION Drawing on our experience with a rational function in Example 2, let's start by graphing f in the viewing rectangle $[-10, 10]$ by $[-10, 10]$. From Figure 10 we have the feeling that we are going to have to zoom in to see some finer detail and also zoom out to see the larger picture. But, as a guide to intelligent zooming, let's first take a close look at the expression for $f(x)$. Because of the factors $(x - 2)^2$ and $(x - 4)^4$ in the denominator, we expect $x = 2$ and $x = 4$ to be the vertical asymptotes. Indeed

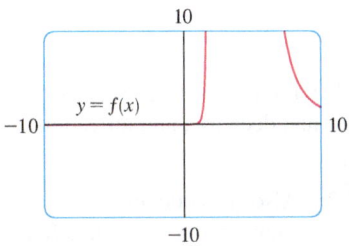

FIGURE 10

$$\lim_{x \to 2} \frac{x^2(x + 1)^3}{(x - 2)^2(x - 4)^4} = \infty \quad \text{and} \quad \lim_{x \to 4} \frac{x^2(x + 1)^3}{(x - 2)^2(x - 4)^4} = \infty$$

To find the horizontal asymptotes, we divide numerator and denominator by x^6:

$$\frac{x^2(x+1)^3}{(x-2)^2(x-4)^4} = \frac{\dfrac{x^2}{x^3} \cdot \dfrac{(x+1)^3}{x^3}}{\dfrac{(x-2)^2}{x^2} \cdot \dfrac{(x-4)^4}{x^4}} = \frac{\dfrac{1}{x}\left(1+\dfrac{1}{x}\right)^3}{\left(1-\dfrac{2}{x}\right)^2\left(1-\dfrac{4}{x}\right)^4}$$

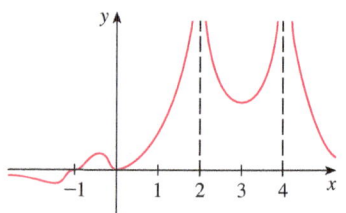

FIGURE 11

This shows that $f(x) \to 0$ as $x \to \pm\infty$, so the x-axis is a horizontal asymptote.

It is also very useful to consider the behavior of the graph near the x-intercepts using an analysis like that in Example 2.6.12. Since x^2 is positive, $f(x)$ does not change sign at 0 and so its graph doesn't cross the x-axis at 0. But, because of the factor $(x+1)^3$, the graph does cross the x-axis at -1 and has a horizontal tangent there. Putting all this information together, but without using derivatives, we see that the curve has to look something like the one in Figure 11.

Now that we know what to look for, we zoom in (several times) to produce the graphs in Figures 12 and 13 and zoom out (several times) to get Figure 14.

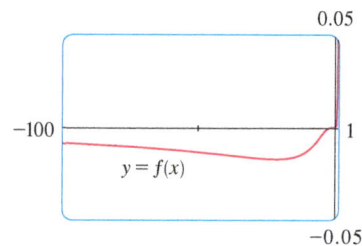

FIGURE 12

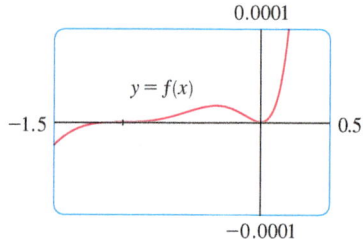

FIGURE 13

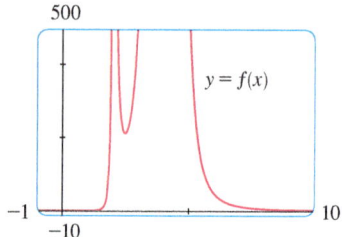

FIGURE 14

We can read from these graphs that the absolute minimum is about -0.02 and occurs when $x \approx -20$. There is also a local maximum ≈ 0.00002 when $x \approx -0.3$ and a local minimum ≈ 211 when $x \approx 2.5$. These graphs also show three inflection points near -35, -5, and -1 and two between -1 and 0. To estimate the inflection points closely we would need to graph f'', but to compute f'' by hand is an unreasonable chore. If you have a computer algebra system, then it's easy to do (see Exercise 15).

We have seen that, for this particular function, *three* graphs (Figures 12, 13, and 14) are necessary to convey all the useful information. The only way to display all these features of the function on a single graph is to draw it by hand. Despite the exaggerations and distortions, Figure 11 does manage to summarize the essential nature of the function. ∎

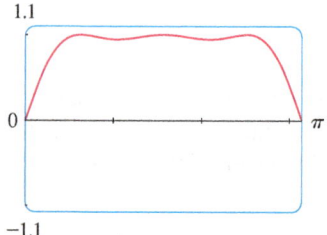

FIGURE 15

EXAMPLE 4 Graph the function $f(x) = \sin(x + \sin 2x)$. For $0 \leq x \leq \pi$, estimate all maximum and minimum values, intervals of increase and decrease, and inflection points.

SOLUTION We first note that f is periodic with period 2π. Also, f is odd and $|f(x)| \leq 1$ for all x. So the choice of a viewing rectangle is not a problem for this function: We start with $[0, \pi]$ by $[-1.1, 1.1]$. (See Figure 15.) It appears that there are three local maximum values and two local minimum values in that window. To confirm

The family of functions

$$f(x) = \sin(x + \sin cx)$$

where c is a constant, occurs in applications to frequency modulation (FM) synthesis. A sine wave is modulated by a wave with a different frequency ($\sin cx$). The case where $c = 2$ is studied in Example 4. Exercise 27 explores another special case.

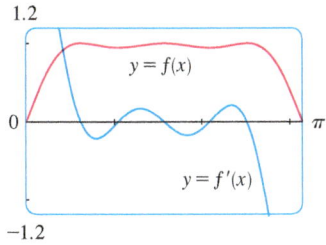

FIGURE 16

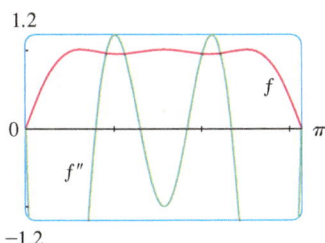

FIGURE 17

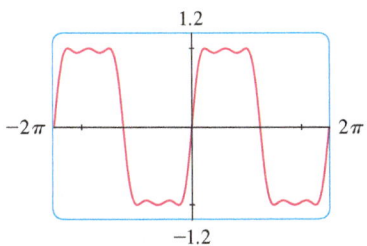

FIGURE 18

this and locate them more accurately, we calculate that

$$f'(x) = \cos(x + \sin 2x) \cdot (1 + 2 \cos 2x)$$

and graph both f and f' in Figure 16.

Using zoom-in and the First Derivative Test, we find the following approximate values:

Intervals of increase:	$(0, 0.6)$, $(1.0, 1.6)$, $(2.1, 2.5)$
Intervals of decrease:	$(0.6, 1.0)$, $(1.6, 2.1)$, $(2.5, \pi)$
Local maximum values:	$f(0.6) \approx 1$, $f(1.6) \approx 1$, $f(2.5) \approx 1$
Local minimum values:	$f(1.0) \approx 0.94$, $f(2.1) \approx 0.94$

The second derivative is

$$f''(x) = -(1 + 2\cos 2x)^2 \sin(x + \sin 2x) - 4 \sin 2x \cos(x + \sin 2x)$$

Graphing both f and f'' in Figure 17, we obtain the following approximate values:

Concave upward on:	$(0.8, 1.3)$, $(1.8, 2.3)$
Concave downward on:	$(0, 0.8)$, $(1.3, 1.8)$, $(2.3, \pi)$
Inflection points:	$(0, 0)$, $(0.8, 0.97)$, $(1.3, 0.97)$, $(1.8, 0.97)$, $(2.3, 0.97)$

Having checked that Figure 15 does indeed represent f accurately for $0 \leq x \leq \pi$, we can state that the extended graph in Figure 18 represents f accurately for $-2\pi \leq x \leq 2\pi$. ■

Our final example is concerned with *families* of functions. This means that the functions in the family are related to each other by a formula that contains one or more arbitrary constants. Each value of the constant gives rise to a member of the family and the idea is to see how the graph of the function changes as the constant changes.

EXAMPLE 5 How does the graph of $f(x) = 1/(x^2 + 2x + c)$ vary as c varies?

SOLUTION The graphs in Figures 19 and 20 (the special cases $c = 2$ and $c = -2$) show two very different-looking curves.

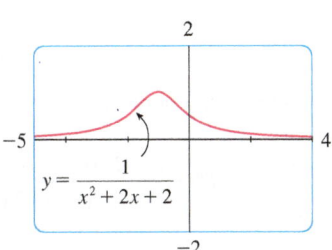

FIGURE 19
$c = 2$

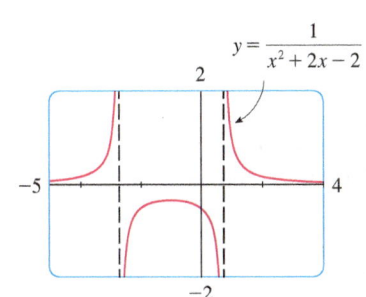

FIGURE 20
$c = -2$

Before drawing any more graphs, let's see what members of this family have in common. Since

$$\lim_{x \to \pm\infty} \frac{1}{x^2 + 2x + c} = 0$$

for any value of c, they all have the x-axis as a horizontal asymptote. A vertical asymptote will occur when $x^2 + 2x + c = 0$. Solving this quadratic equation, we get $x = -1 \pm \sqrt{1 - c}$. When $c > 1$, there is no vertical asymptote (as in Figure 19). When $c = 1$, the graph has a single vertical asymptote $x = -1$ because

$$\lim_{x \to -1} \frac{1}{x^2 + 2x + 1} = \lim_{x \to -1} \frac{1}{(x + 1)^2} = \infty$$

When $c < 1$, there are two vertical asymptotes: $x = -1 \pm \sqrt{1 - c}$ (as in Figure 20)
Now we compute the derivative:

$$f'(x) = -\frac{2x + 2}{(x^2 + 2x + c)^2}$$

This shows that $f'(x) = 0$ when $x = -1$ (if $c \neq 1$), $f'(x) > 0$ when $x < -1$, and $f'(x) < 0$ when $x > -1$. For $c \geq 1$, this means that f increases on $(-\infty, -1)$ and decreases on $(-1, \infty)$. For $c > 1$, there is an absolute maximum value $f(-1) = 1/(c - 1)$. For $c < 1$, $f(-1) = 1/(c - 1)$ is a local maximum value and the intervals of increase and decrease are interrupted at the vertical asymptotes.

Figure 21 is a "slide show" displaying five members of the family, all graphed in the viewing rectangle $[-5, 4]$ by $[-2, 2]$. As predicted, a transition takes place from two vertical asymptotes to one at $c = 1$, and then to none for $c > 1$. As c increases from 1, we see that the maximum point becomes lower; this is explained by the fact that $1/(c - 1) \to 0$ as $c \to \infty$. As c decreases from 1, the vertical asymptotes become more widely separated because the distance between them is $2\sqrt{1 - c}$, which becomes large as $c \to -\infty$. Again, the maximum point approaches the x-axis because $1/(c - 1) \to 0$ as $c \to -\infty$.

TEC See an animation of Figure 21 in Visual 4.6.

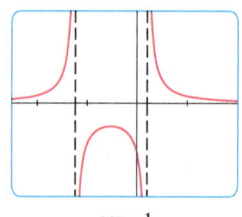
$c = -1$

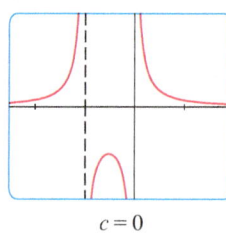
$c = 0$

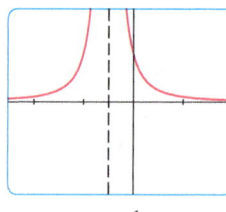
$c = 1$

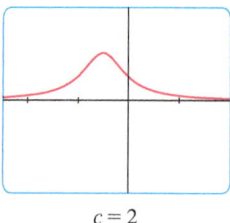
$c = 2$

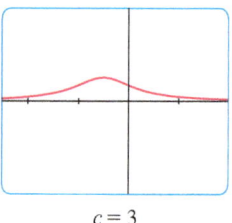
$c = 3$

FIGURE 21
The family of functions
$f(x) = 1/(x^2 + 2x + c)$

There is clearly no inflection point when $c \leq 1$. For $c > 1$ we calculate that

$$f''(x) = \frac{2(3x^2 + 6x + 4 - c)}{(x^2 + 2x + c)^3}$$

and deduce that inflection points occur when $x = -1 \pm \sqrt{3(c - 1)}/3$. So the inflection points become more spread out as c increases and this seems plausible from the last two parts of Figure 21.

4.6 EXERCISES

1–8 Produce graphs of f that reveal all the important aspects of the curve. In particular, you should use graphs of f' and f'' to estimate the intervals of increase and decrease, extreme values, intervals of concavity, and inflection points.

1. $f(x) = x^5 - 5x^4 - x^3 - 28x^2 - 2x$
2. $f(x) = -2x^6 + 5x^5 + 140x^3 - 110x^2$
3. $f(x) = x^6 - 5x^5 + 25x^3 - 6x^2 - 48x$
4. $f(x) = \dfrac{x^4 - x^3 - 8}{x^2 - x - 6}$
5. $f(x) = \dfrac{x}{x^3 + x^2 + 1}$
6. $f(x) = 6 \sin x - x^2$, $-5 \leq x \leq 3$
7. $f(x) = 6 \sin x + \cot x$, $-\pi \leq x \leq \pi$
8. $f(x) = e^x - 0.186x^4$

9–10 Produce graphs of f that reveal all the important aspects of the curve. Estimate the intervals of increase and decrease and intervals of concavity, and use calculus to find these intervals exactly.

9. $f(x) = 1 + \dfrac{1}{x} + \dfrac{8}{x^2} + \dfrac{1}{x^3}$
10. $f(x) = \dfrac{1}{x^8} - \dfrac{2 \times 10^8}{x^4}$

11–12
(a) Graph the function.
(b) Use l'Hospital's Rule to explain the behavior as $x \to 0$.
(c) Estimate the minimum value and intervals of concavity. Then use calculus to find the exact values.

11. $f(x) = x^2 \ln x$
12. $f(x) = xe^{1/x}$

13–14 Sketch the graph by hand using asymptotes and intercepts, but not derivatives. Then use your sketch as a guide to producing graphs (with a graphing device) that display the major features of the curve. Use these graphs to estimate the maximum and minimum values.

13. $f(x) = \dfrac{(x+4)(x-3)^2}{x^4(x-1)}$
14. $f(x) = \dfrac{(2x+3)^2(x-2)^5}{x^3(x-5)^2}$

CAS **15.** If f is the function considered in Example 3, use a computer algebra system to calculate f' and then graph it to confirm that all the maximum and minimum values are as given in the example. Calculate f'' and use it to estimate the intervals of concavity and inflection points.

CAS **16.** If f is the function of Exercise 14, find f' and f'' and use their graphs to estimate the intervals of increase and decrease and concavity of f.

CAS **17–22** Use a computer algebra system to graph f and to find f' and f''. Use graphs of these derivatives to estimate the intervals of increase and decrease, extreme values, intervals of concavity, and inflection points of f.

17. $f(x) = \dfrac{x^3 + 5x^2 + 1}{x^4 + x^3 - x^2 + 2}$
18. $f(x) = \dfrac{x^{2/3}}{1 + x + x^4}$
19. $f(x) = \sqrt{x + 5 \sin x}$, $x \leq 20$
20. $f(x) = x - \tan^{-1}(x^2)$
21. $f(x) = \dfrac{1 - e^{1/x}}{1 + e^{1/x}}$
22. $f(x) = \dfrac{3}{3 + 2 \sin x}$

CAS **23–24** Graph the function using as many viewing rectangles as you need to depict the true nature of the function.

23. $f(x) = \dfrac{1 - \cos(x^4)}{x^8}$
24. $f(x) = e^x + \ln|x - 4|$

CAS **25–26**
(a) Graph the function.
(b) Explain the shape of the graph by computing the limit as $x \to 0^+$ or as $x \to \infty$.
(c) Estimate the maximum and minimum values and then use calculus to find the exact values.
(d) Use a graph of f'' to estimate the x-coordinates of the inflection points.

25. $f(x) = x^{1/x}$
26. $f(x) = (\sin x)^{\sin x}$

27. In Example 4 we considered a member of the family of functions $f(x) = \sin(x + \sin cx)$ that occur in FM synthesis. Here we investigate the function with $c = 3$. Start by graphing f in the viewing rectangle $[0, \pi]$ by $[-1.2, 1.2]$. How many local maximum points do you see? The graph has more than are visible to the naked eye. To discover the hidden maximum and minimum points you will need to examine the graph of f' very carefully. In fact, it helps to look at the graph of f'' at the same time. Find all the maximum and minimum values and inflection points. Then graph f in the viewing rectangle $[-2\pi, 2\pi]$ by $[-1.2, 1.2]$ and comment on symmetry.

28–35 Describe how the graph of f varies as c varies. Graph several members of the family to illustrate the trends that you discover. In particular, you should investigate how maximum and minimum points and inflection points move when c changes. You should also identify any transitional values of c at which the basic shape of the curve changes.

28. $f(x) = x^3 + cx$

29. $f(x) = x^2 + 6x + c/x$ (Trident of Newton)

30. $f(x) = x\sqrt{c^2 - x^2}$

31. $f(x) = e^x + ce^{-x}$

32. $f(x) = \ln(x^2 + c)$

33. $f(x) = \dfrac{cx}{1 + c^2x^2}$

34. $f(x) = \dfrac{\sin x}{c + \cos x}$

35. $f(x) = cx + \sin x$

36. The family of functions $f(t) = C(e^{-at} - e^{-bt})$, where a, b, and C are positive numbers and $b > a$, has been used to model the concentration of a drug injected into the bloodstream at time $t = 0$. Graph several members of this family. What do they have in common? For fixed values of C and a, discover graphically what happens as b increases. Then use calculus to prove what you have discovered.

37. Investigate the family of curves given by $f(x) = xe^{-cx}$, where c is a real number. Start by computing the limits as $x \to \pm\infty$. Identify any transitional values of c where the basic shape changes. What happens to the maximum or minimum points and inflection points as c changes? Illustrate by graphing several members of the family.

38. Investigate the family of curves given by the equation $f(x) = x^4 + cx^2 + x$. Start by determining the transitional value of c at which the number of inflection points changes. Then graph several members of the family to see what shapes are possible. There is another transitional value of c at which the number of critical numbers changes. Try to discover it graphically. Then prove what you have discovered.

39. (a) Investigate the family of polynomials given by the equation $f(x) = cx^4 - 2x^2 + 1$. For what values of c does the curve have minimum points?
    (b) Show that the minimum and maximum points of every curve in the family lie on the parabola $y = 1 - x^2$. Illustrate by graphing this parabola and several members of the family.

40. (a) Investigate the family of polynomials given by the equation $f(x) = 2x^3 + cx^2 + 2x$. For what values of c does the curve have maximum and minimum points?
    (b) Show that the minimum and maximum points of every curve in the family lie on the curve $y = x - x^3$. Illustrate by graphing this curve and several members of the family.

4.7 Optimization Problems

The methods we have learned in this chapter for finding extreme values have practical applications in many areas of life. A businessperson wants to minimize costs and maximize profits. A traveler wants to minimize transportation time. Fermat's Principle in optics states that light follows the path that takes the least time. In this section we solve such problems as maximizing areas, volumes, and profits and minimizing distances, times, and costs.

In solving such practical problems the greatest challenge is often to convert the word problem into a mathematical optimization problem by setting up the function that is to be maximized or minimized. Let's recall the problem-solving principles discussed on page 71 and adapt them to this situation:

Steps In Solving Optimization Problems

1. **Understand the Problem** The first step is to read the problem carefully until it is clearly understood. Ask yourself: What is the unknown? What are the given quantities? What are the given conditions?

2. **Draw a Diagram** In most problems it is useful to draw a diagram and identify the given and required quantities on the diagram.

3. **Introduce Notation** Assign a symbol to the quantity that is to be maximized or minimized (let's call it Q for now). Also select symbols $(a, b, c, \ldots, x, y)$ for other unknown quantities and label the diagram with these symbols. It may help to use initials as suggestive symbols—for example, A for area, h for height, t for time.

4. Express Q in terms of some of the other symbols from Step 3.
5. If Q has been expressed as a function of more than one variable in Step 4, use the given information to find relationships (in the form of equations) among these variables. Then use these equations to eliminate all but one of the variables in the expression for Q. Thus Q will be expressed as a function of *one* variable x, say, $Q = f(x)$. Write the domain of this function in the given context.
6. Use the methods of Sections 4.1 and 4.3 to find the *absolute* maximum or minimum value of f. In particular, if the domain of f is a closed interval, then the Closed Interval Method in Section 4.1 can be used.

EXAMPLE 1 A farmer has 2400 ft of fencing and wants to fence off a rectangular field that borders a straight river. He needs no fence along the river. What are the dimensions of the field that has the largest area?

SOLUTION In order to get a feeling for what is happening in this problem, let's experiment with some specific cases. Figure 1 (not to scale) shows three possible ways of laying out the 2400 ft of fencing.

PS Understand the problem
PS Analogy: Try special cases
PS Draw diagrams

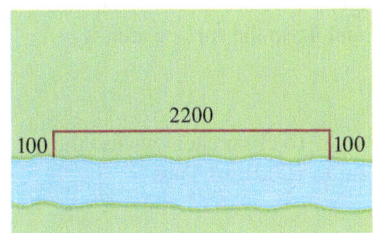
Area = 100 · 2200 = 220,000 ft²

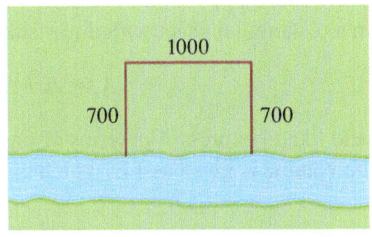
Area = 700 · 1000 = 700,000 ft²

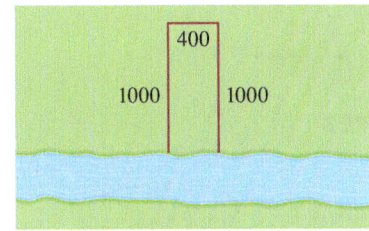
Area = 1000 · 400 = 400,000 ft²

FIGURE 1

We see that when we try shallow, wide fields or deep, narrow fields, we get relatively small areas. It seems plausible that there is some intermediate configuration that produces the largest area.

Figure 2 illustrates the general case. We wish to maximize the area A of the rectangle. Let x and y be the depth and width of the rectangle (in feet). Then we express A in terms of x and y:

$$A = xy$$

PS Introduce notation

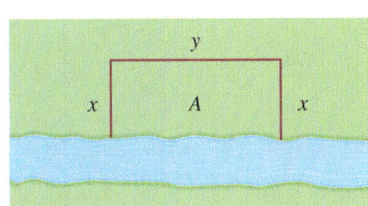

FIGURE 2

We want to express A as a function of just one variable, so we eliminate y by expressing it in terms of x. To do this we use the given information that the total length of the fencing is 2400 ft. Thus

$$2x + y = 2400$$

From this equation we have $y = 2400 - 2x$, which gives

$$A = xy = x(2400 - 2x) = 2400x - 2x^2$$

Note that the largest x can be is 1200 (this uses all the fence for the depth and none for the width) and x can't be negative, so the function that we wish to maximize is

$$A(x) = 2400x - 2x^2 \qquad 0 \leq x \leq 1200$$

The derivative is $A'(x) = 2400 - 4x$, so to find the critical numbers we solve the equation

$$2400 - 4x = 0$$

which gives $x = 600$. The maximum value of A must occur either at this critical number or at an endpoint of the interval. Since $A(0) = 0$, $A(600) = 720{,}000$, and $A(1200) = 0$, the Closed Interval Method gives the maximum value as $A(600) = 720{,}000$.

[Alternatively, we could have observed that $A''(x) = -4 < 0$ for all x, so A is always concave downward and the local maximum at $x = 600$ must be an absolute maximum.]

The corresponding y-value is $y = 2400 - 2(600) = 1200$; so the rectangular field should be 600 ft deep and 1200 ft wide. ∎

EXAMPLE 2 A cylindrical can is to be made to hold 1 L of oil. Find the dimensions that will minimize the cost of the metal to manufacture the can.

SOLUTION Draw the diagram as in Figure 3, where r is the radius and h the height (both in centimeters). In order to minimize the cost of the metal, we minimize the total surface area of the cylinder (top, bottom, and sides). From Figure 4 we see that the sides are made from a rectangular sheet with dimensions $2\pi r$ and h. So the surface area is

$$A = 2\pi r^2 + 2\pi r h$$

FIGURE 3

We would like to express A in terms of one variable, r. To eliminate h we use the fact that the volume is given as 1 L, which is equivalent to 1000 cm³. Thus

$$\pi r^2 h = 1000$$

which gives $h = 1000/(\pi r^2)$. Substitution of this into the expression for A gives

$$A = 2\pi r^2 + 2\pi r \left(\frac{1000}{\pi r^2}\right) = 2\pi r^2 + \frac{2000}{r}$$

We know r must be positive, and there are no limitations on how large r can be. Therefore the function that we want to minimize is

$$A(r) = 2\pi r^2 + \frac{2000}{r} \qquad r > 0$$

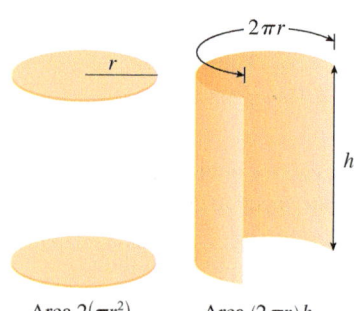

Area $2(\pi r^2)$ Area $(2\pi r)h$

FIGURE 4

To find the critical numbers, we differentiate:

$$A'(r) = 4\pi r - \frac{2000}{r^2} = \frac{4(\pi r^3 - 500)}{r^2}$$

Then $A'(r) = 0$ when $\pi r^3 = 500$, so the only critical number is $r = \sqrt[3]{500/\pi}$.

Since the domain of A is $(0, \infty)$, we can't use the argument of Example 1 concerning endpoints. But we can observe that $A'(r) < 0$ for $r < \sqrt[3]{500/\pi}$ and $A'(r) > 0$ for $r > \sqrt[3]{500/\pi}$, so A is decreasing for *all* r to the left of the critical number and increasing for *all* r to the right. Thus $r = \sqrt[3]{500/\pi}$ must give rise to an *absolute* minimum.

[Alternatively, we could argue that $A(r) \to \infty$ as $r \to 0^+$ and $A(r) \to \infty$ as $r \to \infty$, so there must be a minimum value of $A(r)$, which must occur at the critical number. See Figure 5.]

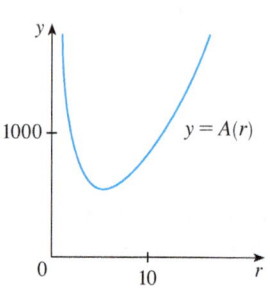

FIGURE 5

In the Applied Project on page 343 we investigate the most economical shape for a can by taking into account other manufacturing costs.

The value of h corresponding to $r = \sqrt[3]{500/\pi}$ is

$$h = \frac{1000}{\pi r^2} = \frac{1000}{\pi (500/\pi)^{2/3}} = 2\sqrt[3]{\frac{500}{\pi}} = 2r$$

Thus, to minimize the cost of the can, the radius should be $\sqrt[3]{500/\pi}$ cm and the height should be equal to twice the radius, namely, the diameter. ∎

NOTE 1 The argument used in Example 2 to justify the absolute minimum is a variant of the First Derivative Test (which applies only to *local* maximum or minimum values) and is stated here for future reference.

> **First Derivative Test for Absolute Extreme Values** Suppose that c is a critical number of a continuous function f defined on an interval.
> (a) If $f'(x) > 0$ for all $x < c$ and $f'(x) < 0$ for all $x > c$, then $f(c)$ is the absolute maximum value of f.
> (b) If $f'(x) < 0$ for all $x < c$ and $f'(x) > 0$ for all $x > c$, then $f(c)$ is the absolute minimum value of f.

TEC Module 4.7 takes you through six additional optimization problems, including animations of the physical situations.

NOTE 2 An alternative method for solving optimization problems is to use implicit differentiation. Let's look at Example 2 again to illustrate the method. We work with the same equations

$$A = 2\pi r^2 + 2\pi rh \qquad \pi r^2 h = 1000$$

but instead of eliminating h, we differentiate both equations implicitly with respect to r:

$$A' = 4\pi r + 2\pi rh' + 2\pi h \qquad \pi r^2 h' + 2\pi rh = 0$$

The minimum occurs at a critical number, so we set $A' = 0$, simplify, and arrive at the equations

$$2r + rh' + h = 0 \qquad rh' + 2h = 0$$

and subtraction gives $2r - h = 0$, or $h = 2r$.

EXAMPLE 3 Find the point on the parabola $y^2 = 2x$ that is closest to the point $(1, 4)$.

SOLUTION The distance between the point $(1, 4)$ and the point (x, y) is

$$d = \sqrt{(x-1)^2 + (y-4)^2}$$

(See Figure 6.) But if (x, y) lies on the parabola, then $x = \tfrac{1}{2}y^2$, so the expression for d becomes

$$d = \sqrt{(\tfrac{1}{2}y^2 - 1)^2 + (y-4)^2}$$

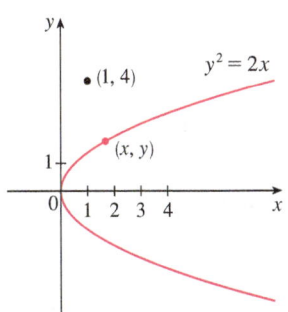

FIGURE 6

(Alternatively, we could have substituted $y = \sqrt{2x}$ to get d in terms of x alone.) Instead of minimizing d, we minimize its square:

$$d^2 = f(y) = (\tfrac{1}{2}y^2 - 1)^2 + (y-4)^2$$

(You should convince yourself that the minimum of d occurs at the same point as the minimum of d^2, but d^2 is easier to work with.) Note that there are no restrictions on y, so the domain is all real numbers. Differentiating, we obtain

$$f'(y) = 2(\tfrac{1}{2}y^2 - 1)y + 2(y - 4) = y^3 - 8$$

so $f'(y) = 0$ when $y = 2$. Observe that $f'(y) < 0$ when $y < 2$ and $f'(y) > 0$ when $y > 2$, so by the First Derivative Test for Absolute Extreme Values, the absolute minimum occurs when $y = 2$. (Or we could simply say that because of the geometric nature of the problem, it's obvious that there is a closest point but not a farthest point.) The corresponding value of x is $x = \tfrac{1}{2}y^2 = 2$. Thus the point on $y^2 = 2x$ closest to $(1, 4)$ is $(2, 2)$. [The distance between the points is $d = \sqrt{f(2)} = \sqrt{5}$.] ∎

EXAMPLE 4 A man launches his boat from point A on a bank of a straight river, 3 km wide, and wants to reach point B, 8 km downstream on the opposite bank, as quickly as possible (see Figure 7). He could row his boat directly across the river to point C and then run to B, or he could row directly to B, or he could row to some point D between C and B and then run to B. If he can row 6 km/h and run 8 km/h, where should he land to reach B as soon as possible? (We assume that the speed of the water is negligible compared with the speed at which the man rows.)

SOLUTION If we let x be the distance from C to D, then the running distance is $|DB| = 8 - x$ and the Pythagorean Theorem gives the rowing distance as $|AD| = \sqrt{x^2 + 9}$. We use the equation

$$\text{time} = \frac{\text{distance}}{\text{rate}}$$

Then the rowing time is $\sqrt{x^2 + 9}/6$ and the running time is $(8 - x)/8$, so the total time T as a function of x is

$$T(x) = \frac{\sqrt{x^2 + 9}}{6} + \frac{8 - x}{8}$$

The domain of this function T is $[0, 8]$. Notice that if $x = 0$, he rows to C and if $x = 8$, he rows directly to B. The derivative of T is

$$T'(x) = \frac{x}{6\sqrt{x^2 + 9}} - \frac{1}{8}$$

Thus, using the fact that $x \geq 0$, we have

$$T'(x) = 0 \iff \frac{x}{6\sqrt{x^2 + 9}} = \frac{1}{8} \iff 4x = 3\sqrt{x^2 + 9}$$

$$\iff 16x^2 = 9(x^2 + 9) \iff 7x^2 = 81$$

$$\iff x = \frac{9}{\sqrt{7}}$$

The only critical number is $x = 9/\sqrt{7}$. To see whether the minimum occurs at this critical number or at an endpoint of the domain $[0, 8]$, we follow the Closed Interval

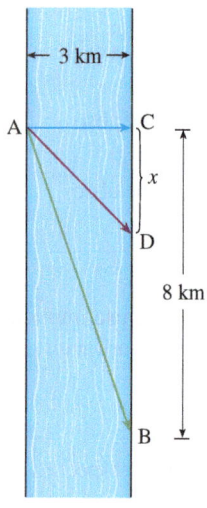

FIGURE 7

Method by evaluating T at all three points:

$$T(0) = 1.5 \qquad T\left(\frac{9}{\sqrt{7}}\right) = 1 + \frac{\sqrt{7}}{8} \approx 1.33 \qquad T(8) = \frac{\sqrt{73}}{6} \approx 1.42$$

Since the smallest of these values of T occurs when $x = 9/\sqrt{7}$, the absolute minimum value of T must occur there. Figure 8 illustrates this calculation by showing the graph of T.

Thus the man should land the boat at a point $9/\sqrt{7}$ km (≈ 3.4 km) downstream from his starting point.

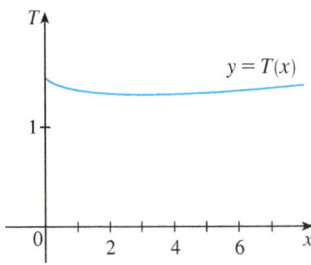

FIGURE 8

EXAMPLE 5 Find the area of the largest rectangle that can be inscribed in a semicircle of radius r.

SOLUTION 1 Let's take the semicircle to be the upper half of the circle $x^2 + y^2 = r^2$ with center the origin. Then the word *inscribed* means that the rectangle has two vertices on the semicircle and two vertices on the x-axis as shown in Figure 9.

Let (x, y) be the vertex that lies in the first quadrant. Then the rectangle has sides of lengths $2x$ and y, so its area is

$$A = 2xy$$

To eliminate y we use the fact that (x, y) lies on the circle $x^2 + y^2 = r^2$ and so $y = \sqrt{r^2 - x^2}$. Thus

$$A = 2x\sqrt{r^2 - x^2}$$

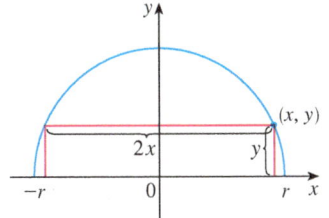

FIGURE 9

The domain of this function is $0 \leq x \leq r$. Its derivative is

$$A' = 2\sqrt{r^2 - x^2} - \frac{2x^2}{\sqrt{r^2 - x^2}} = \frac{2(r^2 - 2x^2)}{\sqrt{r^2 - x^2}}$$

which is 0 when $2x^2 = r^2$, that is, $x = r/\sqrt{2}$ (since $x \geq 0$). This value of x gives a maximum value of A since $A(0) = 0$ and $A(r) = 0$. Therefore the area of the largest inscribed rectangle is

$$A\left(\frac{r}{\sqrt{2}}\right) = 2\frac{r}{\sqrt{2}}\sqrt{r^2 - \frac{r^2}{2}} = r^2$$

SOLUTION 2 A simpler solution is possible if we think of using an angle as a variable. Let θ be the angle shown in Figure 10. Then the area of the rectangle is

$$A(\theta) = (2r\cos\theta)(r\sin\theta) = r^2(2\sin\theta\cos\theta) = r^2\sin 2\theta$$

We know that $\sin 2\theta$ has a maximum value of 1 and it occurs when $2\theta = \pi/2$. So $A(\theta)$ has a maximum value of r^2 and it occurs when $\theta = \pi/4$.

Notice that this trigonometric solution doesn't involve differentiation. In fact, we didn't need to use calculus at all.

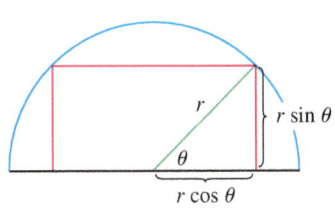

FIGURE 10

■ Applications to Business and Economics

In Section 3.7 we introduced the idea of marginal cost. Recall that if $C(x)$, the **cost function**, is the cost of producing x units of a certain product, then the **marginal cost** is the rate of change of C with respect to x. In other words, the marginal cost function is the derivative, $C'(x)$, of the cost function.

Now let's consider marketing. Let $p(x)$ be the price per unit that the company can charge if it sells x units. Then p is called the **demand function** (or **price function**) and we would expect it to be a decreasing function of x. (More units sold corresponds to a lower price.) If x units are sold and the price per unit is $p(x)$, then the total revenue is

$$R(x) = \text{quantity} \times \text{price} = xp(x)$$

and R is called the **revenue function**. The derivative R' of the revenue function is called the **marginal revenue function** and is the rate of change of revenue with respect to the number of units sold.

If x units are sold, then the total profit is

$$P(x) = R(x) - C(x)$$

and P is called the **profit function**. The **marginal profit function** is P', the derivative of the profit function. In Exercises 59–63 you are asked to use the marginal cost, revenue, and profit functions to minimize costs and maximize revenues and profits.

EXAMPLE 6 A store has been selling 200 flat-screen TVs a week at $350 each. A market survey indicates that for each $10 rebate offered to buyers, the number of TVs sold will increase by 20 a week. Find the demand function and the revenue function. How large a rebate should the store offer to maximize its revenue?

SOLUTION If x is the number of TVs sold per week, then the weekly increase in sales is $x - 200$. For each increase of 20 units sold, the price is decreased by $10. So for each additional unit sold, the decrease in price will be $\frac{1}{20} \times 10$ and the demand function is

$$p(x) = 350 - \tfrac{10}{20}(x - 200) = 450 - \tfrac{1}{2}x$$

The revenue function is

$$R(x) = xp(x) = 450x - \tfrac{1}{2}x^2$$

Since $R'(x) = 450 - x$, we see that $R'(x) = 0$ when $x = 450$. This value of x gives an absolute maximum by the First Derivative Test (or simply by observing that the graph of R is a parabola that opens downward). The corresponding price is

$$p(450) = 450 - \tfrac{1}{2}(450) = 225$$

and the rebate is $350 - 225 = 125$. Therefore, to maximize revenue, the store should offer a rebate of $125. ■

4.7 EXERCISES

1. Consider the following problem: Find two numbers whose sum is 23 and whose product is a maximum.
   (a) Make a table of values, like the one at the right, so that the sum of the numbers in the first two columns is always 23. On the basis of the evidence in your table, estimate the answer to the problem.
   (b) Use calculus to solve the problem and compare with your answer to part (a).

First number	Second number	Product
1	22	22
2	21	42
3	20	60
⋮	⋮	⋮

2. Find two numbers whose difference is 100 and whose product is a minimum.

3. Find two positive numbers whose product is 100 and whose sum is a minimum.

4. The sum of two positive numbers is 16. What is the smallest possible value of the sum of their squares?

5. What is the maximum vertical distance between the line $y = x + 2$ and the parabola $y = x^2$ for $-1 \leq x \leq 2$?

6. What is the minimum vertical distance between the parabolas $y = x^2 + 1$ and $y = x - x^2$?

7. Find the dimensions of a rectangle with perimeter 100 m whose area is as large as possible.

8. Find the dimensions of a rectangle with area 1000 m² whose perimeter is as small as possible.

9. A model used for the yield Y of an agricultural crop as a function of the nitrogen level N in the soil (measured in appropriate units) is
$$Y = \frac{kN}{1 + N^2}$$
where k is a positive constant. What nitrogen level gives the best yield?

10. The rate (in mg carbon/m³/h) at which photosynthesis takes place for a species of phytoplankton is modeled by the function
$$P = \frac{100I}{I^2 + I + 4}$$
where I is the light intensity (measured in thousands of foot-candles). For what light intensity is P a maximum?

11. Consider the following problem: A farmer with 750 ft of fencing wants to enclose a rectangular area and then divide it into four pens with fencing parallel to one side of the rectangle. What is the largest possible total area of the four pens?
    (a) Draw several diagrams illustrating the situation, some with shallow, wide pens and some with deep, narrow pens. Find the total areas of these configurations. Does it appear that there is a maximum area? If so, estimate it.
    (b) Draw a diagram illustrating the general situation. Introduce notation and label the diagram with your symbols.
    (c) Write an expression for the total area.
    (d) Use the given information to write an equation that relates the variables.
    (e) Use part (d) to write the total area as a function of one variable.
    (f) Finish solving the problem and compare the answer with your estimate in part (a).

12. Consider the following problem: A box with an open top is to be constructed from a square piece of cardboard, 3 ft wide, by cutting out a square from each of the four corners and bending up the sides. Find the largest volume that such a box can have.
    (a) Draw several diagrams to illustrate the situation, some short boxes with large bases and some tall boxes with small bases. Find the volumes of several such boxes. Does it appear that there is a maximum volume? If so, estimate it.
    (b) Draw a diagram illustrating the general situation. Introduce notation and label the diagram with your symbols.
    (c) Write an expression for the volume.
    (d) Use the given information to write an equation that relates the variables.
    (e) Use part (d) to write the volume as a function of one variable.
    (f) Finish solving the problem and compare the answer with your estimate in part (a).

13. A farmer wants to fence in an area of 1.5 million square feet in a rectangular field and then divide it in half with a fence parallel to one of the sides of the rectangle. How can he do this so as to minimize the cost of the fence?

14. A box with a square base and open top must have a volume of 32,000 cm³. Find the dimensions of the box that minimize the amount of material used.

15. If 1200 cm² of material is available to make a box with a square base and an open top, find the largest possible volume of the box.

16. A rectangular storage container with an open top is to have a volume of 10 m³. The length of its base is twice the width. Material for the base costs $10 per square meter. Material for the sides costs $6 per square meter. Find the cost of materials for the cheapest such container.

17. Do Exercise 16 assuming the container has a lid that is made from the same material as the sides.

18. A farmer wants to fence in a rectangular plot of land adjacent to the north wall of his barn. No fencing is needed along the barn, and the fencing along the west side of the plot is shared with a neighbor who will split the cost of that portion of the fence. If the fencing costs $20 per linear foot to install and the farmer is not willing to spend more than $5000, find the dimensions for the plot that would enclose the most area.

19. If the farmer in Exercise 18 wants to enclose 8000 square feet of land, what dimensions will minimize the cost of the fence?

20. (a) Show that of all the rectangles with a given area, the one with smallest perimeter is a square.
    (b) Show that of all the rectangles with a given perimeter, the one with greatest area is a square.

21. Find the point on the line $y = 2x + 3$ that is closest to the origin.

22. Find the point on the curve $y = \sqrt{x}$ that is closest to the point (3, 0).

23. Find the points on the ellipse $4x^2 + y^2 = 4$ that are farthest away from the point $(1, 0)$.

24. Find, correct to two decimal places, the coordinates of the point on the curve $y = \sin x$ that is closest to the point $(4, 2)$.

25. Find the dimensions of the rectangle of largest area that can be inscribed in a circle of radius r.

26. Find the area of the largest rectangle that can be inscribed in the ellipse $x^2/a^2 + y^2/b^2 = 1$.

27. Find the dimensions of the rectangle of largest area that can be inscribed in an equilateral triangle of side L if one side of the rectangle lies on the base of the triangle.

28. Find the area of the largest trapezoid that can be inscribed in a circle of radius 1 and whose base is a diameter of the circle.

29. Find the dimensions of the isosceles triangle of largest area that can be inscribed in a circle of radius r.

30. If the two equal sides of an isosceles triangle have length a, find the length of the third side that maximizes the area of the triangle.

31. A right circular cylinder is inscribed in a sphere of radius r. Find the largest possible volume of such a cylinder.

32. A right circular cylinder is inscribed in a cone with height h and base radius r. Find the largest possible volume of such a cylinder.

33. A right circular cylinder is inscribed in a sphere of radius r. Find the largest possible surface area of such a cylinder.

34. A Norman window has the shape of a rectangle surmounted by a semicircle. (Thus the diameter of the semicircle is equal to the width of the rectangle. See Exercise 1.1.62.) If the perimeter of the window is 30 ft, find the dimensions of the window so that the greatest possible amount of light is admitted.

35. The top and bottom margins of a poster are each 6 cm and the side margins are each 4 cm. If the area of printed material on the poster is fixed at 384 cm², find the dimensions of the poster with the smallest area.

36. A poster is to have an area of 180 in² with 1-inch margins at the bottom and sides and a 2-inch margin at the top. What dimensions will give the largest printed area?

37. A piece of wire 10 m long is cut into two pieces. One piece is bent into a square and the other is bent into an equilateral triangle. How should the wire be cut so that the total area enclosed is (a) a maximum? (b) A minimum?

38. Answer Exercise 37 if one piece is bent into a square and the other into a circle.

39. If you are offered one slice from a round pizza (in other words, a sector of a circle) and the slice must have a perimeter of 32 inches, what diameter pizza will reward you with the largest slice?

40. A fence 8 ft tall runs parallel to a tall building at a distance of 4 ft from the building. What is the length of the shortest ladder that will reach from the ground over the fence to the wall of the building?

41. A cone-shaped drinking cup is made from a circular piece of paper of radius R by cutting out a sector and joining the edges CA and CB. Find the maximum capacity of such a cup.

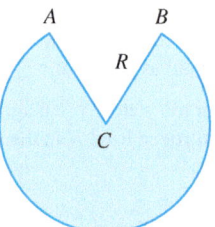

42. A cone-shaped paper drinking cup is to be made to hold 27 cm³ of water. Find the height and radius of the cup that will use the smallest amount of paper.

43. A cone with height h is inscribed in a larger cone with height H so that its vertex is at the center of the base of the larger cone. Show that the inner cone has maximum volume when $h = \frac{1}{3}H$.

44. An object with weight W is dragged along a horizontal plane by a force acting along a rope attached to the object. If the rope makes an angle θ with a plane, then the magnitude of the force is

$$F = \frac{\mu W}{\mu \sin \theta + \cos \theta}$$

where μ is a constant called the coefficient of friction. For what value of θ is F smallest?

45. If a resistor of R ohms is connected across a battery of E volts with internal resistance r ohms, then the power (in watts) in the external resistor is

$$P = \frac{E^2 R}{(R + r)^2}$$

If E and r are fixed but R varies, what is the maximum value of the power?

46. For a fish swimming at a speed v relative to the water, the energy expenditure per unit time is proportional to v^3. It is believed that migrating fish try to minimize the total energy required to swim a fixed distance. If the fish are swimming against a current u $(u < v)$, then the time required to swim a distance L is $L/(v - u)$ and the total energy E required to swim the distance is given by

$$E(v) = av^3 \cdot \frac{L}{v - u}$$

where a is the proportionality constant.
(a) Determine the value of v that minimizes E.
(b) Sketch the graph of E.

Note: This result has been verified experimentally; migrating fish swim against a current at a speed 50% greater than the current speed.

47. In a beehive, each cell is a regular hexagonal prism, open at one end with a trihedral angle at the other end as in the figure. It is believed that bees form their cells in such a way as to minimize the surface area for a given side length and height, thus using the least amount of wax in cell construction. Examination of these cells has shown that the measure of the apex angle θ is amazingly consistent. Based on the geometry of the cell, it can be shown that the surface area S is given by

$$S = 6sh - \tfrac{3}{2}s^2 \cot\theta + (3s^2\sqrt{3}/2)\csc\theta$$

where s, the length of the sides of the hexagon, and h, the height, are constants.
(a) Calculate $dS/d\theta$.
(b) What angle should the bees prefer?
(c) Determine the minimum surface area of the cell (in terms of s and h).

Note: Actual measurements of the angle θ in beehives have been made, and the measures of these angles seldom differ from the calculated value by more than 2°.

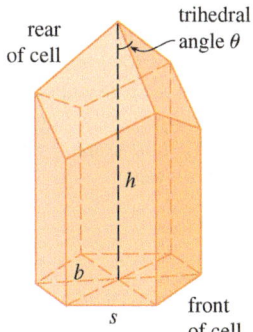

48. A boat leaves a dock at 2:00 PM and travels due south at a speed of 20 km/h. Another boat has been heading due east at 15 km/h and reaches the same dock at 3:00 PM. At what time were the two boats closest together?

49. Solve the problem in Example 4 if the river is 5 km wide and point B is only 5 km downstream from A.

50. A woman at a point A on the shore of a circular lake with radius 2 mi wants to arrive at the point C diametrically opposite A on the other side of the lake in the shortest possible time (see the figure). She can walk at the rate of 4 mi/h and row a boat at 2 mi/h. How should she proceed?

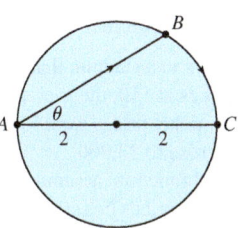

51. An oil refinery is located on the north bank of a straight river that is 2 km wide. A pipeline is to be constructed from the refinery to storage tanks located on the south bank of the river 6 km east of the refinery. The cost of laying pipe is $400,000/km over land to a point P on the north bank and $800,000/km under the river to the tanks. To minimize the cost of the pipeline, where should P be located?

52. Suppose the refinery in Exercise 51 is located 1 km north of the river. Where should P be located?

53. The illumination of an object by a light source is directly proportional to the strength of the source and inversely proportional to the square of the distance from the source. If two light sources, one three times as strong as the other, are placed 10 ft apart, where should an object be placed on the line between the sources so as to receive the least illumination?

54. Find an equation of the line through the point $(3, 5)$ that cuts off the least area from the first quadrant.

55. Let a and b be positive numbers. Find the length of the shortest line segment that is cut off by the first quadrant and passes through the point (a, b).

56. At which points on the curve $y = 1 + 40x^3 - 3x^5$ does the tangent line have the largest slope?

57. What is the shortest possible length of the line segment that is cut off by the first quadrant and is tangent to the curve $y = 3/x$ at some point?

58. What is the smallest possible area of the triangle that is cut off by the first quadrant and whose hypotenuse is tangent to the parabola $y = 4 - x^2$ at some point?

59. (a) If $C(x)$ is the cost of producing x units of a commodity, then the **average cost** per unit is $c(x) = C(x)/x$. Show that if the average cost is a minimum, then the marginal cost equals the average cost.
(b) If $C(x) = 16{,}000 + 200x + 4x^{3/2}$, in dollars, find (i) the cost, average cost, and marginal cost at a production level of 1000 units; (ii) the production level that will minimize the average cost; and (iii) the minimum average cost.

60. (a) Show that if the profit $P(x)$ is a maximum, then the marginal revenue equals the marginal cost.
(b) If $C(x) = 16{,}000 + 500x - 1.6x^2 + 0.004x^3$ is the cost function and $p(x) = 1700 - 7x$ is the demand function, find the production level that will maximize profit.

61. A baseball team plays in a stadium that holds 55,000 spectators. With ticket prices at $10, the average attendance had been 27,000. When ticket prices were lowered to $8, the average attendance rose to 33,000.
(a) Find the demand function, assuming that it is linear.
(b) How should ticket prices be set to maximize revenue?

62. During the summer months Terry makes and sells necklaces on the beach. Last summer he sold the necklaces for $10 each and his sales averaged 20 per day. When he increased the price by $1, he found that the average decreased by two sales per day.
(a) Find the demand function, assuming that it is linear.
(b) If the material for each necklace costs Terry $6, what should the selling price be to maximize his profit?

63. A retailer has been selling 1200 tablet computers a week at $350 each. The marketing department estimates that an additional 80 tablets will sell each week for every $10 that the price is lowered.
(a) Find the demand function.
(b) What should the price be set at in order to maximize revenue?
(c) If the retailer's weekly cost function is
$$C(x) = 35{,}000 + 120x$$
what price should it choose in order to maximize its profit?

64. A company operates 16 oil wells in a designated area. Each pump, on average, extracts 240 barrels of oil daily. The company can add more wells but every added well reduces the average daily ouput of each of the wells by 8 barrels. How many wells should the company add in order to maximize daily production?

65. Show that of all the isosceles triangles with a given perimeter, the one with the greatest area is equilateral.

66. Consider the situation in Exercise 51 if the cost of laying pipe under the river is considerably higher than the cost of laying pipe over land ($400,000/km). You may suspect that in some instances, the minimum distance possible under the river should be used, and P should be located 6 km from the refinery, directly across from the storage tanks. Show that this is *never* the case, no matter what the "under river" cost is.

67. Consider the tangent line to the ellipse $\dfrac{x^2}{a^2} + \dfrac{y^2}{b^2} = 1$ at a point (p, q) in the first quadrant.
(a) Show that the tangent line has x-intercept a^2/p and y-intercept b^2/q.
(b) Show that the portion of the tangent line cut off by the coordinate axes has minimum length $a + b$.
(c) Show that the triangle formed by the tangent line and the coordinate axes has minimum area ab.

68. The frame for a kite is to be made from six pieces of wood. The four exterior pieces have been cut with the lengths indicated in the figure. To maximize the area of the kite, how long should the diagonal pieces be?

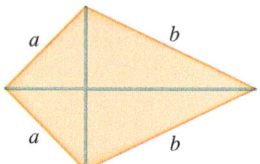

69. A point P needs to be located somewhere on the line AD so that the total length L of cables linking P to the points A, B, and C is minimized (see the figure). Express L as a function of $x = |AP|$ and use the graphs of L and dL/dx to estimate the minimum value of L.

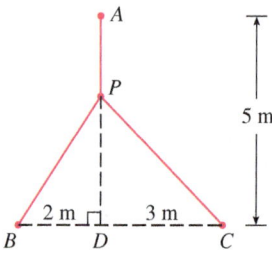

70. The graph shows the fuel consumption c of a car (measured in gallons per hour) as a function of the speed v of the car. At very low speeds the engine runs inefficiently, so initially c decreases as the speed increases. But at high speeds the fuel consumption increases. You can see that $c(v)$ is minimized for this car when $v \approx 30$ mi/h. However, for fuel efficiency, what must be minimized is not the consumption in gallons per hour but rather the fuel consumption in gallons *per mile*. Let's call this consumption G. Using the graph, estimate the speed at which G has its minimum value.

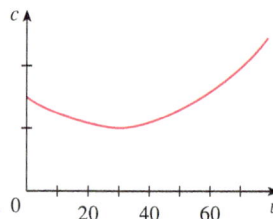

71. Let v_1 be the velocity of light in air and v_2 the velocity of light in water. According to Fermat's Principle, a ray of light will travel from a point A in the air to a point B in the water by a path ACB that minimizes the time taken. Show that
$$\frac{\sin\theta_1}{\sin\theta_2} = \frac{v_1}{v_2}$$

where θ_1 (the angle of incidence) and θ_2 (the angle of refraction) are as shown. This equation is known as Snell's Law.

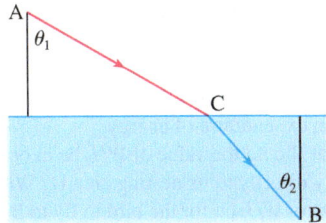

72. Two vertical poles PQ and ST are secured by a rope PRS going from the top of the first pole to a point R on the ground between the poles and then to the top of the second pole as in the figure. Show that the shortest length of such a rope occurs when $\theta_1 = \theta_2$.

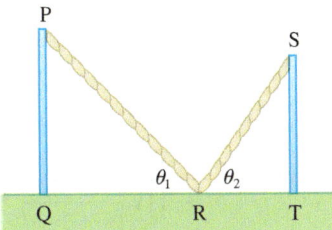

73. The upper right-hand corner of a piece of paper, 12 in. by 8 in., as in the figure, is folded over to the bottom edge. How would you fold it so as to minimize the length of the fold? In other words, how would you choose x to minimize y?

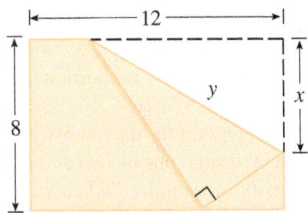

74. A steel pipe is being carried down a hallway 9 ft wide. At the end of the hall there is a right-angled turn into a narrower hallway 6 ft wide. What is the length of the longest pipe that can be carried horizontally around the corner?

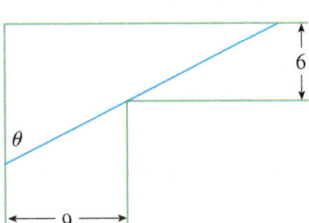

75. An observer stands at a point P, one unit away from a track. Two runners start at the point S in the figure and run along the track. One runner runs three times as fast as the other. Find the maximum value of the observer's angle of sight θ between the runners.

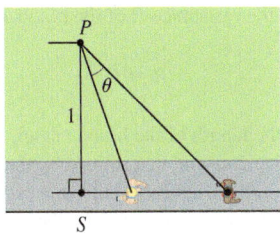

76. A rain gutter is to be constructed from a metal sheet of width 30 cm by bending up one-third of the sheet on each side through an angle θ. How should θ be chosen so that the gutter will carry the maximum amount of water?

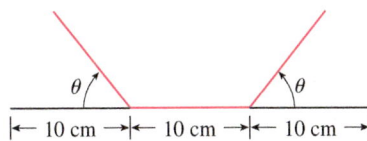

77. Where should the point P be chosen on the line segment AB so as to maximize the angle θ?

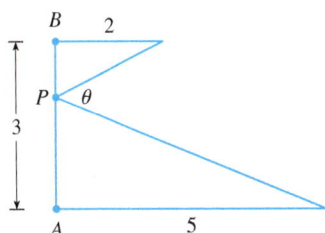

78. A painting in an art gallery has height h and is hung so that its lower edge is a distance d above the eye of an observer (as in the figure). How far from the wall should the observer stand to get the best view? (In other words, where should the observer stand so as to maximize the angle θ subtended at his eye by the painting?)

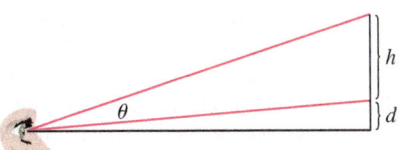

79. Find the maximum area of a rectangle that can be circumscribed about a given rectangle with length L and width W. [Hint: Express the area as a function of an angle θ.]

80. The blood vascular system consists of blood vessels (arteries, arterioles, capillaries, and veins) that convey blood from the heart to the organs and back to the heart. This system should work so as to minimize the energy expended by the heart in pumping the blood. In particular, this energy is reduced when the resistance of the blood is lowered. One of Poiseuille's Laws gives the resistance R of the blood as

$$R = C\frac{L}{r^4}$$

where L is the length of the blood vessel, r is the radius, and C is a positive constant determined by the viscosity of the blood. (Poiseuille established this law experimentally, but it also follows from Equation 8.4.2.) The figure shows a main blood vessel with radius r_1 branching at an angle θ into a smaller vessel with radius r_2.

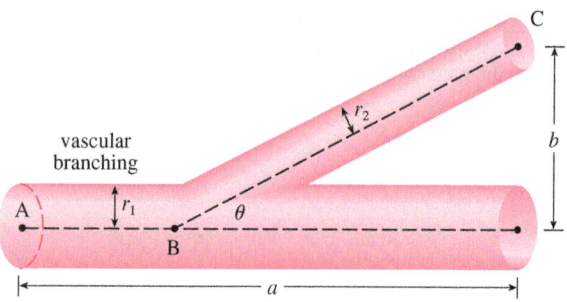

(a) Use Poiseuille's Law to show that the total resistance of the blood along the path ABC is

$$R = C\left(\frac{a - b\cot\theta}{r_1^4} + \frac{b\csc\theta}{r_2^4}\right)$$

where a and b are the distances shown in the figure.

(b) Prove that this resistance is minimized when

$$\cos\theta = \frac{r_2^4}{r_1^4}$$

(c) Find the optimal branching angle (correct to the nearest degree) when the radius of the smaller blood vessel is two-thirds the radius of the larger vessel.

81. Ornithologists have determined that some species of birds tend to avoid flights over large bodies of water during daylight hours. It is believed that more energy is required to fly over water than over land because air generally rises over land and falls over water during the day. A bird with these tendencies is released from an island that is 5 km from the nearest point B on a straight shoreline, flies to a point C on the shoreline, and then flies along the shoreline to its nesting area D. Assume that the bird instinctively chooses a path that will minimize its energy expenditure. Points B and D are 13 km apart.

(a) In general, if it takes 1.4 times as much energy to fly over water as it does over land, to what point C should the bird fly in order to minimize the total energy expended in returning to its nesting area?

(b) Let W and L denote the energy (in joules) per kilometer flown over water and land, respectively. What would a large value of the ratio W/L mean in terms of the bird's flight? What would a small value mean? Determine the ratio W/L corresponding to the minimum expenditure of energy.

(c) What should the value of W/L be in order for the bird to fly directly to its nesting area D? What should the value of W/L be for the bird to fly to B and then along the shore to D?

(d) If the ornithologists observe that birds of a certain species reach the shore at a point 4 km from B, how many times more energy does it take a bird to fly over water than over land?

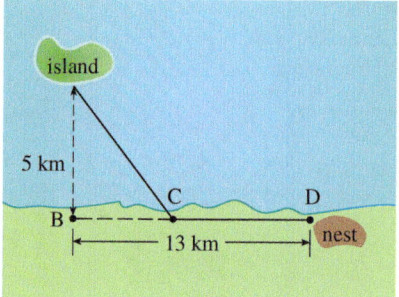

82. Two light sources of identical strength are placed 10 m apart. An object is to be placed at a point P on a line ℓ, parallel to the line joining the light sources and at a distance d meters from it (see the figure). We want to locate P on ℓ so that the intensity of illumination is minimized. We need to use the fact that the intensity of illumination for a single source is directly proportional to the strength of the source and inversely proportional to the square of the distance from the source.

(a) Find an expression for the intensity $I(x)$ at the point P.

(b) If $d = 5$ m, use graphs of $I(x)$ and $I'(x)$ to show that the intensity is minimized when $x = 5$ m, that is, when P is at the midpoint of ℓ.

(c) If $d = 10$ m, show that the intensity (perhaps surprisingly) is *not* minimized at the midpoint.

(d) Somewhere between $d = 5$ m and $d = 10$ m there is a transitional value of d at which the point of minimal illumination abruptly changes. Estimate this value of d by graphical methods. Then find the exact value of d.

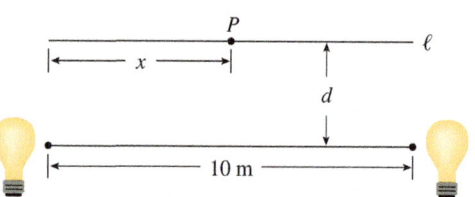

APPLIED PROJECT

THE SHAPE OF A CAN

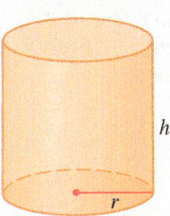

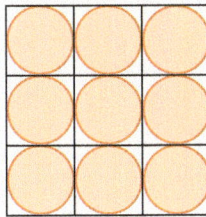
Discs cut from squares

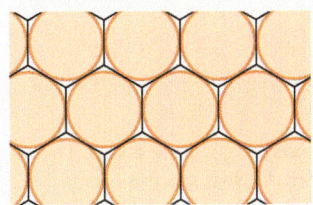
Discs cut from hexagons

In this project we investigate the most economical shape for a can. We first interpret this to mean that the volume V of a cylindrical can is given and we need to find the height h and radius r that minimize the cost of the metal to make the can (see the figure). If we disregard any waste metal in the manufacturing process, then the problem is to minimize the surface area of the cylinder. We solved this problem in Example 4.7.2 and we found that $h = 2r$; that is, the height should be the same as the diameter. But if you go to your cupboard or your supermarket with a ruler, you will discover that the height is usually greater than the diameter and the ratio h/r varies from 2 up to about 3.8. Let's see if we can explain this phenomenon.

1. The material for the cans is cut from sheets of metal. The cylindrical sides are formed by bending rectangles; these rectangles are cut from the sheet with little or no waste. But if the top and bottom discs are cut from squares of side $2r$ (as in the figure), this leaves considerable waste metal, which may be recycled but has little or no value to the can makers. If this is the case, show that the amount of metal used is minimized when

$$\frac{h}{r} = \frac{8}{\pi} \approx 2.55$$

2. A more efficient packing of the discs is obtained by dividing the metal sheet into hexagons and cutting the circular lids and bases from the hexagons (see the figure). Show that if this strategy is adopted, then

$$\frac{h}{r} = \frac{4\sqrt{3}}{\pi} \approx 2.21$$

3. The values of h/r that we found in Problems 1 and 2 are a little closer to the ones that actually occur on supermarket shelves, but they still don't account for everything. If we look more closely at some real cans, we see that the lid and the base are formed from discs with radius larger than r that are bent over the ends of the can. If we allow for this we would increase h/r. More significantly, in addition to the cost of the metal we need to incorporate the manufacturing of the can into the cost. Let's assume that most of the expense is incurred in joining the sides to the rims of the cans. If we cut the discs from hexagons as in Problem 2, then the total cost is proportional to

$$4\sqrt{3}\,r^2 + 2\pi rh + k(4\pi r + h)$$

where k is the reciprocal of the length that can be joined for the cost of one unit area of metal. Show that this expression is minimized when

$$\frac{\sqrt[3]{V}}{k} = \sqrt[3]{\frac{\pi h}{r}} \cdot \frac{2\pi - h/r}{\pi h/r - 4\sqrt{3}}$$

4. Plot $\sqrt[3]{V}/k$ as a function of $x = h/r$ and use your graph to argue that when a can is large or joining is cheap, we should make h/r approximately 2.21 (as in Problem 2). But when the can is small or joining is costly, h/r should be substantially larger.

5. Our analysis shows that large cans should be almost square but small cans should be tall and thin. Take a look at the relative shapes of the cans in a supermarket. Is our conclusion usually true in practice? Are there exceptions? Can you suggest reasons why small cans are not always tall and thin?

APPLIED PROJECT PLANES AND BIRDS: MINIMIZING ENERGY

Small birds like finches alternate between flapping their wings and keeping them folded while gliding (see Figure 1). In this project we analyze this phenomenon and try to determine how frequently a bird should flap its wings. Some of the principles are the same as for fixed-wing aircraft and so we begin by considering how required power and energy depend on the speed of airplanes.[1]

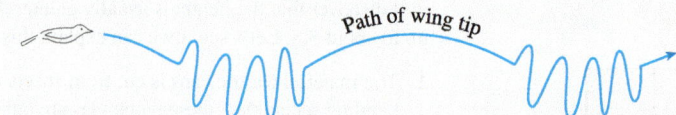

FIGURE 1

1. The power needed to propel an airplane forward at velocity v is

$$P = Av^3 + \frac{BL^2}{v}$$

   where A and B are positive constants specific to the particular aircraft and L is the lift, the upward force supporting the weight of the plane. Find the speed that minimizes the required power.

2. The speed found in Problem 1 minimizes power but a faster speed might use less fuel. The energy needed to propel the airplane a unit distance is $E = P/v$. At what speed is energy minimized?

3. Hows much faster is the speed for minimum energy than the speed for minimum power?

4. In applying the equation of Problem 1 to bird flight we split the term Av^3 into two parts: $A_b v^3$ for the bird's body and $A_w v^3$ for its wings. Let x be the fraction of flying time spent in flapping mode. If m is the bird's mass and all the lift occurs during flapping, then the lift is mg/x and so the power needed during flapping is

$$P_{\text{flap}} = (A_b + A_w)v^3 + \frac{B(mg/x)^2}{v}$$

   The power while wings are folded is $P_{\text{fold}} = A_b v^3$. Show that the average power over an entire flight cycle is

$$\overline{P} = xP_{\text{flap}} + (1-x)P_{\text{fold}} = A_b v^3 + xA_w v^3 + \frac{Bm^2 g^2}{xv}$$

5. For what value of x is the average power a minimum? What can you conclude if the bird flies slowly? What can you conclude if the bird flies faster and faster?

6. The average energy over a cycle is $\overline{E} = \overline{P}/v$. What value of x minimizes $\overline{E}$?

---

1. Adapted from R. McNeill Alexander, *Optima for Animals* (Princeton, NJ: Princeton University Press, 1996.)

4.8 Newton's Method

Suppose that a car dealer offers to sell you a car for $18,000 or for payments of $375 per month for five years. You would like to know what monthly interest rate the dealer is, in effect, charging you. To find the answer, you have to solve the equation

$$\boxed{1} \qquad 48x(1 + x)^{60} - (1 + x)^{60} + 1 = 0$$

(The details are explained in Exercise 41.) How would you solve such an equation?

For a quadratic equation $ax^2 + bx + c = 0$ there is a well-known formula for the solutions. For third- and fourth-degree equations there are also formulas for the solutions, but they are extremely complicated. If f is a polynomial of degree 5 or higher, there is no such formula (see the note on page 211). Likewise, there is no formula that will enable us to find the exact roots of a transcendental equation such as $\cos x = x$.

We can find an *approximate* solution to Equation 1 by plotting the left side of the equation. Using a graphing device, and after experimenting with viewing rectangles, we produce the graph in Figure 1.

We see that in addition to the solution $x = 0$, which doesn't interest us, there is a solution between 0.007 and 0.008. Zooming in shows that the root is approximately 0.0076. If we need more accuracy we could zoom in repeatedly, but that becomes tiresome. A faster alternative is to use a calculator or computer algebra system to solve the equation numerically. If we do so, we find that the root, correct to nine decimal places, is 0.007628603.

How do these devices solve equations? They use a variety of methods, but most of them make some use of **Newton's method**, also called the **Newton-Raphson method**. We will explain how this method works, partly to show what happens inside a calculator or computer, and partly as an application of the idea of linear approximation.

The geometry behind Newton's method is shown in Figure 2. We wish to solve an equation of the form $f(x) = 0$, so the roots of the equation correspond to the x-intercepts of the graph of f. The root that we are trying to find is labeled r in the figure. We start with a first approximation x_1, which is obtained by guessing, or from a rough sketch of the graph of f, or from a computer-generated graph of f. Consider the tangent line L to the curve $y = f(x)$ at the point $(x_1, f(x_1))$ and look at the x-intercept of L, labeled x_2. The idea behind Newton's method is that the tangent line is close to the curve and so its x-intercept, x_2, is close to the x-intercept of the curve (namely, the root r that we are seeking). Because the tangent is a line, we can easily find its x-intercept.

To find a formula for x_2 in terms of x_1 we use the fact that the slope of L is $f'(x_1)$, so its equation is

$$y - f(x_1) = f'(x_1)(x - x_1)$$

Since the x-intercept of L is x_2, we know that the point $(x_2, 0)$ is on the line, and so

$$0 - f(x_1) = f'(x_1)(x_2 - x_1)$$

If $f'(x_1) \neq 0$, we can solve this equation for x_2:

$$x_2 = x_1 - \frac{f(x_1)}{f'(x_1)}$$

We use x_2 as a second approximation to r.

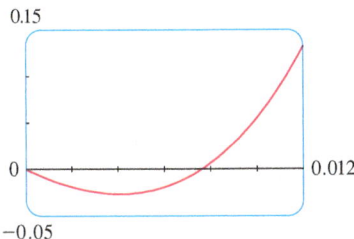

FIGURE 1

Try to solve Equation 1 numerically using your calculator or computer. Some machines are not able to solve it. Others are successful but require you to specify a starting point for the search.

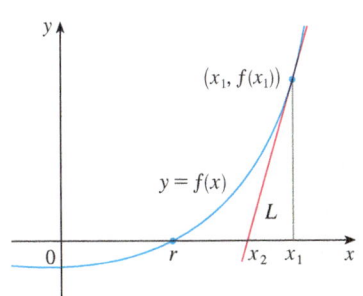

FIGURE 2

346   CHAPTER 4   Applications of Differentiation

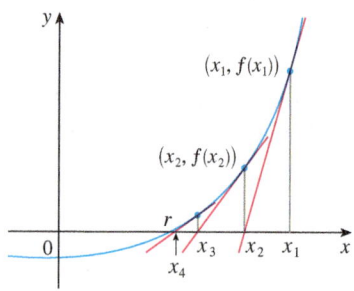

FIGURE 3

Next we repeat this procedure with x_1 replaced by the second approximation x_2, using the tangent line at $(x_2, f(x_2))$. This gives a third approximation:

$$x_3 = x_2 - \frac{f(x_2)}{f'(x_2)}$$

If we keep repeating this process, we obtain a sequence of approximations $x_1, x_2, x_3, x_4, \ldots$ as shown in Figure 3. In general, if the nth approximation is x_n and $f'(x_n) \neq 0$, then the next approximation is given by

$$\boxed{2} \quad x_{n+1} = x_n - \frac{f(x_n)}{f'(x_n)}$$

Sequences were briefly introduced in *A Preview of Calculus* on page 5. A more thorough discussion starts in Section 11.1.

If the numbers x_n become closer and closer to r as n becomes large, then we say that the sequence *converges* to r and we write

$$\lim_{n \to \infty} x_n = r$$

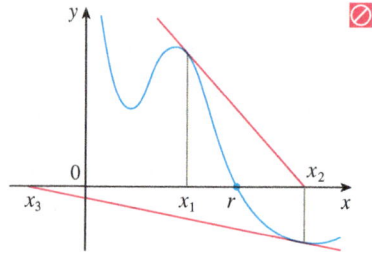

Although the sequence of successive approximations converges to the desired root for functions of the type illustrated in Figure 3, in certain circumstances the sequence may not converge. For example, consider the situation shown in Figure 4. You can see that x_2 is a worse approximation than x_1. This is likely to be the case when $f'(x_1)$ is close to 0. It might even happen that an approximation (such as x_3 in Figure 4) falls outside the domain of f. Then Newton's method fails and a better initial approximation x_1 should be chosen. See Exercises 31–34 for specific examples in which Newton's method works very slowly or does not work at all.

FIGURE 4

EXAMPLE 1 Starting with $x_1 = 2$, find the third approximation x_3 to the root of the equation $x^3 - 2x - 5 = 0$.

SOLUTION We apply Newton's method with

$$f(x) = x^3 - 2x - 5 \quad \text{and} \quad f'(x) = 3x^2 - 2$$

TEC In Module 4.8 you can investigate how Newton's method works for several functions and what happens when you change x_1.

Figure 5 shows the geometry behind the first step in Newton's method in Example 1. Since $f'(2) = 10$, the tangent line to $y = x^3 - 2x - 5$ at $(2, -1)$ has equation $y = 10x - 21$ so its x-intercept is $x_2 = 2.1$.

Newton himself used this equation to illustrate his method and he chose $x_1 = 2$ after some experimentation because $f(1) = -6$, $f(2) = -1$, and $f(3) = 16$. Equation 2 becomes

$$x_{n+1} = x_n - \frac{f(x_n)}{f'(x_n)} = x_n - \frac{x_n^3 - 2x_n - 5}{3x_n^2 - 2}$$

With $n = 1$ we have

$$x_2 = x_1 - \frac{f(x_1)}{f'(x_1)} = x_1 - \frac{x_1^3 - 2x_1 - 5}{3x_1^2 - 2}$$

$$= 2 - \frac{2^3 - 2(2) - 5}{3(2)^2 - 2} = 2.1$$

Then with $n = 2$ we obtain

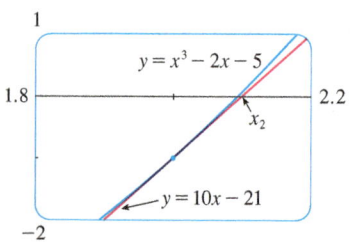

FIGURE 5

$$x_3 = x_2 - \frac{x_2^3 - 2x_2 - 5}{3x_2^2 - 2} = 2.1 - \frac{(2.1)^3 - 2(2.1) - 5}{3(2.1)^2 - 2} \approx 2.0946$$

It turns out that this third approximation $x_3 \approx 2.0946$ is accurate to four decimal places.

Suppose that we want to achieve a given accuracy, say to eight decimal places, using Newton's method. How do we know when to stop? The rule of thumb that is generally used is that we can stop when successive approximations x_n and x_{n+1} agree to eight decimal places. (A precise statement concerning accuracy in Newton's method will be given in Exercise 11.11.39.)

Notice that the procedure in going from n to $n + 1$ is the same for all values of n. (It is called an *iterative* process.) This means that Newton's method is particularly convenient for use with a programmable calculator or a computer.

EXAMPLE 2 Use Newton's method to find $\sqrt[6]{2}$ correct to eight decimal places.

SOLUTION First we observe that finding $\sqrt[6]{2}$ is equivalent to finding the positive root of the equation
$$x^6 - 2 = 0$$
so we take $f(x) = x^6 - 2$. Then $f'(x) = 6x^5$ and Formula 2 (Newton's method) becomes
$$x_{n+1} = x_n - \frac{f(x_n)}{f'(x_n)} = x_n - \frac{x_n^6 - 2}{6x_n^5}$$

If we choose $x_1 = 1$ as the initial approximation, then we obtain
$$x_2 \approx 1.16666667$$
$$x_3 \approx 1.12644368$$
$$x_4 \approx 1.12249707$$
$$x_5 \approx 1.12246205$$
$$x_6 \approx 1.12246205$$

Since x_5 and x_6 agree to eight decimal places, we conclude that
$$\sqrt[6]{2} \approx 1.12246205$$
to eight decimal places.

EXAMPLE 3 Find, correct to six decimal places, the root of the equation $\cos x = x$.

SOLUTION We first rewrite the equation in standard form:
$$\cos x - x = 0$$
Therefore we let $f(x) = \cos x - x$. Then $f'(x) = -\sin x - 1$, so Formula 2 becomes
$$x_{n+1} = x_n - \frac{\cos x_n - x_n}{-\sin x_n - 1} = x_n + \frac{\cos x_n - x_n}{\sin x_n + 1}$$

In order to guess a suitable value for x_1 we sketch the graphs of $y = \cos x$ and $y = x$ in Figure 6. It appears that they intersect at a point whose x-coordinate is somewhat less

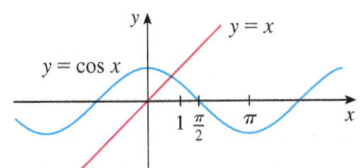

FIGURE 6

than 1, so let's take $x_1 = 1$ as a convenient first approximation. Then, remembering to put our calculator in radian mode, we get

$$x_2 \approx 0.75036387$$
$$x_3 \approx 0.73911289$$
$$x_4 \approx 0.73908513$$
$$x_5 \approx 0.73908513$$

Since x_4 and x_5 agree to six decimal places (eight, in fact), we conclude that the root of the equation, correct to six decimal places, is 0.739085. ■

Instead of using the rough sketch in Figure 6 to get a starting approximation for Newton's method in Example 3, we could have used the more accurate graph that a calculator or computer provides. Figure 7 suggests that we use $x_1 = 0.75$ as the initial approximation. Then Newton's method gives

$$x_2 \approx 0.73911114$$
$$x_3 \approx 0.73908513$$
$$x_4 \approx 0.73908513$$

and so we obtain the same answer as before, but with one fewer step.

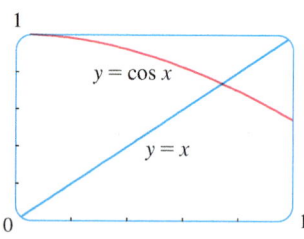

FIGURE 7

4.8 EXERCISES

1. The figure shows the graph of a function f. Suppose that Newton's method is used to approximate the root s of the equation $f(x) = 0$ with initial approximation $x_1 = 6$.
   (a) Draw the tangent lines that are used to find x_2 and x_3, and estimate the numerical values of x_2 and x_3.
   (b) Would $x_1 = 8$ be a better first approximation? Explain.

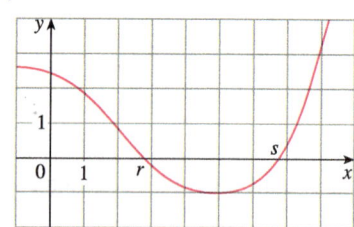

2. Follow the instructions for Exercise 1(a) but use $x_1 = 1$ as the starting approximation for finding the root r.

3. Suppose the tangent line to the curve $y = f(x)$ at the point (2, 5) has the equation $y = 9 - 2x$. If Newton's method is used to locate a root of the equation $f(x) = 0$ and the initial approximation is $x_1 = 2$, find the second approximation x_2.

4. For each initial approximation, determine graphically what happens if Newton's method is used for the function whose graph is shown.

(a) $x_1 = 0$ (b) $x_1 = 1$ (c) $x_1 = 3$
(d) $x_1 = 4$ (e) $x_1 = 5$

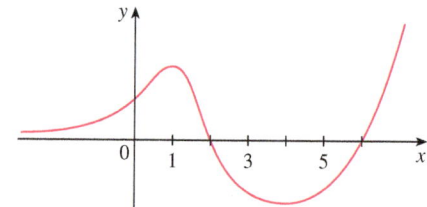

5. For which of the initial approximations $x_1 = a, b, c,$ and d do you think Newton's method will work and lead to the root of the equation $f(x) = 0$?

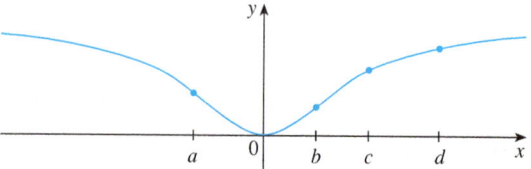

6–8 Use Newton's method with the specified initial approximation x_1 to find x_3, the third approximation to the root of the given equation. (Give your answer to four decimal places.)

6. $2x^3 - 3x^2 + 2 = 0$, $x_1 = -1$

7. $\dfrac{2}{x} - x^2 + 1 = 0$, $x_1 = 2$ **8.** $x^7 + 4 = 0$, $x_1 = -1$

9. Use Newton's method with initial approximation $x_1 = -1$ to find x_2, the second approximation to the root of the equation $x^3 + x + 3 = 0$. Explain how the method works by first graphing the function and its tangent line at $(-1, 1)$.

10. Use Newton's method with initial approximation $x_1 = 1$ to find x_2, the second approximation to the root of the equation $x^4 - x - 1 = 0$. Explain how the method works by first graphing the function and its tangent line at $(1, -1)$.

11–12 Use Newton's method to approximate the given number correct to eight decimal places.

11. $\sqrt[4]{75}$ **12.** $\sqrt[8]{500}$

13–14 (a) Explain how we know that the given equation must have a root in the given interval. (b) Use Newton's method to approximate the root correct to six decimal places.

13. $3x^4 - 8x^3 + 2 = 0$, $[2, 3]$
14. $-2x^5 + 9x^4 - 7x^3 - 11x = 0$, $[3, 4]$

15–16 Use Newton's method to approximate the indicated root of the equation correct to six decimal places.

15. The negative root of $e^x = 4 - x^2$
16. The positive root of $3 \sin x = x$

17–22 Use Newton's method to find all solutions of the equation correct to six decimal places.

17. $3 \cos x = x + 1$ **18.** $\sqrt{x + 1} = x^2 - x$
19. $2^x = 2 - x^2$ **20.** $\ln x = \dfrac{1}{x - 3}$
21. $x^3 = \tan^{-1} x$ **22.** $\sin x = x^2 - 2$

23–28 Use Newton's method to find all the solutions of the equation correct to eight decimal places. Start by drawing a graph to find initial approximations.

23. $-2x^7 - 5x^4 + 9x^3 + 5 = 0$
24. $x^5 - 3x^4 + x^3 - x^2 - x + 6 = 0$
25. $\dfrac{x}{x^2 + 1} = \sqrt{1 - x}$
26. $\cos(x^2 - x) = x^4$
27. $4e^{-x^2} \sin x = x^2 - x + 1$
28. $\ln(x^2 + 2) = \dfrac{3x}{\sqrt{x^2 + 1}}$

29. (a) Apply Newton's method to the equation $x^2 - a = 0$ to derive the following square-root algorithm (used by the ancient Babylonians to compute $\sqrt{a}$):

$$x_{n+1} = \dfrac{1}{2}\left(x_n + \dfrac{a}{x_n}\right)$$

(b) Use part (a) to compute $\sqrt{1000}$ correct to six decimal places.

30. (a) Apply Newton's method to the equation $1/x - a = 0$ to derive the following reciprocal algorithm:

$$x_{n+1} = 2x_n - ax_n^2$$

(This algorithm enables a computer to find reciprocals without actually dividing.)
(b) Use part (a) to compute $1/1.6984$ correct to six decimal places.

31. Explain why Newton's method doesn't work for finding the root of the equation

$$x^3 - 3x + 6 = 0$$

if the initial approximation is chosen to be $x_1 = 1$.

32. (a) Use Newton's method with $x_1 = 1$ to find the root of the equation $x^3 - x = 1$ correct to six decimal places.
(b) Solve the equation in part (a) using $x_1 = 0.6$ as the initial approximation.
(c) Solve the equation in part (a) using $x_1 = 0.57$. (You definitely need a programmable calculator for this part.)
(d) Graph $f(x) = x^3 - x - 1$ and its tangent lines at $x_1 = 1$, 0.6, and 0.57 to explain why Newton's method is so sensitive to the value of the initial approximation.

33. Explain why Newton's method fails when applied to the equation $\sqrt[3]{x} = 0$ with any initial approximation $x_1 \neq 0$. Illustrate your explanation with a sketch.

34. If

$$f(x) = \begin{cases} \sqrt{x} & \text{if } x \geq 0 \\ -\sqrt{-x} & \text{if } x < 0 \end{cases}$$

then the root of the equation $f(x) = 0$ is $x = 0$. Explain why Newton's method fails to find the root no matter which initial approximation $x_1 \neq 0$ is used. Illustrate your explanation with a sketch.

35. (a) Use Newton's method to find the critical numbers of the function

$$f(x) = x^6 - x^4 + 3x^3 - 2x$$

correct to six decimal places.
(b) Find the absolute minimum value of f correct to four decimal places.

36. Use Newton's method to find the absolute maximum value of the function $f(x) = x \cos x$, $0 \leq x \leq \pi$, correct to six decimal places.

37. Use Newton's method to find the coordinates of the inflection point of the curve $y = x^2 \sin x$, $0 \leq x \leq \pi$, correct to six decimal places.

38. Of the infinitely many lines that are tangent to the curve $y = -\sin x$ and pass through the origin, there is one that has the largest slope. Use Newton's method to find the slope of that line correct to six decimal places.

39. Use Newton's method to find the coordinates, correct to six decimal places, of the point on the parabola $y = (x - 1)^2$ that is closest to the origin.

40. In the figure, the length of the chord AB is 4 cm and the length of the arc AB is 5 cm. Find the central angle θ, in radians, correct to four decimal places. Then give the answer to the nearest degree.

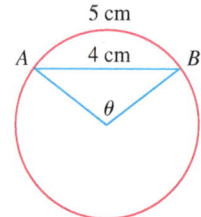

41. A car dealer sells a new car for \$18,000. He also offers to sell the same car for payments of \$375 per month for five years. What monthly interest rate is this dealer charging?

To solve this problem you will need to use the formula for the present value A of an annuity consisting of n equal payments of size R with interest rate i per time period:

$$A = \frac{R}{i}[1 - (1 + i)^{-n}]$$

Replacing i by x, show that

$$48x(1 + x)^{60} - (1 + x)^{60} + 1 = 0$$

Use Newton's method to solve this equation.

42. The figure shows the sun located at the origin and the earth at the point $(1, 0)$. (The unit here is the distance between the centers of the earth and the sun, called an *astronomical unit*: 1 AU $\approx 1.496 \times 10^8$ km.) There are five locations L_1, L_2, L_3, L_4, and L_5 in this plane of rotation of the earth about the sun where a satellite remains motionless with respect to the earth because the forces acting on the satellite (including the gravitational attractions of the earth and the sun) balance each other. These locations are called *libration points*. (A solar research satellite has been placed at one of these libration points.) If m_1 is the mass of the sun, m_2 is the mass of the earth, and $r = m_2/(m_1 + m_2)$, it turns out that the x-coordinate of L_1 is the unique root of the fifth-degree equation

$$p(x) = x^5 - (2 + r)x^4 + (1 + 2r)x^3 - (1 - r)x^2$$
$$+ 2(1 - r)x + r - 1 = 0$$

and the x-coordinate of L_2 is the root of the equation

$$p(x) - 2rx^2 = 0$$

Using the value $r \approx 3.04042 \times 10^{-6}$, find the locations of the libration points (a) L_1 and (b) L_2.

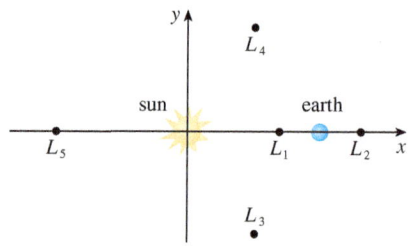

4.9 Antiderivatives

A physicist who knows the velocity of a particle might wish to know its position at a given time. An engineer who can measure the variable rate at which water is leaking from a tank wants to know the amount leaked over a certain time period. A biologist who knows the rate at which a bacteria population is increasing might want to deduce what the size of the population will be at some future time. In each case, the problem is to find a function F whose derivative is a known function f. If such a function F exists, it is called an *antiderivative* of f.

> **Definition** A function F is called an **antiderivative** of f on an interval I if $F'(x) = f(x)$ for all x in I.

For instance, let $f(x) = x^2$. It isn't difficult to discover an antiderivative of f if we keep the Power Rule in mind. In fact, if $F(x) = \frac{1}{3}x^3$, then $F'(x) = x^2 = f(x)$. But the function $G(x) = \frac{1}{3}x^3 + 100$ also satisfies $G'(x) = x^2$. Therefore both F and G are antiderivatives of f. Indeed, any function of the form $H(x) = \frac{1}{3}x^3 + C$, where C is a constant, is an antiderivative of f. The question arises: Are there any others?

To answer this question, recall that in Section 4.2 we used the Mean Value Theorem to prove that if two functions have identical derivatives on an interval, then they must differ by a constant (Corollary 4.2.7). Thus if F and G are any two antiderivatives of f, then

$$F'(x) = f(x) = G'(x)$$

so $G(x) - F(x) = C$, where C is a constant. We can write this as $G(x) = F(x) + C$, so we have the following result.

> **1 Theorem** If F is an antiderivative of f on an interval I, then the most general antiderivative of f on I is
> $$F(x) + C$$
> where C is an arbitrary constant.

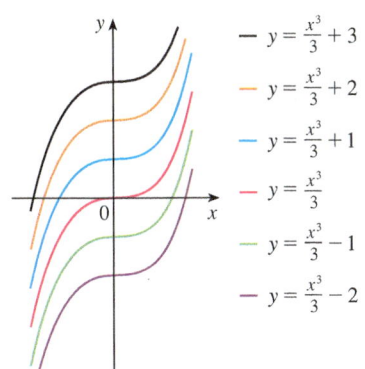

FIGURE 1
Members of the family of antiderivatives of $f(x) = x^2$

Going back to the function $f(x) = x^2$, we see that the general antiderivative of f is $\frac{1}{3}x^3 + C$. By assigning specific values to the constant C, we obtain a family of functions whose graphs are vertical translates of one another (see Figure 1). This makes sense because each curve must have the same slope at any given value of x.

EXAMPLE 1 Find the most general antiderivative of each of the following functions.
(a) $f(x) = \sin x$ (b) $f(x) = 1/x$ (c) $f(x) = x^n$, $n \neq -1$

SOLUTION
(a) If $F(x) = -\cos x$, then $F'(x) = \sin x$, so an antiderivative of $\sin x$ is $-\cos x$. By Theorem 1, the most general antiderivative is $G(x) = -\cos x + C$.
(b) Recall from Section 3.6 that

$$\frac{d}{dx}(\ln x) = \frac{1}{x}$$

So on the interval $(0, \infty)$ the general antiderivative of $1/x$ is $\ln x + C$. We also learned that

$$\frac{d}{dx}(\ln |x|) = \frac{1}{x}$$

for all $x \neq 0$. Theorem 1 then tells us that the general antiderivative of $f(x) = 1/x$ is $\ln |x| + C$ on any interval that doesn't contain 0. In particular, this is true on each of the intervals $(-\infty, 0)$ and $(0, \infty)$. So the general antiderivative of f is

$$F(x) = \begin{cases} \ln x + C_1 & \text{if } x > 0 \\ \ln(-x) + C_2 & \text{if } x < 0 \end{cases}$$

(c) We use the Power Rule to discover an antiderivative of x^n. In fact, if $n \neq -1$, then

$$\frac{d}{dx}\left(\frac{x^{n+1}}{n+1}\right) = \frac{(n+1)x^n}{n+1} = x^n$$

Therefore the general antiderivative of $f(x) = x^n$ is

$$F(x) = \frac{x^{n+1}}{n+1} + C$$

This is valid for $n \geq 0$ since then $f(x) = x^n$ is defined on an interval. If n is negative (but $n \neq -1$), it is valid on any interval that doesn't contain 0.

As in Example 1, every differentiation formula, when read from right to left, gives rise to an antidifferentiation formula. In Table 2 we list some particular antiderivatives. Each formula in the table is true because the derivative of the function in the right column appears in the left column. In particular, the first formula says that the antiderivative of a constant times a function is the constant times the antiderivative of the function. The second formula says that the antiderivative of a sum is the sum of the antiderivatives. (We use the notation $F' = f$, $G' = g$.)

2 Table of Antidifferentiation Formulas

To obtain the most general antiderivative from the particular ones in Table 2, we have to add a constant (or constants), as in Example 1.

Function	Particular antiderivative	Function	Particular antiderivative
$cf(x)$	$cF(x)$	$\sin x$	$-\cos x$
$f(x) + g(x)$	$F(x) + G(x)$	$\sec^2 x$	$\tan x$
x^n ($n \neq -1$)	$\dfrac{x^{n+1}}{n+1}$	$\sec x \tan x$	$\sec x$
$\dfrac{1}{x}$	$\ln|x|$	$\dfrac{1}{\sqrt{1-x^2}}$	$\sin^{-1} x$
e^x	e^x	$\dfrac{1}{1+x^2}$	$\tan^{-1} x$
b^x	$\dfrac{b^x}{\ln b}$	$\cosh x$	$\sinh x$
$\cos x$	$\sin x$	$\sinh x$	$\cosh x$

EXAMPLE 2 Find all functions g such that

$$g'(x) = 4 \sin x + \frac{2x^5 - \sqrt{x}}{x}$$

SOLUTION We first rewrite the given function as follows:

$$g'(x) = 4 \sin x + \frac{2x^5}{x} - \frac{\sqrt{x}}{x} = 4 \sin x + 2x^4 - \frac{1}{\sqrt{x}}$$

Thus we want to find an antiderivative of

$$g'(x) = 4 \sin x + 2x^4 - x^{-1/2}$$

Using the formulas in Table 2 together with Theorem 1, we obtain

$$g(x) = 4(-\cos x) + 2\frac{x^5}{5} - \frac{x^{1/2}}{\frac{1}{2}} + C$$

We often use a capital letter F to represent an antiderivative of a function f. If we begin with derivative notation, f', an antiderivative is f, of course.

$$= -4 \cos x + \tfrac{2}{5} x^5 - 2\sqrt{x} + C$$

In applications of calculus it is very common to have a situation as in Example 2, where it is required to find a function, given knowledge about its derivatives. An equation that involves the derivatives of a function is called a **differential equation**. These will be studied in some detail in Chapter 9, but for the present we can solve some elementary differential equations. The general solution of a differential equation involves an arbitrary constant (or constants) as in Example 2. However, there may be some extra conditions given that will determine the constants and therefore uniquely specify the solution.

EXAMPLE 3 Find f if $f'(x) = e^x + 20(1 + x^2)^{-1}$ and $f(0) = -2$.

SOLUTION The general antiderivative of

$$f'(x) = e^x + \frac{20}{1 + x^2}$$

is

$$f(x) = e^x + 20 \tan^{-1}x + C$$

To determine C we use the fact that $f(0) = -2$:

$$f(0) = e^0 + 20 \tan^{-1} 0 + C = -2$$

Thus we have $C = -2 - 1 = -3$, so the particular solution is

$$f(x) = e^x + 20 \tan^{-1}x - 3$$

Figure 2 shows the graphs of the function f' in Example 3 and its antiderivative f. Notice that $f'(x) > 0$, so f is always increasing. Also notice that when f' has a maximum or minimum, f appears to have an inflection point. So the graph serves as a check on our calculation.

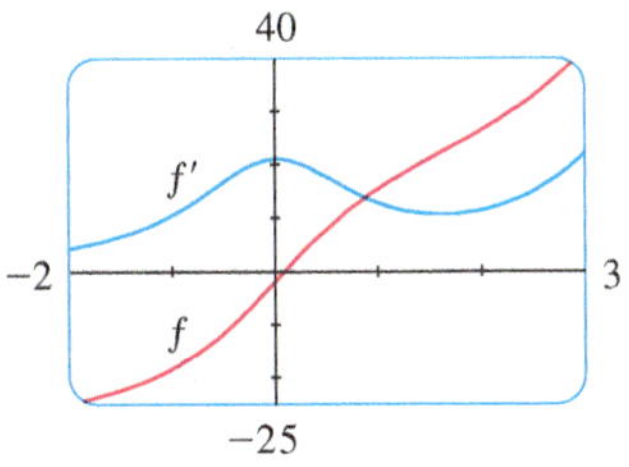

FIGURE 2

EXAMPLE 4 Find f if $f''(x) = 12x^2 + 6x - 4$, $f(0) = 4$, and $f(1) = 1$.

SOLUTION The general antiderivative of $f''(x) = 12x^2 + 6x - 4$ is

$$f'(x) = 12\frac{x^3}{3} + 6\frac{x^2}{2} - 4x + C = 4x^3 + 3x^2 - 4x + C$$

Using the antidifferentiation rules once more, we find that

$$f(x) = 4\frac{x^4}{4} + 3\frac{x^3}{3} - 4\frac{x^2}{2} + Cx + D = x^4 + x^3 - 2x^2 + Cx + D$$

To determine C and D we use the given conditions that $f(0) = 4$ and $f(1) = 1$. Since $f(0) = 0 + D = 4$, we have $D = 4$. Since

$$f(1) = 1 + 1 - 2 + C + 4 = 1$$

we have $C = -3$. Therefore the required function is

$$f(x) = x^4 + x^3 - 2x^2 - 3x + 4$$

If we are given the graph of a function f, it seems reasonable that we should be able to sketch the graph of an antiderivative F. Suppose, for instance, that we are given that $F(0) = 1$. Then we have a place to start, the point $(0, 1)$, and the direction in which we move our pencil is given at each stage by the derivative $F'(x) = f(x)$. In the next example we use the principles of this chapter to show how to graph F even when we don't have a formula for f. This would be the case, for instance, when $f(x)$ is determined by experimental data.

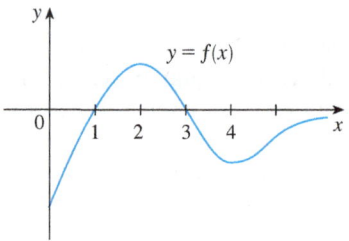

FIGURE 3

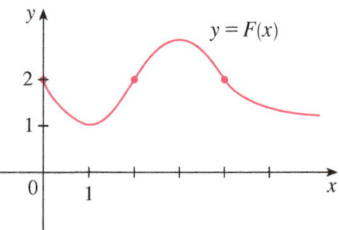

FIGURE 4

EXAMPLE 5 The graph of a function f is given in Figure 3. Make a rough sketch of an antiderivative F, given that $F(0) = 2$.

SOLUTION We are guided by the fact that the slope of $y = F(x)$ is $f(x)$. We start at the point $(0, 2)$ and draw F as an initially decreasing function since $f(x)$ is negative when $0 < x < 1$. Notice that $f(1) = f(3) = 0$, so F has horizontal tangents when $x = 1$ and $x = 3$. For $1 < x < 3$, $f(x)$ is positive and so F is increasing. We see that F has a local minimum when $x = 1$ and a local maximum when $x = 3$. For $x > 3$, $f(x)$ is negative and so F is decreasing on $(3, \infty)$. Since $f(x) \to 0$ as $x \to \infty$, the graph of F becomes flatter as $x \to \infty$. Also notice that $F''(x) = f'(x)$ changes from positive to negative at $x = 2$ and from negative to positive at $x = 4$, so F has inflection points when $x = 2$ and $x = 4$. We use this information to sketch the graph of the antiderivative in Figure 4. ■

■ Rectilinear Motion

Antidifferentiation is particularly useful in analyzing the motion of an object moving in a straight line. Recall that if the object has position function $s = f(t)$, then the velocity function is $v(t) = s'(t)$. This means that the position function is an antiderivative of the velocity function. Likewise, the acceleration function is $a(t) = v'(t)$, so the velocity function is an antiderivative of the acceleration. If the acceleration and the initial values $s(0)$ and $v(0)$ are known, then the position function can be found by antidifferentiating twice.

EXAMPLE 6 A particle moves in a straight line and has acceleration given by $a(t) = 6t + 4$. Its initial velocity is $v(0) = -6$ cm/s and its initial displacement is $s(0) = 9$ cm. Find its position function $s(t)$.

SOLUTION Since $v'(t) = a(t) = 6t + 4$, antidifferentiation gives

$$v(t) = 6\frac{t^2}{2} + 4t + C = 3t^2 + 4t + C$$

Note that $v(0) = C$. But we are given that $v(0) = -6$, so $C = -6$ and

$$v(t) = 3t^2 + 4t - 6$$

Since $v(t) = s'(t)$, s is the antiderivative of v:

$$s(t) = 3\frac{t^3}{3} + 4\frac{t^2}{2} - 6t + D = t^3 + 2t^2 - 6t + D$$

This gives $s(0) = D$. We are given that $s(0) = 9$, so $D = 9$ and the required position function is

$$s(t) = t^3 + 2t^2 - 6t + 9$$
■

An object near the surface of the earth is subject to a gravitational force that produces a downward acceleration denoted by g. For motion close to the ground we may assume that g is constant, its value being about 9.8 m/s² (or 32 ft/s²).

EXAMPLE 7 A ball is thrown upward with a speed of 48 ft/s from the edge of a cliff 432 ft above the ground. Find its height above the ground t seconds later. When does it reach its maximum height? When does it hit the ground?

SOLUTION The motion is vertical and we choose the positive direction to be upward. At time t the distance above the ground is $s(t)$ and the velocity $v(t)$ is decreasing. Therefore the acceleration must be negative and we have

$$a(t) = \frac{dv}{dt} = -32$$

Taking antiderivatives, we have

$$v(t) = -32t + C$$

To determine C we use the given information that $v(0) = 48$. This gives $48 = 0 + C$, so

$$v(t) = -32t + 48$$

The maximum height is reached when $v(t) = 0$, that is, after 1.5 seconds. Since $s'(t) = v(t)$, we antidifferentiate again and obtain

$$s(t) = -16t^2 + 48t + D$$

Using the fact that $s(0) = 432$, we have $432 = 0 + D$ and so

$$s(t) = -16t^2 + 48t + 432$$

The expression for $s(t)$ is valid until the ball hits the ground. This happens when $s(t) = 0$, that is, when

$$-16t^2 + 48t + 432 = 0$$

or, equivalently,

$$t^2 - 3t - 27 = 0$$

Using the quadratic formula to solve this equation, we get

$$t = \frac{3 \pm 3\sqrt{13}}{2}$$

We reject the solution with the minus sign since it gives a negative value for t. Therefore the ball hits the ground after $3(1 + \sqrt{13})/2 \approx 6.9$ seconds. ∎

Figure 5 shows the position function of the ball in Example 7. The graph corroborates the conclusions we reached: The ball reaches its maximum height after 1.5 seconds and hits the ground after about 6.9 seconds.

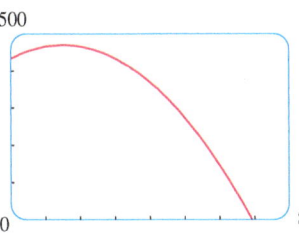

FIGURE 5

4.9 EXERCISES

1–22 Find the most general antiderivative of the function. (Check your answer by differentiation.)

1. $f(x) = 4x + 7$

2. $f(x) = x^2 - 3x + 2$

3. $f(x) = 2x^3 - \frac{2}{3}x^2 + 5x$

4. $f(x) = 6x^5 - 8x^4 - 9x^2$

5. $f(x) = x(12x + 8)$

6. $f(x) = (x - 5)^2$

7. $f(x) = 7x^{2/5} + 8x^{-4/5}$

8. $f(x) = x^{3.4} - 2x^{\sqrt{2}-1}$

9. $f(x) = \sqrt{2}$

10. $f(x) = e^2$

11. $f(x) = 3\sqrt{x} - 2\sqrt[3]{x}$

12. $f(x) = \sqrt[3]{x^2} + x\sqrt{x}$

13. $f(x) = \frac{1}{5} - \frac{2}{x}$

14. $f(t) = \frac{3t^4 - t^3 + 6t^2}{t^4}$

15. $g(t) = \frac{1 + t + t^2}{\sqrt{t}}$

16. $r(\theta) = \sec\theta \tan\theta - 2e^\theta$

17. $h(\theta) = 2\sin\theta - \sec^2\theta$

18. $g(v) = 2\cos v - \frac{3}{\sqrt{1 - v^2}}$

19. $f(x) = 2^x + 4\sinh x$

20. $f(x) = 1 + 2\sin x + 3/\sqrt{x}$

21. $f(x) = \frac{2x^4 + 4x^3 - x}{x^3}, \quad x > 0$

22. $f(x) = \dfrac{2x^2 + 5}{x^2 + 1}$

23–24 Find the antiderivative F of f that satisfies the given condition. Check your answer by comparing the graphs of f and F.

23. $f(x) = 5x^4 - 2x^5$, $F(0) = 4$

24. $f(x) = 4 - 3(1 + x^2)^{-1}$, $F(1) = 0$

25–48 Find f.

25. $f''(x) = 20x^3 - 12x^2 + 6x$

26. $f''(x) = x^6 - 4x^4 + x + 1$

27. $f''(x) = 2x + 3e^x$ **28.** $f''(x) = 1/x^2$

29. $f'''(t) = 12 + \sin t$ **30.** $f'''(t) = \sqrt{t} - 2\cos t$

31. $f'(x) = 1 + 3\sqrt{x}$, $f(4) = 25$

32. $f'(x) = 5x^4 - 3x^2 + 4$, $f(-1) = 2$

33. $f'(t) = 4/(1 + t^2)$, $f(1) = 0$

34. $f'(t) = t + 1/t^3$, $t > 0$, $f(1) = 6$

35. $f'(x) = 5x^{2/3}$, $f(8) = 21$

36. $f'(x) = (x + 1)/\sqrt{x}$, $f(1) = 5$

37. $f'(t) = \sec t (\sec t + \tan t)$, $-\pi/2 < t < \pi/2$, $f(\pi/4) = -1$

38. $f'(t) = 3^t - 3/t$, $f(1) = 2$, $f(-1) = 1$

39. $f''(x) = -2 + 12x - 12x^2$, $f(0) = 4$, $f'(0) = 12$

40. $f''(x) = 8x^3 + 5$, $f(1) = 0$, $f'(1) = 8$

41. $f''(\theta) = \sin \theta + \cos \theta$, $f(0) = 3$, $f'(0) = 4$

42. $f''(t) = t^2 + 1/t^2$, $t > 0$, $f(2) = 3$, $f'(1) = 2$

43. $f''(x) = 4 + 6x + 24x^2$, $f(0) = 3$, $f(1) = 10$

44. $f''(x) = x^3 + \sinh x$, $f(0) = 1$, $f(2) = 2.6$

45. $f''(x) = e^x - 2 \sin x$, $f(0) = 3$, $f(\pi/2) = 0$

46. $f''(t) = \sqrt[3]{t} - \cos t$, $f(0) = 2$, $f(1) = 2$

47. $f''(x) = x^{-2}$, $x > 0$, $f(1) = 0$, $f(2) = 0$

48. $f'''(x) = \cos x$, $f(0) = 1$, $f'(0) = 2$, $f''(0) = 3$

49. Given that the graph of f passes through the point $(2, 5)$ and that the slope of its tangent line at $(x, f(x))$ is $3 - 4x$, find $f(1)$.

50. Find a function f such that $f'(x) = x^3$ and the line $x + y = 0$ is tangent to the graph of f.

51–52 The graph of a function f is shown. Which graph is an antiderivative of f and why?

51. **52.**

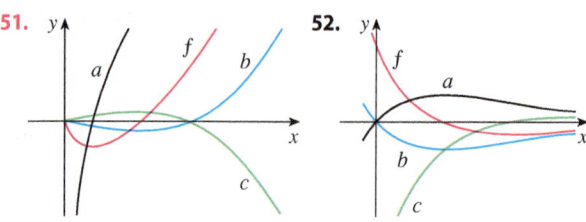

53. The graph of a function is shown in the figure. Make a rough sketch of an antiderivative F, given that $F(0) = 1$.

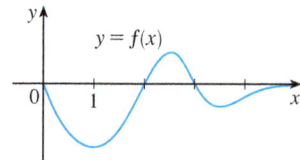

54. The graph of the velocity function of a particle is shown in the figure. Sketch the graph of a position function.

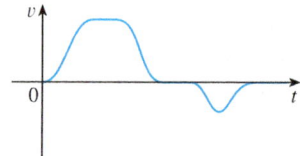

55. The graph of f' is shown in the figure. Sketch the graph of f if f is continuous and $f(0) = -1$.

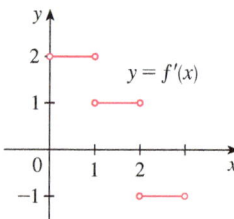

56. (a) Use a graphing device to graph $f(x) = 2x - 3\sqrt{x}$.
(b) Starting with the graph in part (a), sketch a rough graph of the antiderivative F that satisfies $F(0) = 1$.
(c) Use the rules of this section to find an expression for $F(x)$.
(d) Graph F using the expression in part (c). Compare with your sketch in part (b).

57–58 Draw a graph of f and use it to make a rough sketch of the antiderivative that passes through the origin.

57. $f(x) = \dfrac{\sin x}{1 + x^2}$, $-2\pi \le x \le 2\pi$

58. $f(x) = \sqrt{x^4 - 2x^2 + 2} - 2$, $-3 \le x \le 3$

59–64 A particle is moving with the given data. Find the position of the particle.

59. $v(t) = \sin t - \cos t, \quad s(0) = 0$

60. $v(t) = t^2 - 3\sqrt{t}, \quad s(4) = 8$

61. $a(t) = 2t + 1, \quad s(0) = 3, \quad v(0) = -2$

62. $a(t) = 3\cos t - 2\sin t, \quad s(0) = 0, \quad v(0) = 4$

63. $a(t) = 10 \sin t + 3 \cos t, \quad s(0) = 0, \quad s(2\pi) = 12$

64. $a(t) = t^2 - 4t + 6, \quad s(0) = 0, \quad s(1) = 20$

---

65. A stone is dropped from the upper observation deck (the Space Deck) of the CN Tower, 450 m above the ground.
(a) Find the distance of the stone above ground level at time t.
(b) How long does it take the stone to reach the ground?
(c) With what velocity does it strike the ground?
(d) If the stone is thrown downward with a speed of 5 m/s, how long does it take to reach the ground?

66. Show that for motion in a straight line with constant acceleration a, initial velocity v_0, and initial displacement s_0, the displacement after time t is
$$s = \tfrac{1}{2}at^2 + v_0 t + s_0$$

67. An object is projected upward with initial velocity v_0 meters per second from a point s_0 meters above the ground. Show that
$$[v(t)]^2 = v_0^2 - 19.6[s(t) - s_0]$$

68. Two balls are thrown upward from the edge of the cliff in Example 7. The first is thrown with a speed of 48 ft/s and the other is thrown a second later with a speed of 24 ft/s. Do the balls ever pass each other?

69. A stone was dropped off a cliff and hit the ground with a speed of 120 ft/s. What is the height of the cliff?

70. If a diver of mass m stands at the end of a diving board with length L and linear density ρ, then the board takes on the shape of a curve $y = f(x)$, where
$$EIy'' = mg(L - x) + \tfrac{1}{2}\rho g(L - x)^2$$
E and I are positive constants that depend on the material of the board and g (< 0) is the acceleration due to gravity.
(a) Find an expression for the shape of the curve.
(b) Use $f(L)$ to estimate the distance below the horizontal at the end of the board.

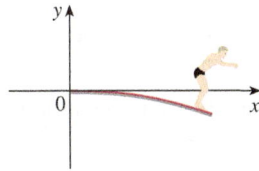

71. A company estimates that the marginal cost (in dollars per item) of producing x items is $1.92 - 0.002x$. If the cost of producing one item is $562, find the cost of producing 100 items.

72. The linear density of a rod of length 1 m is given by $\rho(x) = 1/\sqrt{x}$, in grams per centimeter, where x is measured in centimeters from one end of the rod. Find the mass of the rod.

73. Since raindrops grow as they fall, their surface area increases and therefore the resistance to their falling increases. A raindrop has an initial downward velocity of 10 m/s and its downward acceleration is
$$a = \begin{cases} 9 - 0.9t & \text{if } 0 \le t \le 10 \\ 0 & \text{if } t > 10 \end{cases}$$
If the raindrop is initially 500 m above the ground, how long does it take to fall?

74. A car is traveling at 50 mi/h when the brakes are fully applied, producing a constant deceleration of 22 ft/s². What is the distance traveled before the car comes to a stop?

75. What constant acceleration is required to increase the speed of a car from 30 mi/h to 50 mi/h in 5 seconds?

76. A car braked with a constant deceleration of 16 ft/s², producing skid marks measuring 200 ft before coming to a stop. How fast was the car traveling when the brakes were first applied?

77. A car is traveling at 100 km/h when the driver sees an accident 80 m ahead and slams on the brakes. What constant deceleration is required to stop the car in time to avoid a pileup?

78. A model rocket is fired vertically upward from rest. Its acceleration for the first three seconds is $a(t) = 60t$, at which time the fuel is exhausted and it becomes a freely "falling" body. Fourteen seconds later, the rocket's parachute opens, and the (downward) velocity slows linearly to -18 ft/s in 5 seconds. The rocket then "floats" to the ground at that rate.
(a) Determine the position function s and the velocity function v (for all times t). Sketch the graphs of s and v.
(b) At what time does the rocket reach its maximum height, and what is that height?
(c) At what time does the rocket land?

79. A high-speed bullet train accelerates and decelerates at the rate of 4 ft/s². Its maximum cruising speed is 90 mi/h.
(a) What is the maximum distance the train can travel if it accelerates from rest until it reaches its cruising speed and then runs at that speed for 15 minutes?
(b) Suppose that the train starts from rest and must come to a complete stop in 15 minutes. What is the maximum distance it can travel under these conditions?
(c) Find the minimum time that the train takes to travel between two consecutive stations that are 45 miles apart.
(d) The trip from one station to the next takes 37.5 minutes. How far apart are the stations?

4 REVIEW

CONCEPT CHECK

1. Explain the difference between an absolute maximum and a local maximum. Illustrate with a sketch.

2. (a) What does the Extreme Value Theorem say?
   (b) Explain how the Closed Interval Method works.

3. (a) State Fermat's Theorem.
   (b) Define a critical number of f.

4. (a) State Rolle's Theorem.
   (b) State the Mean Value Theorem and give a geometric interpretation.

5. (a) State the Increasing/Decreasing Test.
   (b) What does it mean to say that f is concave upward on an interval I?
   (c) State the Concavity Test.
   (d) What are inflection points? How do you find them?

6. (a) State the First Derivative Test.
   (b) State the Second Derivative Test.
   (c) What are the relative advantages and disadvantages of these tests?

7. (a) What does l'Hospital's Rule say?
   (b) How can you use l'Hospital's Rule if you have a product $f(x)g(x)$ where $f(x) \to 0$ and $g(x) \to \infty$ as $x \to a$?
   (c) How can you use l'Hospital's Rule if you have a difference $f(x) - g(x)$ where $f(x) \to \infty$ and $g(x) \to \infty$ as $x \to a$?
   (d) How can you use l'Hospital's Rule if you have a power $[f(x)]^{g(x)}$ where $f(x) \to 0$ and $g(x) \to 0$ as $x \to a$?

8. State whether each of the following limit forms is indeterminate. Where possible, state the limit.
   (a) $\dfrac{0}{0}$ (b) $\dfrac{\infty}{\infty}$ (c) $\dfrac{0}{\infty}$ (d) $\dfrac{\infty}{0}$
   (e) $\infty + \infty$ (f) $\infty - \infty$ (g) $\infty \cdot \infty$ (h) $\infty \cdot 0$
   (i) 0^0 (j) 0^∞ (k) ∞^0 (l) 1^∞

9. If you have a graphing calculator or computer, why do you need calculus to graph a function?

10. (a) Given an initial approximation x_1 to a root of the equation $f(x) = 0$, explain geometrically, with a diagram, how the second approximation x_2 in Newton's method is obtained.
    (b) Write an expression for x_2 in terms of x_1, $f(x_1)$, and $f'(x_1)$.
    (c) Write an expression for x_{n+1} in terms of x_n, $f(x_n)$, and $f'(x_n)$.
    (d) Under what circumstances is Newton's method likely to fail or to work very slowly?

11. (a) What is an antiderivative of a function f?
    (b) Suppose F_1 and F_2 are both antiderivatives of f on an interval I. How are F_1 and F_2 related?

TRUE–FALSE QUIZ

Determine whether the statement is true or false. If it is true, explain why. If it is false, explain why or give an example that disproves the statement.

1. If $f'(c) = 0$, then f has a local maximum or minimum at c.

2. If f has an absolute minimum value at c, then $f'(c) = 0$.

3. If f is continuous on (a, b), then f attains an absolute maximum value $f(c)$ and an absolute minimum value $f(d)$ at some numbers c and d in (a, b).

4. If f is differentiable and $f(-1) = f(1)$, then there is a number c such that $|c| < 1$ and $f'(c) = 0$.

5. If $f'(x) < 0$ for $1 < x < 6$, then f is decreasing on $(1, 6)$.

6. If $f''(2) = 0$, then $(2, f(2))$ is an inflection point of the curve $y = f(x)$.

7. If $f'(x) = g'(x)$ for $0 < x < 1$, then $f(x) = g(x)$ for $0 < x < 1$.

8. There exists a function f such that $f(1) = -2$, $f(3) = 0$, and $f'(x) > 1$ for all x.

9. There exists a function f such that $f(x) > 0$, $f'(x) < 0$, and $f''(x) > 0$ for all x.

10. There exists a function f such that $f(x) < 0$, $f'(x) < 0$, and $f''(x) > 0$ for all x.

11. If f and g are increasing on an interval I, then $f + g$ is increasing on I.

12. If f and g are increasing on an interval I, then $f - g$ is increasing on I.

13. If f and g are increasing on an interval I, then fg is increasing on I.

14. If f and g are positive increasing functions on an interval I, then fg is increasing on I.

15. If f is increasing and $f(x) > 0$ on I, then $g(x) = 1/f(x)$ is decreasing on I.

16. If f is even, then f' is even.

17. If f is periodic, then f' is periodic.

18. The most general antiderivative of $f(x) = x^{-2}$ is
$$F(x) = -\frac{1}{x} + C$$

19. If $f'(x)$ exists and is nonzero for all x, then $f(1) \neq f(0)$.

20. If $\lim_{x \to \infty} f(x) = 1$ and $\lim_{x \to \infty} g(x) = \infty$, then
$$\lim_{x \to \infty} [f(x)]^{g(x)} = 1$$

21. $\lim_{x \to 0} \dfrac{x}{e^x} = 1$

EXERCISES

1–6 Find the local and absolute extreme values of the function on the given interval.

1. $f(x) = x^3 - 9x^2 + 24x - 2, \quad [0, 5]$

2. $f(x) = x\sqrt{1-x}, \quad [-1, 1]$

3. $f(x) = \dfrac{3x-4}{x^2+1}, \quad [-2, 2]$

4. $f(x) = \sqrt{x^2 + x + 1}, \quad [-2, 1]$

5. $f(x) = x + 2\cos x, \quad [-\pi, \pi]$

6. $f(x) = x^2 e^{-x}, \quad [-1, 3]$

7–14 Evaluate the limit.

7. $\lim_{x \to 0} \dfrac{e^x - 1}{\tan x}$

8. $\lim_{x \to 0} \dfrac{\tan 4x}{x + \sin 2x}$

9. $\lim_{x \to 0} \dfrac{e^{2x} - e^{-2x}}{\ln(x+1)}$

10. $\lim_{x \to \infty} \dfrac{e^{2x} - e^{-2x}}{\ln(x+1)}$

11. $\lim_{x \to -\infty} (x^2 - x^3) e^{2x}$

12. $\lim_{x \to \pi^-} (x - \pi) \csc x$

13. $\lim_{x \to 1^+} \left(\dfrac{x}{x-1} - \dfrac{1}{\ln x} \right)$

14. $\lim_{x \to (\pi/2)^-} (\tan x)^{\cos x}$

15–17 Sketch the graph of a function that satisfies the given conditions.

15. $f(0) = 0, \quad f'(-2) = f'(1) = f'(9) = 0$,
$\lim_{x \to \infty} f(x) = 0, \quad \lim_{x \to 6} f(x) = -\infty$,
$f'(x) < 0$ on $(-\infty, -2), (1, 6),$ and $(9, \infty)$,
$f'(x) > 0$ on $(-2, 1)$ and $(6, 9)$,
$f''(x) > 0$ on $(-\infty, 0)$ and $(12, \infty)$,
$f''(x) < 0$ on $(0, 6)$ and $(6, 12)$

16. $f(0) = 0, \quad f$ is continuous and even,
$f'(x) = 2x$ if $0 < x < 1, \quad f'(x) = -1$ if $1 < x < 3$,
$f'(x) = 1$ if $x > 3$

17. f is odd, $\quad f'(x) < 0$ for $0 < x < 2$,
$f'(x) > 0$ for $x > 2, \quad f''(x) > 0$ for $0 < x < 3$,
$f''(x) < 0$ for $x > 3, \quad \lim_{x \to \infty} f(x) = -2$

18. The figure shows the graph of the *derivative* f' of a function f.
(a) On what intervals is f increasing or decreasing?
(b) For what values of x does f have a local maximum or minimum?
(c) Sketch the graph of f''.
(d) Sketch a possible graph of f.

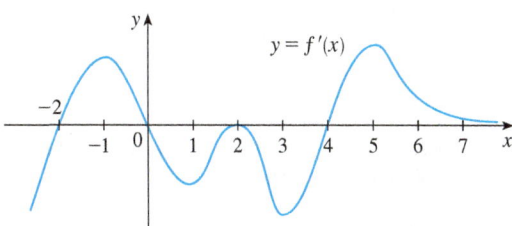

19–34 Use the guidelines of Section 4.5 to sketch the curve.

19. $y = 2 - 2x - x^3$

20. $y = -2x^3 - 3x^2 + 12x + 5$

21. $y = 3x^4 - 4x^3 + 2$

22. $y = \dfrac{x}{1-x^2}$

23. $y = \dfrac{1}{x(x-3)^2}$

24. $y = \dfrac{1}{x^2} - \dfrac{1}{(x-2)^2}$

25. $y = \dfrac{(x-1)^3}{x^2}$

26. $y = \sqrt{1-x} + \sqrt{1+x}$

27. $y = x\sqrt{2+x}$

28. $y = x^{2/3}(x-3)^2$

29. $y = e^x \sin x$, $-\pi \le x \le \pi$

30. $y = 4x - \tan x$, $-\pi/2 < x < \pi/2$

31. $y = \sin^{-1}(1/x)$

32. $y = e^{2x-x^2}$

33. $y = (x - 2)e^{-x}$

34. $y = x + \ln(x^2 + 1)$

35–38 Produce graphs of f that reveal all the important aspects of the curve. Use graphs of f' and f'' to estimate the intervals of increase and decrease, extreme values, intervals of concavity, and inflection points. In Exercise 35 use calculus to find these quantities exactly.

35. $f(x) = \dfrac{x^2 - 1}{x^3}$

36. $f(x) = \dfrac{x^3 + 1}{x^6 + 1}$

37. $f(x) = 3x^6 - 5x^5 + x^4 - 5x^3 - 2x^2 + 2$

38. $f(x) = x^2 + 6.5 \sin x$, $-5 \le x \le 5$

39. Graph $f(x) = e^{-1/x^2}$ in a viewing rectangle that shows all the main aspects of this function. Estimate the inflection points. Then use calculus to find them exactly.

40. (a) Graph the function $f(x) = 1/(1 + e^{1/x})$.
(b) Explain the shape of the graph by computing the limits of $f(x)$ as x approaches ∞, $-\infty$, 0^+, and 0^-.
(c) Use the graph of f to estimate the coordinates of the inflection points.
(d) Use your CAS to compute and graph f''.
(e) Use the graph in part (d) to estimate the inflection points more accurately.

41–42 Use the graphs of f, f', and f'' to estimate the x-coordinates of the maximum and minimum points and inflection points of f.

41. $f(x) = \dfrac{\cos^2 x}{\sqrt{x^2 + x + 1}}$, $-\pi \le x \le \pi$

42. $f(x) = e^{-0.1x} \ln(x^2 - 1)$

43. Investigate the family of functions $f(x) = \ln(\sin x + C)$. What features do the members of this family have in common? How do they differ? For which values of C is f continuous on $(-\infty, \infty)$? For which values of C does f have no graph at all? What happens as $C \to \infty$?

44. Investigate the family of functions $f(x) = cxe^{-cx^2}$. What happens to the maximum and minimum points and the inflection points as c changes? Illustrate your conclusions by graphing several members of the family.

45. Show that the equation $3x + 2\cos x + 5 = 0$ has exactly one real root.

46. Suppose that f is continuous on $[0, 4]$, $f(0) = 1$, and $2 \le f'(x) \le 5$ for all x in $(0, 4)$. Show that $9 \le f(4) \le 21$.

47. By applying the Mean Value Theorem to the function $f(x) = x^{1/5}$ on the interval $[32, 33]$, show that

$$2 < \sqrt[5]{33} < 2.0125$$

48. For what values of the constants a and b is $(1, 3)$ a point of inflection of the curve $y = ax^3 + bx^2$?

49. Let $g(x) = f(x^2)$, where f is twice differentiable for all x, $f'(x) > 0$ for all $x \ne 0$, and f is concave downward on $(-\infty, 0)$ and concave upward on $(0, \infty)$.
(a) At what numbers does g have an extreme value?
(b) Discuss the concavity of g.

50. Find two positive integers such that the sum of the first number and four times the second number is 1000 and the product of the numbers is as large as possible.

51. Show that the shortest distance from the point (x_1, y_1) to the straight line $Ax + By + C = 0$ is

$$\dfrac{|Ax_1 + By_1 + C|}{\sqrt{A^2 + B^2}}$$

52. Find the point on the hyperbola $xy = 8$ that is closest to the point $(3, 0)$.

53. Find the smallest possible area of an isosceles triangle that is circumscribed about a circle of radius r.

54. Find the volume of the largest circular cone that can be inscribed in a sphere of radius r.

55. In $\triangle ABC$, D lies on AB, $CD \perp AB$, $|AD| = |BD| = 4$ cm, and $|CD| = 5$ cm. Where should a point P be chosen on CD so that the sum $|PA| + |PB| + |PC|$ is a minimum?

56. Solve Exercise 55 when $|CD| = 2$ cm.

57. The velocity of a wave of length L in deep water is

$$v = K\sqrt{\dfrac{L}{C} + \dfrac{C}{L}}$$

where K and C are known positive constants. What is the length of the wave that gives the minimum velocity?

58. A metal storage tank with volume V is to be constructed in the shape of a right circular cylinder surmounted by a hemisphere. What dimensions will require the least amount of metal?

59. A hockey team plays in an arena with a seating capacity of 15,000 spectators. With the ticket price set at $12, average attendance at a game has been 11,000. A market survey indicates that for each dollar the ticket price is lowered, average attendance will increase by 1000. How should the owners of the team set the ticket price to maximize their revenue from ticket sales?

60. A manufacturer determines that the cost of making x units of a commodity is
$$C(x) = 1800 + 25x - 0.2x^2 + 0.001x^3$$
and the demand function is $p(x) = 48.2 - 0.03x$.
 (a) Graph the cost and revenue functions and use the graphs to estimate the production level for maximum profit.
 (b) Use calculus to find the production level for maximum profit.
 (c) Estimate the production level that minimizes the average cost.

61. Use Newton's method to find the root of the equation
$$x^5 - x^4 + 3x^2 - 3x - 2 = 0$$
in the interval $[1, 2]$ correct to six decimal places.

62. Use Newton's method to find all solutions of the equation $\sin x = x^2 - 3x + 1$ correct to six decimal places.

63. Use Newton's method to find the absolute maximum value of the function $f(t) = \cos t + t - t^2$ correct to eight decimal places.

64. Use the guidelines in Section 4.5 to sketch the curve $y = x \sin x$, $0 \le x \le 2\pi$. Use Newton's method when necessary.

65–68 Find the most general antiderivative of the function.

65. $f(x) = 4\sqrt{x} - 6x^2 + 3$

66. $g(x) = \dfrac{1}{x} + \dfrac{1}{x^2 + 1}$

67. $f(t) = 2 \sin t - 3e^t$

68. $f(x) = x^{-3} + \cosh x$

69–72 Find f.

69. $f'(t) = 2t - 3 \sin t$, $f(0) = 5$

70. $f'(u) = \dfrac{u^2 + \sqrt{u}}{u}$, $f(1) = 3$

71. $f''(x) = 1 - 6x + 48x^2$, $f(0) = 1$, $f'(0) = 2$

72. $f''(x) = 5x^3 + 6x^2 + 2$, $f(0) = 3$, $f(1) = -2$

73–74 A particle is moving with the given data. Find the position of the particle.

73. $v(t) = 2t - 1/(1 + t^2)$, $s(0) = 1$

74. $a(t) = \sin t + 3 \cos t$, $s(0) = 0$, $v(0) = 2$

75. (a) If $f(x) = 0.1e^x + \sin x$, $-4 \le x \le 4$, use a graph of f to sketch a rough graph of the antiderivative F of f that satisfies $F(0) = 0$.
 (b) Find an expression for $F(x)$.
 (c) Graph F using the expression in part (b). Compare with your sketch in part (a).

76. Investigate the family of curves given by
$$f(x) = x^4 + x^3 + cx^2$$
In particular you should determine the transitional value of c at which the number of critical numbers changes and the transitional value at which the number of inflection points changes. Illustrate the various possible shapes with graphs.

77. A canister is dropped from a helicopter 500 m above the ground. Its parachute does not open, but the canister has been designed to withstand an impact velocity of 100 m/s. Will it burst?

78. In an automobile race along a straight road, car A passed car B twice. Prove that at some time during the race their accelerations were equal. State the assumptions that you make.

79. A rectangular beam will be cut from a cylindrical log of radius 10 inches.
 (a) Show that the beam of maximal cross-sectional area is a square.
 (b) Four rectangular planks will be cut from the four sections of the log that remain after cutting the square

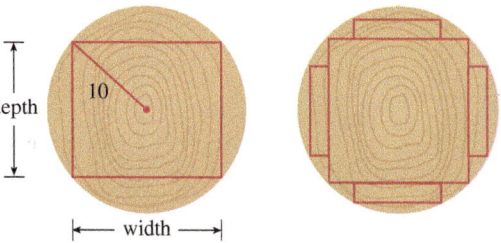

beam. Determine the dimensions of the planks that will have maximal cross-sectional area.

(c) Suppose that the strength of a rectangular beam is proportional to the product of its width and the square of its depth. Find the dimensions of the strongest beam that can be cut from the cylindrical log.

80. If a projectile is fired with an initial velocity v at an angle of inclination θ from the horizontal, then its trajectory, neglecting air resistance, is the parabola

$$y = (\tan\theta)x - \frac{g}{2v^2\cos^2\theta}x^2 \qquad 0 < \theta < \frac{\pi}{2}$$

(a) Suppose the projectile is fired from the base of a plane that is inclined at an angle α, $\alpha > 0$, from the horizontal, as shown in the figure. Show that the range of the projectile, measured up the slope, is given by

$$R(\theta) = \frac{2v^2\cos\theta\,\sin(\theta - \alpha)}{g\cos^2\alpha}$$

(b) Determine θ so that R is a maximum.
(c) Suppose the plane is at an angle α *below* the horizontal. Determine the range R in this case, and determine the angle at which the projectile should be fired to maximize R.

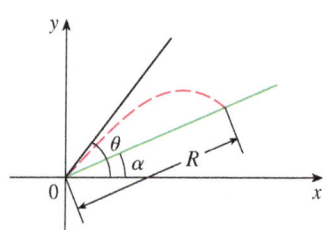

81. If an electrostatic field E acts on a liquid or a gaseous polar dielectric, the net dipole moment P per unit volume is

$$P(E) = \frac{e^E + e^{-E}}{e^E - e^{-E}} - \frac{1}{E}$$

Show that $\lim_{E \to 0^+} P(E) = 0$.

82. If a metal ball with mass m is projected in water and the force of resistance is proportional to the square of the velocity, then the distance the ball travels in time t is

$$s(t) = \frac{m}{c}\ln\cosh\sqrt{\frac{gc}{mt}}$$

where c is a positive constant. Find $\lim_{c \to 0^+} s(t)$.

83. Show that, for $x > 0$,

$$\frac{x}{1 + x^2} < \tan^{-1}x < x$$

84. Sketch the graph of a function f such that $f'(x) < 0$ for all x, $f''(x) > 0$ for $|x| > 1$, $f''(x) < 0$ for $|x| < 1$, and $\lim_{x \to \pm\infty}[f(x) + x] = 0$.

85. A light is to be placed atop a pole of height h feet to illuminate a busy traffic circle, which has a radius of 40 ft. The intensity of illumination I at any point P on the circle is directly proportional to the cosine of the angle θ (see the figure) and inversely proportional to the square of the distance d from the source.

(a) How tall should the light pole be to maximize I?
(b) Suppose that the light pole is h feet tall and that a woman is walking away from the base of the pole at the rate of 4 ft/s. At what rate is the intensity of the light at the point on her back 4 ft above the ground decreasing when she reaches the outer edge of the traffic circle?

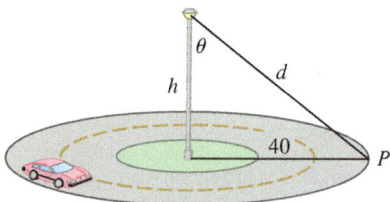

86. Water is flowing at a constant rate into a spherical tank. Let $V(t)$ be the volume of water in the tank and $H(t)$ be the height of the water in the tank at time t.

(a) What are the meanings of $V'(t)$ and $H'(t)$? Are these derivatives positive, negative, or zero?
(b) Is $V''(t)$ positive, negative, or zero? Explain.
(c) Let t_1, t_2, and t_3 be the times when the tank is one-quarter full, half full, and three-quarters full, respectively. Are the values $H''(t_1)$, $H''(t_2)$, and $H''(t_3)$ positive, negative, or zero? Why?

Problems Plus

1. If a rectangle has its base on the *x*-axis and two vertices on the curve $y = e^{-x^2}$, show that the rectangle has the largest possible area when the two vertices are at the points of inflection of the curve.

2. Show that $|\sin x - \cos x| \leq \sqrt{2}$ for all *x*.

3. Does the function $f(x) = e^{10|x-2|-x^2}$ have an absolute maximum? If so, find it. What about an absolute minimum?

4. Show that $x^2 y^2 (4 - x^2)(4 - y^2) \leq 16$ for all numbers *x* and *y* such that $|x| \leq 2$ and $|y| \leq 2$.

5. Show that the inflection points of the curve $y = (\sin x)/x$ lie on the curve $y^2(x^4 + 4) = 4$.

6. Find the point on the parabola $y = 1 - x^2$ at which the tangent line cuts from the first quadrant the triangle with the smallest area.

7. If *a*, *b*, *c*, and *d* are constants such that
$$\lim_{x \to 0} \frac{ax^2 + \sin bx + \sin cx + \sin dx}{3x^2 + 5x^4 + 7x^6} = 8$$
find the value of the sum $a + b + c + d$.

8. Evaluate
$$\lim_{x \to \infty} \frac{(x+2)^{1/x} - x^{1/x}}{(x+3)^{1/x} - x^{1/x}}$$

9. Find the highest and lowest points on the curve $x^2 + xy + y^2 = 12$.

10. Sketch the set of all points (x, y) such that $|x + y| \leq e^x$.

11. If $P(a, a^2)$ is any point on the parabola $y = x^2$, except for the origin, let *Q* be the point where the normal line at *P* intersects the parabola again (see the figure).
    (a) Show that the *y*-coordinate of *Q* is smallest when $a = 1/\sqrt{2}$.
    (b) Show that the line segment *PQ* has the shortest possible length when $a = 1/\sqrt{2}$.

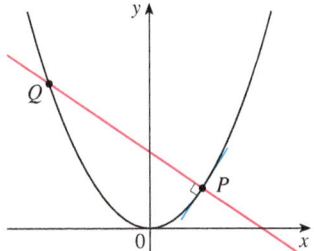

FIGURE FOR PROBLEM 11

12. For what values of *c* does the curve $y = cx^3 + e^x$ have inflection points?

13. An isosceles triangle is circumscribed about the unit circle so that the equal sides meet at the point $(0, a)$ on the *y*-axis (see the figure). Find the value of *a* that minimizes the lengths of the equal sides. (You may be surprised that the result does not give an equilateral triangle.).

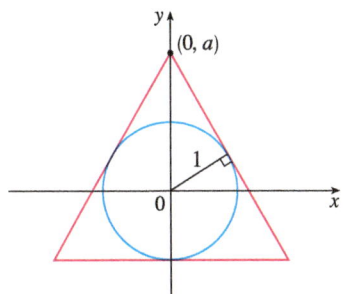

FIGURE FOR PROBLEM 13

14. Sketch the region in the plane consisting of all points (x, y) such that
$$2xy \leq |x - y| \leq x^2 + y^2$$

15. The line $y = mx + b$ intersects the parabola $y = x^2$ in points *A* and *B*. (See the figure.) Find the point *P* on the arc *AOB* of the parabola that maximizes the area of the triangle *PAB*.

16. *ABCD* is a square piece of paper with sides of length 1 m. A quarter-circle is drawn from *B* to *D* with center *A*. The piece of paper is folded along *EF*, with *E* on *AB* and *F* on *AD*, so that *A* falls on the quarter-circle. Determine the maximum and minimum areas that the triangle *AEF* can have.

17. For which positive numbers *a* does the curve $y = a^x$ intersect the line $y = x$?

18. For what value of *a* is the following equation true?
$$\lim_{x \to \infty} \left(\frac{x+a}{x-a}\right)^x = e$$

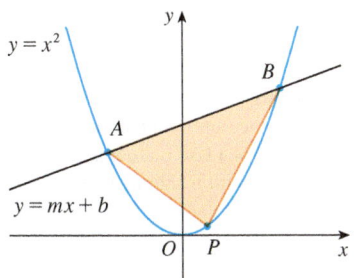

FIGURE FOR PROBLEM 15

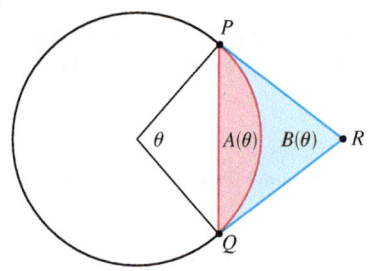

FIGURE FOR PROBLEM 20

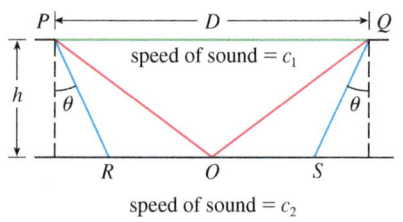

FIGURE FOR PROBLEM 21

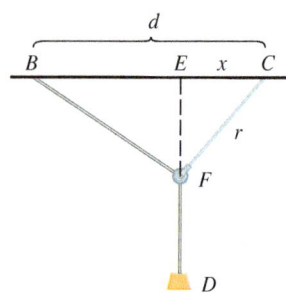

FIGURE FOR PROBLEM 23

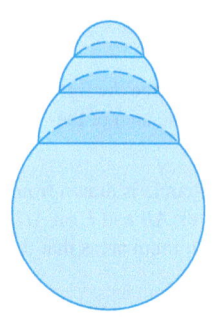

FIGURE FOR PROBLEM 26

19. Let $f(x) = a_1 \sin x + a_2 \sin 2x + \cdots + a_n \sin nx$, where $a_1, a_2, \ldots, a_n$ are real numbers and n is a positive integer. If it is given that $|f(x)| \leq |\sin x|$ for all x, show that
$$|a_1 + 2a_2 + \cdots + na_n| \leq 1$$

20. An arc PQ of a circle subtends a central angle θ as in the figure. Let $A(\theta)$ be the area between the chord PQ and the arc PQ. Let $B(\theta)$ be the area between the tangent lines PR, QR, and the arc. Find
$$\lim_{\theta \to 0^+} \frac{A(\theta)}{B(\theta)}$$

21. The speeds of sound c_1 in an upper layer and c_2 in a lower layer of rock and the thickness h of the upper layer can be determined by seismic exploration if the speed of sound in the lower layer is greater than the speed in the upper layer. A dynamite charge is detonated at a point P and the transmitted signals are recorded at a point Q, which is a distance D from P. The first signal to arrive at Q travels along the surface and takes T_1 seconds. The next signal travels from P to a point R, from R to S in the lower layer, and then to Q, taking T_2 seconds. The third signal is reflected off the lower layer at the midpoint O of RS and takes T_3 seconds to reach Q. (See the figure.)
(a) Express T_1, T_2, and T_3 in terms of D, h, c_1, c_2, and θ.
(b) Show that T_2 is a minimum when $\sin \theta = c_1/c_2$.
(c) Suppose that $D = 1$ km, $T_1 = 0.26$ s, $T_2 = 0.32$ s, and $T_3 = 0.34$ s. Find c_1, c_2, and h.

Note: Geophysicists use this technique when studying the structure of the earth's crust, whether searching for oil or examining fault lines.

22. For what values of c is there a straight line that intersects the curve
$$y = x^4 + cx^3 + 12x^2 - 5x + 2$$
in four distinct points?

23. One of the problems posed by the Marquis de l'Hospital in his calculus textbook *Analyse des Infiniment Petits* concerns a pulley that is attached to the ceiling of a room at a point C by a rope of length r. At another point B on the ceiling, at a distance d from C (where $d > r$), a rope of length ℓ is attached and passed through the pulley at F and connected to a weight W. The weight is released and comes to rest at its equilibrium position D. (See the figure.) As l'Hospital argued, this happens when the distance $|ED|$ is maximized. Show that when the system reaches equilibrium, the value of x is
$$\frac{r}{4d}\left(r + \sqrt{r^2 + 8d^2}\right)$$
Notice that this expression is independent of both W and ℓ.

24. Given a sphere with radius r, find the height of a pyramid of minimum volume whose base is a square and whose base and triangular faces are all tangent to the sphere. What if the base of the pyramid is a regular n-gon? (A regular n-gon is a polygon with n equal sides and angles.) (Use the fact that the volume of a pyramid is $\frac{1}{3}Ah$, where A is the area of the base.)

25. Assume that a snowball melts so that its volume decreases at a rate proportional to its surface area. If it takes three hours for the snowball to decrease to half its original volume, how much longer will it take for the snowball to melt completely?

26. A hemispherical bubble is placed on a spherical bubble of radius 1. A smaller hemispherical bubble is then placed on the first one. This process is continued until n chambers, including the sphere, are formed. (The figure shows the case $n = 4$.) Use mathematical induction to prove that the maximum height of any bubble tower with n chambers is $1 + \sqrt{n}$.

5 Integrals

The photo shows Lake Lanier, which is a reservoir in Georgia, USA. In Exercise 70 in Section 5.4 you will estimate the amount of water that flowed into Lake Lanier during a certain time period.

JRC, Inc. / Alamy

IN CHAPTER 2 we used the tangent and velocity problems to introduce the derivative, which is the central idea in differential calculus. In much the same way, this chapter starts with the area and distance problems and uses them to formulate the idea of a definite integral, which is the basic concept of integral calculus. We will see in Chapters 6 and 8 how to use the integral to solve problems concerning volumes, lengths of curves, population predictions, cardiac output, forces on a dam, work, consumer surplus, and baseball, among many others.

There is a connection between integral calculus and differential calculus. The Fundamental Theorem of Calculus relates the integral to the derivative, and we will see in this chapter that it greatly simplifies the solution of many problems.

5.1 Areas and Distances

Now is a good time to read (or reread) *A Preview of Calculus* (see page 1). It discusses the unifying ideas of calculus and helps put in perspective where we have been and where we are going.

In this section we discover that in trying to find the area under a curve or the distance traveled by a car, we end up with the same special type of limit.

■ The Area Problem

We begin by attempting to solve the *area problem:* Find the area of the region S that lies under the curve $y = f(x)$ from a to b. This means that S, illustrated in Figure 1, is bounded by the graph of a continuous function f [where $f(x) \geq 0$], the vertical lines $x = a$ and $x = b$, and the x-axis.

In trying to solve the area problem we have to ask ourselves: What is the meaning of the word *area*? This question is easy to answer for regions with straight sides. For a rectangle, the area is defined as the product of the length and the width. The area of a triangle is half the base times the height. The area of a polygon is found by dividing it into triangles (as in Figure 2) and adding the areas of the triangles.

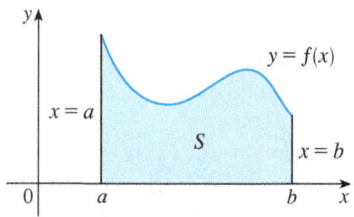

FIGURE 1
$S = \{(x, y) \mid a \leq x \leq b, 0 \leq y \leq f(x)\}$

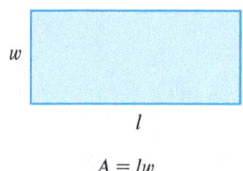

$A = lw$

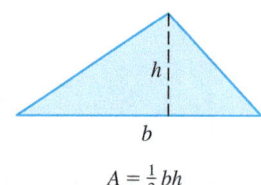

$A = \tfrac{1}{2} bh$

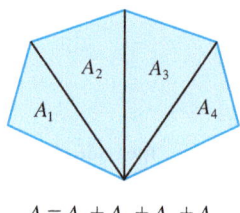

$A = A_1 + A_2 + A_3 + A_4$

FIGURE 2

However, it isn't so easy to find the area of a region with curved sides. We all have an intuitive idea of what the area of a region is. But part of the area problem is to make this intuitive idea precise by giving an exact definition of area.

Recall that in defining a tangent we first approximated the slope of the tangent line by slopes of secant lines and then we took the limit of these approximations. We pursue a similar idea for areas. We first approximate the region S by rectangles and then we take the limit of the areas of these rectangles as we increase the number of rectangles. The following example illustrates the procedure.

EXAMPLE 1 Use rectangles to estimate the area under the parabola $y = x^2$ from 0 to 1 (the parabolic region S illustrated in Figure 3).

SOLUTION We first notice that the area of S must be somewhere between 0 and 1 because S is contained in a square with side length 1, but we can certainly do better than that. Suppose we divide S into four strips S_1, S_2, S_3, and S_4 by drawing the vertical lines $x = \tfrac{1}{4}$, $x = \tfrac{1}{2}$, and $x = \tfrac{3}{4}$ as in Figure 4(a).

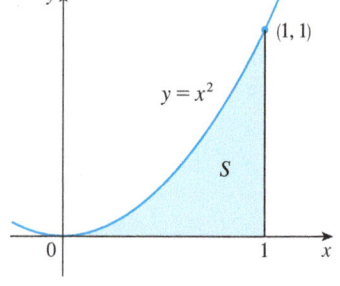

FIGURE 3

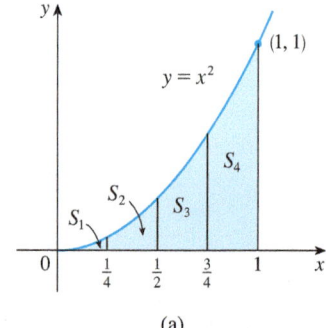

(a)

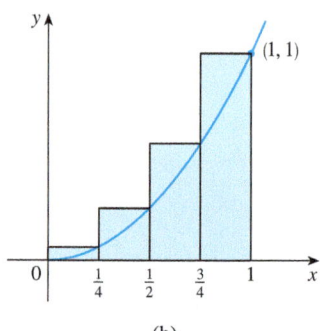

(b)

FIGURE 4

We can approximate each strip by a rectangle that has the same base as the strip and whose height is the same as the right edge of the strip [see Figure 4(b)]. In other words, the heights of these rectangles are the values of the function $f(x) = x^2$ at the *right* endpoints of the subintervals $\left[0, \frac{1}{4}\right]$, $\left[\frac{1}{4}, \frac{1}{2}\right]$, $\left[\frac{1}{2}, \frac{3}{4}\right]$, and $\left[\frac{3}{4}, 1\right]$.

Each rectangle has width $\frac{1}{4}$ and the heights are $\left(\frac{1}{4}\right)^2$, $\left(\frac{1}{2}\right)^2$, $\left(\frac{3}{4}\right)^2$, and 1^2. If we let R_4 be the sum of the areas of these approximating rectangles, we get

$$R_4 = \tfrac{1}{4} \cdot \left(\tfrac{1}{4}\right)^2 + \tfrac{1}{4} \cdot \left(\tfrac{1}{2}\right)^2 + \tfrac{1}{4} \cdot \left(\tfrac{3}{4}\right)^2 + \tfrac{1}{4} \cdot 1^2 = \tfrac{15}{32} = 0.46875$$

From Figure 4(b) we see that the area A of S is less than R_4, so

$$A < 0.46875$$

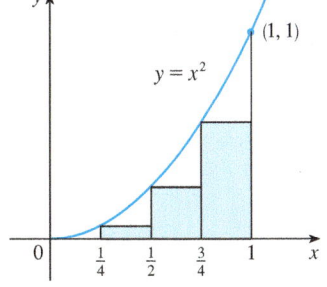

FIGURE 5

Instead of using the rectangles in Figure 4(b) we could use the smaller rectangles in Figure 5 whose heights are the values of f at the *left* endpoints of the subintervals. (The leftmost rectangle has collapsed because its height is 0.) The sum of the areas of these approximating rectangles is

$$L_4 = \tfrac{1}{4} \cdot 0^2 + \tfrac{1}{4} \cdot \left(\tfrac{1}{4}\right)^2 + \tfrac{1}{4} \cdot \left(\tfrac{1}{2}\right)^2 + \tfrac{1}{4} \cdot \left(\tfrac{3}{4}\right)^2 = \tfrac{7}{32} = 0.21875$$

We see that the area of S is larger than L_4, so we have lower and upper estimates for A:

$$0.21875 < A < 0.46875$$

We can repeat this procedure with a larger number of strips. Figure 6 shows what happens when we divide the region S into eight strips of equal width.

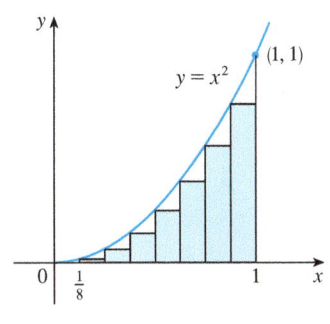
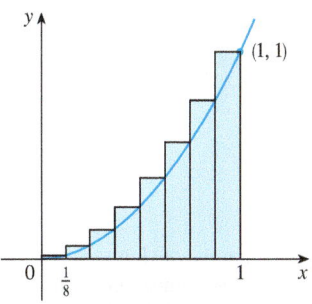

FIGURE 6
Approximating S with eight rectangles

(a) Using left endpoints   (b) Using right endpoints

By computing the sum of the areas of the smaller rectangles (L_8) and the sum of the areas of the larger rectangles (R_8), we obtain better lower and upper estimates for A:

$$0.2734375 < A < 0.3984375$$

So one possible answer to the question is to say that the true area of S lies somewhere between 0.2734375 and 0.3984375.

We could obtain better estimates by increasing the number of strips. The table at the left shows the results of similar calculations (with a computer) using n rectangles whose heights are found with left endpoints (L_n) or right endpoints (R_n). In particular, we see by using 50 strips that the area lies between 0.3234 and 0.3434. With 1000 strips we narrow it down even more: A lies between 0.3328335 and 0.3338335. A good estimate is obtained by averaging these numbers: $A \approx 0.3333335$.

n	L_n	R_n
10	0.2850000	0.3850000
20	0.3087500	0.3587500
30	0.3168519	0.3501852
50	0.3234000	0.3434000
100	0.3283500	0.3383500
1000	0.3328335	0.3338335

From the values in the table in Example 1, it looks as if R_n is approaching $\frac{1}{3}$ as n increases. We confirm this in the next example.

EXAMPLE 2 For the region S in Example 1, show that the sum of the areas of the upper approximating rectangles approaches $\frac{1}{3}$, that is,

$$\lim_{n \to \infty} R_n = \tfrac{1}{3}$$

SOLUTION R_n is the sum of the areas of the n rectangles in Figure 7. Each rectangle has width $1/n$ and the heights are the values of the function $f(x) = x^2$ at the points $1/n, 2/n, 3/n, \ldots, n/n$; that is, the heights are $(1/n)^2, (2/n)^2, (3/n)^2, \ldots, (n/n)^2$. Thus

$$R_n = \frac{1}{n}\left(\frac{1}{n}\right)^2 + \frac{1}{n}\left(\frac{2}{n}\right)^2 + \frac{1}{n}\left(\frac{3}{n}\right)^2 + \cdots + \frac{1}{n}\left(\frac{n}{n}\right)^2$$

$$= \frac{1}{n} \cdot \frac{1}{n^2}(1^2 + 2^2 + 3^2 + \cdots + n^2)$$

$$= \frac{1}{n^3}(1^2 + 2^2 + 3^2 + \cdots + n^2)$$

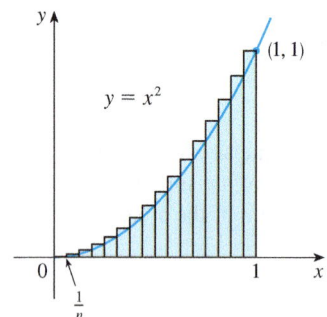

FIGURE 7

Here we need the formula for the sum of the squares of the first n positive integers:

$$\boxed{1} \qquad 1^2 + 2^2 + 3^2 + \cdots + n^2 = \frac{n(n+1)(2n+1)}{6}$$

Perhaps you have seen this formula before. It is proved in Example 5 in Appendix E.
Putting Formula 1 into our expression for R_n, we get

$$R_n = \frac{1}{n^3} \cdot \frac{n(n+1)(2n+1)}{6} = \frac{(n+1)(2n+1)}{6n^2}$$

Thus we have

$$\lim_{n \to \infty} R_n = \lim_{n \to \infty} \frac{(n+1)(2n+1)}{6n^2}$$

$$= \lim_{n \to \infty} \frac{1}{6}\left(\frac{n+1}{n}\right)\left(\frac{2n+1}{n}\right)$$

$$= \lim_{n \to \infty} \frac{1}{6}\left(1 + \frac{1}{n}\right)\left(2 + \frac{1}{n}\right)$$

$$= \frac{1}{6} \cdot 1 \cdot 2 = \frac{1}{3}$$

Here we are computing the limit of the sequence $\{R_n\}$. Sequences and their limits were discussed in *A Preview of Calculus* and will be studied in detail in Section 11.1. The idea is very similar to a limit at infinity (Section 2.6) except that in writing $\lim_{n \to \infty}$ we restrict n to be a positive integer. In particular, we know that

$$\lim_{n \to \infty} \frac{1}{n} = 0$$

When we write $\lim_{n \to \infty} R_n = \frac{1}{3}$ we mean that we can make R_n as close to $\frac{1}{3}$ as we like by taking n sufficiently large.

It can be shown that the lower approximating sums also approach $\frac{1}{3}$, that is,

$$\lim_{n \to \infty} L_n = \tfrac{1}{3}$$

SECTION 5.1 Areas and Distances   369

From Figures 8 and 9 it appears that, as n increases, both L_n and R_n become better and better approximations to the area of S. Therefore we *define* the area A to be the limit of the sums of the areas of the approximating rectangles, that is,

TEC In Visual 5.1 you can create pictures like those in Figures 8 and 9 for other values of n.

$$A = \lim_{n \to \infty} R_n = \lim_{n \to \infty} L_n = \tfrac{1}{3}$$

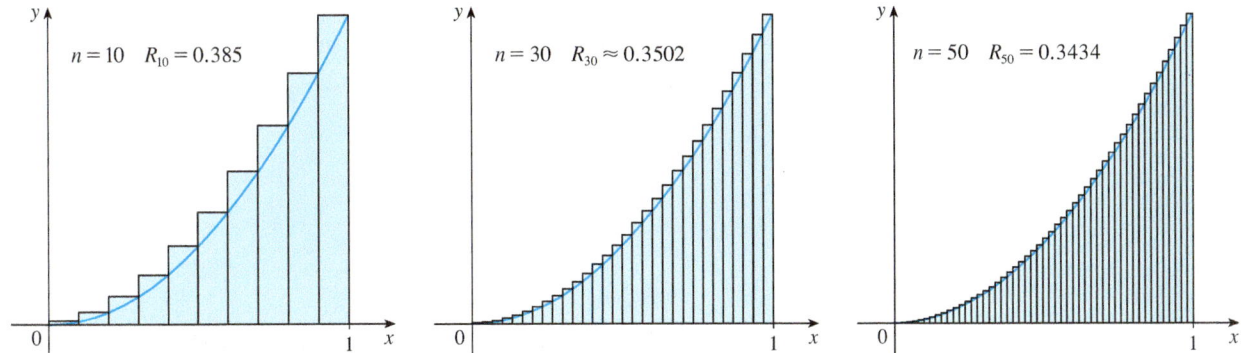

FIGURE 8 Right endpoints produce upper sums because $f(x) = x^2$ is increasing.

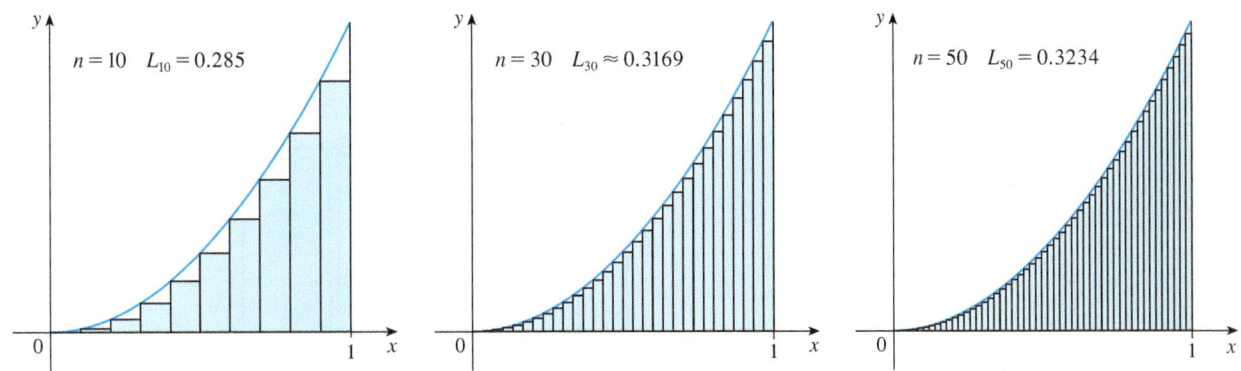

FIGURE 9 Left endpoints produce lower sums because $f(x) = x^2$ is increasing.

Let's apply the idea of Examples 1 and 2 to the more general region S of Figure 1. We start by subdividing S into n strips $S_1, S_2, \ldots, S_n$ of equal width as in Figure 10.

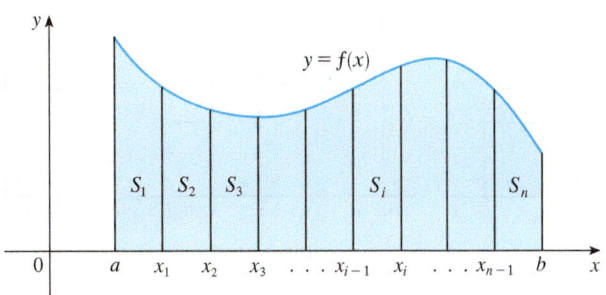

FIGURE 10

The width of the interval $[a, b]$ is $b - a$, so the width of each of the n strips is

$$\Delta x = \frac{b - a}{n}$$

These strips divide the interval $[a, b]$ into n subintervals

$$[x_0, x_1], \quad [x_1, x_2], \quad [x_2, x_3], \quad \ldots, \quad [x_{n-1}, x_n]$$

where $x_0 = a$ and $x_n = b$. The right endpoints of the subintervals are

$$x_1 = a + \Delta x,$$
$$x_2 = a + 2\,\Delta x,$$
$$x_3 = a + 3\,\Delta x,$$
$$\vdots$$

Let's approximate the ith strip S_i by a rectangle with width Δx and height $f(x_i)$, which is the value of f at the right endpoint (see Figure 11). Then the area of the ith rectangle is $f(x_i)\,\Delta x$. What we think of intuitively as the area of S is approximated by the sum of the areas of these rectangles, which is

$$R_n = f(x_1)\,\Delta x + f(x_2)\,\Delta x + \cdots + f(x_n)\,\Delta x$$

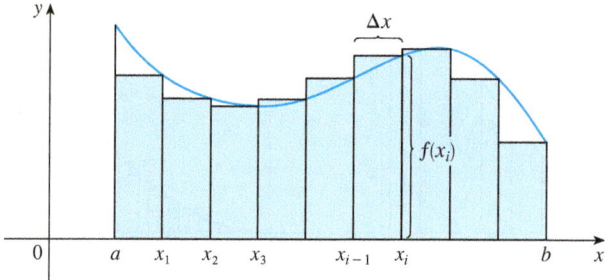

FIGURE 11

Figure 12 shows this approximation for $n = 2, 4, 8,$ and 12. Notice that this approximation appears to become better and better as the number of strips increases, that is, as $n \to \infty$. Therefore we define the area A of the region S in the following way.

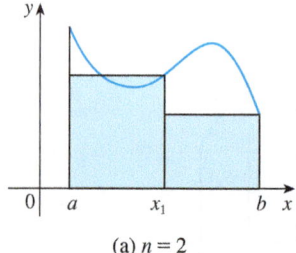
(a) $n = 2$

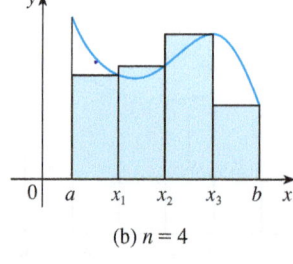
(b) $n = 4$

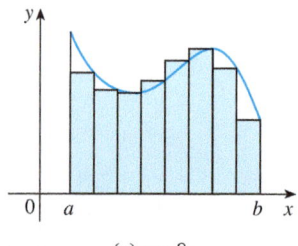
(c) $n = 8$

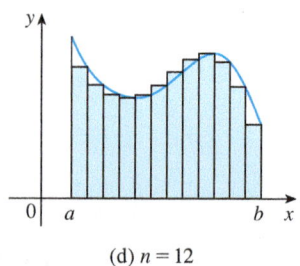
(d) $n = 12$

FIGURE 12

2 Definition The **area** A of the region S that lies under the graph of the continuous function f is the limit of the sum of the areas of approximating rectangles:

$$A = \lim_{n \to \infty} R_n = \lim_{n \to \infty} [f(x_1) \Delta x + f(x_2) \Delta x + \cdots + f(x_n) \Delta x]$$

It can be proved that the limit in Definition 2 always exists, since we are assuming that f is continuous. It can also be shown that we get the same value if we use left endpoints:

$$\boxed{3} \qquad A = \lim_{n \to \infty} L_n = \lim_{n \to \infty} [f(x_0) \Delta x + f(x_1) \Delta x + \cdots + f(x_{n-1}) \Delta x]$$

In fact, instead of using left endpoints or right endpoints, we could take the height of the ith rectangle to be the value of f at *any* number x_i^* in the ith subinterval $[x_{i-1}, x_i]$. We call the numbers $x_1^*, x_2^*, \ldots, x_n^*$ the **sample points**. Figure 13 shows approximating rectangles when the sample points are not chosen to be endpoints. So a more general expression for the area of S is

$$\boxed{4} \qquad A = \lim_{n \to \infty} [f(x_1^*) \Delta x + f(x_2^*) \Delta x + \cdots + f(x_n^*) \Delta x]$$

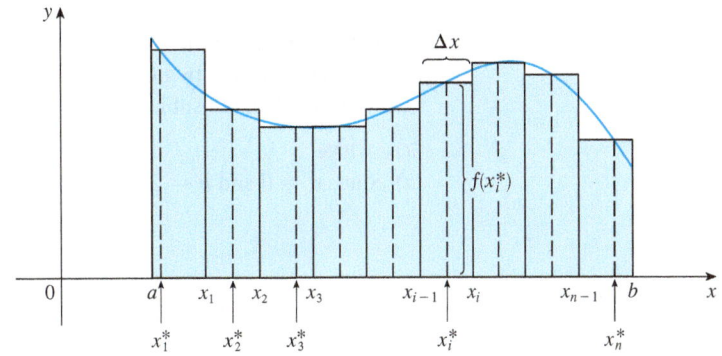

FIGURE 13

NOTE It can be shown that an equivalent definition of area is the following: A is the unique number that is smaller than all the upper sums and bigger than all the lower sums. We saw in Examples 1 and 2, for instance, that the area $\left(A = \frac{1}{3}\right)$ is trapped between all the left approximating sums L_n and all the right approximating sums R_n. The function in those examples, $f(x) = x^2$, happens to be increasing on $[0, 1]$ and so the lower sums arise from left endpoints and the upper sums from right endpoints. (See Figures 8 and 9.) In general, we form **lower** (and **upper**) **sums** by choosing the sample points x_i^* so that $f(x_i^*)$ is the minimum (and maximum) value of f on the ith subinterval. (See Figure 14 and Exercises 7–8.)

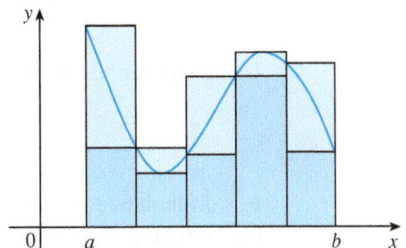

FIGURE 14
Lower sums (short rectangles) and upper sums (tall rectangles)

We often use **sigma notation** to write sums with many terms more compactly. For instance,

$$\sum_{i=1}^{n} f(x_i) \Delta x = f(x_1) \Delta x + f(x_2) \Delta x + \cdots + f(x_n) \Delta x$$

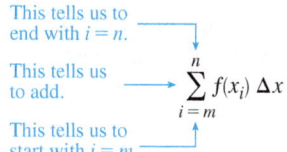

This tells us to end with $i = n$.
This tells us to add.
This tells us to start with $i = m$.

If you need practice with sigma notation, look at the examples and try some of the exercises in Appendix E.

So the expressions for area in Equations 2, 3, and 4 can be written as follows:

$$A = \lim_{n \to \infty} \sum_{i=1}^{n} f(x_i) \Delta x$$

$$A = \lim_{n \to \infty} \sum_{i=1}^{n} f(x_{i-1}) \Delta x$$

$$A = \lim_{n \to \infty} \sum_{i=1}^{n} f(x_i^*) \Delta x$$

We can also rewrite Formula 1 in the following way:

$$\sum_{i=1}^{n} i^2 = \frac{n(n+1)(2n+1)}{6}$$

EXAMPLE 3 Let A be the area of the region that lies under the graph of $f(x) = e^{-x}$ between $x = 0$ and $x = 2$.
(a) Using right endpoints, find an expression for A as a limit. Do not evaluate the limit.
(b) Estimate the area by taking the sample points to be midpoints and using four subintervals and then ten subintervals.

SOLUTION
(a) Since $a = 0$ and $b = 2$, the width of a subinterval is

$$\Delta x = \frac{2 - 0}{n} = \frac{2}{n}$$

So $x_1 = 2/n$, $x_2 = 4/n$, $x_3 = 6/n$, $x_i = 2i/n$, and $x_n = 2n/n$. The sum of the areas of the approximating rectangles is

$$R_n = f(x_1) \Delta x + f(x_2) \Delta x + \cdots + f(x_n) \Delta x$$

$$= e^{-x_1} \Delta x + e^{-x_2} \Delta x + \cdots + e^{-x_n} \Delta x$$

$$= e^{-2/n}\left(\frac{2}{n}\right) + e^{-4/n}\left(\frac{2}{n}\right) + \cdots + e^{-2n/n}\left(\frac{2}{n}\right)$$

According to Definition 2, the area is

$$A = \lim_{n \to \infty} R_n = \lim_{n \to \infty} \frac{2}{n}(e^{-2/n} + e^{-4/n} + e^{-6/n} + \cdots + e^{-2n/n})$$

Using sigma notation we could write

$$A = \lim_{n \to \infty} \frac{2}{n} \sum_{i=1}^{n} e^{-2i/n}$$

It is difficult to evaluate this limit directly by hand, but with the aid of a computer algebra system it isn't hard (see Exercise 30). In Section 5.3 we will be able to find A more easily using a different method.

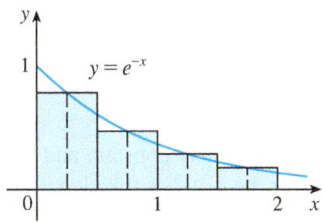

FIGURE 15

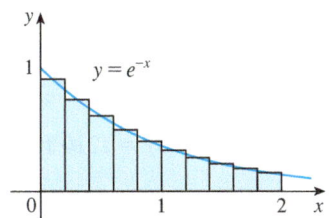

FIGURE 16

(b) With $n = 4$ the subintervals of equal width $\Delta x = 0.5$ are $[0, 0.5]$, $[0.5, 1]$, $[1, 1.5]$, and $[1.5, 2]$. The midpoints of these subintervals are $x_1^* = 0.25$, $x_2^* = 0.75$, $x_3^* = 1.25$, and $x_4^* = 1.75$, and the sum of the areas of the four approximating rectangles (see Figure 15) is

$$M_4 = \sum_{i=1}^{4} f(x_i^*) \Delta x$$

$$= f(0.25) \Delta x + f(0.75) \Delta x + f(1.25) \Delta x + f(1.75) \Delta x$$

$$= e^{-0.25}(0.5) + e^{-0.75}(0.5) + e^{-1.25}(0.5) + e^{-1.75}(0.5)$$

$$= \tfrac{1}{2}(e^{-0.25} + e^{-0.75} + e^{-1.25} + e^{-1.75}) \approx 0.8557$$

So an estimate for the area is

$$A \approx 0.8557$$

With $n = 10$ the subintervals are $[0, 0.2]$, $[0.2, 0.4]$, ..., $[1.8, 2]$ and the midpoints are $x_1^* = 0.1$, $x_2^* = 0.3$, $x_3^* = 0.5$, ..., $x_{10}^* = 1.9$. Thus

$$A \approx M_{10} = f(0.1) \Delta x + f(0.3) \Delta x + f(0.5) \Delta x + \cdots + f(1.9) \Delta x$$

$$= 0.2(e^{-0.1} + e^{-0.3} + e^{-0.5} + \cdots + e^{-1.9}) \approx 0.8632$$

From Figure 16 it appears that this estimate is better than the estimate with $n = 4$. ∎

The Distance Problem

Now let's consider the *distance problem:* Find the distance traveled by an object during a certain time period if the velocity of the object is known at all times. (In a sense this is the inverse problem of the velocity problem that we discussed in Section 2.1.) If the velocity remains constant, then the distance problem is easy to solve by means of the formula

$$\text{distance} = \text{velocity} \times \text{time}$$

But if the velocity varies, it's not so easy to find the distance traveled. We investigate the problem in the following example.

EXAMPLE 4 Suppose the odometer on our car is broken and we want to estimate the distance driven over a 30-second time interval. We take speedometer readings every five seconds and record them in the following table:

Time (s)	0	5	10	15	20	25	30
Velocity (mi/h)	17	21	24	29	32	31	28

In order to have the time and the velocity in consistent units, let's convert the velocity readings to feet per second (1 mi/h $= 5280/3600$ ft/s):

Time (s)	0	5	10	15	20	25	30
Velocity (ft/s)	25	31	35	43	47	45	41

During the first five seconds the velocity doesn't change very much, so we can estimate the distance traveled during that time by assuming that the velocity is constant. If we take the velocity during that time interval to be the initial velocity (25 ft/s), then we obtain the approximate distance traveled during the first five seconds:

$$25 \text{ ft/s} \times 5 \text{ s} = 125 \text{ ft}$$

Similarly, during the second time interval the velocity is approximately constant and we take it to be the velocity when $t = 5$ s. So our estimate for the distance traveled from $t = 5$ s to $t = 10$ s is

$$31 \text{ ft/s} \times 5 \text{ s} = 155 \text{ ft}$$

If we add similar estimates for the other time intervals, we obtain an estimate for the total distance traveled:

$$(25 \times 5) + (31 \times 5) + (35 \times 5) + (43 \times 5) + (47 \times 5) + (45 \times 5) = 1130 \text{ ft}$$

We could just as well have used the velocity at the *end* of each time period instead of the velocity at the beginning as our assumed constant velocity. Then our estimate becomes

$$(31 \times 5) + (35 \times 5) + (43 \times 5) + (47 \times 5) + (45 \times 5) + (41 \times 5) = 1210 \text{ ft}$$

If we had wanted a more accurate estimate, we could have taken velocity readings every two seconds, or even every second. ■

Perhaps the calculations in Example 4 remind you of the sums we used earlier to estimate areas. The similarity is explained when we sketch a graph of the velocity function of the car in Figure 17 and draw rectangles whose heights are the initial velocities for each time interval. The area of the first rectangle is $25 \times 5 = 125$, which is also our estimate for the distance traveled in the first five seconds. In fact, the area of each rectangle can be interpreted as a distance because the height represents velocity and the width represents time. The sum of the areas of the rectangles in Figure 17 is $L_6 = 1130$, which is our initial estimate for the total distance traveled.

In general, suppose an object moves with velocity $v = f(t)$, where $a \leq t \leq b$ and $f(t) \geq 0$ (so the object always moves in the positive direction). We take velocity readings at times $t_0 \, (= a), t_1, t_2, \ldots, t_n \, (= b)$ so that the velocity is approximately constant on each subinterval. If these times are equally spaced, then the time between consecutive readings is $\Delta t = (b - a)/n$. During the first time interval the velocity is approximately $f(t_0)$ and so the distance traveled is approximately $f(t_0) \, \Delta t$. Similarly, the distance traveled during the second time interval is about $f(t_1) \, \Delta t$ and the total distance traveled during the time interval $[a, b]$ is approximately

$$f(t_0) \, \Delta t + f(t_1) \, \Delta t + \cdots + f(t_{n-1}) \, \Delta t = \sum_{i=1}^{n} f(t_{i-1}) \, \Delta t$$

If we use the velocity at right endpoints instead of left endpoints, our estimate for the total distance becomes

$$f(t_1) \, \Delta t + f(t_2) \, \Delta t + \cdots + f(t_n) \, \Delta t = \sum_{i=1}^{n} f(t_i) \, \Delta t$$

The more frequently we measure the velocity, the more accurate our estimates become,

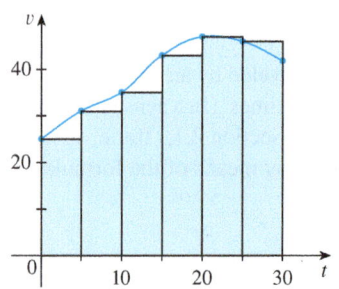

FIGURE 17

so it seems plausible that the *exact* distance d traveled is the *limit* of such expressions:

$$\boxed{5} \quad d = \lim_{n \to \infty} \sum_{i=1}^{n} f(t_{i-1}) \, \Delta t = \lim_{n \to \infty} \sum_{i=1}^{n} f(t_i) \, \Delta t$$

We will see in Section 5.4 that this is indeed true.

Because Equation 5 has the same form as our expressions for area in Equations 2 and 3, it follows that the distance traveled is equal to the area under the graph of the velocity function. In Chapter 6 we will see that other quantities of interest in the natural and social sciences—such as the work done by a variable force or the cardiac output of the heart—can also be interpreted as the area under a curve. So when we compute areas in this chapter, bear in mind that they can be interpreted in a variety of practical ways.

5.1 EXERCISES

1. (a) By reading values from the given graph of f, use five rectangles to find a lower estimate and an upper estimate for the area under the given graph of f from $x = 0$ to $x = 10$. In each case sketch the rectangles that you use.
   (b) Find new estimates using ten rectangles in each case.

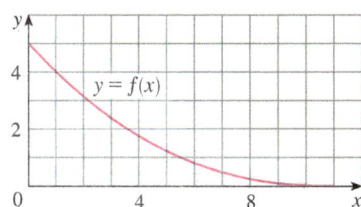

2. (a) Use six rectangles to find estimates of each type for the area under the given graph of f from $x = 0$ to $x = 12$.
      (i) L_6 (sample points are left endpoints)
      (ii) R_6 (sample points are right endpoints)
      (iii) M_6 (sample points are midpoints)
   (b) Is L_6 an underestimate or overestimate of the true area?
   (c) Is R_6 an underestimate or overestimate of the true area?
   (d) Which of the numbers L_6, R_6, or M_6 gives the best estimate? Explain.

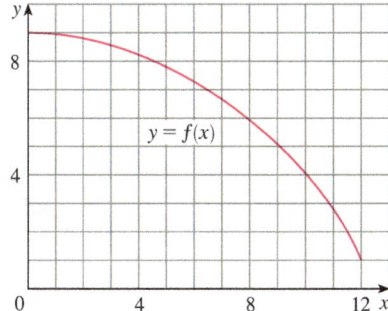

3. (a) Estimate the area under the graph of $f(x) = 1/x$ from $x = 1$ to $x = 2$ using four approximating rectangles and right endpoints. Sketch the graph and the rectangles. Is your estimate an underestimate or an overestimate?
   (b) Repeat part (a) using left endpoints.

4. (a) Estimate the area under the graph of $f(x) = \sin x$ from $x = 0$ to $x = \pi/2$ using four approximating rectangles and right endpoints. Sketch the graph and the rectangles. Is your estimate an underestimate or an overestimate?
   (b) Repeat part (a) using left endpoints.

5. (a) Estimate the area under the graph of $f(x) = 1 + x^2$ from $x = -1$ to $x = 2$ using three rectangles and right endpoints. Then improve your estimate by using six rectangles. Sketch the curve and the approximating rectangles.
   (b) Repeat part (a) using left endpoints.
   (c) Repeat part (a) using midpoints.
   (d) From your sketches in parts (a)–(c), which appears to be the best estimate?

6. (a) Graph the function
   $$f(x) = x - 2 \ln x \qquad 1 \leq x \leq 5$$
   (b) Estimate the area under the graph of f using four approximating rectangles and taking the sample points to be (i) right endpoints and (ii) midpoints. In each case sketch the curve and the rectangles.
   (c) Improve your estimates in part (b) by using eight rectangles.

7. Evaluate the upper and lower sums for $f(x) = 2 + \sin x$, $0 \leq x \leq \pi$, with $n = 2, 4,$ and 8. Illustrate with diagrams like Figure 14.

8. Evaluate the upper and lower sums for $f(x) = 1 + x^2$, $-1 \leq x \leq 1$, with $n = 3$ and 4. Illustrate with diagrams like Figure 14.

9–10 With a programmable calculator (or a computer), it is possible to evaluate the expressions for the sums of areas of approximating rectangles, even for large values of n, using looping. (On a TI use the Is> command or a For-EndFor loop, on a Casio use Isz, on an HP or in BASIC use a FOR-NEXT loop.) Compute the sum of the areas of approximating rectangles using equal subintervals and right endpoints for $n = 10$, 30, 50, and 100. Then guess the value of the exact area.

9. The region under $y = x^4$ from 0 to 1

10. The region under $y = \cos x$ from 0 to $\pi/2$

CAS 11. Some computer algebra systems have commands that will draw approximating rectangles and evaluate the sums of their areas, at least if x_i^* is a left or right endpoint. (For instance, in Maple use leftbox, rightbox, leftsum, and rightsum.)
(a) If $f(x) = 1/(x^2 + 1)$, $0 \le x \le 1$, find the left and right sums for $n = 10$, 30, and 50.
(b) Illustrate by graphing the rectangles in part (a).
(c) Show that the exact area under f lies between 0.780 and 0.791.

CAS 12. (a) If $f(x) = \ln x$, $1 \le x \le 4$, use the commands discussed in Exercise 11 to find the left and right sums for $n = 10$, 30, and 50.
(b) Illustrate by graphing the rectangles in part (a).
(c) Show that the exact area under f lies between 2.50 and 2.59.

13. The speed of a runner increased steadily during the first three seconds of a race. Her speed at half-second intervals is given in the table. Find lower and upper estimates for the distance that she traveled during these three seconds.

t (s)	0	0.5	1.0	1.5	2.0	2.5	3.0
v (ft/s)	0	6.2	10.8	14.9	18.1	19.4	20.2

14. The table shows speedometer readings at 10-second intervals during a 1-minute period for a car racing at the Daytona International Speedway in Florida.
(a) Estimate the distance the race car traveled during this time period using the velocities at the beginning of the time intervals.
(b) Give another estimate using the velocities at the end of the time periods.
(c) Are your estimates in parts (a) and (b) upper and lower estimates? Explain.

Time (s)	Velocity (mi/h)
0	182.9
10	168.0
20	106.6
30	99.8
40	124.5
50	176.1
60	175.6

15. Oil leaked from a tank at a rate of $r(t)$ liters per hour. The rate decreased as time passed and values of the rate at two-hour time intervals are shown in the table. Find lower and upper estimates for the total amount of oil that leaked out.

t (h)	0	2	4	6	8	10
$r(t)$ (L/h)	8.7	7.6	6.8	6.2	5.7	5.3

16. When we estimate distances from velocity data, it is sometimes necessary to use times $t_0, t_1, t_2, t_3, \ldots$ that are not equally spaced. We can still estimate distances using the time periods $\Delta t_i = t_i - t_{i-1}$. For example, on May 7, 1992, the space shuttle *Endeavour* was launched on mission STS-49, the purpose of which was to install a new perigee kick motor in an Intelsat communications satellite. The table, provided by NASA, gives the velocity data for the shuttle between liftoff and the jettisoning of the solid rocket boosters. Use these data to estimate the height above the earth's surface of the *Endeavour*, 62 seconds after liftoff.

Event	Time (s)	Velocity (ft/s)
Launch	0	0
Begin roll maneuver	10	185
End roll maneuver	15	319
Throttle to 89%	20	447
Throttle to 67%	32	742
Throttle to 104%	59	1325
Maximum dynamic pressure	62	1445
Solid rocket booster separation	125	4151

17. The velocity graph of a braking car is shown. Use it to estimate the distance traveled by the car while the brakes are applied.

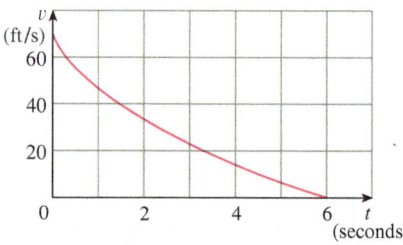

18. The velocity graph of a car accelerating from rest to a speed of 120 km/h over a period of 30 seconds is shown. Estimate the distance traveled during this period.

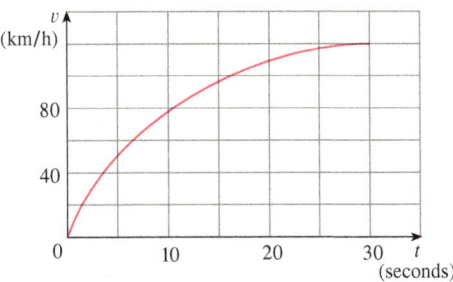

19. In someone infected with measles, the virus level N (measured in number of infected cells per mL of blood plasma) reaches a peak density at about $t = 12$ days (when a rash appears) and then decreases fairly rapidly as a result of immune response. The area under the graph of $N(t)$ from $t = 0$ to $t = 12$ (as shown in the figure) is equal to the total amount of infection needed to develop symptoms (measured in density of infected cells × time). The function N has been modeled by the function

$$f(t) = -t(t-21)(t+1)$$

Use this model with six subintervals and their midpoints to estimate the total amount of infection needed to develop symptoms of measles.

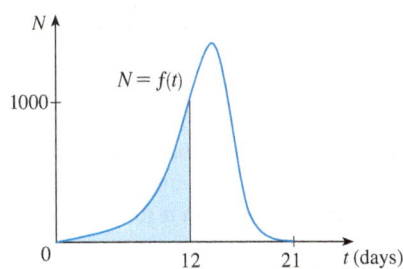

Source: J. M. Heffernan et al., "An In-Host Model of Acute Infection: Measles as a Case Study," *Theoretical Population Biology* 73 (2006): 134–47.

20. The table shows the number of people per day who died from SARS in Singapore at two-week intervals beginning on March 1, 2003.

Date	Deaths per day	Date	Deaths per day
March 1	0.0079	April 26	0.5620
March 15	0.0638	May 10	0.4630
March 29	0.1944	May 24	0.2897
April 12	0.4435		

(a) By using an argument similar to that in Example 4, estimate the number of people who died of SARS in Singapore between March 1 and May 24, 2003, using both left endpoints and right endpoints.
(b) How would you interpret the number of SARS deaths as an area under a curve?

Source: A. Gumel et al., "Modelling Strategies for Controlling SARS Outbreaks," *Proceedings of the Royal Society of London: Series B* 271 (2004): 2223–32.

21–23 Use Definition 2 to find an expression for the area under the graph of f as a limit. Do not evaluate the limit.

21. $f(x) = \dfrac{2x}{x^2 + 1}$, $1 \leq x \leq 3$

22. $f(x) = x^2 + \sqrt{1 + 2x}$, $4 \leq x \leq 7$

23. $f(x) = \sqrt{\sin x}$, $0 \leq x \leq \pi$

24–25 Determine a region whose area is equal to the given limit. Do not evaluate the limit.

24. $\displaystyle\lim_{n \to \infty} \sum_{i=1}^{n} \dfrac{3}{n} \sqrt{1 + \dfrac{3i}{n}}$

25. $\displaystyle\lim_{n \to \infty} \sum_{i=1}^{n} \dfrac{\pi}{4n} \tan \dfrac{i\pi}{4n}$

26. (a) Use Definition 2 to find an expression for the area under the curve $y = x^3$ from 0 to 1 as a limit.
(b) The following formula for the sum of the cubes of the first n integers is proved in Appendix E. Use it to evaluate the limit in part (a).

$$1^3 + 2^3 + 3^3 + \cdots + n^3 = \left[\dfrac{n(n+1)}{2}\right]^2$$

27. Let A be the area under the graph of an increasing continuous function f from a to b, and let L_n and R_n be the approximations to A with n subintervals using left and right endpoints, respectively.
(a) How are A, L_n, and R_n related?
(b) Show that

$$R_n - L_n = \dfrac{b-a}{n}[f(b) - f(a)]$$

Then draw a diagram to illustrate this equation by showing that the n rectangles representing $R_n - L_n$ can be reassembled to form a single rectangle whose area is the right side of the equation.
(c) Deduce that

$$R_n - A < \dfrac{b-a}{n}[f(b) - f(a)]$$

28. If A is the area under the curve $y = e^x$ from 1 to 3, use Exercise 27 to find a value of n such that $R_n - A < 0.0001$.

29. (a) Express the area under the curve $y = x^5$ from 0 to 2 as a limit.
(b) Use a computer algebra system to find the sum in your expression from part (a).
(c) Evaluate the limit in part (a).

30. Find the exact area of the region under the graph of $y = e^{-x}$ from 0 to 2 by using a computer algebra system to evaluate the sum and then the limit in Example 3(a). Compare your answer with the estimate obtained in Example 3(b).

31. Find the exact area under the cosine curve $y = \cos x$ from $x = 0$ to $x = b$, where $0 \leq b \leq \pi/2$. (Use a computer algebra system both to evaluate the sum and compute the limit.) In particular, what is the area if $b = \pi/2$?

32. (a) Let A_n be the area of a polygon with n equal sides inscribed in a circle with radius r. By dividing the polygon into n congruent triangles with central angle $2\pi/n$, show that

$$A_n = \tfrac{1}{2} nr^2 \sin\left(\frac{2\pi}{n}\right)$$

(b) Show that $\lim_{n \to \infty} A_n = \pi r^2$. [*Hint:* Use Equation 3.3.2 on page 191.]

5.2 The Definite Integral

We saw in Section 5.1 that a limit of the form

$$\boxed{1} \quad \lim_{n \to \infty} \sum_{i=1}^{n} f(x_i^*) \, \Delta x = \lim_{n \to \infty} [f(x_1^*) \, \Delta x + f(x_2^*) \, \Delta x + \cdots + f(x_n^*) \, \Delta x]$$

arises when we compute an area. We also saw that it arises when we try to find the distance traveled by an object. It turns out that this same type of limit occurs in a wide variety of situations even when f is not necessarily a positive function. In Chapters 6 and 8 we will see that limits of the form (1) also arise in finding lengths of curves, volumes of solids, centers of mass, force due to water pressure, and work, as well as other quantities. We therefore give this type of limit a special name and notation.

> **2 Definition of a Definite Integral** If f is a function defined for $a \leq x \leq b$, we divide the interval $[a, b]$ into n subintervals of equal width $\Delta x = (b - a)/n$. We let $x_0 \, (= a), x_1, x_2, \ldots, x_n \, (= b)$ be the endpoints of these subintervals and we let $x_1^*, x_2^*, \ldots, x_n^*$ be any **sample points** in these subintervals, so x_i^* lies in the ith subinterval $[x_{i-1}, x_i]$. Then the **definite integral of f from a to b** is
>
> $$\int_a^b f(x) \, dx = \lim_{n \to \infty} \sum_{i=1}^{n} f(x_i^*) \, \Delta x$$
>
> provided that this limit exists and gives the same value for all possible choices of sample points. If it does exist, we say that f is **integrable** on $[a, b]$.

The precise meaning of the limit that defines the integral is as follows:

For every number $\varepsilon > 0$ there is an integer N such that

$$\left| \int_a^b f(x) \, dx - \sum_{i=1}^{n} f(x_i^*) \, \Delta x \right| < \varepsilon$$

for every integer $n > N$ and for every choice of x_i^* in $[x_{i-1}, x_i]$.

NOTE 1 The symbol $\int$ was introduced by Leibniz and is called an **integral sign**. It is an elongated S and was chosen because an integral is a limit of sums. In the notation

$\int_a^b f(x)\,dx$, $f(x)$ is called the **integrand** and a and b are called the **limits of integration**; a is the **lower limit** and b is the **upper limit**. For now, the symbol dx has no meaning by itself; $\int_a^b f(x)\,dx$ is all one symbol. The dx simply indicates that the independent variable is x. The procedure of calculating an integral is called **integration**.

NOTE 2 The definite integral $\int_a^b f(x)\,dx$ is a number; it does not depend on x. In fact, we could use any letter in place of x without changing the value of the integral:

$$\int_a^b f(x)\,dx = \int_a^b f(t)\,dt = \int_a^b f(r)\,dr$$

NOTE 3 The sum

$$\sum_{i=1}^{n} f(x_i^*)\,\Delta x$$

Riemann

Bernhard Riemann received his Ph.D. under the direction of the legendary Gauss at the University of Göttingen and remained there to teach. Gauss, who was not in the habit of praising other mathematicians, spoke of Riemann's "creative, active, truly mathematical mind and gloriously fertile originality." The definition (2) of an integral that we use is due to Riemann. He also made major contributions to the theory of functions of a complex variable, mathematical physics, number theory, and the foundations of geometry. Riemann's broad concept of space and geometry turned out to be the right setting, 50 years later, for Einstein's general relativity theory. Riemann's health was poor throughout his life, and he died of tuberculosis at the age of 39.

that occurs in Definition 2 is called a **Riemann sum** after the German mathematician Bernhard Riemann (1826–1866). So Definition 2 says that the definite integral of an integrable function can be approximated to within any desired degree of accuracy by a Riemann sum.

We know that if f happens to be positive, then the Riemann sum can be interpreted as a sum of areas of approximating rectangles (see Figure 1). By comparing Definition 2 with the definition of area in Section 5.1, we see that the definite integral $\int_a^b f(x)\,dx$ can be interpreted as the area under the curve $y = f(x)$ from a to b. (See Figure 2.)

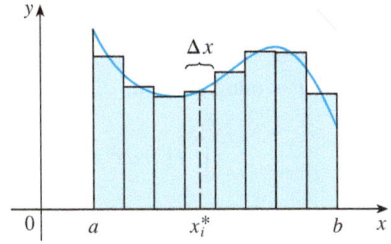

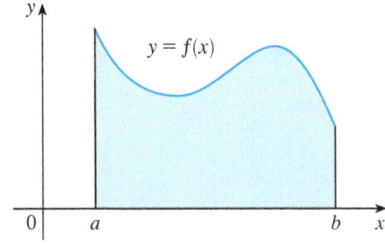

FIGURE 1
If $f(x) \geq 0$, the Riemann sum $\sum f(x_i^*)\,\Delta x$ is the sum of areas of rectangles.

FIGURE 2
If $f(x) \geq 0$, the integral $\int_a^b f(x)\,dx$ is the area under the curve $y = f(x)$ from a to b.

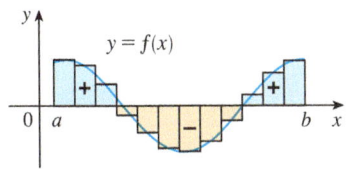

FIGURE 3

$\sum f(x_i^*)\,\Delta x$ is an approximation to the net area.

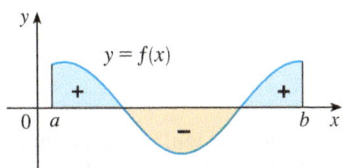

FIGURE 4

$\int_a^b f(x)\,dx$ is the net area.

If f takes on both positive and negative values, as in Figure 3, then the Riemann sum is the sum of the areas of the rectangles that lie above the x-axis and the *negatives* of the areas of the rectangles that lie below the x-axis (the areas of the blue rectangles *minus* the areas of the gold rectangles). When we take the limit of such Riemann sums, we get the situation illustrated in Figure 4. A definite integral can be interpreted as a **net area**, that is, a difference of areas:

$$\int_a^b f(x)\,dx = A_1 - A_2$$

where A_1 is the area of the region above the x-axis and below the graph of f, and A_2 is the area of the region below the x-axis and above the graph of f.

NOTE 4 Although we have defined $\int_a^b f(x)\,dx$ by dividing $[a, b]$ into subintervals of equal width, there are situations in which it is advantageous to work with subintervals of unequal width. For instance, in Exercise 5.1.16, NASA provided velocity data at times that were not equally spaced, but we were still able to estimate the distance traveled. And there are methods for numerical integration that take advantage of unequal subintervals.

If the subinterval widths are $\Delta x_1, \Delta x_2, \ldots, \Delta x_n$, we have to ensure that all these widths approach 0 in the limiting process. This happens if the largest width, max Δx_i, approaches 0. So in this case the definition of a definite integral becomes

$$\int_a^b f(x)\,dx = \lim_{\max \Delta x_i \to 0} \sum_{i=1}^n f(x_i^*)\,\Delta x_i$$

NOTE 5 We have defined the definite integral for an integrable function, but not all functions are integrable (see Exercises 71–72). The following theorem shows that the most commonly occurring functions are in fact integrable. The theorem is proved in more advanced courses.

> **3 Theorem** If f is continuous on $[a, b]$, or if f has only a finite number of jump discontinuities, then f is integrable on $[a, b]$; that is, the definite integral $\int_a^b f(x)\,dx$ exists.

If f is integrable on $[a, b]$, then the limit in Definition 2 exists and gives the same value no matter how we choose the sample points x_i^*. To simplify the calculation of the integral we often take the sample points to be right endpoints. Then $x_i^* = x_i$ and the definition of an integral simplifies as follows.

> **4 Theorem** If f is integrable on $[a, b]$, then
>
> $$\int_a^b f(x)\,dx = \lim_{n \to \infty} \sum_{i=1}^n f(x_i)\,\Delta x$$
>
> where $\quad \Delta x = \dfrac{b - a}{n} \quad$ and $\quad x_i = a + i\,\Delta x$

EXAMPLE 1 Express

$$\lim_{n \to \infty} \sum_{i=1}^n (x_i^3 + x_i \sin x_i)\,\Delta x$$

as an integral on the interval $[0, \pi]$.

SOLUTION Comparing the given limit with the limit in Theorem 4, we see that they will be identical if we choose $f(x) = x^3 + x \sin x$. We are given that $a = 0$ and $b = \pi$. Therefore, by Theorem 4, we have

$$\lim_{n \to \infty} \sum_{i=1}^n (x_i^3 + x_i \sin x_i)\,\Delta x = \int_0^\pi (x^3 1 x \sin x)\,dx \quad \blacksquare$$

Later, when we apply the definite integral to physical situations, it will be important to recognize limits of sums as integrals, as we did in Example 1. When Leibniz chose the notation for an integral, he chose the ingredients as reminders of the limiting process. In general, when we write

$$\lim_{n \to \infty} \sum_{i=1}^{n} f(x_i^*) \Delta x = \int_a^b f(x)\, dx$$

we replace $\lim \Sigma$ by $\int$, x_i^* by x, and Δx by dx.

■ Evaluating Integrals

When we use a limit to evaluate a definite integral, we need to know how to work with sums. The following three equations give formulas for sums of powers of positive integers. Equation 5 may be familiar to you from a course in algebra. Equations 6 and 7 were discussed in Section 5.1 and are proved in Appendix E.

$$\boxed{5} \qquad \sum_{i=1}^{n} i = \frac{n(n+1)}{2}$$

$$\boxed{6} \qquad \sum_{i=1}^{n} i^2 = \frac{n(n+1)(2n+1)}{6}$$

$$\boxed{7} \qquad \sum_{i=1}^{n} i^3 = \left[\frac{n(n+1)}{2}\right]^2$$

The remaining formulas are simple rules for working with sigma notation:

Formulas 8–11 are proved by writing out each side in expanded form. The left side of Equation 9 is

$$ca_1 + ca_2 + \cdots + ca_n$$

The right side is

$$c(a_1 + a_2 + \cdots + a_n)$$

These are equal by the distributive property. The other formulas are discussed in Appendix E.

$$\boxed{8} \qquad \sum_{i=1}^{n} c = nc$$

$$\boxed{9} \qquad \sum_{i=1}^{n} ca_i = c \sum_{i=1}^{n} a_i$$

$$\boxed{10} \qquad \sum_{i=1}^{n} (a_i + b_i) = \sum_{i=1}^{n} a_i + \sum_{i=1}^{n} b_i$$

$$\boxed{11} \qquad \sum_{i=1}^{n} (a_i - b_i) = \sum_{i=1}^{n} a_i - \sum_{i=1}^{n} b_i$$

EXAMPLE 2
(a) Evaluate the Riemann sum for $f(x) = x^3 - 6x$, taking the sample points to be right endpoints and $a = 0$, $b = 3$, and $n = 6$.
(b) Evaluate $\int_0^3 (x^3 - 6x)\, dx$.

SOLUTION
(a) With $n = 6$ the interval width is

$$\Delta x = \frac{b-a}{n} = \frac{3-0}{6} = \frac{1}{2}$$

and the right endpoints are $x_1 = 0.5$, $x_2 = 1.0$, $x_3 = 1.5$, $x_4 = 2.0$, $x_5 = 2.5$, and

$x_6 = 3.0$. So the Riemann sum is

$$R_6 = \sum_{i=1}^{6} f(x_i) \, \Delta x$$

$$= f(0.5) \, \Delta x + f(1.0) \, \Delta x + f(1.5) \, \Delta x + f(2.0) \, \Delta x + f(2.5) \, \Delta x + f(3.0) \, \Delta x$$

$$= \tfrac{1}{2}(-2.875 - 5 - 5.625 - 4 + 0.625 + 9)$$

$$= -3.9375$$

Notice that f is not a positive function and so the Riemann sum does not represent a sum of areas of rectangles. But it does represent the sum of the areas of the blue rectangles (above the x-axis) minus the sum of the areas of the gold rectangles (below the x-axis) in Figure 5.

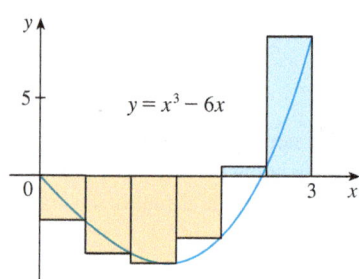

FIGURE 5

(b) With n subintervals we have

$$\Delta x = \frac{b-a}{n} = \frac{3}{n}$$

So $x_0 = 0$, $x_1 = 3/n$, $x_2 = 6/n$, $x_3 = 9/n$, and, in general, $x_i = 3i/n$. Since we are using right endpoints, we can use Theorem 4:

$$\int_0^3 (x^3 - 6x) \, dx = \lim_{n \to \infty} \sum_{i=1}^{n} f(x_i) \, \Delta x = \lim_{n \to \infty} \sum_{i=1}^{n} f\left(\frac{3i}{n}\right) \frac{3}{n}$$

In the sum, n is a constant (unlike i), so we can move $3/n$ in front of the Σ sign.

$$= \lim_{n \to \infty} \frac{3}{n} \sum_{i=1}^{n} \left[\left(\frac{3i}{n}\right)^3 - 6\left(\frac{3i}{n}\right) \right] \quad \text{(Equation 9 with } c = 3/n\text{)}$$

$$= \lim_{n \to \infty} \frac{3}{n} \sum_{i=1}^{n} \left[\frac{27}{n^3} i^3 - \frac{18}{n} i \right]$$

$$= \lim_{n \to \infty} \left[\frac{81}{n^4} \sum_{i=1}^{n} i^3 - \frac{54}{n^2} \sum_{i=1}^{n} i \right] \quad \text{(Equations 11 and 9)}$$

$$= \lim_{n \to \infty} \left\{ \frac{81}{n^4} \left[\frac{n(n+1)}{2} \right]^2 - \frac{54}{n^2} \frac{n(n+1)}{2} \right\} \quad \text{(Equations 7 and 5)}$$

$$= \lim_{n \to \infty} \left[\frac{81}{4} \left(1 + \frac{1}{n}\right)^2 - 27\left(1 + \frac{1}{n}\right) \right]$$

$$= \frac{81}{4} - 27 = -\frac{27}{4} = -6.75$$

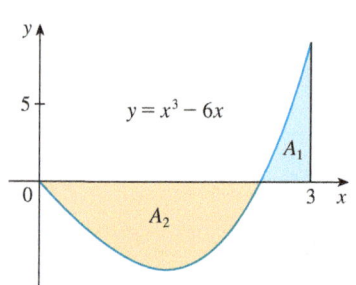

FIGURE 6
$\int_0^3 (x^3 - 6x) \, dx = A_1 - A_2 = -6.75$

This integral can't be interpreted as an area because f takes on both positive and negative values. But it can be interpreted as the difference of areas $A_1 - A_2$, where A_1 and A_2 are shown in Figure 6.

Figure 7 illustrates the calculation by showing the positive and negative terms in the right Riemann sum R_n for $n = 40$. The values in the table show the Riemann sums approaching the exact value of the integral, -6.75, as $n \to \infty$.

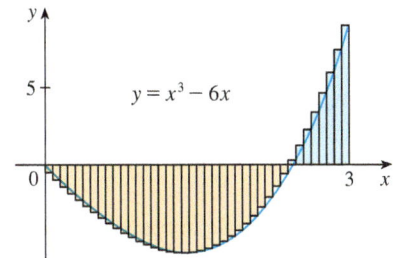

n	R_n
40	-6.3998
100	-6.6130
500	-6.7229
1000	-6.7365
5000	-6.7473

FIGURE 7
$R_{40} \approx -6.3998$

A much simpler method for evaluating the integral in Example 2 will be given in Section 5.4.

EXAMPLE 3
(a) Set up an expression for $\int_1^3 e^x \, dx$ as a limit of sums.
(b) Use a computer algebra system to evaluate the expression.

SOLUTION
(a) Here we have $f(x) = e^x$, $a = 1$, $b = 3$, and
$$\Delta x = \frac{b-a}{n} = \frac{2}{n}$$

So $x_0 = 1$, $x_1 = 1 + 2/n$, $x_2 = 1 + 4/n$, $x_3 = 1 + 6/n$, and
$$x_i = 1 + \frac{2i}{n}$$

From Theorem 4, we get
$$\int_1^3 e^x \, dx = \lim_{n \to \infty} \sum_{i=1}^n f(x_i) \, \Delta x$$
$$= \lim_{n \to \infty} \sum_{i=1}^n f\left(1 + \frac{2i}{n}\right) \frac{2}{n}$$
$$= \lim_{n \to \infty} \frac{2}{n} \sum_{i=1}^n e^{1+2i/n}$$

Because $f(x) = e^x$ is positive, the integral in Example 3 represents the area shown in Figure 8.

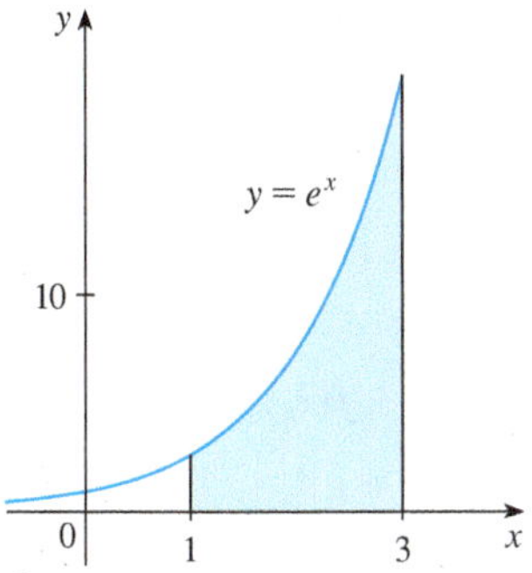

FIGURE 8

A computer algebra system is able to find an explicit expression for this sum because it is a geometric series. The limit could be found using l'Hospital's Rule.

(b) If we ask a computer algebra system to evaluate the sum and simplify, we obtain
$$\sum_{i=1}^n e^{1+2i/n} = \frac{e^{(3n+2)/n} - e^{(n+2)/n}}{e^{2/n} - 1}$$

Now we ask the computer algebra system to evaluate the limit:
$$\int_1^3 e^x \, dx = \lim_{n \to \infty} \frac{2}{n} \cdot \frac{e^{(3n+2)/n} - e^{(n+2)/n}}{e^{2/n} - 1} = e^3 - e$$

We will learn a much easier method for the evaluation of integrals in the next section.

EXAMPLE 4 Evaluate the following integrals by interpreting each in terms of areas.

(a) $\int_0^1 \sqrt{1-x^2}\,dx$
(b) $\int_0^3 (x-1)\,dx$

SOLUTION

(a) Since $f(x) = \sqrt{1-x^2} \geq 0$, we can interpret this integral as the area under the curve $y = \sqrt{1-x^2}$ from 0 to 1. But, since $y^2 = 1 - x^2$, we get $x^2 + y^2 = 1$, which shows that the graph of f is the quarter-circle with radius 1 in Figure 9. Therefore

$$\int_0^1 \sqrt{1-x^2}\,dx = \tfrac{1}{4}\pi(1)^2 = \frac{\pi}{4}$$

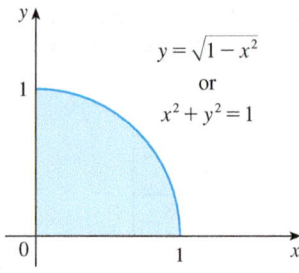

FIGURE 9

(In Section 7.3 we will be able to *prove* that the area of a circle of radius r is πr^2.)

(b) The graph of $y = x - 1$ is the line with slope 1 shown in Figure 10. We compute the integral as the difference of the areas of the two triangles:

$$\int_0^3 (x-1)\,dx = A_1 - A_2 = \tfrac{1}{2}(2 \cdot 2) - \tfrac{1}{2}(1 \cdot 1) = 1.5$$

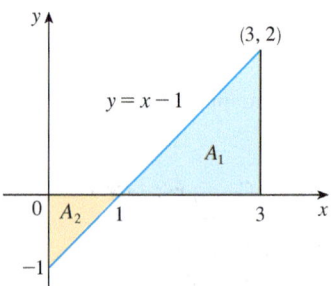

FIGURE 10

■ **The Midpoint Rule**

We often choose the sample point x_i^* to be the right endpoint of the ith subinterval because it is convenient for computing the limit. But if the purpose is to find an *approximation* to an integral, it is usually better to choose x_i^* to be the midpoint of the interval, which we denote by $\bar{x}_i$. Any Riemann sum is an approximation to an integral, but if we use midpoints we get the following approximation.

TEC Module 5.2/7.7 shows how the Midpoint Rule estimates improve as n increases.

> **Midpoint Rule**
>
> $$\int_a^b f(x)\,dx \approx \sum_{i=1}^n f(\bar{x}_i)\,\Delta x = \Delta x\,[f(\bar{x}_1) + \cdots + f(\bar{x}_n)]$$
>
> where $\quad \Delta x = \dfrac{b-a}{n}$
>
> and $\quad \bar{x}_i = \tfrac{1}{2}(x_{i-1} + x_i) = $ midpoint of $[x_{i-1}, x_i]$

EXAMPLE 5 Use the Midpoint Rule with $n = 5$ to approximate $\int_1^2 \dfrac{1}{x}\,dx$.

SOLUTION The endpoints of the five subintervals are 1, 1.2, 1.4, 1.6, 1.8, and 2.0, so the midpoints are 1.1, 1.3, 1.5, 1.7, and 1.9. The width of the subintervals is $\Delta x = (2-1)/5 = \tfrac{1}{5}$, so the Midpoint Rule gives

$$\int_1^2 \frac{1}{x}\,dx \approx \Delta x\,[f(1.1) + f(1.3) + f(1.5) + f(1.7) + f(1.9)]$$

$$= \frac{1}{5}\left(\frac{1}{1.1} + \frac{1}{1.3} + \frac{1}{1.5} + \frac{1}{1.7} + \frac{1}{1.9}\right)$$

$$\approx 0.691908$$

Since $f(x) = 1/x > 0$ for $1 \leq x \leq 2$, the integral represents an area, and the approximation given by the Midpoint Rule is the sum of the areas of the rectangles shown in Figure 11. ■

At the moment we don't know how accurate the approximation in Example 5 is, but in Section 7.7 we will learn a method for estimating the error involved in using the Midpoint Rule. At that time we will discuss other methods for approximating definite integrals.

If we apply the Midpoint Rule to the integral in Example 2, we get the picture in Figure 12. The approximation $M_{40} \approx -6.7563$ is much closer to the true value -6.75 than the right endpoint approximation, $R_{40} \approx -6.3998$, shown in Figure 7.

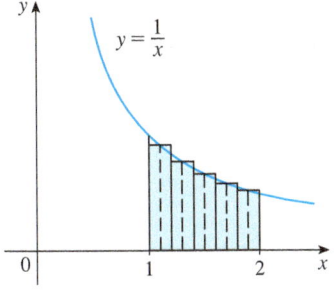

FIGURE 11

TEC In Visual 5.2 you can compare left, right, and midpoint approximations to the integral in Example 2 for different values of n.

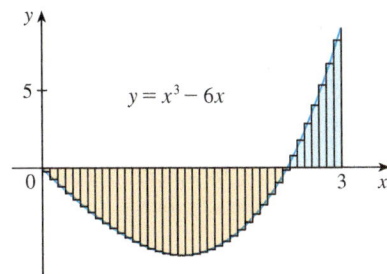

FIGURE 12
$M_{40} \approx -6.7563$

Properties of the Definite Integral

When we defined the definite integral $\int_a^b f(x)\, dx$, we implicitly assumed that $a < b$. But the definition as a limit of Riemann sums makes sense even if $a > b$. Notice that if we reverse a and b, then Δx changes from $(b - a)/n$ to $(a - b)/n$. Therefore

$$\int_b^a f(x)\, dx = -\int_a^b f(x)\, dx$$

If $a = b$, then $\Delta x = 0$ and so

$$\int_a^a f(x)\, dx = 0$$

We now develop some basic properties of integrals that will help us to evaluate integrals in a simple manner. We assume that f and g are continuous functions.

Properties of the Integral

1. $\int_a^b c\, dx = c(b - a)$, where c is any constant

2. $\int_a^b [f(x) + g(x)]\, dx = \int_a^b f(x)\, dx + \int_a^b g(x)\, dx$

3. $\int_a^b cf(x)\, dx = c \int_a^b f(x)\, dx$, where c is any constant

4. $\int_a^b [f(x) - g(x)]\, dx = \int_a^b f(x)\, dx - \int_a^b g(x)\, dx$

Property 1 says that the integral of a constant function $f(x) = c$ is the constant times the length of the interval. If $c > 0$ and $a < b$, this is to be expected because $c(b - a)$ is the area of the shaded rectangle in Figure 13.

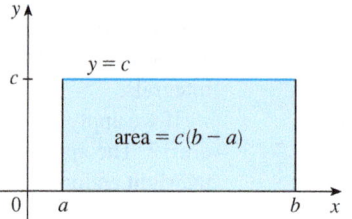

FIGURE 13
$\int_a^b c\, dx = c(b - a)$

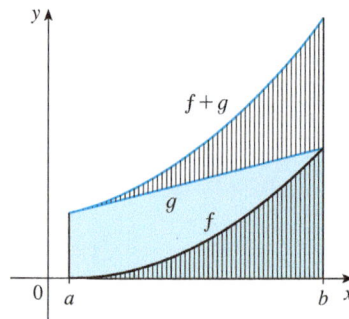

FIGURE 14
$\int_a^b [f(x) + g(x)]\, dx =$
$\int_a^b f(x)\, dx + \int_a^b g(x)\, dx$

Property 2 says that the integral of a sum is the sum of the integrals. For positive functions it says that the area under $f + g$ is the area under f plus the area under g. Figure 14 helps us understand why this is true: In view of how graphical addition works, the corresponding vertical line segments have equal height.

In general, Property 2 follows from Theorem 4 and the fact that the limit of a sum is the sum of the limits:

$$\int_a^b [f(x) + g(x)]\, dx = \lim_{n \to \infty} \sum_{i=1}^{n} [f(x_i) + g(x_i)]\, \Delta x$$

$$= \lim_{n \to \infty} \left[\sum_{i=1}^{n} f(x_i)\, \Delta x + \sum_{i=1}^{n} g(x_i)\, \Delta x \right]$$

$$= \lim_{n \to \infty} \sum_{i=1}^{n} f(x_i)\, \Delta x + \lim_{n \to \infty} \sum_{i=1}^{n} g(x_i)\, \Delta x$$

$$= \int_a^b f(x)\, dx + \int_a^b g(x)\, dx$$

Property 3 seems intuitively reasonable because we know that multiplying a function by a positive number c stretches or shrinks its graph vertically by a factor of c. So it stretches or shrinks each approximating rectangle by a factor c and therefore it has the effect of multiplying the area by c.

Property 3 can be proved in a similar manner and says that the integral of a constant times a function is the constant times the integral of the function. In other words, a constant (but *only* a constant) can be taken in front of an integral sign. Property 4 is proved by writing $f - g = f + (-g)$ and using Properties 2 and 3 with $c = -1$.

EXAMPLE 6 Use the properties of integrals to evaluate $\int_0^1 (4 + 3x^2)\, dx$.

SOLUTION Using Properties 2 and 3 of integrals, we have

$$\int_0^1 (4 + 3x^2)\, dx = \int_0^1 4\, dx + \int_0^1 3x^2\, dx = \int_0^1 4\, dx + 3\int_0^1 x^2\, dx$$

We know from Property 1 that

$$\int_0^1 4\, dx = 4(1 - 0) = 4$$

and we found in Example 5.1.2 that $\int_0^1 x^2\, dx = \tfrac{1}{3}$. So

$$\int_0^1 (4 + 3x^2)\, dx = \int_0^1 4\, dx + 3\int_0^1 x^2\, dx$$

$$= 4 + 3 \cdot \tfrac{1}{3} = 5 \quad\blacksquare$$

The next property tells us how to combine integrals of the same function over adjacent intervals.

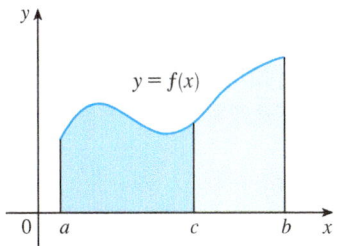

FIGURE 15

5. $$\int_a^c f(x)\,dx + \int_c^b f(x)\,dx = \int_a^b f(x)\,dx$$

This is not easy to prove in general, but for the case where $f(x) \geq 0$ and $a < c < b$ Property 5 can be seen from the geometric interpretation in Figure 15: the area under $y = f(x)$ from a to c plus the area from c to b is equal to the total area from a to b.

EXAMPLE 7 If it is known that $\int_0^{10} f(x)\,dx = 17$ and $\int_0^8 f(x)\,dx = 12$, find $\int_8^{10} f(x)\,dx$.

SOLUTION By Property 5, we have

$$\int_0^8 f(x)\,dx + \int_8^{10} f(x)\,dx = \int_0^{10} f(x)\,dx$$

so

$$\int_8^{10} f(x)\,dx = \int_0^{10} f(x)\,dx - \int_0^8 f(x)\,dx = 17 - 12 = 5$$

Properties 1–5 are true whether $a < b$, $a = b$, or $a > b$. The following properties, in which we compare sizes of functions and sizes of integrals, are true only if $a \leq b$.

Comparison Properties of the Integral

6. If $f(x) \geq 0$ for $a \leq x \leq b$, then $\int_a^b f(x)\,dx \geq 0$.

7. If $f(x) \geq g(x)$ for $a \leq x \leq b$, then $\int_a^b f(x)\,dx \geq \int_a^b g(x)\,dx$.

8. If $m \leq f(x) \leq M$ for $a \leq x \leq b$, then
$$m(b - a) \leq \int_a^b f(x)\,dx \leq M(b - a)$$

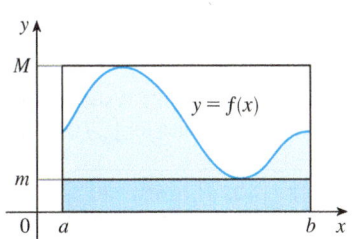

FIGURE 16

If $f(x) \geq 0$, then $\int_a^b f(x)\,dx$ represents the area under the graph of f, so the geometric interpretation of Property 6 is simply that areas are positive. (It also follows directly from the definition because all the quantities involved are positive.) Property 7 says that a bigger function has a bigger integral. It follows from Properties 6 and 4 because $f - g \geq 0$.

Property 8 is illustrated by Figure 16 for the case where $f(x) \geq 0$. If f is continuous, we could take m and M to be the absolute minimum and maximum values of f on the interval $[a, b]$. In this case Property 8 says that the area under the graph of f is greater than the area of the rectangle with height m and less than the area of the rectangle with height M.

PROOF OF PROPERTY 8 Since $m \leq f(x) \leq M$, Property 7 gives

$$\int_a^b m\,dx \leq \int_a^b f(x)\,dx \leq \int_a^b M\,dx$$

Using Property 1 to evaluate the integrals on the left and right sides, we obtain

$$m(b - a) \leq \int_a^b f(x)\,dx \leq M(b - a)$$

Property 8 is useful when all we want is a rough estimate of the size of an integral without going to the bother of using the Midpoint Rule.

EXAMPLE 8 Use Property 8 to estimate $\int_0^1 e^{-x^2}\,dx$.

SOLUTION Because $f(x) = e^{-x^2}$ is a decreasing function on $[0, 1]$, its absolute maximum value is $M = f(0) = 1$ and its absolute minimum value is $m = f(1) = e^{-1}$. Thus, by Property 8,

$$e^{-1}(1 - 0) \le \int_0^1 e^{-x^2}\,dx \le 1(1 - 0)$$

or

$$e^{-1} \le \int_0^1 e^{-x^2}\,dx \le 1$$

Since $e^{-1} \approx 0.3679$, we can write

$$0.367 \le \int_0^1 e^{-x^2}\,dx \le 1$$

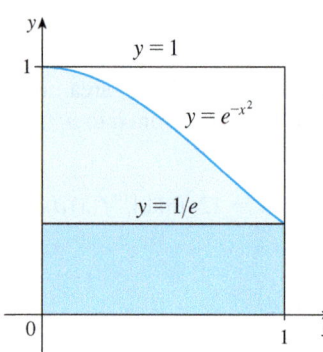

FIGURE 17

The result of Example 8 is illustrated in Figure 17. The integral is greater than the area of the lower rectangle and less than the area of the square.

5.2 EXERCISES

1. Evaluate the Riemann sum for $f(x) = x - 1$, $-6 \le x \le 4$, with five subintervals, taking the sample points to be right endpoints. Explain, with the aid of a diagram, what the Riemann sum represents.

2. If
$$f(x) = \cos x \qquad 0 \le x \le 3\pi/4$$
evaluate the Riemann sum with $n = 6$, taking the sample points to be left endpoints. (Give your answer correct to six decimal places.) What does the Riemann sum represent? Illustrate with a diagram.

3. If $f(x) = x^2 - 4$, $0 \le x \le 3$, find the Riemann sum with $n = 6$, taking the sample points to be midpoints. What does the Riemann sum represent? Illustrate with a diagram.

4. (a) Find the Riemann sum for $f(x) = 1/x$, $1 \le x \le 2$, with four terms, taking the sample points to be right endpoints. (Give your answer correct to six decimal places.) Explain what the Riemann sum represents with the aid of a sketch.
   (b) Repeat part (a) with midpoints as the sample points.

5. The graph of a function f is given. Estimate $\int_0^{10} f(x)\,dx$ using five subintervals with (a) right endpoints, (b) left endpoints, and (c) midpoints.

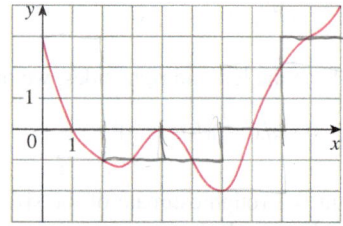

6. The graph of g is shown. Estimate $\int_{-2}^4 g(x)\,dx$ with six subintervals using (a) right endpoints, (b) left endpoints, and (c) midpoints.

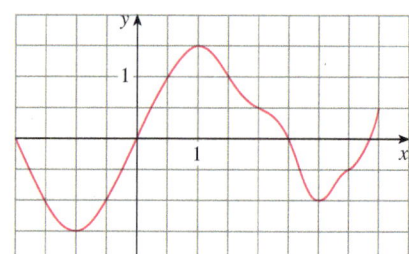

7. A table of values of an increasing function f is shown. Use the table to find lower and upper estimates for $\int_{10}^{30} f(x)\,dx$.

x	10	14	18	22	26	30
$f(x)$	-12	-6	-2	1	3	8

8. The table gives the values of a function obtained from an experiment. Use them to estimate $\int_3^9 f(x)\,dx$ using three equal subintervals with (a) right endpoints, (b) left endpoints, and (c) midpoints. If the function is known to be an increasing function, can you say whether your estimates are less than or greater than the exact value of the integral?

x	3	4	5	6	7	8	9
$f(x)$	-3.4	-2.1	-0.6	0.3	0.9	1.4	1.8

9–12 Use the Midpoint Rule with the given value of n to approximate the integral. Round the answer to four decimal places.

9. $\int_0^8 \sin \sqrt{x}\, dx$, $n = 4$
10. $\int_0^1 \sqrt{x^3 + 1}\, dx$, $n = 5$

11. $\int_0^2 \dfrac{x}{x+1}\, dx$, $n = 5$
12. $\int_0^\pi x \sin^2 x\, dx$, $n = 4$

CAS 13. If you have a CAS that evaluates midpoint approximations and graphs the corresponding rectangles (use `RiemannSum` or `middlesum` and `middlebox` commands in Maple), check the answer to Exercise 11 and illustrate with a graph. Then repeat with $n = 10$ and $n = 20$.

14. With a programmable calculator or computer (see the instructions for Exercise 5.1.9), compute the left and right Riemann sums for the function $f(x) = x/(x + 1)$ on the interval $[0, 2]$ with $n = 100$. Explain why these estimates show that

$$0.8946 < \int_0^2 \dfrac{x}{x+1}\, dx < 0.9081$$

15. Use a calculator or computer to make a table of values of right Riemann sums R_n for the integral $\int_0^\pi \sin x\, dx$ with $n = 5, 10, 50,$ and 100. What value do these numbers appear to be approaching?

16. Use a calculator or computer to make a table of values of left and right Riemann sums L_n and R_n for the integral $\int_0^2 e^{-x^2}\, dx$ with $n = 5, 10, 50,$ and 100. Between what two numbers must the value of the integral lie? Can you make a similar statement for the integral $\int_{-1}^2 e^{-x^2}\, dx$? Explain.

17–20 Express the limit as a definite integral on the given interval.

17. $\displaystyle\lim_{n \to \infty} \sum_{i=1}^n \dfrac{e^{x_i}}{1 + x_i} \Delta x$, $[0, 1]$

18. $\displaystyle\lim_{n \to \infty} \sum_{i=1}^n x_i \sqrt{1 + x_i^3}\, \Delta x$, $[2, 5]$

19. $\displaystyle\lim_{n \to \infty} \sum_{i=1}^n [5(x_i^*)^3 - 4x_i^*]\, \Delta x$, $[2, 7]$

20. $\displaystyle\lim_{n \to \infty} \sum_{i=1}^n \dfrac{x_i^*}{(x_i^*)^2 + 4}\, \Delta x$, $[1, 3]$

21–25 Use the form of the definition of the integral given in Theorem 4 to evaluate the integral.

21. $\int_2^5 (4 - 2x)\, dx$
22. $\int_1^4 (x^2 - 4x + 2)\, dx$

23. $\int_{-2}^0 (x^2 + x)\, dx$
24. $\int_0^2 (2x - x^3)\, dx$

25. $\int_0^1 (x^3 - 3x^2)\, dx$

26. (a) Find an approximation to the integral $\int_0^4 (x^2 - 3x)\, dx$ using a Riemann sum with right endpoints and $n = 8$.
(b) Draw a diagram like Figure 3 to illustrate the approximation in part (a).
(c) Use Theorem 4 to evaluate $\int_0^4 (x^2 - 3x)\, dx$.
(d) Interpret the integral in part (c) as a difference of areas and illustrate with a diagram like Figure 4.

27. Prove that $\int_a^b x\, dx = \dfrac{b^2 - a^2}{2}$.

28. Prove that $\int_a^b x^2\, dx = \dfrac{b^3 - a^3}{3}$.

29–30 Express the integral as a limit of Riemann sums. Do not evaluate the limit.

29. $\int_1^3 \sqrt{4 + x^2}\, dx$
30. $\int_2^5 \left(x^2 + \dfrac{1}{x}\right) dx$

CAS 31–32 Express the integral as a limit of sums. Then evaluate, using a computer algebra system to find both the sum and the limit.

31. $\int_0^\pi \sin 5x\, dx$
32. $\int_2^{10} x^6\, dx$

33. The graph of f is shown. Evaluate each integral by interpreting it in terms of areas.

(a) $\int_0^2 f(x)\, dx$
(b) $\int_0^5 f(x)\, dx$

(c) $\int_5^7 f(x)\, dx$
(d) $\int_0^9 f(x)\, dx$

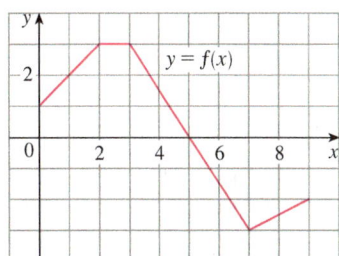

34. The graph of g consists of two straight lines and a semicircle. Use it to evaluate each integral.

(a) $\int_0^2 g(x)\, dx$
(b) $\int_2^6 g(x)\, dx$
(c) $\int_0^7 g(x)\, dx$

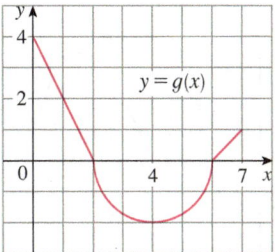

35–40 Evaluate the integral by interpreting it in terms of areas.

35. $\int_{-1}^{2} (1 - x)\, dx$
36. $\int_{0}^{9} \left(\tfrac{1}{3}x - 2\right) dx$
37. $\int_{-3}^{0} \left(1 + \sqrt{9 - x^2}\right) dx$
38. $\int_{-5}^{5} \left(x - \sqrt{25 - x^2}\right) dx$
39. $\int_{-4}^{3} \left|\tfrac{1}{2}x\right| dx$
40. $\int_{0}^{1} |2x - 1|\, dx$

41. Evaluate $\int_{1}^{1} \sqrt{1 + x^4}\, dx$.

42. Given that $\int_{0}^{\pi} \sin^4 x\, dx = \tfrac{3}{8}\pi$, what is $\int_{\pi}^{0} \sin^4 \theta\, d\theta$?

43. In Example 5.1.2 we showed that $\int_{0}^{1} x^2\, dx = \tfrac{1}{3}$. Use this fact and the properties of integrals to evaluate $\int_{0}^{1} (5 - 6x^2)\, dx$.

44. Use the properties of integrals and the result of Example 3 to evaluate $\int_{1}^{3} (2e^x - 1)\, dx$.

45. Use the result of Example 3 to evaluate $\int_{1}^{3} e^{x+2}\, dx$.

46. Use the result of Exercise 27 and the fact that $\int_{0}^{\pi/2} \cos x\, dx = 1$ (from Exercise 5.1.31), together with the properties of integrals, to evaluate $\int_{0}^{\pi/2} (2\cos x - 5x)\, dx$.

47. Write as a single integral in the form $\int_{a}^{b} f(x)\, dx$:
$$\int_{-2}^{2} f(x)\, dx + \int_{2}^{5} f(x)\, dx - \int_{-2}^{-1} f(x)\, dx$$

48. If $\int_{2}^{8} f(x)\, dx = 7.3$ and $\int_{2}^{4} f(x)\, dx = 5.9$, find $\int_{4}^{8} f(x)\, dx$.

49. If $\int_{0}^{9} f(x)\, dx = 37$ and $\int_{0}^{9} g(x)\, dx = 16$, find
$$\int_{0}^{9} [2f(x) + 3g(x)]\, dx$$

50. Find $\int_{0}^{5} f(x)\, dx$ if
$$f(x) = \begin{cases} 3 & \text{for } x < 3 \\ x & \text{for } x \geq 3 \end{cases}$$

51. For the function f whose graph is shown, list the following quantities in increasing order, from smallest to largest, and explain your reasoning.

(A) $\int_{0}^{8} f(x)\, dx$ (B) $\int_{0}^{3} f(x)\, dx$ (C) $\int_{3}^{8} f(x)\, dx$
(D) $\int_{4}^{8} f(x)\, dx$ (E) $f'(1)$

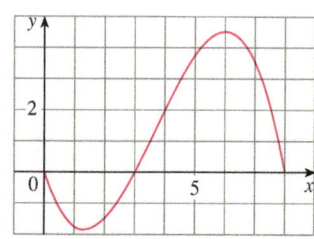

52. If $F(x) = \int_{2}^{x} f(t)\, dt$, where f is the function whose graph is given, which of the following values is largest?
(A) $F(0)$ (B) $F(1)$ (C) $F(2)$
(D) $F(3)$ (E) $F(4)$

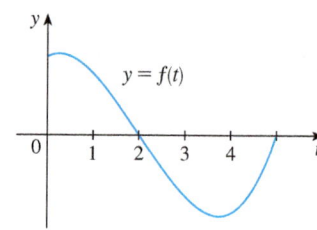

53. Each of the regions A, B, and C bounded by the graph of f and the x-axis has area 3. Find the value of
$$\int_{-4}^{2} [f(x) + 2x + 5]\, dx$$

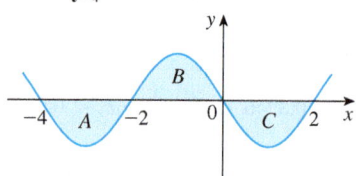

54. Suppose f has absolute minimum value m and absolute maximum value M. Between what two values must $\int_{0}^{2} f(x)\, dx$ lie? Which property of integrals allows you to make your conclusion?

55–58 Use the properties of integrals to verify the inequality without evaluating the integrals.

55. $\int_{0}^{4} (x^2 - 4x + 4)\, dx \geq 0$

56. $\int_{0}^{1} \sqrt{1 + x^2}\, dx \leq \int_{0}^{1} \sqrt{1 + x}\, dx$

57. $2 \leq \int_{-1}^{1} \sqrt{1 + x^2}\, dx \leq 2\sqrt{2}$

58. $\dfrac{\pi}{12} \leq \int_{\pi/6}^{\pi/3} \sin x\, dx \leq \dfrac{\sqrt{3}\,\pi}{12}$

59–64 Use Property 8 to estimate the value of the integral.

59. $\int_{0}^{1} x^3\, dx$
60. $\int_{0}^{3} \dfrac{1}{x + 4}\, dx$
61. $\int_{\pi/4}^{\pi/3} \tan x\, dx$
62. $\int_{0}^{2} (x^3 - 3x + 3)\, dx$
63. $\int_{0}^{2} xe^{-x}\, dx$
64. $\int_{\pi}^{2\pi} (x - 2\sin x)\, dx$

65–66 Use properties of integrals, together with Exercises 27 and 28, to prove the inequality.

65. $\int_{1}^{3} \sqrt{x^4 + 1}\, dx \geq \dfrac{26}{3}$
66. $\int_{0}^{\pi/2} x \sin x\, dx \leq \dfrac{\pi^2}{8}$

67. Which of the integrals $\int_1^2 \arctan x\, dx$, $\int_1^2 \arctan \sqrt{x}\, dx$, and $\int_1^2 \arctan(\sin x)\, dx$ has the largest value? Why?

68. Which of the integrals $\int_0^{0.5} \cos(x^2)\, dx$, $\int_0^{0.5} \cos \sqrt{x}\, dx$ is larger? Why?

69. Prove Property 3 of integrals.

70. (a) If f is continuous on $[a, b]$, show that
$$\left| \int_a^b f(x)\, dx \right| \leq \int_a^b |f(x)|\, dx$$

[Hint: $-|f(x)| \leq f(x) \leq |f(x)|$.]

(b) Use the result of part (a) to show that
$$\left| \int_0^{2\pi} f(x) \sin 2x\, dx \right| \leq \int_0^{2\pi} |f(x)|\, dx$$

71. Let $f(x) = 0$ if x is any rational number and $f(x) = 1$ if x is any irrational number. Show that f is not integrable on $[0, 1]$.

72. Let $f(0) = 0$ and $f(x) = 1/x$ if $0 < x \leq 1$. Show that f is not integrable on $[0, 1]$. [Hint: Show that the first term in the Riemann sum, $f(x_i^*)\, \Delta x$, can be made arbitrarily large.]

73–74 Express the limit as a definite integral.

73. $\displaystyle\lim_{n \to \infty} \sum_{i=1}^{n} \frac{i^4}{n^5}$ [Hint: Consider $f(x) = x^4$.]

74. $\displaystyle\lim_{n \to \infty} \frac{1}{n} \sum_{i=1}^{n} \frac{1}{1 + (i/n)^2}$

75. Find $\int_1^2 x^{-2}\, dx$. Hint: Choose x_i^* to be the geometric mean of x_{i-1} and x_i (that is, $x_i^* = \sqrt{x_{i-1} x_i}$) and use the identity
$$\frac{1}{m(m+1)} = \frac{1}{m} - \frac{1}{m+1}$$

DISCOVERY PROJECT AREA FUNCTIONS

1. (a) Draw the line $y = 2t + 1$ and use geometry to find the area under this line, above the t-axis, and between the vertical lines $t = 1$ and $t = 3$.

(b) If $x > 1$, let $A(x)$ be the area of the region that lies under the line $y = 2t + 1$ between $t = 1$ and $t = x$. Sketch this region and use geometry to find an expression for $A(x)$.

(c) Differentiate the area function $A(x)$. What do you notice?

2. (a) If $x \geq -1$, let
$$A(x) = \int_{-1}^{x} (1 + t^2)\, dt$$

$A(x)$ represents the area of a region. Sketch that region.

(b) Use the result of Exercise 5.2.28 to find an expression for $A(x)$.

(c) Find $A'(x)$. What do you notice?

(d) If $x \geq -1$ and h is a small positive number, then $A(x + h) - A(x)$ represents the area of a region. Describe and sketch the region.

(e) Draw a rectangle that approximates the region in part (d). By comparing the areas of these two regions, show that
$$\frac{A(x + h) - A(x)}{h} \approx 1 + x^2$$

(f) Use part (e) to give an intuitive explanation for the result of part (c).

3. (a) Draw the graph of the function $f(x) = \cos(x^2)$ in the viewing rectangle $[0, 2]$ by $[-1.25, 1.25]$.

(b) If we define a new function g by
$$g(x) = \int_0^x \cos(t^2)\, dt$$

then $g(x)$ is the area under the graph of f from 0 to x [until $f(x)$ becomes negative, at which point $g(x)$ becomes a difference of areas]. Use part (a) to determine the value

of x at which $g(x)$ starts to decrease. [Unlike the integral in Problem 2, it is impossible to evaluate the integral defining g to obtain an explicit expression for $g(x)$.]
(c) Use the integration command on your calculator or computer to estimate $g(0.2)$, $g(0.4)$, $g(0.6)$, ..., $g(1.8)$, $g(2)$. Then use these values to sketch a graph of g.
(d) Use your graph of g from part (c) to sketch the graph of g' using the interpretation of $g'(x)$ as the slope of a tangent line. How does the graph of g' compare with the graph of f?

4. Suppose f is a continuous function on the interval $[a, b]$ and we define a new function g by the equation
$$g(x) = \int_a^x f(t)\, dt$$
Based on your results in Problems 1–3, conjecture an expression for $g'(x)$.

5.3 The Fundamental Theorem of Calculus

The Fundamental Theorem of Calculus is appropriately named because it establishes a connection between the two branches of calculus: differential calculus and integral calculus. Differential calculus arose from the tangent problem, whereas integral calculus arose from a seemingly unrelated problem, the area problem. Newton's mentor at Cambridge, Isaac Barrow (1630–1677), discovered that these two problems are actually closely related. In fact, he realized that differentiation and integration are inverse processes. The Fundamental Theorem of Calculus gives the precise inverse relationship between the derivative and the integral. It was Newton and Leibniz who exploited this relationship and used it to develop calculus into a systematic mathematical method. In particular, they saw that the Fundamental Theorem enabled them to compute areas and integrals very easily without having to compute them as limits of sums as we did in Sections 5.1 and 5.2.

The first part of the Fundamental Theorem deals with functions defined by an equation of the form

$$\boxed{1} \qquad g(x) = \int_a^x f(t)\, dt$$

where f is a continuous function on $[a, b]$ and x varies between a and b. Observe that g depends only on x, which appears as the variable upper limit in the integral. If x is a fixed number, then the integral $\int_a^x f(t)\, dt$ is a definite number. If we then let x vary, the number $\int_a^x f(t)\, dt$ also varies and defines a function of x denoted by $g(x)$.

If f happens to be a positive function, then $g(x)$ can be interpreted as the area under the graph of f from a to x, where x can vary from a to b. (Think of g as the "area so far" function; see Figure 1.)

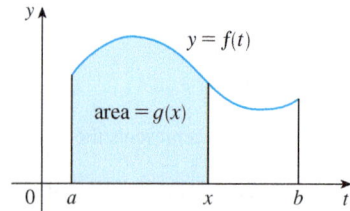

FIGURE 1

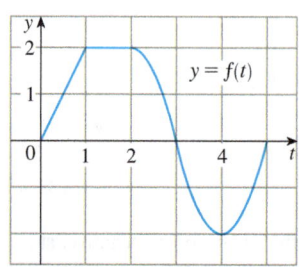

FIGURE 2

EXAMPLE 1 If f is the function whose graph is shown in Figure 2 and $g(x) = \int_0^x f(t)\, dt$, find the values of $g(0)$, $g(1)$, $g(2)$, $g(3)$, $g(4)$, and $g(5)$. Then sketch a rough graph of g.

SOLUTION First we notice that $g(0) = \int_0^0 f(t)\, dt = 0$. From Figure 3 we see that $g(1)$ is the area of a triangle:

$$g(1) = \int_0^1 f(t)\, dt = \tfrac{1}{2}(1 \cdot 2) = 1$$

To find $g(2)$ we add to $g(1)$ the area of a rectangle:

$$g(2) = \int_0^2 f(t)\, dt = \int_0^1 f(t)\, dt + \int_1^2 f(t)\, dt = 1 + (1 \cdot 2) = 3$$

We estimate that the area under f from 2 to 3 is about 1.3, so

$$g(3) = g(2) + \int_2^3 f(t)\, dt \approx 3 + 1.3 = 4.3$$

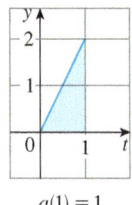
$g(1) = 1$

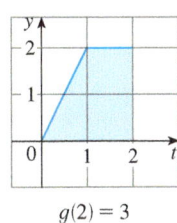
$g(2) = 3$

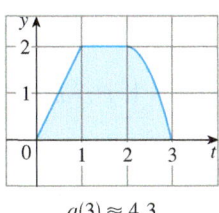
$g(3) \approx 4.3$

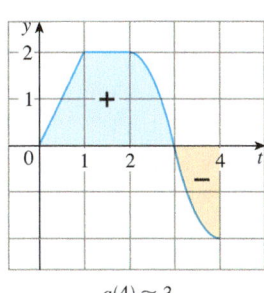
$g(4) \approx 3$

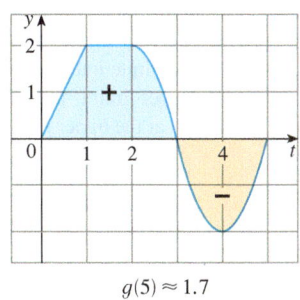
$g(5) \approx 1.7$

FIGURE 3

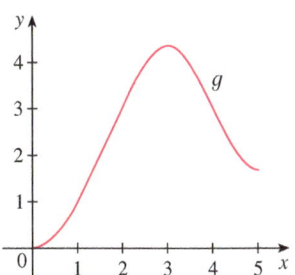

FIGURE 4
$g(x) = \int_0^x f(t)\, dt$

For $t > 3$, $f(t)$ is negative and so we start subtracting areas:

$$g(4) = g(3) + \int_3^4 f(t)\, dt \approx 4.3 + (-1.3) = 3.0$$

$$g(5) = g(4) + \int_4^5 f(t)\, dt \approx 3 + (-1.3) = 1.7$$

We use these values to sketch the graph of g in Figure 4. Notice that, because $f(t)$ is positive for $t < 3$, we keep adding area for $t < 3$ and so g is increasing up to $x = 3$, where it attains a maximum value. For $x > 3$, g decreases because $f(t)$ is negative. ∎

If we take $f(t) = t$ and $a = 0$, then, using Exercise 5.2.27, we have

$$g(x) = \int_0^x t\, dt = \frac{x^2}{2}$$

Notice that $g'(x) = x$, that is, $g' = f$. In other words, if g is defined as the integral of f by Equation 1, then g turns out to be an antiderivative of f, at least in this case. And if we sketch the derivative of the function g shown in Figure 4 by estimating slopes of tangents, we get a graph like that of f in Figure 2. So we suspect that $g' = f$ in Example 1 too.

To see why this might be generally true we consider any continuous function f with $f(x) \geq 0$. Then $g(x) = \int_a^x f(t)\, dt$ can be interpreted as the area under the graph of f from a to x, as in Figure 1.

In order to compute $g'(x)$ from the definition of a derivative we first observe that, for $h > 0$, $g(x + h) - g(x)$ is obtained by subtracting areas, so it is the area under the graph of f from x to $x + h$ (the blue area in Figure 5). For small h you can see from the figure that this area is approximately equal to the area of the rectangle with height $f(x)$ and width h:

$$g(x + h) - g(x) \approx hf(x)$$

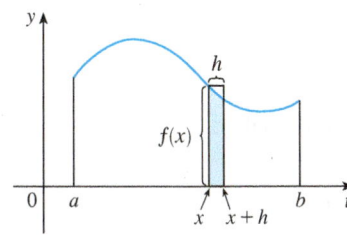

FIGURE 5

so

$$\frac{g(x + h) - g(x)}{h} \approx f(x)$$

Intuitively, we therefore expect that

$$g'(x) = \lim_{h \to 0} \frac{g(x+h) - g(x)}{h} = f(x)$$

The fact that this is true, even when f is not necessarily positive, is the first part of the Fundamental Theorem of Calculus.

We abbreviate the name of this theorem as FTC1. In words, it says that the derivative of a definite integral with respect to its upper limit is the integrand evaluated at the upper limit.

The Fundamental Theorem of Calculus, Part 1 If f is continuous on $[a, b]$, then the function g defined by

$$g(x) = \int_a^x f(t)\,dt \qquad a \leq x \leq b$$

is continuous on $[a, b]$ and differentiable on (a, b), and $g'(x) = f(x)$.

PROOF If x and $x + h$ are in (a, b), then

$$g(x+h) - g(x) = \int_a^{x+h} f(t)\,dt - \int_a^x f(t)\,dt$$

$$= \left(\int_a^x f(t)\,dt + \int_x^{x+h} f(t)\,dt \right) - \int_a^x f(t)\,dt \qquad \text{(by Property 5)}$$

$$= \int_x^{x+h} f(t)\,dt$$

and so, for $h \neq 0$,

$$\boxed{2} \qquad \frac{g(x+h) - g(x)}{h} = \frac{1}{h} \int_x^{x+h} f(t)\,dt$$

For now let's assume that $h > 0$. Since f is continuous on $[x, x + h]$, the Extreme Value Theorem says that there are numbers u and v in $[x, x + h]$ such that $f(u) = m$ and $f(v) = M$, where m and M are the absolute minimum and maximum values of f on $[x, x + h]$. (See Figure 6.)

By Property 8 of integrals, we have

$$mh \leq \int_x^{x+h} f(t)\,dt \leq Mh$$

that is,

$$f(u)h \leq \int_x^{x+h} f(t)\,dt \leq f(v)h$$

Since $h > 0$, we can divide this inequality by h:

$$f(u) \leq \frac{1}{h} \int_x^{x+h} f(t)\,dt \leq f(v)$$

Now we use Equation 2 to replace the middle part of this inequality:

$$\boxed{3} \qquad f(u) \leq \frac{g(x+h) - g(x)}{h} \leq f(v)$$

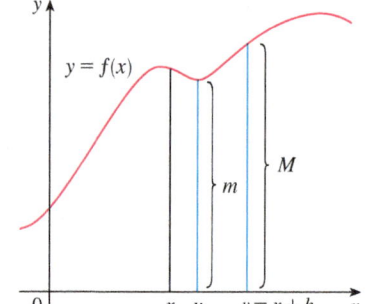

FIGURE 6

TEC Module 5.3 provides visual evidence for FTC1.

Inequality 3 can be proved in a similar manner for the case where $h < 0$. (See Exercise 77.)

Now we let $h \to 0$. Then $u \to x$ and $v \to x$, since u and v lie between x and $x + h$. Therefore

$$\lim_{h \to 0} f(u) = \lim_{u \to x} f(u) = f(x) \quad \text{and} \quad \lim_{h \to 0} f(v) = \lim_{v \to x} f(v) = f(x)$$

because f is continuous at x. We conclude, from (3) and the Squeeze Theorem, that

$$\boxed{4} \qquad g'(x) = \lim_{h \to 0} \frac{g(x + h) - g(x)}{h} = f(x)$$

If $x = a$ or b, then Equation 4 can be interpreted as a one-sided limit. Then Theorem 2.8.4 (modified for one-sided limits) shows that g is continuous on $[a, b]$. ■

Using Leibniz notation for derivatives, we can write FTC1 as

$$\boxed{5} \qquad \frac{d}{dx} \int_a^x f(t) \, dt = f(x)$$

when f is continuous. Roughly speaking, Equation 5 says that if we first integrate f and then differentiate the result, we get back to the original function f.

EXAMPLE 2 Find the derivative of the function $g(x) = \int_0^x \sqrt{1 + t^2} \, dt$.

SOLUTION Since $f(t) = \sqrt{1 + t^2}$ is continuous, Part 1 of the Fundamental Theorem of Calculus gives

$$g'(x) = \sqrt{1 + x^2}$$ ■

EXAMPLE 3 Although a formula of the form $g(x) = \int_a^x f(t) \, dt$ may seem like a strange way of defining a function, books on physics, chemistry, and statistics are full of such functions. For instance, the **Fresnel function**

$$S(x) = \int_0^x \sin(\pi t^2/2) \, dt$$

is named after the French physicist Augustin Fresnel (1788–1827), who is famous for his works in optics. This function first appeared in Fresnel's theory of the diffraction of light waves, but more recently it has been applied to the design of highways.

Part 1 of the Fundamental Theorem tells us how to differentiate the Fresnel function:

$$S'(x) = \sin(\pi x^2/2)$$

This means that we can apply all the methods of differential calculus to analyze S (see Exercise 71).

Figure 7 shows the graphs of $f(x) = \sin(\pi x^2/2)$ and the Fresnel function $S(x) = \int_0^x f(t) \, dt$. A computer was used to graph S by computing the value of this integral for many values of x. It does indeed look as if $S(x)$ is the area under the graph of f from 0 to x [until $x \approx 1.4$ when $S(x)$ becomes a difference of areas]. Figure 8 shows a larger part of the graph of S.

If we now start with the graph of S in Figure 7 and think about what its derivative should look like, it seems reasonable that $S'(x) = f(x)$. [For instance, S is increasing when $f(x) > 0$ and decreasing when $f(x) < 0$.] So this gives a visual confirmation of Part 1 of the Fundamental Theorem of Calculus. ■

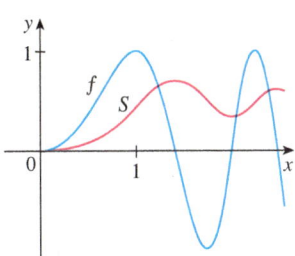

FIGURE 7
$f(x) = \sin(\pi x^2/2)$
$S(x) = \int_0^x \sin(\pi t^2/2) \, dt$

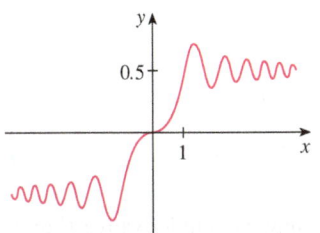

FIGURE 8
The Fresnel function
$S(x) = \int_0^x \sin(\pi t^2/2) \, dt$

EXAMPLE 4 Find $\dfrac{d}{dx} \displaystyle\int_1^{x^4} \sec t \, dt$.

SOLUTION Here we have to be careful to use the Chain Rule in conjunction with FTC1. Let $u = x^4$. Then

$$\begin{aligned}\frac{d}{dx} \int_1^{x^4} \sec t \, dt &= \frac{d}{dx} \int_1^{u} \sec t \, dt \\ &= \frac{d}{du}\left[\int_1^{u} \sec t \, dt\right] \frac{du}{dx} &\text{(by the Chain Rule)} \\ &= \sec u \, \frac{du}{dx} &\text{(by FTC1)} \\ &= \sec(x^4) \cdot 4x^3\end{aligned}$$

In Section 5.2 we computed integrals from the definition as a limit of Riemann sums and we saw that this procedure is sometimes long and difficult. The second part of the Fundamental Theorem of Calculus, which follows easily from the first part, provides us with a much simpler method for the evaluation of integrals.

We abbreviate this theorem as FTC2.

> **The Fundamental Theorem of Calculus, Part 2** If f is continuous on $[a, b]$, then
>
> $$\int_a^b f(x) \, dx = F(b) - F(a)$$
>
> where F is any antiderivative of f, that is, a function such that $F' = f$.

PROOF Let $g(x) = \int_a^x f(t) \, dt$. We know from Part 1 that $g'(x) = f(x)$; that is, g is an antiderivative of f. If F is any other antiderivative of f on $[a, b]$, then we know from Corollary 4.2.7 that F and g differ by a constant:

$$\boxed{6} \qquad F(x) = g(x) + C$$

for $a < x < b$. But both F and g are continuous on $[a, b]$ and so, by taking limits of both sides of Equation 6 (as $x \to a^+$ and $x \to b^-$), we see that it also holds when $x = a$ and $x = b$. So $F(x) = g(x) + C$ for all x in $[a, b]$.

If we put $x = a$ in the formula for $g(x)$, we get

$$g(a) = \int_a^a f(t) \, dt = 0$$

So, using Equation 6 with $x = b$ and $x = a$, we have

$$F(b) - F(a) = [g(b) + C] - [g(a) + C]$$
$$= g(b) - g(a) = g(b) = \int_a^b f(t) \, dt$$

Part 2 of the Fundamental Theorem states that if we know an antiderivative F of f, then we can evaluate $\int_a^b f(x) \, dx$ simply by subtracting the values of F at the endpoints of the interval $[a, b]$. It's very surprising that $\int_a^b f(x) \, dx$, which was defined by a complicated procedure involving all of the values of $f(x)$ for $a \leq x \leq b$, can be found by knowing the values of $F(x)$ at only two points, a and b.

Although the theorem may be surprising at first glance, it becomes plausible if we interpret it in physical terms. If $v(t)$ is the velocity of an object and $s(t)$ is its position at time t, then $v(t) = s'(t)$, so s is an antiderivative of v. In Section 5.1 we considered an object that always moves in the positive direction and made the guess that the area under the velocity curve is equal to the distance traveled. In symbols:

$$\int_a^b v(t)\,dt = s(b) - s(a)$$

That is exactly what FTC2 says in this context.

EXAMPLE 5 Evaluate the integral $\int_1^3 e^x\,dx$.

SOLUTION The function $f(x) = e^x$ is continuous everywhere and we know that an antiderivative is $F(x) = e^x$, so Part 2 of the Fundamental Theorem gives

$$\int_1^3 e^x\,dx = F(3) - F(1) = e^3 - e$$

Compare the calculation in Example 5 with the much harder one in Example 5.2.3.

Notice that FTC2 says we can use *any* antiderivative F of f. So we may as well use the simplest one, namely $F(x) = e^x$, instead of $e^x + 7$ or $e^x + C$. ■

We often use the notation

$$F(x)\Big]_a^b = F(b) - F(a)$$

So the equation of FTC2 can be written as

$$\int_a^b f(x)\,dx = F(x)\Big]_a^b \qquad \text{where} \qquad F' = f$$

Other common notations are $F(x)\big|_a^b$ and $[F(x)]_a^b$.

EXAMPLE 6 Find the area under the parabola $y = x^2$ from 0 to 1.

SOLUTION An antiderivative of $f(x) = x^2$ is $F(x) = \frac{1}{3}x^3$. The required area A is found using Part 2 of the Fundamental Theorem:

In applying the Fundamental Theorem we use a particular antiderivative F of f. It is not necessary to use the most general antiderivative.

$$A = \int_0^1 x^2\,dx = \frac{x^3}{3}\Big]_0^1 = \frac{1^3}{3} - \frac{0^3}{3} = \frac{1}{3}$$

■

If you compare the calculation in Example 6 with the one in Example 5.1.2, you will see that the Fundamental Theorem gives a *much* shorter method.

EXAMPLE 7 Evaluate $\int_3^6 \frac{dx}{x}$.

SOLUTION The given integral is an abbreviation for

$$\int_3^6 \frac{1}{x}\,dx$$

An antiderivative of $f(x) = 1/x$ is $F(x) = \ln|x|$ and, because $3 \leq x \leq 6$, we can write $F(x) = \ln x$. So

$$\int_3^6 \frac{1}{x}\,dx = \ln x\Big]_3^6 = \ln 6 - \ln 3 = \ln \frac{6}{3} = \ln 2$$

■

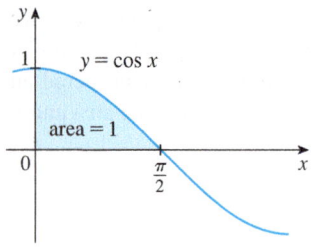

FIGURE 9

EXAMPLE 8 Find the area under the cosine curve from 0 to b, where $0 \leq b \leq \pi/2$.

SOLUTION Since an antiderivative of $f(x) = \cos x$ is $F(x) = \sin x$, we have

$$A = \int_0^b \cos x \, dx = \sin x \Big]_0^b = \sin b - \sin 0 = \sin b$$

In particular, taking $b = \pi/2$, we have proved that the area under the cosine curve from 0 to $\pi/2$ is $\sin(\pi/2) = 1$. (See Figure 9.) ∎

When the French mathematician Gilles de Roberval first found the area under the sine and cosine curves in 1635, this was a very challenging problem that required a great deal of ingenuity. If we didn't have the benefit of the Fundamental Theorem, we would have to compute a difficult limit of sums using obscure trigonometric identities (or a computer algebra system as in Exercise 5.1.31). It was even more difficult for Roberval because the apparatus of limits had not been invented in 1635. But in the 1660s and 1670s, when the Fundamental Theorem was discovered by Barrow and exploited by Newton and Leibniz, such problems became very easy, as you can see from Example 8.

EXAMPLE 9 What is wrong with the following calculation?

$$\int_{-1}^3 \frac{1}{x^2} \, dx = \frac{x^{-1}}{-1}\Big]_{-1}^3 = -\frac{1}{3} - 1 = -\frac{4}{3}$$

SOLUTION To start, we notice that this calculation must be wrong because the answer is negative but $f(x) = 1/x^2 \geq 0$ and Property 6 of integrals says that $\int_a^b f(x) \, dx \geq 0$ when $f \geq 0$. The Fundamental Theorem of Calculus applies to continuous functions. It can't be applied here because $f(x) = 1/x^2$ is not continuous on $[-1, 3]$. In fact, f has an infinite discontinuity at $x = 0$, so

$$\int_{-1}^3 \frac{1}{x^2} \, dx \quad \text{does not exist.}$$
∎

■ Differentiation and Integration as Inverse Processes

We end this section by bringing together the two parts of the Fundamental Theorem.

> **The Fundamental Theorem of Calculus** Suppose f is continuous on $[a, b]$.
>
> **1.** If $g(x) = \int_a^x f(t) \, dt$, then $g'(x) = f(x)$.
>
> **2.** $\int_a^b f(x) \, dx = F(b) - F(a)$, where F is any antiderivative of f, that is, $F' = f$.

We noted that Part 1 can be rewritten as

$$\frac{d}{dx} \int_a^x f(t) \, dt = f(x)$$

which says that if f is integrated and then the result is differentiated, we arrive back at the original function f. Since $F'(x) = f(x)$, Part 2 can be rewritten as

$$\int_a^b F'(x) \, dx = F(b) - F(a)$$

5.3 EXERCISES

1. Explain exactly what is meant by the statement that "differentiation and integration are inverse processes."

2. Let $g(x) = \int_0^x f(t)\, dt$, where f is the function whose graph is shown.
   (a) Evaluate $g(x)$ for $x = 0, 1, 2, 3, 4, 5,$ and 6.
   (b) Estimate $g(7)$.
   (c) Where does g have a maximum value? Where does it have a minimum value?
   (d) Sketch a rough graph of g.

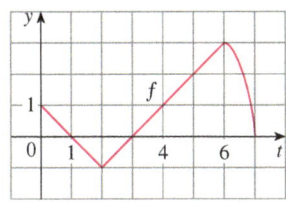

3. Let $g(x) = \int_0^x f(t)\, dt$, where f is the function whose graph is shown.
   (a) Evaluate $g(0), g(1), g(2), g(3),$ and $g(6)$.
   (b) On what interval is g increasing?
   (c) Where does g have a maximum value?
   (d) Sketch a rough graph of g.

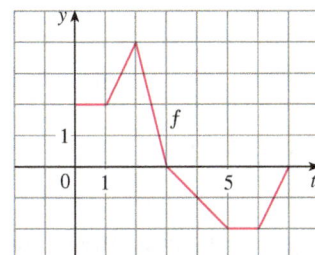

4. Let $g(x) = \int_0^x f(t)\, dt$, where f is the function whose graph is shown.
   (a) Evaluate $g(0)$ and $g(6)$.
   (b) Estimate $g(x)$ for $x = 1, 2, 3, 4,$ and 5.
   (c) On what interval is g increasing?

   (d) Where does g have a maximum value?
   (e) Sketch a rough graph of g.
   (f) Use the graph in part (e) to sketch the graph of $g'(x)$. Compare with the graph of f.

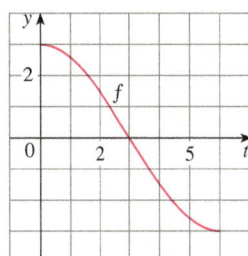

5–6 Sketch the area represented by $g(x)$. Then find $g'(x)$ in two ways: (a) by using Part 1 of the Fundamental Theorem and (b) by evaluating the integral using Part 2 and then differentiating.

5. $g(x) = \int_1^x t^2\, dt$

6. $g(x) = \int_0^x (2 + \sin t)\, dt$

7–18 Use Part 1 of the Fundamental Theorem of Calculus to find the derivative of the function.

7. $g(x) = \int_0^x \sqrt{t + t^3}\, dt$

8. $g(x) = \int_1^x \ln(1 + t^2)\, dt$

9. $g(s) = \int_5^s (t - t^2)^8\, dt$

10. $h(u) = \int_0^u \dfrac{\sqrt{t}}{t + 1}\, dt$

11. $F(x) = \int_x^0 \sqrt{1 + \sec t}\, dt$

$\left[\text{Hint: } \int_x^0 \sqrt{1 + \sec t}\, dt = -\int_0^x \sqrt{1 + \sec t}\, dt\right]$

12. $R(y) = \int_y^2 t^3 \sin t\, dt$

13. $h(x) = \int_1^{e^x} \ln t\, dt$

14. $h(x) = \int_1^{\sqrt{x}} \dfrac{z^2}{z^4 + 1}\, dz$

15. $y = \int_{1}^{3x+2} \dfrac{t}{1+t^3}\,dt$

16. $y = \int_{0}^{x^4} \cos^2\theta\,d\theta$

17. $y = \int_{\sqrt{x}}^{\pi/4} \theta\tan\theta\,d\theta$

18. $y = \int_{\sin x}^{1} \sqrt{1+t^2}\,dt$

19–44 Evaluate the integral.

19. $\int_{1}^{3}(x^2+2x-4)\,dx$

20. $\int_{-1}^{1} x^{100}\,dx$

21. $\int_{0}^{2}\left(\tfrac{4}{5}t^3 - \tfrac{3}{4}t^2 + \tfrac{2}{5}t\right)dt$

22. $\int_{0}^{1}(1-8v^3+16v^7)\,dv$

23. $\int_{1}^{9}\sqrt{x}\,dx$

24. $\int_{1}^{8} x^{-2/3}\,dx$

25. $\int_{\pi/6}^{\pi} \sin\theta\,d\theta$

26. $\int_{-5}^{5} e\,dx$

27. $\int_{0}^{1}(u+2)(u-3)\,du$

28. $\int_{0}^{4}(4-t)\sqrt{t}\,dt$

29. $\int_{1}^{4}\dfrac{2+x^2}{\sqrt{x}}\,dx$

30. $\int_{-1}^{2}(3u-2)(u+1)\,du$

31. $\int_{\pi/6}^{\pi/2} \csc t\cot t\,dt$

32. $\int_{\pi/4}^{\pi/3} \csc^2\theta\,d\theta$

33. $\int_{0}^{1}(1+r)^3\,dr$

34. $\int_{0}^{3}(2\sin x - e^x)\,dx$

35. $\int_{1}^{2}\dfrac{v^3+3v^6}{v^4}\,dv$

36. $\int_{1}^{18}\sqrt{\dfrac{3}{z}}\,dz$

37. $\int_{0}^{1}(x^e + e^x)\,dx$

38. $\int_{0}^{1}\cosh t\,dt$

39. $\int_{1/\sqrt{3}}^{\sqrt{3}}\dfrac{8}{1+x^2}\,dx$

40. $\int_{1}^{3}\dfrac{y^3-2y^2-y}{y^2}\,dy$

41. $\int_{0}^{4} 2^s\,ds$

42. $\int_{1/2}^{1/\sqrt{2}}\dfrac{4}{\sqrt{1-x^2}}\,dx$

43. $\int_{0}^{\pi} f(x)\,dx$ where $f(x) = \begin{cases}\sin x & \text{if } 0 \leq x < \pi/2 \\ \cos x & \text{if } \pi/2 \leq x \leq \pi\end{cases}$

44. $\int_{-2}^{2} f(x)\,dx$ where $f(x) = \begin{cases} 2 & \text{if } -2 \leq x \leq 0 \\ 4 - x^2 & \text{if } 0 < x \leq 2\end{cases}$

45–48 Sketch the region enclosed by the given curves and calculate its area.

45. $y = \sqrt{x}, \quad y = 0, \quad x = 4$

46. $y = x^3, \quad y = 0, \quad x = 1$

47. $y = 4 - x^2, \quad y = 0$

48. $y = 2x - x^2, \quad y = 0$

49–52 Use a graph to give a rough estimate of the area of the region that lies beneath the given curve. Then find the exact area.

49. $y = \sqrt[3]{x}, \quad 0 \leq x \leq 27$

50. $y = x^{-4}, \quad 1 \leq x \leq 6$

51. $y = \sin x, \quad 0 \leq x \leq \pi$

52. $y = \sec^2 x, \quad 0 \leq x \leq \pi/3$

53–54 Evaluate the integral and interpret it as a difference of areas. Illustrate with a sketch.

53. $\int_{-1}^{2} x^3\,dx$

54. $\int_{\pi/6}^{2\pi} \cos x\,dx$

55–58 What is wrong with the equation?

55. $\int_{-2}^{1} x^{-4}\,dx = \left.\dfrac{x^{-3}}{-3}\right]_{-2}^{1} = -\dfrac{3}{8}$

56. $\int_{-1}^{2}\dfrac{4}{x^3}\,dx = \left.-\dfrac{2}{x^2}\right]_{-1}^{2} = \dfrac{3}{2}$

57. $\int_{\pi/3}^{\pi} \sec\theta\tan\theta\,d\theta = \left.\sec\theta\right]_{\pi/3}^{\pi} = -3$

58. $\int_{0}^{\pi} \sec^2 x\,dx = \left.\tan x\right]_{0}^{\pi} = 0$

59–63 Find the derivative of the function.

59. $g(x) = \int_{2x}^{3x}\dfrac{u^2-1}{u^2+1}\,du$

$\left[\text{Hint: } \int_{2x}^{3x} f(u)\,du = \int_{2x}^{0} f(u)\,du + \int_{0}^{3x} f(u)\,du\right]$

60. $g(x) = \int_{1-2x}^{1+2x} t\sin t\,dt$

61. $F(x) = \int_{x}^{x^2} e^{t^2}\,dt$

62. $F(x) = \int_{\sqrt{x}}^{2x} \arctan t\,dt$

63. $y = \int_{\cos x}^{\sin x} \ln(1+2v)\,dv$

64. If $f(x) = \int_{0}^{x}(1-t^2)e^{t^2}\,dt$, on what interval is f increasing?

65. On what interval is the curve

$$y = \int_{0}^{x}\dfrac{t^2}{t^2+t+2}\,dt$$

concave downward?

66. Let $F(x) = \int_1^x f(t)\, dt$, where f is the function whose graph is shown. Where is F concave downward?

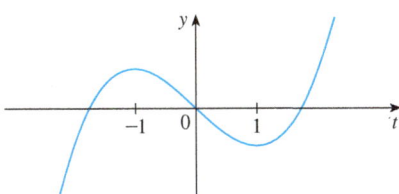

67. Let $F(x) = \int_2^x e^{t^2}\, dt$. Find an equation of the tangent line to the curve $y = F(x)$ at the point with x-coordinate 2.

68. If $f(x) = \int_0^{\sin x} \sqrt{1 + t^2}\, dt$ and $g(y) = \int_3^y f(x)\, dx$, find $g''(\pi/6)$.

69. If $f(1) = 12$, f' is continuous, and $\int_1^4 f'(x)\, dx = 17$, what is the value of $f(4)$?

70. The **error function**
$$\text{erf}(x) = \frac{2}{\sqrt{\pi}} \int_0^x e^{-t^2}\, dt$$
is used in probability, statistics, and engineering.
(a) Show that $\int_a^b e^{-t^2}\, dt = \tfrac{1}{2}\sqrt{\pi}\,[\text{erf}(b) - \text{erf}(a)]$.
(b) Show that the function $y = e^{x^2}\text{erf}(x)$ satisfies the differential equation $y' = 2xy + 2/\sqrt{\pi}$.

71. The Fresnel function S was defined in Example 3 and graphed in Figures 7 and 8.
(a) At what values of x does this function have local maximum values?
(b) On what intervals is the function concave upward?
(c) Use a graph to solve the following equation correct to two decimal places:
$$\int_0^x \sin(\pi t^2/2)\, dt = 0.2$$

72. The **sine integral function**
$$\text{Si}(x) = \int_0^x \frac{\sin t}{t}\, dt$$
is important in electrical engineering. [The integrand $f(t) = (\sin t)/t$ is not defined when $t = 0$, but we know that its limit is 1 when $t \to 0$. So we define $f(0) = 1$ and this makes f a continuous function everywhere.]
(a) Draw the graph of Si.
(b) At what values of x does this function have local maximum values?
(c) Find the coordinates of the first inflection point to the right of the origin.
(d) Does this function have horizontal asymptotes?
(e) Solve the following equation correct to one decimal place:
$$\int_0^x \frac{\sin t}{t}\, dt = 1$$

73–74 Let $g(x) = \int_0^x f(t)\, dt$, where f is the function whose graph is shown.
(a) At what values of x do the local maximum and minimum values of g occur?
(b) Where does g attain its absolute maximum value?
(c) On what intervals is g concave downward?
(d) Sketch the graph of g.

73.

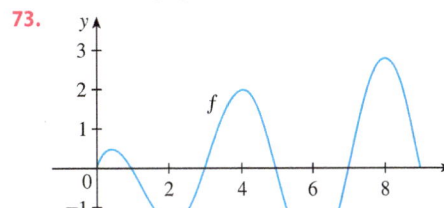

74.

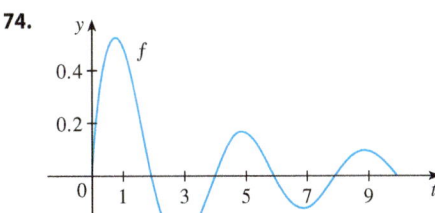

75–76 Evaluate the limit by first recognizing the sum as a Riemann sum for a function defined on $[0, 1]$.

75. $\displaystyle \lim_{n \to \infty} \sum_{i=1}^n \left(\frac{i^4}{n^5} + \frac{i}{n^2} \right)$

76. $\displaystyle \lim_{n \to \infty} \frac{1}{n}\left(\sqrt{\frac{1}{n}} + \sqrt{\frac{2}{n}} + \sqrt{\frac{3}{n}} + \cdots + \sqrt{\frac{n}{n}} \right)$

77. Justify (3) for the case $h < 0$.

78. If f is continuous and g and h are differentiable functions, find a formula for
$$\frac{d}{dx} \int_{g(x)}^{h(x)} f(t)\, dt$$

79. (a) Show that $1 \leq \sqrt{1 + x^3} \leq 1 + x^3$ for $x \geq 0$.
(b) Show that $1 \leq \int_0^1 \sqrt{1 + x^3}\, dx \leq 1.25$.

80. (a) Show that $\cos(x^2) \geq \cos x$ for $0 \leq x \leq 1$.
(b) Deduce that $\int_0^{\pi/6} \cos(x^2)\, dx \geq \tfrac{1}{2}$.

81. Show that
$$0 \leq \int_5^{10} \frac{x^2}{x^4 + x^2 + 1}\, dx \leq 0.1$$
by comparing the integrand to a simpler function.

82. Let

$$f(x) = \begin{cases} 0 & \text{if } x < 0 \\ x & \text{if } 0 \leq x \leq 1 \\ 2 - x & \text{if } 1 < x \leq 2 \\ 0 & \text{if } x > 2 \end{cases}$$

and $\quad g(x) = \int_0^x f(t)\, dt$

(a) Find an expression for $g(x)$ similar to the one for $f(x)$.
(b) Sketch the graphs of f and g.
(c) Where is f differentiable? Where is g differentiable?

83. Find a function f and a number a such that

$$6 + \int_a^x \frac{f(t)}{t^2}\, dt = 2\sqrt{x} \quad \text{for all } x > 0$$

84. The area labeled B is three times the area labeled A. Express b in terms of a.

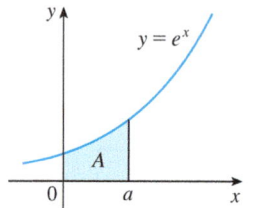 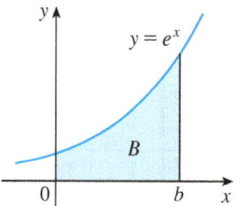

85. A manufacturing company owns a major piece of equipment that depreciates at the (continuous) rate $f = f(t)$, where t is the time measured in months since its last overhaul. Because a fixed cost A is incurred each time the machine is overhauled, the company wants to determine the optimal time T (in months) between overhauls.

(a) Explain why $\int_0^t f(s)\, ds$ represents the loss in value of the machine over the period of time t since the last overhaul.

(b) Let $C = C(t)$ be given by

$$C(t) = \frac{1}{t}\left[A + \int_0^t f(s)\, ds\right]$$

What does C represent and why would the company want to minimize C?

(c) Show that C has a minimum value at the numbers $t = T$ where $C(T) = f(T)$.

86. A high-tech company purchases a new computing system whose initial value is V. The system will depreciate at the rate $f = f(t)$ and will accumulate maintenance costs at the rate $g = g(t)$, where t is the time measured in months. The company wants to determine the optimal time to replace the system.

(a) Let

$$C(t) = \frac{1}{t}\int_0^t [f(s) + g(s)]\, ds$$

Show that the critical numbers of C occur at the numbers t where $C(t) = f(t) + g(t)$.

(b) Suppose that

$$f(t) = \begin{cases} \dfrac{V}{15} - \dfrac{V}{450}t & \text{if } 0 < t \leq 30 \\ 0 & \text{if } t > 30 \end{cases}$$

and $\quad g(t) = \dfrac{Vt^2}{12{,}900} \quad t > 0$

Determine the length of time T for the total depreciation $D(t) = \int_0^t f(s)\, ds$ to equal the initial value V.

(c) Determine the absolute minimum of C on $(0, T]$.

(d) Sketch the graphs of C and $f + g$ in the same coordinate system, and verify the result in part (a) in this case.

5.4 Indefinite Integrals and the Net Change Theorem

We saw in Section 5.3 that the second part of the Fundamental Theorem of Calculus provides a very powerful method for evaluating the definite integral of a function, assuming that we can find an antiderivative of the function. In this section we introduce a notation for antiderivatives, review the formulas for antiderivatives, and use them to evaluate definite integrals. We also reformulate FTC2 in a way that makes it easier to apply to science and engineering problems.

■ Indefinite Integrals

Both parts of the Fundamental Theorem establish connections between antiderivatives and definite integrals. Part 1 says that if f is continuous, then $\int_a^x f(t)\, dt$ is an antiderivative of f. Part 2 says that $\int_a^b f(x)\, dx$ can be found by evaluating $F(b) - F(a)$, where F is an antiderivative of f.

We need a convenient notation for antiderivatives that makes them easy to work with. Because of the relation between antiderivatives and integrals given by the Fundamental

Theorem, the notation $\int f(x)\,dx$ is traditionally used for an antiderivative of f and is called an **indefinite integral**. Thus

$$\int f(x)\,dx = F(x) \quad \text{means} \quad F'(x) = f(x)$$

For example, we can write

$$\int x^2\,dx = \frac{x^3}{3} + C \quad \text{because} \quad \frac{d}{dx}\left(\frac{x^3}{3} + C\right) = x^2$$

So we can regard an indefinite integral as representing an entire *family* of functions (one antiderivative for each value of the constant C).

⊘ You should distinguish carefully between definite and indefinite integrals. A definite integral $\int_a^b f(x)\,dx$ is a *number*, whereas an indefinite integral $\int f(x)\,dx$ is a *function* (or family of functions). The connection between them is given by Part 2 of the Fundamental Theorem: If f is continuous on $[a, b]$, then

$$\int_a^b f(x)\,dx = \left[\int f(x)\,dx\right]_a^b$$

The effectiveness of the Fundamental Theorem depends on having a supply of antiderivatives of functions. We therefore restate the Table of Antidifferentiation Formulas from Section 4.9, together with a few others, in the notation of indefinite integrals. Any formula can be verified by differentiating the function on the right side and obtaining the integrand. For instance,

$$\int \sec^2 x\,dx = \tan x + C \quad \text{because} \quad \frac{d}{dx}(\tan x + C) = \sec^2 x$$

1 Table of Indefinite Integrals

$$\int cf(x)\,dx = c\int f(x)\,dx \qquad \int [f(x) + g(x)]\,dx = \int f(x)\,dx + \int g(x)\,dx$$

$$\int k\,dx = kx + C$$

$$\int x^n\,dx = \frac{x^{n+1}}{n+1} + C \quad (n \neq -1) \qquad \int \frac{1}{x}\,dx = \ln|x| + C$$

$$\int e^x\,dx = e^x + C \qquad \int b^x\,dx = \frac{b^x}{\ln b} + C$$

$$\int \sin x\,dx = -\cos x + C \qquad \int \cos x\,dx = \sin x + C$$

$$\int \sec^2 x\,dx = \tan x + C \qquad \int \csc^2 x\,dx = -\cot x + C$$

$$\int \sec x \tan x\,dx = \sec x + C \qquad \int \csc x \cot x\,dx = -\csc x + C$$

$$\int \frac{1}{x^2+1}\,dx = \tan^{-1} x + C \qquad \int \frac{1}{\sqrt{1-x^2}}\,dx = \sin^{-1} x + C$$

$$\int \sinh x\,dx = \cosh x + C \qquad \int \cosh x\,dx = \sinh x + C$$

Recall from Theorem 4.9.1 that the most general antiderivative *on a given interval* is obtained by adding a constant to a particular antiderivative. **We adopt the convention** that when a formula for a general indefinite integral is given, it is valid only on an interval. Thus we write

$$\int \frac{1}{x^2}\, dx = -\frac{1}{x} + C$$

with the understanding that it is valid on the interval $(0, \infty)$ or on the interval $(-\infty, 0)$. This is true despite the fact that the general antiderivative of the function $f(x) = 1/x^2$, $x \neq 0$, is

$$F(x) = \begin{cases} -\dfrac{1}{x} + C_1 & \text{if } x < 0 \\[4pt] -\dfrac{1}{x} + C_2 & \text{if } x > 0 \end{cases}$$

EXAMPLE 1 Find the general indefinite integral

$$\int (10x^4 - 2\sec^2 x)\, dx$$

SOLUTION Using our convention and Table 1, we have

$$\int (10x^4 - 2\sec^2 x)\, dx = 10 \int x^4\, dx - 2 \int \sec^2 x\, dx$$
$$= 10\,\frac{x^5}{5} - 2\tan x + C$$
$$= 2x^5 - 2\tan x + C$$

You should check this answer by differentiating it.

EXAMPLE 2 Evaluate $\displaystyle \int \frac{\cos \theta}{\sin^2 \theta}\, d\theta$.

SOLUTION This indefinite integral isn't immediately apparent in Table 1, so we use trigonometric identities to rewrite the function before integrating:

$$\int \frac{\cos \theta}{\sin^2 \theta}\, d\theta = \int \left(\frac{1}{\sin \theta}\right)\left(\frac{\cos \theta}{\sin \theta}\right) d\theta$$
$$= \int \csc \theta \cot \theta\, d\theta = -\csc \theta + C$$

EXAMPLE 3 Evaluate $\displaystyle \int_0^3 (x^3 - 6x)\, dx$.

SOLUTION Using FTC2 and Table 1, we have

$$\int_0^3 (x^3 - 6x)\, dx = \frac{x^4}{4} - 6\,\frac{x^2}{2} \bigg]_0^3$$
$$= \left(\tfrac{1}{4} \cdot 3^4 - 3 \cdot 3^2\right) - \left(\tfrac{1}{4} \cdot 0^4 - 3 \cdot 0^2\right)$$
$$= \tfrac{81}{4} - 27 - 0 + 0 = -6.75$$

Compare this calculation with Example 5.2.2(b).

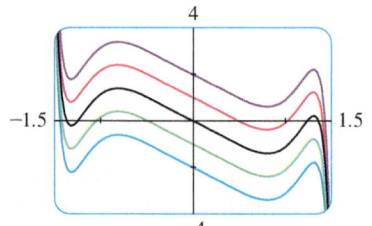

FIGURE 1

The indefinite integral in Example 1 is graphed in Figure 1 for several values of C. Here the value of C is the y-intercept.

SECTION 5.4 Indefinite Integrals and the Net Change Theorem

EXAMPLE 4 Find $\int_0^2 \left(2x^3 - 6x + \frac{3}{x^2 + 1}\right) dx$ and interpret the result in terms of areas.

SOLUTION The Fundamental Theorem gives

$$\int_0^2 \left(2x^3 - 6x + \frac{3}{x^2 + 1}\right) dx = 2\frac{x^4}{4} - 6\frac{x^2}{2} + 3\tan^{-1}x \Big]_0^2$$

$$= \tfrac{1}{2}x^4 - 3x^2 + 3\tan^{-1}x \Big]_0^2$$

$$= \tfrac{1}{2}(2^4) - 3(2^2) + 3\tan^{-1}2 - 0$$

$$= -4 + 3\tan^{-1}2$$

This is the exact value of the integral. If a decimal approximation is desired, we can use a calculator to approximate $\tan^{-1}2$. Doing so, we get

$$\int_0^2 \left(2x^3 - 6x + \frac{3}{x^2 + 1}\right) dx \approx -0.67855$$

Figure 2 shows the graph of the integrand in Example 4. We know from Section 5.2 that the value of the integral can be interpreted as a net area: the sum of the areas labeled with a plus sign minus the area labeled with a minus sign.

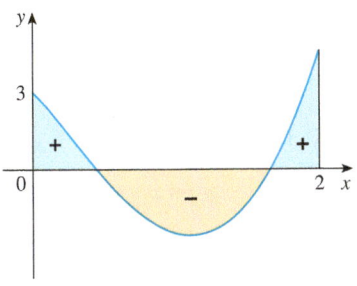

FIGURE 2

EXAMPLE 5 Evaluate $\int_1^9 \frac{2t^2 + t^2\sqrt{t} - 1}{t^2} dt$.

SOLUTION First we need to write the integrand in a simpler form by carrying out the division:

$$\int_1^9 \frac{2t^2 + t^2\sqrt{t} - 1}{t^2} dt = \int_1^9 (2 + t^{1/2} - t^{-2}) dt$$

$$= 2t + \frac{t^{3/2}}{\frac{3}{2}} - \frac{t^{-1}}{-1} \Big]_1^9 = 2t + \tfrac{2}{3}t^{3/2} + \frac{1}{t} \Big]_1^9$$

$$= \left(2 \cdot 9 + \tfrac{2}{3} \cdot 9^{3/2} + \tfrac{1}{9}\right) - \left(2 \cdot 1 + \tfrac{2}{3} \cdot 1^{3/2} + \tfrac{1}{1}\right)$$

$$= 18 + 18 + \tfrac{1}{9} - 2 - \tfrac{2}{3} - 1 = 32\tfrac{4}{9}$$

■ Applications

Part 2 of the Fundamental Theorem says that if f is continuous on $[a, b]$, then

$$\int_a^b f(x) dx = F(b) - F(a)$$

where F is any antiderivative of f. This means that $F' = f$, so the equation can be rewritten as

$$\int_a^b F'(x) dx = F(b) - F(a)$$

We know that $F'(x)$ represents the rate of change of $y = F(x)$ with respect to x and $F(b) - F(a)$ is the change in y when x changes from a to b. [Note that y could, for instance, increase, then decrease, then increase again. Although y might change in both directions, $F(b) - F(a)$ represents the *net* change in y.] So we can reformulate FTC2 in words as follows.

> **Net Change Theorem** The integral of a rate of change is the net change:
> $$\int_a^b F'(x)\,dx = F(b) - F(a)$$

This principle can be applied to all of the rates of change in the natural and social sciences that we discussed in Section 3.7. Here are a few instances of this idea:

- If $V(t)$ is the volume of water in a reservoir at time t, then its derivative $V'(t)$ is the rate at which water flows into the reservoir at time t. So
$$\int_{t_1}^{t_2} V'(t)\,dt = V(t_2) - V(t_1)$$
is the change in the amount of water in the reservoir between time t_1 and time t_2.

- If $[C](t)$ is the concentration of the product of a chemical reaction at time t, then the rate of reaction is the derivative $d[C]/dt$. So
$$\int_{t_1}^{t_2} \frac{d[C]}{dt}\,dt = [C](t_2) - [C](t_1)$$
is the change in the concentration of C from time t_1 to time t_2.

- If the mass of a rod measured from the left end to a point x is $m(x)$, then the linear density is $\rho(x) = m'(x)$. So
$$\int_a^b \rho(x)\,dx = m(b) - m(a)$$
is the mass of the segment of the rod that lies between $x = a$ and $x = b$.

- If the rate of growth of a population is dn/dt, then
$$\int_{t_1}^{t_2} \frac{dn}{dt}\,dt = n(t_2) - n(t_1)$$
is the net change in population during the time period from t_1 to t_2. (The population increases when births happen and decreases when deaths occur. The net change takes into account both births and deaths.)

- If $C(x)$ is the cost of producing x units of a commodity, then the marginal cost is the derivative $C'(x)$. So
$$\int_{x_1}^{x_2} C'(x)\,dx = C(x_2) - C(x_1)$$
is the increase in cost when production is increased from x_1 units to x_2 units.

- If an object moves along a straight line with position function $s(t)$, then its velocity is $v(t) = s'(t)$, so

$$\boxed{2} \quad \int_{t_1}^{t_2} v(t)\,dt = s(t_2) - s(t_1)$$

is the net change of position, or *displacement*, of the particle during the time period from t_1 to t_2. In Section 5.1 we guessed that this was true for the case where the object moves in the positive direction, but now we have proved that it is always true.

- If we want to calculate the distance the object travels during the time interval, we have to consider the intervals when $v(t) \geq 0$ (the particle moves to the right) and also the intervals when $v(t) \leq 0$ (the particle moves to the left). In both cases the distance is computed by integrating $|v(t)|$, the speed. Therefore

$$\boxed{3} \qquad \int_{t_1}^{t_2} |v(t)|\, dt = \text{total distance traveled}$$

Figure 3 shows how both displacement and distance traveled can be interpreted in terms of areas under a velocity curve.

$$\text{displacement} = \int_{t_1}^{t_2} v(t)\, dt = A_1 - A_2 + A_3$$

$$\text{distance} = \int_{t_1}^{t_2} |v(t)|\, dt = A_1 + A_2 + A_3$$

- The acceleration of the object is $a(t) = v'(t)$, so

$$\int_{t_1}^{t_2} a(t)\, dt = v(t_2) - v(t_1)$$

is the change in velocity from time t_1 to time t_2.

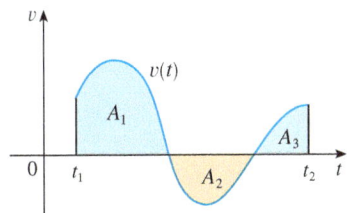

FIGURE 3

EXAMPLE 6 A particle moves along a line so that its velocity at time t is $v(t) = t^2 - t - 6$ (measured in meters per second).
(a) Find the displacement of the particle during the time period $1 \leq t \leq 4$.
(b) Find the distance traveled during this time period.

SOLUTION
(a) By Equation 2, the displacement is

$$s(4) - s(1) = \int_1^4 v(t)\, dt = \int_1^4 (t^2 - t - 6)\, dt$$

$$= \left[\frac{t^3}{3} - \frac{t^2}{2} - 6t \right]_1^4 = -\frac{9}{2}$$

This means that the particle moved 4.5 m toward the left.

(b) Note that $v(t) = t^2 - t - 6 = (t-3)(t+2)$ and so $v(t) \leq 0$ on the interval $[1, 3]$ and $v(t) \geq 0$ on $[3, 4]$. Thus, from Equation 3, the distance traveled is

$$\int_1^4 |v(t)|\, dt = \int_1^3 [-v(t)]\, dt + \int_3^4 v(t)\, dt$$

$$= \int_1^3 (-t^2 + t + 6)\, dt + \int_3^4 (t^2 - t - 6)\, dt$$

$$= \left[-\frac{t^3}{3} + \frac{t^2}{2} + 6t \right]_1^3 + \left[\frac{t^3}{3} - \frac{t^2}{2} - 6t \right]_3^4$$

$$= \frac{61}{6} \approx 10.17 \text{ m}$$

To integrate the absolute value of $v(t)$, we use Property 5 of integrals from Section 5.2 to split the integral into two parts, one where $v(t) \leq 0$ and one where $v(t) \geq 0$.

EXAMPLE 7 Figure 4 shows the power consumption in the city of San Francisco for a day in September (P is measured in megawatts; t is measured in hours starting at midnight). Estimate the energy used on that day.

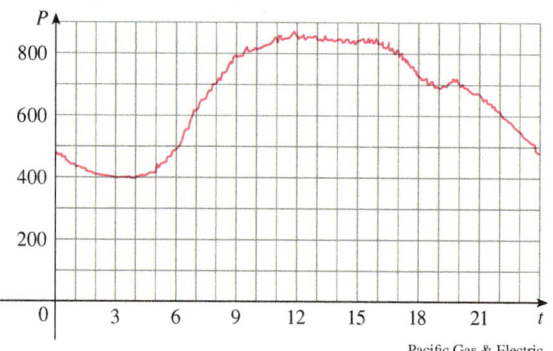

FIGURE 4

Pacific Gas & Electric

SOLUTION Power is the rate of change of energy: $P(t) = E'(t)$. So, by the Net Change Theorem,

$$\int_0^{24} P(t)\,dt = \int_0^{24} E'(t)\,dt = E(24) - E(0)$$

is the total amount of energy used on that day. We approximate the value of the integral using the Midpoint Rule with 12 subintervals and $\Delta t = 2$:

$$\int_0^{24} P(t)\,dt \approx [P(1) + P(3) + P(5) + \cdots + P(21) + P(23)]\,\Delta t$$

$$\approx (440 + 400 + 420 + 620 + 790 + 840 + 850$$

$$+ 840 + 810 + 690 + 670 + 550)(2)$$

$$= 15{,}840$$

The energy used was approximately 15,840 megawatt-hours. ■

A note on units

How did we know what units to use for energy in Example 7? The integral $\int_0^{24} P(t)\,dt$ is defined as the limit of sums of terms of the form $P(t_i^*)\,\Delta t$. Now $P(t_i^*)$ is measured in megawatts and Δt is measured in hours, so their product is measured in megawatt-hours. The same is true of the limit. In general, the unit of measurement for $\int_a^b f(x)\,dx$ is the product of the unit for $f(x)$ and the unit for x.

5.4 EXERCISES

1–4 Verify by differentiation that the formula is correct.

1. $\displaystyle\int \frac{1}{x^2\sqrt{1+x^2}}\,dx = -\frac{\sqrt{1+x^2}}{x} + C$

2. $\displaystyle\int \cos^2 x\,dx = \tfrac{1}{2}x + \tfrac{1}{4}\sin 2x + C$

3. $\displaystyle\int \tan^2 x\,dx = \tan x - x + C$

4. $\displaystyle\int x\sqrt{a+bx}\,dx = \frac{2}{15b^2}(3bx - 2a)(a+bx)^{3/2} + C$

5–18 Find the general indefinite integral.

5. $\displaystyle\int (x^{1.3} + 7x^{2.5})\,dx$

6. $\displaystyle\int \sqrt[4]{x^5}\,dx$

7. $\int (5 + \tfrac{2}{3}x^2 + \tfrac{3}{4}x^3)\, dx$

8. $\int (u^6 - 2u^5 - u^3 + \tfrac{2}{7})\, du$

9. $\int (u + 4)(2u + 1)\, du$

10. $\int \sqrt{t}(t^2 + 3t + 2)\, dt$

11. $\int \dfrac{1 + \sqrt{x} + x}{x}\, dx$

12. $\int \left(x^2 + 1 + \dfrac{1}{x^2 + 1} \right) dx$

13. $\int (\sin x + \sinh x)\, dx$

14. $\int \left(\dfrac{1 + r}{r} \right)^2 dr$

15. $\int (2 + \tan^2\theta)\, d\theta$

16. $\int \sec t (\sec t + \tan t)\, dt$

17. $\int 2^t(1 + 5^t)\, dt$

18. $\int \dfrac{\sin 2x}{\sin x}\, dx$

19–20 Find the general indefinite integral. Illustrate by graphing several members of the family on the same screen.

19. $\int (\cos x + \tfrac{1}{2}x)\, dx$

20. $\int (e^x - 2x^2)\, dx$

21–46 Evaluate the integral.

21. $\int_{-2}^{3}(x^2 - 3)\, dx$

22. $\int_{1}^{2}(4x^3 - 3x^2 + 2x)\, dx$

23. $\int_{-2}^{0}(\tfrac{1}{2}t^4 + \tfrac{1}{4}t^3 - t)\, dt$

24. $\int_{0}^{3}(1 + 6w^2 - 10w^4)\, dw$

25. $\int_{0}^{2}(2x - 3)(4x^2 + 1)\, dx$

26. $\int_{-1}^{1} t(1 - t)^2\, dt$

27. $\int_{0}^{\pi}(5e^x + 3\sin x)\, dx$

28. $\int_{1}^{2}\left(\dfrac{1}{x^2} - \dfrac{4}{x^3} \right) dx$

29. $\int_{1}^{4}\left(\dfrac{4 + 6u}{\sqrt{u}} \right) du$

30. $\int_{0}^{1} \dfrac{4}{1 + p^2}\, dp$

31. $\int_{0}^{1} x(\sqrt[3]{x} + \sqrt[4]{x})\, dx$

32. $\int_{1}^{4} \dfrac{\sqrt{y} - y}{y^2}\, dy$

33. $\int_{1}^{2} \left(\dfrac{x}{2} - \dfrac{2}{x} \right) dx$

34. $\int_{0}^{1}(5x - 5^x)\, dx$

35. $\int_{0}^{1}(x^{10} + 10^x)\, dx$

36. $\int_{0}^{\pi/4} \sec\theta \tan\theta\, d\theta$

37. $\int_{0}^{\pi/4} \dfrac{1 + \cos^2\theta}{\cos^2\theta}\, d\theta$

38. $\int_{0}^{\pi/3} \dfrac{\sin\theta + \sin\theta \tan^2\theta}{\sec^2\theta}\, d\theta$

39. $\int_{1}^{8} \dfrac{2 + t}{\sqrt[3]{t^2}}\, dt$

40. $\int_{-10}^{10} \dfrac{2e^x}{\sinh x + \cosh x}\, dx$

41. $\int_{0}^{\sqrt{3}/2} \dfrac{dr}{\sqrt{1 - r^2}}$

42. $\int_{1}^{2} \dfrac{(x - 1)^3}{x^2}\, dx$

43. $\int_{0}^{1/\sqrt{3}} \dfrac{t^2 - 1}{t^4 - 1}\, dt$

44. $\int_{0}^{2} |2x - 1|\, dx$

45. $\int_{-1}^{2} (x - 2|x|)\, dx$

46. $\int_{0}^{3\pi/2} |\sin x|\, dx$

47. Use a graph to estimate the x-intercepts of the curve $y = 1 - 2x - 5x^4$. Then use this information to estimate the area of the region that lies under the curve and above the x-axis.

48. Repeat Exercise 47 for the curve $y = (x^2 + 1)^{-1} - x^4$.

49. The area of the region that lies to the right of the y-axis and to the left of the parabola $x = 2y - y^2$ (the shaded region in the figure) is given by the integral $\int_{0}^{2}(2y - y^2)\, dy$. (Turn your head clockwise and think of the region as lying below the curve $x = 2y - y^2$ from $y = 0$ to $y = 2$.) Find the area of the region.

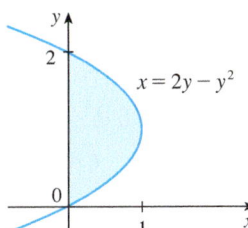

50. The boundaries of the shaded region are the y-axis, the line $y = 1$, and the curve $y = \sqrt[4]{x}$. Find the area of this region by writing x as a function of y and integrating with respect to y (as in Exercise 49).

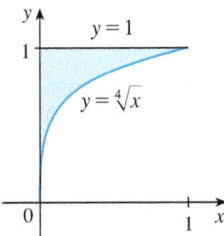

51. If $w'(t)$ is the rate of growth of a child in pounds per year, what does $\int_{5}^{10} w'(t)\, dt$ represent?

52. The current in a wire is defined as the derivative of the charge: $I(t) = Q'(t)$. (See Example 3.7.3.) What does $\int_{a}^{b} I(t)\, dt$ represent?

53. If oil leaks from a tank at a rate of $r(t)$ gallons per minute at time t, what does $\int_0^{120} r(t)\,dt$ represent?

54. A honeybee population starts with 100 bees and increases at a rate of $n'(t)$ bees per week. What does $100 + \int_0^{15} n'(t)\,dt$ represent?

55. In Section 4.7 we defined the marginal revenue function $R'(x)$ as the derivative of the revenue function $R(x)$, where x is the number of units sold. What does $\int_{1000}^{5000} R'(x)\,dx$ represent?

56. If $f(x)$ is the slope of a trail at a distance of x miles from the start of the trail, what does $\int_3^5 f(x)\,dx$ represent?

57. If x is measured in meters and $f(x)$ is measured in newtons, what are the units for $\int_0^{100} f(x)\,dx$?

58. If the units for x are feet and the units for $a(x)$ are pounds per foot, what are the units for da/dx? What units does $\int_2^8 a(x)\,dx$ have?

59–60 The velocity function (in meters per second) is given for a particle moving along a line. Find (a) the displacement and (b) the distance traveled by the particle during the given time interval.

59. $v(t) = 3t - 5$, $\quad 0 \leq t \leq 3$

60. $v(t) = t^2 - 2t - 3$, $\quad 2 \leq t \leq 4$

61–62 The acceleration function (in m/s^2) and the initial velocity are given for a particle moving along a line. Find (a) the velocity at time t and (b) the distance traveled during the given time interval.

61. $a(t) = t + 4$, $\quad v(0) = 5$, $\quad 0 \leq t \leq 10$

62. $a(t) = 2t + 3$, $\quad v(0) = -4$, $\quad 0 \leq t \leq 3$

63. The linear density of a rod of length 4 m is given by $\rho(x) = 9 + 2\sqrt{x}$ measured in kilograms per meter, where x is measured in meters from one end of the rod. Find the total mass of the rod.

64. Water flows from the bottom of a storage tank at a rate of $r(t) = 200 - 4t$ liters per minute, where $0 \leq t \leq 50$. Find the amount of water that flows from the tank during the first 10 minutes.

65. The velocity of a car was read from its speedometer at 10-second intervals and recorded in the table. Use the Midpoint Rule to estimate the distance traveled by the car.

t (s)	v (mi/h)	t (s)	v (mi/h)
0	0	60	56
10	38	70	53
20	52	80	50
30	58	90	47
40	55	100	45
50	51		

66. Suppose that a volcano is erupting and readings of the rate $r(t)$ at which solid materials are spewed into the atmosphere are given in the table. The time t is measured in seconds and the units for $r(t)$ are tonnes (metric tons) per second.

t	0	1	2	3	4	5	6
$r(t)$	2	10	24	36	46	54	60

(a) Give upper and lower estimates for the total quantity $Q(6)$ of erupted materials after six seconds.
(b) Use the Midpoint Rule to estimate $Q(6)$.

67. The marginal cost of manufacturing x yards of a certain fabric is
$$C'(x) = 3 - 0.01x + 0.000006x^2$$
(in dollars per yard). Find the increase in cost if the production level is raised from 2000 yards to 4000 yards.

68. Water flows into and out of a storage tank. A graph of the rate of change $r(t)$ of the volume of water in the tank, in liters per day, is shown. If the amount of water in the tank at time $t = 0$ is 25,000 L, use the Midpoint Rule to estimate the amount of water in the tank four days later.

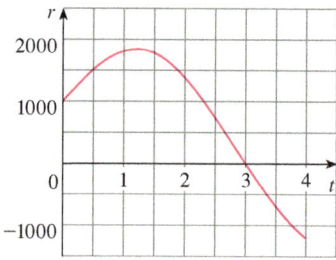

69. The graph of the acceleration $a(t)$ of a car measured in ft/s^2 is shown. Use the Midpoint Rule to estimate the increase in the velocity of the car during the six-second time interval.

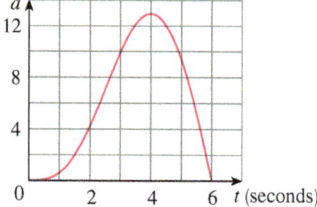

70. Lake Lanier in Georgia, USA, is a reservoir created by Buford Dam on the Chattahoochee River. The table shows the rate of inflow of water, in cubic feet per second, as measured every morning at 7:30 AM by the US Army Corps of Engineers. Use the Midpoint Rule to estimate the amount of water that flowed into Lake Lanier from July 18th, 2013, at 7:30 AM to July 26th at 7:30 AM.

Day	Inflow rate (ft³/s)
July 18	5275
July 19	6401
July 20	2554
July 21	4249
July 22	3016
July 23	3821
July 24	2462
July 25	2628
July 26	3003

71. A bacteria population is 4000 at time $t = 0$ and its rate of growth is $1000 \cdot 2^t$ bacteria per hour after t hours. What is the population after one hour?

72. Shown is the graph of traffic on an Internet service provider's T1 data line from midnight to 8:00 AM. D is the data throughput, measured in megabits per second. Use the Midpoint Rule to estimate the total amount of data transmitted during that time period.

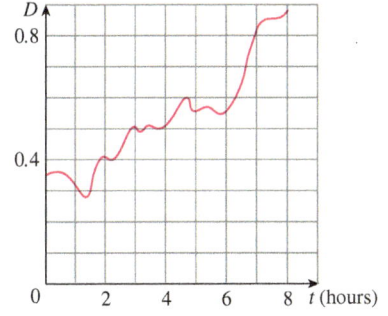

73. Shown is the power consumption in the province of Ontario, Canada, for December 9, 2004 (P is measured in megawatts; t is measured in hours starting at midnight). Using the fact that power is the rate of change of energy, estimate the energy used on that day.

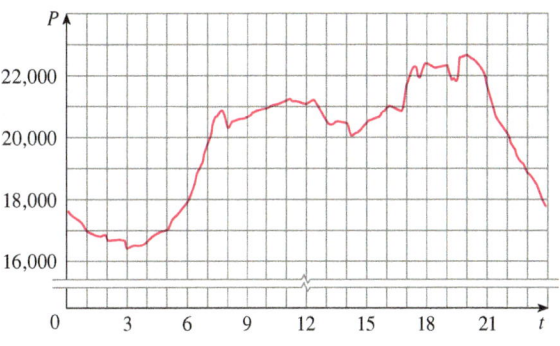

Independent Electricity Market Operator

74. On May 7, 1992, the space shuttle *Endeavour* was launched on mission STS-49, the purpose of which was to install a new perigee kick motor in an Intelsat communications satellite. The table gives the velocity data for the shuttle between liftoff and the jettisoning of the solid rocket boosters.
(a) Use a graphing calculator or computer to model these data by a third-degree polynomial.
(b) Use the model in part (a) to estimate the height reached by the *Endeavour*, 125 seconds after liftoff.

Event	Time (s)	Velocity (ft/s)
Launch	0	0
Begin roll maneuver	10	185
End roll maneuver	15	319
Throttle to 89%	20	447
Throttle to 67%	32	742
Throttle to 104%	59	1325
Maximum dynamic pressure	62	1445
Solid rocket booster separation	125	4151

WRITING PROJECT NEWTON, LEIBNIZ, AND THE INVENTION OF CALCULUS

We sometimes read that the inventors of calculus were Sir Isaac Newton (1642–1727) and Gottfried Wilhelm Leibniz (1646–1716). But we know that the basic ideas behind integration were investigated 2500 years ago by ancient Greeks such as Eudoxus and Archimedes, and methods for finding tangents were pioneered by Pierre Fermat (1601–1665), Isaac Barrow (1630–1677), and others. Barrow—who taught at Cambridge and was a major influence on Newton—was the first to understand the inverse relationship between differentiation and integration. What Newton and Leibniz did was to use this relationship, in the form of the Fundamental Theorem of Calculus, in order to develop calculus into a systematic mathematical discipline. It is in this sense that Newton and Leibniz are credited with the invention of calculus.

Read about the contributions of these men in one or more of the given references and write a report on one of the following three topics. You can include biographical details, but the main thrust of your report should be a description, in some detail, of their methods and notations. In particular, you should consult one of the sourcebooks, which give excerpts from the original publications of Newton and Leibniz, translated from Latin to English.

- The Role of Newton in the Development of Calculus
- The Role of Leibniz in the Development of Calculus
- The Controversy between the Followers of Newton and Leibniz over Priority in the Invention of Calculus

References

1. Carl Boyer and Uta Merzbach, *A History of Mathematics* (New York: Wiley, 1987), Chapter 19.
2. Carl Boyer, *The History of the Calculus and Its Conceptual Development* (New York: Dover, 1959), Chapter V.
3. C. H. Edwards, *The Historical Development of the Calculus* (New York: Springer-Verlag, 1979), Chapters 8 and 9.
4. Howard Eves, *An Introduction to the History of Mathematics,* 6th ed. (New York: Saunders, 1990), Chapter 11.
5. C. C. Gillispie, ed., *Dictionary of Scientific Biography* (New York: Scribner's, 1974). See the article on Leibniz by Joseph Hofmann in Volume VIII and the article on Newton by I. B. Cohen in Volume X.
6. Victor Katz, *A History of Mathematics: An Introduction* (New York: HarperCollins, 1993), Chapter 12.
7. Morris Kline, *Mathematical Thought from Ancient to Modern Times* (New York: Oxford University Press, 1972), Chapter 17.

Sourcebooks

1. John Fauvel and Jeremy Gray, eds., *The History of Mathematics: A Reader* (London: MacMillan Press, 1987), Chapters 12 and 13.
2. D. E. Smith, ed., *A Sourcebook in Mathematics* (New York: Dover, 1959), Chapter V.
3. D. J. Struik, ed., *A Sourcebook in Mathematics,* 1200–1800 (Princeton, NJ: Princeton University Press, 1969), Chapter V.

5.5 The Substitution Rule

Because of the Fundamental Theorem, it's important to be able to find antiderivatives. But our antidifferentiation formulas don't tell us how to evaluate integrals such as

$$\int 2x\sqrt{1 + x^2}\, dx$$

Differentials were defined in Section 3.10. If $u = f(x)$, then
$$du = f'(x)\, dx$$

PS To find this integral we use the problem-solving strategy of *introducing something extra*. Here the "something extra" is a new variable; we change from the variable x to a new variable u. Suppose that we let u be the quantity under the root sign in (1), $u = 1 + x^2$. Then the differential of u is $du = 2x\, dx$. Notice that if the dx in the notation for an integral were to be interpreted as a differential, then the differential $2x\,dx$ would occur in (1) and so, formally, without justifying our calculation, we could write

$$\boxed{2} \qquad \int 2x\sqrt{1+x^2}\,dx = \int \sqrt{1+x^2}\,2x\,dx = \int \sqrt{u}\,du$$

$$= \tfrac{2}{3}u^{3/2} + C = \tfrac{2}{3}(1+x^2)^{3/2} + C$$

But now we can check that we have the correct answer by using the Chain Rule to differentiate the final function of Equation 2:

$$\frac{d}{dx}\left[\tfrac{2}{3}(1+x^2)^{3/2} + C\right] = \tfrac{2}{3} \cdot \tfrac{3}{2}(1+x^2)^{1/2} \cdot 2x = 2x\sqrt{1+x^2}$$

In general, this method works whenever we have an integral that we can write in the form $\int f(g(x))g'(x)\,dx$. Observe that if $F' = f$, then

$$\boxed{3} \qquad \int F'(g(x))g'(x)\,dx = F(g(x)) + C$$

because, by the Chain Rule,

$$\frac{d}{dx}[F(g(x))] = F'(g(x))g'(x)$$

If we make the "change of variable" or "substitution" $u = g(x)$, then from Equation 3 we have

$$\int F'(g(x))g'(x)\,dx = F(g(x)) + C = F(u) + C = \int F'(u)\,du$$

or, writing $F' = f$, we get

$$\int f(g(x))g'(x)\,dx = \int f(u)\,du$$

Thus we have proved the following rule.

$\boxed{4}$ The Substitution Rule If $u = g(x)$ is a differentiable function whose range is an interval I and f is continuous on I, then

$$\int f(g(x))g'(x)\,dx = \int f(u)\,du$$

Notice that the Substitution Rule for integration was proved using the Chain Rule for differentiation. Notice also that if $u = g(x)$, then $du = g'(x)\,dx$, so a way to remember the Substitution Rule is to think of dx and du in (4) as differentials.

Thus the Substitution Rule says: **It is permissible to operate with dx and du after integral signs as if they were differentials.**

EXAMPLE 1 Find $\int x^3 \cos(x^4 + 2)\,dx$.

SOLUTION We make the substitution $u = x^4 + 2$ because its differential is $du = 4x^3\,dx$, which, apart from the constant factor 4, occurs in the integral. Thus, using

$x^3\,dx = \frac{1}{4}\,du$ and the Substitution Rule, we have

$$\int x^3 \cos(x^4 + 2)\,dx = \int \cos u \cdot \tfrac{1}{4}\,du = \tfrac{1}{4}\int \cos u\,du$$

$$= \tfrac{1}{4}\sin u + C$$

$$= \tfrac{1}{4}\sin(x^4 + 2) + C$$

Check the answer by differentiating it.

Notice that at the final stage we had to return to the original variable x. ∎

The idea behind the Substitution Rule is to replace a relatively complicated integral by a simpler integral. This is accomplished by changing from the original variable x to a new variable u that is a function of x. Thus in Example 1 we replaced the integral $\int x^3 \cos(x^4 + 2)\,dx$ by the simpler integral $\tfrac{1}{4}\int \cos u\,du$.

The main challenge in using the Substitution Rule is to think of an appropriate substitution. You should try to choose u to be some function in the integrand whose differential also occurs (except for a constant factor). This was the case in Example 1. If that is not possible, try choosing u to be some complicated part of the integrand (perhaps the inner function in a composite function). Finding the right substitution is a bit of an art. It's not unusual to guess wrong; if your first guess doesn't work, try another substitution.

EXAMPLE 2 Evaluate $\int \sqrt{2x + 1}\,dx$.

SOLUTION 1 Let $u = 2x + 1$. Then $du = 2\,dx$, so $dx = \tfrac{1}{2}\,du$. Thus the Substitution Rule gives

$$\int \sqrt{2x + 1}\,dx = \int \sqrt{u} \cdot \tfrac{1}{2}\,du = \tfrac{1}{2}\int u^{1/2}\,du$$

$$= \tfrac{1}{2} \cdot \frac{u^{3/2}}{3/2} + C = \tfrac{1}{3}u^{3/2} + C$$

$$= \tfrac{1}{3}(2x + 1)^{3/2} + C$$

SOLUTION 2 Another possible substitution is $u = \sqrt{2x + 1}$. Then

$$du = \frac{dx}{\sqrt{2x + 1}} \quad \text{so} \quad dx = \sqrt{2x + 1}\,du = u\,du$$

(Or observe that $u^2 = 2x + 1$, so $2u\,du = 2\,dx$.) Therefore

$$\int \sqrt{2x + 1}\,dx = \int u \cdot u\,du = \int u^2\,du$$

$$= \frac{u^3}{3} + C = \tfrac{1}{3}(2x + 1)^{3/2} + C \quad \blacksquare$$

EXAMPLE 3 Find $\displaystyle\int \frac{x}{\sqrt{1 - 4x^2}}\,dx$.

SOLUTION Let $u = 1 - 4x^2$. Then $du = -8x\,dx$, so $x\,dx = -\tfrac{1}{8}\,du$ and

$$\int \frac{x}{\sqrt{1 - 4x^2}}\,dx = -\tfrac{1}{8}\int \frac{1}{\sqrt{u}}\,du = -\tfrac{1}{8}\int u^{-1/2}\,du$$

$$= -\tfrac{1}{8}(2\sqrt{u}) + C = -\tfrac{1}{4}\sqrt{1 - 4x^2} + C \quad \blacksquare$$

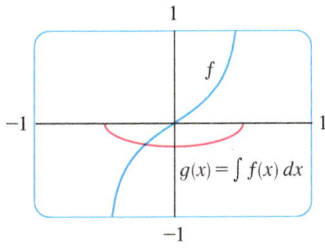

FIGURE 1
$f(x) = \dfrac{x}{\sqrt{1-4x^2}}$

$g(x) = \int f(x)\,dx = -\tfrac{1}{4}\sqrt{1-4x^2}$

The answer to Example 3 could be checked by differentiation, but instead let's check it with a graph. In Figure 1 we have used a computer to graph both the integrand $f(x) = x/\sqrt{1-4x^2}$ and its indefinite integral $g(x) = -\tfrac{1}{4}\sqrt{1-4x^2}$ (we take the case $C = 0$). Notice that $g(x)$ decreases when $f(x)$ is negative, increases when $f(x)$ is positive, and has its minimum value when $f(x) = 0$. So it seems reasonable, from the graphical evidence, that g is an antiderivative of f.

EXAMPLE 4 Calculate $\displaystyle\int e^{5x}\,dx$.

SOLUTION If we let $u = 5x$, then $du = 5\,dx$, so $dx = \tfrac{1}{5}\,du$. Therefore

$$\int e^{5x}\,dx = \tfrac{1}{5}\int e^u\,du = \tfrac{1}{5}e^u + C = \tfrac{1}{5}e^{5x} + C \qquad\blacksquare$$

NOTE With some experience, you might be able to evaluate integrals like those in Examples 1–4 without going to the trouble of making an explicit substitution. By recognizing the pattern in Equation 3, where the integrand on the left side is the product of the derivative of an outer function and the derivative of the inner function, we could work Example 1 as follows:

$$\int x^3 \cos(x^4 + 2)\,dx = \int \cos(x^4+2)\cdot x^3\,dx = \tfrac{1}{4}\int \cos(x^4+2)\cdot(4x^3)\,dx$$

$$= \tfrac{1}{4}\int \cos(x^4+2)\cdot \dfrac{d}{dx}(x^4+2)\,dx = \tfrac{1}{4}\sin(x^4+2) + C$$

Similarly, the solution to Example 4 could be written like this:

$$\int e^{5x}\,dx = \tfrac{1}{5}\int 5e^{5x}\,dx = \tfrac{1}{5}\int \dfrac{d}{dx}(e^{5x})\,dx = \tfrac{1}{5}e^{5x} + C$$

The following example, however, is more complicated and so an explicit substitution is advisable.

EXAMPLE 5 Find $\displaystyle\int \sqrt{1+x^2}\,x^5\,dx$.

SOLUTION An appropriate substitution becomes more obvious if we factor x^5 as $x^4 \cdot x$. Let $u = 1 + x^2$. Then $du = 2x\,dx$, so $x\,dx = \tfrac{1}{2}\,du$. Also $x^2 = u - 1$, so $x^4 = (u-1)^2$:

$$\int \sqrt{1+x^2}\,x^5\,dx = \int \sqrt{1+x^2}\,x^4\cdot x\,dx$$

$$= \int \sqrt{u}\,(u-1)^2 \cdot \tfrac{1}{2}\,du = \tfrac{1}{2}\int \sqrt{u}\,(u^2 - 2u + 1)\,du$$

$$= \tfrac{1}{2}\int (u^{5/2} - 2u^{3/2} + u^{1/2})\,du$$

$$= \tfrac{1}{2}\left(\tfrac{2}{7}u^{7/2} - 2\cdot\tfrac{2}{5}u^{5/2} + \tfrac{2}{3}u^{3/2}\right) + C$$

$$= \tfrac{1}{7}(1+x^2)^{7/2} - \tfrac{2}{5}(1+x^2)^{5/2} + \tfrac{1}{3}(1+x^2)^{3/2} + C \qquad\blacksquare$$

EXAMPLE 6 Calculate $\displaystyle\int \tan x\,dx$.

SOLUTION First we write tangent in terms of sine and cosine:

$$\int \tan x\,dx = \int \dfrac{\sin x}{\cos x}\,dx$$

This suggests that we should substitute $u = \cos x$, since then $du = -\sin x\, dx$ and so $\sin x\, dx = -du$:

$$\int \tan x\, dx = \int \frac{\sin x}{\cos x}\, dx = -\int \frac{1}{u}\, du$$

$$= -\ln|u| + C = -\ln|\cos x| + C \qquad \blacksquare$$

Since $-\ln|\cos x| = \ln(|\cos x|^{-1}) = \ln(1/|\cos x|) = \ln|\sec x|$, the result of Example 6 can also be written as

$$\boxed{\int \tan x\, dx = \ln|\sec x| + C} \qquad \boxed{5}$$

Definite Integrals

When evaluating a *definite* integral by substitution, two methods are possible. One method is to evaluate the indefinite integral first and then use the Fundamental Theorem. For instance, using the result of Example 2, we have

$$\int_0^4 \sqrt{2x+1}\, dx = \int \sqrt{2x+1}\, dx \Big]_0^4$$

$$= \tfrac{1}{3}(2x+1)^{3/2} \Big]_0^4 = \tfrac{1}{3}(9)^{3/2} - \tfrac{1}{3}(1)^{3/2}$$

$$= \tfrac{1}{3}(27 - 1) = \tfrac{26}{3}$$

Another method, which is usually preferable, is to change the limits of integration when the variable is changed.

> This rule says that when using a substitution in a definite integral, we must put everything in terms of the new variable u, not only x and dx but also the limits of integration. The new limits of integration are the values of u that correspond to $x = a$ and $x = b$.

6 The Substitution Rule for Definite Integrals If g' is continuous on $[a, b]$ and f is continuous on the range of $u = g(x)$, then

$$\int_a^b f(g(x))g'(x)\, dx = \int_{g(a)}^{g(b)} f(u)\, du$$

PROOF Let F be an antiderivative of f. Then, by (3), $F(g(x))$ is an antiderivative of $f(g(x))g'(x)$, so by Part 2 of the Fundamental Theorem, we have

$$\int_a^b f(g(x))g'(x)\, dx = F(g(x)) \Big]_a^b = F(g(b)) - F(g(a))$$

But, applying FTC2 a second time, we also have

$$\int_{g(a)}^{g(b)} f(u)\, du = F(u) \Big]_{g(a)}^{g(b)} = F(g(b)) - F(g(a)) \qquad \blacksquare$$

EXAMPLE 7 Evaluate $\int_0^4 \sqrt{2x+1}\, dx$ using (6).

SOLUTION Using the substitution from Solution 1 of Example 2, we have $u = 2x + 1$ and $dx = \tfrac{1}{2}\, du$. To find the new limits of integration we note that

when $x = 0$, $u = 2(0) + 1 = 1$ and when $x = 4$, $u = 2(4) + 1 = 9$

Therefore

$$\int_0^4 \sqrt{2x+1}\, dx = \int_1^9 \tfrac{1}{2}\sqrt{u}\, du$$

$$= \tfrac{1}{2} \cdot \tfrac{2}{3} u^{3/2} \Big]_1^9$$

$$= \tfrac{1}{3}(9^{3/2} - 1^{3/2}) = \tfrac{26}{3}$$

Observe that when using (6) we do *not* return to the variable x after integrating. We simply evaluate the expression in u between the appropriate values of u.

The integral given in Example 8 is an abbreviation for

$$\int_1^2 \frac{1}{(3-5x)^2}\, dx$$

EXAMPLE 8 Evaluate $\displaystyle\int_1^2 \frac{dx}{(3-5x)^2}$.

SOLUTION Let $u = 3 - 5x$. Then $du = -5\, dx$, so $dx = -\tfrac{1}{5}\, du$. When $x = 1$, $u = -2$ and when $x = 2$, $u = -7$. Thus

$$\int_1^2 \frac{dx}{(3-5x)^2} = -\frac{1}{5}\int_{-2}^{-7} \frac{du}{u^2}$$

$$= -\frac{1}{5}\left[-\frac{1}{u}\right]_{-2}^{-7} = \frac{1}{5u}\Big]_{-2}^{-7}$$

$$= \frac{1}{5}\left(-\frac{1}{7} + \frac{1}{2}\right) = \frac{1}{14}$$

Since the function $f(x) = (\ln x)/x$ in Example 9 is positive for $x > 1$, the integral represents the area of the shaded region in Figure 2.

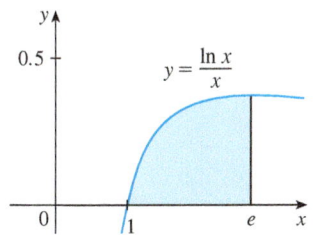

FIGURE 2

EXAMPLE 9 Calculate $\displaystyle\int_1^e \frac{\ln x}{x}\, dx$.

SOLUTION We let $u = \ln x$ because its differential $du = dx/x$ occurs in the integral. When $x = 1$, $u = \ln 1 = 0$; when $x = e$, $u = \ln e = 1$. Thus

$$\int_1^e \frac{\ln x}{x}\, dx = \int_0^1 u\, du = \frac{u^2}{2}\Big]_0^1 = \frac{1}{2}$$

■ Symmetry

The next theorem uses the Substitution Rule for Definite Integrals (6) to simplify the calculation of integrals of functions that possess symmetry properties.

> **7** **Integrals of Symmetric Functions** Suppose f is continuous on $[-a, a]$.
> (a) If f is even [$f(-x) = f(x)$], then $\displaystyle\int_{-a}^a f(x)\, dx = 2\int_0^a f(x)\, dx$.
> (b) If f is odd [$f(-x) = -f(x)$], then $\displaystyle\int_{-a}^a f(x)\, dx = 0$.

PROOF We split the integral in two:

$$\boxed{8} \quad \int_{-a}^a f(x)\, dx = \int_{-a}^0 f(x)\, dx + \int_0^a f(x)\, dx = -\int_0^{-a} f(x)\, dx + \int_0^a f(x)\, dx$$

In the first integral on the far right side we make the substitution $u = -x$. Then $du = -dx$ and when $x = -a$, $u = a$. Therefore

$$-\int_0^{-a} f(x)\,dx = -\int_0^a f(-u)(-du) = \int_0^a f(-u)\,du$$

and so Equation 8 becomes

$$\int_{-a}^a f(x)\,dx = \int_0^a f(-u)\,du + \int_0^a f(x)\,dx$$

(a) If f is even, then $f(-u) = f(u)$ so Equation 9 gives

$$\int_{-a}^a f(x)\,dx = \int_0^a f(u)\,du + \int_0^a f(x)\,dx = 2\int_0^a f(x)\,dx$$

(b) If f is odd, then $f(-u) = -f(u)$ and so Equation 9 gives

$$\int_{-a}^a f(x)\,dx = -\int_0^a f(u)\,du + \int_0^a f(x)\,dx = 0$$

Theorem 7 is illustrated by Figure 3. For the case where f is positive and even, part (a) says that the area under $y = f(x)$ from $-a$ to a is twice the area from 0 to a because of symmetry. Recall that an integral $\int_a^b f(x)\,dx$ can be expressed as the area above the x-axis and below $y = f(x)$ minus the area below the axis and above the curve. Thus part (b) says the integral is 0 because the areas cancel.

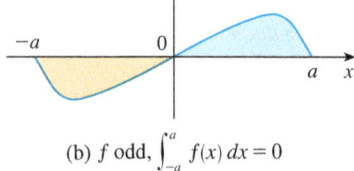

(a) f even, $\int_{-a}^a f(x)\,dx = 2\int_0^a f(x)\,dx$

(b) f odd, $\int_{-a}^a f(x)\,dx = 0$

FIGURE 3

EXAMPLE 10 Since $f(x) = x^6 + 1$ satisfies $f(-x) = f(x)$, it is even and so

$$\int_{-2}^2 (x^6 + 1)\,dx = 2\int_0^2 (x^6 + 1)\,dx$$

$$= 2\left[\tfrac{1}{7}x^7 + x\right]_0^2 = 2\left(\tfrac{128}{7} + 2\right) = \tfrac{284}{7}$$

EXAMPLE 11 Since $f(x) = (\tan x)/(1 + x^2 + x^4)$ satisfies $f(-x) = -f(x)$, it is odd and so

$$\int_{-1}^1 \frac{\tan x}{1 + x^2 + x^4}\,dx = 0$$

5.5 EXERCISES

1–6 Evaluate the integral by making the given substitution.

1. $\int \cos 2x\,dx, \quad u = 2x$

2. $\int xe^{-x^2}\,dx, \quad u = -x^2$

3. $\int x^2\sqrt{x^3 + 1}\,dx, \quad u = x^3 + 1$

4. $\int \sin^2\theta \cos\theta\,d\theta, \quad u = \sin\theta$

5. $\int \dfrac{x^3}{x^4 - 5}\,dx, \quad u = x^4 - 5$

6. $\int \sqrt{2t + 1}\,dt, \quad u = 2t + 1$

7–48 Evaluate the indefinite integral.

7. $\int x\sqrt{1 - x^2}\,dx$

8. $\int x^2 e^{x^3}\,dx$

9. $\int (1 - 2x)^9\,dx$

10. $\int \sin t\sqrt{1 + \cos t}\,dt$

11. $\int \cos(\pi t/2)\,dt$

12. $\int \sec^2 2\theta\,d\theta$

13. $\displaystyle\int \frac{dx}{5-3x}$

14. $\displaystyle\int y^2(4-y^3)^{2/3}\,dy$

15. $\displaystyle\int \cos^3\theta \sin\theta\,d\theta$

16. $\displaystyle\int e^{-5r}\,dr$

17. $\displaystyle\int \frac{e^u}{(1-e^u)^2}\,du$

18. $\displaystyle\int \frac{\sin\sqrt{x}}{\sqrt{x}}\,dx$

19. $\displaystyle\int \frac{a+bx^2}{\sqrt{3ax+bx^3}}\,dx$

20. $\displaystyle\int \frac{z^2}{z^3+1}\,dz$

21. $\displaystyle\int \frac{(\ln x)^2}{x}\,dx$

22. $\displaystyle\int \sin x \sin(\cos x)\,dx$

23. $\displaystyle\int \sec^2\theta \tan^3\theta\,d\theta$

24. $\displaystyle\int x\sqrt{x+2}\,dx$

25. $\displaystyle\int e^x\sqrt{1+e^x}\,dx$

26. $\displaystyle\int \frac{dx}{ax+b}\quad (a\neq 0)$

27. $\displaystyle\int (x^2+1)(x^3+3x)^4\,dx$

28. $\displaystyle\int e^{\cos t}\sin t\,dt$

29. $\displaystyle\int 5^t \sin(5^t)\,dt$

30. $\displaystyle\int \frac{\sec^2 x}{\tan^2 x}\,dx$

31. $\displaystyle\int \frac{(\arctan x)^2}{x^2+1}\,dx$

32. $\displaystyle\int \frac{x}{x^2+4}\,dx$

33. $\displaystyle\int \cos(1+5t)\,dt$

34. $\displaystyle\int \frac{\cos(\pi/x)}{x^2}\,dx$

35. $\displaystyle\int \sqrt{\cot x}\,\csc^2 x\,dx$

36. $\displaystyle\int \frac{2^t}{2^t+3}\,dt$

37. $\displaystyle\int \sinh^2 x \cosh x\,dx$

38. $\displaystyle\int \frac{dt}{\cos^2 t\sqrt{1+\tan t}}$

39. $\displaystyle\int \frac{\sin 2x}{1+\cos^2 x}\,dx$

40. $\displaystyle\int \frac{\sin x}{1+\cos^2 x}\,dx$

41. $\displaystyle\int \cot x\,dx$

42. $\displaystyle\int \frac{\cos(\ln t)}{t}\,dt$

43. $\displaystyle\int \frac{dx}{\sqrt{1-x^2}\,\sin^{-1}x}$

44. $\displaystyle\int \frac{x}{1+x^4}\,dx$

45. $\displaystyle\int \frac{1+x}{1+x^2}\,dx$

46. $\displaystyle\int x^2\sqrt{2+x}\,dx$

47. $\displaystyle\int x(2x+5)^8\,dx$

48. $\displaystyle\int x^3\sqrt{x^2+1}\,dx$

49–52 Evaluate the indefinite integral. Illustrate and check that your answer is reasonable by graphing both the function and its antiderivative (take $C=0$).

49. $\displaystyle\int x(x^2-1)^3\,dx$

50. $\displaystyle\int \tan^2\theta \sec^2\theta\,d\theta$

51. $\displaystyle\int e^{\cos x}\sin x\,dx$

52. $\displaystyle\int \sin x \cos^4 x\,dx$

53–73 Evaluate the definite integral.

53. $\displaystyle\int_0^1 \cos(\pi t/2)\,dt$

54. $\displaystyle\int_0^1 (3t-1)^{50}\,dt$

55. $\displaystyle\int_0^1 \sqrt[3]{1+7x}\,dx$

56. $\displaystyle\int_0^3 \frac{dx}{5x+1}$

57. $\displaystyle\int_0^{\pi/6} \frac{\sin t}{\cos^2 t}\,dt$

58. $\displaystyle\int_{\pi/3}^{2\pi/3} \csc^2\left(\tfrac{1}{2}t\right)dt$

59. $\displaystyle\int_1^2 \frac{e^{1/x}}{x^2}\,dx$

60. $\displaystyle\int_0^1 xe^{-x^2}\,dx$

61. $\displaystyle\int_{-\pi/4}^{\pi/4} (x^3+x^4\tan x)\,dx$

62. $\displaystyle\int_0^{\pi/2} \cos x \sin(\sin x)\,dx$

63. $\displaystyle\int_0^{13} \frac{dx}{\sqrt[3]{(1+2x)^2}}$

64. $\displaystyle\int_0^a x\sqrt{a^2-x^2}\,dx$

65. $\displaystyle\int_0^a x\sqrt{x^2+a^2}\,dx\quad (a>0)$

66. $\displaystyle\int_{-\pi/3}^{\pi/3} x^4\sin x\,dx$

67. $\displaystyle\int_1^2 x\sqrt{x-1}\,dx$

68. $\displaystyle\int_0^4 \frac{x}{\sqrt{1+2x}}\,dx$

69. $\displaystyle\int_e^{e^4} \frac{dx}{x\sqrt{\ln x}}$

70. $\displaystyle\int_0^2 (x-1)e^{(x-1)^2}\,dx$

71. $\displaystyle\int_0^1 \frac{e^z+1}{e^z+z}\,dz$

72. $\displaystyle\int_0^{T/2} \sin(2\pi t/T - \alpha)\,dt$

73. $\displaystyle\int_0^1 \frac{dx}{(1+\sqrt{x})^4}$

74. Verify that $f(x)=\sin\sqrt[3]{x}$ is an odd function and use that fact to show that
$$0 \leq \int_{-2}^3 \sin\sqrt[3]{x}\,dx \leq 1$$

75–76 Use a graph to give a rough estimate of the area of the region that lies under the given curve. Then find the exact area.

75. $y=\sqrt{2x+1},\ 0\leq x\leq 1$

76. $y=2\sin x - \sin 2x,\ 0\leq x\leq \pi$

77. Evaluate $\int_{-2}^2 (x+3)\sqrt{4-x^2}\,dx$ by writing it as a sum of two integrals and interpreting one of those integrals in terms of an area.

78. Evaluate $\int_0^1 x\sqrt{1-x^4}\,dx$ by making a substitution and interpreting the resulting integral in terms of an area.

79. Which of the following areas are equal? Why?

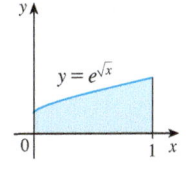
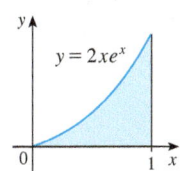
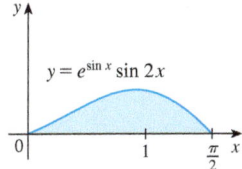

80. A model for the basal metabolism rate, in kcal/h, of a young man is $R(t) = 85 - 0.18\cos(\pi t/12)$, where t is the time in hours measured from 5:00 AM. What is the total basal metabolism of this man, $\int_0^{24} R(t)\,dt$, over a 24-hour time period?

81. An oil storage tank ruptures at time $t = 0$ and oil leaks from the tank at a rate of $r(t) = 100e^{-0.01t}$ liters per minute. How much oil leaks out during the first hour?

82. A bacteria population starts with 400 bacteria and grows at a rate of $r(t) = (450.268)e^{1.12567t}$ bacteria per hour. How many bacteria will there be after three hours?

83. Breathing is cyclic and a full respiratory cycle from the beginning of inhalation to the end of exhalation takes about 5 s. The maximum rate of air flow into the lungs is about 0.5 L/s. This explains, in part, why the function $f(t) = \tfrac{1}{2}\sin(2\pi t/5)$ has often been used to model the rate of air flow into the lungs. Use this model to find the volume of inhaled air in the lungs at time t.

84. The rate of growth of a fish population was modeled by the equation
$$G(t) = \frac{60{,}000e^{-0.6t}}{(1 + 5e^{-0.6t})^2}$$
where t is measured in years and G in kilograms per year. If the biomass was 25,000 kg in the year 2000, what is the predicted biomass for the year 2020?

85. Dialysis treatment removes urea and other waste products from a patient's blood by diverting some of the bloodflow externally through a machine called a dialyzer. The rate at which urea is removed from the blood (in mg/min) is often well described by the equation
$$u(t) = \frac{r}{V}C_0 e^{-rt/V}$$
where r is the rate of flow of blood through the dialyzer (in mL/min), V is the volume of the patient's blood (in mL), and C_0 is the amount of urea in the blood (in mg) at time $t = 0$. Evaluate the integral $\int_0^{30} u(t)\,dt$ and interpret it.

86. Alabama Instruments Company has set up a production line to manufacture a new calculator. The rate of production of these calculators after t weeks is
$$\frac{dx}{dt} = 5000\left(1 - \frac{100}{(t+10)^2}\right)\ \text{calculators/week}$$
(Notice that production approaches 5000 per week as time goes on, but the initial production is lower because of the workers' unfamiliarity with the new techniques.) Find the number of calculators produced from the beginning of the third week to the end of the fourth week.

87. If f is continuous and $\int_0^4 f(x)\,dx = 10$, find $\int_0^2 f(2x)\,dx$.

88. If f is continuous and $\int_0^9 f(x)\,dx = 4$, find $\int_0^3 xf(x^2)\,dx$.

89. If f is continuous on $\mathbb{R}$, prove that
$$\int_a^b f(-x)\,dx = \int_{-b}^{-a} f(x)\,dx$$
For the case where $f(x) \geq 0$ and $0 < a < b$, draw a diagram to interpret this equation geometrically as an equality of areas.

90. If f is continuous on $\mathbb{R}$, prove that
$$\int_a^b f(x+c)\,dx = \int_{a+c}^{b+c} f(x)\,dx$$
For the case where $f(x) \geq 0$, draw a diagram to interpret this equation geometrically as an equality of areas.

91. If a and b are positive numbers, show that
$$\int_0^1 x^a(1-x)^b\,dx = \int_0^1 x^b(1-x)^a\,dx$$

92. If f is continuous on $[0, \pi]$, use the substitution $u = \pi - x$ to show that
$$\int_0^\pi x f(\sin x)\,dx = \frac{\pi}{2}\int_0^\pi f(\sin x)\,dx$$

93. Use Exercise 92 to evaluate the integral
$$\int_0^\pi \frac{x\sin x}{1 + \cos^2 x}\,dx$$

94. (a) If f is continuous, prove that
$$\int_0^{\pi/2} f(\cos x)\,dx = \int_0^{\pi/2} f(\sin x)\,dx$$
(b) Use part (a) to evaluate $\int_0^{\pi/2} \cos^2 x\,dx$ and $\int_0^{\pi/2} \sin^2 x\,dx$.

5 REVIEW

CONCEPT CHECK

Answers to the Concept Check can be found on the back endpapers.

1. (a) Write an expression for a Riemann sum of a function f. Explain the meaning of the notation that you use.
   (b) If $f(x) \geq 0$, what is the geometric interpretation of a Riemann sum? Illustrate with a diagram.
   (c) If $f(x)$ takes on both positive and negative values, what is the geometric interpretation of a Riemann sum? Illustrate with a diagram.

2. (a) Write the definition of the definite integral of a continuous function from a to b.
   (b) What is the geometric interpretation of $\int_a^b f(x)\,dx$ if $f(x) \geq 0$?
   (c) What is the geometric interpretation of $\int_a^b f(x)\,dx$ if $f(x)$ takes on both positive and negative values? Illustrate with a diagram.

3. State the Midpoint Rule.

4. State both parts of the Fundamental Theorem of Calculus.

5. (a) State the Net Change Theorem.
   (b) If $r(t)$ is the rate at which water flows into a reservoir, what does $\int_{t_1}^{t_2} r(t)\,dt$ represent?

6. Suppose a particle moves back and forth along a straight line with velocity $v(t)$, measured in feet per second, and acceleration $a(t)$.
   (a) What is the meaning of $\int_{60}^{120} v(t)\,dt$?
   (b) What is the meaning of $\int_{60}^{120} |v(t)|\,dt$?
   (c) What is the meaning of $\int_{60}^{120} a(t)\,dt$?

7. (a) Explain the meaning of the indefinite integral $\int f(x)\,dx$.
   (b) What is the connection between the definite integral $\int_a^b f(x)\,dx$ and the indefinite integral $\int f(x)\,dx$?

8. Explain exactly what is meant by the statement that "differentiation and integration are inverse processes."

9. State the Substitution Rule. In practice, how do you use it?

TRUE–FALSE QUIZ

Determine whether the statement is true or false. If it is true, explain why. If it is false, explain why or give an example that disproves the statement.

1. If f and g are continuous on $[a, b]$, then
$$\int_a^b [f(x) + g(x)]\,dx = \int_a^b f(x)\,dx + \int_a^b g(x)\,dx$$

2. If f and g are continuous on $[a, b]$, then
$$\int_a^b [f(x)g(x)]\,dx = \left(\int_a^b f(x)\,dx\right)\left(\int_a^b g(x)\,dx\right)$$

3. If f is continuous on $[a, b]$, then
$$\int_a^b 5f(x)\,dx = 5\int_a^b f(x)\,dx$$

4. If f is continuous on $[a, b]$, then
$$\int_a^b xf(x)\,dx = x\int_a^b f(x)\,dx$$

5. If f is continuous on $[a, b]$ and $f(x) \geq 0$, then
$$\int_a^b \sqrt{f(x)}\,dx = \sqrt{\int_a^b f(x)\,dx}$$

6. If f' is continuous on $[1, 3]$, then $\int_1^3 f'(v)\,dv = f(3) - f(1)$.

7. If f and g are continuous and $f(x) \geq g(x)$ for $a \leq x \leq b$, then
$$\int_a^b f(x)\,dx \geq \int_a^b g(x)\,dx$$

8. If f and g are differentiable and $f(x) \geq g(x)$ for $a < x < b$, then $f'(x) \geq g'(x)$ for $a < x < b$.

9. $\int_{-1}^{1} \left(x^5 - 6x^9 + \dfrac{\sin x}{(1 + x^4)^2}\right) dx = 0$

10. $\int_{-5}^{5} (ax^2 + bx + c)\,dx = 2\int_0^5 (ax^2 + c)\,dx$

11. All continuous functions have derivatives.

12. All continuous functions have antiderivatives.

13. $\int_0^3 e^{x^2}\,dx = \int_0^5 e^{x^2}\,dx + \int_5^3 e^{x^2}\,dx$

14. If $\int_0^1 f(x)\,dx = 0$, then $f(x) = 0$ for $0 \leq x \leq 1$.

15. If f is continuous on $[a, b]$, then
$$\frac{d}{dx}\left(\int_a^b f(x)\,dx\right) = f(x)$$

16. $\int_0^2 (x - x^3)\,dx$ represents the area under the curve $y = x - x^3$ from 0 to 2.

17. $\int_{-2}^{1} \dfrac{1}{x^4}\,dx = -\dfrac{3}{8}$

18. If f has a discontinuity at 0, then $\int_{-1}^{1} f(x)\,dx$ does not exist.

EXERCISES

1. Use the given graph of f to find the Riemann sum with six subintervals. Take the sample points to be (a) left endpoints and (b) midpoints. In each case draw a diagram and explain what the Riemann sum represents.

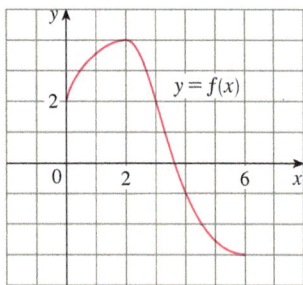

2. (a) Evaluate the Riemann sum for
$$f(x) = x^2 - x \qquad 0 \le x \le 2$$
   with four subintervals, taking the sample points to be right endpoints. Explain, with the aid of a diagram, what the Riemann sum represents.
   (b) Use the definition of a definite integral (with right endpoints) to calculate the value of the integral
$$\int_0^2 (x^2 - x) \, dx$$
   (c) Use the Fundamental Theorem to check your answer to part (b).
   (d) Draw a diagram to explain the geometric meaning of the integral in part (b).

3. Evaluate
$$\int_0^1 \left(x + \sqrt{1 - x^2}\right) dx$$
   by interpreting it in terms of areas.

4. Express
$$\lim_{n \to \infty} \sum_{i=1}^n \sin x_i \, \Delta x$$
   as a definite integral on the interval $[0, \pi]$ and then evaluate the integral.

5. If $\int_0^6 f(x) \, dx = 10$ and $\int_0^4 f(x) \, dx = 7$, find $\int_4^6 f(x) \, dx$.

[CAS] 6. (a) Write $\int_1^5 (x + 2x^5) \, dx$ as a limit of Riemann sums, taking the sample points to be right endpoints. Use a computer algebra system to evaluate the sum and to compute the limit.
   (b) Use the Fundamental Theorem to check your answer to part (a).

7. The figure shows the graphs of f, f', and $\int_0^x f(t) \, dt$. Identify each graph, and explain your choices.

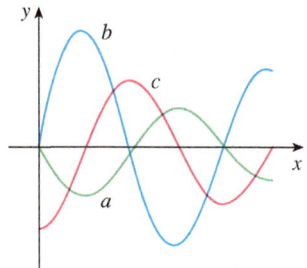

8. Evaluate:
   (a) $\int_0^1 \dfrac{d}{dx}\left(e^{\arctan x}\right) dx$
   (b) $\dfrac{d}{dx} \int_0^1 e^{\arctan x} \, dx$
   (c) $\dfrac{d}{dx} \int_0^x e^{\arctan t} \, dt$

9. The graph of f consists of the three line segments shown. If $g(x) = \int_0^x f(t) \, dt$, find $g(4)$ and $g'(4)$.

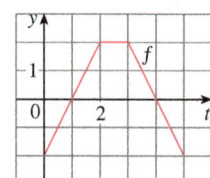

10. If f is the function in Exercise 9, find $g''(4)$.

11–40 Evaluate the integral, if it exists.

11. $\int_1^2 (8x^3 + 3x^2) \, dx$

12. $\int_0^T (x^4 - 8x + 7) \, dx$

13. $\int_0^1 (1 - x^9) \, dx$

14. $\int_0^1 (1 - x)^9 \, dx$

15. $\int_1^9 \dfrac{\sqrt{u} - 2u^2}{u} \, du$

16. $\int_0^1 \left(\sqrt[4]{u} + 1\right)^2 du$

17. $\int_0^1 y(y^2 + 1)^5 \, dy$

18. $\int_0^2 y^2 \sqrt{1 + y^3} \, dy$

19. $\int_1^5 \dfrac{dt}{(t - 4)^2}$

20. $\int_0^1 \sin(3\pi t) \, dt$

21. $\int_0^1 v^2 \cos(v^3) \, dv$

22. $\int_{-1}^1 \dfrac{\sin x}{1 + x^2} \, dx$

23. $\int_{-\pi/4}^{\pi/4} \dfrac{t^4 \tan t}{2 + \cos t} \, dt$

24. $\int_0^1 \dfrac{e^x}{1 + e^{2x}} \, dx$

25. $\int \left(\dfrac{1 - x}{x}\right)^2 dx$

26. $\int_1^{10} \dfrac{x}{x^2 - 4} \, dx$

27. $\int \dfrac{x+2}{\sqrt{x^2+4x}}\, dx$ **28.** $\int \dfrac{\csc^2 x}{1+\cot x}\, dx$

29. $\int \sin \pi t \cos \pi t\, dt$ **30.** $\int \sin x \cos(\cos x)\, dx$

31. $\int \dfrac{e^{\sqrt{x}}}{\sqrt{x}}\, dx$ **32.** $\int \dfrac{\sin(\ln x)}{x}\, dx$

33. $\int \tan x \ln(\cos x)\, dx$ **34.** $\int \dfrac{x}{\sqrt{1-x^4}}\, dx$

35. $\int \dfrac{x^3}{1+x^4}\, dx$ **36.** $\int \sinh(1+4x)\, dx$

37. $\int \dfrac{\sec\theta \tan\theta}{1+\sec\theta}\, d\theta$ **38.** $\int_0^{\pi/4} (1+\tan t)^3 \sec^2 t\, dt$

39. $\int_0^3 |x^2 - 4|\, dx$ **40.** $\int_0^4 |\sqrt{x}-1|\, dx$

41–42 Evaluate the indefinite integral. Illustrate and check that your answer is reasonable by graphing both the function and its antiderivative (take $C=0$).

41. $\int \dfrac{\cos x}{\sqrt{1+\sin x}}\, dx$ **42.** $\int \dfrac{x^3}{\sqrt{x^2+1}}\, dx$

43. Use a graph to give a rough estimate of the area of the region that lies under the curve $y = x\sqrt{x}$, $0 \le x \le 4$. Then find the exact area.

44. Graph the function $f(x) = \cos^2 x \sin x$ and use the graph to guess the value of the integral $\int_0^{2\pi} f(x)\, dx$. Then evaluate the integral to confirm your guess.

45–50 Find the derivative of the function.

45. $F(x) = \int_0^x \dfrac{t^2}{1+t^3}\, dt$ **46.** $F(x) = \int_x^1 \sqrt{t+\sin t}\, dt$

47. $g(x) = \int_0^{x^4} \cos(t^2)\, dt$ **48.** $g(x) = \int_1^{\sin x} \dfrac{1-t^2}{1+t^4}\, dt$

49. $y = \int_{\sqrt{x}}^x \dfrac{e^t}{t}\, dt$ **50.** $y = \int_{2x}^{3x+1} \sin(t^4)\, dt$

51–52 Use Property 8 of integrals to estimate the value of the integral.

51. $\int_1^3 \sqrt{x^2+3}\, dx$ **52.** $\int_3^5 \dfrac{1}{x+1}\, dx$

53–56 Use the properties of integrals to verify the inequality.

53. $\int_0^1 x^2 \cos x\, dx \le \dfrac{1}{3}$ **54.** $\int_{\pi/4}^{\pi/2} \dfrac{\sin x}{x}\, dx \le \dfrac{\sqrt{2}}{2}$

55. $\int_0^1 e^x \cos x\, dx \le e - 1$ **56.** $\int_0^1 x \sin^{-1}x\, dx \le \pi/4$

57. Use the Midpoint Rule with $n = 6$ to approximate $\int_0^3 \sin(x^3)\, dx$.

58. A particle moves along a line with velocity function $v(t) = t^2 - t$, where v is measured in meters per second. Find (a) the displacement and (b) the distance traveled by the particle during the time interval $[0, 5]$.

59. Let $r(t)$ be the rate at which the world's oil is consumed, where t is measured in years starting at $t = 0$ on January 1, 2000, and $r(t)$ is measured in barrels per year. What does $\int_0^8 r(t)\, dt$ represent?

60. A radar gun was used to record the speed of a runner at the times given in the table. Use the Midpoint Rule to estimate the distance the runner covered during those 5 seconds.

t (s)	v (m/s)	t (s)	v (m/s)
0	0	3.0	10.51
0.5	4.67	3.5	10.67
1.0	7.34	4.0	10.76
1.5	8.86	4.5	10.81
2.0	9.73	5.0	10.81
2.5	10.22		

61. A population of honeybees increased at a rate of $r(t)$ bees per week, where the graph of r is as shown. Use the Midpoint Rule with six subintervals to estimate the increase in the bee population during the first 24 weeks.

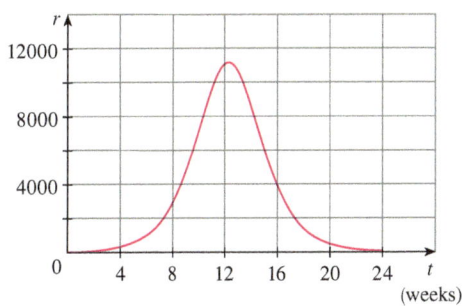

62. Let

$$f(x) = \begin{cases} -x-1 & \text{if } -3 \le x \le 0 \\ -\sqrt{1-x^2} & \text{if } 0 \le x \le 1 \end{cases}$$

Evaluate $\int_{-3}^1 f(x)\, dx$ by interpreting the integral as a difference of areas.

63. If f is continuous and $\int_0^2 f(x)\, dx = 6$, evaluate $\int_0^{\pi/2} f(2 \sin \theta) \cos \theta\, d\theta$.

64. The Fresnel function $S(x) = \int_0^x \sin(\frac{1}{2}\pi t^2)\, dt$ was introduced in Section 5.3. Fresnel also used the function

$$C(x) = \int_0^x \cos(\tfrac{1}{2}\pi t^2)\, dt$$

in his theory of the diffraction of light waves.
(a) On what intervals is C increasing?
(b) On what intervals is C concave upward?
(c) Use a graph to solve the following equation correct to two decimal places:

$$\int_0^x \cos(\tfrac{1}{2}\pi t^2)\, dt = 0.7$$

(d) Plot the graphs of C and S on the same screen. How are these graphs related?

65. Estimate the value of the number c such that the area under the curve $y = \sinh cx$ between $x = 0$ and $x = 1$ is equal to 1.

66. Suppose that the temperature in a long, thin rod placed along the x-axis is initially $C/(2a)$ if $|x| \leq a$ and 0 if $|x| > a$. It can be shown that if the heat diffusivity of the rod is k, then the temperature of the rod at the point x at time t is

$$T(x, t) = \frac{C}{a\sqrt{4\pi kt}} \int_0^a e^{-(x-u)^2/(4kt)}\, du$$

To find the temperature distribution that results from an initial hot spot concentrated at the origin, we need to compute $\lim_{a \to 0} T(x, t)$. Use l'Hospital's Rule to find this limit.

67. If f is a continuous function such that

$$\int_1^x f(t)\, dt = (x - 1)e^{2x} + \int_1^x e^{-t} f(t)\, dt$$

for all x, find an explicit formula for $f(x)$.

68. Suppose h is a function such that $h(1) = -2$, $h'(1) = 2$, $h''(1) = 3$, $h(2) = 6$, $h'(2) = 5$, $h''(2) = 13$, and h'' is continuous everywhere. Evaluate $\int_1^2 h''(u)\, du$.

69. If f' is continuous on $[a, b]$, show that

$$2 \int_a^b f(x) f'(x)\, dx = [f(b)]^2 - [f(a)]^2$$

70. Find

$$\lim_{h \to 0} \frac{1}{h} \int_2^{2+h} \sqrt{1 + t^3}\, dt$$

71. If f is continuous on $[0, 1]$, prove that

$$\int_0^1 f(x)\, dx = \int_0^1 f(1 - x)\, dx$$

72. Evaluate

$$\lim_{n \to \infty} \frac{1}{n}\left[\left(\frac{1}{n}\right)^9 + \left(\frac{2}{n}\right)^9 + \left(\frac{3}{n}\right)^9 + \cdots + \left(\frac{n}{n}\right)^9\right]$$

73. Suppose f is continuous, $f(0) = 0$, $f(1) = 1$, $f'(x) > 0$, and $\int_0^1 f(x)\, dx = \frac{1}{3}$. Find the value of the integral $\int_0^1 f^{-1}(y)\, dy$.

Problems Plus

Before you look at the solution of the following example, cover it up and first try to solve the problem yourself.

EXAMPLE Evaluate $\lim_{x \to 3} \left(\dfrac{x}{x-3} \int_3^x \dfrac{\sin t}{t} dt \right)$.

SOLUTION Let's start by having a preliminary look at the ingredients of the function. What happens to the first factor, $x/(x-3)$, when x approaches 3? The numerator approaches 3 and the denominator approaches 0, so we have

$$\dfrac{x}{x-3} \to \infty \quad \text{as} \quad x \to 3^+ \quad \text{and} \quad \dfrac{x}{x-3} \to -\infty \quad \text{as} \quad x \to 3^-$$

The second factor approaches $\int_3^3 (\sin t)/t \, dt$, which is 0. It's not clear what happens to the function as a whole. (One factor is becoming large while the other is becoming small.) So how do we proceed?

One of the principles of problem solving is *recognizing something familiar.* Is there a part of the function that reminds us of something we've seen before? Well, the integral

$$\int_3^x \dfrac{\sin t}{t} dt$$

has x as its upper limit of integration and that type of integral occurs in Part 1 of the Fundamental Theorem of Calculus:

$$\dfrac{d}{dx} \int_a^x f(t) \, dt = f(x)$$

This suggests that differentiation might be involved.

Once we start thinking about differentiation, the denominator $(x - 3)$ reminds us of something else that should be familiar: One of the forms of the definition of the derivative in Chapter 2 is

$$F'(a) = \lim_{x \to a} \dfrac{F(x) - F(a)}{x - a}$$

and with $a = 3$ this becomes

$$F'(3) = \lim_{x \to 3} \dfrac{F(x) - F(3)}{x - 3}$$

So what is the function F in our situation? Notice that if we define

$$F(x) = \int_3^x \dfrac{\sin t}{t} dt$$

then $F(3) = 0$. What about the factor x in the numerator? That's just a red herring, so let's factor it out and put together the calculation:

$$\lim_{x \to 3} \left(\dfrac{x}{x-3} \int_3^x \dfrac{\sin t}{t} dt \right) = \lim_{x \to 3} x \cdot \lim_{x \to 3} \dfrac{\int_3^x \dfrac{\sin t}{t} dt}{x - 3} = 3 \lim_{x \to 3} \dfrac{F(x) - F(3)}{x - 3}$$

$$= 3F'(3) = 3 \dfrac{\sin 3}{3} = \sin 3 \quad \text{(FTC1)} \quad \blacksquare$$

PS The principles of problem solving are discussed on page 71.

Another approach is to use l'Hospital's Rule

Problems

1. If $x \sin \pi x = \int_0^{x^2} f(t) \, dt$, where f is a continuous function, find $f(4)$.

2. Find the minimum value of the area of the region under the curve $y = x + 1/x$ from $x = a$ to $x = a + 1.5$, for all $a > 0$.

3. If $\int_0^4 e^{(x-2)^4} dx = k$, find the value of $\int_0^4 xe^{(x-2)^4} dx$.

4. (a) Graph several members of the family of functions $f(x) = (2cx - x^2)/c^3$ for $c > 0$ and look at the regions enclosed by these curves and the x-axis. Make a conjecture about how the areas of these regions are related.
   (b) Prove your conjecture in part (a).
   (c) Take another look at the graphs in part (a) and use them to sketch the curve traced out by the vertices (highest points) of the family of functions. Can you guess what kind of curve this is?
   (d) Find an equation of the curve you sketched in part (c).

5. If $f(x) = \int_0^{g(x)} \dfrac{1}{\sqrt{1+t^3}} dt$, where $g(x) = \int_0^{\cos x} [1 + \sin(t^2)] dt$, find $f'(\pi/2)$.

6. If $f(x) = \int_0^x x^2 \sin(t^2) dt$, find $f'(x)$.

7. Evaluate $\displaystyle\lim_{x \to 0} \dfrac{1}{x} \int_0^x (1 - \tan 2t)^{1/t} dt$.

8. The figure shows two regions in the first quadrant: $A(t)$ is the area under the curve $y = \sin(x^2)$ from 0 to t, and $B(t)$ is the area of the triangle with vertices O, P, and $(t, 0)$. Find $\displaystyle\lim_{t \to 0^+} A(t)/B(t)$.

9. Find the interval $[a, b]$ for which the value of the integral $\int_a^b (2 + x - x^2) dx$ is a maximum.

10. Use an integral to estimate the sum $\displaystyle\sum_{i=1}^{10000} \sqrt{i}$.

11. (a) Evaluate $\int_0^n [\![x]\!] dx$, where n is a positive integer.
    (b) Evaluate $\int_a^b [\![x]\!] dx$, where a and b are real numbers with $0 \le a < b$.

12. Find $\dfrac{d^2}{dx^2} \int_0^x \left(\int_1^{\sin t} \sqrt{1 + u^4}\, du \right) dt$.

13. Suppose the coefficients of the cubic polynomial $P(x) = a + bx + cx^2 + dx^3$ satisfy the equation
$$a + \dfrac{b}{2} + \dfrac{c}{3} + \dfrac{d}{4} = 0$$
Show that the equation $P(x) = 0$ has a root between 0 and 1. Can you generalize this result for an nth-degree polynomial?

14. A circular disk of radius r is used in an evaporator and is rotated in a vertical plane. If it is to be partially submerged in the liquid so as to maximize the exposed wetted area of the disk, show that the center of the disk should be positioned at a height $r/\sqrt{1 + \pi^2}$ above the surface of the liquid.

15. Prove that if f is continuous, then $\displaystyle\int_0^x f(u)(x - u) du = \int_0^x \left(\int_0^u f(t) dt \right) du$.

16. The figure shows a parabolic segment, that is, a portion of a parabola cut off by a chord AB. It also shows a point C on the parabola with the property that the tangent line at C is parallel to the chord AB. Archimedes proved that the area of the parabolic segment is $\tfrac{4}{3}$ times the area of the inscribed triangle ABC. Verify Archimedes' result for the parabola $y = 4 - x^2$ and the line $y = x + 2$.

17. Given the point (a, b) in the first quadrant, find the downward-opening parabola that passes through the point (a, b) and the origin such that the area under the parabola is a minimum.

18. The figure shows a region consisting of all points inside a square that are closer to the center than to the sides of the square. Find the area of the region.

19. Evaluate $\displaystyle\lim_{n \to \infty} \left(\dfrac{1}{\sqrt{n}\sqrt{n+1}} + \dfrac{1}{\sqrt{n}\sqrt{n+2}} + \cdots + \dfrac{1}{\sqrt{n}\sqrt{n+n}} \right)$.

20. For any number c, we let $f_c(x)$ be the smaller of the two numbers $(x - c)^2$ and $(x - c - 2)^2$. Then we define $g(c) = \int_0^1 f_c(x) dx$. Find the maximum and minimum values of $g(c)$ if $-2 \le c \le 2$.

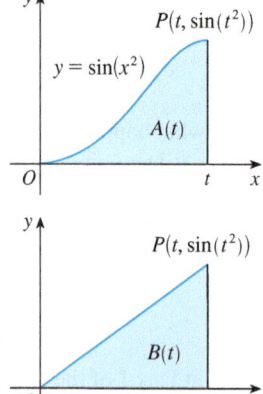

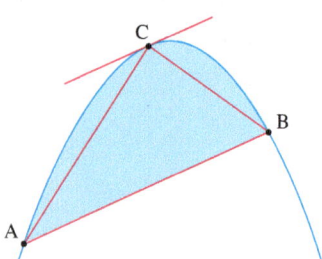

FIGURE FOR PROBLEM 8

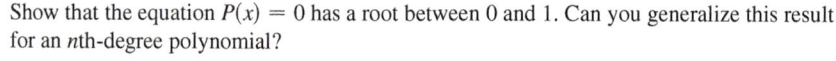

FIGURE FOR PROBLEM 16

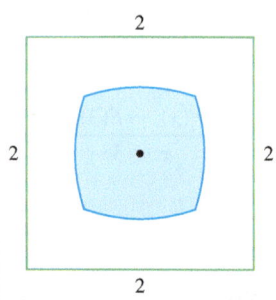

FIGURE FOR PROBLEM 18

6 Applications of Integration

© Richard Paul Kane / Shutterstock.com

When a bat strikes a baseball, the collision lasts only about a thousandth of a second. In the project on page 464, you will use calculus to find the average force on the bat when this happens. Several other applications of calculus to the game of baseball are explored as well.

IN THIS CHAPTER WE EXPLORE some of the applications of the definite integral by using it to compute areas between curves, volumes of solids, and the work done by a varying force. The common theme is the following general method, which is similar to the one we used to find areas under curves: we break up a quantity Q into a large number of small parts. We next approximate each small part by a quantity of the form $f(x_i^*)\,\Delta x$ and thus approximate Q by a Riemann sum. Then we take the limit and express Q as an integral. Finally we evaluate the integral using the Fundamental Theorem of Calculus or the Midpoint Rule.

6.1 Areas Between Curves

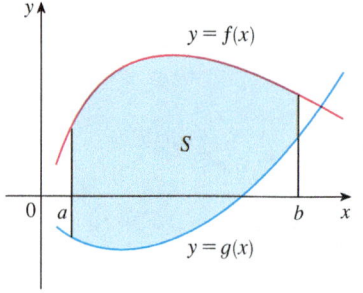

FIGURE 1
$S = \{(x, y) \mid a \leq x \leq b,\ g(x) \leq y \leq f(x)\}$

In Chapter 5 we defined and calculated areas of regions that lie under the graphs of functions. Here we use integrals to find areas of regions that lie between the graphs of two functions.

Consider the region S that lies between two curves $y = f(x)$ and $y = g(x)$ and between the vertical lines $x = a$ and $x = b$, where f and g are continuous functions and $f(x) \geq g(x)$ for all x in $[a, b]$. (See Figure 1.)

Just as we did for areas under curves in Section 5.1, we divide S into n strips of equal width and then we approximate the ith strip by a rectangle with base Δx and height $f(x_i^*) - g(x_i^*)$. (See Figure 2. If we like, we could take all of the sample points to be right endpoints, in which case $x_i^* = x_i$.) The Riemann sum

$$\sum_{i=1}^{n} [f(x_i^*) - g(x_i^*)] \Delta x$$

is therefore an approximation to what we intuitively think of as the area of S.

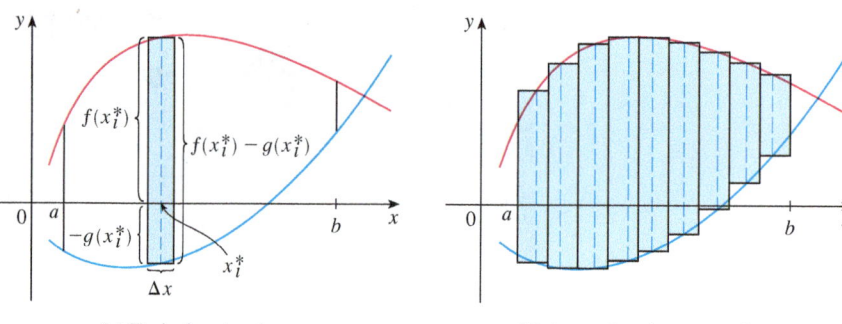

FIGURE 2 (a) Typical rectangle   (b) Approximating rectangles

This approximation appears to become better and better as $n \to \infty$. Therefore we define the **area** A of the region S as the limiting value of the sum of the areas of these approximating rectangles.

1
$$A = \lim_{n \to \infty} \sum_{i=1}^{n} [f(x_i^*) - g(x_i^*)] \Delta x$$

We recognize the limit in (1) as the definite integral of $f - g$. Therefore we have the following formula for area.

2 The area A of the region bounded by the curves $y = f(x)$, $y = g(x)$, and the lines $x = a$, $x = b$, where f and g are continuous and $f(x) \geq g(x)$ for all x in $[a, b]$, is

$$A = \int_a^b [f(x) - g(x)]\, dx$$

Notice that in the special case where $g(x) = 0$, S is the region under the graph of f and our general definition of area (1) reduces to our previous definition (Definition 5.1.2).

SECTION 6.1 Areas Between Curves 429

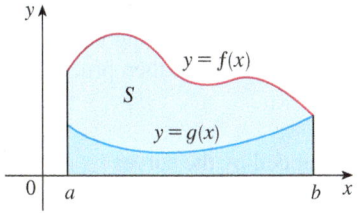

FIGURE 3
$$A = \int_a^b f(x)\,dx - \int_a^b g(x)\,dx$$

In the case where both f and g are positive, you can see from Figure 3 why (2) is true:

$$A = [\text{area under } y = f(x)] - [\text{area under } y = g(x)]$$

$$= \int_a^b f(x)\,dx - \int_a^b g(x)\,dx = \int_a^b [f(x) - g(x)]\,dx$$

EXAMPLE 1 Find the area of the region bounded above by $y = e^x$, bounded below by $y = x$, and bounded on the sides by $x = 0$ and $x = 1$.

SOLUTION The region is shown in Figure 4. The upper boundary curve is $y = e^x$ and the lower boundary curve is $y = x$. So we use the area formula (2) with $f(x) = e^x$, $g(x) = x$, $a = 0$, and $b = 1$:

$$A = \int_0^1 (e^x - x)\,dx = e^x - \tfrac{1}{2}x^2 \Big]_0^1$$

$$= e - \tfrac{1}{2} - 1 = e - 1.5 \qquad \blacksquare$$

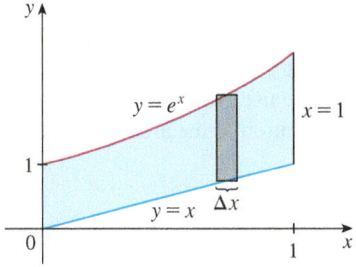

FIGURE 4

In Figure 4 we drew a typical approximating rectangle with width Δx as a reminder of the procedure by which the area is defined in (1). In general, when we set up an integral for an area, it's helpful to sketch the region to identify the top curve y_T, the bottom curve y_B, and a typical approximating rectangle as in Figure 5. Then the area of a typical rectangle is $(y_T - y_B)\,\Delta x$ and the equation

$$A = \lim_{n \to \infty} \sum_{i=1}^n (y_T - y_B)\,\Delta x = \int_a^b (y_T - y_B)\,dx$$

summarizes the procedure of adding (in a limiting sense) the areas of all the typical rectangles.

Notice that in Figure 5 the left-hand boundary reduces to a point, whereas in Figure 3 the right-hand boundary reduces to a point. In the next example both of the side boundaries reduce to a point, so the first step is to find a and b.

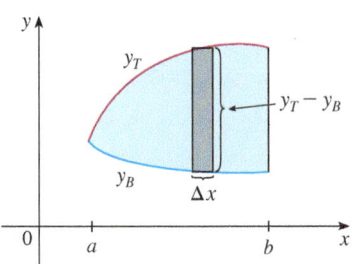

FIGURE 5

EXAMPLE 2 Find the area of the region enclosed by the parabolas $y = x^2$ and $y = 2x - x^2$.

SOLUTION We first find the points of intersection of the parabolas by solving their equations simultaneously. This gives $x^2 = 2x - x^2$, or $2x^2 - 2x = 0$. Thus $2x(x - 1) = 0$, so $x = 0$ or 1. The points of intersection are $(0, 0)$ and $(1, 1)$.

We see from Figure 6 that the top and bottom boundaries are

$$y_T = 2x - x^2 \qquad \text{and} \qquad y_B = x^2$$

The area of a typical rectangle is

$$(y_T - y_B)\,\Delta x = (2x - x^2 - x^2)\,\Delta x$$

and the region lies between $x = 0$ and $x = 1$. So the total area is

$$A = \int_0^1 (2x - 2x^2)\,dx = 2\int_0^1 (x - x^2)\,dx$$

$$= 2\left[\frac{x^2}{2} - \frac{x^3}{3}\right]_0^1 = 2\left(\frac{1}{2} - \frac{1}{3}\right) = \frac{1}{3} \qquad \blacksquare$$

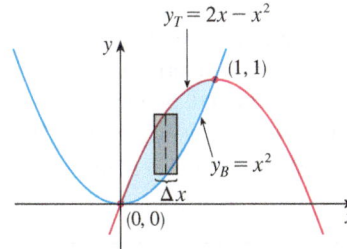

FIGURE 6

Sometimes it's difficult, or even impossible, to find the points of intersection of two curves exactly. As shown in the following example, we can use a graphing calculator or computer to find approximate values for the intersection points and then proceed as before.

EXAMPLE 3 Find the approximate area of the region bounded by the curves $y = x/\sqrt{x^2 + 1}$ and $y = x^4 - x$.

SOLUTION If we were to try to find the exact intersection points, we would have to solve the equation

$$\frac{x}{\sqrt{x^2 + 1}} = x^4 - x$$

This looks like a very difficult equation to solve exactly (in fact, it's impossible), so instead we use a graphing device to draw the graphs of the two curves in Figure 7. One intersection point is the origin. We zoom in toward the other point of intersection and find that $x \approx 1.18$. (If greater accuracy is required, we could use Newton's method or solve numerically on our graphing device.) So an approximation to the area between the curves is

$$A \approx \int_0^{1.18} \left[\frac{x}{\sqrt{x^2 + 1}} - (x^4 - x) \right] dx$$

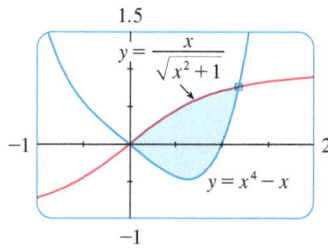

FIGURE 7

To integrate the first term we use the substitution $u = x^2 + 1$. Then $du = 2x\,dx$, and when $x = 1.18$, we have $u \approx 2.39$; when $x = 0$, $u = 1$. So

$$A \approx \tfrac{1}{2} \int_1^{2.39} \frac{du}{\sqrt{u}} - \int_0^{1.18} (x^4 - x)\,dx$$

$$= \sqrt{u}\,\Big]_1^{2.39} - \left[\frac{x^5}{5} - \frac{x^2}{2} \right]_0^{1.18}$$

$$= \sqrt{2.39} - 1 - \frac{(1.18)^5}{5} + \frac{(1.18)^2}{2}$$

$$\approx 0.785 \qquad \blacksquare$$

EXAMPLE 4 Figure 8 shows velocity curves for two cars, A and B, that start side by side and move along the same road. What does the area between the curves represent? Use the Midpoint Rule to estimate it.

SOLUTION We know from Section 5.4 that the area under the velocity curve A represents the distance traveled by car A during the first 16 seconds. Similarly, the area under curve B is the distance traveled by car B during that time period. So the area between these curves, which is the difference of the areas under the curves, is the distance between the cars after 16 seconds. We read the velocities from the graph and convert them to feet per second (1 mi/h = $\tfrac{5280}{3600}$ ft/s).

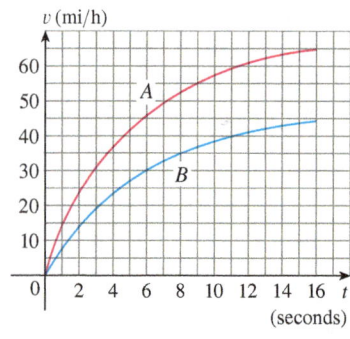

FIGURE 8

t	0	2	4	6	8	10	12	14	16
v_A	0	34	54	67	76	84	89	92	95
v_B	0	21	34	44	51	56	60	63	65
$v_A - v_B$	0	13	20	23	25	28	29	29	30

We use the Midpoint Rule with $n = 4$ intervals, so that $\Delta t = 4$. The midpoints of the intervals are $\bar{t}_1 = 2$, $\bar{t}_2 = 6$, $\bar{t}_3 = 10$, and $\bar{t}_4 = 14$. We estimate the distance between the cars after 16 seconds as follows:

$$\int_0^{16} (v_A - v_B)\, dt \approx \Delta t\, [13 + 23 + 28 + 29]$$

$$= 4(93) = 372 \text{ ft}$$

EXAMPLE 5 Figure 9 is an example of a *pathogenesis curve* for a measles infection. It shows how the disease develops in an individual with no immunity after the measles virus spreads to the bloodstream from the respiratory tract.

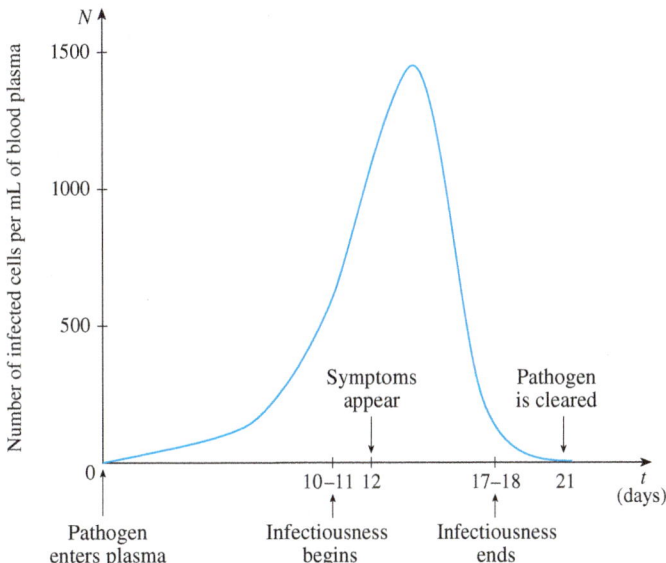

FIGURE 9

Measles pathogenesis curve

Source: J. M. Heffernan et al., "An In-Host Model of Acute Infection: Measles as a Case Study," *Theoretical Population Biology* 73 (2008): 134–47.

The patient becomes infectious to others once the concentration of infected cells becomes great enough, and he or she remains infectious until the immune system manages to prevent further transmission. However, symptoms don't develop until the "amount of infection" reaches a particular threshold. The amount of infection needed to develop symptoms depends on both the concentration of infected cells and time, and corresponds to the area under the pathogenesis curve until symptoms appear. (See Exercise 5.1.19.)

(a) The pathogenesis curve in Figure 9 has been modeled by $f(t) = -t(t - 21)(t + 1)$. If infectiousness begins on day $t_1 = 10$ and ends on day $t_2 = 18$, what are the corresponding concentration levels of infected cells?

(b) The *level of infectiousness* for an infected person is the area between $N = f(t)$ and the line through the points $P_1(t_1, f(t_1))$ and $P_2(t_2, f(t_2))$, measured in (cells/mL) · days. (See Figure 10.) Compute the level of infectiousness for this particular patient.

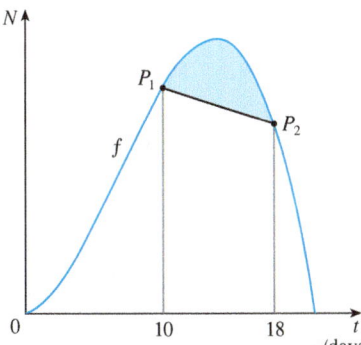

FIGURE 10

SOLUTION

(a) Infectiousness begins when the concentration reaches $f(10) = 1210$ cells/mL and ends when the concentration reduces to $f(18) = 1026$ cells/mL.

(b) The line through P_1 and P_2 has slope $\frac{1026 - 1210}{18 - 10} = -\frac{184}{8} = -23$ and equation $N - 1210 = -23(t - 10) \iff N = -23t + 1440$. The area between f and this line is

$$\int_{10}^{18} [f(t) - (-23t + 1440)]\, dt = \int_{10}^{18} (-t^3 + 20t^2 + 21t + 23t - 1440)\, dt$$

$$= \int_{10}^{18} (-t^3 + 20t^2 + 44t - 1440)\, dt$$

$$= \left[-\frac{t^4}{4} + 20\frac{t^3}{3} + 44\frac{t^2}{2} - 1440t \right]_{10}^{18}$$

$$= -6156 - (-8033\tfrac{1}{3}) \approx 1877$$

Thus the level of infectiousness for this patient is about 1877 (cells/mL) · days. ■

If we are asked to find the area between the curves $y = f(x)$ and $y = g(x)$ where $f(x) \geq g(x)$ for some values of x but $g(x) \geq f(x)$ for other values of x, then we split the given region S into several regions $S_1, S_2, \ldots$ with areas $A_1, A_2, \ldots$ as shown in Figure 11. We then define the area of the region S to be the sum of the areas of the smaller regions $S_1, S_2, \ldots$, that is, $A = A_1 + A_2 + \cdots$. Since

$$|f(x) - g(x)| = \begin{cases} f(x) - g(x) & \text{when } f(x) \geq g(x) \\ g(x) - f(x) & \text{when } g(x) \geq f(x) \end{cases}$$

we have the following expression for A.

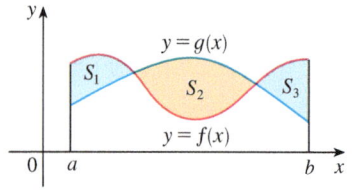

FIGURE 11

3 The area between the curves $y = f(x)$ and $y = g(x)$ and between $x = a$ and $x = b$ is

$$A = \int_a^b |f(x) - g(x)|\, dx$$

When evaluating the integral in (3), however, we must still split it into integrals corresponding to $A_1, A_2, \ldots$.

EXAMPLE 6 Find the area of the region bounded by the curves $y = \sin x$, $y = \cos x$, $x = 0$, and $x = \pi/2$.

SOLUTION The points of intersection occur when $\sin x = \cos x$, that is, when $x = \pi/4$ (since $0 \leq x \leq \pi/2$). The region is sketched in Figure 12.

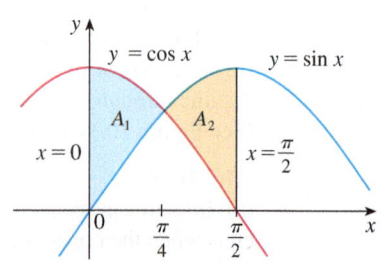

FIGURE 12

Observe that $\cos x \geq \sin x$ when $0 \leq x \leq \pi/4$ but $\sin x \geq \cos x$ when $\pi/4 \leq x \leq \pi/2$. Therefore the required area is

$$A = \int_0^{\pi/2} |\cos x - \sin x|\, dx = A_1 + A_2$$

$$= \int_0^{\pi/4} (\cos x - \sin x)\, dx + \int_{\pi/4}^{\pi/2} (\sin x - \cos x)\, dx$$

$$= \left[\sin x + \cos x\right]_0^{\pi/4} + \left[-\cos x - \sin x\right]_{\pi/4}^{\pi/2}$$

$$= \left(\frac{1}{\sqrt{2}} + \frac{1}{\sqrt{2}} - 0 - 1\right) + \left(-0 - 1 + \frac{1}{\sqrt{2}} + \frac{1}{\sqrt{2}}\right)$$

$$= 2\sqrt{2} - 2$$

In this particular example we could have saved some work by noticing that the region is symmetric about $x = \pi/4$ and so

$$A = 2A_1 = 2 \int_0^{\pi/4} (\cos x - \sin x)\, dx$$

■

Some regions are best treated by regarding x as a function of y. If a region is bounded by curves with equations $x = f(y)$, $x = g(y)$, $y = c$, and $y = d$, where f and g are continuous and $f(y) \geq g(y)$ for $c \leq y \leq d$ (see Figure 13), then its area is

$$A = \int_c^d [f(y) - g(y)]\, dy$$

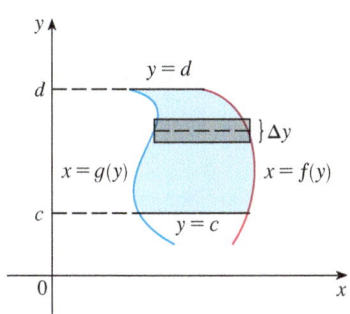

FIGURE 13

If we write x_R for the right boundary and x_L for the left boundary, then, as Figure 14 illustrates, we have

$$A = \int_c^d (x_R - x_L)\, dy$$

Here a typical approximating rectangle has dimensions $x_R - x_L$ and Δy.

EXAMPLE 7 Find the area enclosed by the line $y = x - 1$ and the parabola $y^2 = 2x + 6$.

SOLUTION By solving the two equations we find that the points of intersection are $(-1, -2)$ and $(5, 4)$. We solve the equation of the parabola for x and notice from Figure 15 that the left and right boundary curves are

$$x_L = \tfrac{1}{2}y^2 - 3 \qquad \text{and} \qquad x_R = y + 1$$

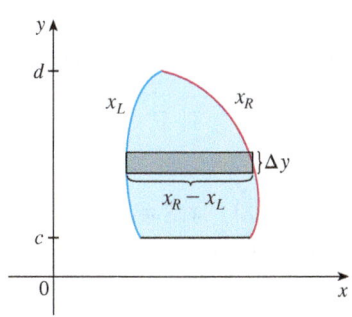

FIGURE 14

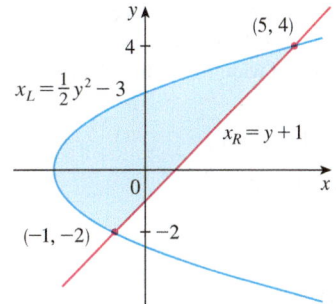

FIGURE 15

We must integrate between the appropriate y-values, $y = -2$ and $y = 4$. Thus

$$A = \int_{-2}^{4} (x_R - x_L)\, dy = \int_{-2}^{4} \left[(y + 1) - \left(\tfrac{1}{2}y^2 - 3\right)\right] dy$$

$$= \int_{-2}^{4} \left(-\tfrac{1}{2}y^2 + y + 4\right) dy$$

$$= -\frac{1}{2}\left(\frac{y^3}{3}\right) + \frac{y^2}{2} + 4y \Bigg]_{-2}^{4}$$

$$= -\tfrac{1}{6}(64) + 8 + 16 - \left(\tfrac{4}{3} + 2 - 8\right) = 18 \quad \blacksquare$$

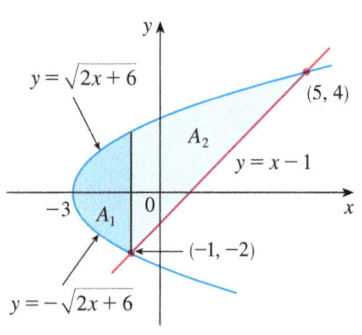

FIGURE 16

NOTE We could have found the area in Example 7 by integrating with respect to x instead of y, but the calculation is much more involved. Because the bottom boundary consists of two different curves, it would have meant splitting the region in two and computing the areas labeled A_1 and A_2 in Figure 16. The method we used in Example 7 is *much* easier.

6.1 EXERCISES

1–4 Find the area of the shaded region.

1.

2.

3.

4.

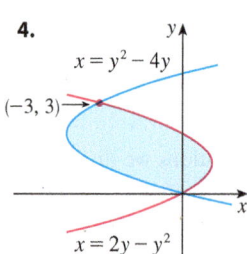

5–12 Sketch the region enclosed by the given curves. Decide whether to integrate with respect to x or y. Draw a typical approximating rectangle and label its height and width. Then find the area of the region.

5. $y = e^x$, $y = x^2 - 1$, $x = -1$, $x = 1$

6. $y = \sin x$, $y = x$, $x = \pi/2$, $x = \pi$

7. $y = (x - 2)^2$, $y = x$

8. $y = x^2 - 4x$, $y = 2x$

9. $y = 1/x$, $y = 1/x^2$, $x = 2$

10. $y = \sin x$, $y = 2x/\pi$, $x \geq 0$

11. $x = 1 - y^2$, $x = y^2 - 1$

12. $4x + y^2 = 12$, $x = y$

13–28 Sketch the region enclosed by the given curves and find its area.

13. $y = 12 - x^2$, $y = x^2 - 6$

14. $y = x^2$, $y = 4x - x^2$

15. $y = \sec^2 x$, $y = 8\cos x$, $-\pi/3 \leq x \leq \pi/3$

16. $y = \cos x$, $y = 2 - \cos x$, $0 \leq x \leq 2\pi$

17. $x = 2y^2$, $x = 4 + y^2$

18. $y = \sqrt{x - 1}$, $x - y = 1$

19. $y = \cos \pi x$, $y = 4x^2 - 1$

20. $x = y^4$, $y = \sqrt{2 - x}$, $y = 0$

21. $y = \tan x$, $y = 2 \sin x$, $-\pi/3 \leq x \leq \pi/3$

22. $y = x^3$, $y = x$

23. $y = \sqrt[3]{2x}$, $y = \tfrac{1}{8}x^2$, $0 \leq x \leq 6$

24. $y = \cos x$, $y = 1 - \cos x$, $0 \leq x \leq \pi$

25. $y = x^4$, $y = 2 - |x|$

26. $y = \sinh x$, $y = e^{-x}$, $x = 0$, $x = 2$

27. $y = 1/x$, $y = x$, $y = \tfrac{1}{4}x$, $x > 0$

28. $y = \tfrac{1}{4}x^2$, $y = 2x^2$, $x + y = 3$, $x \geq 0$

29. The graphs of two functions are shown with the areas of the regions between the curves indicated.
(a) What is the total area between the curves for $0 \leq x \leq 5$?
(b) What is the value of $\int_0^5 [f(x) - g(x)]\, dx$?

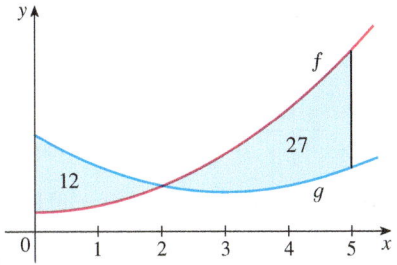

30–32 Sketch the region enclosed by the given curves and find its area.

30. $y = \dfrac{x}{\sqrt{1+x^2}}, \quad y = \dfrac{x}{\sqrt{9-x^2}}, \quad x \geq 0$

31. $y = \dfrac{x}{1+x^2}, \quad y = \dfrac{x^2}{1+x^3}$

32. $y = \dfrac{\ln x}{x}, \quad y = \dfrac{(\ln x)^2}{x}$

33–34 Use calculus to find the area of the triangle with the given vertices.

33. $(0,0), \quad (3,1), \quad (1,2)$

34. $(2,0), \quad (0,2), \quad (-1,1)$

35–36 Evaluate the integral and interpret it as the area of a region. Sketch the region.

35. $\displaystyle\int_0^{\pi/2} |\sin x - \cos 2x|\, dx$ **36.** $\displaystyle\int_{-1}^{1} |3^x - 2^x|\, dx$

37–40 Use a graph to find approximate x-coordinates of the points of intersection of the given curves. Then find (approximately) the area of the region bounded by the curves.

37. $y = x\sin(x^2), \quad y = x^4, \quad x \geq 0$

38. $y = \dfrac{x}{(x^2+1)^2}, \quad y = x^5 - x, \quad x \geq 0$

39. $y = 3x^2 - 2x, \quad y = x^3 - 3x + 4$

40. $y = 1.3^x, \quad y = 2\sqrt{x}$

41–44 Graph the region between the curves and use your calculator to compute the area correct to five decimal places.

41. $y = \dfrac{2}{1+x^4}, \quad y = x^2$ **42.** $y = e^{1-x^2}, \quad y = x^4$

43. $y = \tan^2 x, \quad y = \sqrt{x}$ **44.** $y = \cos x, \quad y = x + 2\sin^4 x$

45. Use a computer algebra system to find the exact area enclosed by the curves $y = x^5 - 6x^3 + 4x$ and $y = x$.

46. Sketch the region in the xy-plane defined by the inequalities $x - 2y^2 \geq 0$, $1 - x - |y| \geq 0$ and find its area.

47. Racing cars driven by Chris and Kelly are side by side at the start of a race. The table shows the velocities of each car (in miles per hour) during the first ten seconds of the race. Use the Midpoint Rule to estimate how much farther Kelly travels than Chris does during the first ten seconds.

t	v_C	v_K	t	v_C	v_K
0	0	0	6	69	80
1	20	22	7	75	86
2	32	37	8	81	93
3	46	52	9	86	98
4	54	61	10	90	102
5	62	71			

48. The widths (in meters) of a kidney-shaped swimming pool were measured at 2-meter intervals as indicated in the figure. Use the Midpoint Rule to estimate the area of the pool.

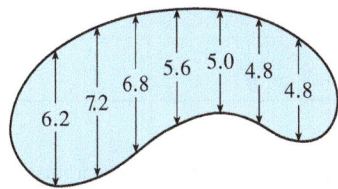

49. A cross-section of an airplane wing is shown. Measurements of the thickness of the wing, in centimeters, at 20-centimeter intervals are 5.8, 20.3, 26.7, 29.0, 27.6, 27.3, 23.8, 20.5, 15.1, 8.7, and 2.8. Use the Midpoint Rule to estimate the area of the wing's cross-section.

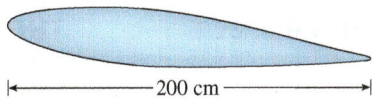

50. If the birth rate of a population is $b(t) = 2200e^{0.024t}$ people per year and the death rate is $d(t) = 1460e^{0.018t}$ people per year, find the area between these curves for $0 \leq t \leq 10$. What does this area represent?

51. In Example 5, we modeled a measles pathogenesis curve by a function f. A patient infected with the measles virus who has some immunity to the virus has a pathogenesis curve that can be modeled by, for instance, $g(t) = 0.9f(t)$.
(a) If the same threshold concentration of the virus is required for infectiousness to begin as in Example 5, on what day does this occur?

(b) Let P_3 be the point on the graph of g where infectiousness begins. It has been shown that infectiousness ends at a point P_4 on the graph of g where the line through P_3, P_4 has the same slope as the line through P_1, P_2 in Example 5(b). On what day does infectiousness end?
(c) Compute the level of infectiousness for this patient.

52. The rates at which rain fell, in inches per hour, in two different locations t hours after the start of a storm are given by $f(t) = 0.73t^3 - 2t^2 + t + 0.6$ and $g(t) = 0.17t^2 - 0.5t + 1.1$. Compute the area between the graphs for $0 \le t \le 2$ and interpret your result in this context.

53. Two cars, A and B, start side by side and accelerate from rest. The figure shows the graphs of their velocity functions.
(a) Which car is ahead after one minute? Explain.
(b) What is the meaning of the area of the shaded region?
(c) Which car is ahead after two minutes? Explain.
(d) Estimate the time at which the cars are again side by side.

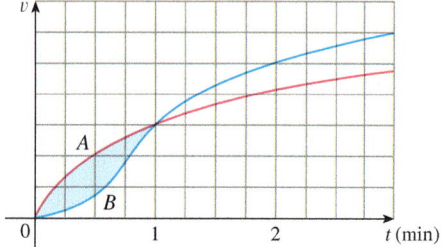

54. The figure shows graphs of the marginal revenue function R' and the marginal cost function C' for a manufacturer. [Recall from Section 4.7 that $R(x)$ and $C(x)$ represent the revenue and cost when x units are manufactured. Assume that R and C are measured in thousands of dollars.] What is the meaning of the area of the shaded region? Use the Midpoint Rule to estimate the value of this quantity.

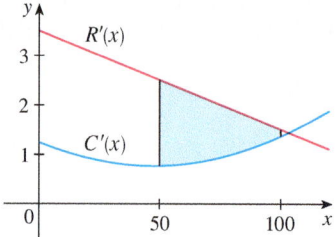

55. The curve with equation $y^2 = x^2(x + 3)$ is called **Tschirnhausen's cubic**. If you graph this curve you will see that part of the curve forms a loop. Find the area enclosed by the loop.

56. Find the area of the region bounded by the parabola $y = x^2$, the tangent line to this parabola at $(1, 1)$, and the x-axis.

57. Find the number b such that the line $y = b$ divides the region bounded by the curves $y = x^2$ and $y = 4$ into two regions with equal area.

58. (a) Find the number a such that the line $x = a$ bisects the area under the curve $y = 1/x^2$, $1 \le x \le 4$.
(b) Find the number b such that the line $y = b$ bisects the area in part (a).

59. Find the values of c such that the area of the region bounded by the parabolas $y = x^2 - c^2$ and $y = c^2 - x^2$ is 576.

60. Suppose that $0 < c < \pi/2$. For what value of c is the area of the region enclosed by the curves $y = \cos x$, $y = \cos(x - c)$, and $x = 0$ equal to the area of the region enclosed by the curves $y = \cos(x - c)$, $x = \pi$, and $y = 0$?

61. For what values of m do the line $y = mx$ and the curve $y = x/(x^2 + 1)$ enclose a region? Find the area of the region.

APPLIED PROJECT THE GINI INDEX

How is it possible to measure the distribution of income among the inhabitants of a given country? One such measure is the *Gini index*, named after the Italian economist Corrado Gini, who first devised it in 1912.

We first rank all households in a country by income and then we compute the percentage of households whose income is at most a given percentage of the country's total income. We define a **Lorenz curve** $y = L(x)$ on the interval $[0, 1]$ by plotting the point $(a/100, b/100)$ on the curve if the bottom $a\%$ of households receive at most $b\%$ of the total income. For instance, in Figure 1 (on page 437) the point $(0.4, 0.12)$ is on the Lorenz curve for the United States in 2010 because the poorest 40% of the population received just 12% of the total income. Likewise, the bottom 80% of the population received 50% of the total income, so the point $(0.8, 0.5)$ lies on the Lorenz curve. (The Lorenz curve is named after the American economist Max Lorenz.)

APPLIED PROJECT The Gini Index

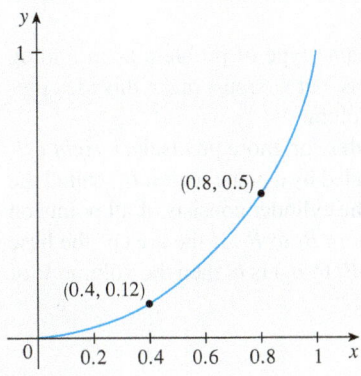

FIGURE 1
Lorenz curve for the US in 2010

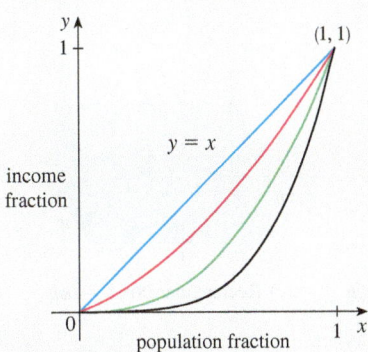

FIGURE 2

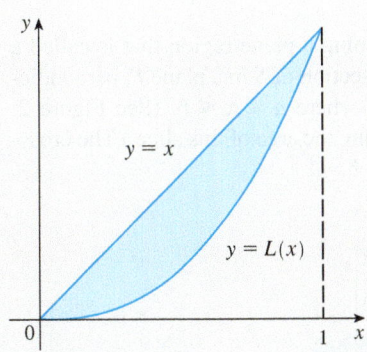

FIGURE 3

Figure 2 shows some typical Lorenz curves. They all pass through the points (0, 0) and (1, 1) and are concave upward. In the extreme case $L(x) = x$, society is perfectly egalitarian: the poorest $a\%$ of the population receives $a\%$ of the total income and so everybody receives the same income. The area between a Lorenz curve $y = L(x)$ and the line $y = x$ measures how much the income distribution differs from absolute equality. The **Gini index** (sometimes called the **Gini coefficient** or the **coefficient of inequality**) is the area between the Lorenz curve and the line $y = x$ (shaded in Figure 3) divided by the area under $y = x$.

1. (a) Show that the Gini index G is twice the area between the Lorenz curve and the line $y = x$, that is,
$$G = 2 \int_0^1 [x - L(x)]\, dx$$

   (b) What is the value of G for a perfectly egalitarian society (everybody has the same income)? What is the value of G for a perfectly totalitarian society (a single person receives all the income)?

2. The following table (derived from data supplied by the US Census Bureau) shows values of the Lorenz function for income distribution in the United States for the year 2010.

x	0.0	0.2	0.4	0.6	0.8	1.0
$L(x)$	0.000	0.034	0.120	0.266	0.498	1.000

   (a) What percentage of the total US income was received by the richest 20% of the population in 2010?

   (b) Use a calculator or computer to fit a quadratic function to the data in the table. Graph the data points and the quadratic function. Is the quadratic model a reasonable fit?

   (c) Use the quadratic model for the Lorenz function to estimate the Gini index for the United States in 2010.

3. The following table gives values for the Lorenz function in the years 1970, 1980, 1990, and 2000. Use the method of Problem 2 to estimate the Gini index for the United States for those years and compare with your answer to Problem 2(c). Do you notice a trend?

x	0.0	0.2	0.4	0.6	0.8	1.0
1970	0.000	0.041	0.149	0.323	0.568	1.000
1980	0.000	0.042	0.144	0.312	0.559	1.000
1990	0.000	0.038	0.134	0.293	0.530	1.000
2000	0.000	0.036	0.125	0.273	0.503	1.000

4. A power model often provides a more accurate fit than a quadratic model for a Lorenz function. If you have a computer with Maple or Mathematica, fit a power function $(y = ax^k)$ to the data in Problem 2 and use it to estimate the Gini index for the United States in 2010. Compare with your answer to parts (b) and (c) of Problem 2.

6.2 Volumes

In trying to find the volume of a solid we face the same type of problem as in finding areas. We have an intuitive idea of what volume means, but we must make this idea precise by using calculus to give an exact definition of volume.

We start with a simple type of solid called a **cylinder** (or, more precisely, a *right cylinder*). As illustrated in Figure 1(a), a cylinder is bounded by a plane region B_1, called the **base**, and a congruent region B_2 in a parallel plane. The cylinder consists of all points on line segments that are perpendicular to the base and join B_1 to B_2. If the area of the base is A and the height of the cylinder (the distance from B_1 to B_2) is h, then the volume V of the cylinder is defined as

$$V = Ah$$

In particular, if the base is a circle with radius r, then the cylinder is a circular cylinder with volume $V = \pi r^2 h$ [see Figure 1(b)], and if the base is a rectangle with length l and width w, then the cylinder is a rectangular box (also called a *rectangular parallelepiped*) with volume $V = lwh$ [see Figure 1(c)].

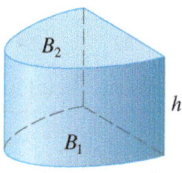
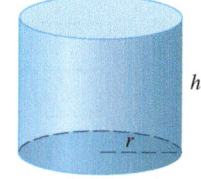
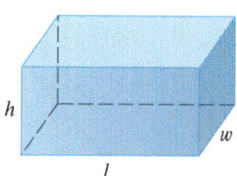

FIGURE 1   (a) Cylinder $V = Ah$ (b) Circular cylinder $V = \pi r^2 h$ (c) Rectangular box $V = lwh$

For a solid S that isn't a cylinder we first "cut" S into pieces and approximate each piece by a cylinder. We estimate the volume of S by adding the volumes of the cylinders. We arrive at the exact volume of S through a limiting process in which the number of pieces becomes large.

We start by intersecting S with a plane and obtaining a plane region that is called a **cross-section** of S. Let $A(x)$ be the area of the cross-section of S in a plane P_x perpendicular to the x-axis and passing through the point x, where $a \leq x \leq b$. (See Figure 2. Think of slicing S with a knife through x and computing the area of this slice.) The cross-sectional area $A(x)$ will vary as x increases from a to b.

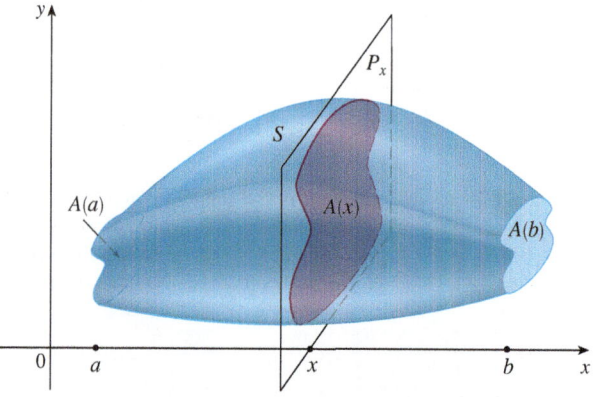

FIGURE 2

Let's divide S into n "slabs" of equal width Δx by using the planes $P_{x_1}, P_{x_2}, \ldots$ to slice the solid. (Think of slicing a loaf of bread.) If we choose sample points x_i^* in $[x_{i-1}, x_i]$, we can approximate the ith slab S_i (the part of S that lies between the planes $P_{x_{i-1}}$ and P_{x_i}) by a cylinder with base area $A(x_i^*)$ and "height" Δx. (See Figure 3.)

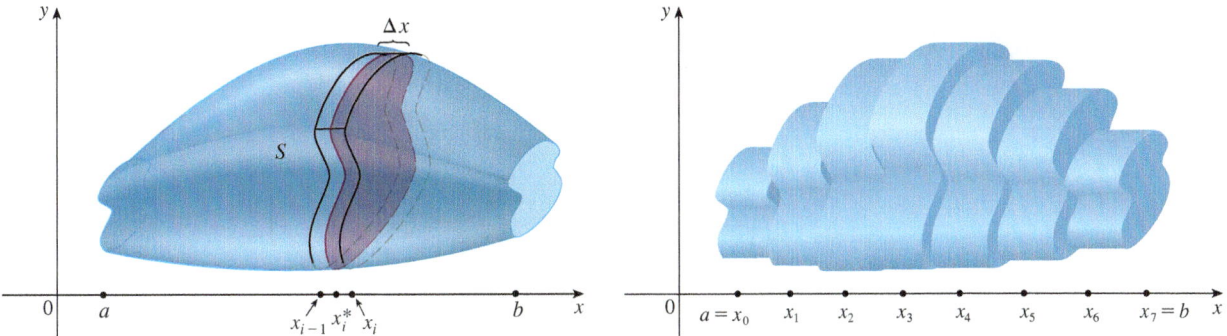

FIGURE 3

The volume of this cylinder is $A(x_i^*) \Delta x$, so an approximation to our intuitive conception of the volume of the ith slab S_i is

$$V(S_i) \approx A(x_i^*) \Delta x$$

Adding the volumes of these slabs, we get an approximation to the total volume (that is, what we think of intuitively as the volume):

$$V \approx \sum_{i=1}^{n} A(x_i^*) \Delta x$$

This approximation appears to become better and better as $n \to \infty$. (Think of the slices as becoming thinner and thinner.) Therefore we *define* the volume as the limit of these sums as $n \to \infty$. But we recognize the limit of Riemann sums as a definite integral and so we have the following definition.

It can be proved that this definition is independent of how S is situated with respect to the x-axis. In other words, no matter how we slice S with parallel planes, we always get the same answer for V.

Definition of Volume Let S be a solid that lies between $x = a$ and $x = b$. If the cross-sectional area of S in the plane P_x, through x and perpendicular to the x-axis, is $A(x)$, where A is a continuous function, then the **volume** of S is

$$V = \lim_{n \to \infty} \sum_{i=1}^{n} A(x_i^*) \Delta x = \int_a^b A(x)\, dx$$

When we use the volume formula $V = \int_a^b A(x)\, dx$, it is important to remember that $A(x)$ is the area of a moving cross-section obtained by slicing through x perpendicular to the x-axis.

Notice that, for a cylinder, the cross-sectional area is constant: $A(x) = A$ for all x. So our definition of volume gives $V = \int_a^b A\, dx = A(b - a)$; this agrees with the formula $V = Ah$.

EXAMPLE 1 Show that the volume of a sphere of radius r is $V = \tfrac{4}{3}\pi r^3$.

SOLUTION If we place the sphere so that its center is at the origin, then the plane P_x intersects the sphere in a circle whose radius (from the Pythagorean Theorem) is

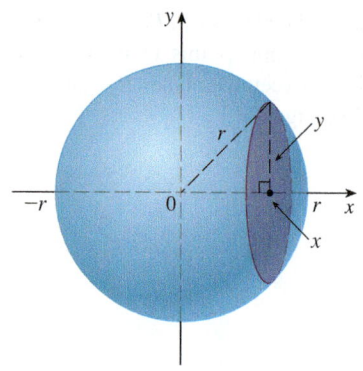

FIGURE 4

$y = \sqrt{r^2 - x^2}$. (See Figure 4.) So the cross-sectional area is

$$A(x) = \pi y^2 = \pi(r^2 - x^2)$$

Using the definition of volume with $a = -r$ and $b = r$, we have

$$V = \int_{-r}^{r} A(x)\, dx = \int_{-r}^{r} \pi(r^2 - x^2)\, dx$$

$$= 2\pi \int_{0}^{r} (r^2 - x^2)\, dx \qquad \text{(The integrand is even.)}$$

$$= 2\pi \left[r^2 x - \frac{x^3}{3} \right]_0^r = 2\pi \left(r^3 - \frac{r^3}{3} \right)$$

$$= \tfrac{4}{3}\pi r^3 \qquad \blacksquare$$

Figure 5 illustrates the definition of volume when the solid is a sphere with radius $r = 1$. From the result of Example 1, we know that the volume of the sphere is $\tfrac{4}{3}\pi$, which is approximately 4.18879. Here the slabs are circular cylinders, or *disks*, and the three parts of Figure 5 show the geometric interpretations of the Riemann sums

$$\sum_{i=1}^{n} A(\bar{x}_i)\, \Delta x = \sum_{i=1}^{n} \pi(1^2 - \bar{x}_i^2)\, \Delta x$$

when $n = 5$, 10, and 20 if we choose the sample points x_i^* to be the midpoints $\bar{x}_i$. Notice that as we increase the number of approximating cylinders, the corresponding Riemann sums become closer to the true volume.

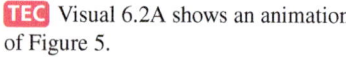

TEC Visual 6.2A shows an animation of Figure 5.

(a) Using 5 disks, $V \approx 4.2726$

(b) Using 10 disks, $V \approx 4.2097$

(c) Using 20 disks, $V \approx 4.1940$

FIGURE 5
Approximating the volume of a sphere with radius 1

EXAMPLE 2 Find the volume of the solid obtained by rotating about the x-axis the region under the curve $y = \sqrt{x}$ from 0 to 1. Illustrate the definition of volume by sketching a typical approximating cylinder.

SOLUTION The region is shown in Figure 6(a). If we rotate about the x-axis, we get the solid shown in Figure 6(b). When we slice through the point x, we get a disk with radius $\sqrt{x}$. The area of this cross-section is

$$A(x) = \pi\left(\sqrt{x}\right)^2 = \pi x$$

and the volume of the approximating cylinder (a disk with thickness Δx) is

$$A(x)\, \Delta x = \pi x\, \Delta x$$

The solid lies between $x = 0$ and $x = 1$, so its volume is

$$V = \int_0^1 A(x)\,dx = \int_0^1 \pi x\,dx = \pi \frac{x^2}{2}\Big]_0^1 = \frac{\pi}{2}$$

Did we get a reasonable answer in Example 2? As a check on our work, let's replace the given region by a square with base [0, 1] and height 1. If we rotate this square, we get a cylinder with radius 1, height 1, and volume $\pi \cdot 1^2 \cdot 1 = \pi$. We computed that the given solid has half this volume. That seems about right.

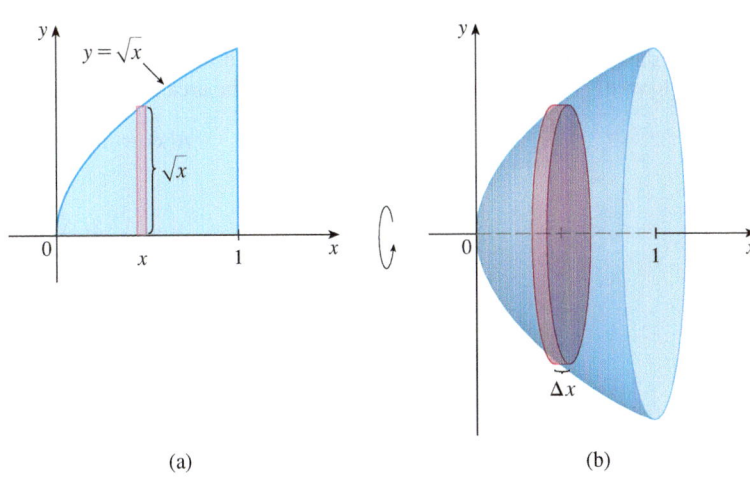

FIGURE 6   (a)   (b)

EXAMPLE 3 Find the volume of the solid obtained by rotating the region bounded by $y = x^3$, $y = 8$, and $x = 0$ about the y-axis.

SOLUTION The region is shown in Figure 7(a) and the resulting solid is shown in Figure 7(b). Because the region is rotated about the y-axis, it makes sense to slice the solid perpendicular to the y-axis (obtaining circular cross-sections) and therefore to integrate with respect to y. If we slice at height y, we get a circular disk with radius x, where $x = \sqrt[3]{y}$. So the area of a cross-section through y is

$$A(y) = \pi x^2 = \pi (\sqrt[3]{y})^2 = \pi y^{2/3}$$

and the volume of the approximating cylinder pictured in Figure 7(b) is

$$A(y)\,\Delta y = \pi y^{2/3}\,\Delta y$$

Since the solid lies between $y = 0$ and $y = 8$, its volume is

$$V = \int_0^8 A(y)\,dy = \int_0^8 \pi y^{2/3}\,dy = \pi \left[\tfrac{3}{5} y^{5/3}\right]_0^8 = \frac{96\pi}{5}$$

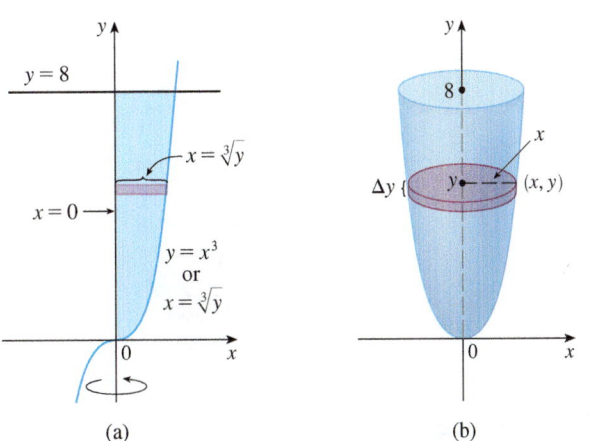

FIGURE 7   (a)   (b)

442   **CHAPTER 6**   Applications of Integration

TEC Visual 6.2B shows how solids of revolution are formed.

EXAMPLE 4 The region $\mathcal{R}$ enclosed by the curves $y = x$ and $y = x^2$ is rotated about the x-axis. Find the volume of the resulting solid.

SOLUTION The curves $y = x$ and $y = x^2$ intersect at the points $(0, 0)$ and $(1, 1)$. The region between them, the solid of rotation, and a cross-section perpendicular to the x-axis are shown in Figure 8. A cross-section in the plane P_x has the shape of a *washer* (an annular ring) with inner radius x^2 and outer radius x, so we find the cross-sectional area by subtracting the area of the inner circle from the area of the outer circle:

$$A(x) = \pi x^2 - \pi(x^2)^2 = \pi(x^2 - x^4)$$

Therefore we have

$$V = \int_0^1 A(x)\, dx = \int_0^1 \pi(x^2 - x^4)\, dx$$

$$= \pi \left[\frac{x^3}{3} - \frac{x^5}{5} \right]_0^1 = \frac{2\pi}{15}$$

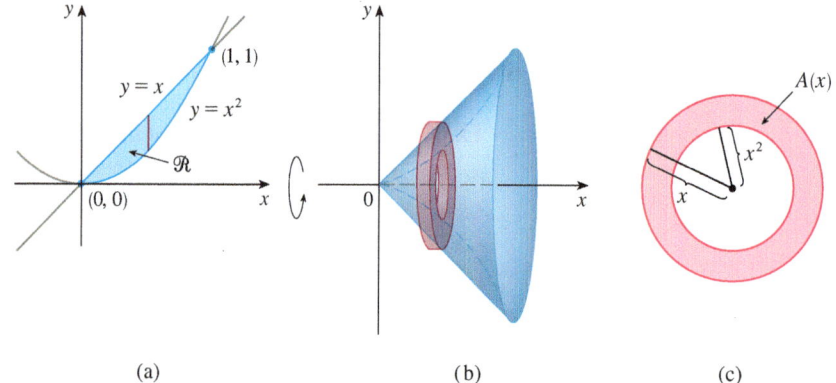

FIGURE 8   (a)   (b)   (c)

EXAMPLE 5 Find the volume of the solid obtained by rotating the region in Example 4 about the line $y = 2$.

SOLUTION The solid and a cross-section are shown in Figure 9. Again the cross-section is a washer, but this time the inner radius is $2 - x$ and the outer radius is $2 - x^2$.

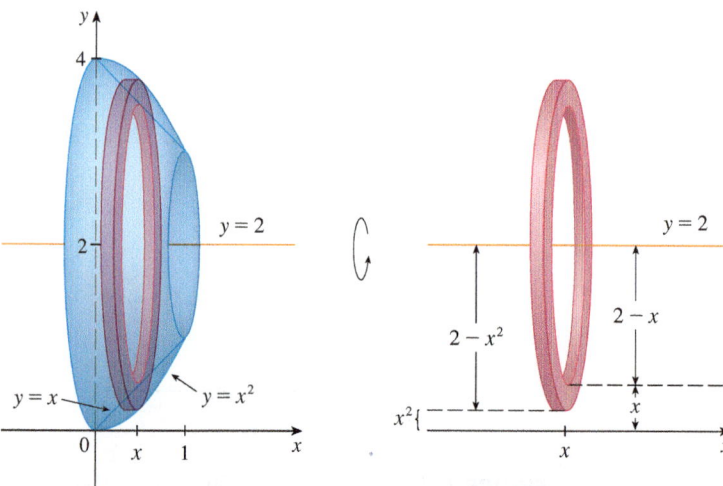

FIGURE 9

The cross-sectional area is

$$A(x) = \pi(2-x^2)^2 - \pi(2-x)^2$$

and so the volume of S is

$$V = \int_0^1 A(x)\,dx$$

$$= \pi \int_0^1 [(2-x^2)^2 - (2-x)^2]\,dx$$

$$= \pi \int_0^1 (x^4 - 5x^2 + 4x)\,dx$$

$$= \pi \left[\frac{x^5}{5} - 5\frac{x^3}{3} + 4\frac{x^2}{2} \right]_0^1$$

$$= \frac{8\pi}{15}$$

■

The solids in Examples 1–5 are all called **solids of revolution** because they are obtained by revolving a region about a line. In general, we calculate the volume of a solid of revolution by using the basic defining formula

$$V = \int_a^b A(x)\,dx \quad \text{or} \quad V = \int_c^d A(y)\,dy$$

and we find the cross-sectional area $A(x)$ or $A(y)$ in one of the following ways:

- If the cross-section is a disk (as in Examples 1–3), we find the radius of the disk (in terms of x or y) and use

$$A = \pi(\text{radius})^2$$

- If the cross-section is a washer (as in Examples 4 and 5), we find the inner radius r_{in} and outer radius r_{out} from a sketch (as in Figures 8, 9, and 10) and compute the area of the washer by subtracting the area of the inner disk from the area of the outer disk:

$$A = \pi(\text{outer radius})^2 - \pi(\text{inner radius})^2$$

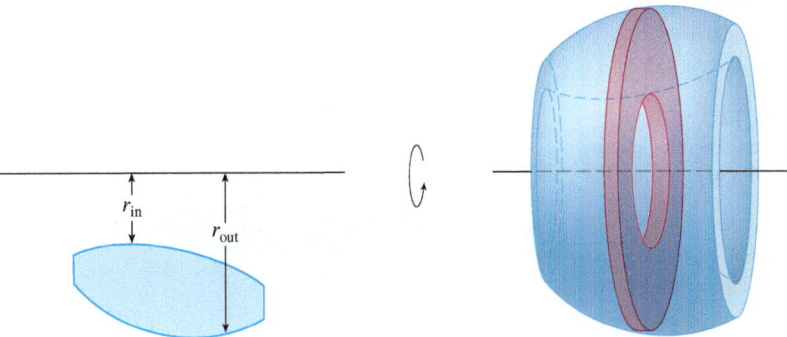

FIGURE 10

The next example gives a further illustration of the procedure.

EXAMPLE 6 Find the volume of the solid obtained by rotating the region in Example 4 about the line $x = -1$.

SOLUTION Figure 11 shows a horizontal cross-section. It is a washer with inner radius $1 + y$ and outer radius $1 + \sqrt{y}$, so the cross-sectional area is

$$A(y) = \pi(\text{outer radius})^2 - \pi(\text{inner radius})^2$$
$$= \pi(1 + \sqrt{y})^2 - \pi(1 + y)^2$$

The volume is

$$V = \int_0^1 A(y)\, dy = \pi \int_0^1 \left[(1 + \sqrt{y})^2 - (1 + y)^2\right] dy$$
$$= \pi \int_0^1 (2\sqrt{y} - y - y^2)\, dy = \pi \left[\frac{4y^{3/2}}{3} - \frac{y^2}{2} - \frac{y^3}{3}\right]_0^1 = \frac{\pi}{2}$$

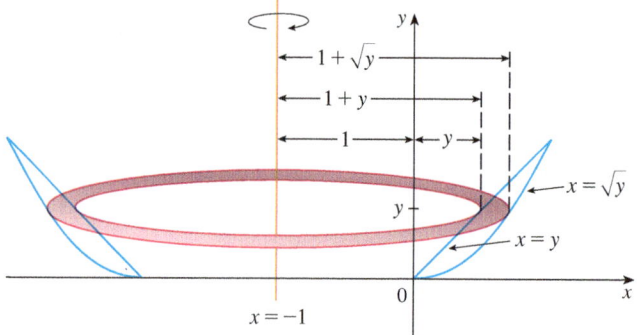

FIGURE 11

We now find the volumes of three solids that are *not* solids of revolution.

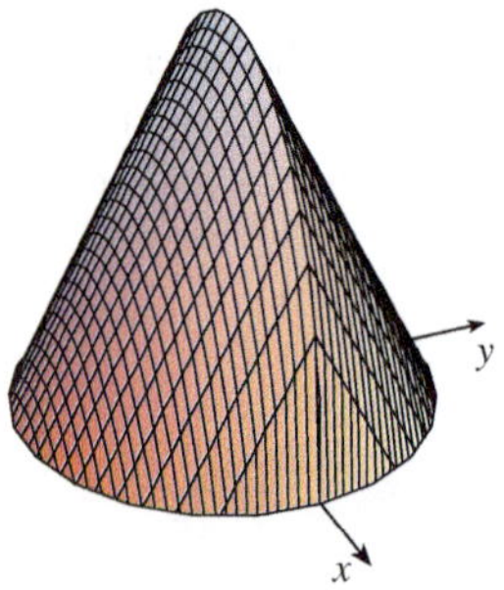

EXAMPLE 7 Figure 12 shows a solid with a circular base of radius 1. Parallel cross-sections perpendicular to the base are equilateral triangles. Find the volume of the solid.

SOLUTION Let's take the circle to be $x^2 + y^2 = 1$. The solid, its base, and a typical cross-section at a distance x from the origin are shown in Figure 13.

Since B lies on the circle, we have $y = \sqrt{1 - x^2}$ and so the base of the triangle ABC is $|AB| = 2y = 2\sqrt{1 - x^2}$. Since the triangle is equilateral, we see from Figure 13(c)

FIGURE 12
Computer-generated picture of the solid in Example 7

TEC Visual 6.2C shows how the solid in Figure 12 is generated.

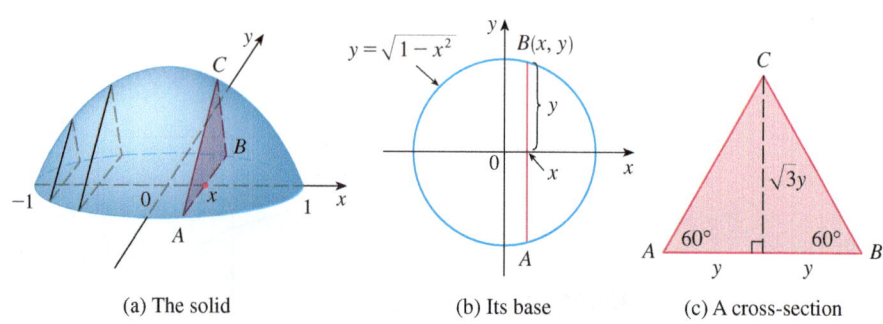

(a) The solid   (b) Its base   (c) A cross-section

FIGURE 13

that its height is $\sqrt{3}\,y = \sqrt{3}\sqrt{1-x^2}$. The cross-sectional area is therefore

$$A(x) = \tfrac{1}{2} \cdot 2\sqrt{1-x^2} \cdot \sqrt{3}\sqrt{1-x^2} = \sqrt{3}\,(1-x^2)$$

and the volume of the solid is

$$V = \int_{-1}^{1} A(x)\,dx = \int_{-1}^{1} \sqrt{3}\,(1-x^2)\,dx$$

$$= 2\int_{0}^{1} \sqrt{3}\,(1-x^2)\,dx = 2\sqrt{3}\left[x - \frac{x^3}{3}\right]_0^1 = \frac{4\sqrt{3}}{3}$$

EXAMPLE 8 Find the volume of a pyramid whose base is a square with side L and whose height is h.

SOLUTION We place the origin O at the vertex of the pyramid and the x-axis along its central axis as in Figure 14. Any plane P_x that passes through x and is perpendicular to the x-axis intersects the pyramid in a square with side of length s, say. We can express s in terms of x by observing from the similar triangles in Figure 15 that

$$\frac{x}{h} = \frac{s/2}{L/2} = \frac{s}{L}$$

and so $s = Lx/h$. [Another method is to observe that the line OP has slope $L/(2h)$ and so its equation is $y = Lx/(2h)$.] Therefore the cross-sectional area is

$$A(x) = s^2 = \frac{L^2}{h^2}x^2$$

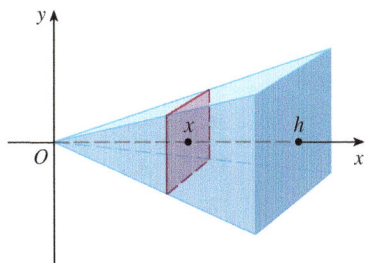
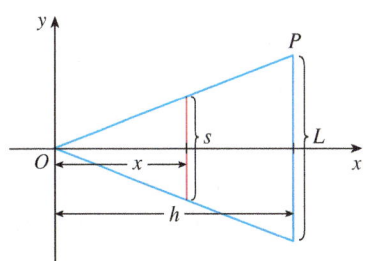

FIGURE 14   **FIGURE 15**

The pyramid lies between $x = 0$ and $x = h$, so its volume is

$$V = \int_0^h A(x)\,dx = \int_0^h \frac{L^2}{h^2}x^2\,dx = \frac{L^2}{h^2}\frac{x^3}{3}\bigg]_0^h = \frac{L^2 h}{3}$$

NOTE We didn't need to place the vertex of the pyramid at the origin in Example 8. We did so merely to make the equations simple. If, instead, we had placed the center of the base at the origin and the vertex on the positive y-axis, as in Figure 16, you can verify that we would have obtained the integral

$$V = \int_0^h \frac{L^2}{h^2}(h-y)^2\,dy = \frac{L^2 h}{3}$$

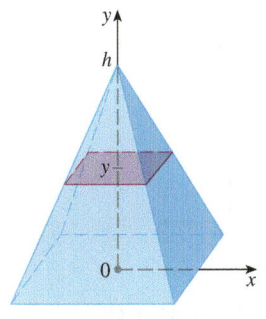

FIGURE 16

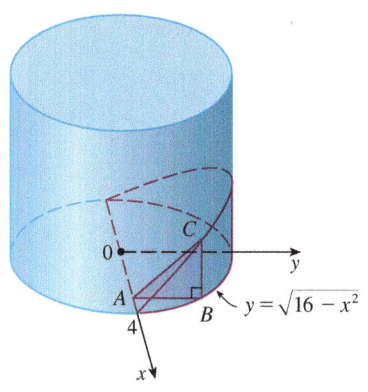

EXAMPLE 9 A wedge is cut out of a circular cylinder of radius 4 by two planes. One plane is perpendicular to the axis of the cylinder. The other intersects the first at an angle of 30° along a diameter of the cylinder. Find the volume of the wedge.

SOLUTION If we place the x-axis along the diameter where the planes meet, then the base of the solid is a semicircle with equation $y = \sqrt{16 - x^2}$, $-4 \leq x \leq 4$. A cross-section perpendicular to the x-axis at a distance x from the origin is a triangle ABC, as shown in Figure 17, whose base is $y = \sqrt{16 - x^2}$ and whose height is $|BC| = y \tan 30° = \sqrt{16 - x^2}/\sqrt{3}$. So the cross-sectional area is

$$A(x) = \tfrac{1}{2}\sqrt{16 - x^2} \cdot \frac{1}{\sqrt{3}}\sqrt{16 - x^2} = \frac{16 - x^2}{2\sqrt{3}}$$

and the volume is

$$V = \int_{-4}^{4} A(x)\,dx = \int_{-4}^{4} \frac{16 - x^2}{2\sqrt{3}}\,dx$$

$$= \frac{1}{\sqrt{3}} \int_{0}^{4} (16 - x^2)\,dx = \frac{1}{\sqrt{3}} \left[16x - \frac{x^3}{3} \right]_{0}^{4}$$

$$= \frac{128}{3\sqrt{3}}$$

For another method see Exercise 64. ∎

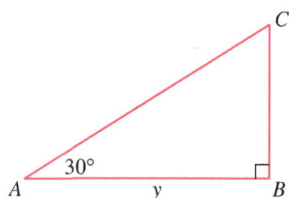

FIGURE 17

6.2 EXERCISES

1–18 Find the volume of the solid obtained by rotating the region bounded by the given curves about the specified line. Sketch the region, the solid, and a typical disk or washer.

1. $y = x + 1$, $y = 0$, $x = 0$, $x = 2$; about the x-axis
2. $y = 1/x$, $y = 0$, $x = 1$, $x = 4$; about the x-axis
3. $y = \sqrt{x - 1}$, $y = 0$, $x = 5$; about the x-axis
4. $y = e^x$, $y = 0$, $x = -1$, $x = 1$; about the x-axis
5. $x = 2\sqrt{y}$, $x = 0$, $y = 9$; about the y-axis
6. $2x = y^2$, $x = 0$, $y = 4$; about the y-axis
7. $y = x^3$, $y = x$, $x \geq 0$; about the x-axis
8. $y = 6 - x^2$, $y = 2$; about the x-axis
9. $y^2 = x$, $x = 2y$; about the y-axis
10. $x = 2 - y^2$, $x = y^4$; about the y-axis
11. $y = x^2$, $x = y^2$; about $y = 1$
12. $y = x^3$, $y = 1$, $x = 2$; about $y = -3$
13. $y = 1 + \sec x$, $y = 3$; about $y = 1$
14. $y = \sin x$, $y = \cos x$, $0 \leq x \leq \pi/4$; about $y = -1$
15. $y = x^3$, $y = 0$, $x = 1$; about $x = 2$
16. $xy = 1$, $y = 0$, $x = 1$, $x = 2$; about $x = -1$
17. $x = y^2$, $x = 1 - y^2$; about $x = 3$
18. $y = x$, $y = 0$, $x = 2$, $x = 4$; about $x = 1$

19–30 Refer to the figure and find the volume generated by rotating the given region about the specified line.

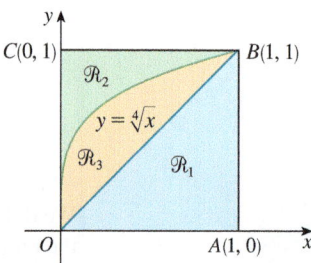

19. $\mathcal{R}_1$ about OA
20. $\mathcal{R}_1$ about OC
21. $\mathcal{R}_1$ about AB
22. $\mathcal{R}_1$ about BC

23. $\mathcal{R}_2$ about OA
24. $\mathcal{R}_2$ about OC
25. $\mathcal{R}_2$ about AB
26. $\mathcal{R}_2$ about BC
27. $\mathcal{R}_3$ about OA
28. $\mathcal{R}_3$ about OC
29. $\mathcal{R}_3$ about AB
30. $\mathcal{R}_3$ about BC

31–34 Set up an integral for the volume of the solid obtained by rotating the region bounded by the given curves about the specified line. Then use your calculator to evaluate the integral correct to five decimal places.

31. $y = e^{-x^2}$, $y = 0$, $x = -1$, $x = 1$
  (a) About the x-axis   (b) About $y = -1$

32. $y = 0$, $y = \cos^2 x$, $-\pi/2 \leq x \leq \pi/2$
  (a) About the x-axis   (b) About $y = 1$

33. $x^2 + 4y^2 = 4$
  (a) About $y = 2$ (b) About $x = 2$

34. $y = x^2$, $x^2 + y^2 = 1$, $y \geq 0$
  (a) About the x-axis   (b) About the y-axis

35–36 Use a graph to find approximate x-coordinates of the points of intersection of the given curves. Then use your calculator to find (approximately) the volume of the solid obtained by rotating about the x-axis the region bounded by these curves.

35. $y = \ln(x^6 + 2)$, $y = \sqrt{3 - x^3}$

36. $y = 1 + xe^{-x^3}$, $y = \arctan x^2$

37–38 Use a computer algebra system to find the exact volume of the solid obtained by rotating the region bounded by the given curves about the specified line.

37. $y = \sin^2 x$, $y = 0$, $0 \leq x \leq \pi$; about $y = -1$

38. $y = x$, $y = xe^{1-x/2}$; about $y = 3$

39–42 Each integral represents the volume of a solid. Describe the solid.

39. $\pi \int_0^\pi \sin x \, dx$
40. $\pi \int_{-1}^1 (1 - y^2)^2 \, dy$
41. $\pi \int_0^1 (y^4 - y^8) \, dy$
42. $\pi \int_1^4 \left[3^2 - (3 - \sqrt{x})^2 \right] dx$

43. A CAT scan produces equally spaced cross-sectional views of a human organ that provide information about the organ otherwise obtained only by surgery. Suppose that a CAT scan of a human liver shows cross-sections spaced 1.5 cm apart. The liver is 15 cm long and the cross-sectional areas, in square centimeters, are 0, 18, 58, 79, 94, 106, 117, 128, 63, 39, and 0. Use the Midpoint Rule to estimate the volume of the liver.

44. A log 10 m long is cut at 1-meter intervals and its cross-sectional areas A (at a distance x from the end of the log) are listed in the table. Use the Midpoint Rule with $n = 5$ to estimate the volume of the log.

x (m)	A (m²)	x (m)	A (m²)
0	0.68	6	0.53
1	0.65	7	0.55
2	0.64	8	0.52
3	0.61	9	0.50
4	0.58	10	0.48
5	0.59		

45. (a) If the region shown in the figure is rotated about the x-axis to form a solid, use the Midpoint Rule with $n = 4$ to estimate the volume of the solid.

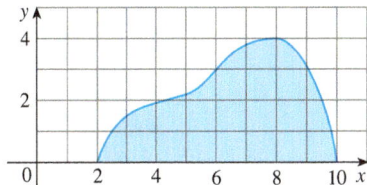

(b) Estimate the volume if the region is rotated about the y-axis. Again use the Midpoint Rule with $n = 4$.

46. (a) A model for the shape of a bird's egg is obtained by rotating about the x-axis the region under the graph of
$$f(x) = (ax^3 + bx^2 + cx + d)\sqrt{1 - x^2}$$
Use a CAS to find the volume of such an egg.
(b) For a red-throated loon, $a = -0.06$, $b = 0.04$, $c = 0.1$, and $d = 0.54$. Graph f and find the volume of an egg of this species.

47–61 Find the volume of the described solid S.

47. A right circular cone with height h and base radius r

48. A frustum of a right circular cone with height h, lower base radius R, and top radius r

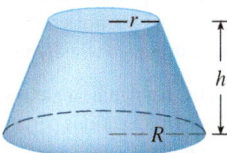

49. A cap of a sphere with radius r and height h

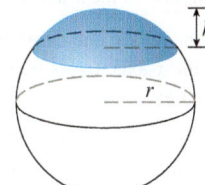

50. A frustum of a pyramid with square base of side b, square top of side a, and height h

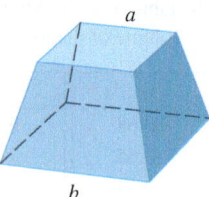

What happens if $a = b$? What happens if $a = 0$?

51. A pyramid with height h and rectangular base with dimensions b and $2b$

52. A pyramid with height h and base an equilateral triangle with side a (a tetrahedron)

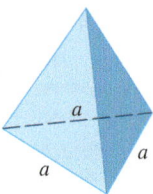

53. A tetrahedron with three mutually perpendicular faces and three mutually perpendicular edges with lengths 3 cm, 4 cm, and 5 cm

54. The base of S is a circular disk with radius r. Parallel cross-sections perpendicular to the base are squares.

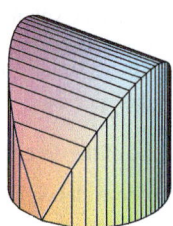

55. The base of S is an elliptical region with boundary curve $9x^2 + 4y^2 = 36$. Cross-sections perpendicular to the x-axis are isosceles right triangles with hypotenuse in the base.

56. The base of S is the triangular region with vertices $(0, 0)$, $(1, 0)$, and $(0, 1)$. Cross-sections perpendicular to the y-axis are equilateral triangles.

57. The base of S is the same base as in Exercise 56, but cross-sections perpendicular to the x-axis are squares.

58. The base of S is the region enclosed by the parabola $y = 1 - x^2$ and the x-axis. Cross-sections perpendicular to the y-axis are squares.

59. The base of S is the same base as in Exercise 58, but cross-sections perpendicular to the x-axis are isosceles triangles with height equal to the base.

60. The base of S is the region enclosed by $y = 2 - x^2$ and the x-axis. Cross-sections perpendicular to the y-axis are quarter-circles.

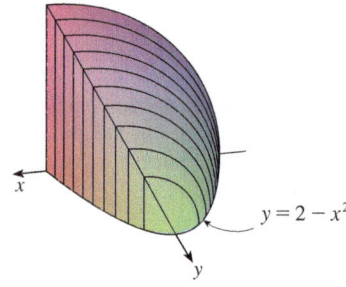

61. The solid S is bounded by circles that are perpendicular to the x-axis, intersect the x-axis, and have centers on the parabola $y = \frac{1}{2}(1 - x^2)$, $-1 \leq x \leq 1$.

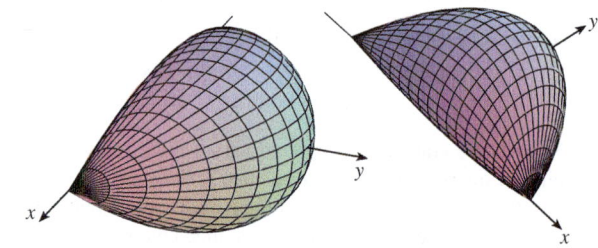

62. The base of S is a circular disk with radius r. Parallel cross-sections perpendicular to the base are isosceles triangles with height h and unequal side in the base.
(a) Set up an integral for the volume of S.
(b) By interpreting the integral as an area, find the volume of S.

63. (a) Set up an integral for the volume of a solid *torus* (the donut-shaped solid shown in the figure) with radii r and R.
(b) By interpreting the integral as an area, find the volume of the torus.

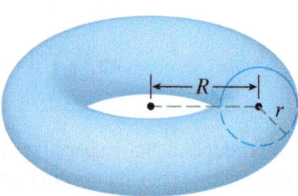

64. Solve Example 9 taking cross-sections to be parallel to the line of intersection of the two planes.

65. (a) Cavalieri's Principle states that if a family of parallel planes gives equal cross-sectional areas for two solids

S_1 and S_2, then the volumes of S_1 and S_2 are equal. Prove this principle.
(b) Use Cavalieri's Principle to find the volume of the oblique cylinder shown in the figure.

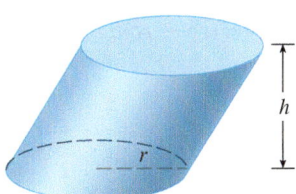

66. Find the volume common to two circular cylinders, each with radius r, if the axes of the cylinders intersect at right angles.

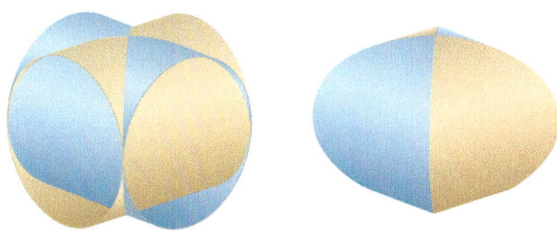

67. Find the volume common to two spheres, each with radius r, if the center of each sphere lies on the surface of the other sphere.

68. A bowl is shaped like a hemisphere with diameter 30 cm. A heavy ball with diameter 10 cm is placed in the bowl and water is poured into the bowl to a depth of h centimeters. Find the volume of water in the bowl.

69. A hole of radius r is bored through the middle of a cylinder of radius $R > r$ at right angles to the axis of the cylinder. Set up, but do not evaluate, an integral for the volume cut out.

70. A hole of radius r is bored through the center of a sphere of radius $R > r$. Find the volume of the remaining portion of the sphere.

71. Some of the pioneers of calculus, such as Kepler and Newton, were inspired by the problem of finding the volumes of wine barrels. (In fact Kepler published a book *Stereometria doliorum* in 1615 devoted to methods for finding the volumes of barrels.) They often approximated the shape of the sides by parabolas.
(a) A barrel with height h and maximum radius R is constructed by rotating about the x-axis the parabola $y = R - cx^2$, $-h/2 \leq x \leq h/2$, where c is a positive constant. Show that the radius of each end of the barrel is $r = R - d$, where $d = ch^2/4$.
(b) Show that the volume enclosed by the barrel is
$$V = \tfrac{1}{3}\pi h\left(2R^2 + r^2 - \tfrac{2}{5}d^2\right)$$

72. Suppose that a region $\mathcal{R}$ has area A and lies above the x-axis. When $\mathcal{R}$ is rotated about the x-axis, it sweeps out a solid with volume V_1. When $\mathcal{R}$ is rotated about the line $y = -k$ (where k is a positive number), it sweeps out a solid with volume V_2. Express V_2 in terms of V_1, k, and A.

6.3 Volumes by Cylindrical Shells

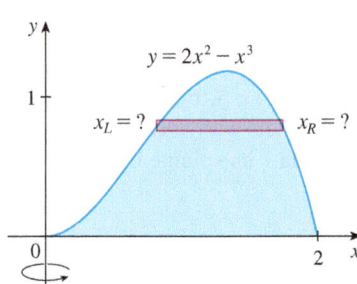

FIGURE 1

Some volume problems are very difficult to handle by the methods of the preceding section. For instance, let's consider the problem of finding the volume of the solid obtained by rotating about the y-axis the region bounded by $y = 2x^2 - x^3$ and $y = 0$. (See Figure 1.) If we slice perpendicular to the y-axis, we get a washer. But to compute the inner radius and the outer radius of the washer, we'd have to solve the cubic equation $y = 2x^2 - x^3$ for x in terms of y; that's not easy.

Fortunately, there is a method, called the **method of cylindrical shells**, that is easier to use in such a case. Figure 2 shows a cylindrical shell with inner radius r_1, outer radius

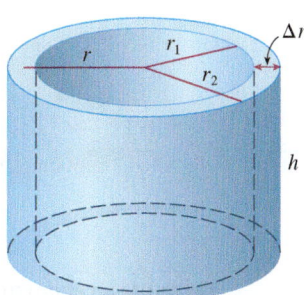

FIGURE 2

450   CHAPTER 6   Applications of Integration

r_2, and height h. Its volume V is calculated by subtracting the volume V_1 of the inner cylinder from the volume V_2 of the outer cylinder:

$$V = V_2 - V_1$$
$$= \pi r_2^2 h - \pi r_1^2 h = \pi (r_2^2 - r_1^2) h$$
$$= \pi (r_2 + r_1)(r_2 - r_1) h$$
$$= 2\pi \frac{r_2 + r_1}{2} h(r_2 - r_1)$$

If we let $\Delta r = r_2 - r_1$ (the thickness of the shell) and $r = \frac{1}{2}(r_2 + r_1)$ (the average radius of the shell), then this formula for the volume of a cylindrical shell becomes

[1]
$$V = 2\pi r h \, \Delta r$$

and it can be remembered as

$$V = [\text{circumference}][\text{height}][\text{thickness}]$$

Now let S be the solid obtained by rotating about the y-axis the region bounded by $y = f(x)$ [where $f(x) \geq 0$], $y = 0$, $x = a$, and $x = b$, where $b > a \geq 0$. (See Figure 3.)

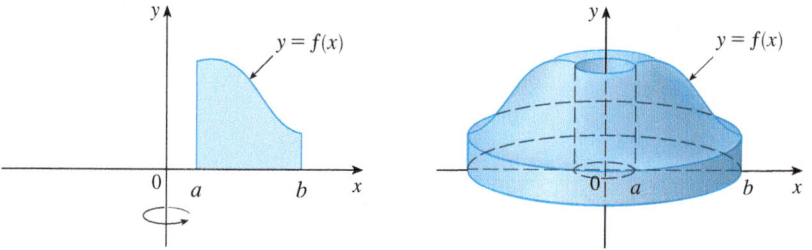

FIGURE 3

We divide the interval $[a, b]$ into n subintervals $[x_{i-1}, x_i]$ of equal width Δx and let $\bar{x}_i$ be the midpoint of the ith subinterval. If the rectangle with base $[x_{i-1}, x_i]$ and height $f(\bar{x}_i)$ is rotated about the y-axis, then the result is a cylindrical shell with average radius $\bar{x}_i$, height $f(\bar{x}_i)$, and thickness Δx (see Figure 4). So by Formula 1 its volume is

$$V_i = (2\pi \bar{x}_i)[f(\bar{x}_i)] \, \Delta x$$

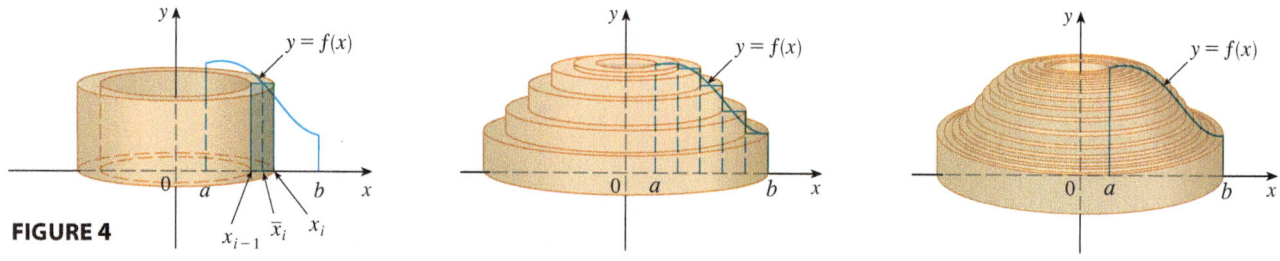

FIGURE 4

Therefore an approximation to the volume V of S is given by the sum of the volumes of these shells:

$$V \approx \sum_{i=1}^{n} V_i = \sum_{i=1}^{n} 2\pi \bar{x}_i f(\bar{x}_i) \, \Delta x$$

This approximation appears to become better as $n \to \infty$. But, from the definition of an integral, we know that

$$\lim_{n \to \infty} \sum_{i=1}^{n} 2\pi \bar{x}_i f(\bar{x}_i) \, \Delta x = \int_{a}^{b} 2\pi x f(x) \, dx$$

Thus the following appears plausible:

> **2** The volume of the solid in Figure 3, obtained by rotating about the y-axis the region under the curve $y = f(x)$ from a to b, is
>
> $$V = \int_{a}^{b} 2\pi x f(x) \, dx \qquad \text{where } 0 \leq a < b$$

The argument using cylindrical shells makes Formula 2 seem reasonable, but later we will be able to prove it (see Exercise 7.1.73).

The best way to remember Formula 2 is to think of a typical shell, cut and flattened as in Figure 5, with radius x, circumference $2\pi x$, height $f(x)$, and thickness Δx or dx:

$$\int_{a}^{b} \underbrace{(2\pi x)}_{\text{circumference}} \underbrace{[f(x)]}_{\text{height}} \underbrace{dx}_{\text{thickness}}$$

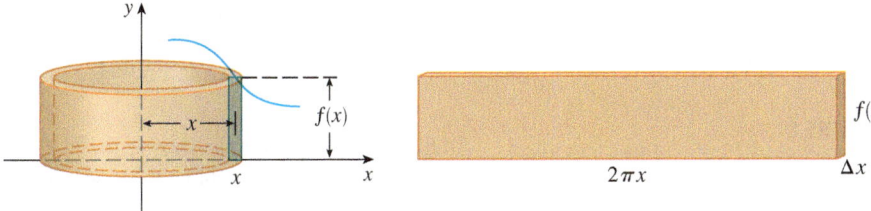

FIGURE 5

This type of reasoning will be helpful in other situations, such as when we rotate about lines other than the y-axis.

EXAMPLE 1 Find the volume of the solid obtained by rotating about the y-axis the region bounded by $y = 2x^2 - x^3$ and $y = 0$.

SOLUTION From the sketch in Figure 6 we see that a typical shell has radius x, circumference $2\pi x$, and height $f(x) = 2x^2 - x^3$. So, by the shell method, the volume is

$$V = \int_{0}^{2} \underbrace{(2\pi x)}_{\text{circumference}} \underbrace{(2x^2 - x^3)}_{\text{height}} \underbrace{dx}_{\text{thickness}}$$

$$= 2\pi \int_{0}^{2} (2x^3 - x^4) \, dx = 2\pi \left[\tfrac{1}{2} x^4 - \tfrac{1}{5} x^5 \right]_{0}^{2}$$

$$= 2\pi \left(8 - \tfrac{32}{5} \right) = \tfrac{16}{5} \pi$$

It can be verified that the shell method gives the same answer as slicing.

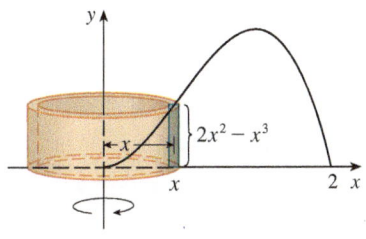

FIGURE 6

TEC Visual 6.3 shows how the solid and shells in Example 1 are formed.

Figure 7 shows a computer-generated picture of the solid whose volume we computed in Example 1.

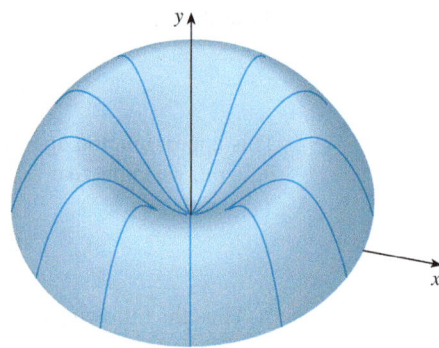

FIGURE 7

NOTE Comparing the solution of Example 1 with the remarks at the beginning of this section, we see that the method of cylindrical shells is much easier than the washer method for this problem. We did not have to find the coordinates of the local maximum and we did not have to solve the equation of the curve for x in terms of y. However, in other examples the methods of the preceding section may be easier.

EXAMPLE 2 Find the volume of the solid obtained by rotating about the y-axis the region between $y = x$ and $y = x^2$.

SOLUTION The region and a typical shell are shown in Figure 8. We see that the shell has radius x, circumference $2\pi x$, and height $x - x^2$. So the volume is

$$V = \int_0^1 (2\pi x)(x - x^2)\,dx = 2\pi \int_0^1 (x^2 - x^3)\,dx$$

$$= 2\pi \left[\frac{x^3}{3} - \frac{x^4}{4}\right]_0^1 = \frac{\pi}{6}$$

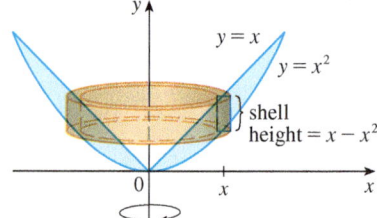

FIGURE 8

As the following example shows, the shell method works just as well if we rotate about the x-axis. We simply have to draw a diagram to identify the radius and height of a shell.

EXAMPLE 3 Use cylindrical shells to find the volume of the solid obtained by rotating about the x-axis the region under the curve $y = \sqrt{x}$ from 0 to 1.

SOLUTION This problem was solved using disks in Example 6.2.2. To use shells we relabel the curve $y = \sqrt{x}$ (in the figure in that example) as $x = y^2$ in Figure 9. For rotation about the x-axis we see that a typical shell has radius y, circumference $2\pi y$, and height $1 - y^2$. So the volume is

$$V = \int_0^1 (2\pi y)(1 - y^2)\,dy = 2\pi \int_0^1 (y - y^3)\,dy$$

$$= 2\pi \left[\frac{y^2}{2} - \frac{y^4}{4}\right]_0^1 = \frac{\pi}{2}$$

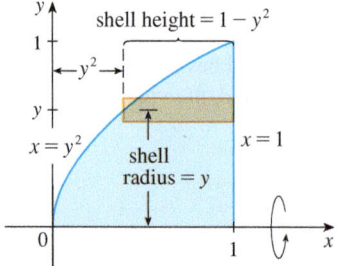

FIGURE 9

In this problem the disk method was simpler.

EXAMPLE 4 Find the volume of the solid obtained by rotating the region bounded by $y = x - x^2$ and $y = 0$ about the line $x = 2$.

SECTION 6.3 Volumes by Cylindrical Shells 453

SOLUTION Figure 10 shows the region and a cylindrical shell formed by rotation about the line $x = 2$. It has radius $2 - x$, circumference $2\pi(2 - x)$, and height $x - x^2$.

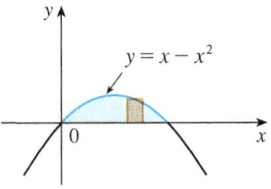
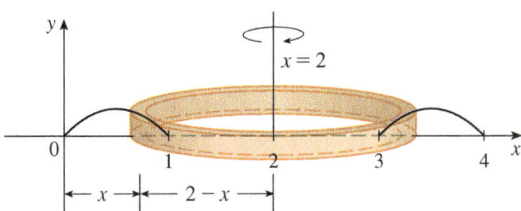

FIGURE 10

The volume of the given solid is

$$V = \int_0^1 2\pi(2 - x)(x - x^2)\, dx$$

$$= 2\pi \int_0^1 (x^3 - 3x^2 + 2x)\, dx$$

$$= 2\pi \left[\frac{x^4}{4} - x^3 + x^2 \right]_0^1 = \frac{\pi}{2}$$

■

■ Disks and Washers versus Cylindrical Shells

When computing the volume of a solid of revolution, how do we know whether to use disks (or washers) or cylindrical shells? There are several considerations to take into account: Is the region more easily described by top and bottom boundary curves of the form $y = f(x)$, or by left and right boundaries $x = g(y)$? Which choice is easier to work with? Are the limits of integration easier to find for one variable versus the other? Does the region require two separate integrals when using x as the variable but only one integral in y? Are we able to evaluate the integral we set up with our choice of variable?

If we decide that one variable is easier to work with than the other, then this dictates which method to use. Draw a sample rectangle in the region, corresponding to a cross-section of the solid. The thickness of the rectangle, either Δx or Δy, corresponds to the integration variable. If you imagine the rectangle revolving, it becomes either a disk (washer) or a shell.

6.3 EXERCISES

1. Let S be the solid obtained by rotating the region shown in the figure about the y-axis. Explain why it is awkward to use slicing to find the volume V of S. Sketch a typical approximating shell. What are its circumference and height? Use shells to find V.

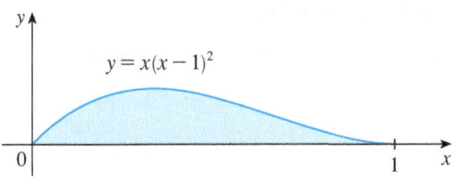

2. Let S be the solid obtained by rotating the region shown in the figure about the y-axis. Sketch a typical cylindrical shell and find its circumference and height. Use shells to find the volume of S. Do you think this method is preferable to slicing? Explain.

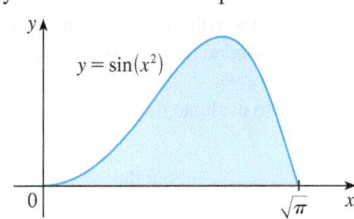

3–7 Use the method of cylindrical shells to find the volume generated by rotating the region bounded by the given curves about the y-axis.

3. $y = \sqrt[3]{x}, \quad y = 0, \quad x = 1$

4. $y = x^3, \quad y = 0, \quad x = 1, \quad x = 2$

5. $y = e^{-x^2}, \quad y = 0, \quad x = 0, \quad x = 1$

6. $y = 4x - x^2, \quad y = x$

7. $y = x^2, \quad y = 6x - 2x^2$

8. Let V be the volume of the solid obtained by rotating about the y-axis the region bounded by $y = \sqrt{x}$ and $y = x^2$. Find V both by slicing and by cylindrical shells. In both cases draw a diagram to explain your method.

9–14 Use the method of cylindrical shells to find the volume of the solid obtained by rotating the region bounded by the given curves about the x-axis.

9. $xy = 1, \quad x = 0, \quad y = 1, \quad y = 3$

10. $y = \sqrt{x}, \quad x = 0, \quad y = 2$

11. $y = x^{3/2}, \quad y = 8, \quad x = 0$

12. $x = -3y^2 + 12y - 9, \quad x = 0$

13. $x = 1 + (y - 2)^2, \quad x = 2$

14. $x + y = 4, \quad x = y^2 - 4y + 4$

15–20 Use the method of cylindrical shells to find the volume generated by rotating the region bounded by the given curves about the specified axis.

15. $y = x^3, \quad y = 8, \quad x = 0; \quad$ about $x = 3$

16. $y = 4 - 2x, \quad y = 0, \quad x = 0; \quad$ about $x = -1$

17. $y = 4x - x^2, \quad y = 3; \quad$ about $x = 1$

18. $y = \sqrt{x}, \quad x = 2y; \quad$ about $x = 5$

19. $x = 2y^2, \quad y \geq 0, \quad x = 2; \quad$ about $y = 2$

20. $x = 2y^2, \quad x = y^2 + 1; \quad$ about $y = -2$

21–26

(a) Set up an integral for the volume of the solid obtained by rotating the region bounded by the given curve about the specified axis.

(b) Use your calculator to evaluate the integral correct to five decimal places.

21. $y = xe^{-x}, \quad y = 0, \quad x = 2; \quad$ about the y-axis

22. $y = \tan x, \quad y = 0, \quad x = \pi/4; \quad$ about $x = \pi/2$

23. $y = \cos^4 x, \quad y = -\cos^4 x, \quad -\pi/2 \leq x \leq \pi/2; \quad$ about $x = \pi$

24. $y = x, \quad y = 2x/(1 + x^3); \quad$ about $x = -1$

25. $x = \sqrt{\sin y}, \quad 0 \leq y \leq \pi, \quad x = 0; \quad$ about $y = 4$

26. $x^2 - y^2 = 7, \quad x = 4; \quad$ about $y = 5$

27. Use the Midpoint Rule with $n = 5$ to estimate the volume obtained by rotating about the y-axis the region under the curve $y = \sqrt{1 + x^3}, 0 \leq x \leq 1$.

28. If the region shown in the figure is rotated about the y-axis to form a solid, use the Midpoint Rule with $n = 5$ to estimate the volume of the solid.

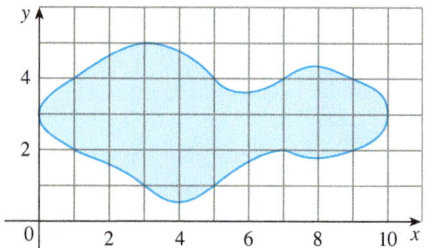

29–32 Each integral represents the volume of a solid. Describe the solid.

29. $\int_0^3 2\pi x^5 \, dx$

30. $\int_1^3 2\pi y \ln y \, dy$

31. $2\pi \int_1^4 \dfrac{y + 2}{y^2} \, dy$

32. $\int_0^1 2\pi (2 - x)(3^x - 2^x) \, dx$

33–34 Use a graph to estimate the x-coordinates of the points of intersection of the given curves. Then use this information and your calculator to estimate the volume of the solid obtained by rotating about the y-axis the region enclosed by these curves.

33. $y = x^2 - 2x, \quad y = \dfrac{x}{x^2 + 1}$

34. $y = e^{\sin x}, \quad y = x^2 - 4x + 5$

CAS 35–36 Use a computer algebra system to find the exact volume of the solid obtained by rotating the region bounded by the given curves about the specified line.

35. $y = \sin^2 x, \quad y = \sin^4 x, \quad 0 \leq x \leq \pi; \quad$ about $x = \pi/2$

36. $y = x^3 \sin x, \quad y = 0, \quad 0 \leq x \leq \pi; \quad$ about $x = -1$

37–43 The region bounded by the given curves is rotated about the specified axis. Find the volume of the resulting solid by any method.

37. $y = -x^2 + 6x - 8$, $y = 0$; about the y-axis

38. $y = -x^2 + 6x - 8$, $y = 0$; about the x-axis

39. $y^2 - x^2 = 1$, $y = 2$; about the x-axis

40. $y^2 - x^2 = 1$, $y = 2$; about the y-axis

41. $x^2 + (y - 1)^2 = 1$; about the y-axis

42. $x = (y - 3)^2$, $x = 4$; about $y = 1$

43. $x = (y - 1)^2$, $x - y = 1$; about $x = -1$

44. Let T be the triangular region with vertices $(0, 0)$, $(1, 0)$, and $(1, 2)$, and let V be the volume of the solid generated when T is rotated about the line $x = a$, where $a > 1$. Express a in terms of V.

45–47 Use cylindrical shells to find the volume of the solid.

45. A sphere of radius r

46. The solid torus of Exercise 6.2.63

47. A right circular cone with height h and base radius r

48. Suppose you make napkin rings by drilling holes with different diameters through two wooden balls (which also have different diameters). You discover that both napkin rings have the same height h, as shown in the figure.
(a) Guess which ring has more wood in it.
(b) Check your guess: Use cylindrical shells to compute the volume of a napkin ring created by drilling a hole with radius r through the center of a sphere of radius R and express the answer in terms of h.

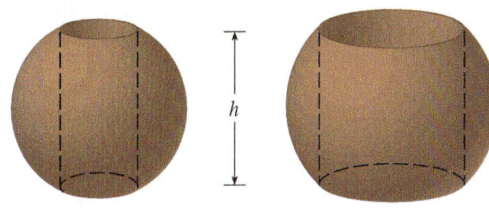

6.4 Work

The term *work* is used in everyday language to mean the total amount of effort required to perform a task. In physics it has a technical meaning that depends on the idea of a *force*. Intuitively, you can think of a force as describing a push or pull on an object—for example, a horizontal push of a book across a table or the downward pull of the earth's gravity on a ball. In general, if an object moves along a straight line with position function $s(t)$, then the **force** F on the object (in the same direction) is given by Newton's Second Law of Motion as the product of its mass m and its acceleration a:

$$\boxed{1} \qquad F = ma = m\frac{d^2s}{dt^2}$$

In the SI metric system, the mass is measured in kilograms (kg), the displacement in meters (m), the time in seconds (s), and the force in newtons (N = kg·m/s²). Thus a force of 1 N acting on a mass of 1 kg produces an acceleration of 1 m/s². In the US Customary system the fundamental unit is chosen to be the unit of force, which is the pound.

In the case of constant acceleration, the force F is also constant and the work done is defined to be the product of the force F and the distance d that the object moves:

$$\boxed{2} \qquad W = Fd \qquad \text{work} = \text{force} \times \text{distance}$$

If F is measured in newtons and d in meters, then the unit for W is a newton-meter, which is called a joule (J). If F is measured in pounds and d in feet, then the unit for W is a foot-pound (ft-lb), which is about 1.36 J.

EXAMPLE 1
(a) How much work is done in lifting a 1.2-kg book off the floor to put it on a desk that is 0.7 m high? Use the fact that the acceleration due to gravity is $g = 9.8$ m/s^2.
(b) How much work is done in lifting a 20-lb weight 6 ft off the ground?

SOLUTION
(a) The force exerted is equal and opposite to that exerted by gravity, so Equation 1 gives
$$F = mg = (1.2)(9.8) = 11.76 \text{ N}$$
and then Equation 2 gives the work done as
$$W = Fd = (11.76 \text{ N})(0.7 \text{ m}) \approx 8.2 \text{ J}$$

(b) Here the force is given as $F = 20$ lb, so the work done is
$$W = Fd = (20 \text{ lb})(6 \text{ ft}) = 120 \text{ ft-lb}$$

Notice that in part (b), unlike part (a), we did not have to multiply by g because we were given the *weight* (which is a force) and not the mass of the object. ∎

Equation 2 defines work as long as the force is constant, but what happens if the force is variable? Let's suppose that the object moves along the x-axis in the positive direction, from $x = a$ to $x = b$, and at each point x between a and b a force $f(x)$ acts on the object, where f is a continuous function. We divide the interval $[a, b]$ into n subintervals with endpoints $x_0, x_1, \ldots, x_n$ and equal width Δx. We choose a sample point x_i^* in the ith subinterval $[x_{i-1}, x_i]$. Then the force at that point is $f(x_i^*)$. If n is large, then Δx is small, and since f is continuous, the values of f don't change very much over the interval $[x_{i-1}, x_i]$. In other words, f is almost constant on the interval and so the work W_i that is done in moving the particle from x_{i-1} to x_i is approximately given by Equation 2:
$$W_i \approx f(x_i^*) \, \Delta x$$

Thus we can approximate the total work by

$$\boxed{3} \qquad W \approx \sum_{i=1}^{n} f(x_i^*) \, \Delta x$$

It seems that this approximation becomes better as we make n larger. Therefore we define the **work done in moving the object from a to b** as the limit of this quantity as $n \to \infty$. Since the right side of (3) is a Riemann sum, we recognize its limit as being a definite integral and so

$$\boxed{4} \qquad W = \lim_{n \to \infty} \sum_{i=1}^{n} f(x_i^*) \, \Delta x = \int_a^b f(x) \, dx$$

EXAMPLE 2 When a particle is located a distance x feet from the origin, a force of $x^2 + 2x$ pounds acts on it. How much work is done in moving it from $x = 1$ to $x = 3$?

SOLUTION
$$W = \int_1^3 (x^2 + 2x) \, dx = \frac{x^3}{3} + x^2 \Big]_1^3 = \frac{50}{3}$$

The work done is $16\frac{2}{3}$ ft-lb. ∎

In the next example we use a law from physics. **Hooke's Law** states that the force required to maintain a spring stretched x units beyond its natural length is proportional to x:

$$f(x) = kx$$

where k is a positive constant called the **spring constant** (see Figure 1). Hooke's Law holds provided that x is not too large.

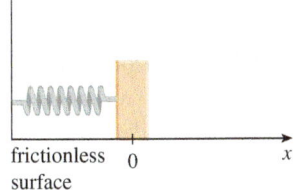

(a) Natural position of spring

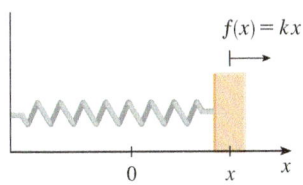

(b) Stretched position of spring

FIGURE 1
Hooke's Law

EXAMPLE 3 A force of 40 N is required to hold a spring that has been stretched from its natural length of 10 cm to a length of 15 cm. How much work is done in stretching the spring from 15 cm to 18 cm?

SOLUTION According to Hooke's Law, the force required to hold the spring stretched x meters beyond its natural length is $f(x) = kx$. When the spring is stretched from 10 cm to 15 cm, the amount stretched is 5 cm = 0.05 m. This means that $f(0.05) = 40$, so

$$0.05k = 40 \qquad k = \frac{40}{0.05} = 800$$

Thus $f(x) = 800x$ and the work done in stretching the spring from 15 cm to 18 cm is

$$W = \int_{0.05}^{0.08} 800x\,dx = 800\,\frac{x^2}{2}\bigg]_{0.05}^{0.08}$$

$$= 400[(0.08)^2 - (0.05)^2] = 1.56 \text{ J}$$

EXAMPLE 4 A 200-lb cable is 100 ft long and hangs vertically from the top of a tall building. How much work is required to lift the cable to the top of the building?

SOLUTION Here we don't have a formula for the force function, but we can use an argument similar to the one that led to Definition 4.

Let's place the origin at the top of the building and the x-axis pointing downward as in Figure 2. We divide the cable into small parts with length Δx. If x_i^* is a point in the ith such interval, then all points in the interval are lifted by approximately the same amount, namely x_i^*. The cable weighs 2 pounds per foot, so the weight of the ith part is $(2 \text{ lb/ft})(\Delta x \text{ ft}) = 2\Delta x$ lb. Thus the work done on the ith part, in foot-pounds, is

$$\underbrace{(2\Delta x)}_{\text{force}} \cdot \underbrace{x_i^*}_{\text{distance}} = 2x_i^* \,\Delta x$$

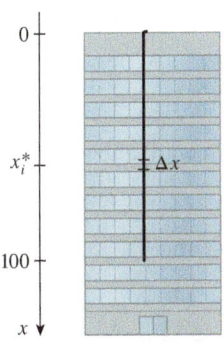

FIGURE 2

If we had placed the origin at the bottom of the cable and the x-axis upward, we would have gotten

$$W = \int_0^{100} 2(100 - x)\,dx$$

which gives the same answer.

We get the total work done by adding all these approximations and letting the number of parts become large (so $\Delta x \to 0$):

$$W = \lim_{n \to \infty} \sum_{i=1}^{n} 2x_i^* \,\Delta x = \int_0^{100} 2x\,dx$$

$$= x^2\Big]_0^{100} = 10{,}000 \text{ ft-lb}$$

EXAMPLE 5 A tank has the shape of an inverted circular cone with height 10 m and base radius 4 m. It is filled with water to a height of 8 m. Find the work required to empty the tank by pumping all of the water to the top of the tank. (The density of water is 1000 kg/m³.)

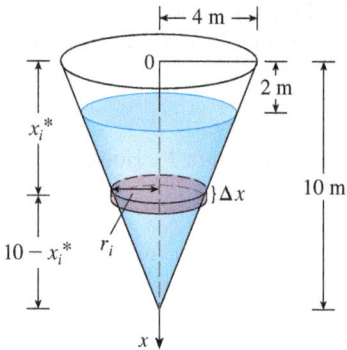

FIGURE 3

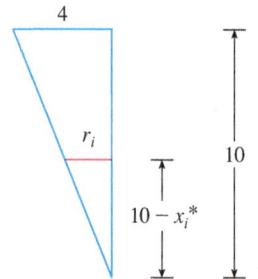

FIGURE 4

SOLUTION Let's measure depths from the top of the tank by introducing a vertical coordinate line as in Figure 3. The water extends from a depth of 2 m to a depth of 10 m and so we divide the interval $[2, 10]$ into n subintervals with endpoints $x_0, x_1, \ldots, x_n$ and choose x_i^* in the ith subinterval. This divides the water into n layers. The ith layer is approximated by a circular cylinder with radius r_i and height Δx. We can compute r_i from similar triangles, using Figure 4, as follows:

$$\frac{r_i}{10 - x_i^*} = \frac{4}{10} \qquad r_i = \tfrac{2}{5}(10 - x_i^*)$$

Thus an approximation to the volume of the ith layer of water is

$$V_i \approx \pi r_i^2 \, \Delta x = \frac{4\pi}{25}(10 - x_i^*)^2 \, \Delta x$$

and so its mass is

$$m_i = \text{density} \times \text{volume}$$

$$\approx 1000 \cdot \frac{4\pi}{25}(10 - x_i^*)^2 \, \Delta x = 160\pi(10 - x_i^*)^2 \, \Delta x$$

The force required to raise this layer must overcome the force of gravity and so

$$F_i = m_i g \approx (9.8)160\pi(10 - x_i^*)^2 \, \Delta x$$

$$= 1568\pi(10 - x_i^*)^2 \, \Delta x$$

Each particle in the layer must travel a distance upward of approximately x_i^*. The work W_i done to raise this layer to the top is approximately the product of the force F_i and the distance x_i^*:

$$W_i \approx F_i x_i^* \approx 1568\pi x_i^*(10 - x_i^*)^2 \, \Delta x$$

To find the total work done in emptying the entire tank, we add the contributions of each of the n layers and then take the limit as $n \to \infty$:

$$W = \lim_{n \to \infty} \sum_{i=1}^{n} 1568\pi x_i^*(10 - x_i^*)^2 \, \Delta x = \int_2^{10} 1568\pi x(10 - x)^2 \, dx$$

$$= 1568\pi \int_2^{10} (100x - 20x^2 + x^3) \, dx = 1568\pi \left[50x^2 - \frac{20x^3}{3} + \frac{x^4}{4} \right]_2^{10}$$

$$= 1568\pi \left(\tfrac{2048}{3} \right) \approx 3.4 \times 10^6 \text{ J}$$ ■

6.4 EXERCISES

1. A 360-lb gorilla climbs a tree to a height of 20 ft. Find the work done if the gorilla reaches that height in
   (a) 10 seconds   (b) 5 seconds

2. How much work is done when a hoist lifts a 200-kg rock to a height of 3 m?

3. A variable force of $5x^{-2}$ pounds moves an object along a straight line when it is x feet from the origin. Calculate the work done in moving the object from $x = 1$ ft to $x = 10$ ft.

4. When a particle is located a distance x meters from the origin, a force of $\cos(\pi x/3)$ newtons acts on it. How much work is done in moving the particle from $x = 1$ to $x = 2$? Interpret your answer by considering the work done from $x = 1$ to $x = 1.5$ and from $x = 1.5$ to $x = 2$.

5. Shown is the graph of a force function (in newtons) that increases to its maximum value and then remains constant. How much work is done by the force in moving an object a distance of 8 m?

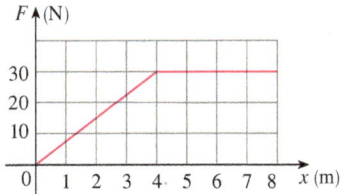

6. The table shows values of a force function $f(x)$, where x is measured in meters and $f(x)$ in newtons. Use the Midpoint Rule to estimate the work done by the force in moving an object from $x = 4$ to $x = 20$.

x	4	6	8	10	12	14	16	18	20
$f(x)$	5	5.8	7.0	8.8	9.6	8.2	6.7	5.2	4.1

7. A force of 10 lb is required to hold a spring stretched 4 in. beyond its natural length. How much work is done in stretching it from its natural length to 6 in. beyond its natural length?

8. A spring has a natural length of 40 cm. If a 60-N force is required to keep the spring compressed 10 cm, how much work is done during this compression? How much work is required to compress the spring to a length of 25 cm?

9. Suppose that 2 J of work is needed to stretch a spring from its natural length of 30 cm to a length of 42 cm.
(a) How much work is needed to stretch the spring from 35 cm to 40 cm?
(b) How far beyond its natural length will a force of 30 N keep the spring stretched?

10. If the work required to stretch a spring 1 ft beyond its natural length is 12 ft-lb, how much work is needed to stretch it 9 in. beyond its natural length?

11. A spring has natural length 20 cm. Compare the work W_1 done in stretching the spring from 20 cm to 30 cm with the work W_2 done in stretching it from 30 cm to 40 cm. How are W_2 and W_1 related?

12. If 6 J of work is needed to stretch a spring from 10 cm to 12 cm and another 10 J is needed to stretch it from 12 cm to 14 cm, what is the natural length of the spring?

13–22 Show how to approximate the required work by a Riemann sum. Then express the work as an integral and evaluate it.

13. A heavy rope, 50 ft long, weighs 0.5 lb/ft and hangs over the edge of a building 120 ft high.
(a) How much work is done in pulling the rope to the top of the building?
(b) How much work is done in pulling half the rope to the top of the building?

14. A thick cable, 60 ft long and weighing 180 lb, hangs from a winch on a crane. Compute in two different ways the work done if the winch winds up 25 ft of the cable.
(a) Follow the method of Example 4.
(b) Write a function for the weight of the remaining cable after x feet has been wound up by the winch. Estimate the amount of work done when the winch pulls up Δx ft of cable.

15. A cable that weighs 2 lb/ft is used to lift 800 lb of coal up a mine shaft 500 ft deep. Find the work done.

16. A chain lying on the ground is 10 m long and its mass is 80 kg. How much work is required to raise one end of the chain to a height of 6 m?

17. A leaky 10-kg bucket is lifted from the ground to a height of 12 m at a constant speed with a rope that weighs 0.8 kg/m. Initially the bucket contains 36 kg of water, but the water leaks at a constant rate and finishes draining just as the bucket reaches the 12-m level. How much work is done?

18. A bucket that weighs 4 lb and a rope of negligible weight are used to draw water from a well that is 80 ft deep. The bucket is filled with 40 lb of water and is pulled up at a rate of 2 ft/s, but water leaks out of a hole in the bucket at a rate of 0.2 lb/s. Find the work done in pulling the bucket to the top of the well.

19. A 10-ft chain weighs 25 lb and hangs from a ceiling. Find the work done in lifting the lower end of the chain to the ceiling so that it's level with the upper end.

20. A circular swimming pool has a diameter of 24 ft, the sides are 5 ft high, and the depth of the water is 4 ft. How much work is required to pump all of the water out over the side? (Use the fact that water weighs 62.5 lb/ft^3.)

21. An aquarium 2 m long, 1 m wide, and 1 m deep is full of water. Find the work needed to pump half of the water out of the aquarium. (Use the fact that the density of water is 1000 kg/m^3.)

22. A spherical water tank, 24 ft in diameter, sits atop a 60 ft tower. The tank is filled by a hose attached to the bottom of the sphere. If a 1.5 horsepower pump is used to deliver water up to the tank, how long will it take to fill the tank? (One horsepower = 550 ft-lb of work per second.)

23–26 A tank is full of water. Find the work required to pump the water out of the spout. In Exercises 25 and 26 use the fact that water weighs 62.5 lb/ft^3.

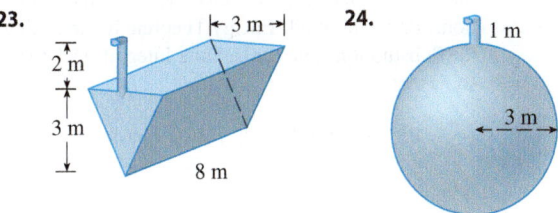

25.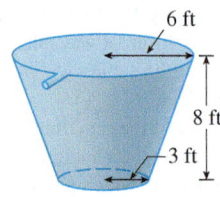
frustum of a cone

26.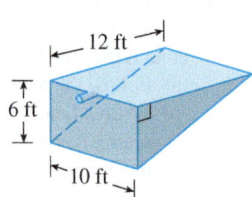

27. Suppose that for the tank in Exercise 23 the pump breaks down after 4.7×10^5 J of work has been done. What is the depth of the water remaining in the tank?

28. Solve Exercise 24 if the tank is half full of oil that has a density of 900 kg/m³.

29. When gas expands in a cylinder with radius r, the pressure at any given time is a function of the volume: $P = P(V)$. The force exerted by the gas on the piston (see the figure) is the product of the pressure and the area: $F = \pi r^2 P$. Show that the work done by the gas when the volume expands from volume V_1 to volume V_2 is

$$W = \int_{V_1}^{V_2} P \, dV$$

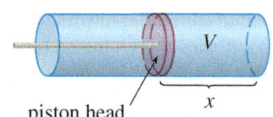
piston head

30. In a steam engine the pressure P and volume V of steam satisfy the equation $PV^{1.4} = k$, where k is a constant. (This is true for adiabatic expansion, that is, expansion in which there is no heat transfer between the cylinder and its surroundings.) Use Exercise 29 to calculate the work done by the engine during a cycle when the steam starts at a pressure of 160 lb/in² and a volume of 100 in³ and expands to a volume of 800 in³.

31. The kinetic energy KE of an object of mass m moving with velocity v is defined as $\text{KE} = \frac{1}{2}mv^2$. If a force $f(x)$ acts on the object, moving it along the x-axis from x_1 to x_2, the *Work-Energy Theorem* states that the net work done is equal to the change in kinetic energy: $\frac{1}{2}mv_2^2 - \frac{1}{2}mv_1^2$, where v_1 is the velocity at x_1 and v_2 is the velocity at x_2.
(a) Let $x = s(t)$ be the position function of the object at time t and $v(t)$, $a(t)$ the velocity and acceleration functions. Prove the Work-Energy Theorem by first using the Substitution Rule for Definite Integrals (5.5.6) to show that

$$W = \int_{x_1}^{x_2} f(x) \, dx = \int_{t_1}^{t_2} f(s(t)) \, v(t) \, dt$$

Then use Newton's Second Law of Motion (force = mass × acceleration) and the substitution $u = v(t)$ to evaluate the integral.
(b) How much work (in ft-lb) is required to hurl a 12-lb bowling ball at 20 mi/h? (*Note:* Divide the weight in pounds by 32 ft/s², the acceleration due to gravity, to find the mass, measured in slugs.)

32. Suppose that when launching an 800-kg roller coaster car an electromagnetic propulsion system exerts a force of $5.7x^2 + 1.5x$ newtons on the car at a distance x meters along the track. Use Exercise 31(a) to find the speed of the car when it has traveled 60 meters.

33. (a) Newton's Law of Gravitation states that two bodies with masses m_1 and m_2 attract each other with a force

$$F = G \frac{m_1 m_2}{r^2}$$

where r is the distance between the bodies and G is the gravitational constant. If one of the bodies is fixed, find the work needed to move the other from $r = a$ to $r = b$.
(b) Compute the work required to launch a 1000-kg satellite vertically to a height of 1000 km. You may assume that the earth's mass is 5.98×10^{24} kg and is concentrated at its center. Take the radius of the earth to be 6.37×10^6 m and $G = 6.67 \times 10^{-11}$ N·m²/kg².

34. The Great Pyramid of King Khufu was built of limestone in Egypt over a 20-year time period from 2580 BC to 2560 BC. Its base is a square with side length 756 ft and its height when built was 481 ft. (It was the tallest manmade structure in the world for more than 3800 years.) The density of the limestone is about 150 lb/ft³.
(a) Estimate the total work done in building the pyramid.
(b) If each laborer worked 10 hours a day for 20 years, for 340 days a year, and did 200 ft-lb/h of work in lifting the limestone blocks into place, about how many laborers were needed to construct the pyramid?

6.5 Average Value of a Function

It is easy to calculate the average value of finitely many numbers $y_1, y_2, \ldots, y_n$:

$$y_{\text{ave}} = \frac{y_1 + y_2 + \cdots + y_n}{n}$$

But how do we compute the average temperature during a day if infinitely many temperature readings are possible? Figure 1 shows the graph of a temperature function $T(t)$, where t is measured in hours and T in °C, and a guess at the average temperature, T_{ave}.

In general, let's try to compute the average value of a function $y = f(x)$, $a \leq x \leq b$. We start by dividing the interval $[a, b]$ into n equal subintervals, each with length $\Delta x = (b - a)/n$. Then we choose points $x_1^*, \ldots, x_n^*$ in successive subintervals and calculate the average of the numbers $f(x_1^*), \ldots, f(x_n^*)$:

$$\frac{f(x_1^*) + \cdots + f(x_n^*)}{n}$$

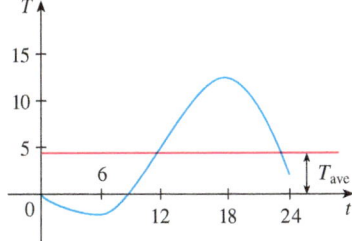

FIGURE 1

(For example, if f represents a temperature function and $n = 24$, this means that we take temperature readings every hour and then average them.) Since $\Delta x = (b - a)/n$, we can write $n = (b - a)/\Delta x$ and the average value becomes

$$\frac{f(x_1^*) + \cdots + f(x_n^*)}{\frac{b-a}{\Delta x}} = \frac{1}{b-a}[f(x_1^*) + \cdots + f(x_n^*)]\Delta x$$

$$= \frac{1}{b-a}[f(x_1^*)\Delta x + \cdots + f(x_n^*)\Delta x]$$

$$= \frac{1}{b-a}\sum_{i=1}^{n} f(x_i^*)\Delta x$$

If we let n increase, we would be computing the average value of a large number of closely spaced values. (For example, we would be averaging temperature readings taken every minute or even every second.) The limiting value is

$$\lim_{n \to \infty} \frac{1}{b-a}\sum_{i=1}^{n} f(x_i^*)\Delta x = \frac{1}{b-a}\int_a^b f(x)\,dx$$

by the definition of a definite integral.

Therefore we define the **average value of f** on the interval $[a, b]$ as

$$f_{\text{ave}} = \frac{1}{b-a}\int_a^b f(x)\,dx$$

For a positive function, we can think of this definition as saying

$$\frac{\text{area}}{\text{width}} = \text{average height}$$

EXAMPLE 1 Find the average value of the function $f(x) = 1 + x^2$ on the interval $[-1, 2]$.

SOLUTION With $a = -1$ and $b = 2$ we have

$$f_{\text{ave}} = \frac{1}{b-a}\int_a^b f(x)\,dx = \frac{1}{2-(-1)}\int_{-1}^{2}(1+x^2)\,dx = \frac{1}{3}\left[x + \frac{x^3}{3}\right]_{-1}^{2} = 2 \quad \blacksquare$$

If $T(t)$ is the temperature at time t, we might wonder if there is a specific time when the temperature is the same as the average temperature. For the temperature function graphed in Figure 1, we see that there are two such times—just before noon and just before midnight. In general, is there a number c at which the value of a function f is exactly equal to the average value of the function, that is, $f(c) = f_{ave}$? The following theorem says that this is true for continuous functions.

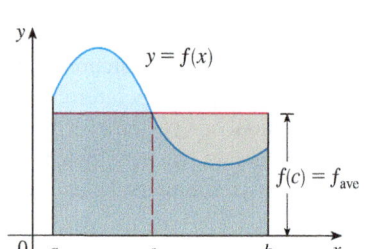

FIGURE 2

You can always chop off the top of a (two-dimensional) mountain at a certain height and use it to fill in the valleys so that the mountain becomes completely flat.

The Mean Value Theorem for Integrals If f is continuous on $[a, b]$, then there exists a number c in $[a, b]$ such that

$$f(c) = f_{ave} = \frac{1}{b-a} \int_a^b f(x)\,dx$$

that is,

$$\int_a^b f(x)\,dx = f(c)(b-a)$$

The Mean Value Theorem for Integrals is a consequence of the Mean Value Theorem for derivatives and the Fundamental Theorem of Calculus. The proof is outlined in Exercise 25.

The geometric interpretation of the Mean Value Theorem for Integrals is that, for *positive* functions f, there is a number c such that the rectangle with base $[a, b]$ and height $f(c)$ has the same area as the region under the graph of f from a to b. (See Figure 2 and the more picturesque interpretation in the margin note.)

EXAMPLE 2 Since $f(x) = 1 + x^2$ is continuous on the interval $[-1, 2]$, the Mean Value Theorem for Integrals says there is a number c in $[-1, 2]$ such that

$$\int_{-1}^2 (1+x^2)\,dx = f(c)[2-(-1)]$$

In this particular case we can find c explicitly. From Example 1 we know that $f_{ave} = 2$, so the value of c satisfies

$$f(c) = f_{ave} = 2$$

Therefore $\quad 1 + c^2 = 2 \quad$ so $\quad c^2 = 1$

So in this case there happen to be two numbers $c = \pm 1$ in the interval $[-1, 2]$ that work in the Mean Value Theorem for Integrals. ■

Examples 1 and 2 are illustrated by Figure 3.

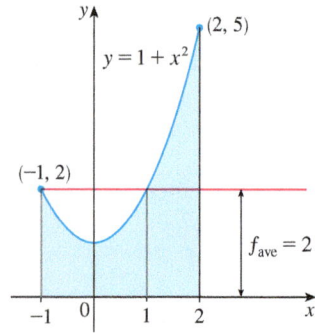

FIGURE 3

EXAMPLE 3 Show that the average velocity of a car over a time interval $[t_1, t_2]$ is the same as the average of its velocities during the trip.

SOLUTION If $s(t)$ is the displacement of the car at time t, then, by definition, the average velocity of the car over the interval is

$$\frac{\Delta s}{\Delta t} = \frac{s(t_2) - s(t_1)}{t_2 - t_1}$$

On the other hand, the average value of the velocity function on the interval is

$$v_{ave} = \frac{1}{t_2 - t_1} \int_{t_1}^{t_2} v(t)\,dt = \frac{1}{t_2 - t_1} \int_{t_1}^{t_2} s'(t)\,dt$$

$$= \frac{1}{t_2 - t_1} [s(t_2) - s(t_1)] \qquad \text{(by the Net Change Theorem)}$$

$$= \frac{s(t_2) - s(t_1)}{t_2 - t_1} = \text{average velocity} \qquad ■$$

6.5 EXERCISES

1–8 Find the average value of the function on the given interval.

1. $f(x) = 3x^2 + 8x$, $[-1, 2]$
2. $f(x) = \sqrt{x}$, $[0, 4]$
3. $g(x) = 3 \cos x$, $[-\pi/2, \pi/2]$
4. $g(t) = \dfrac{t}{\sqrt{3+t^2}}$, $[1, 3]$
5. $f(t) = e^{\sin t} \cos t$, $[0, \pi/2]$
6. $f(x) = x^2/(x^3 + 3)^2$, $[-1, 1]$
7. $h(x) = \cos^4 x \sin x$, $[0, \pi]$
8. $h(u) = (\ln u)/u$, $[1, 5]$

9–12
(a) Find the average value of f on the given interval.
(b) Find c such that $f_{\text{ave}} = f(c)$.
(c) Sketch the graph of f and a rectangle whose area is the same as the area under the graph of f.

9. $f(x) = (x-3)^2$, $[2, 5]$
10. $f(x) = 1/x$, $[1, 3]$
11. $f(x) = 2 \sin x - \sin 2x$, $[0, \pi]$
12. $f(x) = 2xe^{-x^2}$, $[0, 2]$

13. If f is continuous and $\int_1^3 f(x)\, dx = 8$, show that f takes on the value 4 at least once on the interval $[1, 3]$.

14. Find the numbers b such that the average value of $f(x) = 2 + 6x - 3x^2$ on the interval $[0, b]$ is equal to 3.

15. Find the average value of f on $[0, 8]$.

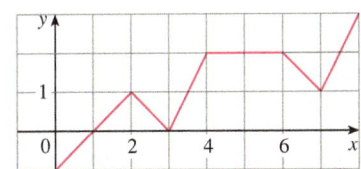

16. The velocity graph of an accelerating car is shown.

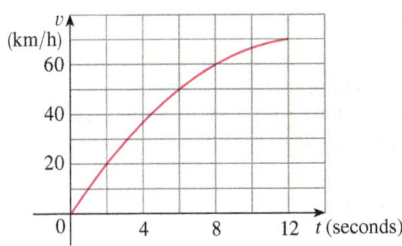

(a) Use the Midpoint Rule to estimate the average velocity of the car during the first 12 seconds.
(b) At what time was the instantaneous velocity equal to the average velocity?

17. In a certain city the temperature (in °F) t hours after 9 AM was modeled by the function
$$T(t) = 50 + 14 \sin \dfrac{\pi t}{12}$$
Find the average temperature during the period from 9 AM to 9 PM.

18. The velocity v of blood that flows in a blood vessel with radius R and length l at a distance r from the central axis is
$$v(r) = \dfrac{P}{4\eta l}(R^2 - r^2)$$
where P is the pressure difference between the ends of the vessel and η is the viscosity of the blood (see Example 3.7.7). Find the average velocity (with respect to r) over the interval $0 \le r \le R$. Compare the average velocity with the maximum velocity.

19. The linear density in a rod 8 m long is $12/\sqrt{x+1}$ kg/m, where x is measured in meters from one end of the rod. Find the average density of the rod.

20. (a) A cup of coffee has temperature 95°C and takes 30 minutes to cool to 61°C in a room with temperature 20°C. Use Newton's Law of Cooling (Section 3.8) to show that the temperature of the coffee after t minutes is
$$T(t) = 20 + 75e^{-kt}$$
where $k \approx 0.02$.
(b) What is the average temperature of the coffee during the first half hour?

21. In Example 3.8.1 we modeled the world population in the second half of the 20th century by the equation $P(t) = 2560e^{0.017185t}$. Use this equation to estimate the average world population during this time period.

22. If a freely falling body starts from rest, then its displacement is given by $s = \tfrac{1}{2}gt^2$. Let the velocity after a time T be v_T. Show that if we compute the average of the velocities with respect to t we get $v_{\text{ave}} = \tfrac{1}{2}v_T$, but if we compute the average of the velocities with respect to s we get $v_{\text{ave}} = \tfrac{2}{3}v_T$.

23. Use the result of Exercise 5.5.83 to compute the average volume of inhaled air in the lungs in one respiratory cycle.

24. Use the diagram to show that if f is concave upward on $[a, b]$, then
$$f_{ave} > f\left(\frac{a+b}{2}\right)$$

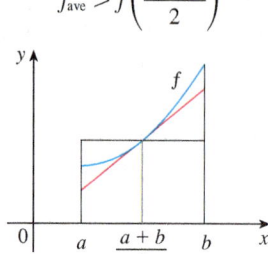

25. Prove the Mean Value Theorem for Integrals by applying the Mean Value Theorem for derivatives (see Section 4.2) to the function $F(x) = \int_a^x f(t)\,dt$.

26. If $f_{ave}[a, b]$ denotes the average value of f on the interval $[a, b]$ and $a < c < b$, show that
$$f_{ave}[a, b] = \frac{c-a}{b-a}\,f_{ave}[a, c] + \frac{b-c}{b-a}\,f_{ave}[c, b]$$

APPLIED PROJECT **CALCULUS AND BASEBALL**

In this project we explore three of the many applications of calculus to baseball. The physical interactions of the game, especially the collision of ball and bat, are quite complex and their models are discussed in detail in a book by Robert Adair, *The Physics of Baseball*, 3d ed. (New York, 2002).

1. It may surprise you to learn that the collision of baseball and bat lasts only about a thousandth of a second. Here we calculate the average force on the bat during this collision by first computing the change in the ball's momentum.

The *momentum* p of an object is the product of its mass m and its velocity v, that is, $p = mv$. Suppose an object, moving along a straight line, is acted on by a force $F = F(t)$ that is a continuous function of time.

(a) Show that the change in momentum over a time interval $[t_0, t_1]$ is equal to the integral of F from t_0 to t_1; that is, show that
$$p(t_1) - p(t_0) = \int_{t_0}^{t_1} F(t)\,dt$$

This integral is called the *impulse* of the force over the time interval.

(b) A pitcher throws a 90-mi/h fastball to a batter, who hits a line drive directly back to the pitcher. The ball is in contact with the bat for 0.001 s and leaves the bat with velocity 110 mi/h. A baseball weighs 5 oz and, in US Customary units, its mass is measured in slugs: $m = w/g$, where $g = 32$ ft/s^2.
  (i) Find the change in the ball's momentum.
  (ii) Find the average force on the bat.

2. In this problem we calculate the work required for a pitcher to throw a 90-mi/h fastball by first considering kinetic energy.

The *kinetic energy* K of an object of mass m and velocity v is given by $K = \tfrac{1}{2}mv^2$. Suppose an object of mass m, moving in a straight line, is acted on by a force $F = F(s)$ that depends on its position s. According to Newton's Second Law
$$F(s) = ma = m\frac{dv}{dt}$$

where a and v denote the acceleration and velocity of the object.

(a) Show that the work done in moving the object from a position s_0 to a position s_1 is equal to the change in the object's kinetic energy; that is, show that
$$W = \int_{s_0}^{s_1} F(s)\,ds = \tfrac{1}{2}mv_1^2 - \tfrac{1}{2}mv_0^2$$

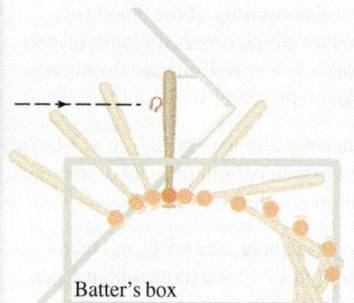

An overhead view of the position of a baseball bat, shown every fiftieth of a second during a typical swing. (Adapted from *The Physics of Baseball*)

where $v_0 = v(s_0)$ and $v_1 = v(s_1)$ are the velocities of the object at the positions s_0 and s_1. *Hint:* By the Chain Rule,

$$m\frac{dv}{dt} = m\frac{dv}{ds}\frac{ds}{dt} = mv\frac{dv}{ds}$$

(b) How many foot-pounds of work does it take to throw a baseball at a speed of 90 mi/h?

3. (a) An outfielder fields a baseball 280 ft away from home plate and throws it directly to the catcher with an initial velocity of 100 ft/s. Assume that the velocity $v(t)$ of the ball after t seconds satisfies the differential equation $dv/dt = -\frac{1}{10}v$ because of air resistance. How long does it take for the ball to reach home plate? (Ignore any vertical motion of the ball.)

(b) The manager of the team wonders whether the ball will reach home plate sooner if it is relayed by an infielder. The shortstop can position himself directly between the outfielder and home plate, catch the ball thrown by the outfielder, turn, and throw the ball to the catcher with an initial velocity of 105 ft/s. The manager clocks the relay time of the shortstop (catching, turning, throwing) at half a second. How far from home plate should the shortstop position himself to minimize the total time for the ball to reach home plate? Should the manager encourage a direct throw or a relayed throw? What if the shortstop can throw at 115 ft/s?

(c) For what throwing velocity of the shortstop does a relayed throw take the same time as a direct throw?

APPLIED PROJECT

CAS WHERE TO SIT AT THE MOVIES

A movie theater has a screen that is positioned 10 ft off the floor and is 25 ft high. The first row of seats is placed 9 ft from the screen and the rows are set 3 ft apart. The floor of the seating area is inclined at an angle of $\alpha = 20°$ above the horizontal and the distance up the incline that you sit is x. The theater has 21 rows of seats, so $0 \leq x \leq 60$. Suppose you decide that the best place to sit is in the row where the angle θ subtended by the screen at your eyes is a maximum. Let's also suppose that your eyes are 4 ft above the floor, as shown in the figure. (In Exercise 4.7.78 we looked at a simpler version of this problem, where the floor is horizontal, but this project involves a more complicated situation and requires technology.)

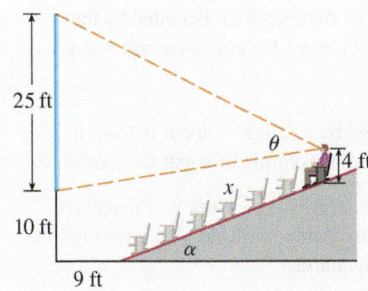

1. Show that

$$\theta = \arccos\left(\frac{a^2 + b^2 - 625}{2ab}\right)$$

where $\qquad a^2 = (9 + x\cos\alpha)^2 + (31 - x\sin\alpha)^2$

and $\qquad b^2 = (9 + x\cos\alpha)^2 + (x\sin\alpha - 6)^2$

2. Use a graph of θ as a function of x to estimate the value of x that maximizes θ. In which row should you sit? What is the viewing angle θ in this row?

3. Use your computer algebra system to differentiate θ and find a numerical value for the root of the equation $d\theta/dx = 0$. Does this value confirm your result in Problem 2?

4. Use the graph of θ to estimate the average value of θ on the interval $0 \leq x \leq 60$. Then use your CAS to compute the average value. Compare with the maximum and minimum values of θ.

6 REVIEW

CONCEPT CHECK

Answers to the Concept Check can be found on the back endpapers.

1. (a) Draw two typical curves $y = f(x)$ and $y = g(x)$, where $f(x) \geq g(x)$ for $a \leq x \leq b$. Show how to approximate the area between these curves by a Riemann sum and sketch the corresponding approximating rectangles. Then write an expression for the exact area.
   (b) Explain how the situation changes if the curves have equations $x = f(y)$ and $x = g(y)$, where $f(y) \geq g(y)$ for $c \leq y \leq d$.

2. Suppose that Sue runs faster than Kathy throughout a 1500-meter race. What is the physical meaning of the area between their velocity curves for the first minute of the race?

3. (a) Suppose S is a solid with known cross-sectional areas. Explain how to approximate the volume of S by a Riemann sum. Then write an expression for the exact volume.
   (b) If S is a solid of revolution, how do you find the cross-sectional areas?

4. (a) What is the volume of a cylindrical shell?
   (b) Explain how to use cylindrical shells to find the volume of a solid of revolution.
   (c) Why might you want to use the shell method instead of slicing?

5. Suppose that you push a book across a 6-meter-long table by exerting a force $f(x)$ at each point from $x = 0$ to $x = 6$. What does $\int_0^6 f(x)\, dx$ represent? If $f(x)$ is measured in newtons, what are the units for the integral?

6. (a) What is the average value of a function f on an interval $[a, b]$?
   (b) What does the Mean Value Theorem for Integrals say? What is its geometric interpretation?

EXERCISES

1–6 Find the area of the region bounded by the given curves.

1. $y = x^2$, $y = 4x - x^2$
2. $y = \sqrt{x}$, $y = -\sqrt[3]{x}$, $y = x - 2$
3. $y = 1 - 2x^2$, $y = |x|$
4. $x + y = 0$, $x = y^2 + 3y$
5. $y = \sin(\pi x/2)$, $y = x^2 - 2x$
6. $y = \sqrt{x}$, $y = x^2$, $x = 2$

7–11 Find the volume of the solid obtained by rotating the region bounded by the given curves about the specified axis.

7. $y = 2x$, $y = x^2$; about the x-axis
8. $x = 1 + y^2$, $y = x - 3$; about the y-axis
9. $x = 0$, $x = 9 - y^2$; about $x = -1$
10. $y = x^2 + 1$, $y = 9 - x^2$; about $y = -1$
11. $x^2 - y^2 = a^2$, $x = a + h$ (where $a > 0, h > 0$); about the y-axis

12–14 Set up, but do not evaluate, an integral for the volume of the solid obtained by rotating the region bounded by the given curves about the specified axis.

12. $y = \tan x$, $y = x$, $x = \pi/3$; about the y-axis
13. $y = \cos^2 x$, $|x| \leq \pi/2$, $y = \frac{1}{4}$; about $x = \pi/2$
14. $y = \sqrt{x}$, $y = x^2$; about $y = 2$

15. Find the volumes of the solids obtained by rotating the region bounded by the curves $y = x$ and $y = x^2$ about the following lines.
    (a) The x-axis   (b) The y-axis   (c) $y = 2$

16. Let $\mathcal{R}$ be the region in the first quadrant bounded by the curves $y = x^3$ and $y = 2x - x^2$. Calculate the following quantities.
    (a) The area of $\mathcal{R}$
    (b) The volume obtained by rotating $\mathcal{R}$ about the x-axis
    (c) The volume obtained by rotating $\mathcal{R}$ about the y-axis

17. Let $\mathcal{R}$ be the region bounded by the curves $y = \tan(x^2)$, $x = 1$, and $y = 0$. Use the Midpoint Rule with $n = 4$ to estimate the following quantities.
    (a) The area of $\mathcal{R}$
    (b) The volume obtained by rotating $\mathcal{R}$ about the x-axis

18. Let $\mathcal{R}$ be the region bounded by the curves $y = 1 - x^2$ and $y = x^6 - x + 1$. Estimate the following quantities.
    (a) The x-coordinates of the points of intersection of the curves
    (b) The area of $\mathcal{R}$
    (c) The volume generated when $\mathcal{R}$ is rotated about the x-axis
    (d) The volume generated when $\mathcal{R}$ is rotated about the y-axis

19–22 Each integral represents the volume of a solid. Describe the solid.

19. $\int_0^{\pi/2} 2\pi x \cos x \, dx$

20. $\int_0^{\pi/2} 2\pi \cos^2 x \, dx$

21. $\int_0^{\pi} \pi (2 - \sin x)^2 \, dx$

22. $\int_0^4 2\pi (6 - y)(4y - y^2) \, dy$

---

23. The base of a solid is a circular disk with radius 3. Find the volume of the solid if parallel cross-sections perpendicular to the base are isosceles right triangles with hypotenuse lying along the base.

24. The base of a solid is the region bounded by the parabolas $y = x^2$ and $y = 2 - x^2$. Find the volume of the solid if the cross-sections perpendicular to the x-axis are squares with one side lying along the base.

25. The height of a monument is 20 m. A horizontal cross-section at a distance x meters from the top is an equilateral triangle with side $\frac{1}{4}x$ meters. Find the volume of the monument.

26. (a) The base of a solid is a square with vertices located at $(1, 0)$, $(0, 1)$, $(-1, 0)$, and $(0, -1)$. Each cross-section perpendicular to the x-axis is a semicircle. Find the volume of the solid.
(b) Show that by cutting the solid of part (a), we can rearrange it to form a cone. Thus compute its volume more simply.

27. A force of 30 N is required to maintain a spring stretched from its natural length of 12 cm to a length of 15 cm. How much work is done in stretching the spring from 12 cm to 20 cm?

28. A 1600-lb elevator is suspended by a 200-ft cable that weighs 10 lb/ft. How much work is required to raise the elevator from the basement to the third floor, a distance of 30 ft?

29. A tank full of water has the shape of a paraboloid of revolution as shown in the figure; that is, its shape is obtained by rotating a parabola about a vertical axis.
(a) If its height is 4 ft and the radius at the top is 4 ft, find the work required to pump the water out of the tank.
(b) After 4000 ft-lb of work has been done, what is the depth of the water remaining in the tank?

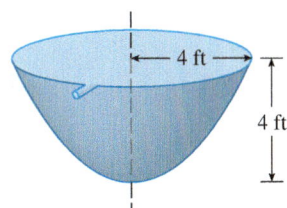

30. A steel tank has the shape of a circular cylinder oriented vertically with diameter 4 m and height 5 m. The tank is currently filled to a level of 3 m with cooking oil that has a density of 920 kg/m³. Compute the work required to pump the oil out through a 1-m spout at the top of the tank.

31. Find the average value of the function $f(t) = \sec^2 t$ on the interval $[0, \pi/4]$.

32. (a) Find the average value of the function $f(x) = 1/\sqrt{x}$ on the interval $[1, 4]$.
(b) Find the value c guaranteed by the Mean Value Theorem for Integrals such that $f_{\text{ave}} = f(c)$.
(c) Sketch the graph of f on $[1, 4]$ and a rectangle whose area is the same as the area under the graph of f.

33. If f is a continuous function, what is the limit as $h \to 0$ of the average value of f on the interval $[x, x + h]$?

34. Let $\mathcal{R}_1$ be the region bounded by $y = x^2$, $y = 0$, and $x = b$, where $b > 0$. Let $\mathcal{R}_2$ be the region bounded by $y = x^2$, $x = 0$, and $y = b^2$.
(a) Is there a value of b such that $\mathcal{R}_1$ and $\mathcal{R}_2$ have the same area?
(b) Is there a value of b such that $\mathcal{R}_1$ sweeps out the same volume when rotated about the x-axis and the y-axis?
(c) Is there a value of b such that $\mathcal{R}_1$ and $\mathcal{R}_2$ sweep out the same volume when rotated about the x-axis?
(d) Is there a value of b such that $\mathcal{R}_1$ and $\mathcal{R}_2$ sweep out the same volume when rotated about the y-axis?

Problems Plus

1. (a) Find a positive continuous function f such that the area under the graph of f from 0 to t is $A(t) = t^3$ for all $t > 0$.
   (b) A solid is generated by rotating about the x-axis the region under the curve $y = f(x)$, where f is a positive function and $x \geq 0$. The volume generated by the part of the curve from $x = 0$ to $x = b$ is b^2 for all $b > 0$. Find the function f.

2. There is a line through the origin that divides the region bounded by the parabola $y = x - x^2$ and the x-axis into two regions with equal area. What is the slope of that line?

3. The figure shows a horizontal line $y = c$ intersecting the curve $y = 8x - 27x^3$. Find the number c such that the areas of the shaded regions are equal.

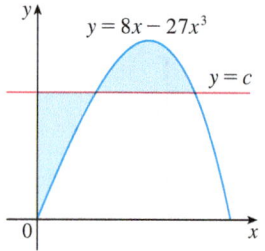

FIGURE FOR PROBLEM 3

4. A cylindrical glass of radius r and height L is filled with water and then tilted until the water remaining in the glass exactly covers its base.
   (a) Determine a way to "slice" the water into parallel rectangular cross-sections and then *set up* a definite integral for the volume of the water in the glass.
   (b) Determine a way to "slice" the water into parallel cross-sections that are trapezoids and then *set up* a definite integral for the volume of the water.
   (c) Find the volume of water in the glass by evaluating one of the integrals in part (a) or part (b).
   (d) Find the volume of the water in the glass from purely geometric considerations.
   (e) Suppose the glass is tilted until the water exactly covers half the base. In what direction can you "slice" the water into triangular cross-sections? Rectangular cross-sections? Cross-sections that are segments of circles? Find the volume of water in the glass.

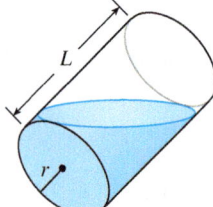 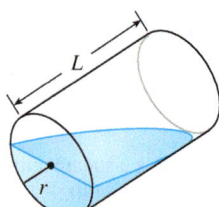

5. (a) Show that the volume of a segment of height h of a sphere of radius r is
$$V = \tfrac{1}{3}\pi h^2(3r - h)$$
   (See the figure.)
   (b) Show that if a sphere of radius 1 is sliced by a plane at a distance x from the center in such a way that the volume of one segment is twice the volume of the other, then x is a solution of the equation
$$3x^3 - 9x + 2 = 0$$
   where $0 < x < 1$. Use Newton's method to find x accurate to four decimal places.

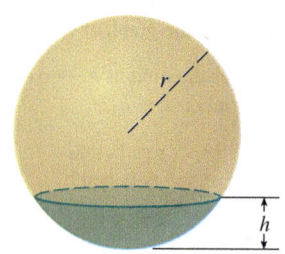

FIGURE FOR PROBLEM 5

   (c) Using the formula for the volume of a segment of a sphere, it can be shown that the depth x to which a floating sphere of radius r sinks in water is a root of the equation
$$x^3 - 3rx^2 + 4r^3 s = 0$$
   where s is the specific gravity of the sphere. Suppose a wooden sphere of radius 0.5 m has specific gravity 0.75. Calculate, to four-decimal-place accuracy, the depth to which the sphere will sink.
   (d) A hemispherical bowl has radius 5 inches and water is running into the bowl at the rate of 0.2 in³/s.
      (i) How fast is the water level in the bowl rising at the instant the water is 3 inches deep?
      (ii) At a certain instant, the water is 4 inches deep. How long will it take to fill the bowl?

6. Archimedes' Principle states that the buoyant force on an object partially or fully submerged in a fluid is equal to the weight of the fluid that the object displaces. Thus, for an

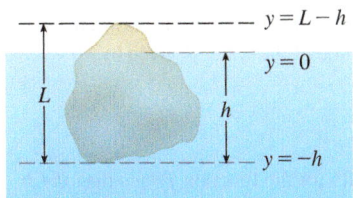

FIGURE FOR PROBLEM 6

object of density ρ_0 floating partly submerged in a fluid of density ρ_f, the buoyant force is given by $F = \rho_f g \int_{-h}^{0} A(y)\, dy$, where g is the acceleration due to gravity and $A(y)$ is the area of a typical cross-section of the object (see the figure). The weight of the object is given by

$$W = \rho_0 g \int_{-h}^{L-h} A(y)\, dy$$

(a) Show that the percentage of the volume of the object above the surface of the liquid is

$$100\,\frac{\rho_f - \rho_0}{\rho_f}$$

(b) The density of ice is 917 kg/m³ and the density of seawater is 1030 kg/m³. What percentage of the volume of an iceberg is above water?

(c) An ice cube floats in a glass filled to the brim with water. Does the water overflow when the ice melts?

(d) A sphere of radius 0.4 m and having negligible weight is floating in a large freshwater lake. How much work is required to completely submerge the sphere? The density of the water is 1000 kg/m³.

7. Water in an open bowl evaporates at a rate proportional to the area of the surface of the water. (This means that the rate of decrease of the volume is proportional to the area of the surface.) Show that the depth of the water decreases at a constant rate, regardless of the shape of the bowl.

8. A sphere of radius 1 overlaps a smaller sphere of radius r in such a way that their intersection is a circle of radius r. (In other words, they intersect in a great circle of the small sphere.) Find r so that the volume inside the small sphere and outside the large sphere is as large as possible.

9. The figure shows a curve C with the property that, for every point P on the middle curve $y = 2x^2$, the areas A and B are equal. Find an equation for C.

10. A paper drinking cup filled with water has the shape of a cone with height h and semi-vertical angle θ. (See the figure.) A ball is placed carefully in the cup, thereby displacing some of the water and making it overflow. What is the radius of the ball that causes the greatest volume of water to spill out of the cup?

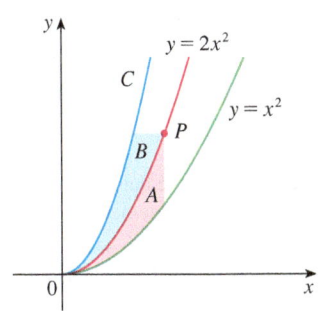

FIGURE FOR PROBLEM 9

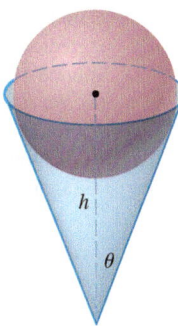

11. A *clepsydra*, or water clock, is a glass container with a small hole in the bottom through which water can flow. The "clock" is calibrated for measuring time by placing markings on the container corresponding to water levels at equally spaced times. Let $x = f(y)$ be continuous on the interval $[0, b]$ and assume that the container is formed by rotating the graph of f about the y-axis. Let V denote the volume of water and h the height of the water level at time t.

(a) Determine V as a function of h.

(b) Show that

$$\frac{dV}{dt} = \pi[f(h)]^2 \frac{dh}{dt}$$

FIGURE FOR PROBLEM 11

(c) Suppose that A is the area of the hole in the bottom of the container. It follows from Torricelli's Law that the rate of change of the volume of the water is given by

$$\frac{dV}{dt} = kA\sqrt{h}$$

where k is a negative constant. Determine a formula for the function f such that dh/dt is a constant C. What is the advantage in having $dh/dt = C$?

12. A cylindrical container of radius r and height L is partially filled with a liquid whose volume is V. If the container is rotated about its axis of symmetry with constant angular speed ω, then the container will induce a rotational motion in the liquid around the same axis. Eventually, the liquid will be rotating at the same angular speed as the container. The surface of the liquid will be convex, as indicated in the figure, because the centrifugal force on the liquid particles increases with the distance from the axis of the container. It can be shown that the surface of the liquid is a paraboloid of revolution generated by rotating the parabola

$$y = h + \frac{\omega^2 x^2}{2g}$$

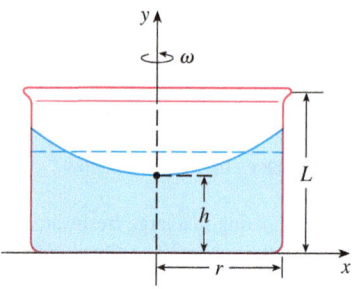

FIGURE FOR PROBLEM 12

about the y-axis, where g is the acceleration due to gravity.
(a) Determine h as a function of ω.
(b) At what angular speed will the surface of the liquid touch the bottom? At what speed will it spill over the top?
(c) Suppose the radius of the container is 2 ft, the height is 7 ft, and the container and liquid are rotating at the same constant angular speed. The surface of the liquid is 5 ft below the top of the tank at the central axis and 4 ft below the top of the tank 1 ft out from the central axis.
  (i) Determine the angular speed of the container and the volume of the fluid.
  (ii) How far below the top of the tank is the liquid at the wall of the container?

13. Suppose the graph of a cubic polynomial intersects the parabola $y = x^2$ when $x = 0$, $x = a$, and $x = b$, where $0 < a < b$. If the two regions between the curves have the same area, how is b related to a?

CAS 14. Suppose we are planning to make a taco from a round tortilla with diameter 8 inches by bending the tortilla so that it is shaped as if it is partially wrapped around a circular cylinder. We will fill the tortilla to the edge (but no more) with meat, cheese, and other ingredients. Our problem is to decide how to curve the tortilla in order to maximize the volume of food it can hold.
(a) We start by placing a circular cylinder of radius r along a diameter of the tortilla and folding the tortilla around the cylinder. Let x represent the distance from the center of the tortilla to a point P on the diameter (see the figure). Show that the cross-sectional area of the filled taco in the plane through P perpendicular to the axis of the cylinder is

$$A(x) = r\sqrt{16 - x^2} - \tfrac{1}{2}r^2 \sin\left(\frac{2}{r}\sqrt{16 - x^2}\right)$$

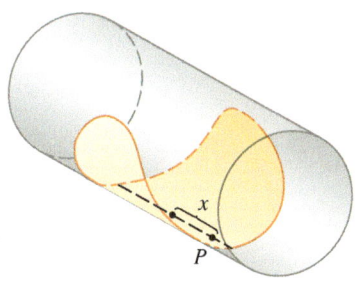

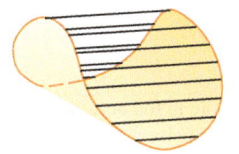

FIGURE FOR PROBLEM 14

and write an expression for the volume of the filled taco.
(b) Determine (approximately) the value of r that maximizes the volume of the taco. (Use a graphical approach with your CAS.)

15. If the tangent at a point P on the curve $y = x^3$ intersects the curve again at Q, let A be the area of the region bounded by the curve and the line segment PQ. Let B be the area of the region defined in the same way starting with Q instead of P. What is the relationship between A and B?

7 Techniques of Integration

The photo shows a screw-worm fly, the first pest effectively eliminated from a region by the sterile insect technique without pesticides. The idea is to introduce into the population sterile males that mate with females but produce no offspring. In Exercise 7.4.67 you will evaluate an integral that relates the female insect population to time.

BECAUSE OF THE FUNDAMENTAL THEOREM of Calculus, we can integrate a function if we know an antiderivative, that is, an indefinite integral. We summarize here the most important integrals that we have learned so far.

$$\int x^n \, dx = \frac{x^{n+1}}{n+1} + C \quad (n \neq -1) \qquad \int \frac{1}{x} \, dx = \ln|x| + C$$

$$\int e^x \, dx = e^x + C \qquad \int b^x \, dx = \frac{b^x}{\ln b} + C$$

$$\int \sin x \, dx = -\cos x + C \qquad \int \cos x \, dx = \sin x + C$$

$$\int \sec^2 x \, dx = \tan x + C \qquad \int \csc^2 x \, dx = -\cot x + C$$

$$\int \sec x \tan x \, dx = \sec x + C \qquad \int \csc x \cot x \, dx = -\csc x + C$$

$$\int \sinh x \, dx = \cosh x + C \qquad \int \cosh x \, dx = \sinh x + C$$

$$\int \tan x \, dx = \ln|\sec x| + C \qquad \int \cot x \, dx = \ln|\sin x| + C$$

$$\int \frac{1}{x^2 + a^2} \, dx = \frac{1}{a} \tan^{-1}\left(\frac{x}{a}\right) + C \qquad \int \frac{1}{\sqrt{a^2 - x^2}} \, dx = \sin^{-1}\left(\frac{x}{a}\right) + C, \quad a > 0$$

In this chapter we develop techniques for using these basic integration formulas to obtain indefinite integrals of more complicated functions. We learned the most important method of integration, the Substitution Rule, in Section 5.5. The other general technique, integration by parts, is presented in Section 7.1. Then we learn methods that are special to particular classes of functions, such as trigonometric functions and rational functions.

Integration is not as straightforward as differentiation; there are no rules that absolutely guarantee obtaining an indefinite integral of a function. Therefore we discuss a strategy for integration in Section 7.5.

7.1 Integration by Parts

Every differentiation rule has a corresponding integration rule. For instance, the Substitution Rule for integration corresponds to the Chain Rule for differentiation. The rule that corresponds to the Product Rule for differentiation is called the rule for *integration by parts*.

The Product Rule states that if f and g are differentiable functions, then

$$\frac{d}{dx}[f(x)g(x)] = f(x)g'(x) + g(x)f'(x)$$

In the notation for indefinite integrals this equation becomes

$$\int [f(x)g'(x) + g(x)f'(x)]\,dx = f(x)g(x)$$

or

$$\int f(x)g'(x)\,dx + \int g(x)f'(x)\,dx = f(x)g(x)$$

We can rearrange this equation as

$$\boxed{1 \quad \int f(x)g'(x)\,dx = f(x)g(x) - \int g(x)f'(x)\,dx}$$

Formula 1 is called the **formula for integration by parts**. It is perhaps easier to remember in the following notation. Let $u = f(x)$ and $v = g(x)$. Then the differentials are $du = f'(x)\,dx$ and $dv = g'(x)\,dx$, so, by the Substitution Rule, the formula for integration by parts becomes

$$\boxed{2 \quad \int u\,dv = uv - \int v\,du}$$

EXAMPLE 1 Find $\int x \sin x\,dx$.

SOLUTION USING FORMULA 1 Suppose we choose $f(x) = x$ and $g'(x) = \sin x$. Then $f'(x) = 1$ and $g(x) = -\cos x$. (For g we can choose *any* antiderivative of g'.) Thus, using Formula 1, we have

$$\int x \sin x\,dx = f(x)g(x) - \int g(x)f'(x)\,dx$$
$$= x(-\cos x) - \int (-\cos x)\,dx$$
$$= -x \cos x + \int \cos x\,dx$$
$$= -x \cos x + \sin x + C$$

It's wise to check the answer by differentiating it. If we do so, we get $x \sin x$, as expected.

SOLUTION USING FORMULA 2 Let

It is helpful to use the pattern:
$$u = \Box \qquad dv = \Box$$
$$du = \Box \qquad v = \Box$$

$$u = x \qquad dv = \sin x \, dx$$

Then
$$du = dx \qquad v = -\cos x$$

and so

$$\int x \sin x \, dx = \int \overbrace{x}^{u} \overbrace{\sin x \, dx}^{dv} = \overbrace{x}^{u} \overbrace{(-\cos x)}^{v} - \int \overbrace{(-\cos x)}^{v} \overbrace{dx}^{du}$$

$$= -x \cos x + \int \cos x \, dx$$

$$= -x \cos x + \sin x + C$$

NOTE Our aim in using integration by parts is to obtain a simpler integral than the one we started with. Thus in Example 1 we started with $\int x \sin x \, dx$ and expressed it in terms of the simpler integral $\int \cos x \, dx$. If we had instead chosen $u = \sin x$ and $dv = x \, dx$, then $du = \cos x \, dx$ and $v = x^2/2$, so integration by parts gives

$$\int x \sin x \, dx = (\sin x) \frac{x^2}{2} - \frac{1}{2} \int x^2 \cos x \, dx$$

Although this is true, $\int x^2 \cos x \, dx$ is a more difficult integral than the one we started with. In general, when deciding on a choice for u and dv, we usually try to choose $u = f(x)$ to be a function that becomes simpler when differentiated (or at least not more complicated) as long as $dv = g'(x) \, dx$ can be readily integrated to give v.

EXAMPLE 2 Evaluate $\int \ln x \, dx$.

SOLUTION Here we don't have much choice for u and dv. Let

$$u = \ln x \qquad dv = dx$$

Then
$$du = \frac{1}{x} dx \qquad v = x$$

Integrating by parts, we get

$$\int \ln x \, dx = x \ln x - \int x \frac{dx}{x}$$

It's customary to write $\int 1 \, dx$ as $\int dx$.

$$= x \ln x - \int dx$$

Check the answer by differentiating it.

$$= x \ln x - x + C$$

Integration by parts is effective in this example because the derivative of the function $f(x) = \ln x$ is simpler than f.

EXAMPLE 3 Find $\int t^2 e^t \, dt$.

SOLUTION Notice that t^2 becomes simpler when differentiated (whereas e^t is unchanged when differentiated or integrated), so we choose

$$u = t^2 \qquad dv = e^t \, dt$$

Then
$$du = 2t \, dt \qquad v = e^t$$

Integration by parts gives

$$\boxed{3} \qquad \int t^2 e^t \, dt = t^2 e^t - 2 \int t e^t \, dt$$

The integral that we obtained, $\int t e^t \, dt$, is simpler than the original integral but is still not obvious. Therefore we use integration by parts a second time, this time with $u = t$ and $dv = e^t \, dt$. Then $du = dt$, $v = e^t$, and

$$\int t e^t \, dt = t e^t - \int e^t \, dt$$
$$= t e^t - e^t + C$$

Putting this in Equation 3, we get

$$\int t^2 e^t \, dt = t^2 e^t - 2 \int t e^t \, dt$$
$$= t^2 e^t - 2(t e^t - e^t + C)$$
$$= t^2 e^t - 2t e^t + 2e^t + C_1 \qquad \text{where } C_1 = -2C$$

EXAMPLE 4 Evaluate $\int e^x \sin x \, dx$.

An easier method, using complex numbers, is given in Exercise 50 in Appendix H.

SOLUTION Neither e^x nor $\sin x$ becomes simpler when differentiated, but we try choosing $u = e^x$ and $dv = \sin x \, dx$ anyway. Then $du = e^x \, dx$ and $v = -\cos x$, so integration by parts gives

$$\boxed{4} \qquad \int e^x \sin x \, dx = -e^x \cos x + \int e^x \cos x \, dx$$

The integral that we have obtained, $\int e^x \cos x \, dx$, is no simpler than the original one, but at least it's no more difficult. Having had success in the preceding example integrating by parts twice, we persevere and integrate by parts again. This time we use $u = e^x$ and $dv = \cos x \, dx$. Then $du = e^x \, dx$, $v = \sin x$, and

$$\boxed{5} \qquad \int e^x \cos x \, dx = e^x \sin x - \int e^x \sin x \, dx$$

At first glance, it appears as if we have accomplished nothing because we have arrived at $\int e^x \sin x \, dx$, which is where we started. However, if we put the expression for $\int e^x \cos x \, dx$ from Equation 5 into Equation 4 we get

$$\int e^x \sin x \, dx = -e^x \cos x + e^x \sin x - \int e^x \sin x \, dx$$

Figure 1 illustrates Example 4 by showing the graphs of $f(x) = e^x \sin x$ and $F(x) = \frac{1}{2}e^x(\sin x - \cos x)$. As a visual check on our work, notice that $f(x) = 0$ when F has a maximum or minimum.

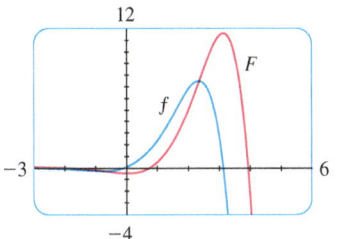

FIGURE 1

This can be regarded as an equation to be solved for the unknown integral. Adding $\int e^x \sin x \, dx$ to both sides, we obtain

$$2 \int e^x \sin x \, dx = -e^x \cos x + e^x \sin x$$

Dividing by 2 and adding the constant of integration, we get

$$\int e^x \sin x \, dx = \tfrac{1}{2}e^x(\sin x - \cos x) + C$$

If we combine the formula for integration by parts with Part 2 of the Fundamental Theorem of Calculus, we can evaluate definite integrals by parts. Evaluating both sides of Formula 1 between a and b, assuming f' and g' are continuous, and using the Fundamental Theorem, we obtain

$$\boxed{6} \quad \int_a^b f(x)g'(x)\,dx = f(x)g(x)\Big]_a^b - \int_a^b g(x)f'(x)\,dx$$

EXAMPLE 5 Calculate $\int_0^1 \tan^{-1} x \, dx$.

SOLUTION Let

$$u = \tan^{-1} x \qquad dv = dx$$

Then

$$du = \frac{dx}{1 + x^2} \qquad v = x$$

So Formula 6 gives

$$\int_0^1 \tan^{-1} x \, dx = x \tan^{-1} x \Big]_0^1 - \int_0^1 \frac{x}{1 + x^2} \, dx$$

$$= 1 \cdot \tan^{-1} 1 - 0 \cdot \tan^{-1} 0 - \int_0^1 \frac{x}{1 + x^2} \, dx$$

$$= \frac{\pi}{4} - \int_0^1 \frac{x}{1 + x^2} \, dx$$

Since $\tan^{-1} x \geq 0$ for $x \geq 0$, the integral in Example 5 can be interpreted as the area of the region shown in Figure 2.

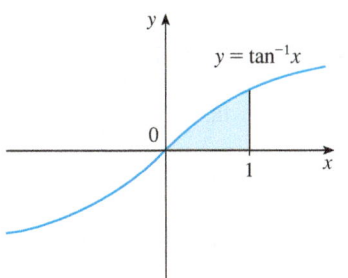

FIGURE 2

To evaluate this integral we use the substitution $t = 1 + x^2$ (since u has another meaning in this example). Then $dt = 2x \, dx$, so $x \, dx = \frac{1}{2} dt$. When $x = 0$, $t = 1$; when $x = 1$, $t = 2$; so

$$\int_0^1 \frac{x}{1 + x^2} \, dx = \tfrac{1}{2} \int_1^2 \frac{dt}{t} = \tfrac{1}{2} \ln |t| \Big]_1^2$$

$$= \tfrac{1}{2}(\ln 2 - \ln 1) = \tfrac{1}{2} \ln 2$$

Therefore

$$\int_0^1 \tan^{-1} x \, dx = \frac{\pi}{4} - \int_0^1 \frac{x}{1 + x^2} \, dx = \frac{\pi}{4} - \frac{\ln 2}{2}$$

Equation 7 is called a *reduction formula* because the exponent n has been reduced to $n - 1$ and $n - 2$.

EXAMPLE 6 Prove the reduction formula

$$\boxed{7} \quad \int \sin^n x \, dx = -\frac{1}{n} \cos x \sin^{n-1} x + \frac{n-1}{n} \int \sin^{n-2} x \, dx$$

where $n \geq 2$ is an integer.

SOLUTION Let

$$u = \sin^{n-1}x \qquad\qquad dv = \sin x\, dx$$

Then $\qquad du = (n-1)\sin^{n-2}x \cos x\, dx \qquad v = -\cos x$

so integration by parts gives

$$\int \sin^n x\, dx = -\cos x \sin^{n-1}x + (n-1)\int \sin^{n-2}x \cos^2 x\, dx$$

Since $\cos^2 x = 1 - \sin^2 x$, we have

$$\int \sin^n x\, dx = -\cos x \sin^{n-1}x + (n-1)\int \sin^{n-2}x\, dx - (n-1)\int \sin^n x\, dx$$

As in Example 4, we solve this equation for the desired integral by taking the last term on the right side to the left side. Thus we have

$$n\int \sin^n x\, dx = -\cos x \sin^{n-1}x + (n-1)\int \sin^{n-2}x\, dx$$

or

$$\int \sin^n x\, dx = -\frac{1}{n}\cos x \sin^{n-1}x + \frac{n-1}{n}\int \sin^{n-2}x\, dx \qquad \blacksquare$$

The reduction formula (7) is useful because by using it repeatedly we could eventually express $\int \sin^n x\, dx$ in terms of $\int \sin x\, dx$ (if n is odd) or $\int (\sin x)^0\, dx = \int dx$ (if n is even).

7.1 EXERCISES

1–2 Evaluate the integral using integration by parts with the indicated choices of u and dv.

1. $\int xe^{2x}\, dx; \quad u = x, \ dv = e^{2x}\, dx$

2. $\int \sqrt{x}\ln x\, dx; \quad u = \ln x, \ dv = \sqrt{x}\, dx$

3–36 Evaluate the integral.

3. $\int x \cos 5x\, dx$

4. $\int ye^{0.2y}\, dy$

5. $\int te^{-3t}\, dt$

6. $\int (x-1)\sin \pi x\, dx$

7. $\int (x^2 + 2x)\cos x\, dx$

8. $\int t^2 \sin \beta t\, dt$

9. $\int \cos^{-1}x\, dx$

10. $\int \ln \sqrt{x}\, dx$

11. $\int t^4 \ln t\, dt$

12. $\int \tan^{-1} 2y\, dy$

13. $\int t \csc^2 t\, dt$

14. $\int x \cosh ax\, dx$

15. $\int (\ln x)^2\, dx$

16. $\int \dfrac{z}{10^z}\, dz$

17. $\int e^{2\theta} \sin 3\theta\, d\theta$

18. $\int e^{-\theta}\cos 2\theta\, d\theta$

19. $\int z^3 e^z\, dz$

20. $\int x \tan^2 x\, dx$

21. $\int \dfrac{xe^{2x}}{(1+2x)^2}\, dx$

22. $\int (\arcsin x)^2\, dx$

23. $\int_0^{1/2} x \cos \pi x\, dx$

24. $\int_0^1 (x^2 + 1)e^{-x}\, dx$

25. $\int_0^2 y \sinh y\, dy$

26. $\int_1^2 w^2 \ln w\, dw$

27. $\int_1^5 \dfrac{\ln R}{R^2}\, dR$

28. $\int_0^{2\pi} t^2 \sin 2t\, dt$

29. $\int_0^{\pi} x \sin x \cos x\, dx$

30. $\int_1^{\sqrt{3}} \arctan(1/x)\, dx$

31. $\displaystyle\int_1^5 \frac{M}{e^M} \, dM$ **32.** $\displaystyle\int_1^2 \frac{(\ln x)^2}{x^3} \, dx$

33. $\displaystyle\int_0^{\pi/3} \sin x \ln(\cos x) \, dx$ **34.** $\displaystyle\int_0^1 \frac{r^3}{\sqrt{4+r^2}} \, dr$

35. $\displaystyle\int_1^2 x^4 (\ln x)^2 \, dx$ **36.** $\displaystyle\int_0^t e^s \sin(t-s) \, ds$

37–42 First make a substitution and then use integration by parts to evaluate the integral.

37. $\displaystyle\int e^{\sqrt{x}} \, dx$ **38.** $\displaystyle\int \cos(\ln x) \, dx$

39. $\displaystyle\int_{\sqrt{\pi/2}}^{\sqrt{\pi}} \theta^3 \cos(\theta^2) \, d\theta$ **40.** $\displaystyle\int_0^{\pi} e^{\cos t} \sin 2t \, dt$

41. $\displaystyle\int x \ln(1+x) \, dx$ **42.** $\displaystyle\int \frac{\arcsin(\ln x)}{x} \, dx$

43–46 Evaluate the indefinite integral. Illustrate, and check that your answer is reasonable, by graphing both the function and its antiderivative (take $C = 0$).

43. $\displaystyle\int xe^{-2x} \, dx$ **44.** $\displaystyle\int x^{3/2} \ln x \, dx$

45. $\displaystyle\int x^3 \sqrt{1+x^2} \, dx$ **46.** $\displaystyle\int x^2 \sin 2x \, dx$

47. (a) Use the reduction formula in Example 6 to show that

$$\int \sin^2 x \, dx = \frac{x}{2} - \frac{\sin 2x}{4} + C$$

  (b) Use part (a) and the reduction formula to evaluate $\int \sin^4 x \, dx$.

48. (a) Prove the reduction formula

$$\int \cos^n x \, dx = \frac{1}{n} \cos^{n-1} x \sin x + \frac{n-1}{n} \int \cos^{n-2} x \, dx$$

  (b) Use part (a) to evaluate $\int \cos^2 x \, dx$.
  (c) Use parts (a) and (b) to evaluate $\int \cos^4 x \, dx$.

49. (a) Use the reduction formula in Example 6 to show that

$$\int_0^{\pi/2} \sin^n x \, dx = \frac{n-1}{n} \int_0^{\pi/2} \sin^{n-2} x \, dx$$

where $n \geq 2$ is an integer.
  (b) Use part (a) to evaluate $\int_0^{\pi/2} \sin^3 x \, dx$ and $\int_0^{\pi/2} \sin^5 x \, dx$.
  (c) Use part (a) to show that, for odd powers of sine,

$$\int_0^{\pi/2} \sin^{2n+1} x \, dx = \frac{2 \cdot 4 \cdot 6 \cdot \,\cdots\, \cdot 2n}{3 \cdot 5 \cdot 7 \cdot \,\cdots\, \cdot (2n+1)}$$

50. Prove that, for even powers of sine,

$$\int_0^{\pi/2} \sin^{2n} x \, dx = \frac{1 \cdot 3 \cdot 5 \cdot \,\cdots\, \cdot (2n-1)}{2 \cdot 4 \cdot 6 \cdot \,\cdots\, \cdot 2n} \frac{\pi}{2}$$

51–54 Use integration by parts to prove the reduction formula.

51. $\displaystyle\int (\ln x)^n \, dx = x(\ln x)^n - n \int (\ln x)^{n-1} \, dx$

52. $\displaystyle\int x^n e^x \, dx = x^n e^x - n \int x^{n-1} e^x \, dx$

53. $\displaystyle\int \tan^n x \, dx = \frac{\tan^{n-1} x}{n-1} - \int \tan^{n-2} x \, dx \quad (n \neq 1)$

54. $\displaystyle\int \sec^n x \, dx = \frac{\tan x \sec^{n-2} x}{n-1} + \frac{n-2}{n-1} \int \sec^{n-2} x \, dx \quad (n \neq 1)$

55. Use Exercise 51 to find $\int (\ln x)^3 \, dx$.

56. Use Exercise 52 to find $\int x^4 e^x \, dx$.

57–58 Find the area of the region bounded by the given curves.

57. $y = x^2 \ln x$, $y = 4 \ln x$ **58.** $y = x^2 e^{-x}$, $y = xe^{-x}$

59–60 Use a graph to find approximate x-coordinates of the points of intersection of the given curves. Then find (approximately) the area of the region bounded by the curves.

59. $y = \arcsin(\tfrac{1}{2} x)$, $y = 2 - x^2$

60. $y = x \ln(x+1)$, $y = 3x - x^2$

61–64 Use the method of cylindrical shells to find the volume generated by rotating the region bounded by the curves about the given axis.

61. $y = \cos(\pi x/2)$, $y = 0$, $0 \leq x \leq 1$;   about the y-axis

62. $y = e^x$, $y = e^{-x}$, $x = 1$;   about the y-axis

63. $y = e^{-x}$, $y = 0$, $x = -1$, $x = 0$;   about $x = 1$

64. $y = e^x$, $x = 0$, $y = 3$;   about the x-axis

65. Calculate the volume generated by rotating the region bounded by the curves $y = \ln x$, $y = 0$, and $x = 2$ about each axis.
  (a) The y-axis   (b) The x-axis

66. Calculate the average value of $f(x) = x \sec^2 x$ on the interval $[0, \pi/4]$.

67. The Fresnel function $S(x) = \int_0^x \sin(\tfrac{1}{2} \pi t^2) \, dt$ was discussed in Example 5.3.3 and is used extensively in the theory of optics. Find $\int S(x) \, dx$. [Your answer will involve $S(x)$.]

68. A rocket accelerates by burning its onboard fuel, so its mass decreases with time. Suppose the initial mass of the rocket at liftoff (including its fuel) is m, the fuel is consumed at rate r, and the exhaust gases are ejected with constant velocity v_e (relative to the rocket). A model for the velocity of the rocket at time t is given by the equation

$$v(t) = -gt - v_e \ln \frac{m - rt}{m}$$

where g is the acceleration due to gravity and t is not too large. If $g = 9.8$ m/s^2, $m = 30{,}000$ kg, $r = 160$ kg/s, and $v_e = 3000$ m/s, find the height of the rocket one minute after liftoff.

69. A particle that moves along a straight line has velocity $v(t) = t^2 e^{-t}$ meters per second after t seconds. How far will it travel during the first t seconds?

70. If $f(0) = g(0) = 0$ and f'' and g'' are continuous, show that

$$\int_0^a f(x) g''(x)\, dx = f(a) g'(a) - f'(a) g(a) + \int_0^a f''(x) g(x)\, dx$$

71. Suppose that $f(1) = 2$, $f(4) = 7$, $f'(1) = 5$, $f'(4) = 3$, and f'' is continuous. Find the value of $\int_1^4 x f''(x)\, dx$.

72. (a) Use integration by parts to show that

$$\int f(x)\, dx = x f(x) - \int x f'(x)\, dx$$

(b) If f and g are inverse functions and f' is continuous, prove that

$$\int_a^b f(x)\, dx = b f(b) - a f(a) - \int_{f(a)}^{f(b)} g(y)\, dy$$

[Hint: Use part (a) and make the substitution $y = f(x)$.]

(c) In the case where f and g are positive functions and $b > a > 0$, draw a diagram to give a geometric interpretation of part (b).

(d) Use part (b) to evaluate $\int_1^e \ln x\, dx$.

73. We arrived at Formula 6.3.2, $V = \int_a^b 2\pi x f(x)\, dx$, by using cylindrical shells, but now we can use integration by parts to prove it using the slicing method of Section 6.2, at least

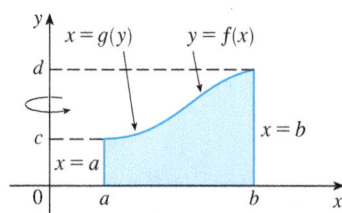

for the case where f is one-to-one and therefore has an inverse function g. Use the figure to show that

$$V = \pi b^2 d - \pi a^2 c - \int_c^d \pi [g(y)]^2\, dy$$

Make the substitution $y = f(x)$ and then use integration by parts on the resulting integral to prove that

$$V = \int_a^b 2\pi x f(x)\, dx$$

74. Let $I_n = \int_0^{\pi/2} \sin^n x\, dx$.

(a) Show that $I_{2n+2} \leq I_{2n+1} \leq I_{2n}$.

(b) Use Exercise 50 to show that

$$\frac{I_{2n+2}}{I_{2n}} = \frac{2n + 1}{2n + 2}$$

(c) Use parts (a) and (b) to show that

$$\frac{2n + 1}{2n + 2} \leq \frac{I_{2n+1}}{I_{2n}} \leq 1$$

and deduce that $\lim_{n \to \infty} I_{2n+1}/I_{2n} = 1$.

(d) Use part (c) and Exercises 49 and 50 to show that

$$\lim_{n \to \infty} \frac{2}{1} \cdot \frac{2}{3} \cdot \frac{4}{3} \cdot \frac{4}{5} \cdot \frac{6}{5} \cdot \frac{6}{7} \cdot \ldots \cdot \frac{2n}{2n - 1} \cdot \frac{2n}{2n + 1} = \frac{\pi}{2}$$

This formula is usually written as an infinite product:

$$\frac{\pi}{2} = \frac{2}{1} \cdot \frac{2}{3} \cdot \frac{4}{3} \cdot \frac{4}{5} \cdot \frac{6}{5} \cdot \frac{6}{7} \cdot \ldots$$

and is called the *Wallis product*.

(e) We construct rectangles as follows. Start with a square of area 1 and attach rectangles of area 1 alternately beside or on top of the previous rectangle (see the figure). Find the limit of the ratios of width to height of these rectangles.

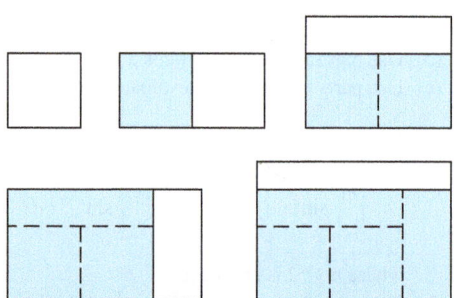

7.2 Trigonometric Integrals

In this section we use trigonometric identities to integrate certain combinations of trigonometric functions. We start with powers of sine and cosine.

EXAMPLE 1 Evaluate $\int \cos^3 x \, dx$.

SOLUTION Simply substituting $u = \cos x$ isn't helpful, since then $du = -\sin x \, dx$. In order to integrate powers of cosine, we would need an extra $\sin x$ factor. Similarly, a power of sine would require an extra $\cos x$ factor. Thus here we can separate one cosine factor and convert the remaining $\cos^2 x$ factor to an expression involving sine using the identity $\sin^2 x + \cos^2 x = 1$:

$$\cos^3 x = \cos^2 x \cdot \cos x = (1 - \sin^2 x) \cos x$$

We can then evaluate the integral by substituting $u = \sin x$, so $du = \cos x \, dx$ and

$$\int \cos^3 x \, dx = \int \cos^2 x \cdot \cos x \, dx = \int (1 - \sin^2 x) \cos x \, dx$$

$$= \int (1 - u^2) \, du = u - \tfrac{1}{3} u^3 + C$$

$$= \sin x - \tfrac{1}{3} \sin^3 x + C \qquad \blacksquare$$

In general, we try to write an integrand involving powers of sine and cosine in a form where we have only one sine factor (and the remainder of the expression in terms of cosine) or only one cosine factor (and the remainder of the expression in terms of sine). The identity $\sin^2 x + \cos^2 x = 1$ enables us to convert back and forth between even powers of sine and cosine.

EXAMPLE 2 Find $\int \sin^5 x \cos^2 x \, dx$.

SOLUTION We could convert $\cos^2 x$ to $1 - \sin^2 x$, but we would be left with an expression in terms of $\sin x$ with no extra $\cos x$ factor. Instead, we separate a single sine factor and rewrite the remaining $\sin^4 x$ factor in terms of $\cos x$:

$$\sin^5 x \cos^2 x = (\sin^2 x)^2 \cos^2 x \sin x = (1 - \cos^2 x)^2 \cos^2 x \sin x$$

Substituting $u = \cos x$, we have $du = -\sin x \, dx$ and so

$$\int \sin^5 x \cos^2 x \, dx = \int (\sin^2 x)^2 \cos^2 x \sin x \, dx$$

$$= \int (1 - \cos^2 x)^2 \cos^2 x \sin x \, dx$$

$$= \int (1 - u^2)^2 u^2 (-du) = -\int (u^2 - 2u^4 + u^6) \, du$$

$$= -\left(\frac{u^3}{3} - 2 \frac{u^5}{5} + \frac{u^7}{7} \right) + C$$

$$= -\tfrac{1}{3} \cos^3 x + \tfrac{2}{5} \cos^5 x - \tfrac{1}{7} \cos^7 x + C \qquad \blacksquare$$

Figure 1 shows the graphs of the integrand $\sin^5 x \cos^2 x$ in Example 2 and its indefinite integral (with $C = 0$). Which is which?

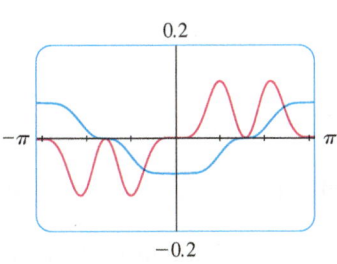

FIGURE 1

In the preceding examples, an odd power of sine or cosine enabled us to separate a single factor and convert the remaining even power. If the integrand contains even powers of both sine and cosine, this strategy fails. In this case, we can take advantage of the following half-angle identities (see Equations 17b and 17a in Appendix D):

$$\sin^2 x = \tfrac{1}{2}(1 - \cos 2x) \quad \text{and} \quad \cos^2 x = \tfrac{1}{2}(1 + \cos 2x)$$

EXAMPLE 3 Evaluate $\int_0^\pi \sin^2 x \, dx$.

SOLUTION If we write $\sin^2 x = 1 - \cos^2 x$, the integral is no simpler to evaluate. Using the half-angle formula for $\sin^2 x$, however, we have

$$\int_0^\pi \sin^2 x \, dx = \tfrac{1}{2} \int_0^\pi (1 - \cos 2x) \, dx$$
$$= \left[\tfrac{1}{2}(x - \tfrac{1}{2} \sin 2x) \right]_0^\pi$$
$$= \tfrac{1}{2}(\pi - \tfrac{1}{2} \sin 2\pi) - \tfrac{1}{2}(0 - \tfrac{1}{2} \sin 0) = \tfrac{1}{2}\pi$$

Notice that we mentally made the substitution $u = 2x$ when integrating $\cos 2x$. Another method for evaluating this integral was given in Exercise 7.1.47. ∎

Example 3 shows that the area of the region shown in Figure 2 is $\pi/2$.

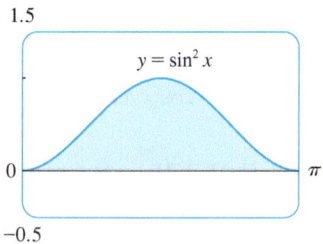

FIGURE 2

EXAMPLE 4 Find $\int \sin^4 x \, dx$.

SOLUTION We could evaluate this integral using the reduction formula for $\int \sin^n x \, dx$ (Equation 7.1.7) together with Example 3 (as in Exercise 7.1.47), but a better method is to write $\sin^4 x = (\sin^2 x)^2$ and use a half-angle formula:

$$\int \sin^4 x \, dx = \int (\sin^2 x)^2 \, dx$$
$$= \int \left(\frac{1 - \cos 2x}{2} \right)^2 dx$$
$$= \tfrac{1}{4} \int (1 - 2\cos 2x + \cos^2 2x) \, dx$$

Since $\cos^2 2x$ occurs, we must use another half-angle formula

$$\cos^2 2x = \tfrac{1}{2}(1 + \cos 4x)$$

This gives

$$\int \sin^4 x \, dx = \tfrac{1}{4} \int \left[1 - 2\cos 2x + \tfrac{1}{2}(1 + \cos 4x) \right] dx$$
$$= \tfrac{1}{4} \int \left(\tfrac{3}{2} - 2\cos 2x + \tfrac{1}{2} \cos 4x \right) dx$$
$$= \tfrac{1}{4}\left(\tfrac{3}{2}x - \sin 2x + \tfrac{1}{8} \sin 4x \right) + C \quad \blacksquare$$

To summarize, we list guidelines to follow when evaluating integrals of the form $\int \sin^m x \cos^n x \, dx$, where $m \geq 0$ and $n \geq 0$ are integers.

Strategy for Evaluating $\int \sin^m x \, \cos^n x \, dx$

(a) If the power of cosine is odd ($n = 2k + 1$), save one cosine factor and use $\cos^2 x = 1 - \sin^2 x$ to express the remaining factors in terms of sine:

$$\int \sin^m x \, \cos^{2k+1} x \, dx = \int \sin^m x \, (\cos^2 x)^k \cos x \, dx$$

$$= \int \sin^m x \, (1 - \sin^2 x)^k \cos x \, dx$$

Then substitute $u = \sin x$.

(b) If the power of sine is odd ($m = 2k + 1$), save one sine factor and use $\sin^2 x = 1 - \cos^2 x$ to express the remaining factors in terms of cosine:

$$\int \sin^{2k+1} x \, \cos^n x \, dx = \int (\sin^2 x)^k \cos^n x \, \sin x \, dx$$

$$= \int (1 - \cos^2 x)^k \cos^n x \, \sin x \, dx$$

Then substitute $u = \cos x$. [Note that if the powers of both sine and cosine are odd, either (a) or (b) can be used.]

(c) If the powers of both sine and cosine are even, use the half-angle identities

$$\sin^2 x = \tfrac{1}{2}(1 - \cos 2x) \qquad \cos^2 x = \tfrac{1}{2}(1 + \cos 2x)$$

It is sometimes helpful to use the identity

$$\sin x \cos x = \tfrac{1}{2} \sin 2x$$

We can use a similar strategy to evaluate integrals of the form $\int \tan^m x \, \sec^n x \, dx$. Since $(d/dx) \tan x = \sec^2 x$, we can separate a $\sec^2 x$ factor and convert the remaining (even) power of secant to an expression involving tangent using the identity $\sec^2 x = 1 + \tan^2 x$. Or, since $(d/dx) \sec x = \sec x \tan x$, we can separate a $\sec x \tan x$ factor and convert the remaining (even) power of tangent to secant.

EXAMPLE 5 Evaluate $\int \tan^6 x \, \sec^4 x \, dx$.

SOLUTION If we separate one $\sec^2 x$ factor, we can express the remaining $\sec^2 x$ factor in terms of tangent using the identity $\sec^2 x = 1 + \tan^2 x$. We can then evaluate the integral by substituting $u = \tan x$ so that $du = \sec^2 x \, dx$:

$$\int \tan^6 x \, \sec^4 x \, dx = \int \tan^6 x \, \sec^2 x \, \sec^2 x \, dx$$

$$= \int \tan^6 x \, (1 + \tan^2 x) \sec^2 x \, dx$$

$$= \int u^6 (1 + u^2) \, du = \int (u^6 + u^8) \, du$$

$$= \frac{u^7}{7} + \frac{u^9}{9} + C$$

$$= \tfrac{1}{7} \tan^7 x + \tfrac{1}{9} \tan^9 x + C$$

EXAMPLE 6 Find $\int \tan^5\theta \sec^7\theta \, d\theta$.

SOLUTION If we separate a $\sec^2\theta$ factor, as in the preceding example, we are left with a $\sec^5\theta$ factor, which isn't easily converted to tangent. However, if we separate a $\sec\theta \tan\theta$ factor, we can convert the remaining power of tangent to an expression involving only secant using the identity $\tan^2\theta = \sec^2\theta - 1$. We can then evaluate the integral by substituting $u = \sec\theta$, so $du = \sec\theta \tan\theta \, d\theta$:

$$\int \tan^5\theta \sec^7\theta \, d\theta = \int \tan^4\theta \sec^6\theta \sec\theta \tan\theta \, d\theta$$

$$= \int (\sec^2\theta - 1)^2 \sec^6\theta \sec\theta \tan\theta \, d\theta$$

$$= \int (u^2 - 1)^2 u^6 \, du$$

$$= \int (u^{10} - 2u^8 + u^6) \, du$$

$$= \frac{u^{11}}{11} - 2\frac{u^9}{9} + \frac{u^7}{7} + C$$

$$= \tfrac{1}{11}\sec^{11}\theta - \tfrac{2}{9}\sec^9\theta + \tfrac{1}{7}\sec^7\theta + C \qquad \blacksquare$$

The preceding examples demonstrate strategies for evaluating integrals of the form $\int \tan^m x \sec^n x \, dx$ for two cases, which we summarize here.

Strategy for Evaluating $\int \tan^m x \sec^n x \, dx$

(a) If the power of secant is even ($n = 2k$, $k \geq 2$), save a factor of $\sec^2 x$ and use $\sec^2 x = 1 + \tan^2 x$ to express the remaining factors in terms of $\tan x$:

$$\int \tan^m x \sec^{2k} x \, dx = \int \tan^m x \, (\sec^2 x)^{k-1} \sec^2 x \, dx$$

$$= \int \tan^m x \, (1 + \tan^2 x)^{k-1} \sec^2 x \, dx$$

Then substitute $u = \tan x$.

(b) If the power of tangent is odd ($m = 2k + 1$), save a factor of $\sec x \tan x$ and use $\tan^2 x = \sec^2 x - 1$ to express the remaining factors in terms of $\sec x$:

$$\int \tan^{2k+1} x \sec^n x \, dx = \int (\tan^2 x)^k \sec^{n-1} x \sec x \tan x \, dx$$

$$= \int (\sec^2 x - 1)^k \sec^{n-1} x \sec x \tan x \, dx$$

Then substitute $u = \sec x$.

For other cases, the guidelines are not as clear-cut. We may need to use identities, integration by parts, and occasionally a little ingenuity. We will sometimes need to be able to integrate $\tan x$ by using the formula established in (5.5.5):

$$\int \tan x \, dx = \ln|\sec x| + C$$

SECTION 7.2 Trigonometric Integrals

We will also need the indefinite integral of secant:

Formula 1 was discovered by James Gregory in 1668. (See his biography on page 198.) Gregory used this formula to solve a problem in constructing nautical tables.

$$\boxed{1} \quad \int \sec x \, dx = \ln |\sec x + \tan x| + C$$

We could verify Formula 1 by differentiating the right side, or as follows. First we multiply numerator and denominator by $\sec x + \tan x$:

$$\int \sec x \, dx = \int \sec x \, \frac{\sec x + \tan x}{\sec x + \tan x} \, dx$$

$$= \int \frac{\sec^2 x + \sec x \tan x}{\sec x + \tan x} \, dx$$

If we substitute $u = \sec x + \tan x$, then $du = (\sec x \tan x + \sec^2 x) \, dx$, so the integral becomes $\int (1/u) \, du = \ln |u| + C$. Thus we have

$$\int \sec x \, dx = \ln |\sec x + \tan x| + C$$

EXAMPLE 7 Find $\int \tan^3 x \, dx$.

SOLUTION Here only $\tan x$ occurs, so we use $\tan^2 x = \sec^2 x - 1$ to rewrite a $\tan^2 x$ factor in terms of $\sec^2 x$:

$$\int \tan^3 x \, dx = \int \tan x \tan^2 x \, dx = \int \tan x (\sec^2 x - 1) \, dx$$

$$= \int \tan x \sec^2 x \, dx - \int \tan x \, dx$$

$$= \frac{\tan^2 x}{2} - \ln |\sec x| + C$$

In the first integral we mentally substituted $u = \tan x$ so that $du = \sec^2 x \, dx$. ∎

If an even power of tangent appears with an odd power of secant, it is helpful to express the integrand completely in terms of $\sec x$. Powers of $\sec x$ may require integration by parts, as shown in the following example.

EXAMPLE 8 Find $\int \sec^3 x \, dx$.

SOLUTION Here we integrate by parts with

$$u = \sec x \qquad\qquad dv = \sec^2 x \, dx$$
$$du = \sec x \tan x \, dx \qquad\qquad v = \tan x$$

Then

$$\int \sec^3 x \, dx = \sec x \tan x - \int \sec x \tan^2 x \, dx$$

$$= \sec x \tan x - \int \sec x (\sec^2 x - 1) \, dx$$

$$= \sec x \tan x - \int \sec^3 x \, dx + \int \sec x \, dx$$

Using Formula 1 and solving for the required integral, we get

$$\int \sec^3 x \, dx = \tfrac{1}{2}(\sec x \tan x + \ln|\sec x + \tan x|) + C$$

Integrals such as the one in the preceding example may seem very special but they occur frequently in applications of integration, as we will see in Chapter 8. Integrals of the form $\int \cot^m x \csc^n x \, dx$ can be found by similar methods because of the identity $1 + \cot^2 x = \csc^2 x$.

Finally, we can make use of another set of trigonometric identities:

> [2] To evaluate the integrals (a) $\int \sin mx \cos nx \, dx$, (b) $\int \sin mx \sin nx \, dx$, or (c) $\int \cos mx \cos nx \, dx$, use the corresponding identity:
>
> (a) $\sin A \cos B = \tfrac{1}{2}[\sin(A - B) + \sin(A + B)]$
>
> (b) $\sin A \sin B = \tfrac{1}{2}[\cos(A - B) - \cos(A + B)]$
>
> (c) $\cos A \cos B = \tfrac{1}{2}[\cos(A - B) + \cos(A + B)]$

These product identities are discussed in Appendix D.

EXAMPLE 9 Evaluate $\int \sin 4x \cos 5x \, dx$.

SOLUTION This integral could be evaluated using integration by parts, but it's easier to use the identity in Equation 2(a) as follows:

$$\int \sin 4x \cos 5x \, dx = \int \tfrac{1}{2}[\sin(-x) + \sin 9x] \, dx$$

$$= \tfrac{1}{2} \int (-\sin x + \sin 9x) \, dx$$

$$= \tfrac{1}{2}\left(\cos x - \tfrac{1}{9} \cos 9x\right) + C$$

7.2 EXERCISES

1–49 Evaluate the integral.

1. $\int \sin^2 x \cos^3 x \, dx$
2. $\int \sin^3 \theta \cos^4 \theta \, d\theta$
3. $\int_0^{\pi/2} \sin^7 \theta \cos^5 \theta \, d\theta$
4. $\int_0^{\pi/2} \sin^5 x \, dx$
5. $\int \sin^5(2t) \cos^2(2t) \, dt$
6. $\int t \cos^5(t^2) \, dt$
7. $\int_0^{\pi/2} \cos^2 \theta \, d\theta$
8. $\int_0^{2\pi} \sin^2(\tfrac{1}{3}\theta) \, d\theta$
9. $\int_0^{\pi} \cos^4(2t) \, dt$
10. $\int_0^{\pi} \sin^2 t \cos^4 t \, dt$
11. $\int_0^{\pi/2} \sin^2 x \cos^2 x \, dx$
12. $\int_0^{\pi/2} (2 - \sin \theta)^2 \, d\theta$
13. $\int \sqrt{\cos \theta} \, \sin^3 \theta \, d\theta$
14. $\int \dfrac{\sin^2(1/t)}{t^2} \, dt$
15. $\int \cot x \cos^2 x \, dx$
16. $\int \tan^2 x \cos^3 x \, dx$
17. $\int \sin^2 x \sin 2x \, dx$
18. $\int \sin x \cos(\tfrac{1}{2}x) \, dx$
19. $\int t \sin^2 t \, dt$
20. $\int x \sin^3 x \, dx$
21. $\int \tan x \sec^3 x \, dx$
22. $\int \tan^2 \theta \sec^4 \theta \, d\theta$
23. $\int \tan^2 x \, dx$
24. $\int (\tan^2 x + \tan^4 x) \, dx$
25. $\int \tan^4 x \sec^6 x \, dx$
26. $\int_0^{\pi/4} \sec^6 \theta \tan^6 \theta \, d\theta$
27. $\int \tan^3 x \sec x \, dx$
28. $\int \tan^5 x \sec^3 x \, dx$
29. $\int \tan^3 x \sec^6 x \, dx$
30. $\int_0^{\pi/4} \tan^4 t \, dt$

31. $\int \tan^5 x \, dx$

32. $\int \tan^2 x \sec x \, dx$

33. $\int x \sec x \tan x \, dx$

34. $\int \dfrac{\sin \phi}{\cos^3 \phi} \, d\phi$

35. $\int_{\pi/6}^{\pi/2} \cot^2 x \, dx$

36. $\int_{\pi/4}^{\pi/2} \cot^3 x \, dx$

37. $\int_{\pi/4}^{\pi/2} \cot^5 \phi \csc^3 \phi \, d\phi$

38. $\int_{\pi/4}^{\pi/2} \csc^4 \theta \cot^4 \theta \, d\theta$

39. $\int \csc x \, dx$

40. $\int_{\pi/6}^{\pi/3} \csc^3 x \, dx$

41. $\int \sin 8x \cos 5x \, dx$

42. $\int \sin 2\theta \sin 6\theta \, d\theta$

43. $\int_0^{\pi/2} \cos 5t \cos 10t \, dt$

44. $\int \sin x \sec^5 x \, dx$

45. $\int_0^{\pi/6} \sqrt{1 + \cos 2x} \, dx$

46. $\int_0^{\pi/4} \sqrt{1 - \cos 4\theta} \, d\theta$

47. $\int \dfrac{1 - \tan^2 x}{\sec^2 x} \, dx$

48. $\int \dfrac{dx}{\cos x - 1}$

49. $\int x \tan^2 x \, dx$

50. If $\int_0^{\pi/4} \tan^6 x \sec x \, dx = I$, express the value of $\int_0^{\pi/4} \tan^8 x \sec x \, dx$ in terms of I.

⚠ 51–54 Evaluate the indefinite integral. Illustrate, and check that your answer is reasonable, by graphing both the integrand and its antiderivative (taking $C = 0$).

51. $\int x \sin^2(x^2) \, dx$

52. $\int \sin^5 x \cos^3 x \, dx$

53. $\int \sin 3x \sin 6x \, dx$

54. $\int \sec^4(\tfrac{1}{2}x) \, dx$

55. Find the average value of the function $f(x) = \sin^2 x \cos^3 x$ on the interval $[-\pi, \pi]$.

56. Evaluate $\int \sin x \cos x \, dx$ by four methods:
    (a) the substitution $u = \cos x$
    (b) the substitution $u = \sin x$
    (c) the identity $\sin 2x = 2 \sin x \cos x$
    (d) integration by parts
    Explain the different appearances of the answers.

57–58 Find the area of the region bounded by the given curves.

57. $y = \sin^2 x, \quad y = \sin^3 x, \quad 0 \leq x \leq \pi$

58. $y = \tan x, \quad y = \tan^2 x, \quad 0 \leq x \leq \pi/4$

⚠ 59–60 Use a graph of the integrand to guess the value of the integral. Then use the methods of this section to prove that your guess is correct.

59. $\int_0^{2\pi} \cos^3 x \, dx$

60. $\int_0^2 \sin 2\pi x \cos 5\pi x \, dx$

61–64 Find the volume obtained by rotating the region bounded by the curves about the given axis.

61. $y = \sin x, \; y = 0, \; \pi/2 \leq x \leq \pi;$ about the x-axis

62. $y = \sin^2 x, \; y = 0, \; 0 \leq x \leq \pi;$ about the x-axis

63. $y = \sin x, \; y = \cos x, \; 0 \leq x \leq \pi/4;$ about $y = 1$

64. $y = \sec x, \; y = \cos x, \; 0 \leq x \leq \pi/3;$ about $y = -1$

65. A particle moves on a straight line with velocity function $v(t) = \sin \omega t \cos^2 \omega t$. Find its position function $s = f(t)$ if $f(0) = 0$.

66. Household electricity is supplied in the form of alternating current that varies from 155 V to -155 V with a frequency of 60 cycles per second (Hz). The voltage is thus given by the equation
$$E(t) = 155 \sin(120 \pi t)$$
where t is the time in seconds. Voltmeters read the RMS (root-mean-square) voltage, which is the square root of the average value of $[E(t)]^2$ over one cycle.
(a) Calculate the RMS voltage of household current.
(b) Many electric stoves require an RMS voltage of 220 V. Find the corresponding amplitude A needed for the voltage $E(t) = A \sin(120 \pi t)$.

67–69 Prove the formula, where m and n are positive integers.

67. $\int_{-\pi}^{\pi} \sin mx \cos nx \, dx = 0$

68. $\int_{-\pi}^{\pi} \sin mx \sin nx \, dx = \begin{cases} 0 & \text{if } m \neq n \\ \pi & \text{if } m = n \end{cases}$

69. $\int_{-\pi}^{\pi} \cos mx \cos nx \, dx = \begin{cases} 0 & \text{if } m \neq n \\ \pi & \text{if } m = n \end{cases}$

70. A *finite Fourier series* is given by the sum
$$f(x) = \sum_{n=1}^{N} a_n \sin nx$$
$$= a_1 \sin x + a_2 \sin 2x + \cdots + a_N \sin Nx$$
Show that the mth coefficient a_m is given by the formula
$$a_m = \dfrac{1}{\pi} \int_{-\pi}^{\pi} f(x) \sin mx \, dx$$

7.3 Trigonometric Substitution

In finding the area of a circle or an ellipse, an integral of the form $\int \sqrt{a^2 - x^2}\, dx$ arises, where $a > 0$. If it were $\int x\sqrt{a^2 - x^2}\, dx$, the substitution $u = a^2 - x^2$ would be effective but, as it stands, $\int \sqrt{a^2 - x^2}\, dx$ is more difficult. If we change the variable from x to θ by the substitution $x = a \sin \theta$, then the identity $1 - \sin^2\theta = \cos^2\theta$ allows us to get rid of the root sign because

$$\sqrt{a^2 - x^2} = \sqrt{a^2 - a^2 \sin^2\theta} = \sqrt{a^2(1 - \sin^2\theta)} = \sqrt{a^2 \cos^2\theta} = a|\cos\theta|$$

Notice the difference between the substitution $u = a^2 - x^2$ (in which the new variable is a function of the old one) and the substitution $x = a \sin \theta$ (the old variable is a function of the new one).

In general, we can make a substitution of the form $x = g(t)$ by using the Substitution Rule in reverse. To make our calculations simpler, we assume that g has an inverse function; that is, g is one-to-one. In this case, if we replace u by x and x by t in the Substitution Rule (Equation 5.5.4), we obtain

$$\int f(x)\, dx = \int f(g(t)) g'(t)\, dt$$

This kind of substitution is called *inverse substitution*.

We can make the inverse substitution $x = a \sin\theta$ provided that it defines a one-to-one function. This can be accomplished by restricting θ to lie in the interval $[-\pi/2, \pi/2]$.

In the following table we list trigonometric substitutions that are effective for the given radical expressions because of the specified trigonometric identities. In each case the restriction on θ is imposed to ensure that the function that defines the substitution is one-to-one. (These are the same intervals used in Section 1.5 in defining the inverse functions.)

Table of Trigonometric Substitutions

Expression	Substitution	Identity
$\sqrt{a^2 - x^2}$	$x = a \sin\theta,\ -\dfrac{\pi}{2} \leq \theta \leq \dfrac{\pi}{2}$	$1 - \sin^2\theta = \cos^2\theta$
$\sqrt{a^2 + x^2}$	$x = a \tan\theta,\ -\dfrac{\pi}{2} < \theta < \dfrac{\pi}{2}$	$1 + \tan^2\theta = \sec^2\theta$
$\sqrt{x^2 - a^2}$	$x = a \sec\theta,\ 0 \leq \theta < \dfrac{\pi}{2}\ \text{or}\ \pi \leq \theta < \dfrac{3\pi}{2}$	$\sec^2\theta - 1 = \tan^2\theta$

EXAMPLE 1 Evaluate $\displaystyle\int \frac{\sqrt{9 - x^2}}{x^2}\, dx$.

SOLUTION Let $x = 3 \sin\theta$, where $-\pi/2 \leq \theta \leq \pi/2$. Then $dx = 3 \cos\theta\, d\theta$ and

$$\sqrt{9 - x^2} = \sqrt{9 - 9\sin^2\theta} = \sqrt{9\cos^2\theta} = 3|\cos\theta| = 3\cos\theta$$

(Note that $\cos\theta \geq 0$ because $-\pi/2 \leq \theta \leq \pi/2$.) Thus the Inverse Substitution Rule gives

$$\int \frac{\sqrt{9-x^2}}{x^2}\,dx = \int \frac{3\cos\theta}{9\sin^2\theta}\,3\cos\theta\,d\theta$$

$$= \int \frac{\cos^2\theta}{\sin^2\theta}\,d\theta = \int \cot^2\theta\,d\theta$$

$$= \int (\csc^2\theta - 1)\,d\theta$$

$$= -\cot\theta - \theta + C$$

Since this is an indefinite integral, we must return to the original variable x. This can be done either by using trigonometric identities to express $\cot\theta$ in terms of $\sin\theta = x/3$ or by drawing a diagram, as in Figure 1, where θ is interpreted as an angle of a right triangle. Since $\sin\theta = x/3$, we label the opposite side and the hypotenuse as having lengths x and 3. Then the Pythagorean Theorem gives the length of the adjacent side as $\sqrt{9-x^2}$, so we can simply read the value of $\cot\theta$ from the figure:

$$\cot\theta = \frac{\sqrt{9-x^2}}{x}$$

(Although $\theta > 0$ in the diagram, this expression for $\cot\theta$ is valid even when $\theta < 0$.) Since $\sin\theta = x/3$, we have $\theta = \sin^{-1}(x/3)$ and so

$$\int \frac{\sqrt{9-x^2}}{x^2}\,dx = -\frac{\sqrt{9-x^2}}{x} - \sin^{-1}\left(\frac{x}{3}\right) + C$$

EXAMPLE 2 Find the area enclosed by the ellipse

$$\frac{x^2}{a^2} + \frac{y^2}{b^2} = 1$$

SOLUTION Solving the equation of the ellipse for y, we get

$$\frac{y^2}{b^2} = 1 - \frac{x^2}{a^2} = \frac{a^2 - x^2}{a^2} \quad \text{or} \quad y = \pm\frac{b}{a}\sqrt{a^2 - x^2}$$

Because the ellipse is symmetric with respect to both axes, the total area A is four times the area in the first quadrant (see Figure 2). The part of the ellipse in the first quadrant is given by the function

$$y = \frac{b}{a}\sqrt{a^2 - x^2} \qquad 0 \leq x \leq a$$

and so

$$\tfrac{1}{4}A = \int_0^a \frac{b}{a}\sqrt{a^2 - x^2}\,dx$$

To evaluate this integral we substitute $x = a\sin\theta$. Then $dx = a\cos\theta\,d\theta$. To change

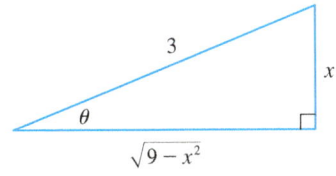

FIGURE 1
$\sin\theta = \dfrac{x}{3}$

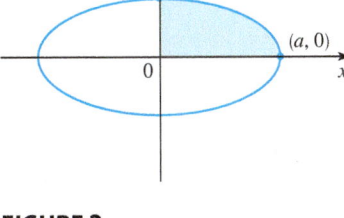

FIGURE 2
$\dfrac{x^2}{a^2} + \dfrac{y^2}{b^2} = 1$

the limits of integration we note that when $x = 0$, $\sin\theta = 0$, so $\theta = 0$; when $x = a$, $\sin\theta = 1$, so $\theta = \pi/2$. Also

$$\sqrt{a^2 - x^2} = \sqrt{a^2 - a^2\sin^2\theta} = \sqrt{a^2\cos^2\theta} = a|\cos\theta| = a\cos\theta$$

since $0 \leq \theta \leq \pi/2$. Therefore

$$A = 4\frac{b}{a}\int_0^a \sqrt{a^2 - x^2}\,dx = 4\frac{b}{a}\int_0^{\pi/2} a\cos\theta \cdot a\cos\theta\,d\theta$$

$$= 4ab\int_0^{\pi/2} \cos^2\theta\,d\theta = 4ab\int_0^{\pi/2} \tfrac{1}{2}(1 + \cos 2\theta)\,d\theta$$

$$= 2ab\left[\theta + \tfrac{1}{2}\sin 2\theta\right]_0^{\pi/2} = 2ab\left(\frac{\pi}{2} + 0 - 0\right) = \pi ab$$

We have shown that the area of an ellipse with semiaxes a and b is πab. In particular, taking $a = b = r$, we have proved the famous formula that the area of a circle with radius r is πr^2. ∎

NOTE Since the integral in Example 2 was a definite integral, we changed the limits of integration and did not have to convert back to the original variable x.

EXAMPLE 3 Find $\int \dfrac{1}{x^2\sqrt{x^2 + 4}}\,dx$.

SOLUTION Let $x = 2\tan\theta$, $-\pi/2 < \theta < \pi/2$. Then $dx = 2\sec^2\theta\,d\theta$ and

$$\sqrt{x^2 + 4} = \sqrt{4(\tan^2\theta + 1)} = \sqrt{4\sec^2\theta} = 2|\sec\theta| = 2\sec\theta$$

So we have

$$\int \frac{dx}{x^2\sqrt{x^2 + 4}} = \int \frac{2\sec^2\theta\,d\theta}{4\tan^2\theta \cdot 2\sec\theta} = \frac{1}{4}\int \frac{\sec\theta}{\tan^2\theta}\,d\theta$$

To evaluate this trigonometric integral we put everything in terms of $\sin\theta$ and $\cos\theta$:

$$\frac{\sec\theta}{\tan^2\theta} = \frac{1}{\cos\theta} \cdot \frac{\cos^2\theta}{\sin^2\theta} = \frac{\cos\theta}{\sin^2\theta}$$

Therefore, making the substitution $u = \sin\theta$, we have

$$\int \frac{dx}{x^2\sqrt{x^2 + 4}} = \frac{1}{4}\int \frac{\cos\theta}{\sin^2\theta}\,d\theta = \frac{1}{4}\int \frac{du}{u^2}$$

$$= \frac{1}{4}\left(-\frac{1}{u}\right) + C = -\frac{1}{4\sin\theta} + C$$

$$= -\frac{\csc\theta}{4} + C$$

We use Figure 3 to determine that $\csc\theta = \sqrt{x^2 + 4}/x$ and so

$$\int \frac{dx}{x^2\sqrt{x^2 + 4}} = -\frac{\sqrt{x^2 + 4}}{4x} + C \qquad \blacksquare$$

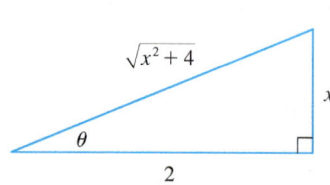

FIGURE 3
$\tan\theta = \dfrac{x}{2}$

EXAMPLE 4 Find $\int \dfrac{x}{\sqrt{x^2+4}}\,dx$.

SOLUTION It would be possible to use the trigonometric substitution $x = 2\tan\theta$ here (as in Example 3). But the direct substitution $u = x^2 + 4$ is simpler, because then $du = 2x\,dx$ and

$$\int \dfrac{x}{\sqrt{x^2+4}}\,dx = \dfrac{1}{2}\int \dfrac{du}{\sqrt{u}} = \sqrt{u} + C = \sqrt{x^2+4} + C$$

NOTE Example 4 illustrates the fact that even when trigonometric substitutions are possible, they may not give the easiest solution. You should look for a simpler method first.

EXAMPLE 5 Evaluate $\int \dfrac{dx}{\sqrt{x^2-a^2}}$, where $a > 0$.

SOLUTION 1 We let $x = a\sec\theta$, where $0 < \theta < \pi/2$ or $\pi < \theta < 3\pi/2$. Then $dx = a\sec\theta\tan\theta\,d\theta$ and

$$\sqrt{x^2-a^2} = \sqrt{a^2(\sec^2\theta - 1)} = \sqrt{a^2\tan^2\theta} = a|\tan\theta| = a\tan\theta$$

Therefore

$$\int \dfrac{dx}{\sqrt{x^2-a^2}} = \int \dfrac{a\sec\theta\tan\theta}{a\tan\theta}\,d\theta = \int \sec\theta\,d\theta = \ln|\sec\theta + \tan\theta| + C$$

The triangle in Figure 4 gives $\tan\theta = \sqrt{x^2-a^2}/a$, so we have

$$\int \dfrac{dx}{\sqrt{x^2-a^2}} = \ln\left|\dfrac{x}{a} + \dfrac{\sqrt{x^2-a^2}}{a}\right| + C$$

$$= \ln|x + \sqrt{x^2-a^2}| - \ln a + C$$

Writing $C_1 = C - \ln a$, we have

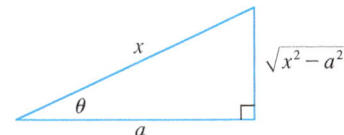

FIGURE 4
$\sec\theta = \dfrac{x}{a}$

$$\boxed{1} \quad \int \dfrac{dx}{\sqrt{x^2-a^2}} = \ln|x + \sqrt{x^2-a^2}| + C_1$$

SOLUTION 2 For $x > 0$ the hyperbolic substitution $x = a\cosh t$ can also be used. Using the identity $\cosh^2 y - \sinh^2 y = 1$, we have

$$\sqrt{x^2-a^2} = \sqrt{a^2(\cosh^2 t - 1)} = \sqrt{a^2\sinh^2 t} = a\sinh t$$

Since $dx = a\sinh t\,dt$, we obtain

$$\int \dfrac{dx}{\sqrt{x^2-a^2}} = \int \dfrac{a\sinh t\,dt}{a\sinh t} = \int dt = t + C$$

Since $\cosh t = x/a$, we have $t = \cosh^{-1}(x/a)$ and

$$\boxed{2} \quad \int \dfrac{dx}{\sqrt{x^2-a^2}} = \cosh^{-1}\left(\dfrac{x}{a}\right) + C$$

Although Formulas 1 and 2 look quite different, they are actually equivalent by Formula 3.11.4.

As Example 6 shows, trigonometric substitution is sometimes a good idea when $(x^2 + a^2)^{n/2}$ occurs in an integral, where n is any integer. The same is true when $(a^2 - x^2)^{n/2}$ or $(x^2 - a^2)^{n/2}$ occur.

NOTE As Example 5 illustrates, hyperbolic substitutions can be used in place of trigonometric substitutions and sometimes they lead to simpler answers. But we usually use trigonometric substitutions because trigonometric identities are more familiar than hyperbolic identities.

EXAMPLE 6 Find $\displaystyle\int_0^{3\sqrt{3}/2} \frac{x^3}{(4x^2 + 9)^{3/2}} \, dx$.

SOLUTION First we note that $(4x^2 + 9)^{3/2} = (\sqrt{4x^2 + 9})^3$ so trigonometric substitution is appropriate. Although $\sqrt{4x^2 + 9}$ is not quite one of the expressions in the table of trigonometric substitutions, it becomes one of them if we make the preliminary substitution $u = 2x$. When we combine this with the tangent substitution, we have $x = \frac{3}{2}\tan\theta$, which gives $dx = \frac{3}{2}\sec^2\theta \, d\theta$ and

$$\sqrt{4x^2 + 9} = \sqrt{9\tan^2\theta + 9} = 3\sec\theta$$

When $x = 0$, $\tan\theta = 0$, so $\theta = 0$; when $x = 3\sqrt{3}/2$, $\tan\theta = \sqrt{3}$, so $\theta = \pi/3$.

$$\int_0^{3\sqrt{3}/2} \frac{x^3}{(4x^2 + 9)^{3/2}} \, dx = \int_0^{\pi/3} \frac{\frac{27}{8}\tan^3\theta}{27\sec^3\theta} \frac{3}{2}\sec^2\theta \, d\theta$$

$$= \frac{3}{16} \int_0^{\pi/3} \frac{\tan^3\theta}{\sec\theta} \, d\theta = \frac{3}{16} \int_0^{\pi/3} \frac{\sin^3\theta}{\cos^2\theta} \, d\theta$$

$$= \frac{3}{16} \int_0^{\pi/3} \frac{1 - \cos^2\theta}{\cos^2\theta} \sin\theta \, d\theta$$

Now we substitute $u = \cos\theta$ so that $du = -\sin\theta \, d\theta$. When $\theta = 0$, $u = 1$; when $\theta = \pi/3$, $u = \frac{1}{2}$. Therefore

$$\int_0^{3\sqrt{3}/2} \frac{x^3}{(4x^2 + 9)^{3/2}} \, dx = -\frac{3}{16} \int_1^{1/2} \frac{1 - u^2}{u^2} \, du$$

$$= \frac{3}{16} \int_1^{1/2} (1 - u^{-2}) \, du = \frac{3}{16} \left[u + \frac{1}{u} \right]_1^{1/2}$$

$$= \frac{3}{16}\left[\left(\tfrac{1}{2} + 2\right) - (1 + 1)\right] = \frac{3}{32} \quad\blacksquare$$

EXAMPLE 7 Evaluate $\displaystyle\int \frac{x}{\sqrt{3 - 2x - x^2}} \, dx$.

SOLUTION We can transform the integrand into a function for which trigonometric substitution is appropriate by first completing the square under the root sign:

$$3 - 2x - x^2 = 3 - (x^2 + 2x) = 3 + 1 - (x^2 + 2x + 1)$$

$$= 4 - (x + 1)^2$$

This suggests that we make the substitution $u = x + 1$. Then $du = dx$ and $x = u - 1$, so

$$\int \frac{x}{\sqrt{3 - 2x - x^2}} \, dx = \int \frac{u - 1}{\sqrt{4 - u^2}} \, du$$

Figure 5 shows the graphs of the integrand in Example 7 and its indefinite integral (with $C = 0$). Which is which?

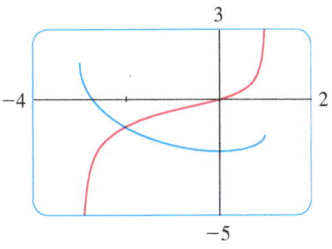

FIGURE 5

We now substitute $u = 2 \sin \theta$, giving $du = 2 \cos \theta \, d\theta$ and $\sqrt{4 - u^2} = 2 \cos \theta$, so

$$\int \frac{x}{\sqrt{3 - 2x - x^2}} \, dx = \int \frac{2 \sin \theta - 1}{2 \cos \theta} \, 2 \cos \theta \, d\theta$$

$$= \int (2 \sin \theta - 1) \, d\theta$$

$$= -2 \cos \theta - \theta + C$$

$$= -\sqrt{4 - u^2} - \sin^{-1}\left(\frac{u}{2}\right) + C$$

$$= -\sqrt{3 - 2x - x^2} - \sin^{-1}\left(\frac{x+1}{2}\right) + C \qquad \blacksquare$$

7.3 EXERCISES

1–3 Evaluate the integral using the indicated trigonometric substitution. Sketch and label the associated right triangle.

1. $\displaystyle\int \frac{dx}{x^2\sqrt{4 - x^2}} \qquad x = 2 \sin \theta$

2. $\displaystyle\int \frac{x^3}{\sqrt{x^2 + 4}} \, dx \qquad x = 2 \tan \theta$

3. $\displaystyle\int \frac{\sqrt{x^2 - 4}}{x} \, dx \qquad x = 2 \sec \theta$

4–30 Evaluate the integral.

4. $\displaystyle\int \frac{x^2}{\sqrt{9 - x^2}} \, dx$

5. $\displaystyle\int \frac{\sqrt{x^2 - 1}}{x^4} \, dx$

6. $\displaystyle\int_0^3 \frac{x}{\sqrt{36 - x^2}} \, dx$

7. $\displaystyle\int_0^a \frac{dx}{(a^2 + x^2)^{3/2}}, \quad a > 0$

8. $\displaystyle\int \frac{dt}{t^2\sqrt{t^2 - 16}}$

9. $\displaystyle\int_2^3 \frac{dx}{(x^2 - 1)^{3/2}}$

10. $\displaystyle\int_0^{2/3} \sqrt{4 - 9x^2} \, dx$

11. $\displaystyle\int_0^{1/2} x \sqrt{1 - 4x^2} \, dx$

12. $\displaystyle\int_0^2 \frac{dt}{\sqrt{4 + t^2}}$

13. $\displaystyle\int \frac{\sqrt{x^2 - 9}}{x^3} \, dx$

14. $\displaystyle\int_0^1 \frac{dx}{(x^2 + 1)^2}$

15. $\displaystyle\int_0^a x^2 \sqrt{a^2 - x^2} \, dx$

16. $\displaystyle\int_{\sqrt{2}/3}^{2/3} \frac{dx}{x^5\sqrt{9x^2 - 1}}$

17. $\displaystyle\int \frac{x}{\sqrt{x^2 - 7}} \, dx$

18. $\displaystyle\int \frac{dx}{[(ax)^2 - b^2]^{3/2}}$

19. $\displaystyle\int \frac{\sqrt{1 + x^2}}{x} \, dx$

20. $\displaystyle\int \frac{x}{\sqrt{1 + x^2}} \, dx$

21. $\displaystyle\int_0^{0.6} \frac{x^2}{\sqrt{9 - 25x^2}} \, dx$

22. $\displaystyle\int_0^1 \sqrt{x^2 + 1} \, dx$

23. $\displaystyle\int \frac{dx}{\sqrt{x^2 + 2x + 5}}$

24. $\displaystyle\int_0^1 \sqrt{x - x^2} \, dx$

25. $\displaystyle\int x^2\sqrt{3 + 2x - x^2} \, dx$

26. $\displaystyle\int \frac{x^2}{(3 + 4x - 4x^2)^{3/2}} \, dx$

27. $\displaystyle\int \sqrt{x^2 + 2x} \, dx$

28. $\displaystyle\int \frac{x^2 + 1}{(x^2 - 2x + 2)^2} \, dx$

29. $\displaystyle\int x\sqrt{1 - x^4} \, dx$

30. $\displaystyle\int_0^{\pi/2} \frac{\cos t}{\sqrt{1 + \sin^2 t}} \, dt$

31. (a) Use trigonometric substitution to show that

$$\int \frac{dx}{\sqrt{x^2 + a^2}} = \ln\left(x + \sqrt{x^2 + a^2}\right) + C$$

(b) Use the hyperbolic substitution $x = a \sinh t$ to show that

$$\int \frac{dx}{\sqrt{x^2 + a^2}} = \sinh^{-1}\left(\frac{x}{a}\right) + C$$

These formulas are connected by Formula 3.11.3.

32. Evaluate

$$\int \frac{x^2}{(x^2 + a^2)^{3/2}} \, dx$$

(a) by trigonometric substitution.
(b) by the hyperbolic substitution $x = a \sinh t$.

33. Find the average value of $f(x) = \sqrt{x^2 - 1}/x$, $1 \leq x \leq 7$.

34. Find the area of the region bounded by the hyperbola $9x^2 - 4y^2 = 36$ and the line $x = 3$.

35. Prove the formula $A = \frac{1}{2}r^2\theta$ for the area of a sector of a circle with radius r and central angle θ. [*Hint:* Assume $0 < \theta < \pi/2$ and place the center of the circle at the origin so it has the equation $x^2 + y^2 = r^2$. Then A is the sum of the area of the triangle POQ and the area of the region PQR in the figure.]

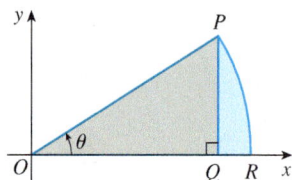

36. Evaluate the integral

$$\int \frac{dx}{x^4\sqrt{x^2 - 2}}$$

Graph the integrand and its indefinite integral on the same screen and check that your answer is reasonable.

37. Find the volume of the solid obtained by rotating about the x-axis the region enclosed by the curves $y = 9/(x^2 + 9)$, $y = 0$, $x = 0$, and $x = 3$.

38. Find the volume of the solid obtained by rotating about the line $x = 1$ the region under the curve $y = x\sqrt{1 - x^2}$, $0 \leq x \leq 1$.

39. (a) Use trigonometric substitution to verify that

$$\int_0^x \sqrt{a^2 - t^2} \, dt = \frac{1}{2}a^2 \sin^{-1}(x/a) + \frac{1}{2}x\sqrt{a^2 - x^2}$$

(b) Use the figure to give trigonometric interpretations of both terms on the right side of the equation in part (a).

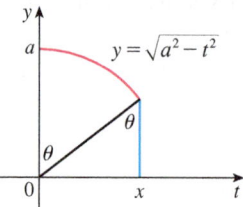

40. The parabola $y = \frac{1}{2}x^2$ divides the disk $x^2 + y^2 \leq 8$ into two parts. Find the areas of both parts.

41. A torus is generated by rotating the circle $x^2 + (y - R)^2 = r^2$ about the x-axis. Find the volume enclosed by the torus.

42. A charged rod of length L produces an electric field at point $P(a, b)$ given by

$$E(P) = \int_{-a}^{L-a} \frac{\lambda b}{4\pi \varepsilon_0 (x^2 + b^2)^{3/2}} \, dx$$

where λ is the charge density per unit length on the rod and ε_0 is the free space permittivity (see the figure). Evaluate the integral to determine an expression for the electric field $E(P)$.

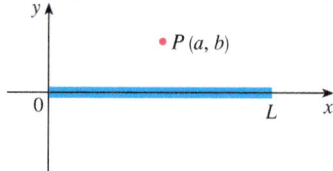

43. Find the area of the crescent-shaped region (called a *lune*) bounded by arcs of circles with radii r and R. (See the figure.)

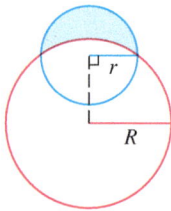

44. A water storage tank has the shape of a cylinder with diameter 10 ft. It is mounted so that the circular cross-sections are vertical. If the depth of the water is 7 ft, what percentage of the total capacity is being used?

7.4 Integration of Rational Functions by Partial Fractions

In this section we show how to integrate any rational function (a ratio of polynomials) by expressing it as a sum of simpler fractions, called *partial fractions*, that we already know how to integrate. To illustrate the method, observe that by taking the fractions $2/(x-1)$ and $1/(x+2)$ to a common denominator we obtain

$$\frac{2}{x-1} - \frac{1}{x+2} = \frac{2(x+2) - (x-1)}{(x-1)(x+2)} = \frac{x+5}{x^2+x-2}$$

If we now reverse the procedure, we see how to integrate the function on the right side of this equation:

$$\int \frac{x+5}{x^2+x-2} \, dx = \int \left(\frac{2}{x-1} - \frac{1}{x+2} \right) dx$$

$$= 2 \ln|x-1| - \ln|x+2| + C$$

To see how the method of partial fractions works in general, let's consider a rational function

$$f(x) = \frac{P(x)}{Q(x)}$$

where P and Q are polynomials. It's possible to express f as a sum of simpler fractions provided that the degree of P is less than the degree of Q. Such a rational function is called *proper*. Recall that if

$$P(x) = a_n x^n + a_{n-1} x^{n-1} + \cdots + a_1 x + a_0$$

where $a_n \neq 0$, then the degree of P is n and we write $\deg(P) = n$.

If f is *improper*, that is, $\deg(P) \geq \deg(Q)$, then we must take the preliminary step of dividing Q into P (by long division) until a remainder $R(x)$ is obtained such that $\deg(R) < \deg(Q)$. The division statement is

$$\boxed{1} \qquad f(x) = \frac{P(x)}{Q(x)} = S(x) + \frac{R(x)}{Q(x)}$$

where S and R are also polynomials.

As the following example illustrates, sometimes this preliminary step is all that is required.

EXAMPLE 1 Find $\int \dfrac{x^3 + x}{x-1} \, dx$.

SOLUTION Since the degree of the numerator is greater than the degree of the denominator, we first perform the long division. This enables us to write

$$\int \frac{x^3+x}{x-1} \, dx = \int \left(x^2 + x + 2 + \frac{2}{x-1} \right) dx$$

$$= \frac{x^3}{3} + \frac{x^2}{2} + 2x + 2\ln|x-1| + C \qquad \blacksquare$$

In the case of an Equation 1 whose denominator is more complicated, the next step is to factor the denominator $Q(x)$ as far as possible. It can be shown that any polynomial Q can be factored as a product of linear factors (of the form $ax + b$) and irreducible quadratic factors (of the form $ax^2 + bx + c$, where $b^2 - 4ac < 0$). For instance, if $Q(x) = x^4 - 16$, we could factor it as

$$Q(x) = (x^2 - 4)(x^2 + 4) = (x - 2)(x + 2)(x^2 + 4)$$

The third step is to express the proper rational function $R(x)/Q(x)$ (from Equation 1) as a sum of **partial fractions** of the form

$$\frac{A}{(ax + b)^i} \quad \text{or} \quad \frac{Ax + B}{(ax^2 + bx + c)^j}$$

A theorem in algebra guarantees that it is always possible to do this. We explain the details for the four cases that occur.

CASE I The denominator $Q(x)$ is a product of distinct linear factors.

This means that we can write

$$Q(x) = (a_1 x + b_1)(a_2 x + b_2) \cdots (a_k x + b_k)$$

where no factor is repeated (and no factor is a constant multiple of another). In this case the partial fraction theorem states that there exist constants $A_1, A_2, \ldots, A_k$ such that

$$\boxed{2} \qquad \frac{R(x)}{Q(x)} = \frac{A_1}{a_1 x + b_1} + \frac{A_2}{a_2 x + b_2} + \cdots + \frac{A_k}{a_k x + b_k}$$

These constants can be determined as in the following example.

EXAMPLE 2 Evaluate $\displaystyle\int \frac{x^2 + 2x - 1}{2x^3 + 3x^2 - 2x}\, dx$.

SOLUTION Since the degree of the numerator is less than the degree of the denominator, we don't need to divide. We factor the denominator as

$$2x^3 + 3x^2 - 2x = x(2x^2 + 3x - 2) = x(2x - 1)(x + 2)$$

Since the denominator has three distinct linear factors, the partial fraction decomposition of the integrand (2) has the form

$$\boxed{3} \qquad \frac{x^2 + 2x - 1}{x(2x - 1)(x + 2)} = \frac{A}{x} + \frac{B}{2x - 1} + \frac{C}{x + 2}$$

Another method for finding A, B, and C is given in the note after this example.

To determine the values of A, B, and C, we multiply both sides of this equation by the product of the denominators, $x(2x - 1)(x + 2)$, obtaining

$$\boxed{4} \qquad x^2 + 2x - 1 = A(2x - 1)(x + 2) + Bx(x + 2) + Cx(2x - 1)$$

Expanding the right side of Equation 4 and writing it in the standard form for polynomials, we get

$$\boxed{5} \qquad x^2 + 2x - 1 = (2A + B + 2C)x^2 + (3A + 2B - C)x - 2A$$

The polynomials in Equation 5 are identical, so their coefficients must be equal. The coefficient of x^2 on the right side, $2A + B + 2C$, must equal the coefficient of x^2 on the left side—namely, 1. Likewise, the coefficients of x are equal and the constant terms are equal. This gives the following system of equations for A, B, and C:

$$2A + B + 2C = 1$$
$$3A + 2B - C = 2$$
$$-2A = -1$$

Solving, we get $A = \frac{1}{2}$, $B = \frac{1}{5}$, and $C = -\frac{1}{10}$, and so

$$\int \frac{x^2 + 2x - 1}{2x^3 + 3x^2 - 2x}\, dx = \int \left(\frac{1}{2}\frac{1}{x} + \frac{1}{5}\frac{1}{2x - 1} - \frac{1}{10}\frac{1}{x + 2} \right) dx$$

$$= \tfrac{1}{2} \ln|x| + \tfrac{1}{10} \ln|2x - 1| - \tfrac{1}{10} \ln|x + 2| + K$$

In integrating the middle term we have made the mental substitution $u = 2x - 1$, which gives $du = 2\, dx$ and $dx = \frac{1}{2}\, du$. ∎

We could check our work by taking the terms to a common denominator and adding them.

Figure 1 shows the graphs of the integrand in Example 2 and its indefinite integral (with $K = 0$). Which is which?

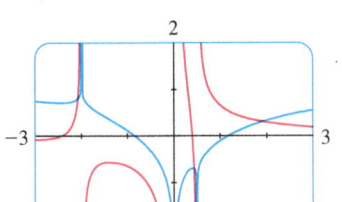

FIGURE 1

NOTE We can use an alternative method to find the coefficients A, B, and C in Example 2. Equation 4 is an identity; it is true for every value of x. Let's choose values of x that simplify the equation. If we put $x = 0$ in Equation 4, then the second and third terms on the right side vanish and the equation then becomes $-2A = -1$, or $A = \frac{1}{2}$. Likewise, $x = \frac{1}{2}$ gives $5B/4 = \frac{1}{4}$ and $x = -2$ gives $10C = -1$, so $B = \frac{1}{5}$ and $C = -\frac{1}{10}$. (You may object that Equation 3 is not valid for $x = 0, \frac{1}{2}$, or -2, so why should Equation 4 be valid for those values? In fact, Equation 4 is true for all values of x, even $x = 0, \frac{1}{2}$, and -2. See Exercise 73 for the reason.)

EXAMPLE 3 Find $\int \dfrac{dx}{x^2 - a^2}$, where $a \neq 0$.

SOLUTION The method of partial fractions gives

$$\frac{1}{x^2 - a^2} = \frac{1}{(x - a)(x + a)} = \frac{A}{x - a} + \frac{B}{x + a}$$

and therefore

$$A(x + a) + B(x - a) = 1$$

Using the method of the preceding note, we put $x = a$ in this equation and get $A(2a) = 1$, so $A = 1/(2a)$. If we put $x = -a$, we get $B(-2a) = 1$, so $B = -1/(2a)$. Thus

$$\int \frac{dx}{x^2 - a^2} = \frac{1}{2a} \int \left(\frac{1}{x - a} - \frac{1}{x + a} \right) dx$$

$$= \frac{1}{2a} \left(\ln|x - a| - \ln|x + a| \right) + C$$

Since $\ln x - \ln y = \ln(x/y)$, we can write the integral as

$$\boxed{6} \quad \int \frac{dx}{x^2 - a^2} = \frac{1}{2a} \ln \left| \frac{x-a}{x+a} \right| + C$$

See Exercises 57–58 for ways of using Formula 6.

CASE II $Q(x)$ **is a product of linear factors, some of which are repeated.**
Suppose the first linear factor $(a_1x + b_1)$ is repeated r times; that is, $(a_1x + b_1)^r$ occurs in the factorization of $Q(x)$. Then instead of the single term $A_1/(a_1x + b_1)$ in Equation 2, we would use

$$\boxed{7} \quad \frac{A_1}{a_1x + b_1} + \frac{A_2}{(a_1x + b_1)^2} + \cdots + \frac{A_r}{(a_1x + b_1)^r}$$

By way of illustration, we could write

$$\frac{x^3 - x + 1}{x^2(x-1)^3} = \frac{A}{x} + \frac{B}{x^2} + \frac{C}{x-1} + \frac{D}{(x-1)^2} + \frac{E}{(x-1)^3}$$

but we prefer to work out in detail a simpler example.

EXAMPLE 4 Find $\int \frac{x^4 - 2x^2 + 4x + 1}{x^3 - x^2 - x + 1} dx$.

SOLUTION The first step is to divide. The result of long division is

$$\frac{x^4 - 2x^2 + 4x + 1}{x^3 - x^2 - x + 1} = x + 1 + \frac{4x}{x^3 - x^2 - x + 1}$$

The second step is to factor the denominator $Q(x) = x^3 - x^2 - x + 1$. Since $Q(1) = 0$, we know that $x - 1$ is a factor and we obtain

$$x^3 - x^2 - x + 1 = (x-1)(x^2 - 1) = (x-1)(x-1)(x+1)$$
$$= (x-1)^2(x+1)$$

Since the linear factor $x - 1$ occurs twice, the partial fraction decomposition is

$$\frac{4x}{(x-1)^2(x+1)} = \frac{A}{x-1} + \frac{B}{(x-1)^2} + \frac{C}{x+1}$$

Multiplying by the least common denominator, $(x-1)^2(x+1)$, we get

$$\boxed{8} \quad 4x = A(x-1)(x+1) + B(x+1) + C(x-1)^2$$
$$= (A + C)x^2 + (B - 2C)x + (-A + B + C)$$

Now we equate coefficients:

$$A + C = 0$$
$$B - 2C = 4$$
$$-A + B + C = 0$$

Another method for finding the coefficients:
Put $x = 1$ in (8): $B = 2$.
Put $x = -1$: $C = -1$.
Put $x = 0$: $A = B + C = 1$.

Solving, we obtain $A = 1$, $B = 2$, and $C = -1$, so

$$\int \frac{x^4 - 2x^2 + 4x + 1}{x^3 - x^2 - x + 1} \, dx = \int \left[x + 1 + \frac{1}{x - 1} + \frac{2}{(x - 1)^2} - \frac{1}{x + 1} \right] dx$$

$$= \frac{x^2}{2} + x + \ln|x - 1| - \frac{2}{x - 1} - \ln|x + 1| + K$$

$$= \frac{x^2}{2} + x - \frac{2}{x - 1} + \ln\left|\frac{x - 1}{x + 1}\right| + K \qquad \blacksquare$$

CASE III $Q(x)$ **contains irreducible quadratic factors, none of which is repeated.**
If $Q(x)$ has the factor $ax^2 + bx + c$, where $b^2 - 4ac < 0$, then, in addition to the partial fractions in Equations 2 and 7, the expression for $R(x)/Q(x)$ will have a term of the form

$$\boxed{9} \qquad \frac{Ax + B}{ax^2 + bx + c}$$

where A and B are constants to be determined. For instance, the function given by $f(x) = x/[(x - 2)(x^2 + 1)(x^2 + 4)]$ has a partial fraction decomposition of the form

$$\frac{x}{(x - 2)(x^2 + 1)(x^2 + 4)} = \frac{A}{x - 2} + \frac{Bx + C}{x^2 + 1} + \frac{Dx + E}{x^2 + 4}$$

The term given in (9) can be integrated by completing the square (if necessary) and using the formula

$$\boxed{10} \qquad \int \frac{dx}{x^2 + a^2} = \frac{1}{a} \tan^{-1}\left(\frac{x}{a}\right) + C$$

EXAMPLE 5 Evaluate $\displaystyle\int \frac{2x^2 - x + 4}{x^3 + 4x} \, dx$.

SOLUTION Since $x^3 + 4x = x(x^2 + 4)$ can't be factored further, we write

$$\frac{2x^2 - x + 4}{x(x^2 + 4)} = \frac{A}{x} + \frac{Bx + C}{x^2 + 4}$$

Multiplying by $x(x^2 + 4)$, we have

$$2x^2 - x + 4 = A(x^2 + 4) + (Bx + C)x$$

$$= (A + B)x^2 + Cx + 4A$$

Equating coefficients, we obtain

$$A + B = 2 \qquad C = -1 \qquad 4A = 4$$

Therefore $A = 1$, $B = 1$, and $C = -1$ and so

$$\int \frac{2x^2 - x + 4}{x^3 + 4x} \, dx = \int \left(\frac{1}{x} + \frac{x - 1}{x^2 + 4} \right) dx$$

In order to integrate the second term we split it into two parts:

$$\int \frac{x-1}{x^2+4}\,dx = \int \frac{x}{x^2+4}\,dx - \int \frac{1}{x^2+4}\,dx$$

We make the substitution $u = x^2 + 4$ in the first of these integrals so that $du = 2x\,dx$. We evaluate the second integral by means of Formula 10 with $a = 2$:

$$\int \frac{2x^2 - x + 4}{x(x^2+4)}\,dx = \int \frac{1}{x}\,dx + \int \frac{x}{x^2+4}\,dx - \int \frac{1}{x^2+4}\,dx$$

$$= \ln|x| + \tfrac{1}{2}\ln(x^2+4) - \tfrac{1}{2}\tan^{-1}(x/2) + K \quad \blacksquare$$

EXAMPLE 6 Evaluate $\displaystyle\int \frac{4x^2 - 3x + 2}{4x^2 - 4x + 3}\,dx$.

SOLUTION Since the degree of the numerator is *not less than* the degree of the denominator, we first divide and obtain

$$\frac{4x^2 - 3x + 2}{4x^2 - 4x + 3} = 1 + \frac{x-1}{4x^2 - 4x + 3}$$

Notice that the quadratic $4x^2 - 4x + 3$ is irreducible because its discriminant is $b^2 - 4ac = -32 < 0$. This means it can't be factored, so we don't need to use the partial fraction technique.

To integrate the given function we complete the square in the denominator:

$$4x^2 - 4x + 3 = (2x - 1)^2 + 2$$

This suggests that we make the substitution $u = 2x - 1$. Then $du = 2\,dx$ and $x = \tfrac{1}{2}(u+1)$, so

$$\int \frac{4x^2 - 3x + 2}{4x^2 - 4x + 3}\,dx = \int \left(1 + \frac{x-1}{4x^2 - 4x + 3}\right) dx$$

$$= x + \tfrac{1}{2}\int \frac{\tfrac{1}{2}(u+1) - 1}{u^2 + 2}\,du = x + \tfrac{1}{4}\int \frac{u - 1}{u^2 + 2}\,du$$

$$= x + \tfrac{1}{4}\int \frac{u}{u^2 + 2}\,du - \tfrac{1}{4}\int \frac{1}{u^2 + 2}\,du$$

$$= x + \tfrac{1}{8}\ln(u^2 + 2) - \tfrac{1}{4}\cdot\frac{1}{\sqrt{2}}\tan^{-1}\left(\frac{u}{\sqrt{2}}\right) + C$$

$$= x + \tfrac{1}{8}\ln(4x^2 - 4x + 3) - \frac{1}{4\sqrt{2}}\tan^{-1}\left(\frac{2x-1}{\sqrt{2}}\right) + C \quad \blacksquare$$

NOTE Example 6 illustrates the general procedure for integrating a partial fraction of the form

$$\frac{Ax + B}{ax^2 + bx + c} \quad \text{where } b^2 - 4ac < 0$$

We complete the square in the denominator and then make a substitution that brings the integral into the form

$$\int \frac{Cu + D}{u^2 + a^2}\,du = C\int \frac{u}{u^2+a^2}\,du + D\int \frac{1}{u^2+a^2}\,du$$

Then the first integral is a logarithm and the second is expressed in terms of $\tan^{-1}$.

CASE IV $Q(x)$ contains a repeated irreducible quadratic factor.

If $Q(x)$ has the factor $(ax^2 + bx + c)^r$, where $b^2 - 4ac < 0$, then instead of the single partial fraction (9), the sum

$$\boxed{11} \quad \frac{A_1 x + B_1}{ax^2 + bx + c} + \frac{A_2 x + B_2}{(ax^2 + bx + c)^2} + \cdots + \frac{A_r x + B_r}{(ax^2 + bx + c)^r}$$

occurs in the partial fraction decomposition of $R(x)/Q(x)$. Each of the terms in (11) can be integrated by using a substitution or by first completing the square if necessary.

EXAMPLE 7 Write out the form of the partial fraction decomposition of the function

$$\frac{x^3 + x^2 + 1}{x(x - 1)(x^2 + x + 1)(x^2 + 1)^3}$$

SOLUTION

$$\frac{x^3 + x^2 + 1}{x(x - 1)(x^2 + x + 1)(x^2 + 1)^3}$$

$$= \frac{A}{x} + \frac{B}{x - 1} + \frac{Cx + D}{x^2 + x + 1} + \frac{Ex + F}{x^2 + 1} + \frac{Gx + H}{(x^2 + 1)^2} + \frac{Ix + J}{(x^2 + 1)^3} \quad \blacksquare$$

It would be extremely tedious to work out by hand the numerical values of the coefficients in Example 7. Most computer algebra systems, however, can find the numerical values very quickly. For instance, the Maple command

```
convert(f, parfrac, x)
```

or the Mathematica command

```
Apart[f]
```

gives the following values:

$A = -1$, $B = \frac{1}{8}$, $C = D = -1$,
$E = \frac{15}{8}$, $F = -\frac{1}{8}$, $G = H = \frac{3}{4}$,
$I = -\frac{1}{2}$, $J = \frac{1}{2}$

EXAMPLE 8 Evaluate $\displaystyle\int \frac{1 - x + 2x^2 - x^3}{x(x^2 + 1)^2}\,dx$.

SOLUTION The form of the partial fraction decomposition is

$$\frac{1 - x + 2x^2 - x^3}{x(x^2 + 1)^2} = \frac{A}{x} + \frac{Bx + C}{x^2 + 1} + \frac{Dx + E}{(x^2 + 1)^2}$$

Multiplying by $x(x^2 + 1)^2$, we have

$$-x^3 + 2x^2 - x + 1 = A(x^2 + 1)^2 + (Bx + C)x(x^2 + 1) + (Dx + E)x$$

$$= A(x^4 + 2x^2 + 1) + B(x^4 + x^2) + C(x^3 + x) + Dx^2 + Ex$$

$$= (A + B)x^4 + Cx^3 + (2A + B + D)x^2 + (C + E)x + A$$

If we equate coefficients, we get the system

$$A + B = 0 \quad C = -1 \quad 2A + B + D = 2 \quad C + E = -1 \quad A = 1$$

which has the solution $A = 1$, $B = -1$, $C = -1$, $D = 1$, and $E = 0$. Thus

$$\int \frac{1 - x + 2x^2 - x^3}{x(x^2 + 1)^2}\,dx = \int \left(\frac{1}{x} - \frac{x + 1}{x^2 + 1} + \frac{x}{(x^2 + 1)^2}\right) dx$$

$$= \int \frac{dx}{x} - \int \frac{x}{x^2 + 1}\,dx - \int \frac{dx}{x^2 + 1} + \int \frac{x\,dx}{(x^2 + 1)^2}$$

$$= \ln|x| - \tfrac{1}{2}\ln(x^2 + 1) - \tan^{-1}x - \frac{1}{2(x^2 + 1)} + K \quad \blacksquare$$

In the second and fourth terms we made the mental substitution $u = x^2 + 1$.

NOTE Example 8 worked out rather nicely because the coefficient E turned out to be 0. In general, we might get a term of the form $1/(x^2 + 1)^2$. One way to integrate such a term is to make the substitution $x = \tan\theta$. Another method is to use the formula in Exercise 72.

Sometimes partial fractions can be avoided when integrating a rational function. For instance, although the integral

$$\int \frac{x^2 + 1}{x(x^2 + 3)}\,dx$$

could be evaluated by using the method of Case III, it's much easier to observe that if $u = x(x^2 + 3) = x^3 + 3x$, then $du = (3x^2 + 3)\,dx$ and so

$$\int \frac{x^2 + 1}{x(x^2 + 3)}\,dx = \tfrac{1}{3}\ln|x^3 + 3x| + C$$

■ Rationalizing Substitutions

Some nonrational functions can be changed into rational functions by means of appropriate substitutions. In particular, when an integrand contains an expression of the form $\sqrt[n]{g(x)}$, then the substitution $u = \sqrt[n]{g(x)}$ may be effective. Other instances appear in the exercises.

EXAMPLE 9 Evaluate $\displaystyle\int \frac{\sqrt{x + 4}}{x}\,dx$.

SOLUTION Let $u = \sqrt{x + 4}$. Then $u^2 = x + 4$, so $x = u^2 - 4$ and $dx = 2u\,du$. Therefore

$$\int \frac{\sqrt{x + 4}}{x}\,dx = \int \frac{u}{u^2 - 4}\,2u\,du = 2\int \frac{u^2}{u^2 - 4}\,du = 2\int\left(1 + \frac{4}{u^2 - 4}\right)du$$

We can evaluate this integral either by factoring $u^2 - 4$ as $(u - 2)(u + 2)$ and using partial fractions or by using Formula 6 with $a = 2$:

$$\int \frac{\sqrt{x + 4}}{x}\,dx = 2\int du + 8\int \frac{du}{u^2 - 4}$$

$$= 2u + 8 \cdot \frac{1}{2 \cdot 2}\ln\left|\frac{u - 2}{u + 2}\right| + C$$

$$= 2\sqrt{x + 4} + 2\ln\left|\frac{\sqrt{x + 4} - 2}{\sqrt{x + 4} + 2}\right| + C \quad \blacksquare$$

7.4 EXERCISES

1–6 Write out the form of the partial fraction decomposition of the function (as in Example 7). Do not determine the numerical values of the coefficients.

1. (a) $\dfrac{4+x}{(1+2x)(3-x)}$ (b) $\dfrac{1-x}{x^3+x^4}$

2. (a) $\dfrac{x-6}{x^2+x-6}$ (b) $\dfrac{x^2}{x^2+x+6}$

3. (a) $\dfrac{1}{x^2+x^4}$ (b) $\dfrac{x^3+1}{x^3-3x^2+2x}$

4. (a) $\dfrac{x^4-2x^3+x^2+2x-1}{x^2-2x+1}$ (b) $\dfrac{x^2-1}{x^3+x^2+x}$

5. (a) $\dfrac{x^6}{x^2-4}$ (b) $\dfrac{x^4}{(x^2-x+1)(x^2+2)^2}$

6. (a) $\dfrac{t^6+1}{t^6+t^3}$ (b) $\dfrac{x^5+1}{(x^2-x)(x^4+2x^2+1)}$

7–38 Evaluate the integral.

7. $\displaystyle\int \dfrac{x^4}{x-1}\,dx$

8. $\displaystyle\int \dfrac{3t-2}{t+1}\,dt$

9. $\displaystyle\int \dfrac{5x+1}{(2x+1)(x-1)}\,dx$

10. $\displaystyle\int \dfrac{y}{(y+4)(2y-1)}\,dy$

11. $\displaystyle\int_0^1 \dfrac{2}{2x^2+3x+1}\,dx$

12. $\displaystyle\int_0^1 \dfrac{x-4}{x^2-5x+6}\,dx$

13. $\displaystyle\int \dfrac{ax}{x^2-bx}\,dx$

14. $\displaystyle\int \dfrac{1}{(x+a)(x+b)}\,dx$

15. $\displaystyle\int_{-1}^{0} \dfrac{x^3-4x+1}{x^2-3x+2}\,dx$

16. $\displaystyle\int_1^2 \dfrac{x^3+4x^2+x-1}{x^3+x^2}\,dx$

17. $\displaystyle\int_1^2 \dfrac{4y^2-7y-12}{y(y+2)(y-3)}\,dy$

18. $\displaystyle\int_1^2 \dfrac{3x^2+6x+2}{x^2+3x+2}\,dx$

19. $\displaystyle\int_0^1 \dfrac{x^2+x+1}{(x+1)^2(x+2)}\,dx$

20. $\displaystyle\int_2^3 \dfrac{x(3-5x)}{(3x-1)(x-1)^2}\,dx$

21. $\displaystyle\int \dfrac{dt}{(t^2-1)^2}$

22. $\displaystyle\int \dfrac{x^4+9x^2+x+2}{x^2+9}\,dx$

23. $\displaystyle\int \dfrac{10}{(x-1)(x^2+9)}\,dx$

24. $\displaystyle\int \dfrac{x^2-x+6}{x^3+3x}\,dx$

25. $\displaystyle\int \dfrac{4x}{x^3+x^2+x+1}\,dx$

26. $\displaystyle\int \dfrac{x^2+x+1}{(x^2+1)^2}\,dx$

27. $\displaystyle\int \dfrac{x^3+4x+3}{x^4+5x^2+4}\,dx$

28. $\displaystyle\int \dfrac{x^3+6x-2}{x^4+6x^2}\,dx$

29. $\displaystyle\int \dfrac{x+4}{x^2+2x+5}\,dx$

30. $\displaystyle\int \dfrac{x^3-2x^2+2x-5}{x^4+4x^2+3}\,dx$

31. $\displaystyle\int \dfrac{1}{x^3-1}\,dx$

32. $\displaystyle\int_0^1 \dfrac{x}{x^2+4x+13}\,dx$

33. $\displaystyle\int_0^1 \dfrac{x^3+2x}{x^4+4x^2+3}\,dx$

34. $\displaystyle\int \dfrac{x^5+x-1}{x^3+1}\,dx$

35. $\displaystyle\int \dfrac{5x^4+7x^2+x+2}{x(x^2+1)^2}\,dx$

36. $\displaystyle\int \dfrac{x^4+3x^2+1}{x^5+5x^3+5x}\,dx$

37. $\displaystyle\int \dfrac{x^2-3x+7}{(x^2-4x+6)^2}\,dx$

38. $\displaystyle\int \dfrac{x^3+2x^2+3x-2}{(x^2+2x+2)^2}\,dx$

39–52 Make a substitution to express the integrand as a rational function and then evaluate the integral.

39. $\displaystyle\int \dfrac{dx}{x\sqrt{x-1}}$

40. $\displaystyle\int \dfrac{dx}{2\sqrt{x+3}+x}$

41. $\displaystyle\int \dfrac{dx}{x^2+x\sqrt{x}}$

42. $\displaystyle\int_0^1 \dfrac{1}{1+\sqrt[3]{x}}\,dx$

43. $\displaystyle\int \dfrac{x^3}{\sqrt[3]{x^2+1}}\,dx$

44. $\displaystyle\int \dfrac{dx}{(1+\sqrt{x})^2}$

45. $\displaystyle\int \dfrac{1}{\sqrt{x}-\sqrt[3]{x}}\,dx$ [Hint: Substitute $u=\sqrt[6]{x}$.]

46. $\displaystyle\int \dfrac{\sqrt{1+\sqrt{x}}}{x}\,dx$

47. $\displaystyle\int \dfrac{e^{2x}}{e^{2x}+3e^x+2}\,dx$

48. $\displaystyle\int \dfrac{\sin x}{\cos^2 x-3\cos x}\,dx$

49. $\displaystyle\int \dfrac{\sec^2 t}{\tan^2 t+3\tan t+2}\,dt$

50. $\displaystyle\int \dfrac{e^x}{(e^x-2)(e^{2x}+1)}\,dx$

51. $\displaystyle\int \dfrac{dx}{1+e^x}$

52. $\displaystyle\int \dfrac{\cosh t}{\sinh^2 t+\sinh^4 t}\,dt$

53–54 Use integration by parts, together with the techniques of this section, to evaluate the integral.

53. $\displaystyle\int \ln(x^2-x+2)\,dx$

54. $\displaystyle\int x\tan^{-1}x\,dx$

55. Use a graph of $f(x)=1/(x^2-2x-3)$ to decide whether $\int_0^2 f(x)\,dx$ is positive or negative. Use the graph to give a rough estimate of the value of the integral and then use partial fractions to find the exact value.

56. Evaluate
$$\int \dfrac{1}{x^2+k}\,dx$$
by considering several cases for the constant k.

57–58 Evaluate the integral by completing the square and using Formula 6.

57. $\displaystyle\int \frac{dx}{x^2 - 2x}$

58. $\displaystyle\int \frac{2x+1}{4x^2 + 12x - 7}\,dx$

59. The German mathematician Karl Weierstrass (1815–1897) noticed that the substitution $t = \tan(x/2)$ will convert any rational function of $\sin x$ and $\cos x$ into an ordinary rational function of t.
   (a) If $t = \tan(x/2)$, $-\pi < x < \pi$, sketch a right triangle or use trigonometric identities to show that
$$\cos\left(\frac{x}{2}\right) = \frac{1}{\sqrt{1+t^2}} \quad \text{and} \quad \sin\left(\frac{x}{2}\right) = \frac{t}{\sqrt{1+t^2}}$$
   (b) Show that
$$\cos x = \frac{1-t^2}{1+t^2} \quad \text{and} \quad \sin x = \frac{2t}{1+t^2}$$
   (c) Show that
$$dx = \frac{2}{1+t^2}\,dt$$

60–63 Use the substitution in Exercise 59 to transform the integrand into a rational function of t and then evaluate the integral.

60. $\displaystyle\int \frac{dx}{1-\cos x}$

61. $\displaystyle\int \frac{1}{3\sin x - 4\cos x}\,dx$

62. $\displaystyle\int_{\pi/3}^{\pi/2} \frac{1}{1+\sin x - \cos x}\,dx$

63. $\displaystyle\int_0^{\pi/2} \frac{\sin 2x}{2+\cos x}\,dx$

64–65 Find the area of the region under the given curve from 1 to 2.

64. $y = \dfrac{1}{x^3 + x}$

65. $y = \dfrac{x^2+1}{3x - x^2}$

66. Find the volume of the resulting solid if the region under the curve $y = 1/(x^2 + 3x + 2)$ from $x = 0$ to $x = 1$ is rotated about (a) the x-axis and (b) the y-axis.

67. One method of slowing the growth of an insect population without using pesticides is to introduce into the population a number of sterile males that mate with fertile females but produce no offspring. (The photo shows a screw-worm fly, the first pest effectively eliminated from a region by this method.)

Let P represent the number of female insects in a population and S the number of sterile males introduced each generation. Let r be the per capita rate of production of females by females, provided their chosen mate is not sterile. Then the female population is related to time t by
$$t = \int \frac{P+S}{P[(r-1)P - S]}\,dP$$
Suppose an insect population with 10,000 females grows at a rate of $r = 1.1$ and 900 sterile males are added initially. Evaluate the integral to give an equation relating the female population to time. (Note that the resulting equation can't be solved explicitly for P.)

68. Factor $x^4 + 1$ as a difference of squares by first adding and subtracting the same quantity. Use this factorization to evaluate $\int 1/(x^4 + 1)\,dx$.

69. (a) Use a computer algebra system to find the partial fraction decomposition of the function
$$f(x) = \frac{4x^3 - 27x^2 + 5x - 32}{30x^5 - 13x^4 + 50x^3 - 286x^2 - 299x - 70}$$
   (b) Use part (a) to find $\int f(x)\,dx$ (by hand) and compare with the result of using the CAS to integrate f directly. Comment on any discrepancy.

CAS 70. (a) Find the partial fraction decomposition of the function
$$f(x) = \frac{12x^5 - 7x^3 - 13x^2 + 8}{100x^6 - 80x^5 + 116x^4 - 80x^3 + 41x^2 - 20x + 4}$$
   (b) Use part (a) to find $\int f(x)\,dx$ and graph f and its indefinite integral on the same screen.
   (c) Use the graph of f to discover the main features of the graph of $\int f(x)\,dx$.

71. The rational number $\tfrac{22}{7}$ has been used as an approximation to the number π since the time of Archimedes. Show that
$$\int_0^1 \frac{x^4(1-x)^4}{1+x^2}\,dx = \frac{22}{7} - \pi$$

72. (a) Use integration by parts to show that, for any positive integer n,
$$\int \frac{dx}{(x^2+a^2)^n}\,dx = \frac{x}{2a^2(n-1)(x^2+a^2)^{n-1}} + \frac{2n-3}{2a^2(n-1)}\int \frac{dx}{(x^2+a^2)^{n-1}}$$
   (b) Use part (a) to evaluate
$$\int \frac{dx}{(x^2+1)^2} \quad \text{and} \quad \int \frac{dx}{(x^2+1)^3}$$

73. Suppose that F, G, and Q are polynomials and

$$\frac{F(x)}{Q(x)} = \frac{G(x)}{Q(x)}$$

for all x except when $Q(x) = 0$. Prove that $F(x) = G(x)$ for all x. [*Hint:* Use continuity.]

74. If f is a quadratic function such that $f(0) = 1$ and

$$\int \frac{f(x)}{x^2(x+1)^3} dx$$

is a rational function, find the value of $f'(0)$.

75. If $a \neq 0$ and n is a positive integer, find the partial fraction decomposition of

$$f(x) = \frac{1}{x^n(x-a)}$$

[*Hint:* First find the coefficient of $1/(x - a)$. Then subtract the resulting term and simplify what is left.]

7.5 Strategy for Integration

As we have seen, integration is more challenging than differentiation. In finding the derivative of a function it is obvious which differentiation formula we should apply. But it may not be obvious which technique we should use to integrate a given function.

Until now individual techniques have been applied in each section. For instance, we usually used substitution in Exercises 5.5, integration by parts in Exercises 7.1, and partial fractions in Exercises 7.4. But in this section we present a collection of miscellaneous integrals in random order and the main challenge is to recognize which technique or formula to use. No hard and fast rules can be given as to which method applies in a given situation, but we give some advice on strategy that you may find useful.

A prerequisite for applying a strategy is a knowledge of the basic integration formulas. In the following table we have collected the integrals from our previous list together with several additional formulas that we have learned in this chapter.

Table of Integration Formulas Constants of integration have been omitted.

1. $\int x^n \, dx = \dfrac{x^{n+1}}{n+1} \quad (n \neq -1)$

2. $\int \dfrac{1}{x} \, dx = \ln|x|$

3. $\int e^x \, dx = e^x$

4. $\int b^x \, dx = \dfrac{b^x}{\ln b}$

5. $\int \sin x \, dx = -\cos x$

6. $\int \cos x \, dx = \sin x$

7. $\int \sec^2 x \, dx = \tan x$

8. $\int \csc^2 x \, dx = -\cot x$

9. $\int \sec x \tan x \, dx = \sec x$

10. $\int \csc x \cot x \, dx = -\csc x$

11. $\int \sec x \, dx = \ln|\sec x + \tan x|$

12. $\int \csc x \, dx = \ln|\csc x - \cot x|$

13. $\int \tan x \, dx = \ln|\sec x|$

14. $\int \cot x \, dx = \ln|\sin x|$

15. $\int \sinh x \, dx = \cosh x$

16. $\int \cosh x \, dx = \sinh x$

17. $\int \dfrac{dx}{x^2 + a^2} = \dfrac{1}{a} \tan^{-1}\left(\dfrac{x}{a}\right)$

18. $\int \dfrac{dx}{\sqrt{a^2 - x^2}} = \sin^{-1}\left(\dfrac{x}{a}\right), \quad a > 0$

*19. $\int \dfrac{dx}{x^2 - a^2} = \dfrac{1}{2a} \ln\left|\dfrac{x-a}{x+a}\right|$

*20. $\int \dfrac{dx}{\sqrt{x^2 \pm a^2}} = \ln\left|x + \sqrt{x^2 \pm a^2}\right|$

Most of these formulas should be memorized. It is useful to know them all, but the ones marked with an asterisk need not be memorized since they are easily derived. Formula 19 can be avoided by using partial fractions, and trigonometric substitutions can be used in place of Formula 20.

Once you are armed with these basic integration formulas, if you don't immediately see how to attack a given integral, you might try the following four-step strategy.

1. **Simplify the Integrand if Possible** Sometimes the use of algebraic manipulation or trigonometric identities will simplify the integrand and make the method of integration obvious. Here are some examples:

$$\int \sqrt{x}\,(1 + \sqrt{x})\,dx = \int (\sqrt{x} + x)\,dx$$

$$\int \frac{\tan\theta}{\sec^2\theta}\,d\theta = \int \frac{\sin\theta}{\cos\theta}\cos^2\theta\,d\theta$$

$$= \int \sin\theta \cos\theta\,d\theta = \tfrac{1}{2}\int \sin 2\theta\,d\theta$$

$$\int (\sin x + \cos x)^2\,dx = \int (\sin^2 x + 2\sin x \cos x + \cos^2 x)\,dx$$

$$= \int (1 + 2\sin x \cos x)\,dx$$

2. **Look for an Obvious Substitution** Try to find some function $u = g(x)$ in the integrand whose differential $du = g'(x)\,dx$ also occurs, apart from a constant factor. For instance, in the integral

$$\int \frac{x}{x^2 - 1}\,dx$$

we notice that if $u = x^2 - 1$, then $du = 2x\,dx$. Therefore we use the substitution $u = x^2 - 1$ instead of the method of partial fractions.

3. **Classify the Integrand According to Its Form** If Steps 1 and 2 have not led to the solution, then we take a look at the form of the integrand $f(x)$.
   (a) *Trigonometric functions.* If $f(x)$ is a product of powers of $\sin x$ and $\cos x$, of $\tan x$ and $\sec x$, or of $\cot x$ and $\csc x$, then we use the substitutions recommended in Section 7.2.
   (b) *Rational functions.* If f is a rational function, we use the procedure of Section 7.4 involving partial fractions.
   (c) *Integration by parts.* If $f(x)$ is a product of a power of x (or a polynomial) and a transcendental function (such as a trigonometric, exponential, or logarithmic function), then we try integration by parts, choosing u and dv according to the advice given in Section 7.1. If you look at the functions in Exercises 7.1, you will see that most of them are the type just described.
   (d) *Radicals.* Particular kinds of substitutions are recommended when certain radicals appear.
       (i) If $\sqrt{\pm x^2 \pm a^2}$ occurs, we use a trigonometric substitution according to the table in Section 7.3.
       (ii) If $\sqrt[n]{ax + b}$ occurs, we use the rationalizing substitution $u = \sqrt[n]{ax + b}$. More generally, this sometimes works for $\sqrt[n]{g(x)}$.

4. Try Again If the first three steps have not produced the answer, remember that there are basically only two methods of integration: substitution and parts.

(a) *Try substitution.* Even if no substitution is obvious (Step 2), some inspiration or ingenuity (or even desperation) may suggest an appropriate substitution.

(b) *Try parts.* Although integration by parts is used most of the time on products of the form described in Step 3(c), it is sometimes effective on single functions. Looking at Section 7.1, we see that it works on $\tan^{-1}x$, $\sin^{-1}x$, and $\ln x$, and these are all inverse functions.

(c) *Manipulate the integrand.* Algebraic manipulations (perhaps rationalizing the denominator or using trigonometric identities) may be useful in transforming the integral into an easier form. These manipulations may be more substantial than in Step 1 and may involve some ingenuity. Here is an example:

$$\int \frac{dx}{1 - \cos x} = \int \frac{1}{1 - \cos x} \cdot \frac{1 + \cos x}{1 + \cos x} dx = \int \frac{1 + \cos x}{1 - \cos^2 x} dx$$

$$= \int \frac{1 + \cos x}{\sin^2 x} dx = \int \left(\csc^2 x + \frac{\cos x}{\sin^2 x} \right) dx$$

(d) *Relate the problem to previous problems.* When you have built up some experience in integration, you may be able to use a method on a given integral that is similar to a method you have already used on a previous integral. Or you may even be able to express the given integral in terms of a previous one. For instance, $\int \tan^2 x \sec x \, dx$ is a challenging integral, but if we make use of the identity $\tan^2 x = \sec^2 x - 1$, we can write

$$\int \tan^2 x \sec x \, dx = \int \sec^3 x \, dx - \int \sec x \, dx$$

and if $\int \sec^3 x \, dx$ has previously been evaluated (see Example 7.2.8), then that calculation can be used in the present problem.

(e) *Use several methods.* Sometimes two or three methods are required to evaluate an integral. The evaluation could involve several successive substitutions of different types, or it might combine integration by parts with one or more substitutions.

In the following examples we indicate a method of attack but do not fully work out the integral.

EXAMPLE 1 $\int \frac{\tan^3 x}{\cos^3 x} dx$

In Step 1 we rewrite the integral:

$$\int \frac{\tan^3 x}{\cos^3 x} dx = \int \tan^3 x \sec^3 x \, dx$$

The integral is now of the form $\int \tan^m x \sec^n x \, dx$ with m odd, so we can use the advice in Section 7.2.

Alternatively, if in Step 1 we had written

$$\int \frac{\tan^3 x}{\cos^3 x} dx = \int \frac{\sin^3 x}{\cos^3 x} \frac{1}{\cos^3 x} dx = \int \frac{\sin^3 x}{\cos^6 x} dx$$

then we could have continued as follows with the substitution $u = \cos x$:

$$\int \frac{\sin^3 x}{\cos^6 x} dx = \int \frac{1 - \cos^2 x}{\cos^6 x} \sin x\, dx = \int \frac{1 - u^2}{u^6}(-du)$$

$$= \int \frac{u^2 - 1}{u^6} du = \int (u^{-4} - u^{-6})\, du$$

EXAMPLE 2 $\int e^{\sqrt{x}}\, dx$

According to (ii) in Step 3(d), we substitute $u = \sqrt{x}$. Then $x = u^2$, so $dx = 2u\, du$ and

$$\int e^{\sqrt{x}}\, dx = 2 \int u e^u\, du$$

The integrand is now a product of u and the transcendental function e^u so it can be integrated by parts.

EXAMPLE 3 $\int \dfrac{x^5 + 1}{x^3 - 3x^2 - 10x}\, dx$

No algebraic simplification or substitution is obvious, so Steps 1 and 2 don't apply here. The integrand is a rational function so we apply the procedure of Section 7.4, remembering that the first step is to divide.

EXAMPLE 4 $\int \dfrac{dx}{x\sqrt{\ln x}}$

Here Step 2 is all that is needed. We substitute $u = \ln x$ because its differential is $du = dx/x$, which occurs in the integral.

EXAMPLE 5 $\int \sqrt{\dfrac{1-x}{1+x}}\, dx$

Although the rationalizing substitution

$$u = \sqrt{\frac{1-x}{1+x}}$$

works here [(ii) in Step 3(d)], it leads to a very complicated rational function. An easier method is to do some algebraic manipulation [either as Step 1 or as Step 4(c)]. Multiplying numerator and denominator by $\sqrt{1-x}$, we have

$$\int \sqrt{\frac{1-x}{1+x}}\, dx = \int \frac{1-x}{\sqrt{1-x^2}}\, dx$$

$$= \int \frac{1}{\sqrt{1-x^2}}\, dx - \int \frac{x}{\sqrt{1-x^2}}\, dx$$

$$= \sin^{-1} x + \sqrt{1-x^2} + C$$

■ Can We Integrate All Continuous Functions?

The question arises: Will our strategy for integration enable us to find the integral of every continuous function? For example, can we use it to evaluate $\int e^{x^2}\, dx$? The answer is No, at least not in terms of the functions that we are familiar with.

The functions that we have been dealing with in this book are called **elementary functions**. These are the polynomials, rational functions, power functions (x^n), exponential functions (b^x), logarithmic functions, trigonometric and inverse trigonometric functions, hyperbolic and inverse hyperbolic functions, and all functions that can be obtained from these by the five operations of addition, subtraction, multiplication, division, and composition. For instance, the function

$$f(x) = \sqrt{\frac{x^2 - 1}{x^3 + 2x - 1}} + \ln(\cosh x) - xe^{\sin 2x}$$

is an elementary function.

If f is an elementary function, then f' is an elementary function but $\int f(x)\,dx$ need not be an elementary function. Consider $f(x) = e^{x^2}$. Since f is continuous, its integral exists, and if we define the function F by

$$F(x) = \int_0^x e^{t^2}\,dt$$

then we know from Part 1 of the Fundamental Theorem of Calculus that

$$F'(x) = e^{x^2}$$

Thus $f(x) = e^{x^2}$ has an antiderivative F, but it has been proved that F is not an elementary function. This means that no matter how hard we try, we will never succeed in evaluating $\int e^{x^2}\,dx$ in terms of the functions we know. (In Chapter 11, however, we will see how to express $\int e^{x^2}\,dx$ as an infinite series.) The same can be said of the following integrals:

$$\int \frac{e^x}{x}\,dx \qquad \int \sin(x^2)\,dx \qquad \int \cos(e^x)\,dx$$

$$\int \sqrt{x^3 + 1}\,dx \qquad \int \frac{1}{\ln x}\,dx \qquad \int \frac{\sin x}{x}\,dx$$

In fact, the majority of elementary functions don't have elementary antiderivatives. You may be assured, though, that the integrals in the following exercises are all elementary functions.

7.5 EXERCISES

1–82 Evaluate the integral.

1. $\int \dfrac{\cos x}{1 - \sin x}\,dx$

2. $\int_0^1 (3x + 1)^{\sqrt{2}}\,dx$

3. $\int_1^4 \sqrt{y}\,\ln y\,dy$

4. $\int \dfrac{\sin^3 x}{\cos x}\,dx$

5. $\int \dfrac{t}{t^4 + 2}\,dt$

6. $\int_0^1 \dfrac{x}{(2x + 1)^3}\,dx$

7. $\int_{-1}^1 \dfrac{e^{\arctan y}}{1 + y^2}\,dy$

8. $\int t \sin t \cos t\,dt$

9. $\int_2^4 \dfrac{x + 2}{x^2 + 3x - 4}\,dx$

10. $\int \dfrac{\cos(1/x)}{x^3}\,dx$

11. $\int \dfrac{1}{x^3\sqrt{x^2 - 1}}\,dx$

12. $\int \dfrac{2x - 3}{x^3 + 3x}\,dx$

13. $\int \sin^5 t \cos^4 t\,dt$

14. $\int \ln(1 + x^2)\,dx$

15. $\int x \sec x \tan x\,dx$

16. $\int_0^{\sqrt{2}/2} \dfrac{x^2}{\sqrt{1 - x^2}}\,dx$

17. $\int_0^\pi t \cos^2 t\,dt$

18. $\int_1^4 \dfrac{e^{\sqrt{t}}}{\sqrt{t}}\,dt$

19. $\int e^{x + e^x}\,dx$

20. $\int e^2\,dx$

21. $\int \arctan \sqrt{x}\,dx$

22. $\int \dfrac{\ln x}{x\sqrt{1 + (\ln x)^2}}\,dx$

23. $\int_0^1 (1 + \sqrt{x})^8 \, dx$

24. $\int (1 + \tan x)^2 \sec x \, dx$

25. $\int_0^1 \dfrac{1 + 12t}{1 + 3t} \, dt$

26. $\int_0^1 \dfrac{3x^2 + 1}{x^3 + x^2 + x + 1} \, dx$

27. $\int \dfrac{dx}{1 + e^x}$

28. $\int \sin \sqrt{at} \, dt$

29. $\int \ln(x + \sqrt{x^2 - 1}) \, dx$

30. $\int_{-1}^2 |e^x - 1| \, dx$

31. $\int \sqrt{\dfrac{1 + x}{1 - x}} \, dx$

32. $\int_1^3 \dfrac{e^{3/x}}{x^2} \, dx$

33. $\int \sqrt{3 - 2x - x^2} \, dx$

34. $\int_{\pi/4}^{\pi/2} \dfrac{1 + 4 \cot x}{4 - \cot x} \, dx$

35. $\int_{-\pi/2}^{\pi/2} \dfrac{x}{1 + \cos^2 x} \, dx$

36. $\int \dfrac{1 + \sin x}{1 + \cos x} \, dx$

37. $\int_0^{\pi/4} \tan^3 \theta \sec^2 \theta \, d\theta$

38. $\int_{\pi/6}^{\pi/3} \dfrac{\sin \theta \cot \theta}{\sec \theta} \, d\theta$

39. $\int \dfrac{\sec \theta \tan \theta}{\sec^2 \theta - \sec \theta} \, d\theta$

40. $\int_0^\pi \sin 6x \cos 3x \, dx$

41. $\int \theta \tan^2 \theta \, d\theta$

42. $\int \dfrac{\tan^{-1} x}{x^2} \, dx$

43. $\int \dfrac{\sqrt{x}}{1 + x^3} \, dx$

44. $\int \sqrt{1 + e^x} \, dx$

45. $\int x^5 e^{-x^3} \, dx$

46. $\int \dfrac{(x - 1)e^x}{x^2} \, dx$

47. $\int x^3(x - 1)^{-4} \, dx$

48. $\int_0^1 x\sqrt{2 - \sqrt{1 - x^2}} \, dx$

49. $\int \dfrac{1}{x\sqrt{4x + 1}} \, dx$

50. $\int \dfrac{1}{x^2\sqrt{4x + 1}} \, dx$

51. $\int \dfrac{1}{x\sqrt{4x^2 + 1}} \, dx$

52. $\int \dfrac{dx}{x(x^4 + 1)}$

53. $\int x^2 \sinh mx \, dx$

54. $\int (x + \sin x)^2 \, dx$

55. $\int \dfrac{dx}{x + x\sqrt{x}}$

56. $\int \dfrac{dx}{\sqrt{x} + x\sqrt{x}}$

57. $\int x\sqrt[3]{x + c} \, dx$

58. $\int \dfrac{x \ln x}{\sqrt{x^2 - 1}} \, dx$

59. $\int \dfrac{dx}{x^4 - 16}$

60. $\int \dfrac{dx}{x^2\sqrt{4x^2 - 1}}$

61. $\int \dfrac{d\theta}{1 + \cos \theta}$

62. $\int \dfrac{d\theta}{1 + \cos^2 \theta}$

63. $\int \sqrt{x} \, e^{\sqrt{x}} \, dx$

64. $\int \dfrac{1}{\sqrt{\sqrt{x} + 1}} \, dx$

65. $\int \dfrac{\sin 2x}{1 + \cos^4 x} \, dx$

66. $\int_{\pi/4}^{\pi/3} \dfrac{\ln(\tan x)}{\sin x \cos x} \, dx$

67. $\int \dfrac{1}{\sqrt{x + 1} + \sqrt{x}} \, dx$

68. $\int \dfrac{x^2}{x^6 + 3x^3 + 2} \, dx$

69. $\int_1^{\sqrt{3}} \dfrac{\sqrt{1 + x^2}}{x^2} \, dx$

70. $\int \dfrac{1}{1 + 2e^x - e^{-x}} \, dx$

71. $\int \dfrac{e^{2x}}{1 + e^x} \, dx$

72. $\int \dfrac{\ln(x + 1)}{x^2} \, dx$

73. $\int \dfrac{x + \arcsin x}{\sqrt{1 - x^2}} \, dx$

74. $\int \dfrac{4^x + 10^x}{2^x} \, dx$

75. $\int \dfrac{dx}{x \ln x - x}$

76. $\int \dfrac{x^2}{\sqrt{x^2 + 1}} \, dx$

77. $\int \dfrac{xe^x}{\sqrt{1 + e^x}} \, dx$

78. $\int \dfrac{1 + \sin x}{1 - \sin x} \, dx$

79. $\int x \sin^2 x \cos x \, dx$

80. $\int \dfrac{\sec x \cos 2x}{\sin x + \sec x} \, dx$

81. $\int \sqrt{1 - \sin x} \, dx$

82. $\int \dfrac{\sin x \cos x}{\sin^4 x + \cos^4 x} \, dx$

83. The functions $y = e^{x^2}$ and $y = x^2 e^{x^2}$ don't have elementary antiderivatives, but $y = (2x^2 + 1)e^{x^2}$ does. Evaluate $\int (2x^2 + 1)e^{x^2} \, dx$.

84. We know that $F(x) = \int_0^x e^{e^t} \, dt$ is a continuous function by FTC1, though it is not an elementary function. The functions

$$\int \dfrac{e^x}{x} \, dx \quad \text{and} \quad \int \dfrac{1}{\ln x} \, dx$$

are not elementary either, but they can be expressed in terms of F. Evaluate the following integrals in terms of F.

(a) $\int_1^2 \dfrac{e^x}{x} \, dx$ (b) $\int_2^3 \dfrac{1}{\ln x} \, dx$

7.6 Integration Using Tables and Computer Algebra Systems

In this section we describe how to use tables and computer algebra systems to integrate functions that have elementary antiderivatives. You should bear in mind, though, that even

Tables of Integrals

Tables of indefinite integrals are very useful when we are confronted by an integral that is difficult to evaluate by hand and we don't have access to a computer algebra system. A relatively brief table of 120 integrals, categorized by form, is provided on the Reference Pages at the back of the book. More extensive tables are available in the *CRC Standard Mathematical Tables and Formulae*, 31st ed. by Daniel Zwillinger (Boca Raton, FL, 2002) (709 entries) or in Gradshteyn and Ryzhik's *Table of Integrals, Series, and Products,* 7e (San Diego, 2007), which contains hundreds of pages of integrals. It should be remembered, however, that integrals do not often occur in exactly the form listed in a table. Usually we need to use the Substitution Rule or algebraic manipulation to transform a given integral into one of the forms in the table.

EXAMPLE 1 The region bounded by the curves $y = \arctan x$, $y = 0$, and $x = 1$ is rotated about the y-axis. Find the volume of the resulting solid.

SOLUTION Using the method of cylindrical shells, we see that the volume is

$$V = \int_0^1 2\pi x \arctan x \, dx$$

In the section of the Table of Integrals titled *Inverse Trigonometric Forms* we locate Formula 92:

$$\int u \tan^{-1} u \, du = \frac{u^2 + 1}{2} \tan^{-1} u - \frac{u}{2} + C$$

The Table of Integrals appears on Reference Pages 6–10 at the back of the book.

So the volume is

$$V = 2\pi \int_0^1 x \tan^{-1} x \, dx = 2\pi \left[\frac{x^2 + 1}{2} \tan^{-1} x - \frac{x}{2} \right]_0^1$$

$$= \pi \left[(x^2 + 1) \tan^{-1} x - x \right]_0^1 = \pi (2 \tan^{-1} 1 - 1)$$

$$= \pi [2(\pi/4) - 1] = \tfrac{1}{2}\pi^2 - \pi \qquad \blacksquare$$

EXAMPLE 2 Use the Table of Integrals to find $\displaystyle\int \frac{x^2}{\sqrt{5 - 4x^2}} \, dx$.

SOLUTION If we look at the section of the table titled *Forms Involving* $\sqrt{a^2 - u^2}$, we see that the closest entry is number 34:

$$\int \frac{u^2}{\sqrt{a^2 - u^2}} \, du = -\frac{u}{2} \sqrt{a^2 - u^2} + \frac{a^2}{2} \sin^{-1}\left(\frac{u}{a}\right) + C$$

This is not exactly what we have, but we will be able to use it if we first make the substitution $u = 2x$:

$$\int \frac{x^2}{\sqrt{5 - 4x^2}} \, dx = \int \frac{(u/2)^2}{\sqrt{5 - u^2}} \frac{du}{2} = \frac{1}{8} \int \frac{u^2}{\sqrt{5 - u^2}} \, du$$

Then we use Formula 34 with $a^2 = 5$ (so $a = \sqrt{5}$):

$$\int \frac{x^2}{\sqrt{5-4x^2}} dx = \frac{1}{8} \int \frac{u^2}{\sqrt{5-u^2}} du = \frac{1}{8}\left(-\frac{u}{2}\sqrt{5-u^2} + \frac{5}{2}\sin^{-1}\frac{u}{\sqrt{5}}\right) + C$$

$$= -\frac{x}{8}\sqrt{5-4x^2} + \frac{5}{16}\sin^{-1}\left(\frac{2x}{\sqrt{5}}\right) + C \qquad \blacksquare$$

EXAMPLE 3 Use the Table of Integrals to evaluate $\int x^3 \sin x \, dx$.

SOLUTION If we look in the section called *Trigonometric Forms*, we see that none of the entries explicitly includes a u^3 factor. However, we can use the reduction formula in entry 84 with $n = 3$:

$$\int x^3 \sin x \, dx = -x^3 \cos x + 3 \int x^2 \cos x \, dx$$

85. $\int u^n \cos u \, du$
$= u^n \sin u - n \int u^{n-1} \sin u \, du$

We now need to evaluate $\int x^2 \cos x \, dx$. We can use the reduction formula in entry 85 with $n = 2$, followed by entry 82:

$$\int x^2 \cos x \, dx = x^2 \sin x - 2 \int x \sin x \, dx$$

$$= x^2 \sin x - 2(\sin x - x \cos x) + K$$

Combining these calculations, we get

$$\int x^3 \sin x \, dx = -x^3 \cos x + 3x^2 \sin x + 6x \cos x - 6 \sin x + C$$

where $C = 3K$. $\qquad \blacksquare$

EXAMPLE 4 Use the Table of Integrals to find $\int x\sqrt{x^2 + 2x + 4} \, dx$.

SOLUTION Since the table gives forms involving $\sqrt{a^2 + x^2}$, $\sqrt{a^2 - x^2}$, and $\sqrt{x^2 - a^2}$, but not $\sqrt{ax^2 + bx + c}$, we first complete the square:

$$x^2 + 2x + 4 = (x + 1)^2 + 3$$

If we make the substitution $u = x + 1$ (so $x = u - 1$), the integrand will involve the pattern $\sqrt{a^2 + u^2}$:

$$\int x\sqrt{x^2 + 2x + 4} \, dx = \int (u - 1)\sqrt{u^2 + 3} \, du$$

$$= \int u\sqrt{u^2 + 3} \, du - \int \sqrt{u^2 + 3} \, du$$

The first integral is evaluated using the substitution $t = u^2 + 3$:

$$\int u\sqrt{u^2 + 3} \, du = \tfrac{1}{2} \int \sqrt{t} \, dt = \tfrac{1}{2} \cdot \tfrac{2}{3} t^{3/2} = \tfrac{1}{3}(u^2 + 3)^{3/2}$$

21. $\int \sqrt{a^2 + u^2} \, du = \frac{u}{2}\sqrt{a^2 + u^2}$
$+ \frac{a^2}{2} \ln\left(u + \sqrt{a^2 + u^2}\right) + C$

For the second integral we use Formula 21 with $a = \sqrt{3}$:

$$\int \sqrt{u^2 + 3} \, du = \frac{u}{2}\sqrt{u^2 + 3} + \tfrac{3}{2} \ln\left(u + \sqrt{u^2 + 3}\right)$$

Therefore

$$\int x\sqrt{x^2 + 2x + 4}\, dx$$

$$= \tfrac{1}{3}(x^2 + 2x + 4)^{3/2} - \frac{x+1}{2}\sqrt{x^2 + 2x + 4} - \tfrac{3}{2}\ln\!\left(x + 1 + \sqrt{x^2 + 2x + 4}\right) + C$$

■

■ Computer Algebra Systems

We have seen that the use of tables involves matching the form of the given integrand with the forms of the integrands in the tables. Computers are particularly good at matching patterns. And just as we used substitutions in conjunction with tables, a CAS can perform substitutions that transform a given integral into one that occurs in its stored formulas. So it isn't surprising that computer algebra systems excel at integration. That doesn't mean that integration by hand is an obsolete skill. We will see that a hand computation sometimes produces an indefinite integral in a form that is more convenient than a machine answer.

To begin, let's see what happens when we ask a machine to integrate the relatively simple function $y = 1/(3x - 2)$. Using the substitution $u = 3x - 2$, an easy calculation by hand gives

$$\int \frac{1}{3x - 2}\, dx = \tfrac{1}{3}\ln|3x - 2| + C$$

whereas Mathematica and Maple both return the answer

$$\tfrac{1}{3}\ln(3x - 2)$$

The first thing to notice is that computer algebra systems omit the constant of integration. In other words, they produce a *particular* antiderivative, not the most general one. Therefore, when making use of a machine integration, we might have to add a constant. Second, the absolute value signs are omitted in the machine answer. That is fine if our problem is concerned only with values of x greater than $\tfrac{2}{3}$. But if we are interested in other values of x, then we need to insert the absolute value symbol.

In the next example we reconsider the integral of Example 4, but this time we ask a machine for the answer.

EXAMPLE 5 Use a computer algebra system to find $\int x\sqrt{x^2 + 2x + 4}\, dx$.

SOLUTION Maple responds with the answer

$$\tfrac{1}{3}(x^2 + 2x + 4)^{3/2} - \tfrac{1}{4}(2x + 2)\sqrt{x^2 + 2x + 4} - \frac{3}{2}\operatorname{arcsinh}\frac{\sqrt{3}}{3}(1 + x)$$

This looks different from the answer we found in Example 4, but it is equivalent because the third term can be rewritten using the identity

This is equation 3.11.3.

$$\operatorname{arcsinh} x = \ln\!\left(x + \sqrt{x^2 + 1}\right)$$

Thus

$$\operatorname{arcsinh}\frac{\sqrt{3}}{3}(1 + x) = \ln\!\left[\frac{\sqrt{3}}{3}(1 + x) + \sqrt{\tfrac{1}{3}(1 + x)^2 + 1}\right]$$

$$= \ln\frac{1}{\sqrt{3}}\left[1 + x + \sqrt{(1 + x)^2 + 3}\right]$$

$$= \ln\frac{1}{\sqrt{3}} + \ln\!\left(x + 1 + \sqrt{x^2 + 2x + 4}\right)$$

The resulting extra term $-\frac{3}{2}\ln(1/\sqrt{3})$ can be absorbed into the constant of integration. Mathematica gives the answer

$$\left(\frac{5}{6} + \frac{x}{6} + \frac{x^2}{3}\right)\sqrt{x^2 + 2x + 4} - \frac{3}{2}\operatorname{arcsinh}\left(\frac{1+x}{\sqrt{3}}\right)$$

Mathematica combined the first two terms of Example 4 (and the Maple result) into a single term by factoring. ∎

EXAMPLE 6 Use a CAS to evaluate $\int x(x^2 + 5)^8 \, dx$.

SOLUTION Maple and Mathematica give the same answer:

$$\tfrac{1}{18}x^{18} + \tfrac{5}{2}x^{16} + 50x^{14} + \tfrac{1750}{3}x^{12} + 4375x^{10} + 21875x^{8} + \tfrac{218750}{3}x^{6} + 156250x^{4} + \tfrac{390625}{2}x^{2}$$

It's clear that both systems must have expanded $(x^2 + 5)^8$ by the Binomial Theorem and then integrated each term.

If we integrate by hand instead, using the substitution $u = x^2 + 5$, we get

The TI-89 also produces this answer.

$$\int x(x^2 + 5)^8 \, dx = \tfrac{1}{18}(x^2 + 5)^9 + C$$

For most purposes, this is a more convenient form of the answer. ∎

EXAMPLE 7 Use a CAS to find $\int \sin^5 x \cos^2 x \, dx$.

SOLUTION In Example 7.2.2 we found that

$$\boxed{1} \qquad \int \sin^5 x \cos^2 x \, dx = -\tfrac{1}{3}\cos^3 x + \tfrac{2}{5}\cos^5 x - \tfrac{1}{7}\cos^7 x + C$$

Maple and the TI-89 report the answer

$$-\tfrac{1}{7}\sin^4 x \cos^3 x - \tfrac{4}{35}\sin^2 x \cos^3 x - \tfrac{8}{105}\cos^3 x$$

whereas Mathematica produces

$$-\tfrac{5}{64}\cos x - \tfrac{1}{192}\cos 3x + \tfrac{3}{320}\cos 5x - \tfrac{1}{448}\cos 7x$$

We suspect that there are trigonometric identities which show that these three answers are equivalent. Indeed, if we ask Maple and Mathematica to simplify their expressions using trigonometric identities, they ultimately produce the same form of the answer as in Equation 1. ∎

7.6 EXERCISES

1–4 Use the indicated entry in the Table of Integrals on the Reference Pages to evaluate the integral.

1. $\int_0^{\pi/2} \cos 5x \cos 2x \, dx$;   entry 80

2. $\int_0^1 \sqrt{x - x^2} \, dx$;   entry 113

3. $\int_1^2 \sqrt{4x^2 - 3} \, dx$;   entry 39

4. $\int_0^1 \tan^3(\pi x/6) \, dx$;   entry 69

5–32 Use the Table of Integrals on Reference Pages 6–10 to evaluate the integral.

5. $\int_0^{\pi/8} \arctan 2x \, dx$

6. $\int_0^2 x^2 \sqrt{4 - x^2} \, dx$

7. $\int \dfrac{\cos x}{\sin^2 x - 9} \, dx$

8. $\int \dfrac{e^x}{4 - e^{2x}} \, dx$

9. $\int \dfrac{\sqrt{9x^2 + 4}}{x^2} \, dx$

10. $\int \dfrac{\sqrt{2y^2 - 3}}{y^2} \, dy$

11. $\int_0^\pi \cos^6 \theta \, d\theta$

12. $\int x \sqrt{2 + x^4} \, dx$

13. $\int \dfrac{\arctan \sqrt{x}}{\sqrt{x}} \, dx$

14. $\int_0^\pi x^3 \sin x \, dx$

15. $\int \dfrac{\coth(1/y)}{y^2} \, dy$

16. $\int \dfrac{e^{3t}}{\sqrt{e^{2t} - 1}} \, dt$

17. $\int y \sqrt{6 + 4y - 4y^2} \, dy$

18. $\int \dfrac{dx}{2x^3 - 3x^2}$

19. $\int \sin^2 x \cos x \ln(\sin x) \, dx$

20. $\int \dfrac{\sin 2\theta}{\sqrt{5 - \sin \theta}} \, d\theta$

21. $\int \dfrac{e^x}{3 - e^{2x}} \, dx$

22. $\int_0^2 x^3 \sqrt{4x^2 - x^4} \, dx$

23. $\int \sec^5 x \, dx$

24. $\int x^3 \arcsin(x^2) \, dx$

25. $\int \dfrac{\sqrt{4 + (\ln x)^2}}{x} \, dx$

26. $\int_0^1 x^4 e^{-x} \, dx$

27. $\int \dfrac{\cos^{-1}(x^{-2})}{x^3} \, dx$

28. $\int \dfrac{dx}{\sqrt{1 - e^{2x}}}$

29. $\int \sqrt{e^{2x} - 1} \, dx$

30. $\int e^t \sin(\alpha t - 3) \, dt$

31. $\int \dfrac{x^4 \, dx}{\sqrt{x^{10} - 2}}$

32. $\int \dfrac{\sec^2 \theta \tan^2 \theta}{\sqrt{9 - \tan^2 \theta}} \, d\theta$

33. The region under the curve $y = \sin^2 x$ from 0 to π is rotated about the x-axis. Find the volume of the resulting solid.

34. Find the volume of the solid obtained when the region under the curve $y = \arcsin x$, $x \geq 0$, is rotated about the y-axis.

35. Verify Formula 53 in the Table of Integrals (a) by differentiation and (b) by using the substitution $t = a + bu$.

36. Verify Formula 31 (a) by differentiation and (b) by substituting $u = a \sin \theta$.

CAS 37–44 Use a computer algebra system to evaluate the integral. Compare the answer with the result of using tables. If the answers are not the same, show that they are equivalent.

37. $\int \sec^4 x \, dx$

38. $\int \csc^5 x \, dx$

39. $\int x^2 \sqrt{x^2 + 4} \, dx$

40. $\int \dfrac{dx}{e^x(3e^x + 2)}$

41. $\int \cos^4 x \, dx$

42. $\int x^2 \sqrt{1 - x^2} \, dx$

43. $\int \tan^5 x \, dx$

44. $\int \dfrac{1}{\sqrt{1 + \sqrt[3]{x}}} \, dx$

CAS 45. (a) Use the table of integrals to evaluate $F(x) = \int f(x) \, dx$, where
$$f(x) = \dfrac{1}{x\sqrt{1 - x^2}}$$

What is the domain of f and F?

(b) Use a CAS to evaluate $F(x)$. What is the domain of the function F that the CAS produces? Is there a discrepancy between this domain and the domain of the function F that you found in part (a)?

CAS 46. Computer algebra systems sometimes need a helping hand from human beings. Try to evaluate
$$\int (1 + \ln x) \sqrt{1 + (x \ln x)^2} \, dx$$

with a computer algebra system. If it doesn't return an answer, make a substitution that changes the integral into one that the CAS *can* evaluate.

DISCOVERY PROJECT CAS PATTERNS IN INTEGRALS

In this project a computer algebra system is used to investigate indefinite integrals of families of functions. By observing the patterns that occur in the integrals of several members of the family, you will first guess, and then prove, a general formula for the integral of any member of the family.

1. (a) Use a computer algebra system to evaluate the following integrals.

  (i) $\int \dfrac{1}{(x + 2)(x + 3)} \, dx$

  (ii) $\int \dfrac{1}{(x + 1)(x + 5)} \, dx$

  (iii) $\int \dfrac{1}{(x + 2)(x - 5)} \, dx$

  (iv) $\int \dfrac{1}{(x + 2)^2} \, dx$

(b) Based on the pattern of your responses in part (a), guess the value of the integral
$$\int \frac{1}{(x+a)(x+b)}\,dx$$
if $a \ne b$. What if $a = b$?

(c) Check your guess by asking your CAS to evaluate the integral in part (b). Then prove it using partial fractions.

2. (a) Use a computer algebra system to evaluate the following integrals.

   (i) $\int \sin x \cos 2x\, dx$ (ii) $\int \sin 3x \cos 7x\, dx$ (iii) $\int \sin 8x \cos 3x\, dx$

   (b) Based on the pattern of your responses in part (a), guess the value of the integral
   $$\int \sin ax \cos bx\, dx$$

   (c) Check your guess with a CAS. Then prove it using the techniques of Section 7.2. For what values of a and b is it valid?

3. (a) Use a computer algebra system to evaluate the following integrals.

   (i) $\int \ln x\, dx$ (ii) $\int x \ln x\, dx$ (iii) $\int x^2 \ln x\, dx$

   (iv) $\int x^3 \ln x\, dx$ (v) $\int x^7 \ln x\, dx$

   (b) Based on the pattern of your responses in part (a), guess the value of
   $$\int x^n \ln x\, dx$$

   (c) Use integration by parts to prove the conjecture that you made in part (b). For what values of n is it valid?

4. (a) Use a computer algebra system to evaluate the following integrals.

   (i) $\int xe^x\, dx$ (ii) $\int x^2 e^x\, dx$ (iii) $\int x^3 e^x\, dx$

   (iv) $\int x^4 e^x\, dx$ (v) $\int x^5 e^x\, dx$

   (b) Based on the pattern of your responses in part (a), guess the value of $\int x^6 e^x\, dx$. Then use your CAS to check your guess.

   (c) Based on the patterns in parts (a) and (b), make a conjecture as to the value of the integral
   $$\int x^n e^x\, dx$$
   when n is a positive integer.

   (d) Use mathematical induction to prove the conjecture you made in part (c).

7.7 Approximate Integration

There are two situations in which it is impossible to find the exact value of a definite integral.

The first situation arises from the fact that in order to evaluate $\int_a^b f(x)\, dx$ using the Fundamental Theorem of Calculus we need to know an antiderivative of f. Sometimes, however, it is difficult, or even impossible, to find an antiderivative (see Section 7.5). For example, it is impossible to evaluate the following integrals exactly:

$$\int_0^1 e^{x^2}\,dx \qquad \int_{-1}^1 \sqrt{1+x^3}\,dx$$

The second situation arises when the function is determined from a scientific experiment through instrument readings or collected data. There may be no formula for the function (see Example 5).

In both cases we need to find approximate values of definite integrals. We already know one such method. Recall that the definite integral is defined as a limit of Riemann sums, so any Riemann sum could be used as an approximation to the integral: If we divide $[a, b]$ into n subintervals of equal length $\Delta x = (b - a)/n$, then we have

$$\int_a^b f(x)\,dx \approx \sum_{i=1}^n f(x_i^*)\,\Delta x$$

where x_i^* is any point in the ith subinterval $[x_{i-1}, x_i]$. If x_i^* is chosen to be the left endpoint of the interval, then $x_i^* = x_{i-1}$ and we have

$$\boxed{1} \qquad \int_a^b f(x)\,dx \approx L_n = \sum_{i=1}^n f(x_{i-1})\,\Delta x$$

If $f(x) \geq 0$, then the integral represents an area and (1) represents an approximation of this area by the rectangles shown in Figure 1(a). If we choose x_i^* to be the right endpoint, then $x_i^* = x_i$ and we have

$$\boxed{2} \qquad \int_a^b f(x)\,dx \approx R_n = \sum_{i=1}^n f(x_i)\,\Delta x$$

[See Figure 1(b).] The approximations L_n and R_n defined by Equations 1 and 2 are called the **left endpoint approximation** and **right endpoint approximation**, respectively.

In Section 5.2 we also considered the case where x_i^* is chosen to be the midpoint $\bar{x}_i$ of the subinterval $[x_{i-1}, x_i]$. Figure 1(c) shows the midpoint approximation M_n, which appears to be better than either L_n or R_n.

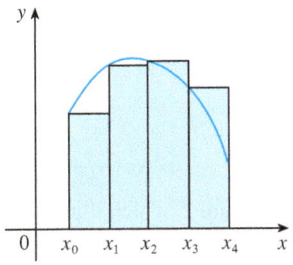

(a) Left endpoint approximation

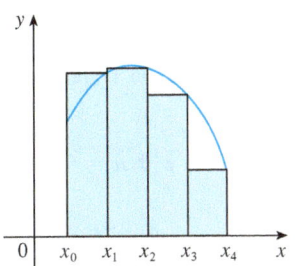

(b) Right endpoint approximation

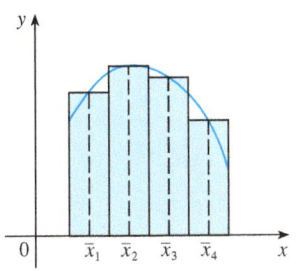

(c) Midpoint approximation

FIGURE 1

Midpoint Rule

$$\int_a^b f(x)\,dx \approx M_n = \Delta x\,[f(\bar{x}_1) + f(\bar{x}_2) + \cdots + f(\bar{x}_n)]$$

where

$$\Delta x = \frac{b-a}{n}$$

and

$$\bar{x}_i = \tfrac{1}{2}(x_{i-1} + x_i) = \text{midpoint of } [x_{i-1}, x_i]$$

Another approximation, called the Trapezoidal Rule, results from averaging the approximations in Equations 1 and 2:

$$\int_a^b f(x)\,dx \approx \frac{1}{2}\left[\sum_{i=1}^n f(x_{i-1})\,\Delta x + \sum_{i=1}^n f(x_i)\,\Delta x\right] = \frac{\Delta x}{2}\left[\sum_{i=1}^n \bigl(f(x_{i-1}) + f(x_i)\bigr)\right]$$

$$= \frac{\Delta x}{2}\bigl[\bigl(f(x_0) + f(x_1)\bigr) + \bigl(f(x_1) + f(x_2)\bigr) + \cdots + \bigl(f(x_{n-1}) + f(x_n)\bigr)\bigr]$$

$$= \frac{\Delta x}{2}[f(x_0) + 2f(x_1) + 2f(x_2) + \cdots + 2f(x_{n-1}) + f(x_n)]$$

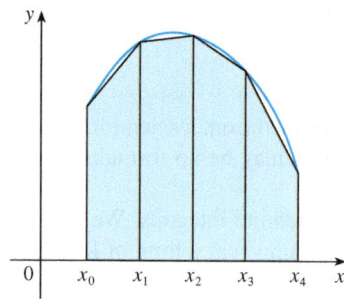

FIGURE 2
Trapezoidal approximation

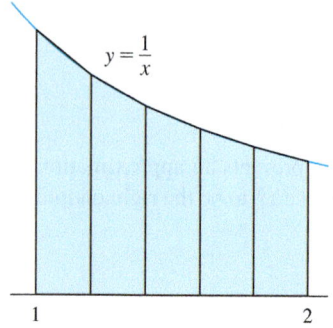

FIGURE 3

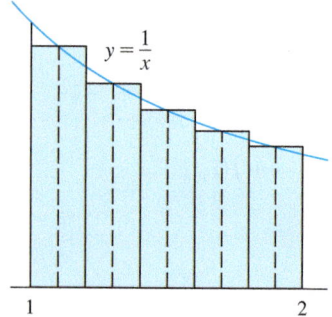

FIGURE 4

$\int_a^b f(x)\,dx = \text{approximation} + \text{error}$

> **Trapezoidal Rule**
>
> $$\int_a^b f(x)\,dx \approx T_n = \frac{\Delta x}{2}[f(x_0) + 2f(x_1) + 2f(x_2) + \cdots + 2f(x_{n-1}) + f(x_n)]$$
>
> where $\Delta x = (b - a)/n$ and $x_i = a + i\,\Delta x$.

The reason for the name Trapezoidal Rule can be seen from Figure 2, which illustrates the case with $f(x) \geq 0$ and $n = 4$. The area of the trapezoid that lies above the ith subinterval is

$$\Delta x \left(\frac{f(x_{i-1}) + f(x_i)}{2}\right) = \frac{\Delta x}{2}[f(x_{i-1}) + f(x_i)]$$

and if we add the areas of all these trapezoids, we get the right side of the Trapezoidal Rule.

EXAMPLE 1 Use (a) the Trapezoidal Rule and (b) the Midpoint Rule with $n = 5$ to approximate the integral $\int_1^2 (1/x)\,dx$.

SOLUTION
(a) With $n = 5$, $a = 1$, and $b = 2$, we have $\Delta x = (2 - 1)/5 = 0.2$, and so the Trapezoidal Rule gives

$$\int_1^2 \frac{1}{x}\,dx \approx T_5 = \frac{0.2}{2}[f(1) + 2f(1.2) + 2f(1.4) + 2f(1.6) + 2f(1.8) + f(2)]$$

$$= 0.1\left(\frac{1}{1} + \frac{2}{1.2} + \frac{2}{1.4} + \frac{2}{1.6} + \frac{2}{1.8} + \frac{1}{2}\right)$$

$$\approx 0.695635$$

This approximation is illustrated in Figure 3.

(b) The midpoints of the five subintervals are 1.1, 1.3, 1.5, 1.7, and 1.9, so the Midpoint Rule gives

$$\int_1^2 \frac{1}{x}\,dx \approx \Delta x\,[f(1.1) + f(1.3) + f(1.5) + f(1.7) + f(1.9)]$$

$$= \frac{1}{5}\left(\frac{1}{1.1} + \frac{1}{1.3} + \frac{1}{1.5} + \frac{1}{1.7} + \frac{1}{1.9}\right)$$

$$\approx 0.691908$$

This approximation is illustrated in Figure 4. ∎

In Example 1 we deliberately chose an integral whose value can be computed explicitly so that we can see how accurate the Trapezoidal and Midpoint Rules are. By the Fundamental Theorem of Calculus,

$$\int_1^2 \frac{1}{x}\,dx = \ln x\Big]_1^2 = \ln 2 = 0.693147\ldots$$

The **error** in using an approximation is defined to be the amount that needs to be added to the approximation to make it exact. From the values in Example 1 we see that the errors in the Trapezoidal and Midpoint Rule approximations for $n = 5$ are

$$E_T \approx -0.002488 \quad \text{and} \quad E_M \approx 0.001239$$

In general, we have

$$E_T = \int_a^b f(x)\,dx - T_n \quad \text{and} \quad E_M = \int_a^b f(x)\,dx - M_n$$

TEC Module 5.2/7.7 allows you to compare approximation methods.

The following tables show the results of calculations similar to those in Example 1, but for $n = 5$, 10, and 20 and for the left and right endpoint approximations as well as the Trapezoidal and Midpoint Rules.

Approximations to $\int_1^2 \frac{1}{x}\,dx$

n	L_n	R_n	T_n	M_n
5	0.745635	0.645635	0.695635	0.691908
10	0.718771	0.668771	0.693771	0.692835
20	0.705803	0.680803	0.693303	0.693069

Corresponding errors

n	E_L	E_R	E_T	E_M
5	-0.052488	0.047512	-0.002488	0.001239
10	-0.025624	0.024376	-0.000624	0.000312
20	-0.012656	0.012344	-0.000156	0.000078

It turns out that these observations are true in most cases.

We can make several observations from these tables:

1. In all of the methods we get more accurate approximations when we increase the value of n. (But very large values of n result in so many arithmetic operations that we have to beware of accumulated round-off error.)

2. The errors in the left and right endpoint approximations are opposite in sign and appear to decrease by a factor of about 2 when we double the value of n.

3. The Trapezoidal and Midpoint Rules are much more accurate than the endpoint approximations.

4. The errors in the Trapezoidal and Midpoint Rules are opposite in sign and appear to decrease by a factor of about 4 when we double the value of n.

5. The size of the error in the Midpoint Rule is about half the size of the error in the Trapezoidal Rule.

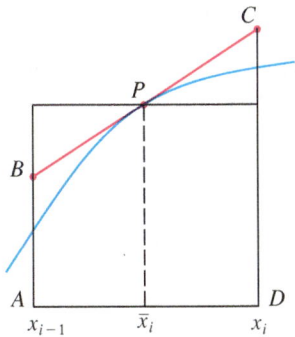

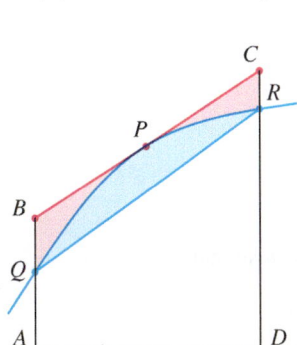

FIGURE 5

Figure 5 shows why we can usually expect the Midpoint Rule to be more accurate than the Trapezoidal Rule. The area of a typical rectangle in the Midpoint Rule is the same as the area of the trapezoid $ABCD$ whose upper side is tangent to the graph at P. The area of this trapezoid is closer to the area under the graph than is the area of the trapezoid $AQRD$ used in the Trapezoidal Rule. [The midpoint error (shaded red) is smaller than the trapezoidal error (shaded blue).]

These observations are corroborated in the following error estimates, which are proved in books on numerical analysis. Notice that Observation 4 corresponds to the n^2 in each denominator because $(2n)^2 = 4n^2$. The fact that the estimates depend on the size of the second derivative is not surprising if you look at Figure 5, because $f''(x)$ measures how much the graph is curved. [Recall that $f''(x)$ measures how fast the slope of $y = f(x)$ changes.]

3 Error Bounds Suppose $|f''(x)| \leq K$ for $a \leq x \leq b$. If E_T and E_M are the errors in the Trapezoidal and Midpoint Rules, then

$$|E_T| \leq \frac{K(b-a)^3}{12n^2} \quad \text{and} \quad |E_M| \leq \frac{K(b-a)^3}{24n^2}$$

Let's apply this error estimate to the Trapezoidal Rule approximation in Example 1. If $f(x) = 1/x$, then $f'(x) = -1/x^2$ and $f''(x) = 2/x^3$. Because $1 \leq x \leq 2$, we have $1/x \leq 1$, so

$$|f''(x)| = \left|\frac{2}{x^3}\right| \leq \frac{2}{1^3} = 2$$

Therefore, taking $K = 2$, $a = 1$, $b = 2$, and $n = 5$ in the error estimate (3), we see that

$$|E_T| \leq \frac{2(2-1)^3}{12(5)^2} = \frac{1}{150} \approx 0.006667$$

K can be any number larger than all the values of $|f''(x)|$, but smaller values of K give better error bounds.

Comparing this error estimate of 0.006667 with the actual error of about 0.002488, we see that it can happen that the actual error is substantially less than the upper bound for the error given by (3).

EXAMPLE 2 How large should we take n in order to guarantee that the Trapezoidal and Midpoint Rule approximations for $\int_1^2 (1/x)\, dx$ are accurate to within 0.0001?

SOLUTION We saw in the preceding calculation that $|f''(x)| \leq 2$ for $1 \leq x \leq 2$, so we can take $K = 2$, $a = 1$, and $b = 2$ in (3). Accuracy to within 0.0001 means that the size of the error should be less than 0.0001. Therefore we choose n so that

$$\frac{2(1)^3}{12n^2} < 0.0001$$

Solving the inequality for n, we get

$$n^2 > \frac{2}{12(0.0001)}$$

or

$$n > \frac{1}{\sqrt{0.0006}} \approx 40.8$$

It's quite possible that a lower value for n would suffice, but 41 is the smallest value for which the error bound formula can guarantee us accuracy to within 0.0001.

Thus $n = 41$ will ensure the desired accuracy.

For the same accuracy with the Midpoint Rule we choose n so that

$$\frac{2(1)^3}{24n^2} < 0.0001 \quad \text{and so} \quad n > \frac{1}{\sqrt{0.0012}} \approx 29 \quad \blacksquare$$

SECTION 7.7 Approximate Integration 519

EXAMPLE 3
(a) Use the Midpoint Rule with $n = 10$ to approximate the integral $\int_0^1 e^{x^2} dx$.
(b) Give an upper bound for the error involved in this approximation.

SOLUTION
(a) Since $a = 0$, $b = 1$, and $n = 10$, the Midpoint Rule gives

$$\int_0^1 e^{x^2} dx \approx \Delta x \left[f(0.05) + f(0.15) + \cdots + f(0.85) + f(0.95) \right]$$

$$= 0.1[e^{0.0025} + e^{0.0225} + e^{0.0625} + e^{0.1225} + e^{0.2025} + e^{0.3025}$$

$$+ e^{0.4225} + e^{0.5625} + e^{0.7225} + e^{0.9025}]$$

$$\approx 1.460393$$

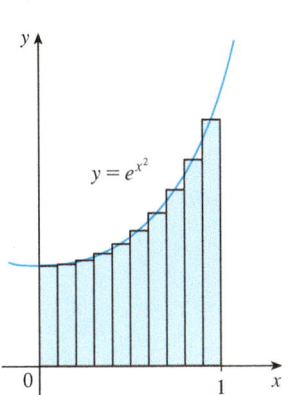

FIGURE 6

Figure 6 illustrates this approximation.

(b) Since $f(x) = e^{x^2}$, we have $f'(x) = 2xe^{x^2}$ and $f''(x) = (2 + 4x^2)e^{x^2}$. Also, since $0 \leq x \leq 1$, we have $x^2 \leq 1$ and so

$$0 \leq f''(x) = (2 + 4x^2)e^{x^2} \leq 6e$$

Error estimates give upper bounds for the error. They are theoretical, worst-case scenarios. The actual error in this case turns out to be about 0.0023.

Taking $K = 6e$, $a = 0$, $b = 1$, and $n = 10$ in the error estimate (3), we see that an upper bound for the error is

$$\frac{6e(1)^3}{24(10)^2} = \frac{e}{400} \approx 0.007$$

■

Simpson's Rule

Another rule for approximate integration results from using parabolas instead of straight line segments to approximate a curve. As before, we divide $[a, b]$ into n subintervals of equal length $h = \Delta x = (b - a)/n$, but this time we assume that n is an *even* number. Then on each consecutive pair of intervals we approximate the curve $y = f(x) \geq 0$ by a parabola as shown in Figure 7. If $y_i = f(x_i)$, then $P_i(x_i, y_i)$ is the point on the curve lying above x_i. A typical parabola passes through three consecutive points P_i, P_{i+1}, and P_{i+2}.

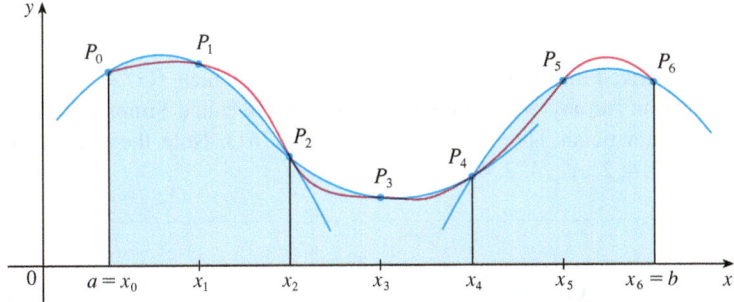

FIGURE 7

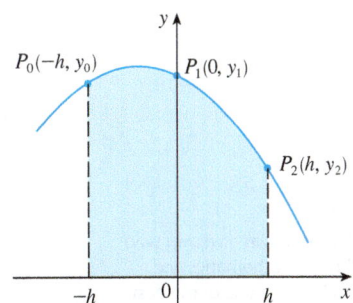

FIGURE 8

To simplify our calculations, we first consider the case where $x_0 = -h$, $x_1 = 0$, and $x_2 = h$. (See Figure 8.) We know that the equation of the parabola through P_0, P_1, and

P_2 is of the form $y = Ax^2 + Bx + C$ and so the area under the parabola from $x = -h$ to $x = h$ is

$$\int_{-h}^{h} (Ax^2 + Bx + C)\, dx = 2\int_{0}^{h} (Ax^2 + C)\, dx = 2\left[A\frac{x^3}{3} + Cx \right]_{0}^{h}$$

$$= 2\left(A\frac{h^3}{3} + Ch \right) = \frac{h}{3}(2Ah^2 + 6C)$$

Here we have used Theorem 5.5.7. Notice that $Ax^2 + C$ is even and Bx is odd.

But, since the parabola passes through $P_0(-h, y_0)$, $P_1(0, y_1)$, and $P_2(h, y_2)$, we have

$$y_0 = A(-h)^2 + B(-h) + C = Ah^2 - Bh + C$$

$$y_1 = C$$

$$y_2 = Ah^2 + Bh + C$$

and therefore

$$y_0 + 4y_1 + y_2 = 2Ah^2 + 6C$$

Thus we can rewrite the area under the parabola as

$$\frac{h}{3}(y_0 + 4y_1 + y_2)$$

Now by shifting this parabola horizontally we do not change the area under it. This means that the area under the parabola through P_0, P_1, and P_2 from $x = x_0$ to $x = x_2$ in Figure 7 is still

$$\frac{h}{3}(y_0 + 4y_1 + y_2)$$

Similarly, the area under the parabola through P_2, P_3, and P_4 from $x = x_2$ to $x = x_4$ is

$$\frac{h}{3}(y_2 + 4y_3 + y_4)$$

If we compute the areas under all the parabolas in this manner and add the results, we get

$$\int_{a}^{b} f(x)\, dx \approx \frac{h}{3}(y_0 + 4y_1 + y_2) + \frac{h}{3}(y_2 + 4y_3 + y_4) + \cdots + \frac{h}{3}(y_{n-2} + 4y_{n-1} + y_n)$$

$$= \frac{h}{3}(y_0 + 4y_1 + 2y_2 + 4y_3 + 2y_4 + \cdots + 2y_{n-2} + 4y_{n-1} + y_n)$$

Although we have derived this approximation for the case in which $f(x) \geq 0$, it is a reasonable approximation for any continuous function f and is called Simpson's Rule after the English mathematician Thomas Simpson (1710–1761). Note the pattern of coefficients: 1, 4, 2, 4, 2, 4, 2, ..., 4, 2, 4, 1.

Simpson

Thomas Simpson was a weaver who taught himself mathematics and went on to become one of the best English mathematicians of the 18th century. What we call Simpson's Rule was actually known to Cavalieri and Gregory in the 17th century, but Simpson popularized it in his book *Mathematical Dissertations* (1743).

Simpson's Rule

$$\int_{a}^{b} f(x)\, dx \approx S_n = \frac{\Delta x}{3}[f(x_0) + 4f(x_1) + 2f(x_2) + 4f(x_3) + \cdots + 2f(x_{n-2}) + 4f(x_{n-1}) + f(x_n)]$$

where n is even and $\Delta x = (b - a)/n$.

EXAMPLE 4 Use Simpson's Rule with $n = 10$ to approximate $\int_1^2 (1/x)\, dx$.

SOLUTION Putting $f(x) = 1/x$, $n = 10$, and $\Delta x = 0.1$ in Simpson's Rule, we obtain

$$\int_1^2 \frac{1}{x}\, dx \approx S_{10}$$

$$= \frac{\Delta x}{3}[f(1) + 4f(1.1) + 2f(1.2) + 4f(1.3) + \cdots + 2f(1.8) + 4f(1.9) + f(2)]$$

$$= \frac{0.1}{3}\left(\frac{1}{1} + \frac{4}{1.1} + \frac{2}{1.2} + \frac{4}{1.3} + \frac{2}{1.4} + \frac{4}{1.5} + \frac{2}{1.6} + \frac{4}{1.7} + \frac{2}{1.8} + \frac{4}{1.9} + \frac{1}{2}\right)$$

$$\approx 0.693150$$

Notice that, in Example 4, Simpson's Rule gives us a *much* better approximation ($S_{10} \approx 0.693150$) to the true value of the integral ($\ln 2 \approx 0.693147\ldots$) than does the Trapezoidal Rule ($T_{10} \approx 0.693771$) or the Midpoint Rule ($M_{10} \approx 0.692835$). It turns out (see Exercise 50) that the approximations in Simpson's Rule are weighted averages of those in the Trapezoidal and Midpoint Rules:

$$S_{2n} = \tfrac{1}{3}T_n + \tfrac{2}{3}M_n$$

(Recall that E_T and E_M usually have opposite signs and $|E_M|$ is about half the size of $|E_T|$.)

In many applications of calculus we need to evaluate an integral even if no explicit formula is known for y as a function of x. A function may be given graphically or as a table of values of collected data. If there is evidence that the values are not changing rapidly, then the Trapezoidal Rule or Simpson's Rule can still be used to find an approximate value for $\int_a^b y\, dx$, the integral of y with respect to x.

EXAMPLE 5 Figure 9 shows data traffic on the link from the United States to SWITCH, the Swiss academic and research network, on February 10, 1998. $D(t)$ is the data throughput, measured in megabits per second (Mb/s). Use Simpson's Rule to estimate the total amount of data transmitted on the link from midnight to noon on that day.

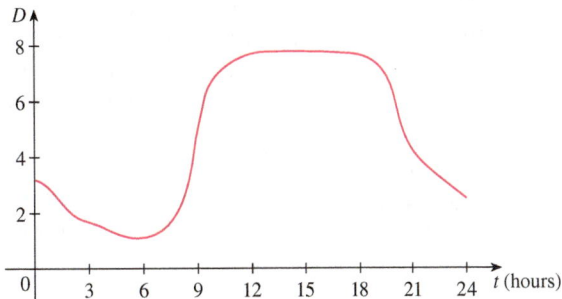

FIGURE 9

SOLUTION Because we want the units to be consistent and $D(t)$ is measured in megabits per second, we convert the units for t from hours to seconds. If we let $A(t)$ be the amount of data (in megabits) transmitted by time t, where t is measured in seconds, then $A'(t) = D(t)$. So, by the Net Change Theorem (see Section 5.4), the total amount of data transmitted by noon (when $t = 12 \times 60^2 = 43{,}200$) is

$$A(43{,}200) = \int_0^{43{,}200} D(t)\, dt$$

We estimate the values of $D(t)$ at hourly intervals from the graph and compile them in the table.

t (hours)	t (seconds)	$D(t)$	t (hours)	t (seconds)	$D(t)$
0	0	3.2	7	25,200	1.3
1	3,600	2.7	8	28,800	2.8
2	7,200	1.9	9	32,400	5.7
3	10,800	1.7	10	36,000	7.1
4	14,400	1.3	11	39,600	7.7
5	18,000	1.0	12	43,200	7.9
6	21,600	1.1			

Then we use Simpson's Rule with $n = 12$ and $\Delta t = 3600$ to estimate the integral:

$$\int_0^{43{,}200} A(t)\, dt \approx \frac{\Delta t}{3}[D(0) + 4D(3600) + 2D(7200) + \cdots + 4D(39{,}600) + D(43{,}200)]$$

$$\approx \frac{3600}{3}[3.2 + 4(2.7) + 2(1.9) + 4(1.7) + 2(1.3) + 4(1.0)$$
$$+ 2(1.1) + 4(1.3) + 2(2.8) + 4(5.7) + 2(7.1) + 4(7.7) + 7.9]$$

$$= 143{,}880$$

Thus the total amount of data transmitted from midnight to noon is about 144,000 megabits, or 144 gigabits. ∎

n	M_n	S_n
4	0.69121989	0.69315453
8	0.69266055	0.69314765
16	0.69302521	0.69314721

n	E_M	E_S
4	0.00192729	−0.00000735
8	0.00048663	−0.00000047
16	0.00012197	−0.00000003

The table in the margin shows how Simpson's Rule compares with the Midpoint Rule for the integral $\int_1^2 (1/x)\, dx$, whose value is about 0.69314718. The second table shows how the error E_S in Simpson's Rule decreases by a factor of about 16 when n is doubled. (In Exercises 27 and 28 you are asked to verify this for two additional integrals.) That is consistent with the appearance of n^4 in the denominator of the following error estimate for Simpson's Rule. It is similar to the estimates given in (3) for the Trapezoidal and Midpoint Rules, but it uses the fourth derivative of f.

4 Error Bound for Simpson's Rule Suppose that $|f^{(4)}(x)| \leq K$ for $a \leq x \leq b$. If E_S is the error involved in using Simpson's Rule, then

$$|E_S| \leq \frac{K(b-a)^5}{180 n^4}$$

EXAMPLE 6 How large should we take n in order to guarantee that the Simpson's Rule approximation for $\int_1^2 (1/x)\, dx$ is accurate to within 0.0001?

SOLUTION If $f(x) = 1/x$, then $f^{(4)}(x) = 24/x^5$. Since $x \geq 1$, we have $1/x \leq 1$ and so

$$|f^{(4)}(x)| = \left|\frac{24}{x^5}\right| \leq 24$$

Many calculators and computer algebra systems have a built-in algorithm that computes an approximation of a definite integral. Some of these machines use Simpson's Rule; others use more sophisticated techniques such as *adaptive numerical integration*. This means that if a function fluctuates much more on a certain part of the interval than it does elsewhere, then that part gets divided into more subintervals. This strategy reduces the number of calculations required to achieve a prescribed accuracy.

Therefore we can take $K = 24$ in (4). Thus, for an error less than 0.0001, we should choose n so that

$$\frac{24(1)^5}{180n^4} < 0.0001$$

This gives

$$n^4 > \frac{24}{180(0.0001)}$$

or

$$n > \frac{1}{\sqrt[4]{0.00075}} \approx 6.04$$

Therefore $n = 8$ (n must be even) gives the desired accuracy. (Compare this with Example 2, where we obtained $n = 41$ for the Trapezoidal Rule and $n = 29$ for the Midpoint Rule.) ∎

EXAMPLE 7
(a) Use Simpson's Rule with $n = 10$ to approximate the integral $\int_0^1 e^{x^2} dx$.
(b) Estimate the error involved in this approximation.

SOLUTION
(a) If $n = 10$, then $\Delta x = 0.1$ and Simpson's Rule gives

$$\int_0^1 e^{x^2} dx \approx \frac{\Delta x}{3}[f(0) + 4f(0.1) + 2f(0.2) + \cdots + 2f(0.8) + 4f(0.9) + f(1)]$$

$$= \frac{0.1}{3}[e^0 + 4e^{0.01} + 2e^{0.04} + 4e^{0.09} + 2e^{0.16} + 4e^{0.25} + 2e^{0.36}$$

$$+ 4e^{0.49} + 2e^{0.64} + 4e^{0.81} + e^1]$$

$$\approx 1.462681$$

(b) The fourth derivative of $f(x) = e^{x^2}$ is

$$f^{(4)}(x) = (12 + 48x^2 + 16x^4)e^{x^2}$$

and so, since $0 \leq x \leq 1$, we have

$$0 \leq f^{(4)}(x) \leq (12 + 48 + 16)e^1 = 76e$$

Figure 10 illustrates the calculation in Example 7. Notice that the parabolic arcs are so close to the graph of $y = e^{x^2}$ that they are practically indistinguishable from it.

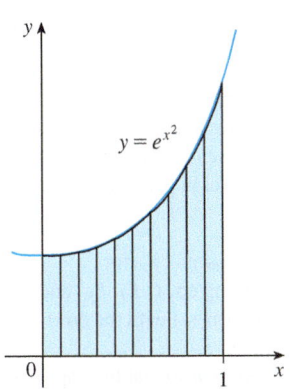

FIGURE 10

Therefore, putting $K = 76e$, $a = 0$, $b = 1$, and $n = 10$ in (4), we see that the error is at most

$$\frac{76e(1)^5}{180(10)^4} \approx 0.000115$$

(Compare this with Example 3.) Thus, correct to three decimal places, we have

$$\int_0^1 e^{x^2} dx \approx 1.463$$

∎

7.7 EXERCISES

1. Let $I = \int_0^4 f(x)\, dx$, where f is the function whose graph is shown.
   (a) Use the graph to find L_2, R_2, and M_2.
   (b) Are these underestimates or overestimates of I?
   (c) Use the graph to find T_2. How does it compare with I?
   (d) For any value of n, list the numbers L_n, R_n, M_n, T_n, and I in increasing order.

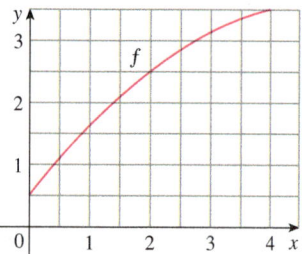

2. The left, right, Trapezoidal, and Midpoint Rule approximations were used to estimate $\int_0^2 f(x)\, dx$, where f is the function whose graph is shown. The estimates were 0.7811, 0.8675, 0.8632, and 0.9540, and the same number of subintervals were used in each case.
   (a) Which rule produced which estimate?
   (b) Between which two approximations does the true value of $\int_0^2 f(x)\, dx$ lie?

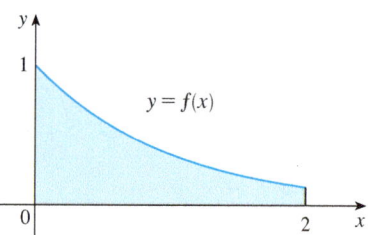

3. Estimate $\int_0^1 \cos(x^2)\, dx$ using (a) the Trapezoidal Rule and (b) the Midpoint Rule, each with $n = 4$. From a graph of the integrand, decide whether your answers are underestimates or overestimates. What can you conclude about the true value of the integral?

4. Draw the graph of $f(x) = \sin(\tfrac{1}{2} x^2)$ in the viewing rectangle $[0, 1]$ by $[0, 0.5]$ and let $I = \int_0^1 f(x)\, dx$.
   (a) Use the graph to decide whether L_2, R_2, M_2, and T_2 underestimate or overestimate I.
   (b) For any value of n, list the numbers L_n, R_n, M_n, T_n, and I in increasing order.
   (c) Compute L_5, R_5, M_5, and T_5. From the graph, which do you think gives the best estimate of I?

5–6 Use (a) the Midpoint Rule and (b) Simpson's Rule to approximate the given integral with the specified value of n. (Round your answers to six decimal places.) Compare your results to the actual value to determine the error in each approximation.

5. $\int_0^2 \dfrac{x}{1+x^2}\, dx$, $n = 10$

6. $\int_0^\pi x \cos x \, dx$, $n = 4$

7–18 Use (a) the Trapezoidal Rule, (b) the Midpoint Rule, and (c) Simpson's Rule to approximate the given integral with the specified value of n. (Round your answers to six decimal places.)

7. $\int_1^2 \sqrt{x^3 - 1}\, dx$, $n = 10$

8. $\int_0^2 \dfrac{1}{1 + x^6}\, dx$, $n = 8$

9. $\int_0^2 \dfrac{e^x}{1 + x^2}\, dx$, $n = 10$

10. $\int_0^{\pi/2} \sqrt[3]{1 + \cos x}\, dx$, $n = 4$

11. $\int_0^4 x^3 \sin x \, dx$, $n = 8$

12. $\int_1^3 e^{1/x}\, dx$, $n = 8$

13. $\int_0^4 \sqrt{y} \cos y \, dy$, $n = 8$

14. $\int_2^3 \dfrac{1}{\ln t}\, dt$, $n = 10$

15. $\int_0^1 \dfrac{x^2}{1 + x^4}\, dx$, $n = 10$

16. $\int_1^3 \dfrac{\sin t}{t}\, dt$, $n = 4$

17. $\int_0^4 \ln(1 + e^x)\, dx$, $n = 8$

18. $\int_0^1 \sqrt{x + x^3}\, dx$, $n = 10$

19. (a) Find the approximations T_8 and M_8 for the integral $\int_0^1 \cos(x^2)\, dx$.
    (b) Estimate the errors in the approximations of part (a).
    (c) How large do we have to choose n so that the approximations T_n and M_n to the integral in part (a) are accurate to within 0.0001?

20. (a) Find the approximations T_{10} and M_{10} for $\int_1^2 e^{1/x}\, dx$.
    (b) Estimate the errors in the approximations of part (a).
    (c) How large do we have to choose n so that the approximations T_n and M_n to the integral in part (a) are accurate to within 0.0001?

21. (a) Find the approximations T_{10}, M_{10}, and S_{10} for $\int_0^\pi \sin x \, dx$ and the corresponding errors E_T, E_M, and E_S.
    (b) Compare the actual errors in part (a) with the error estimates given by (3) and (4).
    (c) How large do we have to choose n so that the approximations T_n, M_n, and S_n to the integral in part (a) are accurate to within 0.00001?

22. How large should n be to guarantee that the Simpson's Rule approximation to $\int_0^1 e^{x^2}\, dx$ is accurate to within 0.00001?

23. The trouble with the error estimates is that it is often very difficult to compute four derivatives and obtain a good upper bound K for $|f^{(4)}(x)|$ by hand. But computer algebra systems have no problem computing $f^{(4)}$ and graphing it, so we can easily find a value for K from a machine graph. This exercise deals with approximations to the integral $I = \int_0^{2\pi} f(x)\, dx$, where $f(x) = e^{\cos x}$.
(a) Use a graph to get a good upper bound for $|f''(x)|$.
(b) Use M_{10} to approximate I.
(c) Use part (a) to estimate the error in part (b).
(d) Use the built-in numerical integration capability of your CAS to approximate I.
(e) How does the actual error compare with the error estimate in part (c)?
(f) Use a graph to get a good upper bound for $|f^{(4)}(x)|$.
(g) Use S_{10} to approximate I.
(h) Use part (f) to estimate the error in part (g).
(i) How does the actual error compare with the error estimate in part (h)?
(j) How large should n be to guarantee that the size of the error in using S_n is less than 0.0001?

24. Repeat Exercise 23 for the integral $\int_{-1}^{1} \sqrt{4-x^3}\, dx$.

25–26 Find the approximations L_n, R_n, T_n, and M_n for $n = 5$, 10, and 20. Then compute the corresponding errors E_L, E_R, E_T, and E_M. (Round your answers to six decimal places. You may wish to use the sum command on a computer algebra system.) What observations can you make? In particular, what happens to the errors when n is doubled?

25. $\int_0^1 xe^x\, dx$ **26.** $\int_1^2 \dfrac{1}{x^2}\, dx$

27–28 Find the approximations T_n, M_n, and S_n for $n = 6$ and 12. Then compute the corresponding errors E_T, E_M, and E_S. (Round your answers to six decimal places. You may wish to use the sum command on a computer algebra system.) What observations can you make? In particular, what happens to the errors when n is doubled?

27. $\int_0^2 x^4\, dx$ **28.** $\int_1^4 \dfrac{1}{\sqrt{x}}\, dx$

29. Estimate the area under the graph in the figure by using (a) the Trapezoidal Rule, (b) the Midpoint Rule, and (c) Simpson's Rule, each with $n = 6$.

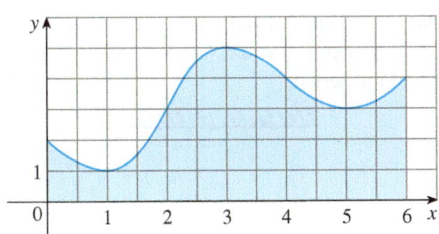

30. The widths (in meters) of a kidney-shaped swimming pool were measured at 2-meter intervals as indicated in the figure. Use Simpson's Rule to estimate the area of the pool.

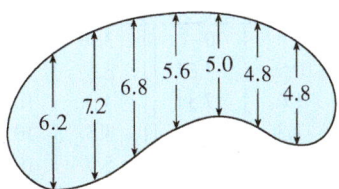

31. (a) Use the Midpoint Rule and the given data to estimate the value of the integral $\int_1^5 f(x)\, dx$.

x	$f(x)$	x	$f(x)$
1.0	2.4	3.5	4.0
1.5	2.9	4.0	4.1
2.0	3.3	4.5	3.9
2.5	3.6	5.0	3.5
3.0	3.8		

(b) If it is known that $-2 \leq f''(x) \leq 3$ for all x, estimate the error involved in the approximation in part (a).

32. (a) A table of values of a function g is given. Use Simpson's Rule to estimate $\int_0^{1.6} g(x)\, dx$.

x	$g(x)$	x	$g(x)$
0.0	12.1	1.0	12.2
0.2	11.6	1.2	12.6
0.4	11.3	1.4	13.0
0.6	11.1	1.6	13.2
0.8	11.7		

(b) If $-5 \leq g^{(4)}(x) \leq 2$ for $0 \leq x \leq 1.6$, estimate the error involved in the approximation in part (a).

33. A graph of the temperature in Boston on August 11, 2013, is shown. Use Simpson's Rule with $n = 12$ to estimate the average temperature on that day.

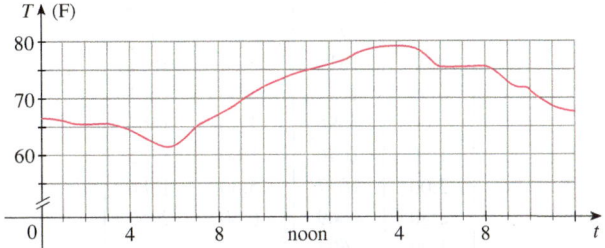

34. A radar gun was used to record the speed of a runner during the first 5 seconds of a race (see the table). Use Simpson's

Rule to estimate the distance the runner covered during those 5 seconds.

t (s)	v (m/s)	t (s)	v (m/s)
0	0	3.0	10.51
0.5	4.67	3.5	10.67
1.0	7.34	4.0	10.76
1.5	8.86	4.5	10.81
2.0	9.73	5.0	10.81
2.5	10.22		

35. The graph of the acceleration $a(t)$ of a car measured in ft/s² is shown. Use Simpson's Rule to estimate the increase in the velocity of the car during the 6-second time interval.

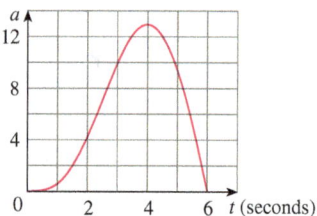

36. Water leaked from a tank at a rate of $r(t)$ liters per hour, where the graph of r is as shown. Use Simpson's Rule to estimate the total amount of water that leaked out during the first 6 hours.

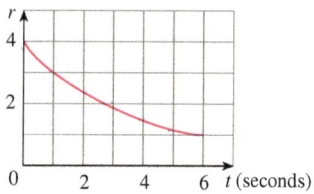

37. The table (supplied by San Diego Gas and Electric) gives the power consumption P in megawatts in San Diego County from midnight to 6:00 AM on a day in December. Use Simpson's Rule to estimate the energy used during that time period. (Use the fact that power is the derivative of energy.)

t	P	t	P
0:00	1814	3:30	1611
0:30	1735	4:00	1621
1:00	1686	4:30	1666
1:30	1646	5:00	1745
2:00	1637	5:30	1886
2:30	1609	6:00	2052
3:00	1604		

38. Shown is the graph of traffic on an Internet service provider's T1 data line from midnight to 8:00 AM. D is the data throughput, measured in megabits per second. Use Simpson's Rule to estimate the total amount of data transmitted during that time period.

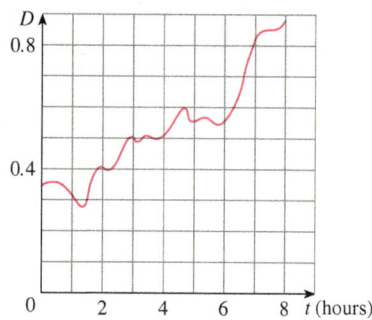

39. Use Simpson's Rule with $n = 8$ to estimate the volume of the solid obtained by rotating the region shown in the figure about (a) the x-axis and (b) the y-axis.

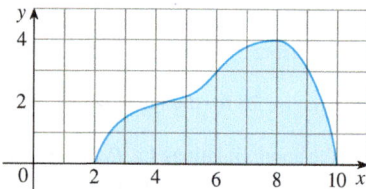

40. The table shows values of a force function $f(x)$, where x is measured in meters and $f(x)$ in newtons. Use Simpson's Rule to estimate the work done by the force in moving an object a distance of 18 m.

x	0	3	6	9	12	15	18
f(x)	9.8	9.1	8.5	8.0	7.7	7.5	7.4

41. The region bounded by the curve $y = 1/(1 + e^{-x})$, the x- and y-axes, and the line $x = 10$ is rotated about the x-axis. Use Simpson's Rule with $n = 10$ to estimate the volume of the resulting solid.

CAS 42. The figure shows a pendulum with length L that makes a maximum angle θ_0 with the vertical. Using Newton's Second Law, it can be shown that the period T (the time for one complete swing) is given by

$$T = 4\sqrt{\frac{L}{g}} \int_0^{\pi/2} \frac{dx}{\sqrt{1 - k^2 \sin^2 x}}$$

where $k = \sin(\tfrac{1}{2}\theta_0)$ and g is the acceleration due to gravity. If $L = 1$ m and $\theta_0 = 42°$, use Simpson's Rule with $n = 10$ to find the period.

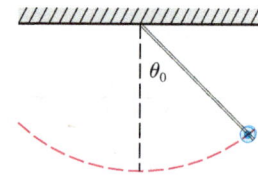

43. The intensity of light with wavelength λ traveling through a diffraction grating with N slits at an angle θ is given by $I(\theta) = N^2 \sin^2 k/k^2$, where $k = (\pi N d \sin\theta)/\lambda$ and d is the distance between adjacent slits. A helium-neon laser with wavelength $\lambda = 632.8 \times 10^{-9}$ m is emitting a narrow band of light, given by $-10^{-6} < \theta < 10^{-6}$, through a grating with 10,000 slits spaced 10^{-4} m apart. Use the Midpoint Rule with $n = 10$ to estimate the total light intensity $\int_{-10^{-6}}^{10^{-6}} I(\theta)\, d\theta$ emerging from the grating.

44. Use the Trapezoidal Rule with $n = 10$ to approximate $\int_0^{20} \cos(\pi x)\, dx$. Compare your result to the actual value. Can you explain the discrepancy?

45. Sketch the graph of a continuous function on [0, 2] for which the Trapezoidal Rule with $n = 2$ is more accurate than the Midpoint Rule.

46. Sketch the graph of a continuous function on [0, 2] for which the right endpoint approximation with $n = 2$ is more accurate than Simpson's Rule.

47. If f is a positive function and $f''(x) < 0$ for $a \leq x \leq b$, show that
$$T_n < \int_a^b f(x)\, dx < M_n$$

48. Show that if f is a polynomial of degree 3 or lower, then Simpson's Rule gives the exact value of $\int_a^b f(x)\, dx$.

49. Show that $\frac{1}{2}(T_n + M_n) = T_{2n}$.

50. Show that $\frac{1}{3}T_n + \frac{2}{3}M_n = S_{2n}$.

7.8 Improper Integrals

In defining a definite integral $\int_a^b f(x)\, dx$ we dealt with a function f defined on a finite interval $[a, b]$ and we assumed that f does not have an infinite discontinuity (see Section 5.2). In this section we extend the concept of a definite integral to the case where the interval is infinite and also to the case where f has an infinite discontinuity in $[a, b]$. In either case the integral is called an *improper* integral. One of the most important applications of this idea, probability distributions, will be studied in Section 8.5.

■ Type 1: Infinite Intervals

Consider the infinite region S that lies under the curve $y = 1/x^2$, above the x-axis, and to the right of the line $x = 1$. You might think that, since S is infinite in extent, its area must be infinite, but let's take a closer look. The area of the part of S that lies to the left of the line $x = t$ (shaded in Figure 1) is

$$A(t) = \int_1^t \frac{1}{x^2}\, dx = -\frac{1}{x}\bigg]_1^t = 1 - \frac{1}{t}$$

FIGURE 1

Notice that $A(t) < 1$ no matter how large t is chosen.

We also observe that

$$\lim_{t \to \infty} A(t) = \lim_{t \to \infty}\left(1 - \frac{1}{t}\right) = 1$$

The area of the shaded region approaches 1 as $t \to \infty$ (see Figure 2), so we say that the area of the infinite region S is equal to 1 and we write

$$\int_1^\infty \frac{1}{x^2}\, dx = \lim_{t \to \infty} \int_1^t \frac{1}{x^2}\, dx = 1$$

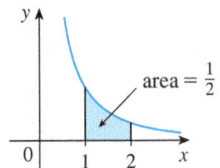
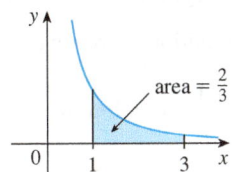
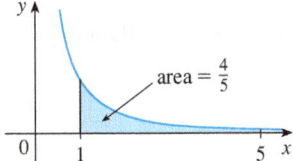
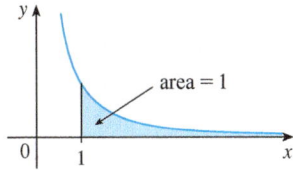

FIGURE 2

528 CHAPTER 7 Techniques of Integration

Using this example as a guide, we define the integral of f (not necessarily a positive function) over an infinite interval as the limit of integrals over finite intervals.

---

[1] Definition of an Improper Integral of Type 1

(a) If $\int_a^t f(x)\, dx$ exists for every number $t \geq a$, then

$$\int_a^\infty f(x)\, dx = \lim_{t \to \infty} \int_a^t f(x)\, dx$$

provided this limit exists (as a finite number).

(b) If $\int_t^b f(x)\, dx$ exists for every number $t \leq b$, then

$$\int_{-\infty}^b f(x)\, dx = \lim_{t \to -\infty} \int_t^b f(x)\, dx$$

provided this limit exists (as a finite number).

The improper integrals $\int_a^\infty f(x)\, dx$ and $\int_{-\infty}^b f(x)\, dx$ are called **convergent** if the corresponding limit exists and **divergent** if the limit does not exist.

(c) If both $\int_a^\infty f(x)\, dx$ and $\int_{-\infty}^a f(x)\, dx$ are convergent, then we define

$$\int_{-\infty}^\infty f(x)\, dx = \int_{-\infty}^a f(x)\, dx + \int_a^\infty f(x)\, dx$$

In part (c) any real number a can be used (see Exercise 76).

Any of the improper integrals in Definition 1 can be interpreted as an area provided that f is a positive function. For instance, in case (a) if $f(x) \geq 0$ and the integral $\int_a^\infty f(x)\, dx$ is convergent, then we define the area of the region $S = \{(x, y) \mid x \geq a, 0 \leq y \leq f(x)\}$ in Figure 3 to be

$$A(S) = \int_a^\infty f(x)\, dx$$

This is appropriate because $\int_a^\infty f(x)\, dx$ is the limit as $t \to \infty$ of the area under the graph of f from a to t.

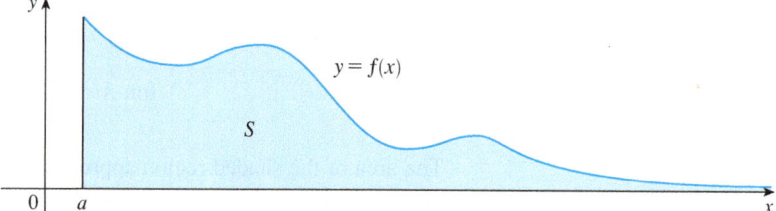

FIGURE 3

EXAMPLE 1 Determine whether the integral $\int_1^\infty (1/x)\, dx$ is convergent or divergent.

SOLUTION According to part (a) of Definition 1, we have

$$\int_1^\infty \frac{1}{x}\, dx = \lim_{t \to \infty} \int_1^t \frac{1}{x}\, dx = \lim_{t \to \infty} \ln|x|\Big]_1^t$$

$$= \lim_{t \to \infty} (\ln t - \ln 1) = \lim_{t \to \infty} \ln t = \infty$$

The limit does not exist as a finite number and so the improper integral $\int_1^\infty (1/x)\,dx$ is divergent.

Let's compare the result of Example 1 with the example given at the beginning of this section:

$$\int_1^\infty \frac{1}{x^2}\,dx \text{ converges} \qquad \int_1^\infty \frac{1}{x}\,dx \text{ diverges}$$

Geometrically, this says that although the curves $y = 1/x^2$ and $y = 1/x$ look very similar for $x > 0$, the region under $y = 1/x^2$ to the right of $x = 1$ (the shaded region in Figure 4) has finite area whereas the corresponding region under $y = 1/x$ (in Figure 5) has infinite area. Note that both $1/x^2$ and $1/x$ approach 0 as $x \to \infty$ but $1/x^2$ approaches 0 faster than $1/x$. The values of $1/x$ don't decrease fast enough for its integral to have a finite value.

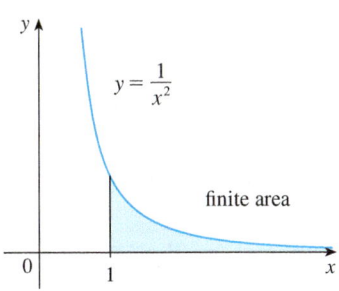

FIGURE 4
$\int_1^\infty (1/x^2)\,dx$ converges

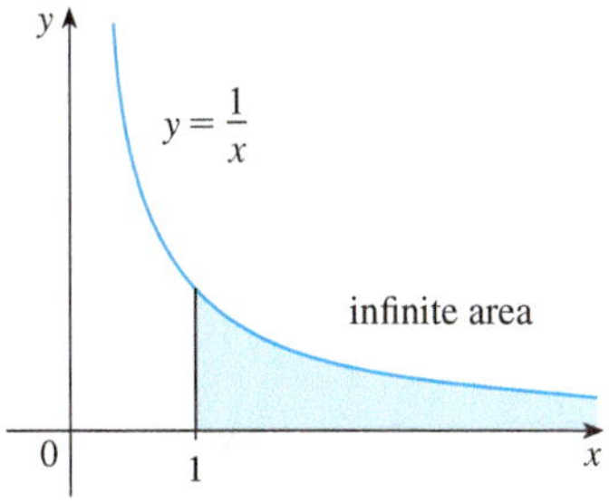

FIGURE 5
$\int_1^\infty (1/x)\,dx$ diverges

EXAMPLE 2 Evaluate $\int_{-\infty}^0 xe^x\,dx$.

SOLUTION Using part (b) of Definition 1, we have

$$\int_{-\infty}^0 xe^x\,dx = \lim_{t \to -\infty} \int_t^0 xe^x\,dx$$

We integrate by parts with $u = x$, $dv = e^x\,dx$ so that $du = dx$, $v = e^x$:

$$\int_t^0 xe^x\,dx = xe^x\Big]_t^0 - \int_t^0 e^x\,dx$$

$$= -te^t - 1 + e^t$$

TEC In Module 7.8 you can investigate visually and numerically whether several improper integrals are convergent or divergent.

We know that $e^t \to 0$ as $t \to -\infty$, and by l'Hospital's Rule we have

$$\lim_{t \to -\infty} te^t = \lim_{t \to -\infty} \frac{t}{e^{-t}} = \lim_{t \to -\infty} \frac{1}{-e^{-t}}$$

$$= \lim_{t \to -\infty} (-e^t) = 0$$

Therefore

$$\int_{-\infty}^0 xe^x\,dx = \lim_{t \to -\infty} (-te^t - 1 + e^t)$$

$$= -0 - 1 + 0 = -1$$

EXAMPLE 3 Evaluate $\int_{-\infty}^{\infty} \frac{1}{1+x^2}\, dx$.

SOLUTION It's convenient to choose $a = 0$ in Definition 1(c):

$$\int_{-\infty}^{\infty} \frac{1}{1+x^2}\, dx = \int_{-\infty}^{0} \frac{1}{1+x^2}\, dx + \int_{0}^{\infty} \frac{1}{1+x^2}\, dx$$

We must now evaluate the integrals on the right side separately:

$$\int_{0}^{\infty} \frac{1}{1+x^2}\, dx = \lim_{t \to \infty} \int_{0}^{t} \frac{dx}{1+x^2} = \lim_{t \to \infty} \tan^{-1}x \Big]_{0}^{t}$$

$$= \lim_{t \to \infty} (\tan^{-1}t - \tan^{-1}0) = \lim_{t \to \infty} \tan^{-1}t = \frac{\pi}{2}$$

$$\int_{-\infty}^{0} \frac{1}{1+x^2}\, dx = \lim_{t \to -\infty} \int_{t}^{0} \frac{dx}{1+x^2} = \lim_{t \to -\infty} \tan^{-1}x \Big]_{t}^{0}$$

$$= \lim_{t \to -\infty} (\tan^{-1}0 - \tan^{-1}t) = 0 - \left(-\frac{\pi}{2}\right) = \frac{\pi}{2}$$

Since both of these integrals are convergent, the given integral is convergent and

$$\int_{-\infty}^{\infty} \frac{1}{1+x^2}\, dx = \frac{\pi}{2} + \frac{\pi}{2} = \pi$$

Since $1/(1+x^2) > 0$, the given improper integral can be interpreted as the area of the infinite region that lies under the curve $y = 1/(1+x^2)$ and above the x-axis (see Figure 6).

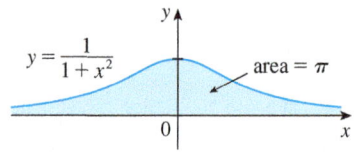

FIGURE 6

EXAMPLE 4 For what values of p is the integral

$$\int_{1}^{\infty} \frac{1}{x^p}\, dx$$

convergent?

SOLUTION We know from Example 1 that if $p = 1$, then the integral is divergent, so let's assume that $p \neq 1$. Then

$$\int_{1}^{\infty} \frac{1}{x^p}\, dx = \lim_{t \to \infty} \int_{1}^{t} x^{-p}\, dx = \lim_{t \to \infty} \frac{x^{-p+1}}{-p+1}\bigg]_{x=1}^{x=t}$$

$$= \lim_{t \to \infty} \frac{1}{1-p}\left[\frac{1}{t^{p-1}} - 1\right]$$

If $p > 1$, then $p - 1 > 0$, so as $t \to \infty$, $t^{p-1} \to \infty$ and $1/t^{p-1} \to 0$. Therefore

$$\int_{1}^{\infty} \frac{1}{x^p}\, dx = \frac{1}{p-1} \quad \text{if } p > 1$$

and so the integral converges. But if $p < 1$, then $p - 1 < 0$ and so

$$\frac{1}{t^{p-1}} = t^{1-p} \to \infty \quad \text{as } t \to \infty$$

and the integral diverges.

SECTION 7.8 Improper Integrals 531

We summarize the result of Example 4 for future reference:

$$\boxed{2} \quad \int_1^\infty \frac{1}{x^p}\, dx \quad \text{is convergent if } p > 1 \text{ and divergent if } p \leq 1.$$

Type 2: Discontinuous Integrands

Suppose that f is a positive continuous function defined on a finite interval $[a, b)$ but has a vertical asymptote at b. Let S be the unbounded region under the graph of f and above the x-axis between a and b. (For Type 1 integrals, the regions extended indefinitely in a horizontal direction. Here the region is infinite in a vertical direction.) The area of the part of S between a and t (the shaded region in Figure 7) is

$$A(t) = \int_a^t f(x)\, dx$$

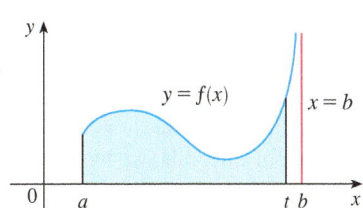

FIGURE 7

If it happens that $A(t)$ approaches a definite number A as $t \to b^-$, then we say that the area of the region S is A and we write

$$\int_a^b f(x)\, dx = \lim_{t \to b^-} \int_a^t f(x)\, dx$$

We use this equation to define an improper integral of Type 2 even when f is not a positive function, no matter what type of discontinuity f has at b.

Parts (b) and (c) of Definition 3 are illustrated in Figures 8 and 9 for the case where $f(x) \geq 0$ and f has vertical asymptotes at a and c, respectively.

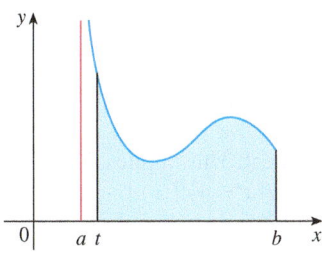

FIGURE 8

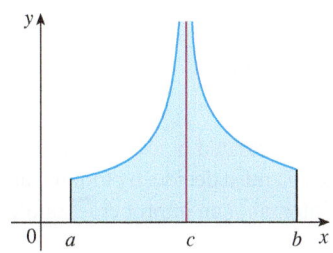

FIGURE 9

$\boxed{3}$ **Definition of an Improper Integral of Type 2**

(a) If f is continuous on $[a, b)$ and is discontinuous at b, then

$$\int_a^b f(x)\, dx = \lim_{t \to b^-} \int_a^t f(x)\, dx$$

if this limit exists (as a finite number).

(b) If f is continuous on $(a, b]$ and is discontinuous at a, then

$$\int_a^b f(x)\, dx = \lim_{t \to a^+} \int_t^b f(x)\, dx$$

if this limit exists (as a finite number).

The improper integral $\int_a^b f(x)\, dx$ is called **convergent** if the corresponding limit exists and **divergent** if the limit does not exist.

(c) If f has a discontinuity at c, where $a < c < b$, and both $\int_a^c f(x)\, dx$ and $\int_c^b f(x)\, dx$ are convergent, then we define

$$\int_a^b f(x)\, dx = \int_a^c f(x)\, dx + \int_c^b f(x)\, dx$$

EXAMPLE 5 Find $\int_2^5 \dfrac{1}{\sqrt{x-2}}\, dx$.

SOLUTION We note first that the given integral is improper because $f(x) = 1/\sqrt{x-2}$ has the vertical asymptote $x = 2$. Since the infinite discontinuity occurs at the left endpoint of $[2, 5]$, we use part (b) of Definition 3:

$$\int_2^5 \frac{dx}{\sqrt{x-2}} = \lim_{t \to 2^+} \int_t^5 \frac{dx}{\sqrt{x-2}} = \lim_{t \to 2^+} 2\sqrt{x-2}\Big]_t^5$$

$$= \lim_{t \to 2^+} 2(\sqrt{3} - \sqrt{t-2}) = 2\sqrt{3}$$

Thus the given improper integral is convergent and, since the integrand is positive, we can interpret the value of the integral as the area of the shaded region in Figure 10. ∎

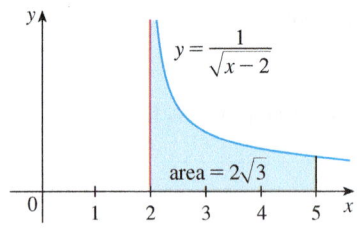

FIGURE 10

EXAMPLE 6 Determine whether $\int_0^{\pi/2} \sec x \, dx$ converges or diverges.

SOLUTION Note that the given integral is improper because $\lim_{x \to (\pi/2)^-} \sec x = \infty$. Using part (a) of Definition 3 and Formula 14 from the Table of Integrals, we have

$$\int_0^{\pi/2} \sec x \, dx = \lim_{t \to (\pi/2)^-} \int_0^t \sec x \, dx = \lim_{t \to (\pi/2)^-} \ln|\sec x + \tan x|\Big]_0^t$$

$$= \lim_{t \to (\pi/2)^-} [\ln(\sec t + \tan t) - \ln 1] = \infty$$

because $\sec t \to \infty$ and $\tan t \to \infty$ as $t \to (\pi/2)^-$. Thus the given improper integral is divergent. ∎

EXAMPLE 7 Evaluate $\int_0^3 \frac{dx}{x-1}$ if possible.

SOLUTION Observe that the line $x = 1$ is a vertical asymptote of the integrand. Since it occurs in the middle of the interval $[0, 3]$, we must use part (c) of Definition 3 with $c = 1$:

$$\int_0^3 \frac{dx}{x-1} = \int_0^1 \frac{dx}{x-1} + \int_1^3 \frac{dx}{x-1}$$

where

$$\int_0^1 \frac{dx}{x-1} = \lim_{t \to 1^-} \int_0^t \frac{dx}{x-1} = \lim_{t \to 1^-} \ln|x-1|\Big]_0^t$$

$$= \lim_{t \to 1^-} (\ln|t-1| - \ln|-1|) = \lim_{t \to 1^-} \ln(1-t) = -\infty$$

because $1 - t \to 0^+$ as $t \to 1^-$. Thus $\int_0^1 dx/(x-1)$ is divergent. This implies that $\int_0^3 dx/(x-1)$ is divergent. [We do not need to evaluate $\int_1^3 dx/(x-1)$.] ∎

⊘ **WARNING** If we had not noticed the asymptote $x = 1$ in Example 7 and had instead confused the integral with an ordinary integral, then we might have made the following erroneous calculation:

$$\int_0^3 \frac{dx}{x-1} = \ln|x-1|\Big]_0^3 = \ln 2 - \ln 1 = \ln 2$$

This is wrong because the integral is improper and must be calculated in terms of limits.

From now on, whenever you meet the symbol $\int_a^b f(x) \, dx$ you must decide, by looking at the function f on $[a, b]$, whether it is an ordinary definite integral or an improper integral.

EXAMPLE 8 $\int_0^1 \ln x \, dx$.

SOLUTION We know that the function $f(x) = \ln x$ has a vertical asymptote at 0 since

$\lim_{x \to 0^+} \ln x = -\infty$. Thus the given integral is improper and we have

$$\int_0^1 \ln x \, dx = \lim_{t \to 0^+} \int_t^1 \ln x \, dx$$

Now we integrate by parts with $u = \ln x$, $dv = dx$, $du = dx/x$, and $v = x$:

$$\int_t^1 \ln x \, dx = x \ln x \Big]_t^1 - \int_t^1 dx$$

$$= 1 \ln 1 - t \ln t - (1 - t) = -t \ln t - 1 + t$$

To find the limit of the first term we use l'Hospital's Rule:

$$\lim_{t \to 0^+} t \ln t = \lim_{t \to 0^+} \frac{\ln t}{1/t} = \lim_{t \to 0^+} \frac{1/t}{-1/t^2} = \lim_{t \to 0^+} (-t) = 0$$

Therefore $\int_0^1 \ln x \, dx = \lim_{t \to 0^+} (-t \ln t - 1 + t) = -0 - 1 + 0 = -1$

Figure 11 shows the geometric interpretation of this result. The area of the shaded region above $y = \ln x$ and below the x-axis is 1.

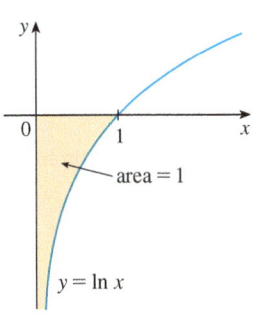

FIGURE 11

A Comparison Test for Improper Integrals

Sometimes it is impossible to find the exact value of an improper integral and yet it is important to know whether it is convergent or divergent. In such cases the following theorem is useful. Although we state it for Type 1 integrals, a similar theorem is true for Type 2 integrals.

> **Comparison Theorem** Suppose that f and g are continuous functions with $f(x) \geq g(x) \geq 0$ for $x \geq a$.
> (a) If $\int_a^\infty f(x) \, dx$ is convergent, then $\int_a^\infty g(x) \, dx$ is convergent.
> (b) If $\int_a^\infty g(x) \, dx$ is divergent, then $\int_a^\infty f(x) \, dx$ is divergent.

We omit the proof of the Comparison Theorem, but Figure 12 makes it seem plausible. If the area under the top curve $y = f(x)$ is finite, then so is the area under the bottom curve $y = g(x)$. And if the area under $y = g(x)$ is infinite, then so is the area under $y = f(x)$. [Note that the reverse is not necessarily true: If $\int_a^\infty g(x) \, dx$ is convergent, $\int_a^\infty f(x) \, dx$ may or may not be convergent, and if $\int_a^\infty f(x) \, dx$ is divergent, $\int_a^\infty g(x) \, dx$ may or may not be divergent.]

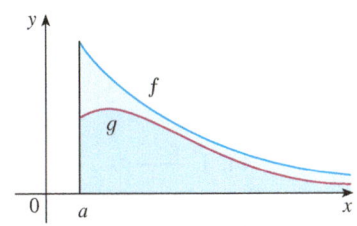

FIGURE 12

EXAMPLE 9 Show that $\int_0^\infty e^{-x^2} dx$ is convergent.

SOLUTION We can't evaluate the integral directly because the antiderivative of e^{-x^2} is not an elementary function (as explained in Section 7.5). We write

$$\int_0^\infty e^{-x^2} dx = \int_0^1 e^{-x^2} dx + \int_1^\infty e^{-x^2} dx$$

and observe that the first integral on the right-hand side is just an ordinary definite integral. In the second integral we use the fact that for $x \geq 1$ we have $x^2 \geq x$, so $-x^2 \leq -x$ and therefore $e^{-x^2} \leq e^{-x}$. (See Figure 13.) The integral of e^{-x} is easy to evaluate:

$$\int_1^\infty e^{-x} dx = \lim_{t \to \infty} \int_1^t e^{-x} dx = \lim_{t \to \infty} (e^{-1} - e^{-t}) = e^{-1}$$

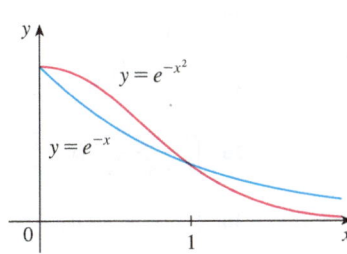

FIGURE 13

534 CHAPTER 7 Techniques of Integration

Table 1

t	$\int_0^t e^{-x^2}\,dx$
1	0.7468241328
2	0.8820813908
3	0.8862073483
4	0.8862269118
5	0.8862269255
6	0.8862269255

Therefore, taking $f(x) = e^{-x}$ and $g(x) = e^{-x^2}$ in the Comparison Theorem, we see that $\int_1^\infty e^{-x^2}\,dx$ is convergent. It follows that $\int_0^\infty e^{-x^2}\,dx$ is convergent. ∎

In Example 9 we showed that $\int_0^\infty e^{-x^2}\,dx$ is convergent without computing its value. In Exercise 72 we indicate how to show that its value is approximately 0.8862. In probability theory it is important to know the exact value of this improper integral, as we will see in Section 8.5; using the methods of multivariable calculus it can be shown that the exact value is $\sqrt{\pi}/2$. Table 1 illustrates the definition of an improper integral by showing how the (computer-generated) values of $\int_0^t e^{-x^2}\,dx$ approach $\sqrt{\pi}/2$ as t becomes large. In fact, these values converge quite quickly because $e^{-x^2} \to 0$ very rapidly as $x \to \infty$.

EXAMPLE 10 The integral $\displaystyle\int_1^\infty \frac{1 + e^{-x}}{x}\,dx$ is divergent by the Comparison Theorem because

$$\frac{1 + e^{-x}}{x} > \frac{1}{x}$$

Table 2

t	$\int_1^t [(1 + e^{-x})/x]\,dx$
2	0.8636306042
5	1.8276735512
10	2.5219648704
100	4.8245541204
1000	7.1271392134
10000	9.4297243064

and $\int_1^\infty (1/x)\,dx$ is divergent by Example 1 [or by (2) with $p = 1$]. ∎

Table 2 illustrates the divergence of the integral in Example 10. It appears that the values are not approaching any fixed number.

7.8 EXERCISES

1. Explain why each of the following integrals is improper.
   (a) $\displaystyle\int_1^2 \frac{x}{x-1}\,dx$ (b) $\displaystyle\int_0^\infty \frac{1}{1+x^3}\,dx$
   (c) $\displaystyle\int_{-\infty}^\infty x^2 e^{-x^2}\,dx$ (d) $\displaystyle\int_0^{\pi/4} \cot x\,dx$

2. Which of the following integrals are improper? Why?
   (a) $\displaystyle\int_0^{\pi/4} \tan x\,dx$ (b) $\displaystyle\int_0^\pi \tan x\,dx$
   (c) $\displaystyle\int_{-1}^1 \frac{dx}{x^2 - x - 2}$ (d) $\displaystyle\int_0^\infty e^{-x^3}\,dx$

3. Find the area under the curve $y = 1/x^3$ from $x = 1$ to $x = t$ and evaluate it for $t = 10$, 100, and 1000. Then find the total area under this curve for $x \geq 1$.

4. (a) Graph the functions $f(x) = 1/x^{1.1}$ and $g(x) = 1/x^{0.9}$ in the viewing rectangles [0, 10] by [0, 1] and [0, 100] by [0, 1].
   (b) Find the areas under the graphs of f and g from $x = 1$ to $x = t$ and evaluate for $t = 10$, 100, 10^4, 10^6, 10^{10}, and 10^{20}.
   (c) Find the total area under each curve for $x \geq 1$, if it exists.

5–40 Determine whether each integral is convergent or divergent. Evaluate those that are convergent.

5. $\displaystyle\int_3^\infty \frac{1}{(x-2)^{3/2}}\,dx$ 6. $\displaystyle\int_0^\infty \frac{1}{\sqrt[4]{1+x}}\,dx$

7. $\displaystyle\int_{-\infty}^0 \frac{1}{3 - 4x}\,dx$ 8. $\displaystyle\int_1^\infty \frac{1}{(2x+1)^3}\,dx$

9. $\displaystyle\int_2^\infty e^{-5p}\,dp$ 10. $\displaystyle\int_{-\infty}^0 2^r\,dr$

11. $\displaystyle\int_0^\infty \frac{x^2}{\sqrt{1+x^3}}\,dx$ 12. $\displaystyle\int_{-\infty}^\infty (y^3 - 3y^2)\,dy$

13. $\displaystyle\int_{-\infty}^\infty xe^{-x^2}\,dx$ 14. $\displaystyle\int_1^\infty \frac{e^{-1/x}}{x^2}\,dx$

15. $\displaystyle\int_0^\infty \sin^2\alpha\,d\alpha$ 16. $\displaystyle\int_0^\infty \sin\theta\, e^{\cos\theta}\,d\theta$

17. $\displaystyle\int_1^\infty \frac{1}{x^2 + x}\,dx$ 18. $\displaystyle\int_2^\infty \frac{dv}{v^2 + 2v - 3}$

19. $\displaystyle\int_{-\infty}^0 ze^{2z}\,dz$ 20. $\displaystyle\int_2^\infty ye^{-3y}\,dy$

21. $\displaystyle\int_1^\infty \frac{\ln x}{x}\,dx$ 22. $\displaystyle\int_1^\infty \frac{\ln x}{x^2}\,dx$

23. $\displaystyle\int_{-\infty}^0 \frac{z}{z^4 + 4}\,dz$ 24. $\displaystyle\int_e^\infty \frac{1}{x(\ln x)^2}\,dx$

25. $\displaystyle\int_0^\infty e^{-\sqrt{y}}\,dy$ 26. $\displaystyle\int_1^\infty \frac{dx}{\sqrt{x} + x\sqrt{x}}$

27. $\displaystyle\int_0^1 \frac{1}{x}\,dx$ 28. $\displaystyle\int_0^5 \frac{1}{\sqrt[3]{5-x}}\,dx$

29. $\displaystyle\int_{-2}^{14} \frac{dx}{\sqrt[4]{x+2}}$ 30. $\displaystyle\int_{-1}^2 \frac{x}{(x+1)^2}\,dx$

31. $\displaystyle\int_{-2}^3 \frac{1}{x^4}\,dx$ 32. $\displaystyle\int_0^1 \frac{dx}{\sqrt{1-x^2}}$

33. $\displaystyle\int_0^9 \frac{1}{\sqrt[3]{x-1}}\,dx$

34. $\displaystyle\int_0^5 \frac{w}{w-2}\,dw$

35. $\displaystyle\int_0^{\pi/2} \tan^2\theta\,d\theta$

36. $\displaystyle\int_0^4 \frac{dx}{x^2-x-2}$

37. $\displaystyle\int_0^1 r\ln r\,dr$

38. $\displaystyle\int_0^{\pi/2} \frac{\cos\theta}{\sqrt{\sin\theta}}\,d\theta$

39. $\displaystyle\int_{-1}^0 \frac{e^{1/x}}{x^3}\,dx$

40. $\displaystyle\int_0^1 \frac{e^{1/x}}{x^3}\,dx$

41–46 Sketch the region and find its area (if the area is finite).

41. $S = \{(x,y) \mid x \geq 1,\ 0 \leq y \leq e^{-x}\}$

42. $S = \{(x,y) \mid x \leq 0,\ 0 \leq y \leq e^x\}$

43. $S = \{(x,y) \mid x \geq 1,\ 0 \leq y \leq 1/(x^3+x)\}$

44. $S = \{(x,y) \mid x \geq 0,\ 0 \leq y \leq xe^{-x}\}$

45. $S = \{(x,y) \mid 0 \leq x < \pi/2,\ 0 \leq y \leq \sec^2 x\}$

46. $S = \{(x,y) \mid -2 < x \leq 0,\ 0 \leq y \leq 1/\sqrt{x+2}\}$

---

47. (a) If $g(x) = (\sin^2 x)/x^2$, use your calculator or computer to make a table of approximate values of $\int_1^t g(x)\,dx$ for $t = 2, 5, 10, 100, 1000$, and $10{,}000$. Does it appear that $\int_1^\infty g(x)\,dx$ is convergent?
(b) Use the Comparison Theorem with $f(x) = 1/x^2$ to show that $\int_1^\infty g(x)\,dx$ is convergent.
(c) Illustrate part (b) by graphing f and g on the same screen for $1 \leq x \leq 10$. Use your graph to explain intuitively why $\int_1^\infty g(x)\,dx$ is convergent.

48. (a) If $g(x) = 1/(\sqrt{x} - 1)$, use your calculator or computer to make a table of approximate values of $\int_2^t g(x)\,dx$ for $t = 5, 10, 100, 1000$, and $10{,}000$. Does it appear that $\int_2^\infty g(x)\,dx$ is convergent or divergent?
(b) Use the Comparison Theorem with $f(x) = 1/\sqrt{x}$ to show that $\int_2^\infty g(x)\,dx$ is divergent.
(c) Illustrate part (b) by graphing f and g on the same screen for $2 \leq x \leq 20$. Use your graph to explain intuitively why $\int_2^\infty g(x)\,dx$ is divergent.

49–54 Use the Comparison Theorem to determine whether the integral is convergent or divergent.

49. $\displaystyle\int_0^\infty \frac{x}{x^3+1}\,dx$

50. $\displaystyle\int_1^\infty \frac{1+\sin^2 x}{\sqrt{x}}\,dx$

51. $\displaystyle\int_1^\infty \frac{x+1}{\sqrt{x^4-x}}\,dx$

52. $\displaystyle\int_0^\infty \frac{\arctan x}{2+e^x}\,dx$

53. $\displaystyle\int_0^1 \frac{\sec^2 x}{x\sqrt{x}}\,dx$

54. $\displaystyle\int_0^\pi \frac{\sin^2 x}{\sqrt{x}}\,dx$

---

55. The integral

$$\int_0^\infty \frac{1}{\sqrt{x}\,(1+x)}\,dx$$

is improper for two reasons: The interval $[0,\infty)$ is infinite and the integrand has an infinite discontinuity at 0. Evaluate it by expressing it as a sum of improper integrals of Type 2 and Type 1 as follows:

$$\int_0^\infty \frac{1}{\sqrt{x}\,(1+x)}\,dx = \int_0^1 \frac{1}{\sqrt{x}\,(1+x)}\,dx + \int_1^\infty \frac{1}{\sqrt{x}\,(1+x)}\,dx$$

56. Evaluate

$$\int_2^\infty \frac{1}{x\sqrt{x^2-4}}\,dx$$

by the same method as in Exercise 55.

57–59 Find the values of p for which the integral converges and evaluate the integral for those values of p.

57. $\displaystyle\int_0^1 \frac{1}{x^p}\,dx$

58. $\displaystyle\int_e^\infty \frac{1}{x(\ln x)^p}\,dx$

59. $\displaystyle\int_0^1 x^p \ln x\,dx$

---

60. (a) Evaluate the integral $\int_0^\infty x^n e^{-x}\,dx$ for $n = 0, 1, 2$, and 3.
(b) Guess the value of $\int_0^\infty x^n e^{-x}\,dx$ when n is an arbitrary positive integer.
(c) Prove your guess using mathematical induction.

61. (a) Show that $\int_{-\infty}^\infty x\,dx$ is divergent.
(b) Show that

$$\lim_{t\to\infty} \int_{-t}^t x\,dx = 0$$

This shows that we can't define

$$\int_{-\infty}^\infty f(x)\,dx = \lim_{t\to\infty}\int_{-t}^t f(x)\,dx$$

62. The *average speed* of molecules in an ideal gas is

$$\bar{v} = \frac{4}{\sqrt{\pi}}\left(\frac{M}{2RT}\right)^{3/2} \int_0^\infty v^3 e^{-Mv^2/(2RT)}\,dv$$

where M is the molecular weight of the gas, R is the gas constant, T is the gas temperature, and v is the molecular speed. Show that

$$\bar{v} = \sqrt{\frac{8RT}{\pi M}}$$

63. We know from Example 1 that the region $\mathcal{R} = \{(x,y) \mid x \geq 1,\ 0 \leq y \leq 1/x\}$ has infinite area. Show that by rotating $\mathcal{R}$ about the x-axis we obtain a solid with finite volume.

64. Use the information and data in Exercise 6.4.33 to find the work required to propel a 1000-kg space vehicle out of the earth's gravitational field.

65. Find the *escape velocity* v_0 that is needed to propel a rocket of mass m out of the gravitational field of a planet with mass M and radius R. Use Newton's Law of Gravitation (see Exercise 6.4.33) and the fact that the initial kinetic energy of $\frac{1}{2}mv_0^2$ supplies the needed work.

66. Astronomers use a technique called *stellar stereography* to determine the density of stars in a star cluster from the observed (two-dimensional) density that can be analyzed from a photograph. Suppose that in a spherical cluster of radius R the density of stars depends only on the distance r from the center of the cluster. If the perceived star density is given by $y(s)$, where s is the observed planar distance from the center of the cluster, and $x(r)$ is the actual density, it can be shown that

$$y(s) = \int_s^R \frac{2r}{\sqrt{r^2 - s^2}} x(r)\, dr$$

If the actual density of stars in a cluster is $x(r) = \frac{1}{2}(R - r)^2$, find the perceived density $y(s)$.

67. A manufacturer of lightbulbs wants to produce bulbs that last about 700 hours but, of course, some bulbs burn out faster than others. Let $F(t)$ be the fraction of the company's bulbs that burn out before t hours, so $F(t)$ always lies between 0 and 1.
(a) Make a rough sketch of what you think the graph of F might look like.
(b) What is the meaning of the derivative $r(t) = F'(t)$?
(c) What is the value of $\int_0^\infty r(t)\, dt$? Why?

68. As we saw in Section 3.8, a radioactive substance decays exponentially: The mass at time t is $m(t) = m(0)e^{kt}$, where $m(0)$ is the initial mass and k is a negative constant. The *mean life* M of an atom in the substance is

$$M = -k \int_0^\infty t e^{kt}\, dt$$

For the radioactive carbon isotope, ^{14}C, used in radiocarbon dating, the value of k is -0.000121. Find the mean life of a ^{14}C atom.

69. In a study of the spread of illicit drug use from an enthusiastic user to a population of N users, the authors model the number of expected new users by the equation

$$\gamma = \int_0^\infty \frac{cN(1 - e^{-kt})}{k} e^{-\lambda t}\, dt$$

where c, k and λ are positive constants. Evaluate this integral to express γ in terms of c, N, k, and λ.

Source: F. Hoppensteadt et al., "Threshold Analysis of a Drug Use Epidemic Model," *Mathematical Biosciences* 53 (1981): 79–87.

70. Dialysis treatment removes urea and other waste products from a patient's blood by diverting some of the bloodflow externally through a machine called a dialyzer. The rate at which urea is removed from the blood (in mg/min) is often well described by the equation

$$u(t) = \frac{r}{V} C_0 e^{-rt/V}$$

where r is the rate of flow of blood through the dialyzer (in mL/min), V is the volume of the patient's blood (in mL), and C_0 is the amount of urea in the blood (in mg) at time $t = 0$. Evaluate the integral $\int_0^\infty u(t)$ and interpret it.

71. Determine how large the number a has to be so that

$$\int_a^\infty \frac{1}{x^2 + 1}\, dx < 0.001$$

72. Estimate the numerical value of $\int_0^\infty e^{-x^2}\, dx$ by writing it as the sum of $\int_0^4 e^{-x^2}\, dx$ and $\int_4^\infty e^{-x^2}\, dx$. Approximate the first integral by using Simpson's Rule with $n = 8$ and show that the second integral is smaller than $\int_4^\infty e^{-4x}\, dx$, which is less than 0.0000001.

73. If $f(t)$ is continuous for $t \geq 0$, the *Laplace transform* of f is the function F defined by

$$F(s) = \int_0^\infty f(t) e^{-st}\, dt$$

and the domain of F is the set consisting of all numbers s for which the integral converges. Find the Laplace transforms of the following functions.
(a) $f(t) = 1$ (b) $f(t) = e^t$ (c) $f(t) = t$

74. Show that if $0 \leq f(t) \leq Me^{at}$ for $t \geq 0$, where M and a are constants, then the Laplace transform $F(s)$ exists for $s > a$.

75. Suppose that $0 \leq f(t) \leq Me^{at}$ and $0 \leq f'(t) \leq Ke^{at}$ for $t \geq 0$, where f' is continuous. If the Laplace transform of $f(t)$ is $F(s)$ and the Laplace transform of $f'(t)$ is $G(s)$, show that

$$G(s) = sF(s) - f(0) \qquad s > a$$

76. If $\int_{-\infty}^\infty f(x)\, dx$ is convergent and a and b are real numbers, show that

$$\int_{-\infty}^a f(x)\, dx + \int_a^\infty f(x)\, dx = \int_{-\infty}^b f(x)\, dx + \int_b^\infty f(x)\, dx$$

77. Show that $\int_0^\infty x^2 e^{-x^2}\, dx = \frac{1}{2} \int_0^\infty e^{-x^2}\, dx$.

78. Show that $\int_0^\infty e^{-x^2}\, dx = \int_0^1 \sqrt{-\ln y}\, dy$ by interpreting the integrals as areas.

79. Find the value of the constant C for which the integral

$$\int_0^\infty \left(\frac{1}{\sqrt{x^2 + 4}} - \frac{C}{x + 2} \right) dx$$

converges. Evaluate the integral for this value of C.

80. Find the value of the constant C for which the integral

$$\int_0^\infty \left(\frac{x}{x^2 + 1} - \frac{C}{3x + 1} \right) dx$$

converges. Evaluate the integral for this value of C.

81. Suppose f is continuous on $[0, \infty)$ and $\lim_{x \to \infty} f(x) = 1$. Is it possible that $\int_0^\infty f(x)\, dx$ is convergent?

82. Show that if $a > -1$ and $b > a + 1$, then the following integral is convergent.

$$\int_0^\infty \frac{x^a}{1 + x^b}\, dx$$

7 REVIEW

CONCEPT CHECK

Answers to the Concept Check can be found on the back endpapers.

1. State the rule for integration by parts. In practice, how do you use it?

2. How do you evaluate $\int \sin^m x \cos^n x \, dx$ if m is odd? What if n is odd? What if m and n are both even?

3. If the expression $\sqrt{a^2 - x^2}$ occurs in an integral, what substitution might you try? What if $\sqrt{a^2 + x^2}$ occurs? What if $\sqrt{x^2 - a^2}$ occurs?

4. What is the form of the partial fraction decomposition of a rational function $P(x)/Q(x)$ if the degree of P is less than the degree of Q and $Q(x)$ has only distinct linear factors? What if a linear factor is repeated? What if $Q(x)$ has an irreducible quadratic factor (not repeated)? What if the quadratic factor is repeated?

5. State the rules for approximating the definite integral $\int_a^b f(x) \, dx$ with the Midpoint Rule, the Trapezoidal Rule, and Simpson's Rule. Which would you expect to give the best estimate? How do you approximate the error for each rule?

6. Define the following improper integrals.
   (a) $\int_a^\infty f(x) \, dx$ (b) $\int_{-\infty}^b f(x) \, dx$ (c) $\int_{-\infty}^\infty f(x) \, dx$

7. Define the improper integral $\int_a^b f(x) \, dx$ for each of the following cases.
   (a) f has an infinite discontinuity at a.
   (b) f has an infinite discontinuity at b.
   (c) f has an infinite discontinuity at c, where $a < c < b$.

8. State the Comparison Theorem for improper integrals.

TRUE-FALSE QUIZ

Determine whether the statement is true or false. If it is true, explain why. If it is false, explain why or give an example that disproves the statement.

1. $\dfrac{x(x^2 + 4)}{x^2 - 4}$ can be put in the form $\dfrac{A}{x+2} + \dfrac{B}{x-2}$.

2. $\dfrac{x^2 + 4}{x(x^2 - 4)}$ can be put in the form $\dfrac{A}{x} + \dfrac{B}{x+2} + \dfrac{C}{x-2}$.

3. $\dfrac{x^2 + 4}{x^2(x - 4)}$ can be put in the form $\dfrac{A}{x^2} + \dfrac{B}{x-4}$.

4. $\dfrac{x^2 - 4}{x(x^2 + 4)}$ can be put in the form $\dfrac{A}{x} + \dfrac{B}{x^2 + 4}$.

5. $\int_0^4 \dfrac{x}{x^2 - 1} \, dx = \tfrac{1}{2} \ln 15$

6. $\int_1^\infty \dfrac{1}{x^{\sqrt{2}}} \, dx$ is convergent.

7. If f is continuous, then $\int_{-\infty}^\infty f(x) \, dx = \lim_{t \to \infty} \int_{-t}^t f(x) \, dx$.

8. The Midpoint Rule is always more accurate than the Trapezoidal Rule.

9. (a) Every elementary function has an elementary derivative.
   (b) Every elementary function has an elementary antiderivative.

10. If f is continuous on $[0, \infty)$ and $\int_1^\infty f(x) \, dx$ is convergent, then $\int_0^\infty f(x) \, dx$ is convergent.

11. If f is a continuous, decreasing function on $[1, \infty)$ and $\lim_{x \to \infty} f(x) = 0$, then $\int_1^\infty f(x) \, dx$ is convergent.

12. If $\int_a^\infty f(x) \, dx$ and $\int_a^\infty g(x) \, dx$ are both convergent, then $\int_a^\infty [f(x) + g(x)] \, dx$ is convergent.

13. If $\int_a^\infty f(x) \, dx$ and $\int_a^\infty g(x) \, dx$ are both divergent, then $\int_a^\infty [f(x) + g(x)] \, dx$ is divergent.

14. If $f(x) \leq g(x)$ and $\int_0^\infty g(x) \, dx$ diverges, then $\int_0^\infty f(x) \, dx$ also diverges.

EXERCISES

Note: Additional practice in techniques of integration is provided in Exercises 7.5.

1–40 Evaluate the integral.

1. $\int_1^2 \dfrac{(x+1)^2}{x} \, dx$

2. $\int_1^2 \dfrac{x}{(x+1)^2} \, dx$

3. $\int \dfrac{e^{\sin x}}{\sec x} \, dx$

4. $\int_0^{\pi/6} t \sin 2t \, dt$

5. $\int \dfrac{dt}{2t^2 + 3t + 1}$

6. $\int_1^2 x^5 \ln x \, dx$

7. $\int_0^{\pi/2} \sin^3 \theta \cos^2 \theta \, d\theta$

8. $\int \dfrac{dx}{\sqrt{e^x - 1}}$

9. $\displaystyle\int \frac{\sin(\ln t)}{t}\, dt$

10. $\displaystyle\int_0^1 \frac{\sqrt{\arctan x}}{1+x^2}\, dx$

11. $\displaystyle\int_1^2 \frac{\sqrt{x^2-1}}{x}\, dx$

12. $\displaystyle\int \frac{e^{2x}}{1+e^{4x}}\, dx$

13. $\displaystyle\int e^{\sqrt[3]{x}}\, dx$

14. $\displaystyle\int \frac{x^2+2}{x+2}\, dx$

15. $\displaystyle\int \frac{x-1}{x^2+2x}\, dx$

16. $\displaystyle\int \frac{\sec^6\theta}{\tan^2\theta}\, d\theta$

17. $\displaystyle\int x\cosh x\, dx$

18. $\displaystyle\int \frac{x^2+8x-3}{x^3+3x^2}\, dx$

19. $\displaystyle\int \frac{x+1}{9x^2+6x+5}\, dx$

20. $\displaystyle\int \tan^5\theta \sec^3\theta\, d\theta$

21. $\displaystyle\int \frac{dx}{\sqrt{x^2-4x}}$

22. $\displaystyle\int \cos\sqrt{t}\, dt$

23. $\displaystyle\int \frac{dx}{x\sqrt{x^2+1}}$

24. $\displaystyle\int e^x \cos x\, dx$

25. $\displaystyle\int \frac{3x^3-x^2+6x-4}{(x^2+1)(x^2+2)}\, dx$

26. $\displaystyle\int x\sin x\cos x\, dx$

27. $\displaystyle\int_0^{\pi/2} \cos^3 x \sin 2x\, dx$

28. $\displaystyle\int \frac{\sqrt[3]{x}+1}{\sqrt[3]{x}-1}\, dx$

29. $\displaystyle\int_{-3}^3 \frac{x}{1+|x|}\, dx$

30. $\displaystyle\int \frac{dx}{e^x\sqrt{1-e^{-2x}}}$

31. $\displaystyle\int_0^{\ln 10} \frac{e^x\sqrt{e^x-1}}{e^x+8}\, dx$

32. $\displaystyle\int_0^{\pi/4} \frac{x\sin x}{\cos^3 x}\, dx$

33. $\displaystyle\int \frac{x^2}{(4-x^2)^{3/2}}\, dx$

34. $\displaystyle\int (\arcsin x)^2\, dx$

35. $\displaystyle\int \frac{1}{\sqrt{x+x^{3/2}}}\, dx$

36. $\displaystyle\int \frac{1-\tan\theta}{1+\tan\theta}\, d\theta$

37. $\displaystyle\int (\cos x + \sin x)^2 \cos 2x\, dx$

38. $\displaystyle\int \frac{2^{\sqrt{x}}}{\sqrt{x}}\, dx$

39. $\displaystyle\int_0^{1/2} \frac{xe^{2x}}{(1+2x)^2}\, dx$

40. $\displaystyle\int_{\pi/4}^{\pi/3} \frac{\sqrt{\tan\theta}}{\sin 2\theta}\, d\theta$

41–50 Evaluate the integral or show that it is divergent.

41. $\displaystyle\int_1^\infty \frac{1}{(2x+1)^3}\, dx$

42. $\displaystyle\int_1^\infty \frac{\ln x}{x^4}\, dx$

43. $\displaystyle\int_2^\infty \frac{dx}{x\ln x}$

44. $\displaystyle\int_2^6 \frac{y}{\sqrt{y-2}}\, dy$

45. $\displaystyle\int_0^4 \frac{\ln x}{\sqrt{x}}\, dx$

46. $\displaystyle\int_0^1 \frac{1}{2-3x}\, dx$

47. $\displaystyle\int_0^1 \frac{x-1}{\sqrt{x}}\, dx$

48. $\displaystyle\int_{-1}^1 \frac{dx}{x^2-2x}$

49. $\displaystyle\int_{-\infty}^\infty \frac{dx}{4x^2+4x+5}$

50. $\displaystyle\int_1^\infty \frac{\tan^{-1}x}{x^2}\, dx$

51–52 Evaluate the indefinite integral. Illustrate and check that your answer is reasonable by graphing both the function and its antiderivative (take $C = 0$).

51. $\displaystyle\int \ln(x^2+2x+2)\, dx$

52. $\displaystyle\int \frac{x^3}{\sqrt{x^2+1}}\, dx$

53. Graph the function $f(x) = \cos^2 x \sin^3 x$ and use the graph to guess the value of the integral $\int_0^{2\pi} f(x)\, dx$. Then evaluate the integral to confirm your guess.

54. (a) How would you evaluate $\int x^5 e^{-2x}\, dx$ by hand? (Don't actually carry out the integration.)
(b) How would you evaluate $\int x^5 e^{-2x}\, dx$ using tables? (Don't actually do it.)
(c) Use a CAS to evaluate $\int x^5 e^{-2x}\, dx$.
(d) Graph the integrand and the indefinite integral on the same screen.

55–58 Use the Table of Integrals on the Reference Pages to evaluate the integral.

55. $\displaystyle\int \sqrt{4x^2-4x-3}\, dx$

56. $\displaystyle\int \csc^5 t\, dt$

57. $\displaystyle\int \cos x\sqrt{4+\sin^2 x}\, dx$

58. $\displaystyle\int \frac{\cot x}{\sqrt{1+2\sin x}}\, dx$

59. Verify Formula 33 in the Table of Integrals (a) by differentiation and (b) by using a trigonometric substitution.

60. Verify Formula 62 in the Table of Integrals.

61. Is it possible to find a number n such that $\int_0^\infty x^n\, dx$ is convergent?

62. For what values of a is $\int_0^\infty e^{ax}\cos x\, dx$ convergent? Evaluate the integral for those values of a.

63–64 Use (a) the Trapezoidal Rule, (b) the Midpoint Rule, and (c) Simpson's Rule with $n = 10$ to approximate the given integral. Round your answers to six decimal places.

63. $\displaystyle\int_2^4 \frac{1}{\ln x}\, dx$

64. $\displaystyle\int_1^4 \sqrt{x}\cos x\, dx$

65. Estimate the errors involved in Exercise 63, parts (a) and (b). How large should n be in each case to guarantee an error of less than 0.00001?

66. Use Simpson's Rule with $n = 6$ to estimate the area under the curve $y = e^x/x$ from $x = 1$ to $x = 4$.

67. The speedometer reading (v) on a car was observed at 1-minute intervals and recorded in the chart. Use Simpson's Rule to estimate the distance traveled by the car.

t (min)	v (mi/h)	t (min)	v (mi/h)
0	40	6	56
1	42	7	57
2	45	8	57
3	49	9	55
4	52	10	56
5	54		

68. A population of honeybees increased at a rate of $r(t)$ bees per week, where the graph of r is as shown. Use Simpson's Rule with six subintervals to estimate the increase in the bee population during the first 24 weeks.

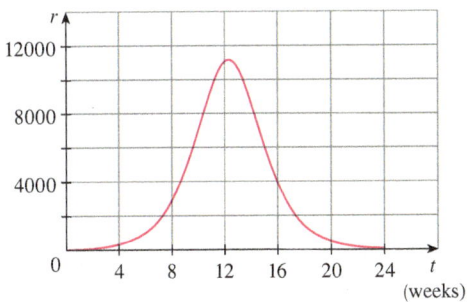

CAS 69. (a) If $f(x) = \sin(\sin x)$, use a graph to find an upper bound for $|f^{(4)}(x)|$.
(b) Use Simpson's Rule with $n = 10$ to approximate $\int_0^\pi f(x)\, dx$ and use part (a) to estimate the error.
(c) How large should n be to guarantee that the size of the error in using S_n is less than 0.00001?

70. Suppose you are asked to estimate the volume of a football. You measure and find that a football is 28 cm long. You use a piece of string and measure the circumference at its widest point to be 53 cm. The circumference 7 cm from each end is 45 cm. Use Simpson's Rule to make your estimate.

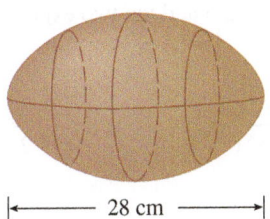

71. Use the Comparison Theorem to determine whether the integral is convergent or divergent.

(a) $\int_1^\infty \dfrac{2 + \sin x}{\sqrt{x}}\, dx$ (b) $\int_1^\infty \dfrac{1}{\sqrt{1 + x^4}}\, dx$

72. Find the area of the region bounded by the hyperbola $y^2 - x^2 = 1$ and the line $y = 3$.

73. Find the area bounded by the curves $y = \cos x$ and $y = \cos^2 x$ between $x = 0$ and $x = \pi$.

74. Find the area of the region bounded by the curves $y = 1/(2 + \sqrt{x})$, $y = 1/(2 - \sqrt{x})$, and $x = 1$.

75. The region under the curve $y = \cos^2 x$, $0 \leq x \leq \pi/2$, is rotated about the x-axis. Find the volume of the resulting solid.

76. The region in Exercise 75 is rotated about the y-axis. Find the volume of the resulting solid.

77. If f' is continuous on $[0, \infty)$ and $\lim_{x \to \infty} f(x) = 0$, show that

$$\int_0^\infty f'(x)\, dx = -f(0)$$

78. We can extend our definition of average value of a continuous function to an infinite interval by defining the average value of f on the interval $[a, \infty)$ to be

$$\lim_{t \to \infty} \frac{1}{t - a} \int_a^t f(x)\, dx$$

(a) Find the average value of $y = \tan^{-1} x$ on the interval $[0, \infty)$.
(b) If $f(x) \geq 0$ and $\int_a^\infty f(x)\, dx$ is divergent, show that the average value of f on the interval $[a, \infty)$ is $\lim_{x \to \infty} f(x)$, if this limit exists.
(c) If $\int_a^\infty f(x)\, dx$ is convergent, what is the average value of f on the interval $[a, \infty)$?
(d) Find the average value of $y = \sin x$ on the interval $[0, \infty)$.

79. Use the substitution $u = 1/x$ to show that

$$\int_0^\infty \frac{\ln x}{1 + x^2}\, dx = 0$$

80. The magnitude of the repulsive force between two point charges with the same sign, one of size 1 and the other of size q, is

$$F = \frac{q}{4\pi \varepsilon_0 r^2}$$

where r is the distance between the charges and ε_0 is a constant. The *potential* V at a point P due to the charge q is defined to be the work expended in bringing a unit charge to P from infinity along the straight line that joins q and P. Find a formula for V.

Problems Plus

Cover up the solution to the example and try it yourself first.

PS The principles of problem solving are discussed on page 71.

EXAMPLE

(a) Prove that if f is a continuous function, then

$$\int_0^a f(x)\, dx = \int_0^a f(a-x)\, dx$$

(b) Use part (a) to show that

$$\int_0^{\pi/2} \frac{\sin^n x}{\sin^n x + \cos^n x}\, dx = \frac{\pi}{4}$$

for all positive numbers n.

SOLUTION

(a) At first sight, the given equation may appear somewhat baffling. How is it possible to connect the left side to the right side? Connections can often be made through one of the principles of problem solving: *introduce something extra*. Here the extra ingredient is a new variable. We often think of introducing a new variable when we use the Substitution Rule to integrate a specific function. But that technique is still useful in the present circumstance in which we have a general function f.

Once we think of making a substitution, the form of the right side suggests that it should be $u = a - x$. Then $du = -dx$. When $x = 0$, $u = a$; when $x = a$, $u = 0$. So

$$\int_0^a f(a-x)\, dx = -\int_a^0 f(u)\, du = \int_0^a f(u)\, du$$

But this integral on the right side is just another way of writing $\int_0^a f(x)\, dx$. So the given equation is proved.

(b) If we let the given integral be I and apply part (a) with $a = \pi/2$, we get

$$I = \int_0^{\pi/2} \frac{\sin^n x}{\sin^n x + \cos^n x}\, dx = \int_0^{\pi/2} \frac{\sin^n(\pi/2 - x)}{\sin^n(\pi/2 - x) + \cos^n(\pi/2 - x)}\, dx$$

A well-known trigonometric identity tells us that $\sin(\pi/2 - x) = \cos x$ and $\cos(\pi/2 - x) = \sin x$, so we get

$$I = \int_0^{\pi/2} \frac{\cos^n x}{\cos^n x + \sin^n x}\, dx$$

Notice that the two expressions for I are very similar. In fact, the integrands have the same denominator. This suggests that we should add the two expressions. If we do so, we get

$$2I = \int_0^{\pi/2} \frac{\sin^n x + \cos^n x}{\sin^n x + \cos^n x}\, dx = \int_0^{\pi/2} 1\, dx = \frac{\pi}{2}$$

Therefore $I = \pi/4$. ∎

The computer graphs in Figure 1 make it seem plausible that all of the integrals in the example have the same value. The graph of each integrand is labeled with the corresponding value of n.

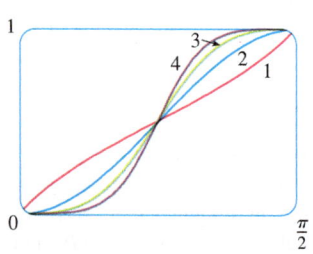

FIGURE 1

Problems

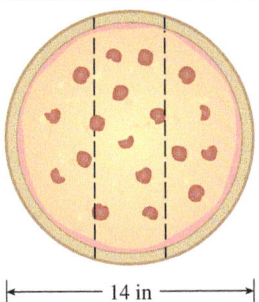

FIGURE FOR PROBLEM 1

1. Three mathematics students have ordered a 14-inch pizza. Instead of slicing it in the traditional way, they decide to slice it by parallel cuts, as shown in the figure. Being mathematics majors, they are able to determine where to slice so that each gets the same amount of pizza. Where are the cuts made?

2. Evaluate
$$\int \frac{1}{x^7 - x}\, dx$$

   The straightforward approach would be to start with partial fractions, but that would be brutal. Try a substitution.

3. Evaluate $\int_0^1 \left(\sqrt[3]{1-x^7} - \sqrt[7]{1-x^3}\right) dx$.

4. The centers of two disks with radius 1 are one unit apart. Find the area of the union of the two disks.

5. An ellipse is cut out of a circle with radius a. The major axis of the ellipse coincides with a diameter of the circle and the minor axis has length $2b$. Prove that the area of the remaining part of the circle is the same as the area of an ellipse with semiaxes a and $a - b$.

6. A man initially standing at the point O walks along a pier pulling a rowboat by a rope of length L. The man keeps the rope straight and taut. The path followed by the boat is a curve called a *tractrix* and it has the property that the rope is always tangent to the curve (see the figure).
   (a) Show that if the path followed by the boat is the graph of the function $y = f(x)$, then
   $$f'(x) = \frac{dy}{dx} = \frac{-\sqrt{L^2 - x^2}}{x}$$
   (b) Determine the function $y = f(x)$.

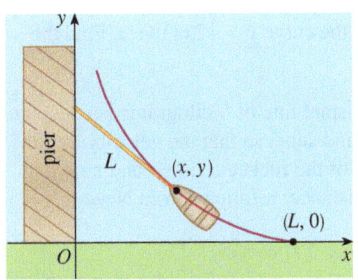

FIGURE FOR PROBLEM 6

7. A function f is defined by
$$f(x) = \int_0^\pi \cos t \, \cos(x - t)\, dt \qquad 0 \leq x \leq 2\pi$$

   Find the minimum value of f.

8. If n is a positive integer, prove that
$$\int_0^1 (\ln x)^n\, dx = (-1)^n n!$$

9. Show that
$$\int_0^1 (1 - x^2)^n\, dx = \frac{2^{2n}(n!)^2}{(2n+1)!}$$

   Hint: Start by showing that if I_n denotes the integral, then
   $$I_{k+1} = \frac{2k+2}{2k+3} I_k$$

10. Suppose that f is a positive function such that f' is continuous.
    (a) How is the graph of $y = f(x) \sin nx$ related to the graph of $y = f(x)$? What happens as $n \to \infty$?

541

(b) Make a guess as to the value of the limit
$$\lim_{n\to\infty} \int_0^1 f(x) \sin nx \, dx$$
based on graphs of the integrand.

(c) Using integration by parts, confirm the guess that you made in part (b). [Use the fact that, since f' is continuous, there is a constant M such that $|f'(x)| \le M$ for $0 \le x \le 1$.]

11. If $0 < a < b$, find
$$\lim_{t\to 0} \left\{ \int_0^1 [bx + a(1-x)]^t \, dx \right\}^{1/t}$$

12. Graph $f(x) = \sin(e^x)$ and use the graph to estimate the value of t such that $\int_t^{t+1} f(x) \, dx$ is a maximum. Then find the exact value of t that maximizes this integral.

13. Evaluate $\int_{-1}^{\infty} \left(\dfrac{x^4}{1 + x^6} \right)^2 dx$.

14. Evaluate $\int \sqrt{\tan x} \, dx$.

15. The circle with radius 1 shown in the figure touches the curve $y = |2x|$ twice. Find the area of the region that lies between the two curves.

16. A rocket is fired straight up, burning fuel at the constant rate of b kilograms per second. Let $v = v(t)$ be the velocity of the rocket at time t and suppose that the velocity u of the exhaust gas is constant. Let $M = M(t)$ be the mass of the rocket at time t and note that M decreases as the fuel burns. If we neglect air resistance, it follows from Newton's Second Law that
$$F = M\frac{dv}{dt} - ub$$
where the force $F = -Mg$. Thus

$$M\frac{dv}{dt} - ub = -Mg$$

Let M_1 be the mass of the rocket without fuel, M_2 the initial mass of the fuel, and $M_0 = M_1 + M_2$. Then, until the fuel runs out at time $t = M_2/b$, the mass is $M = M_0 - bt$.

(a) Substitute $M = M_0 - bt$ into Equation 1 and solve the resulting equation for v. Use the initial condition $v(0) = 0$ to evaluate the constant.

(b) Determine the velocity of the rocket at time $t = M_2/b$. This is called the *burnout velocity*.

(c) Determine the height of the rocket $y = y(t)$ at the burnout time.

(d) Find the height of the rocket at any time t.

FIGURE FOR PROBLEM 15

8 Further Applications of Integration

The Gateway Arch in St. Louis, Missouri, stands 630 feet high and was completed in 1965. The arch was designed by Eero Saarinen using an equation involving the hyperbolic cosine function. In Exercise 8.1.42 you are asked to compute the length of the curve that he used.

© planet5D LLC / Shutterstock.com

WE LOOKED AT SOME APPLICATIONS of integrals in Chapter 6: areas, volumes, work, and average values. Here we explore some of the many other geometric applications of integration—the length of a curve, the area of a surface—as well as quantities of interest in physics, engineering, biology, economics, and statistics. For instance, we will investigate the center of gravity of a plate, the force exerted by water pressure on a dam, the flow of blood from the human heart, and the average time spent on hold during a customer support telephone call.

8.1 Arc Length

FIGURE 1

TEC Visual 8.1 shows an animation of Figure 2.

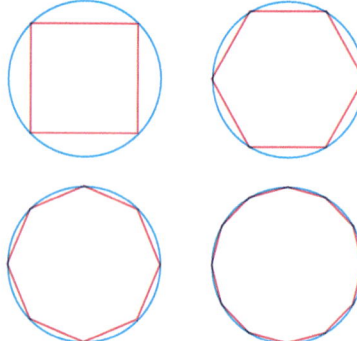

FIGURE 2

What do we mean by the length of a curve? We might think of fitting a piece of string to the curve in Figure 1 and then measuring the string against a ruler. But that might be difficult to do with much accuracy if we have a complicated curve. We need a precise definition for the length of an arc of a curve, in the same spirit as the definitions we developed for the concepts of area and volume.

If the curve is a polygon, we can easily find its length; we just add the lengths of the line segments that form the polygon. (We can use the distance formula to find the distance between the endpoints of each segment.) We are going to define the length of a general curve by first approximating it by a polygon and then taking a limit as the number of segments of the polygon is increased. This process is familiar for the case of a circle, where the circumference is the limit of lengths of inscribed polygons (see Figure 2).

Now suppose that a curve C is defined by the equation $y = f(x)$, where f is continuous and $a \leq x \leq b$. We obtain a polygonal approximation to C by dividing the interval $[a, b]$ into n subintervals with endpoints $x_0, x_1, \ldots, x_n$ and equal width Δx. If $y_i = f(x_i)$, then the point $P_i(x_i, y_i)$ lies on C and the polygon with vertices $P_0, P_1, \ldots, P_n$, illustrated in Figure 3, is an approximation to C.

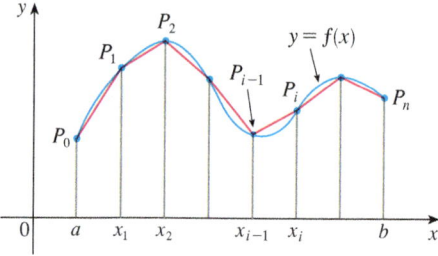

FIGURE 3

The length L of C is approximately the length of this polygon and the approximation gets better as we let n increase. (See Figure 4, where the arc of the curve between P_{i-1} and P_i has been magnified and approximations with successively smaller values of Δx are shown.) Therefore we define the **length** L of the curve C with equation $y = f(x)$, $a \leq x \leq b$, as the limit of the lengths of these inscribed polygons (if the limit exists):

$$\boxed{1} \quad L = \lim_{n \to \infty} \sum_{i=1}^{n} |P_{i-1} P_i|$$

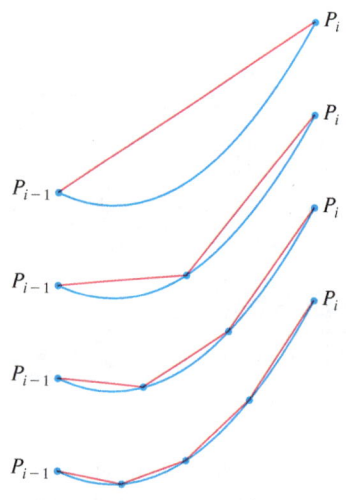

FIGURE 4

Notice that the procedure for defining arc length is very similar to the procedure we used for defining area and volume: We divided the curve into a large number of small parts. We then found the approximate lengths of the small parts and added them. Finally, we took the limit as $n \to \infty$.

The definition of arc length given by Equation 1 is not very convenient for computational purposes, but we can derive an integral formula for L in the case where f has a continuous derivative. [Such a function f is called **smooth** because a small change in x produces a small change in $f'(x)$.]

If we let $\Delta y_i = y_i - y_{i-1}$, then

$$|P_{i-1} P_i| = \sqrt{(x_i - x_{i-1})^2 + (y_i - y_{i-1})^2} = \sqrt{(\Delta x)^2 + (\Delta y_i)^2}$$

By applying the Mean Value Theorem to f on the interval $[x_{i-1}, x_i]$, we find that there is a number x_i^* between x_{i-1} and x_i such that

$$f(x_i) - f(x_{i-1}) = f'(x_i^*)(x_i - x_{i-1})$$

that is,
$$\Delta y_i = f'(x_i^*) \, \Delta x$$

Thus we have

$$|P_{i-1}P_i| = \sqrt{(\Delta x)^2 + (\Delta y_i)^2} = \sqrt{(\Delta x)^2 + [f'(x_i^*) \, \Delta x]^2}$$
$$= \sqrt{1 + [f'(x_i^*)]^2} \sqrt{(\Delta x)^2} = \sqrt{1 + [f'(x_i^*)]^2} \, \Delta x \quad \text{(since } \Delta x > 0\text{)}$$

Therefore, by Definition 1,

$$L = \lim_{n \to \infty} \sum_{i=1}^{n} |P_{i-1}P_i| = \lim_{n \to \infty} \sum_{i=1}^{n} \sqrt{1 + [f'(x_i^*)]^2} \, \Delta x$$

We recognize this expression as being equal to

$$\int_a^b \sqrt{1 + [f'(x)]^2} \, dx$$

by the definition of a definite integral. We know that this integral exists because the function $g(x) = \sqrt{1 + [f'(x)]^2}$ is continuous. Thus we have proved the following theorem:

> **2** **The Arc Length Formula** If f' is continuous on $[a, b]$, then the length of the curve $y = f(x)$, $a \leq x \leq b$, is
>
> $$L = \int_a^b \sqrt{1 + [f'(x)]^2} \, dx$$

If we use Leibniz notation for derivatives, we can write the arc length formula as follows:

> **3**
> $$L = \int_a^b \sqrt{1 + \left(\frac{dy}{dx}\right)^2} \, dx$$

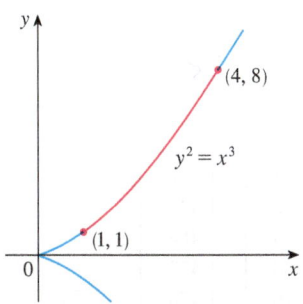

FIGURE 5

EXAMPLE 1 Find the length of the arc of the semicubical parabola $y^2 = x^3$ between the points $(1, 1)$ and $(4, 8)$. (See Figure 5.)

SOLUTION For the top half of the curve we have

$$y = x^{3/2} \qquad \frac{dy}{dx} = \tfrac{3}{2} x^{1/2}$$

and so the arc length formula gives

$$L = \int_1^4 \sqrt{1 + \left(\frac{dy}{dx}\right)^2} \, dx = \int_1^4 \sqrt{1 + \tfrac{9}{4} x} \, dx$$

If we substitute $u = 1 + \tfrac{9}{4} x$, then $du = \tfrac{9}{4} \, dx$. When $x = 1$, $u = \tfrac{13}{4}$; when $x = 4$, $u = 10$.

As a check on our answer to Example 1, notice from Figure 5 that the arc length ought to be slightly larger than the distance from $(1, 1)$ to $(4, 8)$, which is

$$\sqrt{58} \approx 7.615773$$

According to our calculation in Example 1, we have

$$L = \tfrac{1}{27}(80\sqrt{10} - 13\sqrt{13})$$
$$\approx 7.633705$$

Sure enough, this is a bit greater than the length of the line segment.

Therefore

$$L = \tfrac{4}{9}\int_{13/4}^{10} \sqrt{u}\, du = \tfrac{4}{9} \cdot \tfrac{2}{3}u^{3/2}\Big]_{13/4}^{10}$$
$$= \tfrac{8}{27}\left[10^{3/2} - \left(\tfrac{13}{4}\right)^{3/2}\right] = \tfrac{1}{27}(80\sqrt{10} - 13\sqrt{13})$$

If a curve has the equation $x = g(y)$, $c \leq y \leq d$, and $g'(y)$ is continuous, then by interchanging the roles of x and y in Formula 2 or Equation 3, we obtain the following formula for its length:

$$L = \int_c^d \sqrt{1 + [g'(y)]^2}\, dy = \int_c^d \sqrt{1 + \left(\frac{dx}{dy}\right)^2}\, dy$$

EXAMPLE 2 Find the length of the arc of the parabola $y^2 = x$ from $(0, 0)$ to $(1, 1)$.

SOLUTION Since $x = y^2$, we have $dx/dy = 2y$, and Formula 4 gives

$$L = \int_0^1 \sqrt{1 + \left(\frac{dx}{dy}\right)^2}\, dy = \int_0^1 \sqrt{1 + 4y^2}\, dy$$

We make the trigonometric substitution $y = \tfrac{1}{2}\tan\theta$, which gives $dy = \tfrac{1}{2}\sec^2\theta\, d\theta$ and $\sqrt{1 + 4y^2} = \sqrt{1 + \tan^2\theta} = \sec\theta$. When $y = 0$, $\tan\theta = 0$, so $\theta = 0$; when $y = 1$, $\tan\theta = 2$, so $\theta = \tan^{-1}2 = \alpha$, say. Thus

$$L = \int_0^\alpha \sec\theta \cdot \tfrac{1}{2}\sec^2\theta\, d\theta = \tfrac{1}{2}\int_0^\alpha \sec^3\theta\, d\theta$$
$$= \tfrac{1}{2} \cdot \tfrac{1}{2}\Big[\sec\theta\tan\theta + \ln|\sec\theta + \tan\theta|\Big]_0^\alpha \quad \text{(from Example 7.2.8)}$$
$$= \tfrac{1}{4}\Big(\sec\alpha\tan\alpha + \ln|\sec\alpha + \tan\alpha|\Big)$$

(We could have used Formula 21 in the Table of Integrals.) Since $\tan\alpha = 2$, we have $\sec^2\alpha = 1 + \tan^2\alpha = 5$, so $\sec\alpha = \sqrt{5}$ and

$$L = \frac{\sqrt{5}}{2} + \frac{\ln(\sqrt{5} + 2)}{4}$$

Figure 6 shows the arc of the parabola whose length is computed in Example 2, together with polygonal approximations having $n = 1$ and $n = 2$ line segments, respectively. For $n = 1$ the approximate length is $L_1 = \sqrt{2}$, the diagonal of a square. The table shows the approximations L_n that we get by dividing $[0, 1]$ into n equal subintervals. Notice that each time we double the number of sides of the polygon, we get closer to the exact length, which is

$$L = \frac{\sqrt{5}}{2} + \frac{\ln(\sqrt{5} + 2)}{4} \approx 1.478943$$

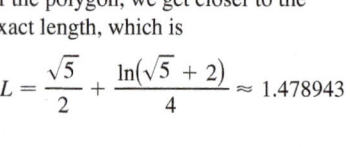

FIGURE 6

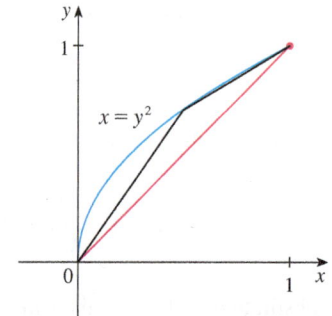

n	L_n
1	1.414
2	1.445
4	1.464
8	1.472
16	1.476
32	1.478
64	1.479

Because of the presence of the square root sign in Formulas 2 and 4, the calculation of an arc length often leads to an integral that is very difficult or even impossible to evaluate explicitly. Thus we sometimes have to be content with finding an approximation to the length of a curve, as in the following example.

EXAMPLE 3
(a) Set up an integral for the length of the arc of the hyperbola $xy = 1$ from the point $(1, 1)$ to the point $(2, \frac{1}{2})$.
(b) Use Simpson's Rule with $n = 10$ to estimate the arc length.

SOLUTION
(a) We have
$$y = \frac{1}{x} \qquad \frac{dy}{dx} = -\frac{1}{x^2}$$

and so the arc length is

$$L = \int_1^2 \sqrt{1 + \left(\frac{dy}{dx}\right)^2} \, dx = \int_1^2 \sqrt{1 + \frac{1}{x^4}} \, dx = \int_1^2 \frac{\sqrt{x^4 + 1}}{x^2} \, dx$$

(b) Using Simpson's Rule (see Section 7.7) with $a = 1$, $b = 2$, $n = 10$, $\Delta x = 0.1$, and $f(x) = \sqrt{1 + 1/x^4}$, we have

$$L = \int_1^2 \sqrt{1 + \frac{1}{x^4}} \, dx$$

$$\approx \frac{\Delta x}{3} [f(1) + 4f(1.1) + 2f(1.2) + 4f(1.3) + \cdots + 2f(1.8) + 4f(1.9) + f(2)]$$

$$\approx 1.1321 \qquad \blacksquare$$

Checking the value of the definite integral with a more accurate approximation produced by a computing device, we see that the approximation using Simpson's Rule is accurate to four decimal places.

■ The Arc Length Function

We will find it useful to have a function that measures the arc length of a curve from a particular starting point to any other point on the curve. Thus if a smooth curve C has the equation $y = f(x)$, $a \le x \le b$, let $s(x)$ be the distance along C from the initial point $P_0(a, f(a))$ to the point $Q(x, f(x))$. Then s is a function, called the **arc length function**, and, by Formula 2,

$$\boxed{5} \qquad s(x) = \int_a^x \sqrt{1 + [f'(t)]^2} \, dt$$

(We have replaced the variable of integration by t so that x does not have two meanings.) We can use Part 1 of the Fundamental Theorem of Calculus to differentiate Equation 5 (since the integrand is continuous):

$$\boxed{6} \qquad \frac{ds}{dx} = \sqrt{1 + [f'(x)]^2} = \sqrt{1 + \left(\frac{dy}{dx}\right)^2}$$

Equation 6 shows that the rate of change of s with respect to x is always at least 1 and is equal to 1 when $f'(x)$, the slope of the curve, is 0. The differential of arc length is

$$\boxed{7} \qquad ds = \sqrt{1 + \left(\frac{dy}{dx}\right)^2} \, dx$$

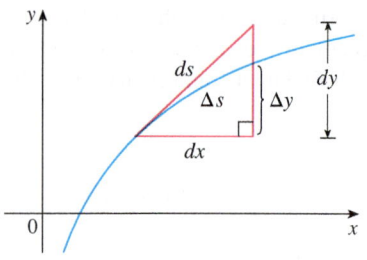

FIGURE 7

and this equation is sometimes written in the symmetric form

$$(ds)^2 = (dx)^2 + (dy)^2 \quad \boxed{8}$$

The geometric interpretation of Equation 8 is shown in Figure 7. It can be used as a mnemonic device for remembering both of the Formulas 3 and 4. If we write $L = \int ds$, then from Equation 8 either we can solve to get (7), which gives (3), or we can solve to get

$$ds = \sqrt{1 + \left(\frac{dx}{dy}\right)^2}\, dy$$

which gives (4).

EXAMPLE 4 Find the arc length function for the curve $y = x^2 - \frac{1}{8}\ln x$ taking $P_0(1, 1)$ as the starting point.

SOLUTION If $f(x) = x^2 - \frac{1}{8}\ln x$, then

$$f'(x) = 2x - \frac{1}{8x}$$

$$1 + [f'(x)]^2 = 1 + \left(2x - \frac{1}{8x}\right)^2 = 1 + 4x^2 - \frac{1}{2} + \frac{1}{64x^2}$$

$$= 4x^2 + \frac{1}{2} + \frac{1}{64x^2} = \left(2x + \frac{1}{8x}\right)^2$$

$$\sqrt{1 + [f'(x)]^2} = 2x + \frac{1}{8x} \quad \text{(since } x > 0\text{)}$$

Thus the arc length function is given by

$$s(x) = \int_1^x \sqrt{1 + [f'(t)]^2}\, dt$$

$$= \int_1^x \left(2t + \frac{1}{8t}\right) dt = t^2 + \frac{1}{8}\ln t \Big]_1^x$$

$$= x^2 + \frac{1}{8}\ln x - 1$$

For instance, the arc length along the curve from $(1, 1)$ to $(3, f(3))$ is

$$s(3) = 3^2 + \frac{1}{8}\ln 3 - 1 = 8 + \frac{\ln 3}{8} \approx 8.1373$$

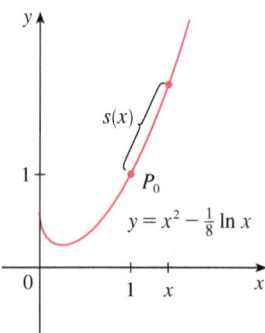

FIGURE 8

Figure 8 shows the interpretation of the arc length function in Example 4. Figure 9 shows the graph of this arc length function. Why is $s(x)$ negative when x is less than 1?

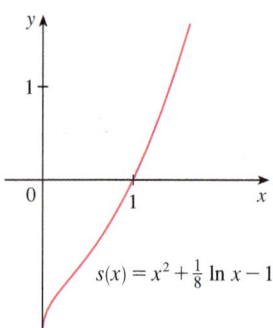

FIGURE 9

8.1 EXERCISES

1. Use the arc length formula (3) to find the length of the curve $y = 2x - 5$, $-1 \leq x \leq 3$. Check your answer by noting that the curve is a line segment and calculating its length by the distance formula.

2. Use the arc length formula to find the length of the curve $y = \sqrt{2 - x^2}$, $0 \leq x \leq 1$. Check your answer by noting that the curve is part of a circle.

3–8 Set up an integral that represents the length of the curve. Then use your calculator to find the length correct to four decimal places.

3. $y = \sin x$, $0 \leq x \leq \pi$
4. $y = xe^{-x}$, $0 \leq x \leq 2$
5. $y = x - \ln x$, $1 \leq x \leq 4$
6. $x = y^2 - 2y$, $0 \leq y \leq 2$
7. $x = \sqrt{y} - y$, $1 \leq y \leq 4$
8. $y^2 = \ln x$, $-1 \leq y \leq 1$

9–20 Find the exact length of the curve.

9. $y = 1 + 6x^{3/2}$, $0 \leq x \leq 1$
10. $36y^2 = (x^2 - 4)^3$, $2 \leq x \leq 3$, $y \geq 0$
11. $y = \dfrac{x^3}{3} + \dfrac{1}{4x}$, $1 \leq x \leq 2$
12. $x = \dfrac{y^4}{8} + \dfrac{1}{4y^2}$, $1 \leq y \leq 2$
13. $x = \tfrac{1}{3}\sqrt{y}\,(y - 3)$, $1 \leq y \leq 9$
14. $y = \ln(\cos x)$, $0 \leq x \leq \pi/3$
15. $y = \ln(\sec x)$, $0 \leq x \leq \pi/4$
16. $y = 3 + \tfrac{1}{2}\cosh 2x$, $0 \leq x \leq 1$
17. $y = \tfrac{1}{4}x^2 - \tfrac{1}{2}\ln x$, $1 \leq x \leq 2$
18. $y = \sqrt{x - x^2} + \sin^{-1}(\sqrt{x})$
19. $y = \ln(1 - x^2)$, $0 \leq x \leq \tfrac{1}{2}$
20. $y = 1 - e^{-x}$, $0 \leq x \leq 2$

21–22 Find the length of the arc of the curve from point P to point Q.

21. $y = \tfrac{1}{2}x^2$, $P\bigl(-1, \tfrac{1}{2}\bigr)$, $Q\bigl(1, \tfrac{1}{2}\bigr)$
22. $x^2 = (y - 4)^3$, $P(1, 5)$, $Q(8, 8)$

23–24 Graph the curve and visually estimate its length. Then use your calculator to find the length correct to four decimal places.

23. $y = x^2 + x^3$, $1 \leq x \leq 2$
24. $y = x + \cos x$, $0 \leq x \leq \pi/2$

25–28 Use Simpson's Rule with $n = 10$ to estimate the arc length of the curve. Compare your answer with the value of the integral produced by a calculator.

25. $y = x \sin x$, $0 \leq x \leq 2\pi$
26. $y = \sqrt[3]{x}$, $1 \leq x \leq 6$
27. $y = \ln(1 + x^3)$, $0 \leq x \leq 5$
28. $y = e^{-x^2}$, $0 \leq x \leq 2$

29. (a) Graph the curve $y = x\sqrt[3]{4 - x}$, $0 \leq x \leq 4$.
    (b) Compute the lengths of inscribed polygons with $n = 1$, 2, and 4 sides. (Divide the interval into equal subintervals.) Illustrate by sketching these polygons (as in Figure 6).
    (c) Set up an integral for the length of the curve.
    (d) Use your calculator to find the length of the curve to four decimal places. Compare with the approximations in part (b).

30. Repeat Exercise 29 for the curve
$$y = x + \sin x \quad 0 \leq x \leq 2\pi$$

31. Use either a computer algebra system or a table of integrals to find the *exact* length of the arc of the curve $y = e^x$ that lies between the points $(0, 1)$ and $(2, e^2)$.

32. Use either a computer algebra system or a table of integrals to find the *exact* length of the arc of the curve $y = x^{4/3}$ that lies between the points $(0, 0)$ and $(1, 1)$. If your CAS has trouble evaluating the integral, make a substitution that changes the integral into one that the CAS can evaluate.

33. Sketch the curve with equation $x^{2/3} + y^{2/3} = 1$ and use symmetry to find its length.

34. (a) Sketch the curve $y^3 = x^2$.
    (b) Use Formulas 3 and 4 to set up two integrals for the arc length from $(0, 0)$ to $(1, 1)$. Observe that one of these is an improper integral and evaluate both of them.
    (c) Find the length of the arc of this curve from $(-1, 1)$ to $(8, 4)$.

35. Find the arc length function for the curve $y = 2x^{3/2}$ with starting point $P_0(1, 2)$.

36. (a) Find the arc length function for the curve $y = \ln(\sin x)$, $0 < x < \pi$, with starting point $(\pi/2, 0)$.
    (b) Graph both the curve and its arc length function on the same screen.

37. Find the arc length function for the curve
$y = \sin^{-1} x + \sqrt{1 - x^2}$ with starting point $(0, 1)$.

38. The arc length function for a curve $y = f(x)$, where f is an increasing function, is $s(x) = \int_0^x \sqrt{3t + 5}\, dt$.
    (a) If f has y-intercept 2, find an equation for f.
    (b) What point on the graph of f is 3 units along the curve from the y-intercept? State your answer rounded to 3 decimal places.

39. For the function $f(x) = \tfrac{1}{4}e^x + e^{-x}$, prove that the arc length on any interval has the same value as the area under the curve.

40. A steady wind blows a kite due west. The kite's height above ground from horizontal position $x = 0$ to $x = 80$ ft is given by $y = 150 - \tfrac{1}{40}(x - 50)^2$. Find the distance traveled by the kite.

41. A hawk flying at 15 m/s at an altitude of 180 m accidentally drops its prey. The parabolic trajectory of the falling prey is described by the equation
$$y = 180 - \dfrac{x^2}{45}$$

until it hits the ground, where y is its height above the ground and x is the horizontal distance traveled in meters. Calculate the distance traveled by the prey from the time it is dropped until the time it hits the ground. Express your answer correct to the nearest tenth of a meter.

42. The Gateway Arch in St. Louis (see the photo on page 543) was constructed using the equation

$$y = 211.49 - 20.96 \cosh 0.03291765x$$

for the central curve of the arch, where x and y are measured in meters and $|x| \leq 91.20$. Set up an integral for the length of the arch and use your calculator to estimate the length correct to the nearest meter.

43. A manufacturer of corrugated metal roofing wants to produce panels that are 28 in. wide and 2 in. high by processing flat sheets of metal as shown in the figure. The profile of the roofing takes the shape of a sine wave. Verify that the sine curve has equation $y = \sin(\pi x/7)$ and find the width w of a flat metal sheet that is needed to make a 28-inch panel. (Use your calculator to evaluate the integral correct to four significant digits.)

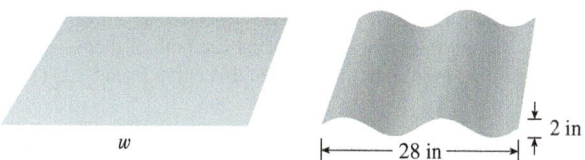

44. (a) The figure shows a telephone wire hanging between two poles at $x = -b$ and $x = b$. It takes the shape of a catenary with equation $y = c + a \cosh(x/a)$. Find the length of the wire.

(b) Suppose two telephone poles are 50 ft apart and the length of the wire between the poles is 51 ft. If the lowest point of the wire must be 20 ft above the ground, how high up on each pole should the wire be attached?

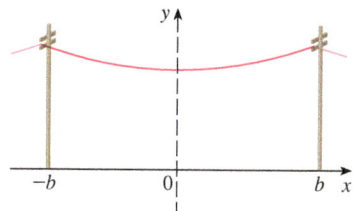

45. Find the length of the curve

$$y = \int_1^x \sqrt{t^3 - 1} \, dt \qquad 1 \leq x \leq 4$$

46. The curves with equations $x^n + y^n = 1$, $n = 4, 6, 8, \ldots$, are called **fat circles**. Graph the curves with $n = 2, 4, 6, 8,$ and 10 to see why. Set up an integral for the length L_{2k} of the fat circle with $n = 2k$. Without attempting to evaluate this integral, state the value of $\lim_{k \to \infty} L_{2k}$.

DISCOVERY PROJECT   ARC LENGTH CONTEST

The curves shown are all examples of graphs of continuous functions f that have the following properties.

1. $f(0) = 0$ and $f(1) = 0$.
2. $f(x) \geq 0$ for $0 \leq x \leq 1$.
3. The area under the graph of f from 0 to 1 is equal to 1.

The lengths L of these curves, however, are different.

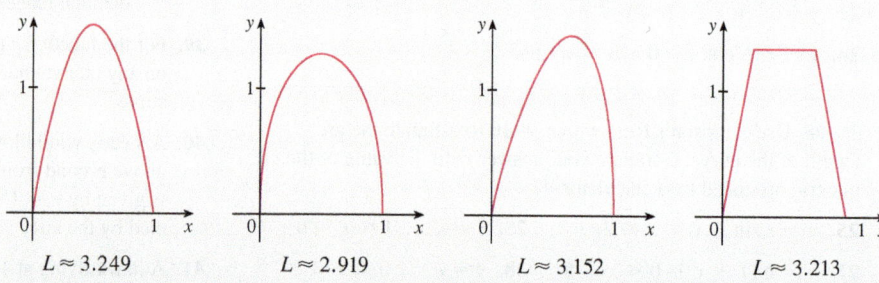

Try to discover formulas for two functions that satisfy the given conditions 1, 2, and 3. (Your graphs might be similar to the ones shown or could look quite different.) Then calculate the arc length of each graph. The winning entry will be the one with the smallest arc length.

8.2 Area of a Surface of Revolution

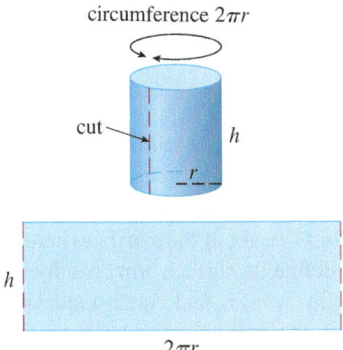

FIGURE 1

A surface of revolution is formed when a curve is rotated about a line. Such a surface is the lateral boundary of a solid of revolution of the type discussed in Sections 6.2 and 6.3.

We want to define the area of a surface of revolution in such a way that it corresponds to our intuition. If the surface area is A, we can imagine that painting the surface would require the same amount of paint as does a flat region with area A.

Let's start with some simple surfaces. The lateral surface area of a circular cylinder with radius r and height h is taken to be $A = 2\pi rh$ because we can imagine cutting the cylinder and unrolling it (as in Figure 1) to obtain a rectangle with dimensions $2\pi r$ and h.

Likewise, we can take a circular cone with base radius r and slant height l, cut it along the dashed line in Figure 2, and flatten it to form a sector of a circle with radius l and central angle $\theta = 2\pi r/l$. We know that, in general, the area of a sector of a circle with radius l and angle θ is $\frac{1}{2}l^2\theta$ (see Exercise 7.3.35) and so in this case the area is

$$A = \tfrac{1}{2}l^2\theta = \tfrac{1}{2}l^2\left(\frac{2\pi r}{l}\right) = \pi rl$$

Therefore we define the lateral surface area of a cone to be $A = \pi rl$.

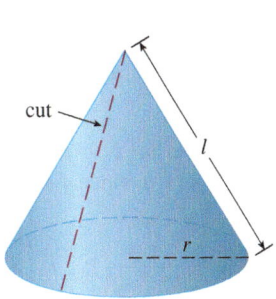
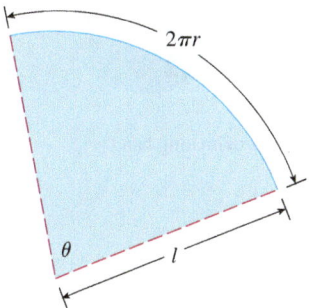

FIGURE 2

What about more complicated surfaces of revolution? If we follow the strategy we used with arc length, we can approximate the original curve by a polygon. When this polygon is rotated about an axis, it creates a simpler surface whose surface area approximates the actual surface area. By taking a limit, we can determine the exact surface area.

The approximating surface, then, consists of a number of *bands*, each formed by rotating a line segment about an axis. To find the surface area, each of these bands can be considered a portion of a circular cone, as shown in Figure 3. The area of the band (or frustum of a cone) with slant height l and upper and lower radii r_1 and r_2 is found by subtracting the areas of two cones:

$$A = \pi r_2(l_1 + l) - \pi r_1 l_1 = \pi[(r_2 - r_1)l_1 + r_2 l]$$

From similar triangles we have

$$\frac{l_1}{r_1} = \frac{l_1 + l}{r_2}$$

which gives

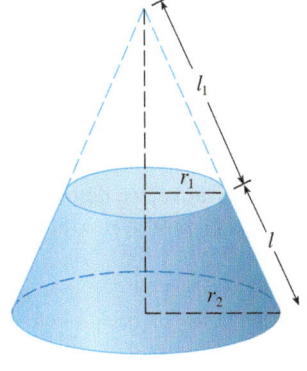

FIGURE 3

$$r_2 l_1 = r_1 l_1 + r_1 l \qquad \text{or} \qquad (r_2 - r_1)l_1 = r_1 l$$

Putting this in Equation 1, we get

$$A = \pi(r_1 l + r_2 l)$$

or

$$\boxed{2 \quad A = 2\pi r l}$$

where $r = \tfrac{1}{2}(r_1 + r_2)$ is the average radius of the band.

Now we apply this formula to our strategy. Consider the surface shown in Figure 4, which is obtained by rotating the curve $y = f(x)$, $a \leq x \leq b$, about the x-axis, where f is positive and has a continuous derivative. In order to define its surface area, we divide the interval $[a, b]$ into n subintervals with endpoints $x_0, x_1, \ldots, x_n$ and equal width Δx, as we did in determining arc length. If $y_i = f(x_i)$, then the point $P_i(x_i, y_i)$ lies on the curve. The part of the surface between x_{i-1} and x_i is approximated by taking the line segment $P_{i-1} P_i$ and rotating it about the x-axis. The result is a band with slant height $l = |P_{i-1} P_i|$ and average radius $r = \tfrac{1}{2}(y_{i-1} + y_i)$ so, by Formula 2, its surface area is

$$2\pi \frac{y_{i-1} + y_i}{2} |P_{i-1} P_i|$$

As in the proof of Theorem 8.1.2, we have

$$|P_{i-1} P_i| = \sqrt{1 + [f'(x_i^*)]^2}\, \Delta x$$

where x_i^* is some number in $[x_{i-1}, x_i]$. When Δx is small, we have $y_i = f(x_i) \approx f(x_i^*)$ and also $y_{i-1} = f(x_{i-1}) \approx f(x_i^*)$, since f is continuous. Therefore

$$2\pi \frac{y_{i-1} + y_i}{2} |P_{i-1} P_i| \approx 2\pi f(x_i^*) \sqrt{1 + [f'(x_i^*)]^2}\, \Delta x$$

and so an approximation to what we think of as the area of the complete surface of revolution is

$$3 \quad \sum_{i=1}^{n} 2\pi f(x_i^*) \sqrt{1 + [f'(x_i^*)]^2}\, \Delta x$$

This approximation appears to become better as $n \to \infty$ and, recognizing (3) as a Riemann sum for the function $g(x) = 2\pi f(x) \sqrt{1 + [f'(x)]^2}$, we have

$$\lim_{n \to \infty} \sum_{i=1}^{n} 2\pi f(x_i^*) \sqrt{1 + [f'(x_i^*)]^2}\, \Delta x = \int_a^b 2\pi f(x) \sqrt{1 + [f'(x)]^2}\, dx$$

Therefore, in the case where f is positive and has a continuous derivative, we define the **surface area** of the surface obtained by rotating the curve $y = f(x)$, $a \leq x \leq b$, about the x-axis as

$$\boxed{4 \quad S = \int_a^b 2\pi f(x) \sqrt{1 + [f'(x)]^2}\, dx}$$

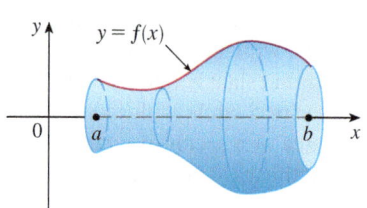

(a) Surface of revolution

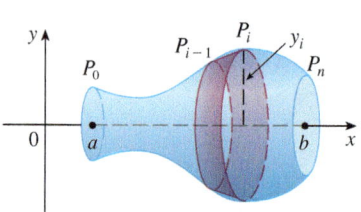

(b) Approximating band

FIGURE 4

With the Leibniz notation for derivatives, this formula becomes

$$\boxed{\;S = \int_a^b 2\pi y \sqrt{1 + \left(\frac{dy}{dx}\right)^2}\, dx\;} \quad [5]$$

If the curve is described as $x = g(y)$, $c \leq y \leq d$, then the formula for surface area becomes

$$\boxed{\;S = \int_c^d 2\pi y \sqrt{1 + \left(\frac{dx}{dy}\right)^2}\, dy\;} \quad [6]$$

and both Formulas 5 and 6 can be summarized symbolically, using the notation for arc length given in Section 8.1, as

$$\boxed{\;S = \int 2\pi y\, ds\;} \quad [7]$$

For rotation about the y-axis, the surface area formula becomes

$$\boxed{\;S = \int 2\pi x\, ds\;} \quad [8]$$

where, as before, we can use either

$$ds = \sqrt{1 + \left(\frac{dy}{dx}\right)^2}\, dx \quad \text{or} \quad ds = \sqrt{1 + \left(\frac{dx}{dy}\right)^2}\, dy$$

These formulas can be remembered by thinking of $2\pi y$ or $2\pi x$ as the circumference of a circle traced out by the point (x, y) on the curve as it is rotated about the x-axis or y-axis, respectively (see Figure 5).

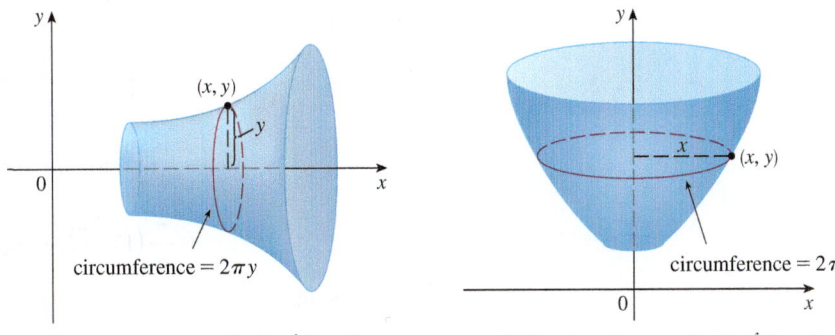

FIGURE 5   (a) Rotation about x-axis: $S = \int 2\pi y\, ds$ (b) Rotation about y-axis: $S = \int 2\pi x\, ds$

554   CHAPTER 8   Further Applications of Integration

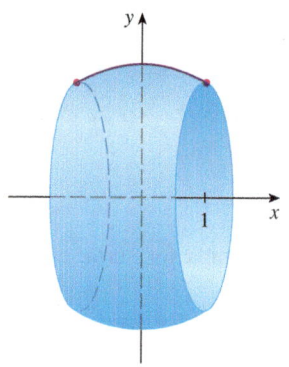

FIGURE 6

Figure 6 shows the portion of the sphere whose surface area is computed in Example 1.

EXAMPLE 1 The curve $y = \sqrt{4 - x^2}$, $-1 \leq x \leq 1$, is an arc of the circle $x^2 + y^2 = 4$. Find the area of the surface obtained by rotating this arc about the x-axis. (The surface is a portion of a sphere of radius 2. See Figure 6.)

SOLUTION We have

$$\frac{dy}{dx} = \tfrac{1}{2}(4 - x^2)^{-1/2}(-2x) = \frac{-x}{\sqrt{4 - x^2}}$$

and so, by Formula 5, the surface area is

$$S = \int_{-1}^{1} 2\pi y \sqrt{1 + \left(\frac{dy}{dx}\right)^2} \, dx$$

$$= 2\pi \int_{-1}^{1} \sqrt{4 - x^2} \sqrt{1 + \frac{x^2}{4 - x^2}} \, dx$$

$$= 2\pi \int_{-1}^{1} \sqrt{4 - x^2} \sqrt{\frac{4 - x^2 + x^2}{4 - x^2}} \, dx$$

$$= 2\pi \int_{-1}^{1} \sqrt{4 - x^2} \, \frac{2}{\sqrt{4 - x^2}} \, dx = 4\pi \int_{-1}^{1} 1 \, dx = 4\pi(2) = 8\pi \quad \blacksquare$$

Figure 7 shows the surface of revolution whose area is computed in Example 2.

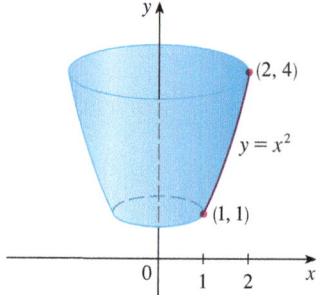

FIGURE 7

EXAMPLE 2 The arc of the parabola $y = x^2$ from $(1, 1)$ to $(2, 4)$ is rotated about the y-axis. Find the area of the resulting surface.

SOLUTION 1 Using

$$y = x^2 \quad \text{and} \quad \frac{dy}{dx} = 2x$$

we have, from Formula 8,

$$S = \int 2\pi x \, ds$$

$$= \int_{1}^{2} 2\pi x \sqrt{1 + \left(\frac{dy}{dx}\right)^2} \, dx$$

$$= 2\pi \int_{1}^{2} x \sqrt{1 + 4x^2} \, dx$$

Substituting $u = 1 + 4x^2$, we have $du = 8x \, dx$. Remembering to change the limits of integration, we have

$$S = 2\pi \int_{5}^{17} \sqrt{u} \cdot \tfrac{1}{8} \, du$$

$$= \frac{\pi}{4} \int_{5}^{17} u^{1/2} \, du = \frac{\pi}{4} \left[\tfrac{2}{3} u^{3/2}\right]_{5}^{17}$$

$$= \frac{\pi}{6} \left(17\sqrt{17} - 5\sqrt{5}\right)$$

SOLUTION 2 Using

$$x = \sqrt{y} \quad \text{and} \quad \frac{dx}{dy} = \frac{1}{2\sqrt{y}}$$

As a check on our answer to Example 2, notice from Figure 7 that the surface area should be close to that of a circular cylinder with the same height and radius halfway between the upper and lower radius of the surface: $2\pi(1.5)(3) \approx 28.27$. We computed that the surface area was

$$\frac{\pi}{6}(17\sqrt{17} - 5\sqrt{5}) \approx 30.85$$

which seems reasonable. Alternatively, the surface area should be slightly larger than the area of a frustum of a cone with the same top and bottom edges. From Equation 2, this is $2\pi(1.5)(\sqrt{10}) \approx 29.80$.

we have

$$S = \int 2\pi x\, ds = \int_1^4 2\pi x \sqrt{1 + \left(\frac{dx}{dy}\right)^2}\, dy$$

$$= 2\pi \int_1^4 \sqrt{y}\sqrt{1 + \frac{1}{4y}}\, dy = 2\pi \int_1^4 \sqrt{y + \tfrac{1}{4}}\, dy = \pi \int_1^4 \sqrt{4y + 1}\, dy$$

$$= \frac{\pi}{4} \int_5^{17} \sqrt{u}\, du \qquad \text{(where } u = 1 + 4y\text{)}$$

$$= \frac{\pi}{6}(17\sqrt{17} - 5\sqrt{5}) \qquad \text{(as in Solution 1)}$$

EXAMPLE 3 Find the area of the surface generated by rotating the curve $y = e^x$, $0 \le x \le 1$, about the x-axis.

SOLUTION Using Formula 5 with

$$y = e^x \qquad \text{and} \qquad \frac{dy}{dx} = e^x$$

Another method: Use Formula 6 with $x = \ln y$.

we have

$$S = \int_0^1 2\pi y \sqrt{1 + \left(\frac{dy}{dx}\right)^2}\, dx = 2\pi \int_0^1 e^x \sqrt{1 + e^{2x}}\, dx$$

$$= 2\pi \int_1^e \sqrt{1 + u^2}\, du \qquad \text{(where } u = e^x\text{)}$$

$$= 2\pi \int_{\pi/4}^{\alpha} \sec^3\theta\, d\theta \qquad \text{(where } u = \tan\theta \text{ and } \alpha = \tan^{-1}e\text{)}$$

Or use Formula 21 in the Table of Integrals.

$$= 2\pi \cdot \tfrac{1}{2}\Big[\sec\theta \tan\theta + \ln|\sec\theta + \tan\theta|\Big]_{\pi/4}^{\alpha} \qquad \text{(by Example 7.2.8)}$$

$$= \pi\Big[\sec\alpha \tan\alpha + \ln(\sec\alpha + \tan\alpha) - \sqrt{2} - \ln(\sqrt{2} + 1)\Big]$$

Since $\tan\alpha = e$, we have $\sec^2\alpha = 1 + \tan^2\alpha = 1 + e^2$ and

$$S = \pi\Big[e\sqrt{1 + e^2} + \ln(e + \sqrt{1 + e^2}) - \sqrt{2} - \ln(\sqrt{2} + 1)\Big]$$

8.2 EXERCISES

1–6
(a) Set up an integral for the area of the surface obtained by rotating the curve about (i) the x-axis and (ii) the y-axis.
(b) Use the numerical integration capability of a calculator to evaluate the surface areas correct to four decimal places.

1. $y = \tan x$, $0 \le x \le \pi/3$
2. $y = x^{-2}$, $1 \le x \le 2$
3. $y = e^{-x^2}$, $-1 \le x \le 1$
4. $x = \ln(2y + 1)$, $0 \le y \le 1$
5. $x = y + y^3$, $0 \le y \le 1$
6. $y = \tan^{-1}x$, $0 \le x \le 2$

7–14 Find the exact area of the surface obtained by rotating the curve about the x-axis.

7. $y = x^3$, $0 \le x \le 2$
8. $y = \sqrt{5 - x}$, $3 \le x \le 5$
9. $y^2 = x + 1$, $0 \le x \le 3$
10. $y = \sqrt{1 + e^x}$, $0 \le x \le 1$
11. $y = \cos(\tfrac{1}{2}x)$, $0 \le x \le \pi$
12. $y = \dfrac{x^3}{6} + \dfrac{1}{2x}$, $\tfrac{1}{2} \le x \le 1$
13. $x = \tfrac{1}{3}(y^2 + 2)^{3/2}$, $1 \le y \le 2$
14. $x = 1 + 2y^2$, $1 \le y \le 2$

15–18 The given curve is rotated about the y-axis. Find the area of the resulting surface.

15. $y = \frac{1}{3}x^{3/2}, \ 0 \leq x \leq 12$

16. $x^{2/3} + y^{2/3} = 1, \ 0 \leq y \leq 1$

17. $x = \sqrt{a^2 - y^2}, \ 0 \leq y \leq a/2$

18. $y = \frac{1}{4}x^2 - \frac{1}{2}\ln x, \ 1 \leq x \leq 2$

19–22 Use Simpson's Rule with $n = 10$ to approximate the area of the surface obtained by rotating the curve about the x-axis. Compare your answer with the value of the integral produced by a calculator.

19. $y = \frac{1}{5}x^5, \ 0 \leq x \leq 5$ **20.** $y = x + x^2, \ 0 \leq x \leq 1$

21. $y = xe^x, \ 0 \leq x \leq 1$ **22.** $y = x \ln x, \ 1 \leq x \leq 2$

CAS **23–24** Use either a CAS or a table of integrals to find the exact area of the surface obtained by rotating the given curve about the x-axis.

23. $y = 1/x, \ 1 \leq x \leq 2$

24. $y = \sqrt{x^2 + 1}, \ 0 \leq x \leq 3$

CAS **25–26** Use a CAS to find the exact area of the surface obtained by rotating the curve about the y-axis. If your CAS has trouble evaluating the integral, express the surface area as an integral in the other variable.

25. $y = x^3, \ 0 \leq y \leq 1$ **26.** $y = \ln(x + 1), \ 0 \leq x \leq 1$

27. If the region $\mathcal{R} = \{(x, y) \mid x \geq 1, \ 0 \leq y \leq 1/x\}$ is rotated about the x-axis, the volume of the resulting solid is finite (see Exercise 7.8.63). Show that the surface area is infinite. (The surface is shown in the figure and is known as **Gabriel's horn**.)

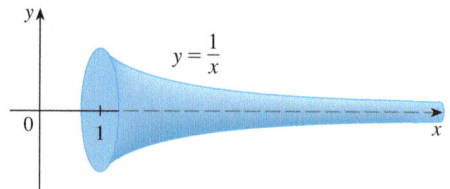

28. If the infinite curve $y = e^{-x}, \ x \geq 0$, is rotated about the x-axis, find the area of the resulting surface.

29. (a) If $a > 0$, find the area of the surface generated by rotating the loop of the curve $3ay^2 = x(a - x)^2$ about the x-axis.
(b) Find the surface area if the loop is rotated about the y-axis.

30. A group of engineers is building a parabolic satellite dish whose shape will be formed by rotating the curve $y = ax^2$ about the y-axis. If the dish is to have a 10-ft diameter and a maximum depth of 2 ft, find the value of a and the surface area of the dish.

31. (a) The ellipse
$$\frac{x^2}{a^2} + \frac{y^2}{b^2} = 1 \quad a > b$$
is rotated about the x-axis to form a surface called an *ellipsoid*, or *prolate spheroid*. Find the surface area of this ellipsoid.
(b) If the ellipse in part (a) is rotated about its minor axis (the y-axis), the resulting ellipsoid is called an *oblate spheroid*. Find the surface area of this ellipsoid.

32. Find the surface area of the torus in Exercise 6.2.63.

33. If the curve $y = f(x), \ a \leq x \leq b$, is rotated about the horizontal line $y = c$, where $f(x) \leq c$, find a formula for the area of the resulting surface.

CAS **34.** Use the result of Exercise 33 to set up an integral to find the area of the surface generated by rotating the curve $y = \sqrt{x}$, $0 \leq x \leq 4$, about the line $y = 4$. Then use a CAS to evaluate the integral.

35. Find the area of the surface obtained by rotating the circle $x^2 + y^2 = r^2$ about the line $y = r$.

36. (a) Show that the surface area of a zone of a sphere that lies between two parallel planes is $S = 2\pi Rh$, where R is the radius of the sphere and h is the distance between the planes. (Notice that S depends only on the distance between the planes and not on their location, provided that both planes intersect the sphere.)
(b) Show that the surface area of a zone of a *cylinder* with radius R and height h is the same as the surface area of the zone of a *sphere* in part (a).

37. Show that if we rotate the curve $y = e^{x/2} + e^{-x/2}$ about the x-axis, the area of the resulting surface is the same value as the enclosed volume for any interval $a \leq x \leq b$.

38. Let L be the length of the curve $y = f(x), \ a \leq x \leq b$, where f is positive and has a continuous derivative. Let S_f be the surface area generated by rotating the curve about the x-axis. If c is a positive constant, define $g(x) = f(x) + c$ and let S_g be the corresponding surface area generated by the curve $y = g(x), \ a \leq x \leq b$. Express S_g in terms of S_f and L.

39. Formula 4 is valid only when $f(x) \geq 0$. Show that when $f(x)$ is not necessarily positive, the formula for surface area becomes
$$S = \int_a^b 2\pi |f(x)| \sqrt{1 + [f'(x)]^2} \ dx$$

DISCOVERY PROJECT ROTATING ON A SLANT

We know how to find the volume of a solid of revolution obtained by rotating a region about a horizontal or vertical line (see Section 6.2). We also know how to find the surface area of a surface of revolution if we rotate a curve about a horizontal or vertical line (see Section 8.2). But what if we rotate about a slanted line, that is, a line that is neither horizontal nor vertical? In this project you are asked to discover formulas for the volume of a solid of revolution and for the area of a surface of revolution when the axis of rotation is a slanted line.

Let C be the arc of the curve $y = f(x)$ between the points $P(p, f(p))$ and $Q(q, f(q))$ and let $\mathcal{R}$ be the region bounded by C, by the line $y = mx + b$ (which lies entirely below C), and by the perpendiculars to the line from P and Q.

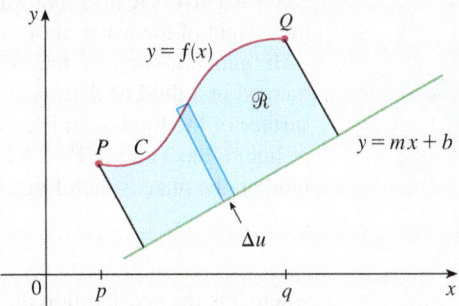

1. Show that the area of $\mathcal{R}$ is

$$\frac{1}{1 + m^2} \int_p^q [f(x) - mx - b][1 + mf'(x)]\, dx$$

[*Hint:* This formula can be verified by subtracting areas, but it will be helpful throughout the project to derive it by first approximating the area using rectangles perpendicular to the line, as shown in the following figure. Use the figure to help express Δu in terms of Δx.]

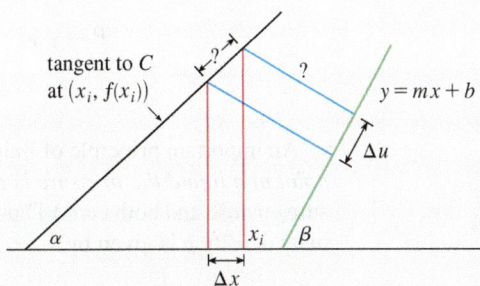

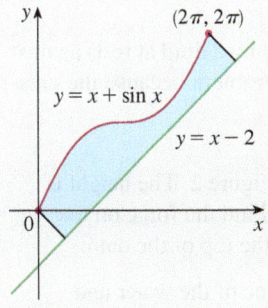

2. Find the area of the region shown in the figure at the left.

3. Find a formula (similar to the one in Problem 1) for the volume of the solid obtained by rotating $\mathcal{R}$ about the line $y = mx + b$.

4. Find the volume of the solid obtained by rotating the region of Problem 2 about the line $y = x - 2$.

5. Find a formula for the area of the surface obtained by rotating C about the line $y = mx + b$.

CAS **6.** Use a computer algebra system to find the exact area of the surface obtained by rotating the curve $y = \sqrt{x}$, $0 \leq x \leq 4$, about the line $y = \frac{1}{2}x$. Then approximate your result to three decimal places.

8.3 Applications to Physics and Engineering

Among the many applications of integral calculus to physics and engineering, we consider two here: force due to water pressure and centers of mass. As with our previous applications to geometry (areas, volumes, and lengths) and to work, our strategy is to break up the physical quantity into a large number of small parts, approximate each small part, add the results (giving a Riemann sum), take the limit, and then evaluate the resulting integral.

■ Hydrostatic Pressure and Force

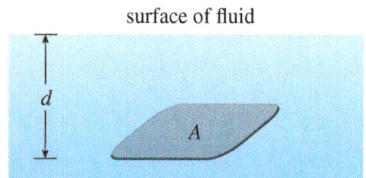

FIGURE 1

Deep-sea divers realize that water pressure increases as they dive deeper. This is because the weight of the water above them increases.

In general, suppose that a thin horizontal plate with area A square meters is submerged in a fluid of density ρ kilograms per cubic meter at a depth d meters below the surface of the fluid as in Figure 1. The fluid directly above the plate (think of a column of liquid) has volume $V = Ad$, so its mass is $m = \rho V = \rho A d$. The force exerted by the fluid on the plate is therefore

$$F = mg = \rho g A d$$

where g is the acceleration due to gravity. The **pressure** P on the plate is defined to be the force per unit area:

$$P = \frac{F}{A} = \rho g d$$

The SI unit for measuring pressure is a newton per square meter, which is called a pascal (abbreviation: $1 \text{ N/m}^2 = 1$ Pa). Since this is a small unit, the kilopascal (kPa) is often used. For instance, because the density of water is $\rho = 1000 \text{ kg/m}^3$, the pressure at the bottom of a swimming pool 2 m deep is

$$P = \rho g d = 1000 \text{ kg/m}^3 \times 9.8 \text{ m/s}^2 \times 2 \text{ m}$$

$$= 19{,}600 \text{ Pa} = 19.6 \text{ kPa}$$

When using US Customary units, we write $P = \rho g d = \delta d$, where $\delta = \rho g$ is the *weight density* (as opposed to ρ, which is the *mass density*). For instance, the weight density of water is $\delta = 62.5 \text{ lb/ft}^3$.

An important principle of fluid pressure is the experimentally verified fact that *at any point in a liquid the pressure is the same in all directions*. (A diver feels the same pressure on nose and both ears.) Thus the pressure in *any* direction at a depth d in a fluid with mass density ρ is given by

$$\boxed{1} \qquad P = \rho g d = \delta d$$

This helps us determine the hydrostatic force (the force exerted by a fluid at rest) against a *vertical* plate or wall or dam. This is not a straightforward problem because the pressure is not constant but increases as the depth increases.

EXAMPLE 1 A dam has the shape of the trapezoid shown in Figure 2. The height is 20 m and the width is 50 m at the top and 30 m at the bottom. Find the force on the dam due to hydrostatic pressure if the water level is 4 m from the top of the dam.

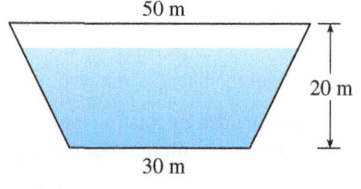

FIGURE 2

SOLUTION We choose a vertical x-axis with origin at the surface of the water and directed downward as in Figure 3(a). The depth of the water is 16 m, so we divide the interval $[0, 16]$ into subintervals of equal length with endpoints x_i and we choose $x_i^* \in [x_{i-1}, x_i]$. The ith horizontal strip of the dam is approximated by a rectangle with height Δx and width w_i, where, from similar triangles in Figure 3(b),

$$\frac{a}{16 - x_i^*} = \frac{10}{20} \quad \text{or} \quad a = \frac{16 - x_i^*}{2} = 8 - \frac{x_i^*}{2}$$

and so
$$w_i = 2(15 + a) = 2\left(15 + 8 - \tfrac{1}{2}x_i^*\right) = 46 - x_i^*$$

If A_i is the area of the ith strip, then

$$A_i \approx w_i \, \Delta x = (46 - x_i^*) \, \Delta x$$

If Δx is small, then the pressure P_i on the ith strip is almost constant and we can use Equation 1 to write

$$P_i \approx 1000 g x_i^*$$

The hydrostatic force F_i acting on the ith strip is the product of the pressure and the area:

$$F_i = P_i A_i \approx 1000 g x_i^* (46 - x_i^*) \, \Delta x$$

Adding these forces and taking the limit as $n \to \infty$, we obtain the total hydrostatic force on the dam:

$$F = \lim_{n \to \infty} \sum_{i=1}^{n} 1000 g x_i^* (46 - x_i^*) \, \Delta x = \int_0^{16} 1000 g x (46 - x) \, dx$$

$$= 1000(9.8) \int_0^{16} (46x - x^2) \, dx = 9800 \left[23x^2 - \frac{x^3}{3} \right]_0^{16}$$

$$\approx 4.43 \times 10^7 \text{ N} \qquad \blacksquare$$

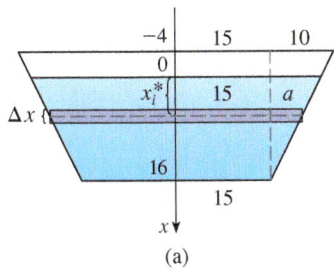

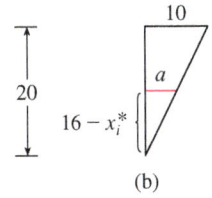

FIGURE 3

EXAMPLE 2 Find the hydrostatic force on one end of a cylindrical drum with radius 3 ft if the drum is submerged in water 10 ft deep.

SOLUTION In this example it is convenient to choose the axes as in Figure 4 so that the origin is placed at the center of the drum. Then the circle has a simple equation, $x^2 + y^2 = 9$. As in Example 1 we divide the circular region into horizontal strips of equal width. From the equation of the circle, we see that the length of the ith strip is $2\sqrt{9 - (y_i^*)^2}$ and so its area is

$$A_i = 2\sqrt{9 - (y_i^*)^2} \, \Delta y$$

Because the weight density of water is $\delta = 62.5$ lb/ft^3, the pressure on this strip is approximately

$$\delta d_i = 62.5(7 - y_i^*)$$

and so the force on the strip is approximately

$$\delta d_i A_i = 62.5(7 - y_i^*) \, 2\sqrt{9 - (y_i^*)^2} \, \Delta y$$

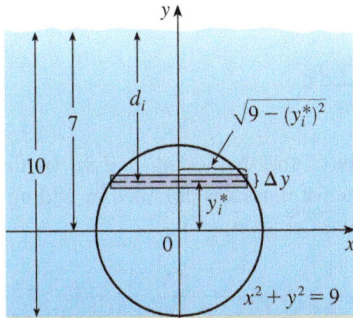

FIGURE 4

The total force is obtained by adding the forces on all the strips and taking the limit:

$$F = \lim_{n \to \infty} \sum_{i=1}^{n} 62.5(7 - y_i^*) \, 2\sqrt{9 - (y_i^*)^2} \, \Delta y$$

$$= 125 \int_{-3}^{3} (7 - y) \sqrt{9 - y^2} \, dy$$

$$= 125 \cdot 7 \int_{-3}^{3} \sqrt{9 - y^2} \, dy - 125 \int_{-3}^{3} y\sqrt{9 - y^2} \, dy$$

The second integral is 0 because the integrand is an odd function (see Theorem 5.5.7). The first integral can be evaluated using the trigonometric substitution $y = 3 \sin \theta$, but it's simpler to observe that it is the area of a semicircular disk with radius 3. Thus

$$F = 875 \int_{-3}^{3} \sqrt{9 - y^2} \, dy = 875 \cdot \tfrac{1}{2}\pi(3)^2$$

$$= \frac{7875\pi}{2} \approx 12{,}370 \text{ lb}$$

■ Moments and Centers of Mass

Our main objective here is to find the point P on which a thin plate of any given shape balances horizontally as in Figure 5. This point is called the **center of mass** (or center of gravity) of the plate.

We first consider the simpler situation illustrated in Figure 6, where two masses m_1 and m_2 are attached to a rod of negligible mass on opposite sides of a fulcrum and at distances d_1 and d_2 from the fulcrum. The rod will balance if

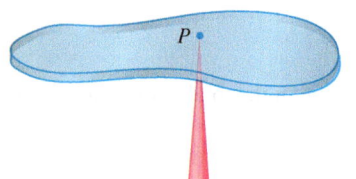

FIGURE 5

$$\boxed{2} \qquad m_1 d_1 = m_2 d_2$$

This is an experimental fact discovered by Archimedes and called the Law of the Lever. (Think of a lighter person balancing a heavier one on a seesaw by sitting farther away from the center.)

Now suppose that the rod lies along the x-axis with m_1 at x_1 and m_2 at x_2 and the center of mass at $\bar{x}$. If we compare Figures 6 and 7, we see that $d_1 = \bar{x} - x_1$ and $d_2 = x_2 - \bar{x}$ and so Equation 2 gives

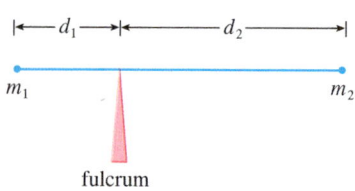

FIGURE 6

$$m_1(\bar{x} - x_1) = m_2(x_2 - \bar{x})$$

$$m_1 \bar{x} + m_2 \bar{x} = m_1 x_1 + m_2 x_2$$

$$\boxed{3} \qquad \bar{x} = \frac{m_1 x_1 + m_2 x_2}{m_1 + m_2}$$

The numbers $m_1 x_1$ and $m_2 x_2$ are called the **moments** of the masses m_1 and m_2 (with respect to the origin), and Equation 3 says that the center of mass $\bar{x}$ is obtained by adding the moments of the masses and dividing by the total mass $m = m_1 + m_2$.

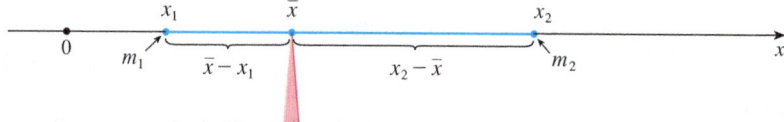

FIGURE 7

In general, if we have a system of n particles with masses $m_1, m_2, \ldots, m_n$ located at the points $x_1, x_2, \ldots, x_n$ on the x-axis, it can be shown similarly that the center of mass of the system is located at

$$\boxed{4} \quad \bar{x} = \frac{\sum_{i=1}^{n} m_i x_i}{\sum_{i=1}^{n} m_i} = \frac{\sum_{i=1}^{n} m_i x_i}{m}$$

where $m = \sum m_i$ is the total mass of the system, and the sum of the individual moments

$$M = \sum_{i=1}^{n} m_i x_i$$

is called the **moment of the system about the origin**. Then Equation 4 could be rewritten as $m\bar{x} = M$, which says that if the total mass were considered as being concentrated at the center of mass $\bar{x}$, then its moment would be the same as the moment of the system.

Now we consider a system of n particles with masses $m_1, m_2, \ldots, m_n$ located at the points $(x_1, y_1), (x_2, y_2), \ldots, (x_n, y_n)$ in the xy-plane as shown in Figure 8. By analogy with the one-dimensional case, we define the **moment of the system about the y-axis** to be

$$\boxed{5} \quad M_y = \sum_{i=1}^{n} m_i x_i$$

and the **moment of the system about the x-axis** as

$$\boxed{6} \quad M_x = \sum_{i=1}^{n} m_i y_i$$

Then M_y measures the tendency of the system to rotate about the y-axis and M_x measures the tendency to rotate about the x-axis.

As in the one-dimensional case, the coordinates $(\bar{x}, \bar{y})$ of the center of mass are given in terms of the moments by the formulas

$$\boxed{7} \quad \bar{x} = \frac{M_y}{m} \qquad \bar{y} = \frac{M_x}{m}$$

where $m = \sum m_i$ is the total mass. Since $m\bar{x} = M_y$ and $m\bar{y} = M_x$, the center of mass $(\bar{x}, \bar{y})$ is the point where a single particle of mass m would have the same moments as the system.

EXAMPLE 3 Find the moments and center of mass of the system of objects that have masses 3, 4, and 8 at the points $(-1, 1)$, $(2, -1)$, and $(3, 2)$, respectively.

SOLUTION We use Equations 5 and 6 to compute the moments:

$$M_y = 3(-1) + 4(2) + 8(3) = 29$$

$$M_x = 3(1) + 4(-1) + 8(2) = 15$$

Since $m = 3 + 4 + 8 = 15$, we use Equations 7 to obtain

$$\bar{x} = \frac{M_y}{m} = \frac{29}{15} \qquad \bar{y} = \frac{M_x}{m} = \frac{15}{15} = 1$$

Thus the center of mass is $\left(1\frac{14}{15}, 1\right)$. (See Figure 9.)

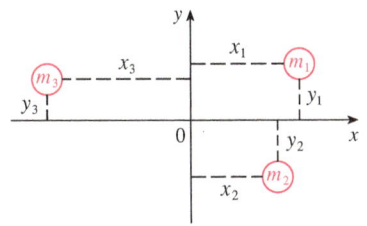

FIGURE 8

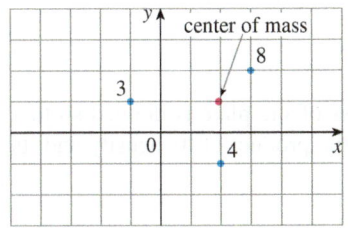

FIGURE 9

Next we consider a flat plate (called a *lamina*) with uniform density ρ that occupies a region $\mathcal{R}$ of the plane. We wish to locate the center of mass of the plate, which is called the **centroid** of $\mathcal{R}$. In doing so we use the following physical principles: The **symmetry principle** says that if $\mathcal{R}$ is symmetric about a line l, then the centroid of $\mathcal{R}$ lies on l. (If $\mathcal{R}$ is reflected about l, then $\mathcal{R}$ remains the same so its centroid remains fixed. But the only fixed points lie on l.) Thus the centroid of a rectangle is its center. Moments should be defined so that if the entire mass of a region is concentrated at the center of mass, then its moments remain unchanged. Also, the moment of the union of two nonoverlapping regions should be the sum of the moments of the individual regions.

Suppose that the region $\mathcal{R}$ is of the type shown in Figure 10(a); that is, $\mathcal{R}$ lies between the lines $x = a$ and $x = b$, above the x-axis, and beneath the graph of f, where f is a continuous function. We divide the interval $[a, b]$ into n subintervals with endpoints x_0, $x_1, \ldots, x_n$ and equal width Δx. We choose the sample point x_i^* to be the midpoint $\bar{x}_i$ of the ith subinterval, that is, $\bar{x}_i = (x_{i-1} + x_i)/2$. This determines the polygonal approximation to $\mathcal{R}$ shown in Figure 10(b). The centroid of the ith approximating rectangle R_i is its center $C_i(\bar{x}_i, \frac{1}{2} f(\bar{x}_i))$. Its area is $f(\bar{x}_i) \Delta x$, so its mass is

$$\rho f(\bar{x}_i) \Delta x$$

The moment of R_i about the y-axis is the product of its mass and the distance from C_i to the y-axis, which is $\bar{x}_i$. Thus

$$M_y(R_i) = [\rho f(\bar{x}_i) \Delta x] \bar{x}_i = \rho \bar{x}_i f(\bar{x}_i) \Delta x$$

Adding these moments, we obtain the moment of the polygonal approximation to $\mathcal{R}$, and then by taking the limit as $n \to \infty$ we obtain the moment of $\mathcal{R}$ itself about the y-axis:

$$M_y = \lim_{n \to \infty} \sum_{i=1}^{n} \rho \bar{x}_i f(\bar{x}_i) \Delta x = \rho \int_a^b x f(x) \, dx$$

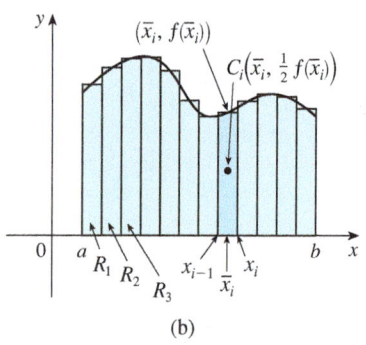

FIGURE 10

In a similar fashion we compute the moment of R_i about the x-axis as the product of its mass and the distance from C_i to the x-axis (which is half the height of R_i):

$$M_x(R_i) = [\rho f(\bar{x}_i) \Delta x] \tfrac{1}{2} f(\bar{x}_i) = \rho \cdot \tfrac{1}{2} [f(\bar{x}_i)]^2 \Delta x$$

Again we add these moments and take the limit to obtain the moment of $\mathcal{R}$ about the x-axis:

$$M_x = \lim_{n \to \infty} \sum_{i=1}^{n} \rho \cdot \tfrac{1}{2} [f(\bar{x}_i)]^2 \Delta x = \rho \int_a^b \tfrac{1}{2} [f(x)]^2 \, dx$$

Just as for systems of particles, the center of mass of the plate is defined so that $m\bar{x} = M_y$ and $m\bar{y} = M_x$. But the mass of the plate is the product of its density and its area:

$$m = \rho A = \rho \int_a^b f(x) \, dx$$

SECTION 8.3 Applications to Physics and Engineering

and so

$$\bar{x} = \frac{M_y}{m} = \frac{\rho \int_a^b xf(x)\,dx}{\rho \int_a^b f(x)\,dx} = \frac{\int_a^b xf(x)\,dx}{\int_a^b f(x)\,dx}$$

$$\bar{y} = \frac{M_x}{m} = \frac{\rho \int_a^b \tfrac{1}{2}[f(x)]^2\,dx}{\rho \int_a^b f(x)\,dx} = \frac{\int_a^b \tfrac{1}{2}[f(x)]^2\,dx}{\int_a^b f(x)\,dx}$$

Notice the cancellation of the ρ's. The location of the center of mass is independent of the density.

In summary, the center of mass of the plate (or the centroid of $\mathcal{R}$) is located at the point $(\bar{x}, \bar{y})$, where

8
$$\bar{x} = \frac{1}{A}\int_a^b xf(x)\,dx \qquad \bar{y} = \frac{1}{A}\int_a^b \tfrac{1}{2}[f(x)]^2\,dx$$

EXAMPLE 4 Find the center of mass of a semicircular plate of radius r.

SOLUTION In order to use (8) we place the semicircle as in Figure 11 so that $f(x) = \sqrt{r^2 - x^2}$ and $a = -r$, $b = r$. Here there is no need to use the formula to calculate $\bar{x}$ because, by the symmetry principle, the center of mass must lie on the y-axis, so $\bar{x} = 0$. The area of the semicircle is $A = \tfrac{1}{2}\pi r^2$, so

$$\bar{y} = \frac{1}{A}\int_{-r}^{r} \tfrac{1}{2}[f(x)]^2\,dx$$

$$= \frac{1}{\tfrac{1}{2}\pi r^2} \cdot \tfrac{1}{2} \int_{-r}^{r} \left(\sqrt{r^2 - x^2}\right)^2 dx$$

$$= \frac{2}{\pi r^2} \int_0^r (r^2 - x^2)\,dx \qquad \text{(since the integrand is even)}$$

$$= \frac{2}{\pi r^2}\left[r^2 x - \frac{x^3}{3}\right]_0^r$$

$$= \frac{2}{\pi r^2}\frac{2r^3}{3} = \frac{4r}{3\pi}$$

The center of mass is located at the point $(0, 4r/(3\pi))$. ∎

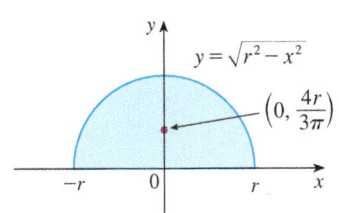

FIGURE 11

EXAMPLE 5 Find the centroid of the region bounded by the curves $y = \cos x$, $y = 0$, $x = 0$, and $x = \pi/2$.

SOLUTION The area of the region is

$$A = \int_0^{\pi/2} \cos x\,dx = \sin x\Big]_0^{\pi/2} = 1$$

so Formulas 8 give

$$\bar{x} = \frac{1}{A} \int_0^{\pi/2} x f(x)\, dx = \int_0^{\pi/2} x \cos x\, dx$$

$$= x \sin x \Big]_0^{\pi/2} - \int_0^{\pi/2} \sin x\, dx \quad \text{(by integration by parts)}$$

$$= \frac{\pi}{2} - 1$$

$$\bar{y} = \frac{1}{A} \int_0^{\pi/2} \tfrac{1}{2}[f(x)]^2\, dx = \tfrac{1}{2} \int_0^{\pi/2} \cos^2 x\, dx$$

$$= \tfrac{1}{4} \int_0^{\pi/2} (1 + \cos 2x)\, dx = \tfrac{1}{4}\Big[x + \tfrac{1}{2} \sin 2x\Big]_0^{\pi/2} = \frac{\pi}{8}$$

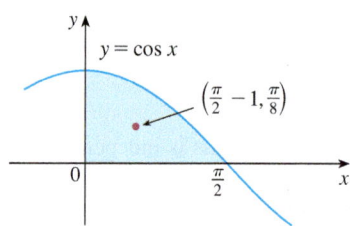

FIGURE 12

The centroid is $\left(\tfrac{1}{2}\pi - 1, \tfrac{1}{8}\pi\right)$ and is shown in Figure 12. ∎

If the region $\mathcal{R}$ lies between two curves $y = f(x)$ and $y = g(x)$, where $f(x) \geq g(x)$, as illustrated in Figure 13, then the same sort of argument that led to Formulas 8 can be used to show that the centroid of $\mathcal{R}$ is $(\bar{x}, \bar{y})$, where

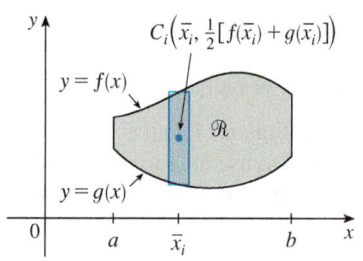

FIGURE 13

$$\boxed{\begin{aligned} \bar{x} &= \frac{1}{A} \int_a^b x[f(x) - g(x)]\, dx \\ \bar{y} &= \frac{1}{A} \int_a^b \tfrac{1}{2}\{[f(x)]^2 - [g(x)]^2\}\, dx \end{aligned}}$$

(9)

(See Exercise 51.)

EXAMPLE 6 Find the centroid of the region bounded by the line $y = x$ and the parabola $y = x^2$.

SOLUTION The region is sketched in Figure 14. We take $f(x) = x$, $g(x) = x^2$, $a = 0$, and $b = 1$ in Formulas 9. First we note that the area of the region is

$$A = \int_0^1 (x - x^2)\, dx = \frac{x^2}{2} - \frac{x^3}{3}\Big]_0^1 = \frac{1}{6}$$

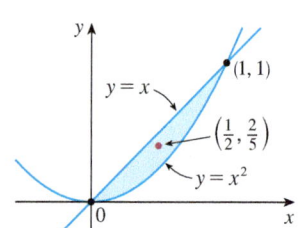

FIGURE 14

Therefore

$$\bar{x} = \frac{1}{A}\int_0^1 x[f(x) - g(x)]\, dx = \frac{1}{\frac{1}{6}}\int_0^1 x(x - x^2)\, dx$$

$$= 6\int_0^1 (x^2 - x^3)\, dx = 6\left[\frac{x^3}{3} - \frac{x^4}{4}\right]_0^1 = \frac{1}{2}$$

$$\bar{y} = \frac{1}{A}\int_0^1 \tfrac{1}{2}\{[f(x)]^2 - [g(x)]^2\}\, dx = \frac{1}{\frac{1}{6}}\int_0^1 \tfrac{1}{2}(x^2 - x^4)\, dx$$

$$= 3\left[\frac{x^3}{3} - \frac{x^5}{5}\right]_0^1 = \frac{2}{5}$$

The centroid is $\left(\tfrac{1}{2}, \tfrac{2}{5}\right)$. ∎

We end this section by showing a surprising connection between centroids and volumes of revolution.

This theorem is named after the Greek mathematician Pappus of Alexandria, who lived in the fourth century AD.

Theorem of Pappus Let $\mathcal{R}$ be a plane region that lies entirely on one side of a line l in the plane. If $\mathcal{R}$ is rotated about l, then the volume of the resulting solid is the product of the area A of $\mathcal{R}$ and the distance d traveled by the centroid of $\mathcal{R}$.

PROOF We give the proof for the special case in which the region lies between $y = f(x)$ and $y = g(x)$ as in Figure 13 and the line l is the y-axis. Using the method of cylindrical shells (see Section 6.3), we have

$$V = \int_a^b 2\pi x [f(x) - g(x)]\, dx$$

$$= 2\pi \int_a^b x[f(x) - g(x)]\, dx$$

$$= 2\pi(\bar{x}A) \quad \text{(by Formulas 9)}$$

$$= (2\pi\bar{x})A = Ad$$

where $d = 2\pi\bar{x}$ is the distance traveled by the centroid during one rotation about the y-axis.

EXAMPLE 7 A torus is formed by rotating a circle of radius r about a line in the plane of the circle that is a distance $R\ (> r)$ from the center of the circle. Find the volume of the torus.

SOLUTION The circle has area $A = \pi r^2$. By the symmetry principle, its centroid is its center and so the distance traveled by the centroid during a rotation is $d = 2\pi R$. Therefore, by the Theorem of Pappus, the volume of the torus is

$$V = Ad = (2\pi R)(\pi r^2) = 2\pi^2 r^2 R$$

The method of Example 7 should be compared with the method of Exercise 6.2.63.

8.3 EXERCISES

1. An aquarium 5 ft long, 2 ft wide, and 3 ft deep is full of water. Find (a) the hydrostatic pressure on the bottom of the aquarium, (b) the hydrostatic force on the bottom, and (c) the hydrostatic force on one end of the aquarium.

2. A tank is 8 m long, 4 m wide, 2 m high, and contains kerosene with density 820 kg/m³ to a depth of 1.5 m. Find (a) the hydrostatic pressure on the bottom of the tank, (b) the hydrostatic force on the bottom, and (c) the hydrostatic force on one end of the tank.

3–11 A vertical plate is submerged (or partially submerged) in water and has the indicated shape. Explain how to approximate the hydrostatic force against one side of the plate by a Riemann sum. Then express the force as an integral and evaluate it.

3.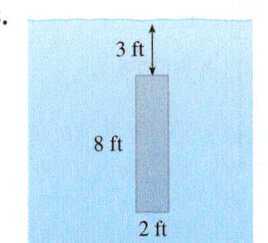

4.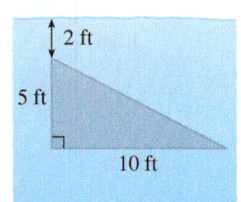

5.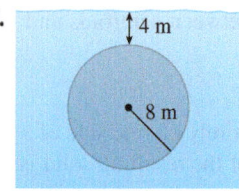

6.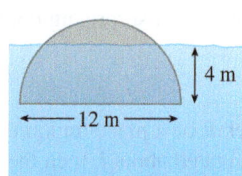

7.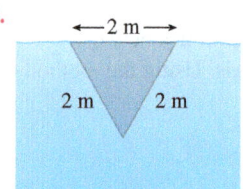

8.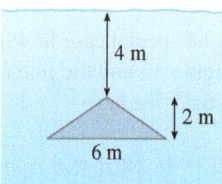

9.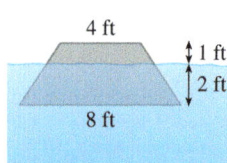

10.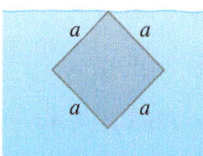

11.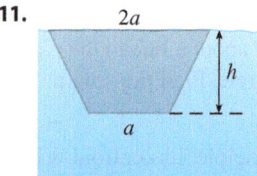

12. A milk truck carries milk with density 64.6 lb/ft³ in a horizontal cylindrical tank with diameter 6 ft.
(a) Find the force exerted by the milk on one end of the tank when the tank is full.
(b) What if the tank is half full?

13. A trough is filled with a liquid of density 840 kg/m³. The ends of the trough are equilateral triangles with sides 8 m long and vertex at the bottom. Find the hydrostatic force on one end of the trough.

14. A vertical dam has a semicircular gate as shown in the figure. Find the hydrostatic force against the gate.

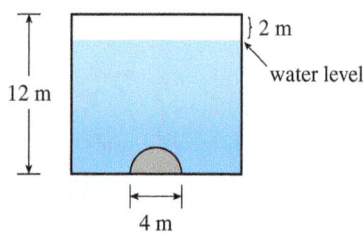

15. A cube with 20-cm-long sides is sitting on the bottom of an aquarium in which the water is one meter deep. Find the hydrostatic force on (a) the top of the cube and (b) one of the sides of the cube.

16. A dam is inclined at an angle of 30° from the vertical and has the shape of an isosceles trapezoid 100 ft wide at the top and 50 ft wide at the bottom and with a slant height of 70 ft. Find the hydrostatic force on the dam when it is full of water.

17. A swimming pool is 20 ft wide and 40 ft long and its bottom is an inclined plane, the shallow end having a depth of 3 ft and the deep end, 9 ft. If the pool is full of water, find the hydrostatic force on (a) the shallow end, (b) the deep end, (c) one of the sides, and (d) the bottom of the pool.

18. Suppose that a plate is immersed vertically in a fluid with density ρ and the width of the plate is $w(x)$ at a depth of x meters beneath the surface of the fluid. If the top of the plate is at depth a and the bottom is at depth b, show that the hydrostatic force on one side of the plate is

$$F = \int_a^b \rho g x w(x)\, dx$$

19. A metal plate was found submerged vertically in seawater, which has density 64 lb/ft³. Measurements of the width of the plate were taken at the indicated depths. Use Simpson's Rule to estimate the force of the water against the plate.

Depth (m)	7.0	7.4	7.8	8.2	8.6	9.0	9.4
Plate width (m)	1.2	1.8	2.9	3.8	3.6	4.2	4.4

20. (a) Use the formula of Exercise 18 to show that

$$F = (\rho g \bar{x}) A$$

where $\bar{x}$ is the x-coordinate of the centroid of the plate and A is its area. This equation shows that the hydrostatic force against a vertical plane region is the same as if the region were horizontal at the depth of the centroid of the region.
(b) Use the result of part (a) to give another solution to Exercise 10.

21–22 Point-masses m_i are located on the x-axis as shown. Find the moment M of the system about the origin and the center of mass $\bar{x}$.

21. $m_1 = 6$ at $x = 10$; $m_2 = 9$ at $x = 30$

22. $m_1 = 12$ at $x = -3$; $m_2 = 15$ at $x = 2$; $m_3 = 20$ at $x = 8$

23–24 The masses m_i are located at the points P_i. Find the moments M_x and M_y and the center of mass of the system.

23. $m_1 = 4$, $m_2 = 2$, $m_3 = 4$;
$P_1(2, -3)$, $P_2(-3, 1)$, $P_3(3, 5)$

24. $m_1 = 5$, $m_2 = 4$, $m_3 = 3$, $m_4 = 6$;
$P_1(-4, 2)$, $P_2(0, 5)$, $P_3(3, 2)$, $P_4(1, -2)$

25–28 Sketch the region bounded by the curves, and visually estimate the location of the centroid. Then find the exact coordinates of the centroid.

25. $y = 2x$, $y = 0$, $x = 1$
26. $y = \sqrt{x}$, $y = 0$, $x = 4$
27. $y = e^x$, $y = 0$, $x = 0$, $x = 1$
28. $y = \sin x$, $y = 0$, $0 \leq x \leq \pi$

29–33 Find the centroid of the region bounded by the given curves.

29. $y = x^2$, $x = y^2$
30. $y = 2 - x^2$, $y = x$
31. $y = \sin x$, $y = \cos x$, $x = 0$, $x = \pi/4$
32. $y = x^3$, $x + y = 2$, $y = 0$
33. $x + y = 2$, $x = y^2$

34–35 Calculate the moments M_x and M_y and the center of mass of a lamina with the given density and shape.

34. $\rho = 4$ **35.** $\rho = 6$

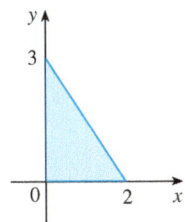 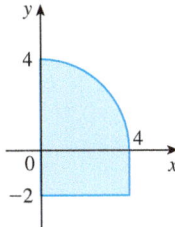

36. Use Simpson's Rule to estimate the centroid of the region shown.

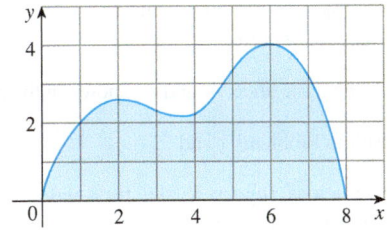

37. Find the centroid of the region bounded by the curves $y = x^3 - x$ and $y = x^2 - 1$. Sketch the region and plot the centroid to see if your answer is reasonable.

38. Use a graph to find approximate x-coordinates of the points of intersection of the curves $y = e^x$ and $y = 2 - x^2$. Then find (approximately) the centroid of the region bounded by these curves.

39. Prove that the centroid of any triangle is located at the point of intersection of the medians. [*Hints*: Place the axes so that the vertices are $(a, 0)$, $(0, b)$, and $(c, 0)$. Recall that a median is a line segment from a vertex to the midpoint of the opposite side. Recall also that the medians intersect at a point two-thirds of the way from each vertex (along the median) to the opposite side.]

40–41 Find the centroid of the region shown, not by integration, but by locating the centroids of the rectangles and triangles (from Exercise 39) and using additivity of moments.

40.   **41.**

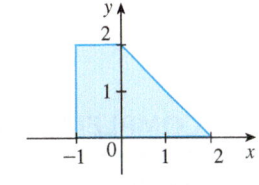 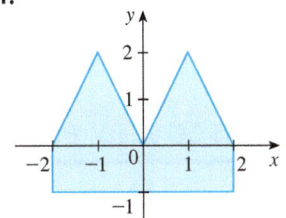

42. A rectangle $\mathcal{R}$ with sides a and b is divided into two parts $\mathcal{R}_1$ and $\mathcal{R}_2$ by an arc of a parabola that has its vertex at one corner of $\mathcal{R}$ and passes through the opposite corner. Find the centroids of both $\mathcal{R}_1$ and $\mathcal{R}_2$.

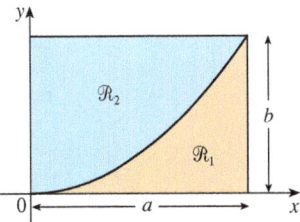

43. If $\bar{x}$ is the x-coordinate of the centroid of the region that lies under the graph of a continuous function f, where $a \leq x \leq b$, show that

$$\int_a^b (cx + d) f(x)\, dx = (c\bar{x} + d) \int_a^b f(x)\, dx$$

44–46 Use the Theorem of Pappus to find the volume of the given solid.

44. A sphere of radius r (Use Example 4.)

45. A cone with height h and base radius r

46. The solid obtained by rotating the triangle with vertices (2, 3), (2, 5), and (5, 4) about the x-axis

47. The centroid of a *curve* can be found by a process similar to the one we used for finding the centroid of a region. If C is a curve with length L, then the centroid is $(\bar{x}, \bar{y})$ where $\bar{x} = (1/L) \int x \, ds$ and $\bar{y} = (1/L) \int y \, ds$. Here we assign appropriate limits of integration, and ds is as defined in Sections 8.1 and 8.2. (The centroid often doesn't lie on the curve itself. If the curve were made of wire and placed on a weightless board, the centroid would be the balance point on the board.) Find the centroid of the quarter-circle $y = \sqrt{16 - x^2}, 0 \le x \le 4$.

48. The *Second Theorem of Pappus* is in the same spirit as Pappus's Theorem on page 565, but for surface area rather than volume: Let C be a curve that lies entirely on one side of a line l in the plane. If C is rotated about l, then the area of the resulting surface is the product of the arc length of C and the distance traveled by the centroid of C (see Exercise 47).

(a) Prove the Second Theorem of Pappus for the case where C is given by $y = f(x), f(x) \ge 0$, and C is rotated about the x-axis.

(b) Use the Second Theorem of Pappus to compute the surface area of the half-sphere obtained by rotating the curve from Exercise 47 about the x-axis. Does your answer agree with the one given by geometric formulas?

49. Use the Second Theorem of Pappus described in Exercise 48 to find the surface area of the torus in Example 7.

50. Let $\mathcal{R}$ be the region that lies between the curves

$$y = x^m \qquad y = x^n \qquad 0 \le x \le 1$$

where m and n are integers with $0 \le n < m$.
(a) Sketch the region $\mathcal{R}$.
(b) Find the coordinates of the centroid of $\mathcal{R}$.
(c) Try to find values of m and n such that the centroid lies *outside* $\mathcal{R}$.

51. Prove Formulas 9.

DISCOVERY PROJECT COMPLEMENTARY COFFEE CUPS

Suppose you have a choice of two coffee cups of the type shown, one that bends outward and one inward, and you notice that they have the same height and their shapes fit together snugly. You wonder which cup holds more coffee. Of course you could fill one cup with water and pour it into the other one but, being a calculus student, you decide on a more mathematical approach. Ignoring the handles, you observe that both cups are surfaces of revolution, so you can think of the coffee as a volume of revolution.

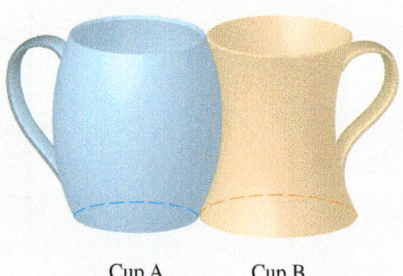
Cup A   Cup B

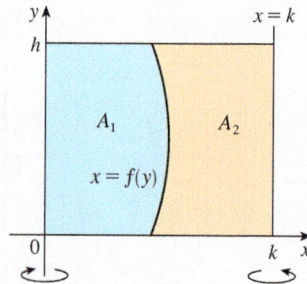

1. Suppose the cups have height h, cup A is formed by rotating the curve $x = f(y)$ about the y-axis, and cup B is formed by rotating the same curve about the line $x = k$. Find the value of k such that the two cups hold the same amount of coffee.

2. What does your result from Problem 1 say about the areas A_1 and A_2 shown in the figure?

3. Use Pappus's Theorem to explain your result in Problems 1 and 2.

4. Based on your own measurements and observations, suggest a value for h and an equation for $x = f(y)$ and calculate the amount of coffee that each cup holds.

8.4 Applications to Economics and Biology

In this section we consider some applications of integration to economics (consumer surplus) and biology (blood flow, cardiac output). Others are described in the exercises.

■ Consumer Surplus

Recall from Section 4.7 that the demand function $p(x)$ is the price that a company has to charge in order to sell x units of a commodity. Usually, selling larger quantities requires lowering prices, so the demand function is a decreasing function. The graph of a typical demand function, called a **demand curve**, is shown in Figure 1. If X is the amount of the commodity that can currently be sold, then $P = p(X)$ is the current selling price.

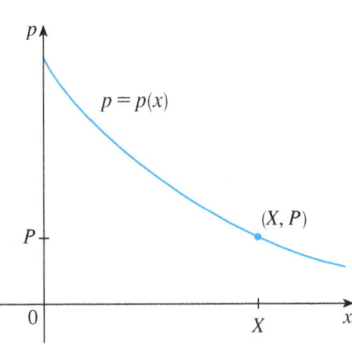

FIGURE 1
A typical demand curve

At a given price, some consumers who buy a good would be willing to pay more; they benefit by not having to. The difference between what a consumer is willing to pay and what the consumer actually pays for a good is called the *consumer surplus*. By finding the total consumer surplus among all purchasers of a good, economists can assess the overall benefit of a market to society.

To determine the total consumer surplus, we look at the demand curve and divide the interval $[0, X]$ into n subintervals, each of length $\Delta x = X/n$, and let $x_i^* = x_i$ be the right endpoint of the ith subinterval, as in Figure 2. According to the demand curve, x_{i-1} units would be purchased at a price of $p(x_{i-1})$ dollars per unit. To increase sales to x_i units, the price would have to be lowered to $p(x_i)$ dollars. In this case, an additional Δx units would be sold (but no more). In general, the consumers who would have paid $p(x_i)$ dollars placed a high value on the product; they would have paid what it was worth to them. So in paying only P dollars they have saved an amount of

$$(\text{savings per unit})(\text{number of units}) = [p(x_i) - P]\Delta x$$

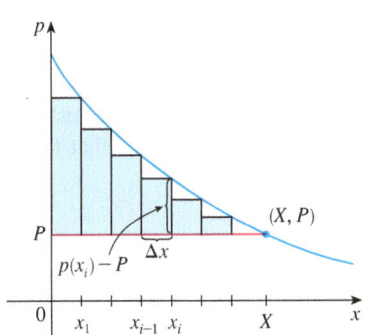

FIGURE 2

Considering similar groups of willing consumers for each of the subintervals and adding the savings, we get the total savings:

$$\sum_{i=1}^{n} [p(x_i) - P]\Delta x$$

(This sum corresponds to the area enclosed by the rectangles in Figure 2.) If we let $n \to \infty$, this Riemann sum approaches the integral

$$\boxed{1} \qquad \int_0^X [p(x) - P]\, dx$$

which economists call the **consumer surplus** for the commodity.

The consumer surplus represents the amount of money saved by consumers in purchasing the commodity at price P, corresponding to an amount demanded of X. Figure 3 shows the interpretation of the consumer surplus as the area under the demand curve and above the line $p = P$.

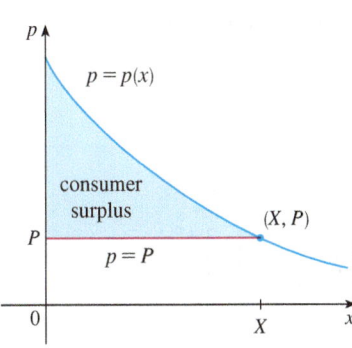

FIGURE 3

EXAMPLE 1 The demand for a product, in dollars, is

$$p = 1200 - 0.2x - 0.0001x^2$$

Find the consumer surplus when the sales level is 500.

SOLUTION Since the number of products sold is $X = 500$, the corresponding price is

$$P = 1200 - (0.2)(500) - (0.0001)(500)^2 = 1075$$

Therefore, from Definition 1, the consumer surplus is

$$\int_0^{500} [p(x) - P]\, dx = \int_0^{500} (1200 - 0.2x - 0.0001x^2 - 1075)\, dx$$

$$= \int_0^{500} (125 - 0.2x - 0.0001x^2)\, dx$$

$$= 125x - 0.1x^2 - (0.0001)\left(\frac{x^3}{3}\right)\Big]_0^{500}$$

$$= (125)(500) - (0.1)(500)^2 - \frac{(0.0001)(500)^3}{3}$$

$$= \$33{,}333.33 \qquad \blacksquare$$

■ Blood Flow

In Example 3.7.7 we discussed the law of laminar flow:

$$v(r) = \frac{P}{4\eta l}(R^2 - r^2)$$

which gives the velocity v of blood that flows along a blood vessel with radius R and length l at a distance r from the central axis, where P is the pressure difference between the ends of the vessel and η is the viscosity of the blood. Now, in order to compute the rate of blood flow, or *flux* (volume per unit time), we consider smaller, equally spaced radii $r_1, r_2, \ldots$. The approximate area of the ring (or washer) with inner radius r_{i-1} and outer radius r_i is

$$2\pi r_i\, \Delta r \qquad \text{where} \quad \Delta r = r_i - r_{i-1}$$

(See Figure 4.) If Δr is small, then the velocity is almost constant throughout this ring and can be approximated by $v(r_i)$. Thus the volume of blood per unit time that flows across the ring is approximately

$$(2\pi r_i\, \Delta r)\, v(r_i) = 2\pi r_i v(r_i)\, \Delta r$$

and the total volume of blood that flows across a cross-section per unit time is about

$$\sum_{i=1}^{n} 2\pi r_i v(r_i)\, \Delta r$$

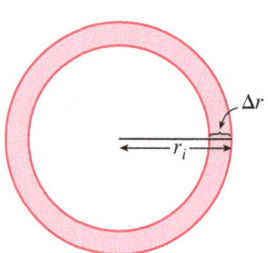

FIGURE 4

This approximation is illustrated in Figure 5. Notice that the velocity (and hence the volume per unit time) increases toward the center of the blood vessel. The approximation gets better as n increases. When we take the limit we get the exact value of the **flux** (or *discharge*), which is the volume of blood that passes a cross-section per unit time:

$$F = \lim_{n \to \infty} \sum_{i=1}^{n} 2\pi r_i v(r_i)\, \Delta r = \int_0^R 2\pi r\, v(r)\, dr$$

$$= \int_0^R 2\pi r\, \frac{P}{4\eta l}(R^2 - r^2)\, dr$$

$$= \frac{\pi P}{2\eta l} \int_0^R (R^2 r - r^3)\, dr = \frac{\pi P}{2\eta l}\left[R^2\, \frac{r^2}{2} - \frac{r^4}{4}\right]_{r=0}^{r=R}$$

$$= \frac{\pi P}{2\eta l}\left[\frac{R^4}{2} - \frac{R^4}{4}\right] = \frac{\pi P R^4}{8\eta l}$$

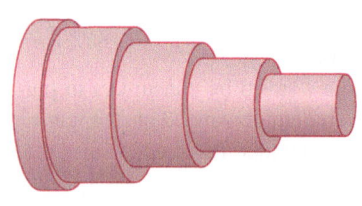

FIGURE 5

The resulting equation

$$F = \frac{\pi P R^4}{8\eta l} \qquad \boxed{2}$$

is called **Poiseuille's Law**; it shows that the flux is proportional to the fourth power of the radius of the blood vessel.

Cardiac Output

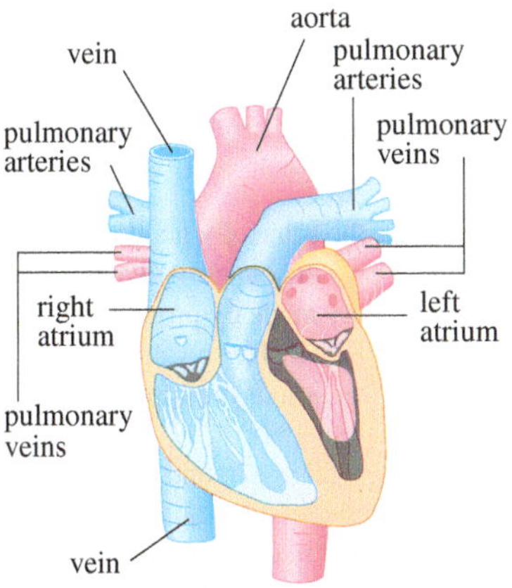

Figure 6 shows the human cardiovascular system. Blood returns from the body through the veins, enters the right atrium of the heart, and is pumped to the lungs through the pulmonary arteries for oxygenation. It then flows back into the left atrium through the pulmonary veins and then out to the rest of the body through the aorta. The **cardiac output** of the heart is the volume of blood pumped by the heart per unit time, that is, the rate of flow into the aorta.

The *dye dilution method* is used to measure the cardiac output. Dye is injected into the right atrium and flows through the heart into the aorta. A probe inserted into the aorta measures the concentration of the dye leaving the heart at equally spaced times over a time interval $[0, T]$ until the dye has cleared. Let $c(t)$ be the concentration of the dye at time t. If we divide $[0, T]$ into subintervals of equal length Δt, then the amount of dye that flows past the measuring point during the subinterval from $t = t_{i-1}$ to $t = t_i$ is approximately

$$(\text{concentration})(\text{volume}) = c(t_i)(F \, \Delta t)$$

where F is the rate of flow that we are trying to determine. Thus the total amount of dye is approximately

$$\sum_{i=1}^{n} c(t_i) F \, \Delta t = F \sum_{i=1}^{n} c(t_i) \, \Delta t$$

and, letting $n \to \infty$, we find that the amount of dye is

$$A = F \int_0^T c(t) \, dt$$

Thus the cardiac output is given by

$$F = \frac{A}{\int_0^T c(t) \, dt} \qquad \boxed{3}$$

where the amount of dye A is known and the integral can be approximated from the concentration readings.

FIGURE 6

t	$c(t)$	t	$c(t)$
0	0	6	6.1
1	0.4	7	4.0
2	2.8	8	2.3
3	6.5	9	1.1
4	9.8	10	0
5	8.9		

EXAMPLE 2 A 5-mg bolus of dye is injected into a right atrium. The concentration of the dye (in milligrams per liter) is measured in the aorta at one-second intervals as shown in the table. Estimate the cardiac output.

SOLUTION Here $A = 5$, $\Delta t = 1$, and $T = 10$. We use Simpson's Rule to approximate the integral of the concentration:

$$\int_0^{10} c(t) \, dt \approx \tfrac{1}{3}[0 + 4(0.4) + 2(2.8) + 4(6.5) + 2(9.8) + 4(8.9)$$
$$+ 2(6.1) + 4(4.0) + 2(2.3) + 4(1.1) + 0]$$
$$\approx 41.87$$

Thus Formula 3 gives the cardiac output to be

$$F = \frac{A}{\int_0^{10} c(t)\,dt} \approx \frac{5}{41.87} \approx 0.12 \text{ L/s} = 7.2 \text{ L/min}$$

8.4 EXERCISES

1. The marginal cost function $C'(x)$ was defined to be the derivative of the cost function. (See Sections 3.7 and 4.7.) The marginal cost of producing x gallons of orange juice is

   $$C'(x) = 0.82 - 0.00003x + 0.000000003x^2$$

   (measured in dollars per gallon). The fixed start-up cost is $C(0) = \$18{,}000$. Use the Net Change Theorem to find the cost of producing the first 4000 gallons of juice.

2. A company estimates that the marginal revenue (in dollars per unit) realized by selling x units of a product is $48 - 0.0012x$. Assuming the estimate is accurate, find the increase in revenue if sales increase from 5000 units to 10,000 units.

3. A mining company estimates that the marginal cost of extracting x tons of copper ore from a mine is $0.6 + 0.008x$, measured in thousands of dollars per ton. Start-up costs are $\$100{,}000$. What is the cost of extracting the first 50 tons of copper? What about the next 50 tons?

4. The demand function for a particular vacation package is $p(x) = 2000 - 46\sqrt{x}$. Find the consumer surplus when the sales level for the packages is 400. Illustrate by drawing the demand curve and identifying the consumer surplus as an area.

5. A demand curve is given by $p = 450/(x + 8)$. Find the consumer surplus when the selling price is $\$10$.

6. The **supply function** $p_S(x)$ for a commodity gives the relation between the selling price and the number of units that manufacturers will produce at that price. For a higher price, manufacturers will produce more units, so p_S is an increasing function of x. Let X be the amount of the commodity currently produced and let $P = p_S(X)$ be the current price. Some producers would be willing to make and sell the commodity for a lower selling price and are therefore receiving more than their minimal price. The excess is called the **producer surplus**. An argument similar to that for consumer surplus shows that the surplus is given by the integral

   $$\int_0^X [P - p_S(x)]\,dx$$

   Calculate the producer surplus for the supply function $p_S(x) = 3 + 0.01x^2$ at the sales level $X = 10$. Illustrate by drawing the supply curve and identifying the producer surplus as an area.

7. If a supply curve is modeled by the equation $p = 125 + 0.002x^2$, find the producer surplus when the selling price is $\$625$.

8. In a purely competitive market, the price of a good is naturally driven to the value where the quantity demanded by consumers matches the quantity made by producers, and the market is said to be in *equilibrium*. These values are the coordinates of the point of intersection of the supply and demand curves.
   (a) Given the demand curve $p = 50 - \frac{1}{20}x$ and the supply curve $p = 20 + \frac{1}{10}x$ for a good, at what quantity and price is the market for the good in equilibrium?
   (b) Find the consumer surplus and the producer surplus when the market is in equilibrium. Illustrate by sketching the supply and demand curves and identifying the surpluses as areas.

9. The sum of consumer surplus and producer surplus is called the *total surplus*; it is one measure economists use as an indicator of the economic health of a society. Total surplus is maximized when the market for a good is in equilibrium.
   (a) The demand function for an electronics company's car stereos is $p(x) = 228.4 - 18x$ and the supply function is $p_S(x) = 27x + 57.4$, where x is measured in thousands. At what quantity is the market for the stereos in equilibrium?
   (b) Compute the maximum total surplus for the stereos.

10. A camera company estimates that the demand function for its new digital camera is $p(x) = 312e^{-0.14x}$ and the supply function is estimated to be $p_S(x) = 26e^{0.2x}$, where x is measured in thousands. Compute the maximum total surplus.

11. A company modeled the demand curve for its product (in dollars) by the equation

    $$p = \frac{800{,}000e^{-x/5000}}{x + 20{,}000}$$

    Use a graph to estimate the sales level when the selling price is $\$16$. Then find (approximately) the consumer surplus for this sales level.

12. A movie theater has been charging $\$10.00$ per person and selling about 500 tickets on a typical weeknight. After surveying their customers, the theater management estimates that for every 50 cents that they lower the price, the number

of moviegoers will increase by 50 per night. Find the demand function and calculate the consumer surplus when the tickets are priced at $8.00.

13. If the amount of capital that a company has at time t is $f(t)$, then the derivative, $f'(t)$, is called the *net investment flow*. Suppose that the net investment flow is $\sqrt{t}$ million dollars per year (where t is measured in years). Find the increase in capital (the *capital formation*) from the fourth year to the eighth year.

14. If revenue flows into a company at a rate of
$f(t) = 9000\sqrt{1 + 2t}$, where t is measured in years and $f(t)$ is measured in dollars per year, find the total revenue obtained in the first four years.

15. If income is continuously collected at a rate of $f(t)$ dollars per year and will be invested at a constant interest rate r (compounded continuously) for a period of T years, then the *future value* of the income is given by $\int_0^T f(t)\, e^{r(T-t)}\, dt$. Compute the future value after 6 years for income received at a rate of $f(t) = 8000\, e^{0.04t}$ dollars per year and invested at 6.2% interest.

16. The *present value* of an income stream is the amount that would need to be invested now to match the future value as described in Exercise 15 and is given by $\int_0^T f(t)\, e^{-rt}\, dt$. Find the present value of the income stream in Exercise 15.

17. *Pareto's Law of Income* states that the number of people with incomes between $x = a$ and $x = b$ is $N = \int_a^b Ax^{-k}\, dx$, where A and k are constants with $A > 0$ and $k > 1$. The average income of these people is
$$\bar{x} = \frac{1}{N}\int_a^b Ax^{1-k}\, dx$$
Calculate $\bar{x}$.

18. A hot, wet summer is causing a mosquito population explosion in a lake resort area. The number of mosquitoes is increasing at an estimated rate of $2200 + 10e^{0.8t}$ per week (where t is measured in weeks). By how much does the mosquito population increase between the fifth and ninth weeks of summer?

19. Use Poiseuille's Law to calculate the rate of flow in a small human artery where we can take $\eta = 0.027$, $R = 0.008$ cm, $l = 2$ cm, and $P = 4000$ dynes/cm^2.

20. High blood pressure results from constriction of the arteries. To maintain a normal flow rate (flux), the heart has to pump harder, thus increasing the blood pressure. Use Poiseuille's Law to show that if R_0 and P_0 are normal values of the radius and pressure in an artery and the constricted values are R and P, then for the flux to remain constant, P and R are related by the equation
$$\frac{P}{P_0} = \left(\frac{R_0}{R}\right)^4$$
Deduce that if the radius of an artery is reduced to three-fourths of its former value, then the pressure is more than tripled.

21. The dye dilution method is used to measure cardiac output with 6 mg of dye. The dye concentrations, in mg/L, are modeled by $c(t) = 20te^{-0.6t}$, $0 \le t \le 10$, where t is measured in seconds. Find the cardiac output.

22. After a 5.5-mg injection of dye, the readings of dye concentration, in mg/L, at two-second intervals are as shown in the table. Use Simpson's Rule to estimate the cardiac output.

t	$c(t)$	t	$c(t)$
0	0.0	10	4.3
2	4.1	12	2.5
4	8.9	14	1.2
6	8.5	16	0.2
8	6.7		

23. The graph of the concentration function $c(t)$ is shown after a 7-mg injection of dye into a heart. Use Simpson's Rule to estimate the cardiac output.

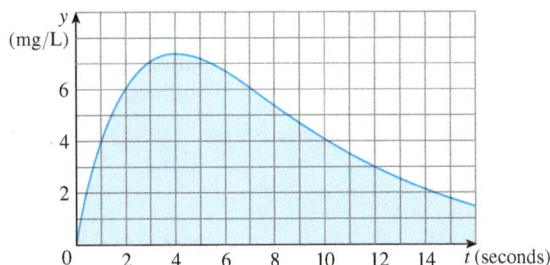

8.5 Probability

Calculus plays a role in the analysis of random behavior. Suppose we consider the cholesterol level of a person chosen at random from a certain age group, or the height of an adult female chosen at random, or the lifetime of a randomly chosen battery of a certain type. Such quantities are called **continuous random variables** because their values actually range over an interval of real numbers, although they might be measured or recorded only to the nearest integer. We might want to know the probability that a blood cholesterol level is greater than 250, or the probability that the height of an adult female is between 60 and 70 inches, or the probability that the battery we are buying lasts between 100 and 200 hours. If X represents the lifetime of that type of battery, we denote this last probability as follows:

$$P(100 \leq X \leq 200)$$

According to the frequency interpretation of probability, this number is the long-run proportion of all batteries of the specified type whose lifetimes are between 100 and 200 hours. Since it represents a proportion, the probability naturally falls between 0 and 1.

Every continuous random variable X has a **probability density function** f. This means that the probability that X lies between a and b is found by integrating f from a to b:

> Note that we always use *intervals* of values when working with probability density functions. We wouldn't, for instance, use a density function to find the probability that X equals a.

$$\boxed{1} \qquad P(a \leq X \leq b) = \int_a^b f(x)\, dx$$

For example, Figure 1 shows the graph of a model for the probability density function f for a random variable X defined to be the height in inches of an adult female in the United States (according to data from the National Health Survey). The probability that the height of a woman chosen at random from this population is between 60 and 70 inches is equal to the area under the graph of f from 60 to 70.

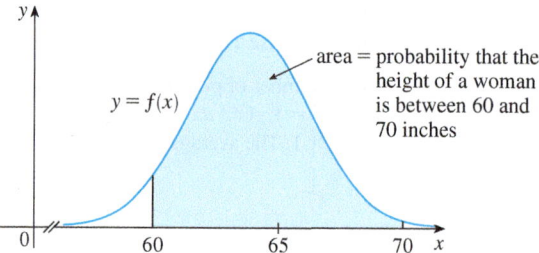

FIGURE 1
Probability density function for the height of an adult female

In general, the probability density function f of a random variable X satisfies the condition $f(x) \geq 0$ for all x. Because probabilities are measured on a scale from 0 to 1, it follows that

$$\boxed{2} \qquad \int_{-\infty}^{\infty} f(x)\, dx = 1$$

EXAMPLE 1 Let $f(x) = 0.006x(10 - x)$ for $0 \leq x \leq 10$ and $f(x) = 0$ for all other values of x.
(a) Verify that f is a probability density function.
(b) Find $P(4 \leq X \leq 8)$.

SOLUTION
(a) For $0 \leq x \leq 10$ we have $0.006x(10 - x) \geq 0$, so $f(x) \geq 0$ for all x. We also need to check that Equation 2 is satisfied:

$$\int_{-\infty}^{\infty} f(x)\, dx = \int_0^{10} 0.006x(10 - x)\, dx = 0.006 \int_0^{10} (10x - x^2)\, dx$$

$$= 0.006\left[5x^2 - \tfrac{1}{3}x^3\right]_0^{10} = 0.006\left(500 - \tfrac{1000}{3}\right) = 1$$

Therefore f is a probability density function.

(b) The probability that X lies between 4 and 8 is

$$P(4 \leq X \leq 8) = \int_4^8 f(x)\,dx = 0.006 \int_4^8 (10x - x^2)\,dx$$

$$= 0.006\left[5x^2 - \tfrac{1}{3}x^3\right]_4^8 = 0.544$$

EXAMPLE 2 Phenomena such as waiting times and equipment failure times are commonly modeled by exponentially decreasing probability density functions. Find the exact form of such a function.

SOLUTION Think of the random variable as being the time you wait on hold before an agent of a company you're telephoning answers your call. So instead of x, let's use t to represent time, in minutes. If f is the probability density function and you call at time $t = 0$, then, from Definition 1, $\int_0^2 f(t)\,dt$ represents the probability that an agent answers within the first two minutes and $\int_4^5 f(t)\,dt$ is the probability that your call is answered during the fifth minute.

It's clear that $f(t) = 0$ for $t < 0$ (the agent can't answer before you place the call). For $t > 0$ we are told to use an exponentially decreasing function, that is, a function of the form $f(t) = Ae^{-ct}$, where A and c are positive constants. Thus

$$f(t) = \begin{cases} 0 & \text{if } t < 0 \\ Ae^{-ct} & \text{if } t \geq 0 \end{cases}$$

We use Equation 2 to determine the value of A:

$$1 = \int_{-\infty}^{\infty} f(t)\,dt = \int_{-\infty}^0 f(t)\,dt + \int_0^{\infty} f(t)\,dt$$

$$= \int_0^{\infty} Ae^{-ct}\,dt = \lim_{x \to \infty} \int_0^x Ae^{-ct}\,dt$$

$$= \lim_{x \to \infty} \left[-\frac{A}{c}e^{-ct}\right]_0^x = \lim_{x \to \infty} \frac{A}{c}(1 - e^{-cx})$$

$$= \frac{A}{c}$$

Therefore $A/c = 1$ and so $A = c$. Thus every exponential density function has the form

$$f(t) = \begin{cases} 0 & \text{if } t < 0 \\ ce^{-ct} & \text{if } t \geq 0 \end{cases}$$

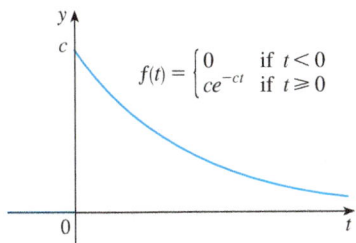

FIGURE 2
An exponential density function

A typical graph is shown in Figure 2.

Average Values

Suppose you're waiting for a company to answer your phone call and you wonder how long, on average, you can expect to wait. Let $f(t)$ be the corresponding density function, where t is measured in minutes, and think of a sample of N people who have called this company. Most likely, none of them had to wait more than an hour, so let's restrict our attention to the interval $0 \leq t \leq 60$. Let's divide that interval into n intervals of length

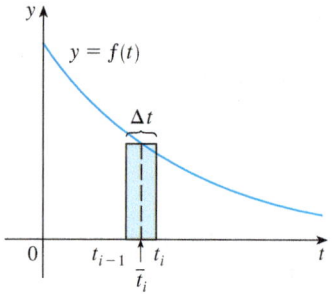

FIGURE 3

Δt and endpoints $0, t_1, t_2, \ldots, t_n = 60$. (Think of Δt as lasting a minute, or half a minute, or 10 seconds, or even a second.) The probability that somebody's call gets answered during the time period from t_{i-1} to t_i is the area under the curve $y = f(t)$ from t_{i-1} to t_i, which is approximately equal to $f(\bar{t}_i)\,\Delta t$. (This is the area of the approximating rectangle in Figure 3, where $\bar{t}_i$ is the midpoint of the interval.)

Since the long-run proportion of calls that get answered in the time period from t_{i-1} to t_i is $f(\bar{t}_i)\,\Delta t$, we expect that, out of our sample of N callers, the number whose call was answered in that time period is approximately $Nf(\bar{t}_i)\,\Delta t$ and the time that each waited is about $\bar{t}_i$. Therefore the total time they waited is the product of these numbers: approximately $\bar{t}_i[Nf(\bar{t}_i)\,\Delta t]$. Adding over all such intervals, we get the approximate total of everybody's waiting times:

$$\sum_{i=1}^{n} N\bar{t}_i\, f(\bar{t}_i)\,\Delta t$$

If we now divide by the number of callers N, we get the approximate *average* waiting time:

$$\sum_{i=1}^{n} \bar{t}_i\, f(\bar{t}_i)\,\Delta t$$

We recognize this as a Riemann sum for the function $tf(t)$. As the time interval shrinks (that is, $\Delta t \to 0$ and $n \to \infty$), this Riemann sum approaches the integral

$$\int_0^{60} tf(t)\,dt$$

This integral is called the *mean waiting time*.

In general, the **mean** of any probability density function f is defined to be

It is traditional to denote the mean by the Greek letter μ (mu).

$$\mu = \int_{-\infty}^{\infty} x f(x)\,dx$$

The mean can be interpreted as the long-run average value of the random variable X. It can also be interpreted as a measure of centrality of the probability density function.

The expression for the mean resembles an integral we have seen before. If $\mathcal{R}$ is the region that lies under the graph of f, we know from Formula 8.3.8 that the x-coordinate of the centroid of $\mathcal{R}$ is

$$\bar{x} = \frac{\int_{-\infty}^{\infty} x f(x)\,dx}{\int_{-\infty}^{\infty} f(x)\,dx} = \int_{-\infty}^{\infty} x f(x)\,dx = \mu$$

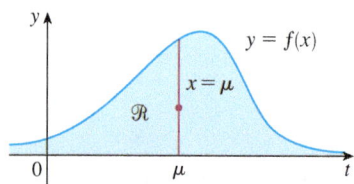

FIGURE 4
$\mathcal{R}$ balances at a point on the line $x = \mu$

because of Equation 2. So a thin plate in the shape of $\mathcal{R}$ balances at a point on the vertical line $x = \mu$. (See Figure 4.)

EXAMPLE 3 Find the mean of the exponential distribution of Example 2:

$$f(t) = \begin{cases} 0 & \text{if } t < 0 \\ ce^{-ct} & \text{if } t \geq 0 \end{cases}$$

SOLUTION According to the definition of a mean, we have

$$\mu = \int_{-\infty}^{\infty} tf(t)\,dt = \int_0^{\infty} tce^{-ct}\,dt$$

To evaluate this integral we use integration by parts, with $u = t$ and $dv = ce^{-ct}\,dt$, so $du = dt$ and $v = -e^{-ct}$:

$$\int_0^\infty tce^{-ct}\,dt = \lim_{x\to\infty} \int_0^x tce^{-ct}\,dt = \lim_{x\to\infty} \left(-te^{-ct}\Big]_0^x + \int_0^x e^{-ct}\,dt \right)$$

$$= \lim_{x\to\infty} \left(-xe^{-cx} + \frac{1}{c} - \frac{e^{-cx}}{c} \right) = \frac{1}{c}$$

The limit of the first term is 0 by l'Hospital's Rule.

The mean is $\mu = 1/c$, so we can rewrite the probability density function as

$$f(t) = \begin{cases} 0 & \text{if } t < 0 \\ \mu^{-1} e^{-t/\mu} & \text{if } t \geq 0 \end{cases}$$

EXAMPLE 4 Suppose the average waiting time for a customer's call to be answered by a company representative is five minutes.
(a) Find the probability that a call is answered during the first minute, assuming that an exponential distribution is appropriate.
(b) Find the probability that a customer waits more than five minutes to be answered.

SOLUTION
(a) We are given that the mean of the exponential distribution is $\mu = 5$ min and so, from the result of Example 3, we know that the probability density function is

$$f(t) = \begin{cases} 0 & \text{if } t < 0 \\ 0.2 e^{-t/5} & \text{if } t \geq 0 \end{cases}$$

where t is measured in minutes. Thus the probability that a call is answered during the first minute is

$$P(0 \leq T \leq 1) = \int_0^1 f(t)\,dt$$

$$= \int_0^1 0.2 e^{-t/5}\,dt = 0.2(-5)e^{-t/5}\Big]_0^1$$

$$= 1 - e^{-1/5} \approx 0.1813$$

So about 18% of customers' calls are answered during the first minute.
(b) The probability that a customer waits more than five minutes is

$$P(T > 5) = \int_5^\infty f(t)\,dt = \int_5^\infty 0.2 e^{-t/5}\,dt$$

$$= \lim_{x\to\infty} \int_5^x 0.2 e^{-t/5}\,dt = \lim_{x\to\infty} (e^{-1} - e^{-x/5})$$

$$= \frac{1}{e} - 0 \approx 0.368$$

About 37% of customers wait more than five minutes before their calls are answered.

Notice the result of Example 4(b): Even though the mean waiting time is 5 minutes, only 37% of callers wait more than 5 minutes. The reason is that some callers have to wait much longer (maybe 10 or 15 minutes), and this brings up the average.

Another measure of centrality of a probability density function is the *median*. That is a number m such that half the callers have a waiting time less than m and the other callers have a waiting time longer than m. In general, the **median** of a probability density function is the number m such that

$$\int_m^\infty f(x)\,dx = \tfrac{1}{2}$$

This means that half the area under the graph of f lies to the right of m. In Exercise 9 you are asked to show that the median waiting time for the company described in Example 4 is approximately 3.5 minutes.

■ Normal Distributions

Many important random phenomena—such as test scores on aptitude tests, heights and weights of individuals from a homogeneous population, annual rainfall in a given location—are modeled by a **normal distribution**. This means that the probability density function of the random variable X is a member of the family of functions

$$\boxed{3} \qquad f(x) = \frac{1}{\sigma\sqrt{2\pi}}\, e^{-(x-\mu)^2/(2\sigma^2)}$$

The standard deviation is denoted by the lowercase Greek letter σ (sigma).

You can verify that the mean for this function is μ. The positive constant σ is called the **standard deviation**; it measures how spread out the values of X are. From the bell-shaped graphs of members of the family in Figure 5, we see that for small values of σ the values of X are clustered about the mean, whereas for larger values of σ the values of X are more spread out. Statisticians have methods for using sets of data to estimate μ and σ.

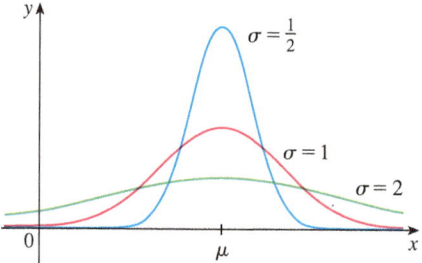

FIGURE 5
Normal distributions

The factor $1/(\sigma\sqrt{2\pi})$ is needed to make f a probability density function. In fact, it can be verified using the methods of multivariable calculus that

$$\int_{-\infty}^\infty \frac{1}{\sigma\sqrt{2\pi}}\, e^{-(x-\mu)^2/(2\sigma^2)}\,dx = 1$$

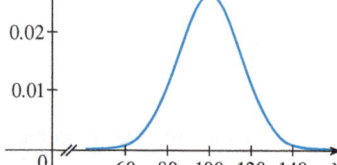

FIGURE 6

EXAMPLE 5 Intelligence Quotient (IQ) scores are distributed normally with mean 100 and standard deviation 15. (Figure 6 shows the corresponding probability density function.)
(a) What percentage of the population has an IQ score between 85 and 115?
(b) What percentage of the population has an IQ above 140?

SOLUTION

(a) Since IQ scores are normally distributed, we use the probability density function given by Equation 3 with $\mu = 100$ and $\sigma = 15$:

$$P(85 \leq X \leq 115) = \int_{85}^{115} \frac{1}{15\sqrt{2\pi}} e^{-(x-100)^2/(2 \cdot 15^2)} dx$$

Recall from Section 7.5 that the function $y = e^{-x^2}$ doesn't have an elementary antiderivative, so we can't evaluate the integral exactly. But we can use the numerical integration capability of a calculator or computer (or the Midpoint Rule or Simpson's Rule) to estimate the integral. Doing so, we find that

$$P(85 \leq X \leq 115) \approx 0.68$$

So about 68% of the population has an IQ score between 85 and 115, that is, within one standard deviation of the mean.

(b) The probability that the IQ score of a person chosen at random is more than 140 is

$$P(X > 140) = \int_{140}^{\infty} \frac{1}{15\sqrt{2\pi}} e^{-(x-100)^2/450} dx$$

To avoid the improper integral we could approximate it by the integral from 140 to 200. (It's quite safe to say that people with an IQ over 200 are extremely rare.) Then

$$P(X > 140) \approx \int_{140}^{200} \frac{1}{15\sqrt{2\pi}} e^{-(x-100)^2/450} dx \approx 0.0038$$

Therefore about 0.4% of the population has an IQ score over 140. ∎

8.5 EXERCISES

1. Let $f(x)$ be the probability density function for the lifetime of a manufacturer's highest quality car tire, where x is measured in miles. Explain the meaning of each integral.

(a) $\int_{30,000}^{40,000} f(x)\, dx$ (b) $\int_{25,000}^{\infty} f(x)\, dx$

2. Let $f(t)$ be the probability density function for the time it takes you to drive to school in the morning, where t is measured in minutes. Express the following probabilities as integrals.
(a) The probability that you drive to school in less than 15 minutes
(b) The probability that it takes you more than half an hour to get to school

3. Let $f(x) = 30x^2(1-x)^2$ for $0 \leq x \leq 1$ and $f(x) = 0$ for all other values of x.
(a) Verify that f is a probability density function.
(b) Find $P(X \leq \frac{1}{3})$.

4. The density function

$$f(x) = \frac{e^{3-x}}{(1+e^{3-x})^2}$$

is an example of a *logistic distribution*.
(a) Verify that f is a probability density function.
(b) Find $P(3 \leq X \leq 4)$.
(c) Graph f. What does the mean appear to be? What about the median?

5. Let $f(x) = c/(1+x^2)$.
(a) For what value of c is f a probability density function?
(b) For that value of c, find $P(-1 < X < 1)$.

6. Let $f(x) = k(3x - x^2)$ if $0 \leq x \leq 3$ and $f(x) = 0$ if $x < 0$ or $x > 3$.
(a) For what value of k is f a probability density function?
(b) For that value of k, find $P(X > 1)$.
(c) Find the mean.

7. A spinner from a board game randomly indicates a real number between 0 and 10. The spinner is fair in the sense that it indicates a number in a given interval with the same probability as it indicates a number in any other interval of the same length.
   (a) Explain why the function
   $$f(x) = \begin{cases} 0.1 & \text{if } 0 \leq x \leq 10 \\ 0 & \text{if } x < 0 \text{ or } x > 10 \end{cases}$$
   is a probability density function for the spinner's values.
   (b) What does your intuition tell you about the value of the mean? Check your guess by evaluating an integral.

8. (a) Explain why the function whose graph is shown is a probability density function.
   (b) Use the graph to find the following probabilities:
      (i) $P(X < 3)$ (ii) $P(3 \leq X \leq 8)$
   (c) Calculate the mean.

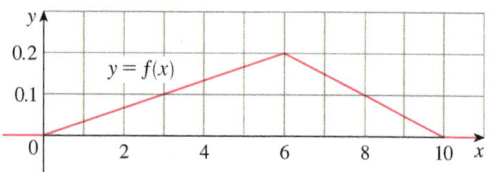

9. Show that the median waiting time for a phone call to the company described in Example 4 is about 3.5 minutes.

10. (a) A type of light bulb is labeled as having an average lifetime of 1000 hours. It's reasonable to model the probability of failure of these bulbs by an exponential density function with mean $\mu = 1000$. Use this model to find the probability that a bulb
    (i) fails within the first 200 hours,
    (ii) burns for more than 800 hours.
    (b) What is the median lifetime of these light bulbs?

11. An online retailer has determined that the average time for credit card transactions to be electronically approved is 1.6 seconds.
    (a) Use an exponential density function to find the probability that a customer waits less than a second for credit card approval.
    (b) Find the probability that a customer waits more than 3 seconds.
    (c) What is the minimum approval time for the slowest 5% of transactions?

12. The time between infection and the display of symptoms for streptococcal sore throat is a random variable whose probabililty density function can be approximated by $f(t) = \frac{1}{15{,}676} t^2 e^{-0.05t}$ if $0 \leq t \leq 150$ and $f(t) = 0$ otherwise (t measured in hours).
    (a) What is the probability that an infected patient will display symptoms within the first 48 hours?
    (b) What is the probability that an infected patient will not display symptoms until after 36 hours?

    Source: Adapted from P. Sartwell, "The Distribution of Incubation Periods of Infectious Disease," *American Journal of Epidemiology* 141 (1995): 386–94.

13. REM sleep is the phase of sleep when most active dreaming occurs. In a study, the amount of REM sleep during the first four hours of sleep was described by a random variable T with probability density function
    $$f(t) = \begin{cases} \frac{1}{1600}t & \text{if } 0 \leq t \leq 40 \\ \frac{1}{20} - \frac{1}{1600}t & \text{if } 40 < t \leq 80 \\ 0 & \text{otherwise} \end{cases}$$
    where t is measured in minutes.
    (a) What is the probability that the amount of REM sleep is between 30 and 60 minutes?
    (b) Find the mean amount of REM sleep.

14. According to the National Health Survey, the heights of adult males in the United States are normally distributed with mean 69.0 inches and standard deviation 2.8 inches.
    (a) What is the probability that an adult male chosen at random is between 65 inches and 73 inches tall?
    (b) What percentage of the adult male population is more than 6 feet tall?

15. The "Garbage Project" at the University of Arizona reports that the amount of paper discarded by households per week is normally distributed with mean 9.4 lb and standard deviation 4.2 lb. What percentage of households throw out at least 10 lb of paper a week?

16. Boxes are labeled as containing 500 g of cereal. The machine filling the boxes produces weights that are normally distributed with standard deviation 12 g.
    (a) If the target weight is 500 g, what is the probability that the machine produces a box with less than 480 g of cereal?
    (b) Suppose a law states that no more than 5% of a manufacturer's cereal boxes can contain less than the stated weight of 500 g. At what target weight should the manufacturer set its filling machine?

17. The speeds of vehicles on a highway with speed limit 100 km/h are normally distributed with mean 112 km/h and standard deviation 8 km/h.
    (a) What is the probability that a randomly chosen vehicle is traveling at a legal speed?
    (b) If police are instructed to ticket motorists driving 125 km/h or more, what percentage of motorists are targeted?

18. Show that the probability density function for a normally distributed random variable has inflection points at $x = \mu \pm \sigma$.

19. For any normal distribution, find the probability that the random variable lies within two standard deviations of the mean.

20. The standard deviation for a random variable with probability density function f and mean μ is defined by

$$\sigma = \left[\int_{-\infty}^{\infty} (x - \mu)^2 f(x)\, dx \right]^{1/2}$$

Find the standard deviation for an exponential density function with mean μ.

21. The hydrogen atom is composed of one proton in the nucleus and one electron, which moves about the nucleus. In the quantum theory of atomic structure, it is assumed that the electron does not move in a well-defined orbit. Instead, it occupies a state known as an *orbital*, which may be thought of as a "cloud" of negative charge surrounding the nucleus. At the state of lowest energy, called the *ground state*, or *1s-orbital*, the shape of this cloud is assumed to be a sphere centered at the nucleus. This sphere is described in terms of the probability density function

$$p(r) = \frac{4}{a_0^3} r^2 e^{-2r/a_0} \qquad r \geq 0$$

where a_0 is the *Bohr radius* ($a_0 \approx 5.59 \times 10^{-11}$ m). The integral

$$P(r) = \int_0^r \frac{4}{a_0^3} s^2 e^{-2s/a_0}\, ds$$

gives the probability that the electron will be found within the sphere of radius r meters centered at the nucleus.
(a) Verify that $p(r)$ is a probability density function.
(b) Find $\lim_{r \to \infty} p(r)$. For what value of r does $p(r)$ have its maximum value?
(c) Graph the density function.
(d) Find the probability that the electron will be within the sphere of radius $4a_0$ centered at the nucleus.
(e) Calculate the mean distance of the electron from the nucleus in the ground state of the hydrogen atom.

8 REVIEW

CONCEPT CHECK

Answers to the Concept Check can be found on the back endpapers.

1. (a) How is the length of a curve defined?
 (b) Write an expression for the length of a smooth curve given by $y = f(x)$, $a \leq x \leq b$.
 (c) What if x is given as a function of y?

2. (a) Write an expression for the surface area of the surface obtained by rotating the curve $y = f(x)$, $a \leq x \leq b$, about the x-axis.
 (b) What if x is given as a function of y?
 (c) What if the curve is rotated about the y-axis?

3. Describe how we can find the hydrostatic force against a vertical wall submersed in a fluid.

4. (a) What is the physical significance of the center of mass of a thin plate?
 (b) If the plate lies between $y = f(x)$ and $y = 0$, where $a \leq x \leq b$, write expressions for the coordinates of the center of mass.

5. What does the Theorem of Pappus say?

6. Given a demand function $p(x)$, explain what is meant by the consumer surplus when the amount of a commodity currently available is X and the current selling price is P. Illustrate with a sketch.

7. (a) What is the cardiac output of the heart?
 (b) Explain how the cardiac output can be measured by the dye dilution method.

8. What is a probability density function? What properties does such a function have?

9. Suppose $f(x)$ is the probability density function for the weight of a female college student, where x is measured in pounds.
 (a) What is the meaning of the integral $\int_0^{130} f(x)\, dx$?
 (b) Write an expression for the mean of this density function.
 (c) How can we find the median of this density function?

10. What is a normal distribution? What is the significance of the standard deviation?

EXERCISES

1–3 Find the length of the curve.

1. $y = 4(x - 1)^{3/2}$, $1 \leq x \leq 4$

2. $y = 2 \ln\left(\sin \tfrac{1}{2}x\right)$, $\pi/3 \leq x \leq \pi$

3. $12x = 4y^3 + 3y^{-1}$, $1 \leq y \leq 3$

4. (a) Find the length of the curve

$$y = \frac{x^4}{16} + \frac{1}{2x^2} \qquad 1 \leq x \leq 2$$

(b) Find the area of the surface obtained by rotating the curve in part (a) about the y-axis.

5. Let C be the arc of the curve $y = 2/(x + 1)$ from the point $(0, 2)$ to $(3, \frac{1}{2})$. Use a calculator or other device to find the value of each of the following, correct to four decimal places.
   (a) The length of C
   (b) The area of the surface obtained by rotating C about the x-axis
   (c) The area of the surface obtained by rotating C about the y-axis

6. (a) The curve $y = x^2$, $0 \leq x \leq 1$, is rotated about the y-axis. Find the area of the resulting surface.
   (b) Find the area of the surface obtained by rotating the curve in part (a) about the x-axis.

7. Use Simpson's Rule with $n = 10$ to estimate the length of the sine curve $y = \sin x$, $0 \leq x \leq \pi$.

8. Use Simpson's Rule with $n = 10$ to estimate the area of the surface obtained by rotating the sine curve in Exercise 7 about the x-axis.

9. Find the length of the curve
$$y = \int_1^x \sqrt{\sqrt{t} - 1}\, dt \qquad 1 \leq x \leq 16$$

10. Find the area of the surface obtained by rotating the curve in Exercise 9 about the y-axis.

11. A gate in an irrigation canal is constructed in the form of a trapezoid 3 ft wide at the bottom, 5 ft wide at the top, and 2 ft high. It is placed vertically in the canal so that the water just covers the gate. Find the hydrostatic force on one side of the gate.

12. A trough is filled with water and its vertical ends have the shape of the parabolic region in the figure. Find the hydrostatic force on one end of the trough.

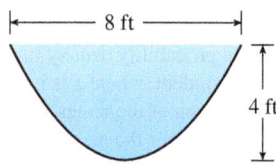

13–14 Find the centroid of the region shown.

13.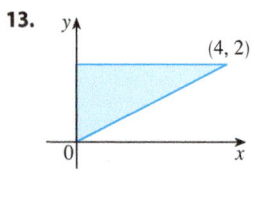

14.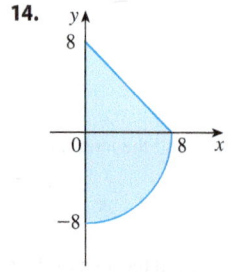

15–16 Find the centroid of the region bounded by the given curves.

15. $y = \frac{1}{2}x$, $y = \sqrt{x}$

16. $y = \sin x$, $y = 0$, $x = \pi/4$, $x = 3\pi/4$

17. Find the volume obtained when the circle of radius 1 with center $(1, 0)$ is rotated about the y-axis.

18. Use the Theorem of Pappus and the fact that the volume of a sphere of radius r is $\frac{4}{3}\pi r^3$ to find the centroid of the semicircular region bounded by the curve $y = \sqrt{r^2 - x^2}$ and the x-axis.

19. The demand function for a commodity is given by
$$p = 2000 - 0.1x - 0.01x^2$$
Find the consumer surplus when the sales level is 100.

20. After a 6-mg injection of dye into a heart, the readings of dye concentration at two-second intervals are as shown in the table. Use Simpson's Rule to estimate the cardiac output.

t	$c(t)$	t	$c(t)$
0	0	14	4.7
2	1.9	16	3.3
4	3.3	18	2.1
6	5.1	20	1.1
8	7.6	22	0.5
10	7.1	24	0
12	5.8		

21. (a) Explain why the function
$$f(x) = \begin{cases} \dfrac{\pi}{20} \sin\left(\dfrac{\pi x}{10}\right) & \text{if } 0 \leq x \leq 10 \\ 0 & \text{if } x < 0 \text{ or } x > 10 \end{cases}$$
is a probability density function.
   (b) Find $P(X < 4)$.
   (c) Calculate the mean. Is the value what you would expect?

22. Lengths of human pregnancies are normally distributed with mean 268 days and standard deviation 15 days. What percentage of pregnancies last between 250 days and 280 days?

23. The length of time spent waiting in line at a certain bank is modeled by an exponential density function with mean 8 minutes.
   (a) What is the probability that a customer is served in the first 3 minutes?
   (b) What is the probability that a customer has to wait more than 10 minutes?
   (c) What is the median waiting time?

Problems Plus

1. Find the area of the region $S = \{(x, y) \mid x \geq 0, \ y \leq 1, \ x^2 + y^2 \leq 4y\}$.

2. Find the centroid of the region enclosed by the loop of the curve $y^2 = x^3 - x^4$.

3. If a sphere of radius r is sliced by a plane whose distance from the center of the sphere is d, then the sphere is divided into two pieces called *segments of one base* (see the first figure). The corresponding surfaces are called *spherical zones of one base*.
   (a) Determine the surface areas of the two spherical zones indicated in the figure.
   (b) Determine the approximate area of the Arctic Ocean by assuming that it is approximately circular in shape, with center at the North Pole and "circumference" at 75° north latitude. Use $r = 3960$ mi for the radius of the earth.
   (c) A sphere of radius r is inscribed in a right circular cylinder of radius r. Two planes perpendicular to the central axis of the cylinder and a distance h apart cut off a *spherical zone of two bases* on the sphere (see the second figure). Show that the surface area of the spherical zone equals the surface area of the region that the two planes cut off on the cylinder.
   (d) The *Torrid Zone* is the region on the surface of the earth that is between the Tropic of Cancer (23.45° north latitude) and the Tropic of Capricorn (23.45° south latitude). What is the area of the Torrid Zone?

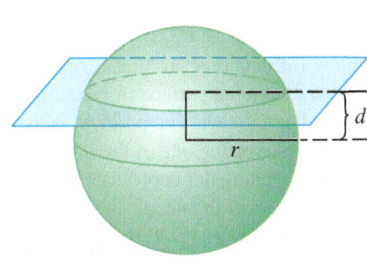
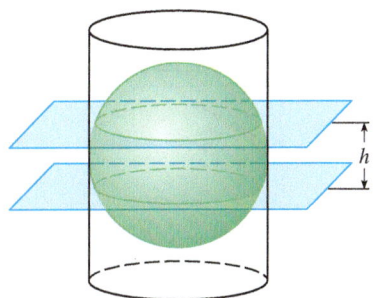

4. (a) Show that an observer at height H above the north pole of a sphere of radius r can see a part of the sphere that has area
$$\frac{2\pi r^2 H}{r + H}$$
   (b) Two spheres with radii r and R are placed so that the distance between their centers is d, where $d > r + R$. Where should a light be placed on the line joining the centers of the spheres in order to illuminate the largest total surface?

5. Suppose that the density of seawater, $\rho = \rho(z)$, varies with the depth z below the surface.
   (a) Show that the hydrostatic pressure is governed by the differential equation
$$\frac{dP}{dz} = \rho(z)g$$
   where g is the acceleration due to gravity. Let P_0 and ρ_0 be the pressure and density at $z = 0$. Express the pressure at depth z as an integral.
   (b) Suppose the density of seawater at depth z is given by $\rho = \rho_0 e^{z/H}$, where H is a positive constant. Find the total force, expressed as an integral, exerted on a vertical circular porthole of radius r whose center is located at a distance $L > r$ below the surface.

6. The figure shows a semicircle with radius 1, horizontal diameter PQ, and tangent lines at P and Q. At what height above the diameter should the horizontal line be placed so as to minimize the shaded area?

7. Let P be a pyramid with a square base of side $2b$ and suppose that S is a sphere with its center on the base of P and S is tangent to all eight edges of P. Find the height of P. Then find the volume of the intersection of S and P.

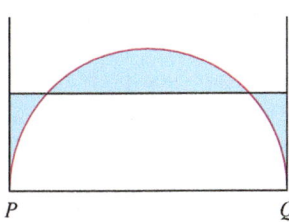

FIGURE FOR PROBLEM 6

8. Consider a flat metal plate to be placed vertically underwater with its top 2 m below the surface of the water. Determine a shape for the plate so that if the plate is divided into any number of horizontal strips of equal height, the hydrostatic force on each strip is the same.

9. A uniform disk with radius 1 m is to be cut by a line so that the center of mass of the smaller piece lies halfway along a radius. How close to the center of the disk should the cut be made? (Express your answer correct to two decimal places.)

10. A triangle with area 30 cm² is cut from a corner of a square with side 10 cm, as shown in the figure. If the centroid of the remaining region is 4 cm from the right side of the square, how far is it from the bottom of the square?

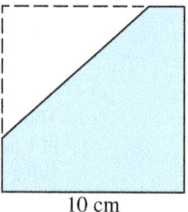

11. In a famous 18th-century problem, known as *Buffon's needle problem*, a needle of length h is dropped onto a flat surface (for example, a table) on which parallel lines L units apart, $L \geq h$, have been drawn. The problem is to determine the probability that the needle will come to rest intersecting one of the lines. Assume that the lines run east-west, parallel to the x-axis in a rectangular coordinate system (as in the figure). Let y be the distance from the "southern" end of the needle to the nearest line to the north. (If the needle's southern end lies on a line, let $y = 0$. If the needle happens to lie east-west, let the "western" end be the "southern" end.) Let θ be the angle that the needle makes with a ray extending eastward from the "southern" end. Then $0 \leq y \leq L$ and $0 \leq \theta \leq \pi$. Note that the needle intersects one of the lines only when $y < h \sin \theta$. The total set of possibilities for the needle can be identified with the rectangular region $0 \leq y \leq L$, $0 \leq \theta \leq \pi$, and the proportion of times that the needle intersects a line is the ratio

$$\frac{\text{area under } y = h \sin \theta}{\text{area of rectangle}}$$

This ratio is the probability that the needle intersects a line. Find the probability that the needle will intersect a line if $h = L$. What if $h = \tfrac{1}{2}L$?

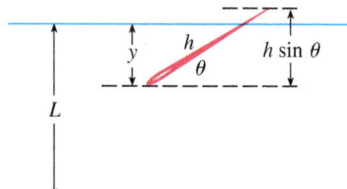
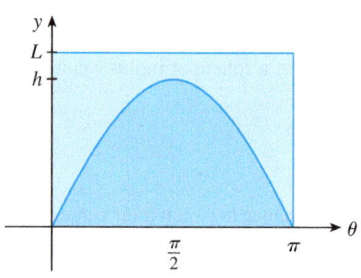

FIGURE FOR PROBLEM 11

12. If the needle in Problem 11 has length $h > L$, it's possible for the needle to intersect more than one line.
    (a) If $L = 4$, find the probability that a needle of length 7 will intersect at least one line. [*Hint:* Proceed as in Problem 11. Define y as before; then the total set of possibilities for the needle can be identified with the same rectangular region $0 \leq y \leq L$, $0 \leq \theta \leq \pi$. What portion of the rectangle corresponds to the needle intersecting a line?]
    (b) If $L = 4$, find the probability that a needle of length 7 will intersect *two* lines.
    (c) If $2L < h \leq 3L$, find a general formula for the probability that the needle intersects three lines.

13. Find the centroid of the region enclosed by the ellipse $x^2 + (x + y + 1)^2 = 1$.

9 Differential Equations

© Dennis Donohue / Shutterstock.com

In the last section of this chapter we use pairs of differential equations to investigate the relationship between populations of predators and prey, such as jaguars and wart hogs, wolves and rabbits, lynx and hares, and ladybugs and aphids.

PERHAPS THE MOST IMPORTANT of all the applications of calculus is to differential equations. When physical scientists or social scientists use calculus, more often than not it is to analyze a differential equation that has arisen in the process of modeling some phenomenon that they are studying. Although it is often impossible to find an explicit formula for the solution of a differential equation, we will see that graphical and numerical approaches provide the needed information.

9.1 Modeling with Differential Equations

Now is a good time to read (or reread) the discussion of mathematical modeling on page 23.

In describing the process of modeling in Section 1.2, we talked about formulating a mathematical model of a real-world problem either through intuitive reasoning about the phenomenon or from a physical law based on evidence from experiments. The mathematical model often takes the form of a *differential equation,* that is, an equation that contains an unknown function and some of its derivatives. This is not surprising because in a real-world problem we often notice that changes occur and we want to predict future behavior on the basis of how current values change. Let's begin by examining several examples of how differential equations arise when we model physical phenomena.

■ Models for Population Growth

One model for the growth of a population is based on the assumption that the population grows at a rate proportional to the size of the population. That is a reasonable assumption for a population of bacteria or animals under ideal conditions (unlimited environment, adequate nutrition, absence of predators, immunity from disease).

Let's identify and name the variables in this model:

$t = $ time   (the independent variable)

$P = $ the number of individuals in the population   (the dependent variable)

The rate of growth of the population is the derivative dP/dt. So our assumption that the rate of growth of the population is proportional to the population size is written as the equation

$$\boxed{1} \qquad \frac{dP}{dt} = kP$$

where k is the proportionality constant. Equation 1 is our first model for population growth; it is a differential equation because it contains an unknown function P and its derivative dP/dt.

Having formulated a model, let's look at its consequences. If we rule out a population of 0, then $P(t) > 0$ for all t. So, if $k > 0$, then Equation 1 shows that $P'(t) > 0$ for all t. This means that the population is always increasing. In fact, as $P(t)$ increases, Equation 1 shows that dP/dt becomes larger. In other words, the growth rate increases as the population increases.

Let's try to think of a solution of Equation 1. This equation asks us to find a function whose derivative is a constant multiple of itself. We know from Chapter 3 that exponential functions have that property. In fact, if we let $P(t) = Ce^{kt}$, then

$$P'(t) = C(ke^{kt}) = k(Ce^{kt}) = kP(t)$$

Thus any exponential function of the form $P(t) = Ce^{kt}$ is a solution of Equation 1. In Section 9.4, we will see that there is no other solution.

Allowing C to vary through all the real numbers, we get the *family* of solutions $P(t) = Ce^{kt}$ whose graphs are shown in Figure 1. But populations have only positive values and so we are interested only in the solutions with $C > 0$. And we are probably

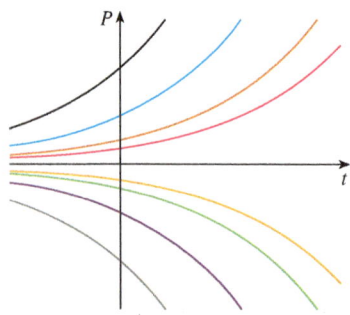

FIGURE 1
The family of solutions of $dP/dt = kP$

concerned only with values of t greater than the initial time $t = 0$. Figure 2 shows the physically meaningful solutions. Putting $t = 0$, we get $P(0) = Ce^{k(0)} = C$, so the constant C turns out to be the initial population, $P(0)$.

Equation 1 is appropriate for modeling population growth under ideal conditions, but we have to recognize that a more realistic model must reflect the fact that a given environment has limited resources. Many populations start by increasing in an exponential manner, but the population levels off when it approaches its *carrying capacity M* (or decreases toward M if it ever exceeds M). For a model to take into account both trends, we make two assumptions:

- $\dfrac{dP}{dt} \approx kP$ if P is small (Initially, the growth rate is proportional to P.)

- $\dfrac{dP}{dt} < 0$ if $P > M$ (P decreases if it ever exceeds M.)

A simple expression that incorporates both assumptions is given by the equation

$$\boxed{2} \qquad \dfrac{dP}{dt} = kP\left(1 - \dfrac{P}{M}\right)$$

Notice that if P is small compared with M, then P/M is close to 0 and so $dP/dt \approx kP$. If $P > M$, then $1 - P/M$ is negative and so $dP/dt < 0$.

Equation 2 is called the *logistic differential equation* and was proposed by the Dutch mathematical biologist Pierre-François Verhulst in the 1840s as a model for world population growth. We will develop techniques that enable us to find explicit solutions of the logistic equation in Section 9.4, but for now we can deduce qualitative characteristics of the solutions directly from Equation 2. We first observe that the constant functions $P(t) = 0$ and $P(t) = M$ are solutions because, in either case, one of the factors on the right side of Equation 2 is zero. (This certainly makes physical sense: If the population is ever either 0 or at the carrying capacity, it stays that way.) These two constant solutions are called **equilibrium solutions**.

If the initial population $P(0)$ lies between 0 and M, then the right side of Equation 2 is positive, so $dP/dt > 0$ and the population increases. But if the population exceeds the carrying capacity ($P > M$), then $1 - P/M$ is negative, so $dP/dt < 0$ and the population decreases. Notice that, in either case, if the population approaches the carrying capacity ($P \to M$), then $dP/dt \to 0$, which means the population levels off. So we expect that the solutions of the logistic differential equation have graphs that look something like the ones in Figure 3. Notice that the graphs move away from the equilibrium solution $P = 0$ and move toward the equilibrium solution $P = M$.

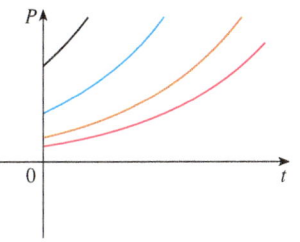

FIGURE 2
The family of solutions $P(t) = Ce^{kt}$ with $C > 0$ and $t \geq 0$

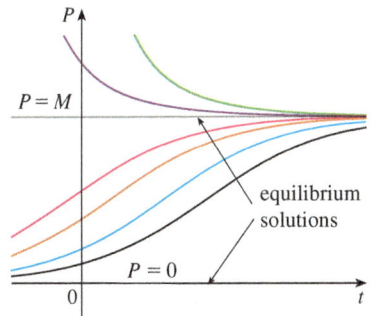

FIGURE 3
Solutions of the logistic equation

A Model for the Motion of a Spring

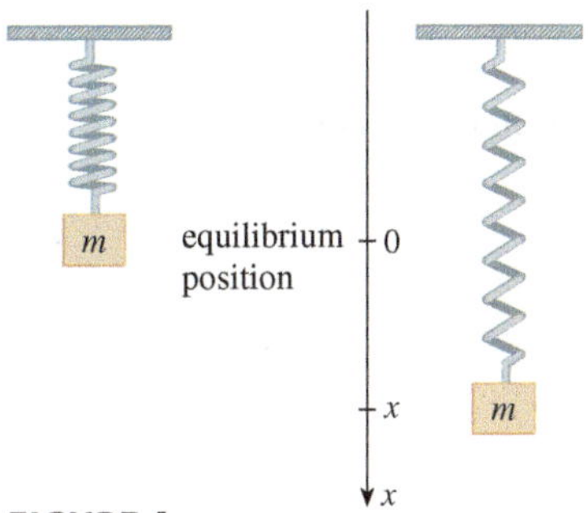

Let's now look at an example of a model from the physical sciences. We consider the motion of an object with mass m at the end of a vertical spring (as in Figure 4). In Section 6.4 we discussed Hooke's Law, which says that if the spring is stretched (or compressed) x units from its natural length, then it exerts a force that is proportional to x:

$$\text{restoring force} = -kx$$

where k is a positive constant (called the *spring constant*). If we ignore any external resisting forces (due to air resistance or friction) then, by Newton's Second Law

FIGURE 4

(force equals mass times acceleration), we have

$$\boxed{3} \qquad m\frac{d^2x}{dt^2} = -kx$$

This is an example of what is called a *second-order differential equation* because it involves second derivatives. Let's see what we can guess about the form of the solution directly from the equation. We can rewrite Equation 3 in the form

$$\frac{d^2x}{dt^2} = -\frac{k}{m}x$$

which says that the second derivative of x is proportional to x but has the opposite sign. We know two functions with this property, the sine and cosine functions. In fact, it turns out that all solutions of Equation 3 can be written as combinations of certain sine and cosine functions (see Exercise 4). This is not surprising; we expect the spring to oscillate about its equilibrium position and so it is natural to think that trigonometric functions are involved.

■ General Differential Equations

In general, a **differential equation** is an equation that contains an unknown function and one or more of its derivatives. The **order** of a differential equation is the order of the highest derivative that occurs in the equation. Thus Equations 1 and 2 are first-order equations and Equation 3 is a second-order equation. In all three of those equations the independent variable is called t and represents time, but in general the independent variable doesn't have to represent time. For example, when we consider the differential equation

$$\boxed{4} \qquad y' = xy$$

it is understood that y is an unknown function of x.

A function f is called a **solution** of a differential equation if the equation is satisfied when $y = f(x)$ and its derivatives are substituted into the equation. Thus f is a solution of Equation 4 if

$$f'(x) = xf(x)$$

for all values of x in some interval.

When we are asked to *solve* a differential equation we are expected to find all possible solutions of the equation. We have already solved some particularly simple differential equations, namely, those of the form

$$y' = f(x)$$

For instance, we know that the general solution of the differential equation

$$y' = x^3$$

is given by

$$y = \frac{x^4}{4} + C$$

where C is an arbitrary constant.

But, in general, solving a differential equation is not an easy matter. There is no systematic technique that enables us to solve all differential equations. In Section 9.2, how-

SECTION 9.1 Modeling with Differential Equations

ever, we will see how to draw rough graphs of solutions even when we have no explicit formula. We will also learn how to find numerical approximations to solutions.

EXAMPLE 1 Show that every member of the family of functions

$$y = \frac{1 + ce^t}{1 - ce^t}$$

is a solution of the differential equation $y' = \frac{1}{2}(y^2 - 1)$.

SOLUTION We use the Quotient Rule to differentiate the expression for y:

$$y' = \frac{(1 - ce^t)(ce^t) - (1 + ce^t)(-ce^t)}{(1 - ce^t)^2}$$

$$= \frac{ce^t - c^2e^{2t} + ce^t + c^2e^{2t}}{(1 - ce^t)^2} = \frac{2ce^t}{(1 - ce^t)^2}$$

The right side of the differential equation becomes

$$\frac{1}{2}(y^2 - 1) = \frac{1}{2}\left[\left(\frac{1 + ce^t}{1 - ce^t}\right)^2 - 1\right]$$

$$= \frac{1}{2}\left[\frac{(1 + ce^t)^2 - (1 - ce^t)^2}{(1 - ce^t)^2}\right]$$

$$= \frac{1}{2}\frac{4ce^t}{(1 - ce^t)^2} = \frac{2ce^t}{(1 - ce^t)^2}$$

Therefore, for every value of c, the given function is a solution of the differential equation. ■

Figure 5 shows graphs of seven members of the family in Example 1. The differential equation shows that if $y \approx \pm 1$, then $y' \approx 0$. That is borne out by the flatness of the graphs near $y = 1$ and $y = -1$.

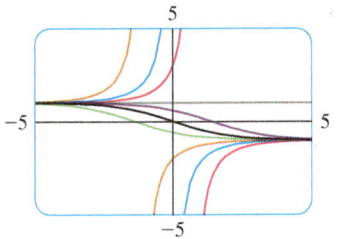

FIGURE 5

When applying differential equations, we are usually not as interested in finding a family of solutions (the *general solution*) as we are in finding a solution that satisfies some additional requirement. In many physical problems we need to find the particular solution that satisfies a condition of the form $y(t_0) = y_0$. This is called an **initial condition**, and the problem of finding a solution of the differential equation that satisfies the initial condition is called an **initial-value problem**.

Geometrically, when we impose an initial condition, we look at the family of solution curves and pick the one that passes through the point (t_0, y_0). Physically, this corresponds to measuring the state of a system at time t_0 and using the solution of the initial-value problem to predict the future behavior of the system.

EXAMPLE 2 Find a solution of the differential equation $y' = \frac{1}{2}(y^2 - 1)$ that satisfies the initial condition $y(0) = 2$.

SOLUTION Substituting the values $t = 0$ and $y = 2$ into the formula

$$y = \frac{1 + ce^t}{1 - ce^t}$$

from Example 1, we get

$$2 = \frac{1 + ce^0}{1 - ce^0} = \frac{1 + c}{1 - c}$$

Solving this equation for c, we get $2 - 2c = 1 + c$, which gives $c = \frac{1}{3}$. So the solution of the initial-value problem is

$$y = \frac{1 + \frac{1}{3}e^t}{1 - \frac{1}{3}e^t} = \frac{3 + e^t}{3 - e^t}$$

9.1 EXERCISES

1. Show that $y = \frac{2}{3}e^x + e^{-2x}$ is a solution of the differential equation $y' + 2y = 2e^x$.

2. Verify that $y = -t \cos t - t$ is a solution of the initial-value problem

$$t \frac{dy}{dt} = y + t^2 \sin t \qquad y(\pi) = 0$$

3. (a) For what values of r does the function $y = e^{rx}$ satisfy the differential equation $2y'' + y' - y = 0$?
(b) If r_1 and r_2 are the values of r that you found in part (a), show that every member of the family of functions $y = ae^{r_1 x} + be^{r_2 x}$ is also a solution.

4. (a) For what values of k does the function $y = \cos kt$ satisfy the differential equation $4y'' = -25y$?
(b) For those values of k, verify that every member of the family of functions $y = A \sin kt + B \cos kt$ is also a solution.

5. Which of the following functions are solutions of the differential equation $y'' + y = \sin x$?
(a) $y = \sin x$
(b) $y = \cos x$
(c) $y = \frac{1}{2}x \sin x$
(d) $y = -\frac{1}{2}x \cos x$

6. (a) Show that every member of the family of functions $y = (\ln x + C)/x$ is a solution of the differential equation $x^2 y' + xy = 1$.
(b) Illustrate part (a) by graphing several members of the family of solutions on a common screen.
(c) Find a solution of the differential equation that satisfies the initial condition $y(1) = 2$.
(d) Find a solution of the differential equation that satisfies the initial condition $y(2) = 1$.

7. (a) What can you say about a solution of the equation $y' = -y^2$ just by looking at the differential equation?
(b) Verify that all members of the family $y = 1/(x + C)$ are solutions of the equation in part (a).
(c) Can you think of a solution of the differential equation $y' = -y^2$ that is not a member of the family in part (b)?
(d) Find a solution of the initial-value problem

$$y' = -y^2 \qquad y(0) = 0.5$$

8. (a) What can you say about the graph of a solution of the equation $y' = xy^3$ when x is close to 0? What if x is large?

(b) Verify that all members of the family $y = (c - x^2)^{-1/2}$ are solutions of the differential equation $y' = xy^3$.
(c) Graph several members of the family of solutions on a common screen. Do the graphs confirm what you predicted in part (a)?
(d) Find a solution of the initial-value problem

$$y' = xy^3 \qquad y(0) = 2$$

9. A population is modeled by the differential equation

$$\frac{dP}{dt} = 1.2P\left(1 - \frac{P}{4200}\right)$$

(a) For what values of P is the population increasing?
(b) For what values of P is the population decreasing?
(c) What are the equilibrium solutions?

10. The Fitzhugh-Nagumo model for the electrical impulse in a neuron states that, in the absence of relaxation effects, the electrical potential in a neuron $v(t)$ obeys the differential equation

$$\frac{dv}{dt} = -v[v^2 - (1 + a)v + a]$$

where a is a positive constant such that $0 < a < 1$.
(a) For what values of v is v unchanging (that is, $dv/dt = 0$)?
(b) For what values of v is v increasing?
(c) For what values of v is v decreasing?

11. Explain why the functions with the given graphs *can't* be solutions of the differential equation

$$\frac{dy}{dt} = e^t(y - 1)^2$$

(a)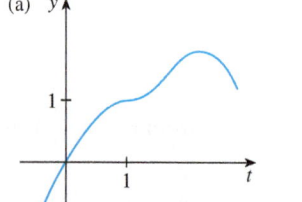

(b)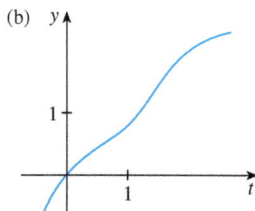

12. The function with the given graph is a solution of one of the following differential equations. Decide which is the correct equation and justify your answer.

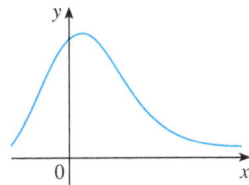

A. $y' = 1 + xy$ **B.** $y' = -2xy$ **C.** $y' = 1 - 2xy$

13. Match the differential equations with the solution graphs labeled I–IV. Give reasons for your choices.

(a) $y' = 1 + x^2 + y^2$ (b) $y' = xe^{-x^2-y^2}$

(c) $y' = \dfrac{1}{1 + e^{x^2+y^2}}$ (d) $y' = \sin(xy)\cos(xy)$

I 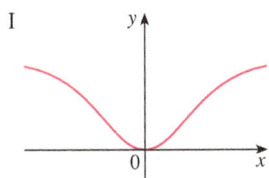   II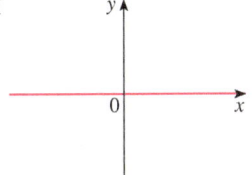

III 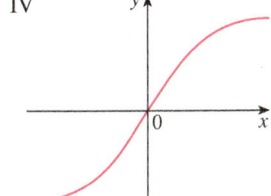   IV

14. Suppose you have just poured a cup of freshly brewed coffee with temperature 95°C in a room where the temperature is 20°C.
  (a) When do you think the coffee cools most quickly? What happens to the rate of cooling as time goes by? Explain.
  (b) Newton's Law of Cooling states that the rate of cooling of an object is proportional to the temperature difference between the object and its surroundings, provided that this difference is not too large. Write a differential equation that expresses Newton's Law of Cooling for this particular situation. What is the initial condition? In view of your answer to part (a), do you think this differential equation is an appropriate model for cooling?
  (c) Make a rough sketch of the graph of the solution of the initial-value problem in part (b).

15. Psychologists interested in learning theory study **learning curves**. A learning curve is the graph of a function $P(t)$, the performance of someone learning a skill as a function of the training time t. The derivative dP/dt represents the rate at which performance improves.
  (a) When do you think P increases most rapidly? What happens to dP/dt as t increases? Explain.
  (b) If M is the maximum level of performance of which the learner is capable, explain why the differential equation

  $$\dfrac{dP}{dt} = k(M - P) \qquad k \text{ a positive constant}$$

  is a reasonable model for learning.
  (c) Make a rough sketch of a possible solution of this differential equation.

16. Von Bertalanffy's equation states that the rate of growth in length of an individual fish is proportional to the difference between the current length L and the asymptotic length L_∞ (in centimeters).
  (a) Write a differential equation that expresses this idea.
  (b) Make a rough sketch of the graph of a solution of a typical initial-value problem for this differential equation.

17. Differential equations have been used extensively in the study of drug dissolution for patients given oral medications. One such equation is the Weibull equation for the concentration $c(t)$ of the drug:

$$\dfrac{dc}{dt} = \dfrac{k}{t^b}(c_s - c)$$

where k and c_s are positive constants and $0 < b < 1$. Verify that

$$c(t) = c_s\left(1 - e^{-\alpha t^{1-b}}\right)$$

is a solution of the Weibull equation for $t > 0$, where $\alpha = k/(1 - b)$. What does the differential equation say about how drug dissolution occurs?

9.2 Direction Fields and Euler's Method

Unfortunately, it's impossible to solve most differential equations in the sense of obtaining an explicit formula for the solution. In this section we show that, despite the absence of an explicit solution, we can still learn a lot about the solution through a graphical approach (direction fields) or a numerical approach (Euler's method).

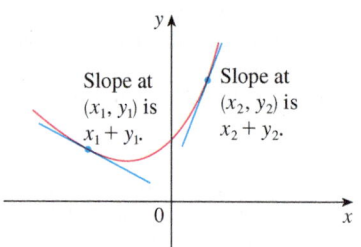

FIGURE 1
A solution of $y' = x + y$

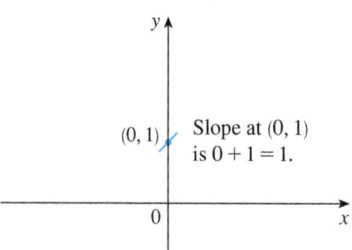

FIGURE 2
Beginning of the solution curve through (0,1)

TEC Module 9.2A shows direction fields and solution curves for a variety of differential equations.

Direction Fields

Suppose we are asked to sketch the graph of the solution of the initial-value problem

$$y' = x + y \qquad y(0) = 1$$

We don't know a formula for the solution, so how can we possibly sketch its graph? Let's think about what the differential equation means. The equation $y' = x + y$ tells us that the slope at any point (x, y) on the graph (called the *solution curve*) is equal to the sum of the x- and y-coordinates of the point (see Figure 1). In particular, because the curve passes through the point $(0, 1)$, its slope there must be $0 + 1 = 1$. So a small portion of the solution curve near the point $(0, 1)$ looks like a short line segment through $(0, 1)$ with slope 1. (See Figure 2.)

As a guide to sketching the rest of the curve, let's draw short line segments at a number of points (x, y) with slope $x + y$. The result is called a *direction field* and is shown in Figure 3. For instance, the line segment at the point $(1, 2)$ has slope $1 + 2 = 3$. The direction field allows us to visualize the general shape of the solution curves by indicating the direction in which the curves proceed at each point.

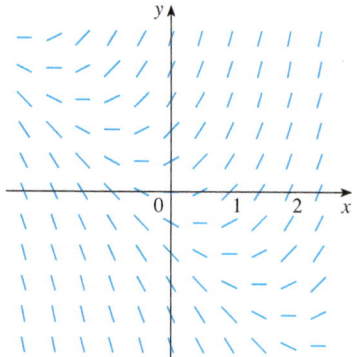

FIGURE 3
Direction field for $y' = x + y$

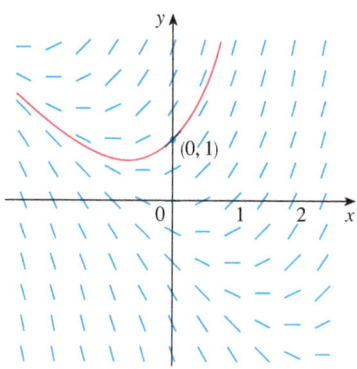

FIGURE 4
The solution curve through (0,1)

Now we can sketch the solution curve through the point $(0, 1)$ by following the direction field as in Figure 4. Notice that we have drawn the curve so that it is parallel to nearby line segments.

In general, suppose we have a first-order differential equation of the form

$$y' = F(x, y)$$

where $F(x, y)$ is some expression in x and y. The differential equation says that the slope of a solution curve at a point (x, y) on the curve is $F(x, y)$. If we draw short line segments with slope $F(x, y)$ at several points (x, y), the result is called a **direction field** (or **slope field**). These line segments indicate the direction in which a solution curve is heading, so the direction field helps us visualize the general shape of these curves.

EXAMPLE 1

(a) Sketch the direction field for the differential equation $y' = x^2 + y^2 - 1$.
(b) Use part (a) to sketch the solution curve that passes through the origin.

SOLUTION

(a) We start by computing the slope at several points in the following chart:

x	-2	-1	0	1	2	-2	-1	0	1	2	...
y	0	0	0	0	0	1	1	1	1	1	...
$y' = x^2 + y^2 - 1$	3	0	-1	0	3	4	1	0	1	4	...

Now we draw short line segments with these slopes at these points. The result is the direction field shown in Figure 5.

(b) We start at the origin and move to the right in the direction of the line segment (which has slope -1). We continue to draw the solution curve so that it moves parallel to the nearby line segments. The resulting solution curve is shown in Figure 6. Returning to the origin, we draw the solution curve to the left as well. ∎

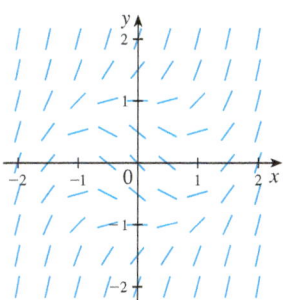

FIGURE 5

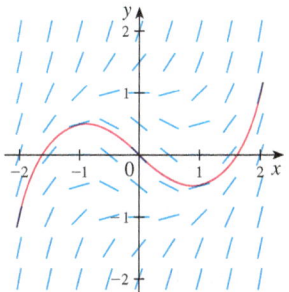

FIGURE 6

The more line segments we draw in a direction field, the clearer the picture becomes. Of course, it's tedious to compute slopes and draw line segments for a huge number of points by hand, but computers are well suited for this task. Figure 7 shows a more detailed, computer-drawn direction field for the differential equation in Example 1. It enables us to draw, with reasonable accuracy, the solution curves with y-intercepts -2, -1, 0, 1, and 2.

Now let's see how direction fields give insight into physical situations. The simple electric circuit shown in Figure 8 contains an electromotive force (usually a battery or generator) that produces a voltage of $E(t)$ volts (V) and a current of $I(t)$ amperes (A) at time t. The circuit also contains a resistor with a resistance of R ohms (Ω) and an inductor with an inductance of L henries (H).

Ohm's Law gives the drop in voltage due to the resistor as RI. The voltage drop due to the inductor is $L(dI/dt)$. One of Kirchhoff's laws says that the sum of the voltage drops is equal to the supplied voltage $E(t)$. Thus we have

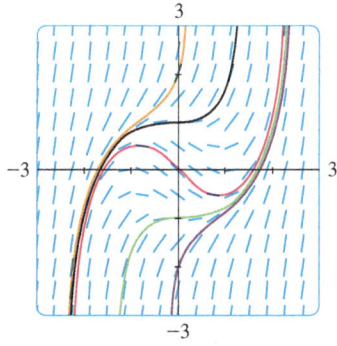

FIGURE 7

$$\boxed{1} \qquad L\frac{dI}{dt} + RI = E(t)$$

which is a first-order differential equation that models the current I at time t.

EXAMPLE 2 Suppose that in the simple circuit of Figure 8 the resistance is 12 Ω, the inductance is 4 H, and a battery gives a constant voltage of 60 V.
(a) Draw a direction field for Equation 1 with these values.
(b) What can you say about the limiting value of the current?
(c) Identify any equilibrium solutions.
(d) If the switch is closed when $t = 0$ so the current starts with $I(0) = 0$, use the direction field to sketch the solution curve.

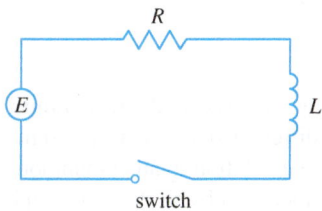

FIGURE 8

SOLUTION

(a) If we put $L = 4$, $R = 12$, and $E(t) = 60$ in Equation 1, we get

$$4\frac{dI}{dt} + 12I = 60 \qquad \text{or} \qquad \frac{dI}{dt} = 15 - 3I$$

The direction field for this differential equation is shown in Figure 9.

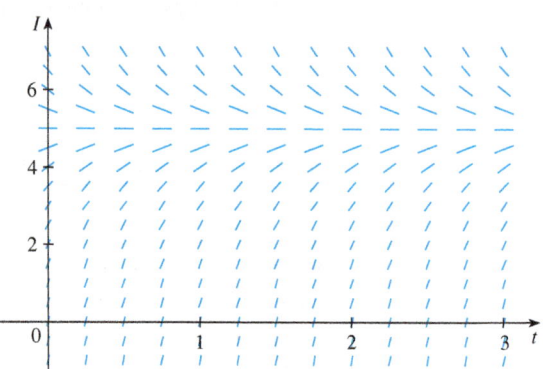

FIGURE 9

(b) It appears from the direction field that all solutions approach the value 5 A, that is,

$$\lim_{t \to \infty} I(t) = 5$$

(c) It appears that the constant function $I(t) = 5$ is an equilibrium solution. Indeed, we can verify this directly from the differential equation $dI/dt = 15 - 3I$. If $I(t) = 5$, then the left side is $dI/dt = 0$ and the right side is $15 - 3(5) = 0$.

(d) We use the direction field to sketch the solution curve that passes through $(0, 0)$, as shown in red in Figure 10.

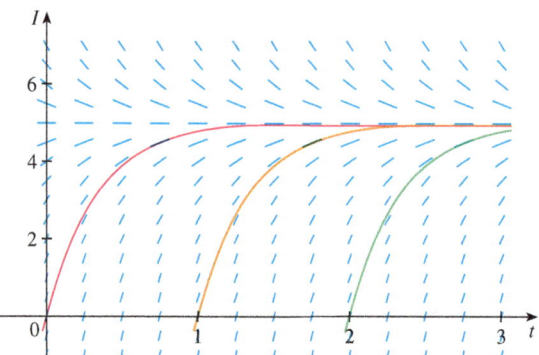

FIGURE 10

Notice from Figure 9 that the line segments along any horizontal line are parallel. That is because the independent variable t does not occur on the right side of the equation $I' = 15 - 3I$. In general, a differential equation of the form

$$y' = f(y)$$

in which the independent variable is missing from the right side, is called **autonomous**. For such an equation, the slopes corresponding to two different points with the same y-coordinate must be equal. This means that if we know one solution to an autonomous differential equation, then we can obtain infinitely many others just by shifting the graph of the known solution to the right or left. In Figure 10 we have shown the solutions that result from shifting the solution curve of Example 2 one and two time units (namely, seconds) to the right. They correspond to closing the switch when $t = 1$ or $t = 2$.

Euler's Method

The basic idea behind direction fields can be used to find numerical approximations to solutions of differential equations. We illustrate the method on the initial-value problem that we used to introduce direction fields:

$$y' = x + y \qquad y(0) = 1$$

The differential equation tells us that $y'(0) = 0 + 1 = 1$, so the solution curve has slope 1 at the point $(0, 1)$. As a first approximation to the solution we could use the linear approximation $L(x) = x + 1$. In other words, we could use the tangent line at $(0, 1)$ as a rough approximation to the solution curve (see Figure 11).

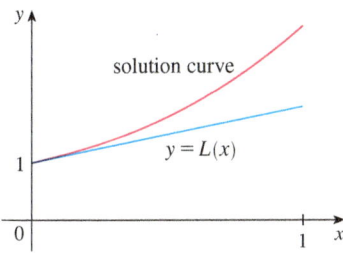

FIGURE 11
First Euler approximation

Euler's idea was to improve on this approximation by proceeding only a short distance along this tangent line and then making a midcourse correction by changing direction as indicated by the direction field. Figure 12 shows what happens if we start out along the tangent line but stop when $x = 0.5$. (This horizontal distance traveled is called the *step size*.) Since $L(0.5) = 1.5$, we have $y(0.5) \approx 1.5$ and we take $(0.5, 1.5)$ as the starting point for a new line segment. The differential equation tells us that $y'(0.5) = 0.5 + 1.5 = 2$, so we use the linear function

$$y = 1.5 + 2(x - 0.5) = 2x + 0.5$$

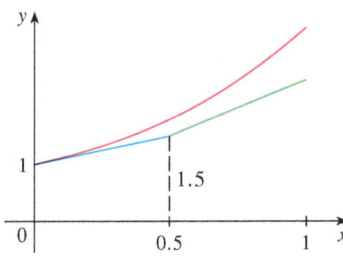

FIGURE 12
Euler approximation with step size 0.5

as an approximation to the solution for $x > 0.5$ (the green segment in Figure 12). If we decrease the step size from 0.5 to 0.25, we get the better Euler approximation shown in Figure 13.

In general, Euler's method says to start at the point given by the initial value and proceed in the direction indicated by the direction field. Stop after a short time, look at the slope at the new location, and proceed in that direction. Keep stopping and changing direction according to the direction field. Euler's method does not produce the exact solution to an initial-value problem—it gives approximations. But by decreasing the step size (and therefore increasing the number of midcourse corrections), we obtain successively better approximations to the exact solution. (Compare Figures 11, 12, and 13.)

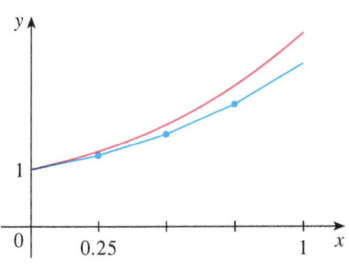

FIGURE 13
Euler approximation with step size 0.25

For the general first-order initial-value problem $y' = F(x, y)$, $y(x_0) = y_0$, our aim is to find approximate values for the solution at equally spaced numbers x_0, $x_1 = x_0 + h$, $x_2 = x_1 + h, \ldots$, where h is the step size. The differential equation tells us that the slope at (x_0, y_0) is $y' = F(x_0, y_0)$, so Figure 14 shows that the approximate value of the solution when $x = x_1$ is

$$y_1 = y_0 + hF(x_0, y_0)$$

Similarly, $$y_2 = y_1 + hF(x_1, y_1)$$

In general, $$y_n = y_{n-1} + hF(x_{n-1}, y_{n-1})$$

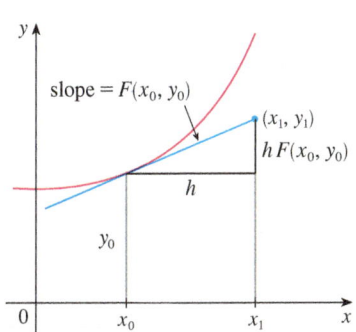

FIGURE 14

Euler's Method Approximate values for the solution of the initial-value problem $y' = F(x, y)$, $y(x_0) = y_0$, with step size h, at $x_n = x_{n-1} + h$, are

$$y_n = y_{n-1} + hF(x_{n-1}, y_{n-1}) \qquad n = 1, 2, 3, \ldots$$

EXAMPLE 3 Use Euler's method with step size 0.1 to construct a table of approximate values for the solution of the initial-value problem

$$y' = x + y \qquad y(0) = 1$$

SOLUTION We are given that $h = 0.1$, $x_0 = 0$, $y_0 = 1$, and $F(x, y) = x + y$. So we have

$$y_1 = y_0 + hF(x_0, y_0) = 1 + 0.1(0 + 1) = 1.1$$

$$y_2 = y_1 + hF(x_1, y_1) = 1.1 + 0.1(0.1 + 1.1) = 1.22$$

$$y_3 = y_2 + hF(x_2, y_2) = 1.22 + 0.1(0.2 + 1.22) = 1.362$$

This means that if $y(x)$ is the exact solution, then $y(0.3) \approx 1.362$.

Proceeding with similar calculations, we get the values in the table:

n	x_n	y_n	n	x_n	y_n
1	0.1	1.100000	6	0.6	1.943122
2	0.2	1.220000	7	0.7	2.197434
3	0.3	1.362000	8	0.8	2.487178
4	0.4	1.528200	9	0.9	2.815895
5	0.5	1.721020	10	1.0	3.187485

Computer software packages that produce numerical approximations to solutions of differential equations use methods that are refinements of Euler's method. Although Euler's method is simple and not as accurate, it is the basic idea on which the more accurate methods are based.

TEC Module 9.2B shows how Euler's method works numerically and visually for a variety of differential equations and step sizes.

For a more accurate table of values in Example 3 we could decrease the step size. But for a large number of small steps the amount of computation is considerable and so we need to program a calculator or computer to carry out these calculations. The following table shows the results of applying Euler's method with decreasing step size to the initial-value problem of Example 3.

Notice that the Euler estimates in the table below seem to be approaching limits, namely, the true values of $y(0.5)$ and $y(1)$. Figure 15 shows graphs of the Euler approximations with step sizes 0.5, 0.25, 0.1, 0.05, 0.02, 0.01, and 0.005. They are approaching the exact solution curve as the step size h approaches 0.

Step size	Euler estimate of $y(0.5)$	Euler estimate of $y(1)$
0.500	1.500000	2.500000
0.250	1.625000	2.882813
0.100	1.721020	3.187485
0.050	1.757789	3.306595
0.020	1.781212	3.383176
0.010	1.789264	3.409628
0.005	1.793337	3.423034
0.001	1.796619	3.433848

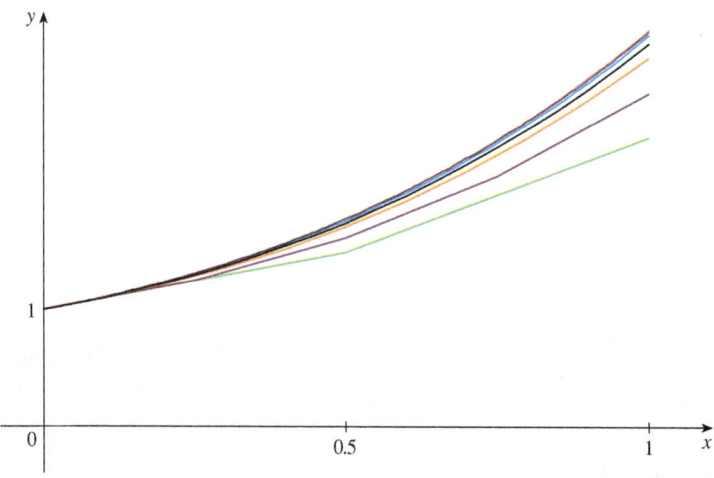

FIGURE 15 Euler approximation approaching the exact solution

Euler

Leonhard Euler (1707–1783) was the leading mathematician of the mid-18th century and the most prolific mathematician of all time. He was born in Switzerland but spent most of his career at the academies of science supported by Catherine the Great in St. Petersburg and Frederick the Great in Berlin. The collected works of Euler (pronounced *Oiler*) fill about 100 large volumes. As the French physicist Arago said, "Euler calculated without apparent effort, as men breathe or as eagles sustain themselves in the air." Euler's calculations and writings were not diminished by raising 13 children or being totally blind for the last 17 years of his life. In fact, when blind, he dictated his discoveries to his helpers from his prodigious memory and imagination. His treatises on calculus and most other mathematical subjects became the standard for mathematics instruction and the equation $e^{i\pi} + 1 = 0$ that he discovered brings together the five most famous numbers in all of mathematics.

EXAMPLE 4 In Example 2 we discussed a simple electric circuit with resistance 12 Ω, inductance 4 H, and a battery with voltage 60 V. If the switch is closed when $t = 0$, we modeled the current I at time t by the initial-value problem

$$\frac{dI}{dt} = 15 - 3I \qquad I(0) = 0$$

Estimate the current in the circuit half a second after the switch is closed.

SOLUTION We use Euler's method with $F(t, I) = 15 - 3I$, $t_0 = 0$, $I_0 = 0$, and step size $h = 0.1$ second:

$$I_1 = 0 + 0.1(15 - 3 \cdot 0) = 1.5$$

$$I_2 = 1.5 + 0.1(15 - 3 \cdot 1.5) = 2.55$$

$$I_3 = 2.55 + 0.1(15 - 3 \cdot 2.55) = 3.285$$

$$I_4 = 3.285 + 0.1(15 - 3 \cdot 3.285) = 3.7995$$

$$I_5 = 3.7995 + 0.1(15 - 3 \cdot 3.7995) = 4.15965$$

So the current after 0.5 s is

$$I(0.5) \approx 4.16 \text{ A}$$

9.2 EXERCISES

1. A direction field for the differential equation $y' = x \cos \pi y$ is shown.
   (a) Sketch the graphs of the solutions that satisfy the given initial conditions.
      (i) $y(0) = 0$ (ii) $y(0) = 0.5$
      (iii) $y(0) = 1$ (iv) $y(0) = 1.6$
   (b) Find all the equilibrium solutions.

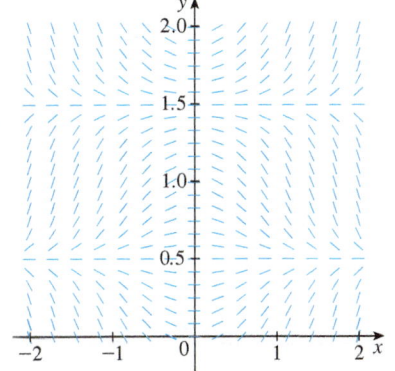

2. A direction field for the differential equation $y' = \tan(\frac{1}{2}\pi y)$ is shown.
   (a) Sketch the graphs of the solutions that satisfy the given initial conditions.
      (i) $y(0) = 1$ (ii) $y(0) = 0.2$
      (iii) $y(0) = 2$ (iv) $y(1) = 3$
   (b) Find all the equilibrium solutions.

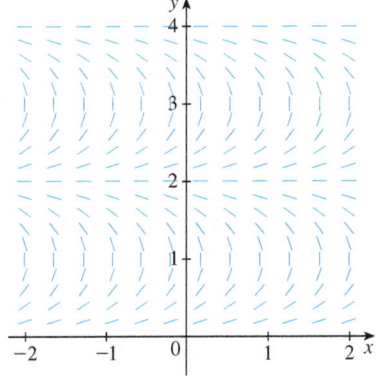

3–6 Match the differential equation with its direction field (labeled I–IV). Give reasons for your answer.

3. $y' = 2 - y$ **4.** $y' = x(2 - y)$

5. $y' = x + y - 1$ **6.** $y' = \sin x \sin y$

I
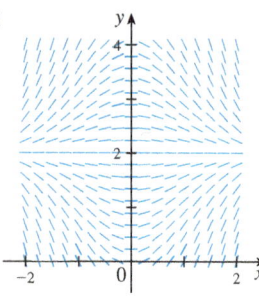

II
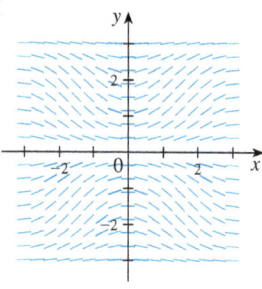

III
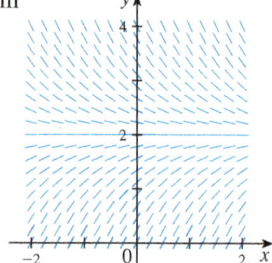

IV
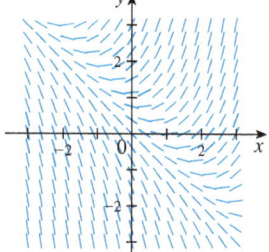

7. Use the direction field labeled I (above) to sketch the graphs of the solutions that satisfy the given initial conditions.
(a) $y(0) = 1$ (b) $y(0) = 2.5$ (c) $y(0) = 3.5$

8. Use the direction field labeled III (above) to sketch the graphs of the solutions that satisfy the given initial conditions.
(a) $y(0) = 1$ (b) $y(0) = 2.5$ (c) $y(0) = 3.5$

9–10 Sketch a direction field for the differential equation. Then use it to sketch three solution curves.

9. $y' = \tfrac{1}{2}y$ **10.** $y' = x - y + 1$

11–14 Sketch the direction field of the differential equation. Then use it to sketch a solution curve that passes through the given point.

11. $y' = y - 2x$, $(1, 0)$ **12.** $y' = xy - x^2$, $(0, 1)$

13. $y' = y + xy$, $(0, 1)$ **14.** $y' = x + y^2$, $(0, 0)$

CAS 15–16 Use a computer algebra system to draw a direction field for the given differential equation. Get a printout and sketch on it the solution curve that passes through $(0, 1)$. Then use the CAS to draw the solution curve and compare it with your sketch.

15. $y' = x^2 y - \tfrac{1}{2}y^2$ **16.** $y' = \cos(x + y)$

CAS 17. Use a computer algebra system to draw a direction field for the differential equation $y' = y^3 - 4y$. Get a printout and sketch on it solutions that satisfy the initial condition $y(0) = c$ for various values of c. For what values of c does $\lim_{t \to \infty} y(t)$ exist? What are the possible values for this limit?

18. Make a rough sketch of a direction field for the autonomous differential equation $y' = f(y)$, where the graph of f is as shown. How does the limiting behavior of solutions depend on the value of $y(0)$?

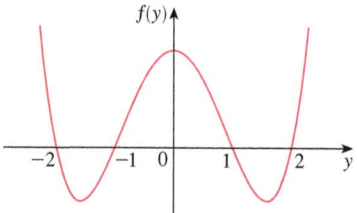

19. (a) Use Euler's method with each of the following step sizes to estimate the value of $y(0.4)$, where y is the solution of the initial-value problem $y' = y$, $y(0) = 1$.
   (i) $h = 0.4$ (ii) $h = 0.2$ (iii) $h = 0.1$
(b) We know that the exact solution of the initial-value problem in part (a) is $y = e^x$. Draw, as accurately as you can, the graph of $y = e^x$, $0 \leq x \leq 0.4$, together with the Euler approximations using the step sizes in part (a). (Your sketches should resemble Figures 11, 12, and 13.) Use your sketches to decide whether your estimates in part (a) are underestimates or overestimates.
(c) The error in Euler's method is the difference between the exact value and the approximate value. Find the errors made in part (a) in using Euler's method to estimate the true value of $y(0.4)$, namely, $e^{0.4}$. What happens to the error each time the step size is halved?

20. A direction field for a differential equation is shown. Draw, with a ruler, the graphs of the Euler approximations to the solution curve that passes through the origin. Use step sizes $h = 1$ and $h = 0.5$. Will the Euler estimates be underestimates or overestimates? Explain.

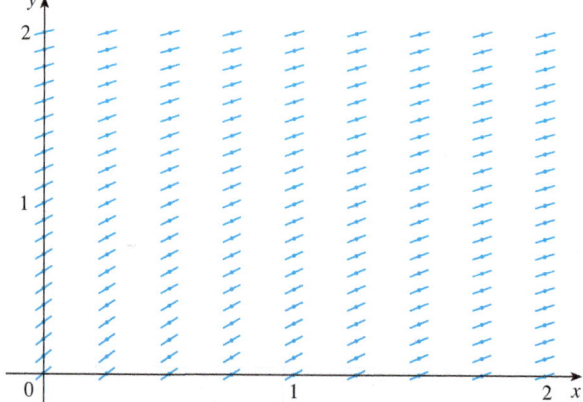

21. Use Euler's method with step size 0.5 to compute the approximate y-values y_1, y_2, y_3, and y_4 of the solution of the initial-value problem $y' = y - 2x$, $y(1) = 0$.

22. Use Euler's method with step size 0.2 to estimate $y(1)$, where $y(x)$ is the solution of the initial-value problem $y' = x^2 y - \frac{1}{2}y^2$, $y(0) = 1$.

23. Use Euler's method with step size 0.1 to estimate $y(0.5)$, where $y(x)$ is the solution of the initial-value problem $y' = y + xy$, $y(0) = 1$.

24. (a) Use Euler's method with step size 0.2 to estimate $y(0.6)$, where $y(x)$ is the solution of the initial-value problem $y' = \cos(x + y)$, $y(0) = 0$.
(b) Repeat part (a) with step size 0.1.

25. (a) Program a calculator or computer to use Euler's method to compute $y(1)$, where $y(x)$ is the solution of the initial-value problem

$$\frac{dy}{dx} + 3x^2 y = 6x^2 \qquad y(0) = 3$$

(i) $h = 1$ (ii) $h = 0.1$
(iii) $h = 0.01$ (iv) $h = 0.001$

(b) Verify that $y = 2 + e^{-x^3}$ is the exact solution of the differential equation.
(c) Find the errors in using Euler's method to compute $y(1)$ with the step sizes in part (a). What happens to the error when the step size is divided by 10?

26. (a) Program your computer algebra system, using Euler's method with step size 0.01, to calculate $y(2)$, where y is the solution of the initial-value problem

$$y' = x^3 - y^3 \qquad y(0) = 1$$

(b) Check your work by using the CAS to draw the solution curve.

27. The figure shows a circuit containing an electromotive force, a capacitor with a capacitance of C farads (F), and a resistor with a resistance of R ohms (Ω). The voltage drop across the capacitor is Q/C, where Q is the charge (in coulombs, C), so in this case Kirchhoff's Law gives

$$RI + \frac{Q}{C} = E(t)$$

But $I = dQ/dt$, so we have

$$R\frac{dQ}{dt} + \frac{1}{C}Q = E(t)$$

Suppose the resistance is 5 Ω, the capacitance is 0.05 F, and a battery gives a constant voltage of 60 V.
(a) Draw a direction field for this differential equation.
(b) What is the limiting value of the charge?
(c) Is there an equilibrium solution?
(d) If the initial charge is $Q(0) = 0$ C, use the direction field to sketch the solution curve.
(e) If the initial charge is $Q(0) = 0$ C, use Euler's method with step size 0.1 to estimate the charge after half a second.

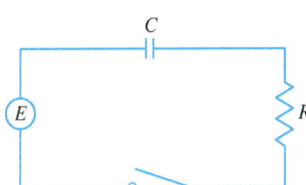

28. In Exercise 9.1.14 we considered a 95°C cup of coffee in a 20°C room. Suppose it is known that the coffee cools at a rate of 1°C per minute when its temperature is 70°C.
(a) What does the differential equation become in this case?
(b) Sketch a direction field and use it to sketch the solution curve for the initial-value problem. What is the limiting value of the temperature?
(c) Use Euler's method with step size $h = 2$ minutes to estimate the temperature of the coffee after 10 minutes.

9.3 Separable Equations

We have looked at first-order differential equations from a geometric point of view (direction fields) and from a numerical point of view (Euler's method). What about the symbolic point of view? It would be nice to have an explicit formula for a solution of a differential equation. Unfortunately, that is not always possible. But in this section we examine a certain type of differential equation that *can* be solved explicitly.

A **separable equation** is a first-order differential equation in which the expression for dy/dx can be factored as a function of x times a function of y. In other words, it can be written in the form

$$\frac{dy}{dx} = g(x)f(y)$$

The name *separable* comes from the fact that the expression on the right side can be "separated" into a function of x and a function of y. Equivalently, if $f(y) \neq 0$, we could write

$$\boxed{1} \qquad \frac{dy}{dx} = \frac{g(x)}{h(y)}$$

where $h(y) = 1/f(y)$. To solve this equation we rewrite it in the differential form

$$h(y)\,dy = g(x)\,dx$$

The technique for solving separable differential equations was first used by James Bernoulli (in 1690) in solving a problem about pendulums and by Leibniz (in a letter to Huygens in 1691). John Bernoulli explained the general method in a paper published in 1694.

so that all y's are on one side of the equation and all x's are on the other side. Then we integrate both sides of the equation:

$$\boxed{2} \qquad \int h(y)\,dy = \int g(x)\,dx$$

Equation 2 defines y implicitly as a function of x. In some cases we may be able to solve for y in terms of x.

We use the Chain Rule to justify this procedure: If h and g satisfy (2), then

$$\frac{d}{dx}\left(\int h(y)\,dy\right) = \frac{d}{dx}\left(\int g(x)\,dx\right)$$

so

$$\frac{d}{dy}\left(\int h(y)\,dy\right)\frac{dy}{dx} = g(x)$$

and

$$h(y)\frac{dy}{dx} = g(x)$$

Thus Equation 1 is satisfied.

EXAMPLE 1

(a) Solve the differential equation $\dfrac{dy}{dx} = \dfrac{x^2}{y^2}$.

(b) Find the solution of this equation that satisfies the initial condition $y(0) = 2$.

SOLUTION

(a) We write the equation in terms of differentials and integrate both sides:

$$y^2\,dy = x^2\,dx$$

$$\int y^2\,dy = \int x^2\,dx$$

$$\tfrac{1}{3}y^3 = \tfrac{1}{3}x^3 + C$$

where C is an arbitrary constant. (We could have used a constant C_1 on the left side and another constant C_2 on the right side. But then we could combine these constants by writing $C = C_2 - C_1$.)

Solving for y, we get

$$y = \sqrt[3]{x^3 + 3C}$$

We could leave the solution like this or we could write it in the form

$$y = \sqrt[3]{x^3 + K}$$

where $K = 3C$. (Since C is an arbitrary constant, so is K.)

Figure 1 shows graphs of several members of the family of solutions of the differential equation in Example 1. The solution of the initial-value problem in part (b) is shown in red.

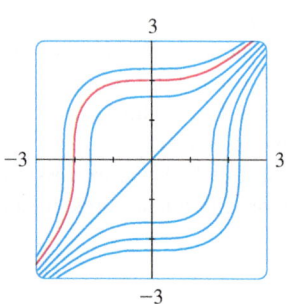

FIGURE 1

(b) If we put $x = 0$ in the general solution in part (a), we get $y(0) = \sqrt[3]{K}$. To satisfy the initial condition $y(0) = 2$, we must have $\sqrt[3]{K} = 2$ and so $K = 8$. Thus the solution of the initial-value problem is

$$y = \sqrt[3]{x^3 + 8}$$

Some computer software can plot curves defined by implicit equations. Figure 2 shows the graphs of several members of the family of solutions of the differential equation in Example 2. As we look at the curves from left to right, the values of C are 3, 2, 1, 0, -1, -2, and -3.

EXAMPLE 2 Solve the differential equation $\dfrac{dy}{dx} = \dfrac{6x^2}{2y + \cos y}$.

SOLUTION Writing the equation in differential form and integrating both sides, we have

$$(2y + \cos y)\,dy = 6x^2\,dx$$

$$\int (2y + \cos y)\,dy = \int 6x^2\,dx$$

$$\boxed{3} \qquad y^2 + \sin y = 2x^3 + C$$

where C is a constant. Equation 3 gives the general solution implicitly. In this case it's impossible to solve the equation to express y explicitly as a function of x.

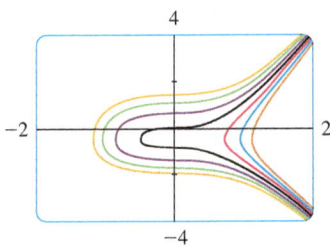

FIGURE 2

EXAMPLE 3 Solve the equation $y' = x^2 y$.

SOLUTION First we rewrite the equation using Leibniz notation:

$$\frac{dy}{dx} = x^2 y$$

If a solution y is a function that satisfies $y(x) \neq 0$ for some x, it follows from a uniqueness theorem for solutions of differential equations that $y(x) \neq 0$ for all x.

If $y \neq 0$, we can rewrite it in differential notation and integrate:

$$\frac{dy}{y} = x^2\,dx \qquad y \neq 0$$

$$\int \frac{dy}{y} = \int x^2\,dx$$

$$\ln |y| = \frac{x^3}{3} + C$$

This equation defines y implicitly as a function of x. But in this case we can solve explicitly for y as follows:

$$|y| = e^{\ln|y|} = e^{(x^3/3)+C} = e^C e^{x^3/3}$$

so

$$y = \pm e^C e^{x^3/3}$$

We can easily verify that the function $y = 0$ is also a solution of the given differential equation. So we can write the general solution in the form

$$y = A e^{x^3/3}$$

where A is an arbitrary constant ($A = e^C$, or $A = -e^C$, or $A = 0$).

Figure 3 shows a direction field for the differential equation in Example 3. Compare it with Figure 4, in which we use the equation $y = Ae^{x^3/3}$ to graph solutions for several values of A. If you use the direction field to sketch solution curves with y-intercepts 5, 2, 1, -1, and -2, they will resemble the curves in Figure 4.

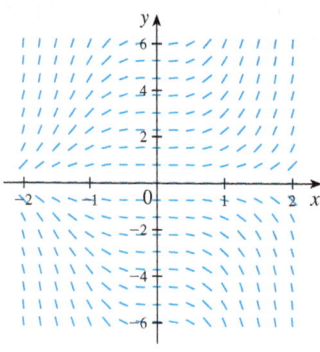

FIGURE 3

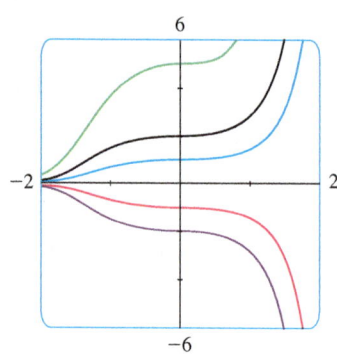

FIGURE 4

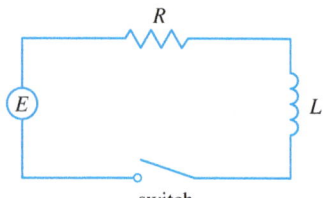

FIGURE 5

EXAMPLE 4 In Section 9.2 we modeled the current $I(t)$ in the electric circuit shown in Figure 5 by the differential equation

$$L\frac{dI}{dt} + RI = E(t)$$

Find an expression for the current in a circuit where the resistance is 12 Ω, the inductance is 4 H, a battery gives a constant voltage of 60 V, and the switch is turned on when $t = 0$. What is the limiting value of the current?

SOLUTION With $L = 4$, $R = 12$, and $E(t) = 60$, the equation becomes

$$4\frac{dI}{dt} + 12I = 60 \quad \text{or} \quad \frac{dI}{dt} = 15 - 3I$$

and the initial-value problem is

$$\frac{dI}{dt} = 15 - 3I \qquad I(0) = 0$$

We recognize this equation as being separable, and we solve it as follows:

$$\int \frac{dI}{15 - 3I} = \int dt \qquad (15 - 3I \neq 0)$$

$$-\tfrac{1}{3} \ln |15 - 3I| = t + C$$

$$|15 - 3I| = e^{-3(t+C)}$$

$$15 - 3I = \pm e^{-3C} e^{-3t} = A e^{-3t}$$

$$I = 5 - \tfrac{1}{3} A e^{-3t}$$

Since $I(0) = 0$, we have $5 - \tfrac{1}{3}A = 0$, so $A = 15$ and the solution is

$$I(t) = 5 - 5e^{-3t}$$

Figure 6 shows how the solution in Example 4 (the current) approaches its limiting value. Comparison with Figure 9.2.10 shows that we were able to draw a fairly accurate solution curve from the direction field.

The limiting current, in amperes, is

$$\lim_{t \to \infty} I(t) = \lim_{t \to \infty} (5 - 5e^{-3t}) = 5 - 5 \lim_{t \to \infty} e^{-3t} = 5 - 0 = 5$$

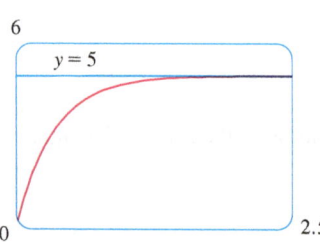

FIGURE 6

Orthogonal Trajectories

An **orthogonal trajectory** of a family of curves is a curve that intersects each curve of the family orthogonally, that is, at right angles (see Figure 7). For instance, each member of the family $y = mx$ of straight lines through the origin is an orthogonal trajectory of the family $x^2 + y^2 = r^2$ of concentric circles with center the origin (see Figure 8). We say that the two families are orthogonal trajectories of each other.

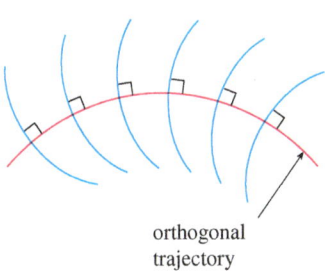
orthogonal trajectory

FIGURE 7

EXAMPLE 5 Find the orthogonal trajectories of the family of curves $x = ky^2$, where k is an arbitrary constant.

SOLUTION The curves $x = ky^2$ form a family of parabolas whose axis of symmetry is the x-axis. The first step is to find a single differential equation that is satisfied by all members of the family. If we differentiate $x = ky^2$, we get

$$1 = 2ky \frac{dy}{dx} \quad \text{or} \quad \frac{dy}{dx} = \frac{1}{2ky}$$

This differential equation depends on k, but we need an equation that is valid for all values of k simultaneously. To eliminate k we note that, from the equation of the given general parabola $x = ky^2$, we have $k = x/y^2$ and so the differential equation can be written as

$$\frac{dy}{dx} = \frac{1}{2ky} = \frac{1}{2\frac{x}{y^2}y} \quad \text{or} \quad \frac{dy}{dx} = \frac{y}{2x}$$

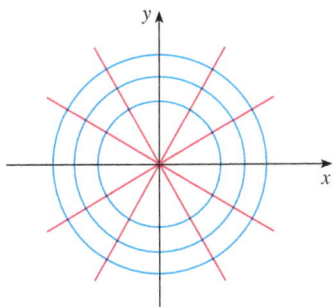

FIGURE 8

This means that the slope of the tangent line at any point (x, y) on one of the parabolas is $y' = y/(2x)$. On an orthogonal trajectory the slope of the tangent line must be the negative reciprocal of this slope. Therefore the orthogonal trajectories must satisfy the differential equation

$$\frac{dy}{dx} = -\frac{2x}{y}$$

This differential equation is separable, and we solve it as follows:

$$\int y \, dy = -\int 2x \, dx$$

$$\frac{y^2}{2} = -x^2 + C$$

$$\boxed{4} \quad x^2 + \frac{y^2}{2} = C$$

where C is an arbitrary positive constant. Thus the orthogonal trajectories are the family of ellipses given by Equation 4 and sketched in Figure 9. ∎

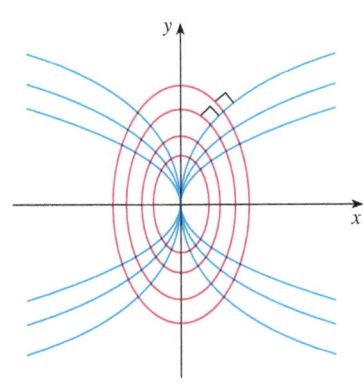

FIGURE 9

Orthogonal trajectories occur in various branches of physics. For example, in an electrostatic field the lines of force are orthogonal to the lines of constant potential. Also, the streamlines in aerodynamics are orthogonal trajectories of the velocity-equipotential curves.

Mixing Problems

A typical mixing problem involves a tank of fixed capacity filled with a thoroughly mixed solution of some substance, such as salt. A solution of a given concentration enters the tank at a fixed rate and the mixture, thoroughly stirred, leaves at a fixed rate, which may differ from the entering rate. If $y(t)$ denotes the amount of substance in the tank at time t, then $y'(t)$ is the rate at which the substance is being added minus the rate at which it is being removed. The mathematical description of this situation often leads to a first-order separable differential equation. We can use the same type of reasoning to model a variety of phenomena: chemical reactions, discharge of pollutants into a lake, injection of a drug into the bloodstream.

EXAMPLE 6 A tank contains 20 kg of salt dissolved in 5000 L of water. Brine that contains 0.03 kg of salt per liter of water enters the tank at a rate of 25 L/min. The solution is kept thoroughly mixed and drains from the tank at the same rate. How much salt remains in the tank after half an hour?

SOLUTION Let $y(t)$ be the amount of salt (in kilograms) after t minutes. We are given that $y(0) = 20$ and we want to find $y(30)$. We do this by finding a differential equation satisfied by $y(t)$. Note that dy/dt is the rate of change of the amount of salt, so

$$\boxed{5} \qquad \frac{dy}{dt} = (\text{rate in}) - (\text{rate out})$$

where (rate in) is the rate at which salt enters the tank and (rate out) is the rate at which salt leaves the tank. We have

$$\text{rate in} = \left(0.03 \, \frac{\text{kg}}{\text{L}}\right)\left(25 \, \frac{\text{L}}{\text{min}}\right) = 0.75 \, \frac{\text{kg}}{\text{min}}$$

The tank always contains 5000 L of liquid, so the concentration at time t is $y(t)/5000$ (measured in kilograms per liter). Since the brine flows out at a rate of 25 L/min, we have

$$\text{rate out} = \left(\frac{y(t)}{5000} \, \frac{\text{kg}}{\text{L}}\right)\left(25 \, \frac{\text{L}}{\text{min}}\right) = \frac{y(t)}{200} \, \frac{\text{kg}}{\text{min}}$$

Thus, from Equation 5, we get

$$\frac{dy}{dt} = 0.75 - \frac{y(t)}{200} = \frac{150 - y(t)}{200}$$

Solving this separable differential equation, we obtain

$$\int \frac{dy}{150 - y} = \int \frac{dt}{200}$$

$$-\ln|150 - y| = \frac{t}{200} + C$$

Since $y(0) = 20$, we have $-\ln 130 = C$, so

$$-\ln|150 - y| = \frac{t}{200} - \ln 130$$

Figure 10 shows the graph of the function $y(t)$ of Example 6. Notice that, as time goes by, the amount of salt approaches 150 kg.

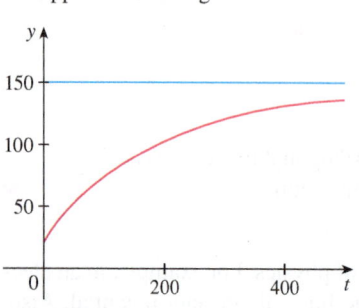

FIGURE 10

SECTION 9.3 Separable Equations

Therefore
$$|150 - y| = 130e^{-t/200}$$

Since $y(t)$ is continuous and $y(0) = 20$ and the right side is never 0, we deduce that $150 - y(t)$ is always positive. Thus $|150 - y| = 150 - y$ and so

$$y(t) = 150 - 130e^{-t/200}$$

The amount of salt after 30 min is

$$y(30) = 150 - 130e^{-30/200} \approx 38.1 \text{ kg}$$

9.3 EXERCISES

1–10 Solve the differential equation.

1. $\dfrac{dy}{dx} = 3x^2 y^2$

2. $\dfrac{dy}{dx} = x\sqrt{y}$

3. $xyy' = x^2 + 1$

4. $y' + xe^y = 0$

5. $(e^y - 1)y' = 2 + \cos x$

6. $\dfrac{du}{dt} = \dfrac{1 + t^4}{ut^2 + u^4 t^2}$

7. $\dfrac{d\theta}{dt} = \dfrac{t \sec \theta}{\theta e^{t^2}}$

8. $\dfrac{dH}{dR} = \dfrac{RH^2 \sqrt{1 + R^2}}{\ln H}$

9. $\dfrac{dp}{dt} = t^2 p - p + t^2 - 1$

10. $\dfrac{dz}{dt} + e^{t+z} = 0$

11–18 Find the solution of the differential equation that satisfies the given initial condition.

11. $\dfrac{dy}{dx} = xe^y, \quad y(0) = 0$

12. $\dfrac{dy}{dx} = \dfrac{x \sin x}{y}, \quad y(0) = -1$

13. $\dfrac{du}{dt} = \dfrac{2t + \sec^2 t}{2u}, \quad u(0) = -5$

14. $x + 3y^2 \sqrt{x^2 + 1}\, \dfrac{dy}{dx} = 0, \quad y(0) = 1$

15. $x \ln x = y\left(1 + \sqrt{3 + y^2}\right) y', \quad y(1) = 1$

16. $\dfrac{dP}{dt} = \sqrt{Pt}, \quad P(1) = 2$

17. $y' \tan x = a + y, \quad y(\pi/3) = a, \quad 0 < x < \pi/2$

18. $\dfrac{dL}{dt} = kL^2 \ln t, \quad L(1) = -1$

19. Find an equation of the curve that passes through the point $(0, 2)$ and whose slope at (x, y) is x/y.

20. Find the function f such that $f'(x) = xf(x) - x$ and $f(0) = 2$.

21. Solve the differential equation $y' = x + y$ by making the change of variable $u = x + y$.

22. Solve the differential equation $xy' = y + xe^{y/x}$ by making the change of variable $v = y/x$.

23. (a) Solve the differential equation $y' = 2x\sqrt{1 - y^2}$.
    (b) Solve the initial-value problem $y' = 2x\sqrt{1 - y^2}$, $y(0) = 0$, and graph the solution.
    (c) Does the initial-value problem $y' = 2x\sqrt{1 - y^2}$, $y(0) = 2$, have a solution? Explain.

24. Solve the equation $e^{-y}y' + \cos x = 0$ and graph several members of the family of solutions. How does the solution curve change as the constant C varies?

25. Solve the initial-value problem $y' = (\sin x)/\sin y$, $y(0) = \pi/2$, and graph the solution (if your CAS does implicit plots).

26. Solve the equation $y' = x\sqrt{x^2 + 1}/(ye^y)$ and graph several members of the family of solutions (if your CAS does implicit plots). How does the solution curve change as the constant C varies?

27–28
(a) Use a computer algebra system to draw a direction field for the differential equation. Get a printout and use it to sketch some solution curves without solving the differential equation.
(b) Solve the differential equation.
(c) Use the CAS to draw several members of the family of solutions obtained in part (b). Compare with the curves from part (a).

27. $y' = y^2$

28. $y' = xy$

29–32 Find the orthogonal trajectories of the family of curves. Use a graphing device to draw several members of each family on a common screen.

29. $x^2 + 2y^2 = k^2$

30. $y^2 = kx^3$

31. $y = \dfrac{k}{x}$

32. $y = \dfrac{1}{x+k}$

33–35 An **integral equation** is an equation that contains an unknown function $y(x)$ and an integral that involves $y(x)$. Solve the given integral equation. [*Hint:* Use an initial condition obtained from the integral equation.]

33. $y(x) = 2 + \displaystyle\int_2^x [t - ty(t)]\, dt$

34. $y(x) = 2 + \displaystyle\int_1^x \dfrac{dt}{ty(t)}, \quad x > 0$

35. $y(x) = 4 + \displaystyle\int_0^x 2t\sqrt{y(t)}\, dt$

---

36. Find a function f such that $f(3) = 2$ and
$$(t^2 + 1)f'(t) + [f(t)]^2 + 1 = 0 \qquad t \neq 1$$

[*Hint:* Use the addition formula for $\tan(x + y)$ on Reference Page 2.]

37. Solve the initial-value problem in Exercise 9.2.27 to find an expression for the charge at time t. Find the limiting value of the charge.

38. In Exercise 9.2.28 we discussed a differential equation that models the temperature of a 95°C cup of coffee in a 20°C room. Solve the differential equation to find an expression for the temperature of the coffee at time t.

39. In Exercise 9.1.15 we formulated a model for learning in the form of the differential equation
$$\dfrac{dP}{dt} = k(M - P)$$

where $P(t)$ measures the performance of someone learning a skill after a training time t, M is the maximum level of performance, and k is a positive constant. Solve this differential equation to find an expression for $P(t)$. What is the limit of this expression?

40. In an elementary chemical reaction, single molecules of two reactants A and B form a molecule of the product C: A + B → C. The law of mass action states that the rate of reaction is proportional to the product of the concentrations of A and B:
$$\dfrac{d[C]}{dt} = k[A][B]$$

(See Example 3.7.4.) Thus, if the initial concentrations are $[A] = a$ moles/L and $[B] = b$ moles/L and we write $x = [C]$, then we have
$$\dfrac{dx}{dt} = k(a - x)(b - x)$$

(a) Assuming that $a \neq b$, find x as a function of t. Use the fact that the initial concentration of C is 0.

(b) Find $x(t)$ assuming that $a = b$. How does this expression for $x(t)$ simplify if it is known that $[C] = \tfrac{1}{2}a$ after 20 seconds?

41. In contrast to the situation of Exercise 40, experiments show that the reaction $H_2 + Br_2 \to 2HBr$ satisfies the rate law
$$\dfrac{d[HBr]}{dt} = k[H_2][Br_2]^{1/2}$$

and so for this reaction the differential equation becomes
$$\dfrac{dx}{dt} = k(a - x)(b - x)^{1/2}$$

where $x = [HBr]$ and a and b are the initial concentrations of hydrogen and bromine.

(a) Find x as a function of t in the case where $a = b$. Use the fact that $x(0) = 0$.

(b) If $a > b$, find t as a function of x. [*Hint:* In performing the integration, make the substitution $u = \sqrt{b - x}$.]

42. A sphere with radius 1 m has temperature 15°C. It lies inside a concentric sphere with radius 2 m and temperature 25°C. The temperature $T(r)$ at a distance r from the common center of the spheres satisfies the differential equation
$$\dfrac{d^2T}{dr^2} + \dfrac{2}{r}\dfrac{dT}{dr} = 0$$

If we let $S = dT/dr$, then S satisfies a first-order differential equation. Solve it to find an expression for the temperature $T(r)$ between the spheres.

43. A glucose solution is administered intravenously into the bloodstream at a constant rate r. As the glucose is added, it is converted into other substances and removed from the bloodstream at a rate that is proportional to the concentration at that time. Thus a model for the concentration $C = C(t)$ of the glucose solution in the bloodstream is
$$\dfrac{dC}{dt} = r - kC$$

where k is a positive constant.

(a) Suppose that the concentration at time $t = 0$ is C_0. Determine the concentration at any time t by solving the differential equation.

(b) Assuming that $C_0 < r/k$, find $\lim_{t \to \infty} C(t)$ and interpret your answer.

44. A certain small country has $10 billion in paper currency in circulation, and each day $50 million comes into the country's banks. The government decides to introduce new currency by having the banks replace old bills with new ones whenever old currency comes into the banks. Let $x = x(t)$ denote the amount of new currency in circulation at time t, with $x(0) = 0$.

(a) Formulate a mathematical model in the form of an initial-value problem that represents the "flow" of the new currency into circulation.

(b) Solve the initial-value problem found in part (a).

(c) How long will it take for the new bills to account for 90% of the currency in circulation?

45. A tank contains 1000 L of brine with 15 kg of dissolved salt. Pure water enters the tank at a rate of 10 L/min. The solution is kept thoroughly mixed and drains from the tank at the same rate. How much salt is in the tank (a) after t minutes and (b) after 20 minutes?

46. The air in a room with volume 180 m³ contains 0.15% carbon dioxide initially. Fresher air with only 0.05% carbon dioxide flows into the room at a rate of 2 m³/min and the mixed air flows out at the same rate. Find the percentage of carbon dioxide in the room as a function of time. What happens in the long run?

47. A vat with 500 gallons of beer contains 4% alcohol (by volume). Beer with 6% alcohol is pumped into the vat at a rate of 5 gal/min and the mixture is pumped out at the same rate. What is the percentage of alcohol after an hour?

48. A tank contains 1000 L of pure water. Brine that contains 0.05 kg of salt per liter of water enters the tank at a rate of 5 L/min. Brine that contains 0.04 kg of salt per liter of water enters the tank at a rate of 10 L/min. The solution is kept thoroughly mixed and drains from the tank at a rate of 15 L/min. How much salt is in the tank (a) after t minutes and (b) after one hour?

49. When a raindrop falls, it increases in size and so its mass at time t is a function of t, namely, $m(t)$. The rate of growth of the mass is $km(t)$ for some positive constant k. When we apply Newton's Law of Motion to the raindrop, we get $(mv)' = gm$, where v is the velocity of the raindrop (directed downward) and g is the acceleration due to gravity. The *terminal velocity* of the raindrop is $\lim_{t \to \infty} v(t)$. Find an expression for the terminal velocity in terms of g and k.

50. An object of mass m is moving horizontally through a medium which resists the motion with a force that is a function of the velocity; that is,
$$m \frac{d^2s}{dt^2} = m \frac{dv}{dt} = f(v)$$
where $v = v(t)$ and $s = s(t)$ represent the velocity and position of the object at time t, respectively. For example, think of a boat moving through the water.
(a) Suppose that the resisting force is proportional to the velocity, that is, $f(v) = -kv$, k a positive constant. (This model is appropriate for small values of v.) Let $v(0) = v_0$ and $s(0) = s_0$ be the initial values of v and s. Determine v and s at any time t. What is the total distance that the object travels from time $t = 0$?
(b) For larger values of v a better model is obtained by supposing that the resisting force is proportional to the square of the velocity, that is, $f(v) = -kv^2$, $k > 0$. (This model was first proposed by Newton.) Let v_0 and s_0 be the initial values of v and s. Determine v and s at any time t. What is the total distance that the object travels in this case?

51. *Allometric growth* in biology refers to relationships between sizes of parts of an organism (skull length and body length, for instance). If $L_1(t)$ and $L_2(t)$ are the sizes of two organs in an organism of age t, then L_1 and L_2 satisfy an allometric law if their specific growth rates are proportional:
$$\frac{1}{L_1} \frac{dL_1}{dt} = k \frac{1}{L_2} \frac{dL_2}{dt}$$
where k is a constant.
(a) Use the allometric law to write a differential equation relating L_1 and L_2 and solve it to express L_1 as a function of L_2.
(b) In a study of several species of unicellular algae, the proportionality constant in the allometric law relating B (cell biomass) and V (cell volume) was found to be $k = 0.0794$. Write B as a function of V.

52. A model for tumor growth is given by the Gompertz equation
$$\frac{dV}{dt} = a(\ln b - \ln V)V$$
where a and b are positive constants and V is the volume of the tumor measured in mm³.
(a) Find a family of solutions for tumor volume as a function of time.
(b) Find the solution that has an initial tumor volume of $V(0) = 1$ mm³.

53. Let $A(t)$ be the area of a tissue culture at time t and let M be the final area of the tissue when growth is complete. Most cell divisions occur on the periphery of the tissue and the number of cells on the periphery is proportional to $\sqrt{A(t)}$. So a reasonable model for the growth of tissue is obtained by assuming that the rate of growth of the area is jointly proportional to $\sqrt{A(t)}$ and $M - A(t)$.
(a) Formulate a differential equation and use it to show that the tissue grows fastest when $A(t) = \frac{1}{3}M$.
(b) Solve the differential equation to find an expression for $A(t)$. Use a computer algebra system to perform the integration.

54. According to Newton's Law of Universal Gravitation, the gravitational force on an object of mass m that has been projected vertically upward from the earth's surface is
$$F = \frac{mgR^2}{(x+R)^2}$$
where $x = x(t)$ is the object's distance above the surface at time t, R is the earth's radius, and g is the acceleration due to gravity. Also, by Newton's Second Law, $F = ma = m(dv/dt)$ and so
$$m \frac{dv}{dt} = -\frac{mgR^2}{(x+R)^2}$$
(a) Suppose a rocket is fired vertically upward with an initial velocity v_0. Let h be the maximum height above

the surface reached by the object. Show that

$$v_0 = \sqrt{\frac{2gRh}{R+h}}$$

[*Hint:* By the Chain Rule, $m(dv/dt) = mv(dv/dx)$.]

(b) Calculate $v_e = \lim_{h \to \infty} v_0$. This limit is called the *escape velocity* for the earth.
(c) Use $R = 3960$ mi and $g = 32$ ft/s² to calculate v_e in feet per second and in miles per second.

APPLIED PROJECT   HOW FAST DOES A TANK DRAIN?

If water (or other liquid) drains from a tank, we expect that the flow will be greatest at first (when the water depth is greatest) and will gradually decrease as the water level decreases. But we need a more precise mathematical description of how the flow decreases in order to answer the kinds of questions that engineers ask: How long does it take for a tank to drain completely? How much water should a tank hold in order to guarantee a certain minimum water pressure for a sprinkler system?

Let $h(t)$ and $V(t)$ be the height and volume of water in a tank at time t. If water drains through a hole with area a at the bottom of the tank, then Torricelli's Law says that

$$\boxed{1} \qquad \frac{dV}{dt} = -a\sqrt{2gh}$$

where g is the acceleration due to gravity. So the rate at which water flows from the tank is proportional to the square root of the water height.

1. (a) Suppose the tank is cylindrical with height 6 ft and radius 2 ft and the hole is circular with radius 1 inch. If we take $g = 32$ ft/s², show that h satisfies the differential equation

$$\frac{dh}{dt} = -\frac{1}{72}\sqrt{h}$$

Problem 2(b) is best done as a classroom demonstration or as a group project with three students in each group: a timekeeper to call out seconds, a bottle keeper to estimate the height every 10 seconds, and a record keeper to record these values.

   (b) Solve this equation to find the height of the water at time t, assuming the tank is full at time $t = 0$.
   (c) How long will it take for the water to drain completely?

2. Because of the rotation and viscosity of the liquid, the theoretical model given by Equation 1 isn't quite accurate. Instead, the model

$$\boxed{2} \qquad \frac{dh}{dt} = k\sqrt{h}$$

is often used and the constant k (which depends on the physical properties of the liquid) is determined from data concerning the draining of the tank.
   (a) Suppose that a hole is drilled in the side of a cylindrical bottle and the height h of the water (above the hole) decreases from 10 cm to 3 cm in 68 seconds. Use Equation 2 to find an expression for $h(t)$. Evaluate $h(t)$ for $t = 10, 20, 30, 40, 50, 60$.
   (b) Drill a 4-mm hole near the bottom of the cylindrical part of a two-liter plastic soft-drink bottle. Attach a strip of masking tape marked in centimeters from 0 to 10, with 0 corresponding to the top of the hole. With one finger over the hole, fill the bottle with water to the 10-cm mark. Then take your finger off the hole and record the values of $h(t)$ for $t = 10, 20, 30, 40, 50, 60$ seconds. (You will probably find that it takes 68 seconds for the level to decrease to $h = 3$ cm.) Compare your data with the values of $h(t)$ from part (a). How well did the model predict the actual values?

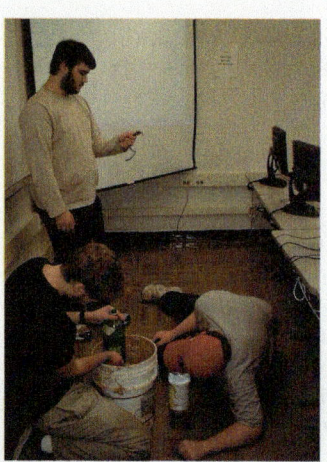

© Richard Le Borne, Dept. Mathematics, Tennessee Technological University

3. In many parts of the world, the water for sprinkler systems in large hotels and hospitals is supplied by gravity from cylindrical tanks on or near the roofs of the buildings. Suppose such a tank has radius 10 ft and the diameter of the outlet is 2.5 inches. An engineer has to guaran-

tee that the water pressure will be at least 2160 lb/ft² for a period of 10 minutes. (When a fire happens, the electrical system might fail and it could take up to 10 minutes for the emergency generator and fire pump to be activated.) What height should the engineer specify for the tank in order to make such a guarantee? (Use the fact that the water pressure at a depth of d feet is $P = 62.5d$. See Section 8.3.)

4. Not all water tanks are shaped like cylinders. Suppose a tank has cross-sectional area $A(h)$ at height h. Then the volume of water up to height h is $V = \int_0^h A(u)\, du$ and so the Fundamental Theorem of Calculus gives $dV/dh = A(h)$. It follows that

$$\frac{dV}{dt} = \frac{dV}{dh}\frac{dh}{dt} = A(h)\frac{dh}{dt}$$

and so Torricelli's Law becomes

$$A(h)\frac{dh}{dt} = -a\sqrt{2gh}$$

(a) Suppose the tank has the shape of a sphere with radius 2 m and is initially half full of water. If the radius of the circular hole is 1 cm and we take $g = 10$ m/s², show that h satisfies the differential equation

$$(4h - h^2)\frac{dh}{dt} = -0.0001\sqrt{20h}$$

(b) How long will it take for the water to drain completely?

APPLIED PROJECT WHICH IS FASTER, GOING UP OR COMING DOWN?

Suppose you throw a ball into the air. Do you think it takes longer to reach its maximum height or to fall back to earth from its maximum height? We will solve the problem in this project, but before getting started, think about that situation and make a guess based on your physical intuition.

In modeling force due to air resistance, various functions have been used, depending on the physical characteristics and speed of the ball. Here we use a linear model, $-pv$, but a quadratic model ($-pv^2$ on the way up and pv^2 on the way down) is another possibility for higher speeds (see Exercise 9.3.50). For a golf ball, experiments have shown that a good model is $-pv^{1.3}$ going up and $p|v|^{1.3}$ coming down. But no matter which force function $-f(v)$ is used [where $f(v) > 0$ for $v > 0$ and $f(v) < 0$ for $v < 0$], the answer to the question remains the same. See F. Brauer, "What Goes Up Must Come Down, Eventually," *American Mathematical Monthly* 108 (2001), pp. 437–440.

1. A ball with mass m is projected vertically upward from the earth's surface with a positive initial velocity v_0. We assume the forces acting on the ball are the force of gravity and a retarding force of air resistance with direction opposite to the direction of motion and with magnitude $p|v(t)|$, where p is a positive constant and $v(t)$ is the velocity of the ball at time t. In both the ascent and the descent, the total force acting on the ball is $-pv - mg$. [During ascent, $v(t)$ is positive and the resistance acts downward; during descent, $v(t)$ is negative and the resistance acts upward.] So, by Newton's Second Law, the equation of motion is

$$mv' = -pv - mg$$

Solve this differential equation to show that the velocity is

$$v(t) = \left(v_0 + \frac{mg}{p}\right)e^{-pt/m} - \frac{mg}{p}$$

2. Show that the height of the ball, until it hits the ground, is

$$y(t) = \left(v_0 + \frac{mg}{p}\right)\frac{m}{p}(1 - e^{-pt/m}) - \frac{mgt}{p}$$

3. Let t_1 be the time that the ball takes to reach its maximum height. Show that

$$t_1 = \frac{m}{p} \ln\left(\frac{mg + pv_0}{mg}\right)$$

Find this time for a ball with mass 1 kg and initial velocity 20 m/s. Assume the air resistance is $\frac{1}{10}$ of the speed.

4. Let t_2 be the time at which the ball falls back to earth. For the particular ball in Problem 3, estimate t_2 by using a graph of the height function $y(t)$. Which is faster, going up or coming down?

5. In general, it's not easy to find t_2 because it's impossible to solve the equation $y(t) = 0$ explicitly. We can, however, use an indirect method to determine whether ascent or descent is faster: we determine whether $y(2t_1)$ is positive or negative. Show that

$$y(2t_1) = \frac{m^2 g}{p^2}\left(x - \frac{1}{x} - 2\ln x\right)$$

where $x = e^{pt_1/m}$. Then show that $x > 1$ and the function

$$f(x) = x - \frac{1}{x} - 2\ln x$$

is increasing for $x > 1$. Use this result to decide whether $y(2t_1)$ is positive or negative. What can you conclude? Is ascent or descent faster?

9.4 Models for Population Growth

In this section we investigate differential equations that are used to model population growth: the law of natural growth, the logistic equation, and several others.

■ The Law of Natural Growth

One of the models for population growth that we considered in Section 9.1 was based on the assumption that the population grows at a rate proportional to the size of the population:

$$\frac{dP}{dt} = kP$$

Is that a reasonable assumption? Suppose we have a population (of bacteria, for instance) with size $P = 1000$ and at a certain time it is growing at a rate of $P' = 300$ bacteria per hour. Now let's take another 1000 bacteria of the same type and put them with the first population. Each half of the combined population was previously growing at a rate of 300 bacteria per hour. We would expect the total population of 2000 to increase at a rate of 600 bacteria per hour initially (provided there's enough room and nutrition). So if we double the size, we double the growth rate. It seems reasonable that the growth rate should be proportional to the size.

In general, if $P(t)$ is the value of a quantity y at time t and if the rate of change of P with respect to t is proportional to its size $P(t)$ at any time, then

$$\boxed{\frac{dP}{dt} = kP} \quad \boxed{1}$$

where k is a constant. Equation 1 is sometimes called the **law of natural growth**. If k is positive, then the population increases; if k is negative, it decreases.

Because Equation 1 is a separable differential equation, we can solve it by the methods of Section 9.3:

$$\int \frac{dP}{P} = \int k\, dt$$

$$\ln |P| = kt + C$$

$$|P| = e^{kt+C} = e^C e^{kt}$$

$$P = Ae^{kt}$$

where A ($= \pm e^C$ or 0) is an arbitrary constant. To see the significance of the constant A, we observe that

$$P(0) = Ae^{k \cdot 0} = A$$

Therefore A is the initial value of the function.

Examples and exercises on the use of (2) are given in Section 3.8.

2 The solution of the initial-value problem

$$\frac{dP}{dt} = kP \qquad P(0) = P_0$$

is

$$P(t) = P_0 e^{kt}$$

Another way of writing Equation 1 is

$$\frac{1}{P} \frac{dP}{dt} = k$$

which says that the **relative growth rate** (the growth rate divided by the population size) is constant. Then (2) says that a population with constant relative growth rate must grow exponentially.

We can account for emigration (or "harvesting") from a population by modifying Equation 1: if the rate of emigration is a constant m, then the rate of change of the population is modeled by the differential equation

3
$$\frac{dP}{dt} = kP - m$$

See Exercise 17 for the solution and consequences of Equation 3.

■ The Logistic Model

As we discussed in Section 9.1, a population often increases exponentially in its early stages but levels off eventually and approaches its carrying capacity because of limited resources. If $P(t)$ is the size of the population at time t, we assume that

$$\frac{dP}{dt} \approx kP \qquad \text{if } P \text{ is small}$$

This says that the growth rate is initially close to being proportional to size. In other words, the relative growth rate is almost constant when the population is small. But we also want to reflect the fact that the relative growth rate decreases as the population P increases and becomes negative if P ever exceeds its **carrying capacity** M, the maximum population that the environment is capable of sustaining in the long run. The simplest expression for the relative growth rate that incorporates these assumptions is

$$\frac{1}{P}\frac{dP}{dt} = k\left(1 - \frac{P}{M}\right)$$

Multiplying by P, we obtain the model for population growth known as the **logistic differential equation**:

4
$$\frac{dP}{dt} = kP\left(1 - \frac{P}{M}\right)$$

Notice from Equation 4 that if P is small compared with M, then P/M is close to 0 and so $dP/dt \approx kP$. However, if $P \to M$ (the population approaches its carrying capacity), then $P/M \to 1$, so $dP/dt \to 0$. We can deduce information about whether solutions increase or decrease directly from Equation 4. If the population P lies between 0 and M, then the right side of the equation is positive, so $dP/dt > 0$ and the population increases. But if the population exceeds the carrying capacity ($P > M$), then $1 - P/M$ is negative, so $dP/dt < 0$ and the population decreases.

Let's start our more detailed analysis of the logistic differential equation by looking at a direction field.

EXAMPLE 1 Draw a direction field for the logistic equation with $k = 0.08$ and carrying capacity $M = 1000$. What can you deduce about the solutions?

SOLUTION In this case the logistic differential equation is

$$\frac{dP}{dt} = 0.08P\left(1 - \frac{P}{1000}\right)$$

A direction field for this equation is shown in Figure 1. We show only the first quadrant because negative populations aren't meaningful and we are interested only in what happens after $t = 0$.

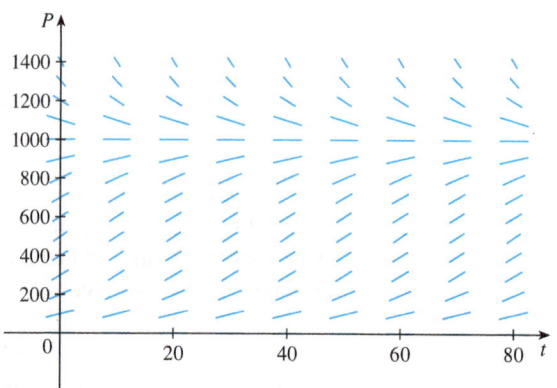

FIGURE 1
Direction field for the logistic equation in Example 1

The logistic equation is autonomous (dP/dt depends only on P, not on t), so the slopes are the same along any horizontal line. As expected, the slopes are positive for $0 < P < 1000$ and negative for $P > 1000$.

The slopes are small when P is close to 0 or 1000 (the carrying capacity). Notice that the solutions move away from the equilibrium solution $P = 0$ and move toward the equilibrium solution $P = 1000$.

In Figure 2 we use the direction field to sketch solution curves with initial populations $P(0) = 100$, $P(0) = 400$, and $P(0) = 1300$. Notice that solution curves that start below $P = 1000$ are increasing and those that start above $P = 1000$ are decreasing. The slopes are greatest when $P \approx 500$ and therefore the solution curves that start below $P = 1000$ have inflection points when $P \approx 500$. In fact we can prove that all solution curves that start below $P = 500$ have an inflection point when P is exactly 500. (See Exercise 13.)

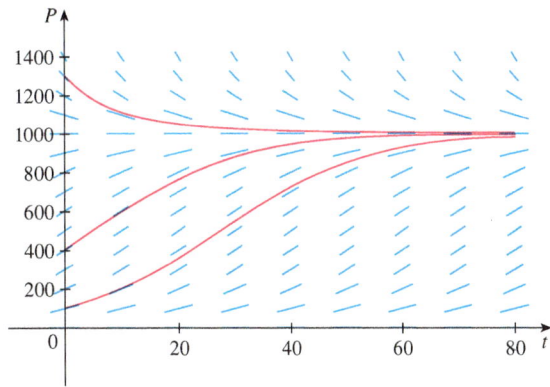

FIGURE 2
Solution curves for the logistic equation in Example 1

The logistic equation (4) is separable and so we can solve it explicitly using the method of Section 9.3. Since

$$\frac{dP}{dt} = kP\left(1 - \frac{P}{M}\right)$$

we have

$$\boxed{5} \qquad \int \frac{dP}{P(1 - P/M)} = \int k\, dt$$

To evaluate the integral on the left side, we write

$$\frac{1}{P(1 - P/M)} = \frac{M}{P(M - P)}$$

Using partial fractions (see Section 7.4), we get

$$\frac{M}{P(M - P)} = \frac{1}{P} + \frac{1}{M - P}$$

This enables us to rewrite Equation 5:

$$\int \left(\frac{1}{P} + \frac{1}{M - P}\right) dP = \int k\, dt$$

$$\ln|P| - \ln|M - P| = kt + C$$

CHAPTER 9 Differential Equations

$$\ln\left|\frac{M-P}{P}\right| = -kt - C$$

$$\left|\frac{M-P}{P}\right| = e^{-kt-C} = e^{-C}e^{-kt}$$

$$\boxed{6} \qquad \frac{M-P}{P} = Ae^{-kt}$$

where $A = \pm e^{-C}$. Solving Equation 6 for P, we get

$$\frac{M}{P} - 1 = Ae^{-kt} \quad \Rightarrow \quad \frac{P}{M} = \frac{1}{1 + Ae^{-kt}}$$

so

$$P = \frac{M}{1 + Ae^{-kt}}$$

We find the value of A by putting $t = 0$ in Equation 6. If $t = 0$, then $P = P_0$ (the initial population), so

$$\frac{M - P_0}{P_0} = Ae^0 = A$$

Thus the solution to the logistic equation is

$$\boxed{7} \qquad P(t) = \frac{M}{1 + Ae^{-kt}} \qquad \text{where } A = \frac{M - P_0}{P_0}$$

Using the expression for $P(t)$ in Equation 7, we see that

$$\lim_{t \to \infty} P(t) = M$$

which is to be expected.

EXAMPLE 2 Write the solution of the initial-value problem

$$\frac{dP}{dt} = 0.08P\left(1 - \frac{P}{1000}\right) \qquad P(0) = 100$$

and use it to find the population sizes $P(40)$ and $P(80)$. At what time does the population reach 900?

SOLUTION The differential equation is a logistic equation with $k = 0.08$, carrying capacity $M = 1000$, and initial population $P_0 = 100$. So Equation 7 gives the population at time t as

$$P(t) = \frac{1000}{1 + Ae^{-0.08t}} \qquad \text{where } A = \frac{1000 - 100}{100} = 9$$

Thus

$$P(t) = \frac{1000}{1 + 9e^{-0.08t}}$$

So the population sizes when $t = 40$ and 80 are

$$P(40) = \frac{1000}{1 + 9e^{-3.2}} \approx 731.6 \qquad P(80) = \frac{1000}{1 + 9e^{-6.4}} \approx 985.3$$

The population reaches 900 when

$$\frac{1000}{1 + 9e^{-0.08t}} = 900$$

Solving this equation for t, we get

$$1 + 9e^{-0.08t} = \tfrac{10}{9}$$

$$e^{-0.08t} = \tfrac{1}{81}$$

$$-0.08t = \ln \tfrac{1}{81} = -\ln 81$$

$$t = \frac{\ln 81}{0.08} \approx 54.9$$

So the population reaches 900 when t is approximately 55. As a check on our work, we graph the population curve in Figure 3 and observe where it intersects the line $P = 900$. The cursor indicates that $t \approx 55$.

Compare the solution curve in Figure 3 with the lowest solution curve we drew from the direction field in Figure 2.

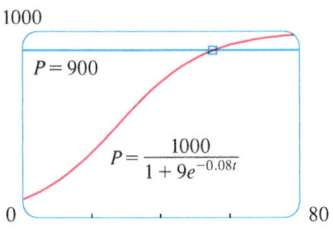

FIGURE 3

Comparison of the Natural Growth and Logistic Models

In the 1930s the biologist G. F. Gause conducted an experiment with the protozoan *Paramecium* and used a logistic equation to model his data. The table gives his daily count of the population of protozoa. He estimated the initial relative growth rate to be 0.7944 and the carrying capacity to be 64.

t (days)	0	1	2	3	4	5	6	7	8	9	10	11	12	13	14	15	16
P (observed)	2	3	22	16	39	52	54	47	50	76	69	51	57	70	53	59	57

EXAMPLE 3 Find the exponential and logistic models for Gause's data. Compare the predicted values with the observed values and comment on the fit.

SOLUTION Given the relative growth rate $k = 0.7944$ and the initial population $P_0 = 2$, the exponential model is

$$P(t) = P_0 e^{kt} = 2e^{0.7944t}$$

Gause used the same value of k for his logistic model. [This is reasonable because $P_0 = 2$ is small compared with the carrying capacity ($M = 64$). The equation

$$\frac{1}{P_0} \frac{dP}{dt} \bigg|_{t=0} = k\left(1 - \frac{2}{64}\right) \approx k$$

shows that the value of k for the logistic model is very close to the value for the exponential model.]

Then the solution of the logistic equation in Equation 7 gives

$$P(t) = \frac{M}{1 + Ae^{-kt}} = \frac{64}{1 + Ae^{-0.7944t}}$$

where $\quad A = \dfrac{M - P_0}{P_0} = \dfrac{64 - 2}{2} = 31$

So $\quad P(t) = \dfrac{64}{1 + 31e^{-0.7944t}}$

We use these equations to calculate the predicted values (rounded to the nearest integer) and compare them in the following table.

t (days)	0	1	2	3	4	5	6	7	8	9	10	11	12	13	14	15	16
P (observed)	2	3	22	16	39	52	54	47	50	76	69	51	57	70	53	59	57
P (logistic model)	2	4	9	17	28	40	51	57	61	62	63	64	64	64	64	64	64
P (exponential model)	2	4	10	22	48	106	...										

We notice from the table and from the graph in Figure 4 that for the first three or four days the exponential model gives results comparable to those of the more sophisticated logistic model. For $t \geq 5$, however, the exponential model is hopelessly inaccurate, but the logistic model fits the observations reasonably well.

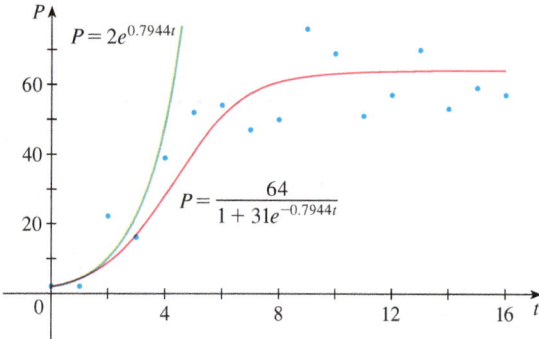

FIGURE 4
The exponential and logistic models for the *Paramecium* data

Many countries that formerly experienced exponential growth are now finding that their rates of population growth are declining and the logistic model provides a better model. The table in the margin shows midyear values of $B(t)$, the population of Belgium, in thousands, at time t, from 1980 to 2012. Figure 5 shows these data points together with a shifted logistic function obtained from a calculator with the ability to fit a logistic function to these points by regression. We see that the logistic model provides a very good fit.

t	$B(t)$	t	$B(t)$
1980	9,847	1998	10,217
1982	9,856	2000	10,264
1984	9,855	2002	10,312
1986	9,862	2004	10,348
1988	9,884	2006	10,379
1990	9,969	2008	10,404
1992	10,046	2010	10,423
1994	10,123	2012	10,438
1996	10,179		

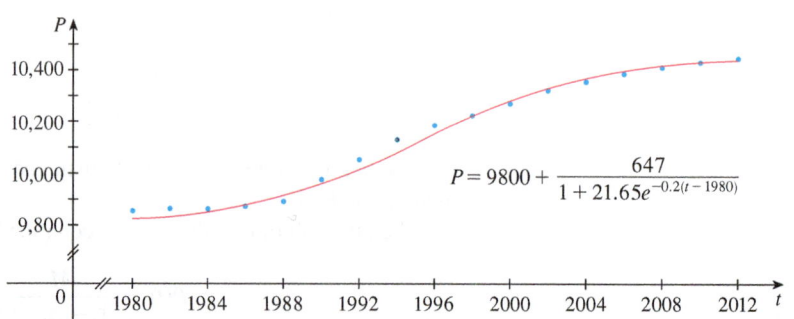

FIGURE 5
Logistic model for the population of Belgium

Other Models for Population Growth

The Law of Natural Growth and the logistic differential equation are not the only equations that have been proposed to model population growth. In Exercise 22 we look at the Gompertz growth function and in Exercises 23 and 24 we investigate seasonal-growth models.

Two of the other models are modifications of the logistic model. The differential equation

$$\frac{dP}{dt} = kP\left(1 - \frac{P}{M}\right) - c$$

has been used to model populations that are subject to harvesting of one sort or another. (Think of a population of fish being caught at a constant rate.) This equation is explored in Exercises 19 and 20.

For some species there is a minimum population level m below which the species tends to become extinct. (Adults may not be able to find suitable mates.) Such populations have been modeled by the differential equation

$$\frac{dP}{dt} = kP\left(1 - \frac{P}{M}\right)\left(1 - \frac{m}{P}\right)$$

where the extra factor, $1 - m/P$, takes into account the consequences of a sparse population (see Exercise 21).

9.4 EXERCISES

1–2 A population grows according to the given logistic equation, where t is measured in weeks.
(a) What is the carrying capacity? What is the value of k?
(b) Write the solution of the equation.
(c) What is the population after 10 weeks?

1. $\dfrac{dP}{dt} = 0.04P\left(1 - \dfrac{P}{1200}\right)$, $P(0) = 60$

2. $\dfrac{dP}{dt} = 0.02P - 0.0004P^2$, $P(0) = 40$

3. Suppose that a population develops according to the logistic equation

$$\frac{dP}{dt} = 0.05P - 0.0005P^2$$

where t is measured in weeks.
(a) What is the carrying capacity? What is the value of k?
(b) A direction field for this equation is shown. Where are the slopes close to 0? Where are they largest? Which solutions are increasing? Which solutions are decreasing?

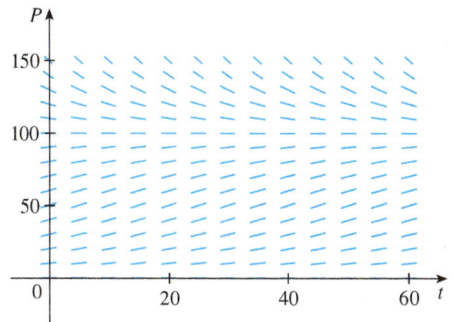

(c) Use the direction field to sketch solutions for initial populations of 20, 40, 60, 80, 120, and 140. What do these solutions have in common? How do they differ? Which solutions have inflection points? At what population levels do they occur?
(d) What are the equilibrium solutions? How are the other solutions related to these solutions?

4. Suppose that a population grows according to a logistic model with carrying capacity 6000 and $k = 0.0015$ per year.
(a) Write the logistic differential equation for these data.

(b) Draw a direction field (either by hand or with a computer algebra system). What does it tell you about the solution curves?

(c) Use the direction field to sketch the solution curves for initial populations of 1000, 2000, 4000, and 8000. What can you say about the concavity of these curves? What is the significance of the inflection points?

(d) Program a calculator or computer to use Euler's method with step size $h = 1$ to estimate the population after 50 years if the initial population is 1000.

(e) If the initial population is 1000, write a formula for the population after t years. Use it to find the population after 50 years and compare with your estimate in part (d).

(f) Graph the solution in part (e) and compare with the solution curve you sketched in part (c).

5. The Pacific halibut fishery has been modeled by the differential equation

$$\frac{dy}{dt} = ky\left(1 - \frac{y}{M}\right)$$

where $y(t)$ is the biomass (the total mass of the members of the population) in kilograms at time t (measured in years), the carrying capacity is estimated to be $M = 8 \times 10^7$ kg, and $k = 0.71$ per year.

(a) If $y(0) = 2 \times 10^7$ kg, find the biomass a year later.
(b) How long will it take for the biomass to reach 4×10^7 kg?

6. Suppose a population $P(t)$ satisfies

$$\frac{dP}{dt} = 0.4P - 0.001P^2 \qquad P(0) = 50$$

where t is measured in years.
(a) What is the carrying capacity?
(b) What is $P'(0)$?
(c) When will the population reach 50% of the carrying capacity?

7. Suppose a population grows according to a logistic model with initial population 1000 and carrying capacity 10,000. If the population grows to 2500 after one year, what will the population be after another three years?

8. The table gives the number of yeast cells in a new laboratory culture.

Time (hours)	Yeast cells	Time (hours)	Yeast cells
0	18	10	509
2	39	12	597
4	80	14	640
6	171	16	664
8	336	18	672

(a) Plot the data and use the plot to estimate the carrying capacity for the yeast population.
(b) Use the data to estimate the initial relative growth rate.

(c) Find both an exponential model and a logistic model for these data.
(d) Compare the predicted values with the observed values, both in a table and with graphs. Comment on how well your models fit the data.
(e) Use your logistic model to estimate the number of yeast cells after 7 hours.

9. The population of the world was about 6.1 billion in 2000. Birth rates around that time ranged from 35 to 40 million per year and death rates ranged from 15 to 20 million per year. Let's assume that the carrying capacity for world population is 20 billion.
(a) Write the logistic differential equation for these data. (Because the initial population is small compared to the carrying capacity, you can take k to be an estimate of the initial relative growth rate.)
(b) Use the logistic model to estimate the world population in the year 2010 and compare with the actual population of 6.9 billion.
(c) Use the logistic model to predict the world population in the years 2100 and 2500.

10. (a) Assume that the carrying capacity for the US population is 800 million. Use it and the fact that the population was 282 million in 2000 to formulate a logistic model for the US population.
(b) Determine the value of k in your model by using the fact that the population in 2010 was 309 million.
(c) Use your model to predict the US population in the years 2100 and 2200.
(d) Use your model to predict the year in which the US population will exceed 500 million.

11. One model for the spread of a rumor is that the rate of spread is proportional to the product of the fraction y of the population who have heard the rumor and the fraction who have not heard the rumor.
(a) Write a differential equation that is satisfied by y.
(b) Solve the differential equation.
(c) A small town has 1000 inhabitants. At 8 AM, 80 people have heard a rumor. By noon half the town has heard it. At what time will 90% of the population have heard the rumor?

12. Biologists stocked a lake with 400 fish and estimated the carrying capacity (the maximal population for the fish of that species in that lake) to be 10,000. The number of fish tripled in the first year.
(a) Assuming that the size of the fish population satisfies the logistic equation, find an expression for the size of the population after t years.
(b) How long will it take for the population to increase to 5000?

13. (a) Show that if P satisfies the logistic equation (4), then

$$\frac{d^2P}{dt^2} = k^2P\left(1 - \frac{P}{M}\right)\left(1 - \frac{2P}{M}\right)$$

(b) Deduce that a population grows fastest when it reaches half its carrying capacity.

14. For a fixed value of M (say $M = 10$), the family of logistic functions given by Equation 7 depends on the initial value P_0 and the proportionality constant k. Graph several members of this family. How does the graph change when P_0 varies? How does it change when k varies?

15. The table gives the midyear population of Japan, in thousands, from 1960 to 2010.

Year	Population	Year	Population
1960	94,092	1990	123,537
1965	98,883	1995	125,327
1970	104,345	2000	126,776
1975	111,573	2005	127,715
1980	116,807	2010	127,579
1985	120,754		

Use a calculator to fit both an exponential function and a logistic function to these data. Graph the data points and both functions, and comment on the accuracy of the models. [*Hint:* Subtract 94,000 from each of the population figures. Then, after obtaining a model from your calculator, add 94,000 to get your final model. It might be helpful to choose $t = 0$ to correspond to 1960 or 1980.]

16. The table gives the midyear population of Norway, in thousands, from 1960 to 2010.

Year	Population	Year	Population
1960	3581	1990	4242
1965	3723	1995	4359
1970	3877	2000	4492
1975	4007	2005	4625
1980	4086	2010	4891
1985	4152		

Use a calculator to fit both an exponential function and a logistic function to these data. Graph the data points and both functions, and comment on the accuracy of the models. [*Hint:* Subtract 3500 from each of the population figures. Then, after obtaining a model from your calculator, add 3500 to get your final model. It might be helpful to choose $t = 0$ to correspond to 1960.]

17. Consider a population $P = P(t)$ with constant relative birth and death rates α and β, respectively, and a constant emigration rate m, where α, β, and m are positive constants. Assume that $\alpha > \beta$. Then the rate of change of the population at time t is modeled by the differential equation

$$\frac{dP}{dt} = kP - m \quad \text{where } k = \alpha - \beta$$

(a) Find the solution of this equation that satisfies the initial condition $P(0) = P_0$.

(b) What condition on m will lead to an exponential expansion of the population?

(c) What condition on m will result in a constant population? A population decline?

(d) In 1847, the population of Ireland was about 8 million and the difference between the relative birth and death rates was 1.6% of the population. Because of the potato famine in the 1840s and 1850s, about 210,000 inhabitants per year emigrated from Ireland. Was the population expanding or declining at that time?

18. Let c be a positive number. A differential equation of the form

$$\frac{dy}{dt} = ky^{1+c}$$

where k is a positive constant, is called a *doomsday equation* because the exponent in the expression ky^{1+c} is larger than the exponent 1 for natural growth.

(a) Determine the solution that satisfies the initial condition $y(0) = y_0$.

(b) Show that there is a finite time $t = T$ (doomsday) such that $\lim_{t \to T^-} y(t) = \infty$.

(c) An especially prolific breed of rabbits has the growth term $ky^{1.01}$. If 2 such rabbits breed initially and the warren has 16 rabbits after three months, then when is doomsday?

19. Let's modify the logistic differential equation of Example 1 as follows:

$$\frac{dP}{dt} = 0.08P\left(1 - \frac{P}{1000}\right) - 15$$

(a) Suppose $P(t)$ represents a fish population at time t, where t is measured in weeks. Explain the meaning of the final term in the equation (-15).

(b) Draw a direction field for this differential equation.

(c) What are the equilibrium solutions?

(d) Use the direction field to sketch several solution curves. Describe what happens to the fish population for various initial populations.

(e) Solve this differential equation explicitly, either by using partial fractions or with a computer algebra system. Use the initial populations 200 and 300. Graph the solutions and compare with your sketches in part (d).

20. Consider the differential equation

$$\frac{dP}{dt} = 0.08P\left(1 - \frac{P}{1000}\right) - c$$

as a model for a fish population, where t is measured in weeks and c is a constant.

(a) Use a CAS to draw direction fields for various values of c.

(b) From your direction fields in part (a), determine the values of c for which there is at least one equilibrium solution. For what values of c does the fish population always die out?

(c) Use the differential equation to prove what you discovered graphically in part (b).

(d) What would you recommend for a limit to the weekly catch of this fish population?

21. There is considerable evidence to support the theory that for some species there is a minimum population m such that the species will become extinct if the size of the population falls below m. This condition can be incorporated into the logistic equation by introducing the factor $(1 - m/P)$. Thus the modified logistic model is given by the differential equation

$$\frac{dP}{dt} = kP\left(1 - \frac{P}{M}\right)\left(1 - \frac{m}{P}\right)$$

(a) Use the differential equation to show that any solution is increasing if $m < P < M$ and decreasing if $0 < P < m$.

(b) For the case where $k = 0.08$, $M = 1000$, and $m = 200$, draw a direction field and use it to sketch several solution curves. Describe what happens to the population for various initial populations. What are the equilibrium solutions?

(c) Solve the differential equation explicitly, either by using partial fractions or with a computer algebra system. Use the initial population P_0.

(d) Use the solution in part (c) to show that if $P_0 < m$, then the species will become extinct. [*Hint:* Show that the numerator in your expression for $P(t)$ is 0 for some value of t.]

22. Another model for a growth function for a limited population is given by the **Gompertz function**, which is a solution of the differential equation

$$\frac{dP}{dt} = c \ln\left(\frac{M}{P}\right) P$$

where c is a constant and M is the carrying capacity.

(a) Solve this differential equation.

(b) Compute $\lim_{t \to \infty} P(t)$.

(c) Graph the Gompertz growth function for $M = 1000$, $P_0 = 100$, and $c = 0.05$, and compare it with the logistic function in Example 2. What are the similarities? What are the differences?

(d) We know from Exercise 13 that the logistic function grows fastest when $P = M/2$. Use the Gompertz differential equation to show that the Gompertz function grows fastest when $P = M/e$.

23. In a **seasonal-growth model**, a periodic function of time is introduced to account for seasonal variations in the rate of growth. Such variations could, for example, be caused by seasonal changes in the availability of food.

(a) Find the solution of the seasonal-growth model

$$\frac{dP}{dt} = kP \cos(rt - \phi) \qquad P(0) = P_0$$

where k, r, and ϕ are positive constants.

(b) By graphing the solution for several values of k, r, and ϕ, explain how the values of k, r, and ϕ affect the solution. What can you say about $\lim_{t \to \infty} P(t)$?

24. Suppose we alter the differential equation in Exercise 23 as follows:

$$\frac{dP}{dt} = kP \cos^2(rt - \phi) \qquad P(0) = P_0$$

(a) Solve this differential equation with the help of a table of integrals or a CAS.

(b) Graph the solution for several values of k, r, and ϕ. How do the values of k, r, and ϕ affect the solution? What can you say about $\lim_{t \to \infty} P(t)$ in this case?

25. Graphs of logistic functions (Figures 2 and 3) look suspiciously similar to the graph of the hyperbolic tangent function (Figure 3.11.3). Explain the similarity by showing that the logistic function given by Equation 7 can be written as

$$P(t) = \tfrac{1}{2}M\left[1 + \tanh\left(\tfrac{1}{2}k(t - c)\right)\right]$$

where $c = (\ln A)/k$. Thus the logistic function is really just a shifted hyperbolic tangent.

9.5 Linear Equations

A first-order **linear** differential equation is one that can be put into the form

$$\boxed{1} \qquad \frac{dy}{dx} + P(x)y = Q(x)$$

where P and Q are continuous functions on a given interval. This type of equation occurs frequently in various sciences, as we will see.

An example of a linear equation is $xy' + y = 2x$ because, for $x \neq 0$, it can be written in the form

$$\boxed{2} \qquad y' + \frac{1}{x}y = 2$$

Notice that this differential equation is not separable because it's impossible to factor the expression for y' as a function of x times a function of y. But we can still solve the equation by noticing, by the Product Rule, that

$$xy' + y = (xy)'$$

and so we can rewrite the equation as

$$(xy)' = 2x$$

If we now integrate both sides of this equation, we get

$$xy = x^2 + C \quad \text{or} \quad y = x + \frac{C}{x}$$

If we had been given the differential equation in the form of Equation 2, we would have had to take the preliminary step of multiplying each side of the equation by x.

It turns out that every first-order linear differential equation can be solved in a similar fashion by multiplying both sides of Equation 1 by a suitable function $I(x)$ called an *integrating factor*. We try to find I so that the left side of Equation 1, when multiplied by $I(x)$, becomes the derivative of the product $I(x)y$:

$$\boxed{3} \qquad I(x)\bigl(y' + P(x)y\bigr) = \bigl(I(x)y\bigr)'$$

If we can find such a function I, then Equation 1 becomes

$$\bigl(I(x)y\bigr)' = I(x)\,Q(x)$$

Integrating both sides, we would have

$$I(x)y = \int I(x)\,Q(x)\,dx + C$$

so the solution would be

$$\boxed{4} \qquad y(x) = \frac{1}{I(x)}\left[\int I(x)\,Q(x)\,dx + C\right]$$

To find such an I, we expand Equation 3 and cancel terms:

$$I(x)y' + I(x)P(x)y = \bigl(I(x)y\bigr)' = I'(x)y + I(x)y'$$

$$I(x)P(x) = I'(x)$$

This is a separable differential equation for I, which we solve as follows:

$$\int \frac{dI}{I} = \int P(x)\,dx$$

$$\ln|I| = \int P(x)\,dx$$

$$I = Ae^{\int P(x)\,dx}$$

where $A = \pm e^C$. We are looking for a particular integrating factor, not the most general one, so we take $A = 1$ and use

$$\boxed{5} \qquad I(x) = e^{\int P(x)\,dx}$$

Thus a formula for the general solution to Equation 1 is provided by Equation 4, where I is given by Equation 5. Instead of memorizing this formula, however, we just remember the form of the integrating factor.

> To solve the linear differential equation $y' + P(x)y = Q(x)$, multiply both sides by the **integrating factor** $I(x) = e^{\int P(x)\,dx}$ and integrate both sides.

EXAMPLE 1 Solve the differential equation $\dfrac{dy}{dx} + 3x^2 y = 6x^2$.

SOLUTION The given equation is linear since it has the form of Equation 1 with $P(x) = 3x^2$ and $Q(x) = 6x^2$. An integrating factor is

$$I(x) = e^{\int 3x^2\,dx} = e^{x^3}$$

Multiplying both sides of the differential equation by e^{x^3}, we get

$$e^{x^3}\frac{dy}{dx} + 3x^2 e^{x^3} y = 6x^2 e^{x^3}$$

or

$$\frac{d}{dx}(e^{x^3} y) = 6x^2 e^{x^3}$$

Integrating both sides, we have

$$e^{x^3} y = \int 6x^2 e^{x^3}\,dx = 2e^{x^3} + C$$

$$y = 2 + Ce^{-x^3}$$

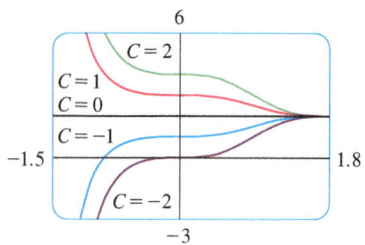

Figure 1 shows the graphs of several members of the family of solutions in Example 1. Notice that they all approach 2 as $x \to \infty$.

FIGURE 1

EXAMPLE 2 Find the solution of the initial-value problem

$$x^2 y' + xy = 1 \qquad x > 0 \qquad y(1) = 2$$

SOLUTION We must first divide both sides by the coefficient of y' to put the differential equation into standard form:

$$\boxed{6} \qquad y' + \frac{1}{x} y = \frac{1}{x^2} \qquad x > 0$$

The integrating factor is

$$I(x) = e^{\int (1/x)\,dx} = e^{\ln x} = x$$

Multiplication of Equation 6 by x gives

$$xy' + y = \frac{1}{x} \qquad \text{or} \qquad (xy)' = \frac{1}{x}$$

SECTION 9.5 Linear Equations

Then
$$xy = \int \frac{1}{x} dx = \ln x + C$$

and so
$$y = \frac{\ln x + C}{x}$$

Since $y(1) = 2$, we have
$$2 = \frac{\ln 1 + C}{1} = C$$

Therefore the solution to the initial-value problem is
$$y = \frac{\ln x + 2}{x}$$

The solution of the initial-value problem in Example 2 is shown in Figure 2.

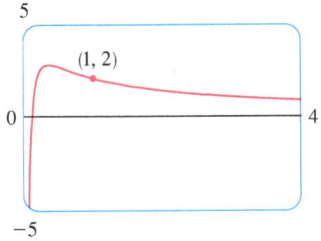

FIGURE 2

EXAMPLE 3 Solve $y' + 2xy = 1$.

SOLUTION The given equation is in the standard form for a linear equation. Multiplying by the integrating factor
$$e^{\int 2x \, dx} = e^{x^2}$$

we get
$$e^{x^2} y' + 2xe^{x^2} y = e^{x^2}$$

or
$$\left(e^{x^2} y\right)' = e^{x^2}$$

Therefore
$$e^{x^2} y = \int e^{x^2} \, dx + C$$

Recall from Section 7.5 that $\int e^{x^2} dx$ can't be expressed in terms of elementary functions. Nonetheless, it's a perfectly good function and we can leave the answer as
$$y = e^{-x^2} \int e^{x^2} \, dx + Ce^{-x^2}$$

Another way of writing the solution is
$$y = e^{-x^2} \int_0^x e^{t^2} \, dt + Ce^{-x^2}$$

(Any number can be chosen for the lower limit of integration.)

Even though the solutions of the differential equation in Example 3 are expressed in terms of an integral, they can still be graphed by a computer algebra system (Figure 3).

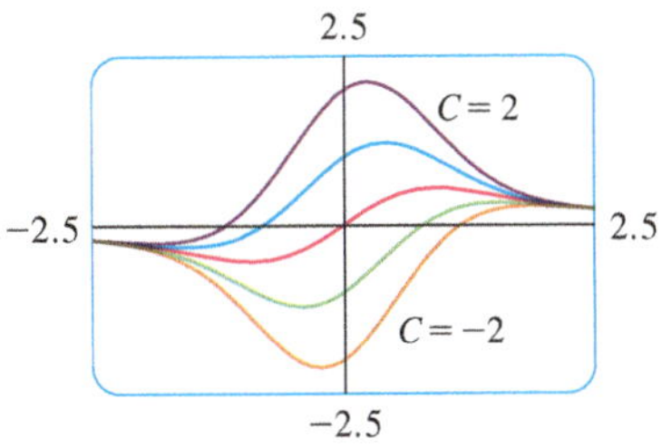

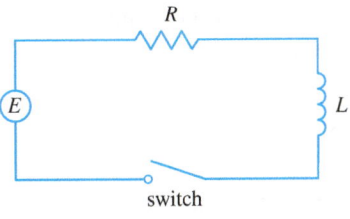

FIGURE 3

Application to Electric Circuits

In Section 9.2 we considered the simple electric circuit shown in Figure 4: An electromotive force (usually a battery or generator) produces a voltage of $E(t)$ volts (V) and a current of $I(t)$ amperes (A) at time t. The circuit also contains a resistor with a resistance of R ohms (Ω) and an inductor with an inductance of L henries (H).

Ohm's Law gives the drop in voltage due to the resistor as RI. The voltage drop due to the inductor is $L(dI/dt)$. One of Kirchhoff's laws says that the sum of the voltage drops is equal to the supplied voltage $E(t)$. Thus we have

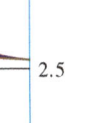

$$L \frac{dI}{dt} + RI = E(t)$$

FIGURE 4

624   CHAPTER 9   Differential Equations

which is a first-order linear differential equation. The solution gives the current I at time t.

EXAMPLE 4 Suppose that in the simple circuit of Figure 4 the resistance is 12 Ω and the inductance is 4 H. If a battery gives a constant voltage of 60 V and the switch is closed when $t = 0$ so the current starts with $I(0) = 0$, find (a) $I(t)$, (b) the current after 1 second, and (c) the limiting value of the current.

SOLUTION

The differential equation in Example 4 is both linear and separable, so an alternative method is to solve it as a separable equation (Example 9.3.4). If we replace the battery by a generator, however, we get an equation that is linear but not separable (Example 5).

(a) If we put $L = 4$, $R = 12$, and $E(t) = 60$ in Equation 7, we obtain the initial-value problem

$$4\frac{dI}{dt} + 12I = 60 \qquad I(0) = 0$$

or

$$\frac{dI}{dt} + 3I = 15 \qquad I(0) = 0$$

Multiplying by the integrating factor $e^{\int 3\,dt} = e^{3t}$, we get

$$e^{3t}\frac{dI}{dt} + 3e^{3t}I = 15e^{3t}$$

$$\frac{d}{dt}(e^{3t}I) = 15e^{3t}$$

$$e^{3t}I = \int 15e^{3t}\,dt = 5e^{3t} + C$$

$$I(t) = 5 + Ce^{-3t}$$

Since $I(0) = 0$, we have $5 + C = 0$, so $C = -5$ and

$$I(t) = 5(1 - e^{-3t})$$

(b) After 1 second the current is

$$I(1) = 5(1 - e^{-3}) \approx 4.75 \text{ A}$$

Figure 5 shows how the current in Example 4 approaches its limiting value.

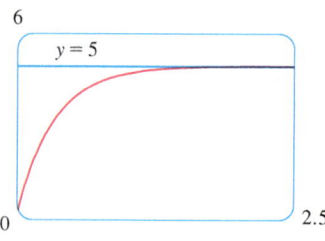

FIGURE 5

(c) The limiting value of the current is given by

$$\lim_{t\to\infty} I(t) = \lim_{t\to\infty} 5(1 - e^{-3t}) = 5 - 5\lim_{t\to\infty} e^{-3t} = 5 - 0 = 5$$

Figure 6 shows the graph of the current when the battery is replaced by a generator.

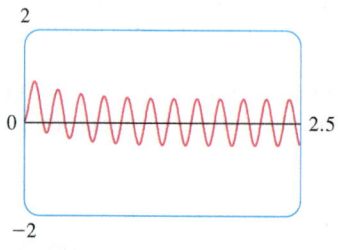

FIGURE 6

EXAMPLE 5 Suppose that the resistance and inductance remain as in Example 4 but, instead of the battery, we use a generator that produces a variable voltage of $E(t) = 60 \sin 30t$ volts. Find $I(t)$.

SOLUTION This time the differential equation becomes

$$4\frac{dI}{dt} + 12I = 60\sin 30t \qquad \text{or} \qquad \frac{dI}{dt} + 3I = 15\sin 30t$$

The same integrating factor e^{3t} gives

$$\frac{d}{dt}(e^{3t}I) = e^{3t}\frac{dI}{dt} + 3e^{3t}I = 15e^{3t}\sin 30t$$

Using Formula 98 in the Table of Integrals, we have

$$e^{3t}I = \int 15e^{3t} \sin 30t \, dt = 15 \frac{e^{3t}}{909} (3 \sin 30t - 30 \cos 30t) + C$$

$$I = \tfrac{5}{101}(\sin 30t - 10 \cos 30t) + Ce^{-3t}$$

Since $I(0) = 0$, we get

$$-\tfrac{50}{101} + C = 0$$

so

$$I(t) = \tfrac{5}{101}(\sin 30t - 10 \cos 30t) + \tfrac{50}{101}e^{-3t}$$

9.5 EXERCISES

1–4 Determine whether the differential equation is linear.

1. $y' + x\sqrt{y} = x^2$
2. $y' - x = y \tan x$
3. $ue^t = t + \sqrt{t}\,\dfrac{du}{dt}$
4. $\dfrac{dR}{dt} + t \cos R = e^{-t}$

5–14 Solve the differential equation.

5. $y' + y = 1$
6. $y' - y = e^x$
7. $y' = x - y$
8. $4x^3y + x^4y' = \sin^3 x$
9. $xy' + y = \sqrt{x}$
10. $2xy' + y = 2\sqrt{x}$
11. $xy' - 2y = x^2$, $x > 0$
12. $y' + 2xy = 1$
13. $t^2\dfrac{dy}{dt} + 3ty = \sqrt{1+t^2}$, $t > 0$
14. $t \ln t \dfrac{dr}{dt} + r = te^t$

15–20 Solve the initial-value problem.

15. $x^2y' + 2xy = \ln x$, $y(1) = 2$
16. $t^3\dfrac{dy}{dt} + 3t^2y = \cos t$, $y(\pi) = 0$
17. $t\dfrac{du}{dt} = t^2 + 3u$, $t > 0$, $u(2) = 4$
18. $xy' + y = x \ln x$, $y(1) = 0$
19. $xy' = y + x^2 \sin x$, $y(\pi) = 0$
20. $(x^2 + 1)\dfrac{dy}{dx} + 3x(y - 1) = 0$, $y(0) = 2$

21–22 Solve the differential equation and use a calculator to graph several members of the family of solutions. How does the solution curve change as C varies?

21. $xy' + 2y = e^x$
22. $xy' = x^2 + 2y$

23. A **Bernoulli differential equation** (named after James Bernoulli) is of the form

$$\frac{dy}{dx} + P(x)y = Q(x)y^n$$

Observe that, if $n = 0$ or 1, the Bernoulli equation is linear. For other values of n, show that the substitution $u = y^{1-n}$ transforms the Bernoulli equation into the linear equation

$$\frac{du}{dx} + (1 - n)P(x)u = (1 - n)Q(x)$$

24–25 Use the method of Exercise 23 to solve the differential equation.

24. $xy' + y = -xy^2$
25. $y' + \dfrac{2}{x}y = \dfrac{y^3}{x^2}$

26. Solve the second-order equation $xy'' + 2y' = 12x^2$ by making the substitution $u = y'$.

27. In the circuit shown in Figure 4, a battery supplies a constant voltage of 40 V, the inductance is 2 H, the resistance is 10 Ω, and $I(0) = 0$.
  (a) Find $I(t)$.
  (b) Find the current after 0.1 seconds.

28. In the circuit shown in Figure 4, a generator supplies a voltage of $E(t) = 40 \sin 60t$ volts, the inductance is 1 H, the resistance is 20 Ω, and $I(0) = 1$ A.
  (a) Find $I(t)$.
  (b) Find the current after 0.1 seconds.
  (c) Use a graphing device to draw the graph of the current function.

29. The figure shows a circuit containing an electromotive force, a capacitor with a capacitance of C farads (F), and a resistor with a resistance of R ohms (Ω). The voltage drop across the capacitor is Q/C, where Q is the charge (in coulombs), so in this case Kirchhoff's Law gives

$$RI + \frac{Q}{C} = E(t)$$

But $I = dQ/dt$ (see Example 3.7.3), so we have

$$R\frac{dQ}{dt} + \frac{1}{C}Q = E(t)$$

Suppose the resistance is 5 Ω, the capacitance is 0.05 F, a battery gives a constant voltage of 60 V, and the initial charge is $Q(0) = 0$ C. Find the charge and the current at time t.

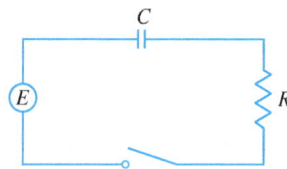

30. In the circuit of Exercise 29, $R = 2$ Ω, $C = 0.01$ F, $Q(0) = 0$, and $E(t) = 10 \sin 60t$. Find the charge and the current at time t.

31. Let $P(t)$ be the performance level of someone learning a skill as a function of the training time t. The graph of P is called a *learning curve*. In Exercise 9.1.15 we proposed the differential equation

$$\frac{dP}{dt} = k[M - P(t)]$$

as a reasonable model for learning, where k is a positive constant. Solve it as a linear differential equation and use your solution to graph the learning curve.

32. Two new workers were hired for an assembly line. Jim processed 25 units during the first hour and 45 units during the second hour. Mark processed 35 units during the first hour and 50 units the second hour. Using the model of Exercise 31 and assuming that $P(0) = 0$, estimate the maximum number of units per hour that each worker is capable of processing.

33. In Section 9.3 we looked at mixing problems in which the volume of fluid remained constant and saw that such problems give rise to separable differentiable equations. (See Example 6 in that section.) If the rates of flow into and out of the system are different, then the volume is not constant and the resulting differential equation is linear but not separable.

A tank contains 100 L of water. A solution with a salt concentration of 0.4 kg/L is added at a rate of 5 L/min. The solution is kept mixed and is drained from the tank at a rate of 3 L/min. If $y(t)$ is the amount of salt (in kilograms) after t minutes, show that y satisfies the differential equation

$$\frac{dy}{dt} = 2 - \frac{3y}{100 + 2t}$$

Solve this equation and find the concentration after 20 minutes.

34. A tank with a capacity of 400 L is full of a mixture of water and chlorine with a concentration of 0.05 g of chlorine per liter. In order to reduce the concentration of chlorine, fresh water is pumped into the tank at a rate of 4 L/s. The mixture is kept stirred and is pumped out at a rate of 10 L/s. Find the amount of chlorine in the tank as a function of time.

35. An object with mass m is dropped from rest and we assume that the air resistance is proportional to the speed of the object. If $s(t)$ is the distance dropped after t seconds, then the speed is $v = s'(t)$ and the acceleration is $a = v'(t)$. If g is the acceleration due to gravity, then the downward force on the object is $mg - cv$, where c is a positive constant, and Newton's Second Law gives

$$m\frac{dv}{dt} = mg - cv$$

(a) Solve this as a linear equation to show that

$$v = \frac{mg}{c}(1 - e^{-ct/m})$$

(b) What is the limiting velocity?
(c) Find the distance the object has fallen after t seconds.

36. If we ignore air resistance, we can conclude that heavier objects fall no faster than lighter objects. But if we take air resistance into account, our conclusion changes. Use the expression for the velocity of a falling object in Exercise 35(a) to find dv/dm and show that heavier objects *do* fall faster than lighter ones.

37. (a) Show that the substitution $z = 1/P$ transforms the logistic differential equation $P' = kP(1 - P/M)$ into the linear differential equation

$$z' + kz = \frac{k}{M}$$

(b) Solve the linear differential equation in part (a) and thus obtain an expression for $P(t)$. Compare with Equation 9.4.7.

38. To account for seasonal variation in the logistic differential equation, we could allow k and M to be functions of t:

$$\frac{dP}{dt} = k(t)P\left(1 - \frac{P}{M(t)}\right)$$

(a) Verify that the substitution $z = 1/P$ transforms this equation into the linear equation

$$\frac{dz}{dt} + k(t)z = \frac{k(t)}{M(t)}$$

(b) Write an expression for the solution of the linear equation in part (a) and use it to show that if the carrying capacity M

is constant, then

$$P(t) = \frac{M}{1 + CMe^{-\int k(t)\,dt}}$$

Deduce that if $\int_0^\infty k(t)\,dt = \infty$, then $\lim_{t\to\infty} P(t) = M$. [This will be true if $k(t) = k_0 + a\cos bt$ with $k_0 > 0$, which describes a positive intrinsic growth rate with a periodic seasonal variation.]

(c) If k is constant but M varies, show that

$$z(t) = e^{-kt}\int_0^t \frac{ke^{ks}}{M(s)}\,ds + Ce^{-kt}$$

and use l'Hospital's Rule to deduce that if $M(t)$ has a limit as $t\to\infty$, then $P(t)$ has the same limit.

9.6 Predator-Prey Systems

We have looked at a variety of models for the growth of a single species that lives alone in an environment. In this section we consider more realistic models that take into account the interaction of two species in the same habitat. We will see that these models take the form of a pair of linked differential equations.

We first consider the situation in which one species, called the *prey*, has an ample food supply and the second species, called the *predators*, feeds on the prey. Examples of prey and predators include rabbits and wolves in an isolated forest, food-fish and sharks, aphids and ladybugs, and bacteria and amoebas. Our model will have two dependent variables and both are functions of time. We let $R(t)$ be the number of prey (using R for rabbits) and $W(t)$ be the number of predators (with W for wolves) at time t.

In the absence of predators, the ample food supply would support exponential growth of the prey, that is,

$$\frac{dR}{dt} = kR \qquad \text{where } k \text{ is a positive constant}$$

In the absence of prey, we assume that the predator population would decline through mortality at a rate proportional to itself, that is,

$$\frac{dW}{dt} = -rW \qquad \text{where } r \text{ is a positive constant}$$

With both species present, however, we assume that the principal cause of death among the prey is being eaten by a predator, and the birth and survival rates of the predators depend on their available food supply, namely, the prey. We also assume that the two species encounter each other at a rate that is proportional to both populations and is therefore proportional to the product RW. (The more there are of either population, the more encounters there are likely to be.) A system of two differential equations that incorporates these assumptions is as follows:

W represents the predators.
R represents the prey.

$$\boxed{1} \qquad \frac{dR}{dt} = kR - aRW \qquad \frac{dW}{dt} = -rW + bRW$$

where k, r, a, and b are positive constants. Notice that the term $-aRW$ decreases the natural growth rate of the prey and the term bRW increases the natural growth rate of the predators.

The Lotka-Volterra equations were proposed as a model to explain the variations in the shark and food-fish populations in the Adriatic Sea by the Italian mathematician Vito Volterra (1860–1940).

The equations in (1) are known as the **predator-prey equations**, or the **Lotka-Volterra equations**. A **solution** of this system of equations is a pair of functions $R(t)$ and $W(t)$ that describe the populations of prey and predators as functions of time. Because the system is coupled (R and W occur in both equations), we can't solve one equation and then the other; we have to solve them simultaneously. Unfortunately, it is usually impossible to find explicit formulas for R and W as functions of t. We can, however, use graphical methods to analyze the equations.

628   **CHAPTER 9**   Differential Equations

EXAMPLE 1 Suppose that populations of rabbits and wolves are described by the Lotka-Volterra equations (1) with $k = 0.08$, $a = 0.001$, $r = 0.02$, and $b = 0.00002$. The time t is measured in months.
(a) Find the constant solutions (called the **equilibrium solutions**) and interpret the answer.
(b) Use the system of differential equations to find an expression for dW/dR.
(c) Draw a direction field for the resulting differential equation in the RW-plane. Then use that direction field to sketch some solution curves.
(d) Suppose that, at some point in time, there are 1000 rabbits and 40 wolves. Draw the corresponding solution curve and use it to describe the changes in both population levels.
(e) Use part (d) to make sketches of R and W as functions of t.

SOLUTION
(a) With the given values of k, a, r, and b, the Lotka-Volterra equations become

$$\frac{dR}{dt} = 0.08R - 0.001RW$$

$$\frac{dW}{dt} = -0.02W + 0.00002RW$$

Both R and W will be constant if both derivatives are 0, that is,

$$R' = R(0.08 - 0.001W) = 0$$

$$W' = W(-0.02 + 0.00002R) = 0$$

One solution is given by $R = 0$ and $W = 0$. (This makes sense: If there are no rabbits or wolves, the populations are certainly not going to increase.) The other constant solution is

$$W = \frac{0.08}{0.001} = 80$$

$$R = \frac{0.02}{0.00002} = 1000$$

So the equilibrium populations consist of 80 wolves and 1000 rabbits. This means that 1000 rabbits are just enough to support a constant wolf population of 80. There are neither too many wolves (which would result in fewer rabbits) nor too few wolves (which would result in more rabbits).

(b) We use the Chain Rule to eliminate t:

$$\frac{dW}{dt} = \frac{dW}{dR} \frac{dR}{dt}$$

so

$$\frac{dW}{dR} = \frac{\frac{dW}{dt}}{\frac{dR}{dt}} = \frac{-0.02W + 0.00002RW}{0.08R - 0.001RW}$$

(c) If we think of W as a function of R, we have the differential equation

$$\frac{dW}{dR} = \frac{-0.02W + 0.00002RW}{0.08R - 0.001RW}$$

We draw the direction field for this differential equation in Figure 1 and we use it to sketch several solution curves in Figure 2. If we move along a solution curve, we observe how the relationship between R and W changes as time passes. Notice that the curves appear to be closed in the sense that if we travel along a curve, we always return to the same point. Notice also that the point $(1000, 80)$ is inside all the solution curves. That point is called an *equilibrium point* because it corresponds to the equilibrium solution $R = 1000$, $W = 80$.

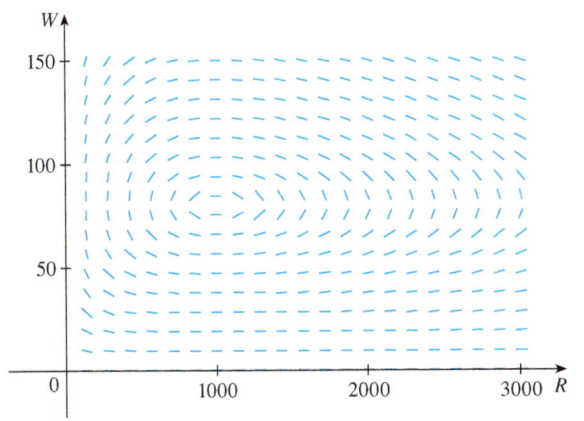

FIGURE 1
Direction field for the predator-prey system

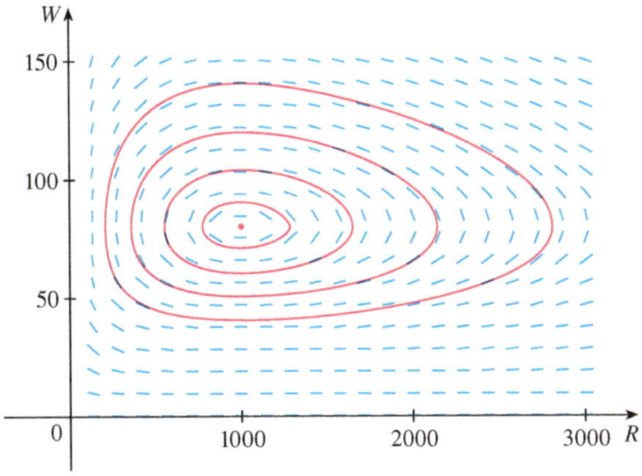

FIGURE 2
Phase portrait of the system

When we represent solutions of a system of differential equations as in Figure 2, we refer to the RW-plane as the **phase plane**, and we call the solution curves **phase trajectories**. So a phase trajectory is a path traced out by solutions (R, W) as time goes by. A **phase portrait** consists of equilibrium points and typical phase trajectories, as shown in Figure 2.

(d) Starting with 1000 rabbits and 40 wolves corresponds to drawing the solution curve through the point $P_0(1000, 40)$. Figure 3 shows this phase trajectory with the direction

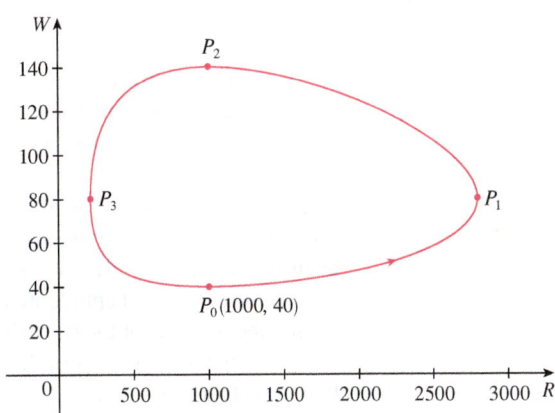

FIGURE 3
Phase trajectory through $(1000, 40)$

field removed. Starting at the point P_0 at time $t = 0$ and letting t increase, do we move clockwise or counterclockwise around the phase trajectory? If we put $R = 1000$ and $W = 40$ in the first differential equation, we get

$$\frac{dR}{dt} = 0.08(1000) - 0.001(1000)(40) = 80 - 40 = 40$$

Since $dR/dt > 0$, we conclude that R is increasing at P_0 and so we move counterclockwise around the phase trajectory.

We see that at P_0 there aren't enough wolves to maintain a balance between the populations, so the rabbit population increases. That results in more wolves and eventually there are so many wolves that the rabbits have a hard time avoiding them. So the number of rabbits begins to decline (at P_1, where we estimate that R reaches its maximum population of about 2800). This means that at some later time the wolf population starts to fall (at P_2, where $R = 1000$ and $W \approx 140$). But this benefits the rabbits, so their population later starts to increase (at P_3, where $W = 80$ and $R \approx 210$). As a consequence, the wolf population eventually starts to increase as well. This happens when the populations return to their initial values of $R = 1000$ and $W = 40$, and the entire cycle begins again.

(e) From the description in part (d) of how the rabbit and wolf populations rise and fall, we can sketch the graphs of $R(t)$ and $W(t)$. Suppose the points P_1, P_2, and P_3 in Figure 3 are reached at times t_1, t_2, and t_3. Then we can sketch graphs of R and W as in Figure 4.

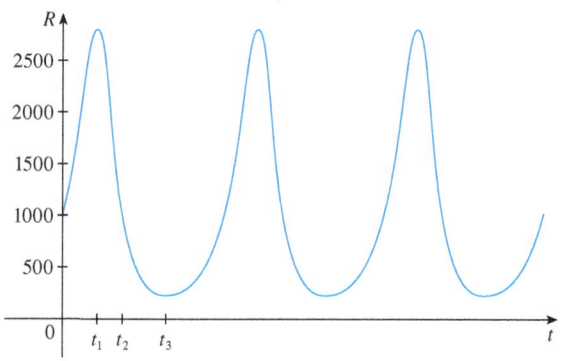 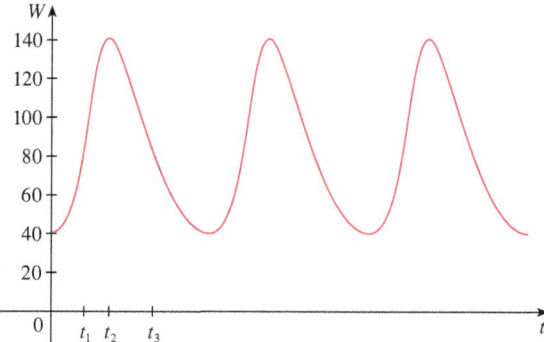

FIGURE 4 Graphs of the rabbit and wolf populations as functions of time

To make the graphs easier to compare, we draw the graphs on the same axes but with different scales for R and W, as in Figure 5 on page 631. Notice that the rabbits reach their maximum populations about a quarter of a cycle before the wolves. ∎

An important part of the modeling process, as we discussed in Section 1.2, is to interpret our mathematical conclusions as real-world predictions and to test the predictions against real data. The Hudson's Bay Company, which started trading in animal furs in Canada in 1670, has kept records that date back to the 1840s. Figure 6 shows graphs of the number of pelts of the snowshoe hare and its predator, the Canada lynx, traded by the company over a 90-year period. You can see that the coupled oscillations in the hare and lynx populations predicted by the Lotka-Volterra model do actually occur and the period of these cycles is roughly 10 years.

SECTION 9.6 Predator-Prey Systems

TEC In Module 9.6 you can change the coefficients in the Lotka-Volterra equations and observe the resulting changes in the phase trajectory and graphs of the rabbit and wolf populations.

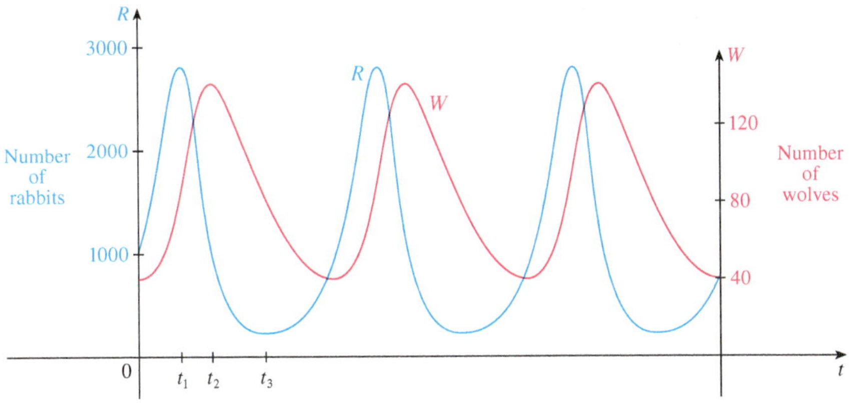

FIGURE 5
Comparison of the rabbit and wolf populations

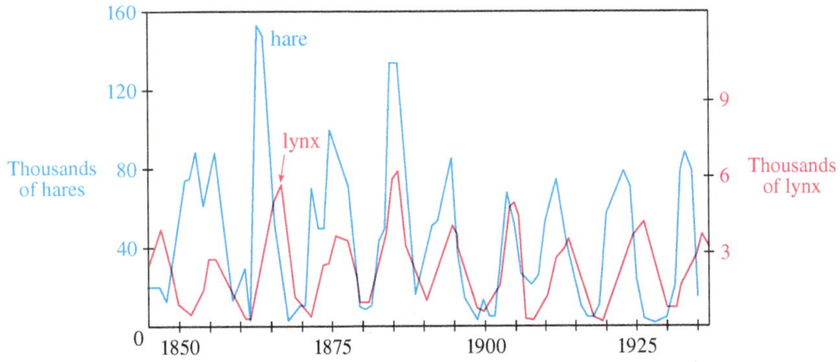

FIGURE 6
Relative abundance of hare and lynx from Hudson's Bay Company records

Although the relatively simple Lotka-Volterra model has had some success in explaining and predicting coupled populations, more sophisticated models have also been proposed. One way to modify the Lotka-Volterra equations is to assume that, in the absence of predators, the prey grow according to a logistic model with carrying capacity M. Then the Lotka-Volterra equations (1) are replaced by the system of differential equations

$$\frac{dR}{dt} = kR\left(1 - \frac{R}{M}\right) - aRW \qquad \frac{dW}{dt} = -rW + bRW$$

This model is investigated in Exercises 11 and 12.

Models have also been proposed to describe and predict population levels of two or more species that compete for the same resources or cooperate for mutual benefit. Such models are explored in Exercises 2–4.

9.6 EXERCISES

1. For each predator-prey system, determine which of the variables, x or y, represents the prey population and which represents the predator population. Is the growth of the prey restricted just by the predators or by other factors as well? Do the predators feed only on the prey or do they have additional food sources? Explain.

   (a) $\dfrac{dx}{dt} = -0.05x + 0.0001xy$

   $\dfrac{dy}{dt} = 0.1y - 0.005xy$

   (b) $\dfrac{dx}{dt} = 0.2x - 0.0002x^2 - 0.006xy$

   $\dfrac{dy}{dt} = -0.015y + 0.00008xy$

2. Each system of differential equations is a model for two species that either compete for the same resources or cooperate for mutual benefit (flowering plants and insect pollinators, for instance). Decide whether each system describes competition or cooperation and explain why it is a reasonable

model. (Ask yourself what effect an increase in one species has on the growth rate of the other.)

(a) $\dfrac{dx}{dt} = 0.12x - 0.0006x^2 + 0.00001xy$

$\dfrac{dy}{dt} = 0.08x + 0.00004xy$

(b) $\dfrac{dx}{dt} = 0.15x - 0.0002x^2 - 0.0006xy$

$\dfrac{dy}{dt} = 0.2y - 0.00008y^2 - 0.0002xy$

3. The system of differential equations

$$\dfrac{dx}{dt} = 0.5x - 0.004x^2 - 0.001xy$$

$$\dfrac{dy}{dt} = 0.4y - 0.001y^2 - 0.002xy$$

is a model for the populations of two species.
(a) Does the model describe cooperation, or competition, or a predator-prey relationship?
(b) Find the equilibrium solutions and explain their significance.

4. Lynx eat snowshoe hares and snowshoe hares eat woody plants like willows. Suppose that, in the absence of hares, the willow population will grow exponentially and the lynx population will decay exponentially. In the absence of lynx and willow, the hare population will decay exponentially. If $L(t)$, $H(t)$, and $W(t)$ represent the populations of these three species at time t, write a system of differential equations as a model for their dynamics. If the constants in your equation are all positive, explain why you have used plus or minus signs.

5–6 A phase trajectory is shown for populations of rabbits (R) and foxes (F).
(a) Describe how each population changes as time goes by.
(b) Use your description to make a rough sketch of the graphs of R and F as functions of time.

5.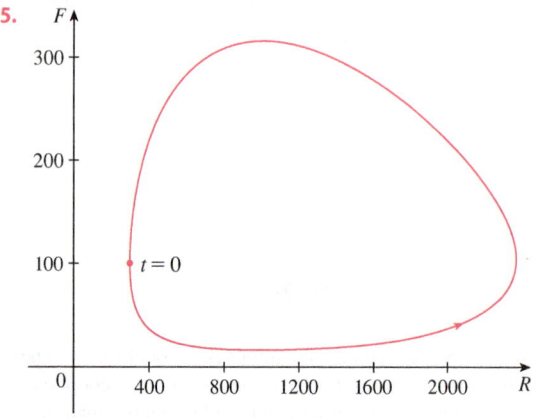

6.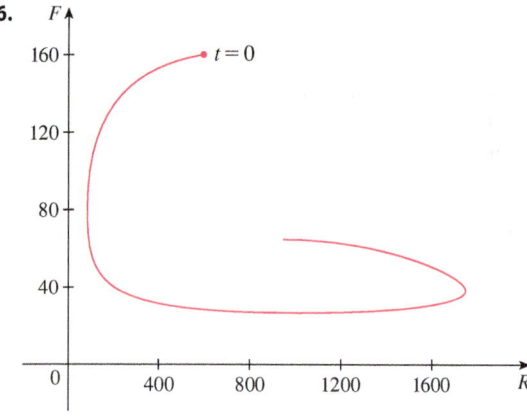

7–8 Graphs of populations of two species are shown. Use them to sketch the corresponding phase trajectory.

7.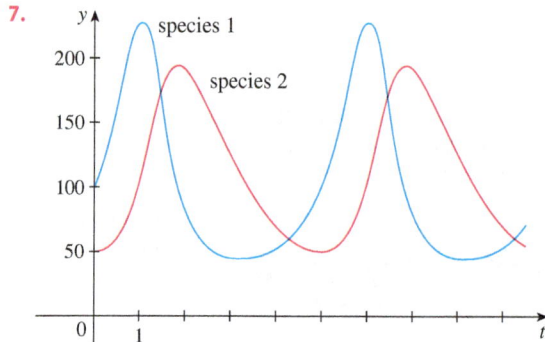

8.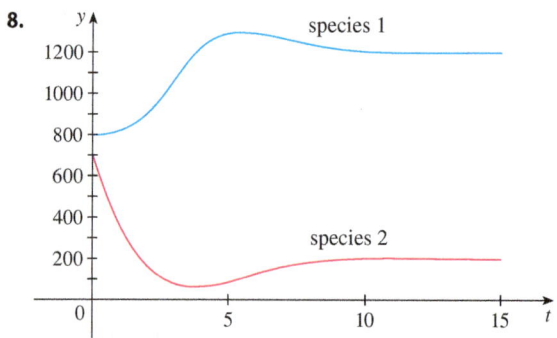

9. In Example 1(b) we showed that the rabbit and wolf populations satisfy the differential equation

$$\dfrac{dW}{dR} = \dfrac{-0.02W + 0.00002RW}{0.08R - 0.001RW}$$

By solving this separable differential equation, show that

$$\frac{R^{0.02}W^{0.08}}{e^{0.00002R}e^{0.001W}} = C$$

where C is a constant.

It is impossible to solve this equation for W as an explicit function of R (or vice versa). If you have a computer algebra system that graphs implicitly defined curves, use this equation and your CAS to draw the solution curve that passes through the point (1000, 40) and compare with Figure 3.

10. Populations of aphids and ladybugs are modeled by the equations

$$\frac{dA}{dt} = 2A - 0.01AL$$

$$\frac{dL}{dt} = -0.5L + 0.0001AL$$

(a) Find the equilibrium solutions and explain their significance.
(b) Find an expression for dL/dA.
(c) The direction field for the differential equation in part (b) is shown. Use it to sketch a phase portrait. What do the phase trajectories have in common?

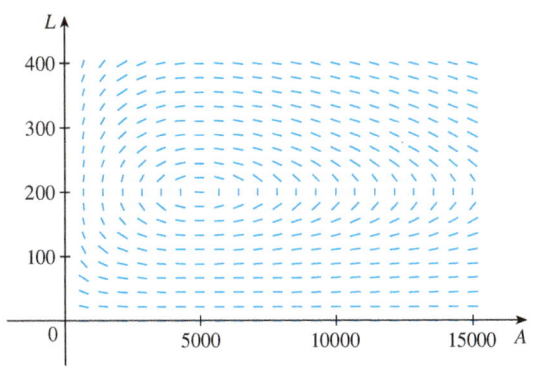

(d) Suppose that at time $t = 0$ there are 1000 aphids and 200 ladybugs. Draw the corresponding phase trajectory and use it to describe how both populations change.
(e) Use part (d) to make rough sketches of the aphid and ladybug populations as functions of t. How are the graphs related to each other?

11. In Example 1 we used Lotka-Volterra equations to model populations of rabbits and wolves. Let's modify those equations as follows:

$$\frac{dR}{dt} = 0.08R(1 - 0.0002R) - 0.001RW$$

$$\frac{dW}{dt} = -0.02W + 0.00002RW$$

(a) According to these equations, what happens to the rabbit population in the absence of wolves?
(b) Find all the equilibrium solutions and explain their significance.
(c) The figure shows the phase trajectory that starts at the point (1000, 40). Describe what eventually happens to the rabbit and wolf populations.

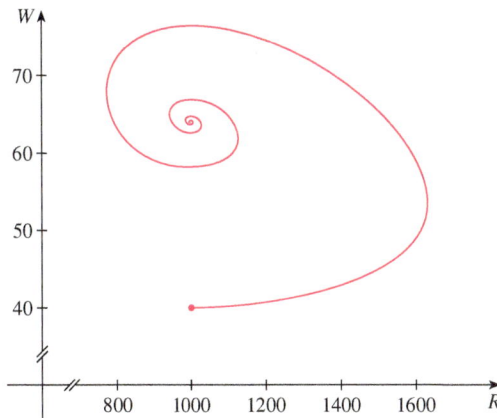

(d) Sketch graphs of the rabbit and wolf populations as functions of time.

CAS 12. In Exercise 10 we modeled populations of aphids and ladybugs with a Lotka-Volterra system. Suppose we modify those equations as follows:

$$\frac{dA}{dt} = 2A(1 - 0.0001A) - 0.01AL$$

$$\frac{dL}{dt} = -0.5L + 0.0001AL$$

(a) In the absence of ladybugs, what does the model predict about the aphids?
(b) Find the equilibrium solutions.
(c) Find an expression for dL/dA.
(d) Use a computer algebra system to draw a direction field for the differential equation in part (c). Then use the direction field to sketch a phase portrait. What do the phase trajectories have in common?
(e) Suppose that at time $t = 0$ there are 1000 aphids and 200 ladybugs. Draw the corresponding phase trajectory and use it to describe how both populations change.
(f) Use part (e) to make rough sketches of the aphid and ladybug populations as functions of t. How are the graphs related to each other?

9 REVIEW

CONCEPT CHECK

Answers to the Concept Check can be found on the back endpapers.

1. (a) What is a differential equation?
   (b) What is the order of a differential equation?
   (c) What is an initial condition?

2. What can you say about the solutions of the equation $y' = x^2 + y^2$ just by looking at the differential equation?

3. What is a direction field for the differential equation $y' = F(x, y)$?

4. Explain how Euler's method works.

5. What is a separable differential equation? How do you solve it?

6. What is a first-order linear differential equation? How do you solve it?

7. (a) Write a differential equation that expresses the law of natural growth. What does it say in terms of relative growth rate?
   (b) Under what circumstances is this an appropriate model for population growth?
   (c) What are the solutions of this equation?

8. (a) Write the logistic differential equation.
   (b) Under what circumstances is this an appropriate model for population growth?

9. (a) Write Lotka-Volterra equations to model populations of food-fish (F) and sharks (S).
   (b) What do these equations say about each population in the absence of the other?

TRUE-FALSE QUIZ

Determine whether the statement is true or false. If it is true, explain why. If it is false, explain why or give an example that disproves the statement.

1. All solutions of the differential equation $y' = -1 - y^4$ are decreasing functions.

2. The function $f(x) = (\ln x)/x$ is a solution of the differential equation $x^2 y' + xy = 1$.

3. The equation $y' = x + y$ is separable.

4. The equation $y' = 3y - 2x + 6xy - 1$ is separable.

5. The equation $e^x y' = y$ is linear.

6. The equation $y' + xy = e^y$ is linear.

7. If y is the solution of the initial-value problem
$$\frac{dy}{dt} = 2y\left(1 - \frac{y}{5}\right) \qquad y(0) = 1$$
then $\lim_{t \to \infty} y = 5$.

EXERCISES

1. (a) A direction field for the differential equation $y' = y(y - 2)(y - 4)$ is shown. Sketch the graphs of the solutions that satisfy the given initial conditions.
   (i) $y(0) = -0.3$ (ii) $y(0) = 1$
   (iii) $y(0) = 3$ (iv) $y(0) = 4.3$
   (b) If the initial condition is $y(0) = c$, for what values of c is $\lim_{t \to \infty} y(t)$ finite? What are the equilibrium solutions?

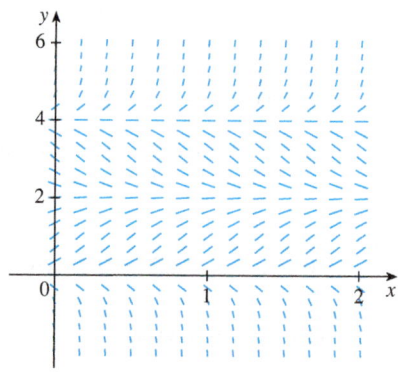

2. (a) Sketch a direction field for the differential equation $y' = x/y$. Then use it to sketch the four solutions that satisfy the initial conditions $y(0) = 1$, $y(0) = -1$, $y(2) = 1$, and $y(-2) = 1$.
(b) Check your work in part (a) by solving the differential equation explicitly. What type of curve is each solution curve?

3. (a) A direction field for the differential equation $y' = x^2 - y^2$ is shown. Sketch the solution of the initial-value problem
$$y' = x^2 - y^2 \qquad y(0) = 1$$
Use your graph to estimate the value of $y(0.3)$.

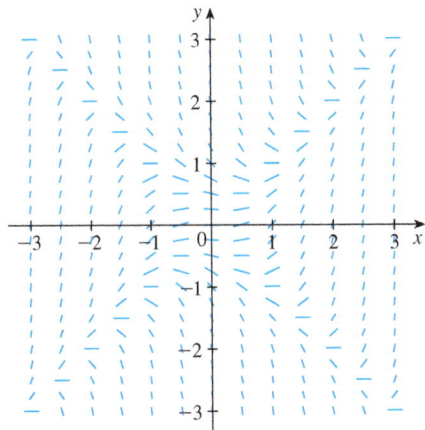

(b) Use Euler's method with step size 0.1 to estimate $y(0.3)$, where $y(x)$ is the solution of the initial-value problem in part (a). Compare with your estimate from part (a).
(c) On what lines are the centers of the horizontal line segments of the direction field in part (a) located? What happens when a solution curve crosses these lines?

4. (a) Use Euler's method with step size 0.2 to estimate $y(0.4)$, where $y(x)$ is the solution of the initial-value problem
$$y' = 2xy^2 \qquad y(0) = 1$$
(b) Repeat part (a) with step size 0.1.
(c) Find the exact solution of the differential equation and compare the value at 0.4 with the approximations in parts (a) and (b).

5–8 Solve the differential equation.

5. $y' = xe^{-\sin x} - y \cos x$

6. $\dfrac{dx}{dt} = 1 - t + x - tx$

7. $2ye^{y^2} y' = 2x + 3\sqrt{x}$

8. $x^2 y' - y = 2x^3 e^{-1/x}$

9–11 Solve the initial-value problem.

9. $\dfrac{dr}{dt} + 2tr = r$, $\quad r(0) = 5$

10. $(1 + \cos x) y' = (1 + e^{-y}) \sin x$, $\quad y(0) = 0$

11. $xy' - y = x \ln x$, $\quad y(1) = 2$

12. Solve the initial-value problem $y' = 3x^2 e^y$, $y(0) = 1$, and graph the solution.

13–14 Find the orthogonal trajectories of the family of curves.

13. $y = ke^x$

14. $y = e^{kx}$

15. (a) Write the solution of the initial-value problem
$$\dfrac{dP}{dt} = 0.1P\left(1 - \dfrac{P}{2000}\right) \qquad P(0) = 100$$
and use it to find the population when $t = 20$.
(b) When does the population reach 1200?

16. (a) The population of the world was 6.1 billion in 2000 and 6.9 billion in 2010. Find an exponential model for these data and use the model to predict the world population in the year 2020.
(b) According to the model in part (a), when will the world population exceed 10 billion?
(c) Use the data in part (a) to find a logistic model for the population. Assume a carrying capacity of 20 billion. Then use the logistic model to predict the population in 2020. Compare with your prediction from the exponential model.
(d) According to the logistic model, when will the world population exceed 10 billion? Compare with your prediction in part (b).

17. The von Bertalanffy growth model is used to predict the length $L(t)$ of a fish over a period of time. If L_∞ is the largest length for a species, then the hypothesis is that the rate of growth in length is proportional to $L_\infty - L$, the length yet to be achieved.
(a) Formulate and solve a differential equation to find an expression for $L(t)$.
(b) For the North Sea haddock it has been determined that $L_\infty = 53$ cm, $L(0) = 10$ cm, and the constant of proportionality is 0.2. What does the expression for $L(t)$ become with these data?

18. A tank contains 100 L of pure water. Brine that contains 0.1 kg of salt per liter enters the tank at a rate of 10 L/min. The solution is kept thoroughly mixed and drains from the tank at the same rate. How much salt is in the tank after 6 minutes?

19. One model for the spread of an epidemic is that the rate of spread is jointly proportional to the number of infected people and the number of uninfected people. In an isolated town of 5000 inhabitants, 160 people have a disease at the beginning of the week and 1200 have it at the end of the week. How long does it take for 80% of the population to become infected?

20. The Brentano-Stevens Law in psychology models the way that a subject reacts to a stimulus. It states that if R represents the reaction to an amount S of stimulus, then the relative rates of increase are proportional:

$$\frac{1}{R}\frac{dR}{dt} = \frac{k}{S}\frac{dS}{dt}$$

where k is a positive constant. Find R as a function of S.

21. The transport of a substance across a capillary wall in lung physiology has been modeled by the differential equation

$$\frac{dh}{dt} = -\frac{R}{V}\left(\frac{h}{k+h}\right)$$

where h is the hormone concentration in the bloodstream, t is time, R is the maximum transport rate, V is the volume of the capillary, and k is a positive constant that measures the affinity between the hormones and the enzymes that assist the process. Solve this differential equation to find a relationship between h and t.

22. Populations of birds and insects are modeled by the equations

$$\frac{dx}{dt} = 0.4x - 0.002xy$$

$$\frac{dy}{dt} = -0.2y + 0.000008xy$$

(a) Which of the variables, x or y, represents the bird population and which represents the insect population? Explain.
(b) Find the equilibrium solutions and explain their significance.
(c) Find an expression for dy/dx.
(d) The direction field for the differential equation in part (c) is shown. Use it to sketch the phase trajectory correspond-

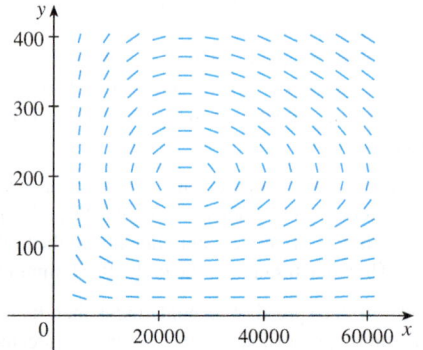

ing to initial populations of 100 birds and 40,000 insects. Then use the phase trajectory to describe how both populations change.
(e) Use part (d) to make rough sketches of the bird and insect populations as functions of time. How are these graphs related to each other?

23. Suppose the model of Exercise 22 is replaced by the equations

$$\frac{dx}{dt} = 0.4x(1 - 0.000005x) - 0.002xy$$

$$\frac{dy}{dt} = -0.2y + 0.000008xy$$

(a) According to these equations, what happens to the insect population in the absence of birds?
(b) Find the equilibrium solutions and explain their significance.
(c) The figure shows the phase trajectory that starts with 100 birds and 40,000 insects. Describe what eventually happens to the bird and insect populations.

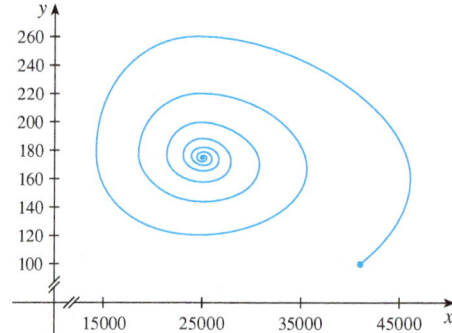

(d) Sketch graphs of the bird and insect populations as functions of time.

24. Barbara weighs 60 kg and is on a diet of 1600 calories per day, of which 850 are used automatically by basal metabolism. She spends about 15 cal/kg/day times her weight doing exercise. If 1 kg of fat contains 10,000 cal and we assume that the storage of calories in the form of fat is 100% efficient, formulate a differential equation and solve it to find her weight as a function of time. Does her weight ultimately approach an equilibrium weight?

Problems Plus

1. Find all functions f such that f' is continuous and
$$[f(x)]^2 = 100 + \int_0^x \{[f(t)]^2 + [f'(t)]^2\}\, dt \qquad \text{for all real } x$$

2. A student forgot the Product Rule for differentiation and made the mistake of thinking that $(fg)' = f'g'$. However, he was lucky and got the correct answer. The function f that he used was $f(x) = e^{x^2}$ and the domain of his problem was the interval $(\tfrac{1}{2}, \infty)$. What was the function g?

3. Let f be a function with the property that $f(0) = 1$, $f'(0) = 1$, and $f(a + b) = f(a)f(b)$ for all real numbers a and b. Show that $f'(x) = f(x)$ for all x and deduce that $f(x) = e^x$.

4. Find all functions f that satisfy the equation
$$\left(\int f(x)\, dx\right)\left(\int \frac{1}{f(x)}\, dx\right) = -1$$

5. Find the curve $y = f(x)$ such that $f(x) \geq 0$, $f(0) = 0$, $f(1) = 1$, and the area under the graph of f from 0 to x is proportional to the $(n + 1)$st power of $f(x)$.

6. A *subtangent* is a portion of the x-axis that lies directly beneath the segment of a tangent line from the point of contact to the x-axis. Find the curves that pass through the point $(c, 1)$ and whose subtangents all have length c.

7. A peach pie is taken out of the oven at 5:00 PM. At that time it is piping hot, 100°C. At 5:10 PM its temperature is 80°C; at 5:20 PM it is 65°C. What is the temperature of the room?

8. Snow began to fall during the morning of February 2 and continued steadily into the afternoon. At noon a snowplow began removing snow from a road at a constant rate. The plow traveled 6 km from noon to 1 PM but only 3 km from 1 PM to 2 PM. When did the snow begin to fall? [*Hints:* To get started, let t be the time measured in hours after noon; let $x(t)$ be the distance traveled by the plow at time t; then the speed of the plow is dx/dt. Let b be the number of hours before noon that it began to snow. Find an expression for the height of the snow at time t. Then use the given information that the rate of removal R (in m³/h) is constant.]

9. A dog sees a rabbit running in a straight line across an open field and gives chase. In a rectangular coordinate system (as shown in the figure), assume:

   (i) The rabbit is at the origin and the dog is at the point $(L, 0)$ at the instant the dog first sees the rabbit.
   (ii) The rabbit runs up the y-axis and the dog always runs straight for the rabbit.
   (iii) The dog runs at the same speed as the rabbit.

   (a) Show that the dog's path is the graph of the function $y = f(x)$, where y satisfies the differential equation
   $$x \frac{d^2 y}{dx^2} = \sqrt{1 + \left(\frac{dy}{dx}\right)^2}$$

   (b) Determine the solution of the equation in part (a) that satisfies the initial conditions $y = y' = 0$ when $x = L$. [*Hint:* Let $z = dy/dx$ in the differential equation and solve the resulting first-order equation to find z; then integrate z to find y.]
   (c) Does the dog ever catch the rabbit?

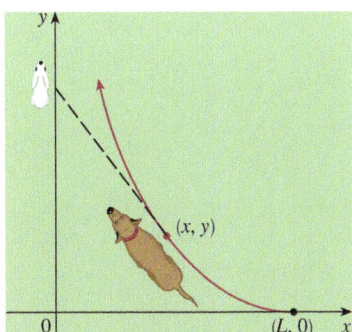

FIGURE FOR PROBLEM 9

10. (a) Suppose that the dog in Problem 9 runs twice as fast as the rabbit. Find a differential equation for the path of the dog. Then solve it to find the point where the dog catches the rabbit.
(b) Suppose the dog runs half as fast as the rabbit. How close does the dog get to the rabbit? What are their positions when they are closest?

11. A planning engineer for a new alum plant must present some estimates to his company regarding the capacity of a silo designed to contain bauxite ore until it is processed into alum. The ore resembles pink talcum powder and is poured from a conveyor at the top of the silo. The silo is a cylinder 100 ft high with a radius of 200 ft. The conveyor carries ore at a rate of $60{,}000\pi$ ft^3/h and the ore maintains a conical shape whose radius is 1.5 times its height.
(a) If, at a certain time t, the pile is 60 ft high, how long will it take for the pile to reach the top of the silo?
(b) Management wants to know how much room will be left in the floor area of the silo when the pile is 60 ft high. How fast is the floor area of the pile growing at that height?
(c) Suppose a loader starts removing the ore at the rate of $20{,}000\pi$ ft^3/h when the height of the pile reaches 90 ft. Suppose, also, that the pile continues to maintain its shape. How long will it take for the pile to reach the top of the silo under these conditions?

12. Find the curve that passes through the point (3, 2) and has the property that if the tangent line is drawn at any point P on the curve, then the part of the tangent line that lies in the first quadrant is bisected at P.

13. Recall that the normal line to a curve at a point P on the curve is the line that passes through P and is perpendicular to the tangent line at P. Find the curve that passes through the point (3, 2) and has the property that if the normal line is drawn at any point on the curve, then the y-intercept of the normal line is always 6.

14. Find all curves with the property that if the normal line is drawn at any point P on the curve, then the part of the normal line between P and the x-axis is bisected by the y-axis.

15. Find all curves with the property that if a line is drawn from the origin to any point (x, y) on the curve, and then a tangent is drawn to the curve at that point and extended to meet the x-axis, the result is an isosceles triangle with equal sides meeting at (x, y).

11 Infinite Sequences and Series

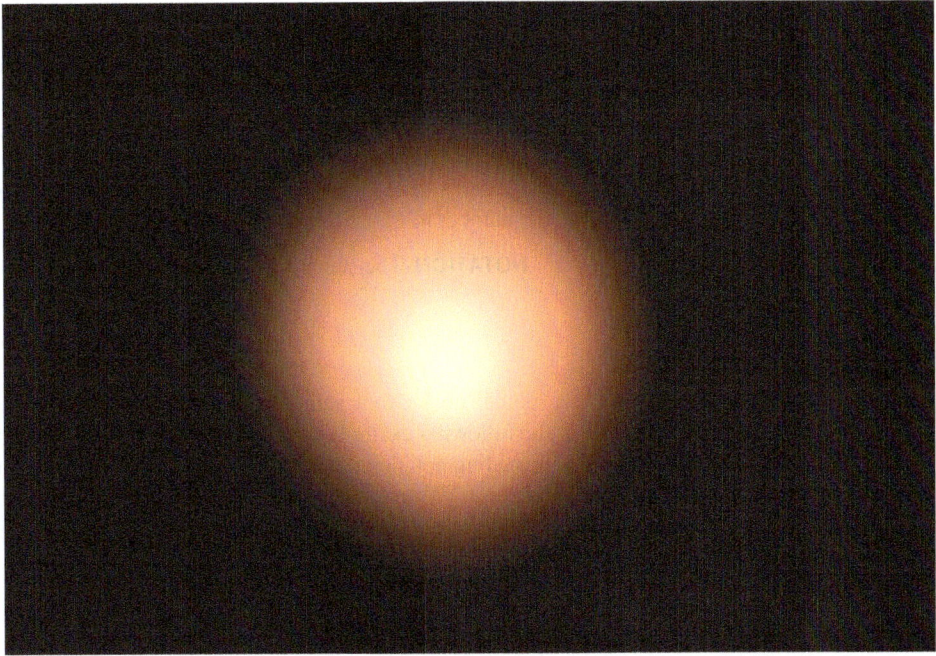

Betelgeuse is a red supergiant star, one of the largest and brightest of the observable stars. In the project on page 783 you are asked to compare the radiation emitted by Betelgeuse with that of other stars.

STScI / NASA / ESA / Galaxy / Galaxy Picture Library / Alamy

INFINITE SEQUENCES AND SERIES WERE introduced briefly in *A Preview of Calculus* in connection with Zeno's paradoxes and the decimal representation of numbers. Their importance in calculus stems from Newton's idea of representing functions as sums of infinite series. For instance, in finding areas he often integrated a function by first expressing it as a series and then integrating each term of the series. We will pursue his idea in Section 11.10 in order to integrate such functions as e^{-x^2}. (Recall that we have previously been unable to do this.) Many of the functions that arise in mathematical physics and chemistry, such as Bessel functions, are defined as sums of series, so it is important to be familiar with the basic concepts of convergence of infinite sequences and series.

Physicists also use series in another way, as we will see in Section 11.11. In studying fields as diverse as optics, special relativity, and electromagnetism, they analyze phenomena by replacing a function with the first few terms in the series that represents it.

11.1 Sequences

A **sequence** can be thought of as a list of numbers written in a definite order:

$$a_1, a_2, a_3, a_4, \ldots, a_n, \ldots$$

The number a_1 is called the *first term*, a_2 is the *second term*, and in general a_n is the *nth term*. We will deal exclusively with infinite sequences and so each term a_n will have a successor a_{n+1}.

Notice that for every positive integer n there is a corresponding number a_n and so a sequence can be defined as a function whose domain is the set of positive integers. But we usually write a_n instead of the function notation $f(n)$ for the value of the function at the number n.

NOTATION The sequence $\{a_1, a_2, a_3, \ldots\}$ is also denoted by

$$\{a_n\} \quad \text{or} \quad \{a_n\}_{n=1}^{\infty}$$

EXAMPLE 1 Some sequences can be defined by giving a formula for the nth term. In the following examples we give three descriptions of the sequence: one by using the preceding notation, another by using the defining formula, and a third by writing out the terms of the sequence. Notice that n doesn't have to start at 1.

(a) $\left\{\dfrac{n}{n+1}\right\}_{n=1}^{\infty}$ $\quad a_n = \dfrac{n}{n+1} \quad$ $\left\{\dfrac{1}{2}, \dfrac{2}{3}, \dfrac{3}{4}, \dfrac{4}{5}, \ldots, \dfrac{n}{n+1}, \ldots\right\}$

(b) $\left\{\dfrac{(-1)^n(n+1)}{3^n}\right\}$ $\quad a_n = \dfrac{(-1)^n(n+1)}{3^n} \quad$ $\left\{-\dfrac{2}{3}, \dfrac{3}{9}, -\dfrac{4}{27}, \dfrac{5}{81}, \ldots, \dfrac{(-1)^n(n+1)}{3^n}, \ldots\right\}$

(c) $\{\sqrt{n-3}\}_{n=3}^{\infty}$ $\quad a_n = \sqrt{n-3}, \; n \geq 3 \quad$ $\{0, 1, \sqrt{2}, \sqrt{3}, \ldots, \sqrt{n-3}, \ldots\}$

(d) $\left\{\cos \dfrac{n\pi}{6}\right\}_{n=0}^{\infty}$ $\quad a_n = \cos \dfrac{n\pi}{6}, \; n \geq 0 \quad$ $\left\{1, \dfrac{\sqrt{3}}{2}, \dfrac{1}{2}, 0, \ldots, \cos \dfrac{n\pi}{6}, \ldots\right\}$ ∎

EXAMPLE 2 Find a formula for the general term a_n of the sequence

$$\left\{\dfrac{3}{5}, -\dfrac{4}{25}, \dfrac{5}{125}, -\dfrac{6}{625}, \dfrac{7}{3125}, \ldots\right\}$$

assuming that the pattern of the first few terms continues.

SOLUTION We are given that

$$a_1 = \dfrac{3}{5} \quad a_2 = -\dfrac{4}{25} \quad a_3 = \dfrac{5}{125} \quad a_4 = -\dfrac{6}{625} \quad a_5 = \dfrac{7}{3125}$$

Notice that the numerators of these fractions start with 3 and increase by 1 whenever we go to the next term. The second term has numerator 4, the third term has numerator 5; in general, the nth term will have numerator $n + 2$. The denominators are the powers of 5, so a_n has denominator 5^n. The signs of the terms are alternately positive and negative, so we need to multiply by a power of -1. In Example 1(b) the factor $(-1)^n$ meant we started with a negative term. Here we want to start with a positive term and so we use $(-1)^{n-1}$ or $(-1)^{n+1}$. Therefore

$$a_n = (-1)^{n-1} \frac{n+2}{5^n}$$

EXAMPLE 3 Here are some sequences that don't have a simple defining equation.
(a) The sequence $\{p_n\}$, where p_n is the population of the world as of January 1 in the year n.
(b) If we let a_n be the digit in the nth decimal place of the number e, then $\{a_n\}$ is a well-defined sequence whose first few terms are

$$\{7, 1, 8, 2, 8, 1, 8, 2, 8, 4, 5, \ldots\}$$

(c) **The Fibonacci sequence** $\{f_n\}$ is defined recursively by the conditions

$$f_1 = 1 \qquad f_2 = 1 \qquad f_n = f_{n-1} + f_{n-2} \qquad n \geq 3$$

Each term is the sum of the two preceding terms. The first few terms are

$$\{1, 1, 2, 3, 5, 8, 13, 21, \ldots\}$$

This sequence arose when the 13th-century Italian mathematician known as Fibonacci solved a problem concerning the breeding of rabbits (see Exercise 83).

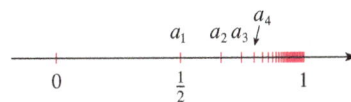

FIGURE 1

A sequence such as the one in Example 1(a), $a_n = n/(n+1)$, can be pictured either by plotting its terms on a number line, as in Figure 1, or by plotting its graph, as in Figure 2. Note that, since a sequence is a function whose domain is the set of positive integers, its graph consists of isolated points with coordinates

$$(1, a_1) \qquad (2, a_2) \qquad (3, a_3) \qquad \ldots \qquad (n, a_n) \qquad \ldots$$

From Figure 1 or Figure 2 it appears that the terms of the sequence $a_n = n/(n+1)$ are approaching 1 as n becomes large. In fact, the difference

$$1 - \frac{n}{n+1} = \frac{1}{n+1}$$

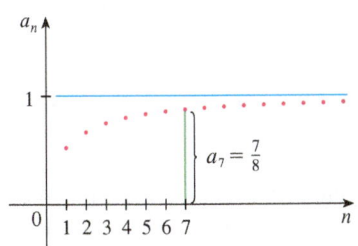

FIGURE 2

can be made as small as we like by taking n sufficiently large. We indicate this by writing

$$\lim_{n \to \infty} \frac{n}{n+1} = 1$$

In general, the notation

$$\lim_{n \to \infty} a_n = L$$

means that the terms of the sequence $\{a_n\}$ approach L as n becomes large. Notice that the following definition of the limit of a sequence is very similar to the definition of a limit of a function at infinity given in Section 2.6.

696 CHAPTER 11 Infinite Sequences and Series

> **1 Definition** A sequence $\{a_n\}$ has the **limit** L and we write
>
> $$\lim_{n \to \infty} a_n = L \quad \text{or} \quad a_n \to L \text{ as } n \to \infty$$
>
> if we can make the terms a_n as close to L as we like by taking n sufficiently large. If $\lim_{n \to \infty} a_n$ exists, we say the sequence **converges** (or is **convergent**). Otherwise, we say the sequence **diverges** (or is **divergent**).

Figure 3 illustrates Definition 1 by showing the graphs of two sequences that have the limit L.

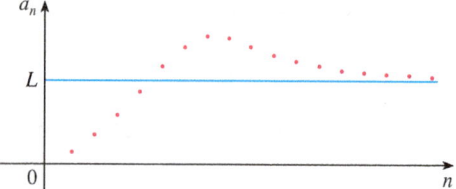 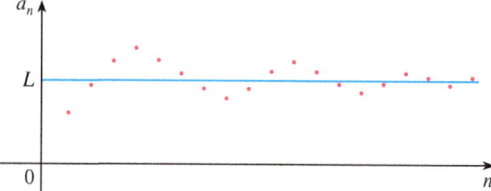

FIGURE 3
Graphs of two sequences with $\lim_{n \to \infty} a_n = L$

A more precise version of Definition 1 is as follows.

Compare this definition with Definition 2.6.7.

> **2 Definition** A sequence $\{a_n\}$ has the **limit** L and we write
>
> $$\lim_{n \to \infty} a_n = L \quad \text{or} \quad a_n \to L \text{ as } n \to \infty$$
>
> if for every $\varepsilon > 0$ there is a corresponding integer N such that
>
> $$\text{if} \quad n > N \quad \text{then} \quad |a_n - L| < \varepsilon$$

Definition 2 is illustrated by Figure 4, in which the terms $a_1, a_2, a_3, \ldots$ are plotted on a number line. No matter how small an interval $(L - \varepsilon, L + \varepsilon)$ is chosen, there exists an N such that all terms of the sequence from a_{N+1} onward must lie in that interval.

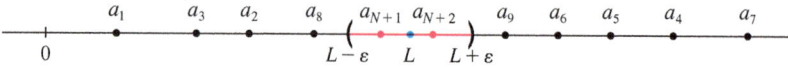

FIGURE 4

Another illustration of Definition 2 is given in Figure 5. The points on the graph of $\{a_n\}$ must lie between the horizontal lines $y = L + \varepsilon$ and $y = L - \varepsilon$ if $n > N$. This picture must be valid no matter how small ε is chosen, but usually a smaller ε requires a larger N.

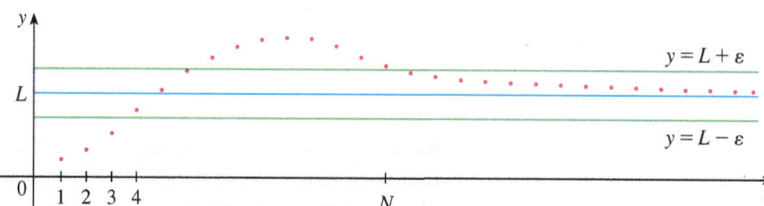

FIGURE 5

If you compare Definition 2 with Definition 2.6.7, you will see that the only difference between $\lim_{n \to \infty} a_n = L$ and $\lim_{x \to \infty} f(x) = L$ is that n is required to be an integer. Thus we have the following theorem, which is illustrated by Figure 6.

3 Theorem If $\lim_{x \to \infty} f(x) = L$ and $f(n) = a_n$ when n is an integer, then $\lim_{n \to \infty} a_n = L$.

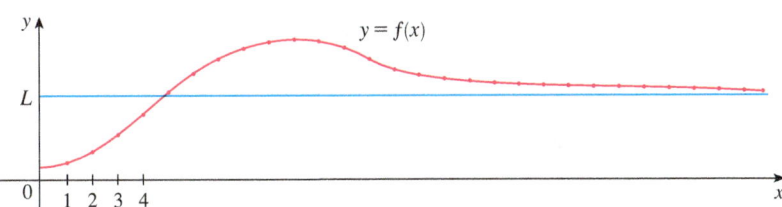

FIGURE 6

In particular, since we know that $\lim_{x \to \infty} (1/x^r) = 0$ when $r > 0$ (Theorem 2.6.5), we have

$$\boxed{4} \quad \lim_{n \to \infty} \frac{1}{n^r} = 0 \quad \text{if } r > 0$$

If a_n becomes large as n becomes large, we use the notation $\lim_{n \to \infty} a_n = \infty$. The following precise definition is similar to Definition 2.6.9.

5 Definition $\lim_{n \to \infty} a_n = \infty$ means that for every positive number M there is an integer N such that

$$\text{if} \quad n > N \quad \text{then} \quad a_n > M$$

If $\lim_{n \to \infty} a_n = \infty$, then the sequence $\{a_n\}$ is divergent but in a special way. We say that $\{a_n\}$ diverges to ∞.

The Limit Laws given in Section 2.3 also hold for the limits of sequences and their proofs are similar.

Limit Laws for Sequences

If $\{a_n\}$ and $\{b_n\}$ are convergent sequences and c is a constant, then

$$\lim_{n \to \infty} (a_n + b_n) = \lim_{n \to \infty} a_n + \lim_{n \to \infty} b_n$$

$$\lim_{n \to \infty} (a_n - b_n) = \lim_{n \to \infty} a_n - \lim_{n \to \infty} b_n$$

$$\lim_{n \to \infty} c a_n = c \lim_{n \to \infty} a_n \qquad \lim_{n \to \infty} c = c$$

$$\lim_{n \to \infty} (a_n b_n) = \lim_{n \to \infty} a_n \cdot \lim_{n \to \infty} b_n$$

$$\lim_{n \to \infty} \frac{a_n}{b_n} = \frac{\lim_{n \to \infty} a_n}{\lim_{n \to \infty} b_n} \quad \text{if } \lim_{n \to \infty} b_n \neq 0$$

$$\lim_{n \to \infty} a_n^p = \left[\lim_{n \to \infty} a_n \right]^p \quad \text{if } p > 0 \text{ and } a_n > 0$$

The Squeeze Theorem can also be adapted for sequences as follows (see Figure 7).

Squeeze Theorem for Sequences

If $a_n \leq b_n \leq c_n$ for $n \geq n_0$ and $\lim_{n \to \infty} a_n = \lim_{n \to \infty} c_n = L$, then $\lim_{n \to \infty} b_n = L$.

Another useful fact about limits of sequences is given by the following theorem, whose proof is left as Exercise 87.

6 Theorem If $\lim_{n \to \infty} |a_n| = 0$, then $\lim_{n \to \infty} a_n = 0$.

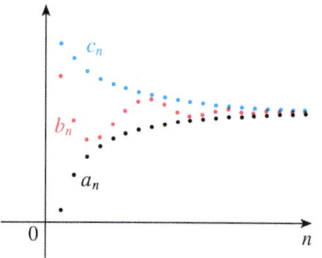

FIGURE 7
The sequence $\{b_n\}$ is squeezed between the sequences $\{a_n\}$ and $\{c_n\}$.

EXAMPLE 4 Find $\lim_{n \to \infty} \dfrac{n}{n+1}$.

SOLUTION The method is similar to the one we used in Section 2.6: Divide numerator and denominator by the highest power of n that occurs in the denominator and then use the Limit Laws.

$$\lim_{n \to \infty} \frac{n}{n+1} = \lim_{n \to \infty} \frac{1}{1 + \dfrac{1}{n}} = \frac{\lim_{n \to \infty} 1}{\lim_{n \to \infty} 1 + \lim_{n \to \infty} \dfrac{1}{n}}$$

$$= \frac{1}{1+0} = 1$$

This shows that the guess we made earlier from Figures 1 and 2 was correct.

Here we used Equation 4 with $r = 1$.

EXAMPLE 5 Is the sequence $a_n = \dfrac{n}{\sqrt{10+n}}$ convergent or divergent?

SOLUTION As in Example 4, we divide numerator and denominator by n:

$$\lim_{n \to \infty} \frac{n}{\sqrt{10+n}} = \lim_{n \to \infty} \frac{1}{\sqrt{\dfrac{10}{n^2} + \dfrac{1}{n}}} = \infty$$

because the numerator is constant and the denominator approaches 0. So $\{a_n\}$ is divergent.

EXAMPLE 6 Calculate $\lim_{n \to \infty} \dfrac{\ln n}{n}$.

SOLUTION Notice that both numerator and denominator approach infinity as $n \to \infty$. We can't apply l'Hospital's Rule directly because it applies not to sequences but to functions of a real variable. However, we can apply l'Hospital's Rule to the related function $f(x) = (\ln x)/x$ and obtain

$$\lim_{x \to \infty} \frac{\ln x}{x} = \lim_{x \to \infty} \frac{1/x}{1} = 0$$

Therefore, by Theorem 3, we have

$$\lim_{n \to \infty} \frac{\ln n}{n} = 0$$

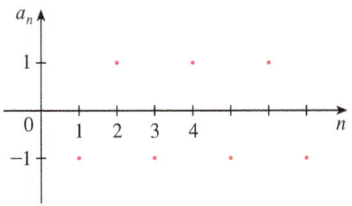

FIGURE 8

The graph of the sequence in Example 8 is shown in Figure 9 and supports our answer.

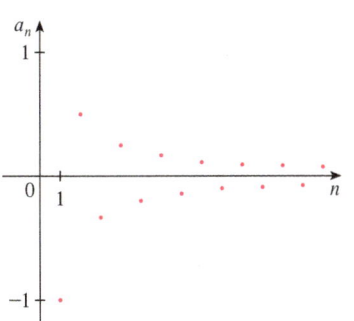

FIGURE 9

Creating Graphs of Sequences
Some computer algebra systems have special commands that enable us to create sequences and graph them directly. With most graphing calculators, however, sequences can be graphed by using parametric equations. For instance, the sequence in Example 10 can be graphed by entering the parametric equations

$$x = t \qquad y = t!/t^t$$

and graphing in dot mode, starting with $t = 1$ and setting the t-step equal to 1. The result is shown in Figure 10.

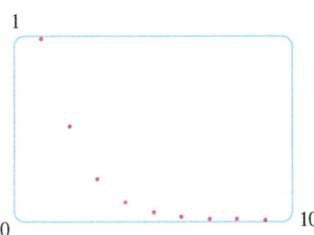

FIGURE 10

EXAMPLE 7 Determine whether the sequence $a_n = (-1)^n$ is convergent or divergent.

SOLUTION If we write out the terms of the sequence, we obtain

$$\{-1, 1, -1, 1, -1, 1, -1, \ldots\}$$

The graph of this sequence is shown in Figure 8. Since the terms oscillate between 1 and -1 infinitely often, a_n does not approach any number. Thus $\lim_{n \to \infty} (-1)^n$ does not exist; that is, the sequence $\{(-1)^n\}$ is divergent.

EXAMPLE 8 Evaluate $\lim_{n \to \infty} \dfrac{(-1)^n}{n}$ if it exists.

SOLUTION We first calculate the limit of the absolute value:

$$\lim_{n \to \infty} \left| \frac{(-1)^n}{n} \right| = \lim_{n \to \infty} \frac{1}{n} = 0$$

Therefore, by Theorem 6,

$$\lim_{n \to \infty} \frac{(-1)^n}{n} = 0$$

The following theorem says that if we apply a continuous function to the terms of a convergent sequence, the result is also convergent. The proof is left as Exercise 88.

> **[7] Theorem** If $\lim_{n \to \infty} a_n = L$ and the function f is continuous at L, then
>
> $$\lim_{n \to \infty} f(a_n) = f(L)$$

EXAMPLE 9 Find $\lim_{n \to \infty} \sin(\pi/n)$.

SOLUTION Because the sine function is continuous at 0, Theorem 7 enables us to write

$$\lim_{n \to \infty} \sin(\pi/n) = \sin\left(\lim_{n \to \infty} (\pi/n)\right) = \sin 0 = 0$$

EXAMPLE 10 Discuss the convergence of the sequence $a_n = n!/n^n$, where $n! = 1 \cdot 2 \cdot 3 \cdot \cdots \cdot n$.

SOLUTION Both numerator and denominator approach infinity as $n \to \infty$ but here we have no corresponding function for use with l'Hospital's Rule ($x!$ is not defined when x is not an integer). Let's write out a few terms to get a feeling for what happens to a_n as n gets large:

$$a_1 = 1 \qquad a_2 = \frac{1 \cdot 2}{2 \cdot 2} \qquad a_3 = \frac{1 \cdot 2 \cdot 3}{3 \cdot 3 \cdot 3}$$

$$\boxed{8} \qquad a_n = \frac{1 \cdot 2 \cdot 3 \cdot \cdots \cdot n}{n \cdot n \cdot n \cdot \cdots \cdot n}$$

It appears from these expressions and the graph in Figure 10 that the terms are decreasing and perhaps approach 0. To confirm this, observe from Equation 8 that

$$a_n = \frac{1}{n} \left(\frac{2 \cdot 3 \cdot \cdots \cdot n}{n \cdot n \cdot \cdots \cdot n} \right)$$

Notice that the expression in parentheses is at most 1 because the numerator is less than (or equal to) the denominator. So

$$0 < a_n \leq \frac{1}{n}$$

We know that $1/n \to 0$ as $n \to \infty$. Therefore $a_n \to 0$ as $n \to \infty$ by the Squeeze Theorem.

EXAMPLE 11 For what values of r is the sequence $\{r^n\}$ convergent?

SOLUTION We know from Section 2.6 and the graphs of the exponential functions in Section 1.4 that $\lim_{x \to \infty} a^x = \infty$ for $a > 1$ and $\lim_{x \to \infty} a^x = 0$ for $0 < a < 1$. Therefore, putting $a = r$ and using Theorem 3, we have

$$\lim_{n \to \infty} r^n = \begin{cases} \infty & \text{if } r > 1 \\ 0 & \text{if } 0 < r < 1 \end{cases}$$

It is obvious that

$$\lim_{n \to \infty} 1^n = 1 \quad \text{and} \quad \lim_{n \to \infty} 0^n = 0$$

If $-1 < r < 0$, then $0 < |r| < 1$, so

$$\lim_{n \to \infty} |r^n| = \lim_{n \to \infty} |r|^n = 0$$

and therefore $\lim_{n \to \infty} r^n = 0$ by Theorem 6. If $r \leq -1$, then $\{r^n\}$ diverges as in Example 7. Figure 11 shows the graphs for various values of r. (The case $r = -1$ is shown in Figure 8.)

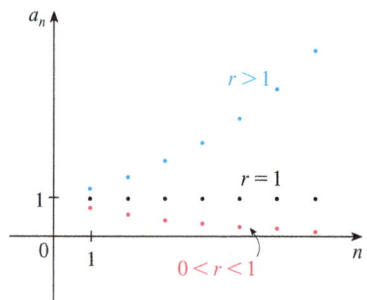
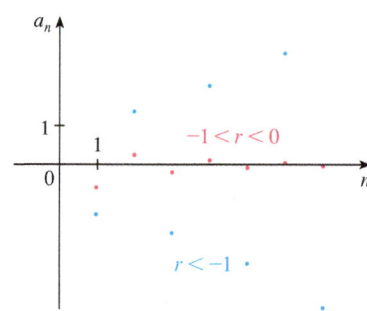

FIGURE 11
The sequence $a_n = r^n$

The results of Example 11 are summarized for future use as follows.

9 The sequence $\{r^n\}$ is convergent if $-1 < r \leq 1$ and divergent for all other values of r.

$$\lim_{n \to \infty} r^n = \begin{cases} 0 & \text{if } -1 < r < 1 \\ 1 & \text{if } r = 1 \end{cases}$$

10 Definition A sequence $\{a_n\}$ is called **increasing** if $a_n < a_{n+1}$ for all $n \geq 1$, that is, $a_1 < a_2 < a_3 < \cdots$. It is called **decreasing** if $a_n > a_{n+1}$ for all $n \geq 1$. A sequence is **monotonic** if it is either increasing or decreasing.

EXAMPLE 12 The sequence $\left\{\dfrac{3}{n+5}\right\}$ is decreasing because

$$\frac{3}{n+5} > \frac{3}{(n+1)+5} = \frac{3}{n+6}$$

The right side is smaller because it has a larger denominator.

and so $a_n > a_{n+1}$ for all $n \geq 1$.

EXAMPLE 13 Show that the sequence $a_n = \dfrac{n}{n^2+1}$ is decreasing.

SOLUTION 1 We must show that $a_{n+1} < a_n$, that is,

$$\frac{n+1}{(n+1)^2+1} < \frac{n}{n^2+1}$$

This inequality is equivalent to the one we get by cross-multiplication:

$$\frac{n+1}{(n+1)^2+1} < \frac{n}{n^2+1} \iff (n+1)(n^2+1) < n[(n+1)^2+1]$$

$$\iff n^3 + n^2 + n + 1 < n^3 + 2n^2 + 2n$$

$$\iff 1 < n^2 + n$$

Since $n \geq 1$, we know that the inequality $n^2 + n > 1$ is true. Therefore $a_{n+1} < a_n$ and so $\{a_n\}$ is decreasing.

SOLUTION 2 Consider the function $f(x) = \dfrac{x}{x^2+1}$:

$$f'(x) = \frac{x^2+1-2x^2}{(x^2+1)^2} = \frac{1-x^2}{(x^2+1)^2} < 0 \quad \text{whenever } x^2 > 1$$

Thus f is decreasing on $(1, \infty)$ and so $f(n) > f(n+1)$. Therefore $\{a_n\}$ is decreasing.

---

11 Definition A sequence $\{a_n\}$ is **bounded above** if there is a number M such that

$$a_n \leq M \quad \text{for all } n \geq 1$$

It is **bounded below** if there is a number m such that

$$m \leq a_n \quad \text{for all } n \geq 1$$

If it is bounded above and below, then $\{a_n\}$ is a **bounded sequence**.

For instance, the sequence $a_n = n$ is bounded below ($a_n > 0$) but not above. The sequence $a_n = n/(n+1)$ is bounded because $0 < a_n < 1$ for all n.

We know that not every bounded sequence is convergent [for instance, the sequence $a_n = (-1)^n$ satisfies $-1 \leq a_n \leq 1$ but is divergent from Example 7] and not every monotonic sequence is convergent ($a_n = n \to \infty$). But if a sequence is both bounded *and* monotonic, then it must be convergent. This fact is proved as Theorem 12, but intuitively you can understand why it is true by looking at Figure 12. If $\{a_n\}$ is increasing and $a_n \leq M$ for all n, then the terms are forced to crowd together and approach some number L.

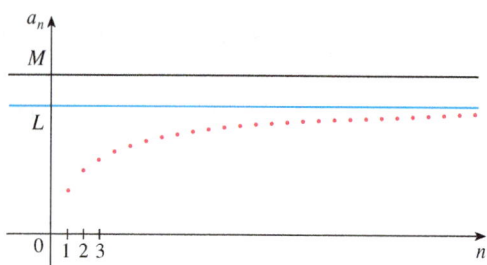

FIGURE 12

The proof of Theorem 12 is based on the **Completeness Axiom** for the set $\mathbb{R}$ of real numbers, which says that if S is a nonempty set of real numbers that has an upper bound M ($x \leq M$ for all x in S), then S has a **least upper bound** b. (This means that b is an upper bound for S, but if M is any other upper bound, then $b \leq M$.) The Completeness Axiom is an expression of the fact that there is no gap or hole in the real number line.

> **12 Monotonic Sequence Theorem** Every bounded, monotonic sequence is convergent.

PROOF Suppose $\{a_n\}$ is an increasing sequence. Since $\{a_n\}$ is bounded, the set $S = \{a_n \mid n \geq 1\}$ has an upper bound. By the Completeness Axiom it has a least upper bound L. Given $\varepsilon > 0$, $L - \varepsilon$ is *not* an upper bound for S (since L is the *least* upper bound). Therefore
$$a_N > L - \varepsilon \quad \text{for some integer } N$$

But the sequence is increasing so $a_n \geq a_N$ for every $n > N$. Thus if $n > N$, we have
$$a_n > L - \varepsilon$$
so
$$0 \leq L - a_n < \varepsilon$$
since $a_n \leq L$. Thus
$$|L - a_n| < \varepsilon \quad \text{whenever } n > N$$
so $\lim_{n \to \infty} a_n = L$.

A similar proof (using the greatest lower bound) works if $\{a_n\}$ is decreasing. ∎

The proof of Theorem 12 shows that a sequence that is increasing and bounded above is convergent. (Likewise, a decreasing sequence that is bounded below is convergent.) This fact is used many times in dealing with infinite series.

EXAMPLE 14 Investigate the sequence $\{a_n\}$ defined by the *recurrence relation*

$$a_1 = 2 \qquad a_{n+1} = \tfrac{1}{2}(a_n + 6) \qquad \text{for } n = 1, 2, 3, \ldots$$

SOLUTION We begin by computing the first several terms:

$$a_1 = 2 \qquad\qquad a_2 = \tfrac{1}{2}(2 + 6) = 4 \qquad\qquad a_3 = \tfrac{1}{2}(4 + 6) = 5$$

$$a_4 = \tfrac{1}{2}(5 + 6) = 5.5 \qquad a_5 = 5.75 \qquad\qquad a_6 = 5.875$$

$$a_7 = 5.9375 \qquad\qquad a_8 = 5.96875 \qquad\qquad a_9 = 5.984375$$

These initial terms suggest that the sequence is increasing and the terms are approaching 6. To confirm that the sequence is increasing, we use mathematical induction to show that $a_{n+1} > a_n$ for all $n \geq 1$. This is true for $n = 1$ because $a_2 = 4 > a_1$. If we assume that it is true for $n = k$, then we have

$$a_{k+1} > a_k$$

so

$$a_{k+1} + 6 > a_k + 6$$

and

$$\tfrac{1}{2}(a_{k+1} + 6) > \tfrac{1}{2}(a_k + 6)$$

Thus

$$a_{k+2} > a_{k+1}$$

We have deduced that $a_{n+1} > a_n$ is true for $n = k + 1$. Therefore the inequality is true for all n by induction.

Next we verify that $\{a_n\}$ is bounded by showing that $a_n < 6$ for all n. (Since the sequence is increasing, we already know that it has a lower bound: $a_n \geq a_1 = 2$ for all n.) We know that $a_1 < 6$, so the assertion is true for $n = 1$. Suppose it is true for $n = k$. Then

$$a_k < 6$$

so

$$a_k + 6 < 12$$

and

$$\tfrac{1}{2}(a_k + 6) < \tfrac{1}{2}(12) = 6$$

Thus

$$a_{k+1} < 6$$

This shows, by mathematical induction, that $a_n < 6$ for all n.

Since the sequence $\{a_n\}$ is increasing and bounded, Theorem 12 guarantees that it has a limit. The theorem doesn't tell us what the value of the limit is. But now that we know $L = \lim_{n \to \infty} a_n$ exists, we can use the given recurrence relation to write

$$\lim_{n \to \infty} a_{n+1} = \lim_{n \to \infty} \tfrac{1}{2}(a_n + 6) = \tfrac{1}{2}\left(\lim_{n \to \infty} a_n + 6\right) = \tfrac{1}{2}(L + 6)$$

Since $a_n \to L$, it follows that $a_{n+1} \to L$ too (as $n \to \infty$, $n + 1 \to \infty$ also). So we have

$$L = \tfrac{1}{2}(L + 6)$$

Solving this equation for L, we get $L = 6$, as we predicted.

11.1 EXERCISES

1. (a) What is a sequence?
   (b) What does it mean to say that $\lim_{n\to\infty} a_n = 8$?
   (c) What does it mean to say that $\lim_{n\to\infty} a_n = \infty$?

2. (a) What is a convergent sequence? Give two examples.
   (b) What is a divergent sequence? Give two examples.

3–12 List the first five terms of the sequence.

3. $a_n = \dfrac{2^n}{2n+1}$

4. $a_n = \dfrac{n^2-1}{n^2+1}$

5. $a_n = \dfrac{(-1)^{n-1}}{5^n}$

6. $a_n = \cos\dfrac{n\pi}{2}$

7. $a_n = \dfrac{1}{(n+1)!}$

8. $a_n = \dfrac{(-1)^n n}{n!+1}$

9. $a_1 = 1,\quad a_{n+1} = 5a_n - 3$

10. $a_1 = 6,\quad a_{n+1} = \dfrac{a_n}{n}$

11. $a_1 = 2,\quad a_{n+1} = \dfrac{a_n}{1+a_n}$

12. $a_1 = 2,\quad a_2 = 1,\quad a_{n+1} = a_n - a_{n-1}$

13–18 Find a formula for the general term a_n of the sequence, assuming that the pattern of the first few terms continues.

13. $\left\{\dfrac{1}{2}, \dfrac{1}{4}, \dfrac{1}{6}, \dfrac{1}{8}, \dfrac{1}{10}, \ldots\right\}$

14. $\left\{4, -1, \dfrac{1}{4}, -\dfrac{1}{16}, \dfrac{1}{64}, \ldots\right\}$

15. $\left\{-3, 2, -\dfrac{4}{3}, \dfrac{8}{9}, -\dfrac{16}{27}, \ldots\right\}$

16. $\{5, 8, 11, 14, 17, \ldots\}$

17. $\left\{\dfrac{1}{2}, -\dfrac{4}{3}, \dfrac{9}{4}, -\dfrac{16}{5}, \dfrac{25}{6}, \ldots\right\}$

18. $\{1, 0, -1, 0, 1, 0, -1, 0, \ldots\}$

19–22 Calculate, to four decimal places, the first ten terms of the sequence and use them to plot the graph of the sequence by hand. Does the sequence appear to have a limit? If so, calculate it. If not, explain why.

19. $a_n = \dfrac{3n}{1+6n}$

20. $a_n = 2 + \dfrac{(-1)^n}{n}$

21. $a_n = 1 + \left(-\dfrac{1}{2}\right)^n$

22. $a_n = 1 + \dfrac{10^n}{9^n}$

23–56 Determine whether the sequence converges or diverges. If it converges, find the limit.

23. $a_n = \dfrac{3+5n^2}{n+n^2}$

24. $a_n = \dfrac{3+5n^2}{1+n}$

25. $a_n = \dfrac{n^4}{n^3-2n}$

26. $a_n = 2 + (0.86)^n$

27. $a_n = 3^n 7^{-n}$

28. $a_n = \dfrac{3\sqrt{n}}{\sqrt{n}+2}$

29. $a_n = e^{-1/\sqrt{n}}$

30. $a_n = \dfrac{4^n}{1+9^n}$

31. $a_n = \sqrt{\dfrac{1+4n^2}{1+n^2}}$

32. $a_n = \cos\left(\dfrac{n\pi}{n+1}\right)$

33. $a_n = \dfrac{n^2}{\sqrt{n^3+4n}}$

34. $a_n = e^{2n/(n+2)}$

35. $a_n = \dfrac{(-1)^n}{2\sqrt{n}}$

36. $a_n = \dfrac{(-1)^{n+1}n}{n+\sqrt{n}}$

37. $\left\{\dfrac{(2n-1)!}{(2n+1)!}\right\}$

38. $\left\{\dfrac{\ln n}{\ln 2n}\right\}$

39. $\{\sin n\}$

40. $a_n = \dfrac{\tan^{-1} n}{n}$

41. $\{n^2 e^{-n}\}$

42. $a_n = \ln(n+1) - \ln n$

43. $a_n = \dfrac{\cos^2 n}{2^n}$

44. $a_n = \sqrt[n]{2^{1+3n}}$

45. $a_n = n\sin(1/n)$

46. $a_n = 2^{-n}\cos n\pi$

47. $a_n = \left(1 + \dfrac{2}{n}\right)^n$

48. $a_n = \sqrt[n]{n}$

49. $a_n = \ln(2n^2+1) - \ln(n^2+1)$

50. $a_n = \dfrac{(\ln n)^2}{n}$

51. $a_n = \arctan(\ln n)$

52. $a_n = n - \sqrt{n+1}\sqrt{n+3}$

53. $\{0, 1, 0, 0, 1, 0, 0, 0, 1, \ldots\}$

54. $\left\{\dfrac{1}{1}, \dfrac{1}{3}, \dfrac{1}{2}, \dfrac{1}{4}, \dfrac{1}{3}, \dfrac{1}{5}, \dfrac{1}{4}, \dfrac{1}{6}, \ldots\right\}$

55. $a_n = \dfrac{n!}{2^n}$

56. $a_n = \dfrac{(-3)^n}{n!}$

57–63 Use a graph of the sequence to decide whether the sequence is convergent or divergent. If the sequence is convergent, guess the value of the limit from the graph and then prove your guess. (See the margin note on page 699 for advice on graphing sequences.)

57. $a_n = (-1)^n \dfrac{n}{n+1}$

58. $a_n = \dfrac{\sin n}{n}$

59. $a_n = \arctan\left(\dfrac{n^2}{n^2+4}\right)$

60. $a_n = \sqrt[n]{3^n + 5^n}$

61. $a_n = \dfrac{n^2 \cos n}{1 + n^2}$

62. $a_n = \dfrac{1 \cdot 3 \cdot 5 \cdot \cdots \cdot (2n-1)}{n!}$

63. $a_n = \dfrac{1 \cdot 3 \cdot 5 \cdot \cdots \cdot (2n-1)}{(2n)^n}$

64. (a) Determine whether the sequence defined as follows is convergent or divergent:
$$a_1 = 1 \qquad a_{n+1} = 4 - a_n \quad \text{for } n \geq 1$$
(b) What happens if the first term is $a_1 = 2$?

65. If $1000 is invested at 6% interest, compounded annually, then after n years the investment is worth $a_n = 1000(1.06)^n$ dollars.
(a) Find the first five terms of the sequence $\{a_n\}$.
(b) Is the sequence convergent or divergent? Explain.

66. If you deposit $100 at the end of every month into an account that pays 3% interest per year compounded monthly, the amount of interest accumulated after n months is given by the sequence
$$I_n = 100\left(\dfrac{1.0025^n - 1}{0.0025} - n\right)$$
(a) Find the first six terms of the sequence.
(b) How much interest will you have earned after two years?

67. A fish farmer has 5000 catfish in his pond. The number of catfish increases by 8% per month and the farmer harvests 300 catfish per month.
(a) Show that the catfish population P_n after n months is given recursively by
$$P_n = 1.08 P_{n-1} - 300 \qquad P_0 = 5000$$
(b) How many catfish are in the pond after six months?

68. Find the first 40 terms of the sequence defined by
$$a_{n+1} = \begin{cases} \tfrac{1}{2} a_n & \text{if } a_n \text{ is an even number} \\ 3a_n + 1 & \text{if } a_n \text{ is an odd number} \end{cases}$$
and $a_1 = 11$. Do the same if $a_1 = 25$. Make a conjecture about this type of sequence.

69. For what values of r is the sequence $\{nr^n\}$ convergent?

70. (a) If $\{a_n\}$ is convergent, show that
$$\lim_{n \to \infty} a_{n+1} = \lim_{n \to \infty} a_n$$
(b) A sequence $\{a_n\}$ is defined by $a_1 = 1$ and $a_{n+1} = 1/(1 + a_n)$ for $n \geq 1$. Assuming that $\{a_n\}$ is convergent, find its limit.

71. Suppose you know that $\{a_n\}$ is a decreasing sequence and all its terms lie between the numbers 5 and 8. Explain why the sequence has a limit. What can you say about the value of the limit?

72–78 Determine whether the sequence is increasing, decreasing, or not monotonic. Is the sequence bounded?

72. $a_n = \cos n$

73. $a_n = \dfrac{1}{2n+3}$

74. $a_n = \dfrac{1-n}{2+n}$

75. $a_n = n(-1)^n$

76. $a_n = 2 + \dfrac{(-1)^n}{n}$

77. $a_n = 3 - 2n e^{-n}$

78. $a_n = n^3 - 3n + 3$

79. Find the limit of the sequence
$$\left\{\sqrt{2},\ \sqrt{2\sqrt{2}},\ \sqrt{2\sqrt{2\sqrt{2}}},\ \ldots\right\}$$

80. A sequence $\{a_n\}$ is given by $a_1 = \sqrt{2}$, $a_{n+1} = \sqrt{2 + a_n}$.
(a) By induction or otherwise, show that $\{a_n\}$ is increasing and bounded above by 3. Apply the Monotonic Sequence Theorem to show that $\lim_{n \to \infty} a_n$ exists.
(b) Find $\lim_{n \to \infty} a_n$.

81. Show that the sequence defined by
$$a_1 = 1 \qquad a_{n+1} = 3 - \dfrac{1}{a_n}$$
is increasing and $a_n < 3$ for all n. Deduce that $\{a_n\}$ is convergent and find its limit.

82. Show that the sequence defined by
$$a_1 = 2 \qquad a_{n+1} = \dfrac{1}{3 - a_n}$$
satisfies $0 < a_n \leq 2$ and is decreasing. Deduce that the sequence is convergent and find its limit.

83. (a) Fibonacci posed the following problem: Suppose that rabbits live forever and that every month each pair produces a new pair which becomes productive at age 2 months. If we start with one newborn pair, how many pairs of rabbits will we have in the nth month? Show that the answer is f_n, where $\{f_n\}$ is the Fibonacci sequence defined in Example 3(c).
(b) Let $a_n = f_{n+1}/f_n$ and show that $a_{n-1} = 1 + 1/a_{n-2}$. Assuming that $\{a_n\}$ is convergent, find its limit.

84. (a) Let $a_1 = a$, $a_2 = f(a)$, $a_3 = f(a_2) = f(f(a)), \ldots,$ $a_{n+1} = f(a_n)$, where f is a continuous function. If $\lim_{n \to \infty} a_n = L$, show that $f(L) = L$.
(b) Illustrate part (a) by taking $f(x) = \cos x$, $a = 1$, and estimating the value of L to five decimal places.

85. (a) Use a graph to guess the value of the limit
$$\lim_{n \to \infty} \frac{n^5}{n!}$$
(b) Use a graph of the sequence in part (a) to find the smallest values of N that correspond to $\varepsilon = 0.1$ and $\varepsilon = 0.001$ in Definition 2.

86. Use Definition 2 directly to prove that $\lim_{n \to \infty} r^n = 0$ when $|r| < 1$.

87. Prove Theorem 6.
[*Hint:* Use either Definition 2 or the Squeeze Theorem.]

88. Prove Theorem 7.

89. Prove that if $\lim_{n \to \infty} a_n = 0$ and $\{b_n\}$ is bounded, then $\lim_{n \to \infty} (a_n b_n) = 0$.

90. Let $a_n = \left(1 + \dfrac{1}{n}\right)^n$.
(a) Show that if $0 \le a < b$, then
$$\frac{b^{n+1} - a^{n+1}}{b - a} < (n + 1)b^n$$
(b) Deduce that $b^n[(n + 1)a - nb] < a^{n+1}$.
(c) Use $a = 1 + 1/(n + 1)$ and $b = 1 + 1/n$ in part (b) to show that $\{a_n\}$ is increasing.
(d) Use $a = 1$ and $b = 1 + 1/(2n)$ in part (b) to show that $a_{2n} < 4$.
(e) Use parts (c) and (d) to show that $a_n < 4$ for all n.
(f) Use Theorem 12 to show that $\lim_{n \to \infty} (1 + 1/n)^n$ exists. (The limit is e. See Equation 3.6.6.)

91. Let a and b be positive numbers with $a > b$. Let a_1 be their arithmetic mean and b_1 their geometric mean:
$$a_1 = \frac{a + b}{2} \qquad b_1 = \sqrt{ab}$$
Repeat this process so that, in general,
$$a_{n+1} = \frac{a_n + b_n}{2} \qquad b_{n+1} = \sqrt{a_n b_n}$$
(a) Use mathematical induction to show that
$$a_n > a_{n+1} > b_{n+1} > b_n$$
(b) Deduce that both $\{a_n\}$ and $\{b_n\}$ are convergent.
(c) Show that $\lim_{n \to \infty} a_n = \lim_{n \to \infty} b_n$. Gauss called the common value of these limits the **arithmetic-geometric mean** of the numbers a and b.

92. (a) Show that if $\lim_{n \to \infty} a_{2n} = L$ and $\lim_{n \to \infty} a_{2n+1} = L$, then $\{a_n\}$ is convergent and $\lim_{n \to \infty} a_n = L$.
(b) If $a_1 = 1$ and
$$a_{n+1} = 1 + \frac{1}{1 + a_n}$$
find the first eight terms of the sequence $\{a_n\}$. Then use part (a) to show that $\lim_{n \to \infty} a_n = \sqrt{2}$. This gives the **continued fraction expansion**
$$\sqrt{2} = 1 + \cfrac{1}{2 + \cfrac{1}{2 + \cdots}}$$

93. The size of an undisturbed fish population has been modeled by the formula
$$p_{n+1} = \frac{bp_n}{a + p_n}$$
where p_n is the fish population after n years and a and b are positive constants that depend on the species and its environment. Suppose that the population in year 0 is $p_0 > 0$.
(a) Show that if $\{p_n\}$ is convergent, then the only possible values for its limit are 0 and $b - a$.
(b) Show that $p_{n+1} < (b/a)p_n$.
(c) Use part (b) to show that if $a > b$, then $\lim_{n \to \infty} p_n = 0$; in other words, the population dies out.
(d) Now assume that $a < b$. Show that if $p_0 < b - a$, then $\{p_n\}$ is increasing and $0 < p_n < b - a$. Show also that if $p_0 > b - a$, then $\{p_n\}$ is decreasing and $p_n > b - a$. Deduce that if $a < b$, then $\lim_{n \to \infty} p_n = b - a$.

LABORATORY PROJECT ▦ LOGISTIC SEQUENCES

A sequence that arises in ecology as a model for population growth is defined by the **logistic difference equation**

$$p_{n+1} = kp_n(1 - p_n)$$

where p_n measures the size of the population of the nth generation of a single species. To keep the numbers manageable, p_n is a fraction of the maximal size of the population, so $0 \leq p_n \leq 1$. Notice that the form of this equation is similar to the logistic differential equation in Section 9.4. The discrete model—with sequences instead of continuous functions—is preferable for modeling insect populations, where mating and death occur in a periodic fashion.

An ecologist is interested in predicting the size of the population as time goes on, and asks these questions: Will it stabilize at a limiting value? Will it change in a cyclical fashion? Or will it exhibit random behavior?

Write a program to compute the first n terms of this sequence starting with an initial population p_0, where $0 < p_0 < 1$. Use this program to do the following.

1. Calculate 20 or 30 terms of the sequence for $p_0 = \frac{1}{2}$ and for two values of k such that $1 < k < 3$. Graph each sequence. Do the sequences appear to converge? Repeat for a different value of p_0 between 0 and 1. Does the limit depend on the choice of p_0? Does it depend on the choice of k?

2. Calculate terms of the sequence for a value of k between 3 and 3.4 and plot them. What do you notice about the behavior of the terms?

3. Experiment with values of k between 3.4 and 3.5. What happens to the terms?

4. For values of k between 3.6 and 4, compute and plot at least 100 terms and comment on the behavior of the sequence. What happens if you change p_0 by 0.001? This type of behavior is called *chaotic* and is exhibited by insect populations under certain conditions.

11.2 Series

The current record for computing a decimal approximation for π was obtained by Shigeru Kondo and Alexander Yee in 2011 and contains more than 10 trillion decimal places.

What do we mean when we express a number as an infinite decimal? For instance, what does it mean to write

$$\pi = 3.14159\ 26535\ 89793\ 23846\ 26433\ 83279\ 50288\ldots$$

The convention behind our decimal notation is that any number can be written as an infinite sum. Here it means that

$$\pi = 3 + \frac{1}{10} + \frac{4}{10^2} + \frac{1}{10^3} + \frac{5}{10^4} + \frac{9}{10^5} + \frac{2}{10^6} + \frac{6}{10^7} + \frac{5}{10^8} + \cdots$$

where the three dots ($\cdots$) indicate that the sum continues forever, and the more terms we add, the closer we get to the actual value of π.

In general, if we try to add the terms of an infinite sequence $\{a_n\}_{n=1}^{\infty}$ we get an expression of the form

$$\boxed{1} \qquad a_1 + a_2 + a_3 + \cdots + a_n + \cdots$$

which is called an **infinite series** (or just a **series**) and is denoted, for short, by the symbol

$$\sum_{n=1}^{\infty} a_n \quad \text{or} \quad \sum a_n$$

Does it make sense to talk about the sum of infinitely many terms?

It would be impossible to find a finite sum for the series

$$1 + 2 + 3 + 4 + 5 + \cdots + n + \cdots$$

because if we start adding the terms we get the cumulative sums 1, 3, 6, 10, 15, 21, ... and, after the nth term, we get $n(n + 1)/2$, which becomes very large as n increases.

However, if we start to add the terms of the series

$$\frac{1}{2} + \frac{1}{4} + \frac{1}{8} + \frac{1}{16} + \frac{1}{32} + \frac{1}{64} + \cdots + \frac{1}{2^n} + \cdots$$

we get $\frac{1}{2}, \frac{3}{4}, \frac{7}{8}, \frac{15}{16}, \frac{31}{32}, \frac{63}{64}, \ldots, 1 - 1/2^n, \ldots$. The table shows that as we add more and more terms, these *partial sums* become closer and closer to 1. (See also Figure 11 in *A Preview of Calculus*, page 6.) In fact, by adding sufficiently many terms of the series we can make the partial sums as close as we like to 1. So it seems reasonable to say that the sum of this infinite series is 1 and to write

$$\sum_{n=1}^{\infty} \frac{1}{2^n} = \frac{1}{2} + \frac{1}{4} + \frac{1}{8} + \frac{1}{16} + \cdots + \frac{1}{2^n} + \cdots = 1$$

n	Sum of first n terms
1	0.50000000
2	0.75000000
3	0.87500000
4	0.93750000
5	0.96875000
6	0.98437500
7	0.99218750
10	0.99902344
15	0.99996948
20	0.99999905
25	0.99999997

We use a similar idea to determine whether or not a general series (1) has a sum. We consider the **partial sums**

$$s_1 = a_1$$
$$s_2 = a_1 + a_2$$
$$s_3 = a_1 + a_2 + a_3$$
$$s_4 = a_1 + a_2 + a_3 + a_4$$

and, in general,

$$s_n = a_1 + a_2 + a_3 + \cdots + a_n = \sum_{i=1}^{n} a_i$$

These partial sums form a new sequence $\{s_n\}$, which may or may not have a limit. If $\lim_{n \to \infty} s_n = s$ exists (as a finite number), then, as in the preceding example, we call it the sum of the infinite series Σa_n.

2 Definition Given a series $\sum_{n=1}^{\infty} a_n = a_1 + a_2 + a_3 + \cdots$, let s_n denote its nth partial sum:

$$s_n = \sum_{i=1}^{n} a_i = a_1 + a_2 + \cdots + a_n$$

If the sequence $\{s_n\}$ is convergent and $\lim_{n \to \infty} s_n = s$ exists as a real number, then the series Σa_n is called **convergent** and we write

$$a_1 + a_2 + \cdots + a_n + \cdots = s \qquad \text{or} \qquad \sum_{n=1}^{\infty} a_n = s$$

The number s is called the **sum** of the series. If the sequence $\{s_n\}$ is divergent, then the series is called **divergent**.

Thus the sum of a series is the limit of the sequence of partial sums. So when we write $\sum_{n=1}^{\infty} a_n = s$, we mean that by adding sufficiently many terms of the series we can get as close as we like to the number s. Notice that

$$\sum_{n=1}^{\infty} a_n = \lim_{n \to \infty} \sum_{i=1}^{n} a_i$$

Compare with the improper integral

$$\int_1^{\infty} f(x)\, dx = \lim_{t \to \infty} \int_1^{t} f(x)\, dx$$

To find this integral we integrate from 1 to t and then let $t \to \infty$. For a series, we sum from 1 to n and then let $n \to \infty$.

EXAMPLE 1 Suppose we know that the sum of the first n terms of the series $\sum_{n=1}^{\infty} a_n$ is

$$s_n = a_1 + a_2 + \cdots + a_n = \frac{2n}{3n + 5}$$

Then the sum of the series is the limit of the sequence $\{s_n\}$:

$$\sum_{n=1}^{\infty} a_n = \lim_{n \to \infty} s_n = \lim_{n \to \infty} \frac{2n}{3n + 5} = \lim_{n \to \infty} \frac{2}{3 + \dfrac{5}{n}} = \frac{2}{3}$$

In Example 1 we were *given* an expression for the sum of the first n terms, but it's usually not easy to *find* such an expression. In Example 2, however, we look at a famous series for which we *can* find an explicit formula for s_n.

EXAMPLE 2 An important example of an infinite series is the **geometric series**

$$a + ar + ar^2 + ar^3 + \cdots + ar^{n-1} + \cdots = \sum_{n=1}^{\infty} ar^{n-1} \qquad a \neq 0$$

Each term is obtained from the preceding one by multiplying it by the **common ratio** r. (We have already considered the special case where $a = \tfrac{1}{2}$ and $r = \tfrac{1}{2}$ on page 708.)

If $r = 1$, then $s_n = a + a + \cdots + a = na \to \pm\infty$. Since $\lim_{n \to \infty} s_n$ doesn't exist, the geometric series diverges in this case.

If $r \neq 1$, we have

$$s_n = a + ar + ar^2 + \cdots + ar^{n-1}$$

and

$$rs_n = \qquad ar + ar^2 + \cdots + ar^{n-1} + ar^n$$

Subtracting these equations, we get

$$s_n - rs_n = a - ar^n$$

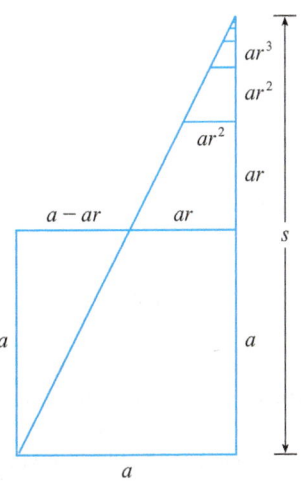

FIGURE 1

Figure 1 provides a geometric demonstration of the result in Example 2. If the triangles are constructed as shown and s is the sum of the series, then, by similar triangles,

$$\frac{s}{a} = \frac{a}{a - ar} \quad \text{so} \quad s = \frac{a}{1 - r}$$

$$\boxed{s_n = \frac{a(1 - r^n)}{1 - r}} \quad \boxed{3}$$

If $-1 < r < 1$, we know from (11.1.9) that $r^n \to 0$ as $n \to \infty$, so

$$\lim_{n \to \infty} s_n = \lim_{n \to \infty} \frac{a(1 - r^n)}{1 - r} = \frac{a}{1 - r} - \frac{a}{1 - r} \lim_{n \to \infty} r^n = \frac{a}{1 - r}$$

Thus when $|r| < 1$ the geometric series is convergent and its sum is $a/(1 - r)$.

If $r \leq -1$ or $r > 1$, the sequence $\{r^n\}$ is divergent by (11.1.9) and so, by Equation 3, $\lim_{n \to \infty} s_n$ does not exist. Therefore the geometric series diverges in those cases. ∎

We summarize the results of Example 2 as follows.

> **4** The geometric series
> $$\sum_{n=1}^{\infty} ar^{n-1} = a + ar + ar^2 + \cdots$$
> is convergent if $|r| < 1$ and its sum is
> $$\sum_{n=1}^{\infty} ar^{n-1} = \frac{a}{1-r} \qquad |r| < 1$$
> If $|r| \geq 1$, the geometric series is divergent.

In words: The sum of a convergent geometric series is

$$\frac{\text{first term}}{1 - \text{common ratio}}$$

EXAMPLE 3 Find the sum of the geometric series
$$5 - \tfrac{10}{3} + \tfrac{20}{9} - \tfrac{40}{27} + \cdots$$

SOLUTION The first term is $a = 5$ and the common ratio is $r = -\tfrac{2}{3}$. Since $|r| = \tfrac{2}{3} < 1$, the series is convergent by (4) and its sum is

$$5 - \frac{10}{3} + \frac{20}{9} - \frac{40}{27} + \cdots = \frac{5}{1 - \left(-\tfrac{2}{3}\right)} = \frac{5}{\tfrac{5}{3}} = 3$$

What do we really mean when we say that the sum of the series in Example 3 is 3? Of course, we can't literally add an infinite number of terms, one by one. But, according to Definition 2, the total sum is the limit of the sequence of partial sums. So, by taking the sum of sufficiently many terms, we can get as close as we like to the number 3. The table shows the first ten partial sums s_n and the graph in Figure 2 shows how the sequence of partial sums approaches 3.

n	s_n
1	5.000000
2	1.666667
3	3.888889
4	2.407407
5	3.395062
6	2.736626
7	3.175583
8	2.882945
9	3.078037
10	2.947975

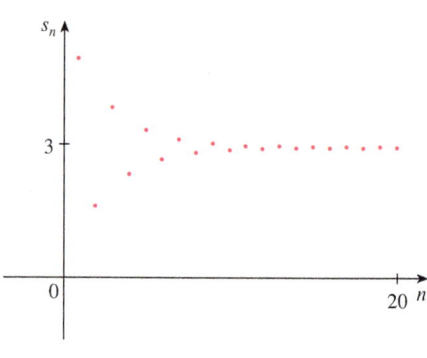

FIGURE 2

EXAMPLE 4 Is the series $\sum_{n=1}^{\infty} 2^{2n}3^{1-n}$ convergent or divergent?

SOLUTION Let's rewrite the nth term of the series in the form ar^{n-1}:

$$\sum_{n=1}^{\infty} 2^{2n}3^{1-n} = \sum_{n=1}^{\infty} (2^2)^n 3^{-(n-1)} = \sum_{n=1}^{\infty} \frac{4^n}{3^{n-1}} = \sum_{n=1}^{\infty} 4\left(\tfrac{4}{3}\right)^{n-1}$$

Another way to identify a and r is to write out the first few terms:

$$4 + \tfrac{16}{3} + \tfrac{64}{9} + \cdots$$

We recognize this series as a geometric series with $a = 4$ and $r = \tfrac{4}{3}$. Since $r > 1$, the series diverges by (4).

EXAMPLE 5 A drug is administered to a patient at the same time every day. Suppose the concentration of the drug is C_n (measured in mg/mL) after the injection on the nth day. Before the injection the next day, only 30% of the drug remains in the bloodstream and the daily dose raises the concentration by 0.2 mg/mL.
(a) Find the concentration after three days.

(b) What is the concentration after the nth dose?
(c) What is the limiting concentration?

SOLUTION
(a) Just before the daily dose of medication is administered, the concentration is reduced to 30% of the preceding day's concentration, that is, $0.3C_n$. With the new dose, the concentration is increased by 0.2 mg/mL and so

$$C_{n+1} = 0.2 + 0.3C_n$$

Starting with $C_0 = 0$ and putting $n = 0, 1, 2$ into this equation, we get

$$C_1 = 0.2 + 0.3C_0 = 0.2$$

$$C_2 = 0.2 + 0.3C_1 = 0.2 + 0.2(0.3) = 0.26$$

$$C_3 = 0.2 + 0.3C_2 = 0.2 + 0.2(0.3) + 0.2(0.3)^2 = 0.278$$

The concentration after three days is 0.278 mg/mL.

(b) After the nth dose the concentration is

$$C_n = 0.2 + 0.2(0.3) + 0.2(0.3)^2 + \cdots + 0.2(0.3)^{n-1}$$

This is a finite geometric series with $a = 0.2$ and $r = 0.3$, so by Formula 3 we have

$$C_n = \frac{0.2[1 - (0.3)^n]}{1 - 0.3} = \frac{2}{7}[1 - (0.3)^n] \text{ mg/mL}$$

(c) Because $0.3 < 1$, we know that $\lim_{n \to \infty} (0.3)^n = 0$. So the limiting concentration is

$$\lim_{n \to \infty} C_n = \lim_{n \to \infty} \frac{2}{7}[1 - (0.3)^n] = \frac{2}{7}(1 - 0) = \frac{2}{7} \text{ mg/mL}$$

EXAMPLE 6 Write the number $2.3\overline{17} = 2.3171717\ldots$ as a ratio of integers.

SOLUTION

$$2.3171717\ldots = 2.3 + \frac{17}{10^3} + \frac{17}{10^5} + \frac{17}{10^7} + \cdots$$

After the first term we have a geometric series with $a = 17/10^3$ and $r = 1/10^2$. Therefore

$$2.3\overline{17} = 2.3 + \frac{\frac{17}{10^3}}{1 - \frac{1}{10^2}} = 2.3 + \frac{\frac{17}{1000}}{\frac{99}{100}}$$

$$= \frac{23}{10} + \frac{17}{990} = \frac{1147}{495}$$

EXAMPLE 7 Find the sum of the series $\sum_{n=0}^{\infty} x^n$, where $|x| < 1$.

SOLUTION Notice that this series starts with $n = 0$ and so the first term is $x^0 = 1$. (With series, we adopt the convention that $x^0 = 1$ even when $x = 0$.)

712 CHAPTER 11 Infinite Sequences and Series

TEC Module 11.2 explores a series that depends on an angle θ in a triangle and enables you to see how rapidly the series converges when θ varies.

Thus

$$\sum_{n=0}^{\infty} x^n = 1 + x + x^2 + x^3 + x^4 + \cdots$$

This is a geometric series with $a = 1$ and $r = x$. Since $|r| = |x| < 1$, it converges and (4) gives

$$\boxed{5} \qquad \sum_{n=0}^{\infty} x^n = \frac{1}{1-x}$$

EXAMPLE 8 Show that the series $\displaystyle\sum_{n=1}^{\infty} \frac{1}{n(n+1)}$ is convergent, and find its sum.

SOLUTION This is not a geometric series, so we go back to the definition of a convergent series and compute the partial sums.

$$s_n = \sum_{i=1}^{n} \frac{1}{i(i+1)} = \frac{1}{1 \cdot 2} + \frac{1}{2 \cdot 3} + \frac{1}{3 \cdot 4} + \cdots + \frac{1}{n(n+1)}$$

We can simplify this expression if we use the partial fraction decomposition

$$\frac{1}{i(i+1)} = \frac{1}{i} - \frac{1}{i+1}$$

(see Section 7.4). Thus we have

Notice that the terms cancel in pairs. This is an example of a **telescoping sum**: Because of all the cancellations, the sum collapses (like a pirate's collapsing telescope) into just two terms.

$$s_n = \sum_{i=1}^{n} \frac{1}{i(i+1)} = \sum_{i=1}^{n} \left(\frac{1}{i} - \frac{1}{i+1} \right)$$

$$= \left(1 - \frac{1}{2} \right) + \left(\frac{1}{2} - \frac{1}{3} \right) + \left(\frac{1}{3} - \frac{1}{4} \right) + \cdots + \left(\frac{1}{n} - \frac{1}{n+1} \right)$$

$$= 1 - \frac{1}{n+1}$$

and so

$$\lim_{n \to \infty} s_n = \lim_{n \to \infty} \left(1 - \frac{1}{n+1} \right) = 1 - 0 = 1$$

Therefore the given series is convergent and

$$\sum_{n=1}^{\infty} \frac{1}{n(n+1)} = 1$$

Figure 3 illustrates Example 8 by showing the graphs of the sequence of terms $a_n = 1/[n(n+1)]$ and the sequence $\{s_n\}$ of partial sums. Notice that $a_n \to 0$ and $s_n \to 1$. See Exercises 78 and 79 for two geometric interpretations of Example 8.

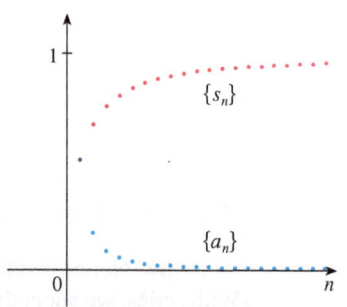

FIGURE 3

EXAMPLE 9 Show that the **harmonic series**

$$\sum_{n=1}^{\infty} \frac{1}{n} = 1 + \frac{1}{2} + \frac{1}{3} + \frac{1}{4} + \cdots$$

is divergent.

SOLUTION For this particular series it's convenient to consider the partial sums s_2, s_4, s_8, s_{16}, s_{32}, ... and show that they become large.

$$s_2 = 1 + \tfrac{1}{2}$$

$$s_4 = 1 + \tfrac{1}{2} + \left(\tfrac{1}{3} + \tfrac{1}{4}\right) > 1 + \tfrac{1}{2} + \left(\tfrac{1}{4} + \tfrac{1}{4}\right) = 1 + \tfrac{2}{2}$$

$$s_8 = 1 + \tfrac{1}{2} + \left(\tfrac{1}{3} + \tfrac{1}{4}\right) + \left(\tfrac{1}{5} + \tfrac{1}{6} + \tfrac{1}{7} + \tfrac{1}{8}\right)$$

$$> 1 + \tfrac{1}{2} + \left(\tfrac{1}{4} + \tfrac{1}{4}\right) + \left(\tfrac{1}{8} + \tfrac{1}{8} + \tfrac{1}{8} + \tfrac{1}{8}\right)$$

$$= 1 + \tfrac{1}{2} + \tfrac{1}{2} + \tfrac{1}{2} = 1 + \tfrac{3}{2}$$

$$s_{16} = 1 + \tfrac{1}{2} + \left(\tfrac{1}{3} + \tfrac{1}{4}\right) + \left(\tfrac{1}{5} + \cdots + \tfrac{1}{8}\right) + \left(\tfrac{1}{9} + \cdots + \tfrac{1}{16}\right)$$

$$> 1 + \tfrac{1}{2} + \left(\tfrac{1}{4} + \tfrac{1}{4}\right) + \left(\tfrac{1}{8} + \cdots + \tfrac{1}{8}\right) + \left(\tfrac{1}{16} + \cdots + \tfrac{1}{16}\right)$$

$$= 1 + \tfrac{1}{2} + \tfrac{1}{2} + \tfrac{1}{2} + \tfrac{1}{2} = 1 + \tfrac{4}{2}$$

Similarly, $s_{32} > 1 + \tfrac{5}{2}$, $s_{64} > 1 + \tfrac{6}{2}$, and in general

$$s_{2^n} > 1 + \frac{n}{2}$$

This shows that $s_{2^n} \to \infty$ as $n \to \infty$ and so $\{s_n\}$ is divergent. Therefore the harmonic series diverges.

> The method used in Example 9 for showing that the harmonic series diverges is due to the French scholar Nicole Oresme (1323–1382).

6 Theorem If the series $\sum_{n=1}^{\infty} a_n$ is convergent, then $\lim_{n \to \infty} a_n = 0$.

PROOF Let $s_n = a_1 + a_2 + \cdots + a_n$. Then $a_n = s_n - s_{n-1}$. Since $\sum a_n$ is convergent, the sequence $\{s_n\}$ is convergent. Let $\lim_{n \to \infty} s_n = s$. Since $n - 1 \to \infty$ as $n \to \infty$, we also have $\lim_{n \to \infty} s_{n-1} = s$. Therefore

$$\lim_{n \to \infty} a_n = \lim_{n \to \infty} (s_n - s_{n-1}) = \lim_{n \to \infty} s_n - \lim_{n \to \infty} s_{n-1} = s - s = 0$$

NOTE 1 With any *series* $\sum a_n$ we associate two *sequences*: the sequence $\{s_n\}$ of its partial sums and the sequence $\{a_n\}$ of its terms. If $\sum a_n$ is convergent, then the limit of the sequence $\{s_n\}$ is s (the sum of the series) and, as Theorem 6 asserts, the limit of the sequence $\{a_n\}$ is 0.

NOTE 2 The converse of Theorem 6 is not true in general. If $\lim_{n \to \infty} a_n = 0$, we cannot conclude that $\sum a_n$ is convergent. Observe that for the harmonic series $\sum 1/n$ we have $a_n = 1/n \to 0$ as $n \to \infty$, but we showed in Example 9 that $\sum 1/n$ is divergent.

7 Test for Divergence If $\lim_{n \to \infty} a_n$ does not exist or if $\lim_{n \to \infty} a_n \neq 0$, then the series $\sum_{n=1}^{\infty} a_n$ is divergent.

The Test for Divergence follows from Theorem 6 because, if the series is not divergent, then it is convergent, and so $\lim_{n \to \infty} a_n = 0$.

EXAMPLE 10 Show that the series $\sum_{n=1}^{\infty} \dfrac{n^2}{5n^2 + 4}$ diverges.

SOLUTION

$$\lim_{n \to \infty} a_n = \lim_{n \to \infty} \frac{n^2}{5n^2 + 4} = \lim_{n \to \infty} \frac{1}{5 + 4/n^2} = \frac{1}{5} \neq 0$$

So the series diverges by the Test for Divergence. ■

NOTE 3 If we find that $\lim_{n \to \infty} a_n \neq 0$, we know that $\Sigma\, a_n$ is divergent. If we find that $\lim_{n \to \infty} a_n = 0$, we know *nothing* about the convergence or divergence of $\Sigma\, a_n$. Remember the warning in Note 2: if $\lim_{n \to \infty} a_n = 0$, the series $\Sigma\, a_n$ might converge or it might diverge.

---

8 Theorem If $\Sigma\, a_n$ and $\Sigma\, b_n$ are convergent series, then so are the series $\Sigma\, ca_n$ (where c is a constant), $\Sigma\, (a_n + b_n)$, and $\Sigma\, (a_n - b_n)$, and

(i) $\displaystyle\sum_{n=1}^{\infty} ca_n = c \sum_{n=1}^{\infty} a_n$

(ii) $\displaystyle\sum_{n=1}^{\infty} (a_n + b_n) = \sum_{n=1}^{\infty} a_n + \sum_{n=1}^{\infty} b_n$

(iii) $\displaystyle\sum_{n=1}^{\infty} (a_n - b_n) = \sum_{n=1}^{\infty} a_n - \sum_{n=1}^{\infty} b_n$

These properties of convergent series follow from the corresponding Limit Laws for Sequences in Section 11.1. For instance, here is how part (ii) of Theorem 8 is proved: Let

$$s_n = \sum_{i=1}^{n} a_i \qquad s = \sum_{n=1}^{\infty} a_n \qquad t_n = \sum_{i=1}^{n} b_i \qquad t = \sum_{n=1}^{\infty} b_n$$

The nth partial sum for the series $\Sigma\, (a_n + b_n)$ is

$$u_n = \sum_{i=1}^{n} (a_i + b_i)$$

and, using Equation 5.2.10, we have

$$\lim_{n \to \infty} u_n = \lim_{n \to \infty} \sum_{i=1}^{n} (a_i + b_i) = \lim_{n \to \infty} \left(\sum_{i=1}^{n} a_i + \sum_{i=1}^{n} b_i \right)$$

$$= \lim_{n \to \infty} \sum_{i=1}^{n} a_i + \lim_{n \to \infty} \sum_{i=1}^{n} b_i$$

$$= \lim_{n \to \infty} s_n + \lim_{n \to \infty} t_n = s + t$$

Therefore $\Sigma\, (a_n + b_n)$ is convergent and its sum is

$$\sum_{n=1}^{\infty} (a_n + b_n) = s + t = \sum_{n=1}^{\infty} a_n + \sum_{n=1}^{\infty} b_n \qquad ■$$

EXAMPLE 11 Find the sum of the series $\sum_{n=1}^{\infty} \left(\frac{3}{n(n+1)} + \frac{1}{2^n} \right)$.

SOLUTION The series $\Sigma \, 1/2^n$ is a geometric series with $a = \frac{1}{2}$ and $r = \frac{1}{2}$, so

$$\sum_{n=1}^{\infty} \frac{1}{2^n} = \frac{\frac{1}{2}}{1 - \frac{1}{2}} = 1$$

In Example 8 we found that

$$\sum_{n=1}^{\infty} \frac{1}{n(n+1)} = 1$$

So, by Theorem 8, the given series is convergent and

$$\sum_{n=1}^{\infty} \left(\frac{3}{n(n+1)} + \frac{1}{2^n} \right) = 3 \sum_{n=1}^{\infty} \frac{1}{n(n+1)} + \sum_{n=1}^{\infty} \frac{1}{2^n}$$

$$= 3 \cdot 1 + 1 = 4 \quad \blacksquare$$

NOTE 4 A finite number of terms doesn't affect the convergence or divergence of a series. For instance, suppose that we were able to show that the series

$$\sum_{n=4}^{\infty} \frac{n}{n^3 + 1}$$

is convergent. Since

$$\sum_{n=1}^{\infty} \frac{n}{n^3 + 1} = \frac{1}{2} + \frac{2}{9} + \frac{3}{28} + \sum_{n=4}^{\infty} \frac{n}{n^3 + 1}$$

it follows that the entire series $\sum_{n=1}^{\infty} n/(n^3 + 1)$ is convergent. Similarly, if it is known that the series $\sum_{n=N+1}^{\infty} a_n$ converges, then the full series

$$\sum_{n=1}^{\infty} a_n = \sum_{n=1}^{N} a_n + \sum_{n=N+1}^{\infty} a_n$$

is also convergent.

11.2 EXERCISES

1. (a) What is the difference between a sequence and a series?
(b) What is a convergent series? What is a divergent series?

2. Explain what it means to say that $\sum_{n=1}^{\infty} a_n = 5$.

3–4 Calculate the sum of the series $\sum_{n=1}^{\infty} a_n$ whose partial sums are given.

3. $s_n = 2 - 3(0.8)^n$

4. $s_n = \frac{n^2 - 1}{4n^2 + 1}$

5–8 Calculate the first eight terms of the sequence of partial sums correct to four decimal places. Does it appear that the series is convergent or divergent?

5. $\sum_{n=1}^{\infty} \frac{1}{n^4 + n^2}$

6. $\sum_{n=1}^{\infty} \frac{1}{\sqrt[3]{n}}$

7. $\sum_{n=1}^{\infty} \sin n$

8. $\sum_{n=1}^{\infty} \frac{(-1)^{n-1}}{n!}$

9–14 Find at least 10 partial sums of the series. Graph both the sequence of terms and the sequence of partial sums on the same screen. Does it appear that the series is convergent or divergent? If it is convergent, find the sum. If it is divergent, explain why.

9. $\sum_{n=1}^{\infty} \frac{12}{(-5)^n}$

10. $\sum_{n=1}^{\infty} \cos n$

11. $\sum_{n=1}^{\infty} \frac{n}{\sqrt{n^2 + 4}}$

12. $\sum_{n=1}^{\infty} \frac{7^{n+1}}{10^n}$

13. $\sum_{n=1}^{\infty} \dfrac{1}{n^2+1}$

14. $\sum_{n=1}^{\infty} \left(\sin \dfrac{1}{n} - \sin \dfrac{1}{n+1}\right)$

15. Let $a_n = \dfrac{2n}{3n+1}$.
  (a) Determine whether $\{a_n\}$ is convergent.
  (b) Determine whether $\sum_{n=1}^{\infty} a_n$ is convergent.

16. (a) Explain the difference between
$$\sum_{i=1}^{n} a_i \quad \text{and} \quad \sum_{j=1}^{n} a_j$$
  (b) Explain the difference between
$$\sum_{i=1}^{n} a_i \quad \text{and} \quad \sum_{i=1}^{n} a_j$$

17–26 Determine whether the geometric series is convergent or divergent. If it is convergent, find its sum.

17. $3 - 4 + \dfrac{16}{3} - \dfrac{64}{9} + \cdots$

18. $4 + 3 + \dfrac{9}{4} + \dfrac{27}{16} + \cdots$

19. $10 - 2 + 0.4 - 0.08 + \cdots$

20. $2 + 0.5 + 0.125 + 0.03125 + \cdots$

21. $\sum_{n=1}^{\infty} 12(0.73)^{n-1}$

22. $\sum_{n=1}^{\infty} \dfrac{5}{\pi^n}$

23. $\sum_{n=1}^{\infty} \dfrac{(-3)^{n-1}}{4^n}$

24. $\sum_{n=0}^{\infty} \dfrac{3^{n+1}}{(-2)^n}$

25. $\sum_{n=1}^{\infty} \dfrac{e^{2n}}{6^{n-1}}$

26. $\sum_{n=1}^{\infty} \dfrac{6 \cdot 2^{2n-1}}{3^n}$

27–42 Determine whether the series is convergent or divergent. If it is convergent, find its sum.

27. $\dfrac{1}{3} + \dfrac{1}{6} + \dfrac{1}{9} + \dfrac{1}{12} + \dfrac{1}{15} + \cdots$

28. $\dfrac{1}{3} + \dfrac{2}{9} + \dfrac{1}{27} + \dfrac{2}{81} + \dfrac{1}{243} + \dfrac{2}{729} + \cdots$

29. $\sum_{n=1}^{\infty} \dfrac{2+n}{1-2n}$

30. $\sum_{k=1}^{\infty} \dfrac{k^2}{k^2-2k+5}$

31. $\sum_{n=1}^{\infty} 3^{n+1} 4^{-n}$

32. $\sum_{n=1}^{\infty} [(-0.2)^n + (0.6)^{n-1}]$

33. $\sum_{n=1}^{\infty} \dfrac{1}{4+e^{-n}}$

34. $\sum_{n=1}^{\infty} \dfrac{2^n + 4^n}{e^n}$

35. $\sum_{k=1}^{\infty} (\sin 100)^k$

36. $\sum_{n=1}^{\infty} \dfrac{1}{1+\left(\frac{2}{3}\right)^n}$

37. $\sum_{n=1}^{\infty} \ln\left(\dfrac{n^2+1}{2n^2+1}\right)$

38. $\sum_{k=0}^{\infty} (\sqrt{2})^{-k}$

39. $\sum_{n=1}^{\infty} \arctan n$

40. $\sum_{n=1}^{\infty} \left(\dfrac{3}{5^n} + \dfrac{2}{n}\right)$

41. $\sum_{n=1}^{\infty} \left(\dfrac{1}{e^n} + \dfrac{1}{n(n+1)}\right)$

42. $\sum_{n=1}^{\infty} \dfrac{e^n}{n^2}$

43–48 Determine whether the series is convergent or divergent by expressing s_n as a telescoping sum (as in Example 8). If it is convergent, find its sum.

43. $\sum_{n=2}^{\infty} \dfrac{2}{n^2-1}$

44. $\sum_{n=1}^{\infty} \ln \dfrac{n}{n+1}$

45. $\sum_{n=1}^{\infty} \dfrac{3}{n(n+3)}$

46. $\sum_{n=4}^{\infty} \left(\dfrac{1}{\sqrt{n}} - \dfrac{1}{\sqrt{n+1}}\right)$

47. $\sum_{n=1}^{\infty} \left(e^{1/n} - e^{1/(n+1)}\right)$

48. $\sum_{n=2}^{\infty} \dfrac{1}{n^3-n}$

49. Let $x = 0.99999\ldots$.
  (a) Do you think that $x < 1$ or $x = 1$?
  (b) Sum a geometric series to find the value of x.
  (c) How many decimal representations does the number 1 have?
  (d) Which numbers have more than one decimal representation?

50. A sequence of terms is defined by
$$a_1 = 1 \qquad a_n = (5-n)a_{n-1}$$
  Calculate $\sum_{n=1}^{\infty} a_n$.

51–56 Express the number as a ratio of integers.

51. $0.\overline{8} = 0.8888\ldots$

52. $0.\overline{46} = 0.46464646\ldots$

53. $2.\overline{516} = 2.516516516\ldots$

54. $10.1\overline{35} = 10.135353535\ldots$

55. $1.234\overline{567}$

56. $5.7\overline{1358}$

57–63 Find the values of x for which the series converges. Find the sum of the series for those values of x.

57. $\sum_{n=1}^{\infty} (-5)^n x^n$

58. $\sum_{n=1}^{\infty} (x+2)^n$

59. $\sum_{n=0}^{\infty} \dfrac{(x-2)^n}{3^n}$

60. $\sum_{n=0}^{\infty} (-4)^n (x-5)^n$

61. $\sum_{n=0}^{\infty} \dfrac{2^n}{x^n}$

62. $\sum_{n=0}^{\infty} \dfrac{\sin^n x}{3^n}$

63. $\sum_{n=0}^{\infty} e^{nx}$

64. We have seen that the harmonic series is a divergent series whose terms approach 0. Show that

$$\sum_{n=1}^{\infty} \ln\left(1 + \frac{1}{n}\right)$$

is another series with this property.

CAS 65–66 Use the partial fraction command on your CAS to find a convenient expression for the partial sum, and then use this expression to find the sum of the series. Check your answer by using the CAS to sum the series directly.

65. $\sum_{n=1}^{\infty} \dfrac{3n^2 + 3n + 1}{(n^2 + n)^3}$ **66.** $\sum_{n=3}^{\infty} \dfrac{1}{n^5 - 5n^3 + 4n}$

67. If the nth partial sum of a series $\sum_{n=1}^{\infty} a_n$ is

$$s_n = \frac{n-1}{n+1}$$

find a_n and $\sum_{n=1}^{\infty} a_n$.

68. If the nth partial sum of a series $\sum_{n=1}^{\infty} a_n$ is $s_n = 3 - n2^{-n}$, find a_n and $\sum_{n=1}^{\infty} a_n$.

69. A doctor prescribes a 100-mg antibiotic tablet to be taken every eight hours. Just before each tablet is taken, 20% of the drug remains in the body.
(a) How much of the drug is in the body just after the second tablet is taken? After the third tablet?
(b) If Q_n is the quantity of the antibiotic in the body just after the nth tablet is taken, find an equation that expresses Q_{n+1} in terms of Q_n.
(c) What quantity of the antibiotic remains in the body in the long run?

70. A patient is injected with a drug every 12 hours. Immediately before each injection the concentration of the drug has been reduced by 90% and the new dose increases the concentration by 1.5 mg/L.
(a) What is the concentration after three doses?
(b) If C_n is the concentration after the nth dose, find a formula for C_n as a function of n.
(c) What is the limiting value of the concentration?

71. A patient takes 150 mg of a drug at the same time every day. Just before each tablet is taken, 5% of the drug remains in the body.
(a) What quantity of the drug is in the body after the third tablet? After the nth tablet?
(b) What quantity of the drug remains in the body in the long run?

72. After injection of a dose D of insulin, the concentration of insulin in a patient's system decays exponentially and so it can be written as De^{-at}, where t represents time in hours and a is a positive constant.
(a) If a dose D is injected every T hours, write an expression for the sum of the residual concentrations just before the $(n+1)$st injection.
(b) Determine the limiting pre-injection concentration.
(c) If the concentration of insulin must always remain at or above a critical value C, determine a minimal dosage D in terms of C, a, and T.

73. When money is spent on goods and services, those who receive the money also spend some of it. The people receiving some of the twice-spent money will spend some of that, and so on. Economists call this chain reaction the *multiplier effect*. In a hypothetical isolated community, the local government begins the process by spending D dollars. Suppose that each recipient of spent money spends $100c\%$ and saves $100s\%$ of the money that he or she receives. The values c and s are called the *marginal propensity to consume* and the *marginal propensity to save* and, of course, $c + s = 1$.
(a) Let S_n be the total spending that has been generated after n transactions. Find an equation for S_n.
(b) Show that $\lim_{n \to \infty} S_n = kD$, where $k = 1/s$. The number k is called the *multiplier*. What is the multiplier if the marginal propensity to consume is 80%?

Note: The federal government uses this principle to justify deficit spending. Banks use this principle to justify lending a large percentage of the money that they receive in deposits.

74. A certain ball has the property that each time it falls from a height h onto a hard, level surface, it rebounds to a height rh, where $0 < r < 1$. Suppose that the ball is dropped from an initial height of H meters.
(a) Assuming that the ball continues to bounce indefinitely, find the total distance that it travels.
(b) Calculate the total time that the ball travels. (Use the fact that the ball falls $\frac{1}{2}gt^2$ meters in t seconds.)
(c) Suppose that each time the ball strikes the surface with velocity v it rebounds with velocity $-kv$, where $0 < k < 1$. How long will it take for the ball to come to rest?

75. Find the value of c if

$$\sum_{n=2}^{\infty} (1 + c)^{-n} = 2$$

76. Find the value of c such that

$$\sum_{n=0}^{\infty} e^{nc} = 10$$

77. In Example 9 we showed that the harmonic series is divergent. Here we outline another method, making use of the fact that $e^x > 1 + x$ for any $x > 0$. (See Exercise 4.3.84.)

If s_n is the nth partial sum of the harmonic series, show that $e^{s_n} > n + 1$. Why does this imply that the harmonic series is divergent?

78. Graph the curves $y = x^n$, $0 \leq x \leq 1$, for $n = 0, 1, 2, 3, 4, \ldots$ on a common screen. By finding the areas between successive curves, give a geometric demonstration of the fact, shown in Example 8, that

$$\sum_{n=1}^{\infty} \frac{1}{n(n+1)} = 1$$

79. The figure shows two circles C and D of radius 1 that touch at P. The line T is a common tangent line; C_1 is the circle that touches C, D, and T; C_2 is the circle that touches C, D, and C_1; C_3 is the circle that touches C, D, and C_2. This procedure can be continued indefinitely and produces an infinite sequence of circles $\{C_n\}$. Find an expression for the diameter of C_n and thus provide another geometric demonstration of Example 8.

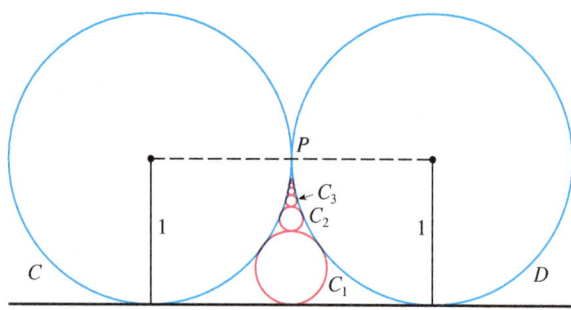

80. A right triangle ABC is given with $\angle A = \theta$ and $|AC| = b$. CD is drawn perpendicular to AB, DE is drawn perpendicular to BC, $EF \perp AB$, and this process is continued indefinitely, as shown in the figure. Find the total length of all the perpendiculars

$$|CD| + |DE| + |EF| + |FG| + \cdots$$

in terms of b and θ.

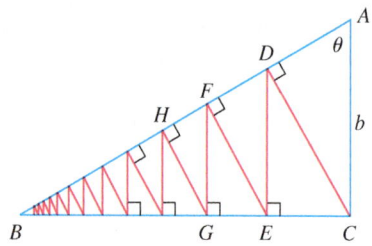

81. What is wrong with the following calculation?

$$0 = 0 + 0 + 0 + \cdots$$
$$= (1 - 1) + (1 - 1) + (1 - 1) + \cdots$$
$$= 1 - 1 + 1 - 1 + 1 - 1 + \cdots$$
$$= 1 + (-1 + 1) + (-1 + 1) + (-1 + 1) + \cdots$$
$$= 1 + 0 + 0 + 0 + \cdots = 1$$

(Guido Ubaldus thought that this proved the existence of God because "something has been created out of nothing.")

82. Suppose that $\sum_{n=1}^{\infty} a_n$ ($a_n \neq 0$) is known to be a convergent series. Prove that $\sum_{n=1}^{\infty} 1/a_n$ is a divergent series.

83. Prove part (i) of Theorem 8.

84. If $\Sigma\, a_n$ is divergent and $c \neq 0$, show that $\Sigma\, ca_n$ is divergent.

85. If $\Sigma\, a_n$ is convergent and $\Sigma\, b_n$ is divergent, show that the series $\Sigma\, (a_n + b_n)$ is divergent. [*Hint:* Argue by contradiction.]

86. If $\Sigma\, a_n$ and $\Sigma\, b_n$ are both divergent, is $\Sigma\, (a_n + b_n)$ necessarily divergent?

87. Suppose that a series $\Sigma\, a_n$ has positive terms and its partial sums s_n satisfy the inequality $s_n \leq 1000$ for all n. Explain why $\Sigma\, a_n$ must be convergent.

88. The Fibonacci sequence was defined in Section 11.1 by the equations

$$f_1 = 1, \quad f_2 = 1, \quad f_n = f_{n-1} + f_{n-2} \quad n \geq 3$$

Show that each of the following statements is true.

(a) $\dfrac{1}{f_{n-1} f_{n+1}} = \dfrac{1}{f_{n-1} f_n} - \dfrac{1}{f_n f_{n+1}}$

(b) $\displaystyle\sum_{n=2}^{\infty} \dfrac{1}{f_{n-1} f_{n+1}} = 1$

(c) $\displaystyle\sum_{n=2}^{\infty} \dfrac{f_n}{f_{n-1} f_{n+1}} = 2$

89. The **Cantor set**, named after the German mathematician Georg Cantor (1845–1918), is constructed as follows. We start with the closed interval $[0, 1]$ and remove the open interval $\left(\tfrac{1}{3}, \tfrac{2}{3}\right)$. That leaves the two intervals $\left[0, \tfrac{1}{3}\right]$ and $\left[\tfrac{2}{3}, 1\right]$ and we remove the open middle third of each. Four intervals remain and again we remove the open middle third of each of them. We continue this procedure indefinitely, at each step removing the open middle third of every interval that remains from the preceding step. The Cantor set consists of the numbers that remain in $[0, 1]$ after all those intervals have been removed.

(a) Show that the total length of all the intervals that are removed is 1. Despite that, the Cantor set contains infinitely many numbers. Give examples of some numbers in the Cantor set.

(b) The **Sierpinski carpet** is a two-dimensional counterpart of the Cantor set. It is constructed by removing the center one-ninth of a square of side 1, then removing the centers of the eight smaller remaining squares, and so on. (The figure shows the first three steps of the construction.) Show that the sum of the areas of the removed squares is 1. This implies that the Sierpinski carpet has area 0.

90. (a) A sequence $\{a_n\}$ is defined recursively by the equation $a_n = \tfrac{1}{2}(a_{n-1} + a_{n-2})$ for $n \geq 3$, where a_1 and a_2 can be any real numbers. Experiment with various values of a_1 and a_2 and use your calculator to guess the limit of the sequence.

(b) Find $\lim_{n\to\infty} a_n$ in terms of a_1 and a_2 by expressing $a_{n+1} - a_n$ in terms of $a_2 - a_1$ and summing a series.

91. Consider the series $\sum_{n=1}^{\infty} n/(n+1)!$.
  (a) Find the partial sums s_1, s_2, s_3, and s_4. Do you recognize the denominators? Use the pattern to guess a formula for s_n.
  (b) Use mathematical induction to prove your guess.
  (c) Show that the given infinite series is convergent, and find its sum.

92. In the figure at the right there are infinitely many circles approaching the vertices of an equilateral triangle, each circle touching other circles and sides of the triangle. If the triangle has sides of length 1, find the total area occupied by the circles.

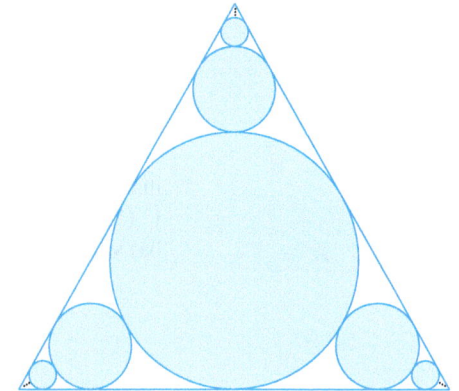

11.3 The Integral Test and Estimates of Sums

In general, it is difficult to find the exact sum of a series. We were able to accomplish this for geometric series and the series $\sum 1/[n(n+1)]$ because in each of those cases we could find a simple formula for the nth partial sum s_n. But usually it isn't easy to discover such a formula. Therefore, in the next few sections, we develop several tests that enable us to determine whether a series is convergent or divergent without explicitly finding its sum. (In some cases, however, our methods will enable us to find good estimates of the sum.) Our first test involves improper integrals.

We begin by investigating the series whose terms are the reciprocals of the squares of the positive integers:

$$\sum_{n=1}^{\infty} \frac{1}{n^2} = \frac{1}{1^2} + \frac{1}{2^2} + \frac{1}{3^2} + \frac{1}{4^2} + \frac{1}{5^2} + \cdots$$

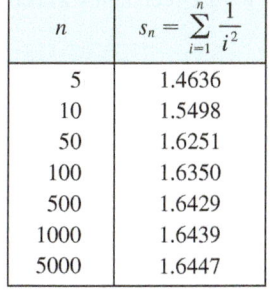

n	$s_n = \sum_{i=1}^{n} \dfrac{1}{i^2}$
5	1.4636
10	1.5498
50	1.6251
100	1.6350
500	1.6429
1000	1.6439
5000	1.6447

There's no simple formula for the sum s_n of the first n terms, but the computer-generated table of approximate values given in the margin suggests that the partial sums are approaching a number near 1.64 as $n \to \infty$ and so it looks as if the series is convergent.

We can confirm this impression with a geometric argument. Figure 1 shows the curve $y = 1/x^2$ and rectangles that lie below the curve. The base of each rectangle is an interval of length 1; the height is equal to the value of the function $y = 1/x^2$ at the right endpoint of the interval.

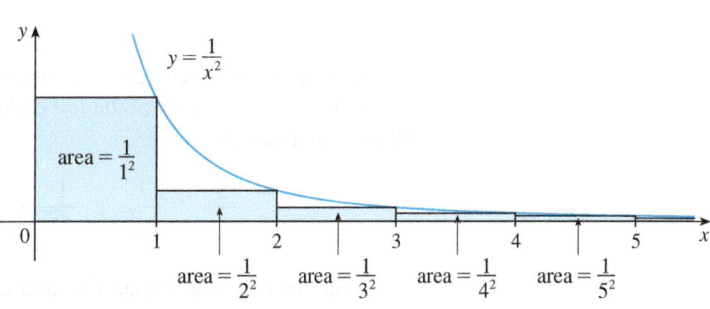

FIGURE 1

So the sum of the areas of the rectangles is

$$\frac{1}{1^2} + \frac{1}{2^2} + \frac{1}{3^2} + \frac{1}{4^2} + \frac{1}{5^2} + \cdots = \sum_{n=1}^{\infty} \frac{1}{n^2}$$

If we exclude the first rectangle, the total area of the remaining rectangles is smaller than the area under the curve $y = 1/x^2$ for $x \geq 1$, which is the value of the integral $\int_1^{\infty} (1/x^2)\, dx$. In Section 7.8 we discovered that this improper integral is convergent and has value 1. So the picture shows that all the partial sums are less than

$$\frac{1}{1^2} + \int_1^{\infty} \frac{1}{x^2}\, dx = 2$$

Thus the partial sums are bounded. We also know that the partial sums are increasing (because all the terms are positive). Therefore the partial sums converge (by the Monotonic Sequence Theorem) and so the series is convergent. The sum of the series (the limit of the partial sums) is also less than 2:

$$\sum_{n=1}^{\infty} \frac{1}{n^2} = \frac{1}{1^2} + \frac{1}{2^2} + \frac{1}{3^2} + \frac{1}{4^2} + \cdots < 2$$

[The exact sum of this series was found by the Swiss mathematician Leonhard Euler (1707–1783) to be $\pi^2/6$, but the proof of this fact is quite difficult. (See Problem 6 in the Problems Plus following Chapter 15.)]

Now let's look at the series

$$\sum_{n=1}^{\infty} \frac{1}{\sqrt{n}} = \frac{1}{\sqrt{1}} + \frac{1}{\sqrt{2}} + \frac{1}{\sqrt{3}} + \frac{1}{\sqrt{4}} + \frac{1}{\sqrt{5}} + \cdots$$

The table of values of s_n suggests that the partial sums aren't approaching a finite number, so we suspect that the given series may be divergent. Again we use a picture for confirmation. Figure 2 shows the curve $y = 1/\sqrt{x}$, but this time we use rectangles whose tops lie *above* the curve.

n	$s_n = \sum_{i=1}^{n} \dfrac{1}{\sqrt{i}}$
5	3.2317
10	5.0210
50	12.7524
100	18.5896
500	43.2834
1000	61.8010
5000	139.9681

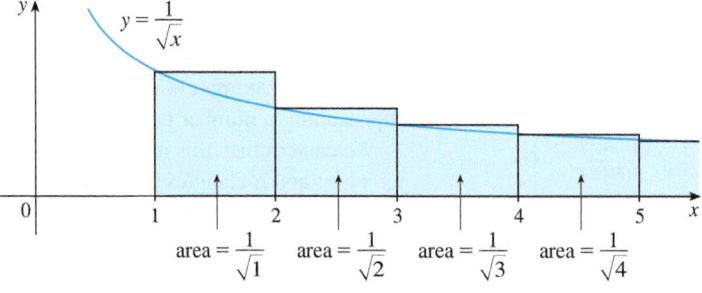

FIGURE 2

The base of each rectangle is an interval of length 1. The height is equal to the value of the function $y = 1/\sqrt{x}$ at the *left* endpoint of the interval. So the sum of the areas of all the rectangles is

$$\frac{1}{\sqrt{1}} + \frac{1}{\sqrt{2}} + \frac{1}{\sqrt{3}} + \frac{1}{\sqrt{4}} + \frac{1}{\sqrt{5}} + \cdots = \sum_{n=1}^{\infty} \frac{1}{\sqrt{n}}$$

This total area is greater than the area under the curve $y = 1/\sqrt{x}$ for $x \geq 1$, which is equal to the integral $\int_1^\infty (1/\sqrt{x})\,dx$. But we know from Section 7.8 that this improper integral is divergent. In other words, the area under the curve is infinite. So the sum of the series must be infinite; that is, the series is divergent.

The same sort of geometric reasoning that we used for these two series can be used to prove the following test. (The proof is given at the end of this section.)

The Integral Test Suppose f is a continuous, positive, decreasing function on $[1, \infty)$ and let $a_n = f(n)$. Then the series $\sum_{n=1}^\infty a_n$ is convergent if and only if the improper integral $\int_1^\infty f(x)\,dx$ is convergent. In other words:

(i) If $\displaystyle\int_1^\infty f(x)\,dx$ is convergent, then $\displaystyle\sum_{n=1}^\infty a_n$ is convergent.

(ii) If $\displaystyle\int_1^\infty f(x)\,dx$ is divergent, then $\displaystyle\sum_{n=1}^\infty a_n$ is divergent.

NOTE When we use the Integral Test, it is not necessary to start the series or the integral at $n = 1$. For instance, in testing the series

$$\sum_{n=4}^\infty \frac{1}{(n-3)^2} \quad \text{we use} \quad \int_4^\infty \frac{1}{(x-3)^2}\,dx$$

Also, it is not necessary that f be *always* decreasing. What is important is that f be *ultimately* decreasing, that is, decreasing for x larger than some number N. Then $\sum_{n=N}^\infty a_n$ is convergent, so $\sum_{n=1}^\infty a_n$ is convergent by Note 4 of Section 11.2.

EXAMPLE 1 Test the series $\displaystyle\sum_{n=1}^\infty \frac{1}{n^2+1}$ for convergence or divergence.

SOLUTION The function $f(x) = 1/(x^2 + 1)$ is continuous, positive, and decreasing on $[1, \infty)$ so we use the Integral Test:

$$\int_1^\infty \frac{1}{x^2+1}\,dx = \lim_{t\to\infty} \int_1^t \frac{1}{x^2+1}\,dx = \lim_{t\to\infty} \tan^{-1}x \Big]_1^t$$

$$= \lim_{t\to\infty}\left(\tan^{-1}t - \frac{\pi}{4}\right) = \frac{\pi}{2} - \frac{\pi}{4} = \frac{\pi}{4}$$

Thus $\int_1^\infty 1/(x^2+1)\,dx$ is a convergent integral and so, by the Integral Test, the series $\sum 1/(n^2+1)$ is convergent. ∎

EXAMPLE 2 For what values of p is the series $\displaystyle\sum_{n=1}^\infty \frac{1}{n^p}$ convergent?

SOLUTION If $p < 0$, then $\lim_{n\to\infty}(1/n^p) = \infty$. If $p = 0$, then $\lim_{n\to\infty}(1/n^p) = 1$. In either case $\lim_{n\to\infty}(1/n^p) \neq 0$, so the given series diverges by the Test for Divergence (11.2.7).

If $p > 0$, then the function $f(x) = 1/x^p$ is clearly continuous, positive, and decreasing on $[1, \infty)$. We found in Chapter 7 [see (7.8.2)] that

$$\int_1^\infty \frac{1}{x^p}\,dx \quad \text{converges if } p > 1 \text{ and diverges if } p \leq 1$$

In order to use the Integral Test we need to be able to evaluate $\int_1^\infty f(x)\,dx$ and therefore we have to be able to find an antiderivative of f. Frequently this is difficult or impossible, so we need other tests for convergence too.

It follows from the Integral Test that the series $\Sigma \, 1/n^p$ converges if $p > 1$ and diverges if $0 < p \leq 1$. (For $p = 1$, this series is the harmonic series discussed in Example 11.2.9.) ∎

The series in Example 2 is called the *p*-series. It is important in the rest of this chapter, so we summarize the results of Example 2 for future reference as follows.

> **1** The *p*-series $\displaystyle\sum_{n=1}^{\infty} \frac{1}{n^p}$ is convergent if $p > 1$ and divergent if $p \leq 1$.

EXAMPLE 3
(a) The series
$$\sum_{n=1}^{\infty} \frac{1}{n^3} = \frac{1}{1^3} + \frac{1}{2^3} + \frac{1}{3^3} + \frac{1}{4^3} + \cdots$$
is convergent because it is a *p*-series with $p = 3 > 1$.

(b) The series
$$\sum_{n=1}^{\infty} \frac{1}{n^{1/3}} = \sum_{n=1}^{\infty} \frac{1}{\sqrt[3]{n}} = 1 + \frac{1}{\sqrt[3]{2}} + \frac{1}{\sqrt[3]{3}} + \frac{1}{\sqrt[3]{4}} + \cdots$$
is divergent because it is a *p*-series with $p = \frac{1}{3} < 1$. ∎

NOTE We should *not* infer from the Integral Test that the sum of the series is equal to the value of the integral. In fact,
$$\sum_{n=1}^{\infty} \frac{1}{n^2} = \frac{\pi^2}{6} \quad \text{whereas} \quad \int_1^{\infty} \frac{1}{x^2} \, dx = 1$$
Therefore, in general,
$$\sum_{n=1}^{\infty} a_n \neq \int_1^{\infty} f(x) \, dx$$

EXAMPLE 4 Determine whether the series $\displaystyle\sum_{n=1}^{\infty} \frac{\ln n}{n}$ converges or diverges.

SOLUTION The function $f(x) = (\ln x)/x$ is positive and continuous for $x > 1$ because the logarithm function is continuous. But it is not obvious whether or not f is decreasing, so we compute its derivative:
$$f'(x) = \frac{(1/x)x - \ln x}{x^2} = \frac{1 - \ln x}{x^2}$$

Thus $f'(x) < 0$ when $\ln x > 1$, that is, when $x > e$. It follows that f is decreasing when $x > e$ and so we can apply the Integral Test:
$$\int_1^{\infty} \frac{\ln x}{x} \, dx = \lim_{t \to \infty} \int_1^t \frac{\ln x}{x} \, dx = \lim_{t \to \infty} \frac{(\ln x)^2}{2} \bigg]_1^t$$
$$= \lim_{t \to \infty} \frac{(\ln t)^2}{2} = \infty$$

Since this improper integral is divergent, the series $\Sigma \, (\ln n)/n$ is also divergent by the Integral Test. ∎

Estimating the Sum of a Series

Suppose we have been able to use the Integral Test to show that a series $\Sigma\, a_n$ is convergent and we now want to find an approximation to the sum s of the series. Of course, any partial sum s_n is an approximation to s because $\lim_{n\to\infty} s_n = s$. But how good is such an approximation? To find out, we need to estimate the size of the **remainder**

$$R_n = s - s_n = a_{n+1} + a_{n+2} + a_{n+3} + \cdots$$

The remainder R_n is the error made when s_n, the sum of the first n terms, is used as an approximation to the total sum.

We use the same notation and ideas as in the Integral Test, assuming that f is decreasing on $[n, \infty)$. Comparing the areas of the rectangles with the area under $y = f(x)$ for $x > n$ in Figure 3, we see that

$$R_n = a_{n+1} + a_{n+2} + \cdots \leq \int_n^\infty f(x)\, dx$$

Similarly, we see from Figure 4 that

$$R_n = a_{n+1} + a_{n+2} + \cdots \geq \int_{n+1}^\infty f(x)\, dx$$

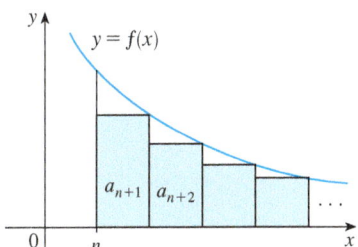

FIGURE 3

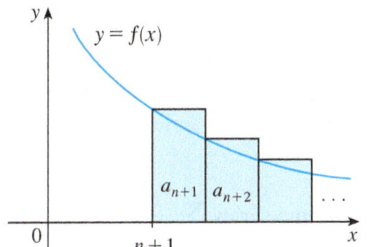

FIGURE 4

So we have proved the following error estimate.

[2] Remainder Estimate for the Integral Test Suppose $f(k) = a_k$, where f is a continuous, positive, decreasing function for $x \geq n$ and $\Sigma\, a_n$ is convergent. If $R_n = s - s_n$, then

$$\int_{n+1}^\infty f(x)\, dx \leq R_n \leq \int_n^\infty f(x)\, dx$$

EXAMPLE 5

(a) Approximate the sum of the series $\Sigma\, 1/n^3$ by using the sum of the first 10 terms. Estimate the error involved in this approximation.
(b) How many terms are required to ensure that the sum is accurate to within 0.0005?

SOLUTION In both parts (a) and (b) we need to know $\int_n^\infty f(x)\, dx$. With $f(x) = 1/x^3$, which satisfies the conditions of the Integral Test, we have

$$\int_n^\infty \frac{1}{x^3}\, dx = \lim_{t\to\infty} \left[-\frac{1}{2x^2} \right]_n^t = \lim_{t\to\infty} \left(-\frac{1}{2t^2} + \frac{1}{2n^2} \right) = \frac{1}{2n^2}$$

(a) Approximating the sum of the series by the 10th partial sum, we have

$$\sum_{n=1}^\infty \frac{1}{n^3} \approx s_{10} = \frac{1}{1^3} + \frac{1}{2^3} + \frac{1}{3^3} + \cdots + \frac{1}{10^3} \approx 1.1975$$

According to the remainder estimate in (2), we have

$$R_{10} \leq \int_{10}^\infty \frac{1}{x^3}\, dx = \frac{1}{2(10)^2} = \frac{1}{200}$$

So the size of the error is at most 0.005.

(b) Accuracy to within 0.0005 means that we have to find a value of n such that $R_n \leq 0.0005$. Since

$$R_n \leq \int_n^\infty \frac{1}{x^3}\,dx = \frac{1}{2n^2}$$

we want

$$\frac{1}{2n^2} < 0.0005$$

Solving this inequality, we get

$$n^2 > \frac{1}{0.001} = 1000 \quad \text{or} \quad n > \sqrt{1000} \approx 31.6$$

We need 32 terms to ensure accuracy to within 0.0005. ∎

If we add s_n to each side of the inequalities in (2), we get

$$\boxed{s_n + \int_{n+1}^\infty f(x)\,dx \leq s \leq s_n + \int_n^\infty f(x)\,dx}$$

[3]

because $s_n + R_n = s$. The inequalities in (3) give a lower bound and an upper bound for s. They provide a more accurate approximation to the sum of the series than the partial sum s_n does.

Although Euler was able to calculate the exact sum of the p-series for $p = 2$, nobody has been able to find the exact sum for $p = 3$. In Example 6, however, we show how to *estimate* this sum.

EXAMPLE 6 Use (3) with $n = 10$ to estimate the sum of the series $\sum_{n=1}^\infty \frac{1}{n^3}$.

SOLUTION The inequalities in (3) become

$$s_{10} + \int_{11}^\infty \frac{1}{x^3}\,dx \leq s \leq s_{10} + \int_{10}^\infty \frac{1}{x^3}\,dx$$

From Example 5 we know that

$$\int_n^\infty \frac{1}{x^3}\,dx = \frac{1}{2n^2}$$

so

$$s_{10} + \frac{1}{2(11)^2} \leq s \leq s_{10} + \frac{1}{2(10)^2}$$

Using $s_{10} \approx 1.197532$, we get

$$1.201664 \leq s \leq 1.202532$$

If we approximate s by the midpoint of this interval, then the error is at most half the length of the interval. So

$$\sum_{n=1}^\infty \frac{1}{n^3} \approx 1.2021 \quad \text{with error} < 0.0005 \quad \blacksquare$$

If we compare Example 6 with Example 5, we see that the improved estimate in (3) can be much better than the estimate $s \approx s_n$. To make the error smaller than 0.0005 we had to use 32 terms in Example 5 but only 10 terms in Example 6.

■ Proof of the Integral Test

We have already seen the basic idea behind the proof of the Integral Test in Figures 1 and 2 for the series $\Sigma\, 1/n^2$ and $\Sigma\, 1/\sqrt{n}$. For the general series $\Sigma\, a_n$, look at Figures 5 and 6. The area of the first shaded rectangle in Figure 5 is the value of f at the right endpoint of $[1, 2]$, that is, $f(2) = a_2$. So, comparing the areas of the shaded rectangles with the area under $y = f(x)$ from 1 to n, we see that

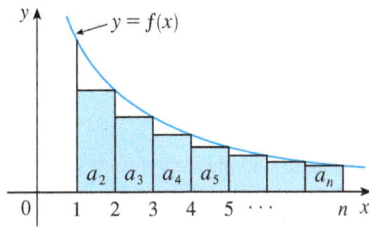

FIGURE 5

$$\boxed{4} \qquad a_2 + a_3 + \cdots + a_n \leq \int_1^n f(x)\, dx$$

(Notice that this inequality depends on the fact that f is decreasing.) Likewise, Figure 6 shows that

$$\boxed{5} \qquad \int_1^n f(x)\, dx \leq a_1 + a_2 + \cdots + a_{n-1}$$

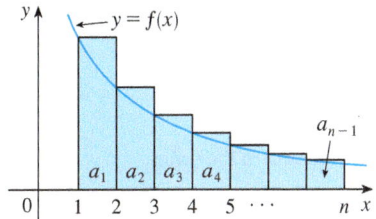

FIGURE 6

(i) If $\int_1^\infty f(x)\, dx$ is convergent, then (4) gives

$$\sum_{i=2}^n a_i \leq \int_1^n f(x)\, dx \leq \int_1^\infty f(x)\, dx$$

since $f(x) \geq 0$. Therefore

$$s_n = a_1 + \sum_{i=2}^n a_i \leq a_1 + \int_1^\infty f(x)\, dx = M, \text{ say}$$

Since $s_n \leq M$ for all n, the sequence $\{s_n\}$ is bounded above. Also

$$s_{n+1} = s_n + a_{n+1} \geq s_n$$

since $a_{n+1} = f(n+1) \geq 0$. Thus $\{s_n\}$ is an increasing bounded sequence and so it is convergent by the Monotonic Sequence Theorem (11.1.12). This means that $\Sigma\, a_n$ is convergent.

(ii) If $\int_1^\infty f(x)\, dx$ is divergent, then $\int_1^n f(x)\, dx \to \infty$ as $n \to \infty$ because $f(x) \geq 0$. But (5) gives

$$\int_1^n f(x)\, dx \leq \sum_{i=1}^{n-1} a_i = s_{n-1}$$

and so $s_{n-1} \to \infty$. This implies that $s_n \to \infty$ and so $\Sigma\, a_n$ diverges. ■

11.3 EXERCISES

1. Draw a picture to show that

$$\sum_{n=2}^{\infty} \frac{1}{n^{1.3}} < \int_1^\infty \frac{1}{x^{1.3}}\, dx$$

What can you conclude about the series?

2. Suppose f is a continuous positive decreasing function for $x \geq 1$ and $a_n = f(n)$. By drawing a picture, rank the following three quantities in increasing order:

$$\int_1^6 f(x)\, dx \qquad \sum_{i=1}^5 a_i \qquad \sum_{i=2}^6 a_i$$

3–8 Use the Integral Test to determine whether the series is convergent or divergent.

3. $\displaystyle\sum_{n=1}^{\infty} n^{-3}$

4. $\displaystyle\sum_{n=1}^{\infty} n^{-0.3}$

5. $\displaystyle\sum_{n=1}^{\infty} \frac{2}{5n-1}$

6. $\displaystyle\sum_{n=1}^{\infty} \frac{1}{(3n-1)^4}$

7. $\displaystyle\sum_{n=1}^{\infty} \frac{n}{n^2+1}$

8. $\displaystyle\sum_{n=1}^{\infty} n^2 e^{-n^3}$

9–26 Determine whether the series is convergent or divergent.

9. $\sum_{n=1}^{\infty} \frac{1}{n^{\sqrt{2}}}$

10. $\sum_{n=3}^{\infty} n^{-0.9999}$

11. $1 + \frac{1}{8} + \frac{1}{27} + \frac{1}{64} + \frac{1}{125} + \cdots$

12. $\frac{1}{5} + \frac{1}{7} + \frac{1}{9} + \frac{1}{11} + \frac{1}{13} + \cdots$

13. $\frac{1}{3} + \frac{1}{7} + \frac{1}{11} + \frac{1}{15} + \frac{1}{19} + \cdots$

14. $1 + \frac{1}{2\sqrt{2}} + \frac{1}{3\sqrt{3}} + \frac{1}{4\sqrt{4}} + \frac{1}{5\sqrt{5}} + \cdots$

15. $\sum_{n=1}^{\infty} \frac{\sqrt{n}+4}{n^2}$

16. $\sum_{n=1}^{\infty} \frac{\sqrt{n}}{1+n^{3/2}}$

17. $\sum_{n=1}^{\infty} \frac{1}{n^2+4}$

18. $\sum_{n=1}^{\infty} \frac{1}{n^2+2n+2}$

19. $\sum_{n=1}^{\infty} \frac{n^3}{n^4+4}$

20. $\sum_{n=3}^{\infty} \frac{3n-4}{n^2-2n}$

21. $\sum_{n=2}^{\infty} \frac{1}{n \ln n}$

22. $\sum_{n=2}^{\infty} \frac{\ln n}{n^2}$

23. $\sum_{k=1}^{\infty} ke^{-k}$

24. $\sum_{k=1}^{\infty} ke^{-k^2}$

25. $\sum_{n=1}^{\infty} \frac{1}{n^2+n^3}$

26. $\sum_{n=1}^{\infty} \frac{n}{n^4+1}$

27–28 Explain why the Integral Test can't be used to determine whether the series is convergent.

27. $\sum_{n=1}^{\infty} \frac{\cos \pi n}{\sqrt{n}}$

28. $\sum_{n=1}^{\infty} \frac{\cos^2 n}{1+n^2}$

29–32 Find the values of p for which the series is convergent.

29. $\sum_{n=2}^{\infty} \frac{1}{n(\ln n)^p}$

30. $\sum_{n=3}^{\infty} \frac{1}{n \ln n \, [\ln(\ln n)]^p}$

31. $\sum_{n=1}^{\infty} n(1+n^2)^p$

32. $\sum_{n=1}^{\infty} \frac{\ln n}{n^p}$

33. The Riemann zeta-function ζ is defined by

$$\zeta(x) = \sum_{n=1}^{\infty} \frac{1}{n^x}$$

and is used in number theory to study the distribution of prime numbers. What is the domain of ζ?

34. Leonhard Euler was able to calculate the exact sum of the p-series with $p = 2$:

$$\zeta(2) = \sum_{n=1}^{\infty} \frac{1}{n^2} = \frac{\pi^2}{6}$$

(See page 720.) Use this fact to find the sum of each series.

(a) $\sum_{n=2}^{\infty} \frac{1}{n^2}$

(b) $\sum_{n=3}^{\infty} \frac{1}{(n+1)^2}$

(c) $\sum_{n=1}^{\infty} \frac{1}{(2n)^2}$

35. Euler also found the sum of the p-series with $p = 4$:

$$\zeta(4) = \sum_{n=1}^{\infty} \frac{1}{n^4} = \frac{\pi^4}{90}$$

Use Euler's result to find the sum of the series.

(a) $\sum_{n=1}^{\infty} \left(\frac{3}{n}\right)^4$

(b) $\sum_{k=5}^{\infty} \frac{1}{(k-2)^4}$

36. (a) Find the partial sum s_{10} of the series $\sum_{n=1}^{\infty} 1/n^4$. Estimate the error in using s_{10} as an approximation to the sum of the series.
(b) Use (3) with $n = 10$ to give an improved estimate of the sum.
(c) Compare your estimate in part (b) with the exact value given in Exercise 35.
(d) Find a value of n so that s_n is within 0.00001 of the sum.

37. (a) Use the sum of the first 10 terms to estimate the sum of the series $\sum_{n=1}^{\infty} 1/n^2$. How good is this estimate?
(b) Improve this estimate using (3) with $n = 10$.
(c) Compare your estimate in part (b) with the exact value given in Exercise 34.
(d) Find a value of n that will ensure that the error in the approximation $s \approx s_n$ is less than 0.001.

38. Find the sum of the series $\sum_{n=1}^{\infty} ne^{-2n}$ correct to four decimal places.

39. Estimate $\sum_{n=1}^{\infty} (2n+1)^{-6}$ correct to five decimal places.

40. How many terms of the series $\sum_{n=2}^{\infty} 1/[n(\ln n)^2]$ would you need to add to find its sum to within 0.01?

41. Show that if we want to approximate the sum of the series $\sum_{n=1}^{\infty} n^{-1.001}$ so that the error is less than 5 in the ninth decimal place, then we need to add more than $10^{11,301}$ terms!

CAS 42. (a) Show that the series $\sum_{n=1}^{\infty} (\ln n)^2/n^2$ is convergent.
(b) Find an upper bound for the error in the approximation $s \approx s_n$.
(c) What is the smallest value of n such that this upper bound is less than 0.05?
(d) Find s_n for this value of n.

43. (a) Use (4) to show that if s_n is the nth partial sum of the harmonic series, then
$$s_n \leq 1 + \ln n$$
(b) The harmonic series diverges, but very slowly. Use part (a) to show that the sum of the first million terms is less than 15 and the sum of the first billion terms is less than 22.

44. Use the following steps to show that the sequence
$$t_n = 1 + \frac{1}{2} + \frac{1}{3} + \cdots + \frac{1}{n} - \ln n$$
has a limit. (The value of the limit is denoted by γ and is called Euler's constant.)
(a) Draw a picture like Figure 6 with $f(x) = 1/x$ and interpret t_n as an area [or use (5)] to show that $t_n > 0$ for all n.
(b) Interpret
$$t_n - t_{n+1} = [\ln(n+1) - \ln n] - \frac{1}{n+1}$$
as a difference of areas to show that $t_n - t_{n+1} > 0$. Therefore $\{t_n\}$ is a decreasing sequence.
(c) Use the Monotonic Sequence Theorem to show that $\{t_n\}$ is convergent.

45. Find all positive values of b for which the series $\sum_{n=1}^{\infty} b^{\ln n}$ converges.

46. Find all values of c for which the following series converges.
$$\sum_{n=1}^{\infty} \left(\frac{c}{n} - \frac{1}{n+1} \right)$$

11.4 The Comparison Tests

In the comparison tests the idea is to compare a given series with a series that is known to be convergent or divergent. For instance, the series

$$\sum_{n=1}^{\infty} \frac{1}{2^n + 1}$$

reminds us of the series $\sum_{n=1}^{\infty} 1/2^n$, which is a geometric series with $a = \frac{1}{2}$ and $r = \frac{1}{2}$ and is therefore convergent. Because the series (1) is so similar to a convergent series, we have the feeling that it too must be convergent. Indeed, it is. The inequality

$$\frac{1}{2^n + 1} < \frac{1}{2^n}$$

shows that our given series (1) has smaller terms than those of the geometric series and therefore all its partial sums are also smaller than 1 (the sum of the geometric series). This means that its partial sums form a bounded increasing sequence, which is convergent. It also follows that the sum of the series is less than the sum of the geometric series:

$$\sum_{n=1}^{\infty} \frac{1}{2^n + 1} < 1$$

Similar reasoning can be used to prove the following test, which applies only to series whose terms are positive. The first part says that if we have a series whose terms are *smaller* than those of a known *convergent* series, then our series is also convergent. The second part says that if we start with a series whose terms are *larger* than those of a known *divergent* series, then it too is divergent.

The Comparison Test Suppose that $\sum a_n$ and $\sum b_n$ are series with positive terms.
(i) If $\sum b_n$ is convergent and $a_n \leq b_n$ for all n, then $\sum a_n$ is also convergent.
(ii) If $\sum b_n$ is divergent and $a_n \geq b_n$ for all n, then $\sum a_n$ is also divergent.

728   **CHAPTER 11**   Infinite Sequences and Series

It is important to keep in mind the distinction between a sequence and a series. A sequence is a list of numbers, whereas a series is a sum. With every series $\Sigma\, a_n$ there are associated two sequences: the sequence $\{a_n\}$ of terms and the sequence $\{s_n\}$ of partial sums.

PROOF

(i) Let $$s_n = \sum_{i=1}^{n} a_i \qquad t_n = \sum_{i=1}^{n} b_i \qquad t = \sum_{n=1}^{\infty} b_n$$

Since both series have positive terms, the sequences $\{s_n\}$ and $\{t_n\}$ are increasing $(s_{n+1} = s_n + a_{n+1} \geq s_n)$. Also $t_n \to t$, so $t_n \leq t$ for all n. Since $a_i \leq b_i$, we have $s_n \leq t_n$. Thus $s_n \leq t$ for all n. This means that $\{s_n\}$ is increasing and bounded above and therefore converges by the Monotonic Sequence Theorem. Thus $\Sigma\, a_n$ converges.

(ii) If $\Sigma\, b_n$ is divergent, then $t_n \to \infty$ (since $\{t_n\}$ is increasing). But $a_i \geq b_i$ so $s_n \geq t_n$. Thus $s_n \to \infty$. Therefore $\Sigma\, a_n$ diverges. ∎

Standard Series for Use with the Comparison Test

In using the Comparison Test we must, of course, have some known series $\Sigma\, b_n$ for the purpose of comparison. Most of the time we use one of these series:

- A p-series $\left[\Sigma\, 1/n^p \text{ converges if } p > 1 \text{ and diverges if } p \leq 1; \text{ see (11.3.1)}\right]$
- A geometric series $\left[\Sigma\, ar^{n-1} \text{ converges if } |r| < 1 \text{ and diverges if } |r| \geq 1; \text{ see (11.2.4)}\right]$

EXAMPLE 1 Determine whether the series $\displaystyle\sum_{n=1}^{\infty} \frac{5}{2n^2 + 4n + 3}$ converges or diverges.

SOLUTION For large n the dominant term in the denominator is $2n^2$, so we compare the given series with the series $\Sigma\, 5/(2n^2)$. Observe that

$$\frac{5}{2n^2 + 4n + 3} < \frac{5}{2n^2}$$

because the left side has a bigger denominator. (In the notation of the Comparison Test, a_n is the left side and b_n is the right side.) We know that

$$\sum_{n=1}^{\infty} \frac{5}{2n^2} = \frac{5}{2} \sum_{n=1}^{\infty} \frac{1}{n^2}$$

is convergent because it's a constant times a p-series with $p = 2 > 1$. Therefore

$$\sum_{n=1}^{\infty} \frac{5}{2n^2 + 4n + 3}$$

is convergent by part (i) of the Comparison Test. ∎

NOTE 1 Although the condition $a_n \leq b_n$ or $a_n \geq b_n$ in the Comparison Test is given for all n, we need verify only that it holds for $n \geq N$, where N is some fixed integer, because the convergence of a series is not affected by a finite number of terms. This is illustrated in the next example.

EXAMPLE 2 Test the series $\displaystyle\sum_{k=1}^{\infty} \frac{\ln k}{k}$ for convergence or divergence.

SOLUTION We used the Integral Test to test this series in Example 11.3.4, but we can also test it by comparing it with the harmonic series. Observe that $\ln k > 1$ for $k \geq 3$ and so

$$\frac{\ln k}{k} > \frac{1}{k} \qquad k \geq 3$$

We know that $\Sigma\, 1/k$ is divergent (p-series with $p = 1$). Thus the given series is divergent by the Comparison Test. ∎

NOTE 2 The terms of the series being tested must be smaller than those of a convergent series or larger than those of a divergent series. If the terms are larger than the terms of a convergent series or smaller than those of a divergent series, then the Comparison Test doesn't apply. Consider, for instance, the series

$$\sum_{n=1}^{\infty} \frac{1}{2^n - 1}$$

The inequality

$$\frac{1}{2^n - 1} > \frac{1}{2^n}$$

is useless as far as the Comparison Test is concerned because $\Sigma\, b_n = \Sigma\, \left(\frac{1}{2}\right)^n$ is convergent and $a_n > b_n$. Nonetheless, we have the feeling that $\Sigma\, 1/(2^n - 1)$ ought to be convergent because it is very similar to the convergent geometric series $\Sigma\, \left(\frac{1}{2}\right)^n$. In such cases the following test can be used.

> **The Limit Comparison Test** Suppose that $\Sigma\, a_n$ and $\Sigma\, b_n$ are series with positive terms. If
>
> $$\lim_{n \to \infty} \frac{a_n}{b_n} = c$$
>
> where c is a finite number and $c > 0$, then either both series converge or both diverge.

Exercises 40 and 41 deal with the cases $c = 0$ and $c = \infty$.

PROOF Let m and M be positive numbers such that $m < c < M$. Because a_n/b_n is close to c for large n, there is an integer N such that

$$m < \frac{a_n}{b_n} < M \qquad \text{when } n > N$$

and so

$$mb_n < a_n < Mb_n \qquad \text{when } n > N$$

If $\Sigma\, b_n$ converges, so does $\Sigma\, Mb_n$. Thus $\Sigma\, a_n$ converges by part (i) of the Comparison Test. If $\Sigma\, b_n$ diverges, so does $\Sigma\, mb_n$ and part (ii) of the Comparison Test shows that $\Sigma\, a_n$ diverges. ∎

EXAMPLE 3 Test the series $\sum_{n=1}^{\infty} \dfrac{1}{2^n - 1}$ for convergence or divergence.

SOLUTION We use the Limit Comparison Test with

$$a_n = \frac{1}{2^n - 1} \qquad b_n = \frac{1}{2^n}$$

and obtain

$$\lim_{n \to \infty} \frac{a_n}{b_n} = \lim_{n \to \infty} \frac{1/(2^n - 1)}{1/2^n} = \lim_{n \to \infty} \frac{2^n}{2^n - 1} = \lim_{n \to \infty} \frac{1}{1 - 1/2^n} = 1 > 0$$

Since this limit exists and $\Sigma\, 1/2^n$ is a convergent geometric series, the given series converges by the Limit Comparison Test. ∎

EXAMPLE 4 Determine whether the series $\displaystyle\sum_{n=1}^{\infty} \frac{2n^2 + 3n}{\sqrt{5 + n^5}}$ converges or diverges.

SOLUTION The dominant part of the numerator is $2n^2$ and the dominant part of the denominator is $\sqrt{n^5} = n^{5/2}$. This suggests taking

$$a_n = \frac{2n^2 + 3n}{\sqrt{5 + n^5}} \qquad b_n = \frac{2n^2}{n^{5/2}} = \frac{2}{n^{1/2}}$$

$$\lim_{n\to\infty} \frac{a_n}{b_n} = \lim_{n\to\infty} \frac{2n^2 + 3n}{\sqrt{5 + n^5}} \cdot \frac{n^{1/2}}{2} = \lim_{n\to\infty} \frac{2n^{5/2} + 3n^{3/2}}{2\sqrt{5 + n^5}}$$

$$= \lim_{n\to\infty} \frac{2 + \dfrac{3}{n}}{2\sqrt{\dfrac{5}{n^5} + 1}} = \frac{2 + 0}{2\sqrt{0 + 1}} = 1$$

Since $\Sigma\, b_n = 2\,\Sigma\, 1/n^{1/2}$ is divergent (p-series with $p = \tfrac{1}{2} < 1$), the given series diverges by the Limit Comparison Test. ∎

Notice that in testing many series we find a suitable comparison series $\Sigma\, b_n$ by keeping only the highest powers in the numerator and denominator.

Estimating Sums

If we have used the Comparison Test to show that a series $\Sigma\, a_n$ converges by comparison with a series $\Sigma\, b_n$, then we may be able to estimate the sum $\Sigma\, a_n$ by comparing remainders. As in Section 11.3, we consider the remainder

$$R_n = s - s_n = a_{n+1} + a_{n+2} + \cdots$$

For the comparison series $\Sigma\, b_n$ we consider the corresponding remainder

$$T_n = t - t_n = b_{n+1} + b_{n+2} + \cdots$$

Since $a_n \leq b_n$ for all n, we have $R_n \leq T_n$. If $\Sigma\, b_n$ is a p-series, we can estimate its remainder T_n as in Section 11.3. If $\Sigma\, b_n$ is a geometric series, then T_n is the sum of a geometric series and we can sum it exactly (see Exercises 35 and 36). In either case we know that R_n is smaller than T_n.

EXAMPLE 5 Use the sum of the first 100 terms to approximate the sum of the series $\Sigma\, 1/(n^3 + 1)$. Estimate the error involved in this approximation.

SOLUTION Since

$$\frac{1}{n^3 + 1} < \frac{1}{n^3}$$

the given series is convergent by the Comparison Test. The remainder T_n for the comparison series $\Sigma\, 1/n^3$ was estimated in Example 11.3.5 using the Remainder Estimate for the Integral Test. There we found that

$$T_n \leq \int_n^{\infty} \frac{1}{x^3}\, dx = \frac{1}{2n^2}$$

Therefore the remainder R_n for the given series satisfies

$$R_n \leq T_n \leq \frac{1}{2n^2}$$

With $n = 100$ we have

$$R_{100} \leq \frac{1}{2(100)^2} = 0.00005$$

Using a programmable calculator or a computer, we find that

$$\sum_{n=1}^{\infty} \frac{1}{n^3 + 1} \approx \sum_{n=1}^{100} \frac{1}{n^3 + 1} \approx 0.6864538$$

with error less than 0.00005. ■

11.4 EXERCISES

1. Suppose Σa_n and Σb_n are series with positive terms and Σb_n is known to be convergent.
   (a) If $a_n > b_n$ for all n, what can you say about Σa_n? Why?
   (b) If $a_n < b_n$ for all n, what can you say about Σa_n? Why?

2. Suppose Σa_n and Σb_n are series with positive terms and Σb_n is known to be divergent.
   (a) If $a_n > b_n$ for all n, what can you say about Σa_n? Why?
   (b) If $a_n < b_n$ for all n, what can you say about Σa_n? Why?

3–32 Determine whether the series converges or diverges.

3. $\sum_{n=1}^{\infty} \frac{1}{n^3 + 8}$

4. $\sum_{n=2}^{\infty} \frac{1}{\sqrt{n} - 1}$

5. $\sum_{n=1}^{\infty} \frac{n+1}{n\sqrt{n}}$

6. $\sum_{n=1}^{\infty} \frac{n-1}{n^3 + 1}$

7. $\sum_{n=1}^{\infty} \frac{9^n}{3 + 10^n}$

8. $\sum_{n=1}^{\infty} \frac{6^n}{5^n - 1}$

9. $\sum_{k=1}^{\infty} \frac{\ln k}{k}$

10. $\sum_{k=1}^{\infty} \frac{k \sin^2 k}{1 + k^3}$

11. $\sum_{k=1}^{\infty} \frac{\sqrt[3]{k}}{\sqrt{k^3 + 4k + 3}}$

12. $\sum_{k=1}^{\infty} \frac{(2k-1)(k^2-1)}{(k+1)(k^2+4)^2}$

13. $\sum_{n=1}^{\infty} \frac{1 + \cos n}{e^n}$

14. $\sum_{n=1}^{\infty} \frac{1}{\sqrt[3]{3n^4 + 1}}$

15. $\sum_{n=1}^{\infty} \frac{4^{n+1}}{3^n - 2}$

16. $\sum_{n=1}^{\infty} \frac{1}{n^n}$

17. $\sum_{n=1}^{\infty} \frac{1}{\sqrt{n^2 + 1}}$

18. $\sum_{n=1}^{\infty} \frac{2}{\sqrt{n} + 2}$

19. $\sum_{n=1}^{\infty} \frac{n+1}{n^3 + n}$

20. $\sum_{n=1}^{\infty} \frac{n^2 + n + 1}{n^4 + n^2}$

21. $\sum_{n=1}^{\infty} \frac{\sqrt{1+n}}{2+n}$

22. $\sum_{n=3}^{\infty} \frac{n+2}{(n+1)^3}$

23. $\sum_{n=1}^{\infty} \frac{5 + 2n}{(1 + n^2)^2}$

24. $\sum_{n=1}^{\infty} \frac{n + 3^n}{n + 2^n}$

25. $\sum_{n=1}^{\infty} \frac{e^n + 1}{ne^n + 1}$

26. $\sum_{n=2}^{\infty} \frac{1}{n\sqrt{n^2 - 1}}$

27. $\sum_{n=1}^{\infty} \left(1 + \frac{1}{n}\right)^2 e^{-n}$

28. $\sum_{n=1}^{\infty} \frac{e^{1/n}}{n}$

29. $\sum_{n=1}^{\infty} \frac{1}{n!}$

30. $\sum_{n=1}^{\infty} \frac{n!}{n^n}$

31. $\sum_{n=1}^{\infty} \sin\left(\frac{1}{n}\right)$

32. $\sum_{n=1}^{\infty} \frac{1}{n^{1+1/n}}$

33–36 Use the sum of the first 10 terms to approximate the sum of the series. Estimate the error.

33. $\sum_{n=1}^{\infty} \frac{1}{5 + n^5}$

34. $\sum_{n=1}^{\infty} \frac{e^{1/n}}{n^4}$

35. $\sum_{n=1}^{\infty} 5^{-n} \cos^2 n$

36. $\sum_{n=1}^{\infty} \frac{1}{3^n + 4^n}$

37. The meaning of the decimal representation of a number $0.d_1d_2d_3\ldots$ (where the digit d_i is one of the numbers 0, 1, 2, $\ldots$, 9) is that

$$0.d_1d_2d_3d_4\ldots = \frac{d_1}{10} + \frac{d_2}{10^2} + \frac{d_3}{10^3} + \frac{d_4}{10^4} + \cdots$$

Show that this series always converges.

38. For what values of p does the series $\sum_{n=2}^{\infty} 1/(n^p \ln n)$ converge?

39. Prove that if $a_n \geq 0$ and $\sum a_n$ converges, then $\sum a_n^2$ also converges.

40. (a) Suppose that $\sum a_n$ and $\sum b_n$ are series with positive terms and $\sum b_n$ is convergent. Prove that if
$$\lim_{n \to \infty} \frac{a_n}{b_n} = 0$$
then $\sum a_n$ is also convergent.
(b) Use part (a) to show that the series converges.
(i) $\sum_{n=1}^{\infty} \frac{\ln n}{n^3}$ (ii) $\sum_{n=1}^{\infty} \frac{\ln n}{\sqrt{n}\, e^n}$

41. (a) Suppose that $\sum a_n$ and $\sum b_n$ are series with positive terms and $\sum b_n$ is divergent. Prove that if
$$\lim_{n \to \infty} \frac{a_n}{b_n} = \infty$$
then $\sum a_n$ is also divergent.

(b) Use part (a) to show that the series diverges.
(i) $\sum_{n=2}^{\infty} \frac{1}{\ln n}$ (ii) $\sum_{n=1}^{\infty} \frac{\ln n}{n}$

42. Give an example of a pair of series $\sum a_n$ and $\sum b_n$ with positive terms where $\lim_{n \to \infty} (a_n/b_n) = 0$ and $\sum b_n$ diverges, but $\sum a_n$ converges. (Compare with Exercise 40.)

43. Show that if $a_n > 0$ and $\lim_{n \to \infty} na_n \neq 0$, then $\sum a_n$ is divergent.

44. Show that if $a_n > 0$ and $\sum a_n$ is convergent, then $\sum \ln(1 + a_n)$ is convergent.

45. If $\sum a_n$ is a convergent series with positive terms, is it true that $\sum \sin(a_n)$ is also convergent?

46. If $\sum a_n$ and $\sum b_n$ are both convergent series with positive terms, is it true that $\sum a_n b_n$ is also convergent?

11.5 Alternating Series

The convergence tests that we have looked at so far apply only to series with positive terms. In this section and the next we learn how to deal with series whose terms are not necessarily positive. Of particular importance are *alternating series*, whose terms alternate in sign.

An **alternating series** is a series whose terms are alternately positive and negative. Here are two examples:

$$1 - \frac{1}{2} + \frac{1}{3} - \frac{1}{4} + \frac{1}{5} - \frac{1}{6} + \cdots = \sum_{n=1}^{\infty} (-1)^{n-1} \frac{1}{n}$$

$$-\frac{1}{2} + \frac{2}{3} - \frac{3}{4} + \frac{4}{5} - \frac{5}{6} + \frac{6}{7} - \cdots = \sum_{n=1}^{\infty} (-1)^n \frac{n}{n+1}$$

We see from these examples that the nth term of an alternating series is of the form

$$a_n = (-1)^{n-1} b_n \quad \text{or} \quad a_n = (-1)^n b_n$$

where b_n is a positive number. (In fact, $b_n = |a_n|$.)

The following test says that if the terms of an alternating series decrease toward 0 in absolute value, then the series converges.

Alternating Series Test If the alternating series
$$\sum_{n=1}^{\infty} (-1)^{n-1} b_n = b_1 - b_2 + b_3 - b_4 + b_5 - b_6 + \cdots \qquad b_n > 0$$
satisfies
(i) $b_{n+1} \leq b_n$ for all n
(ii) $\lim_{n \to \infty} b_n = 0$
then the series is convergent.

Before giving the proof let's look at Figure 1, which gives a picture of the idea behind the proof. We first plot $s_1 = b_1$ on a number line. To find s_2 we subtract b_2, so s_2 is to the left of s_1. Then to find s_3 we add b_3, so s_3 is to the right of s_2. But, since $b_3 < b_2$, s_3 is to the left of s_1. Continuing in this manner, we see that the partial sums oscillate back and forth. Since $b_n \to 0$, the successive steps are becoming smaller and smaller. The even partial sums $s_2, s_4, s_6, \ldots$ are increasing and the odd partial sums $s_1, s_3, s_5, \ldots$ are decreasing. Thus it seems plausible that both are converging to some number s, which is the sum of the series. Therefore we consider the even and odd partial sums separately in the following proof.

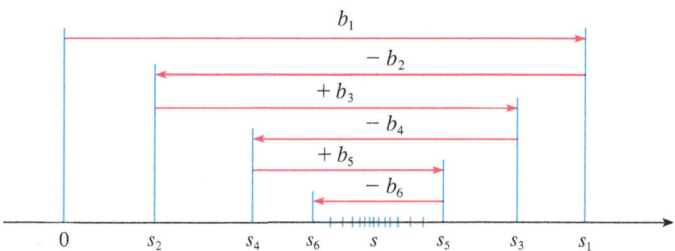

FIGURE 1

PROOF OF THE ALTERNATING SERIES TEST We first consider the even partial sums:

$$s_2 = b_1 - b_2 \geq 0 \qquad \text{since } b_2 \leq b_1$$

$$s_4 = s_2 + (b_3 - b_4) \geq s_2 \qquad \text{since } b_4 \leq b_3$$

In general $\qquad s_{2n} = s_{2n-2} + (b_{2n-1} - b_{2n}) \geq s_{2n-2} \qquad \text{since } b_{2n} \leq b_{2n-1}$

Thus $\qquad 0 \leq s_2 \leq s_4 \leq s_6 \leq \cdots \leq s_{2n} \leq \cdots$

But we can also write

$$s_{2n} = b_1 - (b_2 - b_3) - (b_4 - b_5) - \cdots - (b_{2n-2} - b_{2n-1}) - b_{2n}$$

Every term in parentheses is positive, so $s_{2n} \leq b_1$ for all n. Therefore the sequence $\{s_{2n}\}$ of even partial sums is increasing and bounded above. It is therefore convergent by the Monotonic Sequence Theorem. Let's call its limit s, that is,

$$\lim_{n \to \infty} s_{2n} = s$$

Now we compute the limit of the odd partial sums:

$$\lim_{n \to \infty} s_{2n+1} = \lim_{n \to \infty} (s_{2n} + b_{2n+1})$$

$$= \lim_{n \to \infty} s_{2n} + \lim_{n \to \infty} b_{2n+1}$$

$$= s + 0 \qquad\qquad \text{[by condition (ii)]}$$

$$= s$$

Since both the even and odd partial sums converge to s, we have $\lim_{n \to \infty} s_n = s$ [see Exercise 11.1.92(a)] and so the series is convergent.

Figure 2 illustrates Example 1 by showing the graphs of the terms $a_n = (-1)^{n-1}/n$ and the partial sums s_n. Notice how the values of s_n zigzag across the limiting value, which appears to be about 0.7. In fact, it can be proved that the exact sum of the series is $\ln 2 \approx 0.693$ (see Exercise 36).

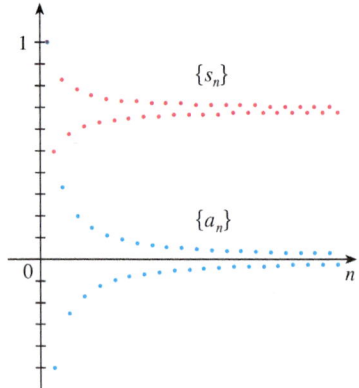

FIGURE 2

Instead of verifying condition (i) of the Alternating Series Test by computing a derivative, we could verify that $b_{n+1} < b_n$ directly by using the technique of Solution 1 of Example 11.1.13.

EXAMPLE 1 The alternating harmonic series

$$1 - \frac{1}{2} + \frac{1}{3} - \frac{1}{4} + \cdots = \sum_{n=1}^{\infty} \frac{(-1)^{n-1}}{n}$$

satisfies

(i) $b_{n+1} < b_n$ because $\dfrac{1}{n+1} < \dfrac{1}{n}$

(ii) $\lim\limits_{n \to \infty} b_n = \lim\limits_{n \to \infty} \dfrac{1}{n} = 0$

so the series is convergent by the Alternating Series Test.

EXAMPLE 2 The series $\sum\limits_{n=1}^{\infty} \dfrac{(-1)^n 3n}{4n - 1}$ is alternating, but

$$\lim_{n \to \infty} b_n = \lim_{n \to \infty} \frac{3n}{4n - 1} = \lim_{n \to \infty} \frac{3}{4 - \dfrac{1}{n}} = \frac{3}{4}$$

so condition (ii) is not satisfied. Instead, we look at the limit of the nth term of the series:

$$\lim_{n \to \infty} a_n = \lim_{n \to \infty} \frac{(-1)^n 3n}{4n - 1}$$

This limit does not exist, so the series diverges by the Test for Divergence.

EXAMPLE 3 Test the series $\sum\limits_{n=1}^{\infty} (-1)^{n+1} \dfrac{n^2}{n^3 + 1}$ for convergence or divergence.

SOLUTION The given series is alternating so we try to verify conditions (i) and (ii) of the Alternating Series Test.

Unlike the situation in Example 1, it is not obvious that the sequence given by $b_n = n^2/(n^3 + 1)$ is decreasing. However, if we consider the related function $f(x) = x^2/(x^3 + 1)$, we find that

$$f'(x) = \frac{x(2 - x^3)}{(x^3 + 1)^2}$$

Since we are considering only positive x, we see that $f'(x) < 0$ if $2 - x^3 < 0$, that is, $x > \sqrt[3]{2}$. Thus f is decreasing on the interval $(\sqrt[3]{2}, \infty)$. This means that $f(n + 1) < f(n)$ and therefore $b_{n+1} < b_n$ when $n \geq 2$. (The inequality $b_2 < b_1$ can be verified directly but all that really matters is that the sequence $\{b_n\}$ is eventually decreasing.)

Condition (ii) is readily verified:

$$\lim_{n \to \infty} b_n = \lim_{n \to \infty} \frac{n^2}{n^3 + 1} = \lim_{n \to \infty} \frac{\dfrac{1}{n}}{1 + \dfrac{1}{n^3}} = 0$$

Thus the given series is convergent by the Alternating Series Test.

Estimating Sums

A partial sum s_n of any convergent series can be used as an approximation to the total sum s, but this is not of much use unless we can estimate the accuracy of the approximation. The error involved in using $s \approx s_n$ is the remainder $R_n = s - s_n$. The next theorem says that for series that satisfy the conditions of the Alternating Series Test, the size of the error is smaller than b_{n+1}, which is the absolute value of the first neglected term.

> **Alternating Series Estimation Theorem** If $s = \Sigma (-1)^{n-1} b_n$, where $b_n > 0$, is the sum of an alternating series that satisfies
>
> (i) $b_{n+1} \leq b_n$ and (ii) $\lim\limits_{n \to \infty} b_n = 0$
>
> then
>
> $$|R_n| = |s - s_n| \leq b_{n+1}$$

You can see geometrically why the Alternating Series Estimation Theorem is true by looking at Figure 1 (on page 733). Notice that $s - s_4 < b_5$, $|s - s_5| < b_6$, and so on. Notice also that s lies between any two consecutive partial sums.

PROOF We know from the proof of the Alternating Series Test that s lies between any two consecutive partial sums s_n and s_{n+1}. (There we showed that s is larger than all the even partial sums. A similar argument shows that s is smaller than all the odd sums.) It follows that

$$|s - s_n| \leq |s_{n+1} - s_n| = b_{n+1} \qquad \blacksquare$$

By definition, $0! = 1$.

EXAMPLE 4 Find the sum of the series $\sum\limits_{n=0}^{\infty} \dfrac{(-1)^n}{n!}$ correct to three decimal places.

SOLUTION We first observe that the series is convergent by the Alternating Series Test because

(i) $\dfrac{1}{(n+1)!} = \dfrac{1}{n!(n+1)} < \dfrac{1}{n!}$

(ii) $0 < \dfrac{1}{n!} < \dfrac{1}{n} \to 0$ so $\dfrac{1}{n!} \to 0$ as $n \to \infty$

To get a feel for how many terms we need to use in our approximation, let's write out the first few terms of the series:

$$s = \frac{1}{0!} - \frac{1}{1!} + \frac{1}{2!} - \frac{1}{3!} + \frac{1}{4!} - \frac{1}{5!} + \frac{1}{6!} - \frac{1}{7!} + \cdots$$

$$= 1 - 1 + \tfrac{1}{2} - \tfrac{1}{6} + \tfrac{1}{24} - \tfrac{1}{120} + \tfrac{1}{720} - \tfrac{1}{5040} + \cdots$$

Notice that $\qquad b_7 = \tfrac{1}{5040} < \tfrac{1}{5000} = 0.0002$

and $\qquad s_6 = 1 - 1 + \tfrac{1}{2} - \tfrac{1}{6} + \tfrac{1}{24} - \tfrac{1}{120} + \tfrac{1}{720} \approx 0.368056$

By the Alternating Series Estimation Theorem we know that

$$|s - s_6| \leq b_7 < 0.0002$$

In Section 11.10 we will prove that $e^x = \sum_{n=0}^{\infty} x^n/n!$ for all x, so what we have obtained in Example 4 is actually an approximation to the number e^{-1}.

This error of less than 0.0002 does not affect the third decimal place, so we have $s \approx 0.368$ correct to three decimal places. ■

11.5 EXERCISES

1. (a) What is an alternating series?
(b) Under what conditions does an alternating series converge?
(c) If these conditions are satisfied, what can you say about the remainder after n terms?

2–20 Test the series for convergence or divergence.

2. $\frac{2}{3} - \frac{2}{5} + \frac{2}{7} - \frac{2}{9} + \frac{2}{11} - \cdots$

3. $-\frac{2}{5} + \frac{4}{6} - \frac{6}{7} + \frac{8}{8} - \frac{10}{9} + \cdots$

4. $\dfrac{1}{\ln 3} - \dfrac{1}{\ln 4} + \dfrac{1}{\ln 5} - \dfrac{1}{\ln 6} + \dfrac{1}{\ln 7} - \cdots$

5. $\displaystyle\sum_{n=1}^{\infty} \frac{(-1)^{n-1}}{3 + 5n}$

6. $\displaystyle\sum_{n=0}^{\infty} \frac{(-1)^{n+1}}{\sqrt{n+1}}$

7. $\displaystyle\sum_{n=1}^{\infty} (-1)^n \frac{3n-1}{2n+1}$

8. $\displaystyle\sum_{n=1}^{\infty} (-1)^n \frac{n^2}{n^2+n+1}$

9. $\displaystyle\sum_{n=1}^{\infty} (-1)^n e^{-n}$

10. $\displaystyle\sum_{n=1}^{\infty} (-1)^n \frac{\sqrt{n}}{2n+3}$

11. $\displaystyle\sum_{n=1}^{\infty} (-1)^{n+1} \frac{n^2}{n^3+4}$

12. $\displaystyle\sum_{n=1}^{\infty} (-1)^{n+1} n e^{-n}$

13. $\displaystyle\sum_{n=1}^{\infty} (-1)^{n-1} e^{2/n}$

14. $\displaystyle\sum_{n=1}^{\infty} (-1)^{n-1} \arctan n$

15. $\displaystyle\sum_{n=0}^{\infty} \frac{\sin\left(n+\frac{1}{2}\right)\pi}{1+\sqrt{n}}$

16. $\displaystyle\sum_{n=1}^{\infty} \frac{n \cos n\pi}{2^n}$

17. $\displaystyle\sum_{n=1}^{\infty} (-1)^n \sin\left(\frac{\pi}{n}\right)$

18. $\displaystyle\sum_{n=1}^{\infty} (-1)^n \cos\left(\frac{\pi}{n}\right)$

19. $\displaystyle\sum_{n=1}^{\infty} (-1)^n \frac{n^n}{n!}$

20. $\displaystyle\sum_{n=1}^{\infty} (-1)^n \left(\sqrt{n+1} - \sqrt{n}\right)$

21–22 Graph both the sequence of terms and the sequence of partial sums on the same screen. Use the graph to make a rough estimate of the sum of the series. Then use the Alternating Series Estimation Theorem to estimate the sum correct to four decimal places.

21. $\displaystyle\sum_{n=1}^{\infty} \frac{(-0.8)^n}{n!}$

22. $\displaystyle\sum_{n=1}^{\infty} (-1)^{n-1} \frac{n}{8^n}$

23–26 Show that the series is convergent. How many terms of the series do we need to add in order to find the sum to the indicated accuracy?

23. $\displaystyle\sum_{n=1}^{\infty} \frac{(-1)^{n+1}}{n^6}$ ($|\text{error}| < 0.00005$)

24. $\displaystyle\sum_{n=1}^{\infty} \frac{\left(-\frac{1}{3}\right)^n}{n}$ ($|\text{error}| < 0.0005$)

25. $\displaystyle\sum_{n=1}^{\infty} \frac{(-1)^{n-1}}{n^2 2^n}$ ($|\text{error}| < 0.0005$)

26. $\displaystyle\sum_{n=1}^{\infty} \left(-\frac{1}{n}\right)^n$ ($|\text{error}| < 0.00005$)

27–30 Approximate the sum of the series correct to four decimal places.

27. $\displaystyle\sum_{n=1}^{\infty} \frac{(-1)^n}{(2n)!}$

28. $\displaystyle\sum_{n=1}^{\infty} \frac{(-1)^{n+1}}{n^6}$

29. $\displaystyle\sum_{n=1}^{\infty} (-1)^n n e^{-2n}$

30. $\displaystyle\sum_{n=1}^{\infty} \frac{(-1)^{n-1}}{n 4^n}$

31. Is the 50th partial sum s_{50} of the alternating series $\sum_{n=1}^{\infty} (-1)^{n-1}/n$ an overestimate or an underestimate of the total sum? Explain.

32–34 For what values of p is each series convergent?

32. $\displaystyle\sum_{n=1}^{\infty} \frac{(-1)^{n-1}}{n^p}$

33. $\displaystyle\sum_{n=1}^{\infty} \frac{(-1)^n}{n+p}$

34. $\displaystyle\sum_{n=2}^{\infty} (-1)^{n-1} \frac{(\ln n)^p}{n}$

35. Show that the series $\Sigma \, (-1)^{n-1} b_n$, where $b_n = 1/n$ if n is odd and $b_n = 1/n^2$ if n is even, is divergent. Why does the Alternating Series Test not apply?

36. Use the following steps to show that
$$\sum_{n=1}^{\infty} \frac{(-1)^{n-1}}{n} = \ln 2$$
Let h_n and s_n be the partial sums of the harmonic and alternating harmonic series.
(a) Show that $s_{2n} = h_{2n} - h_n$.

(b) From Exercise 11.3.44 we have
$$h_n - \ln n \to \gamma \quad \text{as } n \to \infty$$
and therefore
$$h_{2n} - \ln(2n) \to \gamma \quad \text{as } n \to \infty$$
Use these facts together with part (a) to show that $s_{2n} \to \ln 2$ as $n \to \infty$.

11.6 Absolute Convergence and the Ratio and Root Tests

Given any series $\Sigma \, a_n$, we can consider the corresponding series
$$\sum_{n=1}^{\infty} |a_n| = |a_1| + |a_2| + |a_3| + \cdots$$
whose terms are the absolute values of the terms of the original series.

We have convergence tests for series with positive terms and for alternating series. But what if the signs of the terms switch back and forth irregularly? We will see in Example 3 that the idea of absolute convergence sometimes helps in such cases.

1 Definition A series $\Sigma \, a_n$ is called **absolutely convergent** if the series of absolute values $\Sigma \, |a_n|$ is convergent.

Notice that if $\Sigma \, a_n$ is a series with positive terms, then $|a_n| = a_n$ and so absolute convergence is the same as convergence in this case.

EXAMPLE 1 The series
$$\sum_{n=1}^{\infty} \frac{(-1)^{n-1}}{n^2} = 1 - \frac{1}{2^2} + \frac{1}{3^2} - \frac{1}{4^2} + \cdots$$
is absolutely convergent because
$$\sum_{n=1}^{\infty} \left| \frac{(-1)^{n-1}}{n^2} \right| = \sum_{n=1}^{\infty} \frac{1}{n^2} = 1 + \frac{1}{2^2} + \frac{1}{3^2} + \frac{1}{4^2} + \cdots$$
is a convergent p-series ($p = 2$). ■

EXAMPLE 2 We know that the alternating harmonic series
$$\sum_{n=1}^{\infty} \frac{(-1)^{n-1}}{n} = 1 - \frac{1}{2} + \frac{1}{3} - \frac{1}{4} + \cdots$$
is convergent (see Example 11.5.1), but it is not absolutely convergent because the corresponding series of absolute values is
$$\sum_{n=1}^{\infty} \left| \frac{(-1)^{n-1}}{n} \right| = \sum_{n=1}^{\infty} \frac{1}{n} = 1 + \frac{1}{2} + \frac{1}{3} + \frac{1}{4} + \cdots$$
which is the harmonic series (p-series with $p = 1$) and is therefore divergent. ■

738 CHAPTER 11 Infinite Sequences and Series

> **[2] Definition** A series $\sum a_n$ is called **conditionally convergent** if it is convergent but not absolutely convergent.

Example 2 shows that the alternating harmonic series is conditionally convergent. Thus it is possible for a series to be convergent but not absolutely convergent. However, the next theorem shows that absolute convergence implies convergence.

> **[3] Theorem** If a series $\sum a_n$ is absolutely convergent, then it is convergent.

PROOF Observe that the inequality

$$0 \leq a_n + |a_n| \leq 2|a_n|$$

is true because $|a_n|$ is either a_n or $-a_n$. If $\sum a_n$ is absolutely convergent, then $\sum |a_n|$ is convergent, so $\sum 2|a_n|$ is convergent. Therefore, by the Comparison Test, $\sum (a_n + |a_n|)$ is convergent. Then

$$\sum a_n = \sum (a_n + |a_n|) - \sum |a_n|$$

is the difference of two convergent series and is therefore convergent. ∎

EXAMPLE 3 Determine whether the series

$$\sum_{n=1}^{\infty} \frac{\cos n}{n^2} = \frac{\cos 1}{1^2} + \frac{\cos 2}{2^2} + \frac{\cos 3}{3^2} + \cdots$$

is convergent or divergent.

SOLUTION This series has both positive and negative terms, but it is not alternating. (The first term is positive, the next three are negative, and the following three are positive: the signs change irregularly.) We can apply the Comparison Test to the series of absolute values

$$\sum_{n=1}^{\infty} \left| \frac{\cos n}{n^2} \right| = \sum_{n=1}^{\infty} \frac{|\cos n|}{n^2}$$

Since $|\cos n| \leq 1$ for all n, we have

$$\frac{|\cos n|}{n^2} \leq \frac{1}{n^2}$$

We know that $\sum 1/n^2$ is convergent (p-series with $p = 2$) and therefore $\sum |\cos n|/n^2$ is convergent by the Comparison Test. Thus the given series $\sum (\cos n)/n^2$ is absolutely convergent and therefore convergent by Theorem 3. ∎

The following test is very useful in determining whether a given series is absolutely convergent.

Figure 1 shows the graphs of the terms a_n and partial sums s_n of the series in Example 3. Notice that the series is not alternating but has positive and negative terms.

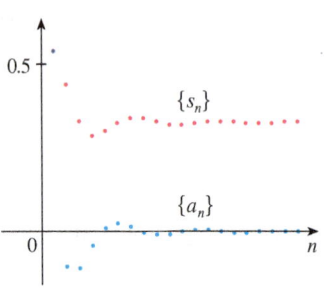

FIGURE 1

The Ratio Test

(i) If $\lim\limits_{n \to \infty} \left| \dfrac{a_{n+1}}{a_n} \right| = L < 1$, then the series $\sum\limits_{n=1}^{\infty} a_n$ is absolutely convergent (and therefore convergent).

(ii) If $\lim\limits_{n \to \infty} \left| \dfrac{a_{n+1}}{a_n} \right| = L > 1$ or $\lim\limits_{n \to \infty} \left| \dfrac{a_{n+1}}{a_n} \right| = \infty$, then the series $\sum\limits_{n=1}^{\infty} a_n$ is divergent.

(iii) If $\lim\limits_{n \to \infty} \left| \dfrac{a_{n+1}}{a_n} \right| = 1$, the Ratio Test is inconclusive; that is, no conclusion can be drawn about the convergence or divergence of $\sum a_n$.

PROOF

(i) The idea is to compare the given series with a convergent geometric series. Since $L < 1$, we can choose a number r such that $L < r < 1$. Since

$$\lim_{n \to \infty} \left| \frac{a_{n+1}}{a_n} \right| = L \quad \text{and} \quad L < r$$

the ratio $|a_{n+1}/a_n|$ will eventually be less than r; that is, there exists an integer N such that

$$\left| \frac{a_{n+1}}{a_n} \right| < r \quad \text{whenever } n \geq N$$

or, equivalently,

$$\boxed{4} \qquad |a_{n+1}| < |a_n| r \quad \text{whenever } n \geq N$$

Putting n successively equal to $N, N+1, N+2, \ldots$ in (4), we obtain

$$|a_{N+1}| < |a_N| r$$
$$|a_{N+2}| < |a_{N+1}| r < |a_N| r^2$$
$$|a_{N+3}| < |a_{N+2}| r < |a_N| r^3$$

and, in general,

$$\boxed{5} \qquad |a_{N+k}| < |a_N| r^k \quad \text{for all } k \geq 1$$

Now the series

$$\sum_{k=1}^{\infty} |a_N| r^k = |a_N| r + |a_N| r^2 + |a_N| r^3 + \cdots$$

is convergent because it is a geometric series with $0 < r < 1$. So the inequality (5), together with the Comparison Test, shows that the series

$$\sum_{n=N+1}^{\infty} |a_n| = \sum_{k=1}^{\infty} |a_{N+k}| = |a_{N+1}| + |a_{N+2}| + |a_{N+3}| + \cdots$$

is also convergent. It follows that the series $\sum_{n=1}^{\infty} |a_n|$ is convergent. (Recall that a finite number of terms doesn't affect convergence.) Therefore $\sum a_n$ is absolutely convergent.

(ii) If $|a_{n+1}/a_n| \to L > 1$ or $|a_{n+1}/a_n| \to \infty$, then the ratio $|a_{n+1}/a_n|$ will eventually be greater than 1; that is, there exists an integer N such that

$$\left|\frac{a_{n+1}}{a_n}\right| > 1 \quad \text{whenever } n \geq N$$

This means that $|a_{n+1}| > |a_n|$ whenever $n \geq N$ and so

$$\lim_{n \to \infty} a_n \neq 0$$

Therefore $\sum a_n$ diverges by the Test for Divergence. ∎

NOTE Part (iii) of the Ratio Test says that if $\lim_{n \to \infty} |a_{n+1}/a_n| = 1$, the test gives no information. For instance, for the convergent series $\sum 1/n^2$ we have

$$\left|\frac{a_{n+1}}{a_n}\right| = \frac{\frac{1}{(n+1)^2}}{\frac{1}{n^2}} = \frac{n^2}{(n+1)^2} = \frac{1}{\left(1 + \frac{1}{n}\right)^2} \to 1 \quad \text{as } n \to \infty$$

whereas for the divergent series $\sum 1/n$ we have

$$\left|\frac{a_{n+1}}{a_n}\right| = \frac{\frac{1}{n+1}}{\frac{1}{n}} = \frac{n}{n+1} = \frac{1}{1 + \frac{1}{n}} \to 1 \quad \text{as } n \to \infty$$

The Ratio Test is usually conclusive if the nth term of the series contains an exponential or a factorial, as we will see in Examples 4 and 5.

Therefore, if $\lim_{n \to \infty} |a_{n+1}/a_n| = 1$, the series $\sum a_n$ might converge or it might diverge. In this case the Ratio Test fails and we must use some other test.

EXAMPLE 4 Test the series $\sum_{n=1}^{\infty} (-1)^n \frac{n^3}{3^n}$ for absolute convergence.

Estimating Sums
In the last three sections we used various methods for estimating the sum of a series—the method depended on which test was used to prove convergence. What about series for which the Ratio Test works? There are two possibilities: If the series happens to be an alternating series, as in Example 4, then it is best to use the methods of Section 11.5. If the terms are all positive, then use the special methods explained in Exercise 46.

SOLUTION We use the Ratio Test with $a_n = (-1)^n n^3/3^n$:

$$\left|\frac{a_{n+1}}{a_n}\right| = \left|\frac{\frac{(-1)^{n+1}(n+1)^3}{3^{n+1}}}{\frac{(-1)^n n^3}{3^n}}\right| = \frac{(n+1)^3}{3^{n+1}} \cdot \frac{3^n}{n^3}$$

$$= \frac{1}{3}\left(\frac{n+1}{n}\right)^3 = \frac{1}{3}\left(1 + \frac{1}{n}\right)^3 \to \frac{1}{3} < 1$$

Thus, by the Ratio Test, the given series is absolutely convergent. ∎

EXAMPLE 5 Test the convergence of the series $\sum_{n=1}^{\infty} \dfrac{n^n}{n!}$.

SOLUTION Since the terms $a_n = n^n/n!$ are positive, we don't need the absolute value signs.

$$\frac{a_{n+1}}{a_n} = \frac{(n+1)^{n+1}}{(n+1)!} \cdot \frac{n!}{n^n} = \frac{(n+1)(n+1)^n}{(n+1)n!} \cdot \frac{n!}{n^n}$$

$$= \left(\frac{n+1}{n}\right)^n = \left(1 + \frac{1}{n}\right)^n \to e \quad \text{as } n \to \infty$$

(see Equation 3.6.6). Since $e > 1$, the given series is divergent by the Ratio Test. ■

NOTE Although the Ratio Test works in Example 5, an easier method is to use the Test for Divergence. Since

$$a_n = \frac{n^n}{n!} = \frac{n \cdot n \cdot n \cdot \cdots \cdot n}{1 \cdot 2 \cdot 3 \cdot \cdots \cdot n} \geq n$$

it follows that a_n does not approach 0 as $n \to \infty$. Therefore the given series is divergent by the Test for Divergence.

The following test is convenient to apply when nth powers occur. Its proof is similar to the proof of the Ratio Test and is left as Exercise 49.

The Root Test

(i) If $\lim\limits_{n \to \infty} \sqrt[n]{|a_n|} = L < 1$, then the series $\sum_{n=1}^{\infty} a_n$ is absolutely convergent (and therefore convergent).

(ii) If $\lim\limits_{n \to \infty} \sqrt[n]{|a_n|} = L > 1$ or $\lim\limits_{n \to \infty} \sqrt[n]{|a_n|} = \infty$, then the series $\sum_{n=1}^{\infty} a_n$ is divergent.

(iii) If $\lim\limits_{n \to \infty} \sqrt[n]{|a_n|} = 1$, the Root Test is inconclusive.

If $\lim_{n \to \infty} \sqrt[n]{|a_n|} = 1$, then part (iii) of the Root Test says that the test gives no information. The series $\sum a_n$ could converge or diverge. (If $L = 1$ in the Ratio Test, don't try the Root Test because L will again be 1. And if $L = 1$ in the Root Test, don't try the Ratio Test because it will fail too.)

EXAMPLE 6 Test the convergence of the series $\sum_{n=1}^{\infty} \left(\dfrac{2n+3}{3n+2}\right)^n$.

SOLUTION

$$a_n = \left(\frac{2n+3}{3n+2}\right)^n$$

$$\sqrt[n]{|a_n|} = \frac{2n+3}{3n+2} = \frac{2 + \dfrac{3}{n}}{3 + \dfrac{2}{n}} \to \frac{2}{3} < 1$$

Thus the given series is absolutely convergent (and therefore convergent) by the Root Test. ■

Rearrangements

The question of whether a given convergent series is absolutely convergent or conditionally convergent has a bearing on the question of whether infinite sums behave like finite sums.

If we rearrange the order of the terms in a finite sum, then of course the value of the sum remains unchanged. But this is not always the case for an infinite series. By a **rearrangement** of an infinite series $\Sigma\, a_n$ we mean a series obtained by simply changing the order of the terms. For instance, a rearrangement of $\Sigma\, a_n$ could start as follows:

$$a_1 + a_2 + a_5 + a_3 + a_4 + a_{15} + a_6 + a_7 + a_{20} + \cdots$$

It turns out that

if $\Sigma\, a_n$ is an absolutely convergent series with sum s,
then any rearrangement of $\Sigma\, a_n$ has the same sum s.

However, any conditionally convergent series can be rearranged to give a different sum. To illustrate this fact let's consider the alternating harmonic series

$$\boxed{6} \qquad 1 - \tfrac{1}{2} + \tfrac{1}{3} - \tfrac{1}{4} + \tfrac{1}{5} - \tfrac{1}{6} + \tfrac{1}{7} - \tfrac{1}{8} + \cdots = \ln 2$$

(See Exercise 11.5.36.) If we multiply this series by $\tfrac{1}{2}$, we get

$$\tfrac{1}{2} - \tfrac{1}{4} + \tfrac{1}{6} - \tfrac{1}{8} + \cdots = \tfrac{1}{2}\ln 2$$

Inserting zeros between the terms of this series, we have

Adding these zeros does not affect the sum of the series; each term in the sequence of partial sums is repeated, but the limit is the same.

$$\boxed{7} \qquad 0 + \tfrac{1}{2} + 0 - \tfrac{1}{4} + 0 + \tfrac{1}{6} + 0 - \tfrac{1}{8} + \cdots = \tfrac{1}{2}\ln 2$$

Now we add the series in Equations 6 and 7 using Theorem 11.2.8:

$$\boxed{8} \qquad 1 + \tfrac{1}{3} - \tfrac{1}{2} + \tfrac{1}{5} + \tfrac{1}{7} - \tfrac{1}{4} + \cdots = \tfrac{3}{2}\ln 2$$

Notice that the series in (8) contains the same terms as in (6) but rearranged so that one negative term occurs after each pair of positive terms. The sums of these series, however, are different. In fact, Riemann proved that

if $\Sigma\, a_n$ is a conditionally convergent series and r is any real number whatsoever, then there is a rearrangement of $\Sigma\, a_n$ that has a sum equal to r.

A proof of this fact is outlined in Exercise 52.

11.6 EXERCISES

1. What can you say about the series $\Sigma\, a_n$ in each of the following cases?

(a) $\displaystyle\lim_{n\to\infty} \left|\frac{a_{n+1}}{a_n}\right| = 8$

(b) $\displaystyle\lim_{n\to\infty} \left|\frac{a_{n+1}}{a_n}\right| = 0.8$

(c) $\displaystyle\lim_{n\to\infty} \left|\frac{a_{n+1}}{a_n}\right| = 1$

2–6 Determine whether the series is absolutely convergent or conditionally convergent.

2. $\displaystyle\sum_{n=1}^{\infty} \frac{(-1)^{n-1}}{\sqrt{n}}$

3. $\displaystyle\sum_{n=0}^{\infty} \frac{(-1)^n}{5n+1}$

4. $\displaystyle\sum_{n=1}^{\infty} \frac{(-1)^n}{n^3+1}$

5. $\sum_{n=1}^{\infty} \dfrac{\sin n}{2^n}$

6. $\sum_{n=1}^{\infty} (-1)^{n-1} \dfrac{n}{n^2+4}$

7–24 Use the Ratio Test to determine whether the series is convergent or divergent.

7. $\sum_{n=1}^{\infty} \dfrac{n}{5^n}$

8. $\sum_{n=1}^{\infty} \dfrac{(-2)^n}{n^2}$

9. $\sum_{n=1}^{\infty} (-1)^{n-1} \dfrac{3^n}{2^n n^3}$

10. $\sum_{n=0}^{\infty} \dfrac{(-3)^n}{(2n+1)!}$

11. $\sum_{k=1}^{\infty} \dfrac{1}{k!}$

12. $\sum_{k=1}^{\infty} k e^{-k}$

13. $\sum_{n=1}^{\infty} \dfrac{10^n}{(n+1)4^{2n+1}}$

14. $\sum_{n=1}^{\infty} \dfrac{n!}{100^n}$

15. $\sum_{n=1}^{\infty} \dfrac{n \pi^n}{(-3)^{n-1}}$

16. $\sum_{n=1}^{\infty} \dfrac{n^{10}}{(-10)^{n+1}}$

17. $\sum_{n=1}^{\infty} \dfrac{\cos(n\pi/3)}{n!}$

18. $\sum_{n=1}^{\infty} \dfrac{n!}{n^n}$

19. $\sum_{n=1}^{\infty} \dfrac{n^{100} 100^n}{n!}$

20. $\sum_{n=1}^{\infty} \dfrac{(2n)!}{(n!)^2}$

21. $1 - \dfrac{2!}{1\cdot 3} + \dfrac{3!}{1\cdot 3\cdot 5} - \dfrac{4!}{1\cdot 3\cdot 5\cdot 7} + \cdots$
$+ (-1)^{n-1} \dfrac{n!}{1\cdot 3\cdot 5\cdot \cdots \cdot (2n-1)} + \cdots$

22. $\dfrac{2}{3} + \dfrac{2\cdot 5}{3\cdot 5} + \dfrac{2\cdot 5\cdot 8}{3\cdot 5\cdot 7} + \dfrac{2\cdot 5\cdot 8\cdot 11}{3\cdot 5\cdot 7\cdot 9} + \cdots$

23. $\sum_{n=1}^{\infty} \dfrac{2\cdot 4\cdot 6 \cdot \cdots \cdot (2n)}{n!}$

24. $\sum_{n=1}^{\infty} (-1)^n \dfrac{2^n n!}{5\cdot 8\cdot 11\cdot \cdots \cdot (3n+2)}$

25–30 Use the Root Test to determine whether the series is convergent or divergent.

25. $\sum_{n=1}^{\infty} \left(\dfrac{n^2+1}{2n^2+1} \right)^n$

26. $\sum_{n=1}^{\infty} \dfrac{(-2)^n}{n^n}$

27. $\sum_{n=2}^{\infty} \dfrac{(-1)^{n-1}}{(\ln n)^n}$

28. $\sum_{n=1}^{\infty} \left(\dfrac{-2n}{n+1} \right)^{5n}$

29. $\sum_{n=1}^{\infty} \left(1 + \dfrac{1}{n} \right)^{n^2}$

30. $\sum_{n=0}^{\infty} (\arctan n)^n$

31–38 Use any test to determine whether the series is absolutely convergent, conditionally convergent, or divergent.

31. $\sum_{n=2}^{\infty} \dfrac{(-1)^n}{\ln n}$

32. $\sum_{n=1}^{\infty} \left(\dfrac{1-n}{2+3n} \right)^n$

33. $\sum_{n=1}^{\infty} \dfrac{(-9)^n}{n 10^{n+1}}$

34. $\sum_{n=1}^{\infty} \dfrac{n 5^{2n}}{10^{n+1}}$

35. $\sum_{n=2}^{\infty} \left(\dfrac{n}{\ln n} \right)^n$

36. $\sum_{n=1}^{\infty} \dfrac{\sin(n\pi/6)}{1 + n\sqrt{n}}$

37. $\sum_{n=1}^{\infty} \dfrac{(-1)^n \arctan n}{n^2}$

38. $\sum_{n=2}^{\infty} \dfrac{(-1)^n}{n \ln n}$

39. The terms of a series are defined recursively by the equations
$$a_1 = 2 \qquad a_{n+1} = \dfrac{5n+1}{4n+3} a_n$$
Determine whether $\Sigma\, a_n$ converges or diverges.

40. A series $\Sigma\, a_n$ is defined by the equations
$$a_1 = 1 \qquad a_{n+1} = \dfrac{2 + \cos n}{\sqrt{n}} a_n$$
Determine whether $\Sigma\, a_n$ converges or diverges.

41–42 Let $\{b_n\}$ be a sequence of positive numbers that converges to $\tfrac{1}{2}$. Determine whether the given series is absolutely convergent.

41. $\sum_{n=1}^{\infty} \dfrac{b_n^n \cos n\pi}{n}$

42. $\sum_{n=1}^{\infty} \dfrac{(-1)^n n!}{n^n b_1 b_2 b_3 \cdots b_n}$

43. For which of the following series is the Ratio Test inconclusive (that is, it fails to give a definite answer)?

(a) $\sum_{n=1}^{\infty} \dfrac{1}{n^3}$

(b) $\sum_{n=1}^{\infty} \dfrac{n}{2^n}$

(c) $\sum_{n=1}^{\infty} \dfrac{(-3)^{n-1}}{\sqrt{n}}$

(d) $\sum_{n=1}^{\infty} \dfrac{\sqrt{n}}{1+n^2}$

44. For which positive integers k is the following series convergent?
$$\sum_{n=1}^{\infty} \dfrac{(n!)^2}{(kn)!}$$

45. (a) Show that $\sum_{n=0}^{\infty} x^n/n!$ converges for all x.
(b) Deduce that $\lim_{n\to\infty} x^n/n! = 0$ for all x.

46. Let $\Sigma\, a_n$ be a series with positive terms and let $r_n = a_{n+1}/a_n$. Suppose that $\lim_{n\to\infty} r_n = L < 1$, so $\Sigma\, a_n$ converges by the Ratio Test. As usual, we let R_n be the remainder after n terms, that is,
$$R_n = a_{n+1} + a_{n+2} + a_{n+3} + \cdots$$

(a) If $\{r_n\}$ is a decreasing sequence and $r_{n+1} < 1$, show, by summing a geometric series, that
$$R_n \le \dfrac{a_{n+1}}{1 - r_{n+1}}$$

(b) If $\{r_n\}$ is an increasing sequence, show that
$$R_n \le \dfrac{a_{n+1}}{1 - L}$$

47. (a) Find the partial sum s_5 of the series $\sum_{n=1}^{\infty} 1/(n2^n)$. Use Exercise 46 to estimate the error in using s_5 as an approximation to the sum of the series.
(b) Find a value of n so that s_n is within 0.00005 of the sum. Use this value of n to approximate the sum of the series.

48. Use the sum of the first 10 terms to approximate the sum of the series
$$\sum_{n=1}^{\infty} \frac{n}{2^n}$$
Use Exercise 46 to estimate the error.

49. Prove the Root Test. [*Hint for part (i):* Take any number r such that $L < r < 1$ and use the fact that there is an integer N such that $\sqrt[n]{|a_n|} < r$ whenever $n \geq N$.]

50. Around 1910, the Indian mathematician Srinivasa Ramanujan discovered the formula
$$\frac{1}{\pi} = \frac{2\sqrt{2}}{9801} \sum_{n=0}^{\infty} \frac{(4n)!(1103 + 26390n)}{(n!)^4 396^{4n}}$$
William Gosper used this series in 1985 to compute the first 17 million digits of π.
(a) Verify that the series is convergent.
(b) How many correct decimal places of π do you get if you use just the first term of the series? What if you use two terms?

51. Given any series $\sum a_n$, we define a series $\sum a_n^+$ whose terms are all the positive terms of $\sum a_n$ and a series $\sum a_n^-$ whose terms are all the negative terms of $\sum a_n$. To be specific, we let
$$a_n^+ = \frac{a_n + |a_n|}{2} \qquad a_n^- = \frac{a_n - |a_n|}{2}$$
Notice that if $a_n > 0$, then $a_n^+ = a_n$ and $a_n^- = 0$, whereas if $a_n < 0$, then $a_n^- = a_n$ and $a_n^+ = 0$.
(a) If $\sum a_n$ is absolutely convergent, show that both of the series $\sum a_n^+$ and $\sum a_n^-$ are convergent.
(b) If $\sum a_n$ is conditionally convergent, show that both of the series $\sum a_n^+$ and $\sum a_n^-$ are divergent.

52. Prove that if $\sum a_n$ is a conditionally convergent series and r is any real number, then there is a rearrangement of $\sum a_n$ whose sum is r. [*Hints:* Use the notation of Exercise 51. Take just enough positive terms a_n^+ so that their sum is greater than r. Then add just enough negative terms a_n^- so that the cumulative sum is less than r. Continue in this manner and use Theorem 11.2.6.]

53. Suppose the series $\sum a_n$ is conditionally convergent.
(a) Prove that the series $\sum n^2 a_n$ is divergent.
(b) Conditional convergence of $\sum a_n$ is not enough to determine whether $\sum na_n$ is convergent. Show this by giving an example of a conditionally convergent series such that $\sum na_n$ converges and an example where $\sum na_n$ diverges.

11.7 Strategy for Testing Series

We now have several ways of testing a series for convergence or divergence; the problem is to decide which test to use on which series. In this respect, testing series is similar to integrating functions. Again there are no hard and fast rules about which test to apply to a given series, but you may find the following advice of some use.

It is not wise to apply a list of the tests in a specific order until one finally works. That would be a waste of time and effort. Instead, as with integration, the main strategy is to classify the series according to its *form*.

1. If the series is of the form $\sum 1/n^p$, it is a *p*-series, which we know to be convergent if $p > 1$ and divergent if $p \leq 1$.

2. If the series has the form $\sum ar^{n-1}$ or $\sum ar^n$, it is a geometric series, which converges if $|r| < 1$ and diverges if $|r| \geq 1$. Some preliminary algebraic manipulation may be required to bring the series into this form.

3. If the series has a form that is similar to a *p*-series or a geometric series, then one of the comparison tests should be considered. In particular, if a_n is a rational function or an algebraic function of n (involving roots of polynomials), then the series should be compared with a *p*-series. Notice that most of the series in Exercises 11.4 have this form. (The value of p should be chosen as in Section 11.4 by keeping only the highest powers of n in the numerator and denominator.) The comparison tests apply only to series with positive terms, but if $\sum a_n$ has some negative terms, then we can apply the Comparison Test to $\sum |a_n|$ and test for absolute convergence.

4. If you can see at a glance that $\lim_{n \to \infty} a_n \neq 0$, then the Test for Divergence should be used.
5. If the series is of the form $\sum (-1)^{n-1} b_n$ or $\sum (-1)^n b_n$, then the Alternating Series Test is an obvious possibility.
6. Series that involve factorials or other products (including a constant raised to the nth power) are often conveniently tested using the Ratio Test. Bear in mind that $|a_{n+1}/a_n| \to 1$ as $n \to \infty$ for all p-series and therefore all rational or algebraic functions of n. Thus the Ratio Test should not be used for such series.
7. If a_n is of the form $(b_n)^n$, then the Root Test may be useful.
8. If $a_n = f(n)$, where $\int_1^\infty f(x)\, dx$ is easily evaluated, then the Integral Test is effective (assuming the hypotheses of this test are satisfied).

In the following examples we don't work out all the details but simply indicate which tests should be used.

EXAMPLE 1 $\displaystyle\sum_{n=1}^{\infty} \frac{n-1}{2n+1}$

Since $a_n \to \frac{1}{2} \neq 0$ as $n \to \infty$, we should use the Test for Divergence.

EXAMPLE 2 $\displaystyle\sum_{n=1}^{\infty} \frac{\sqrt{n^3+1}}{3n^3 + 4n^2 + 2}$

Since a_n is an algebraic function of n, we compare the given series with a p-series. The comparison series for the Limit Comparison Test is $\sum b_n$, where

$$b_n = \frac{\sqrt{n^3}}{3n^3} = \frac{n^{3/2}}{3n^3} = \frac{1}{3n^{3/2}}$$

EXAMPLE 3 $\displaystyle\sum_{n=1}^{\infty} ne^{-n^2}$

Since the integral $\int_1^\infty xe^{-x^2}\, dx$ is easily evaluated, we use the Integral Test. The Ratio Test also works.

EXAMPLE 4 $\displaystyle\sum_{n=1}^{\infty} (-1)^n \frac{n^3}{n^4 + 1}$

Since the series is alternating, we use the Alternating Series Test.

EXAMPLE 5 $\displaystyle\sum_{k=1}^{\infty} \frac{2^k}{k!}$

Since the series involves $k!$, we use the Ratio Test.

EXAMPLE 6 $\displaystyle\sum_{n=1}^{\infty} \frac{1}{2 + 3^n}$

Since the series is closely related to the geometric series $\sum 1/3^n$, we use the Comparison Test.

11.7 EXERCISES

1–38 Test the series for convergence or divergence.

1. $\sum_{n=1}^{\infty} \dfrac{n^2-1}{n^3+1}$

2. $\sum_{n=1}^{\infty} \dfrac{n-1}{n^3+1}$

3. $\sum_{n=1}^{\infty} (-1)^n \dfrac{n^2-1}{n^3+1}$

4. $\sum_{n=1}^{\infty} (-1)^n \dfrac{n^2-1}{n^2+1}$

5. $\sum_{n=1}^{\infty} \dfrac{e^n}{n^2}$

6. $\sum_{n=1}^{\infty} \dfrac{n^{2n}}{(1+n)^{3n}}$

7. $\sum_{n=2}^{\infty} \dfrac{1}{n\sqrt{\ln n}}$

8. $\sum_{n=1}^{\infty} (-1)^{n-1} \dfrac{n^4}{4^n}$

9. $\sum_{n=0}^{\infty} (-1)^n \dfrac{\pi^{2n}}{(2n)!}$

10. $\sum_{n=1}^{\infty} n^2 e^{-n^3}$

11. $\sum_{n=1}^{\infty} \left(\dfrac{1}{n^3} + \dfrac{1}{3^n} \right)$

12. $\sum_{k=1}^{\infty} \dfrac{1}{k\sqrt{k^2+1}}$

13. $\sum_{n=1}^{\infty} \dfrac{3^n n^2}{n!}$

14. $\sum_{n=1}^{\infty} \dfrac{\sin 2n}{1+2^n}$

15. $\sum_{k=1}^{\infty} \dfrac{2^{k-1} 3^{k+1}}{k^k}$

16. $\sum_{n=1}^{\infty} \dfrac{\sqrt{n^4+1}}{n^3+n}$

17. $\sum_{n=1}^{\infty} \dfrac{1 \cdot 3 \cdot 5 \cdots (2n-1)}{2 \cdot 5 \cdot 8 \cdots (3n-1)}$

18. $\sum_{n=2}^{\infty} \dfrac{(-1)^{n-1}}{\sqrt{n}-1}$

19. $\sum_{n=1}^{\infty} (-1)^n \dfrac{\ln n}{\sqrt{n}}$

20. $\sum_{k=1}^{\infty} \dfrac{\sqrt[3]{k}-1}{k(\sqrt{k}+1)}$

21. $\sum_{n=1}^{\infty} (-1)^n \cos(1/n^2)$

22. $\sum_{k=1}^{\infty} \dfrac{1}{2+\sin k}$

23. $\sum_{n=1}^{\infty} \tan(1/n)$

24. $\sum_{n=1}^{\infty} n \sin(1/n)$

25. $\sum_{n=1}^{\infty} \dfrac{n!}{e^{n^2}}$

26. $\sum_{n=1}^{\infty} \dfrac{n^2+1}{5^n}$

27. $\sum_{k=1}^{\infty} \dfrac{k \ln k}{(k+1)^3}$

28. $\sum_{n=1}^{\infty} \dfrac{e^{1/n}}{n^2}$

29. $\sum_{n=1}^{\infty} \dfrac{(-1)^n}{\cosh n}$

30. $\sum_{j=1}^{\infty} (-1)^j \dfrac{\sqrt{j}}{j+5}$

31. $\sum_{k=1}^{\infty} \dfrac{5^k}{3^k+4^k}$

32. $\sum_{n=1}^{\infty} \dfrac{(n!)^n}{n^{4n}}$

33. $\sum_{n=1}^{\infty} \left(\dfrac{n}{n+1} \right)^{n^2}$

34. $\sum_{n=1}^{\infty} \dfrac{1}{n+n\cos^2 n}$

35. $\sum_{n=1}^{\infty} \dfrac{1}{n^{1+1/n}}$

36. $\sum_{n=2}^{\infty} \dfrac{1}{(\ln n)^{\ln n}}$

37. $\sum_{n=1}^{\infty} \left(\sqrt[n]{2} - 1 \right)^n$

38. $\sum_{n=1}^{\infty} \left(\sqrt[n]{2} - 1 \right)$

11.8 Power Series

A **power series** is a series of the form

$$\text{1} \qquad \sum_{n=0}^{\infty} c_n x^n = c_0 + c_1 x + c_2 x^2 + c_3 x^3 + \cdots$$

where x is a variable and the c_n's are constants called the **coefficients** of the series. For each fixed x, the series (1) is a series of constants that we can test for convergence or divergence. A power series may converge for some values of x and diverge for other values of x. The sum of the series is a function

$$f(x) = c_0 + c_1 x + c_2 x^2 + \cdots + c_n x^n + \cdots$$

whose domain is the set of all x for which the series converges. Notice that f resembles a polynomial. The only difference is that f has infinitely many terms.

For instance, if we take $c_n = 1$ for all n, the power series becomes the geometric series

$$\text{2} \qquad \sum_{n=0}^{\infty} x^n = 1 + x + x^2 + \cdots + x^n + \cdots$$

which converges when $-1 < x < 1$ and diverges when $|x| \geq 1$. (See Equation 11.2.5.)

Trigonometric Series
A power series is a series in which each term is a power function. A **trigonometric series**

$$\sum_{n=0}^{\infty} (a_n \cos nx + b_n \sin nx)$$

is a series whose terms are trigonometric functions. This type of series is discussed on the website

www.stewartcalculus.com

Click on *Additional Topics* and then on *Fourier Series*.

In fact if we put $x = \frac{1}{2}$ in the geometric series (2) we get the convergent series

$$\sum_{n=0}^{\infty} \left(\frac{1}{2}\right)^n = 1 + \frac{1}{2} + \frac{1}{4} + \frac{1}{8} + \frac{1}{16} + \cdots$$

but if we put $x = 2$ in (2) we get the divergent series

$$\sum_{n=0}^{\infty} 2^n = 1 + 2 + 4 + 8 + 16 + \cdots$$

More generally, a series of the form

$$\boxed{3} \qquad \sum_{n=0}^{\infty} c_n(x-a)^n = c_0 + c_1(x-a) + c_2(x-a)^2 + \cdots$$

is called a **power series in $(x - a)$** or a **power series centered at a** or a **power series about a**. Notice that in writing out the term corresponding to $n = 0$ in Equations 1 and 3 we have adopted the convention that $(x - a)^0 = 1$ even when $x = a$. Notice also that when $x = a$, all of the terms are 0 for $n \geq 1$ and so the power series (3) always converges when $x = a$.

EXAMPLE 1 For what values of x is the series $\sum_{n=0}^{\infty} n! x^n$ convergent?

SOLUTION We use the Ratio Test. If we let a_n, as usual, denote the nth term of the series, then $a_n = n! x^n$. If $x \neq 0$, we have

Notice that
$(n + 1)! = (n + 1)n(n - 1) \cdot \cdots \cdot 3 \cdot 2 \cdot 1$
$= (n + 1)n!$

$$\lim_{n \to \infty} \left| \frac{a_{n+1}}{a_n} \right| = \lim_{n \to \infty} \left| \frac{(n+1)! x^{n+1}}{n! x^n} \right| = \lim_{n \to \infty} (n + 1)|x| = \infty$$

By the Ratio Test, the series diverges when $x \neq 0$. Thus the given series converges only when $x = 0$. ∎

EXAMPLE 2 For what values of x does the series $\sum_{n=1}^{\infty} \frac{(x-3)^n}{n}$ converge?

SOLUTION Let $a_n = (x - 3)^n/n$. Then

$$\left| \frac{a_{n+1}}{a_n} \right| = \left| \frac{(x-3)^{n+1}}{n+1} \cdot \frac{n}{(x-3)^n} \right|$$

$$= \frac{1}{1 + \dfrac{1}{n}} |x - 3| \to |x - 3| \quad \text{as } n \to \infty$$

By the Ratio Test, the given series is absolutely convergent, and therefore convergent, when $|x - 3| < 1$ and divergent when $|x - 3| > 1$. Now

$$|x - 3| < 1 \iff -1 < x - 3 < 1 \iff 2 < x < 4$$

so the series converges when $2 < x < 4$ and diverges when $x < 2$ or $x > 4$.

The Ratio Test gives no information when $|x - 3| = 1$ so we must consider $x = 2$ and $x = 4$ separately. If we put $x = 4$ in the series, it becomes $\sum 1/n$, the harmonic series, which is divergent. If $x = 2$, the series is $\sum (-1)^n/n$, which converges by the Alternating Series Test. Thus the given power series converges for $2 \leq x < 4$. ∎

Membrane courtesy of National Film Board of Canada

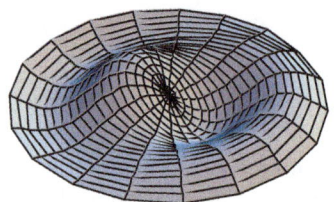

Notice how closely the computer-generated model (which involves Bessel functions and cosine functions) matches the photograph of a vibrating rubber membrane.

We will see that the main use of a power series is that it provides a way to represent some of the most important functions that arise in mathematics, physics, and chemistry. In particular, the sum of the power series in the next example is called a **Bessel function**, after the German astronomer Friedrich Bessel (1784–1846), and the function given in Exercise 35 is another example of a Bessel function. In fact, these functions first arose when Bessel solved Kepler's equation for describing planetary motion. Since that time, these functions have been applied in many different physical situations, including the temperature distribution in a circular plate and the shape of a vibrating drumhead.

EXAMPLE 3 Find the domain of the Bessel function of order 0 defined by

$$J_0(x) = \sum_{n=0}^{\infty} \frac{(-1)^n x^{2n}}{2^{2n}(n!)^2}$$

SOLUTION Let $a_n = (-1)^n x^{2n}/[2^{2n}(n!)^2]$. Then

$$\left| \frac{a_{n+1}}{a_n} \right| = \left| \frac{(-1)^{n+1} x^{2(n+1)}}{2^{2(n+1)}[(n+1)!]^2} \cdot \frac{2^{2n}(n!)^2}{(-1)^n x^{2n}} \right|$$

$$= \frac{x^{2n+2}}{2^{2n+2}(n+1)^2(n!)^2} \cdot \frac{2^{2n}(n!)^2}{x^{2n}}$$

$$= \frac{x^2}{4(n+1)^2} \to 0 < 1 \qquad \text{for all } x$$

Thus, by the Ratio Test, the given series converges for all values of x. In other words, the domain of the Bessel function J_0 is $(-\infty, \infty) = \mathbb{R}$. ∎

Recall that the sum of a series is equal to the limit of the sequence of partial sums. So when we define the Bessel function in Example 3 as the sum of a series we mean that, for every real number x,

$$J_0(x) = \lim_{n \to \infty} s_n(x) \qquad \text{where} \qquad s_n(x) = \sum_{i=0}^{n} \frac{(-1)^i x^{2i}}{2^{2i}(i!)^2}$$

The first few partial sums are

$$s_0(x) = 1$$

$$s_1(x) = 1 - \frac{x^2}{4}$$

$$s_2(x) = 1 - \frac{x^2}{4} + \frac{x^4}{64}$$

$$s_3(x) = 1 - \frac{x^2}{4} + \frac{x^4}{64} - \frac{x^6}{2304}$$

$$s_4(x) = 1 - \frac{x^2}{4} + \frac{x^4}{64} - \frac{x^6}{2304} + \frac{x^8}{147{,}456}$$

Figure 1 shows the graphs of these partial sums, which are polynomials. They are all approximations to the function J_0, but notice that the approximations become better when more terms are included. Figure 2 shows a more complete graph of the Bessel function.

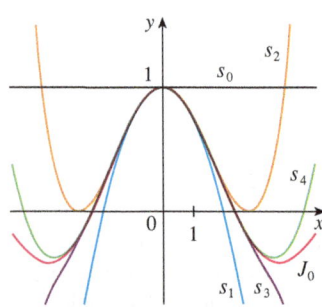

FIGURE 1
Partial sums of the Bessel function J_0

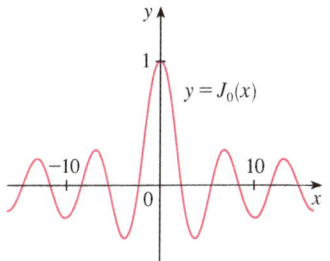

FIGURE 2

For the power series that we have looked at so far, the set of values of x for which the series is convergent has always turned out to be an interval [a finite interval for the geometric series and the series in Example 2, the infinite interval $(-\infty, \infty)$ in Example 3, and a collapsed interval $[0, 0] = \{0\}$ in Example 1]. The following theorem, proved in Appendix F, says that this is true in general.

> **4 Theorem** For a given power series $\sum_{n=0}^{\infty} c_n(x - a)^n$, there are only three possibilities:
>
> (i) The series converges only when $x = a$.
>
> (ii) The series converges for all x.
>
> (iii) There is a positive number R such that the series converges if $|x - a| < R$ and diverges if $|x - a| > R$.

The number R in case (iii) is called the **radius of convergence** of the power series. By convention, the radius of convergence is $R = 0$ in case (i) and $R = \infty$ in case (ii). The **interval of convergence** of a power series is the interval that consists of all values of x for which the series converges. In case (i) the interval consists of just a single point a. In case (ii) the interval is $(-\infty, \infty)$. In case (iii) note that the inequality $|x - a| < R$ can be rewritten as $a - R < x < a + R$. When x is an *endpoint* of the interval, that is, $x = a \pm R$, anything can happen—the series might converge at one or both endpoints or it might diverge at both endpoints. Thus in case (iii) there are four possibilities for the interval of convergence:

$$(a - R, a + R) \qquad (a - R, a + R] \qquad [a - R, a + R) \qquad [a - R, a + R]$$

The situation is illustrated in Figure 3.

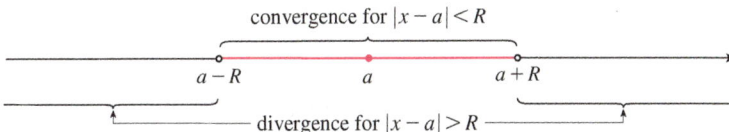

FIGURE 3

We summarize here the radius and interval of convergence for each of the examples already considered in this section.

	Series	Radius of convergence	Interval of convergence
Geometric series	$\sum_{n=0}^{\infty} x^n$	$R = 1$	$(-1, 1)$
Example 1	$\sum_{n=0}^{\infty} n!\, x^n$	$R = 0$	$\{0\}$
Example 2	$\sum_{n=1}^{\infty} \frac{(x - 3)^n}{n}$	$R = 1$	$[2, 4)$
Example 3	$\sum_{n=0}^{\infty} \frac{(-1)^n x^{2n}}{2^{2n}(n!)^2}$	$R = \infty$	$(-\infty, \infty)$

In general, the Ratio Test (or sometimes the Root Test) should be used to determine the radius of convergence R. The Ratio and Root Tests always fail when x is an endpoint of the interval of convergence, so the endpoints must be checked with some other test.

EXAMPLE 4 Find the radius of convergence and interval of convergence of the series

$$\sum_{n=0}^{\infty} \frac{(-3)^n x^n}{\sqrt{n+1}}$$

SOLUTION Let $a_n = (-3)^n x^n / \sqrt{n+1}$. Then

$$\left| \frac{a_{n+1}}{a_n} \right| = \left| \frac{(-3)^{n+1} x^{n+1}}{\sqrt{n+2}} \cdot \frac{\sqrt{n+1}}{(-3)^n x^n} \right| = \left| -3x \sqrt{\frac{n+1}{n+2}} \right|$$

$$= 3 \sqrt{\frac{1 + (1/n)}{1 + (2/n)}} \, |x| \to 3|x| \quad \text{as } n \to \infty$$

By the Ratio Test, the given series converges if $3|x| < 1$ and diverges if $3|x| > 1$. Thus it converges if $|x| < \frac{1}{3}$ and diverges if $|x| > \frac{1}{3}$. This means that the radius of convergence is $R = \frac{1}{3}$.

We know the series converges in the interval $\left(-\frac{1}{3}, \frac{1}{3}\right)$, but we must now test for convergence at the endpoints of this interval. If $x = -\frac{1}{3}$, the series becomes

$$\sum_{n=0}^{\infty} \frac{(-3)^n \left(-\frac{1}{3}\right)^n}{\sqrt{n+1}} = \sum_{n=0}^{\infty} \frac{1}{\sqrt{n+1}} = \frac{1}{\sqrt{1}} + \frac{1}{\sqrt{2}} + \frac{1}{\sqrt{3}} + \frac{1}{\sqrt{4}} + \cdots$$

which diverges. (Use the Integral Test or simply observe that it is a p-series with $p = \frac{1}{2} < 1$.) If $x = \frac{1}{3}$, the series is

$$\sum_{n=0}^{\infty} \frac{(-3)^n \left(\frac{1}{3}\right)^n}{\sqrt{n+1}} = \sum_{n=0}^{\infty} \frac{(-1)^n}{\sqrt{n+1}}$$

which converges by the Alternating Series Test. Therefore the given power series converges when $-\frac{1}{3} < x \leq \frac{1}{3}$, so the interval of convergence is $\left(-\frac{1}{3}, \frac{1}{3}\right]$. ■

EXAMPLE 5 Find the radius of convergence and interval of convergence of the series

$$\sum_{n=0}^{\infty} \frac{n(x+2)^n}{3^{n+1}}$$

SOLUTION If $a_n = n(x+2)^n / 3^{n+1}$, then

$$\left| \frac{a_{n+1}}{a_n} \right| = \left| \frac{(n+1)(x+2)^{n+1}}{3^{n+2}} \cdot \frac{3^{n+1}}{n(x+2)^n} \right|$$

$$= \left(1 + \frac{1}{n}\right) \frac{|x+2|}{3} \to \frac{|x+2|}{3} \quad \text{as } n \to \infty$$

Using the Ratio Test, we see that the series converges if $|x+2|/3 < 1$ and it diverges if $|x+2|/3 > 1$. So it converges if $|x+2| < 3$ and diverges if $|x+2| > 3$. Thus the radius of convergence is $R = 3$.

The inequality $|x + 2| < 3$ can be written as $-5 < x < 1$, so we test the series at the endpoints -5 and 1. When $x = -5$, the series is

$$\sum_{n=0}^{\infty} \frac{n(-3)^n}{3^{n+1}} = \tfrac{1}{3} \sum_{n=0}^{\infty} (-1)^n n$$

which diverges by the Test for Divergence [$(-1)^n n$ doesn't converge to 0]. When $x = 1$, the series is

$$\sum_{n=0}^{\infty} \frac{n(3)^n}{3^{n+1}} = \tfrac{1}{3} \sum_{n=0}^{\infty} n$$

which also diverges by the Test for Divergence. Thus the series converges only when $-5 < x < 1$, so the interval of convergence is $(-5, 1)$. ∎

11.8 EXERCISES

1. What is a power series?

2. (a) What is the radius of convergence of a power series? How do you find it?
(b) What is the interval of convergence of a power series? How do you find it?

3–28 Find the radius of convergence and interval of convergence of the series.

3. $\sum_{n=1}^{\infty} (-1)^n n x^n$

4. $\sum_{n=1}^{\infty} \frac{(-1)^n x^n}{\sqrt[3]{n}}$

5. $\sum_{n=1}^{\infty} \frac{x^n}{2n - 1}$

6. $\sum_{n=1}^{\infty} \frac{(-1)^n x^n}{n^2}$

7. $\sum_{n=0}^{\infty} \frac{x^n}{n!}$

8. $\sum_{n=1}^{\infty} n^n x^n$

9. $\sum_{n=1}^{\infty} \frac{x^n}{n^4 4^n}$

10. $\sum_{n=1}^{\infty} 2^n n^2 x^n$

11. $\sum_{n=1}^{\infty} \frac{(-1)^n 4^n}{\sqrt{n}} x^n$

12. $\sum_{n=1}^{\infty} \frac{(-1)^{n-1}}{n 5^n} x^n$

13. $\sum_{n=1}^{\infty} \frac{n}{2^n(n^2 + 1)} x^n$

14. $\sum_{n=1}^{\infty} \frac{x^{2n}}{n!}$

15. $\sum_{n=0}^{\infty} \frac{(x - 2)^n}{n^2 + 1}$

16. $\sum_{n=1}^{\infty} \frac{(-1)^n}{(2n - 1)2^n} (x - 1)^n$

17. $\sum_{n=2}^{\infty} \frac{(x + 2)^n}{2^n \ln n}$

18. $\sum_{n=1}^{\infty} \frac{\sqrt{n}}{8^n} (x + 6)^n$

19. $\sum_{n=1}^{\infty} \frac{(x - 2)^n}{n^n}$

20. $\sum_{n=1}^{\infty} \frac{(2x - 1)^n}{5^n \sqrt{n}}$

21. $\sum_{n=1}^{\infty} \frac{n}{b^n} (x - a)^n, \quad b > 0$

22. $\sum_{n=2}^{\infty} \frac{b^n}{\ln n} (x - a)^n, \quad b > 0$

23. $\sum_{n=1}^{\infty} n!(2x - 1)^n$

24. $\sum_{n=1}^{\infty} \frac{n^2 x^n}{2 \cdot 4 \cdot 6 \cdots (2n)}$

25. $\sum_{n=1}^{\infty} \frac{(5x - 4)^n}{n^3}$

26. $\sum_{n=2}^{\infty} \frac{x^{2n}}{n(\ln n)^2}$

27. $\sum_{n=1}^{\infty} \frac{x^n}{1 \cdot 3 \cdot 5 \cdots (2n - 1)}$

28. $\sum_{n=1}^{\infty} \frac{n! x^n}{1 \cdot 3 \cdot 5 \cdots (2n - 1)}$

29. If $\sum_{n=0}^{\infty} c_n 4^n$ is convergent, can we conclude that each of the following series is convergent?
(a) $\sum_{n=0}^{\infty} c_n(-2)^n$
(b) $\sum_{n=0}^{\infty} c_n(-4)^n$

30. Suppose that $\sum_{n=0}^{\infty} c_n x^n$ converges when $x = -4$ and diverges when $x = 6$. What can be said about the convergence or divergence of the following series?
(a) $\sum_{n=0}^{\infty} c_n$
(b) $\sum_{n=0}^{\infty} c_n 8^n$
(c) $\sum_{n=0}^{\infty} c_n(-3)^n$
(d) $\sum_{n=0}^{\infty} (-1)^n c_n 9^n$

31. If k is a positive integer, find the radius of convergence of the series

$$\sum_{n=0}^{\infty} \frac{(n!)^k}{(kn)!} x^n$$

32. Let p and q be real numbers with $p < q$. Find a power series whose interval of convergence is
(a) (p, q) (b) $(p, q]$ (c) $[p, q)$ (d) $[p, q]$

33. Is it possible to find a power series whose interval of convergence is $[0, \infty)$? Explain.

34. Graph the first several partial sums $s_n(x)$ of the series $\sum_{n=0}^{\infty} x^n$, together with the sum function $f(x) = 1/(1 - x)$, on a common screen. On what interval do these partial sums appear to be converging to $f(x)$?

35. The function J_1 defined by
$$J_1(x) = \sum_{n=0}^{\infty} \frac{(-1)^n x^{2n+1}}{n!(n+1)!2^{2n+1}}$$
is called the *Bessel function of order 1*.
(a) Find its domain.
(b) Graph the first several partial sums on a common screen.
(c) If your CAS has built-in Bessel functions, graph J_1 on the same screen as the partial sums in part (b) and observe how the partial sums approximate J_1.

36. The function A defined by
$$A(x) = 1 + \frac{x^3}{2 \cdot 3} + \frac{x^6}{2 \cdot 3 \cdot 5 \cdot 6} + \frac{x^9}{2 \cdot 3 \cdot 5 \cdot 6 \cdot 8 \cdot 9} + \cdots$$
is called an *Airy function* after the English mathematician and astronomer Sir George Airy (1801–1892).
(a) Find the domain of the Airy function.
(b) Graph the first several partial sums on a common screen.
(c) If your CAS has built-in Airy functions, graph A on the same screen as the partial sums in part (b) and observe how the partial sums approximate A.

37. A function f is defined by
$$f(x) = 1 + 2x + x^2 + 2x^3 + x^4 + \cdots$$
that is, its coefficients are $c_{2n} = 1$ and $c_{2n+1} = 2$ for all $n \geq 0$. Find the interval of convergence of the series and find an explicit formula for $f(x)$.

38. If $f(x) = \sum_{n=0}^{\infty} c_n x^n$, where $c_{n+4} = c_n$ for all $n \geq 0$, find the interval of convergence of the series and a formula for $f(x)$.

39. Show that if $\lim_{n \to \infty} \sqrt[n]{|c_n|} = c$, where $c \neq 0$, then the radius of convergence of the power series $\sum c_n x^n$ is $R = 1/c$.

40. Suppose that the power series $\sum c_n(x - a)^n$ satisfies $c_n \neq 0$ for all n. Show that if $\lim_{n \to \infty} |c_n/c_{n+1}|$ exists, then it is equal to the radius of convergence of the power series.

41. Suppose the series $\sum c_n x^n$ has radius of convergence 2 and the series $\sum d_n x^n$ has radius of convergence 3. What is the radius of convergence of the series $\sum (c_n + d_n)x^n$?

42. Suppose that the radius of convergence of the power series $\sum c_n x^n$ is R. What is the radius of convergence of the power series $\sum c_n x^{2n}$?

11.9 Representations of Functions as Power Series

In this section we learn how to represent certain types of functions as sums of power series by manipulating geometric series or by differentiating or integrating such a series. You might wonder why we would ever want to express a known function as a sum of infinitely many terms. We will see later that this strategy is useful for integrating functions that don't have elementary antiderivatives, for solving differential equations, and for approximating functions by polynomials. (Scientists do this to simplify the expressions they deal with; computer scientists do this to represent functions on calculators and computers.)

We start with an equation that we have seen before:

$$\boxed{1} \qquad \frac{1}{1-x} = 1 + x + x^2 + x^3 + \cdots = \sum_{n=0}^{\infty} x^n \qquad |x| < 1$$

We first encountered this equation in Example 11.2.7, where we obtained it by observing that the series is a geometric series with $a = 1$ and $r = x$. But here our point of view is different. We now regard Equation 1 as expressing the function $f(x) = 1/(1 - x)$ as a sum of a power series.

A geometric illustration of Equation 1 is shown in Figure 1. Because the sum of a series is the limit of the sequence of partial sums, we have

$$\frac{1}{1-x} = \lim_{n \to \infty} s_n(x)$$

where

$$s_n(x) = 1 + x + x^2 + \cdots + x^n$$

is the nth partial sum. Notice that as n increases, $s_n(x)$ becomes a better approximation to $f(x)$ for $-1 < x < 1$.

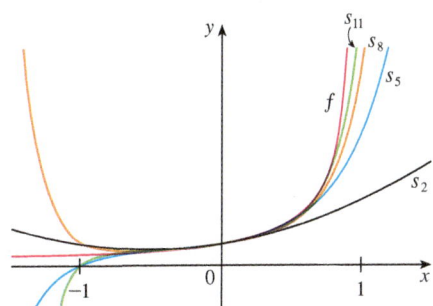

FIGURE 1 $f(x) = \dfrac{1}{1-x}$ and some partial sums

EXAMPLE 1 Express $1/(1 + x^2)$ as the sum of a power series and find the interval of convergence.

SOLUTION Replacing x by $-x^2$ in Equation 1, we have

$$\frac{1}{1+x^2} = \frac{1}{1-(-x^2)} = \sum_{n=0}^{\infty} (-x^2)^n$$

$$= \sum_{n=0}^{\infty} (-1)^n x^{2n} = 1 - x^2 + x^4 - x^6 + x^8 - \cdots$$

Because this is a geometric series, it converges when $|-x^2| < 1$, that is, $x^2 < 1$, or $|x| < 1$. Therefore the interval of convergence is $(-1, 1)$. (Of course, we could have determined the radius of convergence by applying the Ratio Test, but that much work is unnecessary here.) ∎

EXAMPLE 2 Find a power series representation for $1/(x + 2)$.

SOLUTION In order to put this function in the form of the left side of Equation 1, we first factor a 2 from the denominator:

$$\frac{1}{2+x} = \frac{1}{2\left(1+\dfrac{x}{2}\right)} = \frac{1}{2\left[1-\left(-\dfrac{x}{2}\right)\right]}$$

$$= \frac{1}{2}\sum_{n=0}^{\infty}\left(-\frac{x}{2}\right)^n = \sum_{n=0}^{\infty} \frac{(-1)^n}{2^{n+1}} x^n$$

This series converges when $|-x/2| < 1$, that is, $|x| < 2$. So the interval of convergence is $(-2, 2)$. ∎

EXAMPLE 3 Find a power series representation of $x^3/(x + 2)$.

SOLUTION Since this function is just x^3 times the function in Example 2, all we have to do is to multiply that series by x^3:

It's legitimate to move x^3 across the sigma sign because it doesn't depend on n. [Use Theorem 11.2.8(i) with $c = x^3$.]

$$\frac{x^3}{x+2} = x^3 \cdot \frac{1}{x+2} = x^3 \sum_{n=0}^{\infty} \frac{(-1)^n}{2^{n+1}} x^n = \sum_{n=0}^{\infty} \frac{(-1)^n}{2^{n+1}} x^{n+3}$$

$$= \tfrac{1}{2}x^3 - \tfrac{1}{4}x^4 + \tfrac{1}{8}x^5 - \tfrac{1}{16}x^6 + \cdots$$

Another way of writing this series is as follows:

$$\frac{x^3}{x+2} = \sum_{n=3}^{\infty} \frac{(-1)^{n-1}}{2^{n-2}} x^n$$

As in Example 2, the interval of convergence is $(-2, 2)$.

Differentiation and Integration of Power Series

The sum of a power series is a function $f(x) = \sum_{n=0}^{\infty} c_n(x-a)^n$ whose domain is the interval of convergence of the series. We would like to be able to differentiate and integrate such functions, and the following theorem (which we won't prove) says that we can do so by differentiating or integrating each individual term in the series, just as we would for a polynomial. This is called **term-by-term differentiation and integration**.

2 Theorem If the power series $\sum c_n(x-a)^n$ has radius of convergence $R > 0$, then the function f defined by

$$f(x) = c_0 + c_1(x-a) + c_2(x-a)^2 + \cdots = \sum_{n=0}^{\infty} c_n(x-a)^n$$

is differentiable (and therefore continuous) on the interval $(a - R, a + R)$ and

(i) $f'(x) = c_1 + 2c_2(x-a) + 3c_3(x-a)^2 + \cdots = \sum_{n=1}^{\infty} n c_n(x-a)^{n-1}$

(ii) $\int f(x)\, dx = C + c_0(x-a) + c_1 \dfrac{(x-a)^2}{2} + c_2 \dfrac{(x-a)^3}{3} + \cdots$

$\qquad = C + \sum_{n=0}^{\infty} c_n \dfrac{(x-a)^{n+1}}{n+1}$

The radii of convergence of the power series in Equations (i) and (ii) are both R.

In part (ii), $\int c_0\, dx = c_0 x + C_1$ is written as $c_0(x-a) + C$, where $C = C_1 + ac_0$, so all the terms of the series have the same form.

NOTE 1 Equations (i) and (ii) in Theorem 2 can be rewritten in the form

(iii) $\dfrac{d}{dx}\left[\sum_{n=0}^{\infty} c_n(x-a)^n\right] = \sum_{n=0}^{\infty} \dfrac{d}{dx}[c_n(x-a)^n]$

(iv) $\int \left[\sum_{n=0}^{\infty} c_n(x-a)^n\right] dx = \sum_{n=0}^{\infty} \int c_n(x-a)^n\, dx$

We know that, for finite sums, the derivative of a sum is the sum of the derivatives and the integral of a sum is the sum of the integrals. Equations (iii) and (iv) assert that the same is true for infinite sums, provided we are dealing with *power series*. (For other types of series of functions the situation is not as simple; see Exercise 38.)

NOTE 2 Although Theorem 2 says that the radius of convergence remains the same when a power series is differentiated or integrated, this does not mean that the *interval* of convergence remains the same. It may happen that the original series converges at an endpoint, whereas the differentiated series diverges there. (See Exercise 39.)

NOTE 3 The idea of differentiating a power series term by term is the basis for a powerful method for solving differential equations. We will discuss this method in Chapter 17.

EXAMPLE 4 In Example 11.8.3 we saw that the Bessel function

$$J_0(x) = \sum_{n=0}^{\infty} \frac{(-1)^n x^{2n}}{2^{2n}(n!)^2}$$

is defined for all x. Thus, by Theorem 2, J_0 is differentiable for all x and its derivative is found by term-by-term differentiation as follows:

$$J_0'(x) = \sum_{n=0}^{\infty} \frac{d}{dx} \frac{(-1)^n x^{2n}}{2^{2n}(n!)^2} = \sum_{n=1}^{\infty} \frac{(-1)^n 2n x^{2n-1}}{2^{2n}(n!)^2} \quad \blacksquare$$

EXAMPLE 5 Express $1/(1-x)^2$ as a power series by differentiating Equation 1. What is the radius of convergence?

SOLUTION Differentiating each side of the equation

$$\frac{1}{1-x} = 1 + x + x^2 + x^3 + \cdots = \sum_{n=0}^{\infty} x^n$$

we get

$$\frac{1}{(1-x)^2} = 1 + 2x + 3x^2 + \cdots = \sum_{n=1}^{\infty} n x^{n-1}$$

If we wish, we can replace n by $n+1$ and write the answer as

$$\frac{1}{(1-x)^2} = \sum_{n=0}^{\infty} (n+1) x^n$$

According to Theorem 2, the radius of convergence of the differentiated series is the same as the radius of convergence of the original series, namely, $R = 1$. $\quad\blacksquare$

EXAMPLE 6 Find a power series representation for $\ln(1 + x)$ and its radius of convergence.

SOLUTION We notice that the derivative of this function is $1/(1+x)$. From Equation 1 we have

$$\frac{1}{1+x} = \frac{1}{1-(-x)} = 1 - x + x^2 - x^3 + \cdots \quad |x| < 1$$

Integrating both sides of this equation, we get

$$\ln(1+x) = \int \frac{1}{1+x} dx = \int (1 - x + x^2 - x^3 + \cdots) dx$$

$$= x - \frac{x^2}{2} + \frac{x^3}{3} - \frac{x^4}{4} + \cdots + C$$

$$= \sum_{n=1}^{\infty} (-1)^{n-1} \frac{x^n}{n} + C \quad |x| < 1$$

To determine the value of C we put $x = 0$ in this equation and obtain $\ln(1 + 0) = C$.

756   **CHAPTER 11**   Infinite Sequences and Series

Thus $C = 0$ and

$$\ln(1 + x) = x - \frac{x^2}{2} + \frac{x^3}{3} - \frac{x^4}{4} + \cdots = \sum_{n=1}^{\infty} (-1)^{n-1} \frac{x^n}{n} \qquad |x| < 1$$

The radius of convergence is the same as for the original series: $R = 1$. ∎

EXAMPLE 7 Find a power series representation for $f(x) = \tan^{-1}x$.

SOLUTION We observe that $f'(x) = 1/(1 + x^2)$ and find the required series by integrating the power series for $1/(1 + x^2)$ found in Example 1.

$$\tan^{-1}x = \int \frac{1}{1 + x^2} dx = \int (1 - x^2 + x^4 - x^6 + \cdots) dx$$

$$= C + x - \frac{x^3}{3} + \frac{x^5}{5} - \frac{x^7}{7} + \cdots$$

To find C we put $x = 0$ and obtain $C = \tan^{-1} 0 = 0$. Therefore

$$\tan^{-1}x = x - \frac{x^3}{3} + \frac{x^5}{5} - \frac{x^7}{7} + \cdots$$

$$= \sum_{n=0}^{\infty} (-1)^n \frac{x^{2n+1}}{2n + 1}$$

Since the radius of convergence of the series for $1/(1 + x^2)$ is 1, the radius of convergence of this series for $\tan^{-1}x$ is also 1. ∎

> The power series for $\tan^{-1}x$ obtained in Example 7 is called *Gregory's series* after the Scottish mathematician James Gregory (1638–1675), who had anticipated some of Newton's discoveries. We have shown that Gregory's series is valid when $-1 < x < 1$, but it turns out (although it isn't easy to prove) that it is also valid when $x = \pm 1$. Notice that when $x = 1$ the series becomes
>
> $$\frac{\pi}{4} = 1 - \frac{1}{3} + \frac{1}{5} - \frac{1}{7} + \cdots$$
>
> This beautiful result is known as the Leibniz formula for π.

EXAMPLE 8
(a) Evaluate $\int [1/(1 + x^7)] dx$ as a power series.
(b) Use part (a) to approximate $\int_0^{0.5} [1/(1 + x^7)] dx$ correct to within 10^{-7}.

SOLUTION
(a) The first step is to express the integrand, $1/(1 + x^7)$, as the sum of a power series. As in Example 1, we start with Equation 1 and replace x by $-x^7$:

$$\frac{1}{1 + x^7} = \frac{1}{1 - (-x^7)} = \sum_{n=0}^{\infty} (-x^7)^n$$

$$= \sum_{n=0}^{\infty} (-1)^n x^{7n} = 1 - x^7 + x^{14} - \cdots$$

Now we integrate term by term:

$$\int \frac{1}{1 + x^7} dx = \int \sum_{n=0}^{\infty} (-1)^n x^{7n} dx = C + \sum_{n=0}^{\infty} (-1)^n \frac{x^{7n+1}}{7n + 1}$$

$$= C + x - \frac{x^8}{8} + \frac{x^{15}}{15} - \frac{x^{22}}{22} + \cdots$$

> This example demonstrates one way in which power series representations are useful. Integrating $1/(1 + x^7)$ by hand is incredibly difficult. Different computer algebra systems return different forms of the answer, but they are all extremely complicated. (If you have a CAS, try it yourself.) The infinite series answer that we obtain in Example 8(a) is actually much easier to deal with than the finite answer provided by a CAS.

This series converges for $|-x^7| < 1$, that is, for $|x| < 1$.

(b) In applying the Fundamental Theorem of Calculus, it doesn't matter which antiderivative we use, so let's use the antiderivative from part (a) with $C = 0$:

$$\int_0^{0.5} \frac{1}{1+x^7}\, dx = \left[x - \frac{x^8}{8} + \frac{x^{15}}{15} - \frac{x^{22}}{22} + \cdots \right]_0^{1/2}$$

$$= \frac{1}{2} - \frac{1}{8 \cdot 2^8} + \frac{1}{15 \cdot 2^{15}} - \frac{1}{22 \cdot 2^{22}} + \cdots + \frac{(-1)^n}{(7n+1)2^{7n+1}} + \cdots$$

This infinite series is the exact value of the definite integral, but since it is an alternating series, we can approximate the sum using the Alternating Series Estimation Theorem. If we stop adding after the term with $n = 3$, the error is smaller than the term with $n = 4$:

$$\frac{1}{29 \cdot 2^{29}} \approx 6.4 \times 10^{-11}$$

So we have

$$\int_0^{0.5} \frac{1}{1+x^7}\, dx \approx \frac{1}{2} - \frac{1}{8 \cdot 2^8} + \frac{1}{15 \cdot 2^{15}} - \frac{1}{22 \cdot 2^{22}} \approx 0.49951374 \quad \blacksquare$$

11.9 EXERCISES

1. If the radius of convergence of the power series $\sum_{n=0}^{\infty} c_n x^n$ is 10, what is the radius of convergence of the series $\sum_{n=1}^{\infty} n c_n x^{n-1}$? Why?

2. Suppose you know that the series $\sum_{n=0}^{\infty} b_n x^n$ converges for $|x| < 2$. What can you say about the following series? Why?

$$\sum_{n=0}^{\infty} \frac{b_n}{n+1} x^{n+1}$$

3–10 Find a power series representation for the function and determine the interval of convergence.

3. $f(x) = \dfrac{1}{1+x}$

4. $f(x) = \dfrac{5}{1-4x^2}$

5. $f(x) = \dfrac{2}{3-x}$

6. $f(x) = \dfrac{4}{2x+3}$

7. $f(x) = \dfrac{x^2}{x^4+16}$

8. $f(x) = \dfrac{x}{2x^2+1}$

9. $f(x) = \dfrac{x-1}{x+2}$

10. $f(x) = \dfrac{x+a}{x^2+a^2}, \quad a > 0$

11–12 Express the function as the sum of a power series by first using partial fractions. Find the interval of convergence.

11. $f(x) = \dfrac{2x-4}{x^2-4x+3}$

12. $f(x) = \dfrac{2x+3}{x^2+3x+2}$

13. (a) Use differentiation to find a power series representation for
$$f(x) = \frac{1}{(1+x)^2}$$
What is the radius of convergence?
(b) Use part (a) to find a power series for
$$f(x) = \frac{1}{(1+x)^3}$$
(c) Use part (b) to find a power series for
$$f(x) = \frac{x^2}{(1+x)^3}$$

14. (a) Use Equation 1 to find a power series representation for $f(x) = \ln(1-x)$. What is the radius of convergence?
(b) Use part (a) to find a power series for $f(x) = x \ln(1-x)$.
(c) By putting $x = \tfrac{1}{2}$ in your result from part (a), express $\ln 2$ as the sum of an infinite series.

15–20 Find a power series representation for the function and determine the radius of convergence.

15. $f(x) = \ln(5-x)$

16. $f(x) = x^2 \tan^{-1}(x^3)$

17. $f(x) = \dfrac{x}{(1+4x)^2}$

18. $f(x) = \left(\dfrac{x}{2-x}\right)^3$

19. $f(x) = \dfrac{1+x}{(1-x)^2}$

20. $f(x) = \dfrac{x^2+x}{(1-x)^3}$

21–24 Find a power series representation for f, and graph f and several partial sums $s_n(x)$ on the same screen. What happens as n increases?

21. $f(x) = \dfrac{x^2}{x^2 + 1}$

22. $f(x) = \ln(1 + x^4)$

23. $f(x) = \ln\left(\dfrac{1 + x}{1 - x}\right)$

24. $f(x) = \tan^{-1}(2x)$

25–28 Evaluate the indefinite integral as a power series. What is the radius of convergence?

25. $\displaystyle\int \dfrac{t}{1 - t^8}\, dt$

26. $\displaystyle\int \dfrac{t}{1 + t^3}\, dt$

27. $\displaystyle\int x^2 \ln(1 + x)\, dx$

28. $\displaystyle\int \dfrac{\tan^{-1} x}{x}\, dx$

29–32 Use a power series to approximate the definite integral to six decimal places.

29. $\displaystyle\int_0^{0.3} \dfrac{x}{1 + x^3}\, dx$

30. $\displaystyle\int_0^{1/2} \arctan(x/2)\, dx$

31. $\displaystyle\int_0^{0.2} x \ln(1 + x^2)\, dx$

32. $\displaystyle\int_0^{0.3} \dfrac{x^2}{1 + x^4}\, dx$

---

33. Use the result of Example 7 to compute arctan 0.2 correct to five decimal places.

34. Show that the function
$$f(x) = \sum_{n=0}^{\infty} \dfrac{(-1)^n x^{2n}}{(2n)!}$$
is a solution of the differential equation
$$f''(x) + f(x) = 0$$

35. (a) Show that J_0 (the Bessel function of order 0 given in Example 4) satisfies the differential equation
$$x^2 J_0''(x) + x J_0'(x) + x^2 J_0(x) = 0$$
(b) Evaluate $\int_0^1 J_0(x)\, dx$ correct to three decimal places.

36. The Bessel function of order 1 is defined by
$$J_1(x) = \sum_{n=0}^{\infty} \dfrac{(-1)^n x^{2n+1}}{n!(n+1)!\,2^{2n+1}}$$
(a) Show that J_1 satisfies the differential equation
$$x^2 J_1''(x) + x J_1'(x) + (x^2 - 1) J_1(x) = 0$$
(b) Show that $J_0'(x) = -J_1(x)$.

37. (a) Show that the function
$$f(x) = \sum_{n=0}^{\infty} \dfrac{x^n}{n!}$$
is a solution of the differential equation
$$f'(x) = f(x)$$
(b) Show that $f(x) = e^x$.

38. Let $f_n(x) = (\sin nx)/n^2$. Show that the series $\sum f_n(x)$ converges for all values of x but the series of derivatives $\sum f_n'(x)$ diverges when $x = 2n\pi$, n an integer. For what values of x does the series $\sum f_n''(x)$ converge?

39. Let
$$f(x) = \sum_{n=1}^{\infty} \dfrac{x^n}{n^2}$$
Find the intervals of convergence for f, f', and f''.

40. (a) Starting with the geometric series $\sum_{n=0}^{\infty} x^n$, find the sum of the series
$$\sum_{n=1}^{\infty} n x^{n-1} \qquad |x| < 1$$
(b) Find the sum of each of the following series.
  (i) $\displaystyle\sum_{n=1}^{\infty} n x^n, \quad |x| < 1$
  (ii) $\displaystyle\sum_{n=1}^{\infty} \dfrac{n}{2^n}$

(c) Find the sum of each of the following series.
  (i) $\displaystyle\sum_{n=2}^{\infty} n(n-1) x^n, \quad |x| < 1$
  (ii) $\displaystyle\sum_{n=2}^{\infty} \dfrac{n^2 - n}{2^n}$
  (iii) $\displaystyle\sum_{n=1}^{\infty} \dfrac{n^2}{2^n}$

41. Use the power series for $\tan^{-1} x$ to prove the following expression for π as the sum of an infinite series:
$$\pi = 2\sqrt{3} \sum_{n=0}^{\infty} \dfrac{(-1)^n}{(2n+1)\,3^n}$$

42. (a) By completing the square, show that
$$\int_0^{1/2} \dfrac{dx}{x^2 - x + 1} = \dfrac{\pi}{3\sqrt{3}}$$
(b) By factoring $x^3 + 1$ as a sum of cubes, rewrite the integral in part (a). Then express $1/(x^3 + 1)$ as the sum of a power series and use it to prove the following formula for π:
$$\pi = \dfrac{3\sqrt{3}}{4} \sum_{n=0}^{\infty} \dfrac{(-1)^n}{8^n} \left(\dfrac{2}{3n+1} + \dfrac{1}{3n+2} \right)$$

11.10 Taylor and Maclaurin Series

In the preceding section we were able to find power series representations for a certain restricted class of functions. Here we investigate more general problems: Which functions have power series representations? How can we find such representations?

We start by supposing that f is any function that can be represented by a power series

$$\boxed{1} \quad f(x) = c_0 + c_1(x-a) + c_2(x-a)^2 + c_3(x-a)^3 + c_4(x-a)^4 + \cdots \quad |x-a| < R$$

Let's try to determine what the coefficients c_n must be in terms of f. To begin, notice that if we put $x = a$ in Equation 1, then all terms after the first one are 0 and we get

$$f(a) = c_0$$

By Theorem 11.9.2, we can differentiate the series in Equation 1 term by term:

$$\boxed{2} \quad f'(x) = c_1 + 2c_2(x-a) + 3c_3(x-a)^2 + 4c_4(x-a)^3 + \cdots \quad |x-a| < R$$

and substitution of $x = a$ in Equation 2 gives

$$f'(a) = c_1$$

Now we differentiate both sides of Equation 2 and obtain

$$\boxed{3} \quad f''(x) = 2c_2 + 2 \cdot 3c_3(x-a) + 3 \cdot 4c_4(x-a)^2 + \cdots \quad |x-a| < R$$

Again we put $x = a$ in Equation 3. The result is

$$f''(a) = 2c_2$$

Let's apply the procedure one more time. Differentiation of the series in Equation 3 gives

$$\boxed{4} \quad f'''(x) = 2 \cdot 3c_3 + 2 \cdot 3 \cdot 4c_4(x-a) + 3 \cdot 4 \cdot 5c_5(x-a)^2 + \cdots \quad |x-a| < R$$

and substitution of $x = a$ in Equation 4 gives

$$f'''(a) = 2 \cdot 3c_3 = 3!c_3$$

By now you can see the pattern. If we continue to differentiate and substitute $x = a$, we obtain

$$f^{(n)}(a) = 2 \cdot 3 \cdot 4 \cdot \cdots \cdot nc_n = n!c_n$$

Solving this equation for the nth coefficient c_n, we get

$$c_n = \frac{f^{(n)}(a)}{n!}$$

This formula remains valid even for $n = 0$ if we adopt the conventions that $0! = 1$ and $f^{(0)} = f$. Thus we have proved the following theorem.

5 **Theorem** If f has a power series representation (expansion) at a, that is, if

$$f(x) = \sum_{n=0}^{\infty} c_n(x-a)^n \qquad |x-a| < R$$

then its coefficients are given by the formula

$$c_n = \frac{f^{(n)}(a)}{n!}$$

Substituting this formula for c_n back into the series, we see that *if f has a power series expansion at a, then it must be of the following form.*

6
$$f(x) = \sum_{n=0}^{\infty} \frac{f^{(n)}(a)}{n!}(x-a)^n$$

$$= f(a) + \frac{f'(a)}{1!}(x-a) + \frac{f''(a)}{2!}(x-a)^2 + \frac{f'''(a)}{3!}(x-a)^3 + \cdots$$

The series in Equation 6 is called the **Taylor series of the function f at a** (or **about a** or **centered at a**). For the special case $a = 0$ the Taylor series becomes

7
$$f(x) = \sum_{n=0}^{\infty} \frac{f^{(n)}(0)}{n!} x^n = f(0) + \frac{f'(0)}{1!}x + \frac{f''(0)}{2!}x^2 + \cdots$$

This case arises frequently enough that it is given the special name **Maclaurin series**.

NOTE We have shown that *if* f can be represented as a power series about a, then f is equal to the sum of its Taylor series. But there exist functions that are not equal to the sum of their Taylor series. An example of such a function is given in Exercise 84.

EXAMPLE 1 Find the Maclaurin series of the function $f(x) = e^x$ and its radius of convergence.

SOLUTION If $f(x) = e^x$, then $f^{(n)}(x) = e^x$, so $f^{(n)}(0) = e^0 = 1$ for all n. Therefore the Taylor series for f at 0 (that is, the Maclaurin series) is

$$\sum_{n=0}^{\infty} \frac{f^{(n)}(0)}{n!} x^n = \sum_{n=0}^{\infty} \frac{x^n}{n!} = 1 + \frac{x}{1!} + \frac{x^2}{2!} + \frac{x^3}{3!} + \cdots$$

To find the radius of convergence we let $a_n = x^n/n!$. Then

$$\left|\frac{a_{n+1}}{a_n}\right| = \left|\frac{x^{n+1}}{(n+1)!} \cdot \frac{n!}{x^n}\right| = \frac{|x|}{n+1} \to 0 < 1$$

so, by the Ratio Test, the series converges for all x and the radius of convergence is $R = \infty$.

Taylor and Maclaurin
The Taylor series is named after the English mathematician Brook Taylor (1685–1731) and the Maclaurin series is named in honor of the Scottish mathematician Colin Maclaurin (1698–1746) despite the fact that the Maclaurin series is really just a special case of the Taylor series. But the idea of representing particular functions as sums of power series goes back to Newton, and the general Taylor series was known to the Scottish mathematician James Gregory in 1668 and to the Swiss mathematician John Bernoulli in the 1690s. Taylor was apparently unaware of the work of Gregory and Bernoulli when he published his discoveries on series in 1715 in his book *Methodus incrementorum directa et inversa*. Maclaurin series are named after Colin Maclaurin because he popularized them in his calculus textbook *Treatise of Fluxions* published in 1742.

The conclusion we can draw from Theorem 5 and Example 1 is that *if* e^x has a power series expansion at 0, then

$$e^x = \sum_{n=0}^{\infty} \frac{x^n}{n!}$$

So how can we determine whether e^x *does* have a power series representation?

Let's investigate the more general question: under what circumstances is a function equal to the sum of its Taylor series? In other words, if f has derivatives of all orders, when is it true that

$$f(x) = \sum_{n=0}^{\infty} \frac{f^{(n)}(a)}{n!}(x-a)^n$$

As with any convergent series, this means that $f(x)$ is the limit of the sequence of partial sums. In the case of the Taylor series, the partial sums are

$$T_n(x) = \sum_{i=0}^{n} \frac{f^{(i)}(a)}{i!}(x-a)^i$$

$$= f(a) + \frac{f'(a)}{1!}(x-a) + \frac{f''(a)}{2!}(x-a)^2 + \cdots + \frac{f^{(n)}(a)}{n!}(x-a)^n$$

Notice that T_n is a polynomial of degree n called the **nth-degree Taylor polynomial of f at a**. For instance, for the exponential function $f(x) = e^x$, the result of Example 1 shows that the Taylor polynomials at 0 (or Maclaurin polynomials) with $n = 1, 2$, and 3 are

$$T_1(x) = 1 + x \qquad T_2(x) = 1 + x + \frac{x^2}{2!} \qquad T_3(x) = 1 + x + \frac{x^2}{2!} + \frac{x^3}{3!}$$

The graphs of the exponential function and these three Taylor polynomials are drawn in Figure 1.

In general, $f(x)$ is the sum of its Taylor series if

$$f(x) = \lim_{n \to \infty} T_n(x)$$

If we let

$$R_n(x) = f(x) - T_n(x) \quad \text{so that} \quad f(x) = T_n(x) + R_n(x)$$

then $R_n(x)$ is called the **remainder** of the Taylor series. If we can somehow show that $\lim_{n \to \infty} R_n(x) = 0$, then it follows that

$$\lim_{n \to \infty} T_n(x) = \lim_{n \to \infty} [f(x) - R_n(x)] = f(x) - \lim_{n \to \infty} R_n(x) = f(x)$$

We have therefore proved the following theorem.

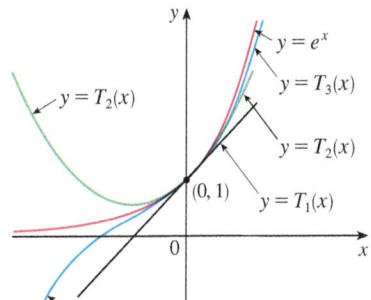

FIGURE 1

As n increases, $T_n(x)$ appears to approach e^x in Figure 1. This suggests that e^x is equal to the sum of its Taylor series.

[8] Theorem If $f(x) = T_n(x) + R_n(x)$, where T_n is the nth-degree Taylor polynomial of f at a and

$$\lim_{n \to \infty} R_n(x) = 0$$

for $|x - a| < R$, then f is equal to the sum of its Taylor series on the interval $|x - a| < R$.

In trying to show that $\lim_{n \to \infty} R_n(x) = 0$ for a specific function f, we usually use the following theorem.

> **9 Taylor's Inequality** If $|f^{(n+1)}(x)| \leq M$ for $|x - a| \leq d$, then the remainder $R_n(x)$ of the Taylor series satisfies the inequality
> $$|R_n(x)| \leq \frac{M}{(n+1)!} |x - a|^{n+1} \qquad \text{for } |x - a| \leq d$$

To see why this is true for $n = 1$, we assume that $|f''(x)| \leq M$. In particular, we have $f''(x) \leq M$, so for $a \leq x \leq a + d$ we have

$$\int_a^x f''(t) \, dt \leq \int_a^x M \, dt$$

An antiderivative of f'' is f', so by Part 2 of the Fundamental Theorem of Calculus, we have

$$f'(x) - f'(a) \leq M(x - a) \qquad \text{or} \qquad f'(x) \leq f'(a) + M(x - a)$$

Thus
$$\int_a^x f'(t) \, dt \leq \int_a^x [f'(a) + M(t - a)] \, dt$$

$$f(x) - f(a) \leq f'(a)(x - a) + M \frac{(x - a)^2}{2}$$

$$f(x) - f(a) - f'(a)(x - a) \leq \frac{M}{2}(x - a)^2$$

But $R_1(x) = f(x) - T_1(x) = f(x) - f(a) - f'(a)(x - a)$. So

$$R_1(x) \leq \frac{M}{2}(x - a)^2$$

A similar argument, using $f''(x) \geq -M$, shows that

$$R_1(x) \geq -\frac{M}{2}(x - a)^2$$

So
$$|R_1(x)| \leq \frac{M}{2}|x - a|^2$$

Although we have assumed that $x > a$, similar calculations show that this inequality is also true for $x < a$.

This proves Taylor's Inequality for the case where $n = 1$. The result for any n is proved in a similar way by integrating $n + 1$ times. (See Exercise 83 for the case $n = 2$.)

NOTE In Section 11.11 we will explore the use of Taylor's Inequality in approximating functions. Our immediate use of it is in conjunction with Theorem 8.

In applying Theorems 8 and 9 it is often helpful to make use of the following fact.

> **10**
> $$\lim_{n \to \infty} \frac{x^n}{n!} = 0 \qquad \text{for every real number } x$$

Formulas for the Taylor Remainder Term

As alternatives to Taylor's Inequality, we have the following formulas for the remainder term. If $f^{(n+1)}$ is continuous on an interval I and $x \in I$, then

$$R_n(x) = \frac{1}{n!} \int_a^x (x - t)^n f^{(n+1)}(t) \, dt$$

This is called the *integral form of the remainder term*. Another formula, called *Lagrange's form of the remainder term*, states that there is a number z between x and a such that

$$R_n(x) = \frac{f^{(n+1)}(z)}{(n+1)!} (x - a)^{n+1}$$

This version is an extension of the Mean Value Theorem (which is the case $n = 0$).

Proofs of these formulas, together with discussions of how to use them to solve the examples of Sections 11.10 and 11.11, are given on the website

www.stewartcalculus.com

Click on *Additional Topics* and then on *Formulas for the Remainder Term in Taylor series*.

This is true because we know from Example 1 that the series $\sum x^n/n!$ converges for all x and so its nth term approaches 0.

EXAMPLE 2 Prove that e^x is equal to the sum of its Maclaurin series.

SOLUTION If $f(x) = e^x$, then $f^{(n+1)}(x) = e^x$ for all n. If d is any positive number and $|x| \le d$, then $|f^{(n+1)}(x)| = e^x \le e^d$. So Taylor's Inequality, with $a = 0$ and $M = e^d$, says that

$$|R_n(x)| \le \frac{e^d}{(n+1)!} |x|^{n+1} \quad \text{for } |x| \le d$$

Notice that the same constant $M = e^d$ works for every value of n. But, from Equation 10, we have

$$\lim_{n\to\infty} \frac{e^d}{(n+1)!} |x|^{n+1} = e^d \lim_{n\to\infty} \frac{|x|^{n+1}}{(n+1)!} = 0$$

It follows from the Squeeze Theorem that $\lim_{n\to\infty} |R_n(x)| = 0$ and therefore $\lim_{n\to\infty} R_n(x) = 0$ for all values of x. By Theorem 8, e^x is equal to the sum of its Maclaurin series, that is,

$$\boxed{e^x = \sum_{n=0}^{\infty} \frac{x^n}{n!} \quad \text{for all } x}$$

[11]

In particular, if we put $x = 1$ in Equation 11, we obtain the following expression for the number e as a sum of an infinite series:

$$\boxed{e = \sum_{n=0}^{\infty} \frac{1}{n!} = 1 + \frac{1}{1!} + \frac{1}{2!} + \frac{1}{3!} + \cdots}$$

[12]

In 1748 Leonhard Euler used Equation 12 to find the value of e correct to 23 digits. In 2010 Shigeru Kondo, again using the series in (12), computed e to more than one trillion decimal places. The special techniques employed to speed up the computation are explained on the website

numbers.computation.free.fr

EXAMPLE 3 Find the Taylor series for $f(x) = e^x$ at $a = 2$.

SOLUTION We have $f^{(n)}(2) = e^2$ and so, putting $a = 2$ in the definition of a Taylor series (6), we get

$$\sum_{n=0}^{\infty} \frac{f^{(n)}(2)}{n!} (x-2)^n = \sum_{n=0}^{\infty} \frac{e^2}{n!} (x-2)^n$$

Again it can be verified, as in Example 1, that the radius of convergence is $R = \infty$. As in Example 2 we can verify that $\lim_{n\to\infty} R_n(x) = 0$, so

[13]
$$e^x = \sum_{n=0}^{\infty} \frac{e^2}{n!} (x-2)^n \quad \text{for all } x$$

We have two power series expansions for e^x, the Maclaurin series in Equation 11 and the Taylor series in Equation 13. The first is better if we are interested in values of x near 0 and the second is better if x is near 2.

764   **CHAPTER 11**   Infinite Sequences and Series

EXAMPLE 4 Find the Maclaurin series for $\sin x$ and prove that it represents $\sin x$ for all x.

SOLUTION We arrange our computation in two columns as follows:

$$f(x) = \sin x \qquad f(0) = 0$$
$$f'(x) = \cos x \qquad f'(0) = 1$$
$$f''(x) = -\sin x \qquad f''(0) = 0$$
$$f'''(x) = -\cos x \qquad f'''(0) = -1$$
$$f^{(4)}(x) = \sin x \qquad f^{(4)}(0) = 0$$

Figure 2 shows the graph of $\sin x$ together with its Taylor (or Maclaurin) polynomials

$$T_1(x) = x$$
$$T_3(x) = x - \frac{x^3}{3!}$$
$$T_5(x) = x - \frac{x^3}{3!} + \frac{x^5}{5!}$$

Notice that, as n increases, $T_n(x)$ becomes a better approximation to $\sin x$.

Since the derivatives repeat in a cycle of four, we can write the Maclaurin series as follows:

$$f(0) + \frac{f'(0)}{1!}x + \frac{f''(0)}{2!}x^2 + \frac{f'''(0)}{3!}x^3 + \cdots$$

$$= x - \frac{x^3}{3!} + \frac{x^5}{5!} - \frac{x^7}{7!} + \cdots = \sum_{n=0}^{\infty} (-1)^n \frac{x^{2n+1}}{(2n+1)!}$$

Since $f^{(n+1)}(x)$ is $\pm \sin x$ or $\pm \cos x$, we know that $|f^{(n+1)}(x)| \leq 1$ for all x. So we can take $M = 1$ in Taylor's Inequality:

$$\boxed{14} \qquad |R_n(x)| \leq \frac{M}{(n+1)!}|x^{n+1}| = \frac{|x|^{n+1}}{(n+1)!}$$

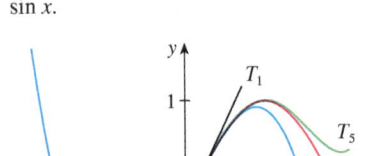

FIGURE 2

By Equation 10 the right side of this inequality approaches 0 as $n \to \infty$, so $|R_n(x)| \to 0$ by the Squeeze Theorem. It follows that $R_n(x) \to 0$ as $n \to \infty$, so $\sin x$ is equal to the sum of its Maclaurin series by Theorem 8. ∎

We state the result of Example 4 for future reference.

$$\boxed{15} \qquad \sin x = x - \frac{x^3}{3!} + \frac{x^5}{5!} - \frac{x^7}{7!} + \cdots$$

$$= \sum_{n=0}^{\infty} (-1)^n \frac{x^{2n+1}}{(2n+1)!} \qquad \text{for all } x$$

EXAMPLE 5 Find the Maclaurin series for $\cos x$.

SOLUTION We could proceed directly as in Example 4, but it's easier to differentiate the Maclaurin series for $\sin x$ given by Equation 15:

$$\cos x = \frac{d}{dx}(\sin x) = \frac{d}{dx}\left(x - \frac{x^3}{3!} + \frac{x^5}{5!} - \frac{x^7}{7!} + \cdots\right)$$

$$= 1 - \frac{3x^2}{3!} + \frac{5x^4}{5!} - \frac{7x^6}{7!} + \cdots = 1 - \frac{x^2}{2!} + \frac{x^4}{4!} - \frac{x^6}{6!} + \cdots$$

The Maclaurin series for e^x, $\sin x$, and $\cos x$ that we found in Examples 2, 4, and 5 were discovered, using different methods, by Newton. These equations are remarkable because they say we know everything about each of these functions if we know all its derivatives at the single number 0.

Since the Maclaurin series for $\sin x$ converges for all x, Theorem 11.9.2 tells us that the differentiated series for $\cos x$ also converges for all x. Thus

$$\boxed{\begin{aligned} \cos x &= 1 - \frac{x^2}{2!} + \frac{x^4}{4!} - \frac{x^6}{6!} + \cdots \\ &= \sum_{n=0}^{\infty} (-1)^n \frac{x^{2n}}{(2n)!} \qquad \text{for all } x \end{aligned}} \quad \boxed{16}$$

EXAMPLE 6 Find the Maclaurin series for the function $f(x) = x \cos x$.

SOLUTION Instead of computing derivatives and substituting in Equation 7, it's easier to multiply the series for $\cos x$ (Equation 16) by x:

$$x \cos x = x \sum_{n=0}^{\infty} (-1)^n \frac{x^{2n}}{(2n)!} = \sum_{n=0}^{\infty} (-1)^n \frac{x^{2n+1}}{(2n)!}$$

EXAMPLE 7 Represent $f(x) = \sin x$ as the sum of its Taylor series centered at $\pi/3$.

SOLUTION Arranging our work in columns, we have

$$f(x) = \sin x \qquad f\left(\frac{\pi}{3}\right) = \frac{\sqrt{3}}{2}$$

$$f'(x) = \cos x \qquad f'\left(\frac{\pi}{3}\right) = \frac{1}{2}$$

$$f''(x) = -\sin x \qquad f''\left(\frac{\pi}{3}\right) = -\frac{\sqrt{3}}{2}$$

$$f'''(x) = -\cos x \qquad f'''\left(\frac{\pi}{3}\right) = -\frac{1}{2}$$

and this pattern repeats indefinitely. Therefore the Taylor series at $\pi/3$ is

$$f\left(\frac{\pi}{3}\right) + \frac{f'\left(\frac{\pi}{3}\right)}{1!}\left(x - \frac{\pi}{3}\right) + \frac{f''\left(\frac{\pi}{3}\right)}{2!}\left(x - \frac{\pi}{3}\right)^2 + \frac{f'''\left(\frac{\pi}{3}\right)}{3!}\left(x - \frac{\pi}{3}\right)^3 + \cdots$$

$$= \frac{\sqrt{3}}{2} + \frac{1}{2 \cdot 1!}\left(x - \frac{\pi}{3}\right) - \frac{\sqrt{3}}{2 \cdot 2!}\left(x - \frac{\pi}{3}\right)^2 - \frac{1}{2 \cdot 3!}\left(x - \frac{\pi}{3}\right)^3 + \cdots$$

The proof that this series represents $\sin x$ for all x is very similar to that in Example 4. [Just replace x by $x - \pi/3$ in (14).] We can write the series in sigma notation if we separate the terms that contain $\sqrt{3}$:

$$\sin x = \sum_{n=0}^{\infty} \frac{(-1)^n \sqrt{3}}{2(2n)!}\left(x - \frac{\pi}{3}\right)^{2n} + \sum_{n=0}^{\infty} \frac{(-1)^n}{2(2n+1)!}\left(x - \frac{\pi}{3}\right)^{2n+1}$$

We have obtained two different series representations for $\sin x$, the Maclaurin series in Example 4 and the Taylor series in Example 7. It is best to use the Maclaurin series for values of x near 0 and the Taylor series for x near $\pi/3$. Notice that the third Taylor polynomial T_3 in Figure 3 is a good approximation to $\sin x$ near $\pi/3$ but not as good near 0. Compare it with the third Maclaurin polynomial T_3 in Figure 2, where the opposite is true.

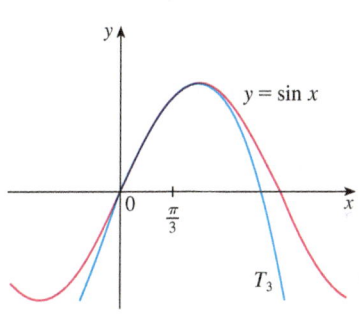

FIGURE 3

The power series that we obtained by indirect methods in Examples 5 and 6 and in Section 11.9 are indeed the Taylor or Maclaurin series of the given functions because Theorem 5 asserts that, no matter how a power series representation $f(x) = \Sigma\, c_n(x - a)^n$ is obtained, it is always true that $c_n = f^{(n)}(a)/n!$. In other words, the coefficients are uniquely determined.

EXAMPLE 8 Find the Maclaurin series for $f(x) = (1 + x)^k$, where k is any real number.

SOLUTION Arranging our work in columns, we have

$$f(x) = (1 + x)^k \qquad\qquad f(0) = 1$$
$$f'(x) = k(1 + x)^{k-1} \qquad\qquad f'(0) = k$$
$$f''(x) = k(k - 1)(1 + x)^{k-2} \qquad\qquad f''(0) = k(k - 1)$$
$$f'''(x) = k(k - 1)(k - 2)(1 + x)^{k-3} \qquad\qquad f'''(0) = k(k - 1)(k - 2)$$
$$\vdots \qquad\qquad\qquad\qquad \vdots$$
$$f^{(n)}(x) = k(k - 1) \cdots (k - n + 1)(1 + x)^{k-n} \qquad f^{(n)}(0) = k(k - 1) \cdots (k - n + 1)$$

Therefore the Maclaurin series of $f(x) = (1 + x)^k$ is

$$\sum_{n=0}^{\infty} \frac{f^{(n)}(0)}{n!} x^n = \sum_{n=0}^{\infty} \frac{k(k - 1) \cdots (k - n + 1)}{n!} x^n$$

This series is called the **binomial series**. Notice that if k is a nonnegative integer, then the terms are eventually 0 and so the series is finite. For other values of k none of the terms is 0 and so we can try the Ratio Test. If the nth term is a_n, then

$$\left| \frac{a_{n+1}}{a_n} \right| = \left| \frac{k(k - 1) \cdots (k - n + 1)(k - n)x^{n+1}}{(n + 1)!} \cdot \frac{n!}{k(k - 1) \cdots (k - n + 1)x^n} \right|$$

$$= \frac{|k - n|}{n + 1} |x| = \frac{\left|1 - \dfrac{k}{n}\right|}{1 + \dfrac{1}{n}} |x| \to |x| \quad \text{as } n \to \infty$$

Thus, by the Ratio Test, the binomial series converges if $|x| < 1$ and diverges if $|x| > 1$. ∎

The traditional notation for the coefficients in the binomial series is

$$\binom{k}{n} = \frac{k(k - 1)(k - 2) \cdots (k - n + 1)}{n!}$$

and these numbers are called the **binomial coefficients**.

The following theorem states that $(1 + x)^k$ is equal to the sum of its Maclaurin series. It is possible to prove this by showing that the remainder term $R_n(x)$ approaches 0, but that turns out to be quite difficult. The proof outlined in Exercise 85 is much easier.

> **17 The Binomial Series** If k is any real number and $|x| < 1$, then
> $$(1+x)^k = \sum_{n=0}^{\infty} \binom{k}{n} x^n = 1 + kx + \frac{k(k-1)}{2!}x^2 + \frac{k(k-1)(k-2)}{3!}x^3 + \cdots$$

Although the binomial series always converges when $|x| < 1$, the question of whether or not it converges at the endpoints, ± 1, depends on the value of k. It turns out that the series converges at 1 if $-1 < k \leq 0$ and at both endpoints if $k \geq 0$. Notice that if k is a positive integer and $n > k$, then the expression for $\binom{k}{n}$ contains a factor $(k - k)$, so $\binom{k}{n} = 0$ for $n > k$. This means that the series terminates and reduces to the ordinary Binomial Theorem when k is a positive integer. (See Reference Page 1.)

EXAMPLE 9 Find the Maclaurin series for the function $f(x) = \dfrac{1}{\sqrt{4-x}}$ and its radius of convergence.

SOLUTION We rewrite $f(x)$ in a form where we can use the binomial series:

$$\frac{1}{\sqrt{4-x}} = \frac{1}{\sqrt{4\left(1 - \frac{x}{4}\right)}} = \frac{1}{2\sqrt{1 - \frac{x}{4}}} = \frac{1}{2}\left(1 - \frac{x}{4}\right)^{-1/2}$$

Using the binomial series with $k = -\tfrac{1}{2}$ and with x replaced by $-x/4$, we have

$$\frac{1}{\sqrt{4-x}} = \frac{1}{2}\left(1 - \frac{x}{4}\right)^{-1/2} = \frac{1}{2}\sum_{n=0}^{\infty} \binom{-\tfrac{1}{2}}{n}\left(-\frac{x}{4}\right)^n$$

$$= \frac{1}{2}\left[1 + \left(-\frac{1}{2}\right)\left(-\frac{x}{4}\right) + \frac{(-\tfrac{1}{2})(-\tfrac{3}{2})}{2!}\left(-\frac{x}{4}\right)^2 + \frac{(-\tfrac{1}{2})(-\tfrac{3}{2})(-\tfrac{5}{2})}{3!}\left(-\frac{x}{4}\right)^3 \right.$$

$$\left. + \cdots + \frac{(-\tfrac{1}{2})(-\tfrac{3}{2})(-\tfrac{5}{2}) \cdots (-\tfrac{1}{2} - n + 1)}{n!}\left(-\frac{x}{4}\right)^n + \cdots \right]$$

$$= \frac{1}{2}\left[1 + \frac{1}{8}x + \frac{1 \cdot 3}{2! 8^2}x^2 + \frac{1 \cdot 3 \cdot 5}{3! 8^3}x^3 + \cdots + \frac{1 \cdot 3 \cdot 5 \cdot \cdots \cdot (2n-1)}{n! 8^n}x^n + \cdots \right]$$

We know from (17) that this series converges when $|-x/4| < 1$, that is, $|x| < 4$, so the radius of convergence is $R = 4$. ∎

We collect in the following table, for future reference, some important Maclaurin series that we have derived in this section and the preceding one.

Table 1

Important Maclaurin Series and Their Radii of Convergence

$$\frac{1}{1-x} = \sum_{n=0}^{\infty} x^n = 1 + x + x^2 + x^3 + \cdots \qquad R = 1$$

$$e^x = \sum_{n=0}^{\infty} \frac{x^n}{n!} = 1 + \frac{x}{1!} + \frac{x^2}{2!} + \frac{x^3}{3!} + \cdots \qquad R = \infty$$

$$\sin x = \sum_{n=0}^{\infty} (-1)^n \frac{x^{2n+1}}{(2n+1)!} = x - \frac{x^3}{3!} + \frac{x^5}{5!} - \frac{x^7}{7!} + \cdots \qquad R = \infty$$

$$\cos x = \sum_{n=0}^{\infty} (-1)^n \frac{x^{2n}}{(2n)!} = 1 - \frac{x^2}{2!} + \frac{x^4}{4!} - \frac{x^6}{6!} + \cdots \qquad R = \infty$$

$$\tan^{-1} x = \sum_{n=0}^{\infty} (-1)^n \frac{x^{2n+1}}{2n+1} = x - \frac{x^3}{3} + \frac{x^5}{5} - \frac{x^7}{7} + \cdots \qquad R = 1$$

$$\ln(1+x) = \sum_{n=1}^{\infty} (-1)^{n-1} \frac{x^n}{n} = x - \frac{x^2}{2} + \frac{x^3}{3} - \frac{x^4}{4} + \cdots \qquad R = 1$$

$$(1+x)^k = \sum_{n=0}^{\infty} \binom{k}{n} x^n = 1 + kx + \frac{k(k-1)}{2!} x^2 + \frac{k(k-1)(k-2)}{3!} x^3 + \cdots \qquad R = 1$$

EXAMPLE 10 Find the sum of the series $\dfrac{1}{1 \cdot 2} - \dfrac{1}{2 \cdot 2^2} + \dfrac{1}{3 \cdot 2^3} - \dfrac{1}{4 \cdot 2^4} + \cdots$.

SOLUTION With sigma notation we can write the given series as

$$\sum_{n=1}^{\infty} (-1)^{n-1} \frac{1}{n \cdot 2^n} = \sum_{n=1}^{\infty} (-1)^{n-1} \frac{\left(\frac{1}{2}\right)^n}{n}$$

Then from Table 1 we see that this series matches the entry for $\ln(1 + x)$ with $x = \frac{1}{2}$. So

$$\sum_{n=1}^{\infty} (-1)^{n-1} \frac{1}{n \cdot 2^n} = \ln\left(1 + \tfrac{1}{2}\right) = \ln \tfrac{3}{2} \qquad \blacksquare$$

TEC Module 11.10/11.11 enables you to see how successive Taylor polynomials approach the original function.

One reason that Taylor series are important is that they enable us to integrate functions that we couldn't previously handle. In fact, in the introduction to this chapter we mentioned that Newton often integrated functions by first expressing them as power series and then integrating the series term by term. The function $f(x) = e^{-x^2}$ can't be integrated by techniques discussed so far because its antiderivative is not an elementary function (see Section 7.5). In the following example we use Newton's idea to integrate this function.

EXAMPLE 11

(a) Evaluate $\int e^{-x^2} dx$ as an infinite series.
(b) Evaluate $\int_0^1 e^{-x^2} dx$ correct to within an error of 0.001.

SOLUTION

(a) First we find the Maclaurin series for $f(x) = e^{-x^2}$. Although it's possible to use the direct method, let's find it simply by replacing x with $-x^2$ in the series for e^x given in

Table 1. Thus, for all values of x,

$$e^{-x^2} = \sum_{n=0}^{\infty} \frac{(-x^2)^n}{n!} = \sum_{n=0}^{\infty} (-1)^n \frac{x^{2n}}{n!} = 1 - \frac{x^2}{1!} + \frac{x^4}{2!} - \frac{x^6}{3!} + \cdots$$

Now we integrate term by term:

$$\int e^{-x^2} dx = \int \left(1 - \frac{x^2}{1!} + \frac{x^4}{2!} - \frac{x^6}{3!} + \cdots + (-1)^n \frac{x^{2n}}{n!} + \cdots\right) dx$$

$$= C + x - \frac{x^3}{3 \cdot 1!} + \frac{x^5}{5 \cdot 2!} - \frac{x^7}{7 \cdot 3!} + \cdots + (-1)^n \frac{x^{2n+1}}{(2n+1)n!} + \cdots$$

This series converges for all x because the original series for e^{-x^2} converges for all x.

(b) The Fundamental Theorem of Calculus gives

$$\int_0^1 e^{-x^2} dx = \left[x - \frac{x^3}{3 \cdot 1!} + \frac{x^5}{5 \cdot 2!} - \frac{x^7}{7 \cdot 3!} + \frac{x^9}{9 \cdot 4!} - \cdots \right]_0^1$$

We can take $C = 0$ in the antiderivative in part (a).

$$= 1 - \tfrac{1}{3} + \tfrac{1}{10} - \tfrac{1}{42} + \tfrac{1}{216} - \cdots$$

$$\approx 1 - \tfrac{1}{3} + \tfrac{1}{10} - \tfrac{1}{42} + \tfrac{1}{216} \approx 0.7475$$

The Alternating Series Estimation Theorem shows that the error involved in this approximation is less than

$$\frac{1}{11 \cdot 5!} = \frac{1}{1320} < 0.001 \qquad \blacksquare$$

Another use of Taylor series is illustrated in the next example. The limit could be found with l'Hospital's Rule, but instead we use a series.

EXAMPLE 12 Evaluate $\lim_{x \to 0} \dfrac{e^x - 1 - x}{x^2}$.

SOLUTION Using the Maclaurin series for e^x, we have

$$\lim_{x \to 0} \frac{e^x - 1 - x}{x^2} = \lim_{x \to 0} \frac{\left(1 + \frac{x}{1!} + \frac{x^2}{2!} + \frac{x^3}{3!} + \cdots\right) - 1 - x}{x^2}$$

Some computer algebra systems compute limits in this way.

$$= \lim_{x \to 0} \frac{\frac{x^2}{2!} + \frac{x^3}{3!} + \frac{x^4}{4!} + \cdots}{x^2}$$

$$= \lim_{x \to 0} \left(\frac{1}{2} + \frac{x}{3!} + \frac{x^2}{4!} + \frac{x^3}{5!} + \cdots \right) = \frac{1}{2}$$

because power series are continuous functions. $\blacksquare$

Multiplication and Division of Power Series

If power series are added or subtracted, they behave like polynomials (Theorem 11.2.8 shows this). In fact, as the following example illustrates, they can also be multiplied and divided like polynomials. We find only the first few terms because the calculations for the later terms become tedious and the initial terms are the most important ones.

EXAMPLE 13 Find the first three nonzero terms in the Maclaurin series for (a) $e^x \sin x$ and (b) $\tan x$.

SOLUTION
(a) Using the Maclaurin series for e^x and $\sin x$ in Table 1, we have

$$e^x \sin x = \left(1 + \frac{x}{1!} + \frac{x^2}{2!} + \frac{x^3}{3!} + \cdots\right)\left(x - \frac{x^3}{3!} + \cdots\right)$$

We multiply these expressions, collecting like terms just as for polynomials:

$$\begin{array}{r}
1 + x + \frac{1}{2}x^2 + \frac{1}{6}x^3 + \cdots \\
\times \quad\quad\quad x \quad\quad - \frac{1}{6}x^3 + \cdots \\
\hline
x + x^2 + \frac{1}{2}x^3 + \frac{1}{6}x^4 + \cdots \\
+ \quad\quad\quad\quad - \frac{1}{6}x^3 - \frac{1}{6}x^4 - \cdots \\
\hline
x + x^2 + \frac{1}{3}x^3 + \cdots
\end{array}$$

Thus $\quad\quad e^x \sin x = x + x^2 + \frac{1}{3}x^3 + \cdots$

(b) Using the Maclaurin series in Table 1, we have

$$\tan x = \frac{\sin x}{\cos x} = \frac{x - \dfrac{x^3}{3!} + \dfrac{x^5}{5!} - \cdots}{1 - \dfrac{x^2}{2!} + \dfrac{x^4}{4!} - \cdots}$$

We use a procedure like long division:

$$\begin{array}{r}
x + \frac{1}{3}x^3 + \frac{2}{15}x^5 + \cdots \\
1 - \frac{1}{2}x^2 + \frac{1}{24}x^4 - \cdots \overline{\smash{)}x - \frac{1}{6}x^3 + \frac{1}{120}x^5 - \cdots} \\
x - \frac{1}{2}x^3 + \frac{1}{24}x^5 - \cdots \\
\hline
\frac{1}{3}x^3 - \frac{1}{30}x^5 + \cdots \\
\frac{1}{3}x^3 - \frac{1}{6}x^5 + \cdots \\
\hline
\frac{2}{15}x^5 + \cdots
\end{array}$$

Thus $\quad\quad \tan x = x + \frac{1}{3}x^3 + \frac{2}{15}x^5 + \cdots$ ∎

Although we have not attempted to justify the formal manipulations used in Example 13, they are legitimate. There is a theorem which states that if both $f(x) = \Sigma\, c_n x^n$ and $g(x) = \Sigma\, b_n x^n$ converge for $|x| < R$ and the series are multiplied as if they were polynomials, then the resulting series also converges for $|x| < R$ and represents $f(x)g(x)$. For division we require $b_0 \neq 0$; the resulting series converges for sufficiently small $|x|$.

11.10 EXERCISES

1. If $f(x) = \sum_{n=0}^{\infty} b_n(x-5)^n$ for all x, write a formula for b_8.

2. The graph of f is shown.

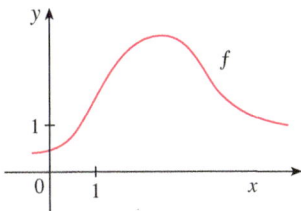

(a) Explain why the series
$$1.6 - 0.8(x-1) + 0.4(x-1)^2 - 0.1(x-1)^3 + \cdots$$
is *not* the Taylor series of f centered at 1.

(b) Explain why the series
$$2.8 + 0.5(x-2) + 1.5(x-2)^2 - 0.1(x-2)^3 + \cdots$$
is *not* the Taylor series of f centered at 2.

3. If $f^{(n)}(0) = (n+1)!$ for $n = 0, 1, 2, \ldots$, find the Maclaurin series for f and its radius of convergence.

4. Find the Taylor series for f centered at 4 if
$$f^{(n)}(4) = \frac{(-1)^n n!}{3^n(n+1)}$$
What is the radius of convergence of the Taylor series?

5–10 Use the definition of a Taylor series to find the first four nonzero terms of the series for $f(x)$ centered at the given value of a.

5. $f(x) = xe^x$, $a = 0$
6. $f(x) = \dfrac{1}{1+x}$, $a = 2$
7. $f(x) = \sqrt[3]{x}$, $a = 8$
8. $f(x) = \ln x$, $a = 1$
9. $f(x) = \sin x$, $a = \pi/6$
10. $f(x) = \cos^2 x$, $a = 0$

11–18 Find the Maclaurin series for $f(x)$ using the definition of a Maclaurin series. [Assume that f has a power series expansion. Do not show that $R_n(x) \to 0$.] Also find the associated radius of convergence.

11. $f(x) = (1-x)^{-2}$
12. $f(x) = \ln(1+x)$
13. $f(x) = \cos x$
14. $f(x) = e^{-2x}$
15. $f(x) = 2^x$
16. $f(x) = x \cos x$
17. $f(x) = \sinh x$
18. $f(x) = \cosh x$

19–26 Find the Taylor series for $f(x)$ centered at the given value of a. [Assume that f has a power series expansion. Do not show that $R_n(x) \to 0$.] Also find the associated radius of convergence.

19. $f(x) = x^5 + 2x^3 + x$, $a = 2$
20. $f(x) = x^6 - x^4 + 2$, $a = -2$
21. $f(x) = \ln x$, $a = 2$
22. $f(x) = 1/x$, $a = -3$
23. $f(x) = e^{2x}$, $a = 3$
24. $f(x) = \cos x$, $a = \pi/2$
25. $f(x) = \sin x$, $a = \pi$
26. $f(x) = \sqrt{x}$, $a = 16$

27. Prove that the series obtained in Exercise 13 represents $\cos x$ for all x.

28. Prove that the series obtained in Exercise 25 represents $\sin x$ for all x.

29. Prove that the series obtained in Exercise 17 represents $\sinh x$ for all x.

30. Prove that the series obtained in Exercise 18 represents $\cosh x$ for all x.

31–34 Use the binomial series to expand the function as a power series. State the radius of convergence.

31. $\sqrt[4]{1-x}$
32. $\sqrt[3]{8+x}$
33. $\dfrac{1}{(2+x)^3}$
34. $(1-x)^{3/4}$

35–44 Use a Maclaurin series in Table 1 to obtain the Maclaurin series for the given function.

35. $f(x) = \arctan(x^2)$
36. $f(x) = \sin(\pi x/4)$
37. $f(x) = x \cos 2x$
38. $f(x) = e^{3x} - e^{2x}$
39. $f(x) = x \cos(\tfrac{1}{2}x^2)$
40. $f(x) = x^2 \ln(1+x^3)$
41. $f(x) = \dfrac{x}{\sqrt{4+x^2}}$
42. $f(x) = \dfrac{x^2}{\sqrt{2+x}}$
43. $f(x) = \sin^2 x$ [*Hint:* Use $\sin^2 x = \tfrac{1}{2}(1 - \cos 2x)$.]
44. $f(x) = \begin{cases} \dfrac{x - \sin x}{x^3} & \text{if } x \neq 0 \\ \tfrac{1}{6} & \text{if } x = 0 \end{cases}$

45–48 Find the Maclaurin series of f (by any method) and its radius of convergence. Graph f and its first few Taylor polynomials on the same screen. What do you notice about the relationship between these polynomials and f?

45. $f(x) = \cos(x^2)$
46. $f(x) = \ln(1+x^2)$
47. $f(x) = xe^{-x}$
48. $f(x) = \tan^{-1}(x^3)$

49. Use the Maclaurin series for $\cos x$ to compute $\cos 5°$ correct to five decimal places.

50. Use the Maclaurin series for e^x to calculate $1/\sqrt[10]{e}$ correct to five decimal places.

51. (a) Use the binomial series to expand $1/\sqrt{1-x^2}$.
(b) Use part (a) to find the Maclaurin series for $\sin^{-1} x$.

52. (a) Expand $1/\sqrt[4]{1+x}$ as a power series.
(b) Use part (a) to estimate $1/\sqrt[4]{1.1}$ correct to three decimal places.

53–56 Evaluate the indefinite integral as an infinite series.

53. $\int \sqrt{1+x^3}\, dx$

54. $\int x^2 \sin(x^2)\, dx$

55. $\int \dfrac{\cos x - 1}{x}\, dx$

56. $\int \arctan(x^2)\, dx$

57–60 Use series to approximate the definite integral to within the indicated accuracy.

57. $\int_0^{1/2} x^3 \arctan x\, dx$ (four decimal places)

58. $\int_0^1 \sin(x^4)\, dx$ (four decimal places)

59. $\int_0^{0.4} \sqrt{1+x^4}\, dx$ ($|\text{error}| < 5 \times 10^{-6}$)

60. $\int_0^{0.5} x^2 e^{-x^2}\, dx$ ($|\text{error}| < 0.001$)

61–65 Use series to evaluate the limit.

61. $\lim_{x\to 0} \dfrac{x - \ln(1+x)}{x^2}$

62. $\lim_{x\to 0} \dfrac{1 - \cos x}{1 + x - e^x}$

63. $\lim_{x\to 0} \dfrac{\sin x - x + \frac{1}{6}x^3}{x^5}$

64. $\lim_{x\to 0} \dfrac{\sqrt{1+x} - 1 - \frac{1}{2}x}{x^2}$

65. $\lim_{x\to 0} \dfrac{x^3 - 3x + 3\tan^{-1} x}{x^5}$

66. Use the series in Example 13(b) to evaluate

$$\lim_{x\to 0} \dfrac{\tan x - x}{x^3}$$

We found this limit in Example 4.4.4 using l'Hospital's Rule three times. Which method do you prefer?

67–72 Use multiplication or division of power series to find the first three nonzero terms in the Maclaurin series for each function.

67. $y = e^{-x^2} \cos x$

68. $y = \sec x$

69. $y = \dfrac{x}{\sin x}$

70. $y = e^x \ln(1+x)$

71. $y = (\arctan x)^2$

72. $y = e^x \sin^2 x$

73–80 Find the sum of the series.

73. $\sum_{n=0}^{\infty} (-1)^n \dfrac{x^{4n}}{n!}$

74. $\sum_{n=0}^{\infty} \dfrac{(-1)^n \pi^{2n}}{6^{2n}(2n)!}$

75. $\sum_{n=1}^{\infty} (-1)^{n-1} \dfrac{3^n}{n\, 5^n}$

76. $\sum_{n=0}^{\infty} \dfrac{3^n}{5^n n!}$

77. $\sum_{n=0}^{\infty} \dfrac{(-1)^n \pi^{2n+1}}{4^{2n+1}(2n+1)!}$

78. $1 - \ln 2 + \dfrac{(\ln 2)^2}{2!} - \dfrac{(\ln 2)^3}{3!} + \cdots$

79. $3 + \dfrac{9}{2!} + \dfrac{27}{3!} + \dfrac{81}{4!} + \cdots$

80. $\dfrac{1}{1\cdot 2} - \dfrac{1}{3\cdot 2^3} + \dfrac{1}{5\cdot 2^5} - \dfrac{1}{7\cdot 2^7} + \cdots$

81. Show that if p is an nth-degree polynomial, then

$$p(x+1) = \sum_{i=0}^{n} \dfrac{p^{(i)}(x)}{i!}$$

82. If $f(x) = (1+x^3)^{30}$, what is $f^{(58)}(0)$?

83. Prove Taylor's Inequality for $n = 2$, that is, prove that if $|f'''(x)| \leq M$ for $|x - a| \leq d$, then

$$|R_2(x)| \leq \dfrac{M}{6}|x-a|^3 \quad \text{for } |x-a| \leq d$$

84. (a) Show that the function defined by

$$f(x) = \begin{cases} e^{-1/x^2} & \text{if } x \neq 0 \\ 0 & \text{if } x = 0 \end{cases}$$

is not equal to its Maclaurin series.

(b) Graph the function in part (a) and comment on its behavior near the origin.

85. Use the following steps to prove (17).
(a) Let $g(x) = \sum_{n=0}^{\infty} \binom{k}{n} x^n$. Differentiate this series to show that

$$g'(x) = \dfrac{kg(x)}{1+x} \quad -1 < x < 1$$

(b) Let $h(x) = (1+x)^{-k} g(x)$ and show that $h'(x) = 0$.
(c) Deduce that $g(x) = (1+x)^k$.

86. In Exercise 10.2.53 it was shown that the length of the ellipse $x = a\sin\theta$, $y = b\cos\theta$, where $a > b > 0$, is

$$L = 4a \int_0^{\pi/2} \sqrt{1 - e^2 \sin^2 \theta}\, d\theta$$

where $e = \sqrt{a^2 - b^2}/a$ is the eccentricity of the ellipse.

Expand the integrand as a binomial series and use the result of Exercise 7.1.50 to express L as a series in powers of the eccentricity up to the term in e^6.

LABORATORY PROJECT — AN ELUSIVE LIMIT

This project deals with the function

$$f(x) = \frac{\sin(\tan x) - \tan(\sin x)}{\arcsin(\arctan x) - \arctan(\arcsin x)}$$

1. Use your computer algebra system to evaluate $f(x)$ for $x = 1, 0.1, 0.01, 0.001,$ and 0.0001. Does it appear that f has a limit as $x \to 0$?
2. Use the CAS to graph f near $x = 0$. Does it appear that f has a limit as $x \to 0$?
3. Try to evaluate $\lim_{x \to 0} f(x)$ with l'Hospital's Rule, using the CAS to find derivatives of the numerator and denominator. What do you discover? How many applications of l'Hospital's Rule are required?
4. Evaluate $\lim_{x \to 0} f(x)$ by using the CAS to find sufficiently many terms in the Taylor series of the numerator and denominator. (Use the command taylor in Maple or Series in Mathematica.)
5. Use the limit command on your CAS to find $\lim_{x \to 0} f(x)$ directly. (Most computer algebra systems use the method of Problem 4 to compute limits.)
6. In view of the answers to Problems 4 and 5, how do you explain the results of Problems 1 and 2?

WRITING PROJECT — HOW NEWTON DISCOVERED THE BINOMIAL SERIES

The Binomial Theorem, which gives the expansion of $(a + b)^k$, was known to Chinese mathematicians many centuries before the time of Newton for the case where the exponent k is a positive integer. In 1665, when he was 22, Newton was the first to discover the infinite series expansion of $(a + b)^k$ when k is a fractional exponent (positive or negative). He didn't publish his discovery, but he stated it and gave examples of how to use it in a letter (now called the *epistola prior*) dated June 13, 1676, that he sent to Henry Oldenburg, secretary of the Royal Society of London, to transmit to Leibniz. When Leibniz replied, he asked how Newton had discovered the binomial series. Newton wrote a second letter, the *epistola posterior* of October 24, 1676, in which he explained in great detail how he arrived at his discovery by a very indirect route. He was investigating the areas under the curves $y = (1 - x^2)^{n/2}$ from 0 to x for $n = 0, 1, 2, 3, 4, \ldots$. These are easy to calculate if n is even. By observing patterns and interpolating, Newton was able to guess the answers for odd values of n. Then he realized he could get the same answers by expressing $(1 - x^2)^{n/2}$ as an infinite series.

Write a report on Newton's discovery of the binomial series. Start by giving the statement of the binomial series in Newton's notation (see the *epistola prior* on page 285 of [4] or page 402 of [2]). Explain why Newton's version is equivalent to Theorem 17 on page 767. Then read Newton's *epistola posterior* (page 287 in [4] or page 404 in [2]) and explain the patterns that Newton discovered in the areas under the curves $y = (1 - x^2)^{n/2}$. Show how he was able to guess the areas under the remaining curves and how he verified his answers. Finally, explain how these discoveries led to the binomial series. The books by Edwards [1] and Katz [3] contain commentaries on Newton's letters.

1. C. H. Edwards, *The Historical Development of the Calculus* (New York: Springer-Verlag, 1979), pp. 178–187.
2. John Fauvel and Jeremy Gray, eds., *The History of Mathematics: A Reader* (London: MacMillan Press, 1987).
3. Victor Katz, *A History of Mathematics: An Introduction* (New York: HarperCollins, 1993), pp. 463–466.
4. D. J. Struik, ed., *A Sourcebook in Mathematics, 1200–1800* (Princeton, NJ: Princeton University Press, 1969).

11.11 Applications of Taylor Polynomials

In this section we explore two types of applications of Taylor polynomials. First we look at how they are used to approximate functions—computer scientists like them because polynomials are the simplest of functions. Then we investigate how physicists and engineers use them in such fields as relativity, optics, blackbody radiation, electric dipoles, the velocity of water waves, and building highways across a desert.

Approximating Functions by Polynomials

Suppose that $f(x)$ is equal to the sum of its Taylor series at a:

$$f(x) = \sum_{n=0}^{\infty} \frac{f^{(n)}(a)}{n!} (x - a)^n$$

In Section 11.10 we introduced the notation $T_n(x)$ for the nth partial sum of this series and called it the nth-degree Taylor polynomial of f at a. Thus

$$T_n(x) = \sum_{i=0}^{n} \frac{f^{(i)}(a)}{i!} (x - a)^i$$

$$= f(a) + \frac{f'(a)}{1!}(x - a) + \frac{f''(a)}{2!}(x - a)^2 + \cdots + \frac{f^{(n)}(a)}{n!}(x - a)^n$$

Since f is the sum of its Taylor series, we know that $T_n(x) \to f(x)$ as $n \to \infty$ and so T_n can be used as an approximation to f: $f(x) \approx T_n(x)$.

Notice that the first-degree Taylor polynomial

$$T_1(x) = f(a) + f'(a)(x - a)$$

is the same as the linearization of f at a that we discussed in Section 3.10. Notice also that T_1 and its derivative have the same values at a that f and f' have. In general, it can be shown that the derivatives of T_n at a agree with those of f up to and including derivatives of order n.

To illustrate these ideas let's take another look at the graphs of $y = e^x$ and its first few Taylor polynomials, as shown in Figure 1. The graph of T_1 is the tangent line to $y = e^x$ at $(0, 1)$; this tangent line is the best linear approximation to e^x near $(0, 1)$. The graph of T_2 is the parabola $y = 1 + x + x^2/2$, and the graph of T_3 is the cubic curve $y = 1 + x + x^2/2 + x^3/6$, which is a closer fit to the exponential curve $y = e^x$ than T_2. The next Taylor polynomial T_4 would be an even better approximation, and so on.

The values in the table give a numerical demonstration of the convergence of the Taylor polynomials $T_n(x)$ to the function $y = e^x$. We see that when $x = 0.2$ the convergence is very rapid, but when $x = 3$ it is somewhat slower. In fact, the farther x is from 0, the more slowly $T_n(x)$ converges to e^x.

When using a Taylor polynomial T_n to approximate a function f, we have to ask the questions: How good an approximation is it? How large should we take n to be in order to achieve a desired accuracy? To answer these questions we need to look at the absolute value of the remainder:

$$|R_n(x)| = |f(x) - T_n(x)|$$

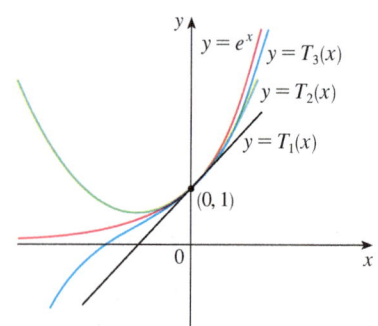

FIGURE 1

	$x = 0.2$	$x = 3.0$
$T_2(x)$	1.220000	8.500000
$T_4(x)$	1.221400	16.375000
$T_6(x)$	1.221403	19.412500
$T_8(x)$	1.221403	20.009152
$T_{10}(x)$	1.221403	20.079665
e^x	1.221403	20.085537

There are three possible methods for estimating the size of the error:

1. If a graphing device is available, we can use it to graph $|R_n(x)|$ and thereby estimate the error.
2. If the series happens to be an alternating series, we can use the Alternating Series Estimation Theorem.
3. In all cases we can use Taylor's Inequality (Theorem 11.10.9), which says that if $|f^{(n+1)}(x)| \leq M$, then
$$|R_n(x)| \leq \frac{M}{(n+1)!}|x-a|^{n+1}$$

EXAMPLE 1
(a) Approximate the function $f(x) = \sqrt[3]{x}$ by a Taylor polynomial of degree 2 at $a = 8$.
(b) How accurate is this approximation when $7 \leq x \leq 9$?

SOLUTION
(a)
$$f(x) = \sqrt[3]{x} = x^{1/3} \qquad f(8) = 2$$
$$f'(x) = \tfrac{1}{3}x^{-2/3} \qquad f'(8) = \tfrac{1}{12}$$
$$f''(x) = -\tfrac{2}{9}x^{-5/3} \qquad f''(8) = \tfrac{1}{144}$$
$$f'''(x) = \tfrac{10}{27}x^{-8/3}$$

Thus the second-degree Taylor polynomial is
$$T_2(x) = f(8) + \frac{f'(8)}{1!}(x-8) + \frac{f''(8)}{2!}(x-8)^2$$
$$= 2 + \tfrac{1}{12}(x-8) - \tfrac{1}{288}(x-8)^2$$

The desired approximation is
$$\sqrt[3]{x} \approx T_2(x) = 2 + \tfrac{1}{12}(x-8) - \tfrac{1}{288}(x-8)^2$$

(b) The Taylor series is not alternating when $x < 8$, so we can't use the Alternating Series Estimation Theorem in this example. But we can use Taylor's Inequality with $n = 2$ and $a = 8$:
$$|R_2(x)| \leq \frac{M}{3!}|x-8|^3$$

where $|f'''(x)| \leq M$. Because $x \geq 7$, we have $x^{8/3} \geq 7^{8/3}$ and so
$$f'''(x) = \frac{10}{27} \cdot \frac{1}{x^{8/3}} \leq \frac{10}{27} \cdot \frac{1}{7^{8/3}} < 0.0021$$

Therefore we can take $M = 0.0021$. Also $7 \leq x \leq 9$, so $-1 \leq x - 8 \leq 1$ and $|x-8| \leq 1$. Then Taylor's Inequality gives
$$|R_2(x)| \leq \frac{0.0021}{3!} \cdot 1^3 = \frac{0.0021}{6} < 0.0004$$

Thus, if $7 \leq x \leq 9$, the approximation in part (a) is accurate to within 0.0004.

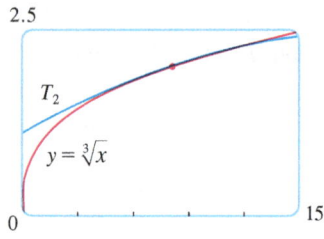

FIGURE 2

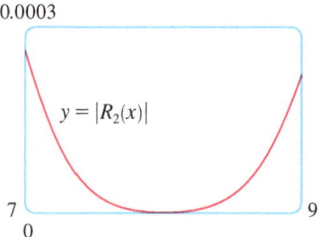

FIGURE 3

Let's use a graphing device to check the calculation in Example 1. Figure 2 shows that the graphs of $y = \sqrt[3]{x}$ and $y = T_2(x)$ are very close to each other when x is near 8. Figure 3 shows the graph of $|R_2(x)|$ computed from the expression

$$|R_2(x)| = |\sqrt[3]{x} - T_2(x)|$$

We see from the graph that

$$|R_2(x)| < 0.0003$$

when $7 \leq x \leq 9$. Thus the error estimate from graphical methods is slightly better than the error estimate from Taylor's Inequality in this case.

EXAMPLE 2
(a) What is the maximum error possible in using the approximation

$$\sin x \approx x - \frac{x^3}{3!} + \frac{x^5}{5!}$$

when $-0.3 \leq x \leq 0.3$? Use this approximation to find $\sin 12°$ correct to six decimal places.
(b) For what values of x is this approximation accurate to within 0.00005?

SOLUTION
(a) Notice that the Maclaurin series

$$\sin x = x - \frac{x^3}{3!} + \frac{x^5}{5!} - \frac{x^7}{7!} + \cdots$$

is alternating for all nonzero values of x, and the successive terms decrease in size because $|x| < 1$, so we can use the Alternating Series Estimation Theorem. The error in approximating $\sin x$ by the first three terms of its Maclaurin series is at most

$$\left|\frac{x^7}{7!}\right| = \frac{|x|^7}{5040}$$

If $-0.3 \leq x \leq 0.3$, then $|x| \leq 0.3$, so the error is smaller than

$$\frac{(0.3)^7}{5040} \approx 4.3 \times 10^{-8}$$

To find $\sin 12°$ we first convert to radian measure:

$$\sin 12° = \sin\left(\frac{12\pi}{180}\right) = \sin\left(\frac{\pi}{15}\right)$$

$$\approx \frac{\pi}{15} - \left(\frac{\pi}{15}\right)^3 \frac{1}{3!} + \left(\frac{\pi}{15}\right)^5 \frac{1}{5!} \approx 0.20791169$$

Thus, correct to six decimal places, $\sin 12° \approx 0.207912$.
(b) The error will be smaller than 0.00005 if

$$\frac{|x|^7}{5040} < 0.00005$$

Solving this inequality for x, we get

$$|x|^7 < 0.252 \quad \text{or} \quad |x| < (0.252)^{1/7} \approx 0.821$$

So the given approximation is accurate to within 0.00005 when $|x| < 0.82$. ∎

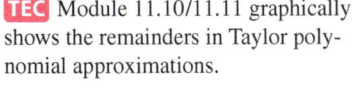

Module 11.10/11.11 graphically shows the remainders in Taylor polynomial approximations.

What if we use Taylor's Inequality to solve Example 2? Since $f^{(7)}(x) = -\cos x$, we have $|f^{(7)}(x)| \leq 1$ and so

$$|R_6(x)| \leq \frac{1}{7!}|x|^7$$

So we get the same estimates as with the Alternating Series Estimation Theorem.

What about graphical methods? Figure 4 shows the graph of

$$|R_6(x)| = |\sin x - (x - \tfrac{1}{6}x^3 + \tfrac{1}{120}x^5)|$$

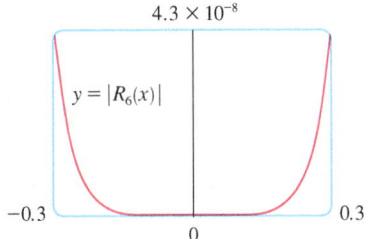

FIGURE 4

and we see from it that $|R_6(x)| < 4.3 \times 10^{-8}$ when $|x| \leq 0.3$. This is the same estimate that we obtained in Example 2. For part (b) we want $|R_6(x)| < 0.00005$, so we graph both $y = |R_6(x)|$ and $y = 0.00005$ in Figure 5. By placing the cursor on the right intersection point we find that the inequality is satisfied when $|x| < 0.82$. Again this is the same estimate that we obtained in the solution to Example 2.

If we had been asked to approximate $\sin 72°$ instead of $\sin 12°$ in Example 2, it would have been wise to use the Taylor polynomials at $a = \pi/3$ (instead of $a = 0$) because they are better approximations to $\sin x$ for values of x close to $\pi/3$. Notice that $72°$ is close to $60°$ (or $\pi/3$ radians) and the derivatives of $\sin x$ are easy to compute at $\pi/3$.

Figure 6 shows the graphs of the Maclaurin polynomial approximations

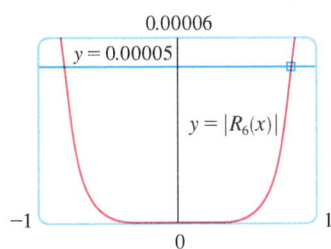

FIGURE 5

$$T_1(x) = x \qquad T_3(x) = x - \frac{x^3}{3!}$$

$$T_5(x) = x - \frac{x^3}{3!} + \frac{x^5}{5!} \qquad T_7(x) = x - \frac{x^3}{3!} + \frac{x^5}{5!} - \frac{x^7}{7!}$$

to the sine curve. You can see that as n increases, $T_n(x)$ is a good approximation to $\sin x$ on a larger and larger interval.

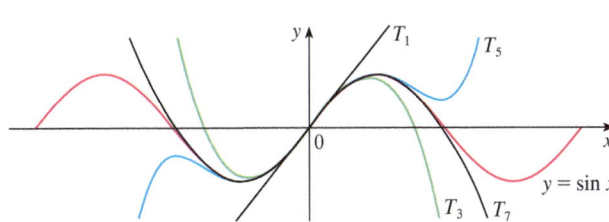

FIGURE 6

One use of the type of calculation done in Examples 1 and 2 occurs in calculators and computers. For instance, when you press the sin or e^x key on your calculator, or when a computer programmer uses a subroutine for a trigonometric or exponential or Bessel function, in many machines a polynomial approximation is calculated. The polynomial is often a Taylor polynomial that has been modified so that the error is spread more evenly throughout an interval.

Applications to Physics

Taylor polynomials are also used frequently in physics. In order to gain insight into an equation, a physicist often simplifies a function by considering only the first two or three terms in its Taylor series. In other words, the physicist uses a Taylor polynomial as an approximation to the function. Taylor's Inequality can then be used to gauge the accuracy of the approximation. The following example shows one way in which this idea is used in special relativity.

EXAMPLE 3 In Einstein's theory of special relativity the mass of an object moving with velocity v is

$$m = \frac{m_0}{\sqrt{1 - v^2/c^2}}$$

where m_0 is the mass of the object when at rest and c is the speed of light. The kinetic energy of the object is the difference between its total energy and its energy at rest:

$$K = mc^2 - m_0 c^2$$

(a) Show that when v is very small compared with c, this expression for K agrees with classical Newtonian physics: $K = \frac{1}{2} m_0 v^2$.
(b) Use Taylor's Inequality to estimate the difference in these expressions for K when $|v| \leq 100$ m/s.

SOLUTION
(a) Using the expressions given for K and m, we get

$$K = mc^2 - m_0 c^2 = \frac{m_0 c^2}{\sqrt{1 - v^2/c^2}} - m_0 c^2 = m_0 c^2 \left[\left(1 - \frac{v^2}{c^2}\right)^{-1/2} - 1 \right]$$

With $x = -v^2/c^2$, the Maclaurin series for $(1 + x)^{-1/2}$ is most easily computed as a binomial series with $k = -\frac{1}{2}$. (Notice that $|x| < 1$ because $v < c$.) Therefore we have

$$(1 + x)^{-1/2} = 1 - \tfrac{1}{2}x + \frac{(-\tfrac{1}{2})(-\tfrac{3}{2})}{2!}x^2 + \frac{(-\tfrac{1}{2})(-\tfrac{3}{2})(-\tfrac{5}{2})}{3!}x^3 + \cdots$$

$$= 1 - \tfrac{1}{2}x + \tfrac{3}{8}x^2 - \tfrac{5}{16}x^3 + \cdots$$

and

$$K = m_0 c^2 \left[\left(1 + \frac{1}{2} \frac{v^2}{c^2} + \frac{3}{8} \frac{v^4}{c^4} + \frac{5}{16} \frac{v^6}{c^6} + \cdots \right) - 1 \right]$$

$$= m_0 c^2 \left(\frac{1}{2} \frac{v^2}{c^2} + \frac{3}{8} \frac{v^4}{c^4} + \frac{5}{16} \frac{v^6}{c^6} + \cdots \right)$$

If v is much smaller than c, then all terms after the first are very small when compared with the first term. If we omit them, we get

$$K \approx m_0 c^2 \left(\frac{1}{2} \frac{v^2}{c^2} \right) = \tfrac{1}{2} m_0 v^2$$

The upper curve in Figure 7 is the graph of the expression for the kinetic energy K of an object with velocity v in special relativity. The lower curve shows the function used for K in classical Newtonian physics. When v is much smaller than the speed of light, the curves are practically identical.

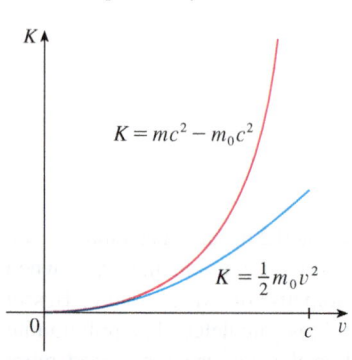

FIGURE 7

(b) If $x = -v^2/c^2$, $f(x) = m_0c^2[(1+x)^{-1/2} - 1]$, and M is a number such that $|f''(x)| \leq M$, then we can use Taylor's Inequality to write

$$|R_1(x)| \leq \frac{M}{2!}x^2$$

We have $f''(x) = \frac{3}{4}m_0c^2(1+x)^{-5/2}$ and we are given that $|v| \leq 100$ m/s, so

$$|f''(x)| = \frac{3m_0c^2}{4(1 - v^2/c^2)^{5/2}} \leq \frac{3m_0c^2}{4(1 - 100^2/c^2)^{5/2}} \quad (=M)$$

Thus, with $c = 3 \times 10^8$ m/s,

$$|R_1(x)| \leq \frac{1}{2} \cdot \frac{3m_0c^2}{4(1 - 100^2/c^2)^{5/2}} \cdot \frac{100^4}{c^4} < (4.17 \times 10^{-10})m_0$$

So when $|v| \leq 100$ m/s, the magnitude of the error in using the Newtonian expression for kinetic energy is at most $(4.2 \times 10^{-10})m_0$. ∎

Another application to physics occurs in optics. Figure 8 is adapted from *Optics*, 4th ed., by Eugene Hecht (San Francisco, 2002), page 153. It depicts a wave from the point source S meeting a spherical interface of radius R centered at C. The ray SA is refracted toward P.

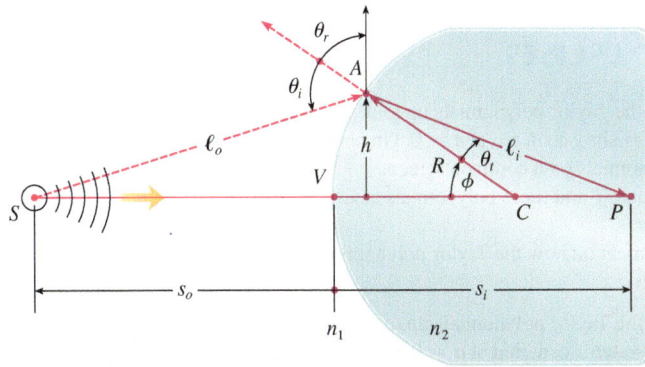

FIGURE 8
Refraction at a spherical interface
Source: Adapted from E. Hecht, *Optics*, 4e (Upper Saddle River, NJ: Pearson Education, 2002).

Using Fermat's principle that light travels so as to minimize the time taken, Hecht derives the equation

$$\boxed{1} \qquad \frac{n_1}{\ell_o} + \frac{n_2}{\ell_i} = \frac{1}{R}\left(\frac{n_2 s_i}{\ell_i} - \frac{n_1 s_o}{\ell_o}\right)$$

where n_1 and n_2 are indexes of refraction and ℓ_o, ℓ_i, s_o, and s_i are the distances indicated in Figure 8. By the Law of Cosines, applied to triangles ACS and ACP, we have

$$\boxed{2} \qquad \ell_o = \sqrt{R^2 + (s_o + R)^2 - 2R(s_o + R)\cos\phi}$$

Here we use the identity
$$\cos(\pi - \phi) = -\cos\phi$$

$$\ell_i = \sqrt{R^2 + (s_i - R)^2 + 2R(s_i - R)\cos\phi}$$

Because Equation 1 is cumbersome to work with, Gauss, in 1841, simplified it by using the linear approximation $\cos\phi \approx 1$ for small values of ϕ. (This amounts to using the Taylor polynomial of degree 1.) Then Equation 1 becomes the following simpler equation [as you are asked to show in Exercise 34(a)]:

$$\boxed{3} \qquad \frac{n_1}{s_o} + \frac{n_2}{s_i} = \frac{n_2 - n_1}{R}$$

The resulting optical theory is known as *Gaussian optics*, or *first-order optics*, and has become the basic theoretical tool used to design lenses.

A more accurate theory is obtained by approximating $\cos\phi$ by its Taylor polynomial of degree 3 (which is the same as the Taylor polynomial of degree 2). This takes into account rays for which ϕ is not so small, that is, rays that strike the surface at greater distances h above the axis. In Exercise 34(b) you are asked to use this approximation to derive the more accurate equation

$$\boxed{4} \qquad \frac{n_1}{s_o} + \frac{n_2}{s_i} = \frac{n_2 - n_1}{R} + h^2 \left[\frac{n_1}{2s_o}\left(\frac{1}{s_o} + \frac{1}{R}\right)^2 + \frac{n_2}{2s_i}\left(\frac{1}{R} - \frac{1}{s_i}\right)^2 \right]$$

The resulting optical theory is known as *third-order optics*.

Other applications of Taylor polynomials to physics and engineering are explored in Exercises 32, 33, 35, 36, 37, and 38, and in the Applied Project on page 783.

11.11 EXERCISES

1. (a) Find the Taylor polynomials up to degree 5 for $f(x) = \sin x$ centered at $a = 0$. Graph f and these polynomials on a common screen.
(b) Evaluate f and these polynomials at $x = \pi/4$, $\pi/2$, and π.
(c) Comment on how the Taylor polynomials converge to $f(x)$.

2. (a) Find the Taylor polynomials up to degree 3 for $f(x) = \tan x$ centered at $a = 0$. Graph f and these polynomials on a common screen.
(b) Evaluate f and these polynomials at $x = \pi/6$, $\pi/4$, and $\pi/3$.
(c) Comment on how the Taylor polynomials converge to $f(x)$.

3–10 Find the Taylor polynomial $T_3(x)$ for the function f centered at the number a. Graph f and T_3 on the same screen.

3. $f(x) = e^x$, $a = 1$

4. $f(x) = \sin x$, $a = \pi/6$

5. $f(x) = \cos x$, $a = \pi/2$

6. $f(x) = e^{-x}\sin x$, $a = 0$

7. $f(x) = \ln x$, $a = 1$

8. $f(x) = x\cos x$, $a = 0$

9. $f(x) = xe^{-2x}$, $a = 0$

10. $f(x) = \tan^{-1}x$, $a = 1$

CAS 11–12 Use a computer algebra system to find the Taylor polynomials T_n centered at a for $n = 2, 3, 4, 5$. Then graph these polynomials and f on the same screen.

11. $f(x) = \cot x$, $a = \pi/4$

12. $f(x) = \sqrt[3]{1 + x^2}$, $a = 0$

13–22
(a) Approximate f by a Taylor polynomial with degree n at the number a.
(b) Use Taylor's Inequality to estimate the accuracy of the approximation $f(x) \approx T_n(x)$ when x lies in the given interval.
(c) Check your result in part (b) by graphing $|R_n(x)|$.

13. $f(x) = 1/x$, $a = 1$, $n = 2$, $0.7 \le x \le 1.3$

14. $f(x) = x^{-1/2}$, $a = 4$, $n = 2$, $3.5 \le x \le 4.5$

15. $f(x) = x^{2/3}$, $a = 1$, $n = 3$, $0.8 \leq x \leq 1.2$

16. $f(x) = \sin x$, $a = \pi/6$, $n = 4$, $0 \leq x \leq \pi/3$

17. $f(x) = \sec x$, $a = 0$, $n = 2$, $-0.2 \leq x \leq 0.2$

18. $f(x) = \ln(1 + 2x)$, $a = 1$, $n = 3$, $0.5 \leq x \leq 1.5$

19. $f(x) = e^{x^2}$, $a = 0$, $n = 3$, $0 \leq x \leq 0.1$

20. $f(x) = x \ln x$, $a = 1$, $n = 3$, $0.5 \leq x \leq 1.5$

21. $f(x) = x \sin x$, $a = 0$, $n = 4$, $-1 \leq x \leq 1$

22. $f(x) = \sinh 2x$, $a = 0$, $n = 5$, $-1 \leq x \leq 1$

23. Use the information from Exercise 5 to estimate $\cos 80°$ correct to five decimal places.

24. Use the information from Exercise 16 to estimate $\sin 38°$ correct to five decimal places.

25. Use Taylor's Inequality to determine the number of terms of the Maclaurin series for e^x that should be used to estimate $e^{0.1}$ to within 0.00001.

26. How many terms of the Maclaurin series for $\ln(1 + x)$ do you need to use to estimate $\ln 1.4$ to within 0.001?

27–29 Use the Alternating Series Estimation Theorem or Taylor's Inequality to estimate the range of values of x for which the given approximation is accurate to within the stated error. Check your answer graphically.

27. $\sin x \approx x - \dfrac{x^3}{6}$ ($|\text{error}| < 0.01$)

28. $\cos x \approx 1 - \dfrac{x^2}{2} + \dfrac{x^4}{24}$ ($|\text{error}| < 0.005$)

29. $\arctan x \approx x - \dfrac{x^3}{3} + \dfrac{x^5}{5}$ ($|\text{error}| < 0.05$)

30. Suppose you know that
$$f^{(n)}(4) = \frac{(-1)^n n!}{3^n(n+1)}$$
and the Taylor series of f centered at 4 converges to $f(x)$ for all x in the interval of convergence. Show that the fifth-degree Taylor polynomial approximates $f(5)$ with error less than 0.0002.

31. A car is moving with speed 20 m/s and acceleration 2 m/s² at a given instant. Using a second-degree Taylor polynomial, estimate how far the car moves in the next second. Would it be reasonable to use this polynomial to estimate the distance traveled during the next minute?

32. The resistivity ρ of a conducting wire is the reciprocal of the conductivity and is measured in units of ohm-meters (Ω-m). The resistivity of a given metal depends on the temperature according to the equation
$$\rho(t) = \rho_{20} e^{\alpha(t-20)}$$
where t is the temperature in °C. There are tables that list the values of α (called the temperature coefficient) and ρ_{20} (the resistivity at 20°C) for various metals. Except at very low temperatures, the resistivity varies almost linearly with temperature and so it is common to approximate the expression for $\rho(t)$ by its first- or second-degree Taylor polynomial at $t = 20$.
(a) Find expressions for these linear and quadratic approximations.
(b) For copper, the tables give $\alpha = 0.0039/$°C and $\rho_{20} = 1.7 \times 10^{-8}$ Ω-m. Graph the resistivity of copper and the linear and quadratic approximations for -250°C $\leq t \leq 1000$°C.
(c) For what values of t does the linear approximation agree with the exponential expression to within one percent?

33. An electric dipole consists of two electric charges of equal magnitude and opposite sign. If the charges are q and $-q$ and are located at a distance d from each other, then the electric field E at the point P in the figure is
$$E = \frac{q}{D^2} - \frac{q}{(D+d)^2}$$
By expanding this expression for E as a series in powers of d/D, show that E is approximately proportional to $1/D^3$ when P is far away from the dipole.

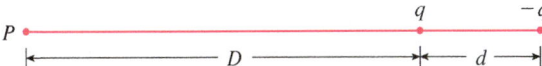

34. (a) Derive Equation 3 for Gaussian optics from Equation 1 by approximating $\cos \phi$ in Equation 2 by its first-degree Taylor polynomial.
(b) Show that if $\cos \phi$ is replaced by its third-degree Taylor polynomial in Equation 2, then Equation 1 becomes Equation 4 for third-order optics. [*Hint:* Use the first two terms in the binomial series for ℓ_o^{-1} and ℓ_i^{-1}. Also, use $\phi \approx \sin \phi$.]

35. If a water wave with length L moves with velocity v across a body of water with depth d, as in the figure on page 782, then
$$v^2 = \frac{gL}{2\pi} \tanh \frac{2\pi d}{L}$$
(a) If the water is deep, show that $v \approx \sqrt{gL/(2\pi)}$.
(b) If the water is shallow, use the Maclaurin series for tanh to show that $v \approx \sqrt{gd}$. (Thus in shallow water the

velocity of a wave tends to be independent of the length of the wave.)

(c) Use the Alternating Series Estimation Theorem to show that if $L > 10d$, then the estimate $v^2 \approx gd$ is accurate to within $0.014gL$.

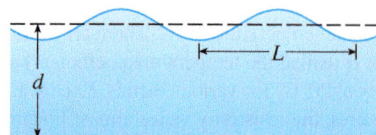

36. A uniformly charged disk has radius R and surface charge density σ as in the figure. The electric potential V at a point P at a distance d along the perpendicular central axis of the disk is

$$V = 2\pi k_e \sigma \left(\sqrt{d^2 + R^2} - d \right)$$

where k_e is a constant (called Coulomb's constant). Show that

$$V \approx \frac{\pi k_e R^2 \sigma}{d} \quad \text{for large } d$$

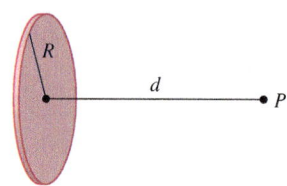

37. If a surveyor measures differences in elevation when making plans for a highway across a desert, corrections must be made for the curvature of the earth.

(a) If R is the radius of the earth and L is the length of the highway, show that the correction is

$$C = R \sec(L/R) - R$$

(b) Use a Taylor polynomial to show that

$$C \approx \frac{L^2}{2R} + \frac{5L^4}{24R^3}$$

(c) Compare the corrections given by the formulas in parts (a) and (b) for a highway that is 100 km long. (Take the radius of the earth to be 6370 km.)

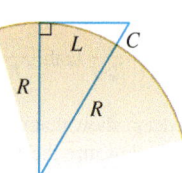

38. The period of a pendulum with length L that makes a maximum angle θ_0 with the vertical is

$$T = 4\sqrt{\frac{L}{g}} \int_0^{\pi/2} \frac{dx}{\sqrt{1 - k^2 \sin^2 x}}$$

where $k = \sin\left(\frac{1}{2}\theta_0\right)$ and g is the acceleration due to gravity. (In Exercise 7.7.42 we approximated this integral using Simpson's Rule.)

(a) Expand the integrand as a binomial series and use the result of Exercise 7.1.50 to show that

$$T = 2\pi \sqrt{\frac{L}{g}} \left[1 + \frac{1^2}{2^2} k^2 + \frac{1^2 3^2}{2^2 4^2} k^4 + \frac{1^2 3^2 5^2}{2^2 4^2 6^2} k^6 + \cdots \right]$$

If θ_0 is not too large, the approximation $T \approx 2\pi \sqrt{L/g}$, obtained by using only the first term in the series, is often used. A better approximation is obtained by using two terms:

$$T \approx 2\pi \sqrt{\frac{L}{g}} \left(1 + \tfrac{1}{4} k^2 \right)$$

(b) Notice that all the terms in the series after the first one have coefficients that are at most $\frac{1}{4}$. Use this fact to compare this series with a geometric series and show that

$$2\pi \sqrt{\frac{L}{g}} \left(1 + \tfrac{1}{4} k^2 \right) \leq T \leq 2\pi \sqrt{\frac{L}{g}} \frac{4 - 3k^2}{4 - 4k^2}$$

(c) Use the inequalities in part (b) to estimate the period of a pendulum with $L = 1$ meter and $\theta_0 = 10°$. How does it compare with the estimate $T \approx 2\pi \sqrt{L/g}$? What if $\theta_0 = 42°$?

39. In Section 4.8 we considered Newton's method for approximating a root r of the equation $f(x) = 0$, and from an initial approximation x_1 we obtained successive approximations $x_2, x_3, \ldots$, where

$$x_{n+1} = x_n - \frac{f(x_n)}{f'(x_n)}$$

Use Taylor's Inequality with $n = 1$, $a = x_n$, and $x = r$ to show that if $f''(x)$ exists on an interval I containing r, x_n, and x_{n+1}, and $|f''(x)| \leq M$, $|f'(x)| \geq K$ for all $x \in I$, then

$$|x_{n+1} - r| \leq \frac{M}{2K} |x_n - r|^2$$

[This means that if x_n is accurate to d decimal places, then x_{n+1} is accurate to about $2d$ decimal places. More precisely, if the error at stage n is at most 10^{-m}, then the error at stage $n+1$ is at most $(M/2K)10^{-2m}$.]

APPLIED PROJECT

RADIATION FROM THE STARS

Any object emits radiation when heated. A *blackbody* is a system that absorbs all the radiation that falls on it. For instance, a matte black surface or a large cavity with a small hole in its wall (like a blast furnace) is a blackbody and emits blackbody radiation. Even the radiation from the sun is close to being blackbody radiation.

Proposed in the late 19th century, the Rayleigh-Jeans Law expresses the energy density of blackbody radiation of wavelength λ as

$$f(\lambda) = \frac{8\pi kT}{\lambda^4}$$

where λ is measured in meters, T is the temperature in kelvins (K), and k is Boltzmann's constant. The Rayleigh-Jeans Law agrees with experimental measurements for long wavelengths but disagrees drastically for short wavelengths. [The law predicts that $f(\lambda) \to \infty$ as $\lambda \to 0^+$ but experiments have shown that $f(\lambda) \to 0$.] This fact is known as the *ultraviolet catastrophe*.

In 1900 Max Planck found a better model (known now as Planck's Law) for blackbody radiation:

$$f(\lambda) = \frac{8\pi hc\lambda^{-5}}{e^{hc/(\lambda kT)} - 1}$$

where λ is measured in meters, T is the temperature (in kelvins), and

$$h = \text{Planck's constant} = 6.6262 \times 10^{-34} \text{ J} \cdot \text{s}$$

$$c = \text{speed of light} = 2.997925 \times 10^8 \text{ m/s}$$

$$k = \text{Boltzmann's constant} = 1.3807 \times 10^{-23} \text{ J/K}$$

1. Use l'Hospital's Rule to show that

$$\lim_{\lambda \to 0^+} f(\lambda) = 0 \quad \text{and} \quad \lim_{\lambda \to \infty} f(\lambda) = 0$$

for Planck's Law. So this law models blackbody radiation better than the Rayleigh-Jeans Law for short wavelengths.

2. Use a Taylor polynomial to show that, for large wavelengths, Planck's Law gives approximately the same values as the Rayleigh-Jeans Law.

3. Graph f as given by both laws on the same screen and comment on the similarities and differences. Use $T = 5700$ K (the temperature of the sun). (You may want to change from meters to the more convenient unit of micrometers: 1 mm $= 10^{-6}$ m.)

4. Use your graph in Problem 3 to estimate the value of λ for which $f(\lambda)$ is a maximum under Planck's Law.

5. Investigate how the graph of f changes as T varies. (Use Planck's Law.) In particular, graph f for the stars Betelgeuse ($T = 3400$ K), Procyon ($T = 6400$ K), and Sirius ($T = 9200$ K), as well as the sun. How does the total radiation emitted (the area under the curve) vary with T? Use the graph to comment on why Sirius is known as a blue star and Betelgeuse as a red star.

11 REVIEW

CONCEPT CHECK

Answers to the Concept Check can be found on the back endpapers.

1. (a) What is a convergent sequence?
   (b) What is a convergent series?
   (c) What does $\lim_{n \to \infty} a_n = 3$ mean?
   (d) What does $\sum_{n=1}^{\infty} a_n = 3$ mean?

2. (a) What is a bounded sequence?
   (b) What is a monotonic sequence?
   (c) What can you say about a bounded monotonic sequence?

3. (a) What is a geometric series? Under what circumstances is it convergent? What is its sum?
   (b) What is a p-series? Under what circumstances is it convergent?

4. Suppose $\Sigma a_n = 3$ and s_n is the nth partial sum of the series. What is $\lim_{n \to \infty} a_n$? What is $\lim_{n \to \infty} s_n$?

5. State the following.
   (a) The Test for Divergence
   (b) The Integral Test
   (c) The Comparison Test
   (d) The Limit Comparison Test
   (e) The Alternating Series Test
   (f) The Ratio Test
   (g) The Root Test

6. (a) What is an absolutely convergent series?
   (b) What can you say about such a series?
   (c) What is a conditionally convergent series?

7. (a) If a series is convergent by the Integral Test, how do you estimate its sum?
   (b) If a series is convergent by the Comparison Test, how do you estimate its sum?
   (c) If a series is convergent by the Alternating Series Test, how do you estimate its sum?

8. (a) Write the general form of a power series.
   (b) What is the radius of convergence of a power series?
   (c) What is the interval of convergence of a power series?

9. Suppose $f(x)$ is the sum of a power series with radius of convergence R.
   (a) How do you differentiate f? What is the radius of convergence of the series for f'?
   (b) How do you integrate f? What is the radius of convergence of the series for $\int f(x)\, dx$?

10. (a) Write an expression for the nth-degree Taylor polynomial of f centered at a.
    (b) Write an expression for the Taylor series of f centered at a.
    (c) Write an expression for the Maclaurin series of f.
    (d) How do you show that $f(x)$ is equal to the sum of its Taylor series?
    (e) State Taylor's Inequality.

11. Write the Maclaurin series and the interval of convergence for each of the following functions.
    (a) $1/(1-x)$ (b) e^x (c) $\sin x$
    (d) $\cos x$ (e) $\tan^{-1} x$ (f) $\ln(1+x)$

12. Write the binomial series expansion of $(1+x)^k$. What is the radius of convergence of this series?

TRUE-FALSE QUIZ

Determine whether the statement is true or false. If it is true, explain why. If it is false, explain why or give an example that disproves the statement.

1. If $\lim_{n \to \infty} a_n = 0$, then Σa_n is convergent.

2. The series $\sum_{n=1}^{\infty} n^{-\sin 1}$ is convergent.

3. If $\lim_{n \to \infty} a_n = L$, then $\lim_{n \to \infty} a_{2n+1} = L$.

4. If $\Sigma c_n 6^n$ is convergent, then $\Sigma c_n(-2)^n$ is convergent.

5. If $\Sigma c_n 6^n$ is convergent, then $\Sigma c_n(-6)^n$ is convergent.

6. If $\Sigma c_n x^n$ diverges when $x = 6$, then it diverges when $x = 10$.

7. The Ratio Test can be used to determine whether $\Sigma 1/n^3$ converges.

8. The Ratio Test can be used to determine whether $\Sigma 1/n!$ converges.

9. If $0 \leq a_n \leq b_n$ and Σb_n diverges, then Σa_n diverges.

10. $\sum_{n=0}^{\infty} \dfrac{(-1)^n}{n!} = \dfrac{1}{e}$

11. If $-1 < \alpha < 1$, then $\lim_{n \to \infty} \alpha^n = 0$.

12. If Σa_n is divergent, then $\Sigma |a_n|$ is divergent.

13. If $f(x) = 2x - x^2 + \tfrac{1}{3}x^3 - \cdots$ converges for all x, then $f'''(0) = 2$.

14. If $\{a_n\}$ and $\{b_n\}$ are divergent, then $\{a_n + b_n\}$ is divergent.

15. If $\{a_n\}$ and $\{b_n\}$ are divergent, then $\{a_n b_n\}$ is divergent.

16. If $\{a_n\}$ is decreasing and $a_n > 0$ for all n, then $\{a_n\}$ is convergent.

17. If $a_n > 0$ and Σa_n converges, then $\Sigma (-1)^n a_n$ converges.

18. If $a_n > 0$ and $\lim_{n \to \infty} (a_{n+1}/a_n) < 1$, then $\lim_{n \to \infty} a_n = 0$.

19. $0.99999\ldots = 1$

20. If $\lim_{n \to \infty} a_n = 2$, then $\lim_{n \to \infty} (a_{n+3} - a_n) = 0$.

21. If a finite number of terms are added to a convergent series, then the new series is still convergent.

22. If $\sum_{n=1}^{\infty} a_n = A$ and $\sum_{n=1}^{\infty} b_n = B$, then $\sum_{n=1}^{\infty} a_n b_n = AB$.

EXERCISES

1–8 Determine whether the sequence is convergent or divergent. If it is convergent, find its limit.

1. $a_n = \dfrac{2 + n^3}{1 + 2n^3}$

2. $a_n = \dfrac{9^{n+1}}{10^n}$

3. $a_n = \dfrac{n^3}{1 + n^2}$

4. $a_n = \cos(n\pi/2)$

5. $a_n = \dfrac{n \sin n}{n^2 + 1}$

6. $a_n = \dfrac{\ln n}{\sqrt{n}}$

7. $\{(1 + 3/n)^{4n}\}$

8. $\{(-10)^n/n!\}$

9. A sequence is defined recursively by the equations $a_1 = 1$, $a_{n+1} = \tfrac{1}{3}(a_n + 4)$. Show that $\{a_n\}$ is increasing and $a_n < 2$ for all n. Deduce that $\{a_n\}$ is convergent and find its limit.

10. Show that $\lim_{n \to \infty} n^4 e^{-n} = 0$ and use a graph to find the smallest value of N that corresponds to $\varepsilon = 0.1$ in the precise definition of a limit.

11–22 Determine whether the series is convergent or divergent.

11. $\sum_{n=1}^{\infty} \dfrac{n}{n^3 + 1}$

12. $\sum_{n=1}^{\infty} \dfrac{n^2 + 1}{n^3 + 1}$

13. $\sum_{n=1}^{\infty} \dfrac{n^3}{5^n}$

14. $\sum_{n=1}^{\infty} \dfrac{(-1)^n}{\sqrt{n+1}}$

15. $\sum_{n=2}^{\infty} \dfrac{1}{n\sqrt{\ln n}}$

16. $\sum_{n=1}^{\infty} \ln\left(\dfrac{n}{3n+1}\right)$

17. $\sum_{n=1}^{\infty} \dfrac{\cos 3n}{1 + (1.2)^n}$

18. $\sum_{n=1}^{\infty} \dfrac{n^{2n}}{(1+2n^2)^n}$

19. $\sum_{n=1}^{\infty} \dfrac{1 \cdot 3 \cdot 5 \cdot \cdots \cdot (2n-1)}{5^n n!}$

20. $\sum_{n=1}^{\infty} \dfrac{(-5)^{2n}}{n^2 9^n}$

21. $\sum_{n=1}^{\infty} (-1)^{n-1} \dfrac{\sqrt{n}}{n+1}$

22. $\sum_{n=1}^{\infty} \dfrac{\sqrt{n+1} - \sqrt{n-1}}{n}$

23–26 Determine whether the series is conditionally convergent, absolutely convergent, or divergent.

23. $\sum_{n=1}^{\infty} (-1)^{n-1} n^{-1/3}$

24. $\sum_{n=1}^{\infty} (-1)^{n-1} n^{-3}$

25. $\sum_{n=1}^{\infty} \dfrac{(-1)^n (n+1) 3^n}{2^{2n+1}}$

26. $\sum_{n=2}^{\infty} \dfrac{(-1)^n \sqrt{n}}{\ln n}$

27–31 Find the sum of the series.

27. $\sum_{n=1}^{\infty} \dfrac{(-3)^{n-1}}{2^{3n}}$

28. $\sum_{n=1}^{\infty} \dfrac{1}{n(n+3)}$

29. $\sum_{n=1}^{\infty} [\tan^{-1}(n+1) - \tan^{-1} n]$

30. $\sum_{n=0}^{\infty} \dfrac{(-1)^n \pi^n}{3^{2n}(2n)!}$

31. $1 - e + \dfrac{e^2}{2!} - \dfrac{e^3}{3!} + \dfrac{e^4}{4!} - \cdots$

32. Express the repeating decimal $4.17326326326\ldots$ as a fraction.

33. Show that $\cosh x \geq 1 + \tfrac{1}{2}x^2$ for all x.

34. For what values of x does the series $\sum_{n=1}^{\infty} (\ln x)^n$ converge?

35. Find the sum of the series $\sum_{n=1}^{\infty} \dfrac{(-1)^{n+1}}{n^5}$ correct to four decimal places.

36. (a) Find the partial sum s_5 of the series $\sum_{n=1}^{\infty} 1/n^6$ and estimate the error in using it as an approximation to the sum of the series.
(b) Find the sum of this series correct to five decimal places.

37. Use the sum of the first eight terms to approximate the sum of the series $\sum_{n=1}^{\infty} (2 + 5^n)^{-1}$. Estimate the error involved in this approximation.

38. (a) Show that the series $\sum_{n=1}^{\infty} \dfrac{n^n}{(2n)!}$ is convergent.
(b) Deduce that $\lim_{n \to \infty} \dfrac{n^n}{(2n)!} = 0$.

39. Prove that if the series $\sum_{n=1}^{\infty} a_n$ is absolutely convergent, then the series

$$\sum_{n=1}^{\infty} \left(\dfrac{n+1}{n}\right) a_n$$

is also absolutely convergent.

40–43 Find the radius of convergence and interval of convergence of the series.

40. $\sum_{n=1}^{\infty} (-1)^n \dfrac{x^n}{n^2 5^n}$

41. $\sum_{n=1}^{\infty} \dfrac{(x+2)^n}{n\,4^n}$

42. $\sum_{n=1}^{\infty} \dfrac{2^n(x-2)^n}{(n+2)!}$

43. $\sum_{n=0}^{\infty} \dfrac{2^n(x-3)^n}{\sqrt{n+3}}$

44. Find the radius of convergence of the series
$$\sum_{n=1}^{\infty} \dfrac{(2n)!}{(n!)^2} x^n$$

45. Find the Taylor series of $f(x) = \sin x$ at $a = \pi/6$.

46. Find the Taylor series of $f(x) = \cos x$ at $a = \pi/3$.

47–54 Find the Maclaurin series for f and its radius of convergence. You may use either the direct method (definition of a Maclaurin series) or known series such as geometric series, binomial series, or the Maclaurin series for e^x, $\sin x$, $\tan^{-1} x$, and $\ln(1+x)$.

47. $f(x) = \dfrac{x^2}{1+x}$

48. $f(x) = \tan^{-1}(x^2)$

49. $f(x) = \ln(4-x)$

50. $f(x) = xe^{2x}$

51. $f(x) = \sin(x^4)$

52. $f(x) = 10^x$

53. $f(x) = 1/\sqrt[4]{16-x}$

54. $f(x) = (1-3x)^{-5}$

55. Evaluate $\int \dfrac{e^x}{x}\,dx$ as an infinite series.

56. Use series to approximate $\int_0^1 \sqrt{1+x^4}\,dx$ correct to two decimal places.

57–58
(a) Approximate f by a Taylor polynomial with degree n at the number a.
(b) Graph f and T_n on a common screen.
(c) Use Taylor's Inequality to estimate the accuracy of the approximation $f(x) \approx T_n(x)$ when x lies in the given interval.
(d) Check your result in part (c) by graphing $|R_n(x)|$.

57. $f(x) = \sqrt{x}$, $\quad a = 1$, $\quad n = 3$, $\quad 0.9 \leq x \leq 1.1$

58. $f(x) = \sec x$, $\quad a = 0$, $\quad n = 2$, $\quad 0 \leq x \leq \pi/6$

59. Use series to evaluate the following limit.
$$\lim_{x \to 0} \dfrac{\sin x - x}{x^3}$$

60. The force due to gravity on an object with mass m at a height h above the surface of the earth is
$$F = \dfrac{mgR^2}{(R+h)^2}$$
where R is the radius of the earth and g is the acceleration due to gravity for an object on the surface of the earth.
(a) Express F as a series in powers of h/R.
(b) Observe that if we approximate F by the first term in the series, we get the expression $F \approx mg$ that is usually used when h is much smaller than R. Use the Alternating Series Estimation Theorem to estimate the range of values of h for which the approximation $F \approx mg$ is accurate to within one percent. (Use $R = 6400$ km.)

61. Suppose that $f(x) = \sum_{n=0}^{\infty} c_n x^n$ for all x.
(a) If f is an odd function, show that
$$c_0 = c_2 = c_4 = \cdots = 0$$
(b) If f is an even function, show that
$$c_1 = c_3 = c_5 = \cdots = 0$$

62. If $f(x) = e^{x^2}$, show that $f^{(2n)}(0) = \dfrac{(2n)!}{n!}$.

Problems Plus

Before you look at the solution of the example, cover it up and first try to solve the problem yourself.

EXAMPLE Find the sum of the series $\sum_{n=0}^{\infty} \dfrac{(x+2)^n}{(n+3)!}$.

SOLUTION The problem-solving principle that is relevant here is *recognizing something familiar*. Does the given series look anything like a series that we already know? Well, it does have some ingredients in common with the Maclaurin series for the exponential function:

$$e^x = \sum_{n=0}^{\infty} \frac{x^n}{n!} = 1 + x + \frac{x^2}{2!} + \frac{x^3}{3!} + \cdots$$

We can make this series look more like our given series by replacing x by $x + 2$:

$$e^{x+2} = \sum_{n=0}^{\infty} \frac{(x+2)^n}{n!} = 1 + (x+2) + \frac{(x+2)^2}{2!} + \frac{(x+2)^3}{3!} + \cdots$$

But here the exponent in the numerator matches the number in the denominator whose factorial is taken. To make that happen in the given series, let's multiply and divide by $(x+2)^3$:

$$\sum_{n=0}^{\infty} \frac{(x+2)^n}{(n+3)!} = \frac{1}{(x+2)^3} \sum_{n=0}^{\infty} \frac{(x+2)^{n+3}}{(n+3)!}$$

$$= (x+2)^{-3} \left[\frac{(x+2)^3}{3!} + \frac{(x+2)^4}{4!} + \cdots \right]$$

We see that the series between brackets is just the series for e^{x+2} with the first three terms missing. So

$$\sum_{n=0}^{\infty} \frac{(x+2)^n}{(n+3)!} = (x+2)^{-3} \left[e^{x+2} - 1 - (x+2) - \frac{(x+2)^2}{2!} \right] \quad \blacksquare$$

Problems

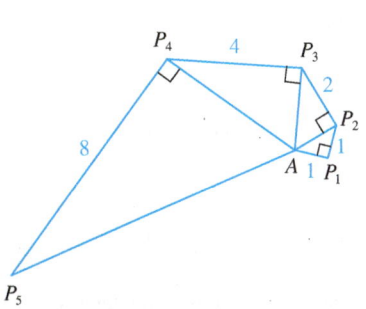

FIGURE FOR PROBLEM 4

1. If $f(x) = \sin(x^3)$, find $f^{(15)}(0)$.

2. A function f is defined by

$$f(x) = \lim_{n \to \infty} \frac{x^{2n} - 1}{x^{2n} + 1}$$

Where is f continuous?

3. (a) Show that $\tan \tfrac{1}{2}x = \cot \tfrac{1}{2}x - 2 \cot x$.
(b) Find the sum of the series

$$\sum_{n=1}^{\infty} \frac{1}{2^n} \tan \frac{x}{2^n}$$

4. Let $\{P_n\}$ be a sequence of points determined as in the figure. Thus $|AP_1| = 1$, $|P_n P_{n+1}| = 2^{n-1}$, and angle $AP_n P_{n+1}$ is a right angle. Find $\lim_{n \to \infty} \angle P_n AP_{n+1}$.

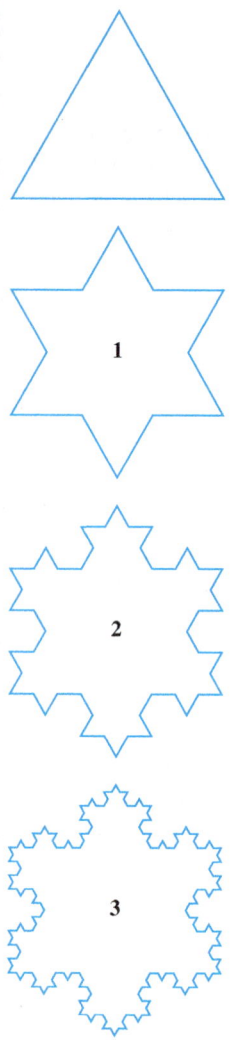

FIGURE FOR PROBLEM 5

5. To construct the **snowflake curve**, start with an equilateral triangle with sides of length 1. Step 1 in the construction is to divide each side into three equal parts, construct an equilateral triangle on the middle part, and then delete the middle part (see the figure). Step 2 is to repeat step 1 for each side of the resulting polygon. This process is repeated at each succeeding step. The snowflake curve is the curve that results from repeating this process indefinitely.
    (a) Let s_n, l_n, and p_n represent the number of sides, the length of a side, and the total length of the nth approximating curve (the curve obtained after step n of the construction), respectively. Find formulas for s_n, l_n, and p_n.
    (b) Show that $p_n \to \infty$ as $n \to \infty$.
    (c) Sum an infinite series to find the area enclosed by the snowflake curve.

    Note: Parts (b) and (c) show that the snowflake curve is infinitely long but encloses only a finite area.

6. Find the sum of the series
$$1 + \frac{1}{2} + \frac{1}{3} + \frac{1}{4} + \frac{1}{6} + \frac{1}{8} + \frac{1}{9} + \frac{1}{12} + \cdots$$
where the terms are the reciprocals of the positive integers whose only prime factors are 2s and 3s.

7. (a) Show that for $xy \neq -1$,
$$\arctan x - \arctan y = \arctan \frac{x - y}{1 + xy}$$
if the left side lies between $-\pi/2$ and $\pi/2$.
    (b) Show that $\arctan \frac{120}{119} - \arctan \frac{1}{239} = \pi/4$.
    (c) Deduce the following formula of John Machin (1680–1751):
$$4 \arctan \tfrac{1}{5} - \arctan \tfrac{1}{239} = \frac{\pi}{4}$$
    (d) Use the Maclaurin series for arctan to show that
$$0.1973955597 < \arctan \tfrac{1}{5} < 0.1973955616$$
    (e) Show that
$$0.004184075 < \arctan \tfrac{1}{239} < 0.004184077$$
    (f) Deduce that, correct to seven decimal places, $\pi \approx 3.1415927$.

    Machin used this method in 1706 to find π correct to 100 decimal places. Recently, with the aid of computers, the value of π has been computed to increasingly greater accuracy. In 2013 Shigeru Kondo and Alexander Yee computed the value of π to more than 12 trillion decimal places!

8. (a) Prove a formula similar to the one in Problem 7(a) but involving arccot instead of arctan.
    (b) Find the sum of the series $\sum_{n=0}^{\infty} \text{arccot}(n^2 + n + 1)$.

9. Use the result of Problem 7(a) to find the sum of the series $\sum_{n=1}^{\infty} \arctan(2/n^2)$.

10. If $a_0 + a_1 + a_2 + \cdots + a_k = 0$, show that
$$\lim_{n \to \infty} \left(a_0 \sqrt{n} + a_1 \sqrt{n+1} + a_2 \sqrt{n+2} + \cdots + a_k \sqrt{n+k} \right) = 0$$
    If you don't see how to prove this, try the problem-solving strategy of *using analogy* (see page 71). Try the special cases $k = 1$ and $k = 2$ first. If you can see how to prove the assertion for these cases, then you will probably see how to prove it in general.

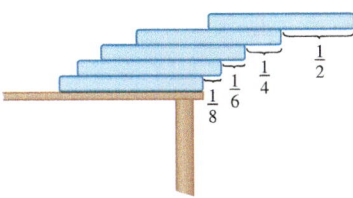

FIGURE FOR PROBLEM 12

11. Find the interval of convergence of $\sum_{n=1}^{\infty} n^3 x^n$ and find its sum.

12. Suppose you have a large supply of books, all the same size, and you stack them at the edge of a table, with each book extending farther beyond the edge of the table than the one beneath it. Show that it is possible to do this so that the top book extends entirely beyond the table. In fact, show that the top book can extend any distance at all beyond the edge of the table if the stack is high enough. Use the following method of stacking: The top book extends half its length beyond the second book. The second book extends a quarter of its length beyond the third. The third extends one-sixth of its length beyond the fourth, and so on. (Try it yourself with a deck of cards.) Consider centers of mass.

13. Find the sum of the series $\sum_{n=2}^{\infty} \ln\left(1 - \frac{1}{n^2}\right)$.

14. If $p > 1$, evaluate the expression

$$\frac{1 + \dfrac{1}{2^p} + \dfrac{1}{3^p} + \dfrac{1}{4^p} + \cdots}{1 - \dfrac{1}{2^p} + \dfrac{1}{3^p} - \dfrac{1}{4^p} + \cdots}$$

15. Suppose that circles of equal diameter are packed tightly in n rows inside an equilateral triangle. (The figure illustrates the case $n = 4$.) If A is the area of the triangle and A_n is the total area occupied by the n rows of circles, show that

$$\lim_{n \to \infty} \frac{A_n}{A} = \frac{\pi}{2\sqrt{3}}$$

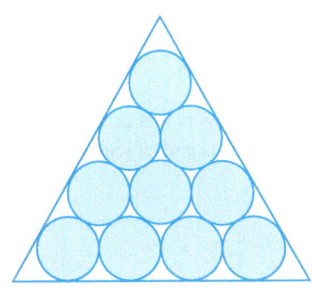

FIGURE FOR PROBLEM 15

16. A sequence $\{a_n\}$ is defined recursively by the equations

$$a_0 = a_1 = 1 \qquad n(n-1)a_n = (n-1)(n-2)a_{n-1} - (n-3)a_{n-2}$$

Find the sum of the series $\sum_{n=0}^{\infty} a_n$.

17. If the curve $y = e^{-x/10} \sin x$, $x \geq 0$, is rotated about the x-axis, the resulting solid looks like an infinite decreasing string of beads.
    (a) Find the exact volume of the nth bead. (Use either a table of integrals or a computer algebra system.)
    (b) Find the total volume of the beads.

18. Starting with the vertices $P_1(0, 1)$, $P_2(1, 1)$, $P_3(1, 0)$, $P_4(0, 0)$ of a square, we construct further points as shown in the figure: P_5 is the midpoint of P_1P_2, P_6 is the midpoint of P_2P_3, P_7 is the midpoint of P_3P_4, and so on. The polygonal spiral path $P_1P_2P_3P_4P_5P_6P_7\ldots$ approaches a point P inside the square.
    (a) If the coordinates of P_n are (x_n, y_n), show that $\frac{1}{2}x_n + x_{n+1} + x_{n+2} + x_{n+3} = 2$ and find a similar equation for the y-coordinates.
    (b) Find the coordinates of P.

19. Find the sum of the series $\sum_{n=1}^{\infty} \frac{(-1)^n}{(2n+1)3^n}$.

20. Carry out the following steps to show that

$$\frac{1}{1 \cdot 2} + \frac{1}{3 \cdot 4} + \frac{1}{5 \cdot 6} + \frac{1}{7 \cdot 8} + \cdots = \ln 2$$

(a) Use the formula for the sum of a finite geometric series (11.2.3) to get an expression for

$$1 - x + x^2 - x^3 + \cdots + x^{2n-2} - x^{2n-1}$$

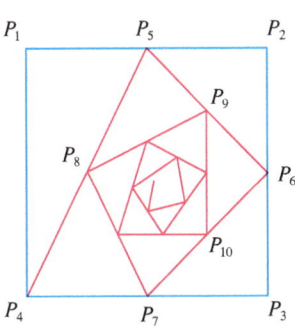

FIGURE FOR PROBLEM 18

(b) Integrate the result of part (a) from 0 to 1 to get an expression for

$$1 - \frac{1}{2} + \frac{1}{3} - \frac{1}{4} + \cdots + \frac{1}{2n-1} - \frac{1}{2n}$$

as an integral.

(c) Deduce from part (b) that

$$\left| \frac{1}{1\cdot 2} + \frac{1}{3\cdot 4} + \frac{1}{5\cdot 6} + \cdots + \frac{1}{(2n-1)(2n)} - \int_0^1 \frac{dx}{1+x} \right| < \int_0^1 x^{2n}\, dx$$

(d) Use part (c) to show that the sum of the given series is $\ln 2$.

21. Find all the solutions of the equation

$$1 + \frac{x}{2!} + \frac{x^2}{4!} + \frac{x^3}{6!} + \frac{x^4}{8!} + \cdots = 0$$

[*Hint:* Consider the cases $x \geq 0$ and $x < 0$ separately.]

22. Right-angled triangles are constructed as in the figure. Each triangle has height 1 and its base is the hypotenuse of the preceding triangle. Show that this sequence of triangles makes indefinitely many turns around P by showing that $\sum \theta_n$ is a divergent series.

23. Consider the series whose terms are the reciprocals of the positive integers that can be written in base 10 notation without using the digit 0. Show that this series is convergent and the sum is less than 90.

24. (a) Show that the Maclaurin series of the function

$$f(x) = \frac{x}{1 - x - x^2} \quad \text{is} \quad \sum_{n=1}^{\infty} f_n x^n$$

where f_n is the nth Fibonacci number, that is, $f_1 = 1$, $f_2 = 1$, and $f_n = f_{n-1} + f_{n-2}$ for $n \geq 3$. [*Hint:* Write $x/(1 - x - x^2) = c_0 + c_1 x + c_2 x^2 + \cdots$ and multiply both sides of this equation by $1 - x - x^2$.]

(b) By writing $f(x)$ as a sum of partial fractions and thereby obtaining the Maclaurin series in a different way, find an explicit formula for the nth Fibonacci number.

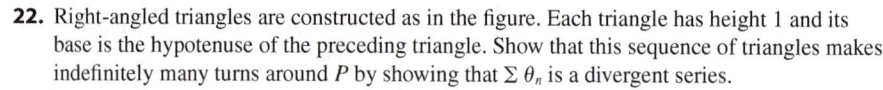

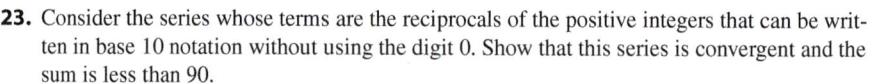

25. Let

$$u = 1 + \frac{x^3}{3!} + \frac{x^6}{6!} + \frac{x^9}{9!} + \cdots$$

$$v = x + \frac{x^4}{4!} + \frac{x^7}{7!} + \frac{x^{10}}{10!} + \cdots$$

$$w = \frac{x^2}{2!} + \frac{x^5}{5!} + \frac{x^8}{8!} + \cdots$$

Show that $u^3 + v^3 + w^3 - 3uvw = 1$.

26. Prove that if $n > 1$, the nth partial sum of the harmonic series is not an integer.

Hint: Let 2^k be the largest power of 2 that is less than or equal to n and let M be the product of all odd integers that are less than or equal to n. Suppose that $s_n = m$, an integer. Then $M2^k s_n = M2^k m$. The right side of this equation is even. Prove that the left side is odd by showing that each of its terms is an even integer, except for the last one.

FIGURE FOR PROBLEM 22

17 Second-Order Differential Equations

The motion of a shock absorber in a motorcycle is described by the differential equations that we solve in Section 17.3.

THE BASIC IDEAS OF DIFFERENTIAL equations were explained in Chapter 9; there we concentrated on first-order equations. In this chapter we study second-order linear differential equations and learn how they can be applied to solve problems concerning the vibrations of springs and the analysis of electric circuits. We will also see how infinite series can be used to solve differential equations.

17.1 Second-Order Linear Equations

A **second-order linear differential equation** has the form

$$\boxed{1} \qquad P(x)\frac{d^2y}{dx^2} + Q(x)\frac{dy}{dx} + R(x)y = G(x)$$

where P, Q, R, and G are continuous functions. We saw in Section 9.1 that equations of this type arise in the study of the motion of a spring. In Section 17.3 we will further pursue this application as well as the application to electric circuits.

In this section we study the case where $G(x) = 0$, for all x, in Equation 1. Such equations are called **homogeneous** linear equations. Thus the form of a second-order linear homogeneous differential equation is

$$\boxed{2} \qquad P(x)\frac{d^2y}{dx^2} + Q(x)\frac{dy}{dx} + R(x)y = 0$$

If $G(x) \neq 0$ for some x, Equation 1 is **nonhomogeneous** and is discussed in Section 17.2.

Two basic facts enable us to solve homogeneous linear equations. The first of these says that if we know two solutions y_1 and y_2 of such an equation, then the **linear combination** $y = c_1 y_1 + c_2 y_2$ is also a solution.

$\boxed{3}$ Theorem If $y_1(x)$ and $y_2(x)$ are both solutions of the linear homogeneous equation (2) and c_1 and c_2 are any constants, then the function

$$y(x) = c_1 y_1(x) + c_2 y_2(x)$$

is also a solution of Equation 2.

PROOF Since y_1 and y_2 are solutions of Equation 2, we have

$$P(x)y_1'' + Q(x)y_1' + R(x)y_1 = 0$$

and

$$P(x)y_2'' + Q(x)y_2' + R(x)y_2 = 0$$

Therefore, using the basic rules for differentiation, we have

$$P(x)y'' + Q(x)y' + R(x)y$$
$$= P(x)(c_1 y_1 + c_2 y_2)'' + Q(x)(c_1 y_1 + c_2 y_2)' + R(x)(c_1 y_1 + c_2 y_2)$$
$$= P(x)(c_1 y_1'' + c_2 y_2'') + Q(x)(c_1 y_1' + c_2 y_2') + R(x)(c_1 y_1 + c_2 y_2)$$
$$= c_1[P(x)y_1'' + Q(x)y_1' + R(x)y_1] + c_2[P(x)y_2'' + Q(x)y_2' + R(x)y_2]$$
$$= c_1(0) + c_2(0) = 0$$

Thus $y = c_1 y_1 + c_2 y_2$ is a solution of Equation 2.

The other fact we need is given by the following theorem, which is proved in more advanced courses. It says that the general solution is a linear combination of two **linearly independent** solutions y_1 and y_2. This means that neither y_1 nor y_2 is a constant multiple of the other. For instance, the functions $f(x) = x^2$ and $g(x) = 5x^2$ are linearly dependent, but $f(x) = e^x$ and $g(x) = xe^x$ are linearly independent.

> **4 Theorem** If y_1 and y_2 are linearly independent solutions of Equation 2 on an interval, and $P(x)$ is never 0, then the general solution is given by
>
> $$y(x) = c_1 y_1(x) + c_2 y_2(x)$$
>
> where c_1 and c_2 are arbitrary constants.

Theorem 4 is very useful because it says that if we know *two* particular linearly independent solutions, then we know *every* solution.

In general, it's not easy to discover particular solutions to a second-order linear equation. But it is always possible to do so if the coefficient functions P, Q, and R are constant functions, that is, if the differential equation has the form

$$\boxed{ay'' + by' + cy = 0} \quad \mathbf{5}$$

where a, b, and c are constants and $a \neq 0$.

It's not hard to think of some likely candidates for particular solutions of Equation 5 if we state the equation verbally. We are looking for a function y such that a constant times its second derivative y'' plus another constant times y' plus a third constant times y is equal to 0. We know that the exponential function $y = e^{rx}$ (where r is a constant) has the property that its derivative is a constant multiple of itself: $y' = re^{rx}$. Furthermore, $y'' = r^2 e^{rx}$. If we substitute these expressions into Equation 5, we see that $y = e^{rx}$ is a solution if

$$ar^2 e^{rx} + bre^{rx} + ce^{rx} = 0$$

or

$$(ar^2 + br + c)e^{rx} = 0$$

But e^{rx} is never 0. Thus $y = e^{rx}$ is a solution of Equation 5 if r is a root of the equation

$$\boxed{ar^2 + br + c = 0} \quad \mathbf{6}$$

Equation 6 is called the **auxiliary equation** (or **characteristic equation**) of the differential equation $ay'' + by' + cy = 0$. Notice that it is an algebraic equation that is obtained from the differential equation by replacing y'' by r^2, y' by r, and y by 1.

Sometimes the roots r_1 and r_2 of the auxiliary equation can be found by factoring. In other cases they are found by using the quadratic formula:

$$\mathbf{7} \qquad r_1 = \frac{-b + \sqrt{b^2 - 4ac}}{2a} \qquad r_2 = \frac{-b - \sqrt{b^2 - 4ac}}{2a}$$

We distinguish three cases according to the sign of the discriminant $b^2 - 4ac$.

CASE I $b^2 - 4ac > 0$

In this case the roots r_1 and r_2 of the auxiliary equation are real and distinct, so $y_1 = e^{r_1 x}$ and $y_2 = e^{r_2 x}$ are two linearly independent solutions of Equation 5. (Note that $e^{r_2 x}$ is not a constant multiple of $e^{r_1 x}$.) Therefore, by Theorem 4, we have the following fact.

> **8** If the roots r_1 and r_2 of the auxiliary equation $ar^2 + br + c = 0$ are real and unequal, then the general solution of $ay'' + by' + cy = 0$ is
> $$y = c_1 e^{r_1 x} + c_2 e^{r_2 x}$$

In Figure 1 the graphs of the basic solutions $f(x) = e^{2x}$ and $g(x) = e^{-3x}$ of the differential equation in Example 1 are shown in blue and red, respectively. Some of the other solutions, linear combinations of f and g, are shown in black.

EXAMPLE 1 Solve the equation $y'' + y' - 6y = 0$.

SOLUTION The auxiliary equation is
$$r^2 + r - 6 = (r - 2)(r + 3) = 0$$

whose roots are $r = 2, -3$. Therefore, by (8), the general solution of the given differential equation is
$$y = c_1 e^{2x} + c_2 e^{-3x}$$

We could verify that this is indeed a solution by differentiating and substituting into the differential equation. ∎

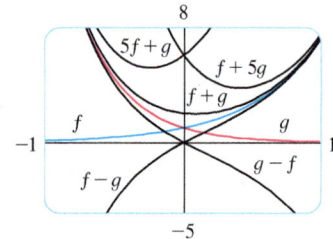

FIGURE 1

EXAMPLE 2 Solve $3 \dfrac{d^2 y}{dx^2} + \dfrac{dy}{dx} - y = 0$.

SOLUTION To solve the auxiliary equation $3r^2 + r - 1 = 0$, we use the quadratic formula:
$$r = \frac{-1 \pm \sqrt{13}}{6}$$

Since the roots are real and distinct, the general solution is
$$y = c_1 e^{(-1+\sqrt{13})x/6} + c_2 e^{(-1-\sqrt{13})x/6}$$
∎

CASE II $b^2 - 4ac = 0$

In this case $r_1 = r_2$; that is, the roots of the auxiliary equation are real and equal. Let's denote by r the common value of r_1 and r_2. Then, from Equations 7, we have

> **9** $$r = -\frac{b}{2a} \qquad \text{so} \qquad 2ar + b = 0$$

We know that $y_1 = e^{rx}$ is one solution of Equation 5. We now verify that $y_2 = xe^{rx}$ is also a solution:

$$ay_2'' + by_2' + cy_2 = a(2re^{rx} + r^2 x e^{rx}) + b(e^{rx} + rxe^{rx}) + cxe^{rx}$$
$$= (2ar + b)e^{rx} + (ar^2 + br + c)xe^{rx}$$
$$= 0(e^{rx}) + 0(xe^{rx}) = 0$$

In the first term, $2ar + b = 0$ by Equations 9; in the second term, $ar^2 + br + c = 0$ because r is a root of the auxiliary equation. Since $y_1 = e^{rx}$ and $y_2 = xe^{rx}$ are linearly independent solutions, Theorem 4 provides us with the general solution.

> **10** If the auxiliary equation $ar^2 + br + c = 0$ has only one real root r, then the general solution of $ay'' + by' + cy = 0$ is
> $$y = c_1 e^{rx} + c_2 x e^{rx}$$

Figure 2 shows the basic solutions $f(x) = e^{-3x/2}$ and $g(x) = xe^{-3x/2}$ in Example 3 and some other members of the family of solutions. Notice that all of them approach 0 as $x \to \infty$.

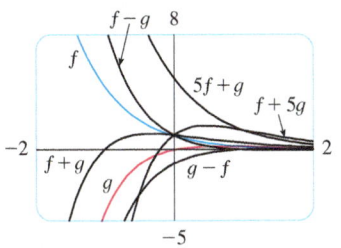

FIGURE 2

EXAMPLE 3 Solve the equation $4y'' + 12y' + 9y = 0$.

SOLUTION The auxiliary equation $4r^2 + 12r + 9 = 0$ can be factored as
$$(2r + 3)^2 = 0$$
so the only root is $r = -\frac{3}{2}$. By (10) the general solution is
$$y = c_1 e^{-3x/2} + c_2 x e^{-3x/2}$$

CASE III $b^2 - 4ac < 0$

In this case the roots r_1 and r_2 of the auxiliary equation are complex numbers. (See Appendix H for information about complex numbers.) We can write
$$r_1 = \alpha + i\beta \qquad r_2 = \alpha - i\beta$$
where α and β are real numbers. [In fact, $\alpha = -b/(2a)$, $\beta = \sqrt{4ac - b^2}/(2a)$.] Then, using Euler's equation
$$e^{i\theta} = \cos\theta + i\sin\theta$$
from Appendix H, we write the solution of the differential equation as
$$y = C_1 e^{r_1 x} + C_2 e^{r_2 x} = C_1 e^{(\alpha + i\beta)x} + C_2 e^{(\alpha - i\beta)x}$$
$$= C_1 e^{\alpha x}(\cos\beta x + i\sin\beta x) + C_2 e^{\alpha x}(\cos\beta x - i\sin\beta x)$$
$$= e^{\alpha x}[(C_1 + C_2)\cos\beta x + i(C_1 - C_2)\sin\beta x]$$
$$= e^{\alpha x}(c_1 \cos\beta x + c_2 \sin\beta x)$$
where $c_1 = C_1 + C_2$, $c_2 = i(C_1 - C_2)$. This gives all solutions (real or complex) of the differential equation. The solutions are real when the constants c_1 and c_2 are real. We summarize the discussion as follows.

> **11** If the roots of the auxiliary equation $ar^2 + br + c = 0$ are the complex numbers $r_1 = \alpha + i\beta$, $r_2 = \alpha - i\beta$, then the general solution of $ay'' + by' + cy = 0$ is
> $$y = e^{\alpha x}(c_1 \cos\beta x + c_2 \sin\beta x)$$

Figure 3 shows the graphs of the solutions in Example 4, $f(x) = e^{3x} \cos 2x$ and $g(x) = e^{3x} \sin 2x$, together with some linear combinations. All solutions approach 0 as $x \to -\infty$.

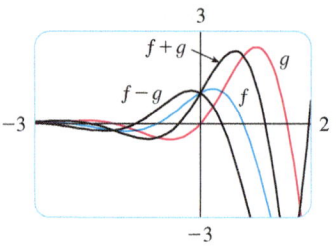

FIGURE 3

EXAMPLE 4 Solve the equation $y'' - 6y' + 13y = 0$.

SOLUTION The auxiliary equation is $r^2 - 6r + 13 = 0$. By the quadratic formula, the roots are

$$r = \frac{6 \pm \sqrt{36-52}}{2} = \frac{6 \pm \sqrt{-16}}{2} = 3 \pm 2i$$

By (11), the general solution of the differential equation is

$$y = e^{3x}(c_1 \cos 2x + c_2 \sin 2x)$$

Initial-Value and Boundary-Value Problems

An **initial-value problem** for the second-order Equation 1 or 2 consists of finding a solution y of the differential equation that also satisfies initial conditions of the form

$$y(x_0) = y_0 \qquad y'(x_0) = y_1$$

where y_0 and y_1 are given constants. If P, Q, R, and G are continuous on an interval and $P(x) \ne 0$ there, then a theorem found in more advanced books guarantees the existence and uniqueness of a solution to this initial-value problem. Examples 5 and 6 illustrate the technique for solving such a problem.

EXAMPLE 5 Solve the initial-value problem

$$y'' + y' - 6y = 0 \qquad y(0) = 1 \qquad y'(0) = 0$$

SOLUTION From Example 1 we know that the general solution of the differential equation is

$$y(x) = c_1 e^{2x} + c_2 e^{-3x}$$

Differentiating this solution, we get

$$y'(x) = 2c_1 e^{2x} - 3c_2 e^{-3x}$$

To satisfy the initial conditions we require that

$$\boxed{12} \qquad y(0) = c_1 + c_2 = 1$$

$$\boxed{13} \qquad y'(0) = 2c_1 - 3c_2 = 0$$

From (13), we have $c_2 = \tfrac{2}{3} c_1$ and so (12) gives

$$c_1 + \tfrac{2}{3} c_1 = 1 \qquad c_1 = \tfrac{3}{5} \qquad c_2 = \tfrac{2}{5}$$

Thus the required solution of the initial-value problem is

$$y = \tfrac{3}{5} e^{2x} + \tfrac{2}{5} e^{-3x}$$

Figure 4 shows the graph of the solution of the initial-value problem in Example 5. Compare with Figure 1.

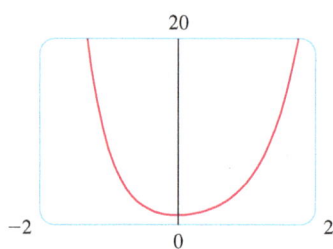

FIGURE 4

EXAMPLE 6 Solve the initial-value problem

$$y'' + y = 0 \qquad y(0) = 2 \qquad y'(0) = 3$$

SOLUTION The auxiliary equation is $r^2 + 1 = 0$, or $r^2 = -1$, whose roots are $\pm i$. Thus $\alpha = 0$, $\beta = 1$, and since $e^{0x} = 1$, the general solution is

$$y(x) = c_1 \cos x + c_2 \sin x$$

Since

$$y'(x) = -c_1 \sin x + c_2 \cos x$$

The solution to Example 6 is graphed in Figure 5. It appears to be a shifted sine curve and, indeed, you can verify that another way of writing the solution is

$$y = \sqrt{13}\sin(x+\phi) \quad \text{where } \tan\phi = \tfrac{2}{3}$$

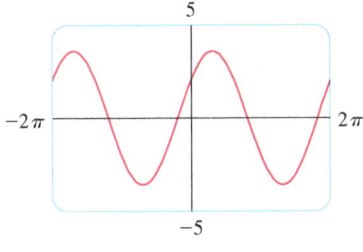

FIGURE 5

the initial conditions become

$$y(0) = c_1 = 2 \qquad y'(0) = c_2 = 3$$

Therefore the solution of the initial-value problem is

$$y(x) = 2\cos x + 3\sin x$$

A **boundary-value problem** for Equation 1 or 2 consists of finding a solution y of the differential equation that also satisfies boundary conditions of the form

$$y(x_0) = y_0 \qquad y(x_1) = y_1$$

In contrast with the situation for initial-value problems, a boundary-value problem does not always have a solution. The method is illustrated in Example 7.

EXAMPLE 7 Solve the boundary-value problem

$$y'' + 2y' + y = 0 \qquad y(0) = 1 \qquad y(1) = 3$$

SOLUTION The auxiliary equation is

$$r^2 + 2r + 1 = 0 \quad \text{or} \quad (r+1)^2 = 0$$

whose only root is $r = -1$. Therefore the general solution is

$$y(x) = c_1 e^{-x} + c_2 x e^{-x}$$

The boundary conditions are satisfied if

$$y(0) = c_1 = 1$$
$$y(1) = c_1 e^{-1} + c_2 e^{-1} = 3$$

The first condition gives $c_1 = 1$, so the second condition becomes

$$e^{-1} + c_2 e^{-1} = 3$$

Solving this equation for c_2 by first multiplying through by e, we get

$$1 + c_2 = 3e \quad \text{so} \quad c_2 = 3e - 1$$

Thus the solution of the boundary-value problem is

$$y = e^{-x} + (3e-1)xe^{-x}$$

Figure 6 shows the graph of the solution of the boundary-value problem in Example 7.

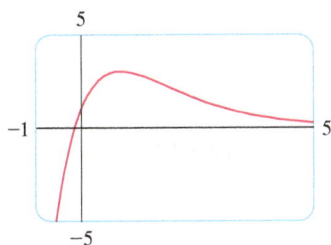

FIGURE 6

Summary: Solutions of $ay'' + by' + c = 0$

Roots of $ar^2 + br + c = 0$	General solution
r_1, r_2 real and distinct	$y = c_1 e^{r_1 x} + c_2 e^{r_2 x}$
$r_1 = r_2 = r$	$y = c_1 e^{rx} + c_2 x e^{rx}$
r_1, r_2 complex: $\alpha \pm i\beta$	$y = e^{\alpha x}(c_1 \cos\beta x + c_2 \sin\beta x)$

17.1 EXERCISES

1–13 Solve the differential equation.

1. $y'' - y' - 6y = 0$
2. $y'' - 6y' + 9y = 0$
3. $y'' + 2y = 0$
4. $y'' + y' - 12y = 0$
5. $4y'' + 4y' + y = 0$
6. $9y'' + 4y = 0$
7. $3y'' = 4y'$
8. $y = y''$
9. $y'' - 4y' + 13y = 0$
10. $3y'' + 4y' - 3y = 0$
11. $2\dfrac{d^2y}{dt^2} + 2\dfrac{dy}{dt} - y = 0$
12. $\dfrac{d^2R}{dt^2} + 6\dfrac{dR}{dt} + 34R = 0$
13. $3\dfrac{d^2V}{dt^2} + 4\dfrac{dV}{dt} + 3V = 0$

14–16 Graph the two basic solutions along with several other solutions of the differential equation. What features do the solutions have in common?

14. $4\dfrac{d^2y}{dx^2} - 4\dfrac{dy}{dx} + y = 0$
15. $\dfrac{d^2y}{dx^2} + 2\dfrac{dy}{dx} + 2y = 0$
16. $2\dfrac{d^2y}{dx^2} + \dfrac{dy}{dx} - y = 0$

17–24 Solve the initial-value problem.

17. $y'' + 3y = 0$, $y(0) = 1$, $y'(0) = 3$
18. $y'' - 2y' - 3y = 0$, $y(0) = 2$, $y'(0) = 2$
19. $9y'' + 12y' + 4y = 0$, $y(0) = 1$, $y'(0) = 0$
20. $3y'' - 2y' - y = 0$, $y(0) = 0$, $y'(0) = -4$
21. $y'' - 6y' + 10y = 0$, $y(0) = 2$, $y'(0) = 3$
22. $4y'' - 20y' + 25y = 0$, $y(0) = 2$, $y'(0) = -3$
23. $y'' - y' - 12y = 0$, $y(1) = 0$, $y'(1) = 1$
24. $4y'' + 4y' + 3y = 0$, $y(0) = 0$, $y'(0) = 1$

25–32 Solve the boundary-value problem, if possible.

25. $y'' + 16y = 0$, $y(0) = -3$, $y(\pi/8) = 2$
26. $y'' + 6y' = 0$, $y(0) = 1$, $y(1) = 0$
27. $y'' + 4y' + 4y = 0$, $y(0) = 2$, $y(1) = 0$
28. $y'' - 8y' + 17y = 0$, $y(0) = 3$, $y(\pi) = 2$
29. $y'' = y'$, $y(0) = 1$, $y(1) = 2$
30. $4y'' - 4y' + y = 0$, $y(0) = 4$, $y(2) = 0$
31. $y'' + 4y' + 20y = 0$, $y(0) = 1$, $y(\pi) = 2$
32. $y'' + 4y' + 20y = 0$, $y(0) = 1$, $y(\pi) = e^{-2\pi}$

33. Let L be a nonzero real number.
    (a) Show that the boundary-value problem $y'' + \lambda y = 0$, $y(0) = 0$, $y(L) = 0$ has only the trivial solution $y = 0$ for the cases $\lambda = 0$ and $\lambda < 0$.
    (b) For the case $\lambda > 0$, find the values of λ for which this problem has a nontrivial solution and give the corresponding solution.

34. If a, b, and c are all positive constants and $y(x)$ is a solution of the differential equation $ay'' + by' + cy = 0$, show that $\lim_{x \to \infty} y(x) = 0$.

35. Consider the boundary-value problem $y'' - 2y' + 2y = 0$, $y(a) = c$, $y(b) = d$.
    (a) If this problem has a unique solution, how are a and b related?
    (b) If this problem has no solution, how are a, b, c, and d related?
    (c) If this problem has infinitely many solutions, how are a, b, c, and d related?

17.2 Nonhomogeneous Linear Equations

In this section we learn how to solve second-order nonhomogeneous linear differential equations with constant coefficients, that is, equations of the form

$$\boxed{1} \qquad ay'' + by' + cy = G(x)$$

where a, b, and c are constants and G is a continuous function. The related homogeneous equation

$$\boxed{2} \qquad ay'' + by' + cy = 0$$

is called the **complementary equation** and plays an important role in the solution of the original nonhomogeneous equation (1).

> **3 Theorem** The general solution of the nonhomogeneous differential equation (1) can be written as
> $$y(x) = y_p(x) + y_c(x)$$
> where y_p is a particular solution of Equation 1 and y_c is the general solution of the complementary Equation 2.

PROOF We verify that if y is any solution of Equation 1, then $y - y_p$ is a solution of the complementary Equation 2. Indeed

$$a(y - y_p)'' + b(y - y_p)' + c(y - y_p) = ay'' - ay_p'' + by' - by_p' + cy - cy_p$$
$$= (ay'' + by' + cy) - (ay_p'' + by_p' + cy_p)$$
$$= G(x) - G(x) = 0$$

This shows that every solution is of the form $y(x) = y_p(x) + y_c(x)$. It is easy to check that every function of this form is a solution. ∎

We know from Section 17.1 how to solve the complementary equation. (Recall that the solution is $y_c = c_1 y_1 + c_2 y_2$, where y_1 and y_2 are linearly independent solutions of Equation 2.) Therefore Theorem 3 says that we know the general solution of the nonhomogeneous equation as soon as we know a particular solution y_p. There are two methods for finding a particular solution: The method of undetermined coefficients is straightforward but works only for a restricted class of functions G. The method of variation of parameters works for every function G but is usually more difficult to apply in practice.

■ The Method of Undetermined Coefficients

We first illustrate the method of undetermined coefficients for the equation

$$ay'' + by' + cy = G(x)$$

where $G(x)$ is a polynomial. It is reasonable to guess that there is a particular solution y_p that is a polynomial of the same degree as G because if y is a polynomial, then $ay'' + by' + cy$ is also a polynomial. We therefore substitute $y_p(x) = $ a polynomial (of the same degree as G) into the differential equation and determine the coefficients.

EXAMPLE 1 Solve the equation $y'' + y' - 2y = x^2$.

SOLUTION The auxiliary equation of $y'' + y' - 2y = 0$ is

$$r^2 + r - 2 = (r - 1)(r + 2) = 0$$

with roots $r = 1, -2$. So the solution of the complementary equation is

$$y_c = c_1 e^x + c_2 e^{-2x}$$

Since $G(x) = x^2$ is a polynomial of degree 2, we seek a particular solution of the form

$$y_p(x) = Ax^2 + Bx + C$$

Then $y_p' = 2Ax + B$ and $y_p'' = 2A$ so, substituting into the given differential equation, we have

$$(2A) + (2Ax + B) - 2(Ax^2 + Bx + C) = x^2$$

or

$$-2Ax^2 + (2A - 2B)x + (2A + B - 2C) = x^2$$

Polynomials are equal when their coefficients are equal. Thus

$$-2A = 1 \qquad 2A - 2B = 0 \qquad 2A + B - 2C = 0$$

The solution of this system of equations is

$$A = -\tfrac{1}{2} \qquad B = -\tfrac{1}{2} \qquad C = -\tfrac{3}{4}$$

A particular solution is therefore

$$y_p(x) = -\tfrac{1}{2}x^2 - \tfrac{1}{2}x - \tfrac{3}{4}$$

and, by Theorem 3, the general solution is

$$y = y_c + y_p = c_1 e^x + c_2 e^{-2x} - \tfrac{1}{2}x^2 - \tfrac{1}{2}x - \tfrac{3}{4}$$

If $G(x)$ (the right side of Equation 1) is of the form Ce^{kx}, where C and k are constants, then we take as a trial solution a function of the same form, $y_p(x) = Ae^{kx}$, because the derivatives of e^{kx} are constant multiples of e^{kx}.

EXAMPLE 2 Solve $y'' + 4y = e^{3x}$.

SOLUTION The auxiliary equation is $r^2 + 4 = 0$ with roots $\pm 2i$, so the solution of the complementary equation is

$$y_c(x) = c_1 \cos 2x + c_2 \sin 2x$$

For a particular solution we try $y_p(x) = Ae^{3x}$. Then $y_p' = 3Ae^{3x}$ and $y_p'' = 9Ae^{3x}$. Substituting into the differential equation, we have

$$9Ae^{3x} + 4(Ae^{3x}) = e^{3x}$$

so $13Ae^{3x} = e^{3x}$ and $A = \tfrac{1}{13}$. Thus a particular solution is

$$y_p(x) = \tfrac{1}{13}e^{3x}$$

and the general solution is

$$y(x) = c_1 \cos 2x + c_2 \sin 2x + \tfrac{1}{13}e^{3x}$$

If $G(x)$ is either $C \cos kx$ or $C \sin kx$, then, because of the rules for differentiating the sine and cosine functions, we take as a trial particular solution a function of the form

$$y_p(x) = A \cos kx + B \sin kx$$

EXAMPLE 3 Solve $y'' + y' - 2y = \sin x$.

SOLUTION We try a particular solution

$$y_p(x) = A \cos x + B \sin x$$

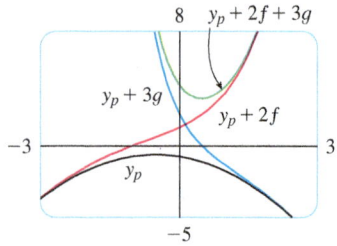

Figure 1 shows four solutions of the differential equation in Example 1 in terms of the particular solution y_p and the functions $f(x) = e^x$ and $g(x) = e^{-2x}$.

FIGURE 1

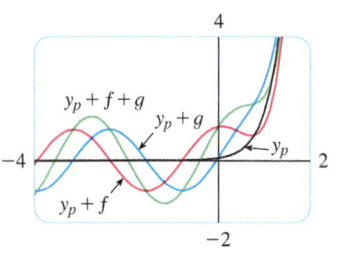

Figure 2 shows solutions of the differential equation in Example 2 in terms of y_p and the functions $f(x) = \cos 2x$ and $g(x) = \sin 2x$. Notice that all solutions approach ∞ as $x \to \infty$ and all solutions (except y_p) resemble sine functions when x is negative.

FIGURE 2

Then $\quad y_p' = -A\sin x + B\cos x \quad\quad y_p'' = -A\cos x - B\sin x$

so substitution in the differential equation gives

$$(-A\cos x - B\sin x) + (-A\sin x + B\cos x) - 2(A\cos x + B\sin x) = \sin x$$

or $\quad\quad\quad\quad\quad\quad\quad\quad (-3A + B)\cos x + (-A - 3B)\sin x = \sin x$

This is true if

$$-3A + B = 0 \quad\quad \text{and} \quad\quad -A - 3B = 1$$

The solution of this system is

$$A = -\tfrac{1}{10} \quad\quad B = -\tfrac{3}{10}$$

so a particular solution is

$$y_p(x) = -\tfrac{1}{10}\cos x - \tfrac{3}{10}\sin x$$

In Example 1 we determined that the solution of the complementary equation is $y_c = c_1 e^x + c_2 e^{-2x}$. Thus the general solution of the given equation is

$$y(x) = c_1 e^x + c_2 e^{-2x} - \tfrac{1}{10}(\cos x + 3\sin x) \quad\blacksquare$$

If $G(x)$ is a product of functions of the preceding types, then we take the trial solution to be a product of functions of the same type. For instance, in solving the differential equation

$$y'' + 2y' + 4y = x\cos 3x$$

we would try

$$y_p(x) = (Ax + B)\cos 3x + (Cx + D)\sin 3x$$

If $G(x)$ is a sum of functions of these types, we use the easily verified *principle of superposition*, which says that if y_{p_1} and y_{p_2} are solutions of

$$ay'' + by' + cy = G_1(x) \quad\quad ay'' + by' + cy = G_2(x)$$

respectively, then $y_{p_1} + y_{p_2}$ is a solution of

$$ay'' + by' + cy = G_1(x) + G_2(x)$$

EXAMPLE 4 Solve $y'' - 4y = xe^x + \cos 2x$.

SOLUTION The auxiliary equation is $r^2 - 4 = 0$ with roots ± 2, so the solution of the complementary equation is $y_c(x) = c_1 e^{2x} + c_2 e^{-2x}$. For the equation $y'' - 4y = xe^x$ we try

$$y_{p_1}(x) = (Ax + B)e^x$$

Then $y_{p_1}' = (Ax + A + B)e^x$, $y_{p_1}'' = (Ax + 2A + B)e^x$, so substitution in the equation gives

$$(Ax + 2A + B)e^x - 4(Ax + B)e^x = xe^x$$

or $\quad\quad\quad\quad\quad\quad\quad\quad (-3Ax + 2A - 3B)e^x = xe^x$

Thus $-3A = 1$ and $2A - 3B = 0$, so $A = -\frac{1}{3}$, $B = -\frac{2}{9}$, and

$$y_{p_1}(x) = \left(-\tfrac{1}{3}x - \tfrac{2}{9}\right)e^x$$

For the equation $y'' - 4y = \cos 2x$, we try

$$y_{p_2}(x) = C \cos 2x + D \sin 2x$$

Substitution gives

$$-4C \cos 2x - 4D \sin 2x - 4(C \cos 2x + D \sin 2x) = \cos 2x$$

or

$$-8C \cos 2x - 8D \sin 2x = \cos 2x$$

Therefore $-8C = 1$, $-8D = 0$, and

$$y_{p_2}(x) = -\tfrac{1}{8} \cos 2x$$

By the superposition principle, the general solution is

$$y = y_c + y_{p_1} + y_{p_2} = c_1 e^{2x} + c_2 e^{-2x} - \left(\tfrac{1}{3}x + \tfrac{2}{9}\right)e^x - \tfrac{1}{8} \cos 2x \quad \blacksquare$$

In Figure 3 we show the particular solution $y_p = y_{p_1} + y_{p_2}$ of the differential equation in Example 4. The other solutions are given in terms of $f(x) = e^{2x}$ and $g(x) = e^{-2x}$.

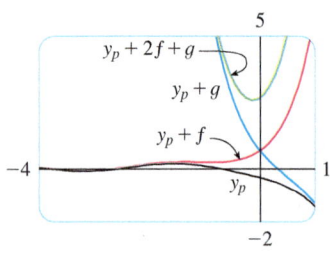

FIGURE 3

Finally we note that the recommended trial solution y_p sometimes turns out to be a solution of the complementary equation and therefore can't be a solution of the nonhomogeneous equation. In such cases we multiply the recommended trial solution by x (or by x^2 if necessary) so that no term in $y_p(x)$ is a solution of the complementary equation.

EXAMPLE 5 Solve $y'' + y = \sin x$.

SOLUTION The auxiliary equation is $r^2 + 1 = 0$ with roots $\pm i$, so the solution of the complementary equation is

$$y_c(x) = c_1 \cos x + c_2 \sin x$$

Ordinarily, we would use the trial solution

$$y_p(x) = A \cos x + B \sin x$$

but we observe that it is a solution of the complementary equation, so instead we try

$$y_p(x) = Ax \cos x + Bx \sin x$$

Then
$$y_p'(x) = A \cos x - Ax \sin x + B \sin x + Bx \cos x$$

$$y_p''(x) = -2A \sin x - Ax \cos x + 2B \cos x - Bx \sin x$$

Substitution in the differential equation gives

$$y_p'' + y_p = -2A \sin x + 2B \cos x = \sin x$$

The graphs of four solutions of the differential equation in Example 5 are shown in Figure 4.

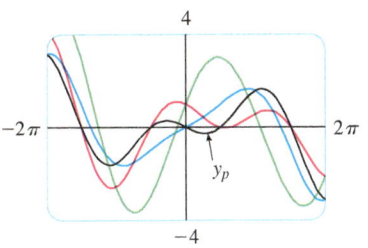

FIGURE 4

so $A = -\frac{1}{2}$, $B = 0$, and
$$y_p(x) = -\tfrac{1}{2}x \cos x$$

The general solution is
$$y(x) = c_1 \cos x + c_2 \sin x - \tfrac{1}{2}x \cos x$$

We summarize the method of undetermined coefficients as follows:

Summary of the Method of Undetermined Coefficients

1. If $G(x) = e^{kx}P(x)$, where P is a polynomial of degree n, then try $y_p(x) = e^{kx}Q(x)$, where $Q(x)$ is an nth-degree polynomial (whose coefficients are determined by substituting in the differential equation).

2. If $G(x) = e^{kx}P(x) \cos mx$ or $G(x) = e^{kx}P(x) \sin mx$, where P is an nth-degree polynomial, then try
$$y_p(x) = e^{kx}Q(x) \cos mx + e^{kx}R(x) \sin mx$$
where Q and R are nth-degree polynomials.

Modification: If any term of y_p is a solution of the complementary equation, multiply y_p by x (or by x^2 if necessary).

EXAMPLE 6 Determine the form of the trial solution for the differential equation $y'' - 4y' + 13y = e^{2x} \cos 3x$.

SOLUTION Here $G(x)$ has the form of part 2 of the summary, where $k = 2$, $m = 3$, and $P(x) = 1$. So, at first glance, the form of the trial solution would be
$$y_p(x) = e^{2x}(A \cos 3x + B \sin 3x)$$

But the auxiliary equation is $r^2 - 4r + 13 = 0$, with roots $r = 2 \pm 3i$, so the solution of the complementary equation is
$$y_c(x) = e^{2x}(c_1 \cos 3x + c_2 \sin 3x)$$

This means that we have to multiply the suggested trial solution by x. So, instead, we use
$$y_p(x) = xe^{2x}(A \cos 3x + B \sin 3x)$$

The Method of Variation of Parameters

Suppose we have already solved the homogeneous equation $ay'' + by' + cy = 0$ and written the solution as

$$\boxed{4} \qquad y(x) = c_1 y_1(x) + c_2 y_2(x)$$

where y_1 and y_2 are linearly independent solutions. Let's replace the constants (or parameters) c_1 and c_2 in Equation 4 by arbitrary functions $u_1(x)$ and $u_2(x)$. We look for a particular solution of the nonhomogeneous equation $ay'' + by' + cy = G(x)$ of the form

$$\boxed{5} \qquad y_p(x) = u_1(x)\, y_1(x) + u_2(x)\, y_2(x)$$

(This method is called **variation of parameters** because we have varied the parameters c_1 and c_2 to make them functions.) Differentiating Equation 5, we get

$$\boxed{6} \qquad y_p' = (u_1' y_1 + u_2' y_2) + (u_1 y_1' + u_2 y_2')$$

Since u_1 and u_2 are arbitrary functions, we can impose two conditions on them. One condition is that y_p is a solution of the differential equation; we can choose the other condition so as to simplify our calculations. In view of the expression in Equation 6, let's impose the condition that

$$\boxed{7} \qquad u_1' y_1 + u_2' y_2 = 0$$

Then

$$y_p'' = u_1' y_1' + u_2' y_2' + u_1 y_1'' + u_2 y_2''$$

Substituting in the differential equation, we get

$$a(u_1' y_1' + u_2' y_2' + u_1 y_1'' + u_2 y_2'') + b(u_1 y_1' + u_2 y_2') + c(u_1 y_1 + u_2 y_2) = G$$

or

$$\boxed{8} \qquad u_1(ay_1'' + by_1' + cy_1) + u_2(ay_2'' + by_2' + cy_2) + a(u_1' y_1' + u_2' y_2') = G$$

But y_1 and y_2 are solutions of the complementary equation, so

$$ay_1'' + by_1' + cy_1 = 0 \quad \text{and} \quad ay_2'' + by_2' + cy_2 = 0$$

and Equation 8 simplifies to

$$\boxed{9} \qquad a(u_1' y_1' + u_2' y_2') = G$$

Equations 7 and 9 form a system of two equations in the unknown functions u_1' and u_2'. After solving this system we may be able to integrate to find u_1 and u_2 and then the particular solution is given by Equation 5.

EXAMPLE 7 Solve the equation $y'' + y = \tan x$, $0 < x < \pi/2$.

SOLUTION The auxiliary equation is $r^2 + 1 = 0$ with roots $\pm i$, so the solution of $y'' + y = 0$ is $y(x) = c_1 \sin x + c_2 \cos x$. Using variation of parameters, we seek a solution of the form

$$y_p(x) = u_1(x) \sin x + u_2(x) \cos x$$

Then

$$y_p' = (u_1' \sin x + u_2' \cos x) + (u_1 \cos x - u_2 \sin x)$$

Set

$$\boxed{10} \qquad u_1' \sin x + u_2' \cos x = 0$$

Then
$$y_p'' = u_1' \cos x - u_2' \sin x - u_1 \sin x - u_2 \cos x$$

For y_p to be a solution we must have

$$\boxed{11} \qquad y_p'' + y_p = u_1' \cos x - u_2' \sin x = \tan x$$

Solving Equations 10 and 11, we get

$$u_1'(\sin^2 x + \cos^2 x) = \cos x \tan x$$

$$u_1' = \sin x \qquad u_1(x) = -\cos x$$

(We seek a particular solution, so we don't need a constant of integration here.) Then, from Equation 10, we obtain

$$u_2' = -\frac{\sin x}{\cos x} u_1' = -\frac{\sin^2 x}{\cos x} = \frac{\cos^2 x - 1}{\cos x} = \cos x - \sec x$$

So
$$u_2(x) = \sin x - \ln(\sec x + \tan x)$$

(Note that $\sec x + \tan x > 0$ for $0 < x < \pi/2$.) Therefore

$$y_p(x) = -\cos x \sin x + [\sin x - \ln(\sec x + \tan x)] \cos x$$

$$= -\cos x \ln(\sec x + \tan x)$$

and the general solution is

$$y(x) = c_1 \sin x + c_2 \cos x - \cos x \ln(\sec x + \tan x) \qquad \blacksquare$$

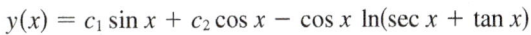

Figure 5 shows four solutions of the differential equation in Example 7.

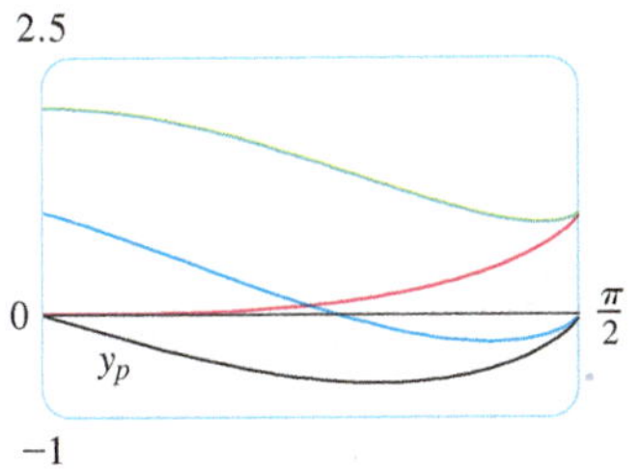

FIGURE 5

17.2 EXERCISES

1–10 Solve the differential equation or initial-value problem using the method of undetermined coefficients.

1. $y'' + 2y' - 8y = 1 - 2x^2$
2. $y'' - 3y' = \sin 2x$
3. $9y'' + y = e^{2x}$
4. $y'' - 2y' + 2y = x + e^x$
5. $y'' - 4y' + 5y = e^{-x}$
6. $y'' - 4y' + 4y = x - \sin x$
7. $y'' - 2y' + 5y = \sin x$, $y(0) = 1$, $y'(0) = 1$
8. $y'' - y = xe^{2x}$, $y(0) = 0$, $y'(0) = 1$
9. $y'' - y' = xe^x$, $y(0) = 2$, $y'(0) = 1$
10. $y'' + y' - 2y = x + \sin 2x$, $y(0) = 1$, $y'(0) = 0$

11–12 Graph the particular solution and several other solutions. What characteristics do these solutions have in common?

11. $y'' + 3y' + 2y = \cos x$
12. $y'' + 4y = e^{-x}$

13–18 Write a trial solution for the method of undetermined coefficients. Do not determine the coefficients.

13. $y'' - y' - 2y = xe^x \cos x$
14. $y'' + 4y = \cos 4x + \cos 2x$
15. $y'' - 3y' + 2y = e^x + \sin x$
16. $y'' + 3y' - 4y = (x^3 + x)e^x$
17. $y'' + 2y' + 10y = x^2 e^{-x} \cos 3x$
18. $y'' + 4y = e^{3x} + x \sin 2x$

19–22 Solve the differential equation using (a) undetermined coefficients and (b) variation of parameters.

19. $4y'' + y = \cos x$ **20.** $y'' - 2y' - 3y = x + 2$

21. $y'' - 2y' + y = e^{2x}$

22. $y'' - y' = e^x$

23–28 Solve the differential equation using the method of variation of parameters.

23. $y'' + y = \sec^2 x$, $0 < x < \pi/2$

24. $y'' + y = \sec^3 x$, $0 < x < \pi/2$

25. $y'' - 3y' + 2y = \dfrac{1}{1 + e^{-x}}$

26. $y'' + 3y' + 2y = \sin(e^x)$

27. $y'' - 2y' + y = \dfrac{e^x}{1 + x^2}$

28. $y'' + 4y' + 4y = \dfrac{e^{-2x}}{x^3}$

17.3 Applications of Second-Order Differential Equations

Second-order linear differential equations have a variety of applications in science and engineering. In this section we explore two of them: the vibration of springs and electric circuits.

■ Vibrating Springs

We consider the motion of an object with mass m at the end of a spring that is either vertical (as in Figure 1) or horizontal on a level surface (as in Figure 2).

In Section 6.4 we discussed Hooke's Law, which says that if the spring is stretched (or compressed) x units from its natural length, then it exerts a force that is proportional to x:

$$\text{restoring force} = -kx$$

where k is a positive constant (called the **spring constant**). If we ignore any external resisting forces (due to air resistance or friction) then, by Newton's Second Law (force equals mass times acceleration), we have

$$\boxed{1} \qquad m\frac{d^2x}{dt^2} = -kx \qquad \text{or} \qquad m\frac{d^2x}{dt^2} + kx = 0$$

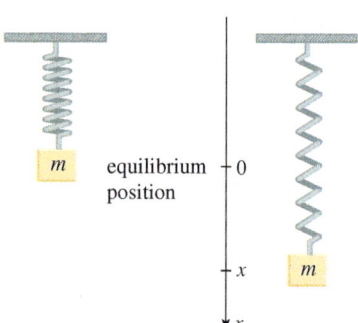

FIGURE 1

This is a second-order linear differential equation. Its auxiliary equation is $mr^2 + k = 0$ with roots $r = \pm \omega i$, where $\omega = \sqrt{k/m}$. Thus the general solution is

$$x(t) = c_1 \cos \omega t + c_2 \sin \omega t$$

which can also be written as

$$x(t) = A\cos(\omega t + \delta)$$

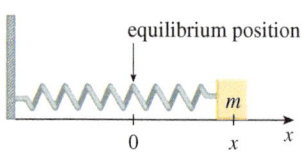

FIGURE 2

where
$$\omega = \sqrt{k/m} \quad \text{(frequency)}$$

$$A = \sqrt{c_1^2 + c_2^2} \quad \text{(amplitude)}$$

$$\cos \delta = \frac{c_1}{A} \qquad \sin \delta = -\frac{c_2}{A} \quad (\delta \text{ is the phase angle})$$

(See Exercise 17.) This type of motion is called **simple harmonic motion**.

EXAMPLE 1 A spring with a mass of 2 kg has natural length 0.5 m. A force of 25.6 N is required to maintain it stretched to a length of 0.7 m. If the spring is stretched to a length of 0.7 m and then released with initial velocity 0, find the position of the mass at any time t.

SOLUTION From Hooke's Law, the force required to stretch the spring is

$$k(0.2) = 25.6$$

so $k = 25.6/0.2 = 128$. Using this value of the spring constant k, together with $m = 2$ in Equation 1, we have

$$2\frac{d^2x}{dt^2} + 128x = 0$$

As in the earlier general discussion, the solution of this equation is

$$\boxed{2} \qquad x(t) = c_1 \cos 8t + c_2 \sin 8t$$

We are given the initial condition that $x(0) = 0.2$. But, from Equation 2, $x(0) = c_1$. Therefore $c_1 = 0.2$. Differentiating Equation 2, we get

$$x'(t) = -8c_1 \sin 8t + 8c_2 \cos 8t$$

Since the initial velocity is given as $x'(0) = 0$, we have $c_2 = 0$ and so the solution is

$$x(t) = 0.2 \cos 8t \qquad \blacksquare$$

Damped Vibrations

We next consider the motion of a spring that is subject to a frictional force (in the case of the horizontal spring of Figure 2) or a damping force (in the case where a vertical spring moves through a fluid as in Figure 3). An example is the damping force supplied by a shock absorber in a car or a bicycle.

We assume that the damping force is proportional to the velocity of the mass and acts in the direction opposite to the motion. (This has been confirmed, at least approximately, by some physical experiments.) Thus

$$\text{damping force} = -c\frac{dx}{dt}$$

where c is a positive constant, called the **damping constant**. Thus, in this case, Newton's Second Law gives

$$m\frac{d^2x}{dt^2} = \text{restoring force} + \text{damping force} = -kx - c\frac{dx}{dt}$$

or

$$\boxed{3} \qquad m\frac{d^2x}{dt^2} + c\frac{dx}{dt} + kx = 0$$

FIGURE 3

Equation 3 is a second-order linear differential equation and its auxiliary equation is $mr^2 + cr + k = 0$. The roots are

$$\boxed{4} \qquad r_1 = \frac{-c + \sqrt{c^2 - 4mk}}{2m} \qquad r_2 = \frac{-c - \sqrt{c^2 - 4mk}}{2m}$$

According to Section 17.1 we need to discuss three cases.

CASE I $c^2 - 4mk > 0$ (overdamping)
In this case r_1 and r_2 are distinct real roots and

$$x = c_1 e^{r_1 t} + c_2 e^{r_2 t}$$

Since c, m, and k are all positive, we have $\sqrt{c^2 - 4mk} < c$, so the roots r_1 and r_2 given by Equations 4 must both be negative. This shows that $x \to 0$ as $t \to \infty$. Typical graphs of x as a function of t are shown in Figure 4. Notice that oscillations do not occur. (It's possible for the mass to pass through the equilibrium position once, but only once.) This is because $c^2 > 4mk$ means that there is a strong damping force (high-viscosity oil or grease) compared with a weak spring or small mass.

CASE II $c^2 - 4mk = 0$ (critical damping)
This case corresponds to equal roots

$$r_1 = r_2 = -\frac{c}{2m}$$

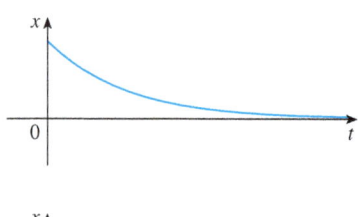
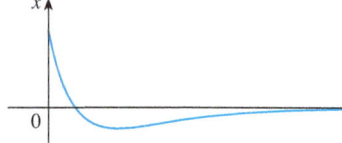

FIGURE 4
Overdamping

and the solution is given by

$$x = (c_1 + c_2 t)e^{-(c/2m)t}$$

It is similar to Case I, and typical graphs resemble those in Figure 4 (see Exercise 12), but the damping is just sufficient to suppress vibrations. Any decrease in the viscosity of the fluid leads to the vibrations of the following case.

CASE III $c^2 - 4mk < 0$ (underdamping)
Here the roots are complex:

$$\left.\begin{matrix} r_1 \\ r_2 \end{matrix}\right\} = -\frac{c}{2m} \pm \omega i$$

where

$$\omega = \frac{\sqrt{4mk - c^2}}{2m}$$

The solution is given by

$$x = e^{-(c/2m)t}(c_1 \cos \omega t + c_2 \sin \omega t)$$

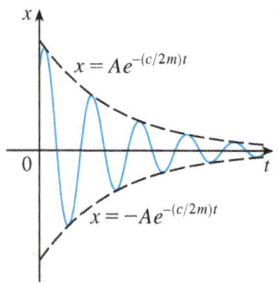

FIGURE 5
Underdamping

We see that there are oscillations that are damped by the factor $e^{-(c/2m)t}$. Since $c > 0$ and $m > 0$, we have $-(c/2m) < 0$ so $e^{-(c/2m)t} \to 0$ as $t \to \infty$. This implies that $x \to 0$ as $t \to \infty$; that is, the motion decays to 0 as time increases. A typical graph is shown in Figure 5.

EXAMPLE 2 Suppose that the spring of Example 1 is immersed in a fluid with damping constant $c = 40$. Find the position of the mass at any time t if it starts from the equilibrium position and is given a push to start it with an initial velocity of 0.6 m/s.

SOLUTION From Example 1, the mass is $m = 2$ and the spring constant is $k = 128$, so the differential equation (3) becomes

$$2 \frac{d^2x}{dt^2} + 40 \frac{dx}{dt} + 128x = 0$$

or

$$\frac{d^2x}{dt^2} + 20 \frac{dx}{dt} + 64x = 0$$

The auxiliary equation is $r^2 + 20r + 64 = (r + 4)(r + 16) = 0$ with roots -4 and -16, so the motion is overdamped and the solution is

$$x(t) = c_1 e^{-4t} + c_2 e^{-16t}$$

We are given that $x(0) = 0$, so $c_1 + c_2 = 0$. Differentiating, we get

$$x'(t) = -4c_1 e^{-4t} - 16c_2 e^{-16t}$$

so

$$x'(0) = -4c_1 - 16c_2 = 0.6$$

Since $c_2 = -c_1$, this gives $12c_1 = 0.6$ or $c_1 = 0.05$. Therefore

$$x = 0.05(e^{-4t} - e^{-16t})$$ ∎

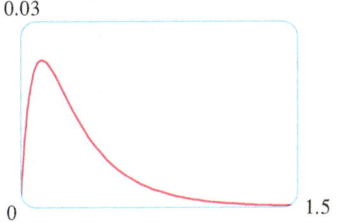

Figure 6 shows the graph of the position function for the overdamped motion in Example 2.

FIGURE 6

Forced Vibrations

Suppose that, in addition to the restoring force and the damping force, the motion of the spring is affected by an external force $F(t)$. Then Newton's Second Law gives

$$m \frac{d^2x}{dt^2} = \text{restoring force} + \text{damping force} + \text{external force}$$

$$= -kx - c \frac{dx}{dt} + F(t)$$

Thus, instead of the homogeneous equation (3), the motion of the spring is now governed by the following nonhomogeneous differential equation:

$$\boxed{5 \qquad m \frac{d^2x}{dt^2} + c \frac{dx}{dt} + kx = F(t)}$$

The motion of the spring can be determined by the methods of Section 17.2.

A commonly occurring type of external force is a periodic force function

$$F(t) = F_0 \cos \omega_0 t \quad \text{where} \quad \omega_0 \neq \omega = \sqrt{k/m}$$

In this case, and in the absence of a damping force ($c = 0$), you are asked in Exercise 9 to use the method of undetermined coefficients to show that

$$\boxed{6} \qquad x(t) = c_1 \cos \omega t + c_2 \sin \omega t + \frac{F_0}{m(\omega^2 - \omega_0^2)} \cos \omega_0 t$$

If $\omega_0 = \omega$, then the applied frequency reinforces the natural frequency and the result is vibrations of large amplitude. This is the phenomenon of **resonance** (see Exercise 10).

Electric Circuits

In Sections 9.3 and 9.5 we were able to use first-order separable and linear equations to analyze electric circuits that contain a resistor and inductor (see Figure 9.3.5 or Figure 9.5.4) or a resistor and capacitor (see Exercise 9.5.29). Now that we know how to solve second-order linear equations, we are in a position to analyze the circuit shown in Figure 7. It contains an electromotive force E (supplied by a battery or generator), a resistor R, an inductor L, and a capacitor C, in series. If the charge on the capacitor at time t is $Q = Q(t)$, then the current is the rate of change of Q with respect to t: $I = dQ/dt$. As in Section 9.5, it is known from physics that the voltage drops across the resistor, inductor, and capacitor are

$$RI \qquad L\frac{dI}{dt} \qquad \frac{Q}{C}$$

respectively. Kirchhoff's voltage law says that the sum of these voltage drops is equal to the supplied voltage:

$$L\frac{dI}{dt} + RI + \frac{Q}{C} = E(t)$$

Since $I = dQ/dt$, this equation becomes

$$\boxed{7} \qquad \boxed{L\frac{d^2Q}{dt^2} + R\frac{dQ}{dt} + \frac{1}{C}Q = E(t)}$$

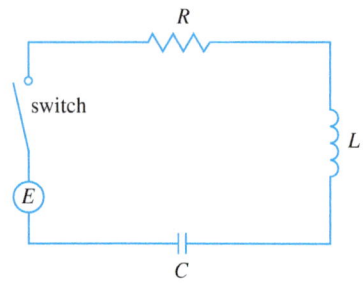

FIGURE 7

which is a second-order linear differential equation with constant coefficients. If the charge Q_0 and the current I_0 are known at time 0, then we have the initial conditions

$$Q(0) = Q_0 \qquad Q'(0) = I(0) = I_0$$

and the initial-value problem can be solved by the methods of Section 17.2.

SECTION 17.3 Applications of Second-Order Differential Equations

A differential equation for the current can be obtained by differentiating Equation 7 with respect to t and remembering that $I = dQ/dt$:

$$L\frac{d^2I}{dt^2} + R\frac{dI}{dt} + \frac{1}{C}I = E'(t)$$

EXAMPLE 3 Find the charge and current at time t in the circuit of Figure 7 if $R = 40\,\Omega$, $L = 1$ H, $C = 16 \times 10^{-4}$ F, $E(t) = 100\cos 10t$, and the initial charge and current are both 0.

SOLUTION With the given values of L, R, C, and $E(t)$, Equation 7 becomes

$$\boxed{8} \qquad \frac{d^2Q}{dt^2} + 40\frac{dQ}{dt} + 625Q = 100\cos 10t$$

The auxiliary equation is $r^2 + 40r + 625 = 0$ with roots

$$r = \frac{-40 \pm \sqrt{-900}}{2} = -20 \pm 15i$$

so the solution of the complementary equation is

$$Q_c(t) = e^{-20t}(c_1 \cos 15t + c_2 \sin 15t)$$

For the method of undetermined coefficients we try the particular solution

$$Q_p(t) = A\cos 10t + B\sin 10t$$

Then
$$Q_p'(t) = -10A\sin 10t + 10B\cos 10t$$
$$Q_p''(t) = -100A\cos 10t - 100B\sin 10t$$

Substituting into Equation 8, we have

$$(-100A\cos 10t - 100B\sin 10t) + 40(-10A\sin 10t + 10B\cos 10t)$$
$$+ 625(A\cos 10t + B\sin 10t) = 100\cos 10t$$

or $\quad (525A + 400B)\cos 10t + (-400A + 525B)\sin 10t = 100\cos 10t$

Equating coefficients, we have

$$525A + 400B = 100 \quad \text{or} \quad 21A + 16B = 4$$
$$-400A + 525B = 0 \quad \text{or} \quad -16A + 21B = 0$$

The solution of this system is $A = \frac{84}{697}$ and $B = \frac{64}{697}$, so a particular solution is

$$Q_p(t) = \frac{1}{697}(84\cos 10t + 64\sin 10t)$$

and the general solution is

$$Q(t) = Q_c(t) + Q_p(t)$$
$$= e^{-20t}(c_1 \cos 15t + c_2 \sin 15t) + \tfrac{4}{697}(21 \cos 10t + 16 \sin 10t)$$

Imposing the initial condition $Q(0) = 0$, we get

$$Q(0) = c_1 + \tfrac{84}{697} = 0 \qquad c_1 = -\tfrac{84}{697}$$

To impose the other initial condition, we first differentiate to find the current:

$$I = \frac{dQ}{dt} = e^{-20t}[(-20c_1 + 15c_2)\cos 15t + (-15c_1 - 20c_2)\sin 15t]$$
$$+ \tfrac{40}{697}(-21 \sin 10t + 16 \cos 10t)$$

$$I(0) = -20c_1 + 15c_2 + \tfrac{640}{697} = 0 \qquad c_2 = -\tfrac{464}{2091}$$

Thus the formula for the charge is

$$Q(t) = \frac{4}{697}\left[\frac{e^{-20t}}{3}(-63 \cos 15t - 116 \sin 15t) + (21 \cos 10t + 16 \sin 10t)\right]$$

and the expression for the current is

$$I(t) = \tfrac{1}{2091}\left[e^{-20t}(-1920 \cos 15t + 13{,}060 \sin 15t) + 120(-21 \sin 10t + 16 \cos 10t)\right] \blacksquare$$

NOTE 1 In Example 3 the solution for $Q(t)$ consists of two parts. Since $e^{-20t} \to 0$ as $t \to \infty$ and both $\cos 15t$ and $\sin 15t$ are bounded functions,

$$Q_c(t) = \tfrac{4}{2091}e^{-20t}(-63 \cos 15t - 116 \sin 15t) \to 0 \qquad \text{as } t \to \infty$$

So, for large values of t,

$$Q(t) \approx Q_p(t) = \tfrac{4}{697}(21 \cos 10t + 16 \sin 10t)$$

and, for this reason, $Q_p(t)$ is called the **steady state solution**. Figure 8 shows how the graph of the steady state solution compares with the graph of Q in this case.

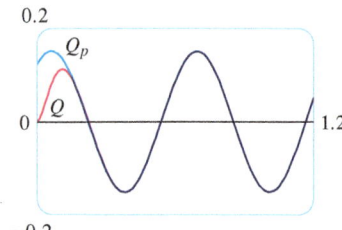

FIGURE 8

$$\boxed{5} \quad m\frac{d^2x}{dt^2} + c\frac{dx}{dt} + kx = F(t)$$

$$\boxed{7} \quad L\frac{d^2Q}{dt^2} + R\frac{dQ}{dt} + \frac{1}{C}Q = E(t)$$

NOTE 2 Comparing Equations 5 and 7, we see that mathematically they are identical. This suggests the analogies given in the following chart between physical situations that, at first glance, are very different.

Spring system		Electric circuit	
x	displacement	Q	charge
dx/dt	velocity	$I = dQ/dt$	current
m	mass	L	inductance
c	damping constant	R	resistance
k	spring constant	$1/C$	elastance
$F(t)$	external force	$E(t)$	electromotive force

17.3 EXERCISES

1. A spring has natural length 0.75 m and a 5-kg mass. A force of 25 N is needed to keep the spring stretched to a length of 1 m. If the spring is stretched to a length of 1.1 m and then released with velocity 0, find the position of the mass after t seconds.

2. A spring with an 8-kg mass is kept stretched 0.4 m beyond its natural length by a force of 32 N. The spring starts at its equilibrium position and is given an initial velocity of 1 m/s. Find the position of the mass at any time t.

3. A spring with a mass of 2 kg has damping constant 14, and a force of 6 N is required to keep the spring stretched 0.5 m beyond its natural length. The spring is stretched 1 m beyond its natural length and then released with zero velocity. Find the position of the mass at any time t.

4. A force of 13 N is needed to keep a spring with a 2-kg mass stretched 0.25 m beyond its natural length. The damping constant of the spring is $c = 8$.
   (a) If the mass starts at the equilibrium position with a velocity of 0.5 m/s, find its position at time t.
   (b) Graph the position function of the mass.

5. For the spring in Exercise 3, find the mass that would produce critical damping.

6. For the spring in Exercise 4, find the damping constant that would produce critical damping.

7. A spring has a mass of 1 kg and its spring constant is $k = 100$. The spring is released at a point 0.1 m above its equilibrium position. Graph the position function for the following values of the damping constant c: 10, 15, 20, 25, 30. What type of damping occurs in each case?

8. A spring has a mass of 1 kg and its damping constant is $c = 10$. The spring starts from its equilibrium position with a velocity of 1 m/s. Graph the position function for the following values of the spring constant k: 10, 20, 25, 30, 40. What type of damping occurs in each case?

9. Suppose a spring has mass m and spring constant k and let $\omega = \sqrt{k/m}$. Suppose that the damping constant is so small that the damping force is negligible. If an external force $F(t) = F_0 \cos \omega_0 t$ is applied, where $\omega_0 \neq \omega$, use the method of undetermined coefficients to show that the motion of the mass is described by Equation 6.

10. As in Exercise 9, consider a spring with mass m, spring constant k, and damping constant $c = 0$, and let $\omega = \sqrt{k/m}$. If an external force $F(t) = F_0 \cos \omega t$ is applied (the applied frequency equals the natural frequency), use the method of undetermined coefficients to show that the motion of the mass is given by
$$x(t) = c_1 \cos \omega t + c_2 \sin \omega t + \frac{F_0}{2m\omega} t \sin \omega t$$

11. Show that if $\omega_0 \neq \omega$, but ω/ω_0 is a rational number, then the motion described by Equation 6 is periodic.

12. Consider a spring subject to a frictional or damping force.
    (a) In the critically damped case, the motion is given by $x = c_1 e^{rt} + c_2 t e^{rt}$. Show that the graph of x crosses the t-axis whenever c_1 and c_2 have opposite signs.
    (b) In the overdamped case, the motion is given by $x = c_1 e^{r_1 t} + c_2 e^{r_2 t}$, where $r_1 > r_2$. Determine a condition on the relative magnitudes of c_1 and c_2 under which the graph of x crosses the t-axis at a positive value of t.

13. A series circuit consists of a resistor with $R = 20\ \Omega$, an inductor with $L = 1$ H, a capacitor with $C = 0.002$ F, and a 12-V battery. If the initial charge and current are both 0, find the charge and current at time t.

14. A series circuit contains a resistor with $R = 24\ \Omega$, an inductor with $L = 2$ H, a capacitor with $C = 0.005$ F, and a 12-V battery. The initial charge is $Q = 0.001$ C and the initial current is 0.
    (a) Find the charge and current at time t.
    (b) Graph the charge and current functions.

15. The battery in Exercise 13 is replaced by a generator producing a voltage of $E(t) = 12 \sin 10t$. Find the charge at time t.

16. The battery in Exercise 14 is replaced by a generator producing a voltage of $E(t) = 12 \sin 10t$.
    (a) Find the charge at time t.
    (b) Graph the charge function.

17. Verify that the solution to Equation 1 can be written in the form $x(t) = A \cos(\omega t + \delta)$.

18. The figure shows a pendulum with length L and the angle θ from the vertical to the pendulum. It can be shown that θ, as a function of time, satisfies the nonlinear differential equation

$$\frac{d^2\theta}{dt^2} + \frac{g}{L}\sin\theta = 0$$

where g is the acceleration due to gravity. For small values of θ we can use the linear approximation $\sin\theta \approx \theta$ and then the differential equation becomes linear.
(a) Find the equation of motion of a pendulum with length 1 m if θ is initially 0.2 rad and the initial angular velocity is $d\theta/dt = 1$ rad/s.
(b) What is the maximum angle from the vertical?
(c) What is the period of the pendulum (that is, the time to complete one back-and-forth swing)?
(d) When will the pendulum first be vertical?
(e) What is the angular velocity when the pendulum is vertical?

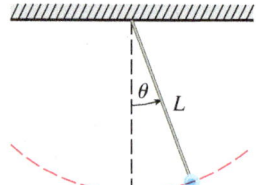

17.4 Series Solutions

Many differential equations can't be solved explicitly in terms of finite combinations of simple familiar functions. This is true even for a simple-looking equation like

$$\boxed{1} \qquad y'' - 2xy' + y = 0$$

But it is important to be able to solve equations such as Equation 1 because they arise from physical problems and, in particular, in connection with the Schrödinger equation in quantum mechanics. In such a case we use the method of power series; that is, we look for a solution of the form

$$y = f(x) = \sum_{n=0}^{\infty} c_n x^n = c_0 + c_1 x + c_2 x^2 + c_3 x^3 + \cdots$$

The method is to substitute this expression into the differential equation and determine the values of the coefficients $c_0, c_1, c_2, \ldots$. This technique resembles the method of undetermined coefficients discussed in Section 17.2.

Before using power series to solve Equation 1, we illustrate the method on the simpler equation $y'' + y = 0$ in Example 1. It's true that we already know how to solve this equation by the techniques of Section 17.1, but it's easier to understand the power series method when it is applied to this simpler equation.

EXAMPLE 1 Use power series to solve the equation $y'' + y = 0$.

SOLUTION We assume there is a solution of the form

$$\boxed{2} \qquad y = c_0 + c_1 x + c_2 x^2 + c_3 x^3 + \cdots = \sum_{n=0}^{\infty} c_n x^n$$

We can differentiate power series term by term, so

$$y' = c_1 + 2c_2 x + 3c_3 x^2 + \cdots = \sum_{n=1}^{\infty} n c_n x^{n-1}$$

$$\boxed{3} \qquad y'' = 2c_2 + 2\cdot 3c_3 x + \cdots = \sum_{n=2}^{\infty} n(n-1) c_n x^{n-2}$$

In order to compare the expressions for y and y'' more easily, we rewrite y'' as follows:

By writing out the first few terms of (4), you can see that it is the same as (3). To obtain (4), we replaced n by $n+2$ and began the summation at 0 instead of 2.

$$\boxed{4} \qquad y'' = \sum_{n=0}^{\infty} (n+2)(n+1)c_{n+2}x^n$$

Substituting the expressions in Equations 2 and 4 into the differential equation, we obtain

$$\sum_{n=0}^{\infty} (n+2)(n+1)c_{n+2}x^n + \sum_{n=0}^{\infty} c_n x^n = 0$$

or

$$\boxed{5} \qquad \sum_{n=0}^{\infty} [(n+2)(n+1)c_{n+2} + c_n]x^n = 0$$

If two power series are equal, then the corresponding coefficients must be equal. Therefore the coefficients of x^n in Equation 5 must be 0:

$$(n+2)(n+1)c_{n+2} + c_n = 0$$

$$\boxed{6} \qquad c_{n+2} = -\frac{c_n}{(n+1)(n+2)} \qquad n = 0, 1, 2, 3, \ldots$$

Equation 6 is called a *recursion relation*. If c_0 and c_1 are known, this equation allows us to determine the remaining coefficients recursively by putting $n = 0, 1, 2, 3, \ldots$ in succession.

Put $n = 0$: $\qquad c_2 = -\dfrac{c_0}{1 \cdot 2}$

Put $n = 1$: $\qquad c_3 = -\dfrac{c_1}{2 \cdot 3}$

Put $n = 2$: $\qquad c_4 = -\dfrac{c_2}{3 \cdot 4} = \dfrac{c_0}{1 \cdot 2 \cdot 3 \cdot 4} = \dfrac{c_0}{4!}$

Put $n = 3$: $\qquad c_5 = -\dfrac{c_3}{4 \cdot 5} = \dfrac{c_1}{2 \cdot 3 \cdot 4 \cdot 5} = \dfrac{c_1}{5!}$

Put $n = 4$: $\qquad c_6 = -\dfrac{c_4}{5 \cdot 6} = -\dfrac{c_0}{4!\, 5 \cdot 6} = -\dfrac{c_0}{6!}$

Put $n = 5$: $\qquad c_7 = -\dfrac{c_5}{6 \cdot 7} = -\dfrac{c_1}{5!\, 6 \cdot 7} = -\dfrac{c_1}{7!}$

By now we see the pattern:

For the even coefficients, $c_{2n} = (-1)^n \dfrac{c_0}{(2n)!}$

For the odd coefficients, $c_{2n+1} = (-1)^n \dfrac{c_1}{(2n+1)!}$

Putting these values back into Equation 2, we write the solution as

$$y = c_0 + c_1 x + c_2 x^2 + c_3 x^3 + c_4 x^4 + c_5 x^5 + \cdots$$

$$= c_0 \left(1 - \frac{x^2}{2!} + \frac{x^4}{4!} - \frac{x^6}{6!} + \cdots + (-1)^n \frac{x^{2n}}{(2n)!} + \cdots \right)$$

$$+ c_1 \left(x - \frac{x^3}{3!} + \frac{x^5}{5!} - \frac{x^7}{7!} + \cdots + (-1)^n \frac{x^{2n+1}}{(2n+1)!} + \cdots \right)$$

$$= c_0 \sum_{n=0}^{\infty} (-1)^n \frac{x^{2n}}{(2n)!} + c_1 \sum_{n=0}^{\infty} (-1)^n \frac{x^{2n+1}}{(2n+1)!}$$

Notice that there are two arbitrary constants, c_0 and c_1. ■

NOTE 1 We recognize the series obtained in Example 1 as being the Maclaurin series for $\cos x$ and $\sin x$. (See Equations 11.10.16 and 11.10.15.) Therefore we could write the solution as

$$y(x) = c_0 \cos x + c_1 \sin x$$

But we are not usually able to express power series solutions of differential equations in terms of known functions.

EXAMPLE 2 Solve $y'' - 2xy' + y = 0$.

SOLUTION We assume there is a solution of the form

$$y = \sum_{n=0}^{\infty} c_n x^n$$

Then

$$y' = \sum_{n=1}^{\infty} n c_n x^{n-1}$$

and

$$y'' = \sum_{n=2}^{\infty} n(n-1) c_n x^{n-2} = \sum_{n=0}^{\infty} (n+2)(n+1) c_{n+2} x^n$$

as in Example 1. Substituting in the differential equation, we get

$$\sum_{n=0}^{\infty} (n+2)(n+1) c_{n+2} x^n - 2x \sum_{n=1}^{\infty} n c_n x^{n-1} + \sum_{n=0}^{\infty} c_n x^n = 0$$

$$\sum_{n=0}^{\infty} (n+2)(n+1) c_{n+2} x^n - \sum_{n=1}^{\infty} 2n c_n x^n + \sum_{n=0}^{\infty} c_n x^n = 0$$

$\sum_{n=1}^{\infty} 2n c_n x^n = \sum_{n=0}^{\infty} 2n c_n x^n$

$$\sum_{n=0}^{\infty} [(n+2)(n+1) c_{n+2} - (2n-1) c_n] x^n = 0$$

This equation is true if the coefficients of x^n are 0:

$$(n+2)(n+1) c_{n+2} - (2n-1) c_n = 0$$

$$\boxed{7} \qquad c_{n+2} = \frac{2n-1}{(n+1)(n+2)} c_n \qquad n = 0, 1, 2, 3, \ldots$$

We solve this recursion relation by putting $n = 0, 1, 2, 3, \ldots$ successively in Equation 7:

Put $n = 0$: $\quad c_2 = \dfrac{-1}{1 \cdot 2} c_0$

Put $n = 1$: $\quad c_3 = \dfrac{1}{2 \cdot 3} c_1$

Put $n = 2$: $\quad c_4 = \dfrac{3}{3 \cdot 4} c_2 = -\dfrac{3}{1 \cdot 2 \cdot 3 \cdot 4} c_0 = -\dfrac{3}{4!} c_0$

Put $n = 3$: $\quad c_5 = \dfrac{5}{4 \cdot 5} c_3 = \dfrac{1 \cdot 5}{2 \cdot 3 \cdot 4 \cdot 5} c_1 = \dfrac{1 \cdot 5}{5!} c_1$

Put $n = 4$: $\quad c_6 = \dfrac{7}{5 \cdot 6} c_4 = -\dfrac{3 \cdot 7}{4! \, 5 \cdot 6} c_0 = -\dfrac{3 \cdot 7}{6!} c_0$

Put $n = 5$: $\quad c_7 = \dfrac{9}{6 \cdot 7} c_5 = \dfrac{1 \cdot 5 \cdot 9}{5! \, 6 \cdot 7} c_1 = \dfrac{1 \cdot 5 \cdot 9}{7!} c_1$

Put $n = 6$: $\quad c_8 = \dfrac{11}{7 \cdot 8} c_6 = -\dfrac{3 \cdot 7 \cdot 11}{8!} c_0$

Put $n = 7$: $\quad c_9 = \dfrac{13}{8 \cdot 9} c_7 = \dfrac{1 \cdot 5 \cdot 9 \cdot 13}{9!} c_1$

In general, the even coefficients are given by

$$c_{2n} = -\dfrac{3 \cdot 7 \cdot 11 \cdot \cdots \cdot (4n-5)}{(2n)!} c_0$$

and the odd coefficients are given by

$$c_{2n+1} = \dfrac{1 \cdot 5 \cdot 9 \cdot \cdots \cdot (4n-3)}{(2n+1)!} c_1$$

The solution is

$$y = c_0 + c_1 x + c_2 x^2 + c_3 x^3 + c_4 x^4 + \cdots$$

$$= c_0 \left(1 - \dfrac{1}{2!} x^2 - \dfrac{3}{4!} x^4 - \dfrac{3 \cdot 7}{6!} x^6 - \dfrac{3 \cdot 7 \cdot 11}{8!} x^8 - \cdots \right)$$

$$+ c_1 \left(x + \dfrac{1}{3!} x^3 + \dfrac{1 \cdot 5}{5!} x^5 + \dfrac{1 \cdot 5 \cdot 9}{7!} x^7 + \dfrac{1 \cdot 5 \cdot 9 \cdot 13}{9!} x^9 + \cdots \right)$$

or

$$\boxed{8} \quad y = c_0 \left(1 - \dfrac{1}{2!} x^2 - \sum_{n=2}^{\infty} \dfrac{3 \cdot 7 \cdot \cdots \cdot (4n-5)}{(2n)!} x^{2n} \right)$$

$$+ c_1 \left(x + \sum_{n=1}^{\infty} \dfrac{1 \cdot 5 \cdot 9 \cdot \cdots \cdot (4n-3)}{(2n+1)!} x^{2n+1} \right)$$

NOTE 2 In Example 2 we had to *assume* that the differential equation had a series solution. But now we could verify directly that the function given by Equation 8 is indeed a solution.

NOTE 3 Unlike the situation of Example 1, the power series that arise in the solution of Example 2 do not define elementary functions. The functions

$$y_1(x) = 1 - \frac{1}{2!}x^2 - \sum_{n=2}^{\infty} \frac{3 \cdot 7 \cdot \cdots \cdot (4n-5)}{(2n)!} x^{2n}$$

and

$$y_2(x) = x + \sum_{n=1}^{\infty} \frac{1 \cdot 5 \cdot 9 \cdot \cdots \cdot (4n-3)}{(2n+1)!} x^{2n+1}$$

are perfectly good functions but they can't be expressed in terms of familiar functions. We can use these power series expressions for y_1 and y_2 to compute approximate values of the functions and even to graph them. Figure 1 shows the first few partial sums $T_0, T_2, T_4, \ldots$ (Taylor polynomials) for $y_1(x)$, and we see how they converge to y_1. In this way we can graph both y_1 and y_2 as in Figure 2.

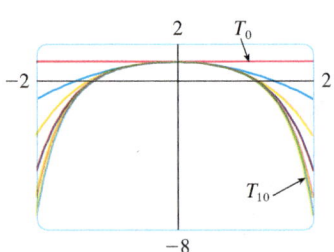

FIGURE 1

NOTE 4 If we were asked to solve the initial-value problem

$$y'' - 2xy' + y = 0 \qquad y(0) = 0 \qquad y'(0) = 1$$

we would observe from Theorem 11.10.5 that

$$c_0 = y(0) = 0 \qquad c_1 = y'(0) = 1$$

This would simplify the calculations in Example 2, since all of the even coefficients would be 0. The solution to the initial-value problem is

$$y(x) = x + \sum_{n=1}^{\infty} \frac{1 \cdot 5 \cdot 9 \cdot \cdots \cdot (4n-3)}{(2n+1)!} x^{2n+1}$$

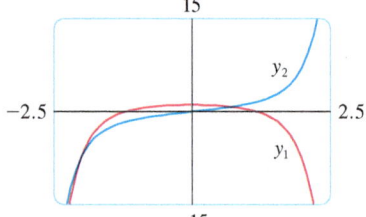

FIGURE 2

17.4 EXERCISES

1–11 Use power series to solve the differential equation.

1. $y' - y = 0$
2. $y' = xy$
3. $y' = x^2 y$
4. $(x - 3)y' + 2y = 0$
5. $y'' + xy' + y = 0$
6. $y'' = y$
7. $(x - 1)y'' + y' = 0$
8. $y'' = xy$
9. $y'' - xy' - y = 0, \quad y(0) = 1, \quad y'(0) = 0$
10. $y'' + x^2 y = 0, \quad y(0) = 1, \quad y'(0) = 0$
11. $y'' + x^2 y' + xy = 0, \quad y(0) = 0, \quad y'(0) = 1$

12. The solution of the initial-value problem

$$x^2 y'' + xy' + x^2 y = 0 \qquad y(0) = 1 \qquad y'(0) = 0$$

is called a Bessel function of order 0.
(a) Solve the initial-value problem to find a power series expansion for the Bessel function.
(b) Graph several Taylor polynomials until you reach one that looks like a good approximation to the Bessel function on the interval $[-5, 5]$.

17 REVIEW

CONCEPT CHECK

Answers to the Concept Check can be found on the back endpapers.

1. (a) Write the general form of a second-order homogeneous linear differential equation with constant coefficients.
   (b) Write the auxiliary equation.
   (c) How do you use the roots of the auxiliary equation to solve the differential equation? Write the form of the solution for each of the three cases that can occur.

2. (a) What is an initial-value problem for a second-order differential equation?
   (b) What is a boundary-value problem for such an equation?

3. (a) Write the general form of a second-order nonhomogeneous linear differential equation with constant coefficients.
   (b) What is the complementary equation? How does it help solve the original differential equation?
   (c) Explain how the method of undetermined coefficients works.
   (d) Explain how the method of variation of parameters works.

4. Discuss two applications of second-order linear differential equations.

5. How do you use power series to solve a differential equation?

TRUE-FALSE QUIZ

Determine whether the statement is true or false. If it is true, explain why. If it is false, explain why or give an example that disproves the statement.

1. If y_1 and y_2 are solutions of $y'' + y = 0$, then $y_1 + y_2$ is also a solution of the equation.

2. If y_1 and y_2 are solutions of $y'' + 6y' + 5y = x$, then $c_1 y_1 + c_2 y_2$ is also a solution of the equation.

3. The general solution of $y'' - y = 0$ can be written as
$$y = c_1 \cosh x + c_2 \sinh x$$

4. The equation $y'' - y = e^x$ has a particular solution of the form
$$y_p = A e^x$$

EXERCISES

1–10 Solve the differential equation.

1. $4y'' - y = 0$
2. $y'' - 2y' + 10y = 0$
3. $y'' + 3y = 0$
4. $y'' + 8y' + 16y = 0$
5. $\dfrac{d^2 y}{dx^2} - 4 \dfrac{dy}{dx} + 5y = e^{2x}$
6. $\dfrac{d^2 y}{dx^2} + \dfrac{dy}{dx} - 2y = x^2$
7. $\dfrac{d^2 y}{dx^2} - 2 \dfrac{dy}{dx} + y = x \cos x$
8. $\dfrac{d^2 y}{dx^2} + 4y = \sin 2x$
9. $\dfrac{d^2 y}{dx^2} - \dfrac{dy}{dx} - 6y = 1 + e^{-2x}$
10. $\dfrac{d^2 y}{dx^2} + y = \csc x$, $0 < x < \pi/2$

11–14 Solve the initial-value problem.

11. $y'' + 6y' = 0$, $y(1) = 3$, $y'(1) = 12$
12. $y'' - 6y' + 25y = 0$, $y(0) = 2$, $y'(0) = 1$
13. $y'' - 5y' + 4y = 0$, $y(0) = 0$, $y'(0) = 1$
14. $9y'' + y = 3x + e^{-x}$, $y(0) = 1$, $y'(0) = 2$

15–16 Solve the boundary-value problem, if possible.

15. $y'' + 4y' + 29y = 0$, $y(0) = 1$, $y(\pi) = -1$
16. $y'' + 4y' + 29y = 0$, $y(0) = 1$, $y(\pi) = -e^{-2\pi}$

17. Use power series to solve the initial-value problem
$$y'' + xy' + y = 0 \quad y(0) = 0 \quad y'(0) = 1$$

18. Use power series to solve the differential equation
$$y'' - xy' - 2y = 0$$

19. A series circuit contains a resistor with $R = 40\ \Omega$, an inductor with $L = 2$ H, a capacitor with $C = 0.0025$ F, and a 12-V battery. The initial charge is $Q = 0.01$ C and the initial current is 0. Find the charge at time t.

20. A spring with a mass of 2 kg has damping constant 16, and a force of 12.8 N keeps the spring stretched 0.2 m beyond its natural length. Find the position of the mass at time t if it starts at the equilibrium position with a velocity of 2.4 m/s.

21. Assume that the earth is a solid sphere of uniform density with mass M and radius $R = 3960$ mi. For a particle of mass m within the earth at a distance r from the earth's center, the gravitational force attracting the particle to the center is

$$F_r = \frac{-GM_r m}{r^2}$$

where G is the gravitational constant and M_r is the mass of the earth within the sphere of radius r.

(a) Show that $F_r = \dfrac{-GMm}{R^3} r$.

(b) Suppose a hole is drilled through the earth along a diameter. Show that if a particle of mass m is dropped from rest at the surface, into the hole, then the distance $y = y(t)$ of the particle from the center of the earth at time t is given by

$$y''(t) = -k^2 y(t)$$

where $k^2 = GM/R^3 = g/R$.

(c) Conclude from part (b) that the particle undergoes simple harmonic motion. Find the period T.

(d) With what speed does the particle pass through the center of the earth?

Appendixes

A Numbers, Inequalities, and Absolute Values

B Coordinate Geometry and Lines

C Graphs of Second-Degree Equations

D Trigonometry

E Sigma Notation

F Proofs of Theorems

G The Logarithm Defined as an Integral

H Complex Numbers

I Answers to Odd-Numbered Exercises

A Numbers, Inequalities, and Absolute Values

Calculus is based on the real number system. We start with the **integers**:

$$\ldots, \quad -3, \quad -2, \quad -1, \quad 0, \quad 1, \quad 2, \quad 3, \quad 4, \quad \ldots$$

Then we construct the **rational numbers**, which are ratios of integers. Thus any rational number r can be expressed as

$$r = \frac{m}{n} \quad \text{where } m \text{ and } n \text{ are integers and } n \neq 0$$

Examples are

$$\tfrac{1}{2} \qquad -\tfrac{3}{7} \qquad 46 = \tfrac{46}{1} \qquad 0.17 = \tfrac{17}{100}$$

(Recall that division by 0 is always ruled out, so expressions like $\tfrac{3}{0}$ and $\tfrac{0}{0}$ are undefined.) Some real numbers, such as $\sqrt{2}$, can't be expressed as a ratio of integers and are therefore called **irrational numbers**. It can be shown, with varying degrees of difficulty, that the following are also irrational numbers:

$$\sqrt{3} \qquad \sqrt{5} \qquad \sqrt[3]{2} \qquad \pi \qquad \sin 1° \qquad \log_{10} 2$$

The set of all real numbers is usually denoted by the symbol $\mathbb{R}$. When we use the word *number* without qualification, we mean "real number."

Every number has a decimal representation. If the number is rational, then the corresponding decimal is repeating. For example,

$$\tfrac{1}{2} = 0.5000\ldots = 0.5\overline{0} \qquad\qquad \tfrac{2}{3} = 0.66666\ldots = 0.\overline{6}$$

$$\tfrac{157}{495} = 0.317171717\ldots = 0.3\overline{17} \qquad\qquad \tfrac{9}{7} = 1.285714285714\ldots = 1.\overline{285714}$$

(The bar indicates that the sequence of digits repeats forever.) On the other hand, if the number is irrational, the decimal is nonrepeating:

$$\sqrt{2} = 1.414213562373095\ldots \qquad\qquad \pi = 3.141592653589793\ldots$$

If we stop the decimal expansion of any number at a certain place, we get an approximation to the number. For instance, we can write

$$\pi \approx 3.14159265$$

where the symbol $\approx$ is read "is approximately equal to." The more decimal places we retain, the better the approximation we get.

The real numbers can be represented by points on a line as in Figure 1. The positive direction (to the right) is indicated by an arrow. We choose an arbitrary reference point O, called the **origin**, which corresponds to the real number 0. Given any convenient unit of measurement, each positive number x is represented by the point on the line a distance of x units to the right of the origin, and each negative number $-x$ is represented by the point x units to the left of the origin. Thus every real number is represented by a point on the line, and every point P on the line corresponds to exactly one real number. The number associated with the point P is called the **coordinate** of P and the line is then called a coordinate line, or a **real number line**, or simply a **real line**. Often we identify the point with its coordinate and think of a number as being a point on the real line.

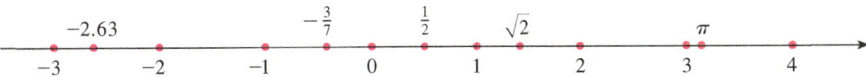

FIGURE 1

The real numbers are ordered. We say *a is less than b* and write $a < b$ if $b - a$ is a positive number. Geometrically this means that a lies to the left of b on the number line. (Equivalently, we say *b is greater than a* and write $b > a$.) The symbol $a \leq b$ (or $b \geq a$) means that either $a < b$ or $a = b$ and is read "*a* is less than or equal to *b*." For instance, the following are true inequalities:

$$7 < 7.4 < 7.5 \qquad -3 > -\pi \qquad \sqrt{2} < 2 \qquad \sqrt{2} \leq 2 \qquad 2 \leq 2$$

In what follows we need to use *set notation*. A **set** is a collection of objects, and these objects are called the **elements** of the set. If S is a set, the notation $a \in S$ means that a is an element of S, and $a \notin S$ means that a is not an element of S. For example, if Z represents the set of integers, then $-3 \in Z$ but $\pi \notin Z$. If S and T are sets, then their **union** $S \cup T$ is the set consisting of all elements that are in S or T (or in both S and T). The **intersection** of S and T is the set $S \cap T$ consisting of all elements that are in both S and T. In other words, $S \cap T$ is the common part of S and T. The empty set, denoted by $\varnothing$, is the set that contains no element.

Some sets can be described by listing their elements between braces. For instance, the set A consisting of all positive integers less than 7 can be written as

$$A = \{1, 2, 3, 4, 5, 6\}$$

We could also write A in *set-builder notation* as

$$A = \{x \mid x \text{ is an integer and } 0 < x < 7\}$$

which is read "A is the set of x such that x is an integer and $0 < x < 7$."

Intervals

Certain sets of real numbers, called **intervals**, occur frequently in calculus and correspond geometrically to line segments. For example, if $a < b$, the **open interval** from a to b consists of all numbers between a and b and is denoted by the symbol (a, b). Using set-builder notation, we can write

$$(a, b) = \{x \mid a < x < b\}$$

FIGURE 2
Open interval (a, b)

Notice that the endpoints of the interval—namely, a and b—are excluded. This is indicated by the round brackets ( ) and by the open dots in Figure 2. The **closed interval** from a to b is the set

$$[a, b] = \{x \mid a \leq x \leq b\}$$

FIGURE 3
Closed interval $[a, b]$

Here the endpoints of the interval are included. This is indicated by the square brackets [ ] and by the solid dots in Figure 3. It is also possible to include only one endpoint in an interval, as shown in Table 1.

We also need to consider infinite intervals such as

$$(a, \infty) = \{x \mid x > a\}$$

This does not mean that ∞ ("infinity") is a number. The notation (a, ∞) stands for the set of all numbers that are greater than a, so the symbol ∞ simply indicates that the interval extends indefinitely far in the positive direction.

1 Table of Intervals

Table 1 lists the nine possible types of intervals. When these intervals are discussed, it is always assumed that $a < b$.

Notation	Set description	Picture
(a, b)	$\{x \mid a < x < b\}$	
$[a, b]$	$\{x \mid a \leq x \leq b\}$	
$[a, b)$	$\{x \mid a \leq x < b\}$	
$(a, b]$	$\{x \mid a < x \leq b\}$	
(a, ∞)	$\{x \mid x > a\}$	
$[a, \infty)$	$\{x \mid x \geq a\}$	
$(-\infty, b)$	$\{x \mid x < b\}$	
$(-\infty, b]$	$\{x \mid x \leq b\}$	
$(-\infty, \infty)$	$\mathbb{R}$ (set of all real numbers)	

Inequalities

When working with inequalities, note the following rules.

2 Rules for Inequalities

1. If $a < b$, then $a + c < b + c$.
2. If $a < b$ and $c < d$, then $a + c < b + d$.
3. If $a < b$ and $c > 0$, then $ac < bc$.
4. If $a < b$ and $c < 0$, then $ac > bc$.
5. If $0 < a < b$, then $1/a > 1/b$.

Rule 1 says that we can add any number to both sides of an inequality, and Rule 2 says that two inequalities can be added. However, we have to be careful with multiplication. Rule 3 says that we can multiply both sides of an inequality by a *positive* number, but Rule 4 says that if we multiply both sides of an inequality by a negative number, then we reverse the direction of the inequality. For example, if we take the inequality $3 < 5$ and multiply by 2, we get $6 < 10$, but if we multiply by -2, we get $-6 > -10$. Finally, Rule 5 says that if we take reciprocals, then we reverse the direction of an inequality (provided the numbers are positive).

EXAMPLE 1 Solve the inequality $1 + x < 7x + 5$.

SOLUTION The given inequality is satisfied by some values of x but not by others. To *solve* an inequality means to determine the set of numbers x for which the inequality is true. This is called the *solution set*.

First we subtract 1 from each side of the inequality (using Rule 1 with $c = -1$):

$$x < 7x + 4$$

Then we subtract $7x$ from both sides (Rule 1 with $c = -7x$):

$$-6x < 4$$

Now we divide both sides by -6 (Rule 4 with $c = -\frac{1}{6}$):

$$x > -\tfrac{4}{6} = -\tfrac{2}{3}$$

These steps can all be reversed, so the solution set consists of all numbers greater than $-\frac{2}{3}$. In other words, the solution of the inequality is the interval $\left(-\frac{2}{3}, \infty\right)$. ∎

EXAMPLE 2 Solve the inequalities $4 \leq 3x - 2 < 13$.

SOLUTION Here the solution set consists of all values of x that satisfy both inequalities. Using the rules given in (2), we see that the following inequalities are equivalent:

$$4 \leq 3x - 2 < 13$$

$$6 \leq 3x < 15 \quad \text{(add 2)}$$

$$2 \leq x < 5 \quad \text{(divide by 3)}$$

Therefore the solution set is $[2, 5)$. ∎

EXAMPLE 3 Solve the inequality $x^2 - 5x + 6 \leq 0$.

SOLUTION First we factor the left side:

$$(x - 2)(x - 3) \leq 0$$

We know that the corresponding equation $(x - 2)(x - 3) = 0$ has the solutions 2 and 3. The numbers 2 and 3 divide the real line into three intervals:

$$(-\infty, 2) \quad (2, 3) \quad (3, \infty)$$

On each of these intervals we determine the signs of the factors. For instance,

$$x \in (-\infty, 2) \;\Rightarrow\; x < 2 \;\Rightarrow\; x - 2 < 0$$

Then we record these signs in the following chart:

Interval	$x - 2$	$x - 3$	$(x - 2)(x - 3)$
$x < 2$	−	−	+
$2 < x < 3$	+	−	−
$x > 3$	+	+	+

Another method for obtaining the information in the chart is to use *test values*. For instance, if we use the test value $x = 1$ for the interval $(-\infty, 2)$, then substitution in $x^2 - 5x + 6$ gives

$$1^2 - 5(1) + 6 = 2$$

A visual method for solving Example 3 is to use a graphing device to graph the parabola $y = x^2 - 5x + 6$ (as in Figure 4) and observe that the curve lies on or below the x-axis when $2 \leq x \leq 3$.

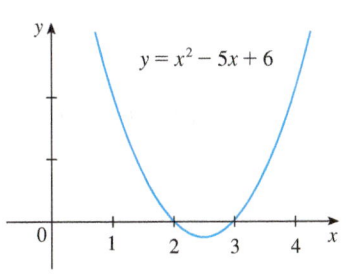

FIGURE 4

The polynomial $x^2 - 5x + 6$ doesn't change sign inside any of the three intervals, so we conclude that it is positive on $(-\infty, 2)$.

Then we read from the chart that $(x - 2)(x - 3)$ is negative when $2 < x < 3$. Thus the solution of the inequality $(x - 2)(x - 3) \leq 0$ is

$$\{x \mid 2 \leq x \leq 3\} = [2, 3]$$

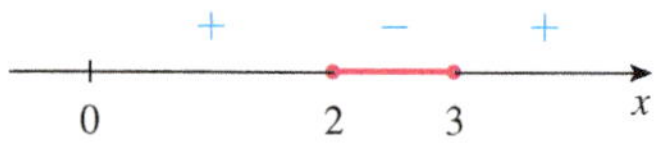

Notice that we have included the endpoints 2 and 3 because we are looking for values of x such that the product is either negative or zero. The solution is illustrated in Figure 5.

FIGURE 5

EXAMPLE 4 Solve $x^3 + 3x^2 > 4x$.

SOLUTION First we take all nonzero terms to one side of the inequality sign and factor the resulting expression:

$$x^3 + 3x^2 - 4x > 0 \quad \text{or} \quad x(x - 1)(x + 4) > 0$$

As in Example 3 we solve the corresponding equation $x(x - 1)(x + 4) = 0$ and use the solutions $x = -4$, $x = 0$, and $x = 1$ to divide the real line into four intervals $(-\infty, -4)$, $(-4, 0)$, $(0, 1)$, and $(1, \infty)$. On each interval the product keeps a constant sign, which we list in the following chart:

Interval	x	$x - 1$	$x + 4$	$x(x - 1)(x + 4)$
$x < -4$	−	−	−	−
$-4 < x < 0$	−	−	+	+
$0 < x < 1$	+	−	+	−
$x > 1$	+	+	+	+

Then we read from the chart that the solution set is

$$\{x \mid -4 < x < 0 \text{ or } x > 1\} = (-4, 0) \cup (1, \infty)$$

The solution is illustrated in Figure 6.

FIGURE 6

Absolute Value

The **absolute value** of a number a, denoted by $|a|$, is the distance from a to 0 on the real number line. Distances are always positive or 0, so we have

$$|a| \geq 0 \qquad \text{for every number } a$$

For example,

$$|3| = 3 \qquad |-3| = 3 \qquad |0| = 0 \qquad |\sqrt{2} - 1| = \sqrt{2} - 1 \qquad |3 - \pi| = \pi - 3$$

In general, we have

Remember that if a is negative, then $-a$ is positive.

3
$$|a| = a \qquad \text{if } a \geq 0$$
$$|a| = -a \qquad \text{if } a < 0$$

APPENDIX A   Numbers, Inequalities, and Absolute Values   A7

EXAMPLE 5 Express $|3x - 2|$ without using the absolute-value symbol.

SOLUTION

$$|3x - 2| = \begin{cases} 3x - 2 & \text{if } 3x - 2 \geq 0 \\ -(3x - 2) & \text{if } 3x - 2 < 0 \end{cases}$$

$$= \begin{cases} 3x - 2 & \text{if } x \geq \frac{2}{3} \\ 2 - 3x & \text{if } x < \frac{2}{3} \end{cases}$$

Recall that the symbol $\sqrt{\ }$ means "the positive square root of." Thus $\sqrt{r} = s$ means $s^2 = r$ and $s \geq 0$. Therefore the equation $\sqrt{a^2} = a$ is not always true. It is true only when $a \geq 0$. If $a < 0$, then $-a > 0$, so we have $\sqrt{a^2} = -a$. In view of (3), we then have the equation

$$\boxed{4} \quad \boxed{\sqrt{a^2} = |a|}$$

which is true for all values of a.

Hints for the proofs of the following properties are given in the exercises.

$\boxed{5}$ **Properties of Absolute Values** Suppose a and b are any real numbers and n is an integer. Then

1. $|ab| = |a||b|$ 2. $\left|\dfrac{a}{b}\right| = \dfrac{|a|}{|b|}$ $(b \neq 0)$ 3. $|a^n| = |a|^n$

For solving equations or inequalities involving absolute values, it's often very helpful to use the following statements.

$\boxed{6}$ Suppose $a > 0$. Then

4. $|x| = a$ if and only if $x = \pm a$
5. $|x| < a$ if and only if $-a < x < a$
6. $|x| > a$ if and only if $x > a$ or $x < -a$

For instance, the inequality $|x| < a$ says that the distance from x to the origin is less than a, and you can see from Figure 7 that this is true if and only if x lies between $-a$ and a.

If a and b are any real numbers, then the distance between a and b is the absolute value of the difference, namely, $|a - b|$, which is also equal to $|b - a|$. (See Figure 8.)

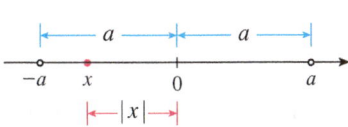

FIGURE 7

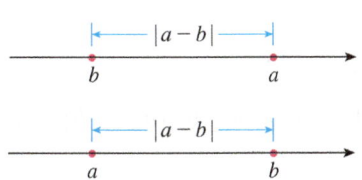

FIGURE 8
Length of a line segment $= |a - b|$

EXAMPLE 6 Solve $|2x - 5| = 3$.

SOLUTION By Property 4 of (6), $|2x - 5| = 3$ is equivalent to

$$2x - 5 = 3 \quad \text{or} \quad 2x - 5 = -3$$

So $2x = 8$ or $2x = 2$. Thus $x = 4$ or $x = 1$.

EXAMPLE 7 Solve $|x - 5| < 2$.

SOLUTION 1 By Property 5 of (6), $|x - 5| < 2$ is equivalent to

$$-2 < x - 5 < 2$$

Therefore, adding 5 to each side, we have

$$3 < x < 7$$

and the solution set is the open interval $(3, 7)$.

FIGURE 9

SOLUTION 2 Geometrically the solution set consists of all numbers x whose distance from 5 is less than 2. From Figure 9 we see that this is the interval $(3, 7)$.

EXAMPLE 8 Solve $|3x + 2| \geq 4$.

SOLUTION By Properties 4 and 6 of (6), $|3x + 2| \geq 4$ is equivalent to

$$3x + 2 \geq 4 \quad \text{or} \quad 3x + 2 \leq -4$$

In the first case $3x \geq 2$, which gives $x \geq \frac{2}{3}$. In the second case $3x \leq -6$, which gives $x \leq -2$. So the solution set is

$$\{x \mid x \leq -2 \text{ or } x \geq \tfrac{2}{3}\} = (-\infty, -2] \cup [\tfrac{2}{3}, \infty)$$

Another important property of absolute value, called the Triangle Inequality, is used frequently not only in calculus but throughout mathematics in general.

> **7 The Triangle Inequality** If a and b are any real numbers, then
>
> $$|a + b| \leq |a| + |b|$$

Observe that if the numbers a and b are both positive or both negative, then the two sides in the Triangle Inequality are actually equal. But if a and b have opposite signs, the left side involves a subtraction and the right side does not. This makes the Triangle Inequality seem reasonable, but we can prove it as follows.

Notice that

$$-|a| \leq a \leq |a|$$

is always true because a equals either $|a|$ or $-|a|$. The corresponding statement for b is

$$-|b| \leq b \leq |b|$$

Adding these inequalities, we get

$$-(|a| + |b|) \leq a + b \leq |a| + |b|$$

If we now apply Properties 4 and 5 (with x replaced by $a + b$ and a by $|a| + |b|$), we obtain

$$|a + b| \leq |a| + |b|$$

which is what we wanted to show.

EXAMPLE 9 If $|x - 4| < 0.1$ and $|y - 7| < 0.2$, use the Triangle Inequality to estimate $|(x + y) - 11|$.

SOLUTION In order to use the given information, we use the Triangle Inequality with $a = x - 4$ and $b = y - 7$:

$$|(x + y) - 11| = |(x - 4) + (y - 7)|$$
$$\leq |x - 4| + |y - 7|$$
$$< 0.1 + 0.2 = 0.3$$

Thus $$|(x + y) - 11| < 0.3$$ ■

A   EXERCISES

1–12 Rewrite the expression without using the absolute-value symbol.

1. $|5 - 23|$
2. $|5| - |-23|$
3. $|-\pi|$
4. $|\pi - 2|$
5. $|\sqrt{5} - 5|$
6. $||-2| - |-3||$
7. $|x - 2|$ if $x < 2$
8. $|x - 2|$ if $x > 2$
9. $|x + 1|$
10. $|2x - 1|$
11. $|x^2 + 1|$
12. $|1 - 2x^2|$

13–38 Solve the inequality in terms of intervals and illustrate the solution set on the real number line.

13. $2x + 7 > 3$
14. $3x - 11 < 4$
15. $1 - x \leq 2$
16. $4 - 3x \geq 6$
17. $2x + 1 < 5x - 8$
18. $1 + 5x > 5 - 3x$
19. $-1 < 2x - 5 < 7$
20. $1 < 3x + 4 \leq 16$
21. $0 \leq 1 - x < 1$
22. $-5 \leq 3 - 2x \leq 9$
23. $4x < 2x + 1 \leq 3x + 2$
24. $2x - 3 < x + 4 < 3x - 2$
25. $(x - 1)(x - 2) > 0$
26. $(2x + 3)(x - 1) \geq 0$
27. $2x^2 + x \leq 1$
28. $x^2 < 2x + 8$
29. $x^2 + x + 1 > 0$
30. $x^2 + x > 1$
31. $x^2 < 3$
32. $x^2 \geq 5$
33. $x^3 - x^2 \leq 0$
34. $(x + 1)(x - 2)(x + 3) \geq 0$
35. $x^3 > x$
36. $x^3 + 3x < 4x^2$
37. $\dfrac{1}{x} < 4$
38. $-3 < \dfrac{1}{x} \leq 1$

39. The relationship between the Celsius and Fahrenheit temperature scales is given by $C = \frac{5}{9}(F - 32)$, where C is the temperature in degrees Celsius and F is the temperature in degrees Fahrenheit. What interval on the Celsius scale corresponds to the temperature range $50 \leq F \leq 95$?

40. Use the relationship between C and F given in Exercise 39 to find the interval on the Fahrenheit scale corresponding to the temperature range $20 \leq C \leq 30$.

41. As dry air moves upward, it expands and in so doing cools at a rate of about 1°C for each 100-m rise, up to about 12 km.
    (a) If the ground temperature is 20°C, write a formula for the temperature at height h.
    (b) What range of temperature can be expected if a plane takes off and reaches a maximum height of 5 km?

42. If a ball is thrown upward from the top of a building 128 ft high with an initial velocity of 16 ft/s, then the height h above the ground t seconds later will be

$$h = 128 + 16t - 16t^2$$

During what time interval will the ball be at least 32 ft above the ground?

43–46 Solve the equation for x.

43. $|2x| = 3$
44. $|3x + 5| = 1$
45. $|x + 3| = |2x + 1|$
46. $\left|\dfrac{2x - 1}{x + 1}\right| = 3$

47–56 Solve the inequality.

47. $|x| < 3$
48. $|x| \geq 3$
49. $|x - 4| < 1$
50. $|x - 6| < 0.1$
51. $|x + 5| \geq 2$
52. $|x + 1| \geq 3$
53. $|2x - 3| \leq 0.4$
54. $|5x - 2| < 6$
55. $1 \leq |x| \leq 4$
56. $0 < |x - 5| < \frac{1}{2}$

57–58 Solve for x, assuming a, b, and c are positive constants.

57. $a(bx - c) \geq bc$

58. $a \leq bx + c < 2a$

59–60 Solve for x, assuming a, b, and c are negative constants.

59. $ax + b < c$

60. $\dfrac{ax + b}{c} \leq b$

61. Suppose that $|x - 2| < 0.01$ and $|y - 3| < 0.04$. Use the Triangle Inequality to show that $|(x + y) - 5| < 0.05$.

62. Show that if $|x + 3| < \frac{1}{2}$, then $|4x + 13| < 3$.

63. Show that if $a < b$, then $a < \dfrac{a + b}{2} < b$.

64. Use Rule 3 to prove Rule 5 of (2).

65. Prove that $|ab| = |a||b|$. [*Hint:* Use Equation 4.]

66. Prove that $\left|\dfrac{a}{b}\right| = \dfrac{|a|}{|b|}$.

67. Show that if $0 < a < b$, then $a^2 < b^2$.

68. Prove that $|x - y| \geq |x| - |y|$. [*Hint:* Use the Triangle Inequality with $a = x - y$ and $b = y$.]

69. Show that the sum, difference, and product of rational numbers are rational numbers.

70. (a) Is the sum of two irrational numbers always an irrational number?
(b) Is the product of two irrational numbers always an irrational number?

B Coordinate Geometry and Lines

Just as the points on a line can be identified with real numbers by assigning them coordinates, as described in Appendix A, so the points in a plane can be identified with ordered pairs of real numbers. We start by drawing two perpendicular coordinate lines that intersect at the origin O on each line. Usually one line is horizontal with positive direction to the right and is called the *x*-axis; the other line is vertical with positive direction upward and is called the *y*-axis.

Any point P in the plane can be located by a unique ordered pair of numbers as follows. Draw lines through P perpendicular to the x- and y-axes. These lines intersect the axes in points with coordinates a and b as shown in Figure 1. Then the point P is assigned the ordered pair (a, b). The first number a is called the **x-coordinate** of P; the second number b is called the **y-coordinate** of P. We say that P is the point with coordinates (a, b), and we denote the point by the symbol $P(a, b)$. Several points are labeled with their coordinates in Figure 2.

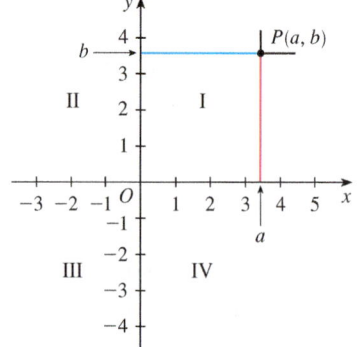

FIGURE 1   **FIGURE 2**

By reversing the preceding process we can start with an ordered pair (a, b) and arrive at the corresponding point P. Often we identify the point P with the ordered pair (a, b) and refer to "the point (a, b)." [Although the notation used for an open interval (a, b) is the same as the notation used for a point (a, b), you will be able to tell from the context which meaning is intended.]

This coordinate system is called the **rectangular coordinate system** or the **Cartesian coordinate system** in honor of the French mathematician René Descartes (1596–1650), even though another Frenchman, Pierre Fermat (1601–1665), invented the principles of analytic geometry at about the same time as Descartes. The plane supplied with this coordinate system is called the **coordinate plane** or the **Cartesian plane** and is denoted by $\mathbb{R}^2$.

The x- and y-axes are called the **coordinate axes** and divide the Cartesian plane into four quadrants, which are labeled I, II, III, and IV in Figure 1. Notice that the first quadrant consists of those points whose x- and y-coordinates are both positive.

EXAMPLE 1 Describe and sketch the regions given by the following sets.
(a) $\{(x, y) \mid x \geq 0\}$ (b) $\{(x, y) \mid y = 1\}$ (c) $\{(x, y) \mid |y| < 1\}$

SOLUTION
(a) The points whose x-coordinates are 0 or positive lie on the y-axis or to the right of it as indicated by the shaded region in Figure 3(a).

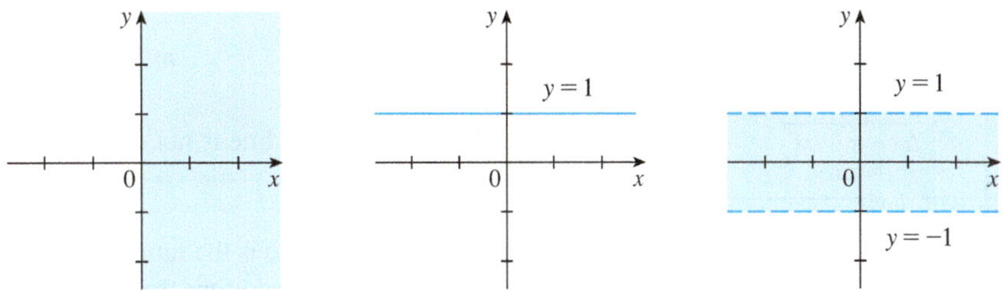

FIGURE 3   (a) $x \geq 0$ (b) $y = 1$ (c) $|y| < 1$

(b) The set of all points with y-coordinate 1 is a horizontal line one unit above the x-axis [see Figure 3(b)].

(c) Recall from Appendix A that

$$|y| < 1 \quad \text{if and only if} \quad -1 < y < 1$$

The given region consists of those points in the plane whose y-coordinates lie between -1 and 1. Thus the region consists of all points that lie between (but not on) the horizontal lines $y = 1$ and $y = -1$. [These lines are shown as dashed lines in Figure 3(c) to indicate that the points on these lines don't lie in the set.] ∎

Recall from Appendix A that the distance between points a and b on a number line is $|a - b| = |b - a|$. Thus the distance between points $P_1(x_1, y_1)$ and $P_3(x_2, y_1)$ on a horizontal line must be $|x_2 - x_1|$ and the distance between $P_2(x_2, y_2)$ and $P_3(x_2, y_1)$ on a vertical line must be $|y_2 - y_1|$. (See Figure 4.)

To find the distance $|P_1P_2|$ between any two points $P_1(x_1, y_1)$ and $P_2(x_2, y_2)$, we note that triangle $P_1P_2P_3$ in Figure 4 is a right triangle, and so by the Pythagorean Theorem we have

$$|P_1P_2| = \sqrt{|P_1P_3|^2 + |P_2P_3|^2} = \sqrt{|x_2 - x_1|^2 + |y_2 - y_1|^2}$$
$$= \sqrt{(x_2 - x_1)^2 + (y_2 - y_1)^2}$$

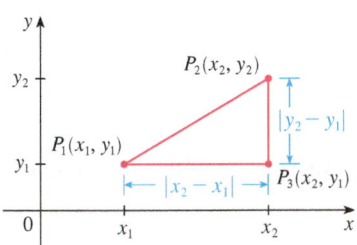

FIGURE 4

1 Distance Formula The distance between the points $P_1(x_1, y_1)$ and $P_2(x_2, y_2)$ is
$$|P_1P_2| = \sqrt{(x_2 - x_1)^2 + (y_2 - y_1)^2}$$

EXAMPLE 2 The distance between $(1, -2)$ and $(5, 3)$ is
$$\sqrt{(5-1)^2 + [3-(-2)]^2} = \sqrt{4^2 + 5^2} = \sqrt{41}$$

Lines

We want to find an equation of a given line L; such an equation is satisfied by the coordinates of the points on L and by no other point. To find the equation of L we use its *slope*, which is a measure of the steepness of the line.

2 Definition The **slope** of a nonvertical line that passes through the points $P_1(x_1, y_1)$ and $P_2(x_2, y_2)$ is
$$m = \frac{\Delta y}{\Delta x} = \frac{y_2 - y_1}{x_2 - x_1}$$

The slope of a vertical line is not defined.

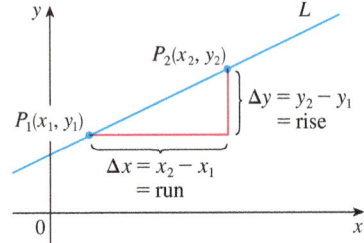

FIGURE 5

Thus the slope of a line is the ratio of the change in y, Δy, to the change in x, Δx. (See Figure 5.) The slope is therefore the rate of change of y with respect to x. The fact that the line is straight means that the rate of change is constant.

Figure 6 shows several lines labeled with their slopes. Notice that lines with positive slope slant upward to the right, whereas lines with negative slope slant downward to the right. Notice also that the steepest lines are the ones for which the absolute value of the slope is largest, and a horizontal line has slope 0.

Now let's find an equation of the line that passes through a given point $P_1(x_1, y_1)$ and has slope m. A point $P(x, y)$ with $x \neq x_1$ lies on this line if and only if the slope of the line through P_1 and P is equal to m; that is,
$$\frac{y - y_1}{x - x_1} = m$$

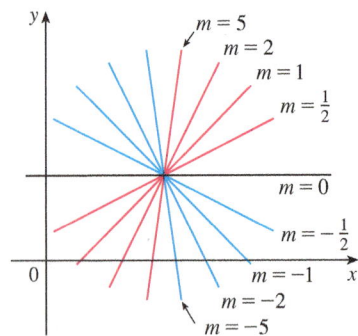

FIGURE 6

This equation can be rewritten in the form
$$y - y_1 = m(x - x_1)$$

and we observe that this equation is also satisfied when $x = x_1$ and $y = y_1$. Therefore it is an equation of the given line.

3 Point-Slope Form of the Equation of a Line An equation of the line passing through the point $P_1(x_1, y_1)$ and having slope m is
$$y - y_1 = m(x - x_1)$$

EXAMPLE 3 Find an equation of the line through $(1, -7)$ with slope $-\frac{1}{2}$.

SOLUTION Using (3) with $m = -\frac{1}{2}$, $x_1 = 1$, and $y_1 = -7$, we obtain an equation of the line as
$$y + 7 = -\tfrac{1}{2}(x - 1)$$
which we can rewrite as
$$2y + 14 = -x + 1 \quad \text{or} \quad x + 2y + 13 = 0 \qquad \blacksquare$$

EXAMPLE 4 Find an equation of the line through the points $(-1, 2)$ and $(3, -4)$.

SOLUTION By Definition 2 the slope of the line is
$$m = \frac{-4 - 2}{3 - (-1)} = -\frac{3}{2}$$

Using the point-slope form with $x_1 = -1$ and $y_1 = 2$, we obtain
$$y - 2 = -\tfrac{3}{2}(x + 1)$$
which simplifies to
$$3x + 2y = 1 \qquad \blacksquare$$

Suppose a nonvertical line has slope m and y-intercept b. (See Figure 7.) This means it intersects the y-axis at the point $(0, b)$, so the point-slope form of the equation of the line, with $x_1 = 0$ and $y_1 = b$, becomes
$$y - b = m(x - 0)$$
This simplifies as follows.

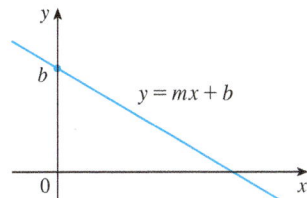

FIGURE 7

4 **Slope-Intercept Form of the Equation of a Line** An equation of the line with slope m and y-intercept b is
$$y = mx + b$$

In particular, if a line is horizontal, its slope is $m = 0$, so its equation is $y = b$, where b is the y-intercept (see Figure 8). A vertical line does not have a slope, but we can write its equation as $x = a$, where a is the x-intercept, because the x-coordinate of every point on the line is a.

Observe that the equation of every line can be written in the form

5
$$Ax + By + C = 0$$

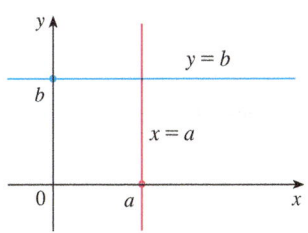

FIGURE 8

because a vertical line has the equation $x = a$ or $x - a = 0$ ($A = 1$, $B = 0$, $C = -a$) and a nonvertical line has the equation $y = mx + b$ or $-mx + y - b = 0$ ($A = -m$, $B = 1$, $C = -b$). Conversely, if we start with a general first-degree equation, that is, an equation of the form (5), where A, B, and C are constants and A and B are not both 0, then we can show that it is the equation of a line. If $B = 0$, the equation becomes $Ax + C = 0$ or $x = -C/A$, which represents a vertical line with x-intercept $-C/A$. If $B \neq 0$, the equation can be rewritten by solving for y:

$$y = -\frac{A}{B}x - \frac{C}{B}$$

and we recognize this as being the slope-intercept form of the equation of a line ($m = -A/B$, $b = -C/B$). Therefore an equation of the form (5) is called a **linear equation** or the **general equation of a line**. For brevity, we often refer to "the line $Ax + By + C = 0$" instead of "the line whose equation is $Ax + By + C = 0$."

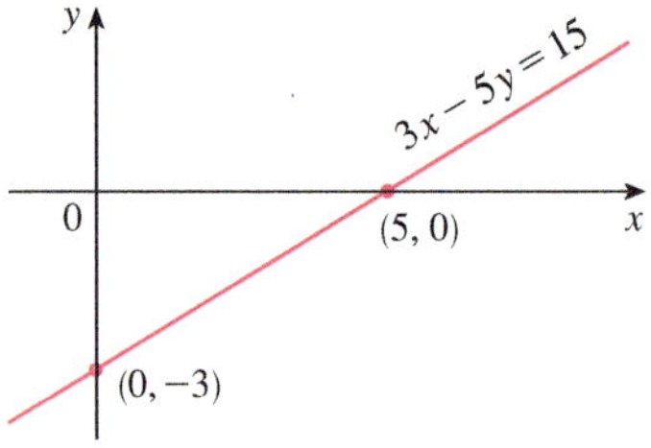

EXAMPLE 5 Sketch the graph of the equation $3x - 5y = 15$.

SOLUTION Since the equation is linear, its graph is a line. To draw the graph, we can simply find two points on the line. It's easiest to find the intercepts. Substituting $y = 0$ (the equation of the x-axis) in the given equation, we get $3x = 15$, so $x = 5$ is the x-intercept. Substituting $x = 0$ in the equation, we see that the y-intercept is -3. This allows us to sketch the graph as in Figure 9. ∎

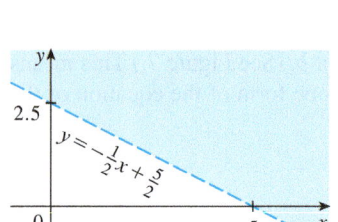

FIGURE 9

EXAMPLE 6 Graph the inequality $x + 2y > 5$.

SOLUTION We are asked to sketch the graph of the set $\{(x, y) \mid x + 2y > 5\}$ and we begin by solving the inequality for y:

$$x + 2y > 5$$
$$2y > -x + 5$$
$$y > -\tfrac{1}{2}x + \tfrac{5}{2}$$

Compare this inequality with the equation $y = -\tfrac{1}{2}x + \tfrac{5}{2}$, which represents a line with slope $-\tfrac{1}{2}$ and y-intercept $\tfrac{5}{2}$. We see that the given graph consists of points whose y-coordinates are *larger* than those on the line $y = -\tfrac{1}{2}x + \tfrac{5}{2}$. Thus the graph is the region that lies *above* the line, as illustrated in Figure 10. ∎

FIGURE 10

Parallel and Perpendicular Lines

Slopes can be used to show that lines are parallel or perpendicular. The following facts are proved, for instance, in *Precalculus: Mathematics for Calculus,* Seventh Edition, by Stewart, Redlin, and Watson (Belmont, CA, 2016).

> **6 Parallel and Perpendicular Lines**
> 1. Two nonvertical lines are parallel if and only if they have the same slope.
> 2. Two lines with slopes m_1 and m_2 are perpendicular if and only if $m_1 m_2 = -1$; that is, their slopes are negative reciprocals:
>
> $$m_2 = -\frac{1}{m_1}$$

EXAMPLE 7 Find an equation of the line through the point $(5, 2)$ that is parallel to the line $4x + 6y + 5 = 0$.

SOLUTION The given line can be written in the form

$$y = -\tfrac{2}{3}x - \tfrac{5}{6}$$

which is in slope-intercept form with $m = -\frac{2}{3}$. Parallel lines have the same slope, so the required line has slope $-\frac{2}{3}$ and its equation in point-slope form is

$$y - 2 = -\tfrac{2}{3}(x - 5)$$

We can write this equation as $2x + 3y = 16$. ∎

EXAMPLE 8 Show that the lines $2x + 3y = 1$ and $6x - 4y - 1 = 0$ are perpendicular.

SOLUTION The equations can be written as

$$y = -\tfrac{2}{3}x + \tfrac{1}{3} \quad \text{and} \quad y = \tfrac{3}{2}x - \tfrac{1}{4}$$

from which we see that the slopes are

$$m_1 = -\tfrac{2}{3} \quad \text{and} \quad m_2 = \tfrac{3}{2}$$

Since $m_1 m_2 = -1$, the lines are perpendicular. ∎

B EXERCISES

1–6 Find the distance between the points.

1. $(1, 1)$, $(4, 5)$
2. $(1, -3)$, $(5, 7)$
3. $(6, -2)$, $(-1, 3)$
4. $(1, -6)$, $(-1, -3)$
5. $(2, 5)$, $(4, -7)$
6. (a, b), (b, a)

7–10 Find the slope of the line through P and Q.

7. $P(1, 5)$, $Q(4, 11)$
8. $P(-1, 6)$, $Q(4, -3)$
9. $P(-3, 3)$, $Q(-1, -6)$
10. $P(-1, -4)$, $Q(6, 0)$

11. Show that the triangle with vertices $A(0, 2)$, $B(-3, -1)$, and $C(-4, 3)$ is isosceles.

12. (a) Show that the triangle with vertices $A(6, -7)$, $B(11, -3)$, and $C(2, -2)$ is a right triangle using the converse of the Pythagorean Theorem.
    (b) Use slopes to show that ABC is a right triangle.
    (c) Find the area of the triangle.

13. Show that the points $(-2, 9)$, $(4, 6)$, $(1, 0)$, and $(-5, 3)$ are the vertices of a square.

14. (a) Show that the points $A(-1, 3)$, $B(3, 11)$, and $C(5, 15)$ are collinear (lie on the same line) by showing that $|AB| + |BC| = |AC|$.
    (b) Use slopes to show that A, B, and C are collinear.

15. Show that $A(1, 1)$, $B(7, 4)$, $C(5, 10)$, and $D(-1, 7)$ are vertices of a parallelogram.

16. Show that $A(1, 1)$, $B(11, 3)$, $C(10, 8)$, and $D(0, 6)$ are vertices of a rectangle.

17–20 Sketch the graph of the equation.

17. $x = 3$
18. $y = -2$
19. $xy = 0$
20. $|y| = 1$

21–36 Find an equation of the line that satisfies the given conditions.

21. Through $(2, -3)$, slope 6
22. Through $(-1, 4)$, slope -3
23. Through $(1, 7)$, slope $\tfrac{2}{3}$
24. Through $(-3, -5)$, slope $-\tfrac{7}{2}$
25. Through $(2, 1)$ and $(1, 6)$
26. Through $(-1, -2)$ and $(4, 3)$
27. Slope 3, y-intercept -2
28. Slope $\tfrac{2}{5}$, y-intercept 4
29. x-intercept 1, y-intercept -3
30. x-intercept -8, y-intercept 6
31. Through $(4, 5)$, parallel to the x-axis
32. Through $(4, 5)$, parallel to the y-axis
33. Through $(1, -6)$, parallel to the line $x + 2y = 6$
34. y-intercept 6, parallel to the line $2x + 3y + 4 = 0$
35. Through $(-1, -2)$, perpendicular to the line $2x + 5y + 8 = 0$
36. Through $(\tfrac{1}{2}, -\tfrac{2}{3})$, perpendicular to the line $4x - 8y = 1$

37–42 Find the slope and y-intercept of the line and draw its graph.

37. $x + 3y = 0$
38. $2x - 5y = 0$

39. $y = -2$
40. $2x - 3y + 6 = 0$
41. $3x - 4y = 12$
42. $4x + 5y = 10$

43–52 Sketch the region in the xy-plane.

43. $\{(x, y) \mid x < 0\}$
44. $\{(x, y) \mid y > 0\}$
45. $\{(x, y) \mid xy < 0\}$
46. $\{(x, y) \mid x \geq 1 \text{ and } y < 3\}$
47. $\{(x, y) \mid |x| \leq 2\}$
48. $\{(x, y) \mid |x| < 3 \text{ and } |y| < 2\}$
49. $\{(x, y) \mid 0 \leq y \leq 4 \text{ and } x \leq 2\}$
50. $\{(x, y) \mid y > 2x - 1\}$
51. $\{(x, y) \mid 1 + x \leq y \leq 1 - 2x\}$
52. $\{(x, y) \mid -x \leq y < \frac{1}{2}(x + 3)\}$

53. Find a point on the y-axis that is equidistant from $(5, -5)$ and $(1, 1)$.

54. Show that the midpoint of the line segment from $P_1(x_1, y_1)$ to $P_2(x_2, y_2)$ is
$$\left(\frac{x_1 + x_2}{2}, \frac{y_1 + y_2}{2} \right)$$

55. Find the midpoint of the line segment joining the given points.
 (a) $(1, 3)$ and $(7, 15)$
 (b) $(-1, 6)$ and $(8, -12)$

56. Find the lengths of the medians of the triangle with vertices $A(1, 0)$, $B(3, 6)$, and $C(8, 2)$. (A median is a line segment from a vertex to the midpoint of the opposite side.)

57. Show that the lines $2x - y = 4$ and $6x - 2y = 10$ are not parallel and find their point of intersection.

58. Show that the lines $3x - 5y + 19 = 0$ and $10x + 6y - 50 = 0$ are perpendicular and find their point of intersection.

59. Find an equation of the perpendicular bisector of the line segment joining the points $A(1, 4)$ and $B(7, -2)$.

60. (a) Find equations for the sides of the triangle with vertices $P(1, 0)$, $Q(3, 4)$, and $R(-1, 6)$.
 (b) Find equations for the medians of this triangle. Where do they intersect?

61. (a) Show that if the x- and y-intercepts of a line are nonzero numbers a and b, then the equation of the line can be put in the form
$$\frac{x}{a} + \frac{y}{b} = 1$$
This equation is called the **two-intercept form** of an equation of a line.
 (b) Use part (a) to find an equation of the line whose x-intercept is 6 and whose y-intercept is -8.

62. A car leaves Detroit at 2:00 PM, traveling at a constant speed west along I-96. It passes Ann Arbor, 40 mi from Detroit, at 2:50 PM.
 (a) Express the distance traveled in terms of the time elapsed.
 (b) Draw the graph of the equation in part (a).
 (c) What is the slope of this line? What does it represent?

C Graphs of Second-Degree Equations

In Appendix B we saw that a first-degree, or linear, equation $Ax + By + C = 0$ represents a line. In this section we discuss second-degree equations such as

$$x^2 + y^2 = 1 \qquad y = x^2 + 1 \qquad \frac{x^2}{9} + \frac{y^2}{4} = 1 \qquad x^2 - y^2 = 1$$

which represent a circle, a parabola, an ellipse, and a hyperbola, respectively.

The graph of such an equation in x and y is the set of all points (x, y) that satisfy the equation; it gives a visual representation of the equation. Conversely, given a curve in the xy-plane, we may have to find an equation that represents it, that is, an equation satisfied by the coordinates of the points on the curve and by no other point. This is the other half of the basic principle of analytic geometry as formulated by Descartes and Fermat. The idea is that if a geometric curve can be represented by an algebraic equation, then the rules of algebra can be used to analyze the geometric problem.

■ Circles

As an example of this type of problem, let's find an equation of the circle with radius r and center (h, k). By definition, the circle is the set of all points $P(x, y)$ whose distance

APPENDIX C Graphs of Second-Degree Equations   A17

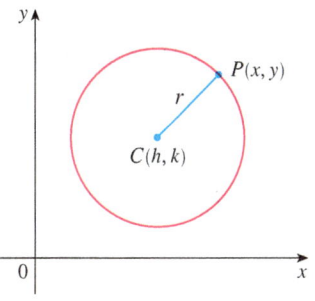

FIGURE 1

from the center $C(h, k)$ is r. (See Figure 1.) Thus P is on the circle if and only if $|PC| = r$. From the distance formula, we have

$$\sqrt{(x - h)^2 + (y - k)^2} = r$$

or equivalently, squaring both sides, we get

$$(x - h)^2 + (y - k)^2 = r^2$$

This is the desired equation.

> **1 Equation of a Circle** An equation of the circle with center (h, k) and radius r is
>
> $$(x - h)^2 + (y - k)^2 = r^2$$
>
> In particular, if the center is the origin $(0, 0)$, the equation is
>
> $$x^2 + y^2 = r^2$$

EXAMPLE 1 Find an equation of the circle with radius 3 and center $(2, -5)$.

SOLUTION From Equation 1 with $r = 3$, $h = 2$, and $k = -5$, we obtain

$$(x - 2)^2 + (y + 5)^2 = 9$$

EXAMPLE 2 Sketch the graph of the equation $x^2 + y^2 + 2x - 6y + 7 = 0$ by first showing that it represents a circle and then finding its center and radius.

SOLUTION We first group the x-terms and y-terms as follows:

$$(x^2 + 2x) + (y^2 - 6y) = -7$$

Then we complete the square within each grouping, adding the appropriate constants (the squares of half the coefficients of x and y) to both sides of the equation:

$$(x^2 + 2x + 1) + (y^2 - 6y + 9) = -7 + 1 + 9$$

or

$$(x + 1)^2 + (y - 3)^2 = 3$$

Comparing this equation with the standard equation of a circle (1), we see that $h = -1$, $k = 3$, and $r = \sqrt{3}$, so the given equation represents a circle with center $(-1, 3)$ and radius $\sqrt{3}$. It is sketched in Figure 2.

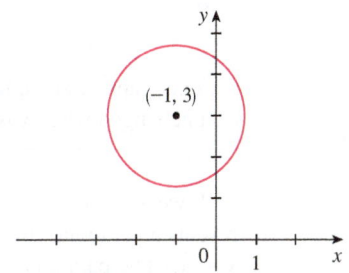

FIGURE 2
$x^2 + y^2 + 2x - 6y + 7 = 0$

Parabolas

The geometric properties of parabolas are reviewed in Section 10.5. Here we regard a parabola as a graph of an equation of the form $y = ax^2 + bx + c$.

EXAMPLE 3 Draw the graph of the parabola $y = x^2$.

SOLUTION We set up a table of values, plot points, and join them by a smooth curve to obtain the graph in Figure 3.

x	$y = x^2$
0	0
$\pm \frac{1}{2}$	$\frac{1}{4}$
± 1	1
± 2	4
± 3	9

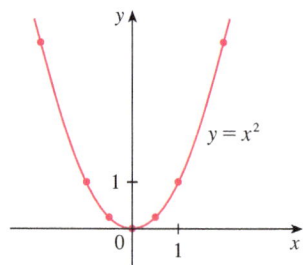

FIGURE 3

Figure 4 shows the graphs of several parabolas with equations of the form $y = ax^2$ for various values of the number a. In each case the *vertex*, the point where the parabola changes direction, is the origin. We see that the parabola $y = ax^2$ opens upward if $a > 0$ and downward if $a < 0$ (as in Figure 5).

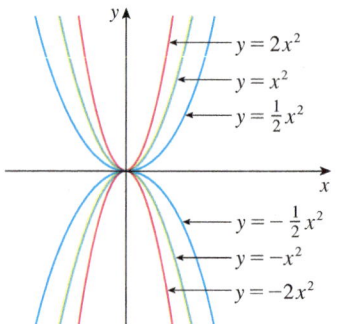

FIGURE 4

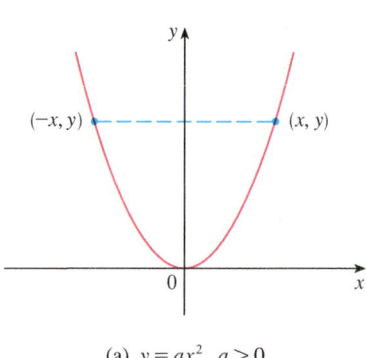

(a) $y = ax^2$, $a > 0$

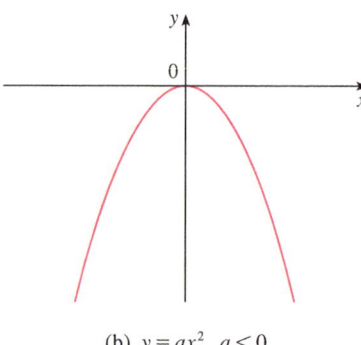

(b) $y = ax^2$, $a < 0$

FIGURE 5

Notice that if (x, y) satisfies $y = ax^2$, then so does $(-x, y)$. This corresponds to the geometric fact that if the right half of the graph is reflected about the y-axis, then the left half of the graph is obtained. We say that the graph is **symmetric with respect to the y-axis**.

> The graph of an equation is symmetric with respect to the y-axis if the equation is unchanged when x is replaced by $-x$.

If we interchange x and y in the equation $y = ax^2$, the result is $x = ay^2$, which also represents a parabola. (Interchanging x and y amounts to reflecting about the diagonal line $y = x$.) The parabola $x = ay^2$ opens to the right if $a > 0$ and to the left if $a < 0$. (See

Figure 6.) This time the parabola is symmetric with respect to the x-axis because if (x, y) satisfies $x = ay^2$, then so does $(x, -y)$.

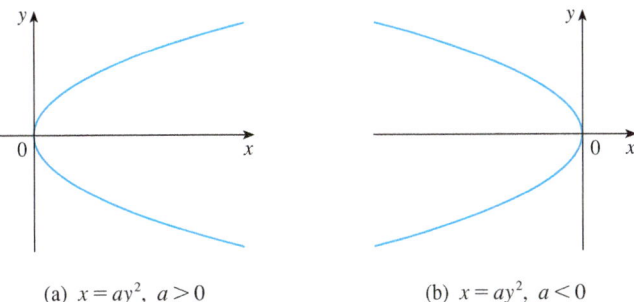

FIGURE 6   (a) $x = ay^2$, $a > 0$ (b) $x = ay^2$, $a < 0$

> The graph of an equation is symmetric with respect to the x-axis if the equation is unchanged when y is replaced by $-y$.

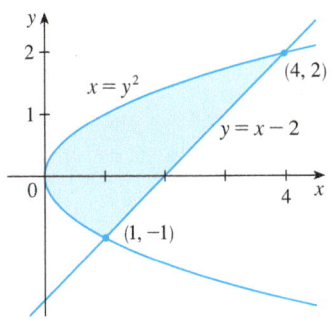

FIGURE 7

EXAMPLE 4 Sketch the region bounded by the parabola $x = y^2$ and the line $y = x - 2$.

SOLUTION First we find the points of intersection by solving the two equations. Substituting $x = y + 2$ into the equation $x = y^2$, we get $y + 2 = y^2$, which gives

$$0 = y^2 - y - 2 = (y - 2)(y + 1)$$

so $y = 2$ or -1. Thus the points of intersection are $(4, 2)$ and $(1, -1)$, and we draw the line $y = x - 2$ passing through these points. We then sketch the parabola $x = y^2$ by referring to Figure 6(a) and having the parabola pass through $(4, 2)$ and $(1, -1)$. The region bounded by $x = y^2$ and $y = x - 2$ means the finite region whose boundaries are these curves. It is sketched in Figure 7. ■

Ellipses

The curve with equation

$$\boxed{2} \qquad \frac{x^2}{a^2} + \frac{y^2}{b^2} = 1$$

where a and b are positive numbers, is called an **ellipse** in standard position. (Geometric properties of ellipses are discussed in Section 10.5.) Observe that Equation 2 is unchanged if x is replaced by $-x$ or y is replaced by $-y$, so the ellipse is symmetric with respect to both axes. As a further aid to sketching the ellipse, we find its intercepts.

> The **x-intercepts** of a graph are the x-coordinates of the points where the graph intersects the x-axis. They are found by setting $y = 0$ in the equation of the graph.
> The **y-intercepts** are the y-coordinates of the points where the graph intersects the y-axis. They are found by setting $x = 0$ in its equation.

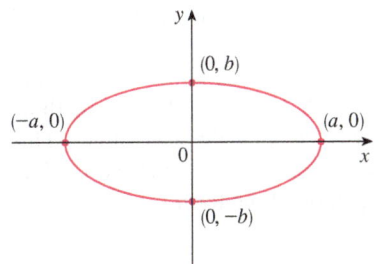

FIGURE 8
$\dfrac{x^2}{a^2} + \dfrac{y^2}{b^2} = 1$

If we set $y = 0$ in Equation 2, we get $x^2 = a^2$ and so the x-intercepts are $\pm a$. Setting $x = 0$, we get $y^2 = b^2$, so the y-intercepts are $\pm b$. Using this information, together with symmetry, we sketch the ellipse in Figure 8. If $a = b$, the ellipse is a circle with radius a.

EXAMPLE 5 Sketch the graph of $9x^2 + 16y^2 = 144$.

SOLUTION We divide both sides of the equation by 144:

$$\frac{x^2}{16} + \frac{y^2}{9} = 1$$

The equation is now in the standard form for an ellipse (2), so we have $a^2 = 16$, $b^2 = 9$, $a = 4$, and $b = 3$. The x-intercepts are ± 4; the y-intercepts are ± 3. The graph is sketched in Figure 9.

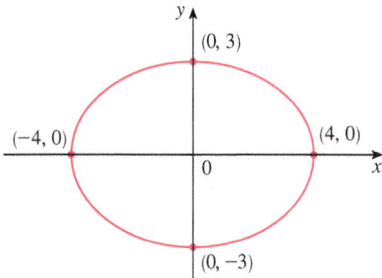

FIGURE 9
$9x^2 + 16y^2 = 144$

Hyperbolas

The curve with equation

$$\frac{x^2}{a^2} - \frac{y^2}{b^2} = 1$$

is called a **hyperbola** in standard position. Again, Equation 3 is unchanged when x is replaced by $-x$ or y is replaced by $-y$, so the hyperbola is symmetric with respect to both axes. To find the x-intercepts we set $y = 0$ and obtain $x^2 = a^2$ and $x = \pm a$. However, if we put $x = 0$ in Equation 3, we get $y^2 = -b^2$, which is impossible, so there is no y-intercept. In fact, from Equation 3 we obtain

$$\frac{x^2}{a^2} = 1 + \frac{y^2}{b^2} \geq 1$$

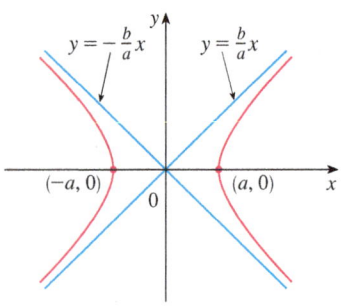

FIGURE 10
The hyperbola $\dfrac{x^2}{a^2} - \dfrac{y^2}{b^2} = 1$

which shows that $x^2 \geq a^2$ and so $|x| = \sqrt{x^2} \geq a$. Therefore we have $x \geq a$ or $x \leq -a$. This means that the hyperbola consists of two parts, called its *branches*. It is sketched in Figure 10.

In drawing a hyperbola it is useful to draw first its *asymptotes*, which are the lines $y = (b/a)x$ and $y = -(b/a)x$ shown in Figure 10. Both branches of the hyperbola approach the asymptotes; that is, they come arbitrarily close to the asymptotes. This involves the idea of a limit, which is discussed in Chapter 2. (See also Exercise 4.5.73.)

By interchanging the roles of x and y we get an equation of the form

$$\frac{y^2}{a^2} - \frac{x^2}{b^2} = 1$$

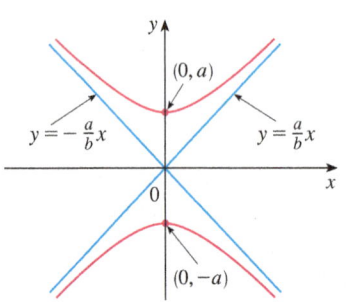

FIGURE 11
The hyperbola $\dfrac{y^2}{a^2} - \dfrac{x^2}{b^2} = 1$

which also represents a hyperbola and is sketched in Figure 11.

EXAMPLE 6 Sketch the curve $9x^2 - 4y^2 = 36$.

SOLUTION Dividing both sides by 36, we obtain

$$\frac{x^2}{4} - \frac{y^2}{9} = 1$$

which is the standard form of the equation of a hyperbola (Equation 3). Since $a^2 = 4$, the x-intercepts are ± 2. Since $b^2 = 9$, we have $b = 3$ and the asymptotes are $y = \pm\frac{3}{2}x$. The hyperbola is sketched in Figure 12.

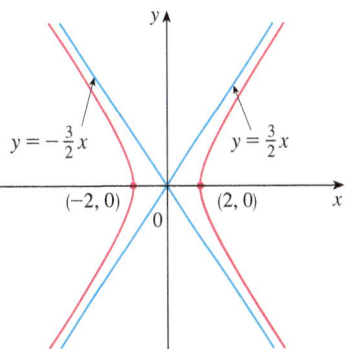

FIGURE 12
The hyperbola $9x^2 - 4y^2 = 36$

If $b = a$, a hyperbola has the equation $x^2 - y^2 = a^2$ (or $y^2 - x^2 = a^2$) and is called an *equilateral hyperbola* [see Figure 13(a)]. Its asymptotes are $y = \pm x$, which are perpendicular. If an equilateral hyperbola is rotated by 45°, the asymptotes become the x- and y-axes, and it can be shown that the new equation of the hyperbola is $xy = k$, where k is a constant [see Figure 13(b)].

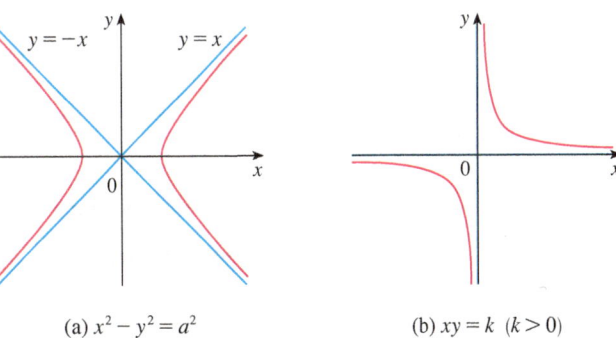

FIGURE 13
Equilateral hyperbolas

(a) $x^2 - y^2 = a^2$ (b) $xy = k$ $(k > 0)$

Shifted Conics

Recall that an equation of the circle with center the origin and radius r is $x^2 + y^2 = r^2$, but if the center is the point (h, k), then the equation of the circle becomes

$$(x - h)^2 + (y - k)^2 = r^2$$

Similarly, if we take the ellipse with equation

$$\boxed{4} \qquad \frac{x^2}{a^2} + \frac{y^2}{b^2} = 1$$

and translate it (shift it) so that its center is the point (h, k), then its equation becomes

$$\boxed{\frac{(x-h)^2}{a^2} + \frac{(y-k)^2}{b^2} = 1}$$ [5]

(See Figure 14.)

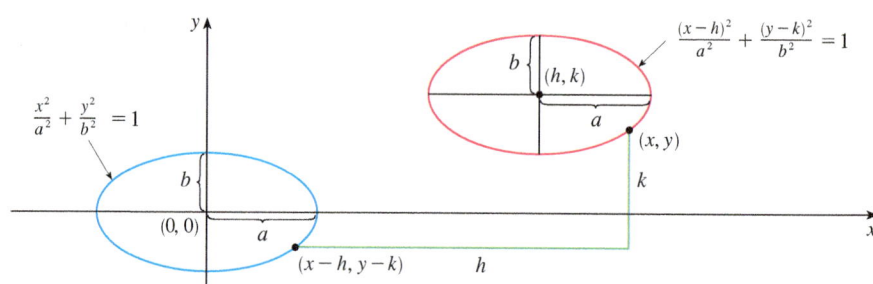

FIGURE 14

Notice that in shifting the ellipse, we replaced x by $x - h$ and y by $y - k$ in Equation 4 to obtain Equation 5. We use the same procedure to shift the parabola $y = ax^2$ so that its vertex (the origin) becomes the point (h, k) as in Figure 15. Replacing x by $x - h$ and y by $y - k$, we see that the new equation is

$$y - k = a(x - h)^2 \quad \text{or} \quad y = a(x - h)^2 + k$$

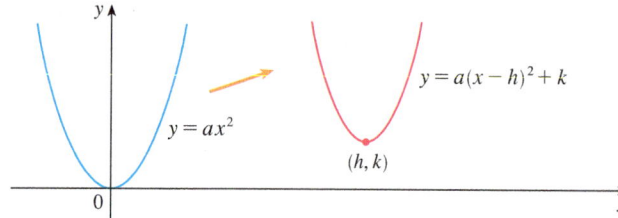

FIGURE 15

EXAMPLE 7 Sketch the graph of the equation $y = 2x^2 - 4x + 1$.

SOLUTION First we complete the square:

$$y = 2(x^2 - 2x) + 1 = 2(x - 1)^2 - 1$$

In this form we see that the equation represents the parabola obtained by shifting $y = 2x^2$ so that its vertex is at the point $(1, -1)$. The graph is sketched in Figure 16.

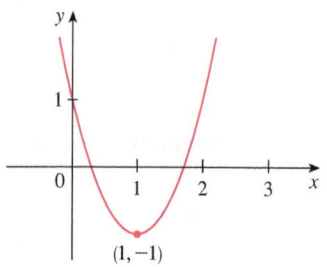

FIGURE 16
$y = 2x^2 - 4x + 1$

EXAMPLE 8 Sketch the curve $x = 1 - y^2$.

SOLUTION This time we start with the parabola $x = -y^2$ (as in Figure 6 with $a = -1$) and shift one unit to the right to get the graph of $x = 1 - y^2$. (See Figure 17.)

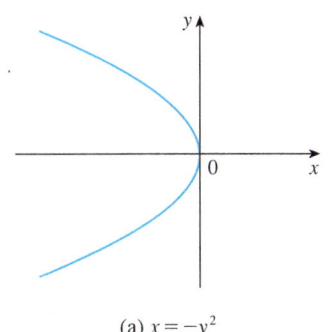
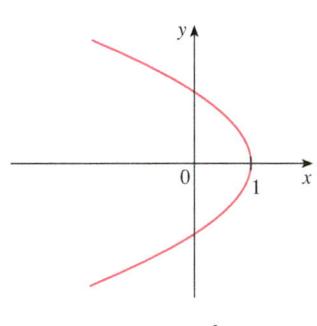

FIGURE 17   (a) $x = -y^2$ (b) $x = 1 - y^2$

C EXERCISES

1–4 Find an equation of a circle that satisfies the given conditions.

1. Center $(3, -1)$, radius 5
2. Center $(-2, -8)$, radius 10
3. Center at the origin, passes through $(4, 7)$
4. Center $(-1, 5)$, passes through $(-4, -6)$

5–9 Show that the equation represents a circle and find the center and radius.

5. $x^2 + y^2 - 4x + 10y + 13 = 0$
6. $x^2 + y^2 + 6y + 2 = 0$
7. $x^2 + y^2 + x = 0$
8. $16x^2 + 16y^2 + 8x + 32y + 1 = 0$
9. $2x^2 + 2y^2 - x + y = 1$

10. Under what condition on the coefficients a, b, and c does the equation $x^2 + y^2 + ax + by + c = 0$ represent a circle? When that condition is satisfied, find the center and radius of the circle.

11–32 Identify the type of curve and sketch the graph. Do not plot points. Just use the standard graphs given in Figures 5, 6, 8, 10, and 11 and shift if necessary.

11. $y = -x^2$
12. $y^2 - x^2 = 1$
13. $x^2 + 4y^2 = 16$
14. $x = -2y^2$
15. $16x^2 - 25y^2 = 400$
16. $25x^2 + 4y^2 = 100$
17. $4x^2 + y^2 = 1$
18. $y = x^2 + 2$
19. $x = y^2 - 1$
20. $9x^2 - 25y^2 = 225$
21. $9y^2 - x^2 = 9$
22. $2x^2 + 5y^2 = 10$
23. $xy = 4$
24. $y = x^2 + 2x$
25. $9(x - 1)^2 + 4(y - 2)^2 = 36$
26. $16x^2 + 9y^2 - 36y = 108$
27. $y = x^2 - 6x + 13$
28. $x^2 - y^2 - 4x + 3 = 0$
29. $x = 4 - y^2$
30. $y^2 - 2x + 6y + 5 = 0$
31. $x^2 + 4y^2 - 6x + 5 = 0$
32. $4x^2 + 9y^2 - 16x + 54y + 61 = 0$

33–34 Sketch the region bounded by the curves.

33. $y = 3x$, $y = x^2$
34. $y = 4 - x^2$, $x - 2y = 2$

35. Find an equation of the parabola with vertex $(1, -1)$ that passes through the points $(-1, 3)$ and $(3, 3)$.

36. Find an equation of the ellipse with center at the origin that passes through the points $\left(1, -10\sqrt{2}/3\right)$ and $\left(-2, 5\sqrt{5}/3\right)$.

37–40 Sketch the graph of the set.

37. $\{(x, y) \mid x^2 + y^2 \leq 1\}$
38. $\{(x, y) \mid x^2 + y^2 > 4\}$
39. $\{(x, y) \mid y \geq x^2 - 1\}$
40. $\{(x, y) \mid x^2 + 4y^2 \leq 4\}$

D Trigonometry

Angles

Angles can be measured in degrees or in radians (abbreviated as rad). The angle given by a complete revolution contains 360°, which is the same as 2π rad. Therefore

$$\boxed{1} \qquad \boxed{\pi \text{ rad} = 180°}$$

and

$$\boxed{2} \qquad 1 \text{ rad} = \left(\frac{180}{\pi}\right)° \approx 57.3° \qquad 1° = \frac{\pi}{180} \text{ rad} \approx 0.017 \text{ rad}$$

EXAMPLE 1
(a) Find the radian measure of 60°. (b) Express $5\pi/4$ rad in degrees.

SOLUTION
(a) From Equation 1 or 2 we see that to convert from degrees to radians we multiply by $\pi/180$. Therefore

$$60° = 60\left(\frac{\pi}{180}\right) = \frac{\pi}{3} \text{ rad}$$

(b) To convert from radians to degrees we multiply by $180/\pi$. Thus

$$\frac{5\pi}{4} \text{ rad} = \frac{5\pi}{4}\left(\frac{180}{\pi}\right) = 225° \qquad \blacksquare$$

In calculus we use radians to measure angles except when otherwise indicated. The following table gives the correspondence between degree and radian measures of some common angles.

Degrees	0°	30°	45°	60°	90°	120°	135°	150°	180°	270°	360°
Radians	0	$\dfrac{\pi}{6}$	$\dfrac{\pi}{4}$	$\dfrac{\pi}{3}$	$\dfrac{\pi}{2}$	$\dfrac{2\pi}{3}$	$\dfrac{3\pi}{4}$	$\dfrac{5\pi}{6}$	π	$\dfrac{3\pi}{2}$	2π

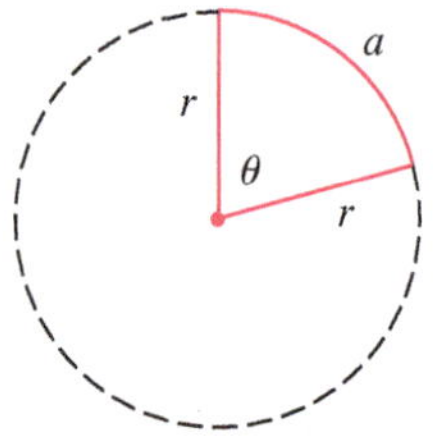

Figure 1 shows a sector of a circle with central angle θ and radius r subtending an arc with length a. Since the length of the arc is proportional to the size of the angle, and since the entire circle has circumference $2\pi r$ and central angle 2π, we have

$$\frac{\theta}{2\pi} = \frac{a}{2\pi r}$$

FIGURE 1

Solving this equation for θ and for a, we obtain

$$\boxed{3} \qquad \boxed{\theta = \frac{a}{r}} \qquad \boxed{a = r\theta}$$

Remember that Equations 3 are valid only when θ is measured in radians.

In particular, putting $a = r$ in Equation 3, we see that an angle of 1 rad is the angle subtended at the center of a circle by an arc equal in length to the radius of the circle (see Figure 2).

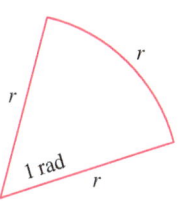

FIGURE 2

EXAMPLE 2
(a) If the radius of a circle is 5 cm, what angle is subtended by an arc of 6 cm?
(b) If a circle has radius 3 cm, what is the length of an arc subtended by a central angle of $3\pi/8$ rad?

SOLUTION
(a) Using Equation 3 with $a = 6$ and $r = 5$, we see that the angle is

$$\theta = \tfrac{6}{5} = 1.2 \text{ rad}$$

(b) With $r = 3$ cm and $\theta = 3\pi/8$ rad, the arc length is

$$a = r\theta = 3\left(\frac{3\pi}{8}\right) = \frac{9\pi}{8} \text{ cm}$$

The **standard position** of an angle occurs when we place its vertex at the origin of a coordinate system and its initial side on the positive x-axis as in Figure 3. A **positive** angle is obtained by rotating the initial side counterclockwise until it coincides with the terminal side. Likewise, **negative** angles are obtained by clockwise rotation as in Figure 4.

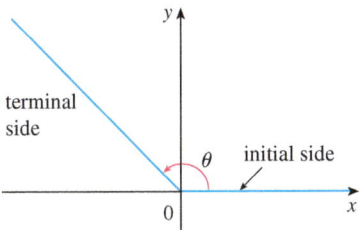

FIGURE 3 $\theta \geq 0$

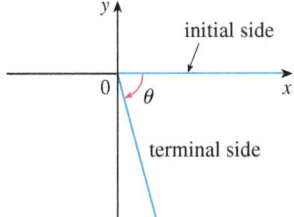

FIGURE 4 $\theta < 0$

Figure 5 shows several examples of angles in standard position. Notice that different angles can have the same terminal side. For instance, the angles $3\pi/4$, $-5\pi/4$, and $11\pi/4$ have the same initial and terminal sides because

$$\frac{3\pi}{4} - 2\pi = -\frac{5\pi}{4} \qquad \frac{3\pi}{4} + 2\pi = \frac{11\pi}{4}$$

and 2π rad represents a complete revolution.

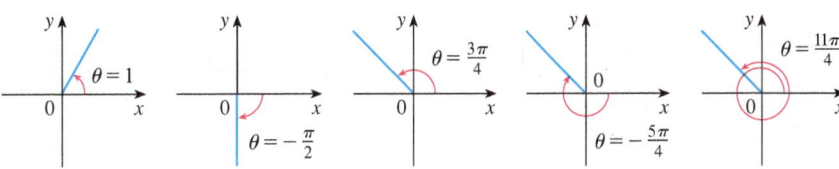

FIGURE 5
Angles in standard position

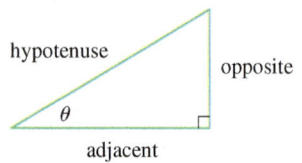

FIGURE 6

The Trigonometric Functions

For an acute angle θ the six trigonometric functions are defined as ratios of lengths of sides of a right triangle as follows (see Figure 6).

$$\boxed{4} \quad \sin\theta = \frac{\text{opp}}{\text{hyp}} \qquad \csc\theta = \frac{\text{hyp}}{\text{opp}}$$

$$\cos\theta = \frac{\text{adj}}{\text{hyp}} \qquad \sec\theta = \frac{\text{hyp}}{\text{adj}}$$

$$\tan\theta = \frac{\text{opp}}{\text{adj}} \qquad \cot\theta = \frac{\text{adj}}{\text{opp}}$$

This definition doesn't apply to obtuse or negative angles, so for a general angle θ in standard position we let $P(x, y)$ be any point on the terminal side of θ and we let r be the distance $|OP|$ as in Figure 7. Then we define

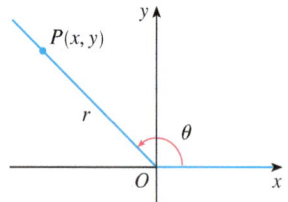

FIGURE 7

$$\boxed{5} \quad \sin\theta = \frac{y}{r} \qquad \csc\theta = \frac{r}{y}$$

$$\cos\theta = \frac{x}{r} \qquad \sec\theta = \frac{r}{x}$$

$$\tan\theta = \frac{y}{x} \qquad \cot\theta = \frac{x}{y}$$

Since division by 0 is not defined, $\tan\theta$ and $\sec\theta$ are undefined when $x = 0$ and $\csc\theta$ and $\cot\theta$ are undefined when $y = 0$. Notice that the definitions in (4) and (5) are consistent when θ is an acute angle.

If θ is a number, the convention is that $\sin\theta$ means the sine of the angle whose *radian* measure is θ. For example, the expression $\sin 3$ implies that we are dealing with an angle of 3 rad. When finding a calculator approximation to this number, we must remember to set our calculator in radian mode, and then we obtain

$$\sin 3 \approx 0.14112$$

If we put $r = 1$ in Definition 5 and draw a unit circle with center the origin and label θ as in Figure 8, then the coordinates of P are $(\cos\theta, \sin\theta)$.

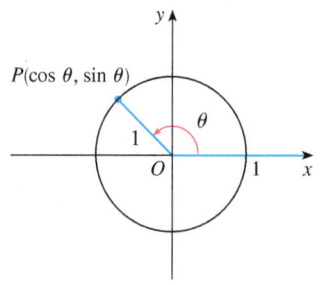

FIGURE 8

If we want to know the sine of the angle 3° we would write $\sin 3°$ and, with our calculator in degree mode, we find that

$$\sin 3° \approx 0.05234$$

The exact trigonometric ratios for certain angles can be read from the triangles in Figure 9. For instance,

$$\sin\frac{\pi}{4} = \frac{1}{\sqrt{2}} \qquad \sin\frac{\pi}{6} = \frac{1}{2} \qquad \sin\frac{\pi}{3} = \frac{\sqrt{3}}{2}$$

$$\cos\frac{\pi}{4} = \frac{1}{\sqrt{2}} \qquad \cos\frac{\pi}{6} = \frac{\sqrt{3}}{2} \qquad \cos\frac{\pi}{3} = \frac{1}{2}$$

$$\tan\frac{\pi}{4} = 1 \qquad \tan\frac{\pi}{6} = \frac{1}{\sqrt{3}} \qquad \tan\frac{\pi}{3} = \sqrt{3}$$

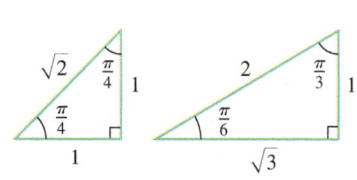

FIGURE 9

The signs of the trigonometric functions for angles in each of the four quadrants can be remembered by means of the rule "All Students Take Calculus" shown in Figure 10.

EXAMPLE 3 Find the exact trigonometric ratios for $\theta = 2\pi/3$.

SOLUTION From Figure 11 we see that a point on the terminal line for $\theta = 2\pi/3$ is $P(-1, \sqrt{3})$. Therefore, taking

$$x = -1 \qquad y = \sqrt{3} \qquad r = 2$$

in the definitions of the trigonometric ratios, we have

$$\sin\frac{2\pi}{3} = \frac{\sqrt{3}}{2} \qquad \cos\frac{2\pi}{3} = -\frac{1}{2} \qquad \tan\frac{2\pi}{3} = -\sqrt{3}$$

$$\csc\frac{2\pi}{3} = \frac{2}{\sqrt{3}} \qquad \sec\frac{2\pi}{3} = -2 \qquad \cot\frac{2\pi}{3} = -\frac{1}{\sqrt{3}}$$

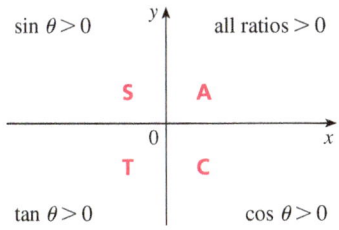

FIGURE 10

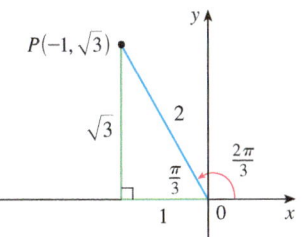

FIGURE 11

The following table gives some values of $\sin\theta$ and $\cos\theta$ found by the method of Example 3.

θ	0	$\frac{\pi}{6}$	$\frac{\pi}{4}$	$\frac{\pi}{3}$	$\frac{\pi}{2}$	$\frac{2\pi}{3}$	$\frac{3\pi}{4}$	$\frac{5\pi}{6}$	π	$\frac{3\pi}{2}$	2π
$\sin\theta$	0	$\frac{1}{2}$	$\frac{1}{\sqrt{2}}$	$\frac{\sqrt{3}}{2}$	1	$\frac{\sqrt{3}}{2}$	$\frac{1}{\sqrt{2}}$	$\frac{1}{2}$	0	-1	0
$\cos\theta$	1	$\frac{\sqrt{3}}{2}$	$\frac{1}{\sqrt{2}}$	$\frac{1}{2}$	0	$-\frac{1}{2}$	$-\frac{1}{\sqrt{2}}$	$-\frac{\sqrt{3}}{2}$	-1	0	1

EXAMPLE 4 If $\cos\theta = \frac{2}{5}$ and $0 < \theta < \pi/2$, find the other five trigonometric functions of θ.

SOLUTION Since $\cos\theta = \frac{2}{5}$, we can label the hypotenuse as having length 5 and the adjacent side as having length 2 in Figure 12. If the opposite side has length x, then the Pythagorean Theorem gives $x^2 + 4 = 25$ and so $x^2 = 21$, $x = \sqrt{21}$. We can now use the diagram to write the other five trigonometric functions:

$$\sin\theta = \frac{\sqrt{21}}{5} \qquad \tan\theta = \frac{\sqrt{21}}{2}$$

$$\csc\theta = \frac{5}{\sqrt{21}} \qquad \sec\theta = \frac{5}{2} \qquad \cot\theta = \frac{2}{\sqrt{21}}$$

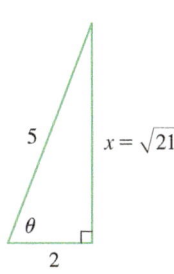

FIGURE 12

EXAMPLE 5 Use a calculator to approximate the value of x in Figure 13.

SOLUTION From the diagram we see that

$$\tan 40° = \frac{16}{x}$$

Therefore

$$x = \frac{16}{\tan 40°} \approx 19.07$$

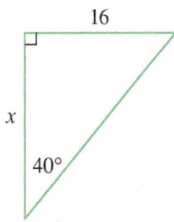

FIGURE 13

Trigonometric Identities

A trigonometric identity is a relationship among the trigonometric functions. The most elementary are the following, which are immediate consequences of the definitions of the trigonometric functions.

$$\boxed{6} \quad \csc\theta = \frac{1}{\sin\theta} \qquad \sec\theta = \frac{1}{\cos\theta} \qquad \cot\theta = \frac{1}{\tan\theta}$$

$$\tan\theta = \frac{\sin\theta}{\cos\theta} \qquad \cot\theta = \frac{\cos\theta}{\sin\theta}$$

For the next identity we refer back to Figure 7. The distance formula (or, equivalently, the Pythagorean Theorem) tells us that $x^2 + y^2 = r^2$. Therefore

$$\sin^2\theta + \cos^2\theta = \frac{y^2}{r^2} + \frac{x^2}{r^2} = \frac{x^2 + y^2}{r^2} = \frac{r^2}{r^2} = 1$$

We have therefore proved one of the most useful of all trigonometric identities:

$$\boxed{7} \quad \sin^2\theta + \cos^2\theta = 1$$

If we now divide both sides of Equation 7 by $\cos^2\theta$ and use Equations 6, we get

$$\boxed{8} \quad \tan^2\theta + 1 = \sec^2\theta$$

Similarly, if we divide both sides of Equation 7 by $\sin^2\theta$, we get

$$\boxed{9} \quad 1 + \cot^2\theta = \csc^2\theta$$

The identities

$$\boxed{10a} \quad \sin(-\theta) = -\sin\theta$$

$$\boxed{10b} \quad \cos(-\theta) = \cos\theta$$

Odd functions and even functions are discussed in Section 1.1.

show that sine is an odd function and cosine is an even function. They are easily proved by drawing a diagram showing θ and $-\theta$ in standard position (see Exercise 39).

Since the angles θ and $\theta + 2\pi$ have the same terminal side, we have

$$\boxed{11} \quad \sin(\theta + 2\pi) = \sin\theta \qquad \cos(\theta + 2\pi) = \cos\theta$$

These identities show that the sine and cosine functions are periodic with period 2π.

The remaining trigonometric identities are all consequences of two basic identities called the **addition formulas**:

12a $$\sin(x + y) = \sin x \cos y + \cos x \sin y$$
12b $$\cos(x + y) = \cos x \cos y - \sin x \sin y$$

The proofs of these addition formulas are outlined in Exercises 85, 86, and 87.

By substituting $-y$ for y in Equations 12a and 12b and using Equations 10a and 10b, we obtain the following **subtraction formulas**:

13a $$\sin(x - y) = \sin x \cos y - \cos x \sin y$$
13b $$\cos(x - y) = \cos x \cos y + \sin x \sin y$$

Then, by dividing the formulas in Equations 12 or Equations 13, we obtain the corresponding formulas for $\tan(x \pm y)$:

14a $$\tan(x + y) = \frac{\tan x + \tan y}{1 - \tan x \tan y}$$
14b $$\tan(x - y) = \frac{\tan x - \tan y}{1 + \tan x \tan y}$$

If we put $y = x$ in the addition formulas (12), we get the **double-angle formulas**:

15a $$\sin 2x = 2 \sin x \cos x$$
15b $$\cos 2x = \cos^2 x - \sin^2 x$$

Then, by using the identity $\sin^2 x + \cos^2 x = 1$, we obtain the following alternate forms of the double-angle formulas for $\cos 2x$:

16a $$\cos 2x = 2 \cos^2 x - 1$$
16b $$\cos 2x = 1 - 2 \sin^2 x$$

If we now solve these equations for $\cos^2 x$ and $\sin^2 x$, we get the following **half-angle formulas**, which are useful in integral calculus:

17a $$\cos^2 x = \frac{1 + \cos 2x}{2}$$
17b $$\sin^2 x = \frac{1 - \cos 2x}{2}$$

Finally, we state the **product formulas**, which can be deduced from Equations 12 and 13:

18a	$\sin x \cos y = \tfrac{1}{2}[\sin(x+y) + \sin(x-y)]$
18b	$\cos x \cos y = \tfrac{1}{2}[\cos(x+y) + \cos(x-y)]$
18c	$\sin x \sin y = \tfrac{1}{2}[\cos(x-y) - \cos(x+y)]$

There are many other trigonometric identities, but those we have stated are the ones used most often in calculus. If you forget any of the identities 13–18, remember that they can all be deduced from Equations 12a and 12b.

EXAMPLE 6 Find all values of x in the interval $[0, 2\pi]$ such that $\sin x = \sin 2x$.

SOLUTION Using the double-angle formula (15a), we rewrite the given equation as

$$\sin x = 2 \sin x \cos x \quad \text{or} \quad \sin x (1 - 2 \cos x) = 0$$

Therefore there are two possibilities:

$$\sin x = 0 \qquad \text{or} \qquad 1 - 2 \cos x = 0$$
$$x = 0, \pi, 2\pi \qquad\qquad \cos x = \tfrac{1}{2}$$
$$\qquad\qquad\qquad\qquad x = \frac{\pi}{3}, \frac{5\pi}{3}$$

The given equation has five solutions: $0, \pi/3, \pi, 5\pi/3$, and 2π. ∎

Graphs of the Trigonometric Functions

The graph of the function $f(x) = \sin x$, shown in Figure 14(a), is obtained by plotting points for $0 \leq x \leq 2\pi$ and then using the periodic nature of the function (from Equation 11) to complete the graph. Notice that the zeros of the sine function occur at the

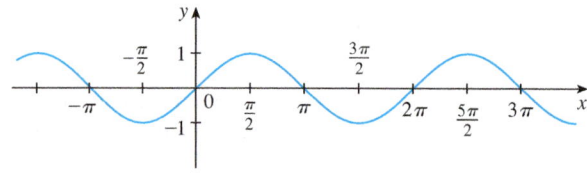

(a) $f(x) = \sin x$

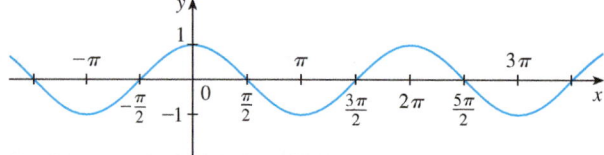

(b) $g(x) = \cos x$

FIGURE 14

integer multiples of π, that is,

$$\sin x = 0 \quad \text{whenever } x = n\pi, \quad n \text{ an integer}$$

Because of the identity

$$\cos x = \sin\left(x + \frac{\pi}{2}\right)$$

(which can be verified using Equation 12a), the graph of cosine is obtained by shifting the graph of sine by an amount $\pi/2$ to the left [see Figure 14(b)]. Note that for both the sine and cosine functions the domain is $(-\infty, \infty)$ and the range is the closed interval $[-1, 1]$. Thus, for all values of x, we have

$$-1 \leq \sin x \leq 1 \qquad -1 \leq \cos x \leq 1$$

The graphs of the remaining four trigonometric functions are shown in Figure 15 and their domains are indicated there. Notice that tangent and cotangent have range $(-\infty, \infty)$, whereas cosecant and secant have range $(-\infty, -1] \cup [1, \infty)$. All four functions are periodic: tangent and cotangent have period π, whereas cosecant and secant have period 2π.

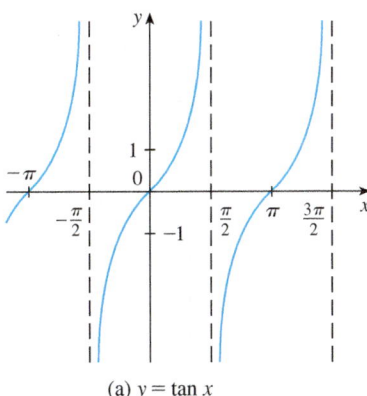

(a) $y = \tan x$

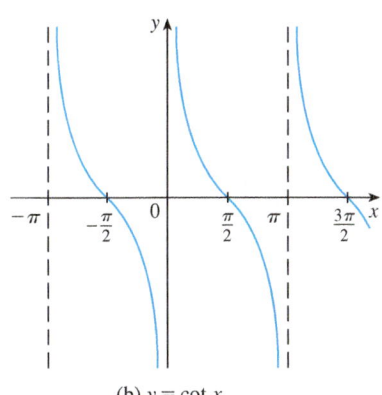

(b) $y = \cot x$

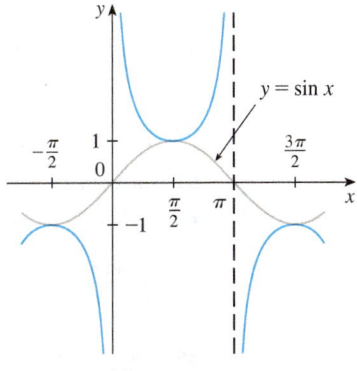

(c) $y = \csc x$

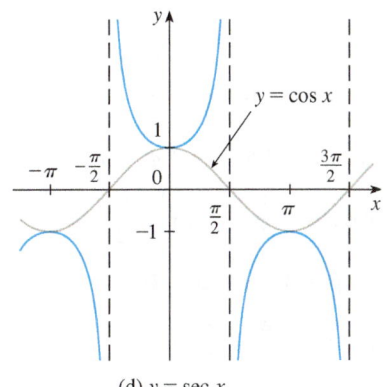

(d) $y = \sec x$

FIGURE 15

D EXERCISES

1–6 Convert from degrees to radians.

1. $210°$
2. $300°$
3. $9°$
4. $-315°$
5. $900°$
6. $36°$

7–12 Convert from radians to degrees.

7. 4π
8. $-\dfrac{7\pi}{2}$
9. $\dfrac{5\pi}{12}$
10. $\dfrac{8\pi}{3}$
11. $-\dfrac{3\pi}{8}$
12. 5

13. Find the length of a circular arc subtended by an angle of $\pi/12$ rad if the radius of the circle is 36 cm.

14. If a circle has radius 10 cm, find the length of the arc subtended by a central angle of $72°$.

15. A circle has radius 1.5 m. What angle is subtended at the center of the circle by an arc 1 m long?

16. Find the radius of a circular sector with angle $3\pi/4$ and arc length 6 cm.

17–22 Draw, in standard position, the angle whose measure is given.

17. $315°$
18. $-150°$
19. $-\dfrac{3\pi}{4}$ rad
20. $\dfrac{7\pi}{3}$ rad
21. 2 rad
22. -3 rad

23–28 Find the exact trigonometric ratios for the angle whose radian measure is given.

23. $\dfrac{3\pi}{4}$
24. $\dfrac{4\pi}{3}$
25. $\dfrac{9\pi}{2}$
26. -5π
27. $\dfrac{5\pi}{6}$
28. $\dfrac{11\pi}{4}$

29–34 Find the remaining trigonometric ratios.

29. $\sin\theta = \dfrac{3}{5}$, $\quad 0 < \theta < \dfrac{\pi}{2}$
30. $\tan\alpha = 2$, $\quad 0 < \alpha < \dfrac{\pi}{2}$
31. $\sec\phi = -1.5$, $\quad \dfrac{\pi}{2} < \phi < \pi$
32. $\cos x = -\dfrac{1}{3}$, $\quad \pi < x < \dfrac{3\pi}{2}$
33. $\cot\beta = 3$, $\quad \pi < \beta < 2\pi$
34. $\csc\theta = -\dfrac{4}{3}$, $\quad \dfrac{3\pi}{2} < \theta < 2\pi$

35–38 Find, correct to five decimal places, the length of the side labeled x.

35.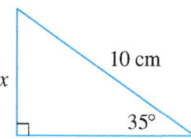
36.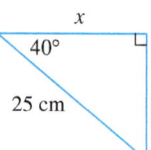
37.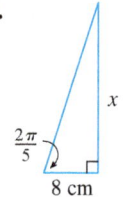
38.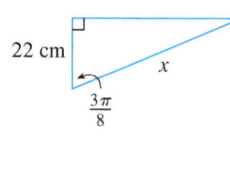

39–41 Prove each equation.

39. (a) Equation 10a (b) Equation 10b
40. (a) Equation 14a (b) Equation 14b
41. (a) Equation 18a (b) Equation 18b
    (c) Equation 18c

42–58 Prove the identity.

42. $\cos\left(\dfrac{\pi}{2} - x\right) = \sin x$
43. $\sin\left(\dfrac{\pi}{2} + x\right) = \cos x$
44. $\sin(\pi - x) = \sin x$
45. $\sin\theta\cot\theta = \cos\theta$
46. $(\sin x + \cos x)^2 = 1 + \sin 2x$
47. $\sec y - \cos y = \tan y \sin y$
48. $\tan^2\alpha - \sin^2\alpha = \tan^2\alpha \sin^2\alpha$
49. $\cot^2\theta + \sec^2\theta = \tan^2\theta + \csc^2\theta$
50. $2\csc 2t = \sec t \csc t$
51. $\tan 2\theta = \dfrac{2\tan\theta}{1 - \tan^2\theta}$
52. $\dfrac{1}{1 - \sin\theta} + \dfrac{1}{1 + \sin\theta} = 2\sec^2\theta$
53. $\sin x \sin 2x + \cos x \cos 2x = \cos x$
54. $\sin^2 x - \sin^2 y = \sin(x+y)\sin(x-y)$
55. $\dfrac{\sin\phi}{1 - \cos\phi} = \csc\phi + \cot\phi$
56. $\tan x + \tan y = \dfrac{\sin(x+y)}{\cos x \cos y}$

57. $\sin 3\theta + \sin \theta = 2 \sin 2\theta \cos \theta$

58. $\cos 3\theta = 4 \cos^3 \theta - 3 \cos \theta$

59–64 If $\sin x = \frac{1}{3}$ and $\sec y = \frac{5}{4}$, where x and y lie between 0 and $\pi/2$, evaluate the expression.

59. $\sin(x + y)$ **60.** $\cos(x + y)$

61. $\cos(x - y)$ **62.** $\sin(x - y)$

63. $\sin 2y$ **64.** $\cos 2y$

65–72 Find all values of x in the interval $[0, 2\pi]$ that satisfy the equation.

65. $2 \cos x - 1 = 0$ **66.** $3 \cot^2 x = 1$

67. $2 \sin^2 x = 1$ **68.** $|\tan x| = 1$

69. $\sin 2x = \cos x$ **70.** $2 \cos x + \sin 2x = 0$

71. $\sin x = \tan x$ **72.** $2 + \cos 2x = 3 \cos x$

73–76 Find all values of x in the interval $[0, 2\pi]$ that satisfy the inequality.

73. $\sin x \leq \frac{1}{2}$ **74.** $2 \cos x + 1 > 0$

75. $-1 < \tan x < 1$ **76.** $\sin x > \cos x$

77–82 Graph the function by starting with the graphs in Figures 14 and 15 and applying the transformations of Section 1.3 where appropriate.

77. $y = \cos\left(x - \dfrac{\pi}{3}\right)$ **78.** $y = \tan 2x$

79. $y = \dfrac{1}{3} \tan\left(x - \dfrac{\pi}{2}\right)$ **80.** $y = 1 + \sec x$

81. $y = |\sin x|$ **82.** $y = 2 + \sin\left(x + \dfrac{\pi}{4}\right)$

---

83. Prove the **Law of Cosines**: If a triangle has sides with lengths a, b, and c, and θ is the angle between the sides with lengths a and b, then
$$c^2 = a^2 + b^2 - 2ab \cos \theta$$

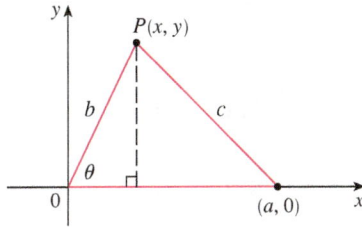

[*Hint:* Introduce a coordinate system so that θ is in standard position, as in the figure. Express x and y in terms of θ and then use the distance formula to compute c.]

84. In order to find the distance $|AB|$ across a small inlet, a point C was located as in the figure and the following measurements were recorded:

$$\angle C = 103° \quad |AC| = 820 \text{ m} \quad |BC| = 910 \text{ m}$$

Use the Law of Cosines from Exercise 83 to find the required distance.

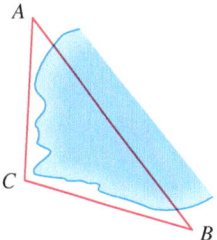

85. Use the figure to prove the subtraction formula
$$\cos(\alpha - \beta) = \cos \alpha \cos \beta + \sin \alpha \sin \beta$$

[*Hint:* Compute c^2 in two ways (using the Law of Cosines from Exercise 83 and also using the distance formula) and compare the two expressions.]

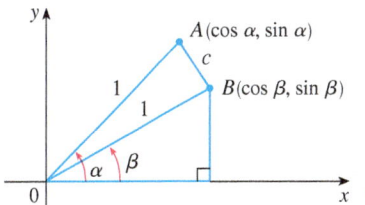

86. Use the formula in Exercise 85 to prove the addition formula for cosine (12b).

87. Use the addition formula for cosine and the identities
$$\cos\left(\frac{\pi}{2} - \theta\right) = \sin \theta \quad \sin\left(\frac{\pi}{2} - \theta\right) = \cos \theta$$
to prove the subtraction formula (13a) for the sine function.

88. Show that the area of a triangle with sides of lengths a and b and with included angle θ is
$$A = \tfrac{1}{2} ab \sin \theta$$

89. Find the area of triangle ABC, correct to five decimal places, if
$$|AB| = 10 \text{ cm} \quad |BC| = 3 \text{ cm} \quad \angle ABC = 107°$$

E Sigma Notation

A convenient way of writing sums uses the Greek letter Σ (capital sigma, corresponding to our letter S) and is called **sigma notation**.

> **1 Definition** If $a_m, a_{m+1}, \ldots, a_n$ are real numbers and m and n are integers such that $m \leq n$, then
> $$\sum_{i=m}^{n} a_i = a_m + a_{m+1} + a_{m+2} + \cdots + a_{n-1} + a_n$$

This tells us to end with $i = n$.
This tells us to add.
$$\sum_{i=m}^{n} a_i$$
This tells us to start with $i = m$.

With function notation, Definition 1 can be written as

$$\sum_{i=m}^{n} f(i) = f(m) + f(m+1) + f(m+2) + \cdots + f(n-1) + f(n)$$

Thus the symbol $\sum_{i=m}^{n}$ indicates a summation in which the letter i (called the **index of summation**) takes on consecutive integer values beginning with m and ending with n, that is, $m, m+1, \ldots, n$. Other letters can also be used as the index of summation.

EXAMPLE 1

(a) $\sum_{i=1}^{4} i^2 = 1^2 + 2^2 + 3^2 + 4^2 = 30$

(b) $\sum_{i=3}^{n} i = 3 + 4 + 5 + \cdots + (n-1) + n$

(c) $\sum_{j=0}^{5} 2^j = 2^0 + 2^1 + 2^2 + 2^3 + 2^4 + 2^5 = 63$

(d) $\sum_{k=1}^{n} \frac{1}{k} = 1 + \frac{1}{2} + \frac{1}{3} + \cdots + \frac{1}{n}$

(e) $\sum_{i=1}^{3} \frac{i-1}{i^2+3} = \frac{1-1}{1^2+3} + \frac{2-1}{2^2+3} + \frac{3-1}{3^2+3} = 0 + \frac{1}{7} + \frac{1}{6} = \frac{13}{42}$

(f) $\sum_{i=1}^{4} 2 = 2 + 2 + 2 + 2 = 8$ ∎

EXAMPLE 2 Write the sum $2^3 + 3^3 + \cdots + n^3$ in sigma notation.

SOLUTION There is no unique way of writing a sum in sigma notation. We could write

$$2^3 + 3^3 + \cdots + n^3 = \sum_{i=2}^{n} i^3$$

or

$$2^3 + 3^3 + \cdots + n^3 = \sum_{j=1}^{n-1} (j+1)^3$$

or

$$2^3 + 3^3 + \cdots + n^3 = \sum_{k=0}^{n-2} (k+2)^3$$ ∎

The following theorem gives three simple rules for working with sigma notation.

[2] Theorem If c is any constant (that is, it does not depend on i), then

(a) $\displaystyle\sum_{i=m}^{n} ca_i = c \sum_{i=m}^{n} a_i$ (b) $\displaystyle\sum_{i=m}^{n} (a_i + b_i) = \sum_{i=m}^{n} a_i + \sum_{i=m}^{n} b_i$

(c) $\displaystyle\sum_{i=m}^{n} (a_i - b_i) = \sum_{i=m}^{n} a_i - \sum_{i=m}^{n} b_i$

PROOF To see why these rules are true, all we have to do is write both sides in expanded form. Rule (a) is just the distributive property of real numbers:

$$ca_m + ca_{m+1} + \cdots + ca_n = c(a_m + a_{m+1} + \cdots + a_n)$$

Rule (b) follows from the associative and commutative properties:

$$(a_m + b_m) + (a_{m+1} + b_{m+1}) + \cdots + (a_n + b_n)$$
$$= (a_m + a_{m+1} + \cdots + a_n) + (b_m + b_{m+1} + \cdots + b_n)$$

Rule (c) is proved similarly. ■

EXAMPLE 3 Find $\displaystyle\sum_{i=1}^{n} 1$.

SOLUTION $\displaystyle\sum_{i=1}^{n} 1 = \underbrace{1 + 1 + \cdots + 1}_{n \text{ terms}} = n$ ■

EXAMPLE 4 Prove the formula for the sum of the first n positive integers:

$$\sum_{i=1}^{n} i = 1 + 2 + 3 + \cdots + n = \frac{n(n+1)}{2}$$

SOLUTION This formula can be proved by mathematical induction (see page 72) or by the following method used by the German mathematician Karl Friedrich Gauss (1777–1855) when he was ten years old.

Write the sum S twice, once in the usual order and once in reverse order:

$$S = 1 + 2 + 3 + \cdots + (n-1) + n$$
$$S = n + (n-1) + (n-2) + \cdots + 2 + 1$$

Adding all columns vertically, we get

$$2S = (n+1) + (n+1) + (n+1) + \cdots + (n+1) + (n+1)$$

On the right side there are n terms, each of which is $n+1$, so

$$2S = n(n+1) \quad \text{or} \quad S = \frac{n(n+1)}{2}$$ ■

EXAMPLE 5 Prove the formula for the sum of the squares of the first n positive integers:

$$\sum_{i=1}^{n} i^2 = 1^2 + 2^2 + 3^2 + \cdots + n^2 = \frac{n(n+1)(2n+1)}{6}$$

APPENDIX E Sigma Notation

SOLUTION 1 Let S be the desired sum. We start with the *telescoping sum* (or *collapsing sum*):

Most terms cancel in pairs.

$$\sum_{i=1}^{n} [(1+i)^3 - i^3] = (2^3 - 1^3) + (3^3 - 2^3) + (4^3 - 3^3) + \cdots + [(n+1)^3 - n^3]$$

$$= (n+1)^3 - 1^3 = n^3 + 3n^2 + 3n$$

On the other hand, using Theorem 2 and Examples 3 and 4, we have

$$\sum_{i=1}^{n} [(1+i)^3 - i^3] = \sum_{i=1}^{n} [3i^2 + 3i + 1] = 3\sum_{i=1}^{n} i^2 + 3\sum_{i=1}^{n} i + \sum_{i=1}^{n} 1$$

$$= 3S + 3\frac{n(n+1)}{2} + n = 3S + \tfrac{3}{2}n^2 + \tfrac{5}{2}n$$

Thus we have

$$n^3 + 3n^2 + 3n = 3S + \tfrac{3}{2}n^2 + \tfrac{5}{2}n$$

Solving this equation for S, we obtain

$$3S = n^3 + \tfrac{3}{2}n^2 + \tfrac{1}{2}n$$

or

$$S = \frac{2n^3 + 3n^2 + n}{6} = \frac{n(n+1)(2n+1)}{6}$$

Principle of Mathematical Induction
Let S_n be a statement involving the positive integer n. Suppose that
1. S_1 is true.
2. If S_k is true, then S_{k+1} is true.
Then S_n is true for all positive integers n.

See pages 72 and 74 for a more thorough discussion of mathematical induction.

SOLUTION 2 Let S_n be the given formula.

1. S_1 is true because $\quad 1^2 = \dfrac{1(1+1)(2 \cdot 1 + 1)}{6}$

2. Assume that S_k is true; that is,

$$1^2 + 2^2 + 3^2 + \cdots + k^2 = \frac{k(k+1)(2k+1)}{6}$$

Then

$$1^2 + 2^2 + 3^2 + \cdots + (k+1)^2 = (1^2 + 2^2 + 3^2 + \cdots + k^2) + (k+1)^2$$

$$= \frac{k(k+1)(2k+1)}{6} + (k+1)^2$$

$$= (k+1)\frac{k(2k+1) + 6(k+1)}{6}$$

$$= (k+1)\frac{2k^2 + 7k + 6}{6}$$

$$= \frac{(k+1)(k+2)(2k+3)}{6}$$

$$= \frac{(k+1)[(k+1)+1][2(k+1)+1]}{6}$$

So S_{k+1} is true.

By the Principle of Mathematical Induction, S_n is true for all n. ∎

We list the results of Examples 3, 4, and 5 together with a similar result for cubes (see Exercises 37–40) as Theorem 3. These formulas are needed for finding areas and evaluating integrals in Chapter 5.

3 Theorem Let c be a constant and n a positive integer. Then

(a) $\displaystyle\sum_{i=1}^{n} 1 = n$ \qquad (b) $\displaystyle\sum_{i=1}^{n} c = nc$

(c) $\displaystyle\sum_{i=1}^{n} i = \frac{n(n+1)}{2}$ \qquad (d) $\displaystyle\sum_{i=1}^{n} i^2 = \frac{n(n+1)(2n+1)}{6}$

(e) $\displaystyle\sum_{i=1}^{n} i^3 = \left[\frac{n(n+1)}{2}\right]^2$

EXAMPLE 6 Evaluate $\displaystyle\sum_{i=1}^{n} i(4i^2 - 3)$.

SOLUTION Using Theorems 2 and 3, we have

$$\sum_{i=1}^{n} i(4i^2 - 3) = \sum_{i=1}^{n} (4i^3 - 3i) = 4\sum_{i=1}^{n} i^3 - 3\sum_{i=1}^{n} i$$

$$= 4\left[\frac{n(n+1)}{2}\right]^2 - 3\frac{n(n+1)}{2}$$

$$= \frac{n(n+1)[2n(n+1) - 3]}{2}$$

$$= \frac{n(n+1)(2n^2 + 2n - 3)}{2}$$

EXAMPLE 7 Find $\displaystyle\lim_{n\to\infty} \sum_{i=1}^{n} \frac{3}{n}\left[\left(\frac{i}{n}\right)^2 + 1\right]$.

The type of calculation in Example 7 arises in Chapter 5 when we compute areas.

SOLUTION

$$\lim_{n\to\infty} \sum_{i=1}^{n} \frac{3}{n}\left[\left(\frac{i}{n}\right)^2 + 1\right] = \lim_{n\to\infty} \sum_{i=1}^{n} \left[\frac{3}{n^3} i^2 + \frac{3}{n}\right]$$

$$= \lim_{n\to\infty} \left[\frac{3}{n^3} \sum_{i=1}^{n} i^2 + \frac{3}{n} \sum_{i=1}^{n} 1\right]$$

$$= \lim_{n\to\infty} \left[\frac{3}{n^3} \cdot \frac{n(n+1)(2n+1)}{6} + \frac{3}{n} \cdot n\right]$$

$$= \lim_{n\to\infty} \left[\frac{1}{2} \cdot \frac{n}{n} \cdot \left(\frac{n+1}{n}\right)\left(\frac{2n+1}{n}\right) + 3\right]$$

$$= \lim_{n\to\infty} \left[\frac{1}{2} \cdot 1\left(1 + \frac{1}{n}\right)\left(2 + \frac{1}{n}\right) + 3\right]$$

$$= \tfrac{1}{2} \cdot 1 \cdot 1 \cdot 2 + 3 = 4$$

E EXERCISES

1–10 Write the sum in expanded form.

1. $\sum_{i=1}^{5} \sqrt{i}$
2. $\sum_{i=1}^{6} \frac{1}{i+1}$
3. $\sum_{i=4}^{6} 3^i$
4. $\sum_{i=4}^{6} i^3$
5. $\sum_{k=0}^{4} \frac{2k-1}{2k+1}$
6. $\sum_{k=5}^{8} x^k$
7. $\sum_{i=1}^{n} i^{10}$
8. $\sum_{j=n}^{n+3} j^2$
9. $\sum_{j=0}^{n-1} (-1)^j$
10. $\sum_{i=1}^{n} f(x_i)\,\Delta x_i$

11–20 Write the sum in sigma notation.

11. $1 + 2 + 3 + 4 + \cdots + 10$
12. $\sqrt{3} + \sqrt{4} + \sqrt{5} + \sqrt{6} + \sqrt{7}$
13. $\frac{1}{2} + \frac{2}{3} + \frac{3}{4} + \frac{4}{5} + \cdots + \frac{19}{20}$
14. $\frac{3}{7} + \frac{4}{8} + \frac{5}{9} + \frac{6}{10} + \cdots + \frac{23}{27}$
15. $2 + 4 + 6 + 8 + \cdots + 2n$
16. $1 + 3 + 5 + 7 + \cdots + (2n-1)$
17. $1 + 2 + 4 + 8 + 16 + 32$
18. $\frac{1}{1} + \frac{1}{4} + \frac{1}{9} + \frac{1}{16} + \frac{1}{25} + \frac{1}{36}$
19. $x + x^2 + x^3 + \cdots + x^n$
20. $1 - x + x^2 - x^3 + \cdots + (-1)^n x^n$

21–35 Find the value of the sum.

21. $\sum_{i=4}^{8} (3i-2)$
22. $\sum_{i=3}^{6} i(i+2)$
23. $\sum_{j=1}^{6} 3^{j+1}$
24. $\sum_{k=0}^{8} \cos k\pi$
25. $\sum_{n=1}^{20} (-1)^n$
26. $\sum_{i=1}^{100} 4$
27. $\sum_{i=0}^{4} (2^i + i^2)$
28. $\sum_{i=-2}^{4} 2^{3-i}$
29. $\sum_{i=1}^{n} 2i$
30. $\sum_{i=1}^{n} (2-5i)$
31. $\sum_{i=1}^{n} (i^2 + 3i + 4)$
32. $\sum_{i=1}^{n} (3 + 2i)^2$
33. $\sum_{i=1}^{n} (i+1)(i+2)$
34. $\sum_{i=1}^{n} i(i+1)(i+2)$

35. $\sum_{i=1}^{n} (i^3 - i - 2)$

36. Find the number n such that $\sum_{i=1}^{n} i = 78$.

37. Prove formula (b) of Theorem 3.

38. Prove formula (e) of Theorem 3 using mathematical induction.

39. Prove formula (e) of Theorem 3 using a method similar to that of Example 5, Solution 1 [start with $(1+i)^4 - i^4$].

40. Prove formula (e) of Theorem 3 using the following method published by Abu Bekr Mohammed ibn Alhusain Alkarchi in about AD 1010. The figure shows a square $ABCD$ in which sides AB and AD have been divided into segments of lengths $1, 2, 3, \ldots, n$. Thus the side of the square has length $n(n+1)/2$ so the area is $[n(n+1)/2]^2$. But the area is also the sum of the areas of the n "gnomons" $G_1, G_2, \ldots, G_n$ shown in the figure. Show that the area of G_i is i^3 and conclude that formula (e) is true.

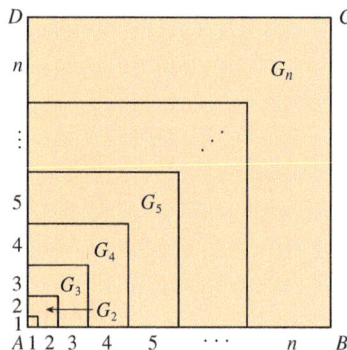

41. Evaluate each telescoping sum.

   (a) $\sum_{i=1}^{n} [i^4 - (i-1)^4]$
   (b) $\sum_{i=1}^{100} (5^i - 5^{i-1})$
   (c) $\sum_{i=3}^{99} \left(\frac{1}{i} - \frac{1}{i+1}\right)$
   (d) $\sum_{i=1}^{n} (a_i - a_{i-1})$

42. Prove the generalized triangle inequality:
$$\left|\sum_{i=1}^{n} a_i\right| \leq \sum_{i=1}^{n} |a_i|$$

43–46 Find the limit.

43. $\lim_{n \to \infty} \sum_{i=1}^{n} \frac{1}{n}\left(\frac{i}{n}\right)^2$
44. $\lim_{n \to \infty} \sum_{i=1}^{n} \frac{1}{n}\left[\left(\frac{i}{n}\right)^3 + 1\right]$
45. $\lim_{n \to \infty} \sum_{i=1}^{n} \frac{2}{n}\left[\left(\frac{2i}{n}\right)^3 + 5\left(\frac{2i}{n}\right)\right]$

46. $\lim_{n \to \infty} \sum_{i=1}^{n} \frac{3}{n} \left[\left(1 + \frac{3i}{n}\right)^3 - 2\left(1 + \frac{3i}{n}\right) \right]$

47. Prove the formula for the sum of a finite geometric series with first term a and common ratio $r \neq 1$:

$$\sum_{i=1}^{n} ar^{i-1} = a + ar + ar^2 + \cdots + ar^{n-1} = \frac{a(r^n - 1)}{r - 1}$$

48. Evaluate $\sum_{i=1}^{n} \frac{3}{2^{i-1}}$.

49. Evaluate $\sum_{i=1}^{n} (2i + 2^i)$.

50. Evaluate $\sum_{i=1}^{m} \left[\sum_{j=1}^{n} (i + j) \right]$.

F Proofs of Theorems

In this appendix we present proofs of several theorems that are stated in the main body of the text. The sections in which they occur are indicated in the margin.

Section 2.3

Limit Laws Suppose that c is a constant and the limits

$$\lim_{x \to a} f(x) = L \quad \text{and} \quad \lim_{x \to a} g(x) = M$$

exist. Then

1. $\lim_{x \to a} [f(x) + g(x)] = L + M$
2. $\lim_{x \to a} [f(x) - g(x)] = L - M$
3. $\lim_{x \to a} [cf(x)] = cL$
4. $\lim_{x \to a} [f(x)g(x)] = LM$
5. $\lim_{x \to a} \frac{f(x)}{g(x)} = \frac{L}{M}$ if $M \neq 0$

PROOF OF LAW 4 Let $\varepsilon > 0$ be given. We want to find $\delta > 0$ such that

$$\text{if} \quad 0 < |x - a| < \delta \quad \text{then} \quad |f(x)g(x) - LM| < \varepsilon$$

In order to get terms that contain $|f(x) - L|$ and $|g(x) - M|$, we add and subtract $Lg(x)$ as follows:

$$|f(x)g(x) - LM| = |f(x)g(x) - Lg(x) + Lg(x) - LM|$$
$$= |[f(x) - L]g(x) + L[g(x) - M]|$$
$$\leq |[f(x) - L]g(x)| + |L[g(x) - M]| \quad \text{(Triangle Inequality)}$$
$$= |f(x) - L||g(x)| + |L||g(x) - M|$$

We want to make each of these terms less than $\varepsilon/2$.

Since $\lim_{x \to a} g(x) = M$, there is a number $\delta_1 > 0$ such that

$$\text{if} \quad 0 < |x - a| < \delta_1 \quad \text{then} \quad |g(x) - M| < \frac{\varepsilon}{2(1 + |L|)}$$

Also, there is a number $\delta_2 > 0$ such that if $0 < |x - a| < \delta_2$, then

$$|g(x) - M| < 1$$

and therefore

$$|g(x)| = |g(x) - M + M| \leq |g(x) - M| + |M| < 1 + |M|$$

Since $\lim_{x \to a} f(x) = L$, there is a number $\delta_3 > 0$ such that

$$\text{if} \quad 0 < |x - a| < \delta_3 \quad \text{then} \quad |f(x) - L| < \frac{\varepsilon}{2(1 + |M|)}$$

Let $\delta = \min\{\delta_1, \delta_2, \delta_3\}$. If $0 < |x - a| < \delta$, then we have $0 < |x - a| < \delta_1$, $0 < |x - a| < \delta_2$, and $0 < |x - a| < \delta_3$, so we can combine the inequalities to obtain

$$|f(x)g(x) - LM| \leq |f(x) - L||g(x)| + |L||g(x) - M|$$

$$< \frac{\varepsilon}{2(1 + |M|)}(1 + |M|) + |L|\frac{\varepsilon}{2(1 + |L|)}$$

$$< \frac{\varepsilon}{2} + \frac{\varepsilon}{2} = \varepsilon$$

This shows that $\lim_{x \to a} [f(x)g(x)] = LM$. ■

PROOF OF LAW 3 If we take $g(x) = c$ in Law 4, we get

$$\lim_{x \to a} [cf(x)] = \lim_{x \to a} [g(x)f(x)] = \lim_{x \to a} g(x) \cdot \lim_{x \to a} f(x)$$

$$= \lim_{x \to a} c \cdot \lim_{x \to a} f(x)$$

$$= c \lim_{x \to a} f(x) \quad \text{(by Law 7)} \quad ■$$

PROOF OF LAW 2 Using Law 1 and Law 3 with $c = -1$, we have

$$\lim_{x \to a} [f(x) - g(x)] = \lim_{x \to a} [f(x) + (-1)g(x)] = \lim_{x \to a} f(x) + \lim_{x \to a} (-1)g(x)$$

$$= \lim_{x \to a} f(x) + (-1) \lim_{x \to a} g(x) = \lim_{x \to a} f(x) - \lim_{x \to a} g(x) \quad ■$$

PROOF OF LAW 5 First let us show that

$$\lim_{x \to a} \frac{1}{g(x)} = \frac{1}{M}$$

To do this we must show that, given $\varepsilon > 0$, there exists $\delta > 0$ such that

$$\text{if} \quad 0 < |x - a| < \delta \quad \text{then} \quad \left|\frac{1}{g(x)} - \frac{1}{M}\right| < \varepsilon$$

Observe that

$$\left|\frac{1}{g(x)} - \frac{1}{M}\right| = \frac{|M - g(x)|}{|Mg(x)|}$$

We know that we can make the numerator small. But we also need to know that the denominator is not small when x is near a. Since $\lim_{x \to a} g(x) = M$, there is a number $\delta_1 > 0$ such that, whenever $0 < |x - a| < \delta_1$, we have

$$|g(x) - M| < \frac{|M|}{2}$$

and therefore

$$|M| = |M - g(x) + g(x)| \leq |M - g(x)| + |g(x)|$$

$$< \frac{|M|}{2} + |g(x)|$$

This shows that

$$\text{if} \quad 0 < |x - a| < \delta_1 \quad \text{then} \quad |g(x)| > \frac{|M|}{2}$$

and so, for these values of x,

$$\frac{1}{|Mg(x)|} = \frac{1}{|M||g(x)|} < \frac{1}{|M|} \cdot \frac{2}{|M|} = \frac{2}{M^2}$$

Also, there exists $\delta_2 > 0$ such that

$$\text{if} \quad 0 < |x - a| < \delta_2 \quad \text{then} \quad |g(x) - M| < \frac{M^2}{2}\varepsilon$$

Let $\delta = \min\{\delta_1, \delta_2\}$. Then, for $0 < |x - a| < \delta$, we have

$$\left| \frac{1}{g(x)} - \frac{1}{M} \right| = \frac{|M - g(x)|}{|Mg(x)|} < \frac{2}{M^2} \frac{M^2}{2}\varepsilon = \varepsilon$$

It follows that $\lim_{x \to a} 1/g(x) = 1/M$. Finally, using Law 4, we obtain

$$\lim_{x \to a} \frac{f(x)}{g(x)} = \lim_{x \to a} \left(f(x) \cdot \frac{1}{g(x)} \right) = \lim_{x \to a} f(x) \lim_{x \to a} \frac{1}{g(x)} = L \cdot \frac{1}{M} = \frac{L}{M} \quad \blacksquare$$

2 Theorem If $f(x) \leq g(x)$ for all x in an open interval that contains a (except possibly at a) and

$$\lim_{x \to a} f(x) = L \quad \text{and} \quad \lim_{x \to a} g(x) = M$$

then $L \leq M$.

PROOF We use the method of proof by contradiction. Suppose, if possible, that $L > M$. Law 2 of limits says that

$$\lim_{x \to a} [g(x) - f(x)] = M - L$$

Therefore, for any $\varepsilon > 0$, there exists $\delta > 0$ such that

$$\text{if} \quad 0 < |x - a| < \delta \quad \text{then} \quad |[g(x) - f(x)] - (M - L)| < \varepsilon$$

In particular, taking $\varepsilon = L - M$ (noting that $L - M > 0$ by hypothesis), we have a number $\delta > 0$ such that

$$\text{if} \quad 0 < |x - a| < \delta \quad \text{then} \quad |[g(x) - f(x)] - (M - L)| < L - M$$

Since $b \leq |b|$ for any number b, we have

$$\text{if} \quad 0 < |x - a| < \delta \quad \text{then} \quad [g(x) - f(x)] - (M - L) < L - M$$

which simplifies to

$$\text{if} \quad 0 < |x - a| < \delta \quad \text{then} \quad g(x) < f(x)$$

But this contradicts $f(x) \leq g(x)$. Thus the inequality $L > M$ must be false. Therefore $L \leq M$. $\blacksquare$

APPENDIX F Proofs of Theorems

3 The Squeeze Theorem If $f(x) \leq g(x) \leq h(x)$ for all x in an open interval that contains a (except possibly at a) and

$$\lim_{x \to a} f(x) = \lim_{x \to a} h(x) = L$$

then

$$\lim_{x \to a} g(x) = L$$

PROOF Let $\varepsilon > 0$ be given. Since $\lim_{x \to a} f(x) = L$, there is a number $\delta_1 > 0$ such that

$$\text{if} \quad 0 < |x - a| < \delta_1 \quad \text{then} \quad |f(x) - L| < \varepsilon$$

that is,

$$\text{if} \quad 0 < |x - a| < \delta_1 \quad \text{then} \quad L - \varepsilon < f(x) < L + \varepsilon$$

Since $\lim_{x \to a} h(x) = L$, there is a number $\delta_2 > 0$ such that

$$\text{if} \quad 0 < |x - a| < \delta_2 \quad \text{then} \quad |h(x) - L| < \varepsilon$$

that is,

$$\text{if} \quad 0 < |x - a| < \delta_2 \quad \text{then} \quad L - \varepsilon < h(x) < L + \varepsilon$$

Let $\delta = \min\{\delta_1, \delta_2\}$. If $0 < |x - a| < \delta$, then $0 < |x - a| < \delta_1$ and $0 < |x - a| < \delta_2$, so

$$L - \varepsilon < f(x) \leq g(x) \leq h(x) < L + \varepsilon$$

In particular,

$$L - \varepsilon < g(x) < L + \varepsilon$$

and so $|g(x) - L| < \varepsilon$. Therefore $\lim_{x \to a} g(x) = L$. ∎

Section 2.5

Theorem If f is a one-to-one continuous function defined on an interval (a, b), then its inverse function f^{-1} is also continuous.

PROOF First we show that if f is both one-to-one and continuous on (a, b), then it must be either increasing or decreasing on (a, b). If it were neither increasing nor decreasing, then there would exist numbers x_1, x_2, and x_3 in (a, b) with $x_1 < x_2 < x_3$ such that $f(x_2)$ does not lie between $f(x_1)$ and $f(x_3)$. There are two possibilities: either (1) $f(x_3)$ lies between $f(x_1)$ and $f(x_2)$ or (2) $f(x_1)$ lies between $f(x_2)$ and $f(x_3)$. (Draw a picture.) In case (1) we apply the Intermediate Value Theorem to the continuous function f to get a number c between x_1 and x_2 such that $f(c) = f(x_3)$. In case (2) the Intermediate Value Theorem gives a number c between x_2 and x_3 such that $f(c) = f(x_1)$. In either case we have contradicted the fact that f is one-to-one.

Let us assume, for the sake of definiteness, that f is increasing on (a, b). We take any number y_0 in the domain of f^{-1} and we let $f^{-1}(y_0) = x_0$; that is, x_0 is the number in (a, b) such that $f(x_0) = y_0$. To show that f^{-1} is continuous at y_0 we take any $\varepsilon > 0$ such that the interval $(x_0 - \varepsilon, x_0 + \varepsilon)$ is contained in the interval (a, b). Since f is increasing, it maps the numbers in the interval $(x_0 - \varepsilon, x_0 + \varepsilon)$ onto the numbers in the interval $(f(x_0 - \varepsilon), f(x_0 + \varepsilon))$ and f^{-1} reverses the correspondence. If we let δ denote the smaller of the numbers $\delta_1 = y_0 - f(x_0 - \varepsilon)$ and $\delta_2 = f(x_0 + \varepsilon) - y_0$, then the interval $(y_0 - \delta, y_0 + \delta)$ is contained in the interval $(f(x_0 - \varepsilon), f(x_0 + \varepsilon))$ and so is mapped into the interval $(x_0 - \varepsilon, x_0 + \varepsilon)$ by f^{-1}. (See the arrow diagram in Figure 1.) We have therefore found a number $\delta > 0$ such that

$$\text{if} \quad |y - y_0| < \delta \quad \text{then} \quad |f^{-1}(y) - f^{-1}(y_0)| < \varepsilon$$

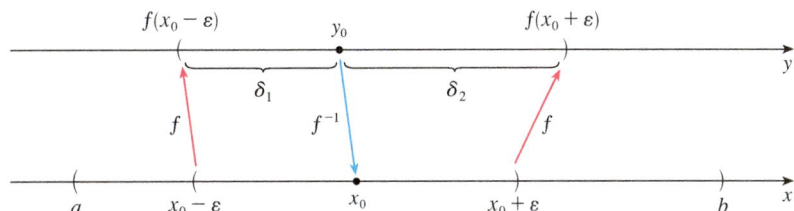

FIGURE 1

This shows that $\lim_{y \to y_0} f^{-1}(y) = f^{-1}(y_0)$ and so f^{-1} is continuous at any number y_0 in its domain. ∎

8 Theorem If f is continuous at b and $\lim_{x \to a} g(x) = b$, then

$$\lim_{x \to a} f(g(x)) = f(b)$$

PROOF Let $\varepsilon > 0$ be given. We want to find a number $\delta > 0$ such that

$$\text{if} \quad 0 < |x - a| < \delta \quad \text{then} \quad |f(g(x)) - f(b)| < \varepsilon$$

Since f is continuous at b, we have

$$\lim_{y \to b} f(y) = f(b)$$

and so there exists $\delta_1 > 0$ such that

$$\text{if} \quad 0 < |y - b| < \delta_1 \quad \text{then} \quad |f(y) - f(b)| < \varepsilon$$

Since $\lim_{x \to a} g(x) = b$, there exists $\delta > 0$ such that

$$\text{if} \quad 0 < |x - a| < \delta \quad \text{then} \quad |g(x) - b| < \delta_1$$

Combining these two statements, we see that whenever $0 < |x - a| < \delta$ we have $|g(x) - b| < \delta_1$, which implies that $|f(g(x)) - f(b)| < \varepsilon$. Therefore we have proved that $\lim_{x \to a} f(g(x)) = f(b)$. ∎

Section 3.3

The proof of the following result was promised when we proved that $\lim_{\theta \to 0} \dfrac{\sin \theta}{\theta} = 1$.

Theorem If $0 < \theta < \pi/2$, then $\theta \leq \tan \theta$.

PROOF Figure 2 shows a sector of a circle with center O, central angle θ, and radius 1. Then

$$|AD| = |OA| \tan \theta = \tan \theta$$

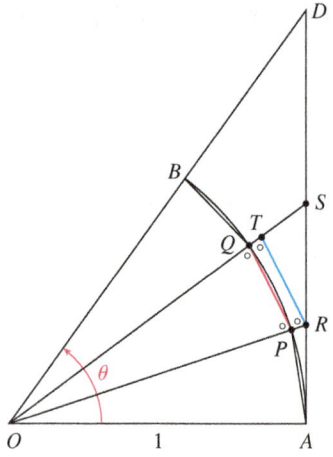

FIGURE 2

We approximate the arc AB by an inscribed polygon consisting of n equal line segments and we look at a typical segment PQ. We extend the lines OP and OQ to meet AD in the points R and S. Then we draw $RT \parallel PQ$ as in Figure 2. Observe that

$$\angle RTO = \angle PQO < 90°$$

and so $\angle RTS > 90°$. Therefore we have

$$|PQ| < |RT| < |RS|$$

If we add n such inequalities, we get

$$L_n < |AD| = \tan \theta$$

where L_n is the length of the inscribed polygon. Thus, by Theorem 2.3.2, we have

$$\lim_{n \to \infty} L_n \leq \tan \theta$$

But the arc length is defined in Equation 8.1.1 as the limit of the lengths of inscribed polygons, so

$$\theta = \lim_{n \to \infty} L_n \leq \tan \theta \qquad \blacksquare$$

Section 4.3

Concavity Test
(a) If $f''(x) > 0$ for all x in I, then the graph of f is concave upward on I.
(b) If $f''(x) < 0$ for all x in I, then the graph of f is concave downward on I.

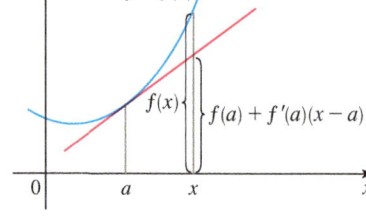

FIGURE 3

PROOF OF (a) Let a be any number in I. We must show that the curve $y = f(x)$ lies above the tangent line at the point $(a, f(a))$. The equation of this tangent is

$$y = f(a) + f'(a)(x - a)$$

So we must show that

$$f(x) > f(a) + f'(a)(x - a)$$

whenever $x \in I$ ($x \neq a$). (See Figure 3.)

First let us take the case where $x > a$. Applying the Mean Value Theorem to f on the interval $[a, x]$, we get a number c, with $a < c < x$, such that

$$\boxed{1} \qquad f(x) - f(a) = f'(c)(x - a)$$

Since $f'' > 0$ on I, we know from the Increasing/Decreasing Test that f' is increasing on I. Thus, since $a < c$, we have

$$f'(a) < f'(c)$$

and so, multiplying this inequality by the positive number $x - a$, we get

$$\boxed{2} \qquad f'(a)(x - a) < f'(c)(x - a)$$

Now we add $f(a)$ to both sides of this inequality:

$$f(a) + f'(a)(x - a) < f(a) + f'(c)(x - a)$$

But from Equation 1 we have $f(x) = f(a) + f'(c)(x - a)$. So this inequality becomes

$$\boxed{3} \qquad f(x) > f(a) + f'(a)(x - a)$$

which is what we wanted to prove.

For the case where $x < a$ we have $f'(c) < f'(a)$, but multiplication by the negative number $x - a$ reverses the inequality, so we get (2) and (3) as before. ∎

Section 4.4

In order to give the promised proof of l'Hospital's Rule, we first need a generalization of the Mean Value Theorem. The following theorem is named after another French mathematician, Augustin-Louis Cauchy (1789–1857).

See the biographical sketch of Cauchy on page 109.

> $\boxed{1}$ **Cauchy's Mean Value Theorem** Suppose that the functions f and g are continuous on $[a, b]$ and differentiable on (a, b), and $g'(x) \neq 0$ for all x in (a, b). Then there is a number c in (a, b) such that
>
> $$\frac{f'(c)}{g'(c)} = \frac{f(b) - f(a)}{g(b) - g(a)}$$

Notice that if we take the special case in which $g(x) = x$, then $g'(c) = 1$ and Theorem 1 is just the ordinary Mean Value Theorem. Furthermore, Theorem 1 can be proved in a similar manner. You can verify that all we have to do is change the function h given by Equation 4.2.4 to the function

$$h(x) = f(x) - f(a) - \frac{f(b) - f(a)}{g(b) - g(a)}[g(x) - g(a)]$$

and apply Rolle's Theorem as before.

> **L'Hospital's Rule** Suppose f and g are differentiable and $g'(x) \neq 0$ on an open interval I that contains a (except possibly at a). Suppose that
>
> $$\lim_{x \to a} f(x) = 0 \quad \text{and} \quad \lim_{x \to a} g(x) = 0$$
>
> or that $\qquad \lim_{x \to a} f(x) = \pm\infty \quad \text{and} \quad \lim_{x \to a} g(x) = \pm\infty$
>
> (In other words, we have an indeterminate form of type $\frac{0}{0}$ or ∞/∞.) Then
>
> $$\lim_{x \to a} \frac{f(x)}{g(x)} = \lim_{x \to a} \frac{f'(x)}{g'(x)}$$
>
> if the limit on the right side exists (or is ∞ or $-\infty$).

PROOF OF L'HOSPITAL'S RULE We are assuming that $\lim_{x \to a} f(x) = 0$ and $\lim_{x \to a} g(x) = 0$. Let

$$L = \lim_{x \to a} \frac{f'(x)}{g'(x)}$$

We must show that $\lim_{x \to a} f(x)/g(x) = L$. Define

$$F(x) = \begin{cases} f(x) & \text{if } x \neq a \\ 0 & \text{if } x = a \end{cases} \qquad G(x) = \begin{cases} g(x) & \text{if } x \neq a \\ 0 & \text{if } x = a \end{cases}$$

Then F is continuous on I since f is continuous on $\{x \in I \mid x \neq a\}$ and

$$\lim_{x \to a} F(x) = \lim_{x \to a} f(x) = 0 = F(a)$$

Likewise, G is continuous on I. Let $x \in I$ and $x > a$. Then F and G are continuous on $[a, x]$ and differentiable on (a, x) and $G' \neq 0$ there (since $F' = f'$ and $G' = g'$). Therefore, by Cauchy's Mean Value Theorem, there is a number y such that $a < y < x$ and

$$\frac{F'(y)}{G'(y)} = \frac{F(x) - F(a)}{G(x) - G(a)} = \frac{F(x)}{G(x)}$$

Here we have used the fact that, by definition, $F(a) = 0$ and $G(a) = 0$. Now, if we let $x \to a^+$, then $y \to a^+$ (since $a < y < x$), so

$$\lim_{x \to a^+} \frac{f(x)}{g(x)} = \lim_{x \to a^+} \frac{F(x)}{G(x)} = \lim_{y \to a^+} \frac{F'(y)}{G'(y)} = \lim_{y \to a^+} \frac{f'(y)}{g'(y)} = L$$

A similar argument shows that the left-hand limit is also L. Therefore

$$\lim_{x \to a} \frac{f(x)}{g(x)} = L$$

This proves l'Hospital's Rule for the case where a is finite.

If a is infinite, we let $t = 1/x$. Then $t \to 0^+$ as $x \to \infty$, so we have

$$\lim_{x \to \infty} \frac{f(x)}{g(x)} = \lim_{t \to 0^+} \frac{f(1/t)}{g(1/t)}$$

$$= \lim_{t \to 0^+} \frac{f'(1/t)(-1/t^2)}{g'(1/t)(-1/t^2)} \qquad \text{(by l'Hospital's Rule for finite } a\text{)}$$

$$= \lim_{t \to 0^+} \frac{f'(1/t)}{g'(1/t)} = \lim_{x \to \infty} \frac{f'(x)}{g'(x)} \qquad \blacksquare$$

Section 11.8

In order to prove Theorem 11.8.4, we first need the following results.

> **Theorem**
>
> **1.** If a power series $\Sigma\, c_n x^n$ converges when $x = b$ (where $b \neq 0$), then it converges whenever $|x| < |b|$.
>
> **2.** If a power series $\Sigma\, c_n x^n$ diverges when $x = d$ (where $d \neq 0$), then it diverges whenever $|x| > |d|$.

PROOF OF 1 Suppose that $\Sigma c_n b^n$ converges. Then, by Theorem 11.2.6, we have $\lim_{n \to \infty} c_n b^n = 0$. According to Definition 11.1.2 with $\varepsilon = 1$, there is a positive integer N such that $|c_n b^n| < 1$ whenever $n \geq N$. Thus, for $n \geq N$, we have

$$|c_n x^n| = \left|\frac{c_n b^n x^n}{b^n}\right| = |c_n b^n| \left|\frac{x}{b}\right|^n < \left|\frac{x}{b}\right|^n$$

If $|x| < |b|$, then $|x/b| < 1$, so $\Sigma |x/b|^n$ is a convergent geometric series. Therefore, by the Comparison Test, the series $\sum_{n=N}^{\infty} |c_n x^n|$ is convergent. Thus the series $\Sigma c_n x^n$ is absolutely convergent and therefore convergent. ■

PROOF OF 2 Suppose that $\Sigma c_n d^n$ diverges. If x is any number such that $|x| > |d|$, then $\Sigma c_n x^n$ cannot converge because, by part 1, the convergence of $\Sigma c_n x^n$ would imply the convergence of $\Sigma c_n d^n$. Therefore $\Sigma c_n x^n$ diverges whenever $|x| > |d|$. ■

Theorem For a power series $\Sigma c_n x^n$ there are only three possibilities:
1. The series converges only when $x = 0$.
2. The series converges for all x.
3. There is a positive number R such that the series converges if $|x| < R$ and diverges if $|x| > R$.

PROOF Suppose that neither case 1 nor case 2 is true. Then there are nonzero numbers b and d such that $\Sigma c_n x^n$ converges for $x = b$ and diverges for $x = d$. Therefore the set $S = \{x \mid \Sigma c_n x^n \text{ converges}\}$ is not empty. By the preceding theorem, the series diverges if $|x| > |d|$, so $|x| \leq |d|$ for all $x \in S$. This says that $|d|$ is an upper bound for the set S. Thus, by the Completeness Axiom (see Section 11.1), S has a least upper bound R. If $|x| > R$, then $x \notin S$, so $\Sigma c_n x^n$ diverges. If $|x| < R$, then $|x|$ is not an upper bound for S and so there exists $b \in S$ such that $b > |x|$. Since $b \in S$, $\Sigma c_n x^n$ converges, so by the preceding theorem $\Sigma c_n x^n$ converges. ■

4 Theorem For a power series $\Sigma c_n (x - a)^n$ there are only three possibilities:
1. The series converges only when $x = a$.
2. The series converges for all x.
3. There is a positive number R such that the series converges if $|x - a| < R$ and diverges if $|x - a| > R$.

PROOF If we make the change of variable $u = x - a$, then the power series becomes $\Sigma c_n u^n$ and we can apply the preceding theorem to this series. In case 3 we have convergence for $|u| < R$ and divergence for $|u| > R$. Thus we have convergence for $|x - a| < R$ and divergence for $|x - a| > R$. ■

Section 14.3

> **Clairaut's Theorem** Suppose f is defined on a disk D that contains the point (a, b). If the functions f_{xy} and f_{yx} are both continuous on D, then $f_{xy}(a, b) = f_{yx}(a, b)$.

PROOF For small values of h, $h \neq 0$, consider the difference

$$\Delta(h) = [f(a + h, b + h) - f(a + h, b)] - [f(a, b + h) - f(a, b)]$$

Notice that if we let $g(x) = f(x, b + h) - f(x, b)$, then

$$\Delta(h) = g(a + h) - g(a)$$

By the Mean Value Theorem, there is a number c between a and $a + h$ such that

$$g(a + h) - g(a) = g'(c)h = h[f_x(c, b + h) - f_x(c, b)]$$

Applying the Mean Value Theorem again, this time to f_x, we get a number d between b and $b + h$ such that

$$f_x(c, b + h) - f_x(c, b) = f_{xy}(c, d)h$$

Combining these equations, we obtain

$$\Delta(h) = h^2 f_{xy}(c, d)$$

If $h \to 0$, then $(c, d) \to (a, b)$, so the continuity of f_{xy} at (a, b) gives

$$\lim_{h \to 0} \frac{\Delta(h)}{h^2} = \lim_{(c, d) \to (a, b)} f_{xy}(c, d) = f_{xy}(a, b)$$

Similarly, by writing

$$\Delta(h) = [f(a + h, b + h) - f(a, b + h)] - [f(a + h, b) - f(a, b)]$$

and using the Mean Value Theorem twice and the continuity of f_{yx} at (a, b), we obtain

$$\lim_{h \to 0} \frac{\Delta(h)}{h^2} = f_{yx}(a, b)$$

It follows that $f_{xy}(a, b) = f_{yx}(a, b)$. ∎

Section 14.4

> **[8] Theorem** If the partial derivatives f_x and f_y exist near (a, b) and are continuous at (a, b), then f is differentiable at (a, b).

PROOF Let

$$\Delta z = f(a + \Delta x, b + \Delta y) - f(a, b)$$

According to (14.4.7), to prove that f is differentiable at (a, b) we have to show that we can write Δz in the form

$$\Delta z = f_x(a, b)\,\Delta x + f_y(a, b)\,\Delta y + \varepsilon_1\,\Delta x + \varepsilon_2\,\Delta y$$

where ε_1 and $\varepsilon_2 \to 0$ as $(\Delta x, \Delta y) \to (0, 0)$.

Referring to Figure 4, we write

$$\boxed{1} \quad \Delta z = [f(a + \Delta x, b + \Delta y) - f(a, b + \Delta y)] + [f(a, b + \Delta y) - f(a, b)]$$

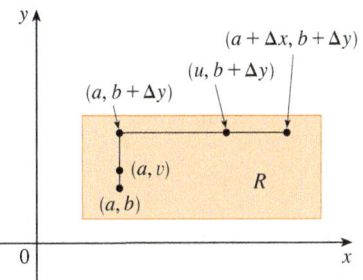

FIGURE 4

Observe that the function of a single variable

$$g(x) = f(x, b + \Delta y)$$

is defined on the interval $[a, a + \Delta x]$ and $g'(x) = f_x(x, b + \Delta y)$. If we apply the Mean Value Theorem to g, we get

$$g(a + \Delta x) - g(a) = g'(u) \Delta x$$

where u is some number between a and $a + \Delta x$. In terms of f, this equation becomes

$$f(a + \Delta x, b + \Delta y) - f(a, b + \Delta y) = f_x(u, b + \Delta y) \Delta x$$

This gives us an expression for the first part of the right side of Equation 1. For the second part we let $h(y) = f(a, y)$. Then h is a function of a single variable defined on the interval $[b, b + \Delta y]$ and $h'(y) = f_y(a, y)$. A second application of the Mean Value Theorem then gives

$$h(b + \Delta y) - h(b) = h'(v) \Delta y$$

where v is some number between b and $b + \Delta y$. In terms of f, this becomes

$$f(a, b + \Delta y) - f(a, b) = f_y(a, v) \Delta y$$

We now substitute these expressions into Equation 1 and obtain

$$\Delta z = f_x(u, b + \Delta y) \Delta x + f_y(a, v) \Delta y$$
$$= f_x(a, b) \Delta x + [f_x(u, b + \Delta y) - f_x(a, b)] \Delta x + f_y(a, b) \Delta y$$
$$\quad + [f_y(a, v) - f_y(a, b)] \Delta y$$
$$= f_x(a, b) \Delta x + f_y(a, b) \Delta y + \varepsilon_1 \Delta x + \varepsilon_2 \Delta y$$

where
$$\varepsilon_1 = f_x(u, b + \Delta y) - f_x(a, b)$$
$$\varepsilon_2 = f_y(a, v) - f_y(a, b)$$

Since $(u, b + \Delta y) \to (a, b)$ and $(a, v) \to (a, b)$ as $(\Delta x, \Delta y) \to (0, 0)$ and since f_x and f_y are continuous at (a, b), we see that $\varepsilon_1 \to 0$ and $\varepsilon_2 \to 0$ as $(\Delta x, \Delta y) \to (0, 0)$.

Therefore f is differentiable at (a, b).

G The Logarithm Defined as an Integral

Our treatment of exponential and logarithmic functions until now has relied on our intuition, which is based on numerical and visual evidence. (See Sections 1.4, 1.5, and 3.1.) Here we use the Fundamental Theorem of Calculus to give an alternative treatment that provides a surer footing for these functions.

Instead of starting with b^x and defining $\log_b x$ as its inverse, this time we start by defining $\ln x$ as an integral and then define the exponential function as its inverse. You should bear in mind that we do not use any of our previous definitions and results concerning exponential and logarithmic functions.

The Natural Logarithm

We first define $\ln x$ as an integral.

> **1 Definition** The **natural logarithmic function** is the function defined by
> $$\ln x = \int_1^x \frac{1}{t}\,dt \qquad x > 0$$

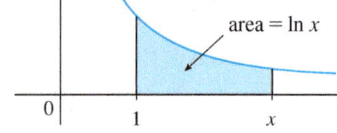

FIGURE 1

The existence of this function depends on the fact that the integral of a continuous function always exists. If $x > 1$, then $\ln x$ can be interpreted geometrically as the area under the hyperbola $y = 1/t$ from $t = 1$ to $t = x$. (See Figure 1.) For $x = 1$, we have
$$\ln 1 = \int_1^1 \frac{1}{t}\,dt = 0$$

For $0 < x < 1$,
$$\ln x = \int_1^x \frac{1}{t}\,dt = -\int_x^1 \frac{1}{t}\,dt < 0$$

and so $\ln x$ is the negative of the area shown in Figure 2.

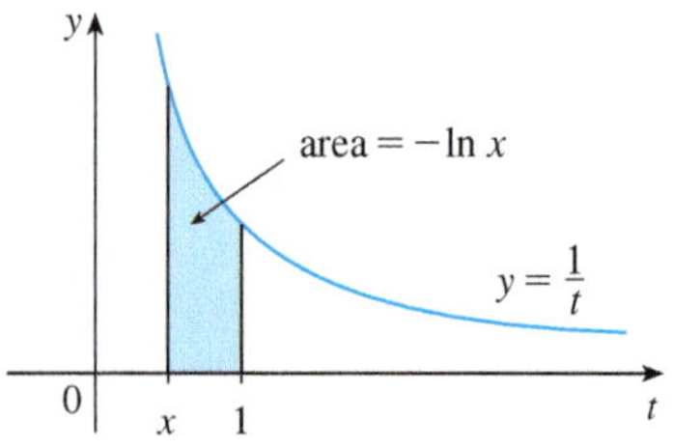

FIGURE 2

EXAMPLE 1

(a) By comparing areas, show that $\tfrac{1}{2} < \ln 2 < \tfrac{3}{4}$.

(b) Use the Midpoint Rule with $n = 10$ to estimate the value of $\ln 2$.

SOLUTION

(a) We can interpret $\ln 2$ as the area under the curve $y = 1/t$ from 1 to 2. From Figure 3 we see that this area is larger than the area of rectangle $BCDE$ and smaller than the area of trapezoid $ABCD$. Thus we have
$$\tfrac{1}{2} \cdot 1 < \ln 2 < 1 \cdot \tfrac{1}{2}\bigl(1 + \tfrac{1}{2}\bigr)$$
$$\tfrac{1}{2} < \ln 2 < \tfrac{3}{4}$$

(b) If we use the Midpoint Rule with $f(t) = 1/t$, $n = 10$, and $\Delta t = 0.1$, we get
$$\ln 2 = \int_1^2 \frac{1}{t}\,dt \approx (0.1)[f(1.05) + f(1.15) + \cdots + f(1.95)]$$
$$= (0.1)\left(\frac{1}{1.05} + \frac{1}{1.15} + \cdots + \frac{1}{1.95}\right) \approx 0.693 \quad\blacksquare$$

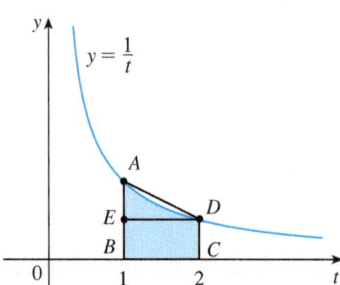

FIGURE 3

Notice that the integral that defines $\ln x$ is exactly the type of integral discussed in Part 1 of the Fundamental Theorem of Calculus (see Section 5.3). In fact, using that theorem, we have

$$\frac{d}{dx}\int_1^x \frac{1}{t}\,dt = \frac{1}{x}$$

and so

$$\boxed{2} \qquad \boxed{\frac{d}{dx}(\ln x) = \frac{1}{x}}$$

We now use this differentiation rule to prove the following properties of the logarithm function.

$\boxed{3}$ **Laws of Logarithms** If x and y are positive numbers and r is a rational number, then

1. $\ln(xy) = \ln x + \ln y$ 2. $\ln\left(\dfrac{x}{y}\right) = \ln x - \ln y$ 3. $\ln(x^r) = r \ln x$

PROOF

1. Let $f(x) = \ln(ax)$, where a is a positive constant. Then, using Equation 2 and the Chain Rule, we have

$$f'(x) = \frac{1}{ax}\frac{d}{dx}(ax) = \frac{1}{ax}\cdot a = \frac{1}{x}$$

Therefore $f(x)$ and $\ln x$ have the same derivative and so they must differ by a constant:

$$\ln(ax) = \ln x + C$$

Putting $x = 1$ in this equation, we get $\ln a = \ln 1 + C = 0 + C = C$. Thus

$$\ln(ax) = \ln x + \ln a$$

If we now replace the constant a by any number y, we have

$$\ln(xy) = \ln x + \ln y$$

2. Using Law 1 with $x = 1/y$, we have

$$\ln\frac{1}{y} + \ln y = \ln\left(\frac{1}{y}\cdot y\right) = \ln 1 = 0$$

and so

$$\ln\frac{1}{y} = -\ln y$$

Using Law 1 again, we have

$$\ln\left(\frac{x}{y}\right) = \ln\left(x\cdot\frac{1}{y}\right) = \ln x + \ln\frac{1}{y} = \ln x - \ln y$$

The proof of Law 3 is left as an exercise. ∎

In order to graph $y = \ln x$, we first determine its limits:

4 (a) $\lim_{x \to \infty} \ln x = \infty$ (b) $\lim_{x \to 0^+} \ln x = -\infty$

PROOF
(a) Using Law 3 with $x = 2$ and $r = n$ (where n is any positive integer), we have $\ln(2^n) = n \ln 2$. Now $\ln 2 > 0$, so this shows that $\ln(2^n) \to \infty$ as $n \to \infty$. But $\ln x$ is an increasing function since its derivative $1/x > 0$. Therefore $\ln x \to \infty$ as $x \to \infty$.

(b) If we let $t = 1/x$, then $t \to \infty$ as $x \to 0^+$. Thus, using (a), we have

$$\lim_{x \to 0^+} \ln x = \lim_{t \to \infty} \ln\left(\frac{1}{t}\right) = \lim_{t \to \infty} (-\ln t) = -\infty$$

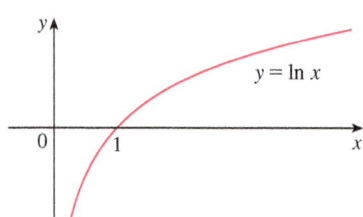

FIGURE 4

If $y = \ln x$, $x > 0$, then

$$\frac{dy}{dx} = \frac{1}{x} > 0 \quad \text{and} \quad \frac{d^2y}{dx^2} = -\frac{1}{x^2} < 0$$

which shows that $\ln x$ is increasing and concave downward on $(0, \infty)$. Putting this information together with (4), we draw the graph of $y = \ln x$ in Figure 4.

Since $\ln 1 = 0$ and $\ln x$ is an increasing continuous function that takes on arbitrarily large values, the Intermediate Value Theorem shows that there is a number where $\ln x$ takes on the value 1. (See Figure 5.) This important number is denoted by e.

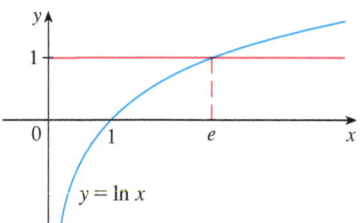

FIGURE 5

5 Definition e is the number such that $\ln e = 1$.

We will show (in Theorem 19) that this definition is consistent with our previous definition of e.

The Natural Exponential Function

Since $\ln$ is an increasing function, it is one-to-one and therefore has an inverse function, which we denote by exp. Thus, according to the definition of an inverse function,

$f^{-1}(x) = y \iff f(y) = x$

6 $\exp(x) = y \iff \ln y = x$

and the cancellation equations are

$f^{-1}(f(x)) = x$
$f(f^{-1}(x)) = x$

7 $\exp(\ln x) = x \quad \text{and} \quad \ln(\exp x) = x$

In particular, we have

$$\exp(0) = 1 \quad \text{since} \quad \ln 1 = 0$$

$$\exp(1) = e \quad \text{since} \quad \ln e = 1$$

We obtain the graph of $y = \exp x$ by reflecting the graph of $y = \ln x$ about the line

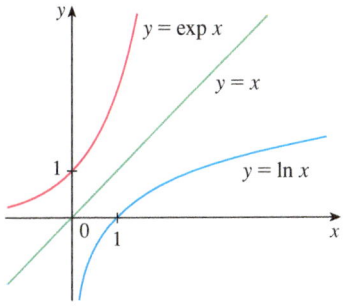

FIGURE 6

$y = x$. (See Figure 6.) The domain of exp is the range of ln, that is, $(-\infty, \infty)$; the range of exp is the domain of ln, that is, $(0, \infty)$.

If r is any rational number, then the third law of logarithms gives

$$\ln(e^r) = r \ln e = r$$

Therefore, by (6), $\quad \exp(r) = e^r$

Thus $\exp(x) = e^x$ whenever x is a rational number. This leads us to define e^x, even for irrational values of x, by the equation

$$e^x = \exp(x)$$

In other words, for the reasons given, we define e^x to be the inverse of the function $\ln x$. In this notation (6) becomes

8
$$e^x = y \iff \ln y = x$$

and the cancellation equations (7) become

9
$$e^{\ln x} = x \qquad x > 0$$

10
$$\ln(e^x) = x \qquad \text{for all } x$$

The natural exponential function $f(x) = e^x$ is one of the most frequently occurring functions in calculus and its applications, so it is important to be familiar with its graph (Figure 7) and its properties (which follow from the fact that it is the inverse of the natural logarithmic function).

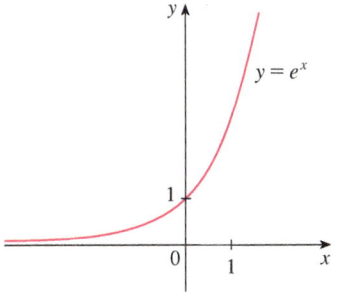

FIGURE 7
The natural exponential function

Properties of the Exponential Function The exponential function $f(x) = e^x$ is an increasing continuous function with domain $\mathbb{R}$ and range $(0, \infty)$. Thus $e^x > 0$ for all x. Also

$$\lim_{x \to -\infty} e^x = 0 \qquad \lim_{x \to \infty} e^x = \infty$$

So the x-axis is a horizontal asymptote of $f(x) = e^x$.

We now verify that f has the other properties expected of an exponential function.

11 Laws of Exponents If x and y are real numbers and r is rational, then

1. $e^{x+y} = e^x e^y$ 2. $e^{x-y} = \dfrac{e^x}{e^y}$ 3. $(e^x)^r = e^{rx}$

PROOF OF LAW 1 Using the first law of logarithms and Equation 10, we have

$$\ln(e^x e^y) = \ln(e^x) + \ln(e^y) = x + y = \ln(e^{x+y})$$

Since ln is a one-to-one function, it follows that $e^x e^y = e^{x+y}$.

Laws 2 and 3 are proved similarly (see Exercises 6 and 7). As we will soon see, Law 3 actually holds when r is any real number. ∎

We now prove the differentiation formula for e^x.

$$\boxed{\frac{d}{dx}(e^x) = e^x}$$ (12)

PROOF The function $y = e^x$ is differentiable because it is the inverse function of $y = \ln x$, which we know is differentiable with nonzero derivative. To find its derivative, we use the inverse function method. Let $y = e^x$. Then $\ln y = x$ and, differentiating this latter equation implicitly with respect to x, we get

$$\frac{1}{y}\frac{dy}{dx} = 1$$

$$\frac{dy}{dx} = y = e^x$$ ∎

■ General Exponential Functions

If $b > 0$ and r is any rational number, then by (9) and (11),

$$b^r = (e^{\ln b})^r = e^{r \ln b}$$

Therefore, even for irrational numbers x, we *define*

$$\boxed{b^x = e^{x \ln b}}$$ (13)

Thus, for instance,

$$2^{\sqrt{3}} = e^{\sqrt{3} \ln 2} \approx e^{1.20} \approx 3.32$$

The function $f(x) = b^x$ is called the **exponential function with base b**. Notice that b^x is positive for all x because e^x is positive for all x.

Definition 13 allows us to extend one of the laws of logarithms. We already know that $\ln(b^r) = r \ln b$ when r is rational. But if we now let r be *any* real number we have, from Definition 13,

$$\ln b^r = \ln(e^{r \ln b}) = r \ln b$$

Thus

$$\ln b^r = r \ln b \qquad \text{for any real number } r$$ (14)

The general laws of exponents follow from Definition 13 together with the laws of exponents for e^x.

> **15 Laws of Exponents** If x and y are real numbers and $a, b > 0$, then
> **1.** $b^{x+y} = b^x b^y$ **2.** $b^{x-y} = b^x/b^y$ **3.** $(b^x)^y = b^{xy}$ **4.** $(ab)^x = a^x b^x$

PROOF

1. Using Definition 13 and the laws of exponents for e^x, we have
$$b^{x+y} = e^{(x+y)\ln b} = e^{x\ln b + y\ln b}$$
$$= e^{x\ln b} e^{y\ln b} = b^x b^y$$

3. Using Equation 14 we obtain
$$(b^x)^y = e^{y\ln(b^x)} = e^{yx\ln b} = e^{xy\ln b} = b^{xy}$$

The remaining proofs are left as exercises.

The differentiation formula for exponential functions is also a consequence of Definition 13:

$$\boxed{\; \frac{d}{dx}(b^x) = b^x \ln b \;} \quad \text{\small 16}$$

PROOF
$$\frac{d}{dx}(b^x) = \frac{d}{dx}(e^{x\ln b}) = e^{x\ln b}\frac{d}{dx}(x\ln b) = b^x \ln b$$

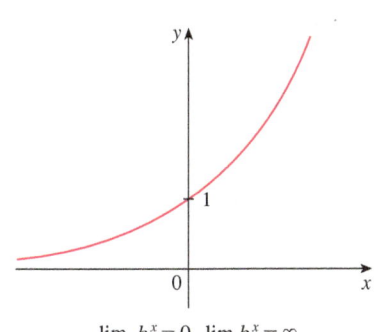

FIGURE 8 $y = b^x$, $b > 1$
$\lim_{x \to -\infty} b^x = 0, \; \lim_{x \to \infty} b^x = \infty$

If $b > 1$, then $\ln b > 0$, so $(d/dx) b^x = b^x \ln b > 0$, which shows that $y = b^x$ is increasing (see Figure 8). If $0 < b < 1$, then $\ln b < 0$ and so $y = b^x$ is decreasing (see Figure 9).

General Logarithmic Functions

If $b > 0$ and $b \neq 1$, then $f(x) = b^x$ is a one-to-one function. Its inverse function is called the **logarithmic function with base b** and is denoted by $\log_b$. Thus

$$\boxed{\; \log_b x = y \iff b^y = x \;} \quad \text{\small 17}$$

In particular, we see that

$$\log_e x = \ln x$$

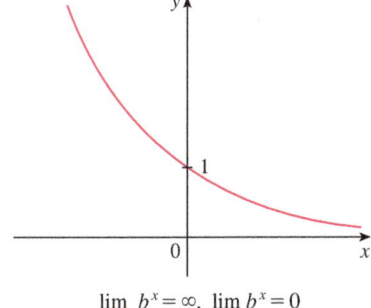

FIGURE 9 $y = b^x$, $0 < b < 1$
$\lim_{x \to -\infty} b^x = \infty, \; \lim_{x \to \infty} b^x = 0$

The laws of logarithms are similar to those for the natural logarithm and can be deduced from the laws of exponents (see Exercise 10).

To differentiate $y = \log_b x$, we write the equation as $b^y = x$. From Equation 14 we have $y \ln b = \ln x$, so

$$\log_b x = y = \frac{\ln x}{\ln b}$$

Since $\ln b$ is a constant, we can differentiate as follows:

$$\frac{d}{dx}(\log_b x) = \frac{d}{dx}\frac{\ln x}{\ln b} = \frac{1}{\ln b}\frac{d}{dx}(\ln x) = \frac{1}{x \ln b}$$

$$\boxed{18 \quad \frac{d}{dx}(\log_b x) = \frac{1}{x \ln b}}$$

The Number e Expressed as a Limit

In this section we defined e as the number such that $\ln e = 1$. The next theorem shows that this is the same as the number e defined in Section 3.1 (see Equation 3.6.5).

$$\boxed{19 \quad e = \lim_{x \to 0}(1 + x)^{1/x}}$$

PROOF Let $f(x) = \ln x$. Then $f'(x) = 1/x$, so $f'(1) = 1$. But, by the definition of derivative,

$$f'(1) = \lim_{h \to 0}\frac{f(1+h) - f(1)}{h} = \lim_{x \to 0}\frac{f(1+x) - f(1)}{x}$$

$$= \lim_{x \to 0}\frac{\ln(1+x) - \ln 1}{x} = \lim_{x \to 0}\frac{1}{x}\ln(1+x) = \lim_{x \to 0}\ln(1+x)^{1/x}$$

Because $f'(1) = 1$, we have

$$\lim_{x \to 0}\ln(1+x)^{1/x} = 1$$

Then, by Theorem 2.5.8 and the continuity of the exponential function, we have

$$e = e^1 = e^{\lim_{x \to 0}\ln(1+x)^{1/x}} = \lim_{x \to 0}e^{\ln(1+x)^{1/x}} = \lim_{x \to 0}(1+x)^{1/x} \quad \blacksquare$$

G EXERCISES

1. (a) By comparing areas, show that

$$\tfrac{1}{3} < \ln 1.5 < \tfrac{5}{12}$$

(b) Use the Midpoint Rule with $n = 10$ to estimate $\ln 1.5$.

2. Refer to Example 1.
(a) Find the equation of the tangent line to the curve $y = 1/t$ that is parallel to the secant line AD.

(b) Use part (a) to show that $\ln 2 > 0.66$.

3. By comparing areas, show that

$$\frac{1}{2} + \frac{1}{3} + \cdots + \frac{1}{n} < \ln n < 1 + \frac{1}{2} + \frac{1}{3} + \cdots + \frac{1}{n-1}$$

4. (a) By comparing areas, show that $\ln 2 < 1 < \ln 3$.
(b) Deduce that $2 < e < 3$.

5. Prove the third law of logarithms. [*Hint:* Start by showing that both sides of the equation have the same derivative.]

6. Prove the second law of exponents for e^x [see (11)].

7. Prove the third law of exponents for e^x [see (11)].

8. Prove the second law of exponents [see (15)].

9. Prove the fourth law of exponents [see (15)].

10. Deduce the following laws of logarithms from (15):
   (a) $\log_b(xy) = \log_b x + \log_b y$
   (b) $\log_b(x/y) = \log_b x - \log_b y$
   (c) $\log_b(x^y) = y \log_b x$

H Complex Numbers

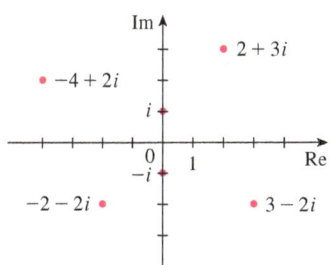

FIGURE 1
Complex numbers as points in the Argand plane

A **complex number** can be represented by an expression of the form $a + bi$, where a and b are real numbers and i is a symbol with the property that $i^2 = -1$. The complex number $a + bi$ can also be represented by the ordered pair (a, b) and plotted as a point in a plane (called the Argand plane) as in Figure 1. Thus the complex number $i = 0 + 1 \cdot i$ is identified with the point $(0, 1)$.

The **real part** of the complex number $a + bi$ is the real number a and the **imaginary part** is the real number b. Thus the real part of $4 - 3i$ is 4 and the imaginary part is -3. Two complex numbers $a + bi$ and $c + di$ are **equal** if $a = c$ and $b = d$, that is, their real parts are equal and their imaginary parts are equal. In the Argand plane the horizontal axis is called the real axis and the vertical axis is called the imaginary axis.

The sum and difference of two complex numbers are defined by adding or subtracting their real parts and their imaginary parts:

$$(a + bi) + (c + di) = (a + c) + (b + d)i$$
$$(a + bi) - (c + di) = (a - c) + (b - d)i$$

For instance,

$$(1 - i) + (4 + 7i) = (1 + 4) + (-1 + 7)i = 5 + 6i$$

The product of complex numbers is defined so that the usual commutative and distributive laws hold:

$$(a + bi)(c + di) = a(c + di) + (bi)(c + di)$$
$$= ac + adi + bci + bdi^2$$

Since $i^2 = -1$, this becomes

$$(a + bi)(c + di) = (ac - bd) + (ad + bc)i$$

EXAMPLE 1

$$(-1 + 3i)(2 - 5i) = (-1)(2 - 5i) + 3i(2 - 5i)$$
$$= -2 + 5i + 6i - 15(-1) = 13 + 11i$$ ∎

Division of complex numbers is much like rationalizing the denominator of a rational expression. For the complex number $z = a + bi$, we define its **complex conjugate** to be $\bar{z} = a - bi$. To find the quotient of two complex numbers we multiply numerator and denominator by the complex conjugate of the denominator.

EXAMPLE 2 Express the number $\dfrac{-1 + 3i}{2 + 5i}$ in the form $a + bi$.

SOLUTION We multiply numerator and denominator by the complex conjugate of $2 + 5i$, namely, $2 - 5i$, and we take advantage of the result of Example 1:

$$\frac{-1 + 3i}{2 + 5i} = \frac{-1 + 3i}{2 + 5i} \cdot \frac{2 - 5i}{2 - 5i} = \frac{13 + 11i}{2^2 + 5^2} = \frac{13}{29} + \frac{11}{29}i$$

The geometric interpretation of the complex conjugate is shown in Figure 2: $\bar{z}$ is the reflection of z in the real axis. We list some of the properties of the complex conjugate in the following box. The proofs follow from the definition and are requested in Exercise 18.

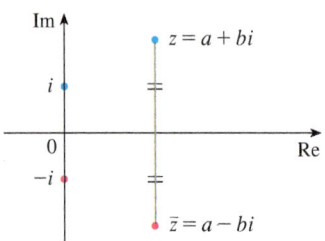

FIGURE 2

> **Properties of Conjugates**
>
> $$\overline{z + w} = \bar{z} + \bar{w} \qquad \overline{zw} = \bar{z}\,\bar{w} \qquad \overline{z^n} = \bar{z}^n$$

The **modulus**, or **absolute value**, $|z|$ of a complex number $z = a + bi$ is its distance from the origin. From Figure 3 we see that if $z = a + bi$, then

$$|z| = \sqrt{a^2 + b^2}$$

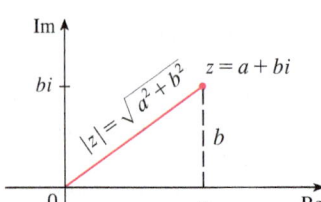

FIGURE 3

Notice that

$$z\bar{z} = (a + bi)(a - bi) = a^2 + abi - abi - b^2i^2 = a^2 + b^2$$

and so

$$z\bar{z} = |z|^2$$

This explains why the division procedure in Example 2 works in general:

$$\frac{z}{w} = \frac{z\bar{w}}{w\bar{w}} = \frac{z\bar{w}}{|w|^2}$$

Since $i^2 = -1$, we can think of i as a square root of -1. But notice that we also have $(-i)^2 = i^2 = -1$ and so $-i$ is also a square root of -1. We say that i is the **principal square root** of -1 and write $\sqrt{-1} = i$. In general, if c is any positive number, we write

$$\sqrt{-c} = \sqrt{c}\,i$$

With this convention, the usual derivation and formula for the roots of the quadratic equation $ax^2 + bx + c = 0$ are valid even when $b^2 - 4ac < 0$:

$$x = \frac{-b \pm \sqrt{b^2 - 4ac}}{2a}$$

EXAMPLE 3 Find the roots of the equation $x^2 + x + 1 = 0$.

SOLUTION Using the quadratic formula, we have

$$x = \frac{-1 \pm \sqrt{1^2 - 4 \cdot 1}}{2} = \frac{-1 \pm \sqrt{-3}}{2} = \frac{-1 \pm \sqrt{3}\,i}{2}$$

We observe that the solutions of the equation in Example 3 are complex conjugates of each other. In general, the solutions of any quadratic equation $ax^2 + bx + c = 0$ with real coefficients a, b, and c are always complex conjugates. (If z is real, $\bar{z} = z$, so z is its own conjugate.)

We have seen that if we allow complex numbers as solutions, then every quadratic equation has a solution. More generally, it is true that every polynomial equation

$$a_n x^n + a_{n-1} x^{n-1} + \cdots + a_1 x + a_0 = 0$$

of degree at least one has a solution among the complex numbers. This fact is known as the Fundamental Theorem of Algebra and was proved by Gauss.

Polar Form

We know that any complex number $z = a + bi$ can be considered as a point (a, b) and that any such point can be represented by polar coordinates (r, θ) with $r \geq 0$. In fact,

$$a = r \cos \theta \qquad b = r \sin \theta$$

as in Figure 4. Therefore we have

$$z = a + bi = (r \cos \theta) + (r \sin \theta)i$$

Thus we can write any complex number z in the form

$$z = r(\cos \theta + i \sin \theta)$$

where $\qquad r = |z| = \sqrt{a^2 + b^2} \qquad$ and $\qquad \tan \theta = \dfrac{b}{a}$

The angle θ is called the **argument** of z and we write $\theta = \arg(z)$. Note that $\arg(z)$ is not unique; any two arguments of z differ by an integer multiple of 2π.

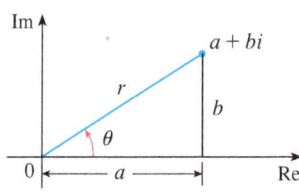

FIGURE 4

EXAMPLE 4 Write the following numbers in polar form.

(a) $z = 1 + i$ \qquad (b) $w = \sqrt{3} - i$

SOLUTION

(a) We have $r = |z| = \sqrt{1^2 + 1^2} = \sqrt{2}$ and $\tan \theta = 1$, so we can take $\theta = \pi/4$. Therefore the polar form is

$$z = \sqrt{2}\left(\cos \frac{\pi}{4} + i \sin \frac{\pi}{4}\right)$$

(b) Here we have $r = |w| = \sqrt{3 + 1} = 2$ and $\tan \theta = -1/\sqrt{3}$. Since w lies in the fourth quadrant, we take $\theta = -\pi/6$ and

$$w = 2\left[\cos\left(-\frac{\pi}{6}\right) + i \sin\left(-\frac{\pi}{6}\right)\right]$$

The numbers z and w are shown in Figure 5.

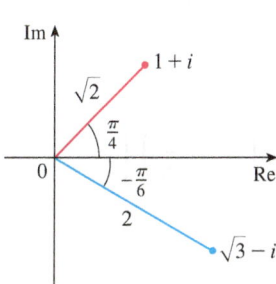

FIGURE 5

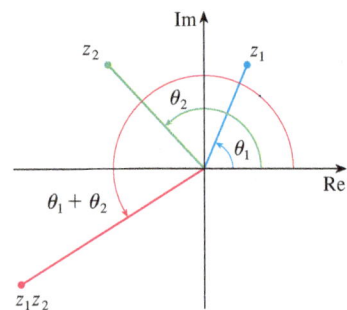

FIGURE 6

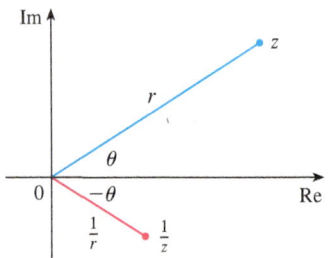

FIGURE 7

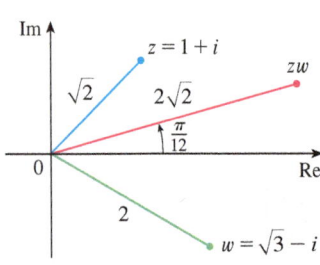

FIGURE 8

The polar form of complex numbers gives insight into multiplication and division. Let

$$z_1 = r_1(\cos\theta_1 + i\sin\theta_1) \qquad z_2 = r_2(\cos\theta_2 + i\sin\theta_2)$$

be two complex numbers written in polar form. Then

$$z_1 z_2 = r_1 r_2 (\cos\theta_1 + i\sin\theta_1)(\cos\theta_2 + i\sin\theta_2)$$
$$= r_1 r_2 [(\cos\theta_1 \cos\theta_2 - \sin\theta_1 \sin\theta_2) + i(\sin\theta_1 \cos\theta_2 + \cos\theta_1 \sin\theta_2)]$$

Therefore, using the addition formulas for cosine and sine, we have

1 $$z_1 z_2 = r_1 r_2 [\cos(\theta_1 + \theta_2) + i\sin(\theta_1 + \theta_2)]$$

This formula says that *to multiply two complex numbers we multiply the moduli and add the arguments*. (See Figure 6.)

A similar argument using the subtraction formulas for sine and cosine shows that *to divide two complex numbers we divide the moduli and subtract the arguments*.

$$\frac{z_1}{z_2} = \frac{r_1}{r_2}[\cos(\theta_1 - \theta_2) + i\sin(\theta_1 - \theta_2)] \qquad z_2 \neq 0$$

In particular, taking $z_1 = 1$ and $z_2 = z$ (and therefore $\theta_1 = 0$ and $\theta_2 = \theta$), we have the following, which is illustrated in Figure 7.

If $z = r(\cos\theta + i\sin\theta)$, then $\dfrac{1}{z} = \dfrac{1}{r}(\cos\theta - i\sin\theta).$

EXAMPLE 5 Find the product of the complex numbers $1 + i$ and $\sqrt{3} - i$ in polar form.

SOLUTION From Example 4 we have

$$1 + i = \sqrt{2}\left(\cos\frac{\pi}{4} + i\sin\frac{\pi}{4}\right)$$

and

$$\sqrt{3} - i = 2\left[\cos\left(-\frac{\pi}{6}\right) + i\sin\left(-\frac{\pi}{6}\right)\right]$$

So, by Equation 1,

$$(1+i)(\sqrt{3}-i) = 2\sqrt{2}\left[\cos\left(\frac{\pi}{4} - \frac{\pi}{6}\right) + i\sin\left(\frac{\pi}{4} - \frac{\pi}{6}\right)\right]$$
$$= 2\sqrt{2}\left(\cos\frac{\pi}{12} + i\sin\frac{\pi}{12}\right)$$

This is illustrated in Figure 8.

Repeated use of Formula 1 shows how to compute powers of a complex number. If

$$z = r(\cos\theta + i\sin\theta)$$

then
$$z^2 = r^2(\cos 2\theta + i\sin 2\theta)$$

and
$$z^3 = zz^2 = r^3(\cos 3\theta + i\sin 3\theta)$$

In general, we obtain the following result, which is named after the French mathematician Abraham De Moivre (1667–1754).

> **2 De Moivre's Theorem** If $z = r(\cos\theta + i\sin\theta)$ and n is a positive integer, then
> $$z^n = [r(\cos\theta + i\sin\theta)]^n = r^n(\cos n\theta + i\sin n\theta)$$

This says that *to take the nth power of a complex number we take the nth power of the modulus and multiply the argument by n.*

EXAMPLE 6 Find $\left(\frac{1}{2} + \frac{1}{2}i\right)^{10}$.

SOLUTION Since $\frac{1}{2} + \frac{1}{2}i = \frac{1}{2}(1 + i)$, it follows from Example 4(a) that $\frac{1}{2} + \frac{1}{2}i$ has the polar form

$$\frac{1}{2} + \frac{1}{2}i = \frac{\sqrt{2}}{2}\left(\cos\frac{\pi}{4} + i\sin\frac{\pi}{4}\right)$$

So by De Moivre's Theorem,

$$\left(\frac{1}{2} + \frac{1}{2}i\right)^{10} = \left(\frac{\sqrt{2}}{2}\right)^{10}\left(\cos\frac{10\pi}{4} + i\sin\frac{10\pi}{4}\right)$$

$$= \frac{2^5}{2^{10}}\left(\cos\frac{5\pi}{2} + i\sin\frac{5\pi}{2}\right) = \frac{1}{32}i$$

De Moivre's Theorem can also be used to find the nth roots of complex numbers. An nth root of the complex number z is a complex number w such that

$$w^n = z$$

Writing these two numbers in trigonometric form as

$$w = s(\cos\phi + i\sin\phi) \quad \text{and} \quad z = r(\cos\theta + i\sin\theta)$$

and using De Moivre's Theorem, we get

$$s^n(\cos n\phi + i\sin n\phi) = r(\cos\theta + i\sin\theta)$$

The equality of these two complex numbers shows that

$$s^n = r \quad \text{or} \quad s = r^{1/n}$$

and
$$\cos n\phi = \cos\theta \quad \text{and} \quad \sin n\phi = \sin\theta$$

From the fact that sine and cosine have period 2π, it follows that

$$n\phi = \theta + 2k\pi \quad \text{or} \quad \phi = \frac{\theta + 2k\pi}{n}$$

Thus
$$w = r^{1/n}\left[\cos\left(\frac{\theta + 2k\pi}{n}\right) + i\sin\left(\frac{\theta + 2k\pi}{n}\right)\right]$$

Since this expression gives a different value of w for $k = 0, 1, 2, \ldots, n - 1$, we have the following.

> **3 Roots of a Complex Number** Let $z = r(\cos\theta + i\sin\theta)$ and let n be a positive integer. Then z has the n distinct nth roots
> $$w_k = r^{1/n}\left[\cos\left(\frac{\theta + 2k\pi}{n}\right) + i\sin\left(\frac{\theta + 2k\pi}{n}\right)\right]$$
> where $k = 0, 1, 2, \ldots, n - 1$.

Notice that each of the nth roots of z has modulus $|w_k| = r^{1/n}$. Thus all the nth roots of z lie on the circle of radius $r^{1/n}$ in the complex plane. Also, since the argument of each successive nth root exceeds the argument of the previous root by $2\pi/n$, we see that the nth roots of z are equally spaced on this circle.

EXAMPLE 7 Find the six sixth roots of $z = -8$ and graph these roots in the complex plane.

SOLUTION In trigonometric form, $z = 8(\cos\pi + i\sin\pi)$. Applying Equation 3 with $n = 6$, we get

$$w_k = 8^{1/6}\left(\cos\frac{\pi + 2k\pi}{6} + i\sin\frac{\pi + 2k\pi}{6}\right)$$

We get the six sixth roots of -8 by taking $k = 0, 1, 2, 3, 4, 5$ in this formula:

$$w_0 = 8^{1/6}\left(\cos\frac{\pi}{6} + i\sin\frac{\pi}{6}\right) = \sqrt{2}\left(\frac{\sqrt{3}}{2} + \frac{1}{2}i\right)$$

$$w_1 = 8^{1/6}\left(\cos\frac{\pi}{2} + i\sin\frac{\pi}{2}\right) = \sqrt{2}\,i$$

$$w_2 = 8^{1/6}\left(\cos\frac{5\pi}{6} + i\sin\frac{5\pi}{6}\right) = \sqrt{2}\left(-\frac{\sqrt{3}}{2} + \frac{1}{2}i\right)$$

$$w_3 = 8^{1/6}\left(\cos\frac{7\pi}{6} + i\sin\frac{7\pi}{6}\right) = \sqrt{2}\left(-\frac{\sqrt{3}}{2} - \frac{1}{2}i\right)$$

$$w_4 = 8^{1/6}\left(\cos\frac{3\pi}{2} + i\sin\frac{3\pi}{2}\right) = -\sqrt{2}\,i$$

$$w_5 = 8^{1/6}\left(\cos\frac{11\pi}{6} + i\sin\frac{11\pi}{6}\right) = \sqrt{2}\left(\frac{\sqrt{3}}{2} - \frac{1}{2}i\right)$$

All these points lie on the circle of radius $\sqrt{2}$ as shown in Figure 9.

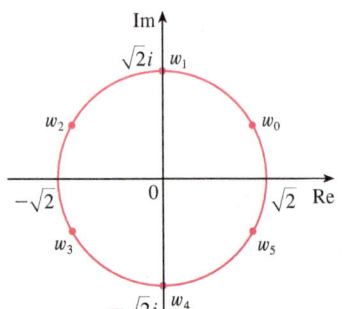

FIGURE 9
The six sixth roots of $z = -8$

Complex Exponentials

We also need to give a meaning to the expression e^z when $z = x + iy$ is a complex number. The theory of infinite series as developed in Chapter 11 can be extended to the case where the terms are complex numbers. Using the Taylor series for e^x (11.10.11) as our guide, we define

$$\boxed{4} \quad e^z = \sum_{n=0}^{\infty} \frac{z^n}{n!} = 1 + z + \frac{z^2}{2!} + \frac{z^3}{3!} + \cdots$$

and it turns out that this complex exponential function has the same properties as the real exponential function. In particular, it is true that

$$\boxed{5} \quad e^{z_1+z_2} = e^{z_1} e^{z_2}$$

If we put $z = iy$, where y is a real number, in Equation 4, and use the facts that

$$i^2 = -1, \quad i^3 = i^2 i = -i, \quad i^4 = 1, \quad i^5 = i, \quad \ldots$$

we get
$$e^{iy} = 1 + iy + \frac{(iy)^2}{2!} + \frac{(iy)^3}{3!} + \frac{(iy)^4}{4!} + \frac{(iy)^5}{5!} + \cdots$$

$$= 1 + iy - \frac{y^2}{2!} - i\frac{y^3}{3!} + \frac{y^4}{4!} + i\frac{y^5}{5!} + \cdots$$

$$= \left(1 - \frac{y^2}{2!} + \frac{y^4}{4!} - \frac{y^6}{6!} + \cdots\right) + i\left(y - \frac{y^3}{3!} + \frac{y^5}{5!} - \cdots\right)$$

$$= \cos y + i \sin y$$

Here we have used the Taylor series for $\cos y$ and $\sin y$ (Equations 11.10.16 and 11.10.15). The result is a famous formula called **Euler's formula**:

$$\boxed{6} \quad e^{iy} = \cos y + i \sin y$$

Combining Euler's formula with Equation 5, we get

$$\boxed{7} \quad e^{x+iy} = e^x e^{iy} = e^x(\cos y + i \sin y)$$

EXAMPLE 8 Evaluate: (a) $e^{i\pi}$ (b) $e^{-1+i\pi/2}$

SOLUTION

(a) From Euler's equation (6) we have

$$e^{i\pi} = \cos \pi + i \sin \pi = -1 + i(0) = -1$$

We could write the result of Example 8(a) as

$$e^{i\pi} + 1 = 0$$

This equation relates the five most famous numbers in all of mathematics: 0, 1, e, i, and π.

(b) Using Equation 7 we get

$$e^{-1+i\pi/2} = e^{-1}\left(\cos \frac{\pi}{2} + i \sin \frac{\pi}{2}\right) = \frac{1}{e}[0 + i(1)] = \frac{i}{e}$$ ∎

Finally, we note that Euler's equation provides us with an easier method of proving De Moivre's Theorem:

$$[r(\cos \theta + i \sin \theta)]^n = (re^{i\theta})^n = r^n e^{in\theta} = r^n(\cos n\theta + i \sin n\theta)$$

APPENDIX H Complex Numbers

H EXERCISES

1–14 Evaluate the expression and write your answer in the form $a + bi$.

1. $(5 - 6i) + (3 + 2i)$
2. $(4 - \frac{1}{2}i) - (9 + \frac{5}{2}i)$
3. $(2 + 5i)(4 - i)$
4. $(1 - 2i)(8 - 3i)$
5. $\overline{12 + 7i}$
6. $\overline{2i(\frac{1}{2} - i)}$
7. $\dfrac{1 + 4i}{3 + 2i}$
8. $\dfrac{3 + 2i}{1 - 4i}$
9. $\dfrac{1}{1 + i}$
10. $\dfrac{3}{4 - 3i}$
11. i^3
12. i^{100}
13. $\sqrt{-25}$
14. $\sqrt{-3}\sqrt{-12}$

15–17 Find the complex conjugate and the modulus of the number.

15. $12 - 5i$
16. $-1 + 2\sqrt{2}\,i$
17. $-4i$

18. Prove the following properties of complex numbers.
   (a) $\overline{z + w} = \bar{z} + \bar{w}$
   (b) $\overline{zw} = \bar{z}\,\bar{w}$
   (c) $\overline{z^n} = \bar{z}^n$, where n is a positive integer
   [*Hint:* Write $z = a + bi$, $w = c + di$.]

19–24 Find all solutions of the equation.

19. $4x^2 + 9 = 0$
20. $x^4 = 1$
21. $x^2 + 2x + 5 = 0$
22. $2x^2 - 2x + 1 = 0$
23. $z^2 + z + 2 = 0$
24. $z^2 + \frac{1}{2}z + \frac{1}{4} = 0$

25–28 Write the number in polar form with argument between 0 and 2π.

25. $-3 + 3i$
26. $1 - \sqrt{3}\,i$
27. $3 + 4i$
28. $8i$

29–32 Find polar forms for zw, z/w, and $1/z$ by first putting z and w into polar form.

29. $z = \sqrt{3} + i$, $w = 1 + \sqrt{3}\,i$
30. $z = 4\sqrt{3} - 4i$, $w = 8i$
31. $z = 2\sqrt{3} - 2i$, $w = -1 + i$
32. $z = 4(\sqrt{3} + i)$, $w = -3 - 3i$

33–36 Find the indicated power using De Moivre's Theorem.

33. $(1 + i)^{20}$
34. $(1 - \sqrt{3}\,i)^5$
35. $(2\sqrt{3} + 2i)^5$
36. $(1 - i)^8$

37–40 Find the indicated roots. Sketch the roots in the complex plane.

37. The eighth roots of 1
38. The fifth roots of 32
39. The cube roots of i
40. The cube roots of $1 + i$

41–46 Write the number in the form $a + bi$.

41. $e^{i\pi/2}$
42. $e^{2\pi i}$
43. $e^{i\pi/3}$
44. $e^{-i\pi}$
45. $e^{2+i\pi}$
46. $e^{\pi+i}$

47. Use De Moivre's Theorem with $n = 3$ to express $\cos 3\theta$ and $\sin 3\theta$ in terms of $\cos\theta$ and $\sin\theta$.

48. Use Euler's formula to prove the following formulas for $\cos x$ and $\sin x$:
$$\cos x = \frac{e^{ix} + e^{-ix}}{2} \qquad \sin x = \frac{e^{ix} - e^{-ix}}{2i}$$

49. If $u(x) = f(x) + ig(x)$ is a complex-valued function of a real variable x and the real and imaginary parts $f(x)$ and $g(x)$ are differentiable functions of x, then the derivative of u is defined to be $u'(x) = f'(x) + ig'(x)$. Use this together with Equation 7 to prove that if $F(x) = e^{rx}$, then $F'(x) = re^{rx}$ when $r = a + bi$ is a complex number.

50. (a) If u is a complex-valued function of a real variable, its indefinite integral $\int u(x)\,dx$ is an antiderivative of u. Evaluate
$$\int e^{(1+i)x}\,dx$$

(b) By considering the real and imaginary parts of the integral in part (a), evaluate the real integrals
$$\int e^x \cos x\,dx \quad \text{and} \quad \int e^x \sin x\,dx$$

(c) Compare with the method used in Example 7.1.4.

Answers to Odd-Numbered Exercises

CHAPTER 1

EXERCISES 1.1 ■ **PAGE 19**

1. Yes
3. (a) 3 (b) -0.2 (c) 0, 3 (d) -0.8
   (e) $[-2, 4], [-1, 3]$ (f) $[-2, 1]$
5. $[-85, 115]$ 7. No
9. Yes, $[-3, 2], [-3, -2) \cup [-1, 3]$
11. (a) $13.8°C$ (b) 1990 (c) 1910, 2005 (d) $[13.5, 14.5]$
13.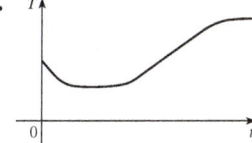
15. (a) 500 MW; 730 MW (b) 4 AM; noon; yes
17.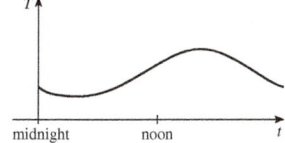
19.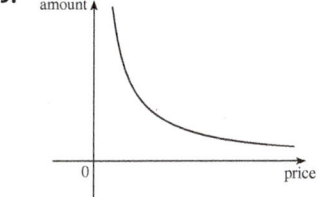
21.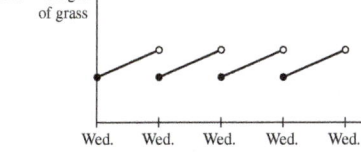
23. (a) 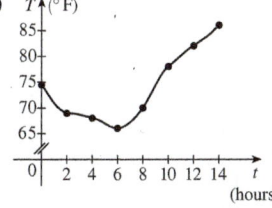   (b) $74°F$
25. $12, 16, 3a^2 - a + 2, 3a^2 + a + 2, 3a^2 + 5a + 4,$
    $6a^2 - 2a + 4, 12a^2 - 2a + 2, 3a^4 - a^2 + 2,$
    $9a^4 - 6a^3 + 13a^2 - 4a + 4, 3a^2 + 6ah + 3h^2 - a - h + 2$
27. $-3 - h$ 29. $-1/(ax)$
31. $(-\infty, -3) \cup (-3, 3) \cup (3, \infty)$ 33. $(-\infty, \infty)$
35. $(-\infty, 0) \cup (5, \infty)$ 37. $[0, 4]$
39. $(-\infty, \infty)$
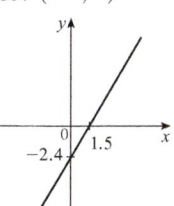
41. $-1, 1, -1$
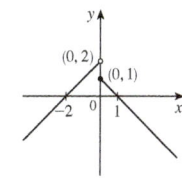
43. $-2, 0, 4$
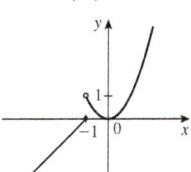
45.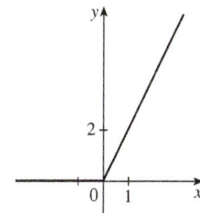
47.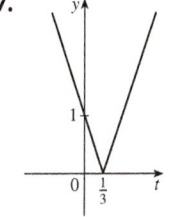
49.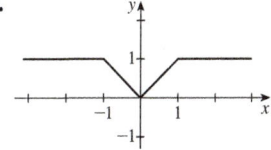
51. $f(x) = \frac{5}{2}x - \frac{11}{2}, 1 \leq x \leq 5$ 53. $f(x) = 1 - \sqrt{-x}$
55. $f(x) = \begin{cases} -x + 3 & \text{if } 0 \leq x \leq 3 \\ 2x - 6 & \text{if } 3 < x \leq 5 \end{cases}$
57. $A(L) = 10L - L^2, 0 < L < 10$
59. $A(x) = \sqrt{3}x^2/4, x > 0$ 61. $S(x) = x^2 + (8/x), x > 0$
63. $V(x) = 4x^3 - 64x^2 + 240x, 0 < x < 6$
65. $F(x) = \begin{cases} 15(40 - x) & \text{if } 0 \leq x < 40 \\ 0 & \text{if } 40 \leq x \leq 65 \\ 15(x - 65) & \text{if } x > 65 \end{cases}$
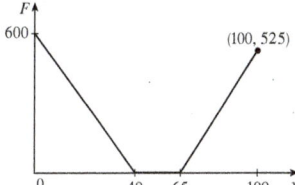

67. (a) 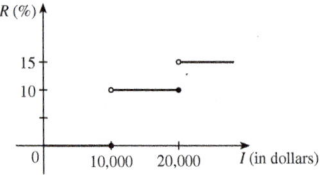 (b) $400, $1900

(c)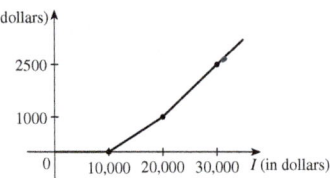

69. f is odd, g is even **71.** (a) $(-5, 3)$ (b) $(-5, -3)$
73. Odd **75.** Neither **77.** Even
79. Even; odd; neither (unless $f = 0$ or $g = 0$)

EXERCISES 1.2 ■ PAGE 33

1. (a) Logarithmic (b) Root (c) Rational
(d) Polynomial, degree 2 (e) Exponential (f) Trigonometric
3. (a) h (b) f (c) g
5. $\{x \mid x \neq \pi/2 + 2n\pi\}$, n an integer

7. (a) $y = 2x + b$,
where b is the y-intercept.

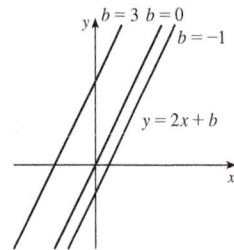

(b) $y = mx + 1 - 2m$,
where m is the slope.

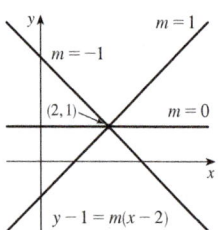

(c) $y = 2x - 3$

9. Their graphs have slope -1.

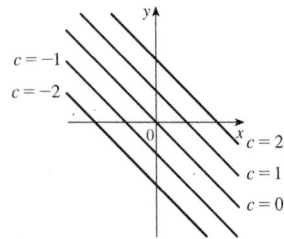

11. $f(x) = -3x(x + 1)(x - 2)$
13. (a) 8.34, change in mg for every 1 year change
(b) 8.34 mg

15. (a)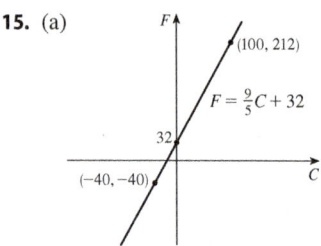

(b) $\frac{9}{5}$, change in °F for every 1°C change; 32, Fahrenheit temperature corresponding to 0°C
17. (a) $T = \frac{1}{6}N + \frac{307}{6}$ (b) $\frac{1}{6}$, change in °F for every chirp per minute change (c) 76°F
19. (a) $P = 0.434d + 15$ (b) 196 ft
21. (a) Cosine (b) Linear
23. (a) 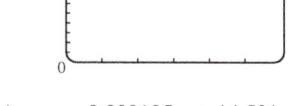 A linear model is appropriate.

(b) $y = -0.000105x + 14.521$

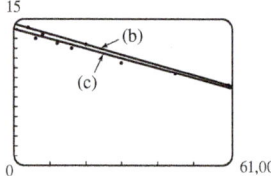

(c) $y = -0.00009979x + 13.951$
(d) About 11.5 per 100 population
(e) About 6% (f) No

25. (a) See graph in part (b).
(b) $y = 1.88074x + 82.64974$

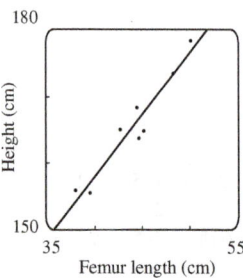

(c) 182.3 cm

27. (a) A linear model is appropriate. See graph in part (b).
(b) $y = 1116.64x + 60{,}188.33$

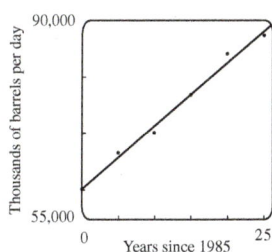

(c) In thousands of barrels per day: 79,171 and 90,338
29. Four times as bright
31. (a) $N = 3.1046A^{0.308}$ (b) 18

EXERCISES 1.3 ■ PAGE 42

1. (a) $y = f(x) + 3$ (b) $y = f(x) - 3$ (c) $y = f(x - 3)$
(d) $y = f(x + 3)$ (e) $y = -f(x)$ (f) $y = f(-x)$
(g) $y = 3f(x)$ (h) $y = \frac{1}{3}f(x)$
3. (a) 3 (b) 1 (c) 4 (d) 5 (e) 2
5. (a)  (b)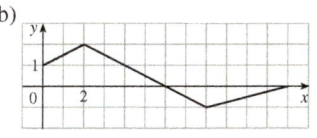

(c) 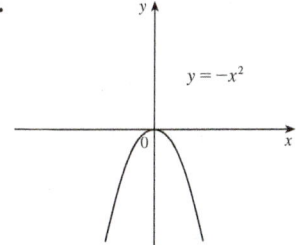 (d)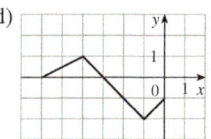

7. $y = -\sqrt{-x^2 - 5x - 4} - 1$

9.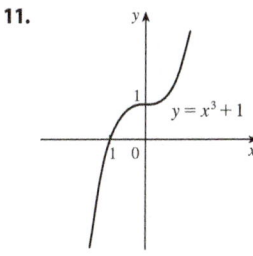

11.

13.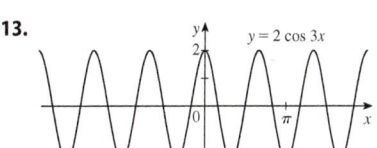

15.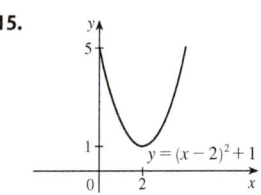

17.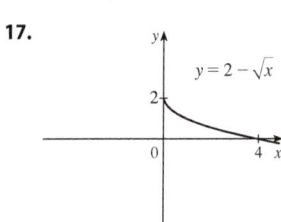

19.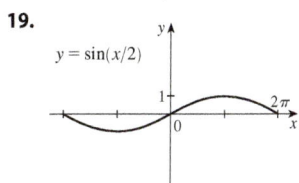

21.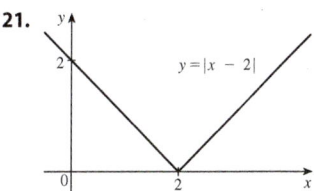

23.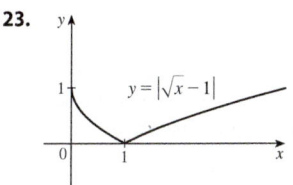

25. $L(t) = 12 + 2\sin\left[\dfrac{2\pi}{365}(t - 80)\right]$
27. $D(t) = 5\cos[(\pi/6)(t - 6.75)] + 7$
29. (a) The portion of the graph of $y = f(x)$ to the right of the y-axis is reflected about the y-axis.
(b)  (c)

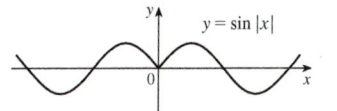 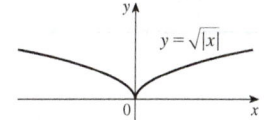

31. (a) $(f+g)(x) = x^3 + 5x^2 - 1, (-\infty, \infty)$
(b) $(f-g)(x) = x^3 - x^2 + 1, (-\infty, \infty)$
(c) $(fg)(x) = 3x^5 + 6x^4 - x^3 - 2x^2, (-\infty, \infty)$
(d) $(f/g)(x) = \dfrac{x^3 + 2x^2}{3x^2 - 1}, \left\{x \mid x \neq \pm\dfrac{1}{\sqrt{3}}\right\}$

33. (a) $(f \circ g)(x) = 3x^2 + 3x + 5, (-\infty, \infty)$
(b) $(g \circ f)(x) = 9x^2 + 33x + 30, (-\infty, \infty)$
(c) $(f \circ f)(x) = 9x + 20, (-\infty, \infty)$
(d) $(g \circ g)(x) = x^4 + 2x^3 + 2x^2 + x, (-\infty, \infty)$

35. (a) $(f \circ g)(x) = \sqrt{4x - 2}, [\tfrac{1}{2}, \infty)$
(b) $(g \circ f)(x) = 4\sqrt{x+1} - 3, [-1, \infty)$
(c) $(f \circ f)(x) = \sqrt{\sqrt{x+1} + 1}, [-1, \infty)$
(d) $(g \circ g)(x) = 16x - 15, (-\infty, \infty)$

37. (a) $(f \circ g)(x) = \dfrac{2x^2 + 6x + 5}{(x+2)(x+1)}, \{x \mid x \neq -2, -1\}$
(b) $(g \circ f)(x) = \dfrac{x^2 + x + 1}{(x+1)^2}, \{x \mid x \neq -1, 0\}$
(c) $(f \circ f)(x) = \dfrac{x^4 + 3x^2 + 1}{x(x^2+1)}, \{x \mid x \neq 0\}$
(d) $(g \circ g)(x) = \dfrac{2x + 3}{3x + 5}, \{x \mid x \neq -2, -\tfrac{5}{3}\}$

39. $(f \circ g \circ h)(x) = 3 \sin(x^2) - 2$
41. $(f \circ g \circ h)(x) = \sqrt{x^6 + 4x^3 + 1}$
43. $g(x) = 2x + x^2, f(x) = x^4$
45. $g(x) = \sqrt[3]{x}, f(x) = x/(1+x)$
47. $g(t) = t^2, f(t) = \sec t \tan t$
49. $h(x) = \sqrt{x}, g(x) = x - 1, f(x) = \sqrt{x}$
51. $h(t) = \cos t, g(t) = \sin t, f(t) = t^2$
53. (a) 4 (b) 3 (c) 0 (d) Does not exist; $f(6) = 6$ is not in the domain of g. (e) 4 (f) -2
55. (a) $r(t) = 60t$ (b) $(A \circ r)(t) = 3600\pi t^2$; the area of the circle as a function of time
57. (a) $s = \sqrt{d^2 + 36}$ (b) $d = 30t$
(c) $(f \circ g)(t) = \sqrt{900t^2 + 36}$; the distance between the lighthouse and the ship as a function of the time elapsed since noon

59. (a) (b)

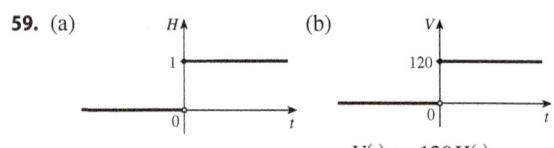

$V(t) = 120H(t)$

(c)

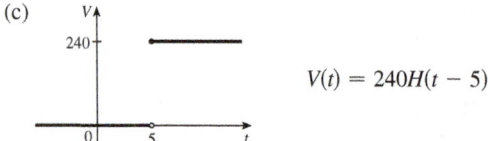

$V(t) = 240H(t - 5)$

61. Yes; $m_1 m_2$
63. (a) $f(x) = x^2 + 6$ (b) $g(x) = x^2 + x - 1$
65. Yes

EXERCISES 1.4 ▪ PAGE 53

1. (a) 4 (b) $x^{-4/3}$
3. (a) $16b^{12}$ (b) $648y^7$
5. (a) $f(x) = b^x, b > 0$ (b) $\mathbb{R}$ (c) $(0, \infty)$
(d) See Figures 4(c), 4(b), and 4(a), respectively.

7.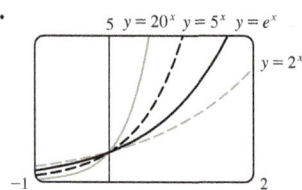
All approach 0 as $x \to -\infty$, all pass through $(0, 1)$, and all are increasing. The larger the base, the faster the rate of increase.

9.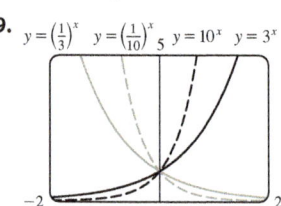
The functions with base greater than 1 are increasing and those with base less than 1 are decreasing. The latter are reflections of the former about the y-axis.

11. 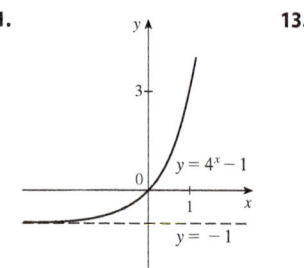 **13.**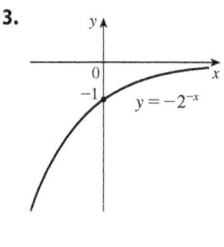

15.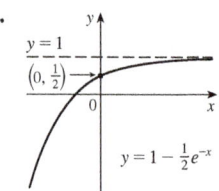

17. (a) $y = e^x - 2$ (b) $y = e^{x-2}$ (c) $y = -e^x$
(d) $y = e^{-x}$ (e) $y = -e^{-x}$
19. (a) $(-\infty, -1) \cup (-1, 1) \cup (1, \infty)$ (b) $(-\infty, \infty)$
21. $f(x) = 3 \cdot 2^x$ **27.** At $x \approx 35.8$
29. (a) See graph in part (c).
(b) $f(t) = 36.89301(1.06614)^t$
(c)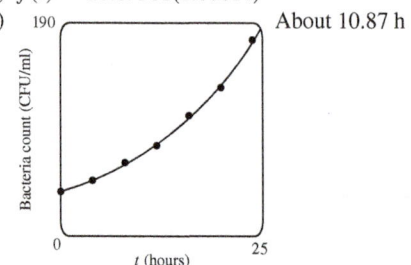
About 10.87 h

31. (a) 25 mg (b) $200 \cdot 2^{-t/5}$ mg
(c) 10.9 mg (d) 38.2 days

33. 3.5 days
35. $P = 2614.086(1.01693)^t$; 5381 million; 8466 million

EXERCISES 1.5 ■ PAGE 66

1. (a) See Definition 1.
(b) It must pass the Horizontal Line Test.
3. No   **5.** No   **7.** Yes   **9.** Yes   **11.** No   **13.** No
15. (a) 6   (b) 3   **17.** 0
19. $F = \frac{9}{5}C + 32$; the Fahrenheit temperature as a function of the Celsius temperature; $[-273.15, \infty)$
21. $f^{-1}(x) = \frac{1}{3}(x-1)^2 - \frac{2}{3}, x \geq 1$
23. $f^{-1}(x) = \frac{1}{2}(1 + \ln x)$ **25.** $y = e^x - 3$
27. $f^{-1}(x) = \frac{1}{4}(x^2 - 3), x \geq 0$

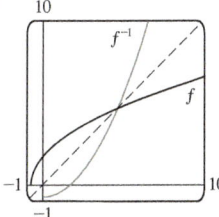

29.

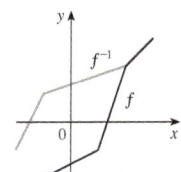

31. (a) $f^{-1}(x) = \sqrt{1 - x^2}, 0 \leq x \leq 1$; f^{-1} and f are the same function.   (b) Quarter-circle in the first quadrant
33. (a) It's defined as the inverse of the exponential function with base b, that is, $\log_b x = y \Longleftrightarrow b^y = x$.
(b) $(0, \infty)$ (c) $\mathbb{R}$ (d) See Figure 11.
35. (a) 5   (b) $\frac{1}{3}$ **37.** (a) 2   (b) $\frac{2}{3}$
39. $\ln 250$ **41.** $\ln \dfrac{\sqrt{x}}{x+1}$

43.

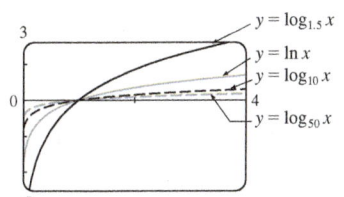

All graphs approach $-\infty$ as $x \to 0^+$, all pass through $(1, 0)$, and all are increasing. The larger the base, the slower the rate of increase.

45. About 1,084,588 mi

47. (a)   (b)

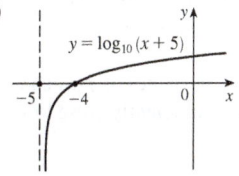

49. (a) $(0, \infty); (-\infty, \infty)$ (b) e^{-2}
(c)

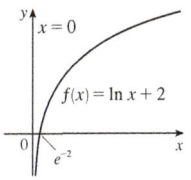

51. (a) $\frac{1}{4}(7 - \ln 6)$ (b) $\frac{1}{3}(e^2 + 10)$
53. (a) $5 + \log_2 3$ or $5 + (\ln 3)/\ln 2$ (b) $\frac{1}{2}(1 + \sqrt{1 + 4e})$
55. (a) $0 < x < 1$ (b) $x > \ln 5$
57. (a) $(\ln 3, \infty)$ (b) $f^{-1}(x) = \ln(e^x + 3)$; $\mathbb{R}$
59. The graph passes the Horizontal Line Test.

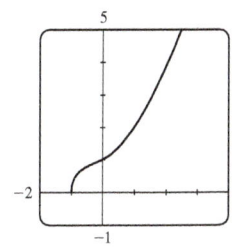

$f^{-1}(x) = -\frac{1}{6}\sqrt[3]{4}\left(\sqrt[3]{D - 27x^2 + 20} - \sqrt[3]{D + 27x^2 - 20} + \sqrt[3]{2}\right)$, where $D = 3\sqrt{3}\sqrt{27x^4 - 40x^2 + 16}$; two of the expressions are complex.
61. (a) $f^{-1}(n) = (3/\ln 2) \ln(n/100)$; the time elapsed when there are n bacteria   (b) After about 26.9 hours
63. (a) π (b) $\pi/6$
65. (a) $\pi/4$ (b) $\pi/2$
67. (a) $5\pi/6$ (b) $\pi/3$
71. $x/\sqrt{1+x^2}$
73.

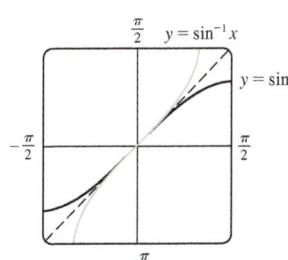

The second graph is the reflection of the first graph about the line $y = x$

75. $\left[-\frac{2}{3}, 0\right], [-\pi/2, \pi/2]$
77. (a) $g^{-1}(x) = f^{-1}(x) - c$ (b) $h^{-1}(x) = (1/c)f^{-1}(x)$

CHAPTER 1 REVIEW ■ PAGE 69

True-False Quiz
1. False   **3.** False   **5.** True   **7.** False   **9.** True
11. False   **13.** False

Exercises
1. (a) 2.7   (b) 2.3, 5.6   (c) $[-6, 6]$ (d) $[-4, 4]$
(e) $[-4, 4]$ (f) No; it fails the Horizontal Line Test.
(g) Odd; its graph is symmetric about the origin.
3. $2a + h - 2$ **5.** $\left(-\infty, \frac{1}{3}\right) \cup \left(\frac{1}{3}, \infty\right), (-\infty, 0) \cup (0, \infty)$
7. $(-6, \infty), \mathbb{R}$

9. (a) Shift the graph 8 units upward.
(b) Shift the graph 8 units to the left.
(c) Stretch the graph vertically by a factor of 2, then shift it 1 unit upward.
(d) Shift the graph 2 units to the right and 2 units downward.
(e) Reflect the graph about the x-axis.
(f) Reflect the graph about the line $y = x$ (assuming f is one-to-one).

11.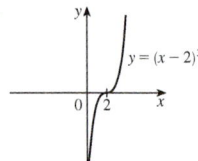

13.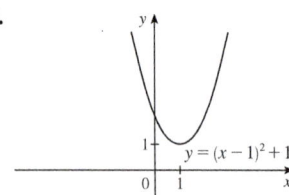

15.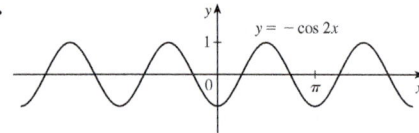

17. (a) Neither (b) Odd (c) Even (d) Neither
19. (a) $(f \circ g)(x) = \ln(x^2 - 9)$, $(-\infty, -3) \cup (3, \infty)$
(b) $(g \circ f)(x) = (\ln x)^2 - 9$, $(0, \infty)$
(c) $(f \circ f)(x) = \ln \ln x$, $(1, \infty)$
(d) $(g \circ g)(x) = (x^2 - 9)^2 - 9$, $(-\infty, \infty)$
21. $y = 0.2493x - 423.4818$; about 77.6 years
23. 1 **25.** (a) 9 (b) 2 (c) $1/\sqrt{3}$ (d) $\tfrac{3}{5}$
27. (a) $\tfrac{1}{16}$ g (b) $m(t) = 2^{-t/4}$
(c) $t(m) = -4 \log_2 m$; the time elapsed when there are m grams of ^{100}Pd
(d) About 26.6 days

PRINCIPLES OF PROBLEM SOLVING ■ PAGE 76

1. $a = 4\sqrt{h^2 - 16}/h$, where a is the length of the altitude and h is the length of the hypotenuse
3. $-\tfrac{7}{3}, 9$
5.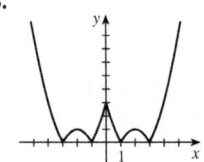
7.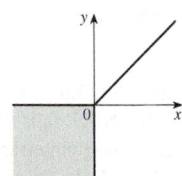

9. (a)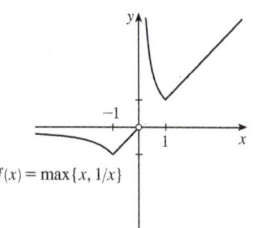

$f(x) = \max\{x, 1/x\}$

(b)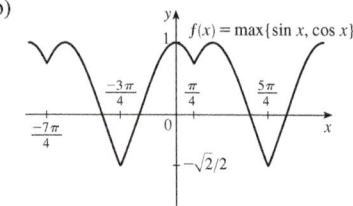

$f(x) = \max\{\sin x, \cos x\}$

(c)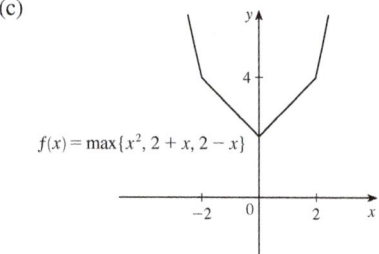

$f(x) = \max\{x^2, 2 + x, 2 - x\}$

11. 5 **13.** $x \in \left[-1, 1 - \sqrt{3}\right) \cup \left(1 + \sqrt{3}, 3\right]$
15. 40 mi/h **19.** $f_n(x) = x^{2^{n+1}}$

CHAPTER 2

EXERCISES 2.1 ■ PAGE 82

1. (a) $-44.4, -38.8, -27.8, -22.2, -16.\overline{6}$
(b) -33.3 (c) $-33\tfrac{1}{3}$
3. (a) (i) 2 (ii) 1.111111 (iii) 1.010101 (iv) 1.001001
(v) 0.666667 (vi) 0.909091 (vii) 0.990099
(viii) 0.999001 (b) 1 (c) $y = x - 3$
5. (a) (i) -32 ft/s (ii) -25.6 ft/s (iii) -24.8 ft/s
(iv) -24.16 ft/s (b) -24 ft/s
7. (a) (i) 29.3 ft/s (ii) 32.7 ft/s (iii) 45.6 ft/s
(iv) 48.75 ft/s (b) 29.7 ft/s
9. (a) 0, 1.7321, -1.0847, -2.7433, 4.3301, -2.8173, 0, -2.1651, -2.6061, -5, 3.4202; no (c) -31.4

EXERCISES 2.2 ■ PAGE 92

1. Yes
3. (a) $\lim_{x \to -3} f(x) = \infty$ means that the values of $f(x)$ can be made arbitrarily large (as large as we please) by taking x sufficiently close to -3 (but not equal to -3).
(b) $\lim_{x \to 4^+} f(x) = -\infty$ means that the values of $f(x)$ can be made arbitrarily large negative by taking x sufficiently close to 4 through values larger than 4.

APPENDIX I Answers to Odd-Numbered Exercises   A71

5. (a) 2 (b) 1 (c) 4 (d) Does not exist (e) 3
7. (a) −1 (b) −2 (c) Does not exist (d) 2 (e) 0
(f) Does not exist (g) 1 (h) 3
9. (a) −∞ (b) ∞ (c) ∞ (d) −∞ (e) ∞
(f) $x = -7, x = -3, x = 0, x = 6$
11. $\lim_{x \to a} f(x)$ exists for all a except $a = -1$.
13. (a) 1 (b) 0 (c) Does not exist
15. 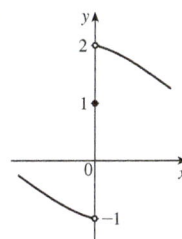 **17.**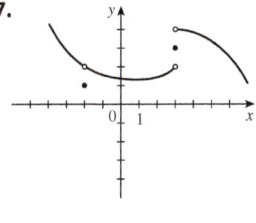

19. $\frac{1}{2}$ **21.** 5 **23.** 0.25 **25.** 1.5 **27.** 1
29. (a) −1.5 **31.** ∞ **33.** ∞ **35.** −∞ **37.** −∞
39. −∞ **41.** ∞ **43.** −∞ **45.** −∞; ∞
47. (a) 2.71828 (b)
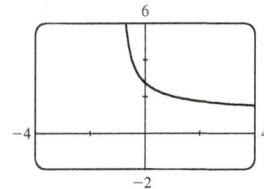

49. (a) 0.998000, 0.638259, 0.358484, 0.158680, 0.038851, 0.008928, 0.001465; 0
(b) 0.000572, −0.000614, −0.000907, −0.000978, −0.000993, −0.001000; −0.001
51. No matter how many times we zoom in toward the origin, the graph appears to consist of almost-vertical lines. This indicates more and more frequent oscillations as $x \to 0$.
53. $x \approx \pm 0.90, \pm 2.24; x = \pm \sin^{-1}(\pi/4), \pm(\pi - \sin^{-1}(\pi/4))$
55. (a) 6 (b) Within 0.0649 of 1

EXERCISES 2.3 ■ PAGE 102

1. (a) −6 (b) −8 (c) 2 (d) −6
(e) Does not exist (f) 0
3. 105 **5.** $\frac{7}{8}$ **7.** 390 **9.** $\frac{3}{2}$ **11.** 4
13. Does not exist **15.** $\frac{6}{5}$ **17.** −10 **19.** $\frac{1}{12}$
21. $\frac{1}{6}$ **23.** $-\frac{1}{9}$ **25.** 1 **27.** $\frac{1}{128}$ **29.** $-\frac{1}{2}$
31. $3x^2$ **33.** (a), (b) $\frac{2}{3}$ **37.** 7 **41.** 6 **43.** −4
45. Does not exist
47. (a) 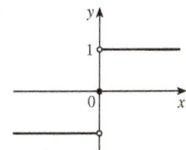 (b) (i) 1
(ii) −1
(iii) Does not exist
(iv) 1

49. (a) (i) 5 (ii) −5 (b) Does not exist
(c)
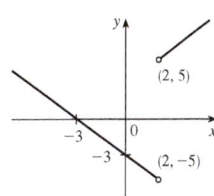

51. 7
53. (a) (i) −2 (ii) Does not exist (iii) −3
(b) (i) $n - 1$ (ii) n (c) a is not an integer.
59. 8 **65.** 15; −1

EXERCISES 2.4 ■ PAGE 113

1. 0.1 (or any smaller positive number)
3. 1.44 (or any smaller positive number)
5. 0.0906 (or any smaller positive number)
7. 0.0219 (or any positive number);
0.011 (or any smaller positive number)
9. (a) 0.01 (or any smaller positive number)
(b) $\lim_{x \to 2^+} \dfrac{1}{\ln(x-1)} = \infty$
11. (a) $\sqrt{1000/\pi}$ cm (b) Within approximately 0.0445 cm
(c) Radius; area; $\sqrt{1000/\pi}$; 1000; 5; ≈0.0445
13. (a) 0.025 (b) 0.0025
35. (a) 0.093 (b) $\delta = (B^{2/3} - 12)/(6B^{1/3}) - 1$, where $B = 216 + 108\varepsilon + 12\sqrt{336 + 324\varepsilon + 81\varepsilon^2}$
41. Within 0.1

EXERCISES 2.5 ■ PAGE 124

1. $\lim_{x \to 4} f(x) = f(4)$
3. (a) −4, −2, 2, 4; $f(-4)$ is not defined and $\lim_{x \to a} f(x)$ does not exist for $a = -2, 2,$ and 4
(b) −4, neither; −2, left; 2, right; 4, right

5.   **7.**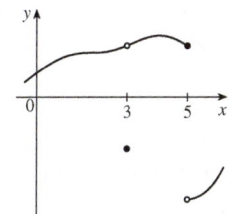

9. (a)

17. $f(-2)$ is undefined.

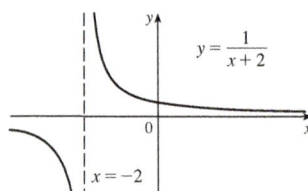

19. $\lim_{x \to -1} f(x)$ does not exist.   **21.** $\lim_{x \to 0} f(x) \neq f(0)$

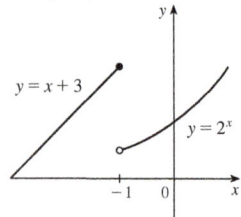 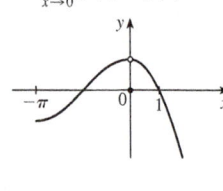

23. Define $f(2) = 3$.   **25.** $(-\infty, \infty)$
27. $(-\infty, \sqrt[3]{2}) \cup (\sqrt[3]{2}, \infty)$ **29.** $[-1, 0]$
31. $(-\infty, -1] \cup (0, \infty)$
33. $x = 0$

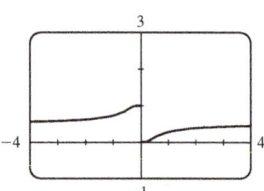

35. 8   **37.** ln 2
41. -1, right   **43.** 0, right; 1, left

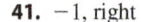
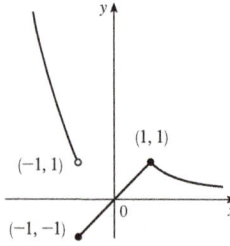 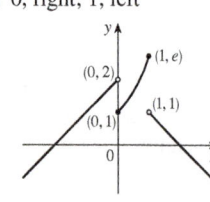

45. $\frac{2}{3}$ **47.** 4
49. (a) $g(x) = x^3 + x^2 + x + 1$ (b) $g(x) = x^2 + x$
57. (b) $(0.86, 0.87)$ **59.** (b) 70.347 **67.** None
69. Yes

EXERCISES 2.6 ■ PAGE 137

1. (a) As x becomes large, $f(x)$ approaches 5.
(b) As x becomes large negative, $f(x)$ approaches 3.
3. (a) -2 (b) 2   (c) ∞ (d) $-\infty$
(e) $x = 1, x = 3, y = -2, y = 2$

5.

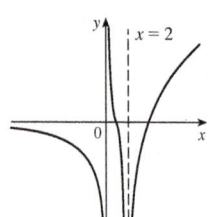

7.

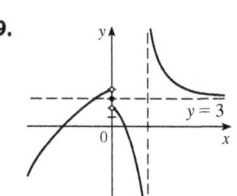

9.

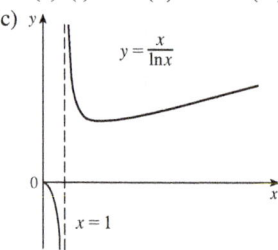

11. 0   **13.** $\frac{2}{5}$ **15.** $\frac{3}{2}$ **17.** 0   **19.** -1 **21.** 4
23. -2 **25.** $\frac{\sqrt{3}}{4}$ **27.** $\frac{1}{6}$ **29.** $\frac{1}{2}(a - b)$ **31.** ∞
33. $-\infty$ **35.** $\pi/2$ **37.** $-\frac{1}{2}$ **39.** 0   **41.** ∞
43. (a) (i) 0   (ii) $-\infty$ (iii) ∞ (b) ∞
(c)

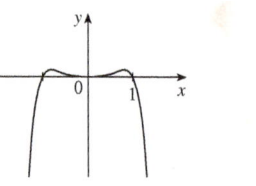

45. (a), (b) $-\frac{1}{2}$ **47.** $y = 4, x = -3$
49. $y = 2; x = -2, x = 1$ **51.** $x = 5$ **53.** $y = 3$
55. (a) 0   (b) $\pm\infty$
57. $f(x) = \dfrac{2 - x}{x^2(x - 3)}$ **59.** (a) $\frac{5}{4}$ (b) 5
61. $-\infty, -\infty$ **63.** $-\infty, \infty$

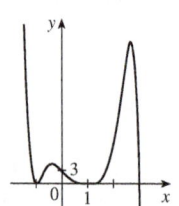 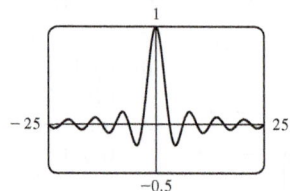

65. (a) 0   (b) An infinite number of times

67. 5

69. (a) v^* (b) 1.2 ≈ 0.47 s

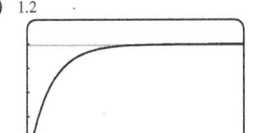

71. $N \geq 15$ **73.** $N \leq -9, N \leq -19$
75. (a) $x > 100$

EXERCISES 2.7 ■ PAGE 148

1. (a) $\dfrac{f(x) - f(3)}{x - 3}$ (b) $\lim\limits_{x \to 3} \dfrac{f(x) - f(3)}{x - 3}$

3. (a) 2 (b) $y = 2x + 1$ (c)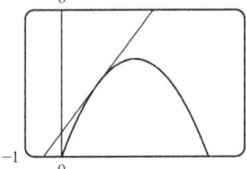

5. $y = -8x + 12$ **7.** $y = \tfrac{1}{2}x + \tfrac{1}{2}$
9. (a) $8a - 6a^2$ (b) $y = 2x + 3, y = -8x + 19$
(c)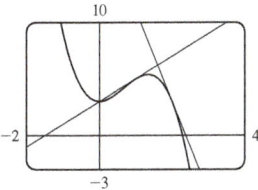

11. (a) Right: $0 < t < 1$ and $4 < t < 6$; left: $2 < t < 3$; standing still: $1 < t < 2$ and $3 < t < 4$
(b)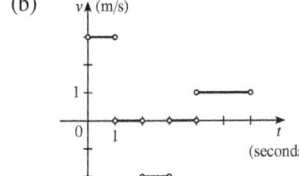

13. -24 ft/s
15. $-2/a^3$ m/s; -2 m/s; $-\tfrac{1}{4}$ m/s; $-\tfrac{2}{27}$ m/s
17. $g'(0), 0, g'(4), g'(2), g'(-2)$
19. (a) 26 (b) No (c) Yes
21. $f(2) = 3; f'(2) = 4$
23.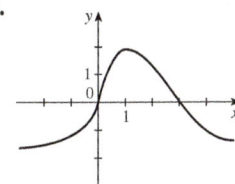

25. 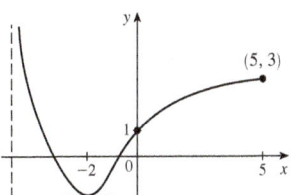 **27.** $y = 3x - 1$

29. (a) $-\tfrac{3}{5}; y = -\tfrac{3}{5}x + \tfrac{16}{5}$ (b)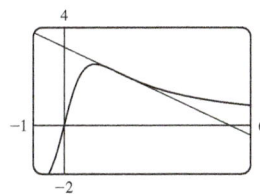

31. $6a - 4$ **33.** $\dfrac{5}{(a+3)^2}$ **35.** $-\dfrac{1}{\sqrt{1-2a}}$
37. $f(x) = \sqrt{x}, a = 9$ **39.** $f(x) = x^6, a = 2$
41. $f(x) = \cos x, a = \pi$ or $f(x) = \cos(\pi + x), a = 0$
43. 32 m/s; 32 m/s
45. 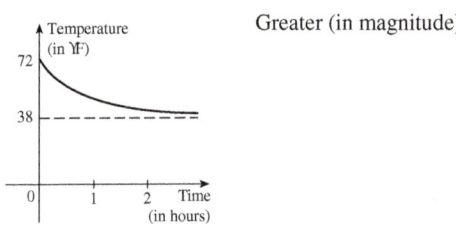 Greater (in magnitude)

47. (a) In (mg/mL)/h: (i) -0.15 (ii) -0.12
(iii) -0.12 (iv) -0.11 (b) -0.12 (mg/mL)/h; After 2 hours, the BAC is decreasing at a rate of 0.12 (mg/mL)/h.
49. (a) 1169.6 thousands of barrels of oil per day per year; oil consumption rose by an average of 1169.6 thousands of barrels of oil per day each year from 1990 to 2005.
(b) 1397.8 thousands of barrels of oil per day per year
51. (a) (i) $20.25/unit (ii) $20.05/unit (b) $20/unit
53. (a) The rate at which the cost is changing per ounce of gold produced; dollars per ounce
(b) When the 800th ounce of gold is produced, the cost of production is $17/oz.
(c) Decrease in the short term; increase in the long term
55. (a) The rate at which daily heating costs change with respect to temperature when the temperature is 58°F; dollars/°F
(b) Negative; If the outside temperature increases, the building should require less heating.
57. (a) The rate at which the oxygen solubility changes with respect to the water temperature; (mg/L)/°C
(b) $S'(16) \approx -0.25$; as the temperature increases past 16°C, the oxygen solubility is decreasing at a rate of 0.25 (mg/L)/°C.
59. Does not exist

61. (a) 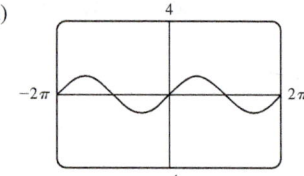 Slope appears to be 1.

(b) 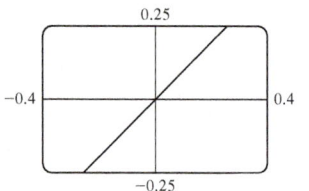 Yes

(c) 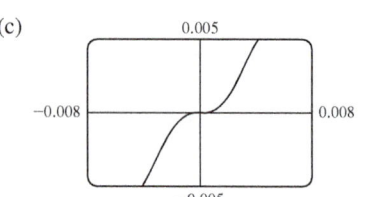 Yes; 0

EXERCISES 2.8 ■ PAGE 160

1. (a) -0.2 (b) 0 (c) 1 (d) 2
(e) 1 (f) 0 (g) -0.2

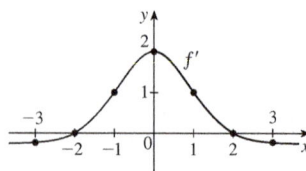

3. (a) II (b) IV (c) I (d) III
5. 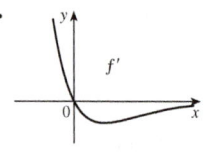 **7.**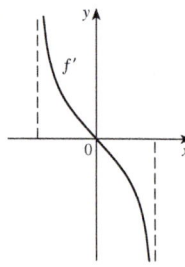

9. 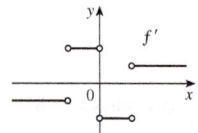 **11.**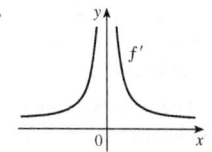

13. (a) The instantaneous rate of change of percentage of full capacity with respect to elapsed time in hours

(b) 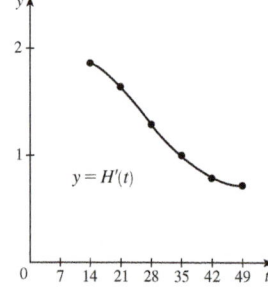 The rate of change of percentage of full capacity is decreasing and approaching 0.

15. 1963 to 1971

17. $f'(x) = e^x$

19. (a) 0, 1, 2, 4 (b) $-1, -2, -4$ (c) $f'(x) = 2x$
21. $f'(x) = 3$, $\mathbb{R}$, $\mathbb{R}$ **23.** $f'(t) = 5t + 6$, $\mathbb{R}$, $\mathbb{R}$
25. $f'(x) = 2x - 6x^2$, $\mathbb{R}$, $\mathbb{R}$
27. $g'(x) = -\dfrac{1}{2\sqrt{9-x}}$, $(-\infty, 9]$, $(-\infty, 9)$
29. $G'(t) = \dfrac{-7}{(3+t)^2}$, $(-\infty, -3) \cup (-3, \infty)$, $(-\infty, -3) \cup (-3, \infty)$
31. $f'(x) = 4x^3$, $\mathbb{R}$, $\mathbb{R}$ **33.** (a) $f'(x) = 4x^3 + 2$
35. (a) The rate at which the unemployment rate is changing, in percent unemployed per year

(b)
t	$U'(t)$	t	$U'(t)$
2003	-0.50	2008	2.35
2004	-0.45	2009	1.90
2005	-0.45	2010	-0.20
2006	-0.25	2011	-0.75
2007	0.60	2012	-0.80

37.
t	14	21	28	35	42	49
$H'(t)$	$\frac{13}{7}$	$\frac{23}{14}$	$\frac{9}{7}$	1	$\frac{11}{14}$	$\frac{5}{7}$

39. (a) The rate at which the percentage of electrical power produced by solar panels is changing, in percentage points per year.
(b) On January 1, 2002, the percentage of electrical power produced by solar panels was increasing at a rate of 3.5 percentage points per year.
41. -4 (corner); 0 (discontinuity)
43. 1 (not defined); 5 (vertical tangent)
45.

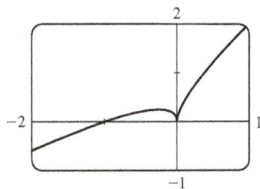

Differentiable at -1; not differentiable at 0

47. $f''(1)$ **49.** $a = f, b = f', c = f''$
51. $a =$ acceleration, $b =$ velocity, $c =$ position
53. $6x + 2$; 6

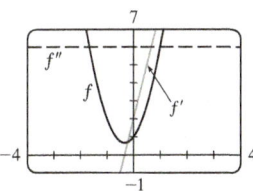

55.

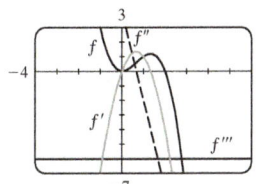

$f'(x) = 4x - 3x^2,$
$f''(x) = 4 - 6x,$
$f'''(x) = -6,$
$f^{(4)}(x) = 0$

57. (a) $\frac{1}{3}a^{-2/3}$

59. $f'(x) = \begin{cases} -1 & \text{if } x < 6 \\ 1 & \text{if } x > 6 \end{cases}$

or $f'(x) = \dfrac{x-6}{|x-6|}$

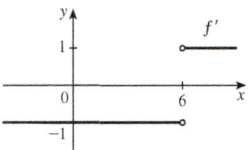

61. (a)   (b) All x
(c) $f'(x) = 2|x|$

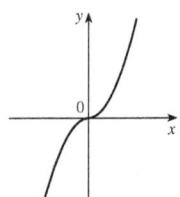

65. (a)   (b)

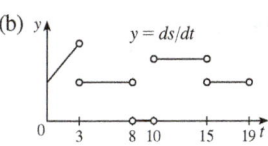

67. $63°$

CHAPTER 2 REVIEW ■ PAGE 166

True-False Quiz
1. False **3.** True **5.** True **7.** False **9.** True
11. True **13.** True **15.** False **17.** True
19. True **21.** False **23.** False **25.** True

Exercises
1. (a) (i) 3 (ii) 0 (iii) Does not exist (iv) 2
(v) ∞ (vi) $-\infty$ (vii) 4 (viii) -1
(b) $y = 4, y = -1$ (c) $x = 0, x = 2$ (d) $-3, 0, 2, 4$
3. 1   **5.** $\frac{3}{2}$ **7.** 3   **9.** ∞ **11.** $\frac{4}{7}$ **13.** $\frac{1}{2}$
15. $-\infty$ **17.** 2   **19.** $\pi/2$ **21.** $x = 0, y = 0$ **23.** 1
29. (a) (i) 3 (ii) 0 (iii) Does not exist
(iv) 0 (v) 0 (vi) 0
(b) At 0 and 3   (c)

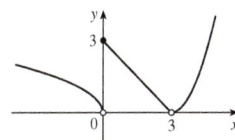

31. $\mathbb{R}$ **35.** (a) -8 (b) $y = -8x + 17$
37. (a) (i) 3 m/s (ii) 2.75 m/s (iii) 2.625 m/s
(iv) 2.525 m/s   (b) 2.5 m/s
39. (a) 10   (b) $y = 10x - 16$
(c)

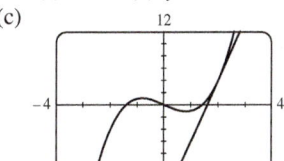

41. (a) The rate at which the cost changes with respect to the interest rate; dollars/(percent per year)
(b) As the interest rate increases past 10%, the cost is increasing at a rate of $1200/(percent per year).
(c) Always positive

43.

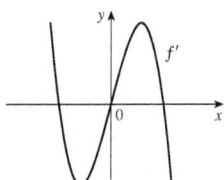

45. (a) $f'(x) = -\frac{5}{2}(3 - 5x)^{-1/2}$ (b) $\left(-\infty, \frac{3}{5}\right], \left(-\infty, \frac{3}{5}\right)$
(c)

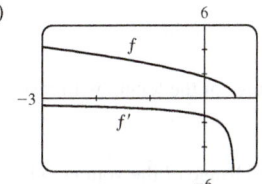

47. -4 (discontinuity), -1 (corner), 2 (discontinuity), 5 (vertical tangent)

APPENDIX I Answers to Odd-Numbered Exercises

49.

51. The rate at which the number of US $20 bills in circulation is changing with respect to time; 0.156 billion bills per year
53. 0

PROBLEMS PLUS ■ PAGE 169

1. $\frac{2}{3}$ **3.** -4 **5.** (a) Does not exist   (b) 1
7. $a = \frac{1}{2} \pm \frac{1}{2}\sqrt{5}$ **9.** $\frac{3}{4}$ **11.** (b) Yes   (c) Yes; no
13. (a) 0   (b) 1   (c) $f'(x) = x^2 + 1$

CHAPTER 3

EXERCISES 3.1 ■ PAGE 180

1. (a) e is the number such that $\lim_{h \to 0} \frac{e^h - 1}{h} = 1$.
(b) 0.99, 1.03; $2.7 < e < 2.8$
3. $f'(x) = 0$ **5.** $f'(x) = 5.2$ **7.** $f'(t) = 6t^2 - 6t - 4$
9. $g'(x) = 2x - 6x^2$ **11.** $g'(t) = -\frac{3}{2}t^{-7/4}$
13. $F'(r) = -15/r^4$ **15.** $R'(a) = 18a + 6$
17. $S'(p) = \frac{1}{2}p^{-1/2} - 1$ **19.** $y' = 3e^x - \frac{4}{3}x^{-4/3}$
21. $h'(u) = 3Au^2 + 2Bu + C$
23. $y' = \frac{3}{2}\sqrt{x} + \frac{2}{\sqrt{x}} - \frac{3}{2x\sqrt{x}}$ **25.** $j'(x) = 2.4x^{1.4}$
27. $G'(q) = -2q^{-2} - 2q^{-3}$ **29.** $f'(v) = -\frac{2}{3}v^{-5/3} - 2e^v$
31. $z' = -10A/y^{11} + Be^y$ **33.** $y = 4x - 1$
35. $y = \frac{1}{2}x + 2$
37. Tangent: $y = 2x + 2$; normal: $y = -\frac{1}{2}x + 2$
39. $y = 3x - 1$ **41.** $f'(x) = 4x^3 - 6x^2 + 2x$
43. (a)   (c) $4x^3 - 9x^2 - 12x + 7$

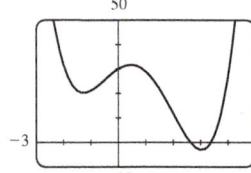
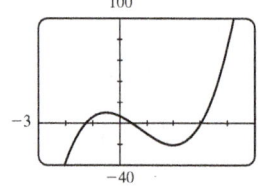

45. $f'(x) = 0.005x^4 - 0.06x^2$, $f''(x) = 0.02x^3 - 0.12x$
47. $f'(x) = 2 - \frac{15}{4}x^{-1/4}$, $f''(x) = \frac{15}{16}x^{-5/4}$
49. (a) $v(t) = 3t^2 - 3$, $a(t) = 6t$ (b) 12 m/s²
(c) $a(1) = 6$ m/s²
51. 1.718; instantaneous rate of change of the length with respect to the age at 12 yr
53. (a) $V = 5.3/P$
(b) -0.00212; instantaneous rate of change of the volume with respect to the pressure at 25°C; m³/kPa
55. $(-2, 21)$, $(1, -6)$ **59.** $y = 3x - 3$, $y = 3x - 7$
61. $y = -2x + 3$
63. $(\pm 2, 4)$ **67.** $P(x) = x^2 - x + 3$

69. $y = \frac{3}{16}x^3 - \frac{9}{4}x + 3$
71. No

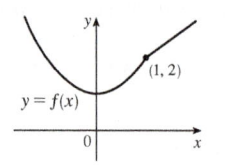
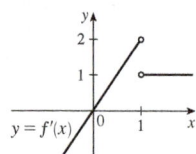

73. (a) Not differentiable at 3 or -3
$$f'(x) = \begin{cases} 2x & \text{if } |x| > 3 \\ -2x & \text{if } |x| < 3 \end{cases}$$
(b)

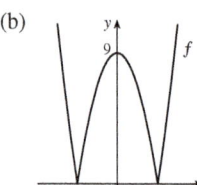
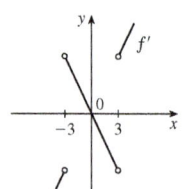

75. $y = 2x^2 - x$ **77.** $a = -\frac{1}{2}, b = 2$ **79.** $-\frac{1}{3}$
81. $m = 4, b = -4$ **83.** 1000   **85.** 3; 1

EXERCISES 3.2 ■ PAGE 188

1. $1 - 2x + 6x^2 - 8x^3$ **3.** $f'(x) = e^x(3x^2 + x - 5)$
5. $y' = \frac{1 - x}{e^x}$ **7.** $g'(x) = \frac{10}{(3 - 4x)^2}$ **9.** $H'(u) = 2u - 1$
11. $F'(y) = 5 + \frac{14}{y^2} + \frac{9}{y^4}$ **13.** $y' = \frac{x(-x^3 - 3x - 2)}{(x^3 - 1)^2}$
15. $y' = \frac{t^4 - 8t^3 + 6t^2 + 9}{(t^2 - 4t + 3)^2}$
17. $y' = e^p\left(1 + \frac{3}{2}\sqrt{p} + p + p\sqrt{p}\right)$ **19.** $y' = \frac{3 - 2\sqrt{s}}{2s^{5/2}}$
21. $f'(t) = \frac{-2t - 3}{3t^{2/3}(t - 3)^2}$ **23.** $f'(x) = \frac{xe^x(x^3 + 2e^x)}{(x^2 + e^x)^2}$
25. $f'(x) = \frac{2cx}{(x^2 + c)^2}$
27. $(x^3 + 3x^2 + 1)e^x$; $(x^3 + 6x^2 + 6x + 1)e^x$
29. $\frac{x(2 + 2e^x - xe^x)}{(1 + e^x)^2}$;
$\frac{2 + 4e^x - 4xe^x - x^2e^x + 2e^{2x} - 4xe^{2x} + x^2e^{2x}}{(1 + e^x)^3}$
31. $y = \frac{2}{3}x - \frac{2}{3}$ **33.** $y = 2x$; $y = -\frac{1}{2}x$
35. (a) $y = \frac{1}{2}x + 1$ (b)

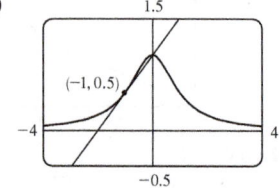

37. (a) $e^x(x^3 + 3x^2 - x - 1)$
(b)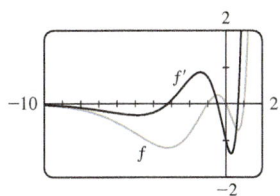

39. (a) $f'(x) = \dfrac{4x}{(x^2 + 1)^2}; f''(x) = \dfrac{4(1 - 3x^2)}{(x^2 + 1)^3}$
(b)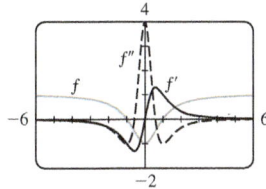

41. $\frac{1}{4}$ **43.** (a) -16 (b) $-\frac{20}{9}$ (c) 20   **45.** 7
47. $y = -2x + 18$ **49.** (a) 0   (b) $-\frac{2}{3}$
51. (a) $y' = xg'(x) + g(x)$ (b) $y' = \dfrac{g(x) - xg'(x)}{[g(x)]^2}$
(c) $y' = \dfrac{xg'(x) - g(x)}{x^2}$
53. Two, $\left(-2 \pm \sqrt{3}, \frac{1}{2}(1 \mp \sqrt{3})\right)$ **55.** 1
57. \$1.627 billion/year
59. $\dfrac{0.0021}{(0.015 + [S])^2}$;
The rate of change of the rate of an enzymatic reaction with respect to the concentration of a substrate S.
61. (c) $3e^{3x}$
63. $f'(x) = (x^2 + 2x)e^x$, $f''(x) = (x^2 + 4x + 2)e^x$,
$f'''(x) = (x^2 + 6x + 6)e^x$, $f^{(4)}(x) = (x^2 + 8x + 12)e^x$,
$f^{(5)}(x) = (x^2 + 10x + 20)e^x$; $f^{(n)}(x) = [x^2 + 2nx + n(n - 1)]e^x$

EXERCISES 3.3 ■ PAGE 196

1. $f'(x) = x^2 \cos x + 2x \sin x$ **3.** $f'(x) = e^x(\cos x - \sin x)$
5. $y' = \sec\theta \,(\sec^2\theta + \tan^2\theta)$
7. $y' = -c \sin t + t(t \cos t + 2 \sin t)$
9. $y' = \dfrac{2 - \tan x + x \sec^2 x}{(2 - \tan x)^2}$ **11.** $f'(\theta) = \dfrac{1}{1 + \cos\theta}$
13. $y' = \dfrac{(t^2 + t)\cos t + \sin t}{(1 + t)^2}$
15. $f'(\theta) = \frac{1}{2}\sin 2\theta + \theta \cos 2\theta$
21. $y = x + 1$ **23.** $y = x - \pi - 1$
25. (a) $y = 2x$ (b) $\frac{3\pi}{2}$
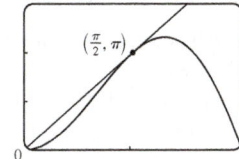

27. (a) $\sec x \tan x - 1$
29. $\theta \cos\theta + \sin\theta$; $2 \cos\theta - \theta \sin\theta$

31. (a) $f'(x) = (1 + \tan x)/\sec x$ (b) $f'(x) = \cos x + \sin x$
33. $(2n + 1)\pi \pm \frac{1}{3}\pi$, n an integer
35. (a) $v(t) = 8 \cos t$, $a(t) = -8 \sin t$
(b) $4\sqrt{3}, -4, -4\sqrt{3}$; to the left
37. 5 ft/rad   **39.** $\frac{5}{3}$ **41.** 3   **43.** $-\frac{3}{4}$
45. $\frac{1}{2}$ **47.** $-\frac{1}{4}$ **49.** $-\sqrt{2}$ **51.** $-\cos x$
53. $A = -\frac{3}{10}, B = -\frac{1}{10}$
55. (a) $\sec^2 x = \dfrac{1}{\cos^2 x}$ (b) $\sec x \tan x = \dfrac{\sin x}{\cos^2 x}$
(c) $\cos x - \sin x = \dfrac{\cot x - 1}{\csc x}$
57. 1

EXERCISES 3.4 ■ PAGE 204

1. $\dfrac{4}{3\sqrt[3]{(1 + 4x)^2}}$ **3.** $\pi \sec^2 \pi x$ **5.** $\dfrac{e^{\sqrt{x}}}{2\sqrt{x}}$
7. $F'(x) = 24x^{11}(5x^3 + 2)^3(5x^3 + 1)$
9. $f'(x) = \dfrac{5}{2\sqrt{5x + 1}}$ **11.** $f'(\theta) = -2\theta \sin(\theta^2)$
13. $y' = xe^{-3x}(2 - 3x)$ **15.** $f'(t) = e^{at}(b \cos bt + a \sin bt)$
17. $f'(x) = (2x - 3)^3(x^2 + x + 1)^4(28x^2 - 12x - 7)$
19. $h'(t) = \frac{2}{3}(t + 1)^{-1/3}(2t^2 - 1)^2(20t^2 + 18t - 1)$
21. $y' = \dfrac{1}{2\sqrt{x}\,(x + 1)^{3/2}}$ **23.** $y' = (\sec^2\theta)\,e^{\tan\theta}$
25. $g'(u) = \dfrac{48u^2(u^3 - 1)^7}{(u^3 + 1)^9}$ **27.** $r'(t) = \dfrac{(\ln 10)10^{2\sqrt{t}}}{\sqrt{t}}$
29. $H'(r) = \dfrac{2(r^2 - 1)^2(r^2 + 3r + 5)}{(2r + 1)^6}$
31. $F'(t) = e^{t \sin 2t}(2t \cos 2t + \sin 2t)$
33. $G'(x) = -C(\ln 4)\dfrac{4^{C/x}}{x^2}$
35. $y' = \dfrac{4e^{2x}}{(1 + e^{2x})^2}\sin\dfrac{1 - e^{2x}}{1 + e^{2x}}$
37. $y' = -2\cos\theta \cot(\sin\theta)\csc^2(\sin\theta)$
39. $f'(t) = -\sec^2(\sec(\cos t))\,\sec(\cos t)\tan(\cos t)\sin t$
41. $f'(t) = 4\sin(e^{\sin^2 t})\cos(e^{\sin^2 t})\,e^{\sin^2 t}\sin t \cos t$
43. $g'(x) = 2r^2 p(\ln a)\,(2ra^{rx} + n)^{p-1}a^{rx}$
45. $y' = \dfrac{-\pi \cos(\tan \pi x)\sec^2(\pi x)\sin\sqrt{\sin(\tan \pi x)}}{2\sqrt{\sin(\tan \pi x)}}$
47. $y' = -3\cos 3\theta \sin(\sin 3\theta)$;
$y'' = -9\cos^2(3\theta)\cos(\sin 3\theta) + 9(\sin 3\theta)\sin(\sin 3\theta)$
49. $y' = \dfrac{-\sec t \tan t}{2\sqrt{1 - \sec t}}$;
$y'' = \dfrac{\sec t\,(3\sec^3 t - 4\sec^2 t - \sec t + 2)}{4(1 - \sec t)^{3/2}}$
51. $y = (\ln 2)x + 1$ **53.** $y = -x + \pi$
55. (a) $y = \frac{1}{2}x + 1$ (b)

57. (a) $f'(x) = \dfrac{2 - 2x^2}{\sqrt{2 - x^2}}$

59. $((\pi/2) + 2n\pi, 3), ((3\pi/2) + 2n\pi, -1)$, n an integer

61. 24 **63.** (a) 30 (b) 36

65. (a) $\tfrac{3}{4}$ (b) Does not exist (c) -2 **67.** $-\tfrac{1}{6}\sqrt{2}$

69. (a) $F'(x) = e^x f'(e^x)$ (b) $G'(x) = e^{f(x)} f'(x)$

71. 120 **73.** 96

77. $-2^{50} \cos 2x$ **79.** $v(t) = \tfrac{5}{2}\pi \cos(10\pi t)$ cm/s

81. (a) $\dfrac{dB}{dt} = \dfrac{7\pi}{54} \cos \dfrac{2\pi t}{5.4}$ (b) 0.16

83. $v(t) = 2e^{-1.5t}(2\pi \cos 2\pi t - 1.5 \sin 2\pi t)$

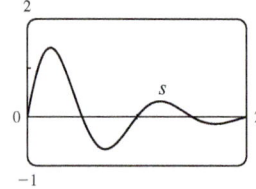 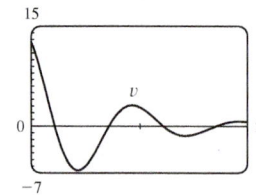

85. (a) 0.0075 (mg/mL)/min (b) 0.0030 (mg/mL)/min

87. dv/dt is the rate of change of velocity with respect to time; dv/ds is the rate of change of velocity with respect to displacement

89. (a) $Q = ab^t$ where $a \approx 100.01244$ and $b \approx 0.000045146$ (b) -670.63 µA

91. (b) The factored form **95.** (b) $-n \cos^{n-1}x \sin[(n+1)x]$

EXERCISES 3.5 ■ PAGE 215

1. (a) $y' = 9x/y$ (b) $y = \pm\sqrt{9x^2 - 1}$, $y' = \pm 9x/\sqrt{9x^2 - 1}$

3. (a) $y' = -\sqrt{y}/\sqrt{x}$ (b) $y = (1 - \sqrt{x})^2$, $y' = 1 - 1/\sqrt{x}$

5. $y' = \dfrac{2y - x}{y - 2x}$ **7.** $y' = -\dfrac{2x(2x^2 + y^2)}{y(2x^2 + 3y)}$

9. $y' = \dfrac{x(x + 2y)}{2x^2y + 4xy^2 + 2y^3 + x^2}$ **11.** $y' = \dfrac{2x + y \sin x}{\cos x - 2y}$

13. $y' = \dfrac{1 - 8x^3\sqrt{x + y}}{8y^3\sqrt{x + y} - 1}$ **15.** $y' = \dfrac{y(y - e^{x/y})}{y^2 - xe^{x/y}}$

17. $y' = \dfrac{1 + x^4y^2 + y^2 + x^4y^4 - 2xy}{x^2 - 2xy - 2x^5y^3}$

19. $y' = -\dfrac{y \cos(xy) + \sin(x + y)}{x \cos(xy) + \sin(x + y)}$ **21.** $-\tfrac{16}{13}$

23. $x' = \dfrac{-2x^4y + x^3 - 6xy^2}{4x^3y^2 - 3x^2y + 2y^3}$ **25.** $y = \tfrac{1}{2}x$

27. $y = \tfrac{3}{4}x - \tfrac{1}{2}$ **29.** $y = x + \tfrac{1}{2}$ **31.** $y = -\tfrac{9}{13}x + \tfrac{40}{13}$

33. (a) $y = \tfrac{9}{2}x - \tfrac{5}{2}$ (b)

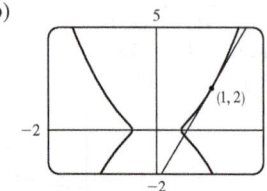

35. $-1/(4y^3)$ **37.** $\dfrac{\cos^2 y \cos x + \sin^2 x \sin y}{\cos^3 y}$ **39.** $1/e^2$

41. (a)

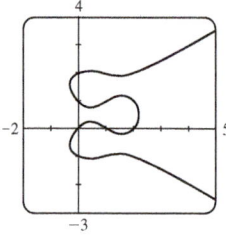

Eight; $x \approx 0.42, 1.58$

(b) $y = -x + 1$, $y = \tfrac{1}{3}x + 2$ (c) $1 \mp \tfrac{1}{3}\sqrt{3}$

43. $\left(\pm\tfrac{5}{4}\sqrt{3}, \pm\tfrac{5}{4}\right)$ **45.** $(x_0 x/a^2) - (y_0 y/b^2) = 1$

49. $y' = \dfrac{2 \tan^{-1} x}{1 + x^2}$ **51.** $y' = \dfrac{1}{\sqrt{-x^2 - x}}$

53. $F'(x) = \dfrac{3}{\sqrt{x^6 - 1}} + \sec^{-1}(x^3)$ **55.** $h'(t) = 0$

57. $y' = \sin^{-1} x$ **59.** $y' = \dfrac{\sqrt{a^2 - b^2}}{a + b \cos x}$

61. $1 - \dfrac{x \arcsin x}{\sqrt{1 - x^2}}$

65. 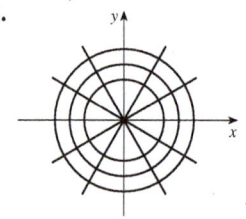 **67.**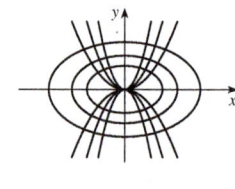

71. (a) $\dfrac{V^3(nb - V)}{PV^3 - n^2aV + 2n^3ab}$ (b) -4.04 L/atm

73. $(\pm\sqrt{3}, 0)$ **75.** $(-1, -1), (1, 1)$ **77.** (b) $\tfrac{3}{2}$

79. (a) 0 (b) $-\tfrac{1}{2}$

EXERCISES 3.6 ■ PAGE 223

1. The differentiation formula is simplest.

3. $f'(x) = \dfrac{\cos(\ln x)}{x}$ **5.** $f'(x) = -\dfrac{1}{x}$

7. $f'(x) = \dfrac{-\sin x}{(1 + \cos x) \ln 10}$ **9.** $g'(x) = \dfrac{1}{x} - 2$

11. $F'(t) = \ln t \left(\ln t \cos t + \dfrac{2 \sin t}{t}\right)$

13. $G'(y) = \dfrac{10}{2y + 1} - \dfrac{y}{y^2 + 1}$ **15.** $F'(s) = \dfrac{1}{s \ln s}$

17. $T'(z) = 2^z \left(\dfrac{1}{z \ln 2} + \ln z\right)$

19. $y' = \dfrac{-x}{1 + x}$ **21.** $y' = \sec^2[\ln(ax + b)] \dfrac{a}{ax + b}$

23. $y' = (2 + \ln x)/(2\sqrt{x})$; $y'' = -\ln x/(4x\sqrt{x})$

25. $y' = \tan x$; $y'' = \sec^2 x$

27. $f'(x) = \dfrac{2x - 1 - (x - 1) \ln(x - 1)}{(x - 1)[1 - \ln(x - 1)]^2}$; $(1, 1 + e) \cup (1 + e, \infty)$

29. $f'(x) = \dfrac{2(x - 1)}{x(x - 2)}$; $(-\infty, 0) \cup (2, \infty)$ **31.** 2

33. $y = 3x - 9$ **35.** $\cos x + 1/x$ **37.** 7

39. $y' = (x^2 + 2)^2(x^4 + 4)^4 \left(\dfrac{4x}{x^2 + 2} + \dfrac{16x^3}{x^4 + 4} \right)$

41. $y' = \sqrt{\dfrac{x-1}{x^4+1}} \left(\dfrac{1}{2x-2} - \dfrac{2x^3}{x^4+1} \right)$

43. $y' = x^x(1 + \ln x)$

45. $y' = x^{\sin x}\left(\dfrac{\sin x}{x} + \cos x \ln x \right)$

47. $y' = (\cos x)^x(-x \tan x + \ln \cos x)$

49. $y' = (\tan x)^{1/x}\left(\dfrac{\sec^2 x}{x \tan x} - \dfrac{\ln \tan x}{x^2} \right)$

51. $y' = \dfrac{2x}{x^2 + y^2 - 2y}$ **53.** $f^{(n)}(x) = \dfrac{(-1)^{n-1}(n-1)!}{(x-1)^n}$

EXERCISES 3.7 ■ PAGE 233

1. (a) $3t^2 - 16t + 24$ (b) 11 ft/s   (c) Never   (d) Always
(e) 72 ft
(f)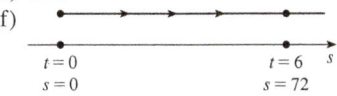
(g) $6t - 16$; -10 ft/s^2
(h) 80
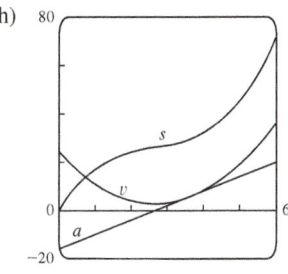
(i) Speeding up when $t > \tfrac{8}{3}$; slowing down when $0 \le t < \tfrac{8}{3}$

3. (a) $(\pi/2)\cos(\pi t/2)$ (b) 0 ft/s
(c) $t = 2n + 1$, t a nonnegative integer
(d) $0 < t < 1$, $3 < t < 5$, $7 < t < 9$, and so on   (e) 6 ft
(f)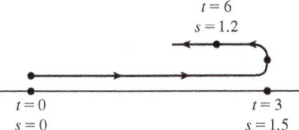
(g) $(-\pi^2/4)\sin(\pi t/2)$; $-\pi^2/4$ ft/s^2

(h)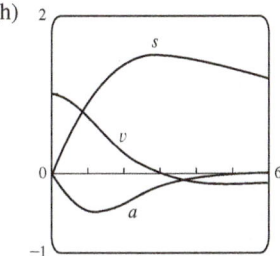

(i) Speeding up when $1 < t < 2$, $3 < t < 4$, and $5 < t < 6$; slowing down when $0 < t < 1$, $2 < t < 3$, and $4 < t < 5$
5. (a) Speeding up when $0 < t < 1$ or $2 < t < 3$; slowing down when $1 < t < 2$

(b) Speeding up when $1 < t < 2$ or $3 < t < 4$; slowing down when $0 < t < 1$ or $2 < t < 3$
7. (a) 4.9 m/s; -14.7 m/s   (b) After 2.5 s   (c) $32\tfrac{5}{8}$ m
(d) ≈ 5.08 s   (e) ≈ -25.3 m/s
9. (a) 7.56 m/s   (b) 6.24 m/s; -6.24 m/s
11. (a) 30 mm^2/mm; the rate at which the area is increasing with respect to side length as x reaches 15 mm
(b) $\Delta A \approx 2x\,\Delta x$
13. (a) (i) 5π (ii) 4.5π (iii) 4.1π
(b) 4π (c) $\Delta A \approx 2\pi r\,\Delta r$
15. (a) 8π ft^2/ft   (b) 16π ft^2/ft   (c) 24π ft^2/ft
The rate increases as the radius increases.
17. (a) 6 kg/m   (b) 12 kg/m   (c) 18 kg/m
At the right end; at the left end
19. (a) 4.75 A   (b) 5 A; $t = \tfrac{2}{3}$ s
23. (a) $dV/dP = -C/P^2$ (b) At the beginning
25. $400(3^t)\ln 3$; ≈ 6850 bacteria/h
27. (a) 16 million/year; 78.5 million/year
(b) $P(t) = at^3 + bt^2 + ct + d$, where $a \approx -0.0002849$, $b \approx 0.5224331$, $c \approx -6.395641$, $d \approx 1720.586$
(c) $P'(t) = 3at^2 + 2bt + c$
(d) 14.16 million/year (smaller); 71.72 million/year (smaller)
(e) $f'(t) = (1.43653 \times 10^9) \cdot (1.01395)^t \ln 1.01395$
(f) 26.25 million/year (larger); 60.28 million/year (smaller)
(g) $P'(85) \approx 76.24$ million/year, $f'(85) = 64.61$ million/year
29. (a) 0.926 cm/s; 0.694 cm/s; 0
(b) 0; -92.6 (cm/s)/cm; -185.2 (cm/s)/cm
(c) At the center; at the edge
31. (a) $C'(x) = 3 + 0.02x + 0.0006x^2$
(b) \$11/pair; the rate at which the cost is changing as the 100th pair of jeans is being produced; the cost of the 101st pair
(c) \$11.07
33. (a) $[xp'(x) - p(x)]/x^2$; the average productivity increases as new workers are added.

35. $\dfrac{dt}{dc} = \dfrac{3\sqrt{9c^2 - 8c} + 9c - 4}{\sqrt{9c^2 - 8c}\,(3c + \sqrt{9c^2 - 8c}\,)}$; the rate of change of duration of dialysis required with respect to the initial urea concentration
37. -0.2436 K/min
39. (a) 0 and 0   (b) $C = 0$
(c) $(0, 0)$, $(500, 50)$; it is possible for the species to coexist.

EXERCISES 3.8 ■ PAGE 242

1. About 235
3. (a) $100(4.2)^t$ (b) ≈ 7409 (c) $\approx 10{,}632$ bacteria/h
(d) $(\ln 100)/(\ln 4.2) \approx 3.2$ h
5. (a) 1508 million, 1871 million   (b) 2161 million
(c) 3972 million; wars in the first half of century, increased life expectancy in second half
7. (a) $Ce^{-0.0005t}$ (b) $-2000 \ln 0.9 \approx 211$ s
9. (a) $100 \times 2^{-t/30}$ mg   (b) ≈ 9.92 mg   (c) ≈ 199.3 years
11. ≈ 2500 years   **13.** Yes; 12.5 billion years
15. (a) $\approx 137°$F   (b) ≈ 116 min
17. (a) $13.\overline{3}°$C   (b) ≈ 67.74 min
19. (a) ≈ 64.5 kPa   (b) ≈ 39.9 kPa

21. (a) (i) $3828.84 (ii) $3840.25 (iii) $3850.08
(iv) $3851.61 (v) $3852.01 (vi) $3852.08
(b) $dA/dt = 0.05A$, $A(0) = 3000$

EXERCISES 3.9 ■ PAGE 249
1. $dV/dt = 3x^2\, dx/dt$ **3.** 48 cm²/s   **5.** $3/(25\pi)$ m/min
7. 128π cm²/min   **9.** (a) 1   (b) 25   **11.** -18
13. (a) The plane's altitude is 1 mi and its speed is 500 mi/h.
(b) The rate at which the distance from the plane to the station is increasing when the plane is 2 mi from the station
(c) 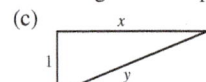   (d) $y^2 = x^2 + 1$
(e) $250\sqrt{3}$ mi/h

15. (a) The height of the pole (15 ft), the height of the man (6 ft), and the speed of the man (5 ft/s)
(b) The rate at which the tip of the man's shadow is moving when he is 40 ft from the pole
(c) 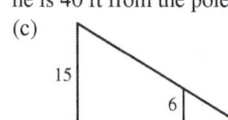   (d) $\dfrac{15}{6} = \dfrac{x+y}{y}$ (e) $\dfrac{25}{3}$ ft/s

17. 65 mi/h   **19.** $837/\sqrt{8674} \approx 8.99$ ft/s
21. -1.6 cm/min   **23.** $\dfrac{720}{13} \approx 55.4$ km/h
25. $(10{,}000 + 800{,}000\pi/9) \approx 2.89 \times 10^5$ cm³/min
27. $\dfrac{10}{3}$ cm/min   **29.** $6/(5\pi) \approx 0.38$ ft/min
31. $150\sqrt{3}$ cm²/min   **33.** 5 m   **35.** $\pi r^2/60$ cm²/min
37. 80 cm³/min   **39.** $\dfrac{107}{810} \approx 0.132$ Ω/s
41. $\sqrt{7}\,\pi/21 \approx 0.396$ m/min
43. (a) 360 ft/s   (b) 0.096 rad/s
45. $\dfrac{10}{9}\pi$ km/min   **47.** $1650/\sqrt{31} \approx 296$ km/h
49. $\dfrac{7}{4}\sqrt{15} \approx 6.78$ m/s

EXERCISES 3.10 ■ PAGE 256
1. $L(x) = 16x + 23$ **3.** $L(x) = \tfrac{1}{4}x + 1$
5. $\sqrt{1-x} \approx 1 - \tfrac{1}{2}x$;
$\sqrt{0.9} \approx 0.95$,
$\sqrt{0.99} \approx 0.995$

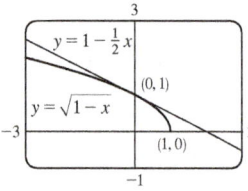

7. $-0.383 < x < 0.516$ **9.** $-0.368 < x < 0.677$
11. (a) $dy = (1 - 4x)e^{-4x}\, dx$ (b) $dy = -\dfrac{2t^3}{\sqrt{1-t^4}}\, dt$
13. (a) $dy = \dfrac{\sec^2\sqrt{t}}{2\sqrt{t}}\, dt$ (b) $dy = \dfrac{-4v}{(1+v^2)^2}\, dv$
15. (a) $dy = \tfrac{1}{10} e^{x/10}\, dx$ (b) 0.01
17. (a) $dy = \dfrac{x}{\sqrt{3+x^2}}\, dx$ (b) -0.05

19. $\Delta y = 1.25$, $dy = 1$

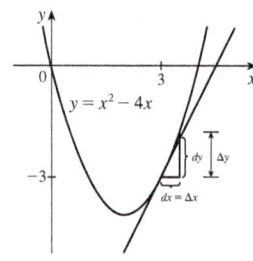

21. $\Delta y \approx 0.34$, $dy = 0.4$

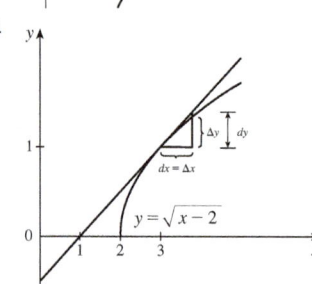

23. 15.968   **25.** $10.00\overline{3}$ **27.** 1.1
33. (a) 270 cm³, 0.01, 1%   (b) 36 cm², $0.00\overline{6}$, $0.\overline{6}$%
35. (a) $84/\pi \approx 27$ cm²; $\tfrac{1}{84} \approx 0.012 = 1.2\%$
(b) $1764/\pi^2 \approx 179$ cm³; $\tfrac{1}{56} \approx 0.018 = 1.8\%$
37. (a) $2\pi rh\, \Delta r$ (b) $\pi(\Delta r)^2 h$
43. (a) 4.8, 5.2   (b) Too large

EXERCISES 3.11 ■ PAGE 264
1. (a) 0   (b) 1   **3.** (a) $\tfrac{13}{5}$ (b) $\tfrac{1}{2}(e^5 + e^{-5}) \approx 74.20995$
5. (a) 1   (b) 0
21. $\text{sech } x = \tfrac{3}{5}$, $\sinh x = \tfrac{4}{3}$, $\text{csch } x = \tfrac{3}{4}$, $\tanh x = \tfrac{4}{5}$, $\coth x = \tfrac{5}{4}$
23. (a) 1   (b) -1 (c) ∞ (d) $-\infty$ (e) 0   (f) 1
(g) ∞ (h) $-\infty$ (i) 0   (j) $\tfrac{1}{2}$
31. $f'(x) = \dfrac{\text{sech}^2\sqrt{x}}{2\sqrt{x}}$ **33.** $h'(x) = 2x\cosh(x^2)$
35. $G'(t) = \dfrac{t^2 + 1}{2t^2}$
37. $y' = 3e^{\cosh 3x} \sinh 3x$
39. $g'(t) = \coth\sqrt{t^2+1} - \dfrac{t^2}{\sqrt{t^2+1}}\,\text{csch}^2\sqrt{t^2+1}$
41. $y' = \dfrac{1}{2\sqrt{x(x-1)}}$
43. $y' = \sinh^{-1}(x/3)$ **45.** $y' = -\csc x$
51. (a) 0.3572   (b) 70.34°
53. (a) 164.50 m   (b) 120 m; 164.13 m
55. (b) $y = 2\sinh 3x - 4\cosh 3x$
57. $\left(\ln\left(1 + \sqrt{2}\right), \sqrt{2}\right)$

CHAPTER 3 REVIEW ■ PAGE 266
True-False Quiz
1. True   **3.** True   **5.** False   **7.** False   **9.** True
11. True   **13.** True   **15.** True

APPENDIX I Answers to Odd-Numbered Exercises

Exercises

1. $4x^7(x+1)^3(3x+2)$ 3. $\frac{3}{2}\sqrt{x} - \frac{1}{2\sqrt{x}} - \frac{1}{\sqrt{x^3}}$
5. $x(\pi x \cos \pi x + 2 \sin \pi x)$
7. $\dfrac{8t^3}{(t^4+1)^2}$ 9. $\dfrac{1+\ln x}{x \ln x}$ 11. $\dfrac{\cos\sqrt{x} - \sqrt{x}\sin\sqrt{x}}{2\sqrt{x}}$
13. $-\dfrac{e^{1/x}(1+2x)}{x^4}$ 15. $\dfrac{2xy - \cos y}{1 - x\sin y - x^2}$
17. $\dfrac{1}{2\sqrt{\arctan x}\,(1+x^2)}$ 19. $\dfrac{1-t^2}{(1+t^2)^2}\sec^2\!\left(\dfrac{t}{1+t^2}\right)$
21. $3^{x\ln x}(\ln 3)(1+\ln x)$ 23. $-(x-1)^{-2}$
25. $\dfrac{2x - y\cos(xy)}{x\cos(xy)+1}$ 27. $\dfrac{2}{(1+2x)\ln 5}$
29. $\cot x - \sin x \cos x$ 31. $\dfrac{4x}{1+16x^2} + \tan^{-1}(4x)$
33. $5 \sec 5x$ 35. $-6x\csc^2(3x^2+5)$
37. $\cos(\tan\sqrt{1+x^3})(\sec^2\sqrt{1+x^3})\dfrac{3x^2}{2\sqrt{1+x^3}}$
39. $2\cos\theta\tan(\sin\theta)\sec^2(\sin\theta)$
41. $\dfrac{(2-x)^4(3x^2 - 55x - 52)}{2\sqrt{x+1}\,(x+3)^8}$ 43. $2x^2\cosh(x^2) + \sinh(x^2)$
45. $3 \tanh 3x$ 47. $\dfrac{\cosh x}{\sqrt{\sinh^2 x - 1}}$
49. $\dfrac{-3\sin(e^{\sqrt{\tan 3x}})e^{\sqrt{\tan 3x}}\sec^2(3x)}{2\sqrt{\tan 3x}}$ 51. $-\frac{4}{27}$
53. $-5x^4/y^{11}$ 57. $y = 2\sqrt{3}x + 1 - \pi\sqrt{3}/3$
59. $y = 2x + 1$ 61. $y = -x + 2;\ y = x + 2$
63. (a) $\dfrac{10 - 3x}{2\sqrt{5-x}}$ (b) $y = \frac{7}{4}x + \frac{1}{4},\ y = -x + 8$
(c)
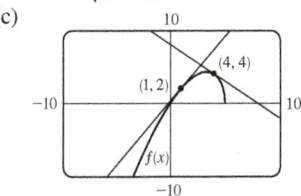
65. $(\pi/4, \sqrt{2}),\ (5\pi/4, -\sqrt{2})$
69. (a) 4   (b) 6   (c) $\frac{7}{9}$ (d) 12
71. $2xg(x) + x^2 g'(x)$ 73. $2g(x)g'(x)$
75. $g'(e^x)e^x$ 77. $g'(x)/g(x)$
79. $\dfrac{f'(x)[g(x)]^2 + g'(x)[f(x)]^2}{[f(x)+g(x)]^2}$
81. $f'(g(\sin 4x))g'(\sin 4x)(\cos 4x)(4)$
83. $(-3, 0)$ 85. $y = -\frac{2}{3}x^2 + \frac{14}{3}x$
87. $v(t) = -Ae^{-ct}[c\cos(\omega t + \delta) + \omega \sin(\omega t + \delta)]$,
$a(t) = Ae^{-ct}[(c^2 - \omega^2)\cos(\omega t + \delta) + 2c\omega\sin(\omega t + \delta)]$

89. (a) $v(t) = 3t^2 - 12;\ a(t) = 6t$ (b) $t > 2;\ 0 \le t < 2$
(c) 23   (d)
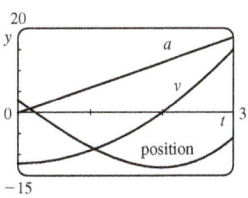
(e) $t > 2;\ 0 < t < 2$
91. 4 kg/m
93. (a) $200(3.24)^t$ (b) $\approx 22{,}040$
(c) $\approx 25{,}910$ bacteria/h   (d) $(\ln 50)/(\ln 3.24) \approx 3.33$ h
95. (a) $C_0 e^{-kt}$ (b) ≈ 100 h   97. $\frac{4}{3}$ cm^2/min
99. 13 ft/s   101. 400 ft/h
103. (a) $L(x) = 1 + x;\ \sqrt[3]{1+3x} \approx 1 + x;\ \sqrt[3]{1.03} \approx 1.01$
(b) $-0.23 < x < 0.40$
105. $12 + \frac{3}{2}\pi \approx 16.7$ cm^2 107. $\frac{1}{32}$ 109. $\frac{1}{4}$ 111. $\frac{1}{8}x^2$

PROBLEMS PLUS ■ PAGE 271

1. $\left(\pm\sqrt{3}/2,\ \frac{1}{4}\right)$ 5. $3\sqrt{2}$ 11. $\left(0, \frac{5}{4}\right)$
13. 3 lines; $(0, 2),\ \left(\frac{4}{3}\sqrt{2}, \frac{2}{3}\right)$ and $\left(\frac{2}{3}\sqrt{2}, \frac{10}{3}\right),\ \left(-\frac{4}{3}\sqrt{2}, \frac{2}{3}\right)$ and $\left(-\frac{2}{3}\sqrt{2}, \frac{10}{3}\right)$
15. (a) $4\pi\sqrt{3}/\sqrt{11}$ rad/s   (b) $40(\cos\theta + \sqrt{8 + \cos^2\theta})$ cm
(c) $-480\pi\sin\theta\,(1 + \cos\theta/\sqrt{8 + \cos^2\theta})$ cm/s
19. $x_T \in (3, \infty),\ y_T \in (2, \infty),\ x_N \in \left(0, \frac{5}{3}\right),\ y_N \in \left(-\frac{5}{2}, 0\right)$
21. (b) (i) $53°$ (or $127°$)   (ii) $63°$ (or $117°$)
23. R approaches the midpoint of the radius AO.
25. $-\sin a$ 27. $2\sqrt{e}$ 31. $(1, -2), (-1, 0)$
33. $\sqrt{29}/58$ 35. $2 + \frac{375}{128}\pi \approx 11.204$ cm^3/min

CHAPTER 4

EXERCISES 4.1 ■ PAGE 283

Abbreviations: abs, absolute; loc, local; max, maximum; min, minimum
1. Abs min: smallest function value on the entire domain of the function; loc min at c: smallest function value when x is near c
3. Abs max at s, abs min at r, loc max at c, loc min at b and r, neither a max nor a min at a and d
5. Abs max $f(4) = 5$, loc max $f(4) = 5$ and $f(6) = 4$, loc min $f(2) = 2$ and $f(1) = f(5) = 3$

7. 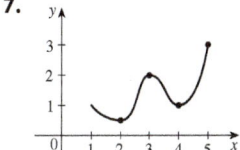   9.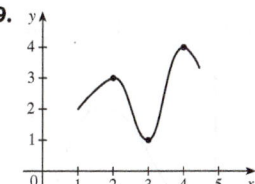

11. (a) 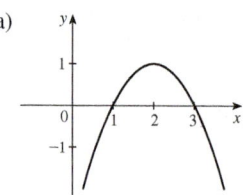 (b)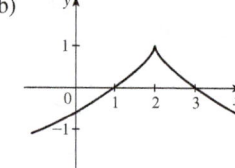

(c)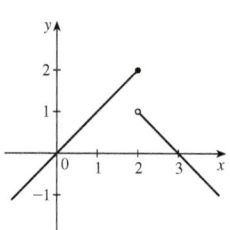

13. (a) 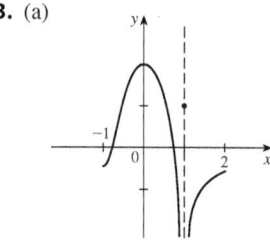 (b)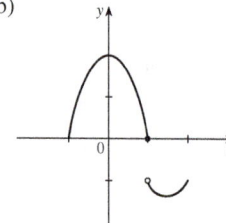

15. Abs max $f(3) = 4$ **17.** Abs max $f(1) = 1$
19. Abs min $f(0) = 0$
21. Abs max $f(\pi/2) = 1$; abs min $f(-\pi/2) = -1$
23. Abs max $f(2) = \ln 2$ **25.** Abs max $f(0) = 1$
27. Abs min $f(1) = -1$ **29.** $\frac{1}{3}$ **31.** $-2, 3$ **33.** 0
35. $0, 2$ **37.** $0, \frac{4}{9}$ **39.** $0, \frac{8}{7}, 4$ **41.** $n\pi$ (n an integer)
43. $0, \frac{2}{3}$ **45.** 10 **47.** $f(2) = 16, f(5) = 7$
49. $f(-1) = 8, f(2) = -19$ **51.** $f(-2) = 33, f(2) = -31$
53. $f(0.2) = 5.2, f(1) = 2$
55. $f(4) = 4 - \sqrt[3]{4}, f(\sqrt{3}/9) = -2\sqrt{3}/9$
57. $f(\pi/6) = \frac{3}{2}\sqrt{3}, f(\pi/2) = 0$
59. $f(e^{1/2}) = 1/(2e), f(\frac{1}{2}) = -4 \ln 2$
61. $f(1) = \ln 3, f(-\frac{1}{2}) = \ln \frac{3}{4}$
63. $f\left(\dfrac{a}{a+b}\right) = \dfrac{a^a b^b}{(a+b)^{a+b}}$
65. (a) 2.19, 1.81 (b) $\frac{6}{25}\sqrt{\frac{3}{5}} + 2, -\frac{6}{25}\sqrt{\frac{3}{5}} + 2$
67. (a) 0.32, 0.00 (b) $\frac{3}{16}\sqrt{3}, 0$
69. 0.177 mg/mL; 21.4 min **71.** $\approx 3.9665°C$
73. About 4.1 months after Jan. 1
75. (a) $r = \frac{2}{3}r_0$ (b) $v = \frac{4}{27}kr_0^3$

(c)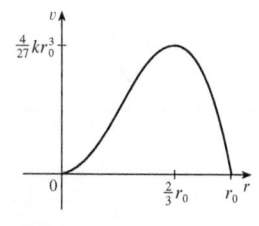

EXERCISES 4.2 ■ PAGE 291

1. 1, 5
3. (a) g is continuous on $[0, 8]$ and differentiable on $(0, 8)$.
(b) 2.1, 6.3 (c) 3.6, 5.6
5. 1 **7.** π
9. f is not differentiable on $(-1, 1)$ **11.** 1
13. $3/\ln 4$ **15.** 1; yes **17.** f is not continous at 3
25. 16 **27.** No **33.** No

EXERCISES 4.3 ■ PAGE 300

1. (a) $(1, 3), (4, 6)$ (b) $(0, 1), (3, 4)$ (c) $(0, 2)$
(d) $(2, 4), (4, 6)$ (e) $(2, 3)$
3. (a) I/D Test (b) Concavity Test
(c) Find points at which the concavity changes.
5. (a) Inc on $(1, 5)$; dec on $(0, 1)$ and $(5, 6)$
(b) Loc max at $x = 5$, loc min at $x = 1$
7. (a) 3, 5 (b) 2, 4, 6 (c) 1, 7
9. (a) Inc on $(-\infty, -1), (3, \infty)$; dec on $(-1, 3)$
(b) Loc max $f(-1) = 9$; loc min $f(3) = -23$
(c) CU on $(1, \infty)$, CD on $(-\infty, 1)$; IP $(1, -7)$
11. (a) Inc on $(-1, 0), (1, \infty)$; dec on $(-\infty, -1), (0, 1)$
(b) Loc max $f(0) = 3$; loc min $f(\pm 1) = 2$
(c) CU on $\left(-\infty, -\sqrt{3}/3\right), \left(\sqrt{3}/3, \infty\right)$;
CD on $\left(-\sqrt{3}/3, \sqrt{3}/3\right)$; IP $\left(\pm\sqrt{3}/3, \frac{22}{9}\right)$
13. (a) Inc on $(0, \pi/4), (5\pi/4, 2\pi)$; dec on $(\pi/4, 5\pi/4)$
(b) Loc max $f(\pi/4) = \sqrt{2}$; loc min $f(5\pi/4) = -\sqrt{2}$
(c) CU on $(3\pi/4, 7\pi/4)$; CD on $(0, 3\pi/4), (7\pi/4, 2\pi)$;
IP $(3\pi/4, 0), (7\pi/4, 0)$
15. (a) Inc on $\left(-\frac{1}{3}\ln 2, \infty\right)$; dec on $\left(-\infty, -\frac{1}{3}\ln 2\right)$
(b) Loc min $f\left(-\frac{1}{3}\ln 2\right) = 2^{-2/3} + 2^{1/3}$ (c) CU on $(-\infty, \infty)$
17. (a) Inc on $(1, \infty)$; dec on $(0, 1)$ (b) Loc min $f(1) = 0$
(c) CU on $(0, \infty)$; No IP
19. Loc max $f(1) = 2$; loc min $f(0) = 1$
21. Loc min $f\left(\frac{1}{16}\right) = -\frac{1}{4}$
23. (a) f has a local maximum at 2.
(b) f has a horizontal tangent at 6.
25. (a) 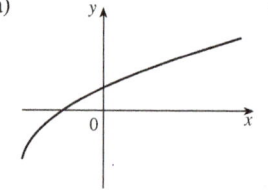 (b)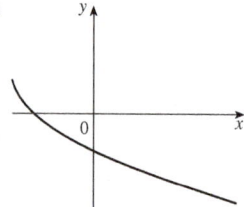

27. 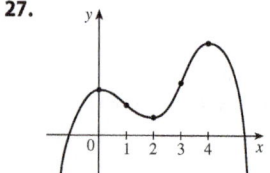 **29.**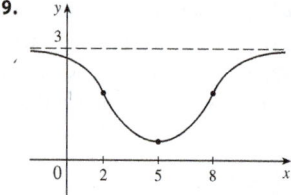

31.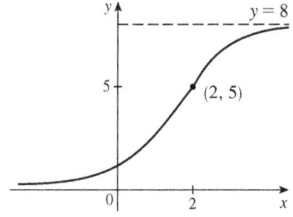

33. (a) No
(b) Yes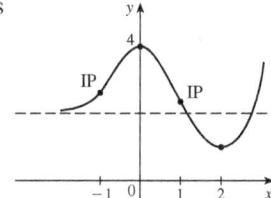

(c)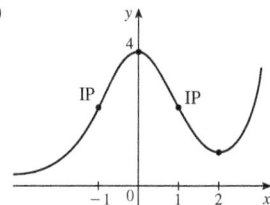

35. (a) Inc on $(0, 2)$, $(4, 6)$, $(8, \infty)$;
dec on $(2, 4)$, $(6, 8)$
(b) Loc max at $x = 2, 6$;
loc min at $x = 4, 8$
(c) CU on $(3, 6)$, $(6, \infty)$;
CD on $(0, 3)$ (d) 3
(e) See graph at right.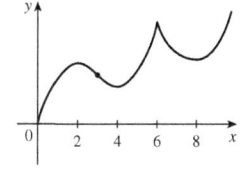

37. (a) Inc on $(-\infty, -2)$, $(2, \infty)$; dec on $(-2, 2)$
(b) Loc max $f(-2) = 18$; loc min $f(2) = -14$
(c) CU on $(0, \infty)$, CD on $(-\infty, 0)$; IP $(0, 2)$
(d)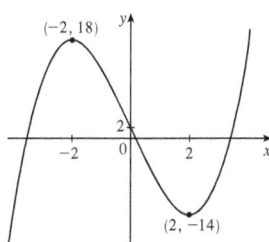

39. (a) Inc on $(-2, 0)$, $(2, \infty)$; dec on $(-\infty, -2)$, $(0, 2)$
(b) Loc max $f(0) = 3$; loc min $f(\pm 2) = -5$
(c) CU on $\left(-\infty, -\dfrac{2}{\sqrt{3}}\right)$, $\left(\dfrac{2}{\sqrt{3}}, \infty\right)$; CD on $\left(-\dfrac{2}{\sqrt{3}}, \dfrac{2}{\sqrt{3}}\right)$;
IPs $\left(\pm\dfrac{2}{\sqrt{3}}, -\dfrac{13}{9}\right)$

(d)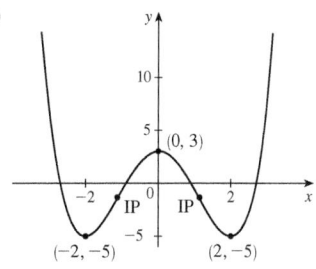

41. (a) Inc on $(-\infty, -2)$, $(0, \infty)$;
dec on $(-2, 0)$
(b) Loc max $h(-2) = 7$;
loc min $h(0) = -1$
(c) CU on $(-1, \infty)$;
CD on $(-\infty, -1)$; IP $(-1, 3)$
(d) See graph at right.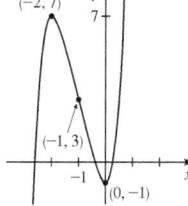

43. (a) Inc on $(-\infty, 4)$;
dec on $(4, 6)$
(b) Loc max $F(4) = 4\sqrt{2}$
(c) CD on $(-\infty, 6)$; No IP
(d) See graph at right.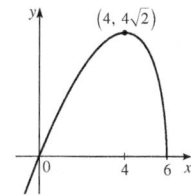

45. (a) Inc on $(-1, \infty)$;
dec on $(-\infty, -1)$
(b) Loc min $C(-1) = -3$
(c) CU on $(-\infty, 0)$, $(2, \infty)$;
CD on $(0, 2)$;
IPs $(0, 0)$, $(2, 6\sqrt[3]{2})$
(d) See graph at right.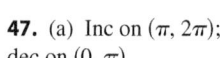

47. (a) Inc on $(\pi, 2\pi)$;
dec on $(0, \pi)$
(b) Loc min $f(\pi) = -1$
(c) CU on $(\pi/3, 5\pi/3)$;
CD on $(0, \pi/3)$, $(5\pi/3, 2\pi)$;
IPs $\left(\pi/3, \tfrac{5}{4}\right)$, $\left(5\pi/3, \tfrac{5}{4}\right)$
(d) See graph at right.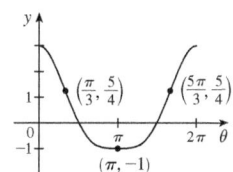

49. (a) VA $x = 0$; HA $y = 1$
(b) Inc on $(0, 2)$;
dec on $(-\infty, 0)$, $(2, \infty)$
(c) Loc max $f(2) = \tfrac{5}{4}$
(d) CU on $(3, \infty)$;
CD on $(-\infty, 0)$, $(0, 3)$; IP $\left(3, \tfrac{11}{9}\right)$
(e) See graph at right.

51. (a) HA $y = 0$
(b) Dec on $(-\infty, \infty)$
(c) None
(d) CU on $(-\infty, \infty)$
(e) See graph at right.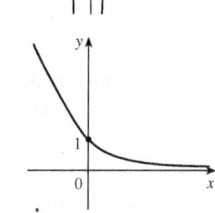

53. (a) HA $y = 0$
(b) Inc on $(-\infty, 0)$,
dec on $(0, \infty)$
(c) Loc max $f(0) = 1$
(d) CU on $(-\infty, -1/\sqrt{2})$,
$(1/\sqrt{2}, \infty)$; CD on $(-1/\sqrt{2}, 1/\sqrt{2})$; IPs $(\pm 1/\sqrt{2}, e^{-1/2})$
(e) See graph at right.

55. (a) VA $x = 0$, $x = e$
(b) Dec on $(0, e)$
(c) None
(d) CU on $(0, 1)$; CD on $(1, e)$;
IP $(1, 0)$
(e) See graph at right.

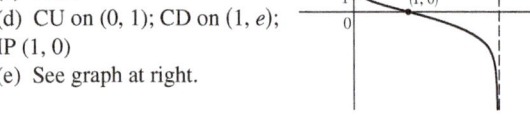

57. $(3, \infty)$
59. (a) Loc and abs max $f(1) = \sqrt{2}$, no min   (b) $\frac{1}{4}(3 - \sqrt{17})$
61. (b) CU on $(0, 0.85)$, $(1.57, 2.29)$; CD on $(0.85, 1.57)$, $(2.29, \pi)$; IP $(0.85, 0.74)$, $(1.57, 0)$, $(2.29, -0.74)$
63. CU on $(-\infty, -0.6)$, $(0.0, \infty)$; CD on $(-0.6, 0.0)$
65. (a) The rate of increase is initially very small, increases to a maximum at $t \approx 8$ h, then decreases toward 0.
(b) When $t = 8$ (c) CU on $(0, 8)$; CD on $(8, 18)$
(d) $(8, 350)$
67. If $D(t)$ is the size of the deficit as a function of time, then at the time of the speech $D'(t) > 0$, but $D''(t) < 0$.
69. $K(3) - K(2)$; CD
71. 28.57 min, when the rate of increase of drug level in the bloodstream is greatest; 85.71 min, when rate of decrease is greatest
73. $f(x) = \frac{1}{9}(2x^3 + 3x^2 - 12x + 7)$
75. (a) $a = 0$, $b = -1$ (b) $y = -x$ at $(0, 0)$

EXERCISES 4.4 ■ PAGE 311

1. (a) Indeterminate   (b) 0   (c) 0
(d) ∞, $-\infty$, or does not exist   (e) Indeterminate
3. (a) $-\infty$ (b) Indeterminate   (c) ∞
5. $\frac{9}{4}$ **7.** 1   **9.** 6   **11.** $-\frac{1}{3}$
13. $-\infty$ **15.** 2   **17.** $\frac{1}{4}$ **19.** 0   **21.** $-\infty$
23. $\frac{8}{5}$ **25.** 3   **27.** $\frac{1}{2}$ **29.** 1   **31.** 1
33. $1/\ln 3$ **35.** 0   **37.** 0   **39.** a/b
41. $\frac{1}{24}$ **43.** π **45.** $\frac{5}{3}$ **47.** 0   **49.** $-2/\pi$
51. $\frac{1}{2}$ **53.** $\frac{1}{2}$ **55.** ∞ **57.** 1   **59.** e^{-2}
61. $1/e$ **63.** 1   **65.** e^4 **67.** e^3 **69.** e^2
71. $\frac{1}{4}$ **75.** 1
77. f has an absolute minimum for $c > 0$. As c increases, the minimum points get farther away from the origin.
81. (a) M; the population should approach its maximum size as time increases   (b) $P_0 e^{kt}$; exponential
83. $\frac{16}{9}a$ **85.** $\frac{1}{2}$ **87.** 56   **91.** (a) 0

EXERCISES 4.5 ■ PAGE 321

Abbreviation: int, intercept; SA, slant asymptote
1. A. $\mathbb{R}$ B. y-int 0; x-int -3, 0
C. None   D. None
E. Inc on $(-\infty, -2)$, $(0, \infty)$;
dec on $(-2, 0)$
F. Loc max $f(-2) = 4$;
loc min $f(0) = 0$
G. CU on $(-1, \infty)$; CD on $(-\infty, -1)$;
IP $(-1, 2)$
H. See graph at right.

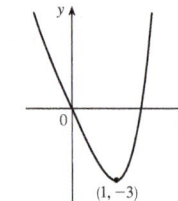

3. A. $\mathbb{R}$ B. y-int 0; x-int 0, $\sqrt[3]{4}$
C. None   D. None
E. Inc on $(1, \infty)$; dec on $(-\infty, 1)$
F. Loc min $f(1) = -3$
G. CU on $(-\infty, \infty)$
H. See graph at right.

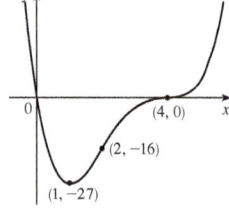

5. A. $\mathbb{R}$ B. y-int 0; x-int 0, 4
C. None   D. None
E. Inc on $(1, \infty)$; dec on $(-\infty, 1)$
F. Loc min $f(1) = -27$
G. CU on $(-\infty, 2)$, $(4, \infty)$;
CD on $(2, 4)$;
IPs $(2, -16)$, $(4, 0)$
H. See graph at right.

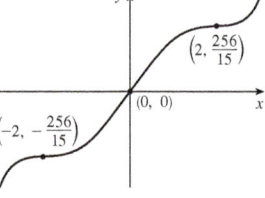

7. A. $\mathbb{R}$ B. y-int 0; x-int 0
C. About $(0, 0)$ D. None
E. Inc on $(-\infty, \infty)$
F. None
G. CU on $(-2, 0)$, $(2, \infty)$;
CD on $(-\infty, -2)$, $(0, 2)$;
IPs $\left(-2, -\frac{256}{15}\right)$, $(0, 0)$, $\left(2, \frac{256}{15}\right)$
H. See graph at right.

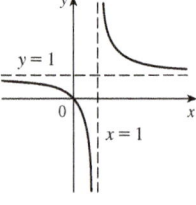

9. A. $\{x \mid x \neq 1\}$ B. y-int 0; x-int 0
C. None   D. VA $x = 1$, HA $y = 1$
E. Dec on $(-\infty, 1)$, $(1, \infty)$
F. None
G. CU on $(1, \infty)$; CD on $(-\infty, 1)$
H. See graph at right.

11. A. $(-\infty, 1) \cup (1, 2) \cup (2, \infty)$
B. y-int 0; x-int 0   C. None
D. HA $y = -1$; VA $x = 2$
E. Inc on $(-\infty, 1)$, $(1, 2)$, $(2, \infty)$
F. None
G. CU on $(-\infty, 1)$, $(1, 2)$;
CD on $(2, \infty)$
H. See graph at right.

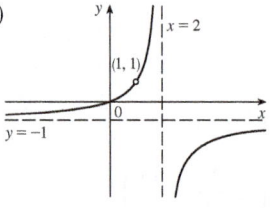

13. A. $(-\infty, -2) \cup (-2, 2) \cup (2, \infty)$ B. y-int 0; x-int 0
C. About $(0, 0)$ D. VA $x = \pm 2$; HA $y = 0$
E. Dec on $(-\infty, -2), (-2, 2), (2, \infty)$
F. No local extrema
G. CU on $(-2, 0), (2, \infty)$;
CD on $(-\infty, -2), (0, 2)$; IP $(0, 0)$
H. See graph at right.

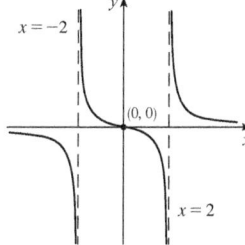

15. A. $\mathbb{R}$ B. y-int 0; x-int 0
C. About y-axis D. HA $y = 1$
E. Inc on $(0, \infty)$; dec on $(-\infty, 0)$
F. Loc min $f(0) = 0$
G. CU on $(-1, 1)$;
CD on $(-\infty, -1), (1, \infty)$; IPs $(\pm 1, \tfrac{1}{4})$
H. See graph at right.

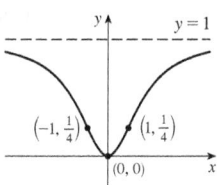

17. A. $(-\infty, 0) \cup (0, \infty)$ B. x-int 1
C. None D. HA $y = 0$; VA $x = 0$
E. Inc on $(0, 2)$;
dec on $(-\infty, 0), (2, \infty)$
F. Loc max $f(2) = \tfrac{1}{4}$
G. CU on $(3, \infty)$;
CD on $(-\infty, 0), (0, 3)$; IP $(3, \tfrac{2}{9})$
H. See graph at right.

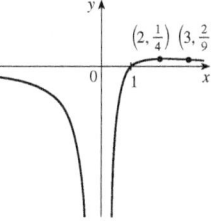

19. A. $(-\infty, -1) \cup (-1, \infty)$
B. y-int 0; x-int 0 C. None
D. VA $x = -1$; HA $y = 1$
E. Inc on $(-\infty, -1), (-1, \infty)$;
F. None
G. CU on $(-\infty, -1), (0, \sqrt[3]{\tfrac{1}{2}})$;
CD on $(-1, 0), (\sqrt[3]{\tfrac{1}{2}}, \infty)$;
IPs $(0, 0), (\sqrt[3]{\tfrac{1}{2}}, \tfrac{1}{3})$
H. See graph at right.

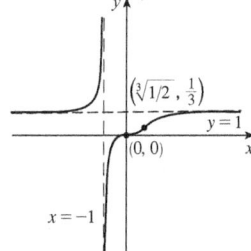

21. A. $[0, \infty)$ B. y-int 0; x-int 0, 3
C. None D. None
E. Inc on $(1, \infty)$; dec on $(0, 1)$
F. Loc min $f(1) = -2$
G. CU on $(0, \infty)$
H. See graph at right.

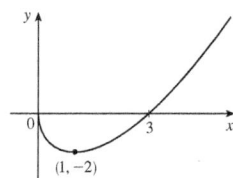

23. A. $(-\infty, -2] \cup [1, \infty)$
B. x-int $-2, 1$ C. None
D. None
E. Inc on $(1, \infty)$; dec on $(-\infty, -2)$
F. None
G. CD on $(-\infty, -2), (1, \infty)$
H. See graph at right.

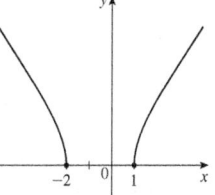

25. A. $\mathbb{R}$ B. y-int 0; x-int 0
C. About $(0, 0)$
D. HA $y = \pm 1$
E. Inc on $(-\infty, \infty)$ F. None
G. CU on $(-\infty, 0)$;
CD on $(0, \infty)$; IP $(0, 0)$
H. See graph at right.

27. A. $[-1, 0) \cup (0, 1]$ B. x-int ± 1 C. About $(0, 0)$
D. VA $x = 0$
E. Dec on $(-1, 0), (0, 1)$
F. None
G. CU on $(-1, -\sqrt{2/3})$,
$(0, \sqrt{2/3})$;
CD on $(-\sqrt{2/3}, 0), (\sqrt{2/3}, 1)$;
IPs $(\pm\sqrt{2/3}, \pm 1/\sqrt{2})$
H. See graph at right.

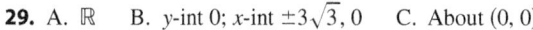

29. A. $\mathbb{R}$ B. y-int 0; x-int $\pm 3\sqrt{3}, 0$ C. About $(0, 0)$
D. None E. Inc on $(-\infty, -1), (1, \infty)$; dec on $(-1, 1)$
F. Loc max $f(-1) = 2$;
loc min $f(1) = -2$
G. CU on $(0, \infty)$;
CD on $(-\infty, 0)$; IP $(0, 0)$
H. See graph at right.

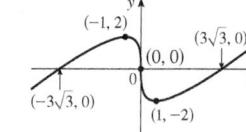

31. A. $\mathbb{R}$ B. y-int -1; x-int ± 1
C. About the y-axis D. None
E. Inc on $(0, \infty)$; dec on $(-\infty, 0)$
F. Loc min $f(0) = -1$
G. CU on $(-1, 1)$;
CD on $(-\infty, -1), (1, \infty)$; IPs $(\pm 1, 0)$
H. See graph at right.

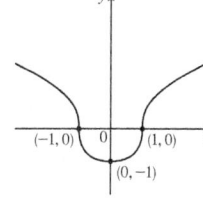

33. A. $\mathbb{R}$ B. y-int 0; x-int $n\pi$ (n an integer)
C. About $(0, 0)$, period 2π D. None
E–G answers for $0 \leq x \leq \pi$:
E. Inc on $(0, \pi/2)$; dec on $(\pi/2, \pi)$ F. Loc max $f(\pi/2) = 1$
G. Let $\alpha = \sin^{-1}\sqrt{2/3}$; CU on $(0, \alpha), (\pi - \alpha, \pi)$;
CD on $(\alpha, \pi - \alpha)$; IPs at $x = 0, \pi, \alpha, \pi - \alpha$
H.

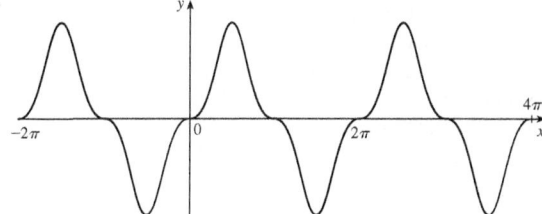

35. A. $(-\pi/2, \pi/2)$ B. y-int 0; x-int 0 C. About y-axis
D. VA $x = \pm\pi/2$
E. Inc on $(0, \pi/2)$;
dec on $(-\pi/2, 0)$
F. Loc min $f(0) = 0$
G. CU on $(-\pi/2, \pi/2)$
H. See graph at right.

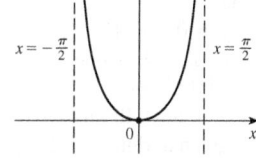

37. A. $[-2\pi, 2\pi]$ B. y-int $\sqrt{3}$; x-int $-4\pi/3, -\pi/3, 2\pi/3, 5\pi/3$ C. Period 2π D. None
E. Inc on $(-2\pi, -11\pi/6), (-5\pi/6, \pi/6), (7\pi/6, 2\pi)$; dec on $(-11\pi/6, -5\pi/6), (\pi/6, 7\pi/6)$
F. Loc max $f(-11\pi/6) = f(\pi/6) = 2$; loc min $f(-5\pi/6) = f(7\pi/6) = -2$
G. CU on $(-4\pi/3, -\pi/3)$, $(2\pi/3, 5\pi/3)$;
CD on $(-2\pi, -4\pi/3)$, $(-\pi/3, 2\pi/3), (5\pi/3, 2\pi)$;
IPs $(-4\pi/3, 0), (-\pi/3, 0)$, $(2\pi/3, 0), (5\pi/3, 0)$
H. See graph at right.

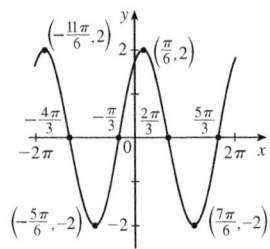

39. A. All reals except $(2n + 1)\pi$ (n an integer)
B. y-int 0; x-int $2n\pi$ C. About the origin, period 2π
D. VA $x = (2n + 1)\pi$ E. Inc on $((2n - 1)\pi, (2n + 1)\pi)$
F. None G. CU on $(2n\pi, (2n + 1)\pi)$; CD on $((2n - 1)\pi, 2n\pi)$; IPs $(2n\pi, 0)$
H.

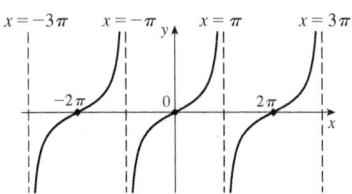

41. A. $\mathbb{R}$ B. y-int $\pi/4$
C. None
D. HA $y = 0, y = \pi/2$
E. Inc on $(-\infty, \infty)$ F. None
G. CU on $(-\infty, 0)$;
CD on $(0, \infty)$; IP $(0, \pi/4)$
H. See graph at right.

43. A. $\mathbb{R}$ B. y-int $\frac{1}{2}$ C. None
D. HA $y = 0, y = 1$
E. Inc on $\mathbb{R}$ F. None
G. CU on $(-\infty, 0)$;
CD on $(0, \infty)$; IP $(0, \frac{1}{2})$
H. See graph at right.

45. A. $(0, \infty)$ B. None
C. None D. VA $x = 0$
E. Inc on $(1, \infty)$; dec on $(0, 1)$
F. Loc min $f(1) = 1$
G. CU on $(0, 2)$; CD on $(2, \infty)$;
IP $(2, \frac{1}{2} + \ln 2)$
H. See graph at right.

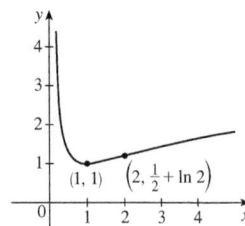

47. A. $\mathbb{R}$ B. y-int $\frac{1}{4}$
C. None
D. HA $y = 0, y = 1$
E. Dec on $\mathbb{R}$ F. None
G. CU on $(\ln \frac{1}{2}, \infty)$;
CD on $(-\infty, \ln \frac{1}{2})$; IP $(\ln \frac{1}{2}, \frac{4}{9})$
H. See graph at right.

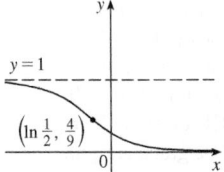

49. A. All x in $(2n\pi, (2n + 1)\pi)$ (n an integer)
B. x-int $\pi/2 + 2n\pi$ C. Period 2π D. VA $x = n\pi$
E. Inc on $(2n\pi, \pi/2 + 2n\pi)$; dec on $(\pi/2 + 2n\pi, (2n + 1)\pi)$
F. Loc max $f(\pi/2 + 2n\pi) = 0$ G. CD on $(2n\pi, (2n + 1)\pi)$
H.

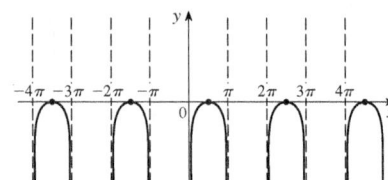

51. A. $(-\infty, 0) \cup (0, \infty)$
B. None C. None
D. VA $x = 0$
E. Inc on $(-\infty, -1), (0, \infty)$;
dec on $(-1, 0)$
F. Loc max $f(-1) = -e$
G. CU on $(0, \infty)$; CD on $(-\infty, 0)$
H. See graph at right.

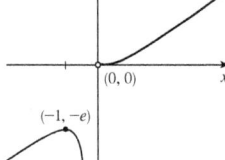

53. A. $\mathbb{R}$ B. y-int 1
C. None D. HA $y = e^{\pm \pi/2}$
E. Inc on $\mathbb{R}$ F. None
G. CU on $(-\infty, \frac{1}{2})$; CD on $(\frac{1}{2}, \infty)$;
IP $(\frac{1}{2}, e^{\arctan(1/2)})$
H. See graph at right.

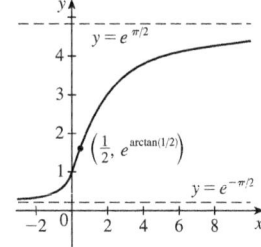

55.

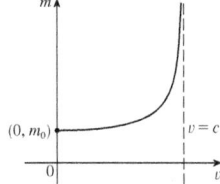

57. (a) When $t = (\ln a)/k$ (b) When $t = (\ln a)/k$
(c)

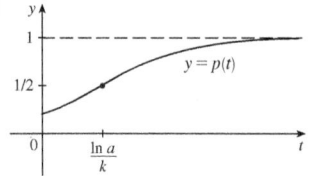

59.

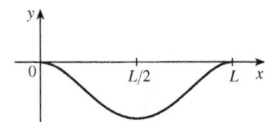

61. $y = x - 1$ **63.** $y = 2x - 3$

65. A. $(-\infty, 1) \cup (1, \infty)$
B. y-int 0; x-int 0   C. None
D. VA $x = 1$; SA $y = x + 1$
E. Inc on $(-\infty, 0)$, $(2, \infty)$;
dec on $(0, 1)$, $(1, 2)$
F. Loc max $f(0) = 0$;
loc min $f(2) = 4$
G. CU on $(1, \infty)$; CD on $(-\infty, 1)$
H. See graph at right.

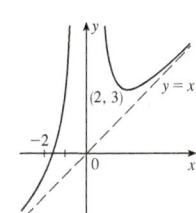

67. A. $(-\infty, 0) \cup (0, \infty)$
B. x-int $-\sqrt[3]{4}$ C. None
D. VA $x = 0$; SA $y = x$
E. Inc on $(-\infty, 0)$, $(2, \infty)$;
dec on $(0, 2)$
F. Loc min $f(2) = 3$
G. CU on $(-\infty, 0)$, $(0, \infty)$
H. See graph at right.

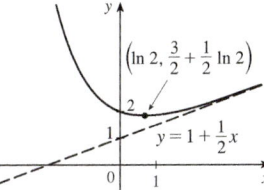

69. A. $\mathbb{R}$ B. y-int 2
C. None
D. SA $y = 1 + \tfrac{1}{2}x$
E. Inc on $(\ln 2, \infty)$;
dec on $(-\infty, \ln 2)$
F. Loc min $f(\ln 2) = \tfrac{3}{2} + \tfrac{1}{2}\ln 2$
G. CU on $(-\infty, \infty)$
H. See graph at right.

71.

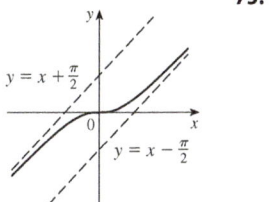

75. VA $x = 0$, asymptotic to $y = x^3$

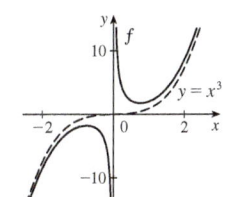

EXERCISES 4.6 ■ PAGE 329

1. Inc on $(-\infty, -1.50)$, $(0.04, 2.62)$, $(2.84, \infty)$; dec on $(-1.50, 0.04)$, $(2.62, 2.84)$; loc max $f(-1.50) \approx 36.47$, $f(2.62) \approx 56.83$; loc min $f(0.04) \approx -0.04$, $f(2.84) \approx 56.73$; CU on $(-0.89, 1.15)$, $(2.74, \infty)$; CD on $(-\infty, -0.89)$, $(1.15, 2.74)$; IPs $(-0.89, 20.90)$, $(1.15, 26.57)$, $(2.74, 56.78)$

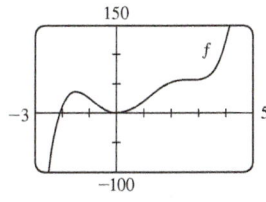
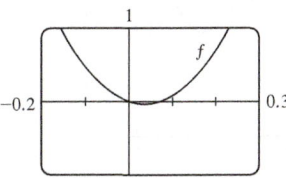
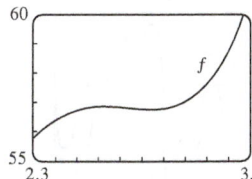

3. Inc on $(-1.31, -0.84)$, $(1.06, 2.50)$, $(2.75, \infty)$; dec on $(-\infty, -1.31)$, $(-0.84, 1.06)$, $(2.50, 2.75)$; loc max $f(-0.84) \approx 23.71$, $f(2.50) \approx -11.02$; loc min $f(-1.31) \approx 20.72$, $f(1.06) \approx -33.12$, $f(2.75) \approx -11.33$; CU on $(-\infty, -1.10)$, $(0.08, 1.72)$, $(2.64, \infty)$; CD on $(-1.10, 0.08)$, $(1.72, 2.64)$; IPs $(-1.10, 22.09)$, $(0.08, -3.88)$, $(1.72, -22.53)$, $(2.64, -11.18)$

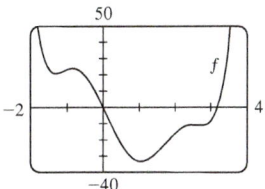

5. Inc on $(-\infty, -1.47)$, $(-1.47, 0.66)$; dec on $(0.66, \infty)$; loc max $f(0.66) \approx 0.38$; CU on $(-\infty, -1.47)$, $(-0.49, 0)$, $(1.10, \infty)$; CD on $(-1.47, -0.49)$, $(0, 1.10)$; IPs $(-0.49, -0.44)$, $(1.10, 0.31)$, $(0, 0)$

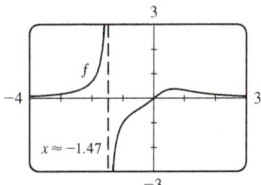

7. Inc on $(-1.40, -0.44)$, $(0.44, 1.40)$; dec on $(-\pi, -1.40)$, $(-0.44, 0)$, $(0, 0.44)$, $(1.40, \pi)$; loc max $f(-0.44) \approx -4.68$, $f(1.40) \approx 6.09$; loc min $f(-1.40) \approx -6.09$, $f(0.44) \approx 4.68$; CU on $(-\pi, -0.77)$, $(0, 0.77)$; CD on $(-0.77, 0)$, $(0.77, \pi)$; IPs $(-0.77, -5.22)$, $(0.77, 5.22)$

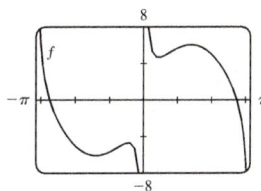

9. Inc on $\left(-8 - \sqrt{61}, -8 + \sqrt{61}\right)$; dec on $\left(-\infty, -8 - \sqrt{61}\right)$, $\left(-8 + \sqrt{61}, 0\right)$, $(0, \infty)$; CU on $\left(-12 - \sqrt{138}, -12 + \sqrt{138}\right)$, $(0, \infty)$; CD on $\left(-\infty, -12 - \sqrt{138}\right)$, $\left(-12 + \sqrt{138}, 0\right)$

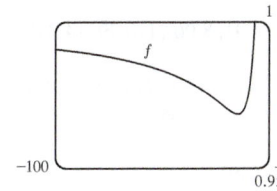
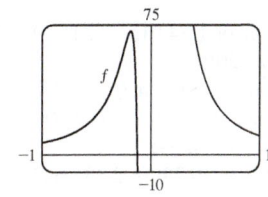

11. (a)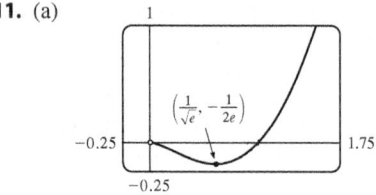

(b) $\lim_{x \to 0^+} f(x) = 0$ (c) Loc min $f(1/\sqrt{e}) = -1/(2e)$;
CD on $(0, e^{-3/2})$; CU on $(e^{-3/2}, \infty)$

13. Loc max $f(-5.6) \approx 0.018$, $f(0.82) \approx -281.5$, $f(5.2) \approx 0.0145$; loc min $f(3) = 0$

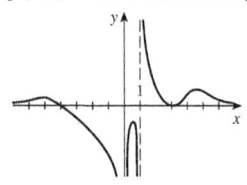 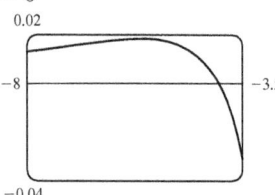

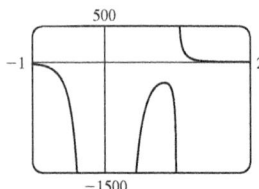 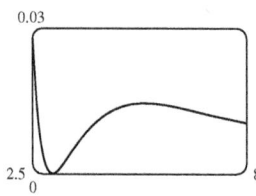

15. $f'(x) = -\dfrac{x(x+1)^2(x^3 + 18x^2 - 44x - 16)}{(x-2)^3(x-4)^5}$

$f''(x) = 2\dfrac{(x+1)(x^6 + 36x^5 + 6x^4 - 628x^3 + 684x^2 + 672x + 64)}{(x-2)^4(x-4)^6}$

CU on $(-35.3, -5.0)$, $(-1, -0.5)$, $(-0.1, 2)$, $(2, 4)$, $(4, \infty)$;
CD on $(-\infty, -35.3)$, $(-5.0, -1)$, $(-0.5, -0.1)$;
IPs $(-35.3, -0.015)$, $(-5.0, -0.005)$, $(-1, 0)$, $(-0.5, 0.00001)$, $(-0.1, 0.0000066)$

17. Inc on $(-9.41, -1.29)$, $(0, 1.05)$;
dec on $(-\infty, -9.41)$, $(-1.29, 0)$, $(1.05, \infty)$;
loc max $f(-1.29) \approx 7.49$, $f(1.05) \approx 2.35$;
loc min $f(-9.41) \approx -0.056$, $f(0) = 0.5$;
CU on $(-13.81, -1.55)$, $(-1.03, 0.60)$, $(1.48, \infty)$;
CD on $(-\infty, -13.81)$, $(-1.55, -1.03)$, $(0.60, 1.48)$;
IPs $(-13.81, -0.05)$, $(-1.55, 5.64)$, $(-1.03, 5.39)$, $(0.60, 1.52)$, $(1.48, 1.93)$

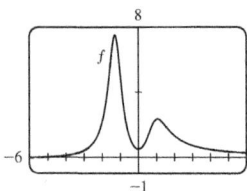 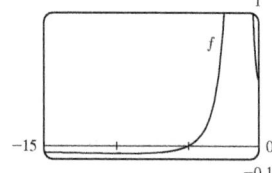

19. Inc on $(-4.91, -4.51)$, $(0, 1.77)$, $(4.91, 8.06)$, $(10.79, 14.34)$, $(17.08, 20)$;
dec on $(-4.51, -4.10)$, $(1.77, 4.10)$, $(8.06, 10.79)$, $(14.34, 17.08)$;
loc max $f(-4.51) \approx 0.62$, $f(1.77) \approx 2.58$, $f(8.06) \approx 3.60$, $f(14.34) \approx 4.39$;
loc min $f(10.79) \approx 2.43$, $f(17.08) \approx 3.49$;

CU on $(9.60, 12.25)$, $(15.81, 18.65)$;
CD on $(-4.91, -4.10)$, $(0, 4.10)$, $(4.91, 9.60)$, $(12.25, 15.81)$, $(18.65, 20)$;
IPs $(9.60, 2.95)$, $(12.25, 3.27)$, $(15.81, 3.91)$, $(18.65, 4.20)$

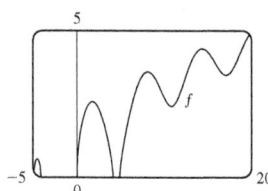

21. Inc on $(-\infty, 0)$, $(0, \infty)$;
CU on $(-\infty, -0.42)$, $(0, 0.42)$;
CD on $(-0.42, 0)$, $(0.42, \infty)$;
IPs $(\mp 0.42, \pm 0.83)$

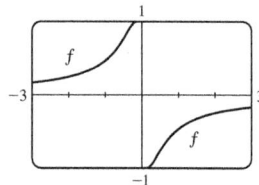

23.

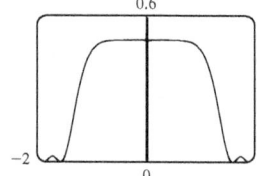 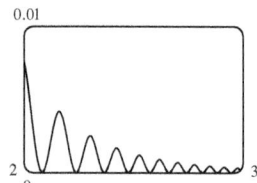

25. (a)

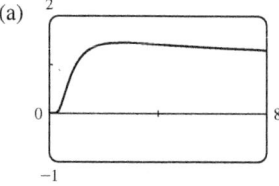

(b) $\lim_{x \to 0^+} x^{1/x} = 0$, $\lim_{x \to \infty} x^{1/x} = 1$
(c) Loc max $f(e) = e^{1/e}$ (d) IPs at $x \approx 0.58, 4.37$

27. Max $f(0.59) \approx 1$, $f(0.68) \approx 1$, $f(1.96) \approx 1$;
min $f(0.64) \approx 0.99996$, $f(1.46) \approx 0.49$, $f(2.73) \approx -0.51$;
IPs $(0.61, 0.99998)$, $(0.66, 0.99998)$, $(1.17, 0.72)$, $(1.75, 0.77)$, $(2.28, 0.34)$

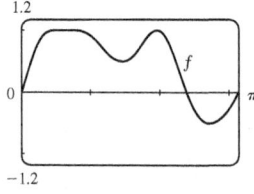

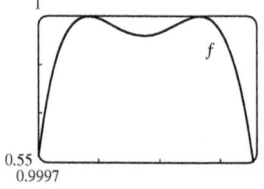 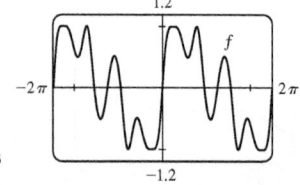

29. For $c < 0$, there is a loc min that moves toward $(-3, -9)$ as c increases. For $0 < c < 8$, there is a loc min that moves toward $(-3, -9)$ and a loc max that moves toward the origin as c decreases. For all $c > 0$, there is a first quadrant loc min that moves toward the origin as c decreases. $c = 0$ is a transitional value that gives the graph of a parabola. For all nonzero c, the y-axis is a VA and there is an IP that moves toward the origin as $|c| \to 0$.

$c \le 0$:

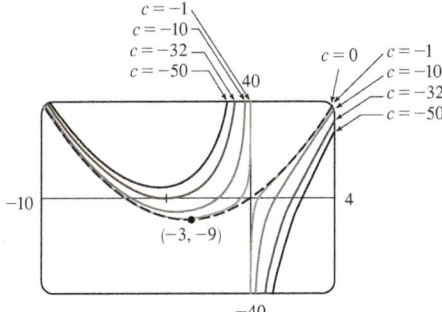

$c \ge 0$:

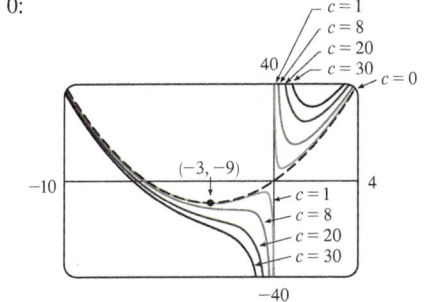

31. For $c < 0$, there is no extreme point and one IP, which decreases along the x-axis. For $c > 0$, there is no IP, and one minimum point.

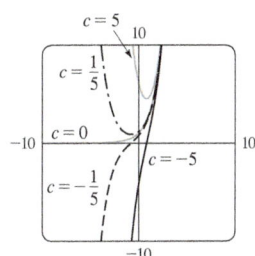

33. For $c > 0$, the maximum and minimum values are always $\pm\frac{1}{2}$, but the extreme points and IPs move closer to the y-axis as c increases. $c = 0$ is a transitional value: when c is replaced by $-c$, the curve is reflected in the x-axis.

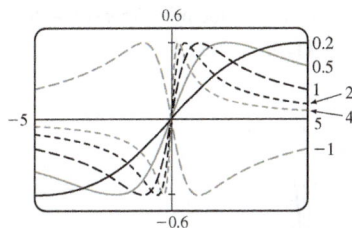

35. For $|c| < 1$, the graph has loc max and min values; for $|c| \ge 1$ it does not. The function increases for $c \ge 1$ and decreases for $c \le -1$. As c changes, the IPs move vertically but not horizontally.

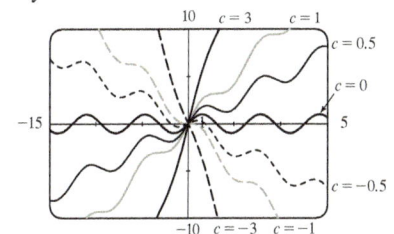

37.

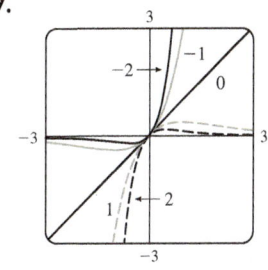

For $c > 0$, $\lim_{x \to \infty} f(x) = 0$ and $\lim_{x \to -\infty} f(x) = -\infty$.
For $c < 0$, $\lim_{x \to \infty} f(x) = \infty$ and $\lim_{x \to -\infty} f(x) = 0$.
As $|c|$ increases, the max and min points and the IPs get closer to the origin.

39. (a) Positive (b)

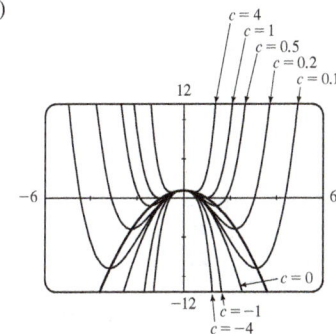

EXERCISES 4.7 ■ PAGE 336

1. (a) 11, 12 (b) 11.5, 11.5 **3.** 10, 10 **5.** $\frac{9}{4}$
7. 25 m by 25 m **9.** $N = 1$
11. (a)

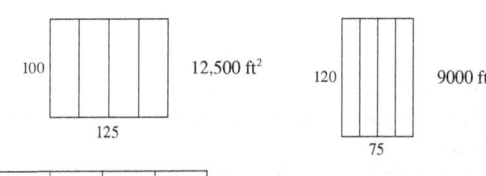

(b)

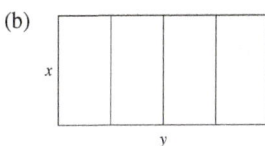

(c) $A = xy$ (d) $5x + 2y = 750$ (e) $A(x) = 375x - \frac{5}{2}x^2$
(f) $14{,}062.5 \text{ ft}^2$
13. 1000 ft by 1500 ft **15.** 4000 cm³ **17.** $191.28
19. $20\sqrt{30}$ ft by $\frac{40}{3}\sqrt{30}$ ft **21.** $\left(-\frac{6}{5}, \frac{3}{5}\right)$
23. $\left(-\frac{1}{3}, \pm\frac{4}{3}\sqrt{2}\right)$ **25.** Square, side $\sqrt{2}\,r$
27. $L/2, \sqrt{3}\,L/4$ **29.** Base $\sqrt{3}\,r$, height $3r/2$
31. $4\pi r^3/(3\sqrt{3})$ **33.** $\pi r^2(1 + \sqrt{5})$ **35.** 24 cm by 36 cm
37. (a) Use all of the wire for the square
(b) $40\sqrt{3}/(9 + 4\sqrt{3})$ m for the square
39. 16 in. **41.** $V = 2\pi R^3/(9\sqrt{3})$ **45.** $E^2/(4r)$
47. (a) $\frac{3}{2}s^2\csc\theta\,(\csc\theta - \sqrt{3}\cot\theta)$ (b) $\cos^{-1}(1/\sqrt{3}) \approx 55°$
(c) $6s\left[h + s/(2\sqrt{2})\right]$
49. Row directly to B **51.** ≈ 4.85 km east of the refinery
53. $10\sqrt[3]{3}/(1 + \sqrt[3]{3})$ ft from the stronger source
55. $(a^{2/3} + b^{2/3})^{3/2}$ **57.** $2\sqrt{6}$
59. (b) (i) $342,491; $342/unit; $390/unit (ii) 400
(iii) $320/unit
61. (a) $p(x) = 19 - \frac{1}{3000}x$ (b) $9.50
63. (a) $p(x) = 500 - \frac{1}{8}x$ (b) $250 (c) $310
69. 9.35 m **73.** $x = 6$ in. **75.** $\pi/6$
77. At a distance $5 - 2\sqrt{5}$ from A **79.** $\frac{1}{2}(L + W)^2$
81. (a) About 5.1 km from B (b) C is close to B; C is close to D; $W/L = \sqrt{25 + x^2}/x$, where $x = |BC|$
(c) ≈ 1.07; no such value (d) $\sqrt{41}/4 \approx 1.6$

EXERCISES 4.8 ■ PAGE 348
1. (a) $x_2 \approx 7.3, x_3 \approx 6.8$ (b) Yes
3. $\frac{9}{2}$ **5.** a, b, c **7.** 1.5215 **9.** -1.25
11. 2.94283096 **13.** (b) 2.630020 **15.** -1.964636
17. $-3.637958, -1.862365, 0.889470$
19. $-1.257691, 0.653483$ **21.** $0, \pm 0.902025$
23. $-1.69312029, -0.74466668, 1.26587094$
25. 0.76682579 **27.** 0.21916368, 1.08422462
29. (b) 31.622777
35. (a) $-1.293227, -0.441731, 0.507854$ (b) -2.0212
37. (1.519855, 2.306964) **39.** (0.410245, 0.347810)
41. 0.76286%

EXERCISES 4.9 ■ PAGE 355
1. $F(x) = 2x^2 + 7x + C$ **3.** $F(x) = \frac{1}{2}x^4 - \frac{2}{9}x^3 + \frac{5}{2}x^2 + C$
5. $F(x) = 4x^3 + 4x^2 + C$ **7.** $F(x) = 5x^{7/5} + 40x^{1/5} + C$
9. $F(x) = \sqrt{2}x + C$ **11.** $F(x) = 2x^{3/2} - \frac{3}{2}x^{4/3} + C$
13. $F(x) = \begin{cases} \frac{1}{5}x - 2\ln|x| + C_1 & \text{if } x < 0 \\ \frac{1}{5}x - 2\ln|x| + C_2 & \text{if } x > 0 \end{cases}$
15. $G(t) = 2t^{1/2} + \frac{2}{3}t^{3/2} + \frac{2}{5}t^{5/2} + C$
17. $H(\theta) = -2\cos\theta - \tan\theta + C_n$ on $(n\pi - \pi/2, n\pi + \pi/2)$, n an integer
19. $F(x) = 2^x/\ln 2 + 4\cosh x + C$
21. $F(x) = x^2 + 4x + 1/x + C, x > 0$
23. $F(x) = x^5 - \frac{1}{3}x^6 + 4$
25. $f(x) = x^5 - x^4 + x^3 + Cx + D$
27. $f(x) = \frac{1}{3}x^3 + 3e^x + Cx + D$

29. $f(t) = 2t^3 + \cos t + Ct^2 + Dt + E$
31. $f(x) = x + 2x^{3/2} + 5$ **33.** $f(t) = 4\arctan t - \pi$
35. $f(x) = 3x^{5/3} - 75$
37. $f(t) = \tan t + \sec t - 2 - \sqrt{2}$
39. $f(x) = -x^2 + 2x^3 - x^4 + 12x + 4$
41. $f(\theta) = -\sin\theta - \cos\theta + 5\theta + 4$
43. $f(x) = 2x^2 + x^3 + 2x^4 + 2x + 3$
45. $f(x) = e^x + 2\sin x - \frac{2}{\pi}(e^{\pi/2} + 4)x + 2$
47. $f(x) = -\ln x + (\ln 2)x - \ln 2$ **49.** 8 **51.** b
53.

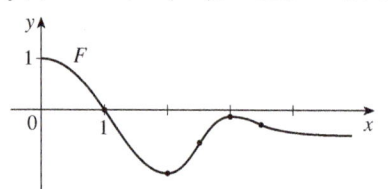

55. **57.**

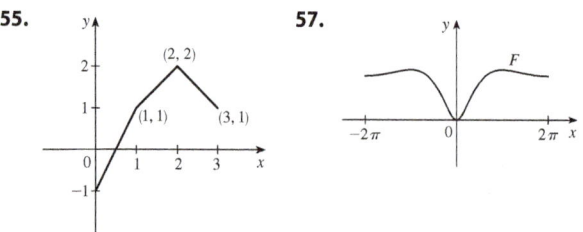

59. $s(t) = 1 - \cos t - \sin t$ **61.** $s(t) = \frac{1}{3}t^3 + \frac{1}{2}t^2 - 2t + 3$
63. $s(t) = -10\sin t - 3\cos t + (6/\pi)t + 3$
65. (a) $s(t) = 450 - 4.9t^2$ (b) $\sqrt{450/4.9} \approx 9.58$ s
(c) $-9.8\sqrt{450/4.9} \approx -93.9$ m/s (d) About 9.09 s
69. 225 ft **71.** $742.08 **73.** $\frac{130}{11} \approx 11.8$ s
75. $\frac{88}{15} \approx 5.87$ ft/s² **77.** 62,500 km/h² ≈ 4.82 m/s²
79. (a) 22.9125 mi (b) 21.675 mi (c) 30 min 33 s
(d) 55.425 mi

CHAPTER 4 REVIEW ■ PAGE 358
True-False Quiz
1. False **3.** False **5.** True **7.** False **9.** True
11. True **13.** False **15.** True **17.** True
19. True **21.** False

Exercises
1. Abs max $f(2) = f(5) = 18$, abs min $f(0) = -2$, loc max $f(2) = 18$, loc min $f(4) = 14$
3. Abs max $f(2) = \frac{2}{5}$, abs and loc min $f\left(-\frac{1}{3}\right) = -\frac{9}{2}$
5. Abs and loc max $f(\pi/6) = \pi/6 + \sqrt{3}$,
abs min $f(-\pi) = -\pi - 2$, loc min $f(5\pi/6) = 5\pi/6 - \sqrt{3}$
7. 1 **9.** 4 **11.** 0 **13.** $\frac{1}{2}$
15.

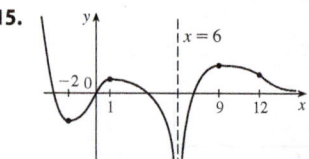

17.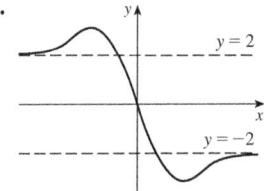

19. A. $\mathbb{R}$ B. y-int 2
C. None   D. None
E. Dec on $(-\infty, \infty)$ F. None
G. CU on $(-\infty, 0)$;
CD on $(0, \infty)$; IP $(0, 2)$
H. See graph at right.

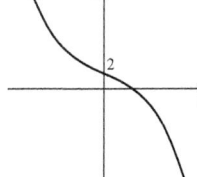

21. A. $\mathbb{R}$ B. y-int 2
C. None   D. None
E. Inc on $(1, \infty)$; dec on $(-\infty, 1)$
F. Loc min $f(1) = 1$
G. CU on $(-\infty, 0)$, $(\frac{2}{3}, \infty)$;
CD on $(0, \frac{2}{3})$; IPs $(0, 2)$, $(\frac{2}{3}, \frac{38}{27})$
H. See graph at right.

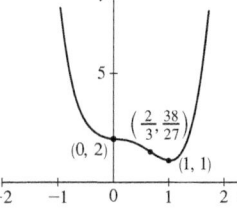

23. A. $\{x \mid x \neq 0, 3\}$
B. None   C. None
D. HA $y = 0$; VA $x = 0$, $x = 3$
E. Inc on $(1, 3)$;
dec on $(-\infty, 0)$, $(0, 1)$, $(3, \infty)$
F. Loc min $f(1) = \frac{1}{4}$
G. CU on $(0, 3)$, $(3, \infty)$;
CD on $(-\infty, 0)$
H. See graph at right.

25. A. $(-\infty, 0) \cup (0, \infty)$
B. x-int 1   C. None
D. VA $x = 0$; SA $y = x - 3$
E. Inc on $(-\infty, -2)$, $(0, \infty)$;
dec on $(-2, 0)$
F. Loc max $f(-2) = -\frac{27}{4}$
G. CU on $(1, \infty)$; CD on $(-\infty, 0)$,
$(0, 1)$; IP $(1, 0)$
H. See graph at right.

27. A. $[-2, \infty)$
B. y-int 0; x-int -2, 0
C. None   D. None
E. Inc on $(-\frac{4}{3}, \infty)$, dec on $(-2, -\frac{4}{3})$
F. Loc min $f(-\frac{4}{3}) = -\frac{4}{9}\sqrt{6}$
G. CU on $(-2, \infty)$
H. See graph at right.

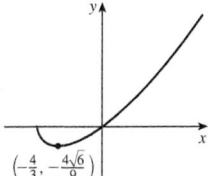

29. A. $[-\pi, \pi]$ B. y-int 0; x-int $-\pi$, 0, π
C. None   D. None
E. Inc on $(-\pi/4, 3\pi/4)$; dec on $(-\pi, -\pi/4)$, $(3\pi/4, \pi)$
F. Loc max $f(3\pi/4) = \frac{1}{2}\sqrt{2}\,e^{3\pi/4}$,
loc min $f(-\pi/4) = -\frac{1}{2}\sqrt{2}\,e^{-\pi/4}$

G. CU on $(-\pi/2, \pi/2)$; CD on $(-\pi, -\pi/2)$, $(\pi/2, \pi)$;
IPs $(-\pi/2, -e^{-\pi/2})$, $(\pi/2, e^{\pi/2})$
H.

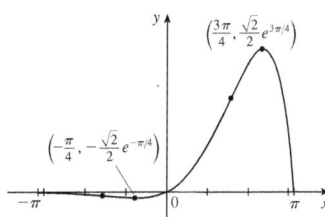

31. A. $\{x \mid |x| \geq 1\}$
B. None   C. About $(0, 0)$
D. HA $y = 0$
E. Dec on $(-\infty, -1)$, $(1, \infty)$
F. None
G. CU on $(1, \infty)$;
CD on $(-\infty, -1)$
H. See graph at right.

33. A. $\mathbb{R}$
B. y-int -2; x-int 2
C. None   D. HA $y = 0$
E. Inc on $(-\infty, 3)$; dec on $(3, \infty)$
F. Loc max $f(3) = e^{-3}$
G. CU on $(4, \infty)$;
CD on $(-\infty, 4)$;
IP $(4, 2e^{-4})$
H. See graph at right.

35. Inc on $(-\sqrt{3}, 0)$, $(0, \sqrt{3})$;
dec on $(-\infty, -\sqrt{3})$, $(\sqrt{3}, \infty)$;
loc max $f(\sqrt{3}) = \frac{2}{9}\sqrt{3}$,
loc min $f(-\sqrt{3}) = -\frac{2}{9}\sqrt{3}$;
CU on $(-\sqrt{6}, 0)$, $(\sqrt{6}, \infty)$;
CD on $(-\infty, -\sqrt{6})$, $(0, \sqrt{6})$;
IPs $(\sqrt{6}, \frac{5}{36}\sqrt{6})$, $(-\sqrt{6}, -\frac{5}{36}\sqrt{6})$

37. Inc on $(-0.23, 0)$, $(1.62, \infty)$; dec on $(-\infty, -0.23)$, $(0, 1.62)$;
loc max $f(0) = 2$; loc min $f(-0.23) \approx 1.96$, $f(1.62) \approx -19.2$;
CU on $(-\infty, -0.12)$, $(1.24, \infty)$;
CD on $(-0.12, 1.24)$; IPs $(-0.12, 1.98)$, $(1.24, -12.1)$

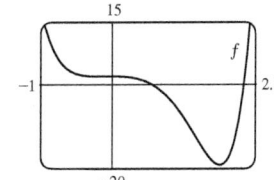 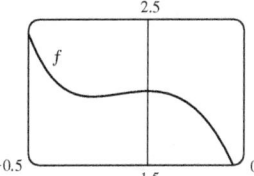

39. 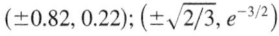 $(\pm 0.82, 0.22)$; $(\pm\sqrt{2/3}, e^{-3/2})$

APPENDIX I Answers to Odd-Numbered Exercises

41. $-2.96, -0.18, 3.01; -1.57, 1.57; -2.16, -0.75, 0.46, 2.21$
43. For $C > -1$, f is periodic with period 2π and has local maxima at $2n\pi + \pi/2$, n an integer. For $C \le -1$, f has no graph. For $-1 < C \le 1$, f has vertical asymptotes. For $C > 1$, f is continuous on $\mathbb{R}$. As C increases, f moves upward and its oscillations become less pronounced.
49. (a) 0   (b) CU on $\mathbb{R}$ **53.** $3\sqrt{3}\,r^2$
55. $4/\sqrt{3}$ cm from D **57.** $L = C$ **59.** \$11.50
61. 1.297383   **63.** 1.16718557
65. $F(x) = \tfrac{8}{3}x^{3/2} - 2x^3 + 3x + C$
67. $F(t) = -2\cos t - 3e^t + C$
69. $f(t) = t^2 + 3\cos t + 2$
71. $f(x) = \tfrac{1}{2}x^2 - x^3 + 4x^4 + 2x + 1$
73. $s(t) = t^2 - \tan^{-1} t + 1$
75. (b) $0.1e^x - \cos x + 0.9$ (c)

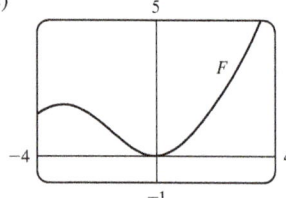

77. No
79. (b) About 8.5 in. by 2 in.   (c) $20/\sqrt{3}$ in. by $20\sqrt{2/3}$ in.
85. (a) $20\sqrt{2} \approx 28$ ft
(b) $\dfrac{dI}{dt} = \dfrac{-480k(h-4)}{[(h-4)^2 + 1600]^{5/2}}$, where k is the constant of proportionality

PROBLEMS PLUS ■ PAGE 363

3. Abs max $f(-5) = e^{45}$, no abs min   **7.** 24
9. $(-2, 4), (2, -4)$ **13.** $(1 + \sqrt{5})/2$ **15.** $(m/2, m^2/4)$
17. $a \le e^{1/e}$
21. (a) $T_1 = D/c_1$, $T_2 = (2h\sec\theta)/c_1 + (D - 2h\tan\theta)/c_2$, $T_3 = \sqrt{4h^2 + D^2}/c_1$
(c) $c_1 \approx 3.85$ km/s, $c_2 \approx 7.66$ km/s, $h \approx 0.42$ km
25. $3/(\sqrt[3]{2} - 1) \approx 11\tfrac{1}{2}$ h

CHAPTER 5

EXERCISES 5.1 ■ PAGE 375

1. (a) $R_5 \approx 12$, $L_5 \approx 22$

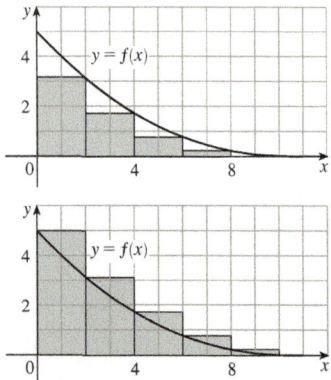

(b) $R_{10} \approx 14.4$, $L_{10} \approx 19.4$

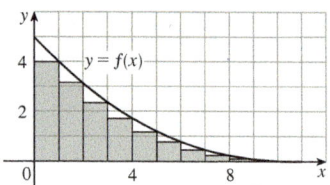

3. (a) 0.6345, underestimate   (b) 0.7595, overestimate

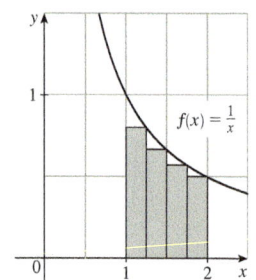 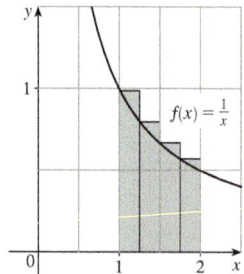

5. (a) 8, 6.875   (b) 5, 5.375

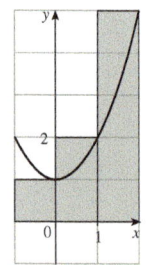 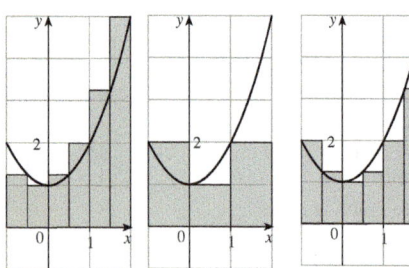

(c) 5.75, 5.9375

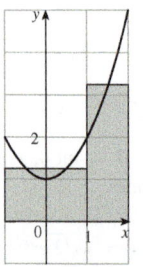 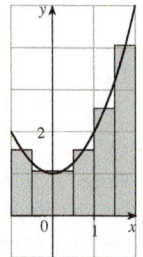

(d) M_6

7. $n = 2$: upper $= 3\pi \approx 9.42$, lower $= 2\pi \approx 6.28$

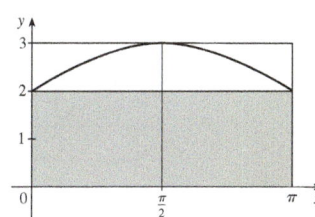

$n = 4$: upper $= (10 + \sqrt{2})(\pi/4) \approx 8.96$,
lower $= (8 + \sqrt{2})(\pi/4) \approx 7.39$

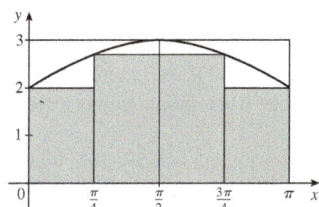

$n = 8$: upper ≈ 8.65, lower ≈ 7.86

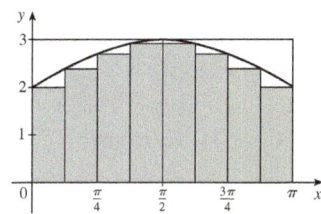

9. 0.2533, 0.2170, 0.2101, 0.2050; 0.2
11. (a) Left: 0.8100, 0.7937, 0.7904;
right: 0.7600, 0.7770, 0.7804
13. 34.7 ft, 44.8 ft **15.** 63.2 L, 70 L **17.** 155 ft
19. 7840 **21.** $\lim\limits_{n\to\infty} \sum\limits_{i=1}^{n} \dfrac{2(1 + 2i/n)}{(1 + 2i/n)^2 + 1} \cdot \dfrac{2}{n}$
23. $\lim\limits_{n\to\infty} \sum\limits_{i=1}^{n} \sqrt{\sin(\pi i/n)} \cdot \dfrac{\pi}{n}$
25. The region under the graph of $y = \tan x$ from 0 to $\pi/4$
27. (a) $L_n < A < R_n$
29. (a) $\lim\limits_{n\to\infty} \dfrac{64}{n^6} \sum\limits_{i=1}^{n} i^5$ (b) $\dfrac{n^2(n+1)^2(2n^2 + 2n - 1)}{12}$
(c) $\dfrac{32}{3}$
31. $\sin b$, 1

EXERCISES 5.2 ■ PAGE 388

1. -10
The Riemann sum represents the sum of the areas of the two rectangles above the x-axis minus the sum of the areas of the three rectangles below the x-axis; that is, the *net area* of the rectangles with respect to the x-axis.

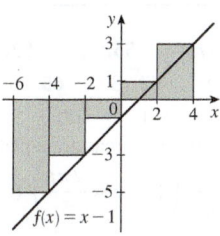

3. $-\dfrac{49}{16}$
The Riemann sum represents the sum of the areas of the two rectangles above the x-axis minus the sum of the areas of the four rectangles below the x-axis.

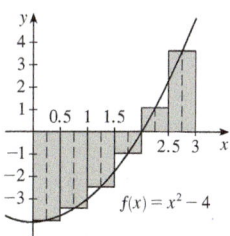

5. (a) 6 (b) 4 (c) 2
7. Lower, $L_5 = -64$; upper, $R_5 = 16$
9. 6.1820 **11.** 0.9071 **13.** 0.9029, 0.9018
15.

n	R_n
5	1.933766
10	1.983524
50	1.999342
100	1.999836

The values of R_n appear to be approaching 2.

17. $\int_0^1 \dfrac{e^x}{1 + x}\,dx$ **19.** $\int_2^7 (5x^3 - 4x)\,dx$
21. -9 **23.** $\dfrac{2}{3}$ **25.** $-\dfrac{3}{4}$
29. $\lim\limits_{n\to\infty} \sum\limits_{i=1}^{n} \sqrt{4 + (1 + 2i/n)} \cdot \dfrac{2}{n}$
31. $\lim\limits_{n\to\infty} \sum\limits_{i=1}^{n} \left(\sin\dfrac{5\pi i}{n}\right)\dfrac{\pi}{n} = \dfrac{2}{5}$
33. (a) 4 (b) 10 (c) -3 (d) 2
35. $\dfrac{3}{2}$ **37.** $3 + \dfrac{9}{4}\pi$ **39.** $\dfrac{25}{4}$ **41.** 0 **43.** 3
45. $e^5 - e^3$ **47.** $\int_{-1}^{5} f(x)\,dx$ **49.** 122
51. B < E < A < D < C **53.** 15
59. $0 \le \int_0^1 x^3\,dx \le 1$ **61.** $\dfrac{\pi}{12} \le \int_{\pi/4}^{\pi/3} \tan x\,dx \le \dfrac{\pi}{12}\sqrt{3}$
63. $0 \le \int_0^2 xe^{-x}\,dx \le 2/e$ **67.** $\int_1^2 \arctan x\,dx$
73. $\int_0^1 x^4\,dx$ **75.** $\dfrac{1}{2}$

EXERCISES 5.3 ■ PAGE 399

1. One process undoes what the other one does. See the Fundamental Theorem of Calculus, page 398.
3. (a) 0, 2, 5, 7, 3
(b) (0, 3)
(c) $x = 3$
(d)

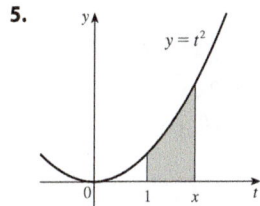

(a), (b) x^2

5.

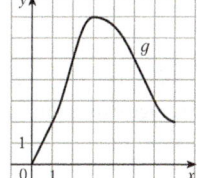

A94 APPENDIX I Answers to Odd-Numbered Exercises

7. $g'(x) = \sqrt{x + x^3}$ **9.** $g'(s) = (s - s^2)^8$
11. $F'(x) = -\sqrt{1 + \sec x}$ **13.** $h'(x) = xe^x$
15. $y' = \dfrac{3(3x+2)}{1+(3x+2)^3}$
17. $y' = -\tfrac{1}{2}\tan\sqrt{x}$ **19.** $\tfrac{26}{3}$ **21.** 2 **23.** $\tfrac{52}{3}$
25. $1 + \sqrt{3}/2$ **27.** $-\tfrac{37}{6}$ **29.** $\tfrac{82}{5}$ **31.** 1 **33.** $\tfrac{15}{4}$
35. $\ln 2 + 7$ **37.** $\dfrac{1}{e+1} + e - 1$ **39.** $4\pi/3$
41. $\dfrac{15}{\ln 2}$ **43.** 0 **45.** $\tfrac{16}{3}$ **47.** $\tfrac{32}{3}$
49. $\tfrac{243}{4}$ **51.** 2
53. 3.75

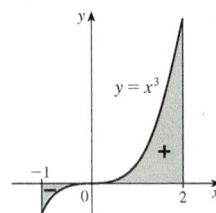

55. The function $f(x) = x^{-4}$ is not continuous on the interval $[-2, 1]$, so FTC2 cannot be applied.
57. The function $f(\theta) = \sec\theta\,\tan\theta$ is not continuous on the interval $[\pi/3, \pi]$, so FTC2 cannot be applied.
59. $g'(x) = \dfrac{-2(4x^2-1)}{4x^2+1} + \dfrac{3(9x^2-1)}{9x^2+1}$
61. $F'(x) = 2xe^{x^4} - e^{x^2}$
63. $y' = \sin x \ln(1 + 2\cos x) + \cos x \ln(1 + 2\sin x)$
65. $(-4, 0)$ **67.** $y = e^4x - 2e^4$ **69.** 29
71. (a) $-2\sqrt{n},\ \sqrt{4n-2},\ n$ an integer > 0
(b) $(0, 1),\ (-\sqrt{4n-1}, -\sqrt{4n-3})$, and $(\sqrt{4n-1}, \sqrt{4n+1})$, n an integer > 0 (c) 0.74

73. (a) Loc max at 1 and 5; loc min at 3 and 7
(b) $x = 9$
(c) $(\tfrac{1}{2}, 2),\ (4, 6),\ (8, 9)$
(d) See graph at right.

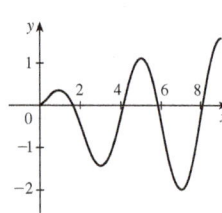

75. $\tfrac{7}{10}$ **83.** $f(x) = x^{3/2},\ a = 9$
85. (b) Average expenditure over $[0, t]$; minimize average expenditure

EXERCISES 5.4 ■ PAGE 408

5. $\tfrac{1}{2.3}x^{2.3} + 2x^{3.5} + C$ **7.** $5x + \tfrac{2}{9}x^3 + \tfrac{3}{16}x^4 + C$
9. $\tfrac{2}{3}u^3 + \tfrac{9}{2}u^2 + 4u + C$ **11.** $\ln|x| + 2\sqrt{x} + x + C$
13. $-\cos x + \cosh x + C$ **15.** $\theta + \tan\theta + C$
17. $\dfrac{2^t}{\ln 2} + \dfrac{10^t}{\ln 10} + C$

19. $\sin x + \tfrac{1}{4}x^2 + C$

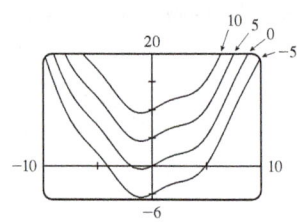

21. $-\tfrac{10}{3}$ **23.** $\tfrac{21}{5}$ **25.** -2 **27.** $5e^\pi + 1$ **29.** 36
31. $\tfrac{55}{63}$ **33.** $\tfrac{3}{4} - 2\ln 2$ **35.** $\dfrac{1}{11} + \dfrac{9}{\ln 10}$
37. $1 + \pi/4$ **39.** $\tfrac{69}{4}$ **41.** $\pi/3$ **43.** $\pi/6$ **45.** -3.5
47. ≈ 1.36 **49.** $\tfrac{4}{3}$
51. The increase in the child's weight (in pounds) between the ages of 5 and 10
53. Number of gallons of oil leaked in the first 2 hours
55. Increase in revenue when production is increased from 1000 to 5000 units
57. Newton-meters (or joules) **59.** (a) $-\tfrac{3}{2}$ m (b) $\tfrac{41}{6}$ m
61. (a) $v(t) = \tfrac{1}{2}t^2 + 4t + 5$ m/s (b) $416\tfrac{2}{3}$ m
63. $46\tfrac{2}{3}$ kg **65.** 1.4 mi **67.** \$58,000 **69.** 39.8 ft/s
71. 5443 bacteria **73.** 4.75×10^5 megawatt-hours

EXERCISES 5.5 ■ PAGE 418

1. $\tfrac{1}{2}\sin 2x + C$ **3.** $\tfrac{2}{9}(x^3+1)^{3/2} + C$
5. $\tfrac{1}{4}\ln|x^4 - 5| + C$ **7.** $-\tfrac{1}{3}(1-x^2)^{3/2} + C$
9. $-\tfrac{1}{20}(1-2x)^{10} + C$ **11.** $(2/\pi)\sin(\pi t/2) + C$
13. $-\tfrac{1}{3}\ln|5 - 3x| + C$
15. $-\tfrac{1}{4}\cos^4\theta + C$ **17.** $\dfrac{1}{1 - e^u} + C$
19. $\tfrac{2}{3}\sqrt{3ax + bx^3} + C$ **21.** $\tfrac{1}{3}(\ln x)^3 + C$ **23.** $\tfrac{1}{4}\tan^4\theta + C$
25. $\tfrac{2}{3}(1+e^x)^{3/2} + C$ **27.** $\tfrac{1}{15}(x^3 + 3x)^5 + C$
29. $-\dfrac{1}{\ln 5}\cos(5^t) + C$ **31.** $\tfrac{1}{3}(\arctan x)^3 + C$
33. $\tfrac{1}{5}\sin(1 + 5t) + C$ **35.** $-\tfrac{2}{3}(\cot x)^{3/2} + C$
37. $\tfrac{1}{3}\sinh^3 x + C$ **39.** $-\ln(1 + \cos^2 x) + C$
41. $\ln|\sin x| + C$ **43.** $\ln|\sin^{-1} x| + C$
45. $\tan^{-1} x + \tfrac{1}{2}\ln(1 + x^2) + C$
47. $\tfrac{1}{40}(2x+5)^{10} - \tfrac{5}{36}(2x+5)^9 + C$
49. $\tfrac{1}{8}(x^2 - 1)^4 + C$ **51.** $-e^{\cos x} + C$

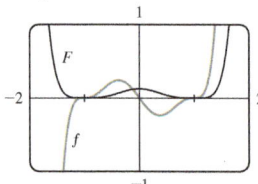 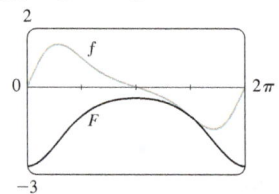

53. $2/\pi$ **55.** $\tfrac{45}{28}$ **57.** $2/\sqrt{3} - 1$ **59.** $e - \sqrt{e}$
61. 0 **63.** 3 **65.** $\tfrac{1}{3}(2\sqrt{2} - 1)a^3$ **67.** $\tfrac{16}{15}$ **69.** 2
71. $\ln(e + 1)$ **73.** $\tfrac{1}{6}$ **75.** $\sqrt{3} - \tfrac{1}{3}$ **77.** 6π
79. All three areas are equal. **81.** ≈ 4512 L
83. $\dfrac{5}{4\pi}\left(1 - \cos\dfrac{2\pi t}{5}\right)$ L

85. $C_0(1 - e^{-30r/V})$; the total amount of urea removed from the blood in the first 30 minutes of dialysis treatment
87. 5   **93.** $\pi^2/4$

CHAPTER 5 REVIEW ■ PAGE 421
True-False Quiz
1. True   **3.** True   **5.** False   **7.** True   **9.** True
11. False   **13.** True   **15.** False   **17.** False

Exercises
1. (a) 8   (b) 5.7

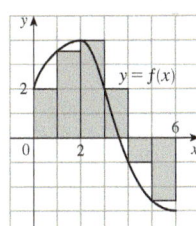 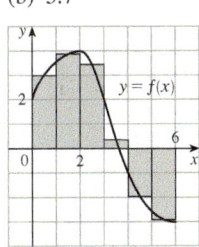

3. $\frac{1}{2} + \pi/4$ **5.** 3   **7.** f is c, f' is b, $\int_0^x f(t)\,dt$ is a
9. 3, 0   **11.** 37   **13.** $\frac{9}{10}$ **15.** -76 **17.** $\frac{21}{4}$
19. Does not exist   **21.** $\frac{1}{3}\sin 1$ **23.** 0
25. $-(1/x) - 2\ln|x| + x + C$
27. $\sqrt{x^2 + 4x} + C$ **29.** $[1/(2\pi)]\sin^2\pi t + C$
31. $2e^{\sqrt{x}} + C$ **33.** $-\frac{1}{2}[\ln(\cos x)]^2 + C$
35. $\frac{1}{4}\ln(1 + x^4) + C$ **37.** $\ln|1 + \sec\theta| + C$ **39.** $\frac{23}{3}$
41. $2\sqrt{1 + \sin x} + C$ **43.** $\frac{64}{5}$ **45.** $F'(x) = x^2/(1 + x^3)$
47. $g'(x) = 4x^3\cos(x^8)$ **49.** $y' = (2e^x - e^{\sqrt{x}})/(2x)$
51. $4 \leq \int_1^3 \sqrt{x^2 + 3}\,dx \leq 4\sqrt{3}$ **57.** 0.280981
59. Number of barrels of oil consumed from Jan. 1, 2000, through Jan. 1, 2008
61. 72,400   **63.** 3   **65.** $c \approx 1.62$
67. $f(x) = e^{2x}(2x - 1)/(1 - e^{-x})$ **73.** $\frac{2}{3}$

PROBLEMS PLUS ■ PAGE 425
1. $\pi/2$ **3.** $2k$ **5.** -1 **7.** e^{-2} **9.** $[-1, 2]$
11. (a) $\frac{1}{2}(n - 1)n$
(b) $\frac{1}{2}[\![b]\!](2b - [\![b]\!] - 1) - \frac{1}{2}[\![a]\!](2a - [\![a]\!] - 1)$
17. $y = -\frac{2b}{a^2}x^2 + \frac{3b}{a}x$ **19.** $2(\sqrt{2} - 1)$

CHAPTER 6

EXERCISES 6.1 ■ PAGE 434
1. $\frac{45}{4} - \ln 8$ **3.** $e - (1/e) + \frac{10}{3}$ **5.** $e - (1/e) + \frac{4}{3}$
7. $\frac{9}{2}$ **9.** $\ln 2 - \frac{1}{2}$ **11.** $\frac{8}{3}$ **13.** 72   **15.** $6\sqrt{3}$
17. $\frac{32}{3}$ **19.** $2/\pi + \frac{2}{3}$ **21.** $2 - 2\ln 2$
23. $\frac{47}{3} - \frac{9}{2}\sqrt[3]{12}$ **25.** $\frac{13}{5}$ **27.** $\ln 2$
29. (a) 39   (b) 15   **31.** $\frac{1}{6}\ln 2$ **33.** $\frac{5}{2}$ **35.** $\frac{3}{2}\sqrt{3} - 1$
37. 0, 0.90; 0.04   **39.** $-1.11, 1.25, 2.86; 8.38$
41. 2.80123   **43.** 0.25142   **45.** $12\sqrt{6} - 9$
47. $117\frac{1}{3}$ ft   **49.** 4232 cm^2

51. (a) Twelfth ($t \approx 11.26$)   (b) Eighteenth ($t \approx 17.18$)
(c) 706
53. (a) Car A   (b) The distance by which A is ahead of B after 1 minute   (c) Car A   (d) $t \approx 2.2$ min
55. $\frac{24}{5}\sqrt{3}$ **57.** $4^{2/3}$ **59.** ± 6
61. $0 < m < 1; m - \ln m - 1$

EXERCISES 6.2 ■ PAGE 446
1. $26\pi/3$

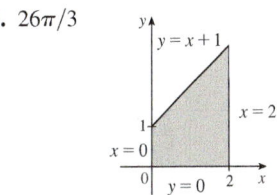 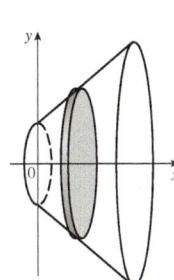

3. 8π

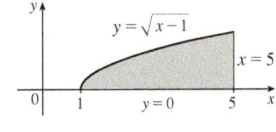 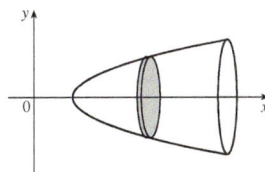

5. 162π

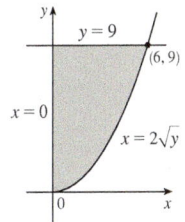 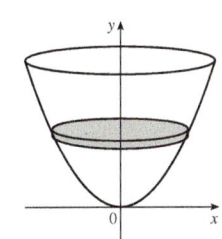

7. $4\pi/21$

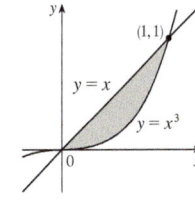 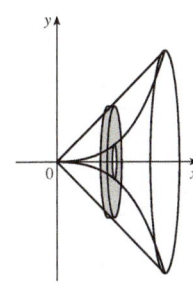

9. $64\pi/15$

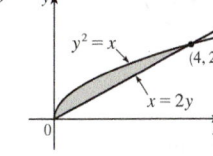 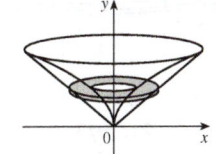

A96 APPENDIX I Answers to Odd-Numbered Exercises

11. $11\pi/30$

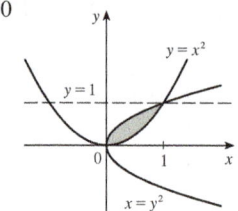
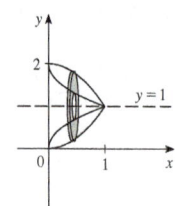

13. $2\pi\left(\tfrac{4}{3}\pi - \sqrt{3}\right)$

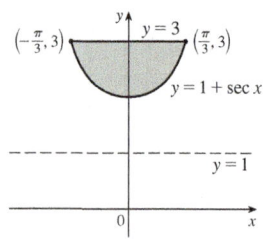
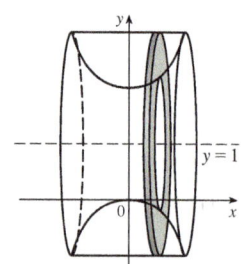

15. $3\pi/5$

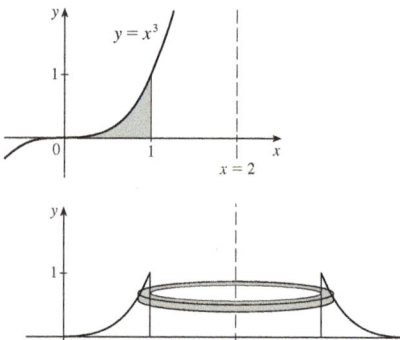

17. $10\sqrt{2}\,\pi/3$

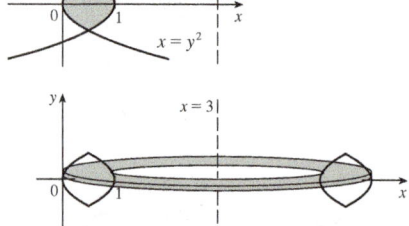

19. $\pi/3$ **21.** $\pi/3$ **23.** $\pi/3$
25. $13\pi/45$ **27.** $\pi/3$ **29.** $17\pi/45$

31. (a) $2\pi \int_0^1 e^{-2x^2}\,dx \approx 3.75825$
(b) $2\pi \int_0^1 \left(e^{-2x^2} + 2e^{-x^2}\right)dx \approx 13.14312$
33. (a) $2\pi \int_0^2 8\sqrt{1 - x^2/4}\,dx \approx 78.95684$
(b) $2\pi \int_0^1 8\sqrt{4 - 4y^2}\,dy \approx 78.95684$
35. $-4.091, -1.467, 1.091; 89.023$ **37.** $\tfrac{11}{8}\pi^2$
39. Solid obtained by rotating the region $0 \leq x \leq \pi$, $0 \leq y \leq \sqrt{\sin x}$ about the x-axis
41. Solid obtained by rotating the region above the x-axis bounded by $x = y^2$ and $x = y^4$ about the y-axis
43. 1110 cm^3 **45.** (a) 196 (b) 838
47. $\tfrac{1}{3}\pi r^2 h$ **49.** $\pi h^2\left(r - \tfrac{1}{3}h\right)$ **51.** $\tfrac{2}{3}b^2 h$
53. 10 cm^3 **55.** 24 **57.** $\tfrac{1}{3}$ **59.** $\tfrac{8}{15}$ **61.** $4\pi/15$
63. (a) $8\pi R \int_0^r \sqrt{r^2 - y^2}\,dy$ (b) $2\pi^2 r^2 R$
65. (b) $\pi r^2 h$ **67.** $\tfrac{5}{12}\pi r^3$ **69.** $8\int_0^r \sqrt{R^2 - y^2}\sqrt{r^2 - y^2}\,dy$

EXERCISES 6.3 ■ **PAGE 453**

1. Circumference $= 2\pi x$, height $= x(x-1)^2$; $\pi/15$

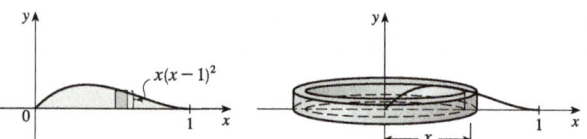

3. $6\pi/7$ **5.** $\pi(1 - 1/e)$ **7.** 8π
9. 4π **11.** 192π **13.** $16\pi/3$
15. $264\pi/5$ **17.** $8\pi/3$ **19.** $13\pi/3$
21. (a) $2\pi \int_0^2 x^2 e^{-x}\,dx$ (b) 4.06300
23. (a) $4\pi \int_{-\pi/2}^{\pi/2} (\pi - x)\cos^4 x\,dx$ (b) 46.50942
25. (a) $\int_0^\pi 2\pi(4 - y)\sqrt{\sin y}\,dy$ (b) 36.57476
27. 3.68
29. Solid obtained by rotating the region $0 \leq y \leq x^4, 0 \leq x \leq 3$ about the y-axis
31. Solid obtained (using shells) by rotating the region $0 \leq x \leq 1/y^2, 1 \leq y \leq 4$ about the line $y = -2$
33. $0, 2.175; 14.450$ **35.** $\tfrac{1}{32}\pi^3$ **37.** 8π
39. $4\sqrt{3}\,\pi$ **41.** $4\pi/3$
43. $117\pi/5$ **45.** $\tfrac{4}{3}\pi r^3$ **47.** $\tfrac{1}{3}\pi r^2 h$

EXERCISES 6.4 ■ **PAGE 458**

1. (a) 7200 ft-lb (b) 7200 ft-lb
3. 4.5 ft-lb **5.** 180 J **7.** $\tfrac{15}{4}$ ft-lb
9. (a) $\tfrac{25}{24} \approx 1.04$ J (b) 10.8 cm **11.** $W_2 = 3W_1$
13. (a) 625 ft-lb (b) $\tfrac{1875}{4}$ ft-lb **15.** $650{,}000$ ft-lb
17. 3857 J **19.** 62.5 ft-lb **21.** 2450 J
23. $\approx 1.06 \times 10^6$ J **25.** $\approx 1.04 \times 10^5$ ft-lb **27.** 2.0 m
31. (b) $161.\overline{3}$ ft-lb
33. (a) $Gm_1 m_2 \left(\dfrac{1}{a} - \dfrac{1}{b}\right)$ (b) $\approx 8.50 \times 10^9$ J

APPENDIX I Answers to Odd-Numbered Exercises

EXERCISES 6.5 ■ PAGE 463

1. 7 **3.** $6/\pi$ **5.** $(2/\pi)(e - 1)$ **7.** $2/(5\pi)$
9. (a) 1 (b) 2, 4 (c)

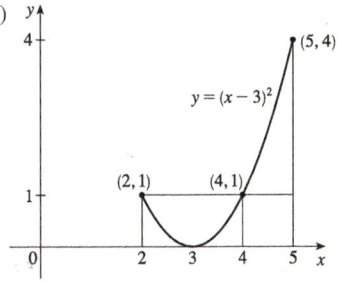

11. (a) $4/\pi$ (b) $\approx 1.24, 2.81$
(c)

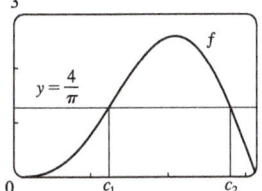

15. $\frac{9}{8}$ **17.** $(50 + 28/\pi)°F \approx 59°F$ **19.** 6 kg/m
21. About 4056 million (or 4 billion) people
23. $5/(4\pi) \approx 0.4$ L

CHAPTER 6 REVIEW ■ PAGE 466

Exercises

1. $\frac{8}{3}$ **3.** $\frac{7}{12}$ **5.** $\frac{4}{3} + 4/\pi$ **7.** $64\pi/15$ **9.** $1656\pi/5$
11. $\frac{4}{3}\pi(2ah + h^2)^{3/2}$ **13.** $\int_{-\pi/3}^{\pi/3} 2\pi(\pi/2 - x)(\cos^2 x - \frac{1}{4})\, dx$
15. (a) $2\pi/15$ (b) $\pi/6$ (c) $8\pi/15$
17. (a) 0.38 (b) 0.87
19. Solid obtained by rotating the region $0 \leq y \leq \cos x$, $0 \leq x \leq \pi/2$ about the y-axis
21. Solid obtained by rotating the region $0 \leq x \leq \pi$, $0 \leq y \leq 2 - \sin x$ about the x-axis
23. 36 **25.** $\frac{125}{3}\sqrt{3}$ m^3 **27.** 3.2 J
29. (a) $8000\pi/3 \approx 8378$ ft-lb (b) 2.1 ft
31. $4/\pi$ **33.** $f(x)$

PROBLEMS PLUS ■ PAGE 468

1. (a) $f(t) = 3t^2$ (b) $f(x) = \sqrt{2x/\pi}$ **3.** $\frac{32}{27}$
5. (b) 0.2261 (c) 0.6736 m
(d) (i) $1/(105\pi) \approx 0.003$ in/s (ii) $370\pi/3$ s ≈ 6.5 min
9. $y = \frac{32}{9}x^2$
11. (a) $V = \int_0^h \pi[f(y)]^2\, dy$
(c) $f(y) = \sqrt{kA/(\pi C)}\, y^{1/4}$. Advantage: the markings on the container are equally spaced.
13. $b = 2a$ **15.** $B = 16A$

CHAPTER 7

EXERCISES 7.1 ■ PAGE 476

1. $\frac{1}{2}xe^{2x} - \frac{1}{4}e^{2x} + C$ **3.** $\frac{1}{5}x \sin 5x + \frac{1}{25}\cos 5x + C$
5. $-\frac{1}{3}te^{-3t} - \frac{1}{9}e^{-3t} + C$
7. $(x^2 + 2x)\sin x + (2x + 2)\cos x - 2\sin x + C$
9. $x \cos^{-1}x - \sqrt{1 - x^2} + C$ **11.** $\frac{1}{5}t^5 \ln t - \frac{1}{25}t^5 + C$
13. $-t \cot t + \ln|\sin t| + C$
15. $x(\ln x)^2 - 2x \ln x + 2x + C$
17. $\frac{1}{13}e^{2\theta}(2 \sin 3\theta - 3 \cos 3\theta) + C$
19. $z^3 e^z - 3z^2 e^z + 6ze^z - 6e^z + C$
21. $\dfrac{e^{2x}}{4(2x + 1)} + C$ **23.** $\dfrac{\pi - 2}{2\pi^2}$
25. $2 \cosh 2 - \sinh 2$ **27.** $\frac{4}{5} - \frac{1}{5}\ln 5$ **29.** $-\pi/4$
31. $2e^{-1} - 6e^{-5}$ **33.** $\frac{1}{2}\ln 2 - \frac{1}{2}$
35. $\frac{32}{5}(\ln 2)^2 - \frac{64}{25}\ln 2 + \frac{62}{125}$
37. $2\sqrt{x}\, e^{\sqrt{x}} - 2e^{\sqrt{x}} + C$ **39.** $-\frac{1}{2} - \pi/4$
41. $\frac{1}{2}(x^2 - 1)\ln(1 + x) - \frac{1}{4}x^2 + \frac{1}{2}x + \frac{3}{4} + C$
43. $-\frac{1}{2}xe^{-2x} - \frac{1}{4}e^{-2x} + C$

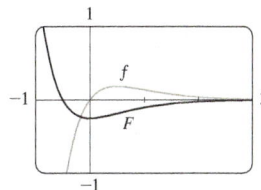

45. $\frac{1}{3}x^2(1 + x^2)^{3/2} - \frac{2}{15}(1 + x^2)^{5/2} + C$

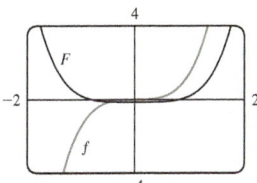

47. (b) $-\frac{1}{4}\cos x \sin^3 x + \frac{3}{8}x - \frac{3}{16}\sin 2x + C$
49. (b) $\frac{2}{3}, \frac{8}{15}$
55. $x[(\ln x)^3 - 3(\ln x)^2 + 6 \ln x - 6] + C$
57. $\frac{16}{3}\ln 2 - \frac{29}{9}$ **59.** $-1.75119, 1.17210; 3.99926$
61. $4 - 8/\pi$ **63.** $2\pi e$
65. (a) $2\pi(2 \ln 2 - \frac{3}{4})$ (b) $2\pi[(\ln 2)^2 - 2\ln 2 + 1]$
67. $xS(x) + \frac{1}{\pi}\cos(\frac{1}{2}\pi x^2) + C$
69. $2 - e^{-t}(t^2 + 2t + 2)$ m **71.** 2

EXERCISES 7.2 ■ PAGE 484

1. $\frac{1}{3}\sin^3 x - \frac{1}{5}\sin^5 x + C$ **3.** $\frac{1}{120}$
5. $-\frac{1}{14}\cos^7(2t) + \frac{1}{5}\cos^5(2t) - \frac{1}{6}\cos^3(2t) + C$
7. $\pi/4$ **9.** $3\pi/8$ **11.** $\pi/16$
13. $\frac{2}{7}(\cos \theta)^{7/2} - \frac{2}{3}(\cos \theta)^{3/2} + C$
15. $\ln|\sin x| - \frac{1}{2}\sin^2 x + C$ **17.** $\frac{1}{2}\sin^4 x + C$
19. $\frac{1}{4}t^2 - \frac{1}{4}t \sin 2t - \frac{1}{8}\cos 2t + C$ **21.** $\frac{1}{3}\sec^3 x + C$
23. $\tan x - x + C$ **25.** $\frac{1}{9}\tan^9 x + \frac{2}{7}\tan^7 x + \frac{1}{5}\tan^5 x + C$
27. $\frac{1}{3}\sec^3 x - \sec x + C$ **29.** $\frac{1}{8}\tan^8 x + \frac{1}{3}\tan^6 x + \frac{1}{4}\tan^4 x + C$
31. $\frac{1}{4}\sec^4 x - \tan^2 x + \ln|\sec x| + C$

33. $x \sec x - \ln|\sec x + \tan x| + C$ **35.** $\sqrt{3} - \tfrac{1}{3}\pi$
37. $\tfrac{22}{105}\sqrt{2} - \tfrac{8}{105}$ **39.** $\ln|\csc x - \cot x| + C$
41. $-\tfrac{1}{6}\cos 3x - \tfrac{1}{26}\cos 13x + C$ **43.** $\tfrac{1}{15}$
45. $\tfrac{1}{2}\sqrt{2}$ **47.** $\tfrac{1}{2}\sin 2x + C$
49. $x \tan x - \ln|\sec x| - \tfrac{1}{2}x^2 + C$
51. $\tfrac{1}{4}x^2 - \tfrac{1}{4}\sin(x^2)\cos(x^2) + C$

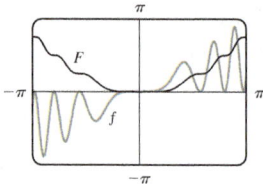

53. $\tfrac{1}{6}\sin 3x - \tfrac{1}{18}\sin 9x + C$

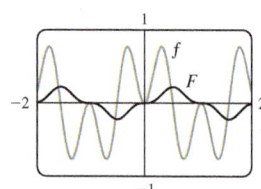

55. 0   **57.** $\tfrac{1}{2}\pi - \tfrac{4}{3}$ **59.** 0   **61.** $\pi^2/4$
63. $\pi(2\sqrt{2} - \tfrac{5}{2})$ **65.** $s = (1 - \cos^3 \omega t)/(3\omega)$

EXERCISES 7.3 ■ PAGE 491

1. $-\dfrac{\sqrt{4 - x^2}}{4x} + C$ **3.** $\sqrt{x^2 - 4} - 2\sec^{-1}\left(\dfrac{x}{2}\right) + C$
5. $\dfrac{1}{3}\dfrac{(x^2 - 1)^{3/2}}{x^3} + C$ **7.** $\dfrac{1}{\sqrt{2}a^2}$
9. $\tfrac{2}{3}\sqrt{3} - \tfrac{3}{4}\sqrt{2}$ **11.** $\tfrac{1}{12}$
13. $\tfrac{1}{6}\sec^{-1}(x/3) - \sqrt{x^2 - 9}/(2x^2) + C$
15. $\tfrac{1}{16}\pi a^4$ **17.** $\sqrt{x^2 - 7} + C$
19. $\ln\left|(\sqrt{1 + x^2} - 1)/x\right| + \sqrt{1 + x^2} + C$ **21.** $\tfrac{9}{500}\pi$
23. $\ln\left|\sqrt{x^2 + 2x + 5} + x + 1\right| + C$
25. $4\sin^{-1}\left(\dfrac{x - 1}{2}\right) + \tfrac{1}{4}(x - 1)^3\sqrt{3 + 2x - x^2}$
$\qquad\qquad\qquad\qquad\qquad - \tfrac{2}{3}(3 + 2x - x^2)^{3/2} + C$
27. $\tfrac{1}{2}(x + 1)\sqrt{x^2 + 2x} - \tfrac{1}{2}\ln\left|x + 1 + \sqrt{x^2 + 2x}\right| + C$
29. $\tfrac{1}{4}\sin^{-1}(x^2) + \tfrac{1}{4}x^2\sqrt{1 - x^4} + C$
33. $\tfrac{1}{6}(\sqrt{48} - \sec^{-1} 7)$ **37.** $\tfrac{3}{8}\pi^2 + \tfrac{3}{4}\pi$
41. $2\pi^2 R r^2$ **43.** $r\sqrt{R^2 - r^2} + \pi r^2/2 - R^2\arcsin(r/R)$

EXERCISES 7.4 ■ PAGE 501

1. (a) $\dfrac{A}{1 + 2x} + \dfrac{B}{3 - x}$ (b) $\dfrac{A}{x} + \dfrac{B}{x^2} + \dfrac{C}{x^3} + \dfrac{D}{1 + x}$
3. (a) $\dfrac{A}{x} + \dfrac{B}{x^2} + \dfrac{Cx + D}{1 + x^2}$ (b) $1 + \dfrac{A}{x} + \dfrac{B}{x - 1} + \dfrac{C}{x - 2}$
5. (a) $x^4 + 4x^2 + 16 + \dfrac{A}{x + 2} + \dfrac{B}{x - 2}$
(b) $\dfrac{Ax + B}{x^2 - x + 1} + \dfrac{Cx + D}{x^2 + 2} + \dfrac{Ex + F}{(x^2 + 2)^2}$
7. $\tfrac{1}{4}x^4 + \tfrac{1}{3}x^3 + \tfrac{1}{2}x^2 + x + \ln|x - 1| + C$

9. $\tfrac{1}{2}\ln|2x + 1| + 2\ln|x - 1| + C$ **11.** $2\ln \tfrac{3}{2}$
13. $a\ln|x - b| + C$ **15.** $\tfrac{5}{2} - \ln 2 - \ln 3$ (or $\tfrac{5}{2} - \ln 6$)
17. $\tfrac{27}{5}\ln 2 - \tfrac{9}{5}\ln 3$ (or $\tfrac{9}{5}\ln \tfrac{8}{3}$)
19. $\tfrac{1}{2} - 5\ln 2 + 3\ln 3$ (or $\tfrac{1}{2} + \ln \tfrac{27}{32}$)
21. $\dfrac{1}{4}\left[\ln|t + 1| - \dfrac{1}{t + 1} - \ln|t - 1| - \dfrac{1}{t - 1}\right] + C$
23. $\ln|x - 1| - \tfrac{1}{2}\ln(x^2 + 9) - \tfrac{1}{3}\tan^{-1}(x/3) + C$
25. $-2\ln|x + 1| + \ln(x^2 + 1) + 2\tan^{-1}x + C$
27. $\tfrac{1}{2}\ln(x^2 + 1) + \tan^{-1}x - \tfrac{1}{2}\tan^{-1}\left(\dfrac{x}{2}\right) + C$
29. $\tfrac{1}{2}\ln(x^2 + 2x + 5) + \tfrac{3}{2}\tan^{-1}\left(\dfrac{x + 1}{2}\right) + C$
31. $\tfrac{1}{3}\ln|x - 1| - \tfrac{1}{6}\ln(x^2 + x + 1) - \dfrac{1}{\sqrt{3}}\tan^{-1}\dfrac{2x + 1}{\sqrt{3}} + C$
33. $\tfrac{1}{4}\ln \tfrac{8}{3}$
35. $2\ln|x| + \tfrac{3}{2}\ln(x^2 + 1) + \tfrac{1}{2}\tan^{-1}x + \dfrac{x}{2(x^2 + 1)} + C$
37. $\tfrac{7}{8}\sqrt{2}\tan^{-1}\left(\dfrac{x - 2}{\sqrt{2}}\right) + \dfrac{3x - 8}{4(x^2 - 4x + 6)} + C$
39. $2\tan^{-1}\sqrt{x - 1} + C$
41. $-2\ln\sqrt{x} - \dfrac{2}{\sqrt{x}} + 2\ln(\sqrt{x} + 1) + C$
43. $\tfrac{3}{10}(x^2 + 1)^{5/3} - \tfrac{3}{4}(x^2 + 1)^{2/3} + C$
45. $2\sqrt{x} + 3\sqrt[3]{x} + 6\sqrt[6]{x} + 6\ln|\sqrt[6]{x} - 1| + C$
47. $\ln\left[\dfrac{(e^x + 2)^2}{(e^x + 1)}\right] + C$
49. $\ln|\tan t + 1| - \ln|\tan t + 2| + C$
51. $x - \ln(e^x + 1) + C$
53. $(x - \tfrac{1}{2})\ln(x^2 - x + 2) - 2x + \sqrt{7}\tan^{-1}\left(\dfrac{2x - 1}{\sqrt{7}}\right) + C$
55. $-\tfrac{1}{2}\ln 3 \approx -0.55$
57. $\tfrac{1}{2}\ln\left|\dfrac{x - 2}{x}\right| + C$ **61.** $\tfrac{1}{5}\ln\left|\dfrac{2\tan(x/2) - 1}{\tan(x/2) + 2}\right| + C$
63. $4\ln \tfrac{2}{3} + 2$ **65.** $-1 + \tfrac{11}{3}\ln 2$
67. $t = \ln\dfrac{10{,}000}{P} + 11\ln\dfrac{P - 9000}{1000}$
69. (a) $\dfrac{24{,}110}{4879}\dfrac{1}{5x + 2} - \dfrac{668}{323}\dfrac{1}{2x + 1} - \dfrac{9438}{80{,}155}\dfrac{1}{3x - 7}$
$\qquad\qquad\qquad\qquad + \dfrac{1}{260{,}015}\dfrac{22{,}098x + 48{,}935}{x^2 + x + 5}$
(b) $\dfrac{4822}{4879}\ln|5x + 2| - \dfrac{334}{323}\ln|2x + 1|$
$\qquad\qquad - \dfrac{3146}{80{,}155}\ln|3x - 7|$
$\qquad + \dfrac{11{,}049}{260{,}015}\ln(x^2 + x + 5) + \dfrac{75{,}772}{260{,}015\sqrt{19}}\tan^{-1}\dfrac{2x + 1}{\sqrt{19}} + C$
The CAS omits the absolute value signs and the constant of integration.
75. $\dfrac{1}{a^n(x - a)} - \dfrac{1}{a^n x} - \dfrac{1}{a^{n-1}x^2} - \cdots - \dfrac{1}{ax^n}$

EXERCISES 7.5 ■ PAGE 507

1. $-\ln(1 - \sin x) + C$ 3. $\frac{32}{3}\ln 2 - \frac{28}{9}$
5. $\frac{1}{2\sqrt{2}}\tan^{-1}\left(\frac{t^2}{\sqrt{2}}\right) + C$ 7. $e^{\pi/4} - e^{-\pi/4}$
9. $\frac{4}{5}\ln 2 + \frac{1}{5}\ln 3$ (or $\frac{1}{5}\ln 48$)   11. $\frac{1}{2}\sec^{-1}x + \frac{\sqrt{x^2-1}}{2x^2} + C$
13. $-\frac{1}{5}\cos^5 t + \frac{2}{7}\cos^7 t - \frac{1}{9}\cos^9 t + C$
15. $x \sec x - \ln|\sec x + \tan x| + C$
17. $\frac{1}{4}\pi^2$ 19. $e^{e^x} + C$ 21. $(x+1)\arctan\sqrt{x} - \sqrt{x} + C$
23. $\frac{4097}{45}$ 25. $4 - \ln 4$ 27. $x - \ln(1 + e^x) + C$
29. $x\ln(x + \sqrt{x^2-1}) - \sqrt{x^2-1} + C$
31. $\sin^{-1}x - \sqrt{1 - x^2} + C$
33. $2\sin^{-1}\left(\frac{x+1}{2}\right) + \frac{x+1}{2}\sqrt{3 - 2x - x^2} + C$
35. 0   37. $\frac{1}{4}$ 39. $\ln|\sec\theta - 1| - \ln|\sec\theta| + C$
41. $\theta\tan\theta - \frac{1}{2}\theta^2 - \ln|\sec\theta| + C$ 43. $\frac{2}{3}\tan^{-1}(x^{3/2}) + C$
45. $-\frac{1}{3}(x^3 + 1)e^{-x^3} + C$
47. $\ln|x-1| - 3(x-1)^{-1} - \frac{3}{2}(x-1)^{-2} - \frac{1}{3}(x-1)^{-3} + C$
49. $\ln\left|\frac{\sqrt{4x+1}-1}{\sqrt{4x+1}+1}\right| + C$ 51. $-\ln\left|\frac{\sqrt{4x^2+1}+1}{2x}\right| + C$
53. $\frac{1}{m}x^2 \cosh mx - \frac{2}{m^2}x\sinh mx + \frac{2}{m^3}\cosh mx + C$
55. $2\ln\sqrt{x} - 2\ln(1 + \sqrt{x}) + C$
57. $\frac{3}{7}(x+c)^{7/3} - \frac{3}{4}c(x+c)^{4/3} + C$
59. $\frac{1}{32}\ln\left|\frac{x-2}{x+2}\right| - \frac{1}{16}\tan^{-1}\left(\frac{x}{2}\right) + C$
61. $\csc\theta - \cot\theta + C$ or $\tan(\theta/2) + C$
63. $2(x - 2\sqrt{x} + 2)e^{\sqrt{x}} + C$
65. $-\tan^{-1}(\cos^2 x) + C$ 67. $\frac{2}{3}[(x+1)^{3/2} - x^{3/2}] + C$
69. $\sqrt{2} - 2/\sqrt{3} + \ln(2 + \sqrt{3}) - \ln(1 + \sqrt{2})$
71. $e^x - \ln(1 + e^x) + C$
73. $-\sqrt{1 - x^2} + \frac{1}{2}(\arcsin x)^2 + C$ 75. $\ln|\ln x - 1| + C$
77. $2(x - 2)\sqrt{1 + e^x} + 2\ln\left|\frac{\sqrt{1+e^x}+1}{\sqrt{1+e^x}-1}\right| + C$
79. $\frac{1}{3}x\sin^3 x + \frac{1}{3}\cos x - \frac{1}{9}\cos^3 x + C$
81. $2\sqrt{1 + \sin x} + C$ 83. $xe^{x^2} + C$

EXERCISES 7.6 ■ PAGE 512

1. $-\frac{5}{21}$ 3. $\sqrt{13} - \frac{3}{2}\ln(4 + \sqrt{13}) - \frac{1}{2} + \frac{3}{2}\ln 3$
5. $\frac{\pi}{8}\arctan\frac{\pi}{4} - \frac{1}{4}\ln(1 + \frac{1}{16}\pi^2)$ 7. $\frac{1}{6}\ln\left|\frac{\sin x - 3}{\sin x + 3}\right| + C$
9. $-\frac{\sqrt{9x^2+4}}{x} + 3\ln(3x + \sqrt{9x^2+4}) + C$
11. $5\pi/16$ 13. $2\sqrt{x}\arctan\sqrt{x} - \ln(1 + x) + C$
15. $-\ln|\sinh(1/y)| + C$
17. $\frac{2y-1}{8}\sqrt{6 + 4y - 4y^2} + \frac{7}{8}\sin^{-1}\left(\frac{2y-1}{\sqrt{7}}\right)$
$\quad - \frac{1}{12}(6 + 4y - 4y^2)^{3/2} + C$

19. $\frac{1}{9}\sin^3 x [3\ln(\sin x) - 1] + C$
21. $\frac{1}{2\sqrt{3}}\ln\left|\frac{e^x + \sqrt{3}}{e^x - \sqrt{3}}\right| + C$
23. $\frac{1}{4}\tan x \sec^3 x + \frac{3}{8}\tan x \sec x + \frac{3}{8}\ln|\sec x + \tan x| + C$
25. $\frac{1}{2}(\ln x)\sqrt{4 + (\ln x)^2} + 2\ln[\ln x + \sqrt{4 + (\ln x)^2}] + C$
27. $-\frac{1}{2}x^{-2}\cos^{-1}(x^{-2}) + \frac{1}{2}\sqrt{1 - x^{-4}} + C$
29. $\sqrt{e^{2x} - 1} - \cos^{-1}(e^{-x}) + C$
31. $\frac{1}{5}\ln|x^5 + \sqrt{x^{10} - 2}| + C$ 33. $\frac{3}{8}\pi^2$
37. $\frac{1}{3}\tan x \sec^2 x + \frac{2}{3}\tan x + C$
39. $\frac{1}{4}x(x^2 + 2)\sqrt{x^2 + 4} - 2\ln(\sqrt{x^2 + 4} + x) + C$
41. $\frac{1}{4}\cos^3 x \sin x + \frac{3}{8}x + \frac{3}{8}\sin x \cos x + C$
43. $-\ln|\cos x| - \frac{1}{2}\tan^2 x + \frac{1}{4}\tan^4 x + C$
45. (a) $-\ln\left|\frac{1 + \sqrt{1-x^2}}{x}\right| + C$;

both have domain $(-1, 0) \cup (0, 1)$

EXERCISES 7.7 ■ PAGE 524

1. (a) $L_2 = 6, R_2 = 12, M_2 \approx 9.6$
(b) L_2 is an underestimate, R_2 and M_2 are overestimates.
(c) $T_2 = 9 < I$ (d) $L_n < T_n < I < M_n < R_n$
3. (a) $T_4 \approx 0.895759$ (underestimate)
(b) $M_4 \approx 0.908907$ (overestimate); $T_4 < I < M_4$
5. (a) $M_{10} \approx 0.806598, E_M \approx -0.001879$
(b) $S_{10} \approx 0.804779, E_S \approx -0.000060$
7. (a) 1.506361   (b) 1.518362   (c) 1.511519
9. (a) 2.660833   (b) 2.664377   (c) 2.663244
11. (a) -7.276910 (b) -4.818251 (c) -5.605350
13. (a) -2.364034 (b) -2.310690 (c) -2.346520
15. (a) 0.243747   (b) 0.243748   (c) 0.243751
17. (a) 8.814278   (b) 8.799212   (c) 8.804229
19. (a) $T_8 \approx 0.902333, M_8 \approx 0.905620$
(b) $|E_T| \leq 0.0078, |E_M| \leq 0.0039$
(c) $n = 71$ for T_n, $n = 50$ for M_n
21. (a) $T_{10} \approx 1.983524, E_T \approx 0.016476$;
$M_{10} \approx 2.008248, E_M \approx -0.008248$;
$S_{10} \approx 2.000110, E_S \approx -0.000110$
(b) $|E_T| \leq 0.025839, |E_M| \leq 0.012919, |E_S| \leq 0.000170$
(c) $n = 509$ for T_n, $n = 360$ for M_n, $n = 22$ for S_n
23. (a) 2.8   (b) 7.954926518   (c) 0.2894
(d) 7.954926521   (e) The actual error is much smaller.
(f) 10.9   (g) 7.953789422   (h) 0.0593
(i) The actual error is smaller.   (j) $n \geq 50$
25.

n	L_n	R_n	T_n	M_n
5	0.742943	1.286599	1.014771	0.992621
10	0.867782	1.139610	1.003696	0.998152
20	0.932967	1.068881	1.000924	0.999538

n	E_L	E_R	E_T	E_M
5	0.257057	-0.286599	-0.014771	0.007379
10	0.132218	-0.139610	-0.003696	0.001848
20	0.067033	-0.068881	-0.000924	0.000462

Observations are the same as after Example 1.

27.

n	T_n	M_n	S_n
6	6.695473	6.252572	6.403292
12	6.474023	6.363008	6.400206

n	E_T	E_M	E_S
6	-0.295473	0.147428	-0.003292
12	-0.074023	0.036992	-0.000206

Observations are the same as after Example 1.
29. (a) 19 (b) 18.6 (c) $18.\overline{6}$
31. (a) 14.4 (b) $\frac{1}{2}$
33. 70.8°F **35.** $37.7\overline{3}$ ft/s **37.** 10,177 megawatt-hours
39. (a) 190 (b) 828
41. 28 **43.** 59.4
45.

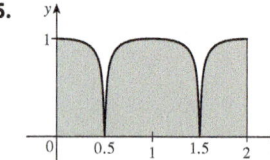

EXERCISES 7.8 ■ PAGE 534
Abbreviations: C, convergent; D, divergent
1. (a), (d) Infinite discontinuity   (b), (c) Infinite interval
3. $\frac{1}{2} - 1/(2t^2)$; 0.495, 0.49995, 0.4999995; 0.5
5. 2 **7.** D **9.** $\frac{1}{5}e^{-10}$ **11.** D **13.** 0 **15.** D
17. ln 2 **19.** $-\frac{1}{4}$ **21.** D **23.** $-\pi/8$ **25.** 2
27. D **29.** $\frac{32}{3}$ **31.** D **33.** $\frac{9}{2}$ **35.** D **37.** $-\frac{1}{4}$
39. $-2/e$
41. $1/e$ **43.** $\frac{1}{2}\ln 2$

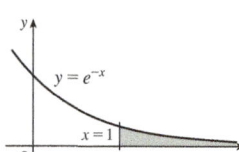
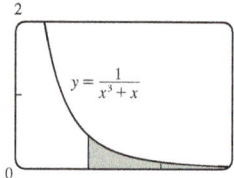

45. Infinite area

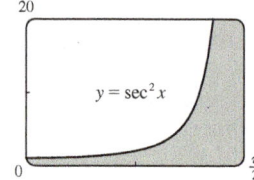

47. (a)

t	$\int_1^t [(\sin^2 x)/x^2]\,dx$
2	0.447453
5	0.577101
10	0.621306
100	0.668479
1,000	0.672957
10,000	0.673407

It appears that the integral is convergent.

(c)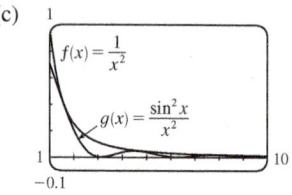

49. C **51.** D **53.** D **55.** π **57.** $p < 1, 1/(1-p)$
59. $p > -1, -1/(p+1)^2$ **63.** π **65.** $\sqrt{2GM/R}$
67. (a)

(b) The rate at which the fraction $F(t)$ increases as t increases
(c) 1; all bulbs burn out eventually

69. $\gamma = \dfrac{cN}{\lambda(k+\lambda)}$ **71.** 1000
73. (a) $F(s) = 1/s, s > 0$ (b) $F(s) = 1/(s-1), s > 1$
(c) $F(s) = 1/s^2, s > 0$
79. $C = 1$; ln 2 **81.** No

CHAPTER 7 REVIEW ■ PAGE 537
True-False Quiz
1. False **3.** False **5.** False **7.** False
9. (a) True (b) False **11.** False **13.** False

Exercises
1. $\frac{7}{2} + \ln 2$ **3.** $e^{\sin x} + C$ **5.** $\ln|2t+1| - \ln|t+1| + C$
7. $\frac{2}{15}$ **9.** $-\cos(\ln t) + C$ **11.** $\sqrt{3} - \frac{1}{3}\pi$
13. $3e^{\sqrt[3]{x}}(x^{2/3} - 2x^{1/3} + 2) + C$
15. $-\frac{1}{2}\ln|x| + \frac{3}{2}\ln|x+2| + C$
17. $x \sinh x - \cosh x + C$
19. $\frac{1}{18}\ln(9x^2 + 6x + 5) + \frac{1}{9}\tan^{-1}\left[\frac{1}{2}(3x+1)\right] + C$
21. $\ln|x - 2 + \sqrt{x^2 - 4x}| + C$
23. $\ln\left|\dfrac{\sqrt{x^2+1}-1}{x}\right| + C$
25. $\frac{3}{2}\ln(x^2 + 1) - 3\tan^{-1}x + \sqrt{2}\tan^{-1}(x/\sqrt{2}) + C$
27. $\frac{2}{5}$ **29.** 0 **31.** $6 - \frac{3}{2}\pi$
33. $\dfrac{x}{\sqrt{4-x^2}} - \sin^{-1}\left(\dfrac{x}{2}\right) + C$
35. $4\sqrt{1+\sqrt{x}} + C$ **37.** $\frac{1}{2}\sin 2x - \frac{1}{8}\cos 4x + C$
39. $\frac{1}{8}e - \frac{1}{4}$ **41.** $\frac{1}{36}$ **43.** D
45. $4\ln 4 - 8$ **47.** $-\frac{4}{3}$ **49.** $\pi/4$
51. $(x+1)\ln(x^2+2x+2) + 2\arctan(x+1) - 2x + C$
53. 0
55. $\frac{1}{4}(2x-1)\sqrt{4x^2-4x-3}$
$\quad - \ln|2x-1+\sqrt{4x^2-4x-3}| + C$
57. $\frac{1}{2}\sin x\sqrt{4+\sin^2 x} + 2\ln(\sin x + \sqrt{4+\sin^2 x}) + C$
61. No
63. (a) 1.925444 (b) 1.920915 (c) 1.922470
65. (a) 0.01348, $n \geq 368$ (b) 0.00674, $n \geq 260$

67. 8.6 mi
69. (a) 3.8   (b) 1.7867, 0.000646   (c) $n \geq 30$
71. (a) D   (b) C
73. 2   **75.** $\frac{3}{16}\pi^2$

PROBLEMS PLUS ■ PAGE 541

1. About 1.85 inches from the center   **3.** 0
7. $f(\pi) = -\pi/2$ **11.** $(b^b a^{-a})^{1/(b-a)} e^{-1}$ **13.** $\frac{1}{8}\pi - \frac{1}{12}$
15. $2 - \sin^{-1}(2/\sqrt{5})$

CHAPTER 8

EXERCISES 8.1 ■ PAGE 548

1. $4\sqrt{5}$ **3.** 3.8202   **5.** 3.4467   **7.** 3.6095
9. $\frac{2}{243}(82\sqrt{82} - 1)$ **11.** $\frac{59}{24}$ **13.** $\frac{32}{3}$
15. $\ln(\sqrt{2} + 1)$ **17.** $\frac{3}{4} + \frac{1}{2}\ln 2$ **19.** $\ln 3 - \frac{1}{2}$
21. $\sqrt{2} + \ln(1 + \sqrt{2})$ **23.** 10.0556
25. 15.498085; 15.374568   **27.** 7.094570; 7.118819
29. (a), (b) 3

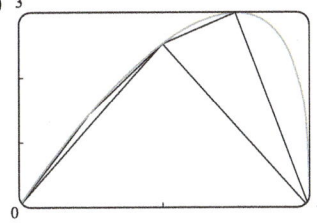

$L_1 = 4$,
$L_2 \approx 6.43$,
$L_4 \approx 7.50$

(c) $\int_0^4 \sqrt{1 + [4(3 - x)/(3(4 - x)^{2/3})]^2}\, dx$ (d) 7.7988
31. $\sqrt{1 + e^4} - \ln(1 + \sqrt{1 + e^4}) + 2 - \sqrt{2} + \ln(1 + \sqrt{2})$
33. 6

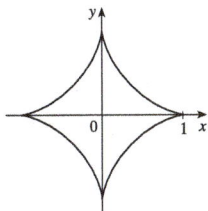

35. $s(x) = \frac{2}{27}[(1 + 9x)^{3/2} - 10\sqrt{10}]$ **37.** $2\sqrt{2}(\sqrt{1 + x} - 1)$
41. 209.1 m   **43.** 29.36 in.   **45.** 12.4

EXERCISES 8.2 ■ PAGE 555

1. (a) (i) $\int_0^{\pi/3} 2\pi \tan x \sqrt{1 + \sec^4 x}\, dx$
(ii) $\int_0^{\pi/3} 2\pi x \sqrt{1 + \sec^4 x}\, dx$ (b) (i) 10.5017   (ii) 7.9353
3. (a) (i) $\int_{-1}^{1} 2\pi e^{-x^2} \sqrt{1 + 4x^2 e^{-2x^2}}\, dx$
(ii) $\int_0^1 2\pi x \sqrt{1 + 4x^2 e^{-2x^2}}\, dx$ (b) (i) 11.0753   (ii) 3.9603
5. (a) (i) $\int_0^1 2\pi y \sqrt{1 + (1 + 3y^2)^2}\, dy$
(ii) $\int_0^1 2\pi(y + y^3)\sqrt{1 + (1 + 3y^2)^2}\, dy$
(b) (i) 8.5302   (ii) 13.5134
7. $\frac{1}{27}\pi(145\sqrt{145} - 1)$ **9.** $\frac{1}{6}\pi(27\sqrt{27} - 5\sqrt{5})$
11. $\pi\sqrt{5} + 4\pi \ln\left(\frac{1 + \sqrt{5}}{2}\right)$ **13.** $\frac{21}{2}\pi$ **15.** $\frac{3712}{15}\pi$

17. πa^2 **19.** 1,230,507   **21.** 24.145807
23. $\frac{1}{4}\pi[4\ln(\sqrt{17} + 4) - 4\ln(\sqrt{2} + 1) - \sqrt{17} + 4\sqrt{2}]$
25. $\frac{1}{6}\pi[\ln(\sqrt{10} + 3) + 3\sqrt{10}]$
29. (a) $\frac{1}{3}\pi a^2$ (b) $\frac{56}{45}\pi\sqrt{3}\, a^2$
31. (a) $2\pi \left[b^2 + \frac{a^2 b \sin^{-1}(\sqrt{a^2 - b^2}/a)}{\sqrt{a^2 - b^2}} \right]$
(b) $2\pi a^2 + \frac{2\pi ab^2}{\sqrt{a^2 - b^2}} \ln \frac{a + \sqrt{a^2 - b^2}}{b}$
33. $\int_a^b 2\pi[c - f(x)]\sqrt{1 + [f'(x)]^2}\, dx$ **35.** $4\pi^2 r^2$
37. Both equal $\pi \int_a^b (e^{x/2} + e^{-x/2})^2\, dx$.

EXERCISES 8.3 ■ PAGE 565

1. (a) 187.5 lb/ft²   (b) 1875 lb   (c) 562.5 lb
3. 7000 lb   **5.** 2.36×10^7 N   **7.** 9.8×10^3 N
9. 889 lb   **11.** $\frac{2}{3}\delta a h^2$ **13.** 5.27×10^5 N
15. (a) 314 N   (b) 353 N
17. (a) 5.63×10^3 lb   (b) 5.06×10^4 lb
(c) 4.88×10^4 lb   (d) 3.03×10^5 lb
19. 4148 lb   **21.** 330; 22
23. 10; 14; (1.4, 1)   **25.** $\left(\frac{2}{3}, \frac{2}{3}\right)$
27. $\left(\frac{1}{e - 1}, \frac{e + 1}{4}\right)$ **29.** $\left(\frac{9}{20}, \frac{9}{20}\right)$
31. $\left(\frac{\pi\sqrt{2} - 4}{4(\sqrt{2} - 1)}, \frac{1}{4(\sqrt{2} - 1)}\right)$ **33.** $\left(\frac{8}{5}, -\frac{1}{2}\right)$
35. $\left(\frac{28}{3(\pi + 2)}, \frac{10}{3(\pi + 2)}\right)$ **37.** $\left(-\frac{1}{5}, -\frac{12}{35}\right)$
41. $\left(0, \frac{1}{12}\right)$ **45.** $\frac{1}{3}\pi r^2 h$ **47.** $\left(\frac{8}{\pi}, \frac{8}{\pi}\right)$
49. $4\pi^2 rR$

EXERCISES 8.4 ■ PAGE 572

1. $21,104   **3.** $140,000; $60,000   **5.** $407.25
7. $166,666.67   **9.** (a) 3800   (b) $324,900
11. 3727; $37,753   **13.** $\frac{2}{3}(16\sqrt{2} - 8) \approx 9.75 million
15. $65,230.48   **17.** $\dfrac{(1 - k)(b^{2-k} - a^{2-k})}{(2 - k)(b^{1-k} - a^{1-k})}$
19. 1.19×10^{-4} cm³/s   **21.** 6.59 L/min   **23.** 5.77 L/min

EXERCISES 8.5 ■ PAGE 579

1. (a) The probability that a randomly chosen tire will have a lifetime between 30,000 and 40,000 miles
(b) The probability that a randomly chosen tire will have a lifetime of at least 25,000 miles
3. (a) $f(x) \geq 0$ for all x and $\int_{-\infty}^{\infty} f(x)\, dx = 1$ (b) $\frac{17}{81}$
5. (a) $1/\pi$ (b) $\frac{1}{2}$
7. (a) $f(x) \geq 0$ for all x and $\int_{-\infty}^{\infty} f(x)\, dx = 1$ (b) 5
11. (a) $\approx 46.5\%$ (b) $\approx 15.3\%$ (c) About 4.8 s
13. $\approx 59.4\%$ (b) 40 min   **15.** $\approx 44\%$
17. (a) 0.0668   (b) $\approx 5.21\%$ **19.** ≈ 0.9545

21. (b) $0; a_0$
(c)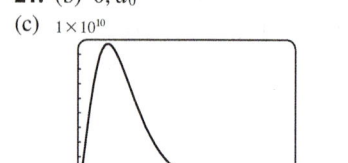
(d) $1 - 41e^{-8} \approx 0.986$ (e) $\tfrac{3}{2}a_0$

CHAPTER 8 REVIEW ■ PAGE 581

Exercises

1. $\tfrac{1}{54}(109\sqrt{109} - 1)$ **3.** $\tfrac{53}{6}$
5. (a) 3.5121 (b) 22.1391 (c) 29.8522
7. 3.8202 **9.** $\tfrac{124}{5}$ **11.** ≈ 458 lb **13.** $(\tfrac{8}{5}, 1)$
15. $(\tfrac{4}{3}, \tfrac{4}{3})$ **17.** $2\pi^2$ **19.** \$7166.67
21. (a) $f(x) \geq 0$ for all x and $\int_{-\infty}^{\infty} f(x)\, dx = 1$
(b) ≈ 0.3455 (c) 5; yes
23. (a) $1 - e^{-3/8} \approx 0.31$ (b) $e^{-5/4} \approx 0.29$
(c) $8 \ln 2 \approx 5.55$ min

PROBLEMS PLUS ■ PAGE 583

1. $\tfrac{2}{3}\pi - \tfrac{1}{2}\sqrt{3}$
3. (a) $2\pi r(r \pm d)$ (b) $\approx 3.36 \times 10^6$ mi^2
(d) $\approx 7.84 \times 10^7$ mi^2
5. (a) $P(z) = P_0 + g\int_0^z \rho(x)\, dx$
(b) $(P_0 - \rho_0 gH)(\pi r^2) + \rho_0 gH e^{L/H} \int_{-r}^{r} e^{x/H} \cdot 2\sqrt{r^2 - x^2}\, dx$
7. Height $\sqrt{2}\, b$, volume $\left(\tfrac{28}{27}\sqrt{6} - 2\right)\pi b^3$ **9.** 0.14 m
11. $2/\pi; 1/\pi$ **13.** $(0, -1)$

CHAPTER 9

EXERCISES 9.1 ■ PAGE 590

3. (a) $\tfrac{1}{2}, -1$ **5.** (d)
7. (a) It must be either 0 or decreasing
(c) $y = 0$ (d) $y = 1/(x + 2)$
9. (a) $0 < P < 4200$ (b) $P > 4200$
(c) $P = 0, P = 4200$
13. (a) III (b) I (c) IV (d) II
15. (a) At the beginning; stays positive, but decreases
(c)

17. It approaches 0 as c approaches c_s.

EXERCISES 9.2 ■ PAGE 597

1. (a)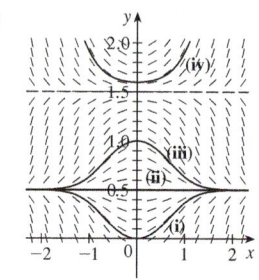

(b) $y = 0.5, y = 1.5$
3. III **5.** IV
7.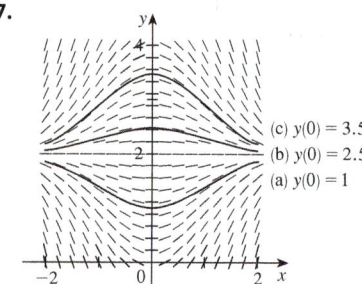

(a) $y(0) = 1$
(b) $y(0) = 2.5$
(c) $y(0) = 3.5$

9.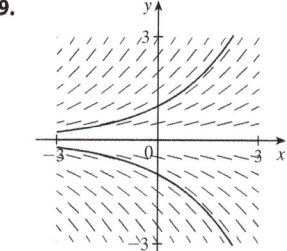

11. 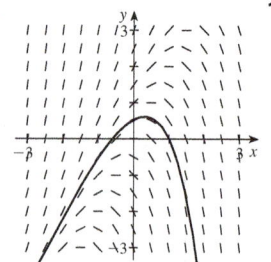 **13.**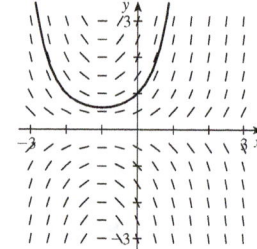

15.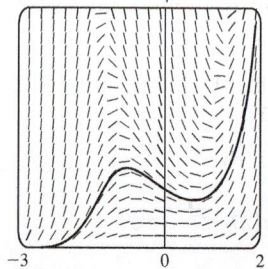

17. 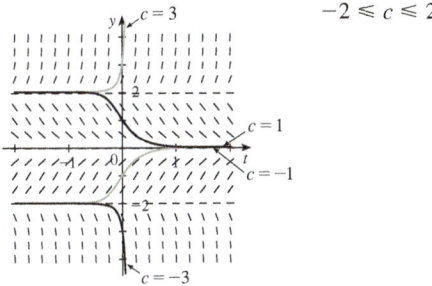 $-2 \le c \le 2; -2, 0, 2$

19. (a) (i) 1.4 (ii) 1.44 (iii) 1.4641

(b) 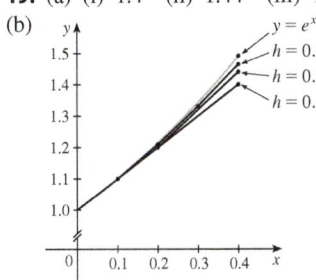 Underestimates

(c) (i) 0.0918 (ii) 0.0518 (iii) 0.0277
It appears that the error is also halved (approximately).

21. $-1, -3, -6.5, -12.25$ **23.** 1.7616

25. (a) (i) 3 (ii) 2.3928 (iii) 2.3701 (iv) 2.3681
(c) (i) -0.6321 (ii) -0.0249 (iii) -0.0022 (iv) -0.0002
It appears that the error is also divided by 10 (approximately).

27. (a), (d) 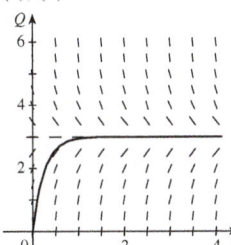   (b) 3
(c) Yes, $Q = 3$
(e) 2.77 C

EXERCISES 9.3 ■ PAGE 605

1. $y = -1/(x^3 + C), y = 0$
3. $y = \pm\sqrt{x^2 + 2\ln|x|} + C$
5. $e^y - y = 2x + \sin x + C$
7. $\theta \sin \theta + \cos \theta = -\frac{1}{2}e^{-t^2} + C$ **9.** $p = Ke^{t^3/3 - t} - 1$
11. $y = -\ln(1 - \frac{1}{2}x^2)$ **13.** $u = -\sqrt{t^2 + \tan t + 25}$
15. $\frac{1}{2}y^2 + \frac{1}{3}(3 + y^2)^{3/2} = \frac{1}{2}x^2 \ln x - \frac{1}{4}x^2 + \frac{41}{12}$
17. $y = \frac{4a}{\sqrt{3}} \sin x - a$
19. $y = \sqrt{x^2 + 4}$ **21.** $y = Ke^x - x - 1$

23. (a) $\sin^{-1} y = x^2 + C$
(b) $y = \sin(x^2), -\sqrt{\pi/2} \le x \le \sqrt{\pi/2}$ (c) No

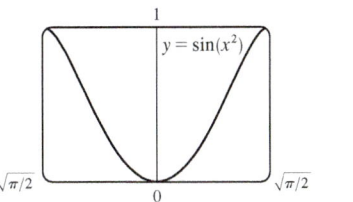

25. $\cos y = \cos x - 1$

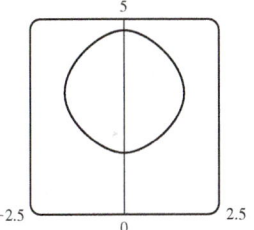

27. (a) 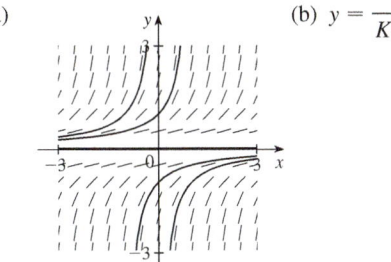   (b) $y = \frac{1}{K - x}$

29. $y = Cx^2$

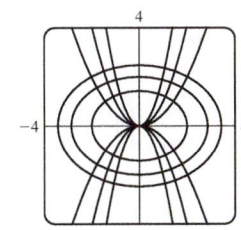

31. $x^2 - y^2 = C$

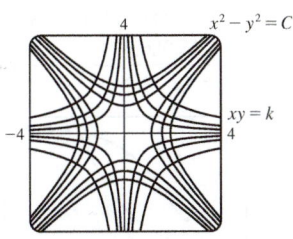

33. $y = 1 + e^{2-x^2/2}$ **35.** $y = (\frac{1}{2}x^2 + 2)^2$
37. $Q(t) = 3 - 3e^{-4t}$; 3   **39.** $P(t) = M - Me^{-kt}$; M
41. (a) $x = a - \dfrac{4}{(kt + 2/\sqrt{a})^2}$

(b) $t = \dfrac{2}{k\sqrt{a-b}}\left(\tan^{-1}\sqrt{\dfrac{b}{a-b}} - \tan^{-1}\sqrt{\dfrac{b-x}{a-b}}\right)$

43. (a) $C(t) = (C_0 - r/k)e^{-kt} + r/k$ (b) r/k; the concentration approaches r/k regardless of the value of C_0
45. (a) $15e^{-t/100}$ kg (b) $15e^{-0.2} \approx 12.3$ kg
47. About 4.9% **49.** g/k
51. (a) $L_1 = KL_2^k$ (b) $B = KV^{0.0794}$
53. (a) $dA/dt = k\sqrt{A}\,(M - A)$ (b) $A(t) = M\left(\dfrac{Ce^{\sqrt{M}\,kt} - 1}{Ce^{\sqrt{M}\,kt} + 1}\right)^2$, where $C = \dfrac{\sqrt{M} + \sqrt{A_0}}{\sqrt{M} - \sqrt{A_0}}$ and $A_0 = A(0)$

EXERCISES 9.4 ■ PAGE 617

1. (a) 1200; 0.04 (b) $P(t) = \dfrac{1200}{1 + 19e^{-0.04t}}$ (c) 87
3. (a) 100; 0.05 (b) Where P is close to 0 or 100; on the line $P = 50$; $0 < P_0 < 100$; $P_0 > 100$
(c)

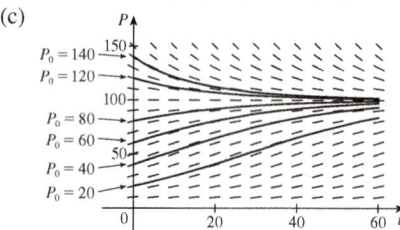

Solutions approach 100; some increase and some decrease, some have an inflection point but others don't; solutions with $P_0 = 20$ and $P_0 = 40$ have inflection points at $P = 50$
(d) $P = 0$, $P = 100$; other solutions move away from $P = 0$ and toward $P = 100$
5. (a) 3.23×10^7 kg (b) ≈ 1.55 years **7.** 9000
9. (a) $\dfrac{dP}{dt} = \dfrac{1}{305}P\left(1 - \dfrac{P}{20}\right)$
(b) 6.24 billion (c) 7.57 billion; 13.87 billion
11. (a) $dy/dt = ky(1 - y)$ (b) $y = \dfrac{y_0}{y_0 + (1 - y_0)e^{-kt}}$
(c) 3:36 PM
15. $P_E(t) = 1909.7761\,(1.0796)^t + 94{,}000$;
$P_L(t) = \dfrac{33{,}086.4394}{1 + 12.3428e^{-0.1657t}} + 94{,}000$

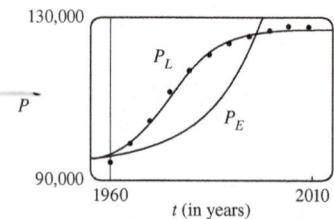

17. (a) $P(t) = \dfrac{m}{k} + \left(P_0 - \dfrac{m}{k}\right)e^{kt}$ (b) $m < kP_0$
(c) $m = kP_0$, $m > kP_0$ (d) Declining
19. (a) Fish are caught at a rate of 15 per week.
(b) See part (d). (c) $P = 250$, $P = 750$

(d)

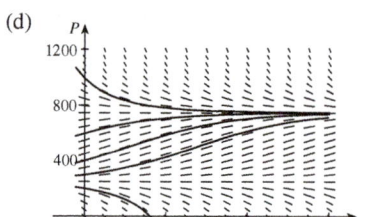

$0 < P_0 < 250$: $P \to 0$;
$P_0 = 250$: $P \to 250$;
$P_0 > 250$: $P \to 750$

(e) $P(t) = \dfrac{250 - 750ke^{t/25}}{1 - ke^{t/25}}$
where $k = \dfrac{1}{11}, -\dfrac{1}{9}$

21. (b)

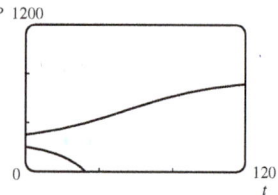

$0 < P_0 < 200$: $P \to 0$;
$P_0 = 200$: $P \to 200$;
$P_0 > 200$: $P \to 1000$

(c) $P(t) = \dfrac{m(M - P_0) + M(P_0 - m)e^{(M-m)(k/M)t}}{M - P_0 + (P_0 - m)e^{(M-m)(k/M)t}}$
23. (a) $P(t) = P_0 e^{(k/r)[\sin(rt - \phi) + \sin \phi]}$ (b) Does not exist

EXERCISES 9.5 ■ PAGE 625

1. No **3.** Yes **5.** $y = 1 + Ce^{-x}$
7. $y = x - 1 + Ce^{-x}$ **9.** $y = \frac{2}{3}\sqrt{x} + C/x$
11. $y = x^2(\ln x + C)$ **13.** $y = \frac{1}{3}t^{-3}(1 + t^2)^{3/2} + Ct^{-3}$
15. $y = \dfrac{1}{x}\ln x - \dfrac{1}{x} + \dfrac{3}{x^2}$ **17.** $u = -t^2 + t^3$
19. $y = -x\cos x - x$
21. $y = \dfrac{(x - 1)e^x + C}{x^2}$

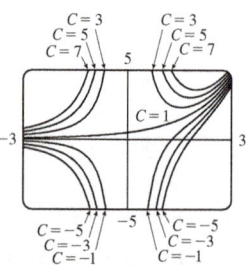

25. $y = \pm\left(Cx^4 + \dfrac{2}{5x}\right)^{-1/2}$
27. (a) $I(t) = 4 - 4e^{-5t}$ (b) $4 - 4e^{-1/2} \approx 1.57$ A
29. $Q(t) = 3(1 - e^{-4t})$, $I(t) = 12e^{-4t}$

31. $P(t) = M + Ce^{-kt}$

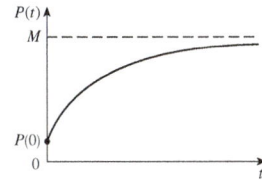

33. $y = \frac{2}{5}(100 + 2t) - 40{,}000(100 + 2t)^{-3/2}$; 0.2275 kg/L
35. (b) mg/c (c) $(mg/c)[t + (m/c)e^{-ct/m}] - m^2g/c^2$
37. (b) $P(t) = \dfrac{M}{1 + MCe^{-kt}}$

EXERCISES 9.6 ■ PAGE 631

1. (a) $x =$ predators, $y =$ prey; growth is restricted only by predators, which feed only on prey.
(b) $x =$ prey, $y =$ predators; growth is restricted by carrying capacity and by predators, which feed only on prey.
3. (a) Competition
(b) (i) $x = 0$, $y = 0$: zero populations
(ii) $x = 0$, $y = 400$: In the absence of an x-population, the y-population stabilizes at 400.
(iii) $x = 125$, $y = 0$: In the absence of a y-population, the x-population stabilizes at 125.
(iv) $x = 50$, $y = 300$: Both populations are stable.
5. (a) The rabbit population starts at about 300, increases to 2400, then decreases back to 300. The fox population starts at 100, decreases to about 20, increases to about 315, decreases to 100, and the cycle starts again.
(b)

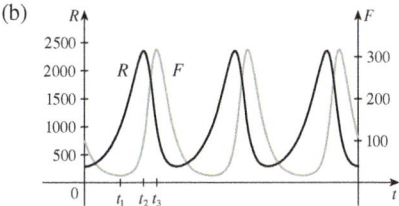

7.

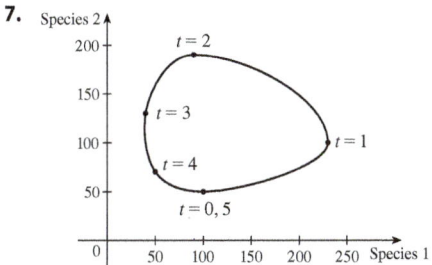

11. (a) Population stabilizes at 5000.
(b) (i) $W = 0$, $R = 0$: Zero populations
(ii) $W = 0$, $R = 5000$: In the absence of wolves, the rabbit population is always 5000.
(iii) $W = 64$, $R = 1000$: Both populations are stable.
(c) The populations stabilize at 1000 rabbits and 64 wolves.

(d)

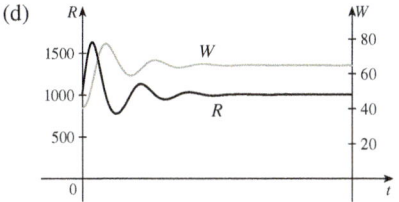

CHAPTER 9 REVIEW ■ PAGE 634

True-False Quiz
1. True   **3.** False   **5.** True   **7.** True

Exercises
1. (a)

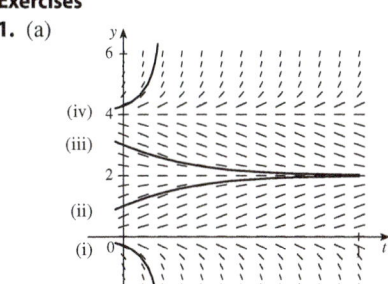

(b) $0 \leq c \leq 4$; $y = 0$, $y = 2$, $y = 4$
3. (a) $y(0.3) \approx 0.8$

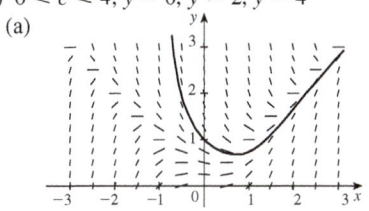

(b) 0.75676
(c) $y = x$ and $y = -x$; there is a loc max or loc min
5. $y = \left(\frac{1}{2}x^2 + C\right)e^{-\sin x}$
7. $y = \pm\sqrt{\ln(x^2 + 2x^{3/2} + C)}$
9. $r(t) = 5e^{t-t^2}$ **11.** $y = \frac{1}{2}x(\ln x)^2 + 2x$ **13.** $x = C - \frac{1}{2}y^2$
15. (a) $P(t) = \dfrac{2000}{1 + 19e^{-0.1t}}$; ≈ 560 (b) $t = -10 \ln \frac{2}{57} \approx 33.5$
17. (a) $L(t) = L_\infty - [L_\infty - L(0)]e^{-kt}$ (b) $L(t) = 53 - 43e^{-0.2t}$
19. 15 days   **21.** $k \ln h + h = (-R/V)t + C$
23. (a) Stabilizes at 200,000
(b) (i) $x = 0$, $y = 0$: Zero populations
(ii) $x = 200{,}000$, $y = 0$: In the absence of birds, the insect population is always 200,000.
(iii) $x = 25{,}000$, $y = 175$: Both populations are stable.
(c) The populations stabilize at 25,000 insects and 175 birds.
(d)

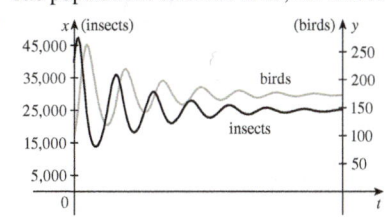

PROBLEMS PLUS ▪ PAGE 637

1. $f(x) = \pm 10 e^x$ **5.** $y = x^{1/n}$ **7.** $20°C$

9. (b) $f(x) = \dfrac{x^2 - L^2}{4L} - \dfrac{L}{2} \ln\left(\dfrac{x}{L}\right)$ (c) No

11. (a) 9.8 h   (b) $31{,}900\pi$ ft^2; 2000π ft^2/h   (c) 5.1 h

13. $x^2 + (y - 6)^2 = 25$ **15.** $y = K/x$, $K \neq 0$

CHAPTER 11

EXERCISES 11.1 ■ PAGE 704

Abbreviations: C, convergent; D, divergent
1. (a) A sequence is an ordered list of numbers. It can also be defined as a function whose domain is the set of positive integers.
(b) The terms a_n approach 8 as n becomes large.
(c) The terms a_n become large as n becomes large.
3. $\frac{2}{3}, \frac{4}{5}, \frac{8}{7}, \frac{16}{9}, \frac{32}{11}$ **5.** $\frac{1}{5}, -\frac{1}{25}, \frac{1}{125}, -\frac{1}{625}, \frac{1}{3125}$ **7.** $\frac{1}{2}, \frac{1}{6}, \frac{1}{24}, \frac{1}{120}, \frac{1}{720}$
9. 1, 2, 7, 32, 157 **11.** $2, \frac{2}{3}, \frac{2}{5}, \frac{2}{7}, \frac{2}{9}$ **13.** $a_n = 1/(2n)$
15. $a_n = -3\left(-\frac{2}{3}\right)^{n-1}$ **17.** $a_n = (-1)^{n+1}\dfrac{n^2}{n+1}$
19. 0.4286, 0.4615, 0.4737, 0.4800, 0.4839, 0.4865, 0.4884, 0.4898, 0.4909, 0.4918; yes; $\frac{1}{2}$
21. 0.5000, 1.2500, 0.8750, 1.0625, 0.9688, 1.0156, 0.9922, 1.0039, 0.9980, 1.0010; yes; 1
23. 5 **25.** D **27.** 0 **29.** 1 **31.** 2
33. D **35.** 0 **37.** 0 **39.** D **41.** 0 **43.** 0
45. 1 **47.** e^2 **49.** ln 2 **51.** $\pi/2$ **53.** D **55.** D
57. D **59.** $\pi/4$ **61.** D **63.** 0

65. (a) 1060, 1123.60, 1191.02, 1262.48, 1338.23 (b) D
67. (b) 5734 **69.** $-1 < r < 1$
71. Convergent by the Monotonic Sequence Theorem; $5 \leq L \leq 8$
73. Decreasing; yes **75.** Not monotonic; no
77. Increasing; yes
79. 2 **81.** $\frac{1}{2}(3 + \sqrt{5})$ **83.** (b) $\frac{1}{2}(1 + \sqrt{5})$
85. (a) 0 (b) 9, 11

EXERCISES 11.2 ▪ PAGE 715

1. (a) A sequence is an ordered list of numbers whereas a series is the *sum* of a list of numbers.
(b) A series is convergent if the sequence of partial sums is a convergent sequence. A series is divergent if it is not convergent.
3. 2
5. 0.5, 0.55, 0.5611, 0.5648, 0.5663, 0.5671, 0.5675, 0.5677; C
7. 1, 1.7937, 2.4871, 3.1170, 3.7018, 4.2521, 4.7749, 5.2749; D
9. $-2.40000, -1.92000,$
$-2.01600, -1.99680,$
$-2.00064, -1.99987,$
$-2.00003, -1.99999,$
$-2.00000, -2.00000;$
convergent, sum $= -2$

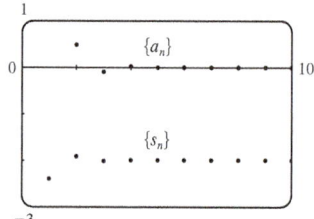

11. 0.44721, 1.15432,
1.98637, 2.88080,
3.80927, 4.75796,
5.71948, 6.68962,
7.66581, 8.64639;
divergent

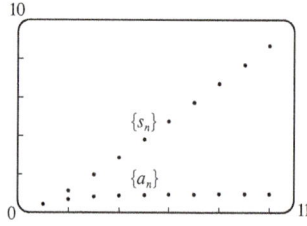

13. 1.00000, 1.33333,
1.50000, 1.60000,
1.66667, 1.71429,
1.75000, 1.77778,
1.80000, 1.81818;
convergent, sum $= 2$

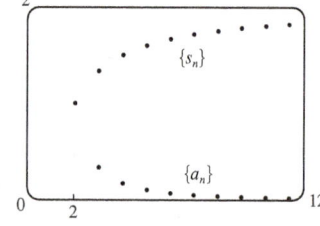

15. (a) Yes (b) No **17.** D **19.** $\frac{25}{3}$ **21.** $\frac{400}{9}$
23. $\frac{1}{7}$ **25.** D **27.** D **29.** D **31.** 9 **33.** D
35. $\dfrac{\sin 100}{1 - \sin 100}$
37. D **39.** D **41.** $e/(e-1)$ **43.** $\frac{3}{2}$ **45.** $\frac{11}{6}$
47. $e - 1$
49. (b) 1 (c) 2 (d) All rational numbers with a terminating decimal representation, except 0
51. $\frac{8}{9}$ **53.** $\frac{838}{333}$ **55.** 45,679/37,000
57. $-\dfrac{1}{5} < x < \dfrac{1}{5}; \dfrac{-5x}{1 + 5x}$

59. $-1 < x < 5; \dfrac{3}{5 - x}$
61. $x > 2$ or $x < -2; \dfrac{x}{x-2}$ **63.** $x < 0; \dfrac{1}{1-e^x}$
65. 1 **67.** $a_1 = 0, a_n = \dfrac{2}{n(n+1)}$ for $n > 1$, sum $= 1$
69. (a) 120 mg; 124 mg
(b) $Q_{n+1} = 100 + 0.20Q_n$ (c) 125 mg
71. (a) 157.875 mg; $\frac{3000}{19}(1 - 0.05^n)$ (b) 157.895 mg
73. (a) $S_n = \dfrac{D(1 - c^n)}{1 - c}$ (b) 5 **75.** $\frac{1}{2}(\sqrt{3} - 1)$
79. $\dfrac{1}{n(n+1)}$ **81.** The series is divergent.
87. $\{s_n\}$ is bounded and increasing.
89. (a) $0, \frac{1}{9}, \frac{2}{9}, \frac{1}{3}, \frac{2}{3}, \frac{7}{9}, \frac{8}{9}, 1$
91. (a) $\frac{1}{2}, \frac{5}{6}, \frac{23}{24}, \frac{119}{120}; \dfrac{(n+1)! - 1}{(n+1)!}$ (c) 1

EXERCISES 11.3 ▪ PAGE 725

1. C

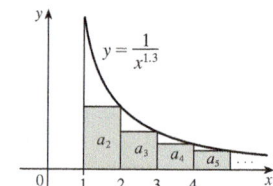

3. C **5.** D **7.** D **9.** C **11.** C **13.** D
15. C **17.** C **19.** D **21.** D **23.** C **25.** C
27. f is neither positive nor decreasing.
29. $p > 1$ **31.** $p < -1$ **33.** $(1, \infty)$
35. (a) $\frac{9}{10}\pi^4$ (b) $\frac{1}{90}\pi^4 - \frac{17}{16}$
37. (a) 1.54977, error ≤ 0.1 (b) 1.64522, error ≤ 0.005
(c) 1.64522 compared to 1.64493 (d) $n > 1000$
39. 0.00145 **45.** $b < 1/e$

EXERCISES 11.4 ▪ PAGE 731

1. (a) Nothing (b) C **3.** C **5.** D **7.** C **9.** D
11. C **13.** C **15.** D **17.** D **19.** C **21.** D
23. C **25.** D **27.** C **29.** C **31.** D
33. 0.1993, error $< 2.5 \times 10^{-5}$
35. 0.0739, error $< 6.4 \times 10^{-8}$
45. Yes

EXERCISES 11.5 ▪ PAGE 736

1. (a) A series whose terms are alternately positive and negative (b) $0 < b_{n+1} \leq b_n$ and $\lim_{n \to \infty} b_n = 0$, where $b_n = |a_n|$ (c) $|R_n| \leq b_{n+1}$
3. D **5.** C **7.** D **9.** C **11.** C **13.** D
15. C **17.** C **19.** D **21.** -0.5507 **23.** 5
25. 5 **27.** -0.4597 **29.** -0.1050
31. An underestimate
33. p is not a negative integer. **35.** $\{b_n\}$ is not decreasing.

EXERCISES 11.6 ■ PAGE 742

Abbreviations: AC, absolutely convergent; CC, conditionally convergent

1. (a) D (b) C (c) May converge or diverge
3. CC **5.** AC **7.** AC **9.** D **11.** AC
13. AC **15.** D **17.** AC **19.** AC **21.** AC
23. D **25.** AC **27.** AC **29.** D **31.** CC
33. AC **35.** D **37.** AC **39.** D **41.** AC
43. (a) and (d)
47. (a) $\frac{661}{960} \approx 0.68854$, error < 0.00521
(b) $n \geq 11$, 0.693109
53. (b) $\sum_{n=2}^{\infty} \frac{(-1)^n}{n \ln n}$; $\sum_{n=1}^{\infty} \frac{(-1)^{n-1}}{n}$

EXERCISES 11.7 ■ PAGE 746

1. D **3.** CC **5.** D **7.** D **9.** C **11.** C
13. C **15.** C **17.** C **19.** C **21.** D **23.** D
25. C **27.** C **29.** C **31.** D
33. C **35.** D **37.** C

EXERCISES 11.8 ■ PAGE 751

1. A series of the form $\sum_{n=0}^{\infty} c_n(x-a)^n$, where x is a variable and a and the c_n's are constants
3. $1, (-1, 1)$ **5.** $1, [-1, 1)$
7. $\infty, (-\infty, \infty)$ **9.** $4, [-4, 4]$
11. $\frac{1}{4}, \left(-\frac{1}{4}, \frac{1}{4}\right]$ **13.** $2, [-2, 2)$
15. $1, [1, 3]$ **17.** $2, [-4, 0)$
19. $\infty, (-\infty, \infty)$ **21.** $b, (a-b, a+b)$ **23.** $0, \{\frac{1}{2}\}$
25. $\frac{1}{5}, \left[\frac{3}{5}, 1\right]$ **27.** $\infty, (-\infty, \infty)$
29. (a) Yes (b) No
31. k^k **33.** No
35. (a) $(-\infty, \infty)$
(b), (c)

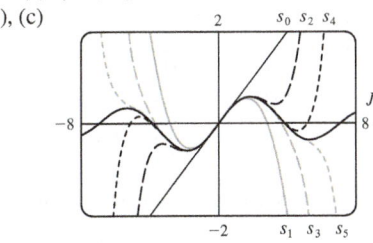

37. $(-1, 1)$, $f(x) = (1+2x)/(1-x^2)$ **41.** 2

EXERCISES 11.9 ■ PAGE 757

1. 10 **3.** $\sum_{n=0}^{\infty} (-1)^n x^n, (-1, 1)$ **5.** $2\sum_{n=0}^{\infty} \frac{1}{3^{n+1}} x^n, (-3, 3)$
7. $\sum_{n=0}^{\infty} \frac{(-1)^n x^{4n+2}}{2^{4n+4}}, (-2, 2)$ **9.** $-\frac{1}{2} - \sum_{n=1}^{\infty} \frac{(-1)^n 3x^n}{2^{n+1}}, (-2, 2)$
11. $\sum_{n=0}^{\infty} \left(-1 - \frac{1}{3^{n+1}}\right) x^n, (-1, 1)$

13. (a) $\sum_{n=0}^{\infty} (-1)^n(n+1)x^n, R = 1$
(b) $\frac{1}{2}\sum_{n=0}^{\infty} (-1)^n(n+2)(n+1)x^n, R = 1$
(c) $\frac{1}{2}\sum_{n=2}^{\infty} (-1)^n n(n-1)x^n, R = 1$
15. $\ln 5 - \sum_{n=1}^{\infty} \frac{x^n}{n5^n}, R = 5$
17. $\sum_{n=0}^{\infty} (-1)^n 4^n (n+1) x^{n+1}, R = \frac{1}{4}$
19. $\sum_{n=0}^{\infty} (2n+1)x^n, R = 1$
21. $\sum_{n=0}^{\infty} (-1)^n x^{2n+2}, R = 1$

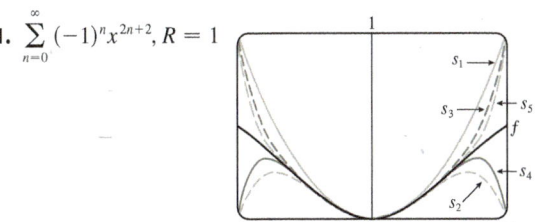

23. $\sum_{n=0}^{\infty} \frac{2x^{2n+1}}{2n+1}, R = 1$

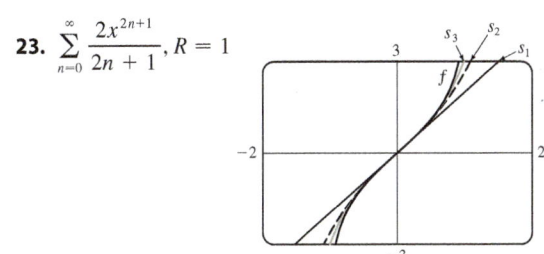

25. $C + \sum_{n=0}^{\infty} \frac{t^{8n+2}}{8n+2}, R = 1$
27. $C + \sum_{n=1}^{\infty} (-1)^n \frac{x^{n+3}}{n(n+3)}, R = 1$
29. 0.044522 **31.** 0.000395
33. 0.19740
35. (b) 0.920 **39.** $[-1, 1], [-1, 1), (-1, 1)$

EXERCISES 11.10 ■ PAGE 771

1. $b_8 = f^{(8)}(5)/8!$ **3.** $\sum_{n=0}^{\infty} (n+1)x^n, R = 1$
5. $x + x^2 + \frac{1}{2}x^3 + \frac{1}{6}x^4$
7. $2 + \frac{1}{12}(x-8) - \frac{1}{288}(x-8)^2 + \frac{5}{20{,}736}(x-8)^3$
9. $\frac{1}{2} + \frac{\sqrt{3}}{2}\left(x - \frac{\pi}{6}\right) - \frac{1}{4}\left(x - \frac{\pi}{6}\right)^2 - \frac{\sqrt{3}}{12}\left(x - \frac{\pi}{6}\right)^3$
11. $\sum_{n=0}^{\infty} (n+1)x^n, R = 1$ **13.** $\sum_{n=0}^{\infty} (-1)^n \frac{x^{2n}}{(2n)!}, R = \infty$

15. $\sum_{n=0}^{\infty} \frac{(\ln 2)^n}{n!} x^n, R = \infty$ **17.** $\sum_{n=0}^{\infty} \frac{x^{2n+1}}{(2n+1)!}, R = \infty$

19. $50 + 105(x-2) + 92(x-2)^2 + 42(x-2)^3 + 10(x-2)^4 + (x-2)^5, R = \infty$

21. $\ln 2 + \sum_{n=1}^{\infty} (-1)^{n+1} \frac{1}{n 2^n} (x-2)^n, R = 2$

23. $\sum_{n=0}^{\infty} \frac{2^n e^6}{n!} (x-3)^n, R = \infty$

25. $\sum_{n=0}^{\infty} \frac{(-1)^{n+1}}{(2n+1)!} (x-\pi)^{2n+1}, R = \infty$

31. $1 - \frac{1}{4}x - \sum_{n=2}^{\infty} \frac{3 \cdot 7 \cdot \cdots \cdot (4n-5)}{4^n \cdot n!} x^n, R = 1$

33. $\sum_{n=0}^{\infty} (-1)^n \frac{(n+1)(n+2)}{2^{n+4}} x^n, R = 2$

35. $\sum_{n=0}^{\infty} (-1)^n \frac{1}{2n+1} x^{4n+2}, R = 1$

37. $\sum_{n=0}^{\infty} (-1)^n \frac{2^{2n}}{(2n)!} x^{2n+1}, R = \infty$

39. $\sum_{n=0}^{\infty} (-1)^n \frac{1}{2^{2n}(2n)!} x^{4n+1}, R = \infty$

41. $\frac{1}{2}x + \sum_{n=1}^{\infty} (-1)^n \frac{1 \cdot 3 \cdot 5 \cdot \cdots \cdot (2n-1)}{n! 2^{3n+1}} x^{2n+1}, R = 2$

43. $\sum_{n=1}^{\infty} (-1)^{n+1} \frac{2^{2n-1}}{(2n)!} x^{2n}, R = \infty$

45. $\sum_{n=0}^{\infty} (-1)^n \frac{1}{(2n)!} x^{4n}, R = \infty$

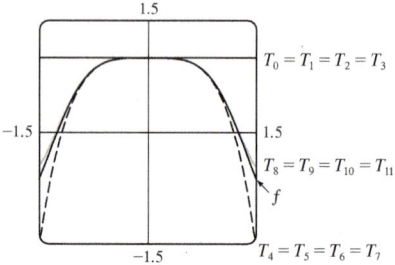

47. $\sum_{n=1}^{\infty} \frac{(-1)^{n-1}}{(n-1)!} x^n, R = \infty$

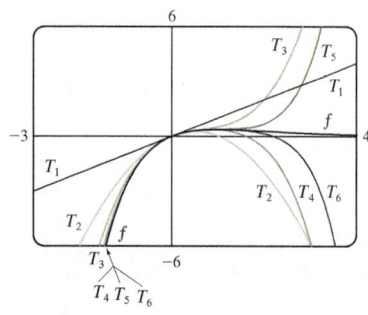

49. 0.99619

51. (a) $1 + \sum_{n=1}^{\infty} \frac{1 \cdot 3 \cdot 5 \cdot \cdots \cdot (2n-1)}{2^n n!} x^{2n}$

(b) $x + \sum_{n=1}^{\infty} \frac{1 \cdot 3 \cdot 5 \cdot \cdots \cdot (2n-1)}{(2n+1) 2^n n!} x^{2n+1}$

53. $C + \sum_{n=0}^{\infty} \binom{1/2}{n} \frac{x^{3n+1}}{3n+1}, R = 1$

55. $C + \sum_{n=1}^{\infty} (-1)^n \frac{1}{2n(2n)!} x^{2n}, R = \infty$

57. 0.0059   **59.** 0.40102   **61.** $\frac{1}{2}$ **63.** $\frac{1}{120}$ **65.** $\frac{3}{5}$

67. $1 - \frac{3}{2}x^2 + \frac{25}{24}x^4$ **69.** $1 + \frac{1}{6}x^2 + \frac{7}{360}x^4$

71. $x - \frac{2}{3}x^4 + \frac{23}{45}x^6$

73. e^{-x^4} **75.** $\ln \frac{8}{5}$

77. $1/\sqrt{2}$ **79.** $e^3 - 1$

EXERCISES 11.11 ■ PAGE 780

1. (a) $T_0(x) = 0, T_1(x) = T_2(x) = x, T_3(x) = T_4(x) = x - \frac{1}{6}x^3$, $T_5(x) = x - \frac{1}{6}x^3 + \frac{1}{120}x^5$

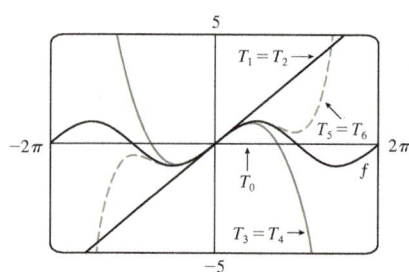

(b)

x	f	T_0	$T_1 = T_2$	$T_3 = T_4$	T_5
$\pi/4$	0.7071	0	0.7854	0.7047	0.7071
$\pi/2$	1	0	1.5708	0.9248	1.0045
π	0	0	3.1416	-2.0261	0.5240

(c) As n increases, $T_n(x)$ is a good approximation to $f(x)$ on a larger and larger interval.

3. $e + e(x-1) + \frac{1}{2}e(x-1)^2 + \frac{1}{6}e(x-1)^3$

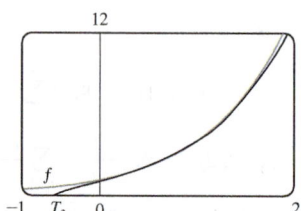

5. $-\left(x - \dfrac{\pi}{2}\right) + \dfrac{1}{6}\left(x - \dfrac{\pi}{2}\right)^3$

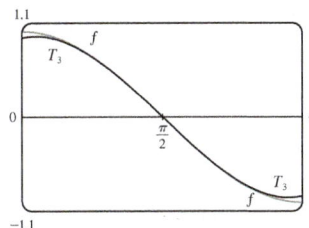

7. $(x - 1) - \dfrac{1}{2}(x - 1)^2 + \dfrac{1}{3}(x - 1)^3$

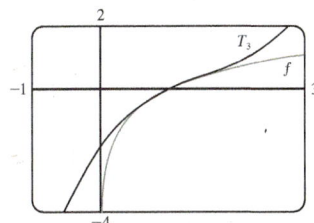

9. $x - 2x^2 + 2x^3$

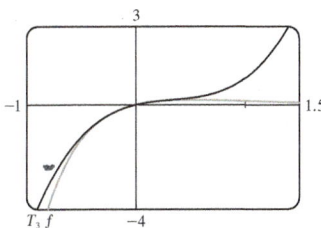

11. $T_5(x) = 1 - 2\left(x - \dfrac{\pi}{4}\right) + 2\left(x - \dfrac{\pi}{4}\right)^2 - \dfrac{8}{3}\left(x - \dfrac{\pi}{4}\right)^3 + \dfrac{10}{3}\left(x - \dfrac{\pi}{4}\right)^4 - \dfrac{64}{15}\left(x - \dfrac{\pi}{4}\right)^5$

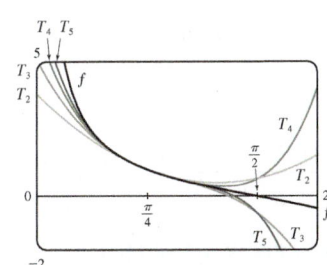

13. (a) $1 - (x - 1) + (x - 1)^2$ (b) $0.006\,482\,7$
15. (a) $1 + \dfrac{2}{3}(x - 1) - \dfrac{1}{9}(x - 1)^2 + \dfrac{4}{81}(x - 1)^3$ (b) $0.000\,097$
17. (a) $1 + \dfrac{1}{2}x^2$ (b) 0.0015
19. (a) $1 + x^2$ (b) $0.000\,06$ **21.** (a) $x^2 - \dfrac{1}{6}x^4$ (b) 0.042

23. 0.17365 **25.** Four **27.** $-1.037 < x < 1.037$
29. $-0.86 < x < 0.86$ **31.** 21 m, no
37. (c) They differ by about 8×10^{-9} km.

CHAPTER 11 REVIEW ■ PAGE 784

True-False Quiz
1. False **3.** True **5.** False **7.** False **9.** False
11. True **13.** True **15.** False **17.** True
19. True **21.** True

Exercises
1. $\dfrac{1}{2}$ **3.** D **5.** 0 **7.** e^{12} **9.** 2 **11.** C
13. C **15.** D **17.** C **19.** C **21.** C **23.** CC
25. AC **27.** $\dfrac{1}{11}$ **29.** $\pi/4$ **31.** e^{-e} **35.** 0.9721
37. $0.189\,762\,24$, error $< 6.4 \times 10^{-7}$
41. $4, [-6, 2]$ **43.** $0.5, [2.5, 3.5)$
45. $\dfrac{1}{2}\displaystyle\sum_{n=0}^{\infty}(-1)^n\left[\dfrac{1}{(2n)!}\left(x - \dfrac{\pi}{6}\right)^{2n} + \dfrac{\sqrt{3}}{(2n + 1)!}\left(x - \dfrac{\pi}{6}\right)^{2n+1}\right]$
47. $\displaystyle\sum_{n=0}^{\infty}(-1)^n x^{n+2}, R = 1$ **49.** $\ln 4 - \displaystyle\sum_{n=1}^{\infty}\dfrac{x^n}{n 4^n}, R = 4$
51. $\displaystyle\sum_{n=0}^{\infty}(-1)^n \dfrac{x^{8n+4}}{(2n + 1)!}, R = \infty$
53. $\dfrac{1}{2} + \displaystyle\sum_{n=1}^{\infty}\dfrac{1 \cdot 5 \cdot 9 \cdot \cdots \cdot (4n - 3)}{n! \, 2^{6n+1}} x^n, R = 16$
55. $C + \ln|x| + \displaystyle\sum_{n=1}^{\infty}\dfrac{x^n}{n \cdot n!}$
57. (a) $1 + \dfrac{1}{2}(x - 1) - \dfrac{1}{8}(x - 1)^2 + \dfrac{1}{16}(x - 1)^3$
(b) 1.5 (c) $0.000\,006$

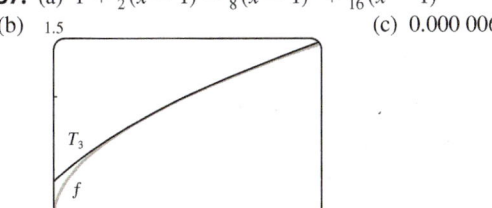

59. $-\dfrac{1}{6}$

PROBLEMS PLUS ■ PAGE 787

1. $15!/5! = 10{,}897{,}286{,}400$
3. (b) 0 if $x = 0$, $(1/x) - \cot x$ if $x \neq k\pi$, k an integer
5. (a) $s_n = 3 \cdot 4^n$, $l_n = 1/3^n$, $p_n = 4^n/3^{n-1}$ (c) $\dfrac{2}{5}\sqrt{3}$
9. $\dfrac{3\pi}{4}$ **11.** $(-1, 1)$, $\dfrac{x^3 + 4x^2 + x}{(1 - x)^4}$ **13.** $\ln \dfrac{1}{2}$
17. (a) $\dfrac{250}{101}\pi(e^{-(n-1)\pi/5} - e^{-n\pi/5})$ (b) $\dfrac{250}{101}\pi$
19. $\dfrac{\pi}{2\sqrt{3}} - 1$
21. $-\left(\dfrac{\pi}{2} - \pi k\right)^2$, where k is a positive integer

APPENDIX I Answers to Odd-Numbered Exercises

CHAPTER 17

EXERCISES 17.1 ■ PAGE 1160

1. $y = c_1 e^{3x} + c_2 e^{-2x}$ 3. $y = c_1 \cos(\sqrt{2}x) + c_2 \sin(\sqrt{2}x)$
5. $y = c_1 e^{-x/2} + c_2 x e^{-x/2}$ 7. $y = c_1 + c_2 e^{4x/3}$
9. $y = e^{2x}(c_1 \cos 3x + c_2 \sin 3x)$
11. $y = c_1 e^{(\sqrt{3}-1)t/2} + c_2 e^{-(\sqrt{3}+1)t/2}$
13. $V = e^{-2t/3}\left[c_1 \cos\left(\dfrac{\sqrt{5}}{3}t\right) + c_2 \sin\left(\dfrac{\sqrt{5}}{3}t\right)\right]$
15. $f(x) = e^{-x} \cos x$, $g(x) = e^{-x} \sin x$. All solution curves approach 0 as $x \to \infty$ and oscillate with amplitudes that become arbitrarily large as $x \to -\infty$.

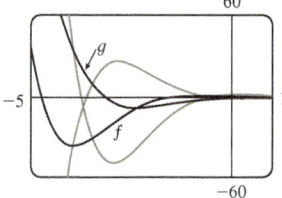

17. $y = \cos(\sqrt{3}x) + \sqrt{3} \sin(\sqrt{3}x)$ 19. $y = e^{-2x/3} + \frac{2}{3}xe^{-2x/3}$
21. $y = e^{3x}(2 \cos x - 3 \sin x)$
23. $y = \frac{1}{7}e^{4x-4} - \frac{1}{7}e^{3-3x}$ 25. $y = -3 \cos 4x + 2 \sin 4x$
27. $y = 2e^{-2x} - 2xe^{-2x}$ 29. $y = \dfrac{e-2}{e-1} + \dfrac{e^x}{e-1}$
31. No solution
33. (b) $\lambda = n^2\pi^2/L^2$, n a positive integer; $y = C \sin(n\pi x/L)$
35. (a) $b - a \neq n\pi$, n any integer
    (b) $b - a = n\pi$ and $\dfrac{c}{d} \neq e^{a-b}\dfrac{\cos a}{\cos b}$ unless $\cos b = 0$, then $\dfrac{c}{d} \neq e^{a-b}\dfrac{\sin a}{\sin b}$
    (c) $b - a = n\pi$ and $\dfrac{c}{d} = e^{a-b}\dfrac{\cos a}{\cos b}$ unless $\cos b = 0$, then $\dfrac{c}{d} = e^{a-b}\dfrac{\sin a}{\sin b}$

EXERCISES 17.2 ■ PAGE 1167

1. $y = c_1 e^{2x} + c_2 e^{-4x} + \frac{1}{4}x^2 + \frac{1}{8}x - \frac{1}{32}$
3. $y = c_1 \cos(\frac{1}{3}x) + c_2 \sin(\frac{1}{3}x) + \frac{1}{37}e^{2x}$
5. $y = e^{2x}(c_1 \cos x + c_2 \sin x) + \frac{1}{10}e^{-x}$
7. $y = e^x\left(\frac{9}{10}\cos 2x - \frac{1}{20}\sin 2x\right) + \frac{1}{10}\cos x + \frac{1}{5}\sin x$
9. $y = e^x\left(\frac{1}{2}x^2 - x + 2\right)$

11.

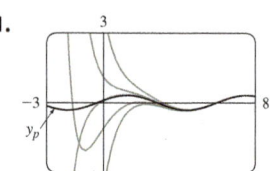

The solutions are all asymptotic to $y_p = \frac{1}{10}\cos x + \frac{3}{10}\sin x$ as $x \to \infty$. Except for y_p, all solutions approach either ∞ or $-\infty$ as $x \to -\infty$.

13. $y_p = (Ax + B)e^x \cos x + (Cx + D)e^x \sin x$
15. $y_p = Axe^x + B\cos x + C\sin x$
17. $y_p = xe^{-x}[(Ax^2 + Bx + C)\cos 3x + (Dx^2 + Ex + F)\sin 3x]$
19. $y = c_1 \cos(\frac{1}{2}x) + c_2 \sin(\frac{1}{2}x) - \frac{1}{3}\cos x$
21. $y = c_1 e^x + c_2 xe^x + e^{2x}$
23. $y = c_1 \sin x + c_2 \cos x + \sin x \ln(\sec x + \tan x) - 1$
25. $y = [c_1 + \ln(1 + e^{-x})]e^x + [c_2 - e^{-x} + \ln(1 + e^{-x})]e^{2x}$
27. $y = e^x\left[c_1 + c_2 x - \frac{1}{2}\ln(1 + x^2) + x \tan^{-1}x\right]$

EXERCISES 17.3 ■ PAGE 1175

1. $x = 0.35 \cos(2\sqrt{5}\,t)$ 3. $x = -\frac{1}{5}e^{-6t} + \frac{6}{5}e^{-t}$ 5. $\frac{49}{12}$ kg

7.

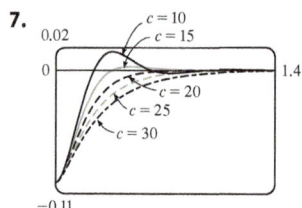

13. $Q(t) = (-e^{-10t}/250)(6 \cos 20t + 3 \sin 20t) + \frac{3}{125}$,
    $I(t) = \frac{3}{5}e^{-10t}\sin 20t$
15. $Q(t) = e^{-10t}\left[\frac{3}{250}\cos 20t - \frac{3}{500}\sin 20t\right] - \frac{3}{250}\cos 10t + \frac{3}{125}\sin 10t$

EXERCISES 17.4 ■ PAGE 1180

1. $c_0 \displaystyle\sum_{n=0}^{\infty} \dfrac{x^n}{n!} = c_0 e^x$ 3. $c_0 \displaystyle\sum_{n=0}^{\infty} \dfrac{x^{3n}}{3^n n!} = c_0 e^{x^3/3}$
5. $c_0 \displaystyle\sum_{n=0}^{\infty} \dfrac{(-1)^n}{2^n n!} x^{2n} + c_1 \displaystyle\sum_{n=0}^{\infty} \dfrac{(-2)^n n!}{(2n+1)!} x^{2n+1}$
7. $c_0 + c_1 \displaystyle\sum_{n=1}^{\infty} \dfrac{x^n}{n} = c_0 - c_1 \ln(1-x)$ for $|x| < 1$
9. $\displaystyle\sum_{n=0}^{\infty} \dfrac{x^{2n}}{2^n n!} = e^{x^2/2}$
11. $x + \displaystyle\sum_{n=1}^{\infty} \dfrac{(-1)^n 2^2 5^2 \cdot \,\cdots\, \cdot (3n-1)^2}{(3n+1)!} x^{3n+1}$

CHAPTER 17 REVIEW ■ PAGE 1181

True-False Quiz
1. True 3. True

Exercises
1. $y = c_1 e^{x/2} + c_2 e^{-x/2}$
3. $y = c_1 \cos(\sqrt{3}x) + c_2 \sin(\sqrt{3}x)$
5. $y = e^{2x}(c_1 \cos x + c_2 \sin x + 1)$
7. $y = c_1 e^x + c_2 xe^x - \frac{1}{2}\cos x - \frac{1}{2}(x+1)\sin x$

9. $y = c_1 e^{3x} + c_2 e^{-2x} - \frac{1}{6} - \frac{1}{5} x e^{-2x}$
11. $y = 5 - 2e^{-6(x-1)}$ **13.** $y = (e^{4x} - e^x)/3$
15. No solution   **17.** $\sum_{n=0}^{\infty} \frac{(-2)^n n!}{(2n+1)!} x^{2n+1}$
19. $Q(t) = -0.02 e^{-10t}(\cos 10t + \sin 10t) + 0.03$
21. (c) $2\pi/k \approx 85$ min   (d) $\approx 17{,}600$ mi/h

APPENDIXES

EXERCISES A ■ PAGE A9

1. 18   **3.** π **5.** $5 - \sqrt{5}$ **7.** $2 - x$
9. $|x+1| = \begin{cases} x+1 & \text{for } x \geq -1 \\ -x-1 & \text{for } x < -1 \end{cases}$ **11.** $x^2 + 1$
13. $(-2, \infty)$ **15.** $[-1, \infty)$
17. $(3, \infty)$ **19.** $(2, 6)$
21. $(0, 1]$ **23.** $[-1, \frac{1}{2})$
25. $(-\infty, 1) \cup (2, \infty)$ **27.** $[-1, \frac{1}{2}]$
29. $(-\infty, \infty)$ **31.** $(-\sqrt{3}, \sqrt{3})$
33. $(-\infty, 1]$ **35.** $(-1, 0) \cup (1, \infty)$
37. $(-\infty, 0) \cup (\frac{1}{4}, \infty)$
39. $10 \leq C \leq 35$ **41.** (a) $T = 20 - 10h, 0 \leq h \leq 12$
(b) $-30°C \leq T \leq 20°C$ **43.** $\pm\frac{3}{2}$ **45.** $2, -\frac{4}{3}$
47. $(-3, 3)$ **49.** $(3, 5)$ **51.** $(-\infty, -7] \cup [-3, \infty)$
53. $[1.3, 1.7]$ **55.** $[-4, -1] \cup [1, 4]$
57. $x \geq (a+b)c/(ab)$ **59.** $x > (c-b)/a$

EXERCISES B ■ PAGE A15

1. 5   **3.** $\sqrt{74}$ **5.** $2\sqrt{37}$ **7.** 2   **9.** $-\frac{9}{2}$
17. 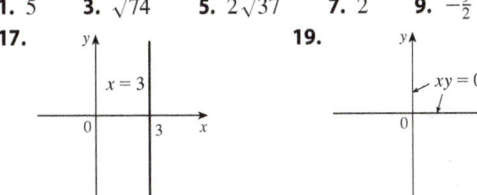   **19.**

21. $y = 6x - 15$ **23.** $2x - 3y + 19 = 0$
25. $5x + y = 11$ **27.** $y = 3x - 2$ **29.** $y = 3x - 3$
31. $y = 5$ **33.** $x + 2y + 11 = 0$ **35.** $5x - 2y + 1 = 0$
37. $m = -\frac{1}{3}$, $b = 0$ **39.** $m = 0$, $b = -2$ **41.** $m = \frac{3}{4}$, $b = -3$

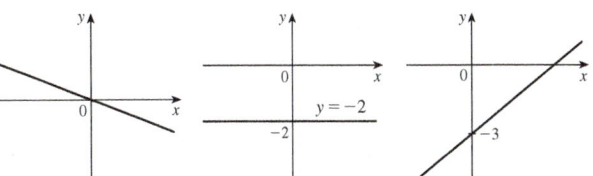

43.   **45.**

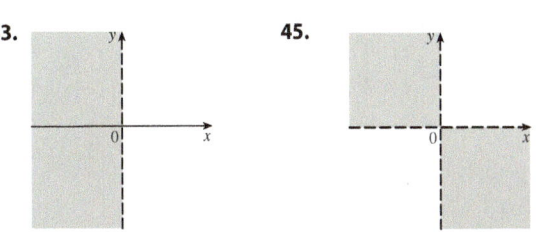

47.   **49.**

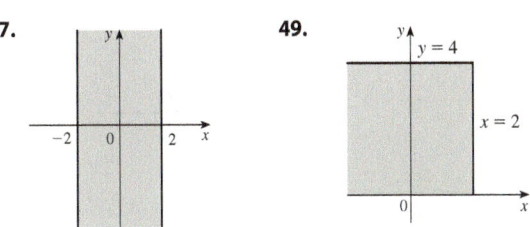

51.

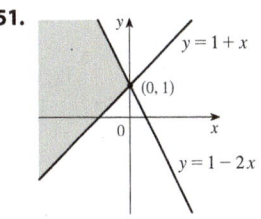

53. $(0, -4)$ **55.** (a) $(4, 9)$ (b) $(3.5, -3)$ **57.** $(1, -2)$
59. $y = x - 3$ **61.** (b) $4x - 3y - 24 = 0$

EXERCISES C ■ PAGE A23

1. $(x-3)^2 + (y+1)^2 = 25$ **3.** $x^2 + y^2 = 65$
5. $(2, -5), 4$ **7.** $(-\frac{1}{2}, 0), \frac{1}{2}$ **9.** $(\frac{1}{4}, -\frac{1}{4}), \sqrt{10}/4$
11. Parabola   **13.** Ellipse

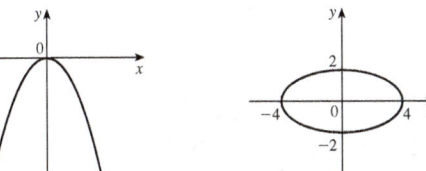

15. Hyperbola

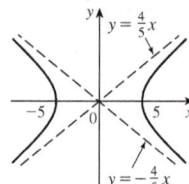

17. Ellipse

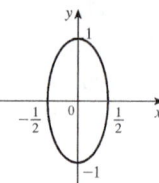

19. Parabola

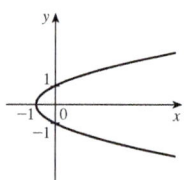

21. Hyperbola

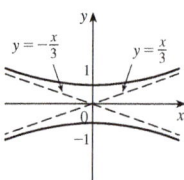

23. Hyperbola

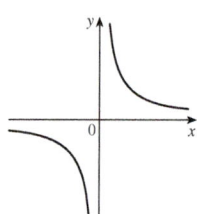

25. Ellipse

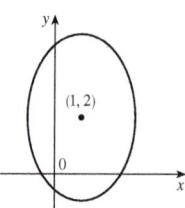

27. Parabola

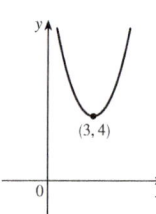

29. Parabola

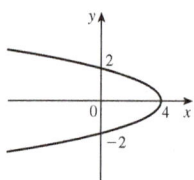

31. Ellipse

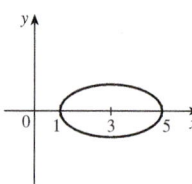

33.

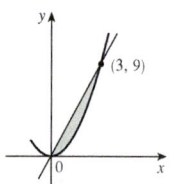

35. $y = x^2 - 2x$

37.

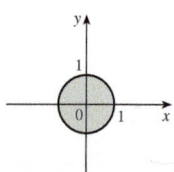

39.

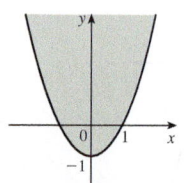

EXERCISES D ■ PAGE A32

1. $7\pi/6$ **3.** $\pi/20$ **5.** 5π **7.** $720°$ **9.** $75°$
11. $-67.5°$ **13.** 3π cm   **15.** $\tfrac{2}{3}$ rad $= (120/\pi)°$

17. 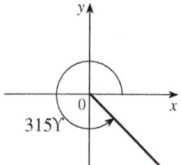   **19.**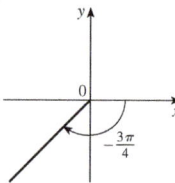

21.

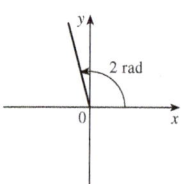

23. $\sin(3\pi/4) = 1/\sqrt{2}$, $\cos(3\pi/4) = -1/\sqrt{2}$, $\tan(3\pi/4) = -1$, $\csc(3\pi/4) = \sqrt{2}$, $\sec(3\pi/4) = -\sqrt{2}$, $\cot(3\pi/4) = -1$
25. $\sin(9\pi/2) = 1$, $\cos(9\pi/2) = 0$, $\csc(9\pi/2) = 1$, $\cot(9\pi/2) = 0$, $\tan(9\pi/2)$ and $\sec(9\pi/2)$ undefined
27. $\sin(5\pi/6) = \tfrac{1}{2}$, $\cos(5\pi/6) = -\sqrt{3}/2$, $\tan(5\pi/6) = -1/\sqrt{3}$, $\csc(5\pi/6) = 2$, $\sec(5\pi/6) = -2/\sqrt{3}$, $\cot(5\pi/6) = -\sqrt{3}$
29. $\cos\theta = \tfrac{4}{5}$, $\tan\theta = \tfrac{3}{4}$, $\csc\theta = \tfrac{5}{3}$, $\sec\theta = \tfrac{5}{4}$, $\cot\theta = \tfrac{4}{3}$
31. $\sin\phi = \sqrt{5}/3$, $\cos\phi = -\tfrac{2}{3}$, $\tan\phi = -\sqrt{5}/2$, $\csc\phi = 3/\sqrt{5}$, $\cot\phi = -2/\sqrt{5}$
33. $\sin\beta = -1/\sqrt{10}$, $\cos\beta = -3/\sqrt{10}$, $\tan\beta = \tfrac{1}{3}$, $\csc\beta = -\sqrt{10}$, $\sec\beta = -\sqrt{10}/3$
35. 5.73576 cm   **37.** 24.62147 cm   **59.** $\tfrac{1}{15}(4 + 6\sqrt{2})$
61. $\tfrac{1}{15}(3 + 8\sqrt{2})$ **63.** $\tfrac{24}{25}$ **65.** $\pi/3, 5\pi/3$
67. $\pi/4, 3\pi/4, 5\pi/4, 7\pi/4$ **69.** $\pi/6, \pi/2, 5\pi/6, 3\pi/2$
71. $0, \pi, 2\pi$ **73.** $0 \leq x \leq \pi/6$ and $5\pi/6 \leq x \leq 2\pi$
75. $0 \leq x < \pi/4$, $3\pi/4 < x < 5\pi/4$, $7\pi/4 < x \leq 2\pi$
77.

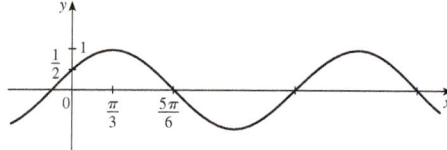

79.

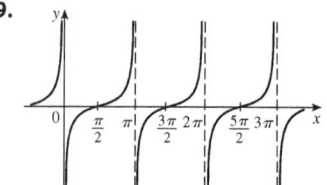

81.

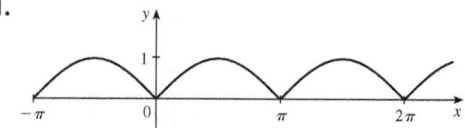

89. 14.34457 cm^2

EXERCISES E ■ PAGE A38

1. $\sqrt{1} + \sqrt{2} + \sqrt{3} + \sqrt{4} + \sqrt{5}$ **3.** $3^4 + 3^5 + 3^6$
5. $-1 + \frac{1}{3} + \frac{3}{5} + \frac{5}{7} + \frac{7}{9}$ **7.** $1^{10} + 2^{10} + 3^{10} + \cdots + n^{10}$
9. $1 - 1 + 1 - 1 + \cdots + (-1)^{n-1}$ **11.** $\sum_{i=1}^{10} i$
13. $\sum_{i=1}^{19} \frac{i}{i+1}$ **15.** $\sum_{i=1}^{n} 2i$ **17.** $\sum_{i=0}^{5} 2^i$ **19.** $\sum_{i=1}^{n} x^i$
21. 80   **23.** 3276   **25.** 0   **27.** 61   **29.** $n(n+1)$
31. $n(n^2 + 6n + 17)/3$ **33.** $n(n^2 + 6n + 11)/3$
35. $n(n^3 + 2n^2 - n - 10)/4$
41. (a) n^4 (b) $5^{100} - 1$ (c) $\frac{97}{300}$ (d) $a_n - a_0$
43. $\frac{1}{3}$ **45.** 14   **49.** $2^{n+1} + n^2 + n - 2$

EXERCISES G ■ PAGE A56

1. (b) 0.405

EXERCISES H ■ PAGE A64

1. $8 - 4i$ **3.** $13 + 18i$ **5.** $12 - 7i$ **7.** $\frac{11}{13} + \frac{10}{13}i$
9. $\frac{1}{2} - \frac{1}{2}i$ **11.** $-i$ **13.** $5i$ **15.** $12 + 5i, 13$
17. $4i, 4$ **19.** $\pm\frac{3}{2}i$ **21.** $-1 \pm 2i$
23. $-\frac{1}{2} \pm (\sqrt{7}/2)i$ **25.** $3\sqrt{2}\,[\cos(3\pi/4) + i\sin(3\pi/4)]$
27. $5\{\cos[\tan^{-1}(\frac{4}{3})] + i\sin[\tan^{-1}(\frac{4}{3})]\}$
29. $4[\cos(\pi/2) + i\sin(\pi/2)]$, $\cos(-\pi/6) + i\sin(-\pi/6)$, $\frac{1}{2}[\cos(-\pi/6) + i\sin(-\pi/6)]$
31. $4\sqrt{2}\,[\cos(7\pi/12) + i\sin(7\pi/12)]$, $(2\sqrt{2})[\cos(13\pi/12) + i\sin(13\pi/12)], \frac{1}{4}[\cos(\pi/6) + i\sin(\pi/6)]$
33. -1024 **35.** $-512\sqrt{3} + 512i$
37. $\pm 1, \pm i, (1/\sqrt{2})(\pm 1 \pm i)$ **39.** $\pm(\sqrt{3}/2) + \frac{1}{2}i, -i$

41. i **43.** $\frac{1}{2} + (\sqrt{3}/2)i$ **45.** $-e^2$
47. $\cos 3\theta = \cos^3\theta - 3\cos\theta\sin^2\theta$,
$\sin 3\theta = 3\cos^2\theta\sin\theta - \sin^3\theta$

Index

RP denotes Reference Page numbers.

Abel, Niels, 211
absolute maximum and minimum, 276, 959, 960, 965
absolute value, 16, A6, A58
absolute value function, 16
absolutely convergent series, 737
acceleration as a rate of change, 159, 224
acceleration of a particle, 871
    components of, 874
    as a vector, 871
Achilles and the tortoise, 5
adaptive numerical integration, 523
addition formulas for sine and cosine, A29
addition of vectors, 798, 801
Airy, Sir George, 752
Airy function, 752
algebraic function, 30
algebraic vector, 800
alternating harmonic series, 734, 737
alternating series, 732
Alternating Series Estimation Theorem, 735
Alternating Series Test, 732
Ampère's Law, 1086
analytic geometry, A10
angle, A24
    between curves, 273
    of deviation, 285
    negative or positive, A25
    between planes, 828
    standard position, A25
    between vectors, 808, 809
angular momentum, 879
angular speed, 872
antiderivative, 350
antidifferentiation formulas, 352
aphelion, 687
apolune, 681
approach path of an aircraft, 208
approximate integration, 514
approximating cylinder, 440
approximating surface, 551
approximation
    by differentials, 254
    to e, 178
    linear, 252, 929, 930, 933

    linear, to a tangent plane, 929
    by the Midpoint Rule, 384, 515
    by Newton's method, 345
    by an nth-degree Taylor polynomial, 258
    quadratic, 258
    by Riemann sums, 379
    by Simpson's Rule, 519, 520
    tangent line, 252
    by Taylor polynomials, 774
    by Taylor's Inequality, 762, 775
    by the Trapezoidal Rule, 516
Archimedes, 411
Archimedes' Principle, 468, 1146
arc length, 544, 861, 862
    of a parametric curve, 652
    of a polar curve, 671
    of a space curve, 861, 862
arc length contest, 550
arc length formula, 545
arc length formula for a space curve, 862
arc length function, 547, 863
arcsine function, 63
area, 2, 366
    of a circle, 488
    under a curve, 366, 371, 378
    between curves, 428, 429
    of an ellipse, 487
    by exhaustion, 2, 97
    enclosed by a parametric curve, 651
    of a plane region, 1099
    in polar coordinates, 658, 669
    of a sector of a circle, 669
    surface, 654, 1026, 1027, 1116, 1117, 1118
    of a surface of a revolution, 551, 557
area function, 391
area problem, 2, 366
argument of a complex number, A59
arithmetic-geometric mean, 706
arrow diagram, 11
astroid, 215, 649
asymptote(s), 316
    in graphing, 316
    horizontal, 128, 316
    of a hyperbola, 678, A20

    slant, 316, 320
    vertical, 90, 316
asymptotic curve, 323
autonomous differential equation, 594
auxiliary equation, 1155
    complex roots of, 1157
    real and distinct roots of, 1156
    real and equal roots of, 1156, 1157
average cost function, 339
average rate of change, 145, 224
average speed of molecules, 535
average value of a function, 461, 575, 997, 1039
average velocity, 4, 81, 143, 224
axes, coordinate, 792, A11
axes of ellipse, A19
axis of a parabola, 674

bacterial growth, 610, 615
Barrow, Isaac, 3, 97, 152, 392, 411
base of a cylinder, 438
base b of a logarithm, 59, A55
    change of, 62
baseball and calculus, 464
basis vectors, 802
Bernoulli, James, 600, 625
Bernoulli, John, 306, 314, 600, 644, 760
Bernoulli differential equation, 625
Bessel, Friedrich, 748
Bessel function, 217, 748, 752
Bézier, Pierre, 657
Bézier curves, 643, 657
binomial coefficients, 766
binomial series, 766
    discovery by Newton, 773
binomial theorem, 173, RP1
binormal vector, 866
blackbody radiation, 783
blood flow, 230, 342, 570
body mass index (BMI), 901, 916
boundary curve, 1134
boundary-value problem, 1159
bounded sequence, 701
bounded set, 965
Boyle's Law, 234
brachistochrone problem, 644

Brahe, Tycho, 875
branches of a hyperbola, 678, A20
Buffon's needle problem, 584
bullet-nose curve, 205

C^1 transformation, 1053
cable (hanging), 259
calculator, graphing, 323, 642, 665
    See also computer algebra system
calculus, 8
    differential, 3
    integral, 2, 3
    invention of, 8, 411
cancellation equations
    for inverse functions, 57
    for inverse trigonometric functions, 58, 64
    for logarithms, 59
cans, minimizing manufacturing cost of, 343
Cantor, Georg, 718
Cantor set, 718
capital formation, 573
cardiac output, 571
cardioid, 215, 662
carrying capacity, 236, 297, 313, 587, 612
Cartesian coordinate system, A11
Cartesian plane, A11
Cassini, Giovanni, 669
CAS. *See* computer algebra system
catenary, 259
Cauchy, Augustin-Louis, 109, 994, A45
Cauchy-Schwarz Inequality, 814
Cauchy's Mean Value Theorem, A45
Cavalieri, 520
Cavalieri's Principle, 448
center of gravity, 560. *See also* center of mass
center of mass, 560
    of a lamina, 1018
    of a plate, 563
    of a solid, 1035
    of a surface, 1124
    of a wire, 1077
centripetal acceleration, 884
centripetal force, 884
centroid of a plane region, 562
centroid of a solid, 1036
Chain Rule, 197, 198, 200
    for several variables, 937, 938, 939, 940
change of base, formula for, 62
change of variables
    in a double integral, 1012, 1052, 1056
    in integration, 412
    in a triple integral, 1042, 1047, 1058, 1059

characteristic equation, 1155
charge, electric, 227, 1016, 1036, 1172
charge density, 1016, 1036
chemical reaction, 227
circle, A16
    area of, 488
    equation of, A16
    in three-dimensional space, 794
circle of curvature, 867
circuit, electric, 1172
circular cylinder, 438
circulation of a velocity field, 1138
cissoid of Diocles, 648, 667
Clairaut, Alexis, 919
Clairaut's Theorem, 919, A48
Clarke, Arthur C., 881
Clarke geosynchronous orbit, 881
clipping planes, 833
closed curve, 1089
closed interval, A3
Closed Interval Method, 281
    for a function of two variables, 966
closed set, 965
closed surface, 1128
Cobb, Charles, 889
Cobb-Douglas production function, 889, 890, 922, 926, 978
    graph of, 891
    level curves for, 896
cochleoid, 690
coefficient(s)
    binomial, 766
    of friction, 197, 284
    of inequality, 437
    of a polynomial, 27
    of a power series, 746
    of static friction, 844
combinations of functions, 40
comets, orbits of, 688
common ratio, 709
comparison properties of the integral, 387
comparison test for improper integrals, 533
Comparison Test for series, 727
Comparison Theorem for integrals, 533
complementary equation, 1161
Completeness Axiom, 702
complex conjugate, A57
complex exponentials, A63
complex number(s), A57
    addition and subtraction of, A57
    argument of, A59
    division of, A57, A60
    equality of, A57
    imaginary part of, A57

    modulus of, A58
    multiplication of, A57, A60
    polar form, A59
    powers of, A61
    principal square root of, A58
    real part of, A57
    roots of, A62
component of b along a, 811
component function, 848, 1069
components of acceleration, 874
components of a vector, 800
composition of functions, 41, 198
    continuity of, 121, 909
    derivative of, 199
compound interest, 241, 313
compressibility, 228
computer algebra system, 86, 511, 642
    for graphing sequences, 699
    for integration, 511, 756
    pitfalls of using, 86
computer algebra system, graphing with
    curve, 323
    function of two variables, 892
    level curves, 897
    parametric equations, 642
    parametric surface, 1114
    partial derivatives, 919
    polar curve, 665
    sequence, 699
    space curve, 851
    vector field, 1070, 1071
concavity, 296
Concavity Test, 296, A44
concentration, 227
conchoid, 645, 667
conditionally convergent series, 738
conductivity (of a substance), 1132
cone, 674, 837
    parametrization of, 1114
conic section, 674, 682
    directrix, 674, 682
    eccentricity, 682
    focus, 674, 676, 682
    polar equation, 684
    shifted, 679, A21
    vertex (vertices), 674
conjugates, properties of, A58
connected region, 1089
conservation of energy, 1093
conservative vector field, 1073, 1090, 1091, 1105
constant force (in work), 455, 811
constant function, 172
Constant Multiple Law of limits, 95
Constant Multiple Rule, 175

constraint, 971, 976
consumer surplus, 569
continued fraction expansion, 706
continuity
    of a function, 114, 849
    of a function of three variables, 909, 910
    of a function of two variables, 907, 908, 910
    on an interval, 117
    from the left, 116
    from the right, 116
    of a vector function, 849
continuous compounding of interest, 241, 313
continuous random variable, 573
contour curves, 893
contour map, 894, 895, 921
convergence
    absolute, 737
    conditional, 738
    of an improper integral, 528, 531
    interval of, 749
    radius of, 749
    of a sequence, 696
    of a series, 708
convergent improper integral, 528, 531
convergent sequence, 696
convergent series, 708
    properties of, 713
conversion of coordinates
    cylindrical to rectangular, 1040
    rectangular to cylindrical, 1040
    rectangular to spherical, 1046
cooling tower, hyperbolic, 839
coordinate axes, 792, A11
coordinate planes, 792
coordinate system, A2
    Cartesian, A11
    cylindrical, 1040
    polar, 658
    retangular, A11
    spherical, 1045
    three-dimensional rectangular, 792, 793
coplanar vectors, 820
Coriolis acceleration, 883
Cornu's spiral, 656
cosine function, A26
    derivative of, 192
    graph of, 31, A31
    power series for, 764, 766
cost function, 231, 335
critical number, 280
critical point(s), 960, 970
critically damped vibration, 1170

cross product, 814, 815
    direction of, 816
    geometric characterization of, 817
    length of, 817, 818
    magnitude of, 817
    properties of, 816, 819
cross-section, 438
cross-section of a surface, 834
cubic function, 28
curl of a vector field, 1103
current, 227
curvature, 657, 864, 865, 875
curvature of a plane parametric curve, 869
curve(s)
    asymptotic, 323
    Bézier, 643, 657
    boundary, 1134
    bullet-nose, 205
    cissoid of Diocles, 667
    closed, 1089
    Cornu's spiral, 656
    demand, 569
    devil's, 215
    dog saddle, 902
    epicycloid, 649
    equipotential, 902
    grid, 1112
    helix, 849
    length of, 544, 861
    level, 893, 897
    long-bow, 691
    monkey saddle, 902
    orientation of, 1080, 1096
    orthogonal, 216
    ovals of Cassini, 669
    parametric, 640, 849
    piecewise-smooth, 1076
    polar, 660
    serpentine, 188
    simple, 1090
    smooth, 544, 863
    space, 849
    strophoid, 673, 691
    swallotail catastrophe, 648
    toroidal spiral, 851
    trochoid, 647
    twisted cubic, 851
    witch of Maria Agnesi, 647
curve fitting, 25
curve-sketching procedure, 315
cusp, 645
cycloid, 643
cylinder, 438, 794, 834
    parabolic, 834
    parametrization of, 1114

cylindrical coordinate system, 1040
    conversion equations for, 1040
    triple integrals in, 1042
cylindrical coordinates, 1040
cylindrical shell, 449

damped vibration, 1169
damping constant, 1169
damping force, 1169, 1171
decay, law of natural, 237
decay, radioactive, 239
decreasing function, 19
decreasing sequence, 700
definite integral, 378, 988
    properties of, 385
    Substitution Rule for, 416
    of a vector function, 859
definite integration
    by parts, 472, 474, 475
    by substitution, 416
degree of a polynomial, 27
del (∇), 949, 951
delta (Δ) notation, 145
demand curve, 336, 569
demand function, 336, 569
De Moivre, Abraham, A61
De Moivre's Theorem, A61
density
    of a lamina, 1016
    linear, 226, 406
    liquid, 558
    mass vs. weight, 558
    of a solid, 1035
dependent variable, 10, 888, 940
derivative(s), 140, 144, 152, 257
    of a composite function, 198
    of a constant function, 172
    directional, 946, 947, 948, 950, 951
    domain of, 153
    of exponential functions, 179, 202, A54, A55
    as a function, 152
    higher, 158
    higher partial, 918
    of hyperbolic functions, 261
    of an integral, 393
    of an inverse function, 218
    of inverse trigonometric functions, 213, 214
    left-hand, 165
    of logarithmic functions, 218, A51, A54
    normal, 1110
    notation, 155
    notation for partial, 914
    partial, 913, 914
    of a polynomial, 172

derivative(s) (*continued*)
  of a power function, 173
  of a power series, 754
  of a product, 183, 184
  of a quotient, 186
  as a rate of change, 140
  right-hand, 165
  second, 158, 858
  second directional, 958
  second partial, 918
  as the slope of a tangent, 141, 146
  third, 159
  of trigonometric functions, 190, 193
  of a vector function, 855, 856, 858
Descartes, René, A11
descent of aircraft, determining start of, 208
determinant, 815
devil's curve, 215
Difference Law of limits, 95
difference of vectors, 799
difference quotient, 12
Difference Rule, 176
differentiable function, 155, 930
differential, 254, 932, 934
differential calculus, 3
differential equation, 181, 237, 353, 585, 586, 588
  autonomous, 594
  Bernoulli, 625
  family of solutions, 586, 589
  first-order, 588
  general solution of, 589
  homogeneous, 1154
  linear, 620
  linearly independent solutions, 1155
  logistic, 612, 707
  nonhomogeneous, 1154, 1160, 1161
  order of, 588
  partial, 920
  second-order, 588, 1154
  separable, 599
  solution of, 588
differentiation, 155
  formulas for, 187, RP5
  formulas for vector functions, 858
  implicit, 208, 209, 917, 942
  logarithmic, 220
  partial, 911, 913, 914, 917
  of a power series, 754
  term-by-term, 754
  of a vector function, 855, 856, 858
differentiation operator, 155
diffusion equation, 926
Direct Substitution Property, 97
directed line segment, 798

direction angles, 810
direction cosines 810
direction field, 592
direction numbers, 825
directional derivative, 946, 947, 948, 950, 951
  maximum value of, 952
  second, 958
directrix, 674, 682
discontinuity, 115, 116
discontinuous function, 115
discontinuous integrand, 531
disk method for approximating volume, 440
dispersion, 286
displacement, 143, 406
displacement vector, 798, 811
distance
  between parallel planes, 830, 833
  between point and line in space, 822
  between point and plane, 822, 829, 830
  between points in a plane, A11
  between points in space, 795
  between real numbers, A7
  between skew lines, 830
distance formula, A12
  in three dimensions, 795
distance problem, 373
divergence
  of an improper integral, 528, 531
  of an infinite series, 708
  of a sequence, 696
  of a vector field, 1106
Divergence, Test for, 713
Divergence Theorem, 1141, 1147
divergent improper integral, 528, 531
divergent sequence, 696
divergent series, 708
division of power series, 770
DNA, helical shape of, 850
dog saddle, 902
domain of a function, 10, 888
domain sketching, 888
Doppler effect, 945
dot product, 800
  in component form, 807
  properties of, 807
  in vector form, 808
double-angle formulas, A29
double helix, 850
double integral(s), 988, 990
  applications of, 1016
  change of variable in, 1012, 1052, 1056
  over general regions, 1001

  Midpoint Rule for, 992
  in polar coordinates, 1010, 1012
  properties of, 1006, 1007
  over rectangles, 988
double Riemann sum, 991
Douglas, Paul, 889
Dumpster design, minimizing cost of, 970
dye dilution method, 571

e (the number), 51, 178, A52
  as a limit, 222, A56
  as a sum of an infinite series, 763
eccentricity, 682
electric charge, 599, 602, 623, 1016, 1036
  on a solid, 1036
electric circuit, 599, 602, 623
  analysis of, 1172
electric current to a flash bulb, 79, 206
electric field (force per unit charge), 1072
electric flux, 1131, 1144
electric force, 1072
elementary function, integrability of, 507
element of a set, A3
ellipse, 215, 676, 682, A19
  area, 487
  directrix, 682
  eccentricity, 682
  foci, 676, 682
  major axis, 676, 687
  minor axis, 676
  polar equation, 684, 687
  reflection property, 677
  rotated, 217
  vertices, 676
ellipsoid, 835, 837
elliptic paraboloid, 836, 837
empirical model, 25
end behavior of a function, 139
endpoint extreme values, 277
energy,
  conservation of, 1093
  kinetic, 464, 1093
  potential, 1093
epicycloid, 649
epitrochoid, 656
equation(s)
  cancellation, 57
  of a circle, A17
  differential. (*see* differential equation)
  diffusion, 926
  of an ellipse, 676, 684, A19
  of a graph, A16, A17
  heat conduction, 925
  of a hyperbola, 67, 679, 684, A20
  Laplace's, 920, 1107
  of a line, A12, A13, A14, A16

of a line in space, 824
of a line through two points, 825
linear, 827, A14
logistic difference, 707
logistic differential, 587, 619
Lotka-Volterra, 627
nth-degree, 211
of a parabola, 674, 684, A18
parametric, 640, 824, 849, 1110
of a plane, 827
of a plane through three points, 828
point-slope, A12
polar, 660, 684
predator-prey, 627
second-degree, A16
slope-intercept, A13
of a space curve, 849
of a sphere, 795
symmetric, 824
two-intercept form, A16
van der Waals, 926
vector, 824, 827
wave, 920
equilateral hyperbola, A21
equilibrium point, 629
equilibrium solution, 587, 628
equipotential curves, 902
equivalent vectors, 798
error
in approximate integration, 516, 517
percentage, 256
relative, 256
in Taylor approximation, 775
error bounds, 518, 522
error estimate
for alternating series, 735
for the Midpoint Rule, 516, 517
for Simpson's Rule, 522
for the Trapezoidal Rule, 516, 517
error function, 401
escape velocity, 535
estimate of the sum of a series, 723, 730, 735, 740
Euclid, 97
Eudoxus, 2, 97, 411
Euler, Leonhard, 52, 597, 720, 726, 763
Euler's formula, A63
Euler's Method, 595
even function, 17, 315
expected values, 1023
exponential decay, 237
exponential function(s), 32, 45, 177, A52, A54, RP4
with base b, A54
derivative, of 179, 202, A55
graphs of, 47, 179

integration of, 383, 413, 768, 769
limits of, 131, A53
power series for, 761
properties of, A53
exponential graph, 46
exponential growth, 237, 615
exponents, laws of, 47, A53, A55
extrapolation, 27
extreme value, 276
Extreme Value Theorem, 278, 965

family
of epicycloids and hypocycloids, 648
of exponential functions, 46
of functions, 29, 327, 328
of parametric curves, 644
of solutions, 586, 589
fat circles, 213, 550
Fermat, Pierre, 3, 152, 279, 411, A11
Fermat's Principle, 340
Fermat's Theorem, 279
Fibonacci, 695, 706
Fibonacci sequence, 695, 706
field
conservative, 1073, 1090, 1091, 1105
electric, 1072
force, 1072
gradient, 956, 1072
gravitational, 1072
incompressible, 1107
irrotational, 1106
scalar, 1069
vector, 1068, 1069
velocity, 1068, 1071
first-degree Taylor polynomial, 970
First Derivative Test, 294
for Absolute Extreme Values, 333
first octant, 792
first-order linear differential equation, 588, 620
first-order optics, 780
fixed point of a function, 170, 292
flash bulb, current to, 79
flow lines, 1074
fluid flow, 1071, 1106, 1107, 1130
flux, 570, 1129, 1131
flux integral, 1129
FM synthesis, 327
foci, 676
focus, 674, 682
of a conic section, 682
of an ellipse, 676, 682
of a hyperbola, 677
of a parabola, 674
folium of Descartes, 209, 691

force, 455
centripetal, 884
constant, 455, 811
exerted by fluid, 558
resultant, 803
torque, 820, 879
force field, 1068, 1072
forced vibrations, 1171
Fourier, Joseph, 233
Fourier series, finite, 485
four-leaved rose, 662
fractions (partial), 493, 494
Frenet-Serret formulas, 870
Fresnel, Augustin, 395
Fresnel function, 395
frustum, 447, 448
Fubini, Guido, 994
Fubini's Theorem, 994, 1030
function(s), 10, 888
absolute value, 16
Airy, 752
algebraic, 30
arc length, 547, 863
arcsine, 64
area, 391
arrow diagram of, 11
average cost, 339
average value of, 461, 575, 997, 1039
Bessel, 217, 748, 752
Cobb-Douglas production, 889, 890, 922, 926, 978
combinations of, 40
component, 848, 1069
composite, 41, 198, 909
constant, 172
continuity of, 114, 849, 907, 908, 909, 910
continuous, 849
cost, 230, 231
cubic, 28
decreasing, 19
demand, 336, 569
derivative of, 144
differentiability of, 155, 930
discontinuous, 115
domain of, 10, 888
elementary, 507
error, 401
even, 17, 315
exponential, 32, 45, 177, A54
extreme values of, 276
family of, 29, 327, 328
fixed point of, 170, 292
Fresnel, 395
Gompertz, 617, 620
gradient of, 936, 950

function(s) (*continued*)
  graph of, 11, 890
  greatest integer, 101
  harmonic, 920, 1110
  Heaviside, 45, 87
  homogeneous, 946
  hyperbolic, 259
  implicit, 209
  increasing, 19
  inverse, 55, 56
  inverse cosine, 64
  inverse hyperbolic, 261
  inverse sine, 63
  inverse tangent, 65
  inverse trigonometric, 63, 64
  joint density, 1021, 1036
  limit of, 83, 105, 904, 909, 910
  linear, 24, 891
  logarithmic, 32, 59, A50, A55
  machine diagram of, 11
  marginal cost, 146, 231, 335, 406
  marginal profit, 336
  marginal revenue, 336
  maximum and minimum values of, 276, 959, 960
  of *n* variables, 898
  natural exponential, 52, A52
  natural logarithmic, 60, A50
  nondifferentiable, 157
  odd, 17, 315
  one-to-one, 56
  periodic, 316
  piecewise defined, 15
  polynomial, 27, 908
  position, 142
  power, 29, 172
  probability density, 574, 1021
  profit, 336
  quadratic, 27
  ramp, 45
  range of, 10, 888
  rational, 30, 493, 908
  reciprocal, 30
  reflected, 37
  representation as a power series, 752
  representations of, 10, 12
  revenue, 336
  root, 29
  of several variables, 898, 909
  shifted, 37
  sine integral, 401
  smooth, 544
  step, 17
  stretched, 37
  tabular, 13
  of three variables, 897, 909
  transformation of, 36
  translation of, 36
  trigonometric, 31, A26
  of two variables, 888
  value of, 10, 11
  vector, 848
Fundamental Theorem of Calculus, 392, 394, 398
  higher-dimensional versions, 1147
  for line integrals, 1087, 1147
  for vector functions, 859

G (gravitational constant), 234, 460
Gabriel's horn, 556
Galileo, 644, 651, 674
Galois, Evariste, 211
Gause, G. F., 615
Gauss, Karl Friedrich, 1141, A35
Gaussian optics, 780
Gauss's Law, 1131, 1144
Gauss's Theorem, 1141
geometric series, 709
geometric vector, 800
geometry of a tetrahedron, 823
geosynchronous orbit, 881
Gibbs, Joseph Willard, 803
Gini, Corrado, 436
Gini coefficient, 437
Gini index, 437
global maximum and minimum, 276
Gompertz function, 617, 620
grad *f*, 949, 951
gradient, 950
gradient vector, 949, 951, 955
  interpretations of, 955
gradient vector field, 956, 1072
graph(s)
  of an equation, A16, A17
  of equations in three dimensions, 793
  of exponential functions, 46, 179, RP4
  of a function, 11
  of a function of two variables, 890
  of logarithmic functions, 60, 62
  of a parametric curve, 640
  of a parametric surface, 1124
  polar, 660, 665
  of power functions, 29, RP3
  of a sequence, 699
  of a surface, 1124
  of trigonometric functions, 31, A30, RP2
graphing calculator, 323, 642, 665
graphing device. *See* computer algebra system
gravitation law, 234, 460
gravitational acceleration, 455
gravitational field, 1072

great circle, 1051
greatest integer function, 101
Green, George, 1096, 1140
Green's identities, 1110
Green's Theorem, 1096, 1140, 1147
  vector forms, 1108
  for a union of simple regions, 1099
Gregory, James, 198, 483, 520, 756, 760
Gregory's series, 756
grid curves, 1112
growth, law of natural, 237, 611
growth rate, 229, 406
  relative, 237, 611

half-angle formulas, A29
half-life, 50, 239
half-space, 898
Hamilton, Sir William Rowan, 815
hare-lynx system, 631
harmonic function, 920, 1110
harmonic series, 713, 722
harmonic series, alternating, 734
heat conduction equation, 925
heat conductivity, 1132
heat flow, 1131
heat index, 911, 931
Heaviside, Oliver, 87
Heaviside function, 45, 87
Hecht, Eugene, 254, 257, 779
helix, 849
hidden line rendering, 834
higher derivatives, 158
higher partial derivatives, 918
homogeneous differential equation, 1154
homogeneous function, 946
Hooke's Law, 457, 1168
horizontal asymptote, 128, 316
horizontal line, equation of, A13
Horizontal Line Test, 56
horizontal plane, 793
Hubble Space Telescope, 282
humidex, 899, 911
Huygens, Christiaan, 644
hydro-turbine optimization, 980
hydrostatic pressure and force, 558
hyperbola, 215, 677, 682, A20
  asymptotes, 678, A20
  branches, 678, A20
  directrix, 682
  eccentricity, 682
  equation, 678, 679, 684, A20
  equilateral, A21
  foci, 677, 682
  polar equation, 684
  reflection property, 682
  vertices, 678

hyperbolic function(s), 259
    derivatives of, 261
    inverse, 261
hyperbolic identities, 260
hyperbolic paraboloid, 836, 837
hyperbolic substitution, 489, 490
hyperboloid, 791, 837
hypersphere, volume of, 1040
hypervolume, 1034
hypocycloid, 648

i (imaginary number), A57
i (standard basis vector), 802
I/D Test, 293
ideal gas law, 236
image of a point, 1053
image of a region, 1053
implicit differentiation, 208, 209, 917, 942
implicit function, 208, 209
Implicit Function Theorem, 942, 943
improper integral, 527
    convergence or divergence of, 528, 531
impulse of a force, 464
incompressible velocity field, 1107
increasing function, 19
increasing sequence, 700
Increasing/Decreasing Test, 293
increment, 145, 933
indefinite integral(s), 403
    table of, 403
independence of path, 1088
independent random variable, 1022
independent variable, 10, 888, 940
indeterminate difference, 309
indeterminate forms of limits, 304
indeterminate power, 310
indeterminate product, 308
index of summation, A34
inequalities, rules for, A4
inertia (moment of), 1019, 1020, 1036, 1086
infinite discontinuity, 116
infinite interval, 527, 528
infinite limit, 89, 112, 132
infinite sequence. *See* sequence
infinite series. *See* series
inflection point, 297
initial condition, 589
initial point
    of a parametric curve, 641
    of a vector, 798
initial-value problem, 589, 1158
inner product, 807
instantaneous rate of change, 80, 145, 224
instantaneous rate of growth, 229
instantaneous rate of reaction, 227

instantaneous velocity, 81, 143, 224
integer, A2
integrable function, 990
integral(s)
    approximations to, 384
    change of variables in, 412, 1012, 1052, 1056, 1058, 1059
    comparison properties of, 387
    conversion to cylindrical coordinates, 1040
    conversion to polar coordinates, 1012
    conversion to spherical coordinates, 1046
    definite, 378, 988
    derivative of, 394
    double (*see* double integral)
    evaluating, 381
    improper, 527
    indefinite, 402
    iterated, 993
    line (*see* line integral)
    patterns in, 513
    properties of, 385
    surface, 1122, 1129
    of symmetric functions, 417
    table of, 471, 503, 509, RP6–10
    triple, 1029, 1030
    units for, 408
integral calculus, 2, 3
Integral Test, 721
integrand, 379
    discontinuous, 531
integration, 379
    approximate, 514
    by computer algebra system, 511
    of exponential functions, 383, 413
    formulas, 471, 503, RP6–10
    indefinite, 402
    limits of, 379
    numerical, 514
    partial, 993, 995
    by partial fractions, 493
    by parts, 472, 473, 474
    of a power series, 754
    of rational functions, 493
    by a rationalizing substitution, 500
    reversing order of, 995, 1006
    over a solid, 1042
    substitution in, 412
    tables, use of, 508
    term-by-term, 754
    of a vector function, 859
intercepts, 315, A19
interest compunded continuously, 241
Intermediate Value Theorem, 123
intermediate variable, 940

interpolation, 27
intersection
    of planes, 828
    of polar graphs, area of, 670
    of sets, A3
    of three cylinders, 1044
interval, A3
interval of convergence, 749
inverse cosine function, 64
inverse function(s), 55, 56
inverse sine function, 63
inverse square laws, 36
inverse tangent function, 65
inverse transformation, 1053
inverse trigonometric functions, 63, 64
irrational number, A2
irrotational vector field, 1106
isobar, 895
isothermal, 893
isothermal compressibility, 228
iterated integral, 993

j (standard basis vector), 802
Jacobi, Carl Gustav Jacob, 1055
Jacobian of a transformation, 1055, 1059
jerk, 160
joint density function, 1021, 1036
joule, 455
jump discontinuity, 116

k (standard basis vector), 802
kampyle of Eudoxus, 215
Kepler, Johannes, 686, 875, 880
Kepler's Laws, 686, 875, 876, 880
kinetic energy, 464, 1093
Kirchhoff's Laws, 593, 1172
Kondo, Shigeru, 763

Lagrange, Joseph-Louis, 288, 289, 972
Lagrange multiplier, 971, 972
lamina, 562, 1016, 1017
Laplace, Pierre, 920, 1107
Laplace operator, 1107
Laplace's equation, 920, 1107
lattice point, 274
law of conservation of angular momentum, 879
Law of Conservation of Energy, 1094
law of cosines, A33
law of gravitation, 460
law of laminar flow, 230, 570
law of natural growth or decay, 237
aw of universal gravitation, 876, 880
laws of exponents, 47, A53, A55
laws of logarithms, 60, A51
learning curve, 591

least squares method, 26, 970
least upper bound, 702
left-hand derivative, 165
left-hand limit, 88, 109
Leibniz, Gottfried Wilhelm, 3, 155, 392, 411, 600, 773
Leibniz notation, 155
lemniscate, 215
length
   of a curve, 544
   of a line segment, A7, A12
   of a parametric curve, 652
   of a polar curve, 671
   of a space curve, 861
   of a vector, 801
level curve(s), 893, 897
level surface, 898
   tangent plane to, 954
l'Hospital, Marquis de, 306, 314
l'Hospital's Rule, 305, 314, A45
   origins of, 314
libration point, 350
limaçon, 665
limit(s), 2, 83
   calculating, 95
   e (the number) as, 222
   of exponential functions, 132, A53
   of a function, 83, 106
   of a function of three variables, 909, 910
   of a function of two variables, 904, 910
   infinite, 89, 112, 132
   at infinity, 126, 127, 132
   of integration, 379
   left-hand, 88, 109
   of logarithmic functions, 91, A50, A52
   one-sided, 88, 109
   precise definitions, 104, 109, 112, 134, 137
   properties of, 95
   properties of, for vector functions, 855
   right-hand, 88, 109
   of a sequence, 5, 368, 696
   involving sine and cosine functions, 190, 191, 192
   of a trigonometric function, 192
   of a vector function, 848
Limit Comparison Test, 729
Limit Laws, 95, A39
   for functions of two variables, 907
   for sequences, 697
line(s) in the plane, 78, A12
   equations of, A12, A13, A14
   horizontal, A13
   normal, 175

   parallel, A14
   perpendicular, A14
   secant, 78, 79
   slope of, A12
   tangent, 78, 79, 141
line(s) in space
   normal, 954
   parametric equations of, 824
   skew, 826
   symmetric equations of, 824
   tangent, 856
   vector equation of, 823, 824
line integral, 1075, 1078
   Fundamental Theorem for, 1087
   for a plane curve, 1075
   with respect to arc length, 1075, 1078, 1080
   with respect to x and y, 1078, 1081
   for a space curve, 1080
   work defined as, 1082
   of vector fields, 1082, 1083, 1084
linear approximation, 252, 929, 930, 933
linear combination, 1154
linear density, 226, 406
linear differential equation, 620, 1154
linear equation, A14
   of a plane, 827
linear function, 24, 891
linear model, 24
linear regression, 26
linearization, 252, 929, 930
linearly independent solutions, 1155
liquid force, 558
Lissajous figure, 642, 648
lithotripsy, 677
local maximum and minimum, 276, 959, 960
logarithm(s), 32, 59
   laws of, 60, A51
   natural, 60, A50
   notation for, 60
logarithmic differentiation, 220
logarithmic function(s), 32, 59, A48
   with base b, 59, A55
   derivatives of, 218, A56
   graphs of, 60, 62
   limits of, 91, A52
   properties of, 59, 60, A51
logistic difference equation, 707
logistic differential equation, 587, 612
logistic model, 587, 611
logistic sequence, 707
long-bow curve, 691
LORAN system, 681
Lorenz curve, 436

Lotka-Volterra equations, 627
lower sum, 371
LZR Racer, 887, 936

machine diagram of a function, 11
Maclaurin, Colin, 760
Maclaurin series, 759, 760
   table of, 768
magnetic field strength of the earth, 921
magnitude of a vector, 801
major axis of ellipse, 676
marginal cost function, 146, 231, 335, 406
marginal productivity, 922
marginal profit function, 336
marginal propensity to consume or save, 717
marginal revenue function, 336
mass
   of a lamina, 1016
   of a solid, 1035
   of a surface, 1124
   of a wire, 1077
mass, center of. *See* center of mass
mathematical induction, 72, 74, 703
   principle of, 72, 74, A36
mathematical model, 13, 23
   for vibration of membrane, 748
maximum and minimum values, 276, 959, 960
mean life of an atom, 536
mean of a probability density function, 576
Mean Value Theorem, 287, 288
   for double integrals, 1063
   for integrals, 462
mean waiting time, 576
median of a probability density function, 578
method of cylindrical shells, 449
method of exhaustion, 2, 97
method of Lagrange multipliers, 971, 972
   with two constraints, 976
method of least squares, 26, 970
method of undetermined coefficients, 1161, 1165
method of variation of parameters, 1165, 1166
midpoint formula, A16
   for points in space, 979
Midpoint Rule, 384, 515
   for double integrals, 992
   error in using, 516
   for triple integrals, 1038
minor axis of ellipse, 676

mixing problems, 604
Möbius, August, 1127
Möbius strip, 1121, 1127
model(s), mathematical, 13, 23
    Cobb-Douglas, for production costs, 889, 890, 922, 926, 978
    comparison of natural growth vs. logistic, 615
    of electric current, 593
    empirical, 25
    exponential, 32, 48
    Gompertz function, 617, 620
    linear, 24
    logarithmic, 32
    polynomial, 28
    for population growth, 237, 586, 617
    power function, 29
    predator-prey, 627
    rational function, 30
    seasonal-growth, 620
    trigonometric, 31, 32
    von Bertalanffy, 635
modeling
    with differential equations, 586
    motion of a spring, 587
    population growth, 48, 237, 586, 611, 617, 635
    vibration of membrane, 748
modulus, A58
moment(s)
    about an axis, 561, 1017
    of inertia, 1019, 1020, 1036, 1086
    of a lamina, 562, 1017, 1018
    of a mass, 561
    about a plane, 1035
    polar, 1020
    second, 1019
    of a solid, 1035
    of a system of particles, 561
momentum of an object, 464
monkey saddle, 902
monotonic sequence, 700
Monotonic Sequence Theorem, 702
motion in space, 870
motion of a projectile, 872
motion of a spring, force affecting
    damping, 1169, 1171
    resonance, 1172
    restoring, 1169, 1171
movie theater seating, 465
multiple integrals. *See* double integral; triple integral(s)
multiplication, scalar, 799, 801
multiplication of power series, 770
multiplier (Lagrange), 971, 972, 975
multiplier effect, 717

natural exponential function, 52, 179, A52
    derivative of, 179, A54
    graph of, 179
    power series for, 760
    properties of, A53
natural growth law, 237, 611
natural logarithm function, 60, A50
    derivative of, 218, A51
    limits of, A51
    properties of, A51
n-dimensional vector, 802
negative angle, A25
negative of a vector, 799
net area, 379
Net Change Theorem, 406
net investment flow, 573
newton (unit of force), 455
Newton, Sir Isaac, 3, 8, 97, 152, 155, 392, 411, 773, 876, 880
Newton's Law of Cooling, 240, 591
Newton's Law of Gravitation, 234, 460, 876, 880, 1071
Newton's method, 345
Newton's Second Law of Motion, 455, 464, 872, 876, 880, 1168
Nicomedes, 645
nondifferentiable function, 157
nonhomogeneous differential equation, 1154, 1160, 1161
nonparallel planes, 828
normal component of acceleration, 874, 875
normal derivative, 1110
normal distribution, 578
normal line, 175
    to a surface, 954
normal plane, 867
normal vector, 827, 866
normally distributed random variable, probability density function of, 1024
nth-degree equation, finding roots of, 211
nth-degree Taylor polynomial, 258, 761
n-tuple, 802
nuclear reactor, cooling towers of, 839
number
    complex, A57
    integer, A2
    irrational, A2
    rational, A2
    real, A2
numerical integration, 514

O (origin), 792
octant, 792

odd function, 17, 315
one-sided limits, 87, 109
one-to-one function, 56
one-to-one transformation, 1053
open interval, A3
open region, 1089
optics
    first-order, 780
    Gaussian, 780
    third-order, 780
optimization problems, 276, 330
orbit of a planet, 876
order of a differential equation, 588
order of integration, reversed, 995, 1006
ordered pair, A10
ordered triple, 792
Oresme, Nicole, 713
orientation of a curve, 1080, 1096
orientation of a surface, 1127
oriented surface, 1127
origin, 792, A2, A10
orthogonal curves, 216
orthogonal projection of a vector, 813
orthogonal surfaces, 959
orthogonal trajectory, 216, 603
orthogonal vectors, 809
osculating circle, 867
osculating plane, 867
Ostrogradsky, Mikhail, 1141
ovals of Cassini, 669
overdamped vibration, 1170

Pappus, Theorem of, 565
Pappus of Alexandria, 565
parabola, 674, 682, A18
    axis, 674
    directrix, 674
    equation, 674, 675
    focus, 674, 682
    polar equation, 684
    reflection property, 274
    vertex, 674
parabolic cylinder, 834
paraboloid, 836, 839
paradoxes of Zeno, 5
parallel lines, A14
parallel planes, 828
parallel vectors, 799, 817
parallelepiped, 438
    volume of, 820
Parallelogram Law, 798, 814
parameter, 640, 824, 849
parametric curve, 640, 849
    arc length of, 652
    area under, 651

parametric equations, 640, 824, 849
    of a line in space, 824
    of a space curve, 849
    of a surface, 1110
    of a trajectory, 873
parametric surface, 1110
    graph of, 1124
    smooth, 1116
    surface area of, 1116, 1117
    surface integral over, 1122, 1123
    tangent plane to, 1115, 1116
parametrization of a space curve, 862
    with respect to arc length, 863
    smooth, 863
paraxial rays, 254
partial derivative(s), 913, 914
    of a function of more than two
        variables, 917
    interpretations of, 915
    notations for, 914
    as a rate of change, 915
    rules for finding, 914
    second, 918
    as slopes of tangent lines, 915
partial differential equation, 920
partial differentiation, 911, 913, 914, 917
partial fractions, 493, 494
partial integration, 472, 473, 474
    for double integrals, 993, 995
partial sum of a series, 708
particle, motion of, 870
parts, integration by, 472, 473, 474
pascal (unit of pressure), 558
path, 1088
patterns in integrals, 513
pendulum, approximating the period
    of, 254, 257
percentage error, 256
perihelion, 687
perilune, 681
period, 316
period of a particle, 884
periodic function, 316
perpendicular lines, A14
perpendicular vectors, 809
phase plane, 629
phase portrait, 629
phase trajectory, 629
piecewise defined function, 15
piecewise-smooth curve, 1076
Planck's Law, 783
plane(s), 826
    angle between, 828
    coordinate, 792
    equation(s) of, 823, 827, 828
    equation of, through three points, 828

horizontal, 793
line of intersection, 829
linear equation of, 827
normal, 867
osculating, 867
parallel, 828
scalar equation of, 827
tangent to a surface, 928, 1115
vector equation of, 827
vertical, 888
plane region of type I, 1002
plane region of type II, 1002
planetary motion, laws of, 686, 875,
    876, 880
planimeter, 1099
point of inflection, 297
point-slope equation of a line, A12
point(s) in space
    coordinates of, 792
    distance between, 794, 795
    projection of, 793
Poiseuille, Jean-Louis-Marie, 230
Poiseuille's Laws, 257, 342, 571
polar axis, 658
polar coordinate system, 658
    area in, 669
    conic sections in, 682
    conversion of double integral to,
        1012, 998
    conversion equations for Cartesian
        coordinates, 659, 660
polar curve, 660
    arc length of, 671
    graph of, 660
    symmetry in, 663
    tangent line to, 663
polar equation(s), 660
    of a conic, 684
    graph of, 660
polar form of a complex number, A59
polar graph, 660
polar moment of inertia, 1020
polar rectangle, 1010
polar region, area of, 669
pole, 658
polynomial, 27
polynomial function, 27
    of two variables, 908
population growth, 48, 237, 610
    of bacteria, 610, 615
    of insects, 502
    models, 586
    world, 49
position function, 142
position vector, 800
positive angle, A25

positive orientation
    of a boundary curve, 1134
    of a closed curve, 1096
    of a surface, 1128
potential, 539
potential energy, 1093
potential function, 1073
pound (unit of force), 455
power, 147
power consumption, approximation
    of, 408
power function(s), 29
    derviative of, 172
Power Law of limits, 96
Power Rule, 173, 174, 200, 221
power series, 746, 747
    coefficients of, 746
    for cosine and sine, 764
    differentiation of, 754
    division of, 770
    for exponential function, 763
    integration of, 754
    interval of convergence, 749
    multiplication of, 770
    radius of convergence, 749
    representations of functions as, 752
predator, 627
predator-prey model, 236, 627
pressure exerted by a fluid, 558
prey, 627
prime notation, 144, 176
principal square root of a complex
    number, A58
principal unit normal vector, 866
principle of mathematical induction,
    72, 74, A36
principle of superposition, 1163
probability, 573, 1021
probability density function, 574, 1021
problem-solving principles, 71
    uses of, 169, 363, 412, 425
producer surplus, 572
product
    cross, 814, 815 (*see also* cross product)
    dot, 807 (*see also* dot product)
    scalar, 807
    scalar triple, 819
    triple, 819
product formulas, A29
Product Law of limits, 95
Product Rule, 183, 184
profit function, 336
projectile, path of, 648, 872
projection, 793, 811
    orthogonal, 813
p-series, 722

quadrant, A11
quadratic approximation, 258, 970
quadratic function, 27
quadric surface(s), 835
    cone, 837
    ellipsoid, 835, 837
    hyperboloid, 837
    paraboloid, 836, 837
    table of graphs, 837
quaternion, 803
Quotient Law of limits, 95
Quotient Rule, 185, 186

radian measure, 190, A24
radiation from stars, 783
radioactive decay, 239
radiocarbon dating, 243
radius of convergence, 749
radius of gyration of a lamina, 1020
rainbow, formation and location of, 285
rainbow angle, 286
ramp function, 45
range of a function, 10, 888
rate of change
    average, 145, 224
    derivative as, 146
    instantaneous, 81, 145, 224
rate of growth, 229, 406
rate of reaction, 147, 227, 406
rates, related, 245
rational function, 30, 493, 908
    continuity of, 118
    integration of, 493
rational number, A2
rationalizing substitution for
    integration, 500
Ratio Test, 739
Rayleigh-Jeans Law, 783
real line, A3
real number, A2
rearrangement of a series, 742
reciprocal function, 30
Reciprocal Rule, 190
rectangular coordinate system, 793, A11
    conversion to cylindrical
        coordinates, 1040
    conversion to spherical
        coordinates, 1046
rectifying plane, 869
rectilinear motion, 354
recursion relation, 1177
reduction formula, 475
reflecting a function, 37
reflection property
    of conics, 273
    of an ellipse, 677

    of a hyperbola, 682
    of a parabola, 273, 274
region
    between two graphs, 428
    connected, 1089
    open, 1089
    plane, of type I or II, 1002, 1003
    simple plane, 1097
    simple solid, 1141
    simply-connected, 1090
    solid (of type 1, 2, or 3), 1031, 1032
    under a graph, 366, 371
regression, linear, 26
related rates, 245
relative error, 256
relative growth rate, 237, 611
relative maximum or minimum, 276
remainder estimates
    for the Alternating Series, 735
    for the Integral Test, 723
remainder of the Taylor series, 761
removable discontinuity, 116
representation(s) of a function, 10, 12, 13
    as a power series, 752
resonance, 1172
restoring force, 1169, 1171
resultant force, 803
revenue function, 336
reversing order of integration, 995, 1006
revolution, solid of, 443
revolution, surface of, 551
Riemann, Georg Bernhard, 379
Riemann sum(s), 379
    double, 991
    triple, 1029
right circular cylinder, 438
right-hand derivative, 165
right-hand limit, 88, 109
right-hand rule, 792, 816
Roberval, Gilles de, 398, 651
rocket stages, determining optimal
    masses for, 979
Rolle, Michel, 287
roller coaster, design of, 182
roller derby, 1052
Rolle's Theorem, 287
root function, 29
Root Law of limits, 97
Root Test, 741
roots of a complex number, A62
roots of an nth-degree equation, 211
rubber membrane, vibration of, 748
ruled surface, 841
ruling of a surface, 834
rumors, rate of spread, 232

saddle point, 961
sample point, 371, 378, 989
satellite dish, parabolic, 839
scalar, 799
scalar equation of a plane, 827
scalar field, 1069
scalar multiple of a vector, 799
scalar product, 807
scalar projection, 811
scalar triple product, 819
    geometric characterization of, 819
scatter plot, 13
seasonal-growth model, 620
secant function, A26
    derivative of, 193
    graph of, A31
secant line, 3, 78, 79, 81
secant vector, 856
second-degree Taylor polynomial, 971
second derivative, 158
    of a vector function, 858
Second Derivative Test, 297
Second Derivatives Test, 961
second directional derivative, 958
second moment of inertia, 1019
second-order differential equation, 588
    boundary-value problem, 1159
    initial-value problem, 1158
    solutions of, 1154, 1159
second partial derivative, 918
sector of a circle, area of, 669
separable differential equation, 599
sequence, 5, 694
    bounded, 701
    convergent, 696
    decreasing, 700
    divergent, 696
    Fibonacci, 695
    graph of, 699
    increasing, 700
    limit of, 5, 368, 696
    logistic, 707
    monotonic, 700
    of partial sums, 708
    term of, 694
series, 6, 707
    absolutely convergent, 737
    alternating, 732
    alternating harmonic, 734, 737, 738
    binomial, 766
    coefficients of, 746
    conditionally convergent, 738
    convergent, 708
    divergent, 708
    geometric, 709
    Gregory's, 756

series (continued)
    harmonic, 713, 722
    infinite, 707
    Maclaurin, 759, 760
    p-, 722
    partial sum of, 708
    power, 746
    rearrangement of, 742
    strategy for testing, 744
    sum of, 6, 708
    Taylor, 759, 760
    term of, 707
    trigonometric, 746
series solution of a differential
        equation, 1176
set, bounded or closed, 965
set notation, A3
serpentine, 188
Shannon index, 969
shell method for approximating
        volume, 449
shift of a function, 37
shifted conics, 679, A21
shock absorber, 1169
Sierpinski carpet, 718
sigma notation, 372, A34
simple curve, 1090
simple harmonic motion, 206, 1168
simple plane region, 1097
simple solid region, 1141
simply-connected region, 1090
Simpson, Thomas, 520, 985
Simpson's Rule, 519, 520
    error bounds for, 522
sine function, A26
    derivative of, 192, 193
    graph of, 31, A31
    power series for, 764
sine integral function, 401
sink, 1145
skew lines, 826
slant asymptote, 316, 320
slope, A12
    of a curve, 141
slope field, 592
slope-intercept equation of a sline, A13
smooth curve, 544, 863
smooth function, 544
smooth parametrization of a space
        curve, 863
smooth surface, 1116
Snell's Law, 341
snowflake curve, 788
solid, 438
solid, volume of, 438, 439, 990, 1031
solid angle, 1151

solid of revolution, 443
    rotated on a slant, 557
    volume of, 445, 451, 557
solid region, 1031, 1141
solution curve, 592
solution of a differential equation, 588
solution of predator-prey equations, 627
source, 1145
space, three-dimensional, 792
space curve, 849
    arc length of, 861, 862
    parametrization of, 851
speed of a particle, 146, 870
sphere
    equation of, 795
    flux across, 1129
    parametrization of, 1113
    surface area of, 1117
spherical coordinate system, 1045
    conversion equations for, 1046
    triple integrals in, 1047
spherical wedge, 1047
spherical zones, 583
spring constant, 457, 587, 1168
Squeeze Theorem, 101, A42
    for sequences, 698
standard basis vectors, 802
    properties of, 818
standard deviation, 578
standard position of an angle, A25
stationary point, 960
steady state solution, 1174
stellar stereography, 536
step function, 17
Stokes, Sir George, 1135, 1140
Stokes' Theorem, 1134, 1140, 1147
strategy
    for integration, 503, 504
    for optimization problems, 330, 331
    for problem solving, 71
    for related rates, 247
    for testing series, 744
    for trigonometric integrals, 481, 482
streamlines, 1074
stretching of a function, 37
strophoid, 673, 691
Substitution Rule, 412, 413, 416
    for definite integrals, 416
subtraction formulas for sine
        and cosine, A29
sum, 371
    of a geometric series, 710
    of an infinite series, 708
    lower, 371
    of partial fractions, 494
    Riemann, 379

    telescoping, 712
    upper, 371
    of vectors, 798, 801
Sum Law of limits, 95
summation notation, A34
Sum Rule, 176
supply function, 572
surface(s)
    closed, 1128
    graph of, 1124
    level, 898
    oriented, 1127
    orthogonal, 959
    parametric, 1111
    positive orientation of, 1128
    quadric, 835
    smooth, 1116

surface area, 552
    of a graph of a function, 1118
    of a parametric surface, 654,
        1116, 1117
    of a sphere, 1117
    of a function of two variables, 1026
surface integral, 1122
    over a parametric surface,
        1122, 1123
    of a vector field, 1128, 1129
surface of revolution, 551
    parametric representation of, 1115
    surface area of, 552
swallowtail catastrophe curve, 648
symmetric equations of a line, 824
symmetric functions, integrals of, 417
symmetry, 17, 315, 417
    in polar graphs, 663
symmetry principle, 562

T and T^{-1} transformations, 1053
table of differentiation
        formulas, 187, RP5
tables of integrals, 503, RP6–10
    use of, 509
tabular function, 13
tangent function, A26
    derivative of, 193
    graph of, 32, A31
tangent line(s), 141
    to a curve, 3, 78, 141
    early methods of finding, 152
    to a parametric curve, 649, 650
    to a polar curve, 663
    to a space curve, 856
    vertical, 158
tangent line approximation, 252

tangent plane
  to a level surface, 954
  to a parametric surface, 1115, 1116
  to a surface, 928
tangent plane approximation, 929, 930
tangent problem, 2, 3, 78, 140
tangent vector, 856
tangential component of acceleration, 874
tautochrone problem, 644
Taylor, Brook, 760
Taylor polynomial, 258, 761, 970
  applications of, 774
Taylor series, 759, 760
Taylor's inequality, 762
techniques of integration, summary, 504
telescoping sum, 712
temperature-humidity index, 899, 911
term of a sequence, 694
term of a series, 707
terminal point of a parametric curve, 641
terminal point of a vector, 798
terminal velocity, 607
term-by-term differentiation and
    integration, 754
Test for Divergence, 713
tests for convergence and divergence
    of series
  Alternating Series Test, 732
  Comparison Test, 727
  Integral Test, 721
  Limit Comparison Test, 729
  Ratio Test, 739
  Root Test, 741
  summary of tests, 744
tetrahedron, 823
third derivative, 159
third-order optics, 780
Thomson, William (Lord Kelvin), 1097,
    1135, 1140
three-dimensional coordinate systems, 792
TNB frame, 866
toroidal spiral, 851
torque, 820, 879
Torricelli, Evangelista, 651
Torricelli's Law, 234
torsion of a space curve, 870
torus, 448, 1122
total differential, 932
total electric charge, 1016, 1036
total fertility rate, 168
trace of a surface, 834
trajectory, parametric equations for, 873
transfer curve, 870, 883
transformation
  of a function, 36
  inverse, 1053

Jacobian of, 1055, 1059
  one-to-one, 1053
  of a root function, 38
translation of a function, 36
Trapezoidal Rule, 516
  error in, 516
tree diagram, 940
trefoil knot, 851, 855
Triangle Inequality, 111, A8
  for vectors, 814
Triangle Law, 798
trigonometric functions, 31, A26
  derivatives of, 190, 193
  graphs of, 31, 32, A30, A31
  integrals of, 403, 479
  inverse, 63
  limits involving, 191, 192
trigonometric identities, A28
trigonometric integrals, 479
  strategy for evaluating, 481, 482
trigonometric series, 746
trigonometric substitutions, 486
  table of, 486
triple integral(s), 1029, 1030
  applications of, 1034
  change of variables in, 1058
  in cylindrical coordinates,
      1040, 1042
  over a general bounded region, 1031
  Midpoint Rule for, 1038
  in spherical coordinates, 1045, 1047
triple product, 819
triple Riemann sum, 1029
trochoid, 647
Tschirnhausen cubic, 215, 436
twisted cubic, 851
type I or type II plane region,
    1002, 1003
type 1, 2, or 3 solid region, 1031, 1032

ultraviolet catastrophe, 783
underdamped vibration, 1170
undetermined coefficients, method of,
    1161, 1165
uniform circular motion, 884
union of sets, A3
unit normal vector, 866
unit tangent vector, 856
unit vector, 803
upper sum, 371

value of a function, 10
van der Waals equation, 217, 926
variable(s)
  change of, 413
  continuous random, 573

dependent, 10, 888, 940
independent, 10, 888, 940
independent random, 1022
intermediate, 940
variables, change of. See change of
    variable(s)
variation of parameters, method of,
    1165, 1166
vascular branching, 342
vector(s), 798
  acceleration, 871
  addition of, 798, 801
  algebraic, 800
  angle between, 808, 809
  basis, 802
  binormal, 866
  combining speed, 806
  components of, 800
  coplanar, 820
  cross product of, 814, 815
  difference, 799
  displacement, 798, 811
  dot product, 807
  equality of, 798
  force, 1071
  geometric representation of, 800
  gradient, 949, 951, 955
  i, **j**, and **k**, 802
  length of, 801
  magnitude of, 801
  multiplication of, 799, 801
  n-dimensional, 802
  normal, 827, 866
  orthogonal, 807
  orthogonal projection of, 813
  parallel, 799, 817
  perpendicular, 807
  position, 800
  properties of, 802
  representation of, 800
  scalar mulitple of, 799, 801
  secant, 856
  standard basis, 802
  subtraction of, 799, 801
  tangent, 856
  three-dimensional, 800
  triple product, 820
  two-dimensional, 800
  unit, 803
  unit normal, 866
  unit tangent, 856
  velocity, 870
  zero, 798
vector equation
  of a line, 824
  of a plane, 827

vector field, 1068, 1069
   component functions, 1069
   conservative, 1073, 1090, 1091, 1105
   curl of, 1103
   divergence of, 1106
   electric flux of, 1131, 1144
   flux of, 1129
   force, 1068, 1072
   gradient, 956, 1072
   gravitational, 1072
   incompressible, 1107
   irrotational, 1106
   line integral of, 1082, 1083, 1084
   surface integral of, 1128, 1129
   velocity, 1068, 1071
vector function, 848
   component functions of, 848
   continuity of, 849
   derivative of, 855, 856, 858
   integration of, 859
   limit of, 848, 855
vector product, 815
     properties of, 816, 819
vector projection, 811
vector triple product, 820
vector-valued function. *See* vector function
velocity, 3, 80, 143, 224, 406
   average, 4, 81, 143, 224
   instantaneous, 81, 143, 224
velocity field, 1071
   airflow, 1068
   ocean currents, 1068
   wind patterns, 1068

velocity gradient, 230
velocity problem, 80, 142
velocity vector, 870
velocity vector field, 1086
Verhulst, Pierre-François, 587
vertex of a parabola, 674
vertical asymptote, 90, 316
vertical line, A13
Vertical Line Test, 15
vertical tangent line, 158
vertical translation of a graph, 37
vertices
   of an ellipse, 676
   of a hyperbola, 678
vibration of a rubber membrane, 748
vibration of a spring, 1168
vibrations, 1168, 1169, 1171
visual representations of a function, 10, 12
volume, 439
   by cross-sections, 438, 439, 570
   by cylindrical shells, 449
   by disks, 440, 443
   by double integrals, 988
   of a hypersphere, 1040
   by polar coordinates, 1012
   of a solid, 438, 990
   of a solid of revolution, 443, 557
   of a solid on a slant, 557
   by triple integrals, 1035
   by washers, 442, 443
Volterra, Vito, 627
von Bertalanffy model, 635

Wallis, John, 3
Wallis product, 478
washer method, 442
wave equation, 920
Weierstrass, Karl, 502
weight (force), 455
wind-chill index, 889
wind patterns in San Francisco Bay area, 1068
witch of Maria Agnesi, 188, 647
work (force), 455, 456, 811
work defined as a line integral, 1082
Wren, Sir Christopher, 654

x-axis, 792, A10
x-coordinate, 792, A10
x-intercept, A13, A19
X-mean, 1023

y-axis, 792, A10
y-coordinate, 792, A10
y-intercept, A13, A19
Y-mean, 1023

z-axis, 792
z-coordinate, 792
Zeno, 5
Zeno's paradoxes, 5
zero vector, 798

REFERENCE page 1

ALGEBRA

Arithmetic Operations

$$a(b+c) = ab + ac \qquad \frac{a}{b} + \frac{c}{d} = \frac{ad+bc}{bd}$$

$$\frac{a+c}{b} = \frac{a}{b} + \frac{c}{b} \qquad \frac{\dfrac{a}{b}}{\dfrac{c}{d}} = \frac{a}{b} \times \frac{d}{c} = \frac{ad}{bc}$$

Exponents and Radicals

$$x^m x^n = x^{m+n} \qquad \frac{x^m}{x^n} = x^{m-n}$$

$$(x^m)^n = x^{mn} \qquad x^{-n} = \frac{1}{x^n}$$

$$(xy)^n = x^n y^n \qquad \left(\frac{x}{y}\right)^n = \frac{x^n}{y^n}$$

$$x^{1/n} = \sqrt[n]{x} \qquad x^{m/n} = \sqrt[n]{x^m} = \left(\sqrt[n]{x}\right)^m$$

$$\sqrt[n]{xy} = \sqrt[n]{x}\sqrt[n]{y} \qquad \sqrt[n]{\frac{x}{y}} = \frac{\sqrt[n]{x}}{\sqrt[n]{y}}$$

Factoring Special Polynomials

$$x^2 - y^2 = (x+y)(x-y)$$
$$x^3 + y^3 = (x+y)(x^2 - xy + y^2)$$
$$x^3 - y^3 = (x-y)(x^2 + xy + y^2)$$

Binomial Theorem

$$(x+y)^2 = x^2 + 2xy + y^2 \qquad (x-y)^2 = x^2 - 2xy + y^2$$
$$(x+y)^3 = x^3 + 3x^2 y + 3xy^2 + y^3$$
$$(x-y)^3 = x^3 - 3x^2 y + 3xy^2 - y^3$$
$$(x+y)^n = x^n + nx^{n-1}y + \frac{n(n-1)}{2}x^{n-2}y^2$$
$$\quad + \cdots + \binom{n}{k} x^{n-k} y^k + \cdots + nxy^{n-1} + y^n$$

where $\displaystyle \binom{n}{k} = \frac{n(n-1)\cdots(n-k+1)}{1 \cdot 2 \cdot 3 \cdot \,\cdots\, \cdot k}$

Quadratic Formula

If $ax^2 + bx + c = 0$, then $x = \dfrac{-b \pm \sqrt{b^2 - 4ac}}{2a}$.

Inequalities and Absolute Value

If $a < b$ and $b < c$, then $a < c$.

If $a < b$, then $a + c < b + c$.

If $a < b$ and $c > 0$, then $ca < cb$.

If $a < b$ and $c < 0$, then $ca > cb$.

If $a > 0$, then

$|x| = a$ means $x = a$ or $x = -a$

$|x| < a$ means $-a < x < a$

$|x| > a$ means $x > a$ or $x < -a$

GEOMETRY

Geometric Formulas

Formulas for area A, circumference C, and volume V:

Triangle
$A = \tfrac{1}{2} bh$
$\quad = \tfrac{1}{2} ab \sin\theta$

Circle
$A = \pi r^2$
$C = 2\pi r$

Sector of Circle
$A = \tfrac{1}{2} r^2 \theta$
$s = r\theta$ (θ in radians)

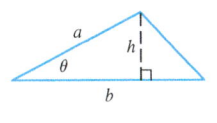 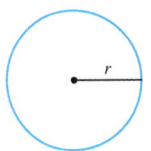 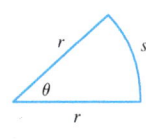

Sphere
$V = \tfrac{4}{3}\pi r^3$
$A = 4\pi r^2$

Cylinder
$V = \pi r^2 h$

Cone
$V = \tfrac{1}{3}\pi r^2 h$
$A = \pi r \sqrt{r^2 + h^2}$

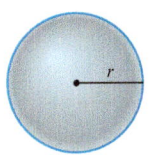 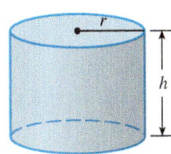 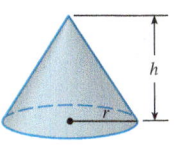

Distance and Midpoint Formulas

Distance between $P_1(x_1, y_1)$ and $P_2(x_2, y_2)$:

$$d = \sqrt{(x_2 - x_1)^2 + (y_2 - y_1)^2}$$

Midpoint of $\overline{P_1 P_2}$: $\left(\dfrac{x_1 + x_2}{2}, \dfrac{y_1 + y_2}{2}\right)$

Lines

Slope of line through $P_1(x_1, y_1)$ and $P_2(x_2, y_2)$:

$$m = \frac{y_2 - y_1}{x_2 - x_1}$$

Point-slope equation of line through $P_1(x_1, y_1)$ with slope m:

$$y - y_1 = m(x - x_1)$$

Slope-intercept equation of line with slope m and y-intercept b:

$$y = mx + b$$

Circles

Equation of the circle with center (h, k) and radius r:

$$(x - h)^2 + (y - k)^2 = r^2$$

REFERENCE page 2

TRIGONOMETRY

Angle Measurement

π radians $= 180°$

$1° = \dfrac{\pi}{180}$ rad $\qquad$ 1 rad $= \dfrac{180°}{\pi}$

$s = r\theta$

(θ in radians)

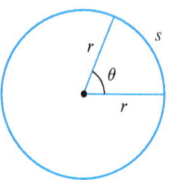

Right Angle Trigonometry

$\sin\theta = \dfrac{\text{opp}}{\text{hyp}} \qquad \csc\theta = \dfrac{\text{hyp}}{\text{opp}}$

$\cos\theta = \dfrac{\text{adj}}{\text{hyp}} \qquad \sec\theta = \dfrac{\text{hyp}}{\text{adj}}$

$\tan\theta = \dfrac{\text{opp}}{\text{adj}} \qquad \cot\theta = \dfrac{\text{adj}}{\text{opp}}$

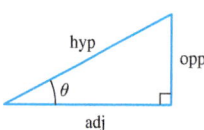

Trigonometric Functions

$\sin\theta = \dfrac{y}{r} \qquad \csc\theta = \dfrac{r}{y}$

$\cos\theta = \dfrac{x}{r} \qquad \sec\theta = \dfrac{r}{x}$

$\tan\theta = \dfrac{y}{x} \qquad \cot\theta = \dfrac{x}{y}$

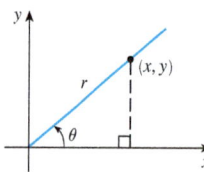

Graphs of Trigonometric Functions

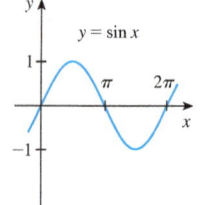
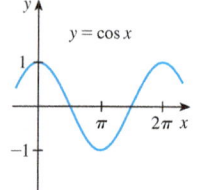
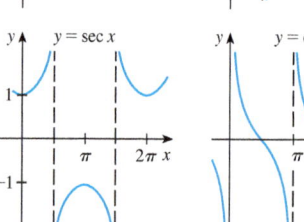

Trigonometric Functions of Important Angles

θ	radians	$\sin\theta$	$\cos\theta$	$\tan\theta$
0°	0	0	1	0
30°	$\pi/6$	$1/2$	$\sqrt{3}/2$	$\sqrt{3}/3$
45°	$\pi/4$	$\sqrt{2}/2$	$\sqrt{2}/2$	1
60°	$\pi/3$	$\sqrt{3}/2$	$1/2$	$\sqrt{3}$
90°	$\pi/2$	1	0	—

Fundamental Identities

$\csc\theta = \dfrac{1}{\sin\theta} \qquad \sec\theta = \dfrac{1}{\cos\theta}$

$\tan\theta = \dfrac{\sin\theta}{\cos\theta} \qquad \cot\theta = \dfrac{\cos\theta}{\sin\theta}$

$\cot\theta = \dfrac{1}{\tan\theta} \qquad \sin^2\theta + \cos^2\theta = 1$

$1 + \tan^2\theta = \sec^2\theta \qquad 1 + \cot^2\theta = \csc^2\theta$

$\sin(-\theta) = -\sin\theta \qquad \cos(-\theta) = \cos\theta$

$\tan(-\theta) = -\tan\theta \qquad \sin\left(\dfrac{\pi}{2} - \theta\right) = \cos\theta$

$\cos\left(\dfrac{\pi}{2} - \theta\right) = \sin\theta \qquad \tan\left(\dfrac{\pi}{2} - \theta\right) = \cot\theta$

The Law of Sines

$\dfrac{\sin A}{a} = \dfrac{\sin B}{b} = \dfrac{\sin C}{c}$

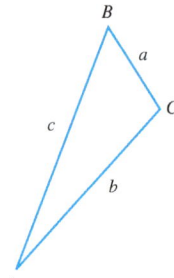

The Law of Cosines

$a^2 = b^2 + c^2 - 2bc\cos A$

$b^2 = a^2 + c^2 - 2ac\cos B$

$c^2 = a^2 + b^2 - 2ab\cos C$

Addition and Subtraction Formulas

$\sin(x + y) = \sin x \cos y + \cos x \sin y$

$\sin(x - y) = \sin x \cos y - \cos x \sin y$

$\cos(x + y) = \cos x \cos y - \sin x \sin y$

$\cos(x - y) = \cos x \cos y + \sin x \sin y$

$\tan(x + y) = \dfrac{\tan x + \tan y}{1 - \tan x \tan y}$

$\tan(x - y) = \dfrac{\tan x - \tan y}{1 + \tan x \tan y}$

Double-Angle Formulas

$\sin 2x = 2\sin x \cos x$

$\cos 2x = \cos^2 x - \sin^2 x = 2\cos^2 x - 1 = 1 - 2\sin^2 x$

$\tan 2x = \dfrac{2\tan x}{1 - \tan^2 x}$

Half-Angle Formulas

$\sin^2 x = \dfrac{1 - \cos 2x}{2} \qquad \cos^2 x = \dfrac{1 + \cos 2x}{2}$

REFERENCE page 3

SPECIAL FUNCTIONS

Power Functions $f(x) = x^a$

(i) $f(x) = x^n$, n a positive integer

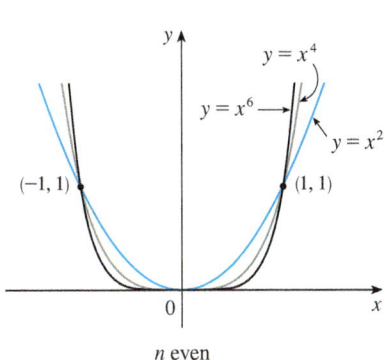

n even

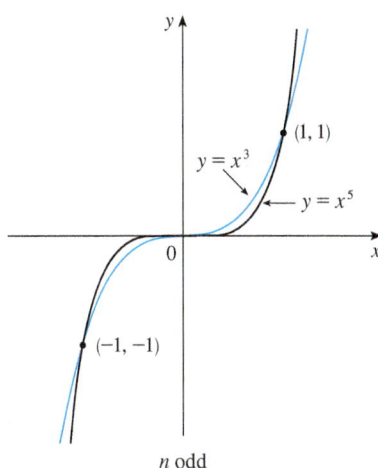

n odd

(ii) $f(x) = x^{1/n} = \sqrt[n]{x}$, n a positive integer

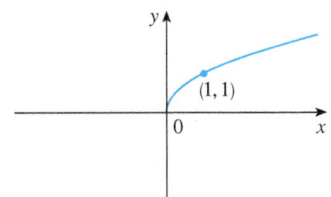

$f(x) = \sqrt{x}$

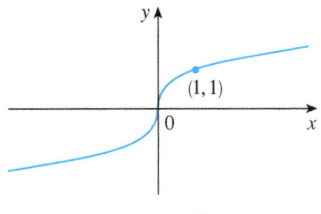

$f(x) = \sqrt[3]{x}$

(iii) $f(x) = x^{-1} = \dfrac{1}{x}$

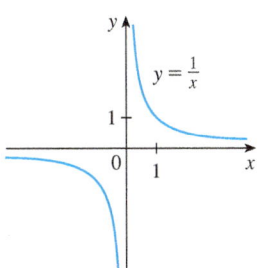

Inverse Trigonometric Functions

$\arcsin x = \sin^{-1}x = y \iff \sin y = x \quad \text{and} \quad -\dfrac{\pi}{2} \leq y \leq \dfrac{\pi}{2}$

$\arccos x = \cos^{-1}x = y \iff \cos y = x \quad \text{and} \quad 0 \leq y \leq \pi$

$\arctan x = \tan^{-1}x = y \iff \tan y = x \quad \text{and} \quad -\dfrac{\pi}{2} < y < \dfrac{\pi}{2}$

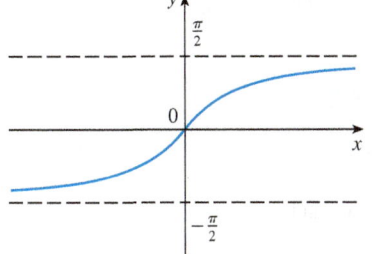

$y = \tan^{-1}x = \arctan x$

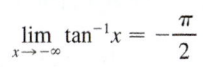

$\lim\limits_{x \to -\infty} \tan^{-1}x = -\dfrac{\pi}{2}$

$\lim\limits_{x \to \infty} \tan^{-1}x = \dfrac{\pi}{2}$

SPECIAL FUNCTIONS

Exponential and Logarithmic Functions

$\log_b x = y \iff b^y = x$

$\ln x = \log_e x, \quad \text{where} \quad \ln e = 1$

$\ln x = y \iff e^y = x$

Cancellation Equations

$\log_b(b^x) = x \qquad b^{\log_b x} = x$

$\ln(e^x) = x \qquad e^{\ln x} = x$

Laws of Logarithms

1. $\log_b(xy) = \log_b x + \log_b y$
2. $\log_b\left(\dfrac{x}{y}\right) = \log_b x - \log_b y$
3. $\log_b(x^r) = r \log_b x$

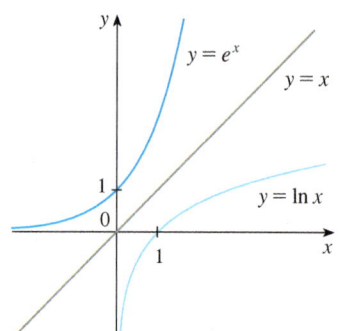

$\lim\limits_{x \to -\infty} e^x = 0 \qquad \lim\limits_{x \to \infty} e^x = \infty$

$\lim\limits_{x \to 0^+} \ln x = -\infty \qquad \lim\limits_{x \to \infty} \ln x = \infty$

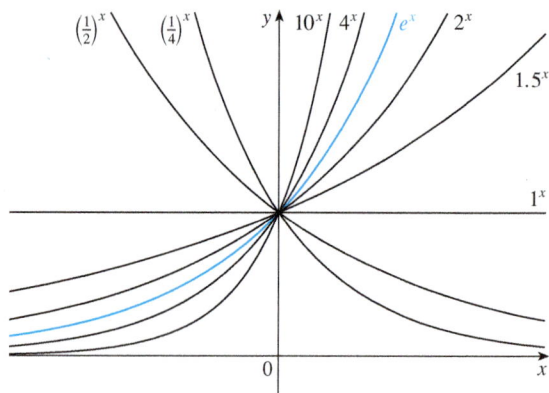

Exponential functions

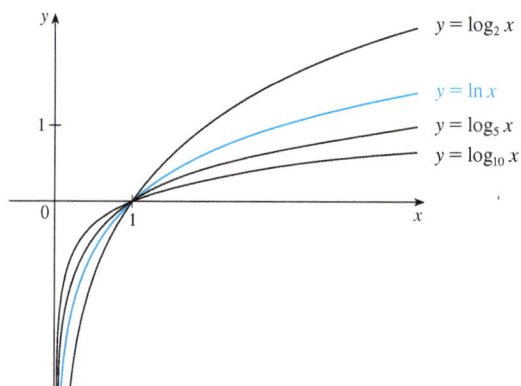

Logarithmic functions

Hyperbolic Functions

$\sinh x = \dfrac{e^x - e^{-x}}{2} \qquad \operatorname{csch} x = \dfrac{1}{\sinh x}$

$\cosh x = \dfrac{e^x + e^{-x}}{2} \qquad \operatorname{sech} x = \dfrac{1}{\cosh x}$

$\tanh x = \dfrac{\sinh x}{\cosh x} \qquad \coth x = \dfrac{\cosh x}{\sinh x}$

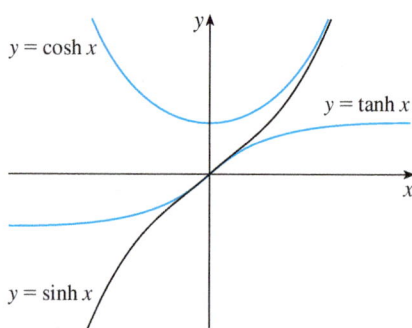

Inverse Hyperbolic Functions

$y = \sinh^{-1} x \iff \sinh y = x \qquad \sinh^{-1} x = \ln\left(x + \sqrt{x^2 + 1}\right)$

$y = \cosh^{-1} x \iff \cosh y = x \text{ and } y \geq 0 \qquad \cosh^{-1} x = \ln\left(x + \sqrt{x^2 - 1}\right)$

$y = \tanh^{-1} x \iff \tanh y = x \qquad \tanh^{-1} x = \tfrac{1}{2} \ln\left(\dfrac{1 + x}{1 - x}\right)$

REFERENCE page 5

DIFFERENTIATION RULES

General Formulas

1. $\dfrac{d}{dx}(c) = 0$

2. $\dfrac{d}{dx}[cf(x)] = cf'(x)$

3. $\dfrac{d}{dx}[f(x) + g(x)] = f'(x) + g'(x)$

4. $\dfrac{d}{dx}[f(x) - g(x)] = f'(x) - g'(x)$

5. $\dfrac{d}{dx}[f(x)g(x)] = f(x)g'(x) + g(x)f'(x)$ (Product Rule)

6. $\dfrac{d}{dx}\left[\dfrac{f(x)}{g(x)}\right] = \dfrac{g(x)f'(x) - f(x)g'(x)}{[g(x)]^2}$ (Quotient Rule)

7. $\dfrac{d}{dx}f(g(x)) = f'(g(x))g'(x)$ (Chain Rule)

8. $\dfrac{d}{dx}(x^n) = nx^{n-1}$ (Power Rule)

Exponential and Logarithmic Functions

9. $\dfrac{d}{dx}(e^x) = e^x$

10. $\dfrac{d}{dx}(b^x) = b^x \ln b$

11. $\dfrac{d}{dx}\ln|x| = \dfrac{1}{x}$

12. $\dfrac{d}{dx}(\log_b x) = \dfrac{1}{x \ln b}$

Trigonometric Functions

13. $\dfrac{d}{dx}(\sin x) = \cos x$

14. $\dfrac{d}{dx}(\cos x) = -\sin x$

15. $\dfrac{d}{dx}(\tan x) = \sec^2 x$

16. $\dfrac{d}{dx}(\csc x) = -\csc x \cot x$

17. $\dfrac{d}{dx}(\sec x) = \sec x \tan x$

18. $\dfrac{d}{dx}(\cot x) = -\csc^2 x$

Inverse Trigonometric Functions

19. $\dfrac{d}{dx}(\sin^{-1} x) = \dfrac{1}{\sqrt{1-x^2}}$

20. $\dfrac{d}{dx}(\cos^{-1} x) = -\dfrac{1}{\sqrt{1-x^2}}$

21. $\dfrac{d}{dx}(\tan^{-1} x) = \dfrac{1}{1+x^2}$

22. $\dfrac{d}{dx}(\csc^{-1} x) = -\dfrac{1}{x\sqrt{x^2-1}}$

23. $\dfrac{d}{dx}(\sec^{-1} x) = \dfrac{1}{x\sqrt{x^2-1}}$

24. $\dfrac{d}{dx}(\cot^{-1} x) = -\dfrac{1}{1+x^2}$

Hyperbolic Functions

25. $\dfrac{d}{dx}(\sinh x) = \cosh x$

26. $\dfrac{d}{dx}(\cosh x) = \sinh x$

27. $\dfrac{d}{dx}(\tanh x) = \text{sech}^2 x$

28. $\dfrac{d}{dx}(\text{csch } x) = -\text{csch } x \coth x$

29. $\dfrac{d}{dx}(\text{sech } x) = -\text{sech } x \tanh x$

30. $\dfrac{d}{dx}(\coth x) = -\text{csch}^2 x$

Inverse Hyperbolic Functions

31. $\dfrac{d}{dx}(\sinh^{-1} x) = \dfrac{1}{\sqrt{1+x^2}}$

32. $\dfrac{d}{dx}(\cosh^{-1} x) = \dfrac{1}{\sqrt{x^2-1}}$

33. $\dfrac{d}{dx}(\tanh^{-1} x) = \dfrac{1}{1-x^2}$

34. $\dfrac{d}{dx}(\text{csch}^{-1} x) = -\dfrac{1}{|x|\sqrt{x^2+1}}$

35. $\dfrac{d}{dx}(\text{sech}^{-1} x) = -\dfrac{1}{x\sqrt{1-x^2}}$

36. $\dfrac{d}{dx}(\coth^{-1} x) = \dfrac{1}{1-x^2}$

TABLE OF INTEGRALS

Basic Forms

1. $\displaystyle\int u\, dv = uv - \int v\, du$

2. $\displaystyle\int u^n\, du = \frac{u^{n+1}}{n+1} + C, \quad n \neq -1$

3. $\displaystyle\int \frac{du}{u} = \ln|u| + C$

4. $\displaystyle\int e^u\, du = e^u + C$

5. $\displaystyle\int b^u\, du = \frac{b^u}{\ln b} + C$

6. $\displaystyle\int \sin u\, du = -\cos u + C$

7. $\displaystyle\int \cos u\, du = \sin u + C$

8. $\displaystyle\int \sec^2 u\, du = \tan u + C$

9. $\displaystyle\int \csc^2 u\, du = -\cot u + C$

10. $\displaystyle\int \sec u \tan u\, du = \sec u + C$

11. $\displaystyle\int \csc u \cot u\, du = -\csc u + C$

12. $\displaystyle\int \tan u\, du = \ln|\sec u| + C$

13. $\displaystyle\int \cot u\, du = \ln|\sin u| + C$

14. $\displaystyle\int \sec u\, du = \ln|\sec u + \tan u| + C$

15. $\displaystyle\int \csc u\, du = \ln|\csc u - \cot u| + C$

16. $\displaystyle\int \frac{du}{\sqrt{a^2 - u^2}} = \sin^{-1}\frac{u}{a} + C, \quad a > 0$

17. $\displaystyle\int \frac{du}{a^2 + u^2} = \frac{1}{a}\tan^{-1}\frac{u}{a} + C$

18. $\displaystyle\int \frac{du}{u\sqrt{u^2 - a^2}} = \frac{1}{a}\sec^{-1}\frac{u}{a} + C$

19. $\displaystyle\int \frac{du}{a^2 - u^2} = \frac{1}{2a}\ln\left|\frac{u+a}{u-a}\right| + C$

20. $\displaystyle\int \frac{du}{u^2 - a^2} = \frac{1}{2a}\ln\left|\frac{u-a}{u+a}\right| + C$

Forms Involving $\sqrt{a^2 + u^2}$, $a > 0$

21. $\displaystyle\int \sqrt{a^2 + u^2}\, du = \frac{u}{2}\sqrt{a^2 + u^2} + \frac{a^2}{2}\ln(u + \sqrt{a^2 + u^2}) + C$

22. $\displaystyle\int u^2\sqrt{a^2 + u^2}\, du = \frac{u}{8}(a^2 + 2u^2)\sqrt{a^2 + u^2} - \frac{a^4}{8}\ln(u + \sqrt{a^2 + u^2}) + C$

23. $\displaystyle\int \frac{\sqrt{a^2 + u^2}}{u}\, du = \sqrt{a^2 + u^2} - a\ln\left|\frac{a + \sqrt{a^2 + u^2}}{u}\right| + C$

24. $\displaystyle\int \frac{\sqrt{a^2 + u^2}}{u^2}\, du = -\frac{\sqrt{a^2 + u^2}}{u} + \ln(u + \sqrt{a^2 + u^2}) + C$

25. $\displaystyle\int \frac{du}{\sqrt{a^2 + u^2}} = \ln(u + \sqrt{a^2 + u^2}) + C$

26. $\displaystyle\int \frac{u^2\, du}{\sqrt{a^2 + u^2}} = \frac{u}{2}\sqrt{a^2 + u^2} - \frac{a^2}{2}\ln(u + \sqrt{a^2 + u^2}) + C$

27. $\displaystyle\int \frac{du}{u\sqrt{a^2 + u^2}} = -\frac{1}{a}\ln\left|\frac{\sqrt{a^2 + u^2} + a}{u}\right| + C$

28. $\displaystyle\int \frac{du}{u^2\sqrt{a^2 + u^2}} = -\frac{\sqrt{a^2 + u^2}}{a^2 u} + C$

29. $\displaystyle\int \frac{du}{(a^2 + u^2)^{3/2}} = \frac{u}{a^2\sqrt{a^2 + u^2}} + C$

TABLE OF INTEGRALS

Forms Involving $\sqrt{a^2 - u^2}$, $a > 0$

30. $\displaystyle\int \sqrt{a^2 - u^2}\, du = \frac{u}{2}\sqrt{a^2 - u^2} + \frac{a^2}{2}\sin^{-1}\frac{u}{a} + C$

31. $\displaystyle\int u^2\sqrt{a^2 - u^2}\, du = \frac{u}{8}(2u^2 - a^2)\sqrt{a^2 - u^2} + \frac{a^4}{8}\sin^{-1}\frac{u}{a} + C$

32. $\displaystyle\int \frac{\sqrt{a^2 - u^2}}{u}\, du = \sqrt{a^2 - u^2} - a\ln\left|\frac{a + \sqrt{a^2 - u^2}}{u}\right| + C$

33. $\displaystyle\int \frac{\sqrt{a^2 - u^2}}{u^2}\, du = -\frac{1}{u}\sqrt{a^2 - u^2} - \sin^{-1}\frac{u}{a} + C$

34. $\displaystyle\int \frac{u^2\, du}{\sqrt{a^2 - u^2}} = -\frac{u}{2}\sqrt{a^2 - u^2} + \frac{a^2}{2}\sin^{-1}\frac{u}{a} + C$

35. $\displaystyle\int \frac{du}{u\sqrt{a^2 - u^2}} = -\frac{1}{a}\ln\left|\frac{a + \sqrt{a^2 - u^2}}{u}\right| + C$

36. $\displaystyle\int \frac{du}{u^2\sqrt{a^2 - u^2}} = -\frac{1}{a^2 u}\sqrt{a^2 - u^2} + C$

37. $\displaystyle\int (a^2 - u^2)^{3/2}\, du = -\frac{u}{8}(2u^2 - 5a^2)\sqrt{a^2 - u^2} + \frac{3a^4}{8}\sin^{-1}\frac{u}{a} + C$

38. $\displaystyle\int \frac{du}{(a^2 - u^2)^{3/2}} = \frac{u}{a^2\sqrt{a^2 - u^2}} + C$

Forms Involving $\sqrt{u^2 - a^2}$, $a > 0$

39. $\displaystyle\int \sqrt{u^2 - a^2}\, du = \frac{u}{2}\sqrt{u^2 - a^2} - \frac{a^2}{2}\ln\left|u + \sqrt{u^2 - a^2}\right| + C$

40. $\displaystyle\int u^2\sqrt{u^2 - a^2}\, du = \frac{u}{8}(2u^2 - a^2)\sqrt{u^2 - a^2} - \frac{a^4}{8}\ln\left|u + \sqrt{u^2 - a^2}\right| + C$

41. $\displaystyle\int \frac{\sqrt{u^2 - a^2}}{u}\, du = \sqrt{u^2 - a^2} - a\cos^{-1}\frac{a}{|u|} + C$

42. $\displaystyle\int \frac{\sqrt{u^2 - a^2}}{u^2}\, du = -\frac{\sqrt{u^2 - a^2}}{u} + \ln\left|u + \sqrt{u^2 - a^2}\right| + C$

43. $\displaystyle\int \frac{du}{\sqrt{u^2 - a^2}} = \ln\left|u + \sqrt{u^2 - a^2}\right| + C$

44. $\displaystyle\int \frac{u^2\, du}{\sqrt{u^2 - a^2}} = \frac{u}{2}\sqrt{u^2 - a^2} + \frac{a^2}{2}\ln\left|u + \sqrt{u^2 - a^2}\right| + C$

45. $\displaystyle\int \frac{du}{u^2\sqrt{u^2 - a^2}} = \frac{\sqrt{u^2 - a^2}}{a^2 u} + C$

46. $\displaystyle\int \frac{du}{(u^2 - a^2)^{3/2}} = -\frac{u}{a^2\sqrt{u^2 - a^2}} + C$

(continued)

TABLE OF INTEGRALS

Forms Involving $a + bu$

47. $\displaystyle\int \frac{u\,du}{a+bu} = \frac{1}{b^2}(a+bu-a\ln|a+bu|) + C$

48. $\displaystyle\int \frac{u^2\,du}{a+bu} = \frac{1}{2b^3}\Big[(a+bu)^2 - 4a(a+bu) + 2a^2\ln|a+bu|\Big] + C$

49. $\displaystyle\int \frac{du}{u(a+bu)} = \frac{1}{a}\ln\left|\frac{u}{a+bu}\right| + C$

50. $\displaystyle\int \frac{du}{u^2(a+bu)} = -\frac{1}{au} + \frac{b}{a^2}\ln\left|\frac{a+bu}{u}\right| + C$

51. $\displaystyle\int \frac{u\,du}{(a+bu)^2} = \frac{a}{b^2(a+bu)} + \frac{1}{b^2}\ln|a+bu| + C$

52. $\displaystyle\int \frac{du}{u(a+bu)^2} = \frac{1}{a(a+bu)} - \frac{1}{a^2}\ln\left|\frac{a+bu}{u}\right| + C$

53. $\displaystyle\int \frac{u^2\,du}{(a+bu)^2} = \frac{1}{b^3}\left(a+bu - \frac{a^2}{a+bu} - 2a\ln|a+bu|\right) + C$

54. $\displaystyle\int u\sqrt{a+bu}\,du = \frac{2}{15b^2}(3bu-2a)(a+bu)^{3/2} + C$

55. $\displaystyle\int \frac{u\,du}{\sqrt{a+bu}} = \frac{2}{3b^2}(bu-2a)\sqrt{a+bu} + C$

56. $\displaystyle\int \frac{u^2\,du}{\sqrt{a+bu}} = \frac{2}{15b^3}(8a^2 + 3b^2u^2 - 4abu)\sqrt{a+bu} + C$

57. $\displaystyle\int \frac{du}{u\sqrt{a+bu}} = \frac{1}{\sqrt{a}}\ln\left|\frac{\sqrt{a+bu}-\sqrt{a}}{\sqrt{a+bu}+\sqrt{a}}\right| + C, \quad \text{if } a > 0$

$\displaystyle\phantom{\int \frac{du}{u\sqrt{a+bu}}} = \frac{2}{\sqrt{-a}}\tan^{-1}\sqrt{\frac{a+bu}{-a}} + C, \quad \text{if } a < 0$

58. $\displaystyle\int \frac{\sqrt{a+bu}}{u}\,du = 2\sqrt{a+bu} + a\int \frac{du}{u\sqrt{a+bu}}$

59. $\displaystyle\int \frac{\sqrt{a+bu}}{u^2}\,du = -\frac{\sqrt{a+bu}}{u} + \frac{b}{2}\int \frac{du}{u\sqrt{a+bu}}$

60. $\displaystyle\int u^n\sqrt{a+bu}\,du = \frac{2}{b(2n+3)}\left[u^n(a+bu)^{3/2} - na\int u^{n-1}\sqrt{a+bu}\,du\right]$

61. $\displaystyle\int \frac{u^n\,du}{\sqrt{a+bu}} = \frac{2u^n\sqrt{a+bu}}{b(2n+1)} - \frac{2na}{b(2n+1)}\int \frac{u^{n-1}\,du}{\sqrt{a+bu}}$

62. $\displaystyle\int \frac{du}{u^n\sqrt{a+bu}} = -\frac{\sqrt{a+bu}}{a(n-1)u^{n-1}} - \frac{b(2n-3)}{2a(n-1)}\int \frac{du}{u^{n-1}\sqrt{a+bu}}$

REFERENCE page 9

TABLE OF INTEGRALS

Trigonometric Forms

63. $\int \sin^2 u \, du = \tfrac{1}{2}u - \tfrac{1}{4}\sin 2u + C$

64. $\int \cos^2 u \, du = \tfrac{1}{2}u + \tfrac{1}{4}\sin 2u + C$

65. $\int \tan^2 u \, du = \tan u - u + C$

66. $\int \cot^2 u \, du = -\cot u - u + C$

67. $\int \sin^3 u \, du = -\tfrac{1}{3}(2 + \sin^2 u)\cos u + C$

68. $\int \cos^3 u \, du = \tfrac{1}{3}(2 + \cos^2 u)\sin u + C$

69. $\int \tan^3 u \, du = \tfrac{1}{2}\tan^2 u + \ln|\cos u| + C$

70. $\int \cot^3 u \, du = -\tfrac{1}{2}\cot^2 u - \ln|\sin u| + C$

71. $\int \sec^3 u \, du = \tfrac{1}{2}\sec u \tan u + \tfrac{1}{2}\ln|\sec u + \tan u| + C$

72. $\int \csc^3 u \, du = -\tfrac{1}{2}\csc u \cot u + \tfrac{1}{2}\ln|\csc u - \cot u| + C$

73. $\int \sin^n u \, du = -\dfrac{1}{n}\sin^{n-1} u \cos u + \dfrac{n-1}{n}\int \sin^{n-2} u \, du$

74. $\int \cos^n u \, du = \dfrac{1}{n}\cos^{n-1} u \sin u + \dfrac{n-1}{n}\int \cos^{n-2} u \, du$

75. $\int \tan^n u \, du = \dfrac{1}{n-1}\tan^{n-1} u - \int \tan^{n-2} u \, du$

76. $\int \cot^n u \, du = \dfrac{-1}{n-1}\cot^{n-1} u - \int \cot^{n-2} u \, du$

77. $\int \sec^n u \, du = \dfrac{1}{n-1}\tan u \sec^{n-2} u + \dfrac{n-2}{n-1}\int \sec^{n-2} u \, du$

78. $\int \csc^n u \, du = \dfrac{-1}{n-1}\cot u \csc^{n-2} u + \dfrac{n-2}{n-1}\int \csc^{n-2} u \, du$

79. $\int \sin au \sin bu \, du = \dfrac{\sin(a-b)u}{2(a-b)} - \dfrac{\sin(a+b)u}{2(a+b)} + C$

80. $\int \cos au \cos bu \, du = \dfrac{\sin(a-b)u}{2(a-b)} + \dfrac{\sin(a+b)u}{2(a+b)} + C$

81. $\int \sin au \cos bu \, du = -\dfrac{\cos(a-b)u}{2(a-b)} - \dfrac{\cos(a+b)u}{2(a+b)} + C$

82. $\int u \sin u \, du = \sin u - u \cos u + C$

83. $\int u \cos u \, du = \cos u + u \sin u + C$

84. $\int u^n \sin u \, du = -u^n \cos u + n \int u^{n-1} \cos u \, du$

85. $\int u^n \cos u \, du = u^n \sin u - n \int u^{n-1} \sin u \, du$

86. $\int \sin^n u \cos^m u \, du = -\dfrac{\sin^{n-1} u \cos^{m+1} u}{n+m} + \dfrac{n-1}{n+m}\int \sin^{n-2} u \cos^m u \, du$
$= \dfrac{\sin^{n+1} u \cos^{m-1} u}{n+m} + \dfrac{m-1}{n+m}\int \sin^n u \cos^{m-2} u \, du$

Inverse Trigonometric Forms

87. $\int \sin^{-1} u \, du = u \sin^{-1} u + \sqrt{1-u^2} + C$

88. $\int \cos^{-1} u \, du = u \cos^{-1} u - \sqrt{1-u^2} + C$

89. $\int \tan^{-1} u \, du = u \tan^{-1} u - \tfrac{1}{2}\ln(1+u^2) + C$

90. $\int u \sin^{-1} u \, du = \dfrac{2u^2-1}{4}\sin^{-1} u + \dfrac{u\sqrt{1-u^2}}{4} + C$

91. $\int u \cos^{-1} u \, du = \dfrac{2u^2-1}{4}\cos^{-1} u - \dfrac{u\sqrt{1-u^2}}{4} + C$

92. $\int u \tan^{-1} u \, du = \dfrac{u^2+1}{2}\tan^{-1} u - \dfrac{u}{2} + C$

93. $\int u^n \sin^{-1} u \, du = \dfrac{1}{n+1}\left[u^{n+1}\sin^{-1} u - \int \dfrac{u^{n+1} \, du}{\sqrt{1-u^2}}\right], \quad n \neq -1$

94. $\int u^n \cos^{-1} u \, du = \dfrac{1}{n+1}\left[u^{n+1}\cos^{-1} u + \int \dfrac{u^{n+1} \, du}{\sqrt{1-u^2}}\right], \quad n \neq -1$

95. $\int u^n \tan^{-1} u \, du = \dfrac{1}{n+1}\left[u^{n+1}\tan^{-1} u - \int \dfrac{u^{n+1} \, du}{1+u^2}\right], \quad n \neq -1$

(continued)

REFERENCE page 10

TABLE OF INTEGRALS

Exponential and Logarithmic Forms

96. $\displaystyle\int ue^{au}\,du = \frac{1}{a^2}(au-1)e^{au} + C$

97. $\displaystyle\int u^n e^{au}\,du = \frac{1}{a}u^n e^{au} - \frac{n}{a}\int u^{n-1}e^{au}\,du$

98. $\displaystyle\int e^{au}\sin bu\,du = \frac{e^{au}}{a^2+b^2}(a\sin bu - b\cos bu) + C$

99. $\displaystyle\int e^{au}\cos bu\,du = \frac{e^{au}}{a^2+b^2}(a\cos bu + b\sin bu) + C$

100. $\displaystyle\int \ln u\,du = u\ln u - u + C$

101. $\displaystyle\int u^n \ln u\,du = \frac{u^{n+1}}{(n+1)^2}[(n+1)\ln u - 1] + C$

102. $\displaystyle\int \frac{1}{u\ln u}\,du = \ln|\ln u| + C$

Hyperbolic Forms

103. $\displaystyle\int \sinh u\,du = \cosh u + C$

104. $\displaystyle\int \cosh u\,du = \sinh u + C$

105. $\displaystyle\int \tanh u\,du = \ln\cosh u + C$

106. $\displaystyle\int \coth u\,du = \ln|\sinh u| + C$

107. $\displaystyle\int \operatorname{sech} u\,du = \tan^{-1}|\sinh u| + C$

108. $\displaystyle\int \operatorname{csch} u\,du = \ln\left|\tanh \tfrac{1}{2}u\right| + C$

109. $\displaystyle\int \operatorname{sech}^2 u\,du = \tanh u + C$

110. $\displaystyle\int \operatorname{csch}^2 u\,du = -\coth u + C$

111. $\displaystyle\int \operatorname{sech} u\tanh u\,du = -\operatorname{sech} u + C$

112. $\displaystyle\int \operatorname{csch} u\coth u\,du = -\operatorname{csch} u + C$

Forms Involving $\sqrt{2au-u^2}$, $a>0$

113. $\displaystyle\int \sqrt{2au-u^2}\,du = \frac{u-a}{2}\sqrt{2au-u^2} + \frac{a^2}{2}\cos^{-1}\left(\frac{a-u}{a}\right) + C$

114. $\displaystyle\int u\sqrt{2au-u^2}\,du = \frac{2u^2 - au - 3a^2}{6}\sqrt{2au-u^2} + \frac{a^3}{2}\cos^{-1}\left(\frac{a-u}{a}\right) + C$

115. $\displaystyle\int \frac{\sqrt{2au-u^2}}{u}\,du = \sqrt{2au-u^2} + a\cos^{-1}\left(\frac{a-u}{a}\right) + C$

116. $\displaystyle\int \frac{\sqrt{2au-u^2}}{u^2}\,du = -\frac{2\sqrt{2au-u^2}}{u} - \cos^{-1}\left(\frac{a-u}{a}\right) + C$

117. $\displaystyle\int \frac{du}{\sqrt{2au-u^2}} = \cos^{-1}\left(\frac{a-u}{a}\right) + C$

118. $\displaystyle\int \frac{u\,du}{\sqrt{2au-u^2}} = -\sqrt{2au-u^2} + a\cos^{-1}\left(\frac{a-u}{a}\right) + C$

119. $\displaystyle\int \frac{u^2\,du}{\sqrt{2au-u^2}} = -\frac{(u+3a)}{2}\sqrt{2au-u^2} + \frac{3a^2}{2}\cos^{-1}\left(\frac{a-u}{a}\right) + C$

120. $\displaystyle\int \frac{du}{u\sqrt{2au-u^2}} = -\frac{\sqrt{2au-u^2}}{au} + C$

CHAPTER 1 CONCEPT CHECK ANSWERS

1. (a) What is a function? What are its domain and range?

A function f is a rule that assigns to each element x in a set D exactly one element, called $f(x)$, in a set E. The domain is the set D and the range is the set of all possible values of $f(x)$ as x varies throughout the domain.

(b) What is the graph of a function?

The graph of a function f consists of all points (x, y) such that $y = f(x)$ and x is in the domain of f.

(c) How can you tell whether a given curve is the graph of a function?

Use the Vertical Line Test: a curve in the xy-plane is the graph of a function of x if and only if no vertical line intersects the curve more than once.

2. Discuss four ways of representing a function. Illustrate your discussion with examples.

A function can be represented verbally, numerically, visually, or algebraically. An example of each is given below.

Verbally: An assignment of students to chairs in a classroom (a description in words)

Numerically: A tax table that assigns an amount of tax to an income (a table of values)

Visually: A graphical history of the Dow Jones average (a graph)

Algebraically: A relationship between the area A and side length s of a square: $A = s^2$ (an explicit formula)

3. (a) What is an even function? How can you tell if a function is even by looking at its graph? Give three examples of an even function.

A function f is even if it satisfies $f(-x) = f(x)$ for every number x in its domain. If the graph of a function is symmetric with respect to the y-axis, then f is even. Examples are $f(x) = x^2$, $f(x) = \cos x$, $f(x) = |x|$.

(b) What is an odd function? How can you tell if a function is odd by looking at its graph? Give three examples of an odd function.

A function f is odd if it satisfies $f(-x) = -f(x)$ for every number x in its domain. If the graph of a function is symmetric with respect to the origin, then f is odd. Examples are $f(x) = x^3$, $f(x) = \sin x$, $f(x) = 1/x$.

4. What is an increasing function?

A function f is increasing on an interval I if $f(x_1) < f(x_2)$ whenever $x_1 < x_2$ in I.

5. What is a mathematical model?

A mathematical model is a mathematical description (often by means of a function or an equation) of a real-world phenomenon. (See the discussion on pages 23–24.)

6. Give an example of each type of function.

(a) Linear function: $f(x) = 2x + 1$, $f(x) = ax + b$

(b) Power function: $f(x) = x^2$, $f(x) = x^n$

(c) Exponential function: $f(x) = 2^x$, $f(x) = b^x$

(d) Quadratic function: $f(x) = x^2 + x + 1$, $f(x) = ax^2 + bx + c$

(e) Polynomial of degree 5: $f(x) = x^5 + 2x^4 - 3x^2 + 7$

(f) Rational function: $f(x) = \dfrac{x}{x+2}$, $f(x) = \dfrac{P(x)}{Q(x)}$

where $P(x)$ and $Q(x)$ are polynomials

7. Sketch by hand, on the same axes, the graphs of the following functions.

(a) $f(x) = x$ **(b)** $g(x) = x^2$
(c) $h(x) = x^3$ **(d)** $j(x) = x^4$

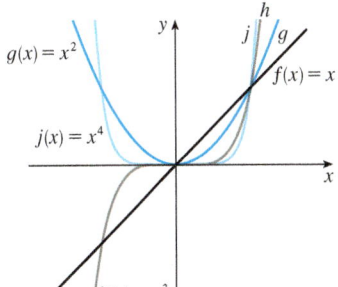

8. Draw, by hand, a rough sketch of the graph of each function.

(a) $y = \sin x$

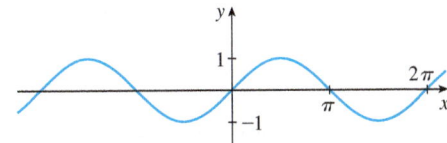

(b) $y = \cos x$

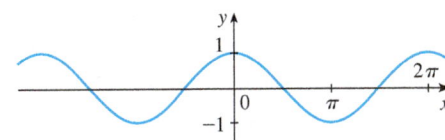

(c) $y = \tan x$

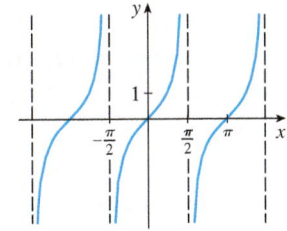

(continued)

CHAPTER 1   CONCEPT CHECK ANSWERS (continued)

(d) $y = 1/x$

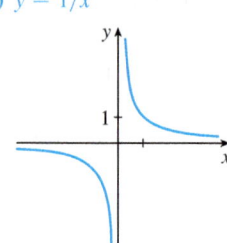

(e) $y = |x|$

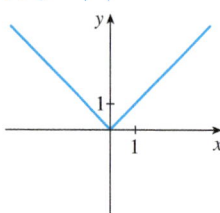

(f) $y = \sqrt{x}$

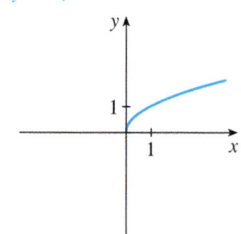

9. Suppose that f has domain A and g has domain B.

(a) What is the domain of $f + g$?

The domain of $f + g$ is the intersection of the domain of f and the domain of g; that is, $A \cap B$.

(b) What is the domain of fg?

The domain of fg is also $A \cap B$.

(c) What is the domain of f/g?

The domain of f/g must exclude values of x that make g equal to 0; that is, $\{x \in A \cap B \mid g(x) \neq 0\}$.

10. How is the composite function $f \circ g$ defined? What is its domain?

The composition of f and g is defined by $(f \circ g)(x) = f(g(x))$. The domain is the set of all x in the domain of g such that $g(x)$ is in the domain of f.

11. Suppose the graph of f is given. Write an equation for each of the graphs that are obtained from the graph of f as follows.

(a) Shift 2 units upward: $y = f(x) + 2$
(b) Shift 2 units downward: $y = f(x) - 2$
(c) Shift 2 units to the right: $y = f(x - 2)$
(d) Shift 2 units to the left: $y = f(x + 2)$
(e) Reflect about the x-axis: $y = -f(x)$
(f) Reflect about the y-axis: $y = f(-x)$
(g) Stretch vertically by a factor of 2: $y = 2f(x)$
(h) Shrink vertically by a factor of 2: $y = \tfrac{1}{2}f(x)$
(i) Stretch horizontally by a factor of 2: $y = f(\tfrac{1}{2}x)$
(j) Shrink horizontally by a factor of 2: $y = f(2x)$

12. Explain what each of the following means and illustrate with a sketch.

(a) $\lim_{x \to a} f(x) = L$ means that the values of $f(x)$ approach L as the values of x approach a (but $x \neq a$).

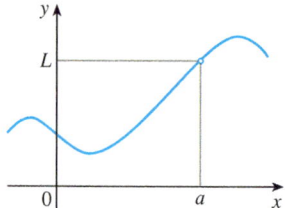

(b) $\lim_{x \to a^+} f(x) = L$ means that the values of $f(x)$ approach L as the values of x approach a through values greater than a.

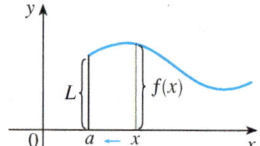

(c) $\lim_{x \to a^-} f(x) = L$ means that the values of $f(x)$ approach L as the values of x approach a through values less than a.

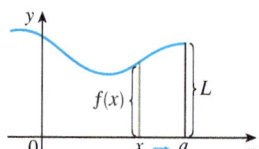

(d) $\lim_{x \to a} f(x) = \infty$ means that the values of $f(x)$ can be made arbitrarily large by taking x sufficiently close to a (but not equal to a).

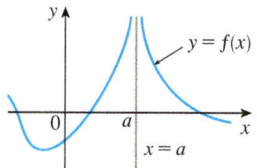

(e) $\lim_{x \to a} f(x) = -\infty$ means that the values of $f(x)$ can be made arbitrarily large negative by taking x sufficiently close to a (but not equal to a).

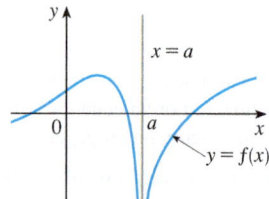

(continued)

CHAPTER 1 CONCEPT CHECK ANSWERS *(continued)*

13. Describe several ways in which a limit can fail to exist. Illustrate with sketches.

In general, the limit of a function fails to exist when the function values do not approach a fixed number. For each of the following functions, the limit fails to exist at $x = 2$.

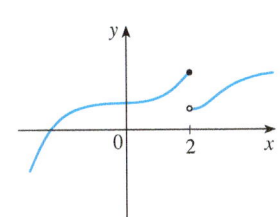
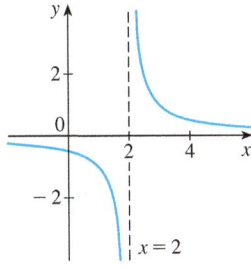

The left and right limits are not equal.

There is an infinite discontinuity.

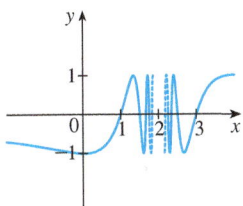

The function values oscillate between 1 and -1 infinitely often.

14. What does it mean to say that the line $x = a$ is a vertical asymptote of the curve $y = f(x)$? Draw curves to illustrate the various possibilities.

It means that the limit of $f(x)$ as x approaches a from one or both sides is positive or negative infinity.

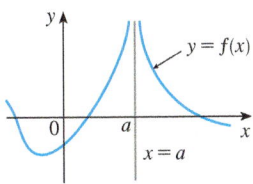
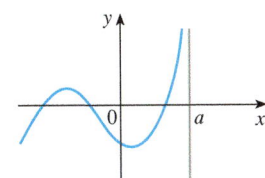

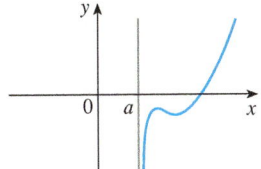
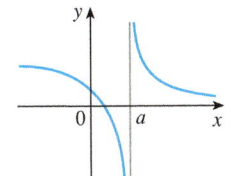

15. State the following Limit Laws.

(a) Sum Law

The limit of a sum is the sum of the limits:
$$\lim_{x \to a} [f(x) + g(x)] = \lim_{x \to a} f(x) + \lim_{x \to a} g(x)$$

(b) Difference Law

The limit of a difference is the difference of the limits:
$$\lim_{x \to a} [f(x) - g(x)] = \lim_{x \to a} f(x) - \lim_{x \to a} g(x)$$

(c) Constant Multiple Law

The limit of a constant times a function is the constant times the limit of the function: $\lim_{x \to a} [cf(x)] = c \lim_{x \to a} f(x)$

(d) Product Law

The limit of a product is the product of the limits:
$$\lim_{x \to a} [f(x)g(x)] = \lim_{x \to a} f(x) \cdot \lim_{x \to a} g(x)$$

(e) Quotient Law

The limit of a quotient is the quotient of the limits, provided that the limit of the denominator is not 0:
$$\lim_{x \to a} \frac{f(x)}{g(x)} = \frac{\lim_{x \to a} f(x)}{\lim_{x \to a} g(x)} \quad \text{if } \lim_{x \to a} g(x) \neq 0$$

(f) Power Law

The limit of a power is the power of the limit:
$$\lim_{x \to a} [f(x)]^n = \left[\lim_{x \to a} f(x)\right]^n \quad \text{(for } n \text{ a positive integer)}$$

(g) Root Law

The limit of a root is the root of the limit:
$$\lim_{x \to a} \sqrt[n]{f(x)} = \sqrt[n]{\lim_{x \to a} f(x)} \quad \text{(for } n \text{ a positive integer)}$$

16. What does the Squeeze Theorem say?

If $f(x) \leq g(x) \leq h(x)$ when x is near a (except possibly at a) and $\lim_{x \to a} f(x) = \lim_{x \to a} h(x) = L$, then $\lim_{x \to a} g(x) = L$. In other words, if $g(x)$ is squeezed between $f(x)$ and $h(x)$ near a, and if f and h have the same limit L at a, then g is forced to have the same limit L at a.

17. (a) What does it mean for f to be continuous at a?

A function f is continuous at a number a if the value of the function at $x = a$ is the same as the limit when x approaches a; that is, $\lim_{x \to a} f(x) = f(a)$.

(b) What does it mean for f to be continuous on the interval $(-\infty, \infty)$? What can you say about the graph of such a function?

A function f is continuous on the interval $(-\infty, \infty)$ if it is continuous at every real number a.

The graph of such a function has no hole or break in it.

(continued)

CHAPTER 1 CONCEPT CHECK ANSWERS (continued)

18. (a) Give examples of functions that are continuous on $[-1, 1]$.

    $f(x) = x^3 - x$, $g(x) = \sqrt{x+2}$, $y = \sin x$, $y = \tan x$, $y = 1/(x-3)$, and $h(x) = |x|$ are all continuous on $[-1, 1]$.

    (b) Give an example of a function that is not continuous on $[0, 1]$.

    $f(x) = \dfrac{1}{x - \frac{1}{2}}$ $\left[f(x) \text{ is not defined at } x = \tfrac{1}{2} \right]$

19. What does the Intermediate Value Theorem say?

    If f is continuous on $[a, b]$ and N is any number between $f(a)$ and $f(b)$ $[f(a) \neq f(b)]$, Then there exists a number c in (a, b) such that $f(c) = N$. In other words, a continuous function takes on every intermediate value between the function values $f(a)$ and $f(b)$.

CHAPTER 2 CONCEPT CHECK ANSWERS

1. Write an expression for the slope of the tangent line to the curve $y = f(x)$ at the point $(a, f(a))$.

   The slope of the tangent line is given by

   $$\lim_{x \to a} \frac{f(x) - f(a)}{x - a} \quad \text{or} \quad \lim_{h \to 0} \frac{f(a + h) - f(a)}{h}$$

2. Suppose an object moves along a straight line with position $f(t)$ at time t. Write an expression for the instantaneous velocity of the object at time $t = a$. How can you interpret this velocity in terms of the graph of f?

   The instantaneous velocity at time $t = a$ is

   $$v(a) = \lim_{h \to 0} \frac{f(a + h) - f(a)}{h}$$

   It is equal to the slope of the tangent line to the graph of f at the point $P(a, f(a))$.

3. If $y = f(x)$ and x changes from x_1 to x_2, write expressions for the following.
   (a) The average rate of change of y with respect to x over the interval $[x_1, x_2]$:

   $$\frac{\Delta y}{\Delta x} = \frac{f(x_2) - f(x_1)}{x_2 - x_1}$$

   (b) The instantaneous rate of change of y with respect to x at $x = x_1$:

   $$\lim_{\Delta x \to 0} \frac{\Delta y}{\Delta x} = \lim_{x_2 \to x_1} \frac{f(x_2) - f(x_1)}{x_2 - x_1}$$

4. Define the derivative $f'(a)$. Discuss two ways of interpreting this number.

   $$f'(a) = \lim_{h \to 0} \frac{f(a + h) - f(a)}{h}$$

   or, equivalently,

   $$f'(a) = \lim_{x \to a} \frac{f(x) - f(a)}{x - a}$$

   The derivative $f'(a)$ is the instantaneous rate of change of $y = f(x)$ (with respect to x) when $x = a$ and also represents the slope of the tangent line to the graph of f at the point $P(a, f(a))$.

5. (a) What does it mean for f to be differentiable at a?

   f is differentiable at a if the derivative $f'(a)$ exists.

   (b) What is the relation between the differentiability and continuity of a function?

   If f is differentiable at a, then f is continuous at a.

   (c) Sketch the graph of a function that is continuous but not differentiable at $a = 2$.

   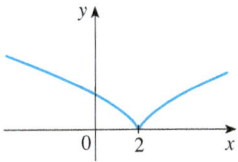

   The graph of f changes direction abruptly at $x = 2$, so f has no tangent line there.

6. Describe several ways in which a function can fail to be differentiable. Illustrate with sketches.

   A function is not differentiable at any value where the graph has a "corner," where the graph has a discontinuity, or where it has a vertical tangent line.

   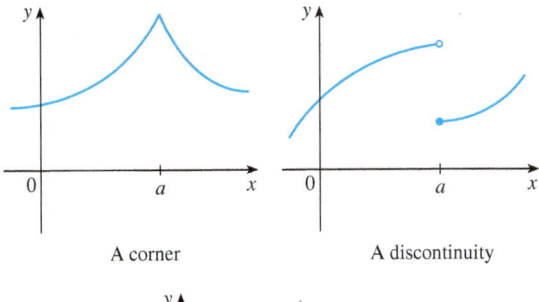

   A corner   A discontinuity

   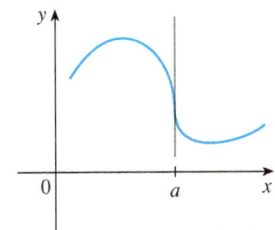

   A vertical tangent

7. What are the second and third derivatives of a function f? If f is the position function of an object, how can you interpret f'' and f'''?

   The second derivative f'' is the derivative of f', and the third derivative f''' is the derivative of f''.

   If f is the postition function of an object, then f' is the velocity function of the object, f'' is the acceleration function, and f''' is the jerk function (the rate of change of acceleration).

(continued)

CHAPTER 2 CONCEPT CHECK ANSWERS (continued)

8. State each differentiation rule both in symbols and in words.

   (a) The Power Rule

   If n is any real number, then $\dfrac{d}{dx}(x^n) = nx^{n-1}$.

   To find the derivative of a variable raised to a constant power, we multiply the expression by the exponent and then subtract one from the exponent.

   (b) The Constant Multiple Rule

   If c is a constant and f is a differentiable function, then
   $$\frac{d}{dx}[cf(x)] = c\frac{d}{dx}f(x)$$

   The derivative of a constant times a function is the constant times the derivative of the function.

   (c) The Sum Rule

   If f and g are both differentiable, then
   $$\frac{d}{dx}[f(x) + g(x)] = \frac{d}{dx}f(x) + \frac{d}{dx}g(x)$$

   The derivative of a sum of functions is the sum of the derivatives.

   (d) The Difference Rule

   If f and g are both differentiable, then
   $$\frac{d}{dx}[f(x) - g(x)] = \frac{d}{dx}f(x) - \frac{d}{dx}g(x)$$

   The derivative of a difference of functions is the difference of the derivatives.

   (e) The Product Rule

   If f and g are both differentiable, then
   $$\frac{d}{dx}[f(x)g(x)] = f(x)\frac{d}{dx}[g(x)] + g(x)\frac{d}{dx}[f(x)]$$

   The derivative of a product of two functions is the first function times the derivative of the second function plus the second function times the derivative of the first function.

   (f) The Quotient Rule

   If f and g are both differentiable, then
   $$\frac{d}{dx}\left[\frac{f(x)}{g(x)}\right] = \frac{g(x)\frac{d}{dx}[f(x)] - f(x)\frac{d}{dx}[g(x)]}{[g(x)]^2}$$

   The derivative of a quotient of functions is the denominator times the derivative of the numerator minus the numerator times the derivative of the denominator, all divided by the square of the denominator.

   (g) The Chain Rule

   If g is differentiable at x and f is differentiable at $g(x)$, then the composite function defined by $F(x) = f(g(x))$ is differentiable at x and F' is given by the product
   $$F'(x) = f'(g(x))g'(x)$$

   The derivative of a composite function is the derivative of the outer function evaluated at the inner function times the derivative of the inner function.

9. State the derivative of each function.

   (a) $y = x^n$: $\quad y' = nx^{n-1}$

   (b) $y = \sin x$: $\quad y' = \cos x$

   (c) $y = \cos x$: $\quad y' = -\sin x$

   (d) $y = \tan x$: $\quad y' = \sec^2 x$

   (e) $y = \csc x$: $\quad y' = -\csc x \cot x$

   (f) $y = \sec x$: $\quad y' = \sec x \tan x$

   (g) $y = \cot x$: $\quad y' = -\csc^2 x$

10. Explain how implicit differentiation works.

    Implicit differentiation consists of differentiating both sides of an equation with respect to x, treating y as a function of x. Then we solve the resulting equation for y'.

11. Give several examples of how the derivative can be interpreted as a rate of change in physics, chemistry, biology, economics, or other sciences.

    In physics, interpretations of the derivative include velocity, linear density, electrical current, power (the rate of change of work), and the rate of radioactive decay. Chemists can use derivatives to measure reaction rates and the compressibility of a substance under pressure. In biology the derivative measures rates of population growth and blood flow. In economics, the derivative measures marginal cost (the rate of change of cost as more items are produced) and marginal profit. Other examples include the rate of heat flow in geology, the rate of performance improvement in psychology, and the rate at which a rumor spreads in sociology.

12. (a) Write an expression for the linearization of f at a.
    $$L(x) = f(a) + f'(a)(x - a)$$

    (b) If $y = f(x)$, write an expression for the differential dy.
    $$dy = f'(x)\,dx$$

    (c) If $dx = \Delta x$, draw a picture showing the geometric meanings of Δy and dy.

    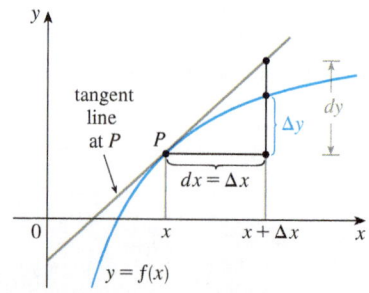

CHAPTER 3 CONCEPT CHECK ANSWERS

1. Explain the difference between an absolute maximum and a local maximum. Illustrate with a sketch.

   The function value $f(c)$ is the absolute maximum value of f if $f(c)$ is the largest function value on the entire domain of f, whereas $f(c)$ is a local maximum value if it is the largest function value when x is near c.

   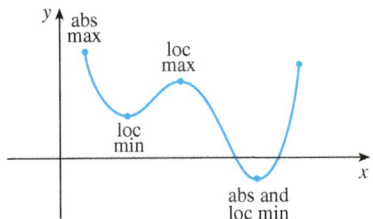

2. What does the Extreme Value Theorem say?

   If f is a continuous function on a closed interval $[a, b]$, then it always attains an absolute maximum and an absolute minimum value on that interval.

3. (a) State Fermat's Theorem.

   If f has a local maximum or minimum at c, and if $f'(c)$ exists, then $f'(c) = 0$.

   (b) Define a critical number of f.

   A critical number of a function f is a number c in the domain of f such that either $f'(c) = 0$ or $f'(c)$ does not exist.

4. Explain how the Closed Interval Method works.

   To find the absolute maximum and minimum values of a continuous function f on a closed interval $[a, b]$, we follow these three steps:
   - Find the critical numbers of f in the interval (a, b) and compute the values of f at these numbers.
   - Find the values of f at the endpoints of the interval.
   - The largest of the values from the previous two steps is the absolute maximum value; the smallest of these values is the absolute minimum value.

5. (a) State Rolle's Theorem.

   Let f be a function that satisfies the following three hypotheses:
   - f is continuous on the closed interval $[a, b]$.
   - f is differentiable on the open interval (a, b).
   - $f(a) = f(b)$

   Then there is a number c in (a, b) such that $f'(c) = 0$.

   (b) State the Mean Value Theorem and give a geometric interpretation.

   If f is continuous on the interval $[a, b]$ and differentiable on (a, b), then there exists a number c between a and b such that
   $$f'(c) = \frac{f(b) - f(a)}{b - a}$$

   Geometrically, the theorem says that there is a point $P(c, f(c))$, where $a < c < b$, on the graph of f where the tangent line is parallel to the secant line that connects $(a, f(a))$ and $(b, f(b))$.

   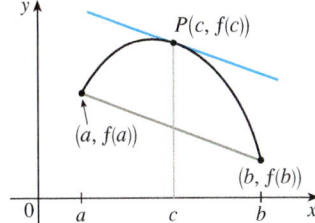

6. (a) State the Increasing/Decreasing Test.

   If $f'(x) > 0$ on an interval, then f is increasing on that interval.

   If $f'(x) < 0$ on an interval, then f is decreasing on that interval.

   (b) What does it mean to say that f is concave upward on an interval I?

   f is concave upward on an interval if the graph of f lies above all of its tangents on that interval.

   (c) State the Concavity Test.

   If $f''(x) > 0$ on an interval, then the graph of f is concave upward on that interval.

   If $f''(x) < 0$ on an interval, then the graph of f is concave downward on that interval.

   (d) What are inflection points? How do you find them?

   Inflection points on the graph of a continuous function f are points where the curve changes from concave upward to concave downward or from concave downward to concave upward. They can be found by determining the values at which the second derivative changes sign.

7. (a) State the First Derivative Test.

   Suppose that c is a critical number of a continuous function f.
   - If f' changes from positive to negative at c, then f has a local maximum at c.
   - If f' changes from negative to positive at c, then f has a local minimum at c.
   - If f' is positive to the left and right of c, or negative to the left and right of c, then f has no local maximum or minimum at c.

   (b) State the Second Derivative Test.

   Suppose f'' is continuous near c.
   - If $f'(c) = 0$ and $f''(c) > 0$, then f has a local minimum at c.
   - If $f'(c) = 0$ and $f''(c) < 0$, then f has a local maximum at c.

(continued)

CHAPTER 3 CONCEPT CHECK ANSWERS (continued)

(c) What are the relative advantages and disadvantages of these tests?

The Second Derivative Test is sometimes easier to use, but it is inconclusive when $f''(c) = 0$ and fails if $f''(c)$ does not exist. In either case the First Derivative Test must be used.

8. Explain the meaning of each of the following statements.

(a) $\lim_{x \to \infty} f(x) = L$ means that the values of $f(x)$ can be made arbitrarily close to L by requiring x to be sufficiently large.

(b) $\lim_{x \to -\infty} f(x) = L$ means that the values of $f(x)$ can be made arbitrarily close to L by requiring x to be sufficiently large negative.

(c) $\lim_{x \to \infty} f(x) = \infty$ means that the values of $f(x)$ can be made arbitrarily large by requiring x to be sufficiently large.

(d) The curve $y = f(x)$ has the horizontal asymptote $y = L$.

The line $y = L$ is called a horizontal asymptote of the curve $y = f(x)$ if either $\lim_{x \to \infty} f(x) = L$ or $\lim_{x \to -\infty} f(x) = L$.

9. If you have a graphing calculator or computer, why do you need calculus to graph a function?

Calculus reveals all the important aspects of a graph, such as local extreme values and inflection points, that can be missed when relying solely on technology. In many cases we can find exact locations of these key points rather than approximations. Using derivatives to identify the behavior of the graph also helps us choose an appropriate viewing window and alerts us to where we may wish to zoom in on a graph.

10. (a) Given an initial approximation x_1 to a root of the equation $f(x) = 0$, explain geometrically, with a diagram, how the second approximation x_2 in Newton's method is obtained.

We find the tangent line L to the graph of $y = f(x)$ at the point $(x_1, f(x_1))$. Then x_2 is the x-intercept of L.

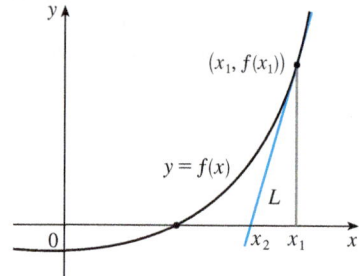

(b) Write an expression for x_2 in terms of x_1, $f(x_1)$, and $f'(x_1)$.

$$x_2 = x_1 - \frac{f(x_1)}{f'(x_1)}$$

(c) Write an expression for x_{n+1} in terms of x_n, $f(x_n)$, and $f'(x_n)$.

$$x_{n+1} = x_n - \frac{f(x_n)}{f'(x_n)}$$

(d) Under what circumstances is Newton's method likely to fail or to work very slowly?

Newton's method is likely to fail or to work very slowly when $f'(x_1)$ is close to 0. It also fails when $f'(x_i)$ is undefined.

11. (a) What is an antiderivative of a function f?

A function F is an antiderivative of f if $F'(x) = f(x)$.

(b) Suppose F_1 and F_2 are both antiderivatives of f on an interval I. How are F_1 and F_2 related?

They are identical or they differ by a constant.

CHAPTER 4   CONCEPT CHECK ANSWERS

1. (a) Write an expression for a Riemann sum of a function f on an interval $[a, b]$. Explain the meaning of the notation that you use.

If f is defined for $a \leq x \leq b$ and we divide the interval $[a, b]$ into n subintervals of equal width Δx, then a Riemann sum of f is

$$\sum_{i=1}^{n} f(x_i^*) \Delta x$$

where x_i^* is a point in the ith subinterval.

(b) If $f(x) \geq 0$, what is the geometric interpretation of a Riemann sum? Illustrate with a diagram.

If f is positive, then a Riemann sum can be interpreted as the sum of areas of approximating rectangles, as shown in the figure.

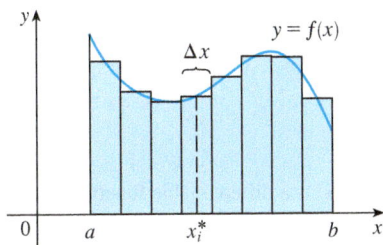

(c) If $f(x)$ takes on both positive and negative values, what is the geometric interpretation of a Riemann sum? Illustrate with a diagram.

If f takes on both positive and negative values then the Riemann sum is the sum of the areas of the rectangles that lie above the x-axis and the negatives of the areas of the rectangles that lie below the x-axis (the areas of the blue rectangles minus the areas of the gray rectangles).

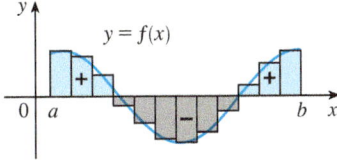

2. (a) Write the definition of the definite integral of a continuous function from a to b.

If f is a continuous function on the interval $[a, b]$, then we divide $[a, b]$ into n subintervals of equal width $\Delta x = (b - a)/n$. We let $x_0 \, (= a), x_1, x_2, \ldots, x_n \, (= b)$ be the endpoints of these subintervals. Then

$$\int_a^b f(x)\, dx = \lim_{n \to \infty} \sum_{i=1}^{n} f(x_i^*)\, \Delta x$$

where x_i^* is any sample point in the ith subinterval $[x_{i-1}, x_i]$.

(b) What is the geometric interpretation of $\int_a^b f(x)\, dx$ if $f(x) \geq 0$?

If f is positive, then $\int_a^b f(x)\, dx$ can be interpreted as the area under the graph of $y = f(x)$ and above the x-axis for $a \leq x \leq b$.

(c) What is the geometric interpretation of $\int_a^b f(x)\, dx$ if $f(x)$ takes on both positive and negative values? Illustrate with a diagram.

In this case $\int_a^b f(x)\, dx$ can be interpreted as a "net area," that is, the area of the region above the x-axis and below the graph of f (labeled "+" in the figure) minus the area of the region below the x-axis and above the graph of f (labeled "−").

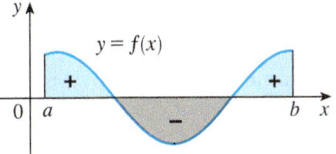

3. State the Midpoint Rule.

If f is a continuous function on the interval $[a, b]$ and we divide $[a, b]$ into n subintervals of equal width $\Delta x = (b - a)/n$, then

$$\int_a^b f(x)\, dx \approx \sum_{i=1}^{n} f(\bar{x}_i)\, \Delta x$$

where $\bar{x}_i = $ midpoint of $[x_{i-1}, x_i] = \tfrac{1}{2}(x_{i-1} + x_i)$.

4. State both parts of the Fundamental Theorem of Calculus.

Suppose f is continuous on $[a, b]$.

Part 1. If $g(x) = \int_a^x f(t)\, dt$, then $g'(x) = f(x)$.

Part 2. $\int_a^b f(x)\, dx = F(b) - F(a)$, where F is any antiderivative of f, that is, $F' = f$.

5. (a) State the Net Change Theorem.

The integral of a rate of change is the net change:

$$\int_a^b F'(x)\, dx = F(b) - F(a)$$

(b) If $r(t)$ is the rate at which water flows into a reservoir, what does $\int_{t_1}^{t_2} r(t)\, dt$ represent?

$\int_{t_1}^{t_2} r(t)\, dt$ represents the change in the amount of water in the reservoir between time t_1 and time t_2.

(continued)

CHAPTER 4   CONCEPT CHECK ANSWERS *(continued)*

6. Suppose a particle moves back and forth along a straight line with velocity $v(t)$, measured in feet per second, and acceleration $a(t)$.

   (a) What is the meaning of $\int_{60}^{120} v(t)\, dt$?

   $\int_{60}^{120} v(t)\, dt$ represents the net change in position (the displacement) of the particle from $t = 60$ s to $t = 120$ s, in other words, in the second minute.

   (b) What is the meaning of $\int_{60}^{120} |v(t)|\, dt$?

   $\int_{60}^{120} |v(t)|\, dt$ represents the total distance traveled by the particle in the second minute.

   (c) What is the meaning of $\int_{60}^{120} a(t)\, dt$?

   $\int_{60}^{120} a(t)\, dt$ represents the change in velocity of the particle in the second minute.

7. (a) Explain the meaning of the indefinite integral $\int f(x)\, dx$.

   The indefinite integral $\int f(x)\, dx$ is another name for an antiderivative of f, so $\int f(x)\, dx = F(x)$ means that $F'(x) = f(x)$.

   (b) What is the connection between the definite integral $\int_a^b f(x)\, dx$ and the indefinite integral $\int f(x)\, dx$?

   The connection is given by Part 2 of the Fundamental Theorem:

$$\int_a^b f(x)\, dx = \Big[\int f(x)\, dx\Big]_a^b$$

   if f is continuous on $[a, b]$.

8. Explain exactly what is meant by the statement that "differentiation and integration are inverse processes."

Part 1 of the Fundamental Theorem of Calculus can be rewritten as

$$\frac{d}{dx}\int_a^x f(t)\, dt = f(x)$$

which says that if f is integrated and then the result is differentiated, we arrive back at the original function f.

Since $F'(x) = f(x)$, Part 2 of the theorem (or, equivalently, the Net Change Theorem) states that

$$\int_a^b F'(x)\, dx = F(b) - F(a)$$

This says that if we take a function F, first differentiate it, and then integrate the result, we arrive back at the original function, but in the form $F(b) - F(a)$.

Also, the indefinite integral $\int f(x)\, dx$ represents an antiderivative of f, so

$$\frac{d}{dx}\int f(x)\, dx = f(x)$$

9. State the Substitution Rule. In practice, how do you use it?

If $u = g(x)$ is a differentiable function and f is continuous on the range of g, then

$$\int f(g(x))\, g'(x)\, dx = \int f(u)\, du$$

In practice, we make the substitutions $u = g(x)$ and $du = g'(x)\, dx$ in the integrand in order to make the integral simpler to evaluate.

CHAPTER 5 CONCEPT CHECK ANSWERS

1. (a) Write an expression for a Riemann sum of a function f. Explain the meaning of the notation that you use.

A Riemann sum of f is

$$\sum_{i=1}^{n} f(x_i^*)\, \Delta x$$

where x_i^* is a point in the ith subinterval $[x_{i-1}, x_i]$ and Δx is the length of the subintervals.

(b) If $f(x) \geq 0$, what is the geometric interpretation of a Riemann sum? Illustrate with a diagram.

If f is positive, then a Riemann sum can be interpreted as the sum of areas of approximating rectangles, as shown in the figure.

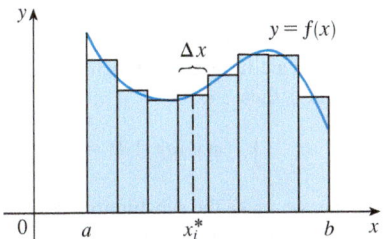

(c) If $f(x)$ takes on both positive and negative values, what is the geometric interpretation of a Riemann sum? Illustrate with a diagram.

If f takes on both positive and negative values then the Riemann sum is the sum of the areas of the rectangles that lie above the x-axis and the negatives of the areas of the rectangles that lie below the x-axis (the areas of the blue rectangles minus the areas of the gray rectangles).

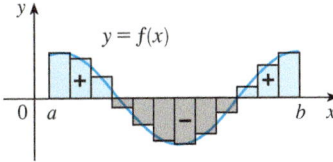

2. (a) Write the definition of the definite integral of a continuous function from a to b.

If f is a continuous function on the interval $[a, b]$, then we divide $[a, b]$ into n subintervals of equal width $\Delta x = (b - a)/n$. We let $x_0\, (= a), x_1, x_2, \ldots, x_n\, (= b)$ be the endpoints of these subintervals. Then

$$\int_a^b f(x)\, dx = \lim_{n \to \infty} \sum_{i=1}^{n} f(x_i^*)\, \Delta x$$

where x_i^* is any sample point in the ith subinterval $[x_{i-1}, x_i]$.

(b) What is the geometric interpretation of $\int_a^b f(x)\, dx$ if $f(x) \geq 0$?

If f is positive, then $\int_a^b f(x)\, dx$ can be interpreted as the area under the graph of $y = f(x)$ and above the x-axis for $a \leq x \leq b$.

(c) What is the geometric interpretation of $\int_a^b f(x)\, dx$ if $f(x)$ takes on both positive and negative values? Illustrate with a diagram.

In this case $\int_a^b f(x)\, dx$ can be interpreted as a "net area," that is, the area of the region above the x-axis and below the graph of f (labeled "+" in the figure) minus the area of the region below the x-axis and above the graph of f (labeled "−").

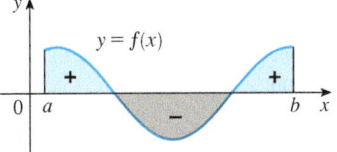

3. State the Midpoint Rule.

If f is a continuous function on the interval $[a, b]$ and we divide $[a, b]$ into n subintervals of equal width $\Delta x = (b - a)/n$, then

$$\int_a^b f(x)\, dx \approx \sum_{i=1}^{n} f(\bar{x}_i)\, \Delta x$$

where $\bar{x}_i =$ midpoint of $[x_{i-1}, x_i] = \frac{1}{2}(x_{i-1} + x_i)$.

4. State both parts of the Fundamental Theorem of Calculus.

Suppose f is continuous on $[a, b]$.

Part 1. If $g(x) = \int_a^x f(t)\, dt$, then $g'(x) = f(x)$.

Part 2. $\int_a^b f(x)\, dx = F(b) - F(a)$, where F is any antiderivative of f, that is, $F' = f$.

5. (a) State the Net Change Theorem.

The integral of a rate of change is the net change:

$$\int_a^b F'(x)\, dx = F(b) - F(a)$$

(b) If $r(t)$ is the rate at which water flows into a reservoir, what does $\int_{t_1}^{t_2} r(t)\, dt$ represent?

$\int_{t_1}^{t_2} r(t)\, dt$ represents the change in the amount of water in the reservoir between time t_1 and time t_2.

6. Suppose a particle moves back and forth along a straight line with velocity $v(t)$, measured in feet per second, and acceleration $a(t)$.

(a) What is the meaning of $\int_{60}^{120} v(t)\, dt$?

$\int_{60}^{120} v(t)\, dt$ represents the net change in position (the displacement) of the particle from $t = 60$ to $t = 120$ seconds, in other words, in the second minute.

(b) What is the meaning of $\int_{60}^{120} |v(t)|\, dt$?

$\int_{60}^{120} |v(t)|\, dt$ represents the total distance traveled by the particle in the second minute.

(continued)

CHAPTER 5   CONCEPT CHECK ANSWERS *(continued)*

(c) What is the meaning of $\int_{60}^{120} a(t)\,dt$?

$\int_{60}^{120} a(t)\,dt$ represents the change in velocity of the particle in the second minute.

7. (a) Explain the meaning of the indefinite integral $\int f(x)\,dx$.

The indefinite integral $\int f(x)\,dx$ is another name for an antiderivative of f, so $\int f(x)\,dx = F(x)$ means that $F'(x) = f(x)$.

(b) What is the connection between the definite integral $\int_a^b f(x)\,dx$ and the indefinite integral $\int f(x)\,dx$?

The connection is given by Part 2 of the Fundamental Theorem:

$$\int_a^b f(x)\,dx = \left[\int f(x)\,dx\right]_a^b$$

if f is continuous on $[a, b]$.

8. Explain exactly what is meant by the statement that "differentiation and integration are inverse processes."

Part 1 of the Fundamental Theorem of Calculus can be rewritten as

$$\frac{d}{dx}\int_a^x f(t)\,dt = f(x)$$

which says that if f is integrated and then the result is differentiated, we arrive back at the original function f.

Since $F'(x) = f(x)$, Part 2 of the theorem (or, equivalently, the Net Change Theorem) states that

$$\int_a^b F'(x)\,dx = F(b) - F(a)$$

This says that if we take a function F, first differentiate it, and then integrate the result, we arrive back at the original function, but in the form $F(b) - F(a)$.

Also, the indefinite integral $\int f(x)\,dx$ represents an antiderivative of f, so

$$\frac{d}{dx}\int f(x)\,dx = f(x)$$

9. State the Substitution Rule. In practice, how do you use it?

If $u = g(x)$ is a differentiable function and f is continuous on the range of g, then

$$\int f(g(x))g'(x)\,dx = \int f(u)\,du$$

In practice, we make the substitutions $u = g(x)$ and $du = g'(x)\,dx$ in the integrand to make the integral simpler to evaluate.

CHAPTER 6 CONCEPT CHECK ANSWERS

1. **(a)** Draw two typical curves $y = f(x)$ and $y = g(x)$, where $f(x) \geq g(x)$ for $a \leq x \leq b$. Show how to approximate the area between these curves by a Riemann sum and sketch the corresponding approximating rectangles. Then write an expression for the exact area.

   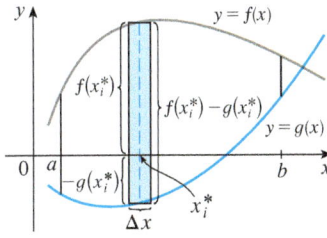

   A Riemann sum that approximates the area between these curves is $\sum_{i=1}^{n} [f(x_i^*) - g(x_i^*)] \Delta x$. A sketch of the corresponding approximating rectangles:

   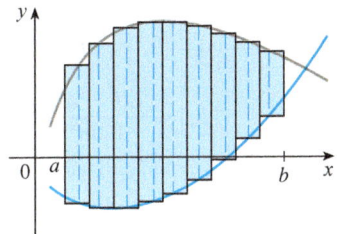

   An expression for the exact area is
   $$\lim_{n \to \infty} \sum_{i=1}^{n} [f(x_i^*) - g(x_i^*)] \Delta x = \int_a^b [f(x) - g(x)] \, dx$$

   (b) Explain how the situation changes if the curves have equations $x = f(y)$ and $x = g(y)$, where $f(y) \geq g(y)$ for $c \leq y \leq d$.

   Instead of using "top minus bottom" and integrating from left to right, we use "right minus left" and integrate from bottom to top: $A = \int_c^d [f(y) - g(y)] \, dy$

   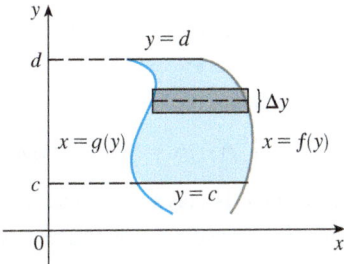

2. Suppose that Sue runs faster than Kathy throughout a 1500-meter race. What is the physical meaning of the area between their velocity curves for the first minute of the race?

   It represents the number of meters by which Sue is ahead of Kathy after 1 minute.

3. **(a)** Suppose S is a solid with known cross-sectional areas. Explain how to approximate the volume of S by a Riemann sum. Then write an expression for the exact volume.

   We slice S into n "slabs" of equal width Δx. The volume of the ith slab is approximately $A(x_i^*) \Delta x$, where x_i^* is a sample point in the ith slab and $A(x_i^*)$ is the cross-sectional area of S at x_i^*. Then the volume of S is approximately $\sum_{i=1}^{n} A(x_i^*) \Delta x$ and the exact volume is
   $$V = \lim_{n \to \infty} \sum_{i=1}^{n} A(x_i^*) \Delta x = \int_a^b A(x) \, dx$$

   (b) If S is a solid of revolution, how do you find the cross-sectional areas?

   If the cross-section is a disk, find the radius in terms of x or y and use $A = \pi(\text{radius})^2$. If the cross-section is a washer, find the inner radius r_{in} and outer radius r_{out} and use $A = \pi(r_{\text{out}}^2) - \pi(r_{\text{in}}^2)$.

4. **(a)** What is the volume of a cylindrical shell?
   $$V = 2\pi r h \, \Delta r = (\text{circumference})(\text{height})(\text{thickness})$$

   (b) Explain how to use cylindrical shells to find the volume of a solid of revolution.

   We approximate the region to be revolved by rectangles, oriented so that revolution forms cylindrical shells rather than disks or washers. For a typical shell, find the circumference and height in terms of x or y and calculate
   $$V = \int_a^b (\text{circumference})(\text{height})(dx \text{ or } dy)$$

   (c) Why might you want to use the shell method instead of slicing?

   Sometimes slicing produces washers or disks whose radii are difficult (or impossible) to find explicitly. On other occasions, the cylindrical shell method leads to an easier integral than slicing does.

5. Suppose that you push a book across a 6-meter-long table by exerting a force $f(x)$ at each point from $x = 0$ to $x = 6$. What does $\int_0^6 f(x) \, dx$ represent? If $f(x)$ is measured in newtons, what are the units for the integral?

   $\int_0^6 f(x) \, dx$ represents the amount of work done. Its units are newton-meters, or joules.

6. **(a)** What is the average value of a function f on an interval $[a, b]$?
   $$f_{\text{ave}} = \frac{1}{b - a} \int_a^b f(x) \, dx$$

   (b) What does the Mean Value Theorem for Integrals say? What is its geometric interpretation?

   If f is continuous on $[a, b]$, then there is a number c in $[a, b]$ at which the value of f is exactly equal to the average value of the function, that is, $f(c) = f_{\text{ave}}$. This means that for positive functions f, there is a number c such that the rectangle with base $[a, b]$ and height $f(c)$ has the same area as the region under the graph of f from a to b.

CHAPTER 7   CONCEPT CHECK ANSWERS

1. State the rule for integration by parts. In practice, how do you use it?

   To integrate $\int f(x)g'(x)\,dx$, let $u = f(x)$ and $v = g(x)$. Then $\int u\,dv = uv - \int v\,du$.

   In practice, try to choose $u = f(x)$ to be a function that becomes simpler when differentiated (or at least not more complicated) at long as $dv = g'(x)\,dx$ can be readily integrated to give v.

2. How do you evaluate $\int \sin^m x \cos^n x\,dx$ if m is odd? What if n is odd? What if m and n are both even?

   If m is odd, use $\sin^2 x = 1 - \cos^2 x$ to write all sine factors except one in terms of cosine. Then substitute $u = \cos x$.

   If n is odd, use $\cos^2 x = 1 - \sin^2 x$ to write all cosine factors except one in terms of sine. Then substitute $u = \sin x$.

   If m and n are even, use the half-angle identities
   $$\sin^2 x = \tfrac{1}{2}(1 - \cos 2x) \qquad \cos^2 x = \tfrac{1}{2}(1 + \cos 2x)$$

3. If the expression $\sqrt{a^2 - x^2}$ occurs in an integral, what substitution might you try? What if $\sqrt{a^2 + x^2}$ occurs? What if $\sqrt{x^2 - a^2}$ occurs?

   If $\sqrt{a^2 - x^2}$ occurs, try $x = a \sin\theta$; if $\sqrt{a^2 + x^2}$ occurs, try $x = a\tan\theta$, and if $\sqrt{x^2 - a^2}$ occurs, try $x = a\sec\theta$.

4. What is the form of the partial fraction decomposition of a rational function $P(x)/Q(x)$ if the degree of P is less than the degree of Q and $Q(x)$ has only distinct linear factors? What if a linear factor is repeated? What if $Q(x)$ has an irreducible quadratic factor (not repeated)? What if the quadratic factor is repeated?

   For distinct linear factors,
   $$\frac{P(x)}{Q(x)} = \frac{A_1}{a_1 x + b_1} + \frac{A_2}{a_2 x + b_2} + \cdots + \frac{A_k}{a_k x + b_k}$$

   If the linear factor $a_1 x + b_1$ is repeated r times, then we must include all the terms
   $$\frac{B_1}{a_1 x + b_1} + \frac{B_2}{(a_1 x + b_1)^2} + \cdots + \frac{B_r}{(a_1 x + b_1)^r}$$

   If $Q(x)$ has an irreducible quadratic factor (not repeated), then we include a term of the form
   $$\frac{Ax + B}{ax^2 + bx + c}$$

   If the irreducible quadratic factor is repeated r times, then we include all the terms
   $$\frac{A_1 x + B_1}{ax^2 + bx + c} + \frac{A_2 x + B_2}{(ax^2 + bx + c)^2} + \cdots + \frac{A_r x + B_r}{(ax^2 + bx + c)^r}$$

5. State the rules for approximating the definite integral $\int_a^b f(x)\,dx$ with the Midpoint Rule, the Trapezoidal Rule, and Simpson's Rule. Which would you expect to give the best estimate? How do you approximate the error for each rule?

   Let $a \le x \le b$, $I = \int_a^b f(x)\,dx$, and $\Delta x = (b - a)/n$.

   Midpoint Rule:
   $$I \approx M_n = \Delta x\,[f(\bar{x}_1) + f(\bar{x}_2) + \cdots + f(\bar{x}_n)]$$
   where $\bar{x}_i$ is the midpoint of $[x_{i-1}, x_i]$.

   Trapezoidal Rule:
   $$I \approx T_n$$
   $$= \frac{\Delta x}{2}[f(x_0) + 2f(x_1) + 2f(x_2) + \cdots + 2f(x_{n-1}) + f(x_n)]$$
   where $x_i = a + i\,\Delta x$.

   Simpson's Rule:
   $$I \approx S_n$$
   $$= \frac{\Delta x}{3}[f(x_0) + 4f(x_1) + 2f(x_2) + 4f(x_3) + \cdots$$
   $$+ 2f(x_{n-2}) + 4f(x_{n-1}) + f(x_n)]$$
   where n is even.

   We would expect the best estimate to be given by Simpson's Rule.

   Suppose $|f''(x)| \le K$ and $|f^{(4)}(x)| \le L$ for $a \le x \le b$. The errors in the Midpoint, Trapezoidal, and Simpson's Rules are given by, respectively,
   $$|E_M| \le \frac{K(b-a)^3}{24n^2} \qquad |E_T| \le \frac{K(b-a)^3}{12n^2}$$
   $$|E_S| \le \frac{L(b-a)^5}{180n^4}$$

6. Define the following improper integrals.

   (a) $\displaystyle\int_a^\infty f(x)\,dx = \lim_{t \to \infty} \int_a^t f(x)\,dx$

   (b) $\displaystyle\int_{-\infty}^b f(x)\,dx = \lim_{t \to -\infty} \int_t^b f(x)\,dx$

   (c) $\displaystyle\int_{-\infty}^\infty f(x)\,dx = \int_{-\infty}^a f(x)\,dx + \int_a^\infty f(x)\,dx$, where a is any real number (assuming that both integrals are convergent)

7. Define the improper integral $\int_a^b f(x)\,dx$ for each of the following cases.

   (a) f has an infinite discontinuity at a.

   If f is continuous on $(a, b]$, then
   $$\int_a^b f(x)\,dx = \lim_{t \to a^+} \int_t^b f(x)\,dx$$
   if this limit exists (as a finite number).

   (b) f has an infinite discontinuity at b.

   If f is continuous on $[a, b)$, then
   $$\int_a^b f(x)\,dx = \lim_{t \to b^-} \int_a^t f(x)\,dx$$
   if this limit exists (as a finite number).

   (c) f has an infinite discontinuity at c, where $a < c < b$.

   If both $\int_a^c f(x)\,dx$ and $\int_c^b f(x)\,dx$ are convergent, then
   $$\int_a^b f(x)\,dx = \int_a^c f(x)\,dx + \int_c^b f(x)\,dx$$

8. State the Comparison Theorem for improper integrals.

   Suppose that f and g are continuous functions with $f(x) \ge g(x) \ge 0$ for $x \ge a$.

   (a) If $\int_a^\infty f(x)\,dx$ is convergent, then $\int_a^\infty g(x)\,dx$ is convergent.

   (b) If $\int_a^\infty g(x)\,dx$ is divergent, then $\int_a^\infty f(x)\,dx$ is divergent.

CHAPTER 8   CONCEPT CHECK ANSWERS

1. **(a) How is the length of a curve defined?**

   We can approximate a curve C by a polygon with vertices P_i along C. The length L of C is defined to be the limit of the lengths of these inscribed polygons:
   $$L = \lim_{n \to \infty} \sum_{i=1}^{n} |P_{i-1}P_i|$$

   (b) Write an expression for the length of a smooth curve given by $y = f(x)$, $a \leq x \leq b$.
   $$L = \int_a^b \sqrt{1 + [f'(x)]^2}\, dx$$

   (c) What if x is given as a function of y?

   If $x = g(y)$, $c \leq y \leq d$, then $L = \int_c^d \sqrt{1 + [g'(y)]^2}\, dy$.

2. **(a) Write an expression for the surface area of the surface obtained by rotating the curve $y = f(x)$, $a \leq x \leq b$, about the x-axis.**
   $$S = \int_a^b 2\pi f(x)\sqrt{1 + [f'(x)]^2}\, dx$$

   (b) What if x is given as a function of y?

   If $x = g(y)$, $c \leq y \leq d$, then $S = \int_c^d 2\pi y\sqrt{1 + [g'(y)]^2}\, dy$.

   (c) What if the curve is rotated about the y-axis?
   $$S = \int_a^b 2\pi x\sqrt{1 + [f'(x)]^2}\, dx$$
   or
   $$S = \int_c^d 2\pi g(y)\sqrt{1 + [g'(y)]^2}\, dy$$

3. **Describe how we can find the hydrostatic force against a vertical wall submersed in a fluid.**

   We divide the wall into horizontal strips of equal height Δx and approximate each by a rectangle with horizontal length $f(x_i)$ at depth x_i. If δ is the weight density of the fluid, then the hydrostatic force is
   $$F = \lim_{n \to \infty} \sum_{i=1}^{n} \delta x_i f(x_i)\, \Delta x = \int_a^b \delta x f(x)\, dx$$

4. **(a) What is the physical significance of the center of mass of a thin plate?**

   The center of mass is the point at which the plate balances horizontally.

   (b) If the plate lies between $y = f(x)$ and $y = 0$, where $a \leq x \leq b$, write expressions for the coordinates of the center of mass.
   $$\bar{x} = \frac{1}{A}\int_a^b x f(x)\, dx \quad \text{and} \quad \bar{y} = \frac{1}{A}\int_a^b \tfrac{1}{2}[f(x)]^2\, dx$$
   where $A = \int_a^b f(x)\, dx$.

5. **What does the Theorem of Pappus say?**

   If a plane region $\mathcal{R}$ that lies entirely on one side of a line ℓ in its plane is rotated about ℓ, then the volume of the resulting solid is the product of the area of $\mathcal{R}$ and the distance traveled by the centroid of $\mathcal{R}$.

6. **Given a demand function $p(x)$, explain what is meant by the consumer surplus when the amount of a commodity currently available is X and the current selling price is P. Illustrate with a sketch.**

   The consumer surplus represents the amount of money saved by consumers in purchasing the commodity at price P [when they were willing to purchase it at price $p(x)$], corresponding to an amount demanded of X.

   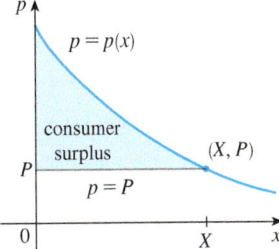

7. **(a) What is the cardiac output of the heart?**

   It is the volume of blood pumped by the heart per unit time, that is, the rate of flow into the aorta.

   (b) Explain how the cardiac output can be measured by the dye dilution method.

   An amount A of dye is injected into part of the heart and its concentration $c(t)$ leaving the heart is measured over a time interval $[0, T]$ until the dye has cleared. The cardiac output is given by $A / \int_0^T c(t)\, dt$.

8. **What is a probability density function? What properties does such a function have?**

   Given a random variable X, its probability density function f is a function such that $\int_a^b f(x)\, dx$ gives the probability that X lies between a and b. The function f has the properties that $f(x) \geq 0$ for all x, and $\int_{-\infty}^{\infty} f(x)\, dx = 1$.

9. **Suppose $f(x)$ is the probability density function for the weight of a female college student, where x is measured in pounds.**

   (a) What is the meaning of the integral $\int_0^{130} f(x)\, dx$?

   It represents the probability that a randomly chosen female college student weighs less than 130 pounds.

   (b) Write an expression for the mean of this density function.
   $$\mu = \int_{-\infty}^{\infty} x f(x)\, dx = \int_0^{\infty} x f(x)\, dx$$
   [since $f(x) = 0$ for $x < 0$]

   (c) How can we find the median of this density function?

   The median of f is the number m such that
   $$\int_m^{\infty} f(x)\, dx = \frac{1}{2}$$

10. **What is a normal distribution? What is the significance of the standard deviation?**

    A normal distribution corresponds to a random variable X that has a probability density function with a bell-shaped graph and equation given by
    $$f(x) = \frac{1}{\sigma\sqrt{2\pi}} e^{-(x-\mu)^2/(2\sigma^2)}$$
    where μ is the mean and the positive constant σ is the standard deviation. σ measures how spread out the values of X are.

CHAPTER 9 CONCEPT CHECK ANSWERS

1. (a) What is a differential equation?

   It is an equation that contains an unknown function and one or more of its derivatives.

   (b) What is the order of a differential equation?

   It is the order of the highest derivative that occurs in the equation.

   (c) What is an initial condition?

   It is a condition of the form $y(t_0) = y_0$.

2. What can you say about the solutions of the equation $y' = x^2 + y^2$ just by looking at the differential equation?

   The equation tells us that the slope of a solution curve at any point (x, y) is $x^2 + y^2$. Note that $x^2 + y^2$ is always positive except at the origin, where $y' = x^2 + y^2 = 0$. Thus there is a horizontal tangent at $(0, 0)$ but nowhere else and the solution curves are increasing everywhere.

3. What is a direction field for the differential equation $y' = F(x, y)$?

   A direction field (or slope field) for the differential equation $y' = F(x, y)$ is a two-dimensional graph consisting of short line segments with slope $F(x, y)$ at point (x, y).

4. Explain how Euler's method works.

   Euler's method says to start at the point given by the initial value and proceed in the direction indicated by the direction field. Stop after a short time, look at the slope at the new location, and proceed in that direction. Keep stopping and changing direction according to the direction field until the approximation is complete.

5. What is a separable differential equation? How do you solve it?

   It is a differential equation in which the expression for dy/dx can be factored as a function of x times a function of y, that is, $dy/dx = g(x) f(y)$. We can solve the equation by rewriting it as $[1/f(y)] \, dy = g(x) \, dx$, integrating both sides, and solving for y.

6. What is a first-order linear differential equation? How do you solve it?

   A first-order linear differential equation is a differential equation that can be put in the form

   $$\frac{dy}{dx} + P(x)y = Q(x)$$

   where P and Q are continuous functions on a given interval. To solve such an equation, we multiply both sides by the integrating factor $I(x) = e^{\int P(x) \, dx}$ to put it in the form $(I(x) y)' = I(x) Q(x)$. We then integrate both sides and solve for y.

7. (a) Write a differential equation that expresses the law of natural growth. What does it say in terms of relative growth rate?

   If $P(t)$ is the value of a quantity y at time t and if the rate of change of P with respect to t is proportional to its size $P(t)$ at any time, then $\dfrac{dP}{dt} = kP$.

   In this case the relative growth rate, $\dfrac{1}{P} \dfrac{dP}{dt}$, is constant.

   (b) Under what circumstances is this an appropriate model for population growth?

   It is an appropriate model under ideal conditions: unlimited environment, adequate nutrition, absence of predators and disease.

   (c) What are the solutions of this equation?

   If $P(0) = P_0$, the initial value, then the solutions are $P(t) = P_0 e^{kt}$.

8. (a) Write the logistic differential equation.

   The logistic differential equation is

   $$\frac{dP}{dt} = kP\left(1 - \frac{P}{M}\right)$$

   where M is the carrying capacity.

   (b) Under what circumstances is this an appropriate model for population growth?

   It is an appropriate model for population growth if the population grows at a rate proportional to the size of the population in the beginning, but eventually levels off and approaches its carrying capacity because of limited resources.

9. (a) Write Lotka-Volterra equations to model populations of food-fish (F) and sharks (S).

   $$\frac{dF}{dt} = kF - aFS \quad \text{and} \quad \frac{dS}{dt} = -rS + bFS$$

   (b) What do these equations say about each population in the absence of the other?

   In the absence of sharks, an ample food supply would support exponential growth of the fish population, that is, $dF/dt = kF$, where k is a positive constant. In the absence of fish, we assume that the shark population would decline at a rate proportional to itself, that is $dS/dt = -rS$, where r is a positive constant.

CHAPTER 11 CONCEPT CHECK ANSWERS

1. **(a)** What is a convergent sequence?

   A convergent sequence $\{a_n\}$ is an ordered list of numbers where $\lim_{n\to\infty} a_n$ exists.

   (b) What is a convergent series?

   A series $\Sigma\, a_n$ is the *sum* of a sequence of numbers. It is convergent if the partial sums $s_n = \sum_{i=1}^{n} a_n$ approach a finite value, that is, $\lim_{n\to\infty} s_n$ exists as a real number.

   (c) What does $\lim_{n\to\infty} a_n = 3$ mean?

   The terms of the sequence $\{a_n\}$ approach 3 as n becomes large.

   (d) What does $\sum_{n=1}^{\infty} a_n = 3$ mean?

   By adding sufficiently many terms of the series, we can make the partial sums as close to 3 as we like.

2. **(a)** What is a bounded sequence?

   A sequence $\{a_n\}$ is bounded if there are numbers m and M such that $m \leq a_n \leq M$ for all $n \geq 1$.

   (b) What is a monotonic sequence?

   A sequence is monotonic if it is either increasing or decreasing for all $n \geq 1$.

   (c) What can you say about a bounded monotonic sequence?

   Every bounded, monotonic sequence is convergent.

3. **(a)** What is a geometric series? Under what circumstances is it convergent? What is its sum?

   A geometric series is of the form
   $$\sum_{n=1}^{\infty} ar^{n-1} = a + ar + ar^2 + \cdots$$
   It is convergent if $|r| < 1$ and its sum is $\dfrac{a}{1-r}$.

   (b) What is a p-series? Under what circumstances is it convergent?

   A p-series is of the form $\sum_{n=1}^{\infty} \dfrac{1}{n^p}$. It is convergent if $p > 1$.

4. Suppose $\Sigma\, a_n = 3$ and s_n is the nth partial sum of the series. What is $\lim_{n\to\infty} a_n$? What is $\lim_{n\to\infty} s_n$?

   If $\Sigma\, a_n = 3$, then $\lim_{n\to\infty} a_n = 0$ and $\lim_{n\to\infty} s_n = 3$.

5. State the following.

   (a) The Test for Divergence

   If $\lim_{n\to\infty} a_n$ does not exist or if $\lim_{n\to\infty} a_n \neq 0$, then the series $\sum_{n=1}^{\infty} a_n$ is divergent.

   (b) The Integral Test

   Suppose f is a continuous, positive, decreasing function on $[1, \infty)$ and let $a_n = f(n)$.
   - If $\int_1^{\infty} f(x)\, dx$ is convergent, then $\sum_{n=1}^{\infty} a_n$ is convergent.
   - If $\int_1^{\infty} f(x)\, dx$ is divergent, then $\sum_{n=1}^{\infty} a_n$ is divergent.

   (c) The Comparison Test

   Suppose that $\Sigma\, a_n$ and $\Sigma\, b_n$ are series with positive terms.
   - If $\Sigma\, b_n$ is convergent and $a_n \leq b_n$ for all n, then $\Sigma\, a_n$ is also convergent.
   - If $\Sigma\, b_n$ is divergent and $a_n \geq b_n$ for all n, then $\Sigma\, a_n$ is also divergent.

   (d) The Limit Comparison Test

   Suppose that $\Sigma\, a_n$ and $\Sigma\, b_n$ are series with positive terms. If $\lim_{n\to\infty} a_n/b_n = c$, where c is a finite number and $c > 0$, then either both series converge or both diverge.

   (e) The Alternating Series Test

   If the alternating series
   $$\sum_{n=1}^{\infty} (-1)^{n-1} b_n = b_1 - b_2 + b_3 - b_4 + b_5 - b_6 + \cdots$$
   where $b_n > 0$ satisfies (i) $b_{n+1} \leq b_n$ for all n and (ii) $\lim_{n\to\infty} b_n = 0$, then the series is convergent.

   (f) The Ratio Test
   - If $\lim_{n\to\infty} \left|\dfrac{a_{n+1}}{a_n}\right| = L < 1$, then the series $\sum_{n=1}^{\infty} a_n$ is absolutely convergent (and therefore convergent).
   - If $\lim_{n\to\infty} \left|\dfrac{a_{n+1}}{a_n}\right| = L > 1$ or $\lim_{n\to\infty} \left|\dfrac{a_{n+1}}{a_n}\right| = \infty$, then the series $\sum_{n=1}^{\infty} a_n$ is divergent.
   - If $\lim_{n\to\infty} \left|\dfrac{a_{n+1}}{a_n}\right| = 1$, the Ratio Test is inconclusive.

   (g) The Root Test
   - If $\lim_{n\to\infty} \sqrt[n]{|a_n|} = L < 1$, then the series $\sum_{n=1}^{\infty} a_n$ is absolutely convergent (and therefore convergent).
   - If $\lim_{n\to\infty} \sqrt[n]{|a_n|} = L > 1$ or $\lim_{n\to\infty} \sqrt[n]{|a_n|} = \infty$, then the series $\sum_{n=1}^{\infty} a_n$ is divergent.
   - If $\lim_{n\to\infty} \sqrt[n]{|a_n|} = 1$, the Root Test is inconclusive.

6. **(a)** What is an absolutely convergent series?

   A series $\Sigma\, a_n$ is called absolutely convergent if the series of absolute values $\Sigma\, |a_n|$ is convergent.

   (b) What can you say about such a series?

   If a series $\Sigma\, a_n$ is absolutely convergent, then it is convergent.

   (c) What is a conditionally convergent series?

   A series $\Sigma\, a_n$ is called conditionally convergent if it is convergent but not absolutely convergent.

(continued)

CHAPTER 11 CONCEPT CHECK ANSWERS *(continued)*

7. (a) If a series is convergent by the Integral Test, how do you estimate its sum?

The sum s can be estimated by the inequality
$$s_n + \int_{n+1}^{\infty} f(x)\,dx \leq s \leq s_n + \int_{n}^{\infty} f(x)\,dx$$
where s_n is the nth partial sum.

(b) If a series is convergent by the Comparison Test, how do you estimate its sum?

We first estimate the remainder for the comparison series. This gives an upper bound for the remainder of the original series (as in Example 11.4.5).

(c) If a series is convergent by the Alternating Series Test, how do you estimate its sum?

We can use a partial sum s_n of an alternating series as an approximation to the total sum. The size of the error is guaranteed to be no more than $|a_{n+1}|$, the absolute value of the first neglected term.

8. (a) Write the general form of a power series.

A power series centered at a is
$$\sum_{n=0}^{\infty} c_n (x-a)^n$$

(b) What is the radius of convergence of a power series?

Given the power series $\sum_{n=0}^{\infty} c_n(x-a)^n$, the radius of convergence is:

(i) 0 if the series converges only when $x = a$,

(ii) ∞ if the series converges for all x, or

(iii) a positive number R such that the series converges if $|x - a| < R$ and diverges if $|x - a| > R$.

(c) What is the interval of convergence of a power series?

The interval of convergence of a power series is the interval that consists of all values of x for which the series converges. Corresponding to the cases in part (b), the interval of convergence is (i) the single point $\{a\}$, (ii) $(-\infty, \infty)$, or (iii) an interval with endpoints $a - R$ and $a + R$ that can contain neither, either, or both of the endpoints.

9. Suppose $f(x)$ is the sum of a power series with radius of convergence R.

(a) How do you differentiate f? What is the radius of convergence of the series for f'?

If $f(x) = \sum_{n=0}^{\infty} c_n(x-a)^n$, then $f'(x) = \sum_{n=1}^{\infty} n c_n (x-a)^{n-1}$ with radius of convergence R.

(b) How do you integrate f? What is the radius of convergence of the series for $\int f(x)\,dx$?

$$\int f(x)\,dx = C + \sum_{n=0}^{\infty} c_n \frac{(x-a)^{n+1}}{n+1}$$ with radius of convergence R.

10. (a) Write an expression for the nth-degree Taylor polynomial of f centered at a.

$$T_n(x) = \sum_{i=0}^{n} \frac{f^{(i)}(a)}{i!} (x-a)^i$$

(b) Write an expression for the Taylor series of f centered at a.

$$\sum_{n=0}^{\infty} \frac{f^{(n)}(a)}{n!} (x-a)^n$$

(c) Write an expression for the Maclaurin series of f.

$$\sum_{n=0}^{\infty} \frac{f^{(n)}(0)}{n!} x^n \quad [a = 0 \text{ in part (b)}]$$

(d) How do you show that $f(x)$ is equal to the sum of its Taylor series?

If $f(x) = T_n(x) + R_n(x)$, where $T_n(x)$ is the nth-degree Taylor polynomial of f and $R_n(x)$ is the remainder of the Taylor series, then we must show that
$$\lim_{n \to \infty} R_n(x) = 0$$

(e) State Taylor's Inequality.

If $|f^{(n+1)}(x)| \leq M$ for $|x-a| \leq d$, then the remainder $R_n(x)$ of the Taylor series satisfies the inequality
$$|R_n(x)| \leq \frac{M}{(n+1)!} |x-a|^{n+1} \quad \text{for } |x-a| \leq d$$

11. Write the Maclaurin series and the interval of convergence for each of the following functions.

(a) $\dfrac{1}{1-x} = \sum_{n=0}^{\infty} x^n$, $R = 1$

(b) $e^x = \sum_{n=0}^{\infty} \dfrac{x^n}{n!}$, $R = \infty$

(c) $\sin x = \sum_{n=0}^{\infty} (-1)^n \dfrac{x^{2n+1}}{(2n+1)!}$, $R = \infty$

(d) $\cos x = \sum_{n=0}^{\infty} (-1)^n \dfrac{x^{2n}}{(2n)!}$, $R = \infty$

(e) $\tan^{-1} x = \sum_{n=0}^{\infty} (-1)^n \dfrac{x^{2n+1}}{2n+1}$, $R = 1$

(f) $\ln(1+x) = \sum_{n=1}^{\infty} (-1)^{n-1} \dfrac{x^n}{n}$, $R = 1$

12. Write the binomial series expansion of $(1+x)^k$. What is the radius of convergence of this series?

If k is any real number and $|x| < 1$, then
$$(1+x)^k = \sum_{n=0}^{\infty} \binom{k}{n} x^n$$
$$= 1 + kx + \frac{k(k-1)}{2!} x^2 + \frac{k(k-1)(k-2)}{3!} x^3 + \cdots$$

The radius of convergence for the binomial series is 1.

CHAPTER 17 CONCEPT CHECK ANSWERS

1. (a) Write the general form of a second-order homogeneous linear differential equation with constant coefficients.

   $$ay'' + by' + cy = 0$$

   where a, b, and c are constants and $a \neq 0$.

   (b) Write the auxiliary equation.

   $$ar^2 + br + c = 0$$

   (c) How do you use the roots of the auxiliary equation to solve the differential equation? Write the form of the solution for each of the three cases that can occur.

   If the auxiliary equation has two distinct real roots r_1 and r_2, the general solution of the differential equation is

   $$y = c_1 e^{r_1 x} + c_2 e^{r_2 x}$$

   If the roots are real and equal, the solution is

   $$y = c_1 e^{rx} + c_2 x e^{rx}$$

   where r is the common root.

   If the roots are complex, we can write $r_1 = \alpha + i\beta$ and $r_2 = \alpha - i\beta$, and the solution is

   $$y = e^{\alpha x}(c_1 \cos \beta x + c_2 \sin \beta x)$$

2. (a) What is an initial-value problem for a second-order differential equation?

   An initial-value problem consists of finding a solution y of the differential equation that also satisfies given conditions $y(x_0) = y_0$ and $y'(x_0) = y_1$, where y_0 and y_1 are constants.

   (b) What is a boundary-value problem for such an equation?

   A boundary-value problem consists of finding a solution y of the differential equation that also satisfies given boundary conditions $y(x_0) = y_0$ and $y(x_1) = y_1$.

3. (a) Write the general form of a second-order nonhomogeneous linear differential equation with constant coefficients.

   $ay'' + by' + cy = G(x)$, where a, b, and c are constants and G is a continuous function.

   (b) What is the complementary equation? How does it help solve the original differential equation?

   The complementary equation is the related homogeneous equation $ay'' + by' + cy = 0$. If we find the general solution y_c of the complementary equation and y_p is any particular solution of the nonhomogeneous differential equation, then the general solution of the original differential equation is $y(x) = y_p(x) + y_c(x)$.

   (c) Explain how the method of undetermined coefficients works.

   To determine a particular solution y_p of $ay'' + by' + cy = G(x)$, we make an initial guess that y_p is a general function of the same type as G. If $G(x)$ is a polynomial, choose y_p to be a general polynomial of the same degree. If $G(x)$ is of the form Ce^{kx}, choose $y_p(x) = Ae^{kx}$. If $G(x)$ is $C \cos kx$ or $C \sin kx$, choose $y_p(x) = A \cos kx + B \sin kx$. If $G(x)$ is a product of functions, choose y_p to be a product of functions of the same type. Some examples are:

$G(x)$	$y_p(x)$
x^2	$Ax^2 + Bx + C$
e^{2x}	Ae^{2x}
$\sin 2x$	$A \cos 2x + B \sin 2x$
xe^{-x}	$(Ax + B)e^{-x}$

   We then substitute y_p, y_p', and y_p'' into the differential equation and determine the coefficients.

   If y_p happens to be a solution of the complementary equation, then multiply the initial trial solution by x (or x^2 if necessary).

   If $G(x)$ is a sum of functions, we find a particular solution for each function and then y_p is the sum of these.

   The general solution of the differential equation is

   $$y(x) = y_p(x) + y_c(x)$$

   (d) Explain how the method of variation of parameters works.

   We write the solution of the complementary equation $ay'' + by' + cy = 0$ as $y_c(x) = c_1 y_1(x) + c_2 y_2(x)$, where y_1 and y_2 are linearly independent solutions. We then take $y_p(x) = u_1(x) y_1(x) + u_2(x) y_2(x)$ as a particular solution, where $u_1(x)$ and $u_2(x)$ are arbitrary functions. After computing y_p', we impose the condition that

   $$u_1' y_1 + u_2' y_2 = 0 \quad (1)$$

   and then compute y_p''. Substituting y_p, y_p', and y_p'' into the original differential equation gives

   $$a(u_1' y_1' + u_2' y_2') = G \quad (2)$$

   We then solve equations (1) and (2) for the unknown functions u_1' and u_2'. If we are able to integrate these functions, then a particular solution is $y_p(x) = u_1(x) y_1(x) + u_2(x) y_2(x)$ and the general solution is $y(x) = y_p(x) + y_c(x)$.

4. Discuss two applications of second-order linear differential equations.

   The motion of an object with mass m at the end of a spring is an example of simple harmonic motion and is described by the second-order linear differential equation

   $$m \frac{d^2 x}{dt^2} + kx = 0$$

   (continued)

CHAPTER 17 CONCEPT CHECK ANSWERS (continued)

where k is the spring constant and x is the distance the spring is stretched (or compressed) from its natural length. If there are external forces acting on the spring, then the differential equation is modified.

Second-order linear differential equations are also used to analyze electrical circuits involving an electromotive force, a resistor, an inductor, and a capacitor in series.

See the discussion in Section 17.3 for additional details.

5. **How do you use power series to solve a differential equation?**

We first assume that the differential equation has a power series solution of the form

$$y = \sum_{n=0}^{\infty} c_n x^n = c_0 + c_1 x + c_2 x^2 + c_3 x^3 + \cdots$$

Differentiating gives

$$y' = \sum_{n=1}^{\infty} n c_n x^{n-1} = \sum_{n=0}^{\infty} (n+1) c_{n+1} x^n$$

and

$$y'' = \sum_{n=2}^{\infty} n(n-1) c_n x^{n-2} = \sum_{n=0}^{\infty} (n+2)(n+1) c_{n+2} x^n$$

We substitute these expressions into the differential equation and equate the coefficients of x^n to find a recursion relation involving the constants c_n. Solving the recursion relation gives a formula for c_n and then

$$y = \sum_{n=0}^{\infty} c_n x^n$$

is the solution of the differential equation.